# Intermediate Algebra

*Fifth Edition*

*Elayn Martin-Gay*

*University of New Orleans*

Boston  Columbus  Hoboken  Indianapolis  New York  San Francisco
Amsterdam  Cape Town  Dubai  London  Madrid  Milan  Munich  Paris  Montréal  Toronto
Delhi  Mexico City  São Paulo  Sydney  Hong Kong  Seoul  Singapore  Taipei  Tokyo

**Editorial Director, Mathematics:** *Christine Hoag*
**Editor-in-Chief:** *Michael Hirsch*
**Acquisitions Editor:** *Mary Beckwith*
**Project Team Lead:** *Christina Lepre*
**Project Manager:** *Lauren Morse*
**Assistant Editor:** *Matthew Summers*
**Editorial Assistant:** *Megan Tripp*
**Executive Development Editor:** *Dawn Nuttall*
**Program Team Lead:** *Karen Wernholm*
**Program Manager:** *Patty Bergin*
**Cover and Illustration Design:** *Tamara Newnam*
**Program Design Lead:** *Heather Scott*
**Executive Content Manager, MathXL:** *Rebecca Williams*
**Senior Content Developer, TestGen:** *John Flanagan*
**Director, Course Production:** *Ruth Berry*
**Media Producer:** *Audra Walsh*
**Senior Marketing Manager:** *Rachel Ross*
**Marketing Assistant:** *Kelly Cross*
**Senior Author Support/Technology Specialist:** *Joe Vetere*
**Senior Procurement Specialist:** *Carol Melville*
**Interior Design, Production Management, Answer Art, and Composition:**
   *Integra Software Services Pvt. Ltd.*
**Text Art:** *Scientific Illustrators*

For permission to use copyrighted material, grateful acknowledgment is made to the copyright holders on page P1 which is hereby made an extension of this copyright page.

PEARSON, ALWAYS LEARNING, and MYMATHLAB are exclusive trademarks in the U.S. and/or other countries owned by Pearson Education, Inc. or its affiliates.

Unless otherwise indicated herein, any third-party trademarks that may appear in this work are the property of their respective owners and any references to third-party trademarks, logos or other trade dress are for demonstrative or descriptive purposes only. Such references are not intended to imply any sponsorship, endorsement, authorization, or promotion of Pearson's products by the owners of such marks or any relationship between the owner and Pearson Education, Inc. or its affiliates, authors, licensees or distributors.

**Library of Congress Cataloging-in-Publication Data**
Martin-Gay, K. Elayn
   Intermediate algebra / Elayn Martin-Gay, University of New Orleans. —5th edition.
    pages cm
   ISBN-13: 978-0-321-97859-2 (alk. paper)
   ISBN-10: 0-321-97859-5 (alk. paper)
1. Algebra—Textbooks.  I. Title.
   QA152.3.M36 2016
   512.9—dc23                                    2014030417

1  2  3  4  5  6  7  8  9  10—CRK—19  18  17  16  15

**PEARSON**
www.pearsonhighered.com

ISBN-10: 0-321-97859-5 (Student Edition)
ISBN-13: 978-0-321-97859-2

This book is dedicated to students everywhere—and we should all be students. After all, is there anyone among us who really knows too much? Take that hint and continue to learn something new every day of your life.

Best of wishes from a fellow student:
Elayn Martin-Gay

# Contents

**Conic Sections   751**

## Appendices

# Preface

*Intermediate Algebra,* **Fifth Edition**, was written to provide a solid foundation in algebra as well as to develop problem-solving skills. Specific care was taken to make sure students have the most up-to-date relevant text preparation for their next mathematics course or for nonmathematical courses that require an understanding of algebraic fundamentals concepts. I have tried to achieve this by writing a user-friendly text that is keyed to objectives and contains many worked-out examples. As suggested by AMATYC and the NCTM Standards (plus Addenda), real-life and real-data applications, data interpretation, conceptual understanding, problem solving, writing, cooperative learning, appropriate use of technology, mental mathematics, number sense, estimation, critical thinking, and geometric concepts are emphasized and integrated throughout the book.

The many factors that contributed to the success of the previous editions have been retained. In preparing the Fifth Edition, I considered comments and suggestions of colleagues, students, and many users of the prior edition throughout the country.

## What's New in the Fifth Edition?

- **The Martin-Gay Program** has been revised and enhanced with a new design in the text and MyMathLab® to actively encourage students to use the text, video program, and Video Organizer as an integrated learning system.

- **The New Video Organizer** is designed to help students take notes and work practice exercises while watching the Interactive Lecture Series videos (available in MyMathLab and on DVD). All content in the Video Organizer is presented in the same order as it is presented in the videos, making it easy for students to create a course notebook and build good study habits.

  — Covers all of the video examples in order.

  — Provides ample space for students to write down key definitions and properties.

  — Includes "Play" and "Pause" button icons to prompt students to follow along with the author for some exercises while they try others on their own.

  The Video Organizer is available in a loose-leaf, notebook-ready format. It is also available for download in MyMathLab.

- **Vocabulary, Readiness & Video Check** questions have been added prior to every section exercise set. These exercises quickly check a student's understanding of new vocabulary words. The **readiness** exercises center on a student's understanding of a concept that is necessary in order to continue to the exercise set. **New video check questions for the Martin-Gay Interactive Lecture videos** are now included in every section for each learning objective. **These exercises are all available for assignment in MyMathLab** and are a great way to assess whether students have viewed and understood the key concepts presented in the videos.

- **New Student Success Tips Videos** are 3- to 5-minute video segments designed to be daily reminders to students to continue practicing and maintaining good organizational and study habits. They are organized in three categories and are available in MyMathLab and the Interactive Lecture Series. The categories are:

  1. Success Tips that apply to any course in college in general, such as Time Management.

  2. Success Tips that apply to any mathematics course. One example is based on understanding that mathematics is a course that requires homework to be completed in a timely fashion.

**3.** Section- or Content-specific Success Tips to help students avoid common mistakes or to better understand concepts that often prove challenging. One example of this type of tip is how to apply the order of operations to simplify an expression.

● **Interactive DVD Lecture Series**, featuring your text author (Elayn Martin-Gay), provides students with active learning at their own pace. The videos offer the following resources and more:

   **A complete lecture for each section of the text** highlights key examples and exercises from the text. "Pop-ups" reinforce key terms, definitions, and concepts.

   **An interface with menu navigation features** allows students to quickly find and focus on the examples and exercises they need to review.

   **Interactive Concept Check** exercises measure students' understanding of key concepts and common trouble spots.

   **New Student Success Tips Videos**.

● **The Interactive DVD Lecture Series** also includes the following resources for test prep:

   **The Chapter Test Prep Videos** help students during their most teachable moment–when they are preparing for a test. This innovation provides step-by-step solutions for the exercises found in each Chapter Test. For the Fifth Edition, the chapter test prep videos are also available on YouTube™. The videos are captioned in English and Spanish.

   **The Practice Final Exam Videos** help students prepare for an end-of-course final. Students can watch full video solutions to each exercise in the Practice Final Exam at the end of this text.

● **The Martin-Gay MyMathLab** course has been updated and revised to provide more exercise coverage, including assignable video check questions and an expanded video program. There are section lecture videos for every section, which students can also access at the specific objective level; Student Success Tips Videos; and an increased number of watch clips at the exercise level to help students while doing homework in MathXL. Suggested homework assignments have been premade for assignment at the instructor's discretion.

● **New MyMathLab Ready to Go courses** (access code required) provide students with all the same great MyMathLab features that you're used to, but make it easier for instructors to get started. Each course includes preassigned homework and quizzes to make creating your course even simpler. Ask your Pearson representative about the details for this particular course or to see a copy of this course.

## Key Pedagogical Features

The following key features have been retained and/or updated for the Fifth Edition of the text:

**Problem-Solving Process**   This is formally introduced in Chapter 2 with a four-step process that is integrated throughout the text. The four steps are **Understand, Translate, Solve,** and **Interpret**. The repeated use of these steps in a variety of examples shows their wide applicability. Reinforcing the steps can increase students' comfort level and confidence in tackling problems.

**Exercise Sets Revised and Updated**   The exercise sets have been carefully examined and extensively revised. Special focus was placed on making sure that even- and odd-numbered exercises are paired and that real-life applications are updated.

**Examples**   Detailed, step-by-step examples were added, deleted, replaced, or updated as needed. Many examples reflect real life. Additional instructional support is provided in the annotated examples.

**Practice Exercises**   Throughout the text, each worked-out example has a parallel Practice exercise. These invite students to be actively involved in the learning process. Students should try each Practice exercise after finishing the corresponding example. Learning by doing will help students grasp ideas before moving on to other concepts. Answers to the Practice exercises are provided at the bottom of each page.

**Helpful Hints**   Helpful Hints contain practical advice on applying mathematical concepts. Strategically placed where students are most likely to need immediate reinforcement, Helpful Hints help students avoid common trouble areas and mistakes.

**Concept Checks**   This feature allows students to gauge their grasp of an idea as it is being presented in the text. Concept Checks stress conceptual understanding at the point of use and help suppress misconceived notions before they start. Answers appear at the bottom of the page. Exercises related to Concept Checks are included in the exercise sets.

**Mixed Practice Exercises**   In the section exercise sets, these exercises require students to determine the problem type and strategy needed to solve it just as they would need to do on a test.

**Integrated Reviews**   This unique, mid-chapter exercise set helps students assimilate new skills and concepts that they have learned separately over several sections. These reviews provide yet another opportunity for students to work with "mixed" exercises as they master the topics.

**Vocabulary Check**   This feature provides an opportunity for students to become more familiar with the use of mathematical terms as they strengthen their verbal skills. These appear at the end of each chapter before the Chapter Highlights. Vocabulary, Readiness & Video exercises provide practice at the section level.

**Chapter Highlights**   Found at the end of every chapter, these contain key definitions and concepts with examples to help students understand and retain what they have learned and help them organize their notes and study for tests.

**Chapter Review**   The end of every chapter contains a comprehensive review of topics introduced in the chapter. The Chapter Review offers exercises keyed to every section in the chapter as well as Mixed Review exercises that are not keyed to sections.

**Chapter Test and Chapter Test Prep Videos**   The Chapter Test is structured to include those problems that involve common student errors. The **Chapter Test Prep Videos** give students instant access to a step-by-step video solution of each exercise in the Chapter Test.

**Cumulative Review**   This review follows every chapter in the text (except Chapter 1). Each odd-numbered exercise contained in the Cumulative Review is an earlier worked example in the text that is referenced in the back of the book along with the answer.

**Writing Exercises**   These exercises occur in almost every exercise set and require students to provide a written response to explain concepts or justify their thinking.

**Applications** Real-world and real-data applications have been thoroughly updated, and many new applications are included. These exercises occur in almost every exercise set and show the relevance of mathematics and help students gradually and continuously develop their problem-solving skills.

**Review Exercises** These exercises occur in each exercise set (except in Chapter 1) and are keyed to earlier sections. They review concepts learned earlier in the text that will be needed in the next section or chapter.

**Exercise Set Resource Icons** Located at the opening of each exercise set, these icons remind students of the resources available for extra practice and support:

See Student Resources descriptions on page xii for details on the individual resources available.

**Exercise Icons** These icons facilitate the assignment of specialized exercises and let students know what resources can support them.

- ◉ DVD Video icon: exercise worked on the Interactive DVD Lecture Series and available in MyMathLab.
- △ Triangle icon: identifies exercises involving geometric concepts.
- ✎ Pencil icon: indicates a written response is needed.
- ▦ Calculator icon: optional exercises intended to be solved using a scientific or graphing calculator.

**Group Activities** Found at the end of each chapter, these activities are for individual or group completion and are usually hands-on or data-based activities that extend the concepts found in the chapter, allowing students to make decisions and interpretations and to think and write about algebra.

**Optional: Calculator Exploration Boxes and Calculator Exercises** The optional Calculator Explorations provide keystrokes and exercises at appropriate points to give students an opportunity to become familiar with these tools. Section exercises that are best completed by using a calculator are identified by ▦ for ease of assignment.

Student and Instructor Resources

## STUDENT RESOURCES

| **Interactive DVD Lecture Series Videos** | **Video Organizer** | **Student Solutions Manual** |
|---|---|---|
| Provides students with active learning at their pace. The videos offer:<br><br>• A complete lecture for each text section. The interface allows easy navigation to examples and exercises students need to review.<br>• Interactive Concept Check exercises<br>• Student Success Tips Videos<br>• Practice Final Exam<br>• Chapter Test Prep Videos | Designed to help students take notes and work practice exercises while watching the Interactive Lecture Series Videos.<br><br>• Covers all of the video examples in order.<br>• Provides ample space for students to write down key definitions and rules.<br>• Includes "Play" and "Pause" button icons to prompt students to follow along with the author for some exercises while they try others on their own.<br><br>Available in loose-leaf, notebook-ready format and in MyMathLab. Answers to exercises available to instructors in MyMathLab. | Provides completely worked-out solutions to the odd-numbered section exercises; all exercises in the Integrated Reviews, Chapter Reviews, Chapter Tests, and Cumulative Reviews |

## INSTRUCTOR RESOURCES

| **Annotated Instructor's Edition** | **Instructor's Resource Manual with Tests and Mini-Lectures** |
|---|---|
| Contains all the content found in the student edition plus the following:<br><br>• Answers to exercises on the same text page<br>• Teaching Tips throughout the text placed at key points<br>• Video Answer Section | • Mini-lectures for each text section<br>• Additional practice worksheets for each section<br>• Several forms of test per chapter—free response and multiple choice<br>• Answers to all items<br><br>**Instructor's Solutions Manual TestGen®** (Available for download from the IRC) |
| Instructor-to-Instructor Videos—available in the Instructor Resources section of the MyMathLab course. | **Online Resources**<br>**MyMathLab®** (access code required)<br><br><br>**MathXL®** (access code required) |

## Acknowledgments

There are many people who helped me develop this text, and I will attempt to thank some of them here. Cindy Trimble was *invaluable* for contributing to the overall accuracy of the text. Dawn Nuttall was *invaluable* for her many suggestions and contributions during the development and writing of this Fifth Edition. Allison Campbell and Lauren Morse provided guidance throughout the production process.

A very special thank you goes to my editor, Mary Beckwith, for being there 24/7/365, as my students say. And my thanks to the staff at Pearson for all their support: Heather Scott, Patty Bergin, Matt Summers, Michelle Renda, Rachel Ross Michael Hirsch, Chris Hoag, and Greg Tobin.

I would like to thank the following reviewers for their input and suggestions:

Rosalie Abraham, *Florida Community College—Jacksonville*
Ana Bacica, *Brazosport College*
Nelson Collins, *Joliet Junior College*
Nancy Desilet, *Carroll Community College*
Elizabeth Eagle, *University of North Carolina—Charlotte*
Dorothy French, *Community College of Philadelphia*
Sharda Gudehithla, *Wilbur Wright College*
Pauline Hall, *Iowa State University*
Debra R. Hill, *University of North Carolina—Charlotte*

Glenn Jablonski, *Triton College*
Sue Kellicut, *Seminole State College*
Jean McArthur, *Joliet Junior College*
Mary T. McMahon, *North Central College*
Owen Mertens, *Missouri State University*
Jeri Rogers, *Seminole State College*
William Stammerman, *Des Moines Area Community College*
Patrick Stevens, *Joliet Junior College*
Arnavaz Taraporevala, *New York City College of Technology*

I would also like to thank the following dedicated group of instructors who participated in our focus groups, Martin-Gay Summits, and our design review for the series. Their feedback and insights have helped to strengthen this edition of the text. These instructors include:

Billie Anderson, *Tyler Junior College*
Cedric Atkins, *Mott Community College*
Lois Beardon, *Schoolcraft College*
Laurel Berry, *Bryant & Stratton College*
John Beyers, *University of Maryland*
Bob Brown, *Community College of Baltimore County–Essex*
Lisa Brown, *Community College of Baltimore County–Essex*
NeKeith Brown, *Richland College*
Gail Burkett, *Palm Beach State College*
Cheryl Cantwell, *Seminole State College*
Ivette Chuca, *El Paso Community College*
Jackie Cohen, *Augusta State College*
Julie Dewan, *Mohawk Valley Community College*
Monette Elizalde, *Palo Alto College*
Kiel Ellis, *Delgado Community College*
Janice Ervin, *Central Piedmont Community College*
Richard Fielding, *Southwestern College*

Dena Frickey, *Delgado Community College*
Cindy Gaddis, *Tyler Junior College*
Gary Garland, *Tarrant County Community College*
Kim Ghiselin, *State College of Florida*
Nita Graham, *St. Louis Community College*
Kim Granger, *St. Louis Community College*
Pauline Hall, *Iowa State University*
Pat Hussey, *Triton College*
Dorothy Johnson, *Lorain County Community College*
Sonya Johnson, *Central Piedmont Community College*
Ann Jones, *Spartanburg Community College*
Irene Jones, *Fullerton College*
Paul Jones, *University of Cincinnati*
Mike Kirby, *Tidewater Community College*
Kathy Kopelousos, *Lewis and Clark Community College*

Tara LaFrance, *Delgado Community College*

John LaMaster, *Indiana Purdue University Fort Wayne*

Nancy Lange, *Inver Hills Community College*

Judy Langer, *Westchester Community College*

Lisa Lindloff, *McLennan Community College*

Sandy Lofstock, *St. Petersburg College*

Kathy Lovelle, *Westchester Community College*

Nicole Mabine, *North Lake College*

Jean McArthur, *Joliet Junior College*

Kevin McCandless, *Evergreen Valley College*

Ena Michael, *State College of Florida*

Daniel Miller, *Niagara County Community College*

Marcia Molle, *Metropolitan Community College*

Carol Murphy, *San Diego Miramar College*

Charlotte Newsom, *Tidewater Community College*

Cao Nguyen, *Central Piedmont Community College*

Greg Nguyen, *Fullerton College*

Eric Oilila, *Jackson Community College*

Linda Padilla, *Joliet Junior College*

Armando Perez, *Laredo Community College*

Davidson Pierre, *State College of Florida*

Marilyn Platt, *Gaston College*

Chris Riola, *Moraine Valley Community College*

Carole Shapero, *Oakton Community College*

Janet Sibol, *Hillsborough Community College*

Anne Smallen, *Mohawk Valley Community College*

Barbara Stoner, *Reading Area Community College*

Jennifer Strehler, *Oakton Community College*

Ellen Stutes, *Louisiana State University Eunice*

Tanomo Taguchi, *Fullerton College*

Robyn Toman, *Anne Arundel Community College*

MaryAnn Tuerk, *Elgin Community College*

Walter Wang, *Baruch College*

Leigh Ann Wheeler, *Greenville Technical Community College*

Darlene Williams, *Delgado Community College*

Valerie Wright, *Central Piedmont Community College*

A special thank you to those students who participated in our design review: Katherine Browne, Mike Bulfin, Nancy Canipe, Ashley Carpenter, Jeff Chojnachi, Roxanne Davis, Mike Dieter, Amy Dombrowski, Kay Herring, Todd Jaycox, Kaleena Levan, Matt Montgomery, Tony Plese, Abigail Polkinghorn, Harley Price, Eli Robinson, Avery Rosen, Robyn Schott, Cynthia Thomas, and Sherry Ward.

*Elayn Martin-Gay*

## About the Author

Elayn Martin-Gay has taught mathematics at the University of New Orleans for more than 25 years. Her numerous teaching awards include the local University Alumni Association's Award for Excellence in Teaching, and Outstanding Developmental Educator at University of New Orleans, presented by the Louisiana Association of Developmental Educators.

Prior to writing textbooks, Elayn Martin-Gay developed an acclaimed series of lecture videos to support developmental mathematics students in their quest for success. These highly successful videos originally served as the foundation material for her texts. Today, the videos are specific to each book in the Martin-Gay series. The author has also created Chapter Test Prep Videos to help students during their most "teachable moment"—as they prepare for a test—along with Instructor-to-Instructor videos that provide teaching tips, hints, and suggestions for each developmental mathematics course, including basic mathematics, prealgebra, beginning algebra, and intermediate algebra.

Elayn is the author of 12 published textbooks as well as multimedia, interactive mathematics, all specializing in developmental mathematics courses. She has also published series in Algebra 1, Algebra 2, and Geometry. She has participated as an author across the broadest range of educational materials: textbooks, videos, tutorial software, and courseware. This provides an opportunity of various combinations for an integrated teaching and learning package that uniquely offers great consistency for the student.

# Applications Index

## Agriculture

## Animals

## Astronomy and space

## Automotive/motor vehicles

## Aviation

## Business

## Chemistry/physics

# Real Numbers and Algebraic Expressions

## A Selection of Resources for Success in This Mathematics Course

**Textbook**

**Instructor**

**MyMathLab and MathXL**

**Video Organizer**

**Interactive Lecture Series**

For more information about the resources illustrated above, read Section 1.1.

In arithmetic, we add, subtract, multiply, divide, raise to powers, and take roots of numbers. In algebra, we add, subtract, multiply, divide, raise to powers, and take roots of variables. Letters, such as $x$, that represent numbers are called variables. Understanding algebraic expressions made up of combinations of variables and numbers depends on your understanding of arithmetic expressions. This chapter reviews the arithmetic operations on real numbers and the corresponding algebraic expressions. After this review, we will be prepared to explore how widely useful these algebraic expressions are for problem solving in diverse situations.

## 1.1 Study Skill Tips for Success in Mathematics

Before reading Section 1.1, you might want to ask yourself a few questions.

**1.** When you took your last math course, were you organized? Were your notes and materials from that course easy to find, or were they disorganized and hard to find—if you saved them at all?

**2.** Were you satisfied—really satisfied—with your performance in that course? In other words, do you feel that your outcome represented your best effort?

If the answer is "no" to these questions, then it is time to make a change. Changing to or resuming good study skill habits is not a process you can start and stop as you please. It is something that you must remember and practice each and every day. To begin, continue reading this section.

### Objective A Getting Ready for This Course

Now that you have decided to take this course, remember that a *positive attitude* will make all the difference in the world. Your belief that you can succeed is just as important as your commitment to this course. Make sure you are ready for this course by having the time and positive attitude that it takes to succeed.

Make sure that you are familiar with the way that this course is being taught. Is it a traditional course, in which you have a printed textbook and meet with an instructor? Is it taught totally online, and your textbook is electronic and you e-mail your instructor? Or is your course structured somewhere in between these two methods? (Not all of the tips that follow will apply to all forms of instruction.)

Also make sure that you have scheduled your math course for a time that will give you the best chance for success. For example, if you are also working, you may want to check with your employer to make sure that your work hours will not conflict with your course schedule.

On the day of your first class period, double-check your schedule and allow yourself extra time to arrive on time in case of traffic problems or difficulty locating your classroom. Make sure that you are aware of and bring all necessary class materials.

### Objective B General Tips for Success

Below are some general tips that will increase your chance for success in a mathematics class. Many of these tips will also help you in other courses you may be taking.

*Most important! Organize your class materials.* In the next couple pages, many ideas will be presented to help you organize your class materials—notes, any handouts, completed homework, previous tests, and so on. In general, you MUST have these materials organized. All of them will be valuable references throughout your course and when studying for upcoming tests and the final exam. One way to make sure you can locate these materials when you need them is to use a three-ring binder. This binder should be used solely for your mathematics class and should be brought to each and every class or lab. This way, any material can be immediately inserted in a section of this binder and will be there when you need it.

*Form study groups and/or exchange names and e-mail addresses.* Depending on how your course is taught, you may want to keep in contact with your fellow students. One way of doing this is to form a study group—whether in person or through the Internet. Also, you may want to ask if anyone is interested in exchanging e-mail addresses or any other form of contact.

> **Helpful Hint**
>
> **MyMathLab® and MathXL®**
> When assignments are turned in online, keep a hard copy of your complete written work. You will need to refer to your written work to be able to ask questions and to study for tests later.

*Choose to attend all class periods.* If possible, sit near the front of the classroom. This way, you will see and hear the presentation better. It may also be easier for you to participate in classroom activities.

*Do your homework.* You've probably heard the phrase "practice makes perfect" in relation to music and sports. It also applies to mathematics. You will find that the more time you spend solving mathematics exercises, the easier the process becomes. Be sure to schedule enough time to complete your assignments before the due date assigned by your instructor.

*Check your work.* Review the steps you took while working a problem. Learn to check your answers in the original exercises. You may also compare your answers with the "Answers to Selected Exercises" section in the back of the book. If you have made a mistake, try to figure out what went wrong. Then correct your mistake. If you can't find what went wrong, **don't** erase your work or throw it away. Show your work to your instructor, a tutor in a math lab, or a classmate. It is easier for someone to find where you had trouble if he or she looks at your original work.

*Learn from your mistakes and be patient with yourself.* Everyone, even your instructor, makes mistakes. (That definitely includes me—Elayn Martin-Gay.) Use your errors to learn and to become a better math student. The key is finding and understanding your errors.

Was your mistake a careless one, or did you make it because you can't read your own math writing? If so, try to work more slowly or write more neatly and make a conscious effort to carefully check your work.

Did you make a mistake because you don't understand a concept? Take the time to review the concept or ask questions to better understand it.

Did you skip too many steps? Skipping steps or trying to do too many steps mentally may lead to preventable mistakes.

*Know how to get help if you need it.* It's all right to ask for help. In fact, it's a good idea to ask for help whenever there is something that you don't understand. Make sure you know when your instructor has office hours and how to find his or her office. Find out whether math tutoring services are available on your campus. Check on the hours, location, and requirements of the tutoring service.

*Don't be afraid to ask questions.* You are not the only person in class with questions. Other students are normally grateful that someone has spoken up.

*Turn in assignments on time.* This way, you can be sure that you will not lose points for being late. Show every step of a problem and be neat and organized. Also be sure that you understand which problems are assigned for homework. If allowed, you can always double-check the assignment with another student in your class.

## Objective C Knowing and Using Your Text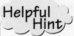

Flip through the pages of this text or view the e-text pages on a computer screen. Start noticing examples, exercise sets, end-of-chapter material, and so on. Every text is organized in some manner. Learn the way this text is organized by reading about and then finding an example in your text of each type of resource listed below. Finding and using these resources throughout your course will increase your chance of success.

- *Practice Exercises.* Each example in every section has a parallel Practice exercise. As you read a section, try each Practice exercise after you've finished the corresponding example. Answers are at the bottom of the page. This "learn-by-doing" approach will help you grasp ideas before you move on to other concepts.

- *Symbols at the Beginning of an Exercise Set.* If you need help with a particular section, the symbols listed at the beginning of each exercise set will remind you of the resources available.

- *Objectives.* The main section of exercises in each exercise set is referenced by an objective, such as **A** or **B**, and also an example(s). There is also often a section of exercises titled "Mixed Practice," which is referenced by two or more objectives or sections. These are mixed exercises written to prepare you for your next exam. Use all of this referencing if you have trouble completing an assignment from the exercise set.

**Helpful Hint**

**MyMathLab®** and **MathXL®**
If you are doing your homework online, you can work and re-work those exercises that you struggle with until you master them. Try working through all the assigned exercises twice before the due date.

**Helpful Hint**

**MyMathLab®** and **MathXL®**
If you are completing your homework online, it's important to work each exercise on paper before submitting the answer. That way, you can check your work and follow your steps to find and correct any mistakes.

**Helpful Hint**

**MyMathLab®** and **MathXL®**
Be aware of assignments and due dates set by your instructor. Don't wait until the last minute to submit work online.

- *Icons (Symbols).* Make sure that you understand the meaning of the icons that are beside many exercises. ▶ tells you that the corresponding exercise may be viewed on the video Lecture Series that corresponds to that section. ⬉ tells you that this exercise is a writing exercise in which you should answer in complete sentences. △ tells you that the exercise involves geometry.

- *Integrated Reviews.* Found in the middle of each chapter, these reviews offer you a chance to practice—in one place—the many concepts that you have learned separately over several sections.

- *End-of-Chapter Opportunities.* There are many opportunities at the end of each chapter to help you understand the concepts of the chapter.

  **Vocabulary Checks** contain key vocabulary terms introduced in the chapter.

  **Chapter Highlights** contain chapter summaries and examples.

  **Chapter Reviews** contain review problems. The first part is organized section by section, and the second part contains a set of mixed exercises.

  **Chapter Tests** are sample tests to help you prepare for an exam. The Chapter Test Prep Videos found in the Interactive Lecture Series, MyMathLab, and YouTube provide the video solution to each question on each Chapter Test.

  **Cumulative Reviews** start at Chapter 2 and are reviews consisting of material from the beginning of the book to the end of that particular chapter.

- *Student Resources in Your Textbook.* You will find a **Student Resources** section at the back of this textbook. It contains the following to help you study and prepare for tests:

  **Study Skill Builders** contain study skills advice. To increase your chance for success in the course, read these study tips, and answer the questions.

  **Bigger Picture—Study Guide Outline** provides you with a study guide outline of the course, with examples.

  **Practice Final** provides you with a Practice Final Exam to help you prepare for a final.

  **Answers to Selected Exercises** – described below.

- *Resources to Check Your Work.* The **Answers to Selected Exercises** section provides answers to all odd-numbered section exercises and to all Integrated Review, Chapter Test, and Cumulative Review exercises. Answers to the margin Practice exercises are written at the bottom of the page.

## Objective D Knowing and Using Video and Notebook Organizer Resources ▶

### Video Resources

Below is a list of video resources that are all made by me—the author of your text, Elayn Martin-Gay. By making these videos, I can be sure that the methods presented are consistent with those in the text.

- *Interactive DVD Lecture Series.* Exercises marked with a ▶ are fully worked out by the author on the DVDs and within MyMathLab. The Lecture Series provides Section Lecture Videos that contain approximately 20 minutes of instruction per section and are organized by objective.

- *Chapter Test Prep Videos.* These videos provide solutions to all of the Chapter Test exercises worked out by the author. They can be found in MyMathLab, the Interactive Lecture Series, and YouTube. This supplement is very helpful before a test or exam.

- *Student Success Tips.* These video segments are about 3 minutes long and are daily reminders to help you continue practicing and maintaining good organizational and study habits.

**Helpful Hint**

**MyMathLab®**

In MyMathLab, you have access to the following video resources:

- Lecture Videos for each section
- Chapter Test Prep Videos
- Final Exam Videos

Use these videos provided by the author to prepare for class, review, and study for tests.

- *Final Exam Videos.* These video segments provide solutions to all of the Practice Final Exam exercises worked out by the author. These videos can be found within MyMathLab and the Interactive Lecture Series.

## Notebook Organizer Resources

The resources below are in three-ring notebook ready form. They are to be inserted in a three-ring binder and completed. Both resources are numbered according to the sections in your text to which they refer.

- *Video Organizer.* This organizer is closely tied to the Interactive Lecture (Video) Series. Each section should be completed while watching the lecture video on the same section. Once completed, you will have a set of notes to accompany the Lecture (Video) Series section by section.
- *Student Organizer.* This organizer helps you study effectively through note-taking hints, practice, and homework while referencing examples in the text and examples in the Lecture Series.

## Objective E Getting Help

If you have trouble completing assignments or understanding the mathematics, get help as soon as you need it! This tip is presented as an objective on its own because it is so important. In mathematics, usually the material presented in one section builds on your understanding of the previous section. This means that if you don't understand the concepts covered during a class period, there is a good chance that you will not understand the concepts covered during the next class period. If this happens to you, get help as soon as you can.

Where can you get help? Many suggestions have been made in this section on where to get help, and now it is up to you to get it. Try your instructor, a tutoring center, or a math lab, or you may want to form a study group with fellow classmates. If you do decide to see your instructor or go to a tutoring center, make sure that you have a neat notebook and are ready with your questions.

> **Helpful Hint**
>
> **MyMathLab® and MathXL®**
>
> - Use the **Help Me Solve This** button to get step-by-step help for the exercise you are working. You will need to work an additional exercise of the same type before you can get credit for having worked it correctly.
> - Use the **Video** button to view a video clip of the author working a similar exercise.

## Objective F Preparing for and Taking an Exam

Make sure that you allow yourself plenty of time to prepare for a test. If you think that you are a little "math anxious," it may be that you are not preparing for a test in a way that will ensure success. The way that you prepare for a test in mathematics is important. To prepare for a test:

1. Review your previous homework assignments.
2. Review any notes from class and section-level quizzes you have taken. (If this is a final exam, also review chapter tests you have taken.)
3. Review concepts and definitions by reading the Chapter Highlights at the end of each chapter.
4. Practice working out exercises by completing the Chapter Review found at the end of each chapter. (If this is a final exam, go through a Cumulative Review. There is one found at the end of each chapter except Chapter 1. Choose the review found at the end of the latest chapter that you have covered in your course.) *Don't stop here!*
5. It is important that you place yourself in conditions similar to test conditions to find out how you will perform. In other words, as soon as you feel that you know the material, get a few blank sheets of paper and take a sample test. There is a Chapter Test available at the end of each chapter, or you can work selected problems from the Chapter Review. Your instructor may also provide you with a review sheet. During this sample test, do not use your notes or your textbook.

> **Helpful Hint**
>
> **MyMathLab® and MathXL®**
>
> Review your written work for previous assignments. Then go back and re-work previous assignments. Open a previous assignment, and click **Similar Exercise** to generate new exercises. Re-work the exercises until you fully understand them and can work them without help features.

Then check your sample test. If your sample test is the Chapter Test in the text, don't forget that the video solutions are in MyMathLab, the Interactive Lecture Series, and YouTube. If you are not satisfied with the results, study the areas that you are weak in and try again.

6. On the day of the test, allow yourself plenty of time to arrive at where you will be taking your exam.

When taking your test:

1. Read the directions on the test carefully.

2. Read each problem carefully as you take the test. Make sure that you answer the question asked.

3. Watch your time and pace yourself so that you can attempt each problem on your test.

4. If you have time, check your work and answers.

5. Do not turn your test in early. If you have extra time, spend it double-checking your work.

## Objective G  Managing Your Time

As a college student, you know the demands that classes, homework, work, and family place on your time. Some days you probably wonder how you'll ever get everything done. One key to managing your time is developing a schedule. Here are some hints for making a schedule:

1. Make a list of all of your weekly commitments for the term. Include classes, work, regular meetings, extracurricular activities, and so on. You may also find it helpful to list such things as laundry, regular workouts, grocery shopping, and so on.

2. Next, estimate the time needed for each item on the list. Also make a note of how often you will need to do each item. Don't forget to include time estimates for the reading, studying, and homework you do outside of your classes. You may want to ask your instructor for help estimating the time needed.

3. In the exercise set that follows, you are asked to block out a typical week on the schedule grid given. Start with items with fixed time slots like classes and work.

4. Next, include the items on your list with flexible time slots. Think carefully about how best to schedule items such as study time.

5. Don't fill up every time slot on the schedule. Remember that you need to allow time for eating, sleeping, and relaxing! You should also allow a little extra time in case some items take longer than planned.

6. If you find that your weekly schedule is too full for you to handle, you may need to make some changes in your workload, classload, or in other areas of your life. You may want to talk to your advisor, manager or supervisor at work, or someone in your college's academic counseling center for help with such decisions.

## 1.1  Exercise Set  MyMathLab®

1. What is your instructor's name?

2. What are your instructor's office location and office hours?

3. What is the best way to contact your instructor?

4. Do you have the name and contact information of at least one other student in class?

5. Will your instructor allow you to use a calculator in this class?

6. Why is it important that you write step-by-step solutions to homework exercises and keep a hard copy of all work submitted?

**7.** Is there a tutoring service available on campus? If so, what are its hours? What services are available?

**8.** Have you attempted this course before? If so, write down ways that you might improve your chances of success during this next attempt.

**9.** List some steps that you can take if you begin having trouble understanding the material or completing an assignment. If you are completing your homework in MyMathLab® and MathXL®, list the resources you can use for help.

**10.** How many hours of studying does your instructor advise for each hour of instruction?

**11.** What does the ╲ icon in this text mean?

**12.** What does the △ icon in this text mean?

**13.** What does the ◗ icon in this text mean?

**14.** Search the minor columns in your text. What are Practice exercises?

**15.** When might be the best time to work a Practice exercise?

**16.** Where are the answers to Practice exercises?

**17.** What answers are contained in this text, and where are they?

**18.** What are Study Skill Tips of the Day, and where are they?

**19.** What and where are Integrated Reviews?

**20.** How many times is it suggested that you work through the homework exercises in MathXL® before the submission deadline?

**21.** How far in advance of the assigned due date is it suggested that homework be submitted online? Why?

**22.** Chapter Highlights are found at the end of each chapter. Find the Chapter 1 Highlights and explain how you might use it and how it might be helpful.

**23.** Chapter Reviews are found at the end of each chapter. Find the Chapter 1 Review and explain how you might use it and how it might be helpful.

**24.** Chapter Tests are found at the end of each chapter. Find the Chapter 1 Test and explain how you might use it and how it might be helpful when preparing for an exam on Chapter 1. Include how the Chapter Test Prep Videos may help. If you are working in MyMathLab® and MathXL®, how can you use previous homework assignments to study?

**25.** What is the Video Organizer? Explain the contents and how it might be used.

**26.** What is the Student Organizer? Explain the contents and how it might be used.

**27.** Read or reread objective **G** and fill out the schedule grid on the next page.

|  | Monday | Tuesday | Wednesday | Thursday | Friday | Saturday | Sunday |
|---|---|---|---|---|---|---|---|
| 4:00 a.m. |  |  |  |  |  |  |  |
| 5:00 a.m. |  |  |  |  |  |  |  |
| 6:00 a.m. |  |  |  |  |  |  |  |
| 7:00 a.m. |  |  |  |  |  |  |  |
| 8:00 a.m. |  |  |  |  |  |  |  |
| 9:00 a.m. |  |  |  |  |  |  |  |
| 10:00 a.m. |  |  |  |  |  |  |  |
| 11:00 a.m. |  |  |  |  |  |  |  |
| 12:00 p.m. |  |  |  |  |  |  |  |
| 1:00 p.m. |  |  |  |  |  |  |  |
| 2:00 p.m. |  |  |  |  |  |  |  |
| 3:00 p.m. |  |  |  |  |  |  |  |
| 4:00 p.m. |  |  |  |  |  |  |  |
| 5:00 p.m. |  |  |  |  |  |  |  |
| 6:00 p.m. |  |  |  |  |  |  |  |
| 7:00 p.m. |  |  |  |  |  |  |  |
| 8:00 p.m. |  |  |  |  |  |  |  |
| 9:00 p.m. |  |  |  |  |  |  |  |
| 10:00 p.m. |  |  |  |  |  |  |  |
| 11:00 p.m. |  |  |  |  |  |  |  |
| Midnight |  |  |  |  |  |  |  |
| 1:00 a.m. |  |  |  |  |  |  |  |
| 2:00 a.m. |  |  |  |  |  |  |  |
| 3:00 a.m. |  |  |  |  |  |  |  |

## **1.2**  Algebraic Expressions and Sets of Numbers

### Objectives

**A** Identify Natural Numbers, Whole Numbers, Integers, Rational Numbers, and Irrational Real Numbers.

**B** Write Phrases as Algebraic Expressions.

Recall that letters that represent numbers are called **variables.** An **algebraic expression** (or simply **expression**) is formed by numbers and variables connected by the operations of addition, subtraction, multiplication, division, raising to powers, or taking roots. For example,

$$2x, \qquad \frac{x+5}{6}, \qquad \sqrt{y} - 1.6, \qquad \text{and} \qquad z^3$$

are algebraic expressions or, more simply, expressions. (Recall that the expression $2x$ means $2 \cdot x$.)

Algebraic expressions occur often during problem solving. For example, the average cost to own and operate a car in the United States for 2013 was $0.608 per mile.

The expression $0.608m$ gives the total cost to operate a car annually for $m$ miles. (*Source:* AAA)

To find the cost of driving a car for 12,000 miles, for example, we replace the variable $m$ with 12,000 and perform the indicated operation. This process is called **evaluating** an expression, and the result is called the **value** of the expression for the given replacement value.

In our example, when $m = 12{,}000$,

$$0.608m = 0.608(12{,}000) = 7296$$

Thus, it costs $7296 to own and operate a car for 12,000 miles in 2013.

When evaluating an expression to solve a problem, we often need to think about the kind of number that is appropriate for the solution. For example, if we are asked to determine the maximum number of parking spaces for a parking lot to be constructed, an answer of $98\frac{1}{10}$ is not appropriate because $\frac{1}{10}$ of a parking space is not realistic.

We practice writing algebraic expressions in Objective **B** and evaluating algebraic expressions in Section 1.5.

## Objective A Identifying Common Sets of Numbers

Let's review some common sets of numbers and their graphs on a **number line.** To construct a number line, we draw a line and label a point 0 with which we associate the number 0. This point is called the **origin.** If we choose a point to the right of 0 and label it 1, the distance from 0 to 1 is called the **unit distance** and can be used to locate more points. The **positive numbers** lie to the right of the origin, and the **negative numbers** lie to the left of the origin. The number 0 is neither positive nor negative.

✓**Concept Check** Use the definitions of positive numbers, negative numbers, and zero to describe the meaning of *nonnegative numbers*.

A number is **graphed** on a number line by shading the point on the number line that corresponds to the number. Some common sets of numbers and their graphs are shown next.

### Identifying Numbers

Natural numbers: $\{1, 2, 3, \dots\}$

Whole numbers: $\{0, 1, 2, 3, \dots\}$

Integers: $\{\dots, -3, -2, -1, 0, 1, 2, 3, \dots\}$

Each listing of three dots, . . . , is called an **ellipsis** and means to continue in the same pattern.

A **set** is a collection of objects. The objects of a set are called its **members** or **elements.** When the elements of a set are listed, such as those displayed in the box on the previous page, the set is written in **roster** form. A set can also be written in **set builder notation,** which describes the members of the set but does not list them. The following set is written in set builder notation.

$$\{x \mid x \text{ is a natural number less than 3}\}$$

The set of all $x$     such that     $x$ is a natural number less than 3

This same set written in roster form is $\{1, 2\}$.

A set that contains *no* elements is called the **empty set** or **null set,** symbolized by $\{\ \}$ or $\varnothing$.

$$\{x \mid x \text{ is a month with 32 days}\} \text{ is } \varnothing \text{ or } \{\ \}$$

because no month has 32 days. The set has no elements.

## Practice 1–2

Write each set in roster form.

**1.** $\{x \mid x$ is a whole number between 0 and 4$\}$

**2.** $\{x \mid x$ is a natural number greater than 80$\}$

**Examples**   Write each set in roster form.

**1.** $\{x \mid x$ is a whole number between 1 and 6$\}$
   $\{2, 3, 4, 5\}$

**2.** $\{x \mid x$ is a natural number greater than 100$\}$
   $\{101, 102, 103, \dots\}$

■ Work Practice 1–2

The symbol $\in$ is used to denote that an element is in a particular set. The symbol $\in$ is read as "is an element of." For example, the true statement "3 is an element of $\{1, 2, 3, 4, 5\}$" can be written in symbols as

$$3 \in \{1, 2, 3, 4, 5\}$$

The symbol $\notin$ is read as "is not an element of." In symbols, we write the true statement "$p$ is not an element of $\{a, 5, g, j, q\}$" as

$$p \notin \{a, 5, g, j, q\}$$

## Practice 3–4

Determine whether each statement is true or false.

**3.** $0 \in \{x \mid x$ is a natural number$\}$

**4.** $9 \notin \{4, 6, 8, 10\}$

**Examples**   Determine whether each statement is true or false.

**3.** $3 \in \{x \mid x$ is a natural number$\}$    True, since 3 is a natural number and therefore an element of the set.

**4.** $7 \notin \{1, 2, 3\}$    True, since 7 is not an element of the set $\{1, 2, 3\}$.

■ Work Practice 3–4

We can use set builder notation to describe three other common sets of numbers.

## Identifying Numbers

Real numbers: $\{x \mid x$ corresponds to a point on the number line$\}$

Rational numbers: $\left\{ \dfrac{a}{b} \,\middle|\, a \text{ and } b \text{ are integers and } b \neq 0 \right\}$

Irrational numbers: $\{x \mid x$ is a real number and $x$ is not a rational number$\}$

Notice that every integer is also a rational number since each integer can be written as the quotient of itself and 1:

$$3 = \frac{3}{1}, \qquad 0 = \frac{0}{1}, \qquad -8 = \frac{-8}{1}$$

Not every rational number, however, is an integer. The rational number $\frac{2}{3}$, for example, is not an integer. Some square roots are rational numbers, and some are irrational numbers. For example, $\sqrt{2}$, $\sqrt{3}$, and $\sqrt{7}$ are irrational numbers, but $\sqrt{25}$ is a rational number because $\sqrt{25} = 5 = \frac{5}{1}$. The number $\pi$ is an irrational number. To help you make the distinction between rational and irrational numbers, here are a few examples of each. (Note: We study square roots further in Section 1.4 and in Chapter 7.)

| Rational Numbers | | Irrational Numbers |
|---|---|---|
| Number | Equivalent Quotient of Integers, $\frac{a}{b}$ | |
| $-\frac{2}{3}$ $=$ | $\frac{-2}{3}$ or $\frac{2}{-3}$ | $\sqrt{5}$ |
| $\sqrt{36}$ $=$ | $\frac{6}{1}$ | $\frac{\sqrt{6}}{7}$ |
| $5$ $=$ | $\frac{5}{1}$ | $-\sqrt{3}$ |
| $0$ $=$ | $\frac{0}{1}$ | $\pi$ |
| $1.2$ $=$ | $\frac{12}{10}$ | $\frac{2}{\sqrt{3}}$ |
| $3\frac{7}{8}$ $=$ | $\frac{31}{8}$ | |

Some rational and irrational numbers are graphed below.

Every rational number can be written as a decimal that either repeats or terminates. For example,

$$\frac{1}{2} = 0.5 \qquad\qquad \frac{5}{4} = 1.25$$

$$\frac{2}{3} = 0.6666666\ldots = 0.\overline{6} \qquad \frac{1}{11} = 0.090909\ldots = 0.\overline{09}$$

An irrational number written as a decimal neither terminates nor repeats. When we perform calculations with irrational numbers, we often use decimal approximations that have been rounded. For example, consider the following irrational numbers along with a four-decimal-place approximation of each:

$$\pi \approx 3.1416 \qquad \sqrt{2} \approx 1.4142$$

Earlier we mentioned that every integer is also a rational number. In other words, all the elements of the set of integers are also elements of the set of rational numbers. When this happens, we say that the set of integers, set $I$, is a **subset** of the set of rational numbers, set $Q$. In symbols,

$$\underbrace{I \subseteq Q}_{\text{is a subset of}} \qquad \textit{Note:} \text{ The symbol } \nsubseteq \text{ reads } \textbf{is not a subset of}. \text{ For example, } Q \nsubseteq I.$$

The natural numbers, whole numbers, integers, rational numbers, and irrational numbers are each a subset of the set of real numbers. The relationships among these sets of numbers are shown in the following diagram.

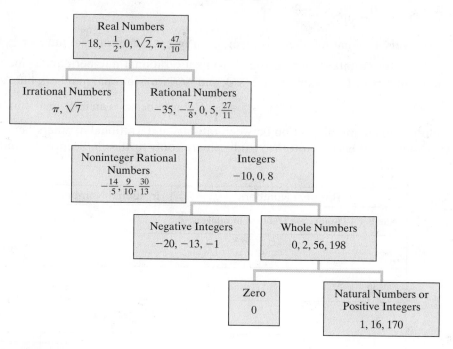

Copyright 2015 Pearson Education, Inc.

## Practice 5–7

Determine whether each statement is true or false.

**5.** 0 is a real number.

**6.** Every integer is a rational number.

**7.** $\sqrt{3}$ is a rational number.

**Examples** Determine whether each statement is true or false.

**5.** 3 is a real number.

True. Every whole number is a real number.

**6.** Every rational number is an integer.

False. The number $\frac{2}{3}$, for example, is a rational number, but it is not an integer.

**7.** $\frac{1}{5}$ is an irrational number.

False. The number $\frac{1}{5}$ is a rational number since it is in the form $\frac{a}{b}$ with $a$ and $b$ integers and $b \neq 0$.

■ **Work Practice 5–7**

## Objective B  Writing Phrases as Algebraic Expressions

Often, solving problems involves translating a phrase into an algebraic expression. The following is a list of key words and phrases and their translations.

| Selected Key Words/Phrases and Their Translations | | | |
|---|---|---|---|
| **Addition** | **Subtraction** | **Multiplication** | **Division** |
| sum | difference of | product | quotient |
| plus | minus | times | divide |
| added to | subtracted from | multiply | into |
| more than | less than | twice | ratio |
| increased by | decreased by | of | |
| total | less | | |

**Examples**  Write each phrase as an algebraic expression. Use the variable $x$ to represent each unknown number.

8. Eight times a number                                          $8 \cdot x$ or $8x$
9. Three more than eight times a number                          $8x + 3$
10. The quotient of a number and $-7$                             $x \div -7$ or $\dfrac{x}{-7}$
11. One and six-tenths subtracted from twice a number            $2x - 1.6$ or $2x - 1\dfrac{6}{10}$
12. Six less than a number                                       $x - 6$
13. Twice the sum of four and a number                           $2(4 + x)$

■ Work Practice 8–13

**Practice 8–13**

Write each phrase as an algebraic expression. Use the variable $x$ to represent each unknown number.

8. Seventeen times a number
9. Five more than six times a number
10. The quotient of six and a number
11. One-fourth subtracted from three times a number
12. Eleven less than a number
13. Three times the difference of a number and ten

**Answers**

**8.** $17 \cdot x$ or $17x$   **9.** $6x + 5$   **10.** $6 \div x$ or $\dfrac{6}{x}$

**11.** $3x - \dfrac{1}{4}$   **12.** $x - 11$   **13.** $3(x - 10)$

## Vocabulary, Readiness & Video Check

*Use the choices below to fill in each blank.*

| | | | |
|---|---|---|---|
| whole numbers | integers | rational number | empty |
| natural numbers | value | irrational number | set |
| set builder notation | expression | variables | roster form |

1. Letters that represent numbers are called _____.
2. Finding the _____ of an expression means evaluating the expression.
3. A set that contains no elements is called the _____ set.
4. A(n) _____ is formed by numbers and variables connected by operations such as addition, subtraction, multiplication, division, raising to powers, and/or taking roots.
5. The _____ are $\{1, 2, 3, \ldots\}$.
6. The _____ are $\{0, 1, 2, 3, \ldots\}$.
7. The _____ are $\{\ldots, -3, -2, -1, 0, 1, 2, 3, \ldots\}$.
8. The number $\sqrt{5}$ is a(n) _____.
9. The number $\dfrac{5}{7}$ is a(n) _____.
10. A(n) _____ is a collection of objects.
11. When the elements of a set are listed such as $\{1, 3, 5\}$, the set is written in _____.
12. When the elements of a set are described but not listed, such as $\{x \mid x$ is an odd number between 1 and 5$\}$, the set is written in _____.

*Martin-Gay Interactive Videos*  *Watch the section lecture video and answer the following questions.*

**Objective A** 13. Based on the lecture before ▣ Example 1, what is the relationship among real numbers, rational numbers, and irrational numbers? ▶

**Objective B** 14. From ▣ Example 5, why must we be careful when translating phrases dealing with subtraction? ▶

*See Video 1.2* 🍎

## 1.2 Exercise Set MyMathLab® ▶

**Objective A** *Write each set in roster form. See Examples 1 and 2.*

▶ **1.** $\{x \,|\, x$ is a natural number less than 6$\}$

**2.** $\{x \,|\, x$ is a natural number greater than 6$\}$

**3.** $\{x \,|\, x$ is a natural number between 10 and 17$\}$

**4.** $\{x \,|\, x$ is an odd natural number$\}$

**5.** $\{x \,|\, x$ is a whole number that is not a natural number$\}$

**6.** $\{x \,|\, x$ is a natural number less than 1$\}$

**7.** $\{x \,|\, x$ is an even whole number less than 9$\}$

**8.** $\{x \,|\, x$ is an odd whole number less than 9$\}$

*List the elements of the set* $\left\{ 3, 0, \sqrt{7}, \sqrt{36}, \dfrac{2}{5}, -134 \right\}$ *that are also elements of the given set. See Examples 3 through 7.*

▶ **9.** Whole numbers

**10.** Integers

**11.** Natural numbers

**12.** Rational numbers

**13.** Irrational numbers

**14.** Real numbers

*Place* $\in$ *or* $\notin$ *in the space provided to make each statement true. See Examples 3 through 7.*

▶ **15.** $-11$ ___ $\{x \,|\, x$ is an integer$\}$

**16.** $-6$ ___ $\{2, 4, 6, \dots\}$

**17.** $0$ ___ $\{x \,|\, x$ is a positive integer$\}$

**18.** $12$ ___ $\{1, 2, 3, \dots\}$

▶ **19.** $12$ ___ $\{1, 3, 5, \dots\}$

**20.** $\dfrac{1}{2}$ ___ $\{x \,|\, x$ is an irrational number$\}$

*Determine whether each statement is true or false. See Examples 5 through 7.*

**21.** Every whole number is a real number.

**22.** Every irrational number is a real number.

**23.** Some real numbers are irrational numbers.

**24.** Some real numbers are whole numbers.

**25.** Every whole number is a natural number.

**26.** Every irrational number is a rational number.

**27.** $\frac{1}{2}$ is a real number.

**28.** $-\frac{4}{5}$ is a real number.

**Mixed Practice** *Determine whether each statement is true or false. See Examples 3 through 7.*

**29.** $0 \in \{x \mid x \text{ is an integer}\}$

**30.** $0 \in \{x \mid x \text{ is a whole number}\}$

**31.** $\sqrt{7} \notin \{x \mid x \text{ is an irrational number}\}$

**32.** $-\frac{7}{11} \notin \{x \mid x \text{ is a rational number}\}$

**33.** $\{x \mid x \text{ is a day of the week starting with the letter B}\}$ is $\varnothing$.

**34.** $\{x \mid x \text{ is a rational number and an irrational number}\}$ is $\varnothing$.

**35.** Some real numbers are integers.

**36.** Every integer is a real number.

**37.** Some rational numbers are irrational numbers.

**38.** Every real number is also an integer.

**Objective** **B** *Write each phrase as an algebraic expression. Use the variable x to represent each unknown number. See Examples 8 through 13.*

**39.** Twice a number

**40.** Six times a number

▶ **41.** Ten less than a number

**42.** A number minus seven

**43.** The sum of a number and two

**44.** The difference of twenty-five and a number

**45.** A number divided by eleven

**46.** The quotient of a number and thirteen

**47.** Four subtracted from a number

**48.** Seventeen subtracted from a number

**49.** A number plus two and three-tenths

**50.** Fifteen and seven-tenths plus a number

**51.** A number less than one and one-third

**52.** Two and three-fourths less than a number

**53.** Nine times a number

**54.** Nine minus a number

**55.** Nine added to a number

**56.** Nine divided by a number

**57.** Five more than twice a number

**58.** One more than six times a number

**59.** Twelve minus three times a number

**60.** Four subtracted from three times a number

**61.** One plus twice a number

**62.** Three less than twice a number

**63.** Ten subtracted from five times a number

**64.** Four minus three times a number

**65.** The quotient of five and the difference of four and a number

**66.** The quotient of four and the sum of a number and one

**67.** Twice the sum of a number and three

**68.** Eight times the difference of a number and nine

*For Exercises 69 through 72, fill in the blank so that each is a true statement. There are many possible correct answers.*

**69.** _____ $\in \{1, 3, 5, 7\}$

**70.** _____ $\notin \{1, 3, 5, 7\}$

**71.** _____ $\subseteq \{1, 3, 5, 7\}$

**72.** _____ $\not\subseteq \{1, 3, 5, 7\}$

**73.** Name a whole number that is not a natural number.

**74.** Name a rational number that does not simplify to an integer.

## Concept Extensions

*The bar graph shows the life expectancy at birth for all Americans born in the years shown. Use the graph to calculate the increase in life expectancy over each ten-year period shown. (The first row has been completed for you.)*

**U.S. Life Expectancy at Birth**

Source: U.S. Center for Health Statistics

| Year | Increase in Life Expectancy (in years) from 10 Years Earlier |
|------|---------------------------------|
| 1950 | 2.9 |
| **75.** 1960 | |
| **76.** 1970 | |
| **77.** 1980 | |
| **78.** 1990 | |
| **79.** 2000 | |
| **80.** 2010 | |

**81.** In your own words, explain why every natural number is also a rational number but not every rational number is a natural number.

**82.** In your own words, explain why every irrational number is a real number but not every real number is an irrational number.

## 1.3 Equations, Inequalities, and Properties of Real Numbers

### Objective A Writing Sentences as Equations

When writing sentences as equations, we use the symbol = to translate the phrase **"is equal to."** All of the following key words and phrases also mean equality.

**Equality**

| | | | |
|---|---|---|---|
| equals | is/was | represents | is the same as |
| gives | yields | amounts to | is equal to |

**Examples** Write each sentence as an equation.

**1.** The sum of $x$ and 5 is 20

$$x + 5 = 20$$

**2.** The difference of 8 and $x$ is the same as the product of 2 and $x$.

$$8 - x = 2 \cdot x$$

**3.** The quotient of $z$ and 9 amounts to 9 plus $z$.

$$z \div 9 = 9 + z$$

or

$$\frac{z}{9} = 9 + z$$

Work Practice 1–3

**Objectives**

**A** Write Sentences as Equations.

**B** Use Inequality Symbols.

**C** Find the Opposite, or Additive Inverse, and the Reciprocal, or Multiplicative Inverse, of a Number.

**D** Identify and Use the Commutative, Associative, and Distributive Properties.

**Practice 1–3**

Write each sentence as an equation.

**1.** The difference of $x$ and 7 is 45.

**2.** The product of 5 and $x$ amounts to the sum of $x$ and 14.

**3.** The quotient of $y$ and 23 is the same as 20 subtracted from $y$.

**Answers**

**1.** $x - 7 = 45$  **2.** $5 \cdot x = x + 14$

**3.** $y \div 23 = y - 20$ or $\frac{y}{23} = y - 20$

## Objective B Using Inequality Symbols

If we want to write in symbols that two numbers are not equal, we can use the symbol $\neq$, which means **"is not equal to."** For example,

$$3 \neq 2$$

Graphing two numbers on a number line gives us a way to compare two numbers. For any two real numbers graphed on a number line, the number to the left is less than the number to the right. This means that the number to the right is greater than the number to the left.

The symbol $<$ means **"is less than."** Since $-4$ is to the left of $-1$ on the number line, we write $-4 < -1$. The symbol $>$ means **"is greater than."** Since $-1$ is to the right of $-4$ on the number line, we write $-1 > -4$.

$$-4 < -1 \text{ or } -1 > -4$$

```
 ←─┼──┼──┼──┼──┼──┼──┼──┼──┼──┼──→
   -5 -4 -3 -2 -1  0  1  2  3  4  5
```

Notice that since $-4 < -1$, then we also know that $-1 > -4$. This is true for any two numbers, say, $a$ and $b$.

If $a < b$, then also $b > a$. For example, since $-4 < -1$, then $-1 > -4$.

**Examples** Insert $<$, $>$, or $=$ between each pair of numbers to form a true statement.

**4.** $-5$ $\quad$ $5$ $\qquad$ $-5$ is to the left of $5$ on a number line, so $-5 < 5$.

**5.** $3$ $\quad$ $-7$ $\qquad$ $3$ is to the right of $-7$, so $3 > -7$.

**6.** $-16$ $\quad$ $-6$ $\qquad$ $-16$ is to the left of $-6$, so $-16 < -6$.

**7.** $-2.5$ $\quad$ $-2.1$ $\qquad$ $-2.5$ is to the left of $-2.1$, so $-2.5 < -2.1$.

**8.** $6.36$ $\quad$ $6.360$ $\qquad$ The true statement is $6.36 = 6.360$.

**9.** $4.3$ $\quad$ $-5.2$ $\qquad$ $4.3$ is to the right of $-5.2$, so $4.3 > -5.2$.

**10.** $\dfrac{5}{8}$ $\quad$ $\dfrac{3}{8}$ $\qquad$ The denominators are the same, so $\dfrac{5}{8} > \dfrac{3}{8}$ since $5 > 3$.

**11.** $\dfrac{2}{3}$ $\quad$ $\dfrac{3}{4}$ $\qquad$ By dividing, we see that $\dfrac{2}{3} = 0.666\ldots$ and $\dfrac{3}{4} = 0.75$.

$\qquad$ Thus $\dfrac{2}{3} < \dfrac{3}{4}$ since $0.666\ldots < 0.75$.

■ Work Practice 4–11

### Practice 4–11

Insert $<$, $>$, or $=$ between each pair of numbers to form a true statement.

**4.** $7$ $\quad$ $-7$

**5.** $-1$ $\quad$ $11$

**6.** $-10$ $\quad$ $-12$

**7.** $-3.25$ $\quad$ $-3.025$

**8.** $7.206$ $\quad$ $7.2060$

**9.** $18.6$ $\quad$ $-14.2$

**10.** $\dfrac{4}{7}$ $\quad$ $\dfrac{5}{7}$

**11.** $\dfrac{3}{8}$ $\quad$ $\dfrac{1}{3}$

**Helpful Hint**

When inserting the $>$ or $<$ symbol, think of the symbols as arrowheads that "point" toward the smaller number when the statement is true.

In addition to $<$ and $>$, there are the inequality symbols $\leq$ and $\geq$. The symbol $\leq$ means **"is less than or equal to,"** and the symbol $\geq$ means **"is greater than or equal to."**

**Answers**

**4.** $>$  **5.** $<$  **6.** $>$  **7.** $<$  **8.** $=$
**9.** $>$  **10.** $<$  **11.** $>$

**Examples**    Determine whether each statement is true or false.

**12.** $-9 \leq 7$    True, since $-9 < 7$ is true.
**13.** $-5 \leq -5$    True, since $-5 = -5$ is true.
**14.** $-5 \geq -5$    True, since $-5 = -5$ is true.
**15.** $-24 \geq -20$    False, since neither $-24 > -20$ nor $-24 = -20$ is true.

▪ Work Practice 12–15

**Practice 12–15**

Determine whether each statement is true or false.

**12.** $-11 \leq 16$
**13.** $-7 \leq -7$
**14.** $-7 \geq -7$
**15.** $-25 \leq -30$

## Objective C  Finding Opposites and Reciprocals ▶

Of all the real numbers, two of them stand out as extraordinary: 0 and 1. Zero is the only real number that can be added to *any* real number and result in the same real number. Also, 1 is the only real number that can be multiplied by *any* real number and result in the same real number. This is why 0 is called the **additive identity** and 1 is called the **multiplicative identity.**

---

### Identity Properties

For any real number $a$,

Identity Property of 0:  $a + 0 = 0 + a = a$

Also,

Identity Property of 1:  $a \cdot 1 = 1 \cdot a = a$

---

We use the identity property of 1 when we say that $x$, for example, means $1 \cdot x$ or $1x$. We also use this property when we write equivalent expressions. For example,

$$\underbrace{\frac{2}{3} = \frac{2}{3} \cdot 1}_{\text{identity property of 1}} = \frac{2}{3} \cdot \frac{5}{5} = \frac{10}{15} \quad \frac{5}{5} \text{ is another name for 1.}$$

Two numbers whose sum is the additive identity 0 are called **opposites** or **additive inverses** of each other. Each real number has a unique opposite.

---

### Opposites or Additive Inverses

If $a$ is a real number, then the unique **opposite,** or **additive inverse,** of $a$ is written as $-a$ and the following is true:

$$a + (-a) = 0 \quad (\text{Also, } -a + a = 0)$$

---

On a number line, we picture a real number and its opposite as being the same distance from 0 but on opposite sides of 0.

The opposite of 6 is $-6$.

The opposite of $\frac{2}{3}$ is $-\frac{2}{3}$.

The opposite of $-4$ is 4.

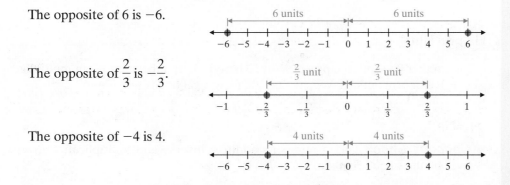

**Answers**
**12.** true  **13.** true  **14.** true  **15.** false

We stated that the opposite or additive inverse of a number $a$ is $-a$. This means that the opposite of $-4$ is $-(-4)$. But we stated also that the opposite of $-4$ is 4. This means that $-(-4) = 4$, and in general, we have the following property.

### Double Negative Property

For every real number $a$, we have $-(-a) = a$.

---

**Examples**    Find the opposite, or additive inverse, of each number.

**16.** 8      The opposite of 8 is $-8$.

**17.** $-\dfrac{1}{5}$      The opposite of $-\dfrac{1}{5}$ is $-\left(-\dfrac{1}{5}\right)$ or $\dfrac{1}{5}$.

**18.** 0      The opposite of 0 is $-0$, or 0.

**19.** $-3.5$      The opposite of $-3.5$ is $-(-3.5)$ or 3.5.

■ Work Practice 16–19

Two numbers whose product is 1 are called **reciprocals** or **multiplicative inverses** of each other. Just as each real number has a unique opposite, each nonzero real number also has a unique reciprocal.

### Reciprocals or Multiplicative Inverses

If $a$ is a nonzero real number, then its **reciprocal, or multiplicative inverse,** is $\dfrac{1}{a}$ and the following is true:

$$a \cdot \frac{1}{a} = 1 \qquad \left(\text{Also,} \frac{1}{a} \cdot a = 1\right)$$

---

**Examples**    Find the reciprocal, or multiplicative inverse, of each number.

**20.** 11      The reciprocal of 11 is $\dfrac{1}{11}$.

**21.** $-9$      The reciprocal of $-9$ is $-\dfrac{1}{9}$.

**22.** $\dfrac{7}{4}$      The reciprocal of $\dfrac{7}{4}$ is $\dfrac{4}{7}$   $\left(\text{since } \dfrac{7}{4} \cdot \dfrac{4}{7} = 1\right)$.

■ Work Practice 20–22

---

### Helpful Hint

The number 0 has no reciprocal. Why? There is no number that when multiplied by 0 gives a product of 1. We say that the reciprocal of 0 is **undefined.**

---

✔**Concept Check** Can a number's additive inverse and multiplicative inverse ever be the same? Explain.

---

**Practice 16–19**

Find the opposite, or additive inverse, of each number.

**16.** 7     **17.** $\dfrac{2}{13}$

**18.** $-\dfrac{5}{7}$     **19.** $-4.7$

**Practice 20–22**

Find the reciprocal, or multiplicative inverse, of each number.

**20.** 13

**21.** $-5$

**22.** $\dfrac{2}{3}$

**Answers**

**16.** $-7$   **17.** $-\dfrac{2}{13}$   **18.** $\dfrac{5}{7}$   **19.** 4.7

**20.** $\dfrac{1}{13}$   **21.** $-\dfrac{1}{5}$   **22.** $\dfrac{3}{2}$

✔**Concept Check Answer**

no; answers may vary

## Objective D  Using the Commutative, Associative, and Distributive Properties ▶

In addition to these special real numbers, all real numbers have certain properties that allow us to write equivalent expressions—that is, expressions that have the same value. These properties will be especially useful in Chapter 2 when we solve equations.

The **commutative properties** state that the order in which two real numbers are added or multiplied does not affect their sum or product.

### Commutative Properties

For any real numbers $a$ and $b$,

Addition: $a + b = b + a$
Multiplication: $a \cdot b = b \cdot a$

For example,

$7 + 11 = 18$ and $11 + 7 = 18$    Addition
$7 \cdot 11 = 77$ and $11 \cdot 7 = 77$    Multiplication

The **associative properties** state that regrouping numbers that are added or multiplied does not affect their sum or product.

### Associative Properties

For real numbers $a, b,$ and $c$,

Addition: $(a + b) + c = a + (b + c)$
Multiplication: $(a \cdot b) \cdot c = a \cdot (b \cdot c)$

For example,

$(2 + 3) + 7 = 5 + 7 = 12$    Addition
$2 + (3 + 7) = 2 + 10 = 12$
$(2 \cdot 3) \cdot 7 = 6 \cdot 7 = 42$    Multiplication
$2 \cdot (3 \cdot 7) = 2 \cdot 21 = 42$

**Example 23**   Use the commutative property of addition to write an expression equivalent to $7x + 5$.

**Solution:** $7x + 5 = 5 + 7x$

▪ Work Practice 23

**Example 24**   Use the associative property of multiplication to write an expression equivalent to $4 \cdot (9y)$. Then simplify this equivalent expression.

**Solution:** $4 \cdot (9y) = (4 \cdot 9)y = 36y$

▪ Work Practice 24

**Practice 23**

Use the commutative property of addition to write an expression equivalent to $9 + 4x$.

**Practice 24**

Use the associative property of multiplication to write an expression equivalent to $5 \cdot (6x)$. Then simplify this equivalent expression.

**Answers**
**23.** $4x + 9$   **24.** $(5 \cdot 6)x = 30x$

The **distributive property** states that multiplication distributes over addition. In Section 1.4, we learn that subtraction is defined in terms of addition. Because of this, we can also say that multiplication distributes over subtraction.

### Distributive Properties

For real numbers $a$, $b$, and $c$,

$$a(b + c) = ab + ac$$

Also,

$$a(b - c) = ab - ac$$

For example,

$$3(6 + 2) = 3(8) = 24$$
$$3(6 + 2) = 3(6) + 3(2) = 18 + 6 = 24$$

**Helpful Hint**    The distributive property also applies to sums or differences of more than two terms. For example,

$$a(b + c + d) = ab + ac + ad$$

**Practice 25–27**

Use the distributive property to multiply.

**25.** $7(4x - y)$

**26.** $8(3 + x)$

**27.** $5x(y - 4)$

**Answers**
**25.** $28x - 7y$   **26.** $24 + 8x$
**27.** $5xy - 20x$

✔ **Concept Check Answer**

no; $6(2a)(3b) = 6(6ab) = 36ab$

**Examples**    Use the distributive property to multiply.

**25.** $3(2x - y) = 3 \cdot 2x - 3 \cdot y = 6x - 3y$

**26.** $4(2y + 5) = 4 \cdot 2y + 4 \cdot 5 = 8y + 20$

**27.** $0.7x(y - 2) = 0.7x \cdot y - 0.7x \cdot 2 = 0.7xy - 1.4x$

🔲 **Work Practice 25–27**

✔**Concept Check**  Is the statement below true? Why or why not?
$$6(2a)(3b) = 6(2a) \cdot 6(3b)$$

## Vocabulary, Readiness & Video Check

*Complete the table by filling in the symbols.*

|     | Symbol | Meaning |     | Symbol | Meaning |
| --- | --- | --- | --- | --- | --- |
| **1.** |  | is less than | **2.** |  | is greater than |
| **3.** |  | is not equal to | **4.** |  | is equal to |
| **5.** |  | is greater than or equal to | **6.** |  | is less than or equal to |

*Use the choices below to fill in each blank.*

distributive        $-a$        commutative

associative        $\dfrac{1}{a}$        $a$

**7.** The opposite of nonzero number $a$ is _____.

**8.** The reciprocal of nonzero number $a$ is _____.

**9.** The _____ property has to do with "order."

**10.** The _____ property has to do with "grouping."

**11.** $a(b + c) = ab + ac$ illustrates the _____ property.

**12.** For every real number $a$, $-(-a) = $ _____ .

*Martin-Gay Interactive Videos*      Watch the section lecture video and answer the following questions.

See Video 1.3

**Objective** **A** **13.** In the lecture before ⊞ Example 1, what 6 symbols are discussed that can be used to compare two numbers? ▶

**Objective** **B** **14.** In ⊞ Example 2, explain why the two numbers are equal. ▶

**Objective** **C** **15.** Complete these statements based on the lecture given before ⊞ Example 4. *Reciprocal* is the same as _____ inverse, and *opposite* is the same as _____ inverse. ▶

**Objective** **D** **16.** The commutative and associative properties are discussed in ⊞ Examples 7 and 8 and the lecture before. What's the one word used again and again to describe the commutative property? The associative property? ▶

## 1.3   Exercise Set   MyMathLab® ▶

**Objective** **A**   *Write each sentence as an equation. See Examples 1 through 3.*

**1.** The sum of 10 and $x$ is $-12$.

**2.** The difference of $y$ and 3 amounts to 12.

▶ **3.** Twice $x$ plus 5 is the same as $-14$.

**4.** Three more than the product of 4 and $c$ is 7.

**5.** The quotient of $n$ and 5 is 4 times $n$.

**6.** The quotient of 8 and $y$ is 3 more than $y$.

**7.** The difference of $z$ and one-half is the same as the product of $z$ and one-half.

**8.** Five added to one-fourth $q$ is the same as 4 more than $q$.

**Objective** **B**   *Insert $<$, $>$, or $=$ between each pair of numbers to form a true statement. See Examples 4 through 11.*

**9.** 0      $-2$

**10.** $-5$      0

**11.** $-16$      $-17$

**12.** $-14$      $-24$

**13.** $\dfrac{12}{3}$      $\dfrac{12}{2}$

**14.** $\dfrac{20}{5}$      $\dfrac{20}{4}$

▶ **15.** 7.4      7.40

**16.** $\dfrac{12}{4}$      $\dfrac{15}{5}$

**17.** 8.6      $-3.5$

**18.** $-4.7$      3.8

▶ **19.** $\dfrac{7}{11}$      $\dfrac{9}{11}$

**20.** $\dfrac{9}{20}$      $\dfrac{3}{20}$

**21.** $\dfrac{1}{2}$      $\dfrac{5}{8}$

**22.** $\dfrac{3}{4}$      $\dfrac{7}{8}$

**23.** $-7.9$      $-7.09$

**24.** $-13.07$      $-13.7$

*Determine whether each statement is true or false. See Examples 12 through 15.*

**25.** $-6 \le 0$

**26.** $0 \le -4$

**27.** $-3 \ge -3$

**28.** $-8 \le -8$

**29.** $-14 \ge -1$

**30.** $-14 \le -1$

**31.** $-3 \le -3$

**32.** $-8 \ge -8$

## Objective C  *Write the opposite (or additive inverse) of each number. See Examples 16 through 19.*

**33.** $9$

**34.** $15$

**35.** $-6.2$

**36.** $-7.8$

**37.** $\dfrac{4}{7}$

**38.** $\dfrac{9}{5}$

**39.** $-\dfrac{5}{11}$

**40.** $-\dfrac{14}{3}$

**41.** $0$

**42.** $10.3$

*Write the reciprocal (or multiplicative inverse) of each number if one exists. See Examples 20 through 22.*

**43.** $5$

**44.** $9$

**45.** $-10$

**46.** $-4$

**47.** $-\dfrac{1}{4}$

**48.** $\dfrac{1}{9}$

**49.** $0$

**50.** $\dfrac{0}{6}$

**51.** $\dfrac{19}{16}$

**52.** $-\dfrac{23}{5}$

*Fill in the chart. See Examples 16 through 22.*

|  | **Number** | **Opposite** | **Reciprocal** |
|---|---|---|---|
| **53.** | 25 |  |  |
| **54.** | 7 |  |  |
| **55.** |  | 8 |  |
| **56.** |  |  | $-\dfrac{1}{6}$ |
| **57.** | $-\dfrac{1}{7}$ |  |  |
| **58.** | $\dfrac{1}{11}$ |  |  |
| **59.** | 0 |  |  |
| **60.** | 1 |  |  |
| **61.** |  |  | $\dfrac{8}{7}$ |
| **62.** |  | $\dfrac{36}{13}$ |  |

## Objective D  *Use a commutative property to write an equivalent expression. See Example 23.*

**63.** $7x + y$

**64.** $3a + 2b$

**65.** $z \cdot w$

**66.** $r \cdot s$

**67.** $\dfrac{1}{3} \cdot \dfrac{x}{5}$

**68.** $\dfrac{x}{2} \cdot \dfrac{9}{10}$

*Use an associative property to write an equivalent expression. See Example 24.*

**▶ 69.** $5 \cdot (7x)$

**70.** $3 \cdot (10z)$

**71.** $(x + 1.2) + y$

**72.** $5q + (2r + s)$

**73.** $(14z) \cdot y$

**74.** $(9.2x) \cdot y$

*Use the distributive property to multiply. See Examples 25 through 27.*

**▶ 75.** $3(x + 5)$

**76.** $7(y + 2)$

**77.** $4(z - 6)$

**78.** $2(7 - y)$

**79.** $8(2a + b)$

**80.** $9(c + 7d)$

**81.** $6x(y - 4)$

**82.** $11y(z - 2)$

**83.** $0.4(2x + 5y)$

**84.** $0.5(3a - 4b)$

**85.** $\frac{1}{2}(4x - 9y)$

**86.** $\frac{1}{3}(4x + 9y)$

**87.** $2(6x + 5y + 2z)$

**88.** $5(3a + b + 9c)$

*Complete each statement to illustrate the given property. See Examples 23 through 27.*

**89.** $3x + 6 = $ _____ Commutative property of addition

**90.** $8 + 0 = $ _____ Additive identity property

**91.** $\frac{2}{3} + \left(-\frac{2}{3}\right) = $ _____ Additive inverse property

**92.** $4(x + 3) = $ _____ Distributive property

**93.** $7 \cdot 1 = $ _____ Multiplicative identity property

**94.** $0 + 5.4 = $ _____ Additive identity property

**95.** $10(2y) = $ _____ Associative property of multiplication

**96.** $9y + (x + 3z) = $ _____ Associative property of addition

## Concept Extensions

*In each statement, a property of real numbers has been incorrectly applied. Correct the right-hand side of each statement. See the last Concept Check in this section.*

**97.** $3(x + 4) = 3x + 4$

**98.** $4 + 8y = 4y + 8$

**99.** $5(7y) = (5 \cdot 7)(5 \cdot y)$

**100.** $5(3a)(7b) = 5(3a) \cdot 5(7b)$

*Solve.*

**101.** Name the only real number that is its own opposite, and explain why this is so.

**102.** Name the only real number that has no reciprocal, and explain why this is so.

**103.** Is division commutative? Explain why or why not.

**104.** Is subtraction commutative? Explain why or why not.

**105.** Evaluate $24 \div (6 \div 3)$ and $(24 \div 6) \div 3$. Use these two expressions and discuss whether division is associative.

**106.** Evaluate $12 - (5 - 3)$ and $(12 - 5) - 3$. Use these two expressions and discuss whether subtraction is associative.

*To demonstrate the distributive property geometrically, represent the area of the larger rectangle in two ways: first as width times length and second as the sum of the areas of the smaller rectangles.*

*Example:*

Area of larger rectangle: $x(y + z)$
Area of larger rectangle: $xy + xz$
Thus: $x(y + z) = xy + xz$

**107.**

**108.**

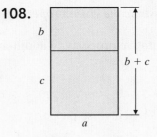

---

## 1.4  Operations on Real Numbers

### Objectives

**A** Find the Absolute Value of a Number.

**B** Add and Subtract Real Numbers.

**C** Multiply and Divide Real Numbers.

**D** Simplify Expressions Containing Exponents.

**E** Find Roots of Numbers.

### Objective A  Finding the Absolute Value of a Number

In Section 1.3, we used a number line to compare two real numbers. A number line can also be used to visualize distance, which leads to the concept of absolute value. The **absolute value** of a number is the distance between the number and 0 on a number line. The symbol for absolute value is $|\ \ |$. For example, since $-4$ and $4$ are both 4 units from 0 on the number line, each has an absolute value of 4.

$|-4| = 4$ and $|4| = 4$

An equivalent definition of the absolute value of a real number $a$ is given next.

## Absolute Value

The absolute value of $a$, written as $|a|$, is

$$|a| = \begin{cases} a \text{ if } a \text{ is } 0 \text{ or a positive number} \\ -a \text{ if } a \text{ is a negative number} \end{cases}$$

        ↑
the opposite of

**Examples**  Find each absolute value.

1. $|3| = 3$
2. $|0| = 0$
3. $\left|-\dfrac{1}{7}\right| = \underset{\uparrow}{-}\left(-\dfrac{1}{7}\right) = \dfrac{1}{7}$

    the opposite of

4. $-|2.7| = -2.7$
5. $-|-8| = -8$  Since $|-8|$ is 8, we have $-|-8| = -8$.

▪ **Work Practice 1–5**

**Helpful Hint**

Since distance is always positive or zero, the absolute value of a number is always positive or zero.

✓**Concept Check**  Explain how you know that $|14| = -14$ is a false statement.

## Objective B  Adding and Subtracting Real Numbers ▶

When solving problems, we often need to add real numbers. For example, if the New Orleans Saints lose 5 yards in one play, then lose another 7 yards in the next play, their total loss may be described by $-5 + (-7)$.

The addition of two real numbers may be summarized by the following.

### Adding Real Numbers

1. To add two numbers with the *same* sign, add their absolute values and attach their common sign.
2. To add two numbers with *different* signs, subtract the smaller absolute value from the larger absolute value and attach the sign of the number with the larger absolute value.

For example, to add $-5 + (-7)$, we first add their absolute values.

$$|-5| = 5, \quad |-7| = 7, \quad \text{and} \quad 5 + 7 = 12$$

Next, we attach their common negative sign.

$$-5 + (-7) = -12$$

(This represents a total loss of 12 yards for the New Orleans Saints.)

To find $-4 + 3$, we first subtract their absolute values. (Subtract smaller absolute value from larger absolute value.)

**Practice 1–5**

Find each absolute value.

1. $|7|$      2. $|-20|$
3. $\left|-\dfrac{4}{9}\right|$      4. $-|5.9|$
5. $-|-3|$

**Answers**

1. 7  2. 20  3. $\dfrac{4}{9}$  4. $-5.9$  5. $-3$

✓**Concept Check Answer**

$|14| = 14$ since the absolute value of a number is the distance between the number and 0 and distance cannot be negative

$$|-4| = 4, \quad |3| = 3, \quad \text{and} \quad 4 - 3 = 1$$

Next, we attach the sign of the number with the larger absolute value.

$$-4 + 3 = -1$$

## Practice 6–11

Add.

**6.** $-7 + (-10)$

**7.** $8 + (-12)$

**8.** $-14 + 20$

**9.** $-4.6 + (-1.9)$

**10.** $-\dfrac{2}{3} + \dfrac{1}{6}$

**11.** $-\dfrac{1}{7} + \dfrac{1}{2}$

### Examples  Add.

**6.** $-3 + (-11) = -(3 + 11) = -14$    Add their absolute values, or $3 + 11 = 14$. Then attach the common negative sign.

**7.** $3 + (-7) = -4$    Subtract their absolute values, or $7 - 3 = 4$. Since $-7$ has the larger absolute value, the answer is $-4$.

**8.** $-10 + 15 = 5$

**9.** $-8.3 + (-1.9) = -10.2$

**10.** $-\dfrac{1}{4} + \dfrac{1}{2} = -\dfrac{1}{4} + \dfrac{1}{2} \cdot \dfrac{2}{2} = -\dfrac{1}{4} + \dfrac{2}{4} = \dfrac{1}{4}$

**11.** $-\dfrac{2}{3} + \dfrac{3}{7} = -\dfrac{2}{3} \cdot \dfrac{7}{7} + \dfrac{3}{7} \cdot \dfrac{3}{3} = -\dfrac{14}{21} + \dfrac{9}{21} = -\dfrac{5}{21}$

■ Work Practice 6–11

Subtraction of two real numbers may be defined in terms of addition.

### Subtracting Real Numbers

If $a$ and $b$ are real numbers, then the difference of $a$ and $b$, written $a - b$, is defined by

$$a - b = a + (-b)$$

In other words, to subtract a second real number from a first, we add the first number and the opposite of the second number.

## Practice 12–15

Subtract.

**12.** $7 - 14$

**13.** $-10 - (-2)$

**14.** $13.3 - (-8.9)$

**15.** $-\dfrac{1}{3} - \dfrac{1}{2}$

### Examples  Subtract.

Add the opposite.

**12.** $2 - 8 = 2 + (-8) = -6$

Add the opposite.

**13.** $-8 - (-1) = -8 + (1) = -7$

**14.** $10.7 - (-9.8) = 10.7 + 9.8 = 20.5$

**15.** $-\dfrac{2}{3} - \dfrac{1}{4} = -\dfrac{2}{3} + \left(-\dfrac{1}{4}\right) = -\dfrac{2}{3} \cdot \dfrac{4}{4} + \left(-\dfrac{1}{4} \cdot \dfrac{3}{3}\right) = -\dfrac{8}{12} + \left(-\dfrac{3}{12}\right) = -\dfrac{11}{12}$

■ Work Practice 12–15

## Practice 16–17

Simplify each expression.

**16.** $18 + 3 - 4$

**17.** $-3 - 11 + 7$

To add or subtract three or more real numbers, we add or subtract from left to right.

### Examples  Simplify each expression.

**16.** $11 + 2 - 7 = 13 - 7 = 13 + (-7) = 6$

**17.** $-5 - 4 + 2 = -5 + (-4) + 2 = -9 + 2 = -7$

■ Work Practice 16–17

### Answers

**6.** $-17$   **7.** $-4$   **8.** $6$   **9.** $-6.5$   **10.** $-\dfrac{1}{2}$

**11.** $\dfrac{5}{14}$   **12.** $-7$   **13.** $-8$   **14.** $22.2$

**15.** $-\dfrac{5}{6}$   **16.** $17$   **17.** $-7$

# Objective C Multiplying and Dividing Real Numbers ▶

To discover sign patterns when you multiply real numbers, recall that multiplication by a positive integer is the same as repeated addition. For example,

$$3(2) = 2 + 2 + 2 = 6$$
$$3(-2) = (-2) + (-2) + (-2) = -6$$

Notice here that $3(-2) = -6$. This illustrates that the product of two numbers with different signs is negative. We summarize sign patterns for multiplying any two real numbers as follows.

### Multiplying Two Real Numbers

1. The product of two numbers with the *same* sign is positive.
2. The product of two numbers with *different* signs is negative.

Also recall that the product of zero and any real number $a$ is zero.

### Product Property of 0

$$0 \cdot a = 0 \qquad \text{Also,} \qquad a \cdot 0 = 0$$

**Examples** Multiply.

**18.** $-8(-1) = 8$

**19.** $2\left(-\dfrac{1}{6}\right) = \dfrac{2}{1} \cdot \left(-\dfrac{1}{6}\right) = -\dfrac{2}{6} = -\dfrac{1}{3}$

**20.** $-1.2(0.3) = -0.36$

**21.** $7(-6) = -42$

**22.** $-\dfrac{1}{3}\left(-\dfrac{1}{2}\right) = \dfrac{1}{6}$

**23.** $(-4.6)(-2.5) = 11.5$

**24.** $0(-6) = 0$

■ Work Practice 18–24

Recall that $\dfrac{8}{4} = 2$ because $2 \cdot 4 = 8$. Likewise, $\dfrac{8}{-4} = -2$ because $(-2)(-4) = 8$. Also, $\dfrac{-8}{4} = -2$ because $(-2)4 = -8$, and $\dfrac{-8}{-4} = 2$ because $2(-4) = -8$. From these examples, we can see that the sign patterns for division are the same as for multiplication.

### Dividing Two Real Numbers

1. The quotient of two numbers with the *same* sign is positive.
2. The quotient of two numbers with *different* signs is negative.

Notice from the previous reasoning that we cannot divide by 0. Why? If $\dfrac{5}{0}$ did exist, it would equal a number such that the number times 0 would equal 5. There is no such number, so we cannot define division by 0. We say, for example, that $\dfrac{5}{0}$ is **undefined.**

**Practice 18–24**

Multiply.

**18.** $-4(-2)$

**19.** $5\left(-\dfrac{1}{10}\right)$

**20.** $-3.2(0.1)$

**21.** $8(-6)$

**22.** $-\dfrac{2}{5}\left(-\dfrac{1}{3}\right)$

**23.** $(-1.3)(-1.5)$

**24.** $0(-10)$

**Answers**
**18.** 8  **19.** $-\dfrac{1}{2}$  **20.** $-0.32$  **21.** $-48$
**22.** $\dfrac{2}{15}$  **23.** $1.95$  **24.** 0

## Practice 25–30

Divide.

**25.** $\dfrac{45}{-9}$

**26.** $\dfrac{-16}{-4}$

**27.** $\dfrac{25}{-5}$

**28.** $\dfrac{-3}{0}$

**29.** $\dfrac{0}{-3}$

**30.** $\dfrac{-10}{-40}$

**Examples**  Divide.

**25.** $\dfrac{20}{-4} = -5$

**26.** $\dfrac{-9}{-3} = 3$

**27.** $\dfrac{-40}{10} = -4$

**28.** $\dfrac{-8}{0}$ is undefined.

**29.** $\dfrac{0}{-8} = 0$

**30.** $\dfrac{-10}{-80} = \dfrac{1}{8}$

◼ Work Practice 25–30

With sign rules for division, we can understand why the positioning of the negative sign in a fraction does not change the value of the fraction. For example,

$$\frac{-12}{3} = -4, \qquad \frac{12}{-3} = -4, \qquad \text{and} \qquad -\frac{12}{3} = -4$$

Since all these fractions equal $-4$, we can say that

$$\frac{-12}{3} = \frac{12}{-3} = -\frac{12}{3}$$

In general, the following holds true:

If $a$ and $b$ are real numbers and $b \neq 0$, then

$$\frac{a}{-b} = \frac{-a}{b} = -\frac{a}{b}$$

Also recall that division by a nonzero real number $b$ is the same as multiplication by its reciprocal $\dfrac{1}{b}$. In other words,

$$a \div b = a \cdot \frac{1}{b}$$

## Practice 31–32

Divide.

**31.** $-\dfrac{3}{4} \div \left(-\dfrac{3}{8}\right)$

**32.** $-\dfrac{1}{11} \div \dfrac{2}{7}$

**Examples**  Divide.

**31.** $-\dfrac{1}{10} \div \left(-\dfrac{2}{5}\right) = -\dfrac{1}{10} \cdot \left(-\dfrac{5}{2}\right) = \dfrac{5}{20} = \dfrac{1}{4}$

**32.** $-\dfrac{1}{4} \div \dfrac{3}{7} = -\dfrac{1}{4} \cdot \dfrac{7}{3} = -\dfrac{7}{12}$

◼ Work Practice 31–32

## Objective D  Simplifying Expressions Containing Exponents ▶

Recall that when two numbers are multiplied, each number is called a **factor** of the product. For example, in $3 \cdot 5 = 15$, the 3 and 5 are factors.

### Answers

**25.** $-5$  **26.** 4  **27.** $-5$  **28.** undefined

**29.** 0  **30.** $\dfrac{1}{4}$  **31.** 2  **32.** $-\dfrac{7}{22}$

A natural number *exponent* is a shorthand notation for repeated multiplication of the same factor. This repeated factor is called the **base,** and the number of times it is used as a factor is indicated by the **exponent.** For example,

$$4^3 = \underbrace{4 \cdot 4 \cdot 4}_{\text{4 is a factor 3 times.}} = 64$$

exponent ↓   base ↑

## Exponents

If $a$ is a real number and $n$ is a natural number, then the **nth power of $a$,** or **$a$ raised to the nth power,** written as **$a^n$,** is the product of $n$ factors, each of which is $a$.

$$a^n = \underbrace{a \cdot a \cdot a \cdot a \cdots \cdots a}_{a \text{ is a factor } n \text{ times.}}$$

exponent ↓   base ↑

It is not necessary to write an exponent of 1. For example, 3 is assumed to be $3^1$.

**Examples**   Find the value of each expression.

**33.** $3^2 = 3 \cdot 3 = 9$

**34.** $-5^2 = -(5 \cdot 5) = -25$

**35.** $-5^3 = -(5 \cdot 5 \cdot 5) = -125$

**36.** $\left(\dfrac{1}{2}\right)^4 = \left(\dfrac{1}{2}\right)\left(\dfrac{1}{2}\right)\left(\dfrac{1}{2}\right)\left(\dfrac{1}{2}\right) = \dfrac{1}{16}$

**37.** $(-5)^2 = (-5)(-5) = 25$

**38.** $(-5)^3 = (-5)(-5)(-5) = -125$

◼ Work Practice 33–38

**Practice 33–38**

Simplify each expression.

**33.** $4^2$

**34.** $-3^2$

**35.** $-3^3$

**36.** $\left(\dfrac{1}{3}\right)^4$

**37.** $(-3)^2$

**38.** $(-3)^3$

✓**Concept Check** When $(-8.2)^7$ is evaluated, will the value be positive or negative? How can you tell without making any calculations?

**Helpful Hint**

Be very careful when finding the value of expressions such as $-5^2$ and $(-5)^2$.

$$-5^2 = -(5 \cdot 5) = -25 \qquad \text{and} \qquad (-5)^2 = (-5)(-5) = 25$$

Without parentheses, the base to square is 5, not $-5$.

## Objective E  Finding the Root of a Number

The opposite of squaring a number is taking the **square root** of a number. For example, since the square of 4, or $4^2$, is 16, we say that a square root of 16 is 4. The notation $\sqrt{a}$ is used to denote the **positive,** or **principal square root** of a nonnegative number $a$. We then have in symbols that

$$\sqrt{16} = 4$$

**Answers**

**33.** 16  **34.** $-9$  **35.** $-27$  **36.** $\dfrac{1}{81}$

**37.** 9  **38.** $-27$

✓**Concept Check Answer**

negative; the exponent is an odd number

**Practice 39–41**

Find each root.

**39.** $\sqrt{36}$

**40.** $\sqrt{4}$

**41.** $\sqrt{\dfrac{1}{49}}$

---

**Examples** Find each root.

**39.** $\sqrt{9} = 3$, since 3 is positive and $3^2 = 9$.

**40.** $\sqrt{25} = 5$, since $5^2 = 25$.

**41.** $\sqrt{\dfrac{1}{4}} = \dfrac{1}{2}$, since $\left(\dfrac{1}{2}\right)^2 = \dfrac{1}{4}$.

■ Work Practice 39–41

We can find roots other than square roots. Since 2 cubed, written as $2^3$, is 8, we say that the **cube root** of 8 is 2. This is written as

$$\sqrt[3]{8} = 3$$

Also, since $3^4 = 81$ and 3 is positive,

$$\sqrt[4]{81} = 3$$

**Practice 42–44**

Find each root.

**42.** $\sqrt[3]{64}$

**43.** $\sqrt[5]{1}$

**44.** $\sqrt[5]{243}$

---

**Examples** Find each root.

**42.** $\sqrt[3]{27} = 3$, since $3^3 = 27$.

**43.** $\sqrt[5]{1} = 1$, since $1^5 = 1$.

**44.** $\sqrt[4]{16} = 2$, since 2 is positive and $2^4 = 16$.

■ Work Practice 42–44

Of course, as mentioned in Section 1.2, not all roots simplify to rational numbers. We study radicals further in Chapter 7.

**Answers**

**39.** 6  **40.** 2  **41.** $\dfrac{1}{7}$  **42.** 4
**43.** 1  **44.** 3

---

## Vocabulary, Readiness & Video Check

*Choose the fraction(s) equivalent to the given fraction. (There may sometimes be more than one correct choice.)*

**1.** $-\dfrac{1}{7}$     **a.** $\dfrac{-1}{-7}$     **b.** $\dfrac{-1}{7}$     **c.** $\dfrac{1}{-7}$     **d.** $\dfrac{1}{7}$

**2.** $\dfrac{-x}{y}$     **a.** $\dfrac{x}{-y}$     **b.** $-\dfrac{x}{y}$     **c.** $\dfrac{x}{y}$     **d.** $\dfrac{-x}{-y}$

**3.** $\dfrac{5}{-(x+y)}$     **a.** $\dfrac{5}{(x+y)}$     **b.** $\dfrac{-5}{(x+y)}$     **c.** $\dfrac{-5}{-(x+y)}$     **d.** $-\dfrac{5}{(x+y)}$

**4.** $-\dfrac{(y+z)}{3y}$     **a.** $\dfrac{-(y+z)}{3y}$     **b.** $\dfrac{-(y+z)}{-3y}$     **c.** $\dfrac{(y+z)}{3y}$     **d.** $\dfrac{(y+z)}{-3y}$

**5.** $\dfrac{-9x}{-2y}$     **a.** $\dfrac{-9x}{2y}$     **b.** $\dfrac{9x}{2y}$     **c.** $\dfrac{9x}{-2y}$     **d.** $-\dfrac{9x}{2y}$

**6.** $\dfrac{-a}{-b}$     **a.** $\dfrac{a}{b}$     **b.** $\dfrac{a}{-b}$     **c.** $\dfrac{-a}{b}$     **d.** $-\dfrac{a}{b}$

*Use the choices below to fill in each blank. Some choices may be used more than once and some used not at all.*

| | | | | | |
|---|---|---|---|---|---|
| exponent | undefined | base | $\dfrac{-a}{-b}$ | $\dfrac{a}{b}$ | absolute value |
| square root | reciprocal | 0 | $\dfrac{-a}{b}$ | $\dfrac{a}{-b}$ | |

**7.** $0 \cdot a =$ _____ .

**8.** $\dfrac{0}{4}$ simplifies to _____ while $\dfrac{4}{0}$ is _____ .

**9.** The _____ of the nonzero number $b$ is $\dfrac{1}{b}$.

**10.** The fraction $-\dfrac{a}{b} =$ _____ $=$ _____ .

**11.** A(n) _____ is a shorthand notation for repeated multiplication of the same number.

**12.** In $(-5)^2$, the 2 is the _____ and the $-5$ is the _____ .

**13.** The opposite of squaring a number is taking the _____ of a number.

**14.** The _____ of a number is that number's distance from 0 on a number line.

---

*Martin-Gay Interactive Videos*    *Watch the section lecture video and answer the following questions.*

*See Video 1.4*

**Objective A 15.** Based on the lecture before ⊞ Example 1, complete the following statements. The absolute value of a number is that number's _____ from zero on a number line. Or, more formally, $|a| = a$ if $a$ is _____ or a _____ number. Also, $|a| = -a$ if $a$ is a _____ number. ▶

**Objective B 16.** From ⊞ Example 7, we learn that subtraction of two real numbers may be defined in terms of what operation? ▶

**Objective C 17.** From ⊞ Examples 8 and 9, explain the significance of the signs of numbers when multiplying and dividing two real numbers. ▶

**Objective D 18.** Explain the significance of the use of parentheses when comparing ⊞ Examples 12 and 13. ▶

**Objective E 19.** From ⊞ Example 14, what does the $\sqrt{\phantom{x}}$ notation mean? ▶

---

**1.4 Exercise Set** MyMathLab® ▶

**Objective A** *Find each absolute value. See Examples 1 through 5.*

**1.** $|2|$      **2.** $|8|$      **3.** $|-4|$      **4.** $|-6|$      **5.** $|0|$

**6.** $|-1|$      **7.** $-|3|$      **8.** $-|11|$      **9.** $-\left|-\dfrac{2}{9}\right|$      **10.** $-\left|-\dfrac{5}{13}\right|$

**Objective B** *Add or subtract as indicated. See Examples 6 through 17.*

**11.** $-3 + 8$      **12.** $12 + (-7)$      ▶ **13.** $-14 + (-10)$      **14.** $-5 + (-9)$

**15.** $-4.3 - 6.7$        **16.** $-8.2 - (-6.6)$        **17.** $15 - 18$        ▶**18.** $13 - 17$

**19.** $\dfrac{11}{15} - \left(-\dfrac{3}{5}\right)$        **20.** $\dfrac{7}{10} - \left(-\dfrac{4}{5}\right)$        **21.** $19 - 10 - 11$        **22.** $-13 - 4 + 9$

**23.** $-14 - 7$        **24.** $-6 - 31$        **25.** $-\dfrac{4}{5} - \left(-\dfrac{3}{10}\right)$        **26.** $-\dfrac{5}{2} - \left(-\dfrac{2}{3}\right)$

**27.** Subtract 14 from 8.            **28.** Subtract 9 from $-3$.

**Objective** **C**  *Multiply or divide as indicated. See Examples 18 through 32.*

▶**29.** $-5 \cdot 12$        **30.** $-3 \cdot 8$        **31.** $-8(-10)$        **32.** $-4(-11)$        **33.** $-17 \cdot 0$

**34.** $-5 \cdot 0$        **35.** $0(-1)$        **36.** $0(-34)$        **37.** $\dfrac{-9}{3}$        **38.** $\dfrac{-20}{5}$

▶**39.** $\dfrac{-12}{-4}$        **40.** $\dfrac{-36}{-6}$        **41.** $3\left(-\dfrac{1}{18}\right)$        **42.** $5\left(-\dfrac{1}{50}\right)$        **43.** $(-0.7)(-0.8)$

**44.** $(-0.9)(-0.5)$    **45.** $9.1 \div (-1.3)$    **46.** $22.5 \div (-2.5)$    **47.** $-4(-2)(-1)$    **48.** $-5(-3)(-2)$

**Objective** **D**  *Find the value of each expression. See Examples 33 through 38.*

▶**49.** $-7^2$        ▶**50.** $(-7)^2$        **51.** $(-6)^2$        **52.** $-6^2$

**53.** $(-2)^3$        **54.** $-2^3$        **55.** $\left(-\dfrac{1}{3}\right)^3$        **56.** $\left(-\dfrac{1}{2}\right)^4$

**Objective** **E**  *Find each root. See Examples 39 through 44.*

▶**57.** $\sqrt{49}$        **58.** $\sqrt{81}$        **59.** $\sqrt{64}$        **60.** $\sqrt{100}$        **61.** $\sqrt{\dfrac{1}{9}}$

**62.** $\sqrt{\dfrac{1}{25}}$        **63.** $\sqrt[3]{64}$        ▶**64.** $\sqrt[4]{81}$        **65.** $\sqrt{\dfrac{4}{25}}$        **66.** $\sqrt{\dfrac{4}{81}}$

**Objectives** **A** **B** **C** **D** **E** **Mixed Practice** *Perform the indicated operations. See Examples 1 through 44.*

**67.** $-4 + 7$        **68.** $-9 + 15$        **69.** $-9 + (-3)$        **70.** $-17 + (-2)$

**71.** $6(-3)$        **72.** $5(-4)$        **73.** $-9 \cdot 8$        **74.** $-6 \cdot 7$

**75.** $(-11)^2$

**76.** $-11^2$

**77.** $\dfrac{16}{-2}$

**78.** $\dfrac{35}{-7}$

**79.** $-18 \div 6$

**80.** $-42 \div 6$

**81.** $-\dfrac{2}{7} \cdot \left(-\dfrac{1}{6}\right)$

**82.** $\dfrac{5}{9} \cdot \left(-\dfrac{3}{5}\right)$

**83.** $-2(-3.6)$

**84.** $-5(-4.2)$

**85.** $-4 - (-19)$

**86.** $-5 - (-17)$

**87.** $6.3 - 18.5$

**88.** $15.9 - 21.7$

▶ **89.** $\dfrac{0}{-2}$

**90.** $\dfrac{0}{-11}$

▶ **91.** $\dfrac{-2}{0}$

**92.** $\dfrac{-22}{0}$

**93.** $\sqrt{\dfrac{1}{16}}$

**94.** $\sqrt{\dfrac{1}{49}}$

**95.** $\dfrac{-6}{7} \div 2$

**96.** $\dfrac{-9}{13} \div (-3)$

**97.** $\dfrac{-5.2}{-1.3}$

**98.** $\dfrac{-4.2}{-1.4}$

**99.** $-25 \div (-5)$

**100.** $-88 \div (-11)$

**101.** $\sqrt[4]{16}$

**102.** $\sqrt[3]{1}$

**103.** $\sqrt[5]{32}$

**104.** $\sqrt[3]{125}$

**105.** $-\dfrac{1}{6} \div \dfrac{9}{10}$

**106.** $\dfrac{4}{7} \div \left(-\dfrac{1}{8}\right)$

**107.** $-\dfrac{2}{3} \cdot \left(\dfrac{6}{4}\right)$

**108.** $\dfrac{5}{6} \cdot \left(\dfrac{-12}{15}\right)$

**109.** $\dfrac{3}{5} \div \left(-\dfrac{2}{5}\right)$

**110.** $\dfrac{2}{7} \div \left(-\dfrac{1}{14}\right)$

**111.** $16 - 8 - 9$

**112.** $-14 - 3 + 6$

**113.** $-5 + (-7) - 10$

**114.** $-8 + (-10) - 6$

**115.** $-7(-1)(5)$

**116.** $-6(2)(-3)$

**117.** $-6(-5)(0)$

**118.** $4(-3)(0)$

**119.** $-4^4$

**120.** $(-4)^4$

## Concept Extensions

**121.** Explain why $-3^2$ and $(-3)^2$ simplify to different numbers.

**122.** Explain why $-3^3$ and $(-3)^3$ simplify to the same number.

*Answer each statement true or false.*

**123.** $-7 - 1 < -7(-1)$

**124.** $\dfrac{-100}{-5} > -100 + (-5)$

*Each circle below represents a whole, or 1. Determine the unknown fractional part of each circle.*

**125.**

**126.**

**127.** Most of Mauna Kea, a volcano on Hawaii, lies below sea level. If this volcano begins at 5998 meters below sea level and then rises 10,203 meters, find the height of the volcano above sea level.

**128.** The highest point on land on Earth is the top of Mt. Everest, in the Himalayas, at an elevation of 29,028 feet above sea level. The lowest point on land is the Dead Sea, between Israel and Jordan, at 1312 feet below sea level. Find the difference in elevations. (*Source:* National Geographic Society)

*A fair game is one in which each team or player has the same chance of winning. Suppose that a game consists of three players taking turns spinning a spinner. If the spinner lands on yellow, player 1 gets a point. If the spinner lands on red, player 2 gets a point, and if the spinner lands on blue, player 3 gets a point. After 12 spins, the player with the most points wins. Use this information to answer Exercises 129 through 134.*

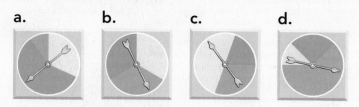

**a.**  **b.**  **c.**  **d.**

**129.** Which spinner(s) lead to a fair game?

**130.** Which spinner(s) would lead to an unfair game?

**131.** If you are player 2 and want to win the game, which spinner would you choose?

**132.** If you are player 1 and want to lose the game, which spinner would you choose?

**133.** Is it possible for the game to end in a three-way tie? If so, list the possible ending scores.

**134.** Is it possible for the game to end in a two-way tie? If so, list the possible ending scores.

*Use a calculator to approximate each square root. Round to four decimal places.*

**135.** $\sqrt{273}$    **136.** $\sqrt{10}$    **137.** $\sqrt{19.6}$    **138.** $\sqrt{7.9}$

*Investment firms often advertise their gains and losses in the form of bar graphs such as the one that follows. This graph shows investment risk over time for common stocks by showing average annual compound returns for 1 year, 5 years, 10 years, 15 years, and 20 years. For example, after 1 year, the annual compound return in percent for an investor is anywhere from a gain of 56% to a loss of 64%. Use this graph to answer Exercises 139 through 143.*

**Investment Risk Over Time**

Source: Yahoo Finance; mutual funds

**139.** A person investing in common stocks may expect at most an average annual gain of what percent after 15 years?

**140.** A person investing in common stocks may expect to lose at most an average per year of what percent after 5 years?

**141.** Find the difference in percent of the highest average annual return and the lowest average annual return after 20 years.

**142.** Find the difference in percent of the highest average annual return and the lowest average annual return after 5 years.

**143.** Do you think that the type of investment shown in the figure is recommended for short-term investments or long-term investments? Explain your answer.

# Real Numbers

*Write each set in roster form.*

**1.** $\{x \mid x \text{ is a natural number less than } 4\}$

**2.** $\{x \mid x \text{ is an odd whole number less than } 6\}$

**3.** $\{x \mid x \text{ is an even natural number greater than } 7\}$

**4.** $\{x \mid x \text{ is a whole number between } 10 \text{ and } 15\}$

*Write each phrase as an algebraic expression. Let x represent the unknown number.*

**5.** Twice the difference of a number and three

**6.** The quotient of six and the sum of a number and ten

*Insert $<$, $>$, or $=$ between each pair of numbers to form a true statement.*

**7.** $-4 \quad -6$

**8.** $8.6 \quad 8.600$

**9.** $\dfrac{9}{10} \quad \dfrac{11}{10}$

**10.** $-6.1 \quad -6.01$

*Write each sentence as an equation.*

**11.** The product of 5 and $x$ is the same as 20.

**12.** The sum of $a$ and 12 amounts to 14.

**13.** The quotient of $y$ and 10 is the same as the product of $y$ and 10.

**14.** The sum of $x$ and negative 1 equals the difference of $x$ and 1.

*Perform each indicated operation.*

**15.** $-4 + 7$

**16.** $-11 + 20$

**17.** $-4(7)$

**18.** $-11(20)$

**19.** $-8 - (-13)$

**20.** $-12 - 16$

**21.** $\dfrac{-20}{-4}$

**22.** $\dfrac{-18}{6}$

**23.** $-5^2$

**24.** $(-5)^2$

**25.** $-6 - 1 + 20$

**26.** $18 - 4 - 19$

**27.** $\dfrac{0}{-3}$

**28.** $\dfrac{5}{0}$

**29.** $-4(3)(2)$

**30.** $-5(-1)(6)$

1. _____
2. _____
3. _____
4. _____
5. _____
6. _____
7. _____
8. _____
9. _____
10. _____
11. _____
12. _____
13. _____
14. _____
15. _____
16. _____
17. _____
18. _____
19. _____
20. _____
21. _____
22. _____
23. _____
24. _____
25. _____
26. _____
27. _____
28. _____
29. _____
30. _____

31. _____

32. _____

33. _____

34. _____

35. _____

36. _____

37. _____

38. _____

39. _____

40. _____

41. _____

42. _____

43. _____

44. _____

45. _____

46. _____

47. _____

48. _____

49. _____

50. _____

51. _____

52. _____

53. _____

54. _____

55. _____

56. _____

57. _____

58. _____

**31.** $-\dfrac{1}{2} \cdot \dfrac{6}{7}$

**32.** $\dfrac{4}{5} \cdot \left(-\dfrac{1}{8}\right)$

**33.** $\dfrac{3}{10} - \dfrac{4}{5}$

**34.** $-\dfrac{2}{3} - \dfrac{1}{4}$

**35.** $\dfrac{1.6}{-0.2}$

**36.** $\dfrac{-4.8}{16}$

**37.** $6.7 - (-1.3)$

**38.** $-4.6 + 9$

**39.** $\dfrac{1}{2} + \left(-\dfrac{7}{8}\right)$

**40.** $\dfrac{1}{2} \div \left(-\dfrac{7}{8}\right)$

**41.** $\sqrt{49}$

**42.** $\sqrt[3]{27}$

**43.** $-2^2$

**44.** $(-2)^3$

**45.** $\sqrt{\dfrac{1}{81}}$

**46.** $\sqrt{\dfrac{1}{100}}$

*Fill in the chart.*

| Number | Opposite | Reciprocal |
|---|---|---|
| **47.** $-6$ | | |
| **48.** $4$ | | |
| **49.** | | $\dfrac{7}{5}$ |
| **50.** | $\dfrac{7}{30}$ | |

*Use the distributive property to multiply.*

**51.** $9(m + 5)$

**52.** $11(7 + r)$

**53.** $3(2y - 3x)$

**54.** $8(4m - 7n)$

**55.** $0.2(3a + 7)$

**56.** $0.6(2n + 5)$

**57.** $\dfrac{1}{5}(10x - 19y + 20)$

**58.** $\dfrac{1}{2}(10x - 19y + 20)$

# 1.5 Order of Operations and Algebraic Expressions

In this section, we review order of operations and algebraic expressions.

## Objective A Using the Order of Operations

The expression $3 + 2 \cdot 30$ represents the total number of USB flash drives shown.

Expressions containing more than one operation are written to follow a particular agreed-upon **order of operations.** For example, when we write $3 + 2 \cdot 30$, we mean to multiply first and then add.

### Order of Operations

Simplify expressions using the following order.

1. If grouping symbols such as parentheses are present, simplify expressions within those first, starting with the innermost set.
2. Evaluate exponential expressions, roots, and absolute values.
3. Perform multiplications or divisions in order from left to right.
4. Perform additions or subtractions in order from left to right.

### Helpful Hint

Fraction bars, radical signs, and absolute value bars can sometimes be used as grouping symbols. For example,

|  | Fraction Bar | Radical Sign | Absolute Value Bars |
|---|---|---|---|
| **Grouping Symbol** | $\dfrac{-1 - 7}{6 - 11}$ | $\sqrt{15 + 1}$ | $\lvert -7.2 - \sqrt{4} \rvert$ |
| **Not Grouping Symbol** | $-\dfrac{8}{9}$ | $\sqrt{9}$ | $\lvert -3.2 \rvert$ |

---

**Example 1** Simplify: $3 + 2 \cdot 30$

**Solution:** First we multiply; then we add.

$$3 + 2 \cdot 30 = 3 + 60 = 63$$

 Work Practice 1

## Objectives

A Use the Order of Operations.

B Identify and Evaluate Algebraic Expressions.

C Identify Like Terms and Simplify Algebraic Expressions.

**Practice 1**

Simplify: $15 - 2 \cdot 5$

**Practice 2**

Simplify: $2 + 5(2 - 6)^2$

**Example 2** Simplify: $1 + 2(1 - 4)^2$

**Solution:** Remember order of operations so that you are *not* tempted to add 1 and 2 first. Unless there are grouping symbols, addition is last in order of operations.

$$1 + 2(1 - 4)^2 = 1 + 2(-3)^2 \quad \text{Simplify inside parentheses first.}$$
$$= 1 + 2(9) \quad \text{Write } (-3)^2 \text{ as 9.}$$
$$= 1 + 18 \quad \text{Multiply.}$$
$$= 19 \quad \text{Add.}$$

■ Work Practice 2

**Practice 3**

Simplify: $\dfrac{|-3|^2 + 5}{\sqrt{9} - 10}$

**Example 3** Simplify: $\dfrac{|-2|^3 + 1}{-7 - \sqrt{4}}$

**Solution:** Here, the fraction bar serves as a grouping symbol. We simplify the numerator and the denominator separately. Then we divide.

$$\frac{|-2|^3 + 1}{-7 - \sqrt{4}} = \frac{2^3 + 1}{-7 - 2} \quad \text{Write } |-2| \text{ as 2 and } \sqrt{4} \text{ as 2.}$$
$$= \frac{8 + 1}{-9} \quad \text{Write } 2^3 \text{ as 8; simplify the denominator.}$$
$$= \frac{9}{-9} = -1 \quad \text{Simplify the numerator; then divide.}$$

■ Work Practice 3

Besides parentheses, other symbols used for grouping expressions are brackets [ ], braces { }, radical signs, and absolute value bars. Brackets [ ] and braces { } are commonly used when we group expressions that already contain parentheses.

**Practice 4**

Simplify:
$7 - [2(1 - 3) + 5(10 - 12)]$

**Helpful Hint** When grouping symbols occur within grouping symbols, remember to perform operations within the innermost set first.

**Example 4** Simplify: $3 - [6(4 - 6) + 2(5 - 9)]$

**Solution:**

$$3 - [6(4 - 6) + 2(5 - 9)] = 3 - [6(-2) + 2(-4)] \quad \text{Simplify within the innermost sets of parentheses.}$$
$$= 3 - [-12 + (-8)]$$
$$= 3 - [-20]$$
$$= 23$$

■ Work Practice 4

**Practice 5**

Simplify: $\dfrac{|9 - 16| + 5^2}{-3\sqrt{8 + 1} + (-4)^2}$

**Example 5** Simplify: $\dfrac{-5\sqrt{30 - 5} + (-2)^2}{4^2 + |7 - 10|}$

**Solution:** Here, the fraction bar, radical sign, and absolute value bars serve as grouping symbols. Thus, we simplify within the radical sign and absolute value bars first, remembering to calculate above and below the fraction bar separately.

$$\frac{-5\sqrt{30 - 5} + (-2)^2}{4^2 + |7 - 10|} = \frac{-5\sqrt{25} + (-2)^2}{4^2 + |-3|} = \frac{-5 \cdot 5 + 4}{16 + 3} = \frac{-25 + 4}{16 + 3}$$
$$= \frac{-21}{19} \text{ or } -\frac{21}{19}$$

 Work Practice 5

**Answers**

**2.** 82  **3.** −2  **4.** 21  **5.** $\dfrac{32}{7}$

✓**Concept Check** True or false? If two different people use the order of operations to simplify a numerical expression and neither makes a calculation error, it is not possible that they each obtain a different result. Explain.

## Objective B Evaluating Algebraic Expressions

Recall from Section 1.2 that an algebraic expression is formed by numbers and variables connected by the operations of addition, subtraction, multiplication, division, raising to powers, or taking roots. Also, if numbers are substituted for the variables in an algebraic expression and the operations performed, the result is called the **value** of the expression for the given replacement values. This entire process is called **evaluating an expression.**

**Example 6** Evaluate each expression when $x = 4$ and $y = -3$.

**a.** $3x - 7y$      **b.** $-2y^2$      **c.** $\dfrac{\sqrt{x}}{y} - \dfrac{y}{x}$

**Solution:** For each expression, replace $x$ with 4 and $y$ with $-3$.

**a.** $3x - 7y = 3 \cdot 4 - 7(-3)$     Let $x = 4$ and $y = -3$.

$\phantom{3x - 7y} = 12 - (-21)$     Multiply.

$\phantom{3x - 7y} = 12 + 21$     Write as an addition.

$\phantom{3x - 7y} = 33$     Add.

**b.** $-2y^2 = -2(-3)^2$     Let $y = -3$.

$\phantom{-2y^2} = -2(9)$     Write $(-3)^2$ as 9.

$\phantom{-2y^2} = -18$     Multiply.

**c.** $\dfrac{\sqrt{x}}{y} - \dfrac{y}{x} = \dfrac{\sqrt{4}}{-3} - \dfrac{-3}{4}$

$\phantom{\dfrac{\sqrt{x}}{y} - \dfrac{y}{x}} = -\dfrac{2}{3} + \dfrac{3}{4}$     Write $\sqrt{4}$ as 2.

$\phantom{\dfrac{\sqrt{x}}{y} - \dfrac{y}{x}} = -\dfrac{2}{3} \cdot \dfrac{4}{4} + \dfrac{3}{4} \cdot \dfrac{3}{3}$     The LCD is 12.

$\phantom{\dfrac{\sqrt{x}}{y} - \dfrac{y}{x}} = -\dfrac{8}{12} + \dfrac{9}{12}$     Write each fraction with a denominator of 12.

$\phantom{\dfrac{\sqrt{x}}{y} - \dfrac{y}{x}} = \dfrac{1}{12}$     Add.

◼ Work Practice 6

> **Helpful Hint** In $-2(-3)^2$, the exponent 2 goes with the base of $-3$ only.

**Example 7** Converting Degrees Fahrenheit to Degrees Celsius

The algebraic expression $\dfrac{5(x - 32)}{9}$ represents the equivalent temperature in degrees Celsius when $x$ is degrees Fahrenheit. Complete the following table by evaluating this expression at the given values of $x$.

| Degrees Fahrenheit | $x$ | $-4$ | $10$ | $32$ |
|---|---|---|---|---|
| Degrees Celsius | $\dfrac{5(x - 32)}{9}$ | | | |

*(Continued on next page)*

---

### Practice 6

Evaluate each expression when $x = 9$ and $y = -4$.

**a.** $2x - 6y$

**b.** $-3y^2$

**c.** $\dfrac{\sqrt{x}}{y} + \dfrac{y}{x}$

### Practice 7

Use the algebraic expression given in Example 7 to complete the following table.

| Degrees Fahrenheit | $x$ | $-13$ | $0$ | $41$ |
|---|---|---|---|---|
| Degrees Celsius | $\dfrac{5(x - 32)}{9}$ | | | |

**Answers**

6. **a.** 42   **b.** $-48$   **c.** $-\dfrac{43}{36}$

7. $-25; -\dfrac{160}{9}$ or $-17\dfrac{7}{9}; 5$

✓**Concept Check Answer**

true; answers may vary

**Solution:**   To complete the table, we evaluate $\dfrac{5(x-32)}{9}$ at each given replacement value.

When $x = -4$,

$$\frac{5(x-32)}{9} = \frac{5(-4-32)}{9} = \frac{5(-36)}{9} = -20$$

When $x = 10$,

$$\frac{5(x-32)}{9} = \frac{5(10-32)}{9} = \frac{5(-22)}{9} = \frac{-110}{9} \text{ or } -12\frac{2}{9}$$

When $x = 32$,

$$\frac{5(x-32)}{9} = \frac{5(32-32)}{9} = \frac{5 \cdot 0}{9} = 0$$

The completed table is as follows:

| Degrees Fahrenheit | $x$ | $-4$ | 10 | 32 |
|---|---|---|---|---|
| Degrees Celsius | $\dfrac{5(x-32)}{9}$ | $-20$ | $\dfrac{-110}{9}$ or $-12\dfrac{2}{9}$ | 0 |

Thus, $-4°$F is equivalent to $-20°$C, $10°$F is equivalent to $-\dfrac{110}{9}°$C or $-12\dfrac{2}{9}°$C, and $32°$F is equivalent to $0°$C.

■ **Work Practice 7**

## Objective C  Simplifying Algebraic Expressions by Combining Like Terms ▶

Often, an algebraic expression may be **simplified** by removing grouping symbols and combining any like terms. The **terms** of an expression are the addends of the expression. For example, in the expression $3x^2 + 4x$, the terms are $3x^2$ and $4x$.

| Expression | Terms |
|---|---|
| $-2x + y$ | $-2x, y$ |
| $3x^2 - \dfrac{y}{5} + 7$ | $3x^2, \quad -\dfrac{y}{5}, \quad 7$ |

**Helpful Hint**

The expression $3x^2 - \dfrac{y}{5} + 7$ can be written as $3x^2 + \left(-\dfrac{y}{5}\right) + 7$ so

↑ ↑ ↑

that the addends (terms) are $\qquad 3x^2 \quad -\dfrac{y}{5} \quad 7.$

Terms with the same variable(s) raised to the same power(s) are called **like terms.** We can add or subtract like terms by using the distributive property. This process is called **combining like terms.**

> **Examples** Simplify by combining like terms.
>
> **8.** $3x - 5x + 4 = (3 - 5)x + 4$   Use the distributive property.
> $= -2x + 4$
>
> **9.** $y + 3y = 1y + 3y = (1 + 3)y$
> $= 4y$
>
> **10.** $7x + 9x - 6 - 10 = (7 + 9)x + (-6 - 10)$
> $= 16x - 16$

■ Work Practice 8–10

**Practice 8–10**

Simplify by combining like terms.
**8.** $9x - 15x + 7$
**9.** $8y + y$
**10.** $4x + 12x - 9 - 10$

The associative and commutative properties may sometimes be needed to rearrange and group like terms when we simplify expressions.

> **Examples** Simplify.
>
> **11.** $-7x + 5 + 3x - 2 = -7x + 3x + 5 - 2$   Use the commutative property.
> $= (-7 + 3)x + (5 - 2)$   Use the distributive property.
> $= -4x + 3$   Simplify.
>
> **12.** $3y - 2y - 5 - 7 + y = 3y - 2y + y - 5 - 7$   Use the commutative property.
> $= (3 - 2 + 1)y + (-5 - 7)$   Use the distributive property.
> $= 2y - 12$   Simplify.

■ Work Practice 11–12

**Practice 11–12**

Simplify.
**11.** $-4x + 7 - 5x - 8$
**12.** $5y - 6y + 2 - 11 + 2y$

> **Examples** Simplify by using the distributive property to multiply and then combining like terms.
>
> **13.** $-2(x + 3) + 7 = -2(x) + (-2)(3) + 7 = -2x - 6 + 7 = -2x + 1$
>
> **14.** $7x + 3 - 5(x - 4) = 7x + 3 - 5x + 20$   Use the distributive property.
> $= 2x + 23$   Combine like terms.
>
> **15.** $(2.1x - 5.6) - (-x - 5.3) = (2.1x - 5.6) - 1(-x - 5.3)$
> $= 2.1x - 5.6 + 1x + 5.3$   Use the distributive property.
> $= 3.1x - 0.3$   Combine like terms.
>
> **16.** $\frac{1}{2}(4a - 6b) - \frac{1}{3}(9a + 12b - 1) + \frac{1}{4}$
>
> $= 2a - 3b - 3a - 4b + \frac{1}{3} + \frac{1}{4}$   Use the distributive property.
>
> $= -a - 7b + \frac{7}{12}$   Combine like terms.

■ Work Practice 13–16

**Practice 13–16**

Simplify by using the distributive property to multiply and then combining like terms.
**13.** $-3(y + 1) + 4$
**14.** $8x + 2 - 4(x - 9)$
**15.** $(3.2x - 4.1) - (-x + 7.6)$
**16.** $\frac{1}{5}(15m - 40n)$
$- \frac{1}{4}(8m - 4n + 1) + \frac{1}{5}$

**Answers**
**8.** $-6x + 7$ **9.** $9y$ **10.** $16x - 19$
**11.** $-9x - 1$ **12.** $y - 9$ **13.** $-3y + 1$
**14.** $4x + 38$ **15.** $4.2x - 11.7$
**16.** $m - 7n - \frac{1}{20}$

✔**Concept Check** Find and correct the error in the following:

$\cancel{x - 4(x - 5) = x - 4x - 20}$
$\cancel{= -3x - 20}$

✔**Concept Check Answer**
$x - 4(x - 5) = x - 4x + 20$
$= -3x + 20$

# Vocabulary, Readiness & Video Check

*Use the choices below to fill in each blank. Not all choices will be used.*

| | | |
|---|---|---|
| like | terms | 1 |
| unlike | combining | 9 |

**1.** Terms with the same variable(s) raised to the same powers are called _____ terms.

**2.** The _____ of an expression are the addends of the expression.

**3.** The process of adding or subtracting like terms is called _____ like terms.

**4.** Using order of operations, $9 \div 3 \cdot 3 = $ _____.

*Martin-Gay Interactive Videos*     *Watch the section lecture video and answer the following questions.*

**Objective A** **5.** In the lecture before ⊞ Example 1, what reason is given for needing the order of operations? ▶

**Objective B** **6.** In ⊞ Example 4, why can you not first add the 3 and 2 in the numerator? ▶

**Objective C** **7.** From ⊞ Examples 5–7, how do we simplify algebraic expressions? If the expression contains parentheses, what property might we apply first? ▶

*See Video 1.5* 🍎

## 1.5 Exercise Set MyMathLab® ▶

**Objective A** *Simplify each expression. See Examples 1 through 5.*

**1.** $3(5 - 7)^4$

**2.** $7(3 - 8)^2$

▶**3.** $-3^2 + 2^3$

**4.** $-5^2 - 2^4$

**5.** $\dfrac{3.1 - (-1.4)}{-0.5}$

**6.** $\dfrac{4.2 - (-8.2)}{-0.4}$

**7.** $|3.6 - 7.2| + |3.6 + 7.2|$

**8.** $|8.6 - 1.9| - |2.1 + 5.3|$

**9.** $(-3)^2 + 2^3$

**10.** $(-15)^2 - 2^4$

▶**11.** $-8 \div 4 \cdot 2$

**12.** $-20 \div 5 \cdot 4$

**13.** $4[8 - (2 - 4)]$

**14.** $3[11 - (1 - 3)]$

**15.** $-8\left(-\dfrac{3}{4}\right) - 8$

**16.** $-10\left(-\dfrac{2}{5}\right) - 10$

**17.** $2 - [(7 - 6) + (9 - 19)]$

**18.** $8 - [(4 - 7) + (8 - 1)]$

**19.** $\dfrac{(-9 + 6)(-1^2)}{-2 - 2}$

**20.** $\dfrac{(-1 - 2)(-3^2)}{-6 - 3}$

**21.** $\left(\sqrt[3]{8}\right)(-4) - \left(\sqrt{9}\right)(-5)$

**22.** $\left(\sqrt[3]{27}\right)(-5) - \left(\sqrt{25}\right)(-3)$

**23.** $12 + \{6 - [5 - 2(-5)]\}$

**24.** $18 + \{9 - [1 - 6(-3)]\}$

**25.** $25 - [(3 - 5) + (14 - 18)]^2$

**26.** $10 - [(4 - 5)^2 + (12 - 14)]^4$

▶**27.** $\dfrac{(3 - \sqrt{9}) - (-5 - 1.3)}{-3}$

**28.** $\dfrac{-\sqrt{16} - (6 - 2.4)}{-2}$

**29.** $\dfrac{|3-9|-|-5|}{-3}$

**30.** $\dfrac{|-14|-|2-7|}{-15}$

**31.** $\dfrac{3(-2+1)}{5} - \dfrac{-7(2-4)}{1-(-2)}$

**32.** $\dfrac{-1-2}{2(-3)+10} - \dfrac{2(-5)}{-1(8)+1}$

**33.** $\dfrac{\frac{1}{3}\cdot 9 - 7}{3 + \frac{1}{2}\cdot 4}$

**34.** $\dfrac{\frac{1}{5}\cdot 20 - 6}{10 + \frac{1}{4}\cdot 12}$

**35.** $3\{-2 + 5[1 - 2(-2 + 5)]\}$

**36.** $2\{-1 + 3[7 - 4(-10 + 12)]\}$

**37.** $\dfrac{-4\sqrt{80+1} + (-4)^2}{3^3 + |-2(3)|}$

**38.** $\dfrac{(-2)^4 + 3\sqrt{120-20}}{4^3 + |5(-1)|}$

▦ **39.** $-150(3.25 - 1.68)$

▦ **40.** $-290(9.61 - 6.27)$

▦ **41.** $\left(\dfrac{5.6 - 8.4}{1.9 - 2.7}\right)^2$

▦ **42.** $\left(\dfrac{9.4 - 10.8}{8.7 - 7.9}\right)^2$

**Objective B** *Evaluate each expression when $x = 9$ and $y = -2$. See Example 6.*

**43.** $9x - 6y$

**44.** $4x - 10y$

**45.** $-3y^2$

**46.** $-7y^2$

**47.** $\dfrac{\sqrt{x}}{y} - \dfrac{y}{x}$

**48.** $\dfrac{y}{2x} - \dfrac{\sqrt{x}}{3y}$

▶ **49.** $\dfrac{3 + 2|x - y|}{x + 2y}$

**50.** $\dfrac{5 + 2|y - x|}{x + 6y}$

(*Hint:* Remember order of operations.)

**51.** $\dfrac{y^3 + \sqrt{x - 5}}{|4x - y|}$

**52.** $\dfrac{y^2 + \sqrt{x + 7}}{|3x - y|}$

*Complete each table. See Example 7.*

△ **53.** The algebraic expression $8 + 2y$ represents the perimeter of a rectangle with width 4 and length $y$.

**a.** Complete the table by evaluating this expression at the given values of $y$.

| Length | $y$ | 5 | 7 | 10 | 100 |
|---|---|---|---|---|---|
| Perimeter | $8 + 2y$ | | | | |

**b.** Use the results of the table in (a) to answer the following question. As the width of a rectangle remains the same and the length increases, does the perimeter increase or decrease? Explain how you arrived at your answer.

△ **54.** The algebraic expression $\pi r^2$ represents the area of a circle with radius $r$.

**a.** Complete the table by evaluating this expression at the given values of $r$. (Use 3.14 for $\pi$.)

| Radius | $r$ | 2 | 3 | 7 | 10 |
|---|---|---|---|---|---|
| Area | $\pi r^2$ | | | | |

**b.** As the radius of a circle increases, does its area increase or decrease? Explain your answer.

**55.** The algebraic expression $\dfrac{100x + 5000}{x}$ represents the cost per bookshelf (in dollars) of producing $x$ bookshelves.

**a.** Complete the table.

| Number of Bookshelves | $x$ | 10 | 100 | 1000 |
|---|---|---|---|---|
| Cost per Bookshelf | $\dfrac{100x + 5000}{x}$ | | | |

**b.** As the number of bookshelves manufactured increases, does the cost per bookshelf increase or decrease? Why do you think that this is so?

**56.** If $C$ is degrees Celsius, the algebraic expression $1.8C + 32$ represents the equivalent temperature in degrees Fahrenheit.

°C

**a.** Complete the table.

| Degrees Celsius | $C$ | −10 | 0 | 50 |
|---|---|---|---|---|
| Degrees Fahrenheit | $1.8C + 32$ | | | |

**b.** As degrees Celsius increase, do degrees Fahrenheit increase or decrease? Explain your answer.

*Solve.*

**57.** The average price for an ounce of gold in the United States for a month in 2013 was $1275.86. The algebraic expression $1275.86z$ gives the average cost for $z$ ounces of gold during this period. Find the cost if 6.5 ounces of gold had been purchased during this time. (*Source:* Max Keiser Financial War Reports).

**58.** The Comet Ison, which unfortunately did not survive its trip around our Sun in November 2013, was traveling at a speed of 377.75 kilometers per second as it approached the Sun. At this speed, the algebraic expression $377.75t$ gives the total distance covered in $t$ seconds. Find the distance covered by the comet in 5 seconds. (*Source:* NASA)

## Objective C  *Simplify. See Examples 8 through 12.*

**59.** $6x + 2x$

**60.** $9y - 11y$

**61.** $19y - y$

**62.** $14x - x$

**63.** $9x - 8 - 10x$

**64.** $14x - 1 - 20x$

**65.** $-9 + 4x + 18 - 10x$

**66.** $5y - 14 + 7y - 20y$

**67.** $3a - 4b + a - 9b$

**68.** $11x - y + 11x - 6y$

**69.** $x - y + x - y$

**70.** $a - b + 3a - 3b$

**71.** $1.5x + 2.3 - 0.7x - 5.9$

**72.** $6.3y - 9.7 + 2.2y - 11.1$

**73.** $\dfrac{3}{4}b - \dfrac{1}{2} + \dfrac{1}{6}b - \dfrac{2}{3}$

**74.** $\frac{7}{8}a - \frac{11}{12} - \frac{1}{2}a + \frac{5}{6}$

**75.** $8ab - 4.6 - 11ab - 8.2$

**76.** $4mn - 6.01 - 6mn - 8.1$

*Simplify. See Examples 13 through 16.*

**77.** $2(3x + 7)$

**78.** $4(5y + 12)$

**79.** $5k - (3k - 10)$

**80.** $-11c - (4 - 2c)$

**81.** $(3x + 4) - (6x - 1)$

**82.** $(8 - 5y) - (4 - 3y)$

▶**83.** $3(xy - 2) + xy + 15 - x^2$

**84.** $4(yz + 3) - 7yz + 1 + y^2$

**85.** $-(n + 5) + (5n - 3)$

**86.** $-(8 - t) + (2t - 6)$

**87.** $4(6n - 3) - 3(8n + 4)$

**88.** $5(2z - 6) + 10(3 - z)$

▶**89.** $\frac{1}{2}(10x - 2) - \frac{1}{6}(60x - 5y)$

**90.** $\frac{1}{4}(8x - 4) - \frac{1}{5}(20x - 6y)$

**91.** $3x - 2(x - 5) + x$

**92.** $7n + 3(2n - 6) - n$

**93.** $\frac{1}{6}(24a - 18b) - \frac{1}{7}(7a - 21b - 2) - \frac{1}{5}$

**94.** $\frac{1}{3}(6x - 33y) - \frac{1}{8}(24x - 40y + 1) - \frac{1}{3}$

**95.** $5.7a + 1.7 - 3(2.1a - 0.6)$

**96.** $6.2b + 5.1 - 2(4.2b - 0.1)$

**97.** $-4[6(2t + 1) - (9 + 10t)]$

**98.** $-5[3(4x + 2) - (13 + 8x)]$

**99.** $-1.2(5.7x - 3.6) + 8.75x$

**100.** $5.8(-9.6 - 31.2y) - 18.65$

**101.** $8.1z + 7.3(z + 5.2) - 6.85$

**102.** $6.5y - 4.4(1.8y - 3.3) + 10.95$

## Concept Extensions

*Insert parentheses so that, when simplified, each expression is equal to the given number.*

**103.** $2 + 7 \cdot 1 + 3; \quad 36$

**104.** $6 - 5 \cdot 2 + 2; \quad -6$

**105.** Write an algebraic expression that simplifies to $-3x - 1$.

**106.** Write an algebraic expression that simplifies to $-7x - 4$.

*Find and correct the error in each expression. See the last Concept Check in this section.*

**107.** $(3x + 2) - (5x + 7) = 3x + 2 - 5x + 7$
$$= -2x + 9$$

**108.** $2 + 3(7 - 6x) = 5(7 - 6x)$
$$= 35 - 30x$$

*The following graph is called a broken-line graph, or simply a line graph. This particular graph shows the past, present, and future predicted U.S. population over 65. Just as with a bar graph, to find the population over 65 for a particular year, read the height of the corresponding point. To read the height, follow the point horizontally to the left until you reach the vertical axis. Use this graph to answer Exercises 109 through 114.*

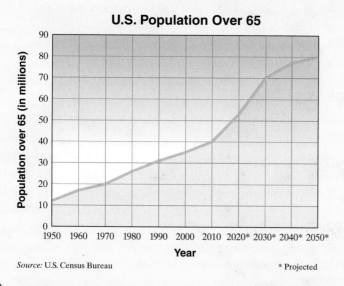

**U.S. Population Over 65**

Source: U.S. Census Bureau

\* Projected

**109.** Estimate the population over 65 in the year 1970.

**110.** Estimate the predicted population over 65 in the year 2050.

**111.** Estimate the predicted population over 65 in the year 2030.

**112.** Estimate the population over 65 in the year 2000.

**113.** Is the population over 65 increasing as time passes or decreasing? Explain how you arrived at your answer.

**114.** The percent of Americans over 65 in 1950 was 8.1%. The percent of Americans over 65 in 2050 is expected to be 2.5 times the percent over 65 in 1950. Estimate the percent of Americans expected to be over age 65 in 2050.

*Each circle below represents a whole, or 1. Determine the unknown fractional part of each circle.*

**115.**

**116.**

*Find the value of the expression when $x_1 = 2$, $x_2 = 4$, $y_1 = -3$, $y_2 = 2$.*

**117.** $\dfrac{y_2 - y_1}{x_2 - x_1}$

**118.** $\sqrt{(x_2 - x_1)^2 + (y_2 - y_1)^2}$

*Simplify. Round each result to the nearest ten-thousandth.*

**119.** $\dfrac{-1.682 - 17.895}{(-7.102)(-4.691)}$

**120.** $\dfrac{(-5.161)(3.222)}{7.955 - 19.676}$

## 1.6 Exponents and Scientific Notation

### Objective A  Using the Product Rule

Recall from Section 1.4 that exponents may be used to write repeated factors in a more compact form. As we have seen in the previous sections, exponents can be used when the repeated factor is a number or a variable. For example,

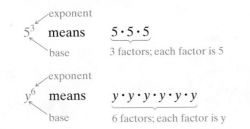

$5^3$ means $\underbrace{5 \cdot 5 \cdot 5}_{\text{3 factors; each factor is 5}}$

exponent
base

$y^6$ means $\underbrace{y \cdot y \cdot y \cdot y \cdot y \cdot y}_{\text{6 factors; each factor is y}}$

exponent
base

Expressions that contain exponents such as $5^3$ and $y^6$ are called **exponential expressions.**

Exponential expressions can be multiplied, divided, added, subtracted, and themselves raised to powers. In this section, we review operations on exponential expressions.

We review multiplication first. To multiply $x^2$ by $x^3$, we use the definition of $a^n$:

$$x^2 \cdot x^3 = \underbrace{(x \cdot x)(x \cdot x \cdot x)}_{x \text{ is a factor 5 times.}}$$

$$= x^5$$

Notice that the result is exactly the same if we add the exponents.

$$x^2 \cdot x^3 = x^{2+3} = x^5$$

This suggests the following rule.

> **Product Rule for Exponents**
>
> If $m$ and $n$ are integers and $a$ is a real number, then
>
> $$a^m \cdot a^n = a^{m+n}$$

In other words, the *product* of exponential expressions with a common base is the common base raised to a power equal to the *sum* of the exponents of the factors.

**Examples**  Use the product rule to simplify.

1. $2^2 \cdot 2^5 = 2^{2+5} = 2^7$
2. $x^7 \cdot x^3 = x^{7+3} = x^{10}$
3. $y \cdot y^2 \cdot y^4 = (y^1 \cdot y^2) \cdot y^4$
   $= y^3 \cdot y^4$
   $= y^7$

■ Work Practice 1–3

**Examples**  Use the product rule to simplify.

4. $(3x^6)(5x) = 3(5)x^6x^1 = 15x^7$  Use properties of multiplication to group like bases.
5. $(-2.4x^3p^2)(4xp^{10}) = -2.4(4)x^3x^1p^2p^{10} = -9.6x^4p^{12}$

■ Work Practice 4–5

### Objectives

A  Use the Product Rule for Exponents.

B  Simplify Expressions Raised to the Zero Power.

C  Use the Quotient Rule for Exponents.

D  Simplify Expressions Raised to Negative Powers.

E  Simplify Exponential Expressions Containing Variables in the Exponent.

F  Convert Between Scientific Notation and Standard Notation.

**Practice 1–3**

Use the product rule to simplify.
1. $5^2 \cdot 5^6$   2. $x^5 \cdot x^9$   3. $y \cdot y^4 \cdot y^3$

**Practice 4–5**

Use the product rule to simplify.
4. $(7y^5)(6y)$
5. $(-3x^2y^7)(5xy^6)$

**Answers**
1. $5^8$   2. $x^{14}$   3. $y^8$   4. $42y^6$
5. $-15x^3y^{13}$

## Objective B  Simplifying Expressions Raised to the Zero Power ▶

The definition of $a^n$ does not include the possibility that $n$ might be 0. But if it did, then, by the product rule,

$$a^0 \cdot a^n = a^{0+n} = a^n = 1 \cdot a^n$$

From this, we reasonably define that $a^0 = 1$, as long as $a$ does not equal 0.

### Zero Exponent

If $a$ does not equal 0, then $a^0 = 1$.

**Examples**    Evaluate each expression.

**6.** $7^0 = 1$

**7.** $-7^0 = -7^0 = -1$    Without parentheses, only 7 is raised to the 0 power.

**8.** $(2x + 5)^0 = 1$

**9.** $2x^0 = 2(1) = 2$

▪ Work Practice 6–9

## Objective C  Using the Quotient Rule ▶

To find quotients of exponential expressions, we again begin with the definition of $a^n$ to simplify $\dfrac{x^9}{x^2}$:

$$\frac{x^9}{x^2} = \frac{x \cdot x \cdot x \cdot x \cdot x \cdot x \cdot x \cdot x \cdot x}{x \cdot x} = \frac{x}{x} \cdot \frac{x}{x} \cdot \frac{x^7}{1} = 1 \cdot 1 \cdot x^7 = x^7$$

(Assume for the rest of this section that denominators containing variables are not 0.) Notice that the result is exactly the same if we subtract the exponents.

$$\frac{x^9}{x^2} = x^{9-2} = x^7$$

This suggests the following rule.

### Quotient Rule for Exponents

If $a$ is a nonzero real number and $m$ and $n$ are integers, then

$$\frac{a^m}{a^n} = a^{m-n}$$

In other words, the *quotient* of exponential expressions with a common base is the common base raised to a power equal to the *difference* of the exponents.

**Examples**    Use the quotient rule to simplify.

**10.** $\dfrac{5^8}{5^2} = 5^{8-2} = 5^6$

**11.** $\dfrac{x^7}{x^4} = x^{7-4} = x^3$

**12.** $\dfrac{20x^6}{4x^5} = 5x^{6-5} = 5x^1$, or $5x$

**13.** $\dfrac{12y^{10}z^7}{14y^8z^7} = \dfrac{6}{7}y^{10-8} \cdot z^{7-7} = \dfrac{6}{7}y^2 \cdot z^0 = \dfrac{6}{7}y^2 \cdot 1 = \dfrac{6}{7}y^2$ or $\dfrac{6y^2}{7}$

▪ Work Practice 10–13

---

**Practice 6–9**

Evaluate each expression.

**6.** $8^0$

**7.** $-14^0$

**8.** $(y - 3)^0$

**9.** $5x^0$

**Practice 10–13**

Use the quotient rule to simplify.

**10.** $\dfrac{6^{10}}{6^2}$    **11.** $\dfrac{y^6}{y^2}$

**12.** $\dfrac{36x^5}{9x}$    **13.** $\dfrac{10a^7b^9}{15a^5b^9}$

**Answers**

**6.** 1  **7.** −1  **8.** 1  **9.** 5  **10.** $6^8$

**11.** $y^4$  **12.** $4x^4$  **13.** $\dfrac{2}{3}a^2$

# Objective D Simplifying Expressions Raised to Negative Powers ▶

When the exponent of the denominator is larger than the exponent of the numerator, applying the quotient rule gives a negative exponent. For example,

$$\frac{x^3}{x^5} = x^{3-5} = x^{-2}$$

However, using the definition of $a^n$ gives us

$$\frac{x^3}{x^5} = \frac{x \cdot x \cdot x}{x \cdot x \cdot x \cdot x \cdot x} = 1 \cdot 1 \cdot 1 \cdot \frac{1}{x^2} = \frac{1}{x^2}$$

From this, we reasonably define $x^{-2} = \dfrac{1}{x^2}$ or, in general, $a^{-n} = \dfrac{1}{a^n}$.

> ## Negative Exponents
>
> If $a$ is a real number other than 0 and $n$ is a positive integer, then
>
> $$a^{-n} = \frac{1}{a^n}$$

**Examples** Simplify and write with positive exponents only.

**14.** $5^{-2} = \dfrac{1}{5^2} = \dfrac{1}{25}$

**15.** $2x^{-3} = 2 \cdot \dfrac{1}{x^3} = \dfrac{2}{x^3}$  *Without parentheses, only $x$ is raised to the $-3$ power.*

**16.** $(3x)^{-1} = \dfrac{1}{(3x)^1} = \dfrac{1}{3x}$  *With parentheses, both 3 and $x$ are raised to the $-1$ power.*

**17.** $2^{-1} + 3^{-2} = \dfrac{1}{2^1} + \dfrac{1}{3^2} = \dfrac{1}{2} + \dfrac{1}{9} = \dfrac{9}{18} + \dfrac{2}{18} = \dfrac{11}{18}$

**18.** $\dfrac{1}{t^{-5}} = \dfrac{1}{\frac{1}{t^5}} = 1 \div \dfrac{1}{t^5} = 1 \cdot \dfrac{t^5}{1} = t^5$

■ Work Practice 14–18

> ### Helpful Hint
>
> Notice that when a factor containing an exponent is moved from the numerator to the denominator or from the denominator to the numerator, the sign of its exponent changes.
>
> $$x^{-3} = \frac{1}{x^3} \qquad 5^{-2} = \frac{1}{5^2} = \frac{1}{25}$$
>
> $$\frac{1}{t^{-5}} = t^5 \qquad \frac{1}{2^{-3}} = 2^3 = 8$$

**Examples** Simplify and write using positive exponents only.

**19.** $\dfrac{m^5}{m^{15}} = m^{5-15} = m^{-10} = \dfrac{1}{m^{10}}$

**20.** $\dfrac{3^3}{3^6} = 3^{3-6} = 3^{-3} = \dfrac{1}{3^3} = \dfrac{1}{27}$

*(Continued on next page)*

## Practice 14–18

Simplify and write with positive exponents only.

**14.** $7^{-2}$

**15.** $5x^{-4}$

**16.** $(2x)^{-1}$

**17.** $3^{-1} + 2^{-2}$

**18.** $\dfrac{1}{y^{-4}}$

## Practice 19–25

Simplify and write using positive exponents only.

**19.** $\dfrac{x^3}{x^{10}}$  **20.** $\dfrac{4^2}{4^5}$  **21.** $y^{-10} \cdot y^3$

**22.** $\dfrac{q^5}{q^{-4}}$  **23.** $\dfrac{5^{-4}}{5^{-2}}$  **24.** $\dfrac{10x^{-8}y^5}{20xy^{-5}}$

**25.** $\dfrac{(4x^{-1})(x^5)}{x^7}$

**Answers**

**14.** $\dfrac{1}{49}$  **15.** $\dfrac{5}{x^4}$  **16.** $\dfrac{1}{2x}$  **17.** $\dfrac{7}{12}$

**18.** $y^4$  **19.** $\dfrac{1}{x^7}$  **20.** $\dfrac{1}{64}$  **21.** $\dfrac{1}{y^7}$

**22.** $q^9$  **23.** $\dfrac{1}{25}$  **24.** $\dfrac{y^{10}}{2x^9}$  **25.** $\dfrac{4}{x^3}$

**21.** $x^{-9} \cdot x^2 = x^{-9+2} = x^{-7} = \dfrac{1}{x^7}$

**22.** $\dfrac{5p^4}{p^{-3}} = 5 \cdot p^{4-(-3)} = 5p^7$

**23.** $\dfrac{2^{-3}}{2^{-1}} = 2^{-3-(-1)} = 2^{-2} = \dfrac{1}{2^2} = \dfrac{1}{4}$

**24.** $\dfrac{2x^{-7}y^2}{10xy^{-5}} = \dfrac{x^{-7-1} \cdot y^{2-(-5)}}{5} = \dfrac{x^{-8}y^7}{5} = \dfrac{y^7}{5x^8}$

**25.** $\dfrac{(3x^{-3})(x^2)}{x^6} = \dfrac{3x^{-3+2}}{x^6} = \dfrac{3x^{-1}}{x^6} = 3x^{-1-6} = 3x^{-7} = \dfrac{3}{x^7}$

■ **Work Practice 19–25**

✔**Concept Check**  Find and correct the error in the following:

$$\dfrac{y^{-6}}{y^{-2}} = y^{-6-2} = y^{-8} = \dfrac{1}{y^8}$$

## Objective  E  Simplifying with Variables in the Exponent

  Simplify. Assume that $a$ and $t$ are nonzero integers and that $x$ is not 0.

**26.** $x^{2a} \cdot x^3 = x^{2a+3}$     Use the product rule.

**27.** $\dfrac{x^{2t-1}}{x^{t-5}} = x^{(2t-1)-(t-5)}$     Use the quotient rule.

$\qquad\quad = x^{2t-1-t+5} = x^{t+4}$

■ **Work Practice 26–27**

## Objective  F  Converting Between Scientific Notation and Standard Notation

Very large and very small numbers occur frequently in nature. For example, the distance between Earth and the Sun is approximately 150,000,000 kilometers. A helium atom has a diameter of 0.000000022 centimeters. It can be tedious to write these very large and very small numbers in standard notation like this. **Scientific notation** is a convenient shorthand notation for writing very large and very small numbers.

Helium atom
0.000000022 cm

150,000,000 km

### Scientific Notation

A positive number is written in **scientific notation** if it is written as the product of a number $a$, where $1 \le a < 10$, and an integer power $n$ of 10: $a \times 10^n$.

For example,

$2.03 \times 10^2 \qquad 7.362 \times 10^7 \qquad 8.1 \times 10^{-5}$

---

**Practice 26–27**

Simplify. Assume that $n$ and $m$ are nonzero integers and that $x$ is not 0.

**26.** $x^{3m} \cdot x^n$

**27.** $\dfrac{x^{2m-2}}{x^{m-6}}$

**Answers**

**26.** $x^{3m+n}$   **27.** $x^{m+4}$

✔**Concept Check Answer**

$\dfrac{y^{-6}}{y^{-2}} = y^{-6-(-2)} = y^{-4} = \dfrac{1}{y^4}$

## Writing a Number in Scientific Notation

**Step 1:** Move the decimal point in the original number until the new number has a value between 1 and 10.

**Step 2:** Count the number of decimal places the decimal point was moved in Step 1. If the original number is 10 or greater, the count is positive. If the original number is less than 1, the count is negative.

**Step 3:** Write the product of the new number from Step 1 and 10 raised to an exponent equal to the count found in Step 2.

**Example 28**   Write 730,000 in scientific notation.

**Solution:**

**Step 1:** Move the decimal point until the number is between 1 and 10.

730,000.

**Step 2:** The decimal point is moved 5 places and the original number is 10 or greater, so the count is positive 5.

**Step 3:** $730,000 = 7.3 \times 10^5$

■ Work Practice 28

**Example 29**   Write 0.00000104 in scientific notation.

**Solution:**

**Step 1:** Move the decimal point until the number is between 1 and 10.

0.00000104

**Step 2:** The decimal point is moved 6 places and the original number is less than 1, so the count is $-6$.

**Step 3:** $0.00000104 = 1.04 \times 10^{-6}$

■ Work Practice 29

To write a scientific notation number in standard form, we reverse the preceding steps.

## Writing a Scientific Notation Number in Standard Notation

Move the decimal point in the number the same number of places as the exponent on 10. If the exponent is positive, move the decimal point to the right. If the exponent is negative, move the decimal point to the left.

**Examples**   Write each number in standard notation.

**30.** $7.7 \times 10^8 = 770,000,000$   Since the exponent is positive, move the decimal point 8 places to the right. Add zeros as needed.

**31.** $1.025 \times 10^{-3} = 0.001025$   Since the exponent is negative, move the decimal point 3 places to the left. Add zeros as needed.

■ Work Practice 30–31

✓**Concept Check** Which of the following numbers have values that are less than 1?

   **a.** $3.5 \times 10^{-5}$   **b.** $3.5 \times 10^5$   **c.** $-3.5 \times 10^5$   **d.** $-3.5 \times 10^{-5}$

**Practice 28**

Write 1,760,000 in scientific notation.

**Practice 29**

Write 0.00028 in scientific notation.

**Practice 30–31**

Write each number in standard notation.
**30.** $8.6 \times 10^7$
**31.** $3.022 \times 10^{-4}$

**Answers**
**28.** $1.76 \times 10^6$   **29.** $2.8 \times 10^{-4}$
**30.** 86,000,000   **31.** 0.0003022

✓**Concept Check Answer**
**a, c, d**

## Calculator Explorations

Multiply 5,000,000 by 700,000 on your calculator. The display should read $\boxed{3.5 \quad 12}$ or $\boxed{3.5\ E\ 12}$, which is the product written in scientific notation. Both these notations mean $3.5 \times 10^{12}$.

To enter a number written in scientific notation on a calculator, find the key marked $\boxed{EE}$. (On some calculators, this key may be marked $\boxed{EXP}$.)
To enter $7.26 \times 10^{13}$, press the keys

$$\boxed{7.26} \quad \boxed{EE} \quad \boxed{13}$$

The display will read $\boxed{7.26 \quad 13}$ or $\boxed{7.26\ E\ 13}$.

*Use your calculator to perform each operation indicated.*

**1.** Multiply $3 \times 10^{11}$ and $2 \times 10^{32}$.

**2.** Divide $6 \times 10^{14}$ by $3 \times 10^{9}$.

**3.** Multiply $5.2 \times 10^{23}$ and $7.3 \times 10^{4}$.

**4.** Divide $4.38 \times 10^{41}$ by $3 \times 10^{17}$.

## Vocabulary, Readiness & Video Check

*State the base of the exponent 5 in each expression.*

**1.** $9x^5$     **2.** $yz^5$     **3.** $-3^5$     **4.** $(-3)^5$     **5.** $(y^7)^5$     **6.** $9 \cdot 2^5$

*Write each expression with positive exponents.*

**7.** $5x^{-1}y^{-2}$     **8.** $7xy^{-4}$     **9.** $a^2b^{-1}c^{-5}$     **10.** $a^{-4}b^2c^{-6}$     **11.** $\dfrac{y^{-2}}{x^{-4}}$     **12.** $\dfrac{x^{-7}}{z^{-3}}$

### Martin-Gay Interactive Videos

*Watch the section lecture video and answer the following questions.*

See Video 1.6

**Objective A**   **13.** Why are we reminded that multiplication is commutative and associative during the simplifying of Example 2?

**Objective B**   **14.** Example 3 does not contain parentheses yet a discussion of parentheses is an important part of evaluating. Explain.

**Objective C**   **15.** When applying the quotient rule in Example 5, how do you know which exponents to subtract?

**Objective D**   **16.** What important reminder is made at the beginning of Example 6?

**Objective E**   **17.** In Example 9, what rule for exponents is used to simplify the expression?

**Objective F**   **18.** From Examples 10 and 11, explain how the direction of the movement of the decimal point in Step 1 might suggest another way to determine the sign of the exponent on 10.

## 1.6 Exercise Set MyMathLab® ▶

**Objective A** *Use the product rule to simplify each expression. See Examples 1 through 5.*

**1.** $4^2 \cdot 4^3$

**2.** $3^3 \cdot 3^5$

**▶3.** $x^5 \cdot x^3$

**4.** $a^2 \cdot a^9$

**5.** $m \cdot m^7 \cdot m^6$

**6.** $n \cdot n^{10} \cdot n^{12}$

**7.** $(4xy)(-5x)$

**8.** $(-7xy)(7y)$

**▶9.** $(-4x^3p^2)(4y^3x^3)$

**10.** $(-6a^2c^3)(-3ab^3)$

**Objective B** *Evaluate each expression. See Examples 6 through 9.*

**▶11.** $-8^0$

**12.** $(-9)^0$

**13.** $(4x + 5)^0$

**14.** $(3x - 1)^0$

**15.** $-x^0$

**16.** $-5x^0$

**▶17.** $4x^0 + 5$

**18.** $8x^0 + 1$

**Objective C** *Use the quotient rule to simplify. See Examples 10 through 13.*

**19.** $\dfrac{a^5}{a^2}$

**20.** $\dfrac{x^9}{x^4}$

**21.** $-\dfrac{26z^{11}}{2z^7}$

**22.** $-\dfrac{16x^5}{8x}$

**▶23.** $\dfrac{x^9y^6}{x^8y^6}$

**24.** $\dfrac{a^{12}b^2}{a^9b}$

**25.** $\dfrac{12x^4y^7}{9xy^5}$

**26.** $\dfrac{24a^{10}b^{11}}{10ab^3}$

**27.** $\dfrac{-36a^5b^7c^{10}}{6ab^3c^4}$

**28.** $\dfrac{49a^3bc^{14}}{-7abc^8}$

**Objective D** *Simplify and write using positive exponents only. See Examples 14 through 25.*

**▶29.** $4^{-2}$

**30.** $2^{-3}$

**31.** $(-3)^{-3}$

**32.** $(-6)^{-2}$

**33.** $\dfrac{x^7}{x^{15}}$

**34.** $\dfrac{z}{z^3}$

**35.** $5a^{-4}$

**36.** $10b^{-1}$

**37.** $\dfrac{x^{-7}}{y^{-2}}$

**38.** $\dfrac{p^{-13}}{q^{-3}}$

**39.** $\dfrac{x^{-2}}{x^5}$

**40.** $\dfrac{z^{-12}}{z^{10}}$

**41.** $\dfrac{8r^4}{2r^{-4}}$

**42.** $\dfrac{3s^3}{15s^{-3}}$

**43.** $\dfrac{x^{-9}x^4}{x^{-5}}$

**44.** $\dfrac{y^{-7}y}{y^8}$

**45.** $\dfrac{2a^{-6}b^2}{18ab^{-5}}$

**46.** $\dfrac{18ab^{-6}}{3a^{-3}b^6}$

**47.** $\dfrac{(24x^8)(x)}{20x^{-7}}$

**48.** $\dfrac{(30z^2)(z^5)}{55z^{-4}}$

## Objectives A B C D Mixed Practice *Simplify and write using positive exponents only. See Examples 1 through 25.*

**49.** $-7x^3 \cdot 20x^9$

**50.** $-3y \cdot -9y^4$

**51.** $x^7 \cdot x^8 \cdot x$

**52.** $y^6 \cdot y \cdot y^9$

**53.** $2x^3 \cdot 5x^7$

**54.** $-3z^4 \cdot 10z^7$

**55.** $(5x)^0 + 5x^0$

**56.** $4y^0 - (4y)^0$

**57.** $\dfrac{z^{12}}{z^{15}}$

**58.** $\dfrac{x^{11}}{x^{20}}$

**59.** $3^0 - 3t^0$

**60.** $4^0 + 4x^0$

**61.** $\dfrac{y^{-3}}{y^{-7}}$

**62.** $\dfrac{y^{-6}}{y^{-9}}$

**63.** $4^{-1} + 3^{-2}$

**64.** $1^{-3} - 4^{-2}$

**65.** $3x^{-1}$

**66.** $(4x)^{-1}$

**67.** $\dfrac{r^4}{r^{-4}}$

**68.** $\dfrac{x^{-5}}{x^3}$

**69.** $\dfrac{x^{-7}y^{-2}}{x^2y^2}$

**70.** $\dfrac{a^{-5}b^7}{a^{-2}b^{-3}}$

**71.** $(-4x^2y)(3x^4)(-2xy^5)$

**72.** $(-6a^4b)(2b^3)(-3ab^6)$

**73.** $2^{-4} \cdot x$

**74.** $5^{-2} \cdot y$

**75.** $\dfrac{5^{17}}{5^{13}}$

**76.** $\dfrac{10^{25}}{10^{23}}$

**77.** $\dfrac{8^{-7}}{8^{-6}}$

**78.** $\dfrac{13^{-10}}{13^{-9}}$

**79.** $\dfrac{9^{-5}a^4}{9^{-3}a^{-1}}$

**80.** $\dfrac{11^{-9}b^3}{11^{-7}b^{-4}}$

**81.** $\dfrac{14x^{-2}yz^{-4}}{2xyz}$

**82.** $\dfrac{30x^{-7}yz^{-14}}{3xyz}$

## Objective E *Simplify. Assume that variables in the exponents represent nonzero integers and that x, y, and z are not 0. See Examples 26 and 27.*

**83.** $x^5 \cdot x^{7a}$

**84.** $y^{2p} \cdot y^{9p}$

**85.** $\dfrac{x^{3t-1}}{x^t}$

**86.** $\dfrac{y^{4p-2}}{y^{3p}}$

**87.** $x^{4a} \cdot x^7$

**88.** $x^{9y} \cdot x^{-7y}$

**89.** $\dfrac{z^{6x}}{z^7}$

**90.** $\dfrac{y^6}{y^{4z}}$

**91.** $\dfrac{x^{3t} \cdot x^{4t-1}}{x^t}$

**92.** $\dfrac{z^{5x} \cdot z^{x-7}}{z^x}$

## Objective F *Write each number in scientific notation. See Examples 28 and 29.*

**93.** 31,250,000

**94.** 678,000

**95.** 0.016

**96.** 0.007613

**97.** 67,413

**98.** 36,800,000

**99.** 0.0125

**100.** 0.00084

**101.** 0.000053

**102.** 98,700,000,000

*Write each number in scientific notation.*

**103.** The approximate distance between Jupiter and the sun is 778,300,000 kilometers. (*Source:* National Space Data Center)

**104.** For the 2013 Major League Baseball season, the payroll of the World Series Champion Boston Red Sox was approximately $140,660,000. (*Source:* ESPN)

**105.** The exoplanet identified as WASP-17b is a "hot Jupiter," or a massive gaseous world that orbits close to its host star. This exoplanet is especially interesting because it is one of the few where water has been detected in its atmosphere. WASP-17b orbits a star about 1000 light years away in the constellation Scorpius. Thus, WASP-17b is about 6,000,000,000,000,000 miles from Earth. Write the distance of WASP-17b from Earth in scientific notation.

**106.** By the end of November 2013, Nintendo had sold 101,570,000 Wii units. (*Source:* Nintendo)

**107.** Lake Mead, created from the Colorado River by the Hoover Dam, has a capacity of 124,000,000,000 cubic feet of water. (*Source:* U.S. Bureau of Reclamation)

**108.** The temperature of the core of the Sun is about 27,000,000°F.

**109.** A pulsar is a rotating neutron star that gives off sharp, regular pulses of radio waves. For one particular pulsar, the rate of pulses is every 0.001 second.

△ **110.** To convert from cubic inches to cubic meters, multiply by 0.0000164.

*Write each number in standard notation. See Examples 30 and 31.*

▶ **111.** $3.6 \times 10^{-9}$     **112.** $2.7 \times 10^{-5}$     ▶ **113.** $9.3 \times 10^7$     **114.** $6.378 \times 10^8$     **115.** $1.278 \times 10^6$

**116.** $7.6 \times 10^4$     **117.** $7.35 \times 10^{12}$     **118.** $1.66 \times 10^{-5}$     **119.** $4.03 \times 10^{-7}$     **120.** $8.007 \times 10^8$

*Write each number in standard notation.*

**121.** The estimated world population in 1 A.D. was $3.0 \times 10^8$. (*Source: World Almanac and Book of Facts*)

**122.** There are $3.949 \times 10^6$ miles of highways, roads, and streets in the United States. (*Source:* Bureau of Transportation Statistics)

**123.** Chiricahua National Monument in Arizona contains $9.5 \times 10^{-3}$ square kilometers of privately-owned land.

**124.** Hagerman Fossil Beds National Monument in Idaho contains $6.68 \times 10^{-2}$ square kilometers of privately-owned land.

**125.** In December 2013, Coca-Cola averaged $1.55 \times 10^5$ followers on Twitter. (*Source:* The Coca-Cola Company)

**126.** In 2013, McDonald's served approximately $6.8 \times 10^7$ customers per day. (*Source:* McDonald's)

## Concept Extensions

**127.** Explain how to convert a number from standard notation to scientific notation.

**128.** Explain how to convert a number from scientific notation to standard notation.

**129.** Explain why $(-5)^0$ simplifies to 1 but $-5^0$ simplifies to $-1$.

**130.** Explain why both $4x^0 - 3y^0$ and $(4x - 3y)^0$ simplify to 1.

---

## 1.7 More Work with Exponents and Scientific Notation

### Objectives

**A** Use the Power Rules for Exponents.

**B** Use Exponent Rules and Definitions to Simplify Exponential Expressions.

**C** Simplify Exponential Expressions Containing Variables in the Exponent.

**D** Use Scientific Notation to Compute.

### Objective A  Using the Power Rules

The volume of the cube shown whose side measures $x^2$ units is $(x^2)^3$ cubic units. To simplify an expression such as $(x^2)^3$, we use the definition of $a^n$:

$$(x^2)^3 = \underbrace{(x^2)(x^2)(x^2)}_{x^2 \text{ is a factor 3 times.}} = x^{2+2+2} = x^6$$

$x^2$ units

Notice that the result is exactly the same if the exponents are multiplied.

$$(x^2)^3 = x^{2 \cdot 3} = x^6$$

This suggests that an expression raised to a power that is then all raised to another power is equal to the original expression raised to the product of the powers. Two additional rules for exponents are given in the following box.

## Power Rule and Power of a Product or Quotient Rules for Exponents

If $a$ and $b$ are real numbers and $m$ and $n$ are integers, then

$(a^m)^n = a^{m \cdot n}$     Power rule

$(ab)^m = a^m b^m$     Power of a product

$\left(\dfrac{a}{b}\right)^n = \dfrac{a^n}{b^n}$     $(b \neq 0)$     Power of a quotient

**Examples**   Use the power rule to simplify each expression. Write each answer using positive exponents only.

**1.** $(x^5)^7 = x^{5 \cdot 7} = x^{35}$

**2.** $(2^2)^3 = 2^{2 \cdot 3} = 2^6 = 64$

**3.** $(5^{-1})^2 = 5^{-1 \cdot 2} = 5^{-2} = \dfrac{1}{5^2} = \dfrac{1}{25}$

**4.** $(y^{-3})^{-4} = y^{-3(-4)} = y^{12}$

■ **Work Practice 1–4**

**Examples**   Use the power rules to simplify each expression. Write each answer using positive exponents only.

**5.** $(5x^2)^3 = 5^3 \cdot (x^2)^3 = 5^3 \cdot x^{2 \cdot 3} = 125x^6$

**6.** $\left(\dfrac{2}{3}\right)^3 = \dfrac{2^3}{3^3} = \dfrac{8}{27}$

**7.** $\left(\dfrac{3p^4}{q^5}\right)^2 = \dfrac{(3p^4)^2}{(q^5)^2} = \dfrac{3^2 \cdot (p^4)^2}{(q^5)^2} = \dfrac{9p^8}{q^{10}}$

**8.** $\left(\dfrac{2^{-3}}{y}\right)^{-2} = \dfrac{(2^{-3})^{-2}}{y^{-2}}$

$= \dfrac{2^6}{y^{-2}} = 64y^2$   Use the negative exponent rule.

**9.** $(x^{-5}y^2z^{-1})^7 = (x^{-5})^7 \cdot (y^2)^7 \cdot (z^{-1})^7$

$= x^{-35}y^{14}z^{-7} = \dfrac{y^{14}}{x^{35}z^7}$

■ **Work Practice 5–9**

## Objective B  Using Exponent Rules to Simplify Expressions ▶

In the next few examples, we practice the use of several of the rules and definitions for exponents. The following is a summary of these rules and definitions.

### Practice 1–4

Use the power rule to simplify each expression. Write each answer using positive exponents only.

**1.** $(y^2)^8$   **2.** $(3^3)^2$

**3.** $(6^2)^{-1}$   **4.** $(x^{-5})^{-7}$

### Practice 5–9

Use the power rules to simplify each expression. Write each answer using positive exponents only.

**5.** $(3x^4)^3$   **6.** $\left(\dfrac{4}{5}\right)^2$

**7.** $\left(\dfrac{4m^5}{n^3}\right)^3$   **8.** $\left(\dfrac{2^{-1}}{y}\right)^{-3}$

**9.** $(a^{-4}b^3c^{-2})^6$

**Answers**

**1.** $y^{16}$   **2.** $729$   **3.** $\dfrac{1}{36}$   **4.** $x^{35}$   **5.** $27x^{12}$

**6.** $\dfrac{16}{25}$   **7.** $\dfrac{64m^{15}}{n^9}$   **8.** $8y^3$   **9.** $\dfrac{b^{18}}{a^{24}c^{12}}$

## Summary of Rules for Exponents

If $a$ and $b$ are real numbers and $m$ and $n$ are integers, then

| | | |
|---|---|---|
| Product rule | $a^m \cdot a^n = a^{m+n}$ | |
| Zero exponent | $a^0 = 1$ | $(a \neq 0)$ |
| Negative exponent | $a^{-n} = \dfrac{1}{a^n}$ | $(a \neq 0)$ |
| Quotient rule | $\dfrac{a^m}{a^n} = a^{m-n}$ | $(a \neq 0)$ |
| Power rule | $(a^m)^n = a^{m \cdot n}$ | |
| Power of a product | $(ab)^m = a^m \cdot b^m$ | |
| Power of a quotient | $\left(\dfrac{a}{b}\right)^n = \dfrac{a^n}{b^n}$ | $(b \neq 0)$ |

### Practice 10–13

Simplify each expression. Write each answer using positive exponents only.

**10.** $(7xy^{-2})^{-2}$

**11.** $\left(\dfrac{y^{-7}}{y^{-10}}\right)^{-3}$

**12.** $\left(\dfrac{3}{5}\right)^{-2}$

**13.** $\dfrac{6^{-2}x^{-4}y^{10}}{x^2 y^{-6}}$

**Examples** Simplify each expression. Write each answer using positive exponents only.

**10.** $(2x^0y^{-3})^{-2} = 2^{-2}(x^0)^{-2}(y^{-3})^{-2}$

$= 2^{-2}x^0 y^6$

$= \dfrac{1(y^6)}{2^2}$  Write $x^0$ as 1.

$= \dfrac{y^6}{4}$

**11.** $\left(\dfrac{x^{-5}}{x^{-2}}\right)^{-3} = \dfrac{(x^{-5})^{-3}}{(x^{-2})^{-3}} = \dfrac{x^{15}}{x^6} = x^{15-6} = x^9$

**12.** $\left(\dfrac{2}{7}\right)^{-2} = \dfrac{2^{-2}}{7^{-2}} = \dfrac{7^2}{2^2} = \dfrac{49}{4}$

**13.** $\dfrac{5^{-2}x^{-3}y^{11}}{x^2 y^{-5}} = 5^{-2}x^{-3-2}y^{11-(-5)} = 5^{-2}x^{-5}y^{16} = \dfrac{y^{16}}{5^2 x^5} = \dfrac{y^{16}}{25x^5}$

■ Work Practice 10–13

### Practice 14–15

Simplify each expression. Write each answer using positive exponents only.

**14.** $\left(\dfrac{4a^3b^2}{b^{-6}c}\right)^{-2}$

**15.** $\left(\dfrac{4x^3}{3y^{-1}}\right)^3\left(\dfrac{y^{-2}}{3x^{-1}}\right)^{-1}$

**Examples** Simplify each expression. Write each answer using positive exponents only.

**14.** $\left(\dfrac{3x^2y}{y^{-9}z}\right)^{-2} = \left(\dfrac{3x^2y^{10}}{z}\right)^{-2} = \dfrac{3^{-2}x^{-4}y^{-20}}{z^{-2}} = \dfrac{z^2}{3^2x^4y^{20}} = \dfrac{z^2}{9x^4y^{20}}$

**15.** $\left(\dfrac{3a^2}{2x^{-1}}\right)^3\left(\dfrac{x^{-3}}{4a^{-2}}\right)^{-1} = \dfrac{27a^6}{8x^{-3}} \cdot \dfrac{x^3}{4^{-1}a^2}$

$= \dfrac{27 \cdot 4 \cdot a^6 \cdot x^3 \cdot x^3}{8 \cdot a^2} = \dfrac{27a^4x^6}{2}$

■ Work Practice 14–15

**Answers**

**10.** $\dfrac{y^4}{49x^2}$  **11.** $\dfrac{1}{y^9}$  **12.** $\dfrac{25}{9}$  **13.** $\dfrac{y^{16}}{36x^6}$

**14.** $\dfrac{c^2}{16a^6b^{16}}$  **15.** $\dfrac{64x^8y^5}{9}$

## Objective C Simplifying with Variables in the Exponent

### Examples  Simplify. Assume that $a$ and $b$ are integers and that $x$ and $y$ are not 0.

**16.** $x^{-b}(2x^b)^2 = x^{-b}2^2x^{2b} = 4x^{-b+2b} = 4x^b$

**17.** $\dfrac{(y^{3a})^2}{y^{a-6}} = \dfrac{y^{2(3a)}}{y^{a-6}} = \dfrac{y^{6a}}{y^{a-6}} = y^{6a-(a-6)} = y^{6a-a+6} = y^{5a+6}$

■ Work Practice 16–17

### Practice 16–17

Simplify. Assume that $m$ and $n$ are integers and that $x$ and $y$ are not 0.

**16.** $x^{-n}(3x^n)^2$    **17.** $\dfrac{(y^{2m})^2}{y^{m-3}}$

## Objective D Using Scientific Notation to Compute ▶

To perform operations on numbers written in scientific notation, we use the properties of exponents.

### Examples  Perform each indicated operation. Write each answer in scientific notation.

**18.** $(8.1 \times 10^5)(5 \times 10^{-7}) = 8.1 \times 5 \times 10^5 \times 10^{-7}$
$$= 40.5 \times 10^{-2}$$
$$= (4.05 \times 10^1) \times 10^{-2}$$
$$= 4.05 \times 10^{-1}$$

**19.** $\dfrac{1.2 \times 10^4}{3 \times 10^{-2}} = \left(\dfrac{1.2}{3}\right)\left(\dfrac{10^4}{10^{-2}}\right) = 0.4 \times 10^{4-(-2)}$
$$= 0.4 \times 10^6 = (4 \times 10^{-1}) \times 10^6 = 4 \times 10^5$$

■ Work Practice 18–19

### Practice 18–19

Perform each indicated operation. Write each answer in scientific notation.

**18.** $(9.6 \times 10^6)(4 \times 10^{-8})$

**19.** $\dfrac{4.2 \times 10^7}{7 \times 10^{-3}}$

### Example 20  Use scientific notation to simplify: $\dfrac{2000 \times 0.000021}{700}$

**Solution:**

$$\dfrac{2000 \times 0.000021}{700} = \dfrac{(2 \times 10^3)(2.1 \times 10^{-5})}{7 \times 10^2} = \dfrac{2(2.1)}{7} \cdot \dfrac{10^3 \cdot 10^{-5}}{10^2}$$
$$= 0.6 \times 10^{-4}$$
$$= (6 \times 10^{-1}) \times 10^{-4}$$
$$= 6 \times 10^{-5}$$

■ Work Practice 20

### Practice 20

Use scientific notation to simplify:

$$\dfrac{3000 \times 0.000012}{400}$$

**Answers**
**16.** $9x^n$  **17.** $y^{3m+3}$  **18.** $3.84 \times 10^{-1}$
**19.** $6 \times 10^9$  **20.** $9 \times 10^{-5}$

## Vocabulary, Readiness & Video Check

*Simplify. See Examples 1 through 4, and Section 1.6.*

**1.** $(x^4)^5$    **2.** $(5^6)^2$    **3.** $x^4 \cdot x^5$    **4.** $x^7 \cdot x^8$    **5.** $(y^6)^7$

**6.** $(x^3)^4$    **7.** $(z^4)^9$    **8.** $(z^3)^7$    **9.** $(z^{-6})^{-3}$    **10.** $(y^{-4})^{-2}$

*Martin-Gay Interactive Videos*    Watch the section lecture video and answer the following questions

**Objective A** 11. Based on ▣ Example 1 and the lecture before, what are the differences between the power rule and product rule for exponents? ⊙

**Objective B** 12. Name all the rules and definitions used to simplify ▣ Example 5. ⊙

**Objective C** 13. In ▣ Example 6, to remove the parentheses in the numerator, were the exponents added or multiplied? ⊙

**Objective D** 14. In the ▣ Example 7 multiplication problem, why is the first product not the final answer? How can we tell that this first product is not written in scientific notation? ⊙

*See Video 1.7* 🍎

## 1.7 Exercise Set MyMathLab® ⊙

**Objective A** *Simplify. Write each answer using positive exponents only. See Examples 1 through 9.*

**1.** $(3^{-1})^2$

**2.** $(2^{-2})^2$

⊙**3.** $(x^4)^{-9}$

**4.** $(y^7)^{-3}$

**5.** $(3x^2y^3)^2$

**6.** $(4x^3yz)^2$

**7.** $\left(\dfrac{2x^5}{y^{-3}}\right)^4$

**8.** $\left(\dfrac{3a^{-4}}{b^7}\right)^3$

**9.** $(2a^2bc^{-3})^{-6}$

**10.** $(6x^{-6}y^7z^0)^{-2}$

⊙**11.** $\left(\dfrac{x^7y^{-3}}{z^{-4}}\right)^{-5}$

**12.** $\left(\dfrac{a^{-2}b^{-5}}{c^{-11}}\right)^{-6}$

**13.** $(-2^{-2}y^{-1})^{-3}$

**14.** $(-4^{-6}y^{-6})^{-4}$

**Objective B** *Simplify. Write each answer using positive exponents only. See Examples 10 through 15.*

**15.** $\left(\dfrac{a^{-4}}{a^{-5}}\right)^{-2}$

**16.** $\left(\dfrac{x^{-9}}{x^{-4}}\right)^{-3}$

**17.** $\left(\dfrac{6p^6}{p^{12}}\right)^2$

**18.** $\left(\dfrac{4p^6}{p^9}\right)^3$

**19.** $(-8y^3xa^{-2})^{-3}$

**20.** $(-5y^0x^2a^3)^{-3}$

**21.** $\left(\dfrac{3}{4}\right)^{-3}$

**22.** $\left(\dfrac{5}{8}\right)^{-2}$

⊙**23.** $\left(\dfrac{2a^{-2}b^5}{4a^2b^7}\right)^{-2}$

**24.** $\left(\dfrac{5x^7y^4}{10x^3y^{-2}}\right)^{-3}$

**25.** $\left(\dfrac{x^{-2}y^{-2}}{a^{-3}}\right)^{-7}$

**26.** $\left(\dfrac{x^{-1}y^{-2}}{z^{-3}}\right)^{-5}$

**Objectives A B** **Mixed Practice** *Simplify. Write each answer using positive exponents only. See Examples 1 through 15.*

**27.** $(y^{-5})^2$

**28.** $(z^{-2})^{13}$

**29.** $(5^{-1})^3$

**30.** $(8^2)^{-1}$

**31.** $(x^7)^{-9}$

**32.** $(y^{-4})^5$

**33.** $\left(\dfrac{x^4}{y^{-3}}\right)^{-5}$

**34.** $\left(\dfrac{a^{-3}}{b^{-6}}\right)^{-4}$

**35.** $(4x^2)^2$

**36.** $(-8x^3)^2$

**37.** $\left(\dfrac{4^{-4}}{y^3x}\right)^{-2}$

**38.** $\left(\dfrac{7^{-3}}{ab^2}\right)^{-2}$

**39.** $\left(\dfrac{2x^{-3}}{y^{-1}}\right)^{-3}$

**40.** $\left(\dfrac{n^5}{2m^{-2}}\right)^{-4}$

**41.** $\dfrac{4^{-1}x^2yz}{x^{-2}yz^3}$

**42.** $\dfrac{8^{-2}x^{-3}y^{11}}{x^2y^{-5}}$

**43.** $\left(\dfrac{3x^5}{6x^4}\right)^4$

**44.** $\left(\dfrac{8^{-3}}{y^2}\right)^{-2}$

**45.** $\dfrac{(y^3)^{-4}}{y^3}$

**46.** $\dfrac{2(y^3)^{-3}}{y^{-3}}$

**47.** $\dfrac{3^{-2}a^{-5}b^6}{4^{-2}a^{-7}b^{-3}}$

**48.** $\dfrac{2^{-3}m^{-4}n^{-5}}{5^{-2}m^{-5}n}$

**49.** $(4x^6y^5)^{-2}(6x^4y^3)$

**50.** $(5x^2y^4)^{-2}(3x^9y^4)$

**51.** $x^6(x^6bc)^{-6}$

**52.** $y^2(y^2bx)^{-4}$

**53.** $\dfrac{2^{-3}x^2y^{-5}}{5^{-2}x^7y^{-1}}$

**54.** $\dfrac{7^{-1}a^{-3}b^5}{a^2b^{-2}}$

**55.** $\left(\dfrac{2x^2}{y^4}\right)^3\left(\dfrac{2x^5}{y}\right)^{-2}$

**56.** $\left(\dfrac{3z^{-2}}{y}\right)^2\left(\dfrac{9y^{-4}}{z^{-3}}\right)^{-1}$

**Objective C** *Simplify. Assume that variables in the exponents represent integers and that all other variables are not 0. See Examples 16 and 17.*

**57.** $(x^{3a+6})^3$

**58.** $(x^{2b+7})^2$

**59.** $\dfrac{x^{4a}(x^{4a})^3}{x}$

**60.** $\dfrac{x^{-5y+2}x^{2y}}{x}$

**61.** $(b^{5x-2})^2$

**62.** $(c^{2a+3})^3$

**63.** $\dfrac{(y^{2a})^8}{y^{a-3}}$

**64.** $\dfrac{(y^{4a})^7}{y^{2a-1}}$

**65.** $\left(\dfrac{2x^{3t}}{x^{2t-1}}\right)^4$

**66.** $\left(\dfrac{3y^{5a}}{y^{-a+1}}\right)^2$

**67.** $\dfrac{25x^{2a+b}y^{2a-b}}{5x^{a-b}y^{a+b}}$

**68.** $\dfrac{16x^{3a-b}y^{4a+b}}{2x^{a-2b}y^{a+3b}}$

**Objective D** *Perform each indicated operation. Write each answer in scientific notation. See Examples 18 through 20.*

**69.** $(5 \times 10^{11})(2.9 \times 10^{-3})$

**70.** $(3.6 \times 10^{-12})(6 \times 10^9)$

**71.** $(2 \times 10^5)^3$

**72.** $(3 \times 10^{-7})^3$

**73.** $\dfrac{3.6 \times 10^{-4}}{9 \times 10^2}$

**74.** $\dfrac{1.2 \times 10^9}{2 \times 10^{-5}}$

**75.** $\dfrac{0.0069}{0.023}$

**76.** $\dfrac{0.00048}{0.0016}$

**77.** $\dfrac{18{,}200 \times 100}{91{,}000}$

**78.** $\dfrac{0.0003 \times 0.0024}{0.0006 \times 20}$

**79.** $\dfrac{6000 \times 0.006}{0.009 \times 400}$

**80.** $\dfrac{0.00016 \times 300}{0.064 \times 100}$

**81.** $\dfrac{0.00064 \times 2000}{16{,}000}$

**82.** $\dfrac{0.00072 \times 0.003}{0.00024}$

**83.** $\dfrac{66{,}000 \times 0.001}{0.002 \times 0.003}$

**84.** $\dfrac{0.0007 \times 11{,}000}{0.001 \times 0.0001}$

**85.** $\dfrac{9.24 \times 10^{15}}{(2.2 \times 10^{-2})(1.2 \times 10^{-5})}$

**86.** $\dfrac{(2.6 \times 10^{-3})(4.8 \times 10^{-4})}{1.3 \times 10^{-12}}$

*Solve.*

**87.** To convert from square inches to square meters, multiply by $6.452 \times 10^{-4}$. The area of the following square is $4 \times 10^{-2}$ square inches. Convert this area to square meters.

$4 \times 10^{-2}$ sq in.

**88.** To convert from cubic inches to cubic meters, multiply by $1.64 \times 10^{-5}$. A grain of salt is in the shape of a cube. If an average size of a grain of salt is $3.8 \times 10^{-6}$ cubic inches, convert this volume to cubic meters.

$3.8 \times 10^{-6}$ cu in.

## Concept Extensions

*Solve.*

**89.** Each side of the cube shown is $\dfrac{2x^{-2}}{y}$ meters. Find its volume.

$\dfrac{2x^{-2}}{y}$ m

**90.** The lot shown is in the shape of a parallelogram with base $\dfrac{3x^{-1}}{y^{-3}}$ feet and height $5x^{-7}$ feet. Find its area.

$5x^{-7}$ ft

$\dfrac{3x^{-1}}{y^{-3}}$ ft

**91.** The density $D$ of an object is equivalent to the quotient of its mass $M$ and volume $V$. Thus $D = \dfrac{M}{V}$. Express in scientific notation the density of an object whose mass is 500,000 pounds and whose volume is 250 cubic feet.

**92.** The density of ordinary water is $3.12 \times 10^{-2}$ tons per cubic foot. The volume of water in the largest of the Great Lakes, Lake Superior, is $4.269 \times 10^{14}$ cubic feet. Use the formula $D = \dfrac{M}{V}$ (see Exercise 91) to find the mass (in tons) of the water in Lake Superior. Express your answer in scientific notation. (*Source:* National Ocean Service)

**93.** Is there a number $a$ such that $a^{-1} = a^1$? If so, give the value of $a$.

**94.** Is there a number $a$ such that $a^{-2}$ is a negative number? If so, give the value of $a$.

**95.** Explain whether $0.4 \times 10^{-5}$ is written in scientific notation.

**96.** In your own words, explain why $a^0 = 1$ (as long as $a$ is not 0).

**97.** The estimated population of the United States in 2014 was $3.17 \times 10^8$ people. The land area of the United States is $3.536 \times 10^6$ square miles. Find the population density (number of people per square mile) for the United States in 2014. Round to the nearest whole. (*Source:* U.S. News & World Report)

**98.** The estimated population of China in 2014 was $1.39 \times 10^9$ people. The land area was $3.826 \times 10^6$ square miles. Find the population density (number of people per square mile) for China in 2014. Round to the nearest whole. (*Source:* World Population Review)

**99.** In June, 2014, the value of goods and services imported into the United States was $2.37 \times 10^{11}$ dollars. The estimated population of the United States was $3.17 \times 10^8$ people. Find the average value of imports per person in the United States for October 2013. Round to the nearest dollar. (*Source:* U.S. Census Bureau, Bureau of Economic Analysis)

**100.** In 2014, the population of the European Union was $5.07 \times 10^8$ people. At the same time, the population of the United States was $3.17 \times 10^8$ people. How many times greater was the population of the European Union than the population of the United States? Round to the nearest tenth. (*Source:* European Commission: Eurostats)

---

## Chapter 1 Group Activity

### Geometry Investigations

#### Sections 1.2 through 1.6

Recall that the perimeter of a figure is the distance around the outside of the figure. For a rectangle with length $l$ and width $w$, the perimeter of the rectangle is given by the expression $2l + 2w$.

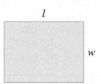

Area is a measure of the surface of a region. For example, we measure a plot of land or the floor space of a home by area. For a rectangle with length $l$ and width $w$, the area of the rectangle is given by the expression $lw$.

A circular cylinder can be formed by rolling a rectangle into a tube. The surface area of the cylinder (excluding the two ends of the cylinder) is the same as the area of the rectangle used to form the cylinder. Recall that volume is a measure of the space inside a three-dimensional region. The volume of a circular cylinder with height $h$ and radius $r$ is given by the expression $\pi r^2 h$.

### Group Activity

1. Work together to discover whether two rectangles with the same perimeter always have the same area. Explain your results. Give examples.

2. Do figures with the same surface area always have the same volume? To see, take two $8\frac{1}{2}$-by-11-inch sheets of paper and construct two cylinders using the following figures as a guide. Verify that both cylinders have the same surface area. Measure the height and radius of each resulting cylinder. Then find the volume of each cylinder to the nearest tenth of a cubic inch. Explain your results.

## Chapter 1 Vocabulary Check

*Fill in each blank with one of the words or phrases listed below.*

| | | | |
|---|---|---|---|
| distributive | absolute value | inequality | algebraic expression |
| real | opposite | commutative | exponent |
| reciprocals | associative | whole | variable |

**1.** A(n) _____ is formed by numbers and variables connected by the operations of addition, subtraction, multiplication, division, raising to powers, and/or taking roots.

**2.** The _____ of a number $a$ is $-a$.

**3.** $3(x - 6) = 3x - 18$ by the _____ property.

**4.** The _____ of a number is the distance between that number and 0 on a number line.

**5.** A(n) _____ is a shorthand notation for repeated multiplication of the same factor.

**6.** A letter that represents a number is called a(n) _____.

**7.** The symbols $<$ and $>$ are called _____ symbols.

**8.** If $a$ is not 0, then $a$ and $\frac{1}{a}$ are called _____.

**9.** $A + B = B + A$ by the _____ property of addition.

**10.** $(A + B) + C = A + (B + C)$ by the _____ property of addition.

**11.** The numbers $0, 1, 2, 3, \ldots$ are called _____ numbers.

**12.** If a number corresponds to a point on a number line, we know that number is a _____ number.

> **Helpful Hint**
>
> ▶ Are you preparing for your test? Don't forget to take the Chapter 1 Test on page 74. Then check your answers at the back of the text and use the Chapter Test Prep Videos to see the fully worked-out solutions to any of the exercises you want to review.

# 1 Chapter Highlights

| Definitions and Concepts | Examples |
|---|---|
| **Section 1.2    Algebraic Expressions and Sets of Numbers** | |
| Letters that represent numbers are called **variables**. | **Variables:** $\quad x, \quad a, \quad m, \quad y$ |
| An **algebraic expression** is formed by numbers and variables connected by the operations of addition, subtraction, multiplication, division, raising to powers, or taking roots. | **Algebraic Expressions:** $$7y, \quad -3, \quad \frac{x^2 - 9}{-2} + 14x, \quad \sqrt{3} + \sqrt{m}$$ |
| **Natural numbers:** $\{1, 2, 3, \ldots\}$ | List the elements of the set $\left\{-9.6, -5, -\sqrt{2}, 0, \frac{2}{5}, 101\right\}$ that belong to the set of |
| **Whole numbers:** $\{0, 1, 2, 3, \ldots\}$ | **Natural numbers:** $\quad 101$ |
| **Integers:** $\{\ldots, -3, -2, -1, 0, 1, 2, 3, \ldots\}$ | **Whole numbers:** $\quad 0, 101$ |
| Each listing of three dots is called an **ellipsis**. | **Integers:** $\quad -5, 0, 101$ |
| The objects of a set are called its members or **elements**. | Write the set $\{x \mid x \text{ is an integer between } -2 \text{ and } 5\}$ in roster form. |
| **Set builder notation** describes the elements of a set but does not list them. | $$\{-1, 0, 1, 2, 3, 4\}$$ |
| **Rational numbers:** $\left\{\frac{a}{b} \middle\| a \text{ and } b \text{ are integers and } b \neq 0\right\}$ | **Rational numbers:** $\quad -9.6, -5, 0, \frac{2}{5}, 101$ |
| **Irrational numbers:** $\{x \mid x \text{ is a real number and } x \text{ is not a rational number}\}$ | **Irrational numbers:** $\quad -\sqrt{2}$ |

| Definitions and Concepts | Examples |
|---|---|

### Section 1.2   Algebraic Expressions and Sets of Numbers (*continued*)

**Real numbers:**   $\{x \mid x$ corresponds to a point on the

If all the elements of set $A$ are also in set $B$, we say that set $A$ is a **subset** of set $B$.

Real numbers   $-9.6, -5, -\sqrt{2}, 0, \dfrac{2}{5}, 101$

The set of integers is a subset of the set of rational numbers.

### Section 1.3   Equations, Inequalities, and Properties of Real Numbers

**Symbols**

| $=$ | is equal to |
|---|---|
| $\neq$ | is not equal to |
| $>$ | is greater than |
| $<$ | is less than |
| $\geq$ | is greater than or equal to |
| $\leq$ | is less than or equal to |

$-5 = -5$
$-5 \neq -3$
$1.7 > 1.2$
$-1.7 < -1.2$
$\dfrac{5}{3} \geq \dfrac{5}{3}$
$-\dfrac{1}{2} \leq \dfrac{1}{2}$

**Identity**

$a + 0 = a \qquad 0 + a = a$
$a \cdot 1 = a \qquad 1 \cdot a = a$

$3 + 0 = 3 \qquad\qquad 0 + 3 = 3$
$-1.8 \cdot 1 = -1.8 \qquad 1 \cdot -1.8 = -1.8$

**Inverse**

$a + (-a) = 0 \qquad -a + a = 0$
$a \cdot \dfrac{1}{a} = 1 \qquad \dfrac{1}{a} \cdot a = 1; a \neq 0$

$7 + (-7) = 0 \qquad -7 + 7 = 0$
$5 \cdot \dfrac{1}{5} = 1 \qquad \dfrac{1}{5} \cdot 5 = 1$

**Commutative**

$a + b = b + a$
$a \cdot b = b \cdot a$

$x + 7 = 7 + x$
$9 \cdot y = y \cdot 9$

**Associative**

$(a + b) + c = a + (b + c)$
$(a \cdot b) \cdot c = a \cdot (b \cdot c)$

$(3 + 1) + 10 = 3 + (1 + 10)$
$(3 \cdot 1) \cdot 10 = 3 \cdot (1 \cdot 10)$

**Distributive**

$a(b + c) = ab + ac$

$6(x + 5) = 6 \cdot x + 6 \cdot 5$
$\qquad\qquad = 6x + 30$

### Section 1.4   Operations on Real Numbers

**Absolute Value**

$$|a| = \begin{cases} a \text{ if } a \text{ is 0 or a positive number} \\ -a \text{ if } a \text{ is a negative number} \end{cases}$$

$|3| = 3, \quad |0| = 0, \quad |-7.2| = 7.2$

**Adding Real Numbers**

1. To add two numbers with the same sign, add their absolute values and attach their common sign.
2. To add two numbers with different signs, subtract the smaller absolute value from the larger absolute value and attach the sign of the number with the larger absolute value.

$$\dfrac{2}{7} + \dfrac{1}{7} = \dfrac{3}{7}$$
$-5 + (-2.6) = -7.6$
$-18 + 6 = -12$
$20.8 + (-10.2) = 10.6$

**Subtracting Real Numbers**

$a - b = a + (-b)$

$18 - 21 = 18 + (-21) = -3$

(*continued*)

| Definitions and Concepts | Examples |
|---|---|

### Section 1.4   Operations on Real Numbers (*continued*)

**Multiplying and Dividing Real Numbers**

The product or quotient of two numbers with the same sign is positive.

$$(-8)(-4) = 32 \qquad \frac{-8}{-4} = 2$$

$$8 \cdot 4 = 32 \qquad \frac{8}{4} = 2$$

The product or quotient of two numbers with different signs is negative.

$$-17 \cdot 2 = -34 \qquad \frac{-14}{2} = -7$$

$$4(-1.6) = -6.4 \qquad \frac{22}{-2} = -11$$

A natural number **exponent** is a shorthand notation for repeated multiplication of the same factor.

$$3^4 = 3 \cdot 3 \cdot 3 \cdot 3 = 81$$

The notation $\sqrt{a}$ is used to denote the **positive,** or **principal square root** of a nonnegative number $a$.

$$\sqrt{a} = b \text{ if } b^2 = a \text{ and } b \text{ is positive}$$

$$\sqrt{49} = 7$$

Also,

$$\sqrt[3]{a} = b \text{ if } b^3 = a$$

$$\sqrt[4]{a} = b \text{ if } b^4 = a \text{ and } b \text{ is positive.}$$

$$\sqrt[3]{64} = 4$$

$$\sqrt[4]{16} = 2$$

### Section 1.5   Order of Operations and Algebraic Expressions

**Order of Operations**

Simplify expressions using the order that follows.

1. If grouping symbols such as parentheses are present, simplify expressions within those first, starting with the innermost set.

2. Evaluate exponential expressions, roots, and absolute values.

3. Multiply or divide in order from left to right.

4. Add or subtract in order from left to right.

Simplify:  $\dfrac{42 - 2(3^2 - \sqrt{16})}{-8}$

$$\frac{42 - 2(3^2 - \sqrt{16})}{-8} = \frac{42 - 2(9 - 4)}{-8}$$

$$= \frac{42 - 2(5)}{-8}$$

$$= \frac{42 - 10}{-8}$$

$$= \frac{32}{-8} = -4$$

To **evaluate** an algebraic expression containing variables, substitute the given numbers for the variables and simplify. The result is called the **value** of the expression.

Evaluate:  $2.7x$ when $x = 3$

$$2.7x = 2.7(3)$$

$$= 8.1$$

### Section 1.6   Exponents and Scientific Notation

**Product Rule**   $a^m \cdot a^n = a^{m+n}$

$$x^2 \cdot x^3 = x^5$$

**Zero Exponent**   $a^0 = 1 \quad (a \neq 0)$

$$7^0 = 1, \, (-10)^0 = 1$$

**Quotient Rule**   $\dfrac{a^m}{a^n} = a^{m-n} \quad (a \neq 0)$

$$\frac{y^{10}}{y^4} = y^{10-4} = y^6$$

**Negative Exponent**   $a^{-n} = \dfrac{1}{a^n} \quad (a \neq 0)$

$$3^{-2} = \frac{1}{3^2} = \frac{1}{9}, \frac{x^{-5}}{x^{-7}} = x^{-5-(-7)} = x^2$$

A positive number is written in **scientific notation** if it is written as the product of a number $a$, where $1 \leq a < 10$, and an integer power of 10: $a \times 10^n$.

Numbers written in scientific notation:

$$568,000 = 5.68 \times 10^5$$

$$0.0002117 = 2.117 \times 10^{-4}$$

| Definitions and Concepts | Examples |
|---|---|

**Section 1.7   More Work with Exponents and Scientific Notation**

| | |
|---|---|
| **Power Rules** $$(a^m)^n = a^{m \cdot n}$$ $$(ab)^m = a^m b^m$$ $$\left(\frac{a}{b}\right)^n = \frac{a^n}{b^n} \quad (b \neq 0)$$ | $$(7^8)^2 = 7^{16}$$ $$(2y)^3 = 2^3 y^3 = 8y^3$$ $$\left(\frac{5x^{-3}}{x^2}\right)^{-2} = \frac{5^{-2}x^6}{x^{-4}}$$ $$= 5^{-2} \cdot x^{6-(-4)}$$ $$= \frac{x^{10}}{5^2}, \text{ or } \frac{x^{10}}{25}$$ |

## Chapter 1   Review

**(1.2)** *Write each as an expression. Use x to represent the unknown number.*

1. The quotient of a number and seven

2. The product of a number and seven

3. Four times the sum of a number and ten

4. The difference of three times a number and nine

*Write each set in roster form.*

5. $\{x \mid x \text{ is an odd integer between } -2 \text{ and } 4\}$

6. $\{x \mid x \text{ is an even integer between } -3 \text{ and } 7\}$

7. $\{x \mid x \text{ is a negative whole number}\}$

8. $\{x \mid x \text{ is a natural number that is not a rational number}\}$

9. $\{x \mid x \text{ is a whole number greater than } 5\}$

10. $\{x \mid x \text{ is an integer less than } 3\}$

*Determine whether each statement is true or false if $A = \{6, 10, 12\}$, $B = \{5, 9, 11\}$, $C = \{\ldots, -3, -2, -1, 0, 1, 2, 3, \ldots\}$, $D = \{2, 4, 6, \ldots, 16\}$, $E = \{x \mid x \text{ is a rational number}\}$, $F = \{\ \}$, $G = \{x \mid x \text{ is an irrational number}\}$, and $H = \{x \mid x \text{ is a real number}\}$.*

11. $10 \in D$

12. $59 \in B$

13. $\sqrt{169} \notin A$

14. $0 \notin F$

15. $\pi \in E$

16. $\pi \in H$

17. $\sqrt{4} \in G$

18. $-9 \in C$

*List the elements of the set $\left\{5, -\frac{2}{3}, \frac{8}{2}, \sqrt{9}, 0.3, \sqrt{7}, 1\frac{5}{8}, -1, \pi\right\}$ that are also elements of each given set.*

19. Whole numbers

20. Natural numbers

21. Rational numbers

22. Irrational numbers

23. Real numbers

24. Integers

**(1.3)** *Write each statement as an equation.*

25. Twelve is the product of $x$ and negative 4.

26. The sum of $n$ and twice $n$ is negative fifteen.

**27.** Four times the sum of $y$ and three is $-1$.

**28.** The difference of $t$ and five, multiplied by six is four.

**29.** Seven subtracted from $z$ is six.

**30.** Ten less than the product of $x$ and nine is five.

**31.** The difference of $x$ and 5 is the same as 12.

**32.** The opposite of four is equal to the product of $y$ and seven.

**33.** Two-thirds is equal to twice the sum of $n$ and one-fourth.

**34.** The sum of $t$ and six amounts to negative twelve.

*Find the opposite, or additive inverse, of each number.*

**35.** $-\dfrac{3}{4}$       **36.** 0.6       **37.** 0       **38.** $-1$

*Find the reciprocal, or multiplicative inverse, of each number.*

**39.** $-\dfrac{3}{4}$       **40.** $\dfrac{1}{5}$       **41.** 0       **42.** $-1$

*Name each property illustrated.*

**43.** $(M + 5) + P = M + (5 + P)$

**44.** $5(3x - 4) = 15x - 20$

**45.** $(-4) + 4 = 0$

**46.** $(3 + x) + 7 = 7 + (3 + x)$

**47.** $(XY)Z = X(YZ)$

**48.** $\left(-\dfrac{3}{5}\right) \cdot \left(-\dfrac{5}{3}\right) = 1$

**49.** $T \cdot 1 = T$

**50.** $(x + y) + z = (y + x) + z$

**51.** $A + 0 = A$

**52.** $8 \cdot 1 = 8$

*Complete each equation using the given property.*

**53.** $5(x - 3z) = $ _____ Distributive property

**54.** $(7 + y) + (3 + x) = $ _____ Commutative property

**55.** $0 = $ _____ Additive inverse property

**56.** $1 = $ _____ Multiplicative inverse property

**57.** $[(3.4)(0.7)]5 = $ _____ Associative property

**58.** $7 = $ _____ Additive identity property

*Insert $<$, $>$, or $=$ to make each statement true.*

**59.** $-9$    $-12$       **60.** $0$    $-6$       **61.** $-3$    $-1$       **62.** $7$    $|-7|$

**63.** $-5$    $-(-5)$       **64.** $-(-2)$    $-2$

*Use the distributive property to multiply.*

**65.** $-3(2x - 7y)$

**66.** $-9(10a + 4b)$

**67.** $\dfrac{1}{2}(18m - 8n + 1)$

**68.** $\dfrac{1}{3}(12x - 33y + 2)$

**(1.4)** *Simplify.*

**69.** $-7 + 3$

**70.** $-10 + (-25)$

**71.** $5(-0.4)$

**72.** $(-3.1)(-0.1)$

**73.** $-7 - (-15)$

**74.** $9 - (-4.3)$

**75.** $\sqrt{16} - 2^3$

**76.** $\sqrt[3]{27} - 5^2$

**77.** $(-24) \div 0$

**78.** $0 \div (-45)$

**79.** $(-36) \div (-9)$

**80.** $60 \div (-12)$

**81.** $-\dfrac{4}{5} - \left(-\dfrac{2}{3}\right)$

**82.** $\dfrac{5}{4} - \left(-\dfrac{3}{4}\right)$

△ **83.** Determine the unknown fractional part.

**84.** The Bertha Rogers gas well in Washita County, Oklahoma, is the deepest well in the United States. From the surface, this now-capped well extends 31,441 feet into the earth. The elevation of the nearby Cordell Municipal Airport is 1589 feet above sea level. Assuming that the surface elevation of the well is the same as at the Cordell Municipal Airport, find the elevation relative to sea level of the *bottom* of the Bertha Rogers gas well. *(Sources: U.S. Geological Survey, Oklahoma Department of Transportation)*

**(1.5)** *Simplify.*

**85.** $-5 + 7 - 3 - (-10)$

**86.** $8 - (-3) + (-4) + 6$

**87.** $3(4 - 5)^4$

**88.** $6(7 - 10)^2$

**89.** $\left(-\dfrac{8}{15}\right) \cdot \left(-\dfrac{2}{3}\right)^2$

**90.** $\left(-\dfrac{3}{4}\right)^2 \cdot \left(-\dfrac{10}{21}\right)$

**91.** $-\dfrac{6}{15} \div \dfrac{8}{25}$

**92.** $\dfrac{4}{9} \div \left(-\dfrac{8}{45}\right)$

**93.** $-\dfrac{3}{8} + 3(2) \div 6$

**94.** $5(-2) - (-3) - \dfrac{1}{6} + \dfrac{2}{3}$

**95.** $|2^3 - 3^2| - |5 - 7|$

**96.** $|5^2 - 2^2| + |9 \div (-3)|$

**97.** $(2^3 - 3^2) - (5 - 7)$

**98.** $(5^2 - 2^2) + [9 \div (-3)]$

**99.** $\dfrac{(8 - 10)^3 - (-4)^2}{2 + 8(2) \div 4}$

**100.** $\dfrac{(2 + 4)^2 + (-1)^5}{12 \div 2 \cdot 3 - 3}$

**101.** $\dfrac{(4 - 9) + 4 - 9}{10 - 12 \div 4 \cdot 8}$

**102.** $\dfrac{3 - 7 - (7 - 3)}{15 + 30 \div 6 \cdot 2}$

**103.** $\dfrac{\sqrt{25}}{4 + 3 \cdot 7}$

**104.** $\dfrac{\sqrt{64}}{24 - 8 \cdot 2}$

△ *The algebraic expression 2πr represents the circumference of (distance around) a circle of radius r.*

△ **105.** Complete the table by evaluating the expression at the given values of $r$. (Use 3.14 for $\pi$.)

| Radius | $r$ | 1 | 10 | 100 |
|---|---|---|---|---|
| Circumference | $2\pi r$ | | | |

**106.** As the radius of a circle increases, does the circumference of the circle increase or decrease?

*Simplify.*

**107.** $14x - 3 - 11x - 10$

**108.** $81y + 19 - y - 20$

**109.** $7a - 3(2a - y) + 4y - 6$

**110.** $9b - 4(8b - x) + 9x - 10$

**111.** $\dfrac{1}{5}(15m - 5n) - (3m - 5) + 2n$

**112.** $-(2x - 9) + \dfrac{1}{4}(8x + 4y) - 13$

**(1.6)** *Evaluate.*

**113.** $(-2)^2$

**114.** $(-3)^4$

**115.** $-2^2$

**116.** $-3^4$

**117.** $8^0$

**118.** $-9^0$

**119.** $-4^{-2}$

**120.** $(-4)^{-2}$

*Simplify each expression. Write each answer with positive exponents only. Assume that variables in the exponents represent integers and that all other variables are not 0.*

**121.** $-xy^2 \cdot y^3 \cdot xy^2z$

**122.** $(-4xy)(-3xy^2b)$

**123.** $a^{-14} \cdot a^5$

**124.** $\dfrac{a^{16}}{a^{17}}$

**125.** $\dfrac{x^{-7}}{x^4}$

**126.** $\dfrac{9a(a^{-3})}{18a^{15}}$

**127.** $\dfrac{y^{6p-3}}{y^{6p+2}}$

**128.** $\dfrac{4y^{3x-3}}{2y^{2x+4}}$

*Write each number in scientific notation.*

**129.** 36,890,000

**130.** 0.000362

*Write each number without exponents.*

**131.** $1.678 \times 10^{-6}$

**132.** $4.1 \times 10^5$

**(1.7)** *Simplify. Write each answer with positive exponents only.*

**133.** $(8^5)^3$

**134.** $\left(\dfrac{a}{4}\right)^2$

**135.** $(3x)^3$

**136.** $(-4x)^{-2}$

**137.** $\left(\dfrac{6x}{5}\right)^2$      **138.** $(8^6)^{-3}$      **139.** $\left(\dfrac{4}{3}\right)^{-2}$      **140.** $(-2x^3)^{-3}$

**141.** $\left(\dfrac{8p^6}{4p^4}\right)^{-2}$      **142.** $(-3x^{-2}y^2)^3$      **143.** $\left(\dfrac{x^{-5}y^{-3}}{z^3}\right)^{-5}$      **144.** $\dfrac{4^{-1}x^3yz}{x^{-2}yx^4}$

**145.** $(5xyz)^{-4}(x^{-2})^{-3}$      **146.** $\dfrac{2(3yz)^{-3}}{y^{-3}}$

*Simplify each expression.*

**147.** $x^{4a}(3x^{5a})^3$      **148.** $(3x^{2a+b}y^{-3b})^2$

## Mixed Review

*Complete the table.*

| | Number | Opposite of Number | Reciprocal of Number |
|---|---|---|---|
| **149.** | $-\dfrac{3}{4}$ | | |
| **150.** | | $-5$ | |

*Simplify. If necessary, write answers with positive exponents only.*

**151.** $-2\left(5x + \dfrac{1}{2}\right) + 7.1$      **152.** $\sqrt{36} \div 2 \cdot 3$

**153.** $-\dfrac{7}{11} - \left(-\dfrac{1}{11}\right)$      **154.** $10 - (-1) + (-2) + 6$

**155.** $\left(-\dfrac{2}{3}\right)^3 \div \dfrac{10}{9}$      **156.** $\dfrac{(3-5)^2 + (-1)^3}{1 + 2(3 - (-1))^2}$

**157.** $\dfrac{1}{3}(9x - 3y) - (4x - 1) + 4y$      **158.** $-5^{-2}$

**159.** $\left(\dfrac{5x^7}{10x^{-3}}\right)^{-3}$      **160.** $(-5a^{-2}bc)^{-2} \cdot (3a^{-3}b^2c^3)^2$

# Chapter 1

Test    Step-by-step test solutions are found on the Chapter Test Prep Videos. Where available:    MyMathLab® or You Tube™

**Answers**

*Determine whether each statement is true or false.*

**1.** $-2.3 > -2.33$

**2.** $-6^2 = (-6)^2$

**3.** $(-2)(-3)(0) = \dfrac{(-4)}{0}$

**4.** All rational numbers are integers.

**5.** All natural numbers are integers.

**6.** Write the set in roster form. $\{x \mid x \text{ is a whole number less than } 2\}$

*Simplify.*

**7.** $5 - 12 \div 3(2)$

**8.** $(4 - 9)^3 - |-4 - 6|^2$

**9.** $[3|4 - 5|^5 - (-9)] \div (-6)$

**10.** $\dfrac{6(7 - 9)^3 + (-2)}{(-2)(-5)(-5)}$

**11.** $\dfrac{(4 - \sqrt{16}) - (-7 - 20)}{-2(1 - 4)^2}$

*Evaluate each expression when $q = 4$, $r = -2$, and $t = 1$.*

**12.** $q^2 - r^2$

**13.** $\dfrac{5t - 3q}{3r - 1}$

**14.** The algebraic expression $5.75x$ represents the total cost for $x$ adults to attend the theater.

   **a.** Complete the table that follows.

   **b.** As the number of adults increases, does the total cost increase or decrease?

| Adults | $x$ | 1 | 3 | 10 | 20 |
|---|---|---|---|---|---|
| Total Cost | $5.75x$ | | | | |

*Write each statement as an equation.*

**15.** Three times the quotient of $n$ and five is the opposite of $n$.

**16.** Twenty is equal to six subtracted from twice $x$.

**17.** Negative two is equal to $x$ divided by the sum of $x$ and five.

**Answers**

1. _____
2. _____
3. _____
4. _____
5. _____
6. _____
7. _____
8. _____
9. _____
10. _____
11. _____
12. _____
13. _____
14. a. _____
   b. _____
15. _____
16. _____
17. _____

*Name each property illustrated.*

**18.** $6(x - 4) = 6x - 24$

**19.** $(4 + x) + z = 4 + (x + z)$

**20.** $(-7) + 7 = 0$

**21.** Find the reciprocal and opposite of $-\dfrac{7}{11}$.

*Simplify.*

**22.** $9x + 12y - 3x - 6.2 + 20$

**23.** $\dfrac{1}{3}(15x - 27y) - 2(3x - y) - 4y$

*Simplify. Write answers using positive exponents only.*

**24.** $(-9x)^{-2}$

**25.** $\dfrac{6^{-1}a^2b^{-3}}{3^{-2}a^{-5}b^2}$

**26.** $\left(\dfrac{-xy^{-5}z}{xy^3}\right)^{-5}$

**27.** $(-6a^{-5}b^{12})^{-2}(3ab^5)^2$

**28.** $\dfrac{(x^w)^2}{(x^{4w})^{-2}}$

*Write each number in scientific notation.*

**29.** 630,000,000

**30.** 0.01200

**31.** Write $5.0 \times 10^{-6}$ in standard notation.

*Use scientific notation to find the quotient. Express the quotient in scientific notation.*

**32.** $\dfrac{(0.00012)(144,000)}{0.0003}$

**33.** $\dfrac{(0.0024)(0.00012)}{0.00032}$

**34.** In 2013, the United States Postal Service handled approximately 22,600,000,000 pieces of first-class mail. Write this number in scientific notation. (*Source:* United States Postal Service)

**18.** _____

**19.** _____

**20.** _____

**21.** _____

**22.** _____

**23.** _____

**24.** _____

**25.** _____

**26.** _____

**27.** _____

**28.** _____

**29.** _____

**30.** _____

**31.** _____

**32.** _____

**33.** _____

**34.** _____

# Equations, Inequalities, and Problem Solving

Mathematics is a tool for solving problems in such diverse fields as transportation, engineering, economics, medicine, business, and biology. We solve problems using mathematics by modeling real-world phenomena with mathematical equations or inequalities. Our ability to solve problems using mathematics, then, depends in part on our ability to solve equations and inequalities. In this chapter, we solve linear equations and inequalities in one variable and graph their solutions on number lines.

## Use of Tablets Is Increasing

The widespread use of tablets began with Apple's introduction of the iPad in 2010, but did you know that the development of these devices began as early as 1888? This early device used a sheet of paper as a display and a pen attached to electromagnetic actuators. By the 1980s, early versions of the tablets we use today were commercially produced.

Today's tablets have cellular or wireless Internet capabilities, and each newer model offers better displays and faster processing. There are even tablets that talk to you. Who knows what the next great innovations will be?

In Section 2.1, objective A, and Section 2.2, Exercises 37 and 38, you will find the projected increase in the number of tablet users worldwide (in millions of users) as well as the percent increase. (*Source:* Forrester Research: World Tablet Adoption Forecast 2012–2017)

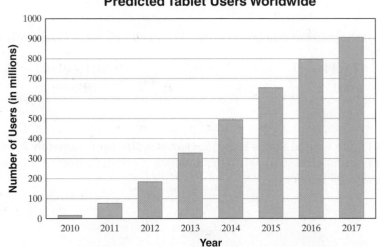

**Predicted Tablet Users Worldwide**

### Objective A Deciding Whether a Number Is a Solution of an Equation

An **equation** is a statement that two expressions are equal. To solve problems, we need to be able to solve equations. In this section, we will solve a special type of equation called a *linear equation in one variable*.

Linear equations model many real-life problems. For example, we can use a linear equation to calculate the increase in the number (in millions) of tablet users worldwide.

Tablets are smaller and lighter than desktop computers and often just as powerful. There are many types of tablets available, and the technology will continue to expand. Predicted numbers of worldwide tablet users for various years are shown below.

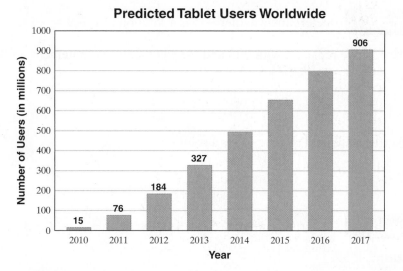

**Predicted Tablet Users Worldwide**

To find the increase in the number of tablet users worldwide from 2013 to 2017, for example, we can use the equation below.

| In words: | Increase in tablet users | is | tablet users in 2017 | minus | tablet users in 2013 |
|---|---|---|---|---|---|
| Translate: | $x$ | $=$ | 906 | $-$ | 327 |

Since our variable $x$ (increase in the number of tablet users worldwide) is by itself on one side of the equation, we can find the value of $x$ by simplifying the right side.

$x = 579$

The increase in the number of tablet users worldwide from 2013 to 2017 is predicted to be 579 million.

The equation, $x = 906 - 327$, like every other equation, is a statement that two expressions are equal. Oftentimes, the unknown variable is not by itself on one side of the equation. In these cases, we will use properties of equality to write equivalent equations so that a solution many be found. This is called **solving the equation.** In this section, we concentrate on solving equations such as this one, called **linear equations** in one variable. Linear equations are also called **first-degree equations** since the exponent on the variable is 1.

---

**Linear Equations in One Variable—Examples**

$3x = -15$     $7 - y = 3y$     $4n - 9n + 6 = 0$     $z = -2$

---

**Objectives**

**A** Decide Whether a Number Is a Solution of an Equation.

**B** Solve Linear Equations Using Properties of Equality.

**C** Solve Linear Equations That Can Be Simplified by Combining Like Terms.

**D** Solve Linear Equations Containing Fractions or Decimals.

**E** Recognize Identities and Equations with No Solution.

> **Linear Equation in One Variable**
>
> A **linear equation in one variable** is an equation that can be written in the form
>
> $$ax + b = c$$
>
> where $a$, $b$, and $c$ are real numbers and $a \neq 0$.

When a variable in an equation is replaced by a number and the resulting equation is true, then that number is called a **solution** of the equation. For example, 1 is a solution of the equation $3x + 4 = 7$ since $3(1) + 4 = 7$ is a true statement. But 2 is not a solution of this equation since $3(2) + 4 = 7$ is *not* a true statement. The **solution set** of an equation is the set of solutions of the equation. For example, the solution set of $3x + 4 = 7$ is $\{1\}$.

**Practice 1**

Determine whether $-7$ is a solution of $14 - x = 21$.

**Example 1**  Determine whether $-15$ is a solution of $x - 9 = -24$.

**Solution:**  We replace $x$ with $-15$ and see whether a true statement results.

$$x - 9 = -24$$
$$-15 - 9 \overset{?}{=} -24 \quad \text{Replace } x \text{ with } -15.$$
$$-24 = -24 \quad \text{True}$$

Since a true statement results, $-15$ is a solution.

■ **Work Practice 1**

**Practice 2**

Determine whether 8 is a solution of $x - 10 = 2x - 14$.

**Example 2**  Determine whether 5 is a solution of $2x - 3 = x + 3$.

**Solution:**

$$2x - 3 = x + 3$$
$$2 \cdot 5 - 3 \overset{?}{=} 5 + 3 \quad \text{Replace } x \text{ with } 5.$$
$$7 = 8 \quad \text{False}$$

Since a false statement results, 5 is not a solution.

■ **Work Practice 2**

## Objective B  Using the Properties of Equality ▶

To **solve an equation** is to find the solution set of an equation. Equations with the same solution set are called **equivalent equations**. For example,

$$3x + 4 = 7 \qquad 3x = 3 \qquad x = 1$$

are equivalent equations because they all have the same solution set, namely, $\{1\}$. To solve an equation in $x$, we start with the given equation and write a series of simpler equivalent equations until we obtain an equation of the form

$$x = \textbf{number}$$

**Answers**

**1.** $-7$ is a solution
**2.** 8 is not a solution

To write equivalent equations, we use two important properties.

> ### Addition Property of Equality
>
> If $a$, $b$, and $c$, are real numbers, then
>
> $$a = b \quad \text{and} \quad a + c = b + c$$
>
> are equivalent equations.

> ### Multiplication Property of Equality
>
> If $c \neq 0$, then
>
> $$a = b \quad \text{and} \quad ac = bc$$
>
> are equivalent equations.

The **addition property of equality** guarantees that the same number may be added to both sides of an equation and the result is an equivalent equation. Recall that we define subtraction in terms of addition.

$$7 - 10 = 7 + (-10) = -3$$

This means that the addition property also says we can *subtract* the same number from both sides and the result is an equivalent equation.

The **multiplication property of equality** guarantees that both sides of an equation may be multiplied by the same nonzero number and the result is an equivalent equation. Recall that we define division in terms of multiplication. This means that the multiplication property also says we can *divide* both sides by the same nonzero number and the result is an equivalent equation.

For example, to solve $2x + 5 = 9$, we use the addition and multiplication properties of equality to get $x$ alone—that is, to write an equivalent equation of the form

$$x = \text{number}$$

We will do this in the next example.

**Example 3** Solve: $2x + 5 = 9$

**Solution:** First we use the addition property of equality and subtract 5 from both sides.

$$2x + 5 = 9$$
$$2x + 5 - 5 = 9 - 5 \quad \text{Subtract 5 from both sides.}$$
$$2x = 4 \quad \text{Simplify.}$$

Now we use the multiplication property of equality and divide both sides by 2.

$$\frac{2x}{2} = \frac{4}{2} \quad \text{Divide both sides by 2.}$$
$$x = 2 \quad \text{Simplify.}$$

**Check:** To check, we replace $x$ in the original equation with 2.

$$2x + 5 = 9 \quad \text{Original equation}$$
$$2(2) + 5 \stackrel{?}{=} 9 \quad \text{Replace } x \text{ with 2.}$$
$$4 + 5 \stackrel{?}{=} 9$$
$$9 = 9 \quad \text{True}$$

The solution is 2, or the solution set is $\{2\}$.

📖 **Work Practice 3**

**Practice 3**

Solve: $3x + 6 = 21$

**Answer**

**3.** $\{5\}$

## Practice 4

Solve: $4.5 = 3 + 2.5x$

**Example 4**   Solve:   $0.6 = 2 - 3.5c$

**Solution:**   We use both the addition property and the multiplication property of equality.

$$0.6 = 2 - 3.5c$$
$$0.6 - 2 = 2 - 3.5c - 2 \quad \text{Subtract 2 from both sides.}$$
$$-1.4 = -3.5c \quad \text{Simplify.}$$
$$\frac{-1.4}{-3.5} = \frac{-3.5c}{-3.5} \quad \text{Divide both sides by } -3.5.$$
$$0.4 = c \quad \text{Simplify } \frac{-1.4}{-3.5}.$$

**Check:**

$$0.6 = 2 - 3.5c$$
$$0.6 \overset{?}{=} 2 - 3.5(0.4) \quad \text{Replace } c \text{ with } 0.4.$$
$$0.6 \overset{?}{=} 2 - 1.4 \quad \text{Multiply.}$$
$$0.6 = 0.6 \quad \text{True}$$

The solution is 0.4, or the solution set is $\{0.4\}$.

■ **Work Practice 4**

## Objective C  Solving Linear Equations by Combining Like Terms ▶

Often, an equation can be simplified by removing any grouping symbols and combining any like terms.

## Practice 5

Solve:

$-2x + 2 - 3x = 8x + 20 - 7x$

**Example 5**   Solve:   $-4x - 1 + 5x = 9x + 3 - 7x$

**Solution:**   First we simplify both sides of this equation by combining like terms. Then let's get variable terms on the same side of the equation by using the addition property of equality to subtract $2x$ from both sides. Next, we use this same property to add 1 to both sides of the equation. These steps are shown below.

$$-4x - 1 + 5x = 9x + 3 - 7x$$
$$x - 1 = 2x + 3 \quad \text{Combine like terms.}$$
$$x - 1 - 2x = 2x + 3 - 2x \quad \text{Subtract } 2x \text{ from both sides.}$$
$$-x - 1 = 3 \quad \text{Simplify.}$$
$$-x - 1 + 1 = 3 + 1 \quad \text{Add 1 to both sides.}$$
$$-x = 4 \quad \text{Simplify.}$$

Notice that this equation is not solved for $x$ since we have $-x$, or $-1x$, not $x$. To get $x$ alone, we divide both sides by $-1$.

$$\frac{-x}{-1} = \frac{4}{-1} \quad \text{Divide both sides by } -1.$$
$$x = -4 \quad \text{Simplify.}$$

Check to see that the solution set is $\{-4\}$.

■ **Work Practice 5**

If an equation contains parentheses, we use the distributive property to remove them.

**Answers**

**4.** $\{0.6\}$   **5.** $\{-3\}$

**Example 6**   Solve: $3(x - 3) = 5x - 9$

**Solution:**   First we use the distributive property.

$$3(x - 3) = 5x - 9$$
$$3x - 9 = 5x - 9 \quad \text{Use the distributive property.}$$

Next we get variable terms on the same side of the equation by using the addition property of equality.

$$3x - 9 - 5x = 5x - 9 - 5x \quad \text{Subtract } 5x \text{ from both sides.}$$
$$-2x - 9 = -9 \quad \text{Simplify.}$$
$$-2x - 9 + 9 = -9 + 9 \quad \text{Add 9 to both sides.}$$
$$-2x = 0 \quad \text{Simplify.}$$
$$\frac{-2x}{-2} = \frac{0}{-2} \quad \text{Divide both sides by } -2.$$
$$x = 0$$

Check to see that $\{0\}$ is the solution set.

■ **Work Practice 6**

## Objective D   Solving Linear Equations Containing Fractions or Decimals

If an equation contains fractions, we can first clear the equation of fractions by multiplying both sides of the equation by the *least common denominator* (LCD) of all fractions in the equation.

**Example 7**   Solve: $\dfrac{y}{3} - \dfrac{y}{4} = \dfrac{1}{6}$

**Solution:**   First we clear the equation of fractions by multiplying both sides of the equation by 12, the LCD of the denominators 3, 4, and 6.

$$\frac{y}{3} - \frac{y}{4} = \frac{1}{6}$$
$$12\left(\frac{y}{3} - \frac{y}{4}\right) = 12\left(\frac{1}{6}\right) \quad \text{Multiply both sides by the LCD, 12.}$$
$$12\left(\frac{y}{3}\right) - 12\left(\frac{y}{4}\right) = 2 \quad \text{Use the distributive property.}$$
$$4y - 3y = 2 \quad \text{Simplify.}$$
$$y = 2 \quad \text{Simplify.}$$

**Check:**   To check, we replace $y$ with 2 in the original equation.

$$\frac{y}{3} - \frac{y}{4} = \frac{1}{6} \quad \text{Original equation}$$
$$\frac{2}{3} - \frac{2}{4} \overset{?}{=} \frac{1}{6} \quad \text{Replace } y \text{ with 2.}$$
$$\frac{8}{12} - \frac{6}{12} \overset{?}{=} \frac{1}{6} \quad \text{Write fractions with the LCD.}$$
$$\frac{2}{12} \overset{?}{=} \frac{1}{6} \quad \text{Subtract.}$$
$$\frac{1}{6} = \frac{1}{6} \quad \text{True}$$

Since a true statement results, the solution set is $\{2\}$.

■ **Work Practice 7**

**Practice 6**

Solve: $4(x - 2) = 6x - 10$

**Practice 7**

Solve: $\dfrac{x}{6} - \dfrac{x}{8} = \dfrac{1}{8}$

As a general guideline, the following steps may be used to solve a linear equation in one variable.

> ## Solving a Linear Equation in One Variable
>
> **Step 1:** Clear the equation of fractions or decimals by multiplying both sides of the equation by an appropriate nonzero number.
>
> **Step 2:** Use the distributive property to remove grouping symbols such as parentheses.
>
> **Step 3:** Combine like terms on each side of the equation.
>
> **Step 4:** Use the addition property of equality to rewrite the equation as an equivalent equation, with variable terms on one side and numbers on the other side.
>
> **Step 5:** Use the multiplication property of equality to get the variable alone.
>
> **Step 6:** Check the proposed solution in the original equation.

**Practice 8**

Solve:

$$\frac{x-1}{3} + \frac{2}{3} = x - \frac{2x+3}{9}$$

**Helpful Hint**

When we multiply both sides of an equation by a number, the distributive property tells us that each term of the equation is multiplied by the number.

**Example 8**  Solve: $\dfrac{x+5}{2} + \dfrac{1}{2} = \dfrac{1}{8}(15x + 3)$

**Solution:**  To begin, we multiply both sides of the equation by 8, the LCD of 2 and 8. This will clear the equation of fractions.

$$8\left(\frac{x+5}{2} + \frac{1}{2}\right) = 8\left[\frac{1}{8}(15x + 3)\right] \qquad \text{Multiply both sides by 8.}$$

$$8\left(\frac{x+5}{2}\right) + 8\left(\frac{1}{2}\right) = 8 \cdot \frac{1}{8}(15x + 3) \qquad \text{Use the distributive property.}$$

$$4(x + 5) + 4 = 15x + 3 \qquad \text{Multiply.}$$

$$4x + 20 + 4 = 15x + 3 \qquad \text{Use the distributive property to remove parentheses.}$$

$$4x + 24 = 15x + 3 \qquad \text{Combine like terms.}$$

$$4x - 15x = 3 - 24 \qquad \text{Subtract } 15x \text{ and } 24 \text{ from both sides.}$$

$$-11x = -21 \qquad \text{Simplify.}$$

$$\frac{-11x}{-11} = \frac{-21}{-11} \qquad \text{Divide both sides by } -11.$$

$$x = \frac{21}{11} \qquad \text{Simplify.}$$

To check, verify that replacing $x$ with $\dfrac{21}{11}$ makes the original equation true. The solution set is $\left\{\dfrac{21}{11}\right\}$.

■ **Work Practice 8**

If an equation contains decimals, you may want to first clear the equation of decimals by multiplying by an appropriate power of 10.

**Answer**

8. $\left\{\dfrac{3}{2}\right\}$

**Example 9**　Solve:　$0.3x + 0.1 = 0.27x - 0.02$

**Solution:**　To clear this equation of decimals, we multiply both sides of the equation by 100. Recall that multiplying a number by 100 moves its decimal point two places to the right.

$$100(0.3x + 0.1) = 100(0.27x - 0.02)$$

| | |
|---|---|
| $100(0.3x) + 100(0.1) = 100(0.27x) - 100(0.02)$ | Use the distributive property. |
| $30x + 10 = 27x - 2$ | Multiply. |
| $30x - 27x = -2 - 10$ | Subtract $27x$ and $10$ from both sides. |
| $3x = -12$ | Simplify. |
| $\dfrac{3x}{3} = \dfrac{-12}{3}$ | Divide both sides by 3. |
| $x = -4$ | Simplify. |

Check to see that the solution set is $\{-4\}$.

■ **Work Practice 9**

✔**Concept Check**　Explain what is wrong with the following:

$$3x - 5 = 16$$
$$3x = 11$$
$$\frac{3x}{3} = \frac{11}{3}$$
$$x = \frac{11}{3}$$

## Objective E  Recognizing Identities and Equations with No Solution ▶

So far, each linear equation that we have solved has had a single solution. We will now look at two other types of equations: *contradictions* and *identities*.

An equation in one variable that has no solution is called a **contradiction,** and an equation in one variable that has every number (for which the equation is defined) as a solution is called an **identity.** The next examples show how to recognize contradictions and identities.

**Example 10**　Solve:　$3x + 5 = 3(x + 2)$

**Solution:**　First we use the distributive property to remove parentheses.

| | |
|---|---|
| $3x + 5 = 3(x + 2)$ | |
| $3x + 5 = 3x + 6$ | Use the distributive property. |
| $3x + 5 - 3x = 3x + 6 - 3x$ | Subtract $3x$ from both sides. |
| $5 = 6$ | False. |

The equation $5 = 6$ is a false statement no matter what value the variable $x$ might have. Thus the original equation has no solution. Its solution set is written either as $\{\ \}$ or $\varnothing$. This equation is a contradiction.

■ **Work Practice 10**

**Practice 9**

Solve:

$$0.2x + 0.1 = 0.12x - 0.06$$

**Practice 10**

Solve:　$5x - 1 = 5(x + 3)$

**Answers**

**9.** $\{-2\}$　**10.** $\varnothing$

✔**Concept Check Answer**

Add 5 instead of subtracting.

$$3x - 5 = 16$$
$$3x = 21$$
$$x = 7$$

Therefore the correct solution set is $\{7\}$.

**Practice 11**

Solve:

$-4(x - 1) = -4x - 9 + 13$

**Example 11** Solve: $6x - 4 = 2 + 6(x - 1)$

**Solution:** First we use the distributive property to remove parentheses.

$$6x - 4 = 2 + 6(x - 1)$$
$$6x - 4 = 2 + 6x - 6 \qquad \text{Use the distributive property.}$$
$$6x - 4 = 6x - 4 \qquad \text{Combine the terms.}$$

At this point we might notice that both sides of the equation are the same, so replacing $x$ by any real number gives a true statement. Thus, the solution set of this equation is the set of real numbers, and the equation is an identity. Continuing to "solve" $6x - 4 = 6x - 4$, we eventually arrive at the same conclusion.

$$6x - 4 + 4 = 6x - 4 + 4 \qquad \text{Add 4 to both sides.}$$
$$6x = 6x \qquad \text{Simplify.}$$
$$6x - 6x = 6x - 6x \qquad \text{Subtract } 6x \text{ from both sides.}$$
$$0 = 0 \qquad \text{True}$$

Since $0 = 0$ is a true statement for every value of $x$, the solution set is the set of all real numbers, which can be written as $\{x \mid x \text{ is a real number}\}$. The equation is called an identity.

■ **Work Practice 11**

**Answer**

**11.** $\{x \mid x \text{ is a real number}\}$

# Vocabulary, Readiness & Video Check

*Use the choices below to fill in the blanks. Not all choices will be used.*

multiplication        value          like

addition             solution        equivalent

1. Equations with the same solution set are called _____ equations.

2. A value for the variable in an equation that makes the equation a true statement is called a(n) _____ of the equation.

3. By the _____ property of equality, $y = -3$ and $y - 7 = -3 - 7$ are equivalent equations.

4. By the _____ property of equality, $2y = -3$ and $\dfrac{2y}{2} = \dfrac{-3}{2}$ are equivalent equations.

*Identify each as an equation or an expression.*

**5.** $\frac{1}{3}x - 5$

_____

**6.** $2(x - 3) = 7$

_____

**7.** $\frac{5}{9}x + \frac{1}{3} = \frac{2}{9} - x$

_____

**8.** $\frac{5}{9}x + \frac{1}{3} - \frac{2}{9} - x$

_____

*By inspection, decide which equations have no solution and which equations have all real numbers as solutions.*

**9.** $2x + 3 = 2x + 3$

**10.** $2x + 1 = 2x + 3$

**11.** $5x - 2 = 5x - 7$

**12.** $5x - 3 = 5x - 3$

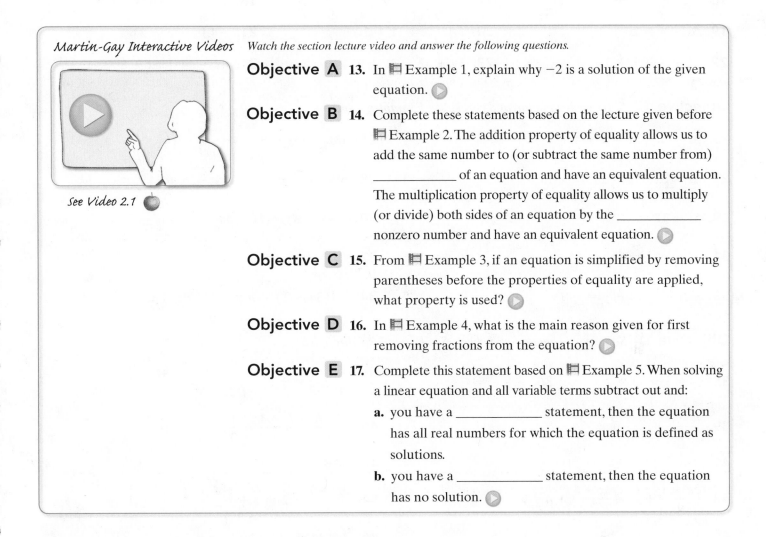

**Martin-Gay Interactive Videos**

*See Video 2.1*

*Watch the section lecture video and answer the following questions.*

**Objective A** **13.** In ⊞ Example 1, explain why −2 is a solution of the given equation. ▶

**Objective B** **14.** Complete these statements based on the lecture given before ⊞ Example 2. The addition property of equality allows us to add the same number to (or subtract the same number from) _____ of an equation and have an equivalent equation. The multiplication property of equality allows us to multiply (or divide) both sides of an equation by the _____ nonzero number and have an equivalent equation. ▶

**Objective C** **15.** From ⊞ Example 3, if an equation is simplified by removing parentheses before the properties of equality are applied, what property is used? ▶

**Objective D** **16.** In ⊞ Example 4, what is the main reason given for first removing fractions from the equation? ▶

**Objective E** **17.** Complete this statement based on ⊞ Example 5. When solving a linear equation and all variable terms subtract out and:

**a.** you have a _____ statement, then the equation has all real numbers for which the equation is defined as solutions.

**b.** you have a _____ statement, then the equation has no solution. ▶

## 2.1 Exercise Set MyMathLab® ▶

**Objective A** *Determine whether each number is a solution of the given equation. See Examples 1 and 2.*

**1.** Is −24 a solution of $\frac{x}{-6} = 4$?

**2.** Is 15 a solution of $\frac{x}{-3} = -5$?

**3.** Is −3 a solution of $x - 17 = 20$?

**4.** Is −8 a solution of $x - 10 = -2$?

▶**5.** Is −2 a solution of $5 + 3x = -1$?

**6.** Is −1 a solution of $6 - 2x = 4$?

**7.** Is 5 a solution of $x - 7 = x + 2$?

**8.** Is 5 a solution of $x - 1 = x - 1$?

**9.** Is 5 a solution of $4(x - 3) = 12$?

**10.** Is 12 a solution of $5(x - 6) = 30$?

**11.** Is $-8$ a solution of $4x - 2 = 5x + 6$?

**12.** Is 2 a solution of $7x + 1 = 6x - 1$?

**Objective** **B**  *Solve each equation and check. See Examples 3 and 4.*

**13.** $-5x = -30$

**14.** $-2x = 18$

**15.** $-10 = x + 12$

**16.** $-25 = y + 30$

**17.** $x - 2.8 = 1.9$

**18.** $y - 8.6 = -6.3$

**19.** $5x - 4 = 26 + 2x$

**20.** $5y - 3 = 11 + 3y$

▶ **21.** $-4.1 - 7z = 3.6$

**22.** $10.3 - 6x = -2.3$

**23.** $5y + 12 = 2y - 3$

**24.** $4x + 14 = 6x + 8$

**Objective** **C**  *Solve each equation and check. See Examples 5 and 6.*

**25.** $3x - 4 - 5x = x + 4 + x$

**26.** $13x - 15x + 8 = 4x + 2 - 24$

**27.** $8x - 5x + 3 = x - 7 + 10$

**28.** $6 + 3x + x = -x + 8 - 26 + 24$

**29.** $5x + 12 = 2(2x + 7)$

**30.** $2(4x + 3) = 7x + 5$

**31.** $3(x - 6) = 5x$

**32.** $6x = 4(x - 5)$

▶ **33.** $-2(5y - 1) - y = -4(y - 3)$

**34.** $-4(3n - 2) - n = -11(n - 1)$

**Objective** **D**  *Solve each equation and check. See Examples 7 through 9.*

**35.** $\dfrac{x}{2} + \dfrac{x}{3} = \dfrac{3}{4}$

**36.** $\dfrac{x}{2} + \dfrac{x}{5} = \dfrac{5}{4}$

**37.** $\dfrac{3t}{4} - \dfrac{t}{2} = 1$

**38.** $\dfrac{4r}{5} - 7 = \dfrac{r}{10}$

**39.** $\dfrac{n - 3}{4} + \dfrac{n + 5}{7} = \dfrac{5}{14}$

**40.** $\dfrac{2 + h}{9} + \dfrac{h - 1}{3} = \dfrac{1}{3}$

**41.** $0.6x - 10 = 1.4x - 14$

**42.** $0.3x + 2.4 = 0.1x + 4$

**43.** $\dfrac{3x - 1}{9} + x = \dfrac{3x + 1}{3} + 4$

**44.** $\dfrac{2z + 7}{8} - 2 = z + \dfrac{z - 1}{2}$

**45.** $1.5(4 - x) = 1.3(2 - x)$

**46.** $2.4(2x + 3) = -0.1(2x + 78)$

**Objective** **E**  *Solve each equation. See Examples 10 and 11.*

**47.** $4(n + 3) = 2(6 + 2n)$

**48.** $6(4n + 4) = 8(3 + 3n)$

▶ **49.** $3(x + 1) + 5 = 3x + 2$

**50.** $4(x + 2) + 4 = 4x - 8$

**51.** $2(x - 8) + x = 3(x - 6) + 2$

**52.** $5(x - 4) + x = 6(x - 2) - 8$

**53.** $4(x + 5) = 3(x - 4) + x$

**54.** $9(x - 2) = 8(x - 3) + x$

## Objectives B C D E Mixed Practice *Solve each equation. See Examples 3 through 11.*

**55.** $\dfrac{3}{8} + \dfrac{b}{3} = \dfrac{5}{12}$

**56.** $\dfrac{a}{2} + \dfrac{7}{4} = 5$

**57.** $x - 10 = -6x - 10$

**58.** $4x - 7 = 2x - 7$

**59.** $5(x - 2) + 2x = 7(x + 4) - 38$

**60.** $3x + 2(x + 4) = 5(x + 1) + 3$

**61.** $y + 0.2 = 0.6(y + 3)$

**62.** $-(w + 0.2) = 0.3(4 - w)$

**63.** $\dfrac{1}{4}(a + 2) = \dfrac{1}{6}(5 - a)$

**64.** $\dfrac{1}{3}(8 + 2c) = \dfrac{1}{5}(3c - 5)$

**65.** $2y + 5(y - 4) = 4y - 2(y - 10)$

**66.** $9c - 3(6 - 5c) = c - 2(3c + 9)$

**67.** $6x - 2(x - 3) = 4(x + 1) + 4$

**68.** $10x - 2(x + 4) = 8(x - 2) + 6$

**69.** $\dfrac{m - 4}{3} - \dfrac{3m - 1}{5} = 1$

**70.** $\dfrac{n + 1}{8} - \dfrac{2 - n}{3} = \dfrac{5}{6}$

**71.** $8x - 12 - 3x = 9x - 7$

**72.** $10y - 18 - 4y = 12y - 13$

**73.** $-(3x - 5) - (2x - 6) + 1 = -5(x - 1) - (3x + 2) + 3$

**74.** $-4(2x - 3) - (10x + 7) - 2 = -(12x - 5) - (4x + 9) - 1$

**75.** $\dfrac{1}{3}(y + 4) + 6 = \dfrac{1}{4}(3y - 1) - 2$

**76.** $\dfrac{1}{5}(2y - 1) - 2 = \dfrac{1}{2}(3y - 5) + 3$

**77.** $2[7 - 5(1 - n)] + 8n = -16 + 3[6(n + 1) - 3n]$

**78.** $3[8 - 4(n - 2)] + 5n = -20 + 2[5(1 - n) - 6n]$

## Review

*Translate each phrase into an expression. Use the variable x to represent each unknown number. See Section 1.2.*

**79.** The quotient of 8 and a number

**80.** The sum of 8 and a number

**81.** The product of 8 and a number

**82.** The difference of 8 and a number

**83.** Five subtracted from twice a number

**84.** Two more than three times a number

## Concept Extensions

*Find the error in each proposed solution. Then correct the proposed solution. See the Concept Check in this section.*

**85.** $2x + 19 = 13$

$2x = 32$

$\dfrac{2x}{2} = \dfrac{32}{2}$

$x = 16$

**86.** $-3(x - 4) = 10$

$-3x - 12 = 10$

$-3x = 22$

$\dfrac{-3x}{-3} = \dfrac{22}{-3}$

$x = -\dfrac{22}{3}$

**87.** $9x + 1.6 = 4x + 0.4$

$5x = 1.2$

$\dfrac{5x}{5} = \dfrac{1.2}{5}$

$x = 0.24$

**88.** $\dfrac{x}{3} + 7 = \dfrac{5x}{3}$

$x + 7 = 5x$

$7 = 4x$

$\dfrac{7}{4} = \dfrac{4x}{4}$

$\dfrac{7}{4} = x$

*Solve.*

**89. a.** Simplify the expression $4(x + 1) + 1$.

**b.** Solve the equation $4(x + 1) + 1 = -7$.

**c.** Explain the difference between solving an equation for a variable and simplifying an expression.

**90.** Explain why the multiplication property of equality does not include multiplying both sides of an equation by 0. (*Hint:* Write down a false statement and then multiply both sides by 0. Is the result true or false? What does this mean?)

**91.** In your own words, explain why the equation $x + 7 = x + 6$ has no solution while the solution set of the equation $x + 7 = x + 7$ contains all real numbers.

**92.** In your own words, explain why the equation $x = -x$ has one solution—namely, 0—while the solution set of the equation $x = x$ is all real numbers.

*For Exercises 93–96, find the value of K such that the equations are equivalent.*

**93.** $3.2x + 4 = 5.4x - 7$

$3.2x = 5.4x + K$

**94.** $-7.6y - 10 = -1.1y + 12$

$-7.6y = -1.1y + K$

**95.** $\dfrac{7}{11}x + 9 = \dfrac{3}{11}x - 14$

$\dfrac{7}{11}x = \dfrac{3}{11}x + K$

**96.** $\dfrac{x}{6} + 4 = \dfrac{x}{3}$

$x + K = 2x$

**97.** Write a linear equation in $x$ whose only solution is 5.

**98.** Write an equation in $x$ that has no solution.

*Solve and check.*

**99.** $-9.112y = -47.537304$

**100.** $2.86z - 8.1258 = -3.75$

**101.** $x(x - 6) + 7 = x(x + 1)$

**102.** $7x^2 + 2x - 3 = 6x(x + 4) + x^2$

## 2.2 An Introduction to Problem Solving

### Objective A Writing Algebraic Expressions

In order to prepare for problem solving, we practice writing algebraic expressions that can be simplified.

Our first example involves consecutive integers and perimeter. Recall that *consecutive integers* are integers that follow one another in order. Study the examples of consecutive integers, consecutive even integers, and consecutive odd integers and their representations.

**Objectives**

**A** Write Algebraic Expressions That Can Be Simplified.

**B** Apply the Steps for Problem Solving.

> **Helpful Hint**
>
> You may want to begin this section by studying key words and phrases and their translations in Sections 1.2, objective B, and 1.3, objective A.

**Example 1** Write the following as algebraic expressions. Then simplify.

**a.** The sum of three consecutive integers, if $x$ is the first consecutive integer.

**b.** The perimeter of the triangle with sides of length $x$, $5x$, and $6x - 3$.

**Solution:**

**a.** Recall that if $x$ is the first integer, then the next consecutive integer is 1 more, or $x + 1$, and the next consecutive integer is 1 more than $x + 1$, or $x + 1 + 1$, or $x + 2$.

| In words: | first integer | plus | next consecutive integer | plus | next consecutive integer |
|---|---|---|---|---|---|

Translate: $\quad x \quad + \quad (x + 1) \quad + \quad (x + 2)$

Then $x + (x + 1) + (x + 2) = x + x + 1 + x + 2$

$\qquad\qquad\qquad\qquad\qquad = 3x + 3$    Simplify by combining like terms.

**b.** The perimeter of a triangle is the sum of the lengths of the sides.

In words:   side   +   side   +   side

Translate: $\quad x \quad + \quad 5x \quad + (6x - 3)$

Then $x + 5x + (6x - 3) = x + 5x + 6x - 3$

$\qquad\qquad\qquad\qquad\qquad = 12x - 3$   Simplify.

**Work Practice 1**

### Practice 1

Write the following as algebraic expressions. Then simplify.

**a.** The sum of three consecutive even integers, if $x$ is the first even integer.

**b.** The perimeter of the triangle with sides of length $x$, $2x + 1$, and $4x$.

## Practice 2

Athletes from all over the world participated in the 2012 Summer Olympics in London, England. The number of athletes from the United States was 120 more than the athletes from Australia, and the number of athletes from Great Britain was 271 fewer than twice the number of athletes from Australia. Write the sum of the number of athletes competing from these countries in the 2012 Summer Olympics as an algebraic expression. Let $x$ be the number of athletes from Australia. (In Exercise 30, we will find the exact number of athletes from each country.) (*Source:* Theguardian.com)

**Example 2** Writing Algebraic Expressions Representing Metropolitan Regions

The three most populous metropolitan regions in the United States are the cities and surrounding regions of New York City, Los Angeles, and Chicago, in that order. The population of the New York City area is 2 million more than double the population of the Chicago area. The population of the Los Angeles area is 2.4 million less than twice the population of the Chicago area. Write the sum of the populations of three metropolitan regions as an algebraic expression. Let $x$ be the population of the Chicago area (in millions). (*Source:* worldatlas.com)

### Solution:

If $x$ = the population of Chicago (in millions) then

$2x + 2$ = the population of New York City (in millions) and

$2x - 2.4$ = the population of Los Angeles (in millions)

In words:  | population of Chicago | population of New York City | population of Los Angeles |

Translate:       $x$       +       $(2x + 2)$       +       $(2x - 2.4)$

Then $x + (2x + 2) + (2x - 2.4) = x + 2x + 2 + 2x - 2.4$

$$= 5x - 0.4$$

In Exercise 29, we will find the actual population of these metropolitan regions.

■ Work Practice 2

## Objective B  Solving Problems ▶

Our main purpose for studying algebra is to solve problems. The following problem-solving strategy will be used throughout this text and may also be used to solve real-life problems that occur outside the mathematics classroom.

### General Strategy for Problem Solving

1. UNDERSTAND the problem. During this step, become comfortable with the problem. Some ways of doing this are:

   Read and reread the problem.

   Choose a variable to represent the unknown.

   Construct a drawing if necessary.

   Propose a solution and check. Pay careful attention to how you check your proposed solution. This will help when writing an equation to model the problem.

2. TRANSLATE the problem into an equation.

3. SOLVE the equation.

4. INTERPRET the results: If possible, check to see whether your answer is reasonable. Then *check* the proposed solution in the stated problem and *state* your conclusion.

**Answer**

**2.** $x$ = Australia,
$x + 120$ = United States,
$2x - 271$ = Great Britain;
sum = $4x - 151$

Let's review this strategy by solving a problem involving unknown numbers.

**Example 3** Finding Unknown Numbers

Find three numbers such that the second number is 3 more than twice the first number and the third number is four times the first number. The sum of the three numbers is 164.

**Solution:**

1. UNDERSTAND the problem. First let's read and reread the problem and then propose a solution. For example, if the first number is 25, then the second number is 3 more than twice 25, or 53. The third number is four times 25, or 100. The sum of 25, 53, and 100 is 178, not the required sum, but we have gained some valuable information about the problem. First, we know that the first number is less than 25 since our guess led to a sum greater than the required sum. Also, we have gained some information as to how to model the problem.

   Next let's assign a variable and use this variable to represent any other unknown quantities. If we let

$$x = \text{the first number, then}$$

$$\underbrace{2x}_{} + \overset{\uparrow}{3} = \text{the second number}$$

$$\underset{\substack{\uparrow \\ \text{twice the second number}}}{\overset{\text{3 more than}}{}}$$

$$4x = \text{the third number}$$

2. TRANSLATE the problem into an equation. To do so, we use the fact that the sum of the numbers is 164. First let's write this relationship in words and then translate to an equation.

| In words: | first number | added to | second number | added to | third number | is | 164 |
|-----------|--------------|----------|---------------|----------|--------------|-----|-----|
| | ↓ | ↓ | ↓ | ↓ | ↓ | ↓ | ↓ |
| Translate: | $x$ | $+$ | $(2x + 3)$ | $+$ | $4x$ | $=$ | $164$ |

3. SOLVE the equation.

$$x + (2x + 3) + 4x = 164$$
$$x + 2x + 4x + 3 = 164 \quad \text{Remove parentheses.}$$
$$7x + 3 = 164 \quad \text{Combine like terms.}$$
$$7x = 161 \quad \text{Subtract 3 from both sides.}$$
$$x = 23 \quad \text{Divide both sides by 7.}$$

4. INTERPRET. Here, we *check* our work and *state* the solution. Recall that if the first number $x = 23$, then the second number $2x + 3 = 2 \cdot 23 + 3 = 49$, and the third number $4x = 4 \cdot 23 = 92$.

**Check:** Is the second number 3 more than twice the first number? Yes, since 3 more than twice 23 is $46 + 3$, or 49. Is the third number four times the first number? Yes, since $4 \cdot 23 = 92$. Also, their sum, $23 + 49 + 92 = 164$, is the required sum.

**State:** The three numbers are 23, 49, and 92.

■ Work Practice 3

Many of today's rates and statistics are given as percents. Interest rates, tax rates, nutrition labeling, and percent of households in a given category are just a few examples. Before we practice solving problems containing percents, let's take a moment to review the meaning of percent and how to find a percent of a number.

**Practice 3**

One number is three times the first number. A third number is 50 more than the first number. If their sum is 235, find the three numbers.

**Helpful Hint** The purpose of guessing a solution is not to guess correctly but to gain confidence and to help understand the problem and how to model it.

**Answer**
**3.** 37, 87, and 111

The word *percent* means *per hundred,* and the symbol % is used to denote percent. This means that 23% is 23 per one hundred, or $\frac{23}{100}$. Also,

$$41\% = \frac{41}{100} = 0.41$$

To find a percent of a number, we multiply.

$$16\% \text{ of } 25 = 16\% \cdot 25 = 0.16 \cdot 25 = 4$$

Thus, 16% of 25 is 4.

Study the table below. It will help you become more familiar with finding percents.

| Percent | Meaning/Shortcut | Example |
|---------|------------------|---------|
| 50% | $\frac{1}{2}$ or half of a number | 50% of 60 is 30. |
| 25% | $\frac{1}{4}$ or a quarter of a number | 25% of 60 is 15. |
| 10% | 0.1 or $\frac{1}{10}$ of a number (move the decimal point 1 place to the left) | 10% of 60 is 6.0 or 6. |
| 1% | 0.01 or $\frac{1}{100}$ of a number (move the decimal point 2 places to the left) | 1% of 60 is 0.60 or 0.6. |
| 100% | 1 or all of a number | 100% of 60 is 60. |
| 200% | 2 or double a number | 200% of 60 is 120. |

✓**Concept Check** Suppose you are finding 112% of a number $x$. Which of the following is a correct description of the result? Explain.

**a.** The result is less than $x$.

**b.** The result is equal to $x$.

**c.** The result is greater than $x$.

Next, we solve a problem containing a percent.

**Example 4** Finding the Original Price of a Computer

Suppose that a computer store just announced an 8% decrease in the price of a particular computer model. If this computer sells for $2162 after the decrease, find its original price.

**Solution:**

**1. UNDERSTAND.** Read and reread the problem. Recall that a percent decrease means a percent of the original price. Let's guess that the original price of the computer is $2500. The amount of decrease is then 8% of $2500, or $(0.08)(\$2500) = \$200$. This means that the new price of the computer is the

---

**Practice 4**

The price of a home was just decreased by 6%. If the decreased price is $83,660, find the original price of the home.

**Answer**

**4.** $89,000

✓**Concept Check Answer**

c

original price minus the decrease, or $2500 − $200 = $2300. Our guess is incorrect, but we now have an idea of how to model this problem. In our model, we will let $x$ = the original price of the computer.

**2.** TRANSLATE.

| In words: | original price of computer | minus | 8% of original price | is | new price |
|---|---|---|---|---|---|
| | ↓ | ↓ | ↓ | ↓ | ↓ |
| Translate: | $x$ | − | $0.08x$ | = | 2162 |

**3.** SOLVE the equation.

$$x - 0.08x = 2162$$
$$0.92x = 2162 \qquad \text{Combine like terms.}$$
$$x = \frac{2162}{0.92} = 2350 \qquad \text{Divide both sides by 0.92.}$$

**4.** INTERPRET.

**Check:**  The amount $2350 is a reasonable price for a computer. If the original price of the computer was $2350, the new price is

$$\$2350 - (0.08)(\$2350) = \$2350 - \$188$$
$$= \$2162 \qquad \text{The given new price}$$

**State:**  The original price of the computer was $2350.

■ Work Practice 4

---

⚠ **Example 5**  Finding the Lengths of a Triangle's Sides

A pennant in the shape of an isosceles triangle is to be constructed for the Slidell High School Athletic Club and sold at a fundraiser. The company manufacturing the pennant charges according to perimeter, and the athletic club has determined that a perimeter of 149 centimeters should make a nice profit. If each equal side of the triangle is twice the length of the third side, increased by 12 centimeters, find the lengths of the sides of the triangular pennant.

**Solution:**

**1.** UNDERSTAND. Read and reread the problem. Recall that the perimeter of a triangle is the distance around. Let's guess that the third side of the triangular pennant is 20 centimeters. This means that each equal side is twice 20 centimeters, increased by 12 centimeters, or $2(20) + 12 = 52$ centimeters.

This gives a perimeter of $20 + 52 + 52 = 124$ centimeters. Our guess is incorrect, but we now have a better understanding of how to model this problem. Now we let

$$x = \text{the third side of the triangle, then}$$
$$2x + 12 = \text{the first side}$$
$$2x + 12 = \text{the second side}$$

*(Continued on next page)*

---

**Practice 5**

A rectangle has a perimeter of 106 meters. Its length is 5 meters more than twice its width. Find the length and the width of the rectangle.

**Answer**
**5.** length: 37 m, width: 16 m

**2. TRANSLATE.**

In words:

| first side | + | second side | + | third side | = | 149 |

Translate: $(2x + 12) + (2x + 12) + x = 149$

**3. SOLVE the equation.**

$$(2x + 12) + (2x + 12) + x = 149$$
$$2x + 2x + x + 12 + 12 = 149 \quad \text{Remove parentheses.}$$
$$5x + 24 = 149 \quad \text{Combine like terms.}$$
$$5x = 125 \quad \text{Subtract 24 from both sides.}$$
$$x = 25 \quad \text{Divide both sides by 5.}$$

**4. INTERPRET.** If the third side is $25$ centimeters, then the first side is $2(25) + 12 = 62$ centimeters and the second side is 62 centimeters also.

**Check:** The first and second sides are each twice 25 centimeters increased by 12 centimeters, or 62 centimeters. Also, the perimeter is $25 + 62 + 62 = 149$ centimeters, the required perimeter.

**State:** The dimensions of the triangle are 62 centimeters, 62 centimeters, and 25 centimeters.

🔳 **Work Practice 5**

**Practice 6**

Find three consecutive integers whose sum is 378.

### Example 6  Finding Consecutive Integers

Kelsey was helping her friend Benji study for an exam. Kelsey told Benji that her three latest art history quiz scores happened to be three consecutive even integers whose sum is 264. Help Benji find the scores.

**Solution:**

**1. UNDERSTAND.** Read and reread the problem. Since we are looking for consecutive even integers, let

$x = $ the first integer. Then
$x + 2 = $ the next consecutive even integer.
$x + 4 = $ "   "   "   "   "

**2. TRANSLATE.**

In words:

| first integer | + | next even integer | + | next even integer | = | 264 |

Translate: $x + (x + 2) + (x + 4) = 264$

**3. SOLVE.**

$$x + (x + 2) + (x + 4) = 264$$
$$3x + 6 = 264 \quad \text{Combine like terms.}$$
$$3x = 258 \quad \text{Subtract 6 from both sides.}$$
$$x = 86 \quad \text{Divide both sides by 3.}$$

**4. INTERPRET.** If $x = 86$, then $x + 2 = 86 + 2$ or 88, and $x + 4 = 86 + 4$ or 90.

**Check:** The numbers 86, 88, and 90 are three consecutive even integers. Their sum is 264, the required sum.

**State:** Kelsey's art history quiz scores are 86, 88, and 90.

**Answer**
**6.** 125, 126, and 127

🔳 **Work Practice 6**

## Vocabulary, Readiness & Video Check

*Fill in each blank with <, >, or =. (Assume that the unknown number is a positive number.)*

**1.** 130% of a number _____ the number.

**2.** 70% of a number _____ the number.

**3.** 100% of a number _____ the number.

**4.** 200% of a number _____ the number.

*Complete the table. The first row has been completed for you.*

|   | First Integer | All Described Integers |
|---|---|---|
| Three consecutive integers | 18 | 18, 19, 20 |
| **5.** Four consecutive integers | 31 | |
| **6.** Three consecutive odd integers | 31 | |
| **7.** Three consecutive even integers | 18 | |
| **8.** Four consecutive even integers | 92 | |
| **9.** Three consecutive integers | $y$ | |
| **10.** Three consecutive even integers | $z$ ($z$ is even) | |
| **11.** Four consecutive integers | $p$ | |
| **12.** Three consecutive odd integers | $s$ ($s$ is odd) | |

*Martin-Gay Interactive Videos*     *Watch the section lecture video and answer the following questions.*

*See Video 2.2*

**Objective A** **13.** In ⊞ Example 2, once the phrase is translated, what property is used to remove parentheses? ▶

**Objective B** **14.** The equation for ⊞ Example 3 is solved and the result is $x = 45$. Why is 45 not the complete solution to the application? ▶

## 2.2 Exercise Set MyMathLab® ▶

**Objective A** *Write the following as algebraic expressions. Then simplify. See Examples 1 and 2.*

△ **1.** The perimeter of the square with side length $y$.

△ **2.** The perimeter of the rectangle with length $x$ and width $x - 5$.

▶ **3.** The sum of three consecutive integers if the first is $z$.

**4.** The sum of three consecutive odd integers if the first integer is $x$.

▶ **5.** The total amount of money (in cents) in $x$ nickels, $(x + 3)$ dimes, and $2x$ quarters. (*Hint:* The value of a nickel is 5 cents, the value of a dime is 10 cents, and the value of a quarter is 25 cents.)

**6.** The total amount of money (in cents) in $y$ quarters, $7y$ dimes, and $(2y - 1)$ nickels. (Use the hint for Exercise 5.)

△ **7.** A piece of land along Bayou Liberty is to be fenced and subdivided as shown so that each rectangle has the same dimensions. Express the total amount of fencing needed as an algebraic expression in $x$.

**8.** A flooded piece of land near the Mississippi River in New Orleans is to be surveyed and divided into 4 rectangles of equal dimension. Express the total amount of fencing needed as an algebraic expression in $x$.

△ **9.** Write the perimeter of the floor plan shown as an algebraic expression in $x$.

**10.** Write the perimeter of the floor plan shown as an algebraic expression in $x$.

## Objective B   *Solve. For Exercises 11 and 12, the solutions have been started for you. See Example 3.*

**11.** Four times the difference of a number and 2 is the same as 2, increased by four times the number, plus twice the number. Find the number.

Start the solution:

1. UNDERSTAND the problem. Reread it as many times as needed.
2. TRANSLATE into an equation. (Fill in the blanks below.)

| Four times | the difference of a number and 2 | is the same as | 2 | increased by | four times the number | plus | twice the number |
|---|---|---|---|---|---|---|---|
| ↓ | ↓ | ↓ | ↓ | ↓ | ↓ | ↓ | ↓ |
| 4 · | $(x - 2)$ | = | 2 | ___ | ___ | ___ | ___ |

Finish with:
3. SOLVE and 4. INTERPRET

**12.** Twice the sum of a number and 3 is the same as five times the number, minus 1, minus four times the number. Find the number.

Start the solution:

1. UNDERSTAND the problem. Reread it as many times as needed.
2. TRANSLATE into an equation. (Fill in the blanks below.)

| Twice | the sum of a number and 3 | is the same as | five times the number | minus | 1 | minus | four times the number |
|---|---|---|---|---|---|---|---|
| ↓ | ↓ | ↓ | ↓ | ↓ | ↓ | ↓ | ↓ |
| 2 | $(x + 3)$ | = | ___ | ___ | 1 | ___ | ___ |

Finish with:
3. SOLVE and 4. INTERPRET

▶ **13.** A second number is five times a first number. A third number is 100 more than the first number. If the sum of the three numbers is 415, find the numbers.

**14.** A second number is 6 less than a first number. A third number is twice the first number. If the sum of the three numbers is 306, find the numbers.

*Solve. See Examples 4 through 6.*

**15.** The United States consists of 2271 million acres of land. Approximately 29% of this land is federally owned. Find the number of acres that are not federally owned. (*Source:* U.S. General Services Administration)

**16.** The state of Nevada contains the most federally owned acres of land in the United States. If 90% of the state's 70 million acres of land is federally owned, find the number of acres that are not federally owned. (*Source:* U.S. General Services Administration)

**17.** In 2012, a total of 3836 earthquakes occurred in the United States. Of these, 92.1% were minor tremors with magnitudes of 3.9 or less on the Richter scale. How many minor earthquakes occurred in the United States in 2012? Round to the nearest whole. (*Source:* U.S. Geological Survey National Earthquake Information Center)

**18.** Of the 939 tornadoes that occurred in the United States during 2012, 21.9% occurred during the month of April. How many tornadoes occurred in the United States during April 2012? Round to the nearest whole. (*Source:* Storm Prediction Center)

**19.** In a recent survey, 15% of online shoppers in the United States say that they prefer to do business only with large, well-known retailers. In a group of 1500 online shoppers, how many are willing to do business with any size retailers? (*Source:* Inc.com)

**20.** In 2013, the restaurant industry employed 9% of the U.S. workforce. If there are estimated to be 145 million Americans in the workforce, how many people are employed by the restaurant industry? Round to the nearest tenth. (*Sources:* National Restaurant Association, U.S. Bureau of Labor Statistics)

*The following graph is called a circle graph or a pie chart. The circle represents a whole or, in this case, 100%. This particular graph shows the number of minutes per day that people use e-mail at work. Use this graph to answer Exercises 21 through 24.*

**Time Spent on E-mail at Work**

Source: Pew Internet & American Life Project

**21.** What percent of e-mail users at work spend less than 15 minutes on e-mail per day?

**22.** Among e-mail users at work, what is the most common time spent on e-mail per day?

**23.** If it were estimated that a large company has 4633 employees, how many of these would you expect to be using e-mail more than 3 hours per day?

**24.** If it were estimated that a medium-size company has 250 employees, how many of these would you expect to be using e-mail between 2 and 3 hours per day?

**25.** Daytona International Speedway in Florida has 37,000 more grandstand seats than twice the number of grandstand seats at Darlington Motor Raceway in South Carolina. Together, these two race tracks seat 220,000 NASCAR fans. How many seats does each race track have? (*Source:* NASCAR)

**26.** For the 2013–2014 National Hockey League season, the payroll for the New York Rangers was $3,005,000 less than the payroll for the Boston Bruins. The total payroll for these two teams was $129,675,000. What was the payroll for each of these teams for the 2013–2014 NHL season? (*Source: USA Today*)

△ **27.** The perimeter of the triangle in Example 1b in this section is 483 feet. Find the length of each side.

△ **28.** The perimeter of the triangle in Practice 1b in this section is 130.5 meters. Find its dimensions.

**29.** The sum of the populations of the metropolitan regions of New York, Los Angeles, and Chicago is 48.6 million. Use this information and Example 2 in this section to find the population of each metropolitan region. (*Source:* worldatlas.com)

**30.** The total number of athletes from the United States, Australia, and Great Britain at the 2012 Summer Olympics in London was 1505. Use this information and Practice 2 in this section to find the number of athletes from each of these medal-winning countries. (*Source:* Theguardian.com)

*Use the diagrams to find the unknown measures of angles or lengths of sides. Recall that the sum of the angle measures of a triangle is 180°.*

**31.** $4x°$, $(x + 6)°$, $x°$

**32.** $3x°$, $(x + 10)°$, $x°$

**33.** $(4x)$ m, $(5x + 1)$ m, $(5x + 3)$ m
Perimeter is 102 meters.

**34.** $(2x)$ cm, $(3.5x)$ cm, $(3x + 7)$ cm
Perimeter is 75 centimeters.

**35.** $(2.5x - 9)$ in., Isoceles Trapezoid, $x$ in., $x$ in., $1.5x$ in.
Perimeter is 99 inches.

**36.** $(9.2x - 3)$ ft, $7.3x$ ft, Parallelogram, $7.3x$ ft, $(9.2x - 3)$ ft
Perimeter is 324 feet.

*Many companies predict the growth or decline of various technologies. The following data are based on information from Forrester Research, a global research firm. Notice that the first table is the predicted increase in the number of tablet users worldwide (in millions) and the second is the predicted percent increase in the number of tablet users worldwide.*

**37.** Use the middle column in the table to find the predicted number of tablet users worldwide for each year.

| Year | Increase in Tablet Users Worldwide | Predicted Number |
|------|-----------------------------------|-----------------|
| 2014 | $x + 44$ | |
| 2015 | $1.5x - 21$ | |
| 2016 | $2x - 104$ | |
| Total | 1944 million | |

**38.** Use the middle column in the table to find the predicted percent increase in the number of tablet users worldwide for each year.

| Year | Percent Increase in Tablet Users Worldwide since 2013 | Predicted Percent Increase |
|------|-------------------------------------------------------|---------------------------|
| 2014 | $x$ | |
| 2015 | $2x - 2$ | |
| 2016 | $3x - 10$ | |
| Total | 294% | |

*Solve.*

**39.** The occupations of biomedical engineers, event planners, and interpreters are among the 10 with the largest growth from 2010 to 2020. The number of event planners will grow four times the number of biomedical engineer jobs. The number of interpreters will grow 1 thousand less than three times the number of biomedical engineer jobs. If the total growth of these three jobs is predicted to be 63 thousand, find the predicted growth of each job. (*Source:* U.S. Department of Labor, Bureau of Labor Statistics)

**40.** The job categories of utilities, manufacturing, and agriculture are among the 10 job categories with the largest decline from 2010 to 2020. The number of jobs in manufacturing is predicted to decline 1 thousand more than twice the number of jobs in utilities. The number of jobs in agriculture is predicted to decline by 24 thousand less than double the number of jobs in utilities. If the total decline of these three job categories is 157 thousand, find the predicted decline of each job category. (*Source:* U.S. Department of Labor, Bureau of Labor Statistics)

**41.** AT&T Stadium, home of the Dallas Cowboys of the NFL, seats approximately 11,500 more fans than does Levi's Stadium, home of the San Francisco 49ers. Soldier Field, home of the Chicago Bears, seats 7000 fewer fans than Levi's Stadium. If the total seats in these three stadiums is 210,000, how many seats are in each of the three stadiums?

**42.** The B767-300ER aircraft has 88 more seats than the B737-200 aircraft. The F-100 has 32 fewer seats than the B737-200 aircraft. If their total number of seats is 413, find the number of seats for each aircraft. (*Source:* Air Transport Association of America)

**43.** A new fax machine was recently purchased for an office in Hopedale for $464.40 including tax. If the tax rate in Hopedale is 8%, find the price of the fax machine before tax.

**44.** A premedical student at a local university was complaining that she had just paid $158.60 for her human anatomy book, including tax. Find the price of the book before taxes if the tax rate at this university is 9%.

**45.** The median compensation for presidents of U.S. public research universities was $441,400 for the 2012–2013 school year. Calculate the salary of a university president who received a 4.5% raise for the 2013–2014 school year.

**46.** In 2013, the population of Japan was 127.3 million people. This represented a decrease in population of 0.2% from 2012. What was the population of Japan in 2012? Round to the nearest tenth of a million. (*Source:* Population Reference Bureau)

**47.** The sum of three consecutive integers is 228. Find the integers.

**48.** The sum of three consecutive odd integers is 327. Find the integers.

*Recall that the sum of the angle measures of a triangle is 180°.*

△ **49.** Find the measures of the angles of a triangle if the measure of one angle is twice the measure of a second angle and the third angle measures 3 times the second angle decreased by 12.

△ **50.** Find the angles of an isoceles triangle whose two base angles are equal and whose third angle is 10° less than three times a base angle.

**51.** Two frames are needed with the same perimeter: one frame in the shape of a square and one in the shape of an equilateral triangle. Each side of the triangle is 6 centimeters longer than each side of the square. Find the length of a side of each frame. (An equilateral triangle has sides that are the same length.)

**52.** Two frames are needed with the same perimeter: one frame in the shape of a square and one in the shape of a regular pentagon. Each side of the square is 7 inches longer than each side of the pentagon. Find the length of a side of each frame. (A regular polygon has sides that are the same length.)

**53.** Cisco Systems, a networking company giant, recently announced it would be cutting 6500 jobs. This is equivalent to 9% of Cisco's workforce. Find the size of Cisco's workforce prior to this round of layoffs. Round to the nearest whole. (*Source:* MSN.com)

**54.** Dana, an auto parts supplier headquartered in Toledo, Ohio, recently announced it would be cutting 11,000 jobs worldwide. This is equivalent to 15% of Dana's workforce. Find the size of Dana's workforce prior to this round of job layoffs. Round to the nearest whole. (*Source:* Dana Corporation)

**55.** The ZIP codes of three Nevada locations—Fallon, Fernley, and Gardnerville Ranchos—are three consecutive even integers. If twice the first integer added to the third is 268,222, find each ZIP code.

**56.** During a recent year, the average SAT scores in math for the states of Alabama, Louisiana, and Michigan were 3 consecutive integers. If the sum of the first integer, second integer, and three times the third integer is 2637, find each score.

*Recall that two angles are complements of each other if their sum is 90°. Two angles are supplements of each other if their sum is 180°. Find the measure of each angle.*

**57.** One angle is three times its supplement increased by 20°. Find the measures of the two supplementary angles.

**58.** One angle is twice its complement increased by 30°. Find the measure of the two complementary angles.

**59.** Incandescent, fluorescent, and halogen bulbs are lasting longer today than ever before. On average, the number of bulb hours for a fluorescent bulb is 25 times the number of bulb hours for a halogen bulb. The number of bulb hours for an incandescent bulb is 2500 less than the halogen bulb. If the total number of bulb hours for the three types of bulbs is 105,500, find the number of bulb hours for each type. (*Source: Popular Science* magazine)

**60.** Singapore, Hong Kong, and the United Kingdom (UK) are the top three countries that have the greatest penetration rates (number of cellular subscriptions per 100 inhabitants) in the world. Hong Kong has a 20.7 greater penetration rate than Singapore. The UK has a penetration rate 7.2 less than Singapore. (*Source:* This Week in Consumer Electronics)

   **a.** If the sum of the penetration rates is 390.6, find the number of inhabitants per 100 in each of these countries with cellular subscriptions.

   **b.** Explain, in your own words, how the penetration rate of cellular subscriptions in these three countries is greater than the number of inhabitants.

**61.** During the 2013 Major League Baseball season, the numbers of home runs hit by Alfonso Soriano of the New York Yankees, Edwin Encarnacion of the Toronto Blue Jays, and Evan Longoria of the Tampa Bay Rays were three consecutive even integers. Of these three players, Encarnacion hit the most home runs and Longoria hit the fewest home runs. The total number of home runs hit by these three players was 102. How many home runs did each player hit during the 2013 season? (*Source:* Major League Baseball)

**62.** In the 2012 Summer Olympics, Italy won more medals than Australia, which won more medals than the Netherlands. It the numbers of medals won by these three countries are three consecutive integers whose sum is 21, find the number of medals won by each. (*Source:* Olympic.org)

## Review

*Find the value of each expression for the given values. See Section 1.5.*

**63.** $4ab - 3bc$; $a = -5$, $b = -8$, and $c = 2$

**64.** $ab + 6bc$; $a = 0$, $b = -1$, and $c = 9$

**65.** $n^2 - m^2$; $n = -3$ and $m = -8$

**66.** $2n^2 + 3m^2$; $n = -2$ and $m = 7$

**67.** $P + PRT$; $P = 3000$, $R = 0.0325$, and $T = 2$

**68.** $\frac{1}{3} lwh$; $l = 37.8$, $w = 5.6$, and $h = 7.9$

## Concept Extensions

**69.** For Exercise 38, the percents have a sum of 294%. Is this possible? Why or why not?

**70.** In your own words, explain the differences in the tables for Exercises 37 and 38.

*Use the figure below to help you answer Exercises 71 and 72.*

**71.** Find an angle such that its supplement is equal to 10 times its complement.

**72.** Find an angle such that its supplement is equal to twice its complement increased by 50°.

*The average annual number of cigarettes smoked by an American adult continues to decline. For the years 2000–2012, the equation $y = -88x + 2105$ approximates this data. Here, $x$ is the number of years after 2000, and $y$ is the average annual number of cigarettes smoked. (Source: United States Department of Agriculture: Economic Research Service)*

**73.** If this trend continues, find the year in which the average annual number of cigarettes smoked is zero. To do this, let $y = 0$ and solve for $x$.

**74.** Predict the average annual number of cigarettes smoked by an American adult in 2015. To do so, let $x = 15$ (since $2015 - 2000 = 15$) and find $y$.

**75.** Predict the average annual number of cigarettes smoked by an American adult in 2020. To do so, let $x = 20$ (since $2020 - 2000 = 20$) and find $y$.

**76.** Use the results of Exercise 74 to predict the average *daily* number of cigarettes smoked by an American adult in 2015. Round to the nearest whole. Do you think this number represents the average daily number of cigarettes smoked by an adult smoker? Why or why not?

*Solve.*

**77.** Determine whether there are three consecutive integers such that their sum is three times the second integer.

**78.** Determine whether there are two consecutive odd integers such that 7 times the first exceeds 5 times the second by 54.

*To break even in a manufacturing business, income or revenue R must equal the cost of production C. Use this information to answer Exercises 79 through 82.*

**79.** The cost $C$ to produce $x$ number of skateboards is $C = 100 + 20x$. The skateboards are sold wholesale for $24 each, so revenue $R$ is given by $R = 24x$. Find how many skateboards the manufacturer needs to produce and sell to break even. (*Hint:* Set the cost expression equal to the revenue expression and solve for $x$.)

**80.** The revenue $R$ from selling $x$ number of computer boards is given by $R = 60x$, and the cost $C$ of producing them is given by $C = 50x + 5000$. Find how many boards must be sold to break even. Find how much money is needed to produce the break-even number of boards.

**81.** In your own words, explain what happens if a company makes and sells fewer products than the break-even number.

**82.** In your own words, explain what happens if more products than the break-even number are made and sold.

---

### 2.3  Formulas and Problem Solving

**Objectives**

**A** Solve a Formula for a Specified Variable.

**B** Use Formulas to Solve Problems.

#### Objective A  Solving Formulas for Specified Variables

Solving problems that we encounter in the real world sometimes requires us to express relationships among measured quantities. A **formula** is an equation that describes a known relationship among measured phenomena, such as time, area, and gravity. Some examples of formulas follow.

| Formula | Meaning |
|---|---|
| $I = Prt$ | Interest = principal $\cdot$ rate $\cdot$ time |
| $A = lw$ | Area of a rectangle = length $\cdot$ width |
| $d = rt$ | Distance = rate $\cdot$ time |
| $C = 2\pi r$ | Circumference of a circle = $2 \cdot \pi \cdot$ radius |
| $V = lwh$ | Volume of a rectangular solid = length $\cdot$ width $\cdot$ height |

Other formulas are listed on an inside cover of this text. Notice that the formula for the volume of a rectangular solid, $V = lwh$, is solved for $V$ since $V$ is by itself on one side of the equation with no $V$s on the other side of the equation. Suppose that the volume of a rectangular solid is known as well as its width and its length, and we wish to find its height. One way to find its height is to begin by solving the formula $V = lwh$ for $h$.

**Example 1** Solve $V = lwh$ for $h$.

**Solution:** To solve $V = lwh$ for $h$, we want to get $h$ alone on one side of the equation. To do so, we divide both sides of the equation by $lw$.

$$V = lwh$$

$$\frac{V}{lw} = \frac{lw\,h}{lw} \qquad \text{Divide both sides by } lw.$$

$$\frac{V}{lw} = h \text{ or } h = \frac{V}{lw} \qquad \text{Simplify.}$$

Thus, we see that to find the height of a rectangular solid, we divide its volume by the product of its length and its width.

■ Work Practice 1

The following steps may be used to solve formulas and equations for a specified variable.

> ### Solving an Equation for a Specified Variable
>
> **Step 1:** Clear the equation of fractions or decimals by multiplying each side of the equation by an appropriate nonzero number.
>
> **Step 2:** Use the distributive property to remove grouping symbols such as parentheses.
>
> **Step 3:** Combine like terms on each side of the equation.
>
> **Step 4:** Use the addition property of equality to rewrite the equation as an equivalent equation with terms containing the specified variable on one side and all other terms on the other side.
>
> **Step 5:** Use the distributive property and the multiplication property of equality to get the specified variable alone.

**Example 2** Solve $3y - 2x = 7$ for $y$.

**Solution:** This is a linear equation in two variables. Often an equation such as this is solved for $y$ to reveal some properties about the graph of this equation, which we will learn more about in Chapter 3. Since there are no fractions or grouping symbols, we begin with Step 4 and get the term containing the specified variable $y$ on one side by adding $2x$ to both sides of the equation.

$$3y - 2x = 7$$

$$3y - 2x + 2x = 7 + 2x \qquad \text{Add } 2x \text{ to both sides.}$$

$$3y = 7 + 2x$$

*(Continued on next page)*

**Practice 1**

Solve $I = Prt$ for $P$.

**Practice 2**

Solve $2y + 5x = 10$ for $y$.

**Answers**

**1.** $P = \dfrac{I}{rt}$  **2.** $y = \dfrac{10 - 5x}{2}$ or

$y = 5 - \dfrac{5}{2}x$

To solve for $y$, we divide both sides by 3.

$$\frac{3y}{3} = \frac{7 + 2x}{3} \quad \text{Divide both sides by 3.}$$

$$y = \frac{7 + 2x}{3} \quad \text{or} \quad y = \frac{7}{3} + \frac{2}{3}x$$

■ Work Practice 2

**Practice 3**

Solve $A = \frac{1}{2}(B + b)h$ for $B$.

**Example 3**  Solve $A = \frac{1}{2}(B + b)h$ for $b$.

**Solution:**   Since this formula for finding the area of a trapezoid contains fractions, we begin by multiplying both sides of the equation by the LCD, 2.

> **Helpful Hint**   Remember that we may get the specified variable alone on either side of the equation.

$$A = \frac{1}{2}(B + b)h$$

$$2 \cdot A = 2 \cdot \frac{1}{2}(B + b)h \quad \text{Multiply both sides by 2.}$$

$$2A = (B + b)h \quad \text{Simplify.}$$

$$2A = Bh + bh \quad \text{Use the distributive property.}$$

$$2A - Bh = bh \quad \text{Get the term containing } b \text{ alone by subtracting } Bh \text{ from both sides.}$$

$$\frac{2A - Bh}{h} = \frac{bh}{h} \quad \text{Divide both sides by } h.$$

$$\frac{2A - Bh}{h} = b \quad \text{or} \quad b = \frac{2A - Bh}{h} \quad \text{or} \quad b = \frac{2A}{h} - B$$

■ Work Practice 3

## Objective B  Using Formulas to Solve Problems

In this section, we also solve problems that can be modeled by known formulas. We use the same problem-solving steps that were introduced in the previous section.

Formulas are very useful in problem solving. For example, the compound interest formula

$$A = P\left(1 + \frac{r}{n}\right)^{nt}$$

is used by banks to compute the amount $A$ in an account that pays compound interest. The variable $P$ represents the principal or amount invested in the account, $r$ is the annual rate of interest, $t$ is the time in years, and $n$ is the number of times compounded per year.

**Answer**

**3.** $B = \dfrac{2A - bh}{h}$ or $B = \dfrac{2A}{h} - b$

**Example 4** Finding the Amount in a Savings Account

Marial Callier just received an inheritance of $10,000 and plans to place all the money in a savings account that pays 5% compounded quarterly to help her son go to college in 3 years. How much money will be in the account in 3 years?

**Solution:**

1. **UNDERSTAND.** Read and reread the problem. The appropriate formula needed to solve this problem is the compound interest formula

$$A = P\left(1 + \frac{r}{n}\right)^{nt}$$

Make sure that you understand the meaning of all the variables in this formula:

$A$ = amount in the account after $t$ years

$P$ = principal or amount invested

$t$ = time in years

$r$ = annual rate of interest

$n$ = number of times compounded per year

2. **TRANSLATE.** Use the compound interest formula and let $P = \$10,000$, $r = 5\% = 0.05$, $t = 3$ years, and $n = 4$ since the account is compounded quarterly, or 4 times a year.

Formula: $\quad A = P\left(1 + \frac{r}{n}\right)^{nt}$

Substitute: $\quad A = 10,000\left(1 + \frac{0.05}{4}\right)^{4\cdot3}$

3. **SOLVE.** We simplify the right side of the equation.

$$A = 10,000\left(1 + \frac{0.05}{4}\right)^{4\cdot3}$$

$A = 10,000(1.0125)^{12\cdot}$     Simplify $1 + \dfrac{0.05}{4}$ and write $4 \cdot 3$ as 12.

$A \approx 10,000(1.160754518)$    Approximate $(1.0125)^{12}$.

$A \approx 11,607.55$         Multiply and round to two decimal places.

4. **INTERPRET.**

**Check:** Repeat your calculations to make sure that you made no error. Notice that $11,607.55 is a reasonable amount to have in the account after 3 years.

**State:** In 3 years, the account will contain $11,607.55.

◼ Work Practice 4

**Example 5** Finding Cycling Time

The fastest average speed by a cyclist across the continental United States is 15.4 mph, by Pete Penseyres. If he traveled a total distance of about 3107.5 miles at this speed, find his time cycling. Write the time in days, hours, and minutes. (*Source: Guinness World Records*)

**Solution:**

1. **UNDERSTAND.** Read and reread the problem. The appropriate formula needed is the distance formula

$d = rt$     where

$d$ = distance traveled    $r$ = rate    and    $t$ = time

(*Continued on next page*)

**Practice 4**

If $5000 is invested in an account paying 4% compounded monthly, determine how much money will be in the account in 2 years. Use the formula from Example 4.

**Practice 5**

Nearly 4800 cyclists from 36 U.S. states and 6 countries rode in the Pan-Massachusetts Challenge recently to raise money for cancer research and treatment. If the riders of a certain team traveled their 192-mile route at an average speed of 7.5 miles per hour, find the time they spent cycling. Write the answer in hours and minutes.

**Answers**

**4.** $5415.71    **5.** 25 hours, 36 minutes

**2. TRANSLATE.** Use the distance formula and let $d = 3107.5$ miles and $r = 15.4$ mph.

$$d = rt$$
$$3107.5 = 15.4t$$

**3. SOLVE.**

$$\frac{3107.5}{15.4} = \frac{15.4t}{15.4} \qquad \text{Divide both sides by 15.4.}$$
$$201.79 \approx t$$

The time is approximately 201.79 hours. Since there are 24 hours in a day, we divide 201.79 by 24 and find that the time is approximately 8.41 days. Now, let's convert the decimal part of 8.41 days back to hours. To do this, multiply 0.41 by 24 and the result is 9.84 hours. Next, we convert the decimal part of 9.84 hours to minutes by multiplying by 60 since there are 60 minutes in an hour. We have $0.84 \cdot 60 \approx 50$ minutes rounded to the nearest whole. The time is then approximately

8 days, 9 hours, 50 minutes.

**4. INTERPRET.**

**Check:**   Repeat your calculations to make sure that an error was not made.

**State:**   Pete Penseyres's cycling time was approximately 8 days, 9 hours, 50 minutes.

■ Work Practice 5

## Vocabulary, Readiness & Video Check

*Solve each equation for the specified variable. See Examples 1 through 3.*

**1.** $2x + y = 5$ for $y$

**2.** $7x - y = 3$ for $y$

**3.** $a - 5b = 8$ for $a$

**4.** $7r + s = 10$ for $s$

**5.** $5j + k - h = 6$ for $k$

**6.** $w - 4y + z = 0$ for $z$

*Martin-Gay Interactive Videos*

See Video 2.3

*Watch the section lecture video and answer the following questions.*

**Objective A** **7.** Based on the lecture before ⊞ Example 1, what two things does solving an equation for a specific variable mean? ▶

**Objective B** **8.** As the solution is checked at the end of ⊞ Example 3, why do you think it is mentioned to be especially careful that you use the correct formula when solving problems? ▶

## 2.3 Exercise Set MyMathLab® ▶

**Objective A** *Solve each equation for the specified variable or value. See Examples 1 through 3.*

**1.** $d = rt$ for $t$

**2.** $W = gh$ for $g$

**3.** $I = Prt$ for $r$

△ **4.** $V = lwh$ for $w$

△ **5.** $P = a + b + c$ for $c$

△ **6.** $a^2 + b^2 = c^2$ for $b^2$

▶ **7.** $9x - 4y = 16$ for $y$

**8.** $2x + 3y = 17$ for $y$

△ **9.** $P = 2l + 2w$ for $l$

△ **10.** $P = 2l + 2w$ for $w$

**11.** $E = I(r + R)$ for $r$

**12.** $A = P(1 + rt)$ for $t$

**13.** $5x + 4y = 20$ for $y$

**14.** $-9x - 5y = 18$ for $y$

△ **15.** $S = 2LW + 2LH + 2WH$ for $H$

△ **16.** $S = 2\pi r^2 + 2\pi rh$ for $h$

△ **17.** $C = 2\pi r$ for $r$

△ **18.** $A = \pi r^2$ for $\pi$

▶ **19.** $s = \dfrac{n}{2}(a + L)$ for $L$

**20.** $F = \dfrac{9}{5}C + 32$ for $C$

**Objective B** *Solve. For Exercises 25 and 26, the solutions have been started for you. Round all dollar amounts to two decimal places. See Examples 4 and 5.*

**21.** Use the compound interest formula to complete the table and find the balance $A$ if $3500 is invested at an annual rate of 3% for 10 years and compounded $n$ times a year.

| $n$ | 1 | 2 | 4 | 12 | 365 |
|-----|---|---|---|----|-----|
| $A$ |   |   |   |    |     |

**22.** Use the compound interest formula to complete the table and find the balance $A$ if $5000 is invested at an annual rate of 6% for 15 years and compounded $n$ times a year.

| $n$ | 1 | 2 | 4 | 12 | 365 |
|-----|---|---|---|----|-----|
| $A$ |   |   |   |    |     |

**23.** A principal of $6000 is invested in an account paying an annual rate of 4%. Find the amount in the account after 5 years if the account is compounded

**a.** semiannually.

**b.** quarterly.

**c.** monthly.

**24.** A principal of $25,000 is invested in an account paying an annual rate of 5%. Find the amount in the account after 2 years if the account is compounded

**a.** semiannually.

**b.** quarterly.

**c.** monthly.

**25.** Omaha, Nebraska, is about 90 miles from Lincoln, Nebraska. Irania Schmidt must go to the law library in Lincoln to get a document for the law firm she works for. Find how long it takes her to drive *round-trip* if she averages 50 mph.

**Start the solution:**

**1.** UNDERSTAND the problem. Reread it as many times as needed.

**2.** TRANSLATE into an equation. (Fill in the blanks below.) Here, we simply use the formula $d = r \cdot t$.

| distance (round-trip) | equals | rate or speed | · | time |
|:---:|:---:|:---:|:---:|:---:|
| ↓ | ↓ | ↓ | ↓ | ↓ |
| ____ | = | ____ | · | $t$ |

Finish with:

**3.** SOLVE and **4.** INTERPRET

**26.** It took the Selby family $5\frac{1}{2}$ hours round-trip to drive from their house to their beach house 154 miles away. Find their average speed.

**Start the solution:**

**1.** UNDERSTAND the problem. Reread it as many times as needed.

**2.** TRANSLATE into an equation. (Fill in the blanks below.) Here, we simply use the formula $d = r \cdot t$.

| distance (round-trip) | equals | rate or speed | · | time |
|:---:|:---:|:---:|:---:|:---:|
| ↓ | ↓ | ↓ | ↓ | ↓ |
| ____ | = | $r$ | · | ____ |

Finish with:

**3.** SOLVE and **4.** INTERPRET

**27.** One day's high temperature in Phoenix, Arizona, was recorded as 104°F. Write 104°F as degrees Celsius. [Use the formula $C = \frac{5}{9}(F - 32)$.]

**28.** One year's low temperature in Nome, Alaska, was recorded as −15°C. Write −15°C as degrees Fahrenheit. (Use the formula $F = \frac{9}{5}C + 32$.)

△ **29.** A package of floor tiles contains 24 one-foot-square tiles. Find how many packages should be bought to cover a square ballroom floor whose side measures 64 feet.

△ **30.** One-foot-square ceiling tiles are sold in packages of 50. Find how many packages must be bought for a rectangular ceiling 18 feet by 12 feet.

64 ft    64 ft

▶ **31.** If the area of a triangular kite is 18 square feet and its base is 4 feet, find the height of the kite.

△

4 ft

height

▦ **32.** Bailey, Ethan, Avery, and Mia would like to go to Disneyland in 3 years. The total cost should be $4500. If each invests $1000 in a savings account paying 5.5% interest, compounded semiannually, will they have enough in 3 years?

△ **33.** A gallon of latex paint can cover 500 square feet. Find how many gallon containers of paint should be bought to paint two coats on the walls of a rectangular room whose dimensions are 14 feet by 16 feet. (Assume 8-foot ceilings and disregard any openings such as windows or doors.)

△ **34.** A gallon of enamel paint can cover 300 square feet. Find how many gallon containers of paint should be bought to paint three coats on a wall that measures 21 feet by 8 feet.

△ **35.** A portion of the external tank of the Space Shuttle *Endeavour* is a liquid hydrogen tank. If the ends of the tank are hemispheres, find the volume of the tank. To do so, answer parts **(a)** through **(c)**. (*Source:* NASA/Kennedy Space Center)

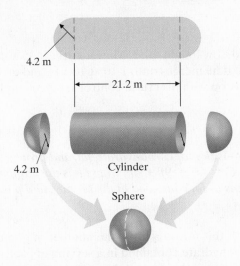

a. Find the volume of the cylinder shown. Round to two decimal places.
b. Find the volume of the sphere shown. Round to two decimal places.
c. Add the results of parts **(a)** and **(b)**. This sum is the approximate volume of the tank.

△ **36.** In 1945, Arthur C. Clarke, a scientist and science-fiction writer, predicted that an artificial satellite placed at a height of 22,248 miles directly above the equator would orbit the globe at the same speed with which Earth was rotating. This belt along the equator is known as the Clarke belt. Use the formula for the circumference of a circle and approximate the "length" of the Clarke belt. Round to the nearest whole mile. (*Hint:* Recall that the radius of Earth is approximately 4000 miles.)

**37.** Amelia Earhart was the first woman to fly solo nonstop coast to coast, setting the women's nonstop transcontinental speed record. She traveled 2447.8 miles in 19 hours 5 minutes. Find the average speed of her flight in miles per hour. (Change 19 hours 5 minutes into hours and use the formula $d = rt$.) Round to the nearest tenth of a mile per hour.

△ **38.** The *Endeavour* Space Shuttle has a cargo bay that is in the shape of a cylinder whose length is 18.3 meters and whose diameter is 4.6 meters. Find its volume.

△ **39.** The deepest hole in the ocean floor is beneath the Pacific Ocean and is called Hole 504B. It is located off the coast of Ecuador. Scientists are drilling it to learn more about Earth's history. Currently, the hole is in the shape of a cylinder whose volume is approximately 3800 cubic feet and whose length is 1.3 miles. Find the radius of the hole to the nearest hundredth of a foot. (*Hint:* Make sure the same units of measurement are used.)

**40.** The deepest man-made hole is called the Kola Superdeep Borehole. It is approximately 8 miles deep and is located near a small Russian town in the Arctic Circle. If it takes 7.5 hours to remove the drill from the bottom of the hole, find the rate that the drill can be retrieved in feet per second. Round to the nearest tenth. (*Hint:* Write 8 miles as feet and 7.5 hours as seconds, and then use the formula $d = rt$.)

△ **41.** Eartha is the world's largest globe. It is located at the headquarters of DeLorme, a mapmaking company in Yarmouth, Maine. Eartha is 41.125 feet in diameter. Find its exact circumference (distance around) and then approximate its circumference using 3.14 for $\pi$. (*Source:* DeLorme)

△ **42.** Eartha is in the shape of a sphere. Its radius is about 20.6 feet. Approximate Eartha's volume to the nearest cubic foot. Use the approximation 3.14 for $\pi$. (*Source:* DeLorme)

**43.** Find *how much interest* $10,000 earns in 2 years in a certificate of deposit paying 8.5% interest compounded quarterly.

**44.** Find how long it takes Mark to drive 135 miles on I-10 if he merges onto I-10 at 10 A.M. and drives nonstop with his cruise control set on 60 mph.

*The calorie count of a serving of food can be computed based on its composition of carbohydrate, fat, and protein. The calorie count C for a serving of food can be computed using the formula $C = 4h + 9f + 4p$, where h is the number of grams of carbohydrate contained in the serving, f is the number of grams of fat contained in the serving, and p is the number of grams of protein contained in the serving.*

**45.** Solve this formula for $f$, the number of grams of fat contained in a serving of food.

**46.** Solve this formula for $h$, the number of grams of carbohydrate contained in a serving of food.

**47.** A serving of cashews contains 14 grams of fat, 7 grams of carbohydrate, and 6 grams of protein. How many calories are in this serving of cashews?

**48.** A serving of chocolate candies contains 9 grams of fat, 30 grams of carbohydrate, and 2 grams of protein. How many calories are in this serving of chocolate candies?

**49.** A serving of raisins contains 130 calories and 31 grams of carbohydrate. If raisins are a fat-free food, how much protein is provided by this serving of raisins?

**50.** A serving of yogurt contains 120 calories, 21 grams of carbohydrate, and 5 grams of protein. How much fat is provided by this serving of yogurt? Round to the nearest tenth of a gram.

## Review

*Determine which numbers in the set $\{-3, -2, -1, 0, 1, 2, 3\}$ are solutions of each inequality. See Sections 1.3 and 2.1.*

**51.** $x < 0$

**52.** $x > 1$

**53.** $x + 5 \leq 6$

**54.** $x - 3 \geq -7$

**55.** In your own words, explain what real numbers are solutions of $x < 0$.

**56.** In your own words, explain what real numbers are solutions of $x > 1$.

## Concept Extensions

Solar system distances are so great that units other than miles or kilometers are often used. For example, the astronomical unit (AU) is the average distance between Earth and the Sun, or 92,900,000 miles. Use this information to convert each planet's distance in miles from the Sun to astronomical units. Round to three decimal places. The planet Mercury's AU from the Sun has been completed for you. (Source: National Space Science Data Center)

| Planet | Miles from the Sun | AU from the Sun |
|---|---|---|
| Mercury | 36 million | 0.388 |
| 57. Venus | 67.2 million | |
| 59. Earth | 92.9 million | |
| 61. Mars | 141.5 million | |
| 63. Jupiter | 483.3 million | |

| Planet | Miles from the Sun | AU from the Sun |
|---|---|---|
| 58. Saturn | 886.1 million | |
| 60. Uranus | 1783 million | |
| 62. Neptune | 2793 million | |
| 64. Pluto (dwarf planet) | 3670 million | |

65. To borrow money at a rate of $r$, which loan plan should you choose—one compounding 4 times a year or 12 times a year? Explain your choice.

66. If you are investing money in a savings account paying a rate of $r$, which account should you choose—an account compounded 4 times a year or 12 times a year? Explain your choice.

67. To solve the formula $W = gh - 3gt^2$ for $g$, explain why it is a good idea to first factor $g$ from the terms on the right side of the equation. Then perform this step and solve for $g$.

68. An orbit such as Clarke's belt in Exercise 36 is called a geostationary orbit. In your own words, why do you think that communications satellites are placed in geostationary orbits?

The measure of the chance or likelihood of an event occurring is its **probability**. A formula basic to the study of probability is the formula for the probability of an event when all the outcomes are equally likely. This formula is

$$\text{Probability of an event} = \frac{\text{number of ways that the event can occur}}{\text{number of possible outcomes}}$$

For example, to find the probability that a single spin on the spinner will result in red, notice first that the spinner is divided into 8 parts, so there are 8 possible outcomes. Next, notice that there is only one sector of the spinner colored red, so the number of ways that the spinner can land on red is 1. Then this probability denoted by P(red) is

$$P(\text{red}) = \frac{1}{8}$$

Find each probability in simplest form.

69. $P(\text{green})$

70. $P(\text{yellow})$

71. $P(\text{black})$

72. $P(\text{blue})$

73. $P(\text{green or blue})$

74. $P(\text{black or yellow})$

75. $P(\text{red, green, or black})$

76. $P(\text{yellow, blue, or black})$

77. $P(\text{white})$

78. $P(\text{red, yellow, green, blue, or black})$

79. From the probability formula, what do you think is always the probability of an event that is impossible occuring?

80. What do you think is always the probability of an event that is sure to occur?

## Objectives

**A** Use Interval Notation.

**B** Solve Linear Inequalities Using the Addition Property of Inequality.

**C** Solve Linear Inequalities Using the Multiplication Property of Inequality.

**D** Solve Linear Inequalities Using Both Properties of Inequality.

**E** Solve Problems That Can Be Modeled by Linear Inequalities.

Relationships among measurable quantities are not always described by equations. For example, suppose that a salesperson earns a base salary of $600 per month plus a commission of 20% of sales. Suppose we want to find the minimum amount of sales needed to receive a total income of *at least* $1500 per month. Here, the phrase "at least" implies that an income of $1500 *or more* is acceptable. In symbols, we can write

income $\geq 1500$

This is an example of an inequality, which we will solve in Example 12.

A *linear inequality* in one variable is similar to a linear equation in one variable except that the equality symbol is replaced with an inequality symbol, such as $<, >, \leq,$ or $\geq$.

### Linear Inequalities in One Variable—Examples

| $3x + 5 \geq 4$ | $2y < 0$ | $3(x - 4) > 5x$ | $\dfrac{x}{3} \leq 5$ |
|:---:|:---:|:---:|:---:|
| ↑ | ↑ | ↑ | ↑ |
| is greater than or equal to | is less than | is greater than | is less than or equal to |

### Linear Inequality in One Variable

A **linear inequality in one variable** is an inequality that can be written in the form

$ax + b < c$

where $a, b,$ and $c$ are real numbers and $a \neq 0$.

In this section, when we make definitions, state properties, or list steps about an inequality containing the symbol $<$, we mean that the definition, property, or steps apply to inequalities containing the symbols $>, \leq,$ and $\geq$ also.

## Objective A Using Interval Notation

A **solution** of an inequality is a value of the variable that makes the inequality a true statement. The **solution set** of an inequality is the set of all solutions. Notice that the solution set of the inequality $x > 2$, for example, contains all numbers greater than 2. Its graph is an interval on the number line since an infinite number of values satisfy the variable. If we use open/closed-circle notation, the graph of $\{x \mid x > 2\}$ looks like:

$\{x \mid x > 2\}$

In this text, a different graphing notation will be used to help us understand **interval notation.** Instead of an open circle, we use a parenthesis; instead of a closed circle, we use a bracket. With this new notation, the graph of $\{x \mid x > 2\}$ now looks like:

and can be represented in interval notation as $(2, \infty)$. The symbol $\infty$ is read "infinity" and indicates that the interval includes *all* numbers greater than 2. The left parenthesis indicates that 2 *is not* included in the interval. Using a left bracket, [, would indicate that 2 *is* included in the interval. The following table shows three equivalent ways to describe an interval: in set notation, as a graph, and in interval notation.

| Set Notation | Graph | Interval Notation |
|---|---|---|
| $\{x \mid x < a\}$ | 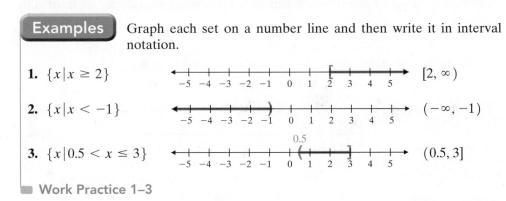 | $(-\infty, a)$ |
| $\{x \mid x > a\}$ | | $(a, \infty)$ |
| $\{x \mid x \leq a\}$ | | $(-\infty, a]$ |
| $\{x \mid x \geq a\}$ | | $[a, \infty)$ |
| $\{x \mid a < x < b\}$ | | $(a, b)$ |
| $\{x \mid a \leq x \leq b\}$ | | $[a, b]$ |
| $\{x \mid a < x \leq b\}$ | | $(a, b]$ |
| $\{x \mid a \leq x < b\}$ | | $[a, b)$ |
| $\{x \mid x \text{ is a real number}\}$ | | $(-\infty, \infty)$ |

**Helpful Hint**

Notice that a parenthesis is always used to enclose $\infty$ and $-\infty$.

**Examples** Graph each set on a number line and then write it in interval notation.

**1.** $\{x \mid x \geq 2\}$ 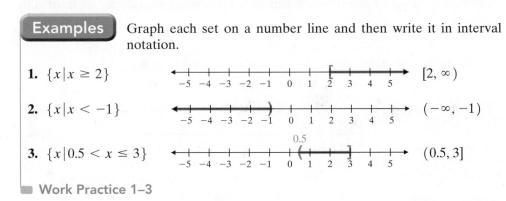 $[2, \infty)$

**2.** $\{x \mid x < -1\}$ $(-\infty, -1)$

**3.** $\{x \mid 0.5 < x \leq 3\}$ $(0.5, 3]$

**Work Practice 1–3**

**✓Concept Check** Explain what is wrong with writing the interval $(5, \infty]$.

## Objective B  Using the Addition Property of Inequality

To solve a linear inequality, we use a process similar to the one used to solve a linear equation. We use properties of inequalities to write equivalent inequalities until the variable is alone on one side of the inequality.

**Addition Property of Inequality**

If $a, b,$ and $c$ are real numbers, then

$$a < b \quad \text{and} \quad a + c < b + c$$

are equivalent inequalities.

In other words, we may add the same real number to both sides of an inequality, and the resulting inequality will have the same solution set. This property also allows us to subtract the same real number from both sides.

### Practice 4

Solve: $x + 3 < 1$. Graph the solution set.

**Example 4** Solve: $x - 2 < 5$. Graph the solution set.

**Solution:**

$$x - 2 < 5$$
$$x - 2 + 2 < 5 + 2 \quad \text{Add 2 to both sides.}$$
$$x < 7 \quad \text{Simplify.}$$

The solution set is $\{x \mid x < 7\}$, which in interval notation is $(-\infty, 7)$. The graph of the solution set is

■ Work Practice 4

### Helpful Hint

In Example 4, the solution set is $\{x \mid x < 7\}$. This means that *all* numbers less than 7 are solutions. For example, $6.9, 0, -\pi, 1$, and $-56.7$ are solutions, just to name a few. To see this, replace $x$ in $x - 2 < 5$ with each of these numbers and see that the result is a true inequality.

### Practice 5

Solve: $3x - 4 < 4x$. Graph the solution set.

**Example 5** Solve: $4x - 2 < 5x$. Graph the solution set.

**Solution:** To get $x$ alone on one side of the inequality, we subtract $4x$ from both sides.

$$4x - 2 < 5x$$
$$4x - 2 - 4x < 5x - 4x \quad \text{Subtract } 4x \text{ from both sides.}$$
$$-2 < x \quad \text{or} \quad x > -2 \quad \text{Simplify.}$$

The solution set is $\{x \mid x > -2\}$, which in interval notation is $(-2, \infty)$. The graph is

■ Work Practice 5

### Helpful Hint

Don't forget that $-2 < x$ means the same as $x > -2$.

### Practice 6

Solve: $5x - 1 \geq 4x + 4$. Graph the solution set.

**Example 6** Solve: $3x + 4 \geq 2x - 6$. Graph the solution set.

**Solution:**

$$3x + 4 \geq 2x - 6$$
$$3x + 4 - 2x \geq 2x - 6 - 2x \quad \text{Subtract } 2x \text{ from both sides.}$$
$$x + 4 \geq -6 \quad \text{Combine like terms.}$$
$$x + 4 - 4 \geq -6 - 4 \quad \text{Subtract 4 from both sides.}$$
$$x \geq -10 \quad \text{Simplify.}$$

The solution set is $\{x \mid x \geq -10\}$, which in interval notation is $[-10, \infty)$. The graph of the solution set is

■ Work Practice 6

**Answers**

4. $\{x \mid x < -2\}, (-\infty, -2)$

5. $\{x \mid x > -4\}, (-4, -\infty)$

6. $\{x \mid x \geq 5\}, [5, \infty)$

## Objective C Using the Multiplication Property of Inequality ▶

Next, we introduce and use the multiplication property of inequality to solve linear inequalities. To understand this property, let's start with the true statement $-3 < 7$ and multiply both sides by 2.

$$-3 < 7$$
$$2(-3) < 2(7) \quad \text{Multiply both sides by 2.}$$
$$-6 < 14 \quad \text{True}$$

The statement remains true.

Notice what happens if both sides of $-3 < 7$ are multiplied by $-2$.

$$-3 < 7$$
$$-2(-3) < -2(7) \quad \text{Multiply both sides by 2.}$$
$$6 < -14 \quad \text{False}$$

The inequality $6 < -14$ is a false statement. However, *if the direction of the inequality sign is reversed*, the result is

$$6 > -14 \quad \text{True}$$

These examples suggest the following property.

---

### Multiplication Property of Inequality

If $a, b,$ and $c$ are real numbers and $c$ is **positive,** then

$$a < b \text{ and } ac < bc$$

are equivalent inequalities.
If $a, b,$ and $c$ are real numbers and $c$ is **negative,** then

$$a < b \text{ and } ac > bc$$

are equivalent inequalities.

---

In other words, we may multiply both sides of an inequality by the same positive real number, and the result is an equivalent inequality. We may also multiply both sides of an inequality by the same *negative number* and *reverse the direction of the inequality symbol,* and the result is an equivalent inequality. The multiplication property holds for division also since division is defined in terms of multiplication.

**Helpful Hint** Whenever both sides of an inequality are multiplied or divided by a negative number, the direction of the inequality symbol *must be* reversed to form an equivalent inequality.

**Example 7** Solve: $\frac{1}{4}x \le \frac{3}{2}$. Graph the solution set.

**Solution:**

$$\frac{1}{4}x \le \frac{3}{2}$$
$$4 \cdot \frac{1}{4}x \le 4 \cdot \frac{3}{2} \quad \text{Multiply both sides by 4.}$$
$$x \le 6 \quad \text{Simplify.}$$

**Helpful Hint** The inequality symbol is the same since we are multiplying by a *positive* number.

The solution set is $\{x \mid x \le 6\}$, which in interval notation is $(-\infty, 6]$. The graph of this solution set is

■ **Work Practice 7**

**Practice 7**

Solve: $\frac{1}{6}x \le \frac{2}{3}$. Graph the solution set.

**Answer**

**7.** $\{x \mid x \le 4\}, (-\infty, 4]$

## Practice 8

Solve: $-1.1x < 5.5$. Graph the solution set.

---

**Example 8**   Solve: $-2.3x < 6.9$. Graph the solution set.

**Solution:**

$$-2.3x < 6.9$$

$$\frac{-2.3x}{-2.3} > \frac{6.9}{-2.3} \quad \text{Divide both sides by } -2.3 \text{ and reverse the inequality symbol.}$$

$$x > -3 \quad \text{Simplify.}$$

**Helpful Hint**   The inequality symbol is *reversed* since we divided by a *negative* number.

The solution set is $\{x \mid x > -3\}$, which is $(-3, \infty)$ in interval notation. The graph of the solution set is

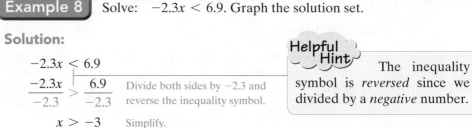

■ Work Practice 8

---

✓**Concept Check** In which of the following inequalities must the inequality symbol be reversed during the solution process? (If necessary, assume that variable terms are moved to the left and constants are moved to the right.)

**a.** $-2x > 7$          **c.** $-x + 4 + 3x < 5$

**d.** $2x - 3 > 10$          **d.** $-x + 4 < 5$

## Objective D   Using Both Properties of Inequality ▶

Many problems require us to use both properties of inequality. To solve linear inequalities in general, we follow steps similar to those for solving linear equations.

> ### Solving a Linear Inequality in One Variable
>
> **Step 1:** Clear the equation of fractions or decimals by multiplying both sides of the inequality by an appropriate nonzero number.
>
> **Step 2:** Use the distributive property to remove grouping symbols such as parentheses.
>
> **Step 3:** Combine like terms on each side of the inequality.
>
> **Step 4:** Use the addition property of inequality to rewrite the inequality as an equivalent inequality with variable terms on one side and numbers on the other side.
>
> **Step 5:** Use the multiplication property of inequality to get the variable alone on one side of the inequality.

## Practice 9

Solve:
$-(2x - 6) \le 4(2x - 4) + 2$.
Graph the solution set.

**Answers**

**8.** $\{x \mid x > -5\}, (-5, \infty)$

**9.** $[2, \infty)$

✓ **Concept Check Answers**

**a, d**

---

**Example 9**   Solve: $-(x - 3) + 2 \le 3(2x - 5) + x$. Graph the solution set.

**Solution:**

$$-(x - 3) + 2 \le 3(2x - 5) + x$$

$$-x + 3 + 2 \le 6x - 15 + x \quad \text{Apply the distributive property.}$$

$$5 - x \le 7x - 15 \quad \text{Combine like terms.}$$

$$5 - x + x \le 7x - 15 + x \quad \text{Add } x \text{ to both sides.}$$

$$5 \le 8x - 15 \quad \text{Combine like terms.}$$

$$5 + 15 \le 8x - 15 + 15 \quad \text{Add 15 to both sides.}$$

$$20 \le 8x \quad \text{Combine like terms.}$$

$$\frac{20}{8} \le \frac{8x}{8}$$    Divide both sides by 8.

$$\frac{5}{2} \le x, \quad \text{or } x \ge \frac{5}{2} \quad \text{Simplify.}$$

The solution set written in interval notation is $\left[\frac{5}{2}, \infty\right)$, and its graph is

■ **Work Practice 9**

**Helpful Hint**    In Example 9, don't forget that $\frac{5}{2} \le x$ means the same as $x \ge \frac{5}{2}$.

---

**Example 10**    Solve: $\frac{2}{5}(x - 6) \ge x - 1$. Write the solution set in interval notation.

**Solution:**

$$\frac{2}{5}(x - 6) \ge x - 1$$

$$5\left[\frac{2}{5}(x - 6)\right] \ge 5(x - 1) \quad \text{Multiply both sides by 5 to eliminate fractions.}$$

$$2(x - 6) \ge 5(x - 1)$$

$$2x - 12 \ge 5x - 5 \quad \text{Use the distributive property.}$$

$$-3x - 12 \ge -5 \quad \text{Subtract } 5x \text{ from both sides.}$$

$$-3x \ge 7 \quad \text{Add 12 to both sides.}$$

$$\frac{-3x}{-3} \le \frac{7}{-3} \quad \text{Divide both sides by } -3 \text{ and reverse the inequality symbol.}$$

$$x \le -\frac{7}{3} \quad \text{Simplify.}$$

The solution set is $\left(-\infty, -\frac{7}{3}\right]$.

■ **Work Practice 10**

**Practice 10**

Solve: $\frac{3}{4}(x + 2) \ge x - 6$.

Write the solution set in interval notation.

---

**Example 11**    Solve: $2(x + 3) > 2x + 1$.

**Solution:**

$$2(x + 3) > 2x + 1$$

$$2x + 6 > 2x + 1 \quad \text{Distribute on the left side.}$$

$$2x + 6 - 2x > 2x + 1 - 2x \quad \text{Subtract } 2x \text{ from both sides.}$$

$$6 > 1 \quad \text{Simplify.}$$

$6 > 1$ is a true statement for all values of $x$, so this inequality and the original inequality are true for all numbers. The solution set is $\{x \mid x \text{ is a real number}\}$, or $(-\infty, \infty)$ in interval notation, and its graph is

■ **Work Practice 11**

**Practice 11**

Solve: $5(x - 3) < 5x + 2$.

**Helpful Hint**    If variables subtract out and a false statement occurs, such as $6 < 1$, then the original inequality has no solution.

**Answers**
**10.** $(-\infty, 30]$   **11.** $(-\infty, \infty)$

## Objective E  Linear Inequalities and Problem Solving ▶

Problems containing words such as "at least," "at most," "between," "no more than," and "no less than" usually indicate that an inequality is to be solved instead of an equation. In solving applications involving linear inequalities, we use the same four-step strategy as when we solved applications involving linear equations.

### Practice 12

A salesperson earns $1000 a month plus a commission of 15% of sales. Find the minimum amount of sales needed to receive a total income of at least $4000 per month.

### Example 12  Calculating Income with Commission

A salesperson earns $600 per month plus a commission of 20% of sales. Find the minimum amount of sales needed to receive a total income of at least $1500 per month.

**Solution:**

1.  UNDERSTAND. Read and reread the problem. Let

    $x$ = amount of sales

2.  TRANSLATE. Since the income is to be at least $1500, this means that we want the income to be greater than or equal to $1500. To write an inequality, notice that the salesperson's income consists of $600 plus a commission (20% of sales).

    In words:   600   +   commission of 20% of sales   ≥   1500

    Translate:  600   +        0.20x              ≥   1500

3.  SOLVE the inequality for $x$.

$$600 + 0.20x \geq 1500$$
$$600 + 0.20x - 600 \geq 1500 - 600$$
$$0.20x \geq 900$$
$$x \geq 4500$$

4.  INTERPRET.

**Check:**  The income for sales of $4500 is

$$600 + 0.20(4500), \text{ or } 1500$$

Thus, if sales are greater than or equal to $4500, income is greater than or equal to $1500.

**State:**  The minimum amount of sales needed for the salesperson to earn at least $1500 per month is $4500.

■ Work Practice 12

### Practice 13

Use the formula given in Example 13 to predict when the consumption of cigarettes will be less than 100 billion per year.

### Example 13  Finding the Annual Consumption

In the United States, the annual consumption of cigarettes is declining. The consumption $c$ in billions of cigarettes per year since 2000 can be approximated by the formula $c = -13.5t + 440$, where $t$ is the number of years after 2000. Use this formula to predict when the consumption of cigarettes will be less than 200 billion per year. (*Source:* United States Department of Agriculture: Economic Research Service)

**Answers**

**12.** $20,000  **13.** after the year 2025

**Solution:**

1. **UNDERSTAND.** Read and reread the problem. To become familiar with the given formula, let's find the cigarette consumption after 15 years, which would be the year $2000 + 15 = 2015$. To do so, we substitute 15 for $t$ in the given formula.

$$c = -13.5(15) + 440 = 237.5$$

Thus, in 2015, we predict cigarette consumption to be about 237.5 billion.

Variables have already been assigned in the given formula. For review, they are

$c =$ the annual consumption of cigarettes in the United States in billions of cigarettes

$t =$ the number of years after 2000

2. **TRANSLATE.** We are looking for the first year that the consumption of cigarettes $c$ is less than 200. Since we are finding years $t$, we substitute the expression in the formula given for $c$, or

$$-13.5t + 440 < 200$$

3. **SOLVE** the inequality.

$$-13.5t + 440 < 200$$
$$-13.5t < -240$$
$$t > \text{approximately } 17.77$$

4. **INTERPRET.**

**Check:** Substitute a number greater than 17.77 and see that $c$ is less than 200.

**State:** The annual consumption of cigarettes will be less than 200 billion more than 17.77 years after 2000, or approximately in $2018(2000 + 18 = 2018)$.

■ **Work Practice 13**

## Vocabulary, Readiness & Video Check

*Match each graph with the interval notation that describes it.*

**1.**

a. $(-5, \infty)$     b. $(-5, -\infty)$

c. $(\infty, -5)$     d. $(-\infty, -5)$

**2.**

a. $(-\infty, -11]$     b. $(-11, \infty)$

c. $[-11, \infty)$     d. $(-\infty, -11)$

**3.**

a. $\left[\dfrac{7}{4}, -2.5\right)$     b. $\left(-2.5, \dfrac{7}{4}\right]$

c. $\left[-2.5, \dfrac{7}{4}\right)$     d. $\left(\dfrac{7}{4}, -2.5\right)$

**4.**

a. $\left[-\dfrac{10}{3}, 0.2\right)$     b. $\left(0.2, -\dfrac{10}{3}\right]$

c. $\left(-\dfrac{10}{3}, 0.2\right]$     d. $\left[0.2, -\dfrac{10}{3}\right)$

*Use the choices below to fill in each blank. Not all choices will be used.*

$(-\infty, -0.4)$    $(-\infty, -0.4]$    $[-0.4, \infty)$    $(-0.4, \infty)$    $(\infty, -0.4]$

**5.** The set $\{x \mid x \geq -0.4\}$ written in interval notation is _____.

**6.** The set $\{x \mid x < -0.4\}$ written in interval notation is _____.

**7.** The set $\{x \mid x \leq -0.4\}$ written in interval notation is _____.

**8.** The set $\{x \mid x > -0.4\}$ written in interval notation is _____.

*Each inequality below is solved by dividing both sides by the coefficient of x. Will the inequality symbol be reversed during this solution process?*

**9.** $2x > -15$        **10.** $-2x \leq 15$        **11.** $-2x < -15$        **12.** $-x \geq 23$

*Martin-Gay Interactive Videos*    *Watch the section lecture video and answer the following questions.*

**Objective A** **13.** Using ⊞ Example 1 as a reference, explain how the graph of the solution set of an inequality can help you write the solution set in interval notation. ▶

**Objective B** **14.** From the lecture before ⊞ Example 3, explain the addition property of inequality in your own words. What equality property does it closely resemble? ▶

*See Video 2.4* 🍎

**Objective C** **15.** Based on the lecture before ⊞ Example 4, complete the following statement. If you multiply or divide both sides of an inequality by the _____ nonzero negative number, you must _____ the direction of the inequality symbol. ▶

**Objective D** **16.** To solve ⊞ Example 6 using the given steps, what first step do we perform? ▶

**Objective E** **17.** What words or phrases in the ⊞ Example 7 statement tell you this is an inequality application (besides the last line telling you to use an inequality)? ▶

**2.4** **Exercise Set** MyMathLab® ▶

**Objectives A** *Graph the solution set of each inequality on a number line and then write it in interval notation. See Examples 1 through 3.*

▶**1.** $\{x \mid x < -3\}$

−5 −4 −3 −2 −1  0  1  2  3  4  5

**2.** $\{x \mid x > 5\}$

−5 −4 −3 −2 −1  0  1  2  3  4  5

**3.** $\{x \mid x \geq 0.3\}$

−5 −4 −3 −2 −1  0  1  2  3  4  5

**4.** $\{x \mid x < -0.2\}$

−5 −4 −3 −2 −1  0  1  2  3  4  5

**5.** $\{x \mid -7 \leq x\}$

−10 −8 −6 −4 −2  0  2  4  6  8  10

**6.** $\{x \mid -7 \geq x\}$

−10 −8 −6 −4 −2  0  2  4  6  8  10

**7.** $\{x \mid -2 < x < 5\}$

**8.** $\{x \mid -5 \leq x \leq -1\}$

**9.** $\{x \mid 5 \geq x > -1\}$

**10.** $\{x \mid -3 > x \geq -7\}$

**Objective** **B**  *Solve. Graph the solution set and then write it in interval notation. See Examples 4 through 6.*

**11.** $x - 7 \geq -9$

**12.** $x + 2 \leq -1$

**13.** $7x < 6x + 1$

**14.** $11x < 10x + 5$

**15.** $8x - 7 \leq 7x - 5$

**16.** $7x - 1 \geq 6x - 1$

**Objective** **C**  *Solve. Graph the solution set and then write it in interval notation. See Examples 7 and 8.*

**17.** $\dfrac{3}{4}x \geq 6$

**18.** $\dfrac{5}{6}x \geq 5$

**19.** $5x < -23.5$

**20.** $4x > -11.2$

**21.** $-3x \geq 9$

**22.** $-4x \geq 8$

**Objective** **D**  *Solve. Write the solution set using interval notation. See Examples 9 through 11.*

**23.** $-2x + 7 \geq 9$

**24.** $8 - 5x \leq 23$

**25.** $15 + 2x \geq 4x - 7$

**26.** $20 + x < 6x - 15$

**27.** $4(2x + 1) > 4$

**28.** $6(2 - 3x) \geq 12$

**29.** $3(x - 5) < 2(2x - 1)$

**30.** $5(x + 4) \leq 4(2x + 3)$

**31.** $\dfrac{5x + 1}{7} - \dfrac{2x - 6}{4} \geq -4$

**32.** $\dfrac{1 - 2x}{3} + \dfrac{3x + 7}{7} > 1$

**33.** $-3(2x - 1) < -4[2 + 3(x + 2)]$

**34.** $-2(4x + 2) > -5[1 + 2(x - 1)]$

**Objectives** B C D **Mixed Practice** *Solve. Write the solution set using interval notation. See Examples 4 through 11.*

**35.** $x + 9 < 3$

**36.** $x - 9 < -12$

**37.** $-x < -4$

**38.** $-x > -2$

▷ **39.** $-7x \leq 3.5$

**40.** $-6x \leq 4.2$

**41.** $\frac{1}{2} + \frac{2}{3} \geq \frac{x}{6}$

**42.** $\frac{3}{4} - \frac{2}{3} \geq \frac{x}{6}$

**43.** $-5x + 4 \leq -4(x - 1)$

**44.** $-6x + 2 < -3(x + 4)$

**45.** $\frac{3}{4}(x - 7) \geq x + 2$

**46.** $\frac{4}{5}(x + 1) \leq x + 1$

**47.** $0.8x + 0.6x \geq 4.2$

**48.** $0.7x - x > 0.45$

**49.** $4(x - 6) + 2x - 4 \geq 3(x - 7) + 10x$

**50.** $7(2x + 3) + 4x \leq 7 + 5(3x - 4) + x$

**51.** $14 - (5x - 6) \geq -6(x + 1) - 5$

**52.** $13y - (9y + 2) \leq 5(y - 6) + 10$

**53.** $\frac{1}{2}(3x - 4) \leq \frac{3}{4}(x - 6) + 1$

**54.** $\frac{2}{3}(x + 3) < \frac{1}{6}(2x - 8) + 2$

**55.** $0.4(4x - 3) < 1.2(x + 2)$

**56.** $0.2(8x - 2) < 1.2(x - 3)$

**57.** $\frac{2}{5}x - \frac{1}{4} \leq \frac{3}{10}x - \frac{4}{5}$

**58.** $\frac{7}{12}x - \frac{1}{3} \leq \frac{3}{8}x - \frac{5}{6}$

**59.** $\frac{-x + 2}{2} - \frac{1 - 5x}{8} < -1$

**60.** $\frac{3 - 4x}{6} - \frac{1 - 2x}{12} \leq -2$

**61.** $\frac{x + 5}{5} - \frac{3 + x}{8} \geq -\frac{3}{10}$

**62.** $\frac{x - 4}{2} - \frac{x - 2}{3} > \frac{5}{6}$

**63.** $\frac{x + 3}{12} + \frac{x - 5}{15} < \frac{2}{3}$

**64.** $\frac{3x + 2}{18} - \frac{1 + 2x}{6} \leq -\frac{1}{2}$

**Objective** E  *Solve. See Examples 12 and 13. For Exercises 65 through 72, **a.** answer with an inequality, and **b.** in your own words, explain the meaning of your answer to part **a**.*

*Exercises 65 and 66 are written to help you get started. Exercises 67 and 68 are written to help you write and solve the inequality.*

**65.** Shureka Washburn has scores of 72, 67, 82, and 79 on her algebra tests.
   **a.** Use an inequality to find the scores she must make on the final exam to pass the course with an average of 77 or higher, given that the final exam counts as two tests.
   **b.** In your own words, explain the meaning of your answer to part **a**.

**66.** In a Winter Olympics 5000-meter speed-skating event, Hans Holden scored times of 6.85, 7.04, and 6.92 minutes on his first three trials.
   **a.** Use an inequality to find the times he can score on his last trial so that his average time is under 7.0 minutes.
   **b.** In your own words, explain the meaning of your answer to part **a**.

**67.** A clerk must use the elevator to move boxes of paper. The elevator's maximum weight limit is 1500 pounds. If each box of paper weighs 66 pounds and the clerk weighs 147 pounds, use an inequality to find the number of whole boxes she can move on the elevator at one time.

**Start the solution:**

**1.** UNDERSTAND the problem. Reread it as many times as needed.

**2.** TRANSLATE into an inequality. (Fill in the blanks below.)

Let $x$ = number of boxes

| Clerk's weight | + | number of boxes | times | weight of each box | ≤ | elevator maximum weight |
|---|---|---|---|---|---|---|
| ↓ | ↓ | ↓ | ↓ | ↓ | ↓ | ↓ |
| ____ | + | $x$ | · | ____ | ≤ | ____ |

Finish with:

**3.** SOLVE and **4.** INTERPRET

**68.** To mail a large envelope first class, the U.S. Post Office charges 90 cents for the first ounce and 20 cents per ounce for each additional ounce. Use an inequality to find the number of whole ounces that can be mailed for no more than $3.00

**Start the solution:**

**1.** UNDERSTAND the problem. Reread it as many times as needed.

**2.** TRANSLATE into an inequality. (Fill in the blanks below.)

Let $x$ = number of additional ounces (after first ounce)

| Price of first ounce | + | number of additional ounces | times | price per additional ounce | ≤ | 300 cents |
|---|---|---|---|---|---|---|
| ↓ | ↓ | ↓ | ↓ | ↓ | ↓ | ↓ |
| ____ | + | $x$ | · | ____ | ≤ | ____ |

Finish with:

**3.** SOLVE and **4.** INTERPRET

**69.** A small plane's maximum takeoff weight is 2000 pounds or less. Six passengers weigh an average of 160 pounds each. Use an inequality to find the luggage and cargo weights the plane can carry.

**70.** A shopping mall parking garage charges $1 for the first half-hour and 60 cents for each additional half-hour. Use an inequality to find how long you can park if you have only $4.00 in cash.

**71.** Northeast Telephone Company offers two billing plans for local calls.

Plan 1: $25 per month for unlimited calls
Plan 2: $13 per month plus $0.06 per call

Use an inequality to find the number of monthly calls for which plan 1 is more economical than plan 2.

**72.** A car rental company offers two subcompact rental plans.

Plan A: $36 per day and unlimited mileage
Plan B: $24 per day plus $0.15 per mile

Use an inequality to find the number of daily miles for which plan A is more economical than plan B.

**73.** At room temperature, glass used in windows actually has some properties of a liquid. It has a very slow, viscous flow. (Viscosity is the property of a fluid that resists internal flow. For example, lemonade flows more easily than fudge syrup. Fudge syrup has a higher viscosity than lemonade.) Glass does not become a true liquid until temperatures are greater than or equal to 500°C. Find the Fahrenheit temperatures for which glass is a liquid. (Use the formula $F = \frac{9}{5}C + 32$.)

**74.** Stibnite is a silvery white mineral with a metallic luster. It is one of the few minerals that melts easily in a match flame or at temperatures of approximately 977°F or greater. Find the Celsius temperatures for which stibnite melts. [Use the formula $C = \frac{5}{9}(F - 32)$.]

**75.** Although beginning salaries vary greatly according to your field of study, the equation

$$s = 651.2t + 27{,}821$$

can be used to approximate and to predict average beginning salaries for candidates with bachelor's degrees. The variable $s$ is the starting salary and $t$ is the number of years after 1989.

**a.** Approximate when beginning salaries for candidates were greater than $35,000.

**b.** Determine the year you plan to graduate from college. Use this year to find the corresponding value of $t$ and approximate your beginning salary.

**76. a.** Use the formula in Example 13 to estimate the years that the consumption of cigarettes will be less than 50 billion per year.

**b.** Use your answer to part **a** to describe the limitations of your answer.

*The average consumption per person per year of whole milk w can be approximated by the equation*

$$w = -1.9t + 66.6$$

*where t is the number of years after 2000 and w is measured in pounds. The average consumption of skim milk s per person per year can be approximated by the equation*

$$s = -0.25t + 28.7$$

*where t is the number of years after 2000 and s is measured in pounds. The consumption of whole milk is shown on the graph in blue and the consumption of skim milk is shown on the graph in red. Use this information to answer Exercises 77 through 86.*

**Whole Milk versus Skim Milk**

*Source:* Based on data from U.S. Department of Agriculture: Economic Research Service

**77.** Is the consumption of whole milk increasing or decreasing over time? Explain how you arrived at your answer.

**78.** Is the consumption of skim milk increasing or decreasing over time? Explain how you arrived at your answer.

**79.** Predict the consumption of whole milk per person in 2020. (*Hint:* Find the value of $t$ that corresponds to 2020.)

**80.** Predict the consumption of skim milk per person in 2020. (*Hint:* Find the value of $t$ that corresponds to 2020.)

**81.** Determine when the consumption of whole milk will be less than 40 pounds per person per year.

**82.** For 2000 through 2012, the consumption of whole milk was greater than the consumption of skim milk. Explain how this can be determined from the graph.

**83.** Both lines have a negative slope, that is, the amount of each type of milk consumed per person per year is decreasing as time goes on. However, the amount of whole milk being consumed is decreasing faster than the amount of skim milk being consumed. Explain how this could be.

**84.** Do you think it is possible that the consumption per person of whole milk will eventually be the same as the consumption of skim milk? Explain your answer.

**85.** The consumption of skim milk will be greater than the consumption of whole milk when $s > w$.
   **a.** Find when this will occur by substituting the given equivalent expression for $w$ and the given equivalent expression for $s$ and solving for $t$. (Round to the nearest hundredth.)
   **b.** Estimate to the nearest whole the first year when this will occur.

**86.** How will the two lines in the graph appear if the consumption of whole milk is the same as the consumption of skim milk?

## Review

*List or describe the integers that make both inequalities true. See Section 1.3.*

**87.** $x \geq 0$ and $x \leq 7$

**88.** $x < 5$ and $x > 1$

**89.** $x < 6$ and $x < -5$

**90.** $x \geq -2$ and $x \geq 2$

*Solve each equation for x. See Section 2.1.*

**91.** $3x - 12 = 3$

**92.** $2x - 6 = 4$

**93.** $-5x - 4 = -x - 4$

**94.** $-x + 7 = 5x - 6$

## Concept Extensions

*Each row of the table shows three equivalent ways of describing an interval. Complete this table by filling in the equivalent descriptions. The first row has been completed for you.*

| | Set Notation | Graph | Interval Notation |
|---|---|---|---|
| | $\{x \mid x < -3\}$ | ⟵──────)──────⟶ −3 | $(-\infty, -3)$ |
| **95.** | | ⟵──[──────⟶ 2 | |
| **96.** | | ⟵──(──────⟶ −4 | |
| **97.** | $\{x \mid x < 0\}$ | ⟵──────────⟶ 0 | |
| **98.** | $\{x \mid x \leq 5\}$ | ⟵──────────⟶ 5 | |
| **99.** | | ⟵──────────⟶ −2   1.5 | $(-2, 1.5]$ |
| **100.** | | ⟵──────────⟶ −3.7   4 | $[-3.7, 4)$ |

*Will the inequality symbol be reversed during the solution process for Exercises 101–104? Assume that the variable term remains on the left side of the inequality.*

**101.** $-3x \leq 14$

**102.** $3x > -14$

**103.** $-x - 17 \geq -23$

**104.** $-3x < -14$

**105.** Solve: $2x - 3 = 5$

**106.** Write an inequality whose solution set is $\{x \mid x \leq 2\}$.

**107.** Solve: $2x - 3 > 5$

**108.** Solve: $2x - 3 < 5$

**109.** Read the equations and inequalities for Exercises 105, 107, and 108 and their solutions. In your own words, write down your thoughts.

**110.** Write an inequality whose solution set is $\{x \mid x \leq -2\}$.

**111.** When graphing the solution set of an inequality, explain how you know whether to use a parenthesis or a bracket.

**112.** Explain what is wrong with the interval notation $(-6, -\infty)$.

**113.** Explain how solving a linear inequality is similar to solving a linear equation.

**114.** Explain how solving a linear inequality is different from solving a linear equation.

# Linear Equations and Inequalities

*Solve each equation or inequality.*

**1.** $-4x = 20$

**2.** $-4x < 20$

**3.** $\dfrac{3x}{4} \geq 2$

**4.** $5x + 3 \geq 2 + 4x$

**5.** $6(y - 4) = 3(y - 8)$

**6.** $-4x \leq \dfrac{2}{5}$

**7.** $-3x \geq \dfrac{1}{2}$

**8.** $5(y + 4) = 4(y + 5)$

**9.** $7x < 7(x - 2)$

**10.** $\dfrac{-5x + 11}{2} \leq 7$

**11.** $-5x + 1.5 = -19.5$

**12.** $-5x + 4 = -26$

**13.** $5 + 2x - x = -x + 3 - 14$

**14.** $12x + 14 < 11x - 2$

**15.** _____

**16.** _____

**17.** _____

**18.** _____

**19.** _____

**20.** _____

**21.** _____

**22.** _____

**23.** _____

**24.** _____

**25.** _____

**26.** _____

**27.** _____

**28.** _____

**15.** $\dfrac{x}{5} - \dfrac{x}{4} = \dfrac{x-2}{2}$

**16.** $12x - 12 = 8(x - 1)$

**17.** $2(x - 3) > 70$

**18.** $-3x - 4.7 = 11.8$

**19.** $-2(b - 4) - (3b - 1) = 5b + 3$

**20.** $8(x + 3) < 7(x + 5) + x$

**21.** $\dfrac{3t + 1}{8} = \dfrac{5 + 2t}{7} + 2$

**22.** $4(x - 6) - x = 8(x - 3) - 5x$

**23.** $\dfrac{x + 3}{12} + \dfrac{x - 5}{15} < \dfrac{2}{3}$

**24.** $\dfrac{y}{3} + \dfrac{y}{5} = \dfrac{y + 3}{10}$

**25.** $5(x - 6) + 2x > 3(2x - 1) - 4$

**26.** $14(x - 1) - 7x \le 2(3x - 6) + 4$

**27.** $\dfrac{1}{4}(3x + 2) - x \ge \dfrac{3}{8}(x - 5) + 2$

**28.** $\dfrac{1}{3}(x - 10) - 4x > \dfrac{5}{6}(2x + 1) - 1$

Two inequalities joined by the words **and** or **or** are called **compound inequalities**.

**Compound Inequalities**

$x + 3 < 8$ and $x > 2$

$\dfrac{2x}{3} \geq 5$ or $-x + 10 < 7$

**Objectives**

**A** Find the Intersection of Two Sets.

**B** Solve Compound Inequalities Containing "and."

**C** Find the Union of Two Sets.

**D** Solve Compound Inequalities Containing "or."

## Objective A Finding the Intersection of Two Sets

The solution set of a compound inequality formed by the word **and** is the *intersection* of the solution sets of the two inequalities. We use the symbol ∩ to denote "intersection."

---

### Intersection of Two Sets

The **intersection** of two sets, $A$ and $B$, is the set of all elements common to both sets. $A$ intersect $B$ is denoted by $A \cap B$.

$A \cap B$

---

**Example 1** Find the intersection: $\{2, 4, 6, 8\} \cap \{3, 4, 5, 6\}$

**Solution:** The numbers 4 and 6 are in both sets. The intersection is $\{4, 6\}$.

■ Work Practice 1

**Practice 1**

Find the intersection:
$\{1, 2, 3, 4, 5\} \cap \{3, 4, 5, 6\}$

## Objective B Solving Compound Inequalities Containing "and"

A value is a **solution** of a compound inequality formed by the word **and** if it is a solution of *both* inequalities. For example, the solution set of the compound inequality $x \leq 5$ and $x \geq 3$ contains all values of $x$ that make the inequality $x \leq 5$ a true statement **and** the inequality $x \geq 3$ a true statement. The first graph shown below is the graph of $x \leq 5$, the second graph is the graph of $x \geq 3$, and the third graph shows the intersection of the two graphs. The third graph is the graph of $x \leq 5$ **and** $x \geq 3$.

$\{x \mid x \leq 5\}$       [number line from −4 to 6, shaded left to 5]     $(-\infty, 5]$

$\{x \mid x \geq 3\}$       [number line from −4 to 6, shaded right from 3]     $[3, \infty)$

$\{x \mid x \leq 5 \text{ and } x \geq 3\}$, also $\{x \mid 3 \leq x \leq 5\}$ (see below)    [number line from −4 to 6, shaded 3 to 5]     $[3, 5]$

Since $x \geq 3$ is the same as $3 \leq x$, the compound inequality $3 \leq x$ and $x \leq 5$ can be written in a more compact form as $3 \leq x \leq 5$. The solution set $\{x \mid 3 \leq x \leq 5\}$ includes all numbers that are greater than or equal to 3 and at the same time less than or equal to 5.

In interval notation, the set $\{x \mid x \leq 5 \text{ and } x \geq 3\}$ or $\{x \mid 3 \leq x \leq 5\}$ is written as $[3, 5]$.

**Answer**
1. $\{3, 4, 5\}$

129

## Helpful Hint

Don't forget that some compound inequalities containing "and" can be written in a more compact form.

| Compound Inequality | Compact Form | Interval Notation |
|---|---|---|
| $2 \leq x$ and $x \leq 6$ | $2 \leq x \leq 6$ | $[2, 6]$ |

Graph:

**Practice 2**

Solve: $x + 5 < 9$ and
$3x - 1 < 2$

**Example 2**   Solve:   $x - 7 < 2$ and $2x + 1 < 9$

**Solution:**   First we solve each inequality separately.

$$x - 7 < 2 \quad and \quad 2x + 1 < 9$$
$$x < 9 \quad and \quad 2x < 8$$
$$x < 9 \quad and \quad x < 4$$

Now we can graph the two intervals on two number lines and find their intersection.

$\{x \mid x < 9\}$ ⟶ $(-\infty, 9)$

$\{x \mid x < 4\}$ ⟶ $(-\infty, 4)$

$\{x \mid x < 9 \text{ and } x < 4\}$
$= \{x \mid x < 4\}$ ⟶ $(-\infty, 4)$

The solution set is $(-\infty, 4)$.

■ Work Practice 2

**Practice 3**

Solve: $4x \geq 0$ and $2x + 4 \leq 2$

**Example 3**   Solve:   $2x \geq 0$ and $4x - 1 \leq -9$

**Solution:**   First we solve each inequality separately.

$$2x \geq 0 \quad and \quad 4x - 1 \leq -9$$
$$x \geq 0 \quad and \quad 4x \leq -8$$
$$x \geq 0 \quad and \quad x \leq -2$$

Now we can graph the two intervals and find their intersection.

$\{x \mid x \geq 0\}$ ⟶ $[0, \infty)$

$\{x \mid x \leq -2\}$ ⟶ $(-\infty, -2]$

$\{x \mid x \geq 0 \text{ and } x \leq -2\} = \varnothing$

There is no number that is greater than or equal to 0 **and** less than or equal to $-2$. The solution set is $\varnothing$.

■ Work Practice 3

## Helpful Hint

Example 3 shows that some compound inequalities have no solution. Also, some have all real numbers as solutions.

To solve a compound inequality like $2 < 4 - x < 7$, we get $x$ alone in the middle. Since a compound inequality such as this one is really two inequalities in one statement, we must perform the same operation to all three parts of the inequality. For example:

$$2 < 4 - x < 7 \quad \text{means} \quad 2 < 4 - x \quad and \quad 4 - x < 7.$$

**Answers**

**2.** $(-\infty, 1)$   **3.** $\varnothing$

**Example 4** Solve: $2 < 4 - x < 7$

**Solution:** To get $x$ alone, we first subtract 4 from all three parts.

$$2 < 4 - x < 7$$
$$2 - 4 < 4 - x - 4 < 7 - 4 \quad \text{Subtract 4 from all three parts.}$$
$$-2 < -x < 3 \quad \text{Simplify.}$$
$$\frac{-2}{-1} > \frac{-x}{-1} > \frac{3}{-1} \quad \text{Divide all three parts by } -1 \text{ and reverse the inequality symbols.}$$
$$2 > x > -3$$

This is equivalent to $-3 < x < 2$, and its graph is shown.

The solution set in interval notation is $(-3, 2)$.

■ **Work Practice 4**

**Practice 4**

Solve: $5 < 1 - x < 9$

**Helpful Hint** Don't forget to reverse both inequality symbols.

**Example 5** Solve: $-1 \le \frac{2}{3}x + 5 < 2$

**Solution:** First we clear the inequality of fractions by multiplying all three parts by the LCD, 3.

$$-1 \le \frac{2}{3}x + 5 < 2$$
$$3(-1) \le 3\left(\frac{2}{3}x + 5\right) < 3(2) \quad \text{Multiply all three parts by the LCD, 3.}$$
$$-3 \le 2x + 15 < 6 \quad \text{Use the distributive property and multiply.}$$
$$-3 - 15 \le 2x + 15 - 15 < 6 - 15 \quad \text{Subtract 15 from all three parts.}$$
$$-18 \le 2x < -9 \quad \text{Simplify.}$$
$$\frac{-18}{2} \le \frac{2x}{2} < \frac{-9}{2} \quad \text{Divide all three parts by 2.}$$
$$-9 \le x < -\frac{9}{2} \quad \text{Simplify.}$$

The graph of the solution is shown.

The solution set in interval notation is $\left[-9, -\frac{9}{2}\right)$.

■ **Work Practice 5**

**Practice 5**

Solve: $-2 < \frac{3}{4}x + 2 \le 5$

## Objective C Finding the Union of Two Sets ▶

The solution set of a compound inequality formed by the word **or** is the **union** of the solution sets of the two inequalities. We use the symbol $\cup$ to denote "union."

**Union of Two Sets**

The **union** of two sets, $A$ and $B$, is the set of elements that belong to *either* of the sets. $A$ union $B$ is denoted by $A \cup B$.

**Helpful Hint** The word "either" in this definition means "one or the other or both."

**Answers**

**4.** $(-8, -4)$  **5.** $\left(-\frac{16}{3}, 4\right]$

**Practice 6**

Find the union:
$\{1, 2, 3, 4, 5\} \cup \{3, 4, 5, 6\}$

**Example 6**   Find the union:   $\{2, 4, 6, 8\} \cup \{3, 4, 5, 6\}$

**Solution:**   The numbers that are in either set are $\{2, 3, 4, 5, 6, 8\}$. This set is the union.

■ Work Practice 6

## Objective D  Solving Compound Inequalities Containing "or" ▶

A value of $x$ is a solution of a compound inequality formed by the word **or** if it is a solution of **either** inequality. For example, the solution set of the compound inequality $x \le 1$ **or** $x \ge 3$ contains all numbers that make the inequality $x \le 1$ a true statement **or** the inequality $x \ge 3$ a true statement. In other words, the solution of such an inequality is the *union* of the solutions of the individual inequalities.

| $\{x \mid x \le 1\}$ | | $(-\infty, 1]$ |
|---|---|---|
| $\{x \mid x \ge 3\}$ | | $[3, \infty)$ |
| $\{x \mid x \le 1$ or $x \ge 3\}$ | | $(-\infty, 1] \cup [3, \infty)$ |

In interval notation, the set $\{x \mid x \le 1 \text{ or } x \ge 3\}$ is written as $(-\infty, 1] \cup [3, \infty)$.

**Practice 7**

Solve:
$3x - 2 \ge 10 \text{ or } x - 6 \le -4$

**Example 7**   Solve:   $5x - 3 \le 10 \text{ or } x + 1 \ge 5$

**Solution:**   First we solve each inequality separately.

$$5x - 3 \le 10 \quad \text{or} \quad x + 1 \ge 5$$
$$5x \le 13 \quad \text{or} \quad x \ge 4$$
$$x \le \frac{13}{5} \quad \text{or} \quad x \ge 4$$

Now we can graph each interval and find their union.

| $\left\{x \mid x \le \dfrac{13}{5}\right\}$ | | $\left(-\infty, \dfrac{13}{5}\right]$ |
|---|---|---|
| $\{x \mid x \ge 4\}$ | | $[4, \infty)$ |
| $\left\{x \mid x \le \dfrac{13}{5} \text{ or } x \ge 4\right\}$ | | $\left(-\infty, \dfrac{13}{5}\right] \cup [4, \infty)$ |

The solution set is $\left(-\infty, \dfrac{13}{5}\right] \cup [4, \infty)$.

■ Work Practice 7

**Practice 8**

Solve:  $x - 7 \le -1$ or
$2x - 6 \ge 2$

**Example 8**   Solve:   $-2x - 5 < -3 \text{ or } 6x < 0$

**Solution:**   First we solve each inequality separately.

$$-2x - 5 < -3 \quad \text{or} \quad 6x < 0$$
$$-2x < 2 \quad \text{or} \quad x < 0$$
$$x > -1 \quad \text{or} \quad x < 0$$

**Answers**

**6.** $\{1, 2, 3, 4, 5, 6\}$

**7.** $(-\infty, 2] \cup [4, \infty)$   **8.** $(-\infty, \infty)$

Now we can graph each interval and find their union.

$\{x \mid x > -1\}$                                        $(-1, \infty)$

$\{x \mid x < 0\}$                                        $(-\infty, 0)$

$\{x \mid x > -1 \text{ or } x < 0\}$                      $(-\infty, \infty)$
= all real numbers

The solution set is $(-\infty, \infty)$.

■ Work Practice 8

✓**Concept Check** Which of the following is *not* a correct way to represent the set of all numbers between $-3$ and $5$?

  **a.** $\{x \mid -3 < x < 5\}$
  **b.** $-3 < x \text{ or } x < 5$
  **c.** $(-3, 5)$
  **d.** $x > -3 \text{ and } x < 5$

✓**Concept Check Answer**
**b** is not correct

## Vocabulary, Readiness & Video Check

*Use the choices below to fill in each blank. Some choices may be used more than once.*

| | | |
|---|---|---|
| or | $\cup$ | $\varnothing$ |
| and | $\cap$ | compound |

**1.** Two inequalities joined by the words "and" or "or" are called _____ inequalities.

**2.** The word _____ means intersection.

**3.** The word _____ means union.

**4.** The symbol _____ represents intersection.

**5.** The symbol _____ represents union.

**6.** The symbol _____ is the empty set.

**7.** The inequality $-2 \le x < 1$ means $-2 \le x$ _____ $x < 1$.

**8.** $\{x \mid x < 0 \text{ and } x > 0\} = $ _____

*Martin-Gay Interactive Videos*

*Watch the section lecture video and answer the following questions.*

**Objective A**  **9.** Based on ▤ Example 1 and the lecture before, complete the following statement. For an element to be in the intersection of sets $A$ and $B$, the element must be in set $A$ _____ in set $B$. ▶

**Objective B**  **10.** In ▤ Example 2, how can using three number lines help us find the solution to this "and" compound inequality? ▶

**Objective C**  **11.** Based on ▤ Example 4 and the lecture before, complete the following statement. For an element to be in the union of sets $A$ and $B$, the element must be in set $A$ _____ in set $B$. ▶

**Objective D**  **12.** In ▤ Example 5, how can using three number lines help us find the solution to this "or" compound inequality? ▶

*See Video 2.5* 🍎

## 2.5 Exercise Set MyMathLab® ▶

**Objectives** **A** **C** **Mixed Practice** *If A = {x | x is an even integer}, B = {x | x is an odd integer},*
*C = {2, 3, 4, 5}, and D = {4, 5, 6, 7}, list the elements of each set. See Examples 1 and 6.*

▶**1.** $C \cup D$

**2.** $C \cap D$

▶**3.** $A \cap D$

**4.** $A \cup D$

**5.** $A \cup B$

**6.** $A \cap B$

**7.** $B \cap D$

**8.** $B \cup D$

**9.** $B \cup C$

**10.** $B \cap C$

**11.** $A \cap C$

**12.** $A \cup C$

**Objective** **B** *Solve each compound inequality. Graph the two inequalities on the first two number lines and the solution set on the third number line. Write the solutions in interval notation. See Examples 2 and 3.*

**13.** $x < 1$ *and* $x > -3$

**14.** $x \le 0$ *and* $x \ge -2$

**15.** $x \le -3$ *and* $x \ge -2$

**16.** $x < 2$ *and* $x > 4$

**17.** $x < -1$ *and* $x < 1$

**18.** $x \ge -4$ *and* $x > 1$

*Solve each compound inequality. Write solutions in interval notation. See Examples 2 and 3.*

**19.** $x + 1 \geq 7$ *and* $3x - 1 \geq 5$

**20.** $x + 2 \geq 3$ *and* $5x - 1 \geq 9$

**21.** $4x + 2 \leq -10$ *and* $2x \leq 0$

**22.** $2x + 4 > 0$ *and* $4x > 0$

**23.** $-2x < -8$ *and* $x - 5 < 5$

**24.** $-7x \leq -21$ *and* $x - 20 \leq -15$

*Solve each compound inequality. See Examples 4 and 5.*

**25.** $5 < x - 6 < 11$

**26.** $-2 \leq x + 3 \leq 0$

**27.** $-2 \leq 3x - 5 \leq 7$

**28.** $1 < 4 + 2x < 7$

**29.** $1 \leq \dfrac{2}{3}x + 3 \leq 4$

**30.** $-2 < \dfrac{1}{2}x - 5 < 1$

**31.** $-5 \leq \dfrac{-3x + 1}{4} \leq 2$

**32.** $-4 \leq \dfrac{-2x + 5}{3} \leq 1$

**Objective D** *Solve each compound inequality. Graph the two given inequalities on the first two number lines and the solution set on the third number line. Write the solutions in interval notation. See Examples 7 and 8.*

**33.** $x < 4$ *or* $x < 5$

**34.** $x < 0$ *or* $x < 1$

**35.** $x \leq -4$ *or* $x \geq 1$

**36.** $x \geq -3$ *or* $x \leq -4$

**37.** $x > 0$ *or* $x < 3$

**38.** $x \geq -2$ *or* $x \leq 2$

*Solve each compound inequality. Write answers in interval notation. See Examples 7 and 8.*

**39.** $-2x \leq -4 \ or \ 5x - 20 \geq 5$

**40.** $-5x \leq 10 \ or \ 3x - 5 \geq 1$

**41.** $x + 4 < 0 \ or \ 6x > -12$

**42.** $x + 9 < 0 \ or \ 4x > -12$

**43.** $3(x - 1) < 12 \ or \ x + 7 > 10$

**44.** $5(x - 1) \leq -5 \ or \ 5 - x \leq 11$

## Objectives B D Mixed Practice *Solve each compound inequality. Write solutions in interval notation.*

**45.** $x < \frac{2}{3} \ and \ x > -\frac{1}{2}$

**46.** $x < \frac{5}{7} \ and \ x < 1$

**47.** $x < \frac{2}{3} \ or \ x > -\frac{1}{2}$

**48.** $x < \frac{5}{7} \ or \ x < 1$

**49.** $0 \leq 2x - 3 \leq 9$

**50.** $3 < 5x + 1 < 11$

**51.** $\frac{1}{2} < x - \frac{3}{4} < 2$

**52.** $\frac{2}{3} < x + \frac{1}{2} < 4$

**53.** $x + 3 \geq 3 \ and \ x + 3 \leq 2$

**54.** $2x - 1 \geq 3 \ and \ -x > 2$

**55.** $3x \geq 5 \ or \ -\frac{5}{8}x - 6 > 1$

**56.** $\frac{3}{8}x + 1 \leq 0 \ or \ -2x < -4$

**57.** $0 < \frac{5 - 2x}{3} < 5$

**58.** $-2 < \frac{-2x - 1}{3} < 2$

**59.** $-6 < 3(x - 2) \leq 8$

**60.** $-5 < 2(x + 4) \leq 8$

**61.** $-x + 5 > 6 \ and \ 1 + 2x \leq -5$

**62.** $5x \leq 0 \ and \ -x + 5 < 8$

▶ **63.** $3x + 2 \leq 5 \ or \ 7x > 29$

**64.** $-x < 7 \ or \ 3x + 1 < -20$

**65.** $5 - x > 7 \ and \ 2x + 3 \geq 13$

**66.** $-2x < -6 \ and \ 1 - x > -2$

**67.** $-\frac{1}{2} \leq \frac{4x - 1}{6} < \frac{5}{6}$

**68.** $-\frac{1}{2} \leq \frac{3x - 1}{10} < \frac{1}{2}$

**69.** $\frac{1}{15} < \frac{8 - 3x}{15} < \frac{4}{5}$

**70.** $-\frac{1}{4} < \frac{6 - x}{12} < -\frac{1}{6}$

**71.** $0.3 < 0.2x - 0.9 < 1.5$

**72.** $-0.7 < 0.4x + 0.8 < 0.5$

## Review

*Evaluate. See Section 1.5.*

**73.** $|-7| - |19|$

**74.** $|-7 - 19|$

**75.** $-(-6) - |-10|$

**76.** $|-4| - (-4) + |-20|$

*Find by inspection all values for x that make each equation true.*

**77.** $|x| = 7$

**78.** $|x| = 5$

**79.** $|x| = 0$

**80.** $|x| = -2$

## Concept Extensions

*Use the graph to answer Exercises 81 and 82.*

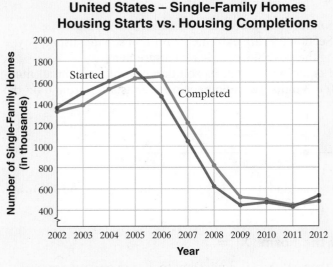

**United States – Single-Family Homes
Housing Starts vs. Housing Completions**

*Source:* U.S. Census Bureau

**81.** For which years were the number of single-family housing starts greater than 1500 and the number of single-family home completions greater than 1500?

**82.** For which years were the number of single-family housing starts less than 1000 or the number of single-family housing completions greater than 1500?

**83.**  In your own words, describe how to find the union of two sets.

**84.** In your own words, describe how to find the intersection of two sets.

*To solve a compound inequality such as x − 6 < 3x < 2x + 5, we solve*

$$x - 6 < 3x \quad and \quad 3x < 2x + 5$$

*Use this information to solve the inequalities in Exercises 85 through 88.*

**85.** $x - 6 < 3x < 2x + 5$

**86.** $2x - 3 < 3x + 1 < 4x - 5$

**87.** $x + 3 < 2x + 1 < 4x + 6$

**88.** $-3(x - 2) \leq 3 - 2x \leq 10 - 3x$

*The formula for converting Fahrenheit temperatures to Celsius temperatures is $C = \dfrac{5}{9}(F - 32)$. Use this formula for Exercises 89 and 90.*

**89.** During a recent year, the temperatures in Chicago ranged from −29° to 35° Celsius. Use a compound inequality to convert these temperatures to Fahrenheit temperatures.

**90.** In Oslo, the average temperature ranges from −10° to 18° Celsius. Use a compound inequality to convert these temperatures to Fahrenheit temperatures.

*Solve.*

**91.** Christian D'Angelo has scores of 68, 65, 75, and 78 on his algebra tests. Use a compound inequality to find the scores he can make on his final exam to receive a C in the course. The final exam counts as two tests, and a C is received if the final course average is from 70 to 79.

**92.** Wendy Wood has scores of 80, 90, 82, and 75 on her chemistry tests. Use a compound inequality to find the range of scores she can make on her final exam to receive a B in the course. The final exam counts as two tests, and a B is received if the final course average is from 80 to 89.

## Objective

**A** Solve Absolute Value Equations.

## Objective A Solving Absolute Value Equations

In Chapter 1, we defined the absolute value of a number as its distance from 0 on a number line.

$|-2| = 2$ and $|3| = 3$

In this section, we concentrate on solving equations containing the absolute value of a variable or a variable expression. Examples of absolute value equations are

$$|x| = 3 \qquad -5 = |2y + 7| \qquad |z - 6.7| = |3z + 1.2|$$

Since distance and absolute value are so closely related, absolute value equations and inequalities (see Section 2.7) are extremely useful in solving distance-type problems, such as calculating the possible error in a measurement.

For the absolute value equation $|x| = 3$, its solution set will contain all numbers whose distance from 0 is 3 units. Two numbers are 3 units away from 0 on the number line: 3 and $-3$.

Thus, the solution set of the equation $|x| = 3$ is $\{3, -3\}$. This suggests the following:

---

### Solving Equations of the Form $|X| = a$

If $a$ is a positive number, then $|X| = a$ is equivalent to $X = a$ or $X = -a$.

---

**Practice 1**

Solve: $|q| = 5$.

**Example 1** Solve: $|p| = 2$.

**Solution:** Since 2 is positive, $|p| = 2$ is equivalent to $p = 2$ or $p = -2$.

To check, let $p = 2$ and then $p = -2$ in the original equation.

$$
\begin{array}{ll}
|p| = 2 & \text{Original equation} \\
|2| = 2 & \text{Let } p = 2. \\
2 = 2 & \text{True}
\end{array}
\qquad
\begin{array}{ll}
|p| = 2 & \text{Original equation} \\
|-2| = 2 & \text{Let } p = -2. \\
2 = 2 & \text{True}
\end{array}
$$

The solutions are 2 and $-2$ or the solution set is $\{2, -2\}$.

■ Work Practice 1

If the expression inside the absolute value bars is more complicated than a single variable, we can still apply the property for solving absolute value equations above.

---

**Helpful Hint**

For the equation $|X| = a$ in the box above, $X$ can be a single variable or a variable expression.

---

**Answer**

**1.** $\{-5, 5\}$

**Example 2** Solve: $|5w + 3| = 7$.

**Solution:** Here the expression inside the absolute value bars is $5w + 3$. If we think of the expression $5w + 3$ as $X$ in the property for solving absolute value equations, we see that $|X| = 7$ is equivalent to

$$X = 7 \quad \text{or} \quad X = -7$$

Then substitute $5w + 3$ for $X$, and we have

$$5w + 3 = 7 \quad \text{or} \quad 5w + 3 = -7$$

Solve these two equations for $w$.

$$5w + 3 = 7 \quad \text{or} \quad 5w + 3 = -7$$
$$5w = 4 \quad \text{or} \quad 5w = -10$$
$$w = \frac{4}{5} \quad \text{or} \quad w = -2$$

**Check:** To check, let $w = -2$ and then $w = \frac{4}{5}$ in the original equation.

| Let $w = -2$ | Let $w = \frac{4}{5}$ |
|---|---|
| $|5(-2) + 3| = 7$ | $\left|5\left(\frac{4}{5}\right) + 3\right| = 7$ |
| $|-10 + 3| = 7$ | $|4 + 3| = 7$ |
| $|-7| = 7$ | $|7| = 7$ |
| $7 = 7$   True | $7 = 7$   True |

Both solutions check, and the solutions are $-2$ and $\frac{4}{5}$ or the solution set is $\left\{-2, \frac{4}{5}\right\}$.

■ **Work Practice 2**

**Example 3** Solve: $\left|\frac{x}{2} - 1\right| = 11$.

**Solution:** $\left|\frac{x}{2} - 1\right| = 11$ is equivalent to

$$\frac{x}{2} - 1 = 11 \quad \text{or} \quad \frac{x}{2} - 1 = -11$$
$$2\left(\frac{x}{2} - 1\right) = 2(11) \quad \text{or} \quad 2\left(\frac{x}{2} - 1\right) = 2(-11) \quad \text{Clear fractions.}$$
$$x - 2 = 22 \quad \text{or} \quad x - 2 = -22 \quad \text{Apply the distributive property.}$$
$$x = 24 \quad \text{or} \quad x = -20$$

The solutions are 24 and $-20$, or the solution set is $\{24, -20\}$.

■ **Work Practice 3**

To apply the absolute value rule, first make sure that the absolute value expression is isolated.

**Helpful Hint**

If the equation has a single absolute value expression containing variables, isolate the absolute value expression first.

**Practice 2**

Solve: $|2x - 3| = 5$.

**Practice 3**

Solve: $\left|\frac{x}{5} + 1\right| = 15$.

**Answers**

**2.** $\{-1, 4\}$   **3.** $\{-80, 70\}$

**Practice 4**

Solve: $|3x| + 8 = 14.$

**Example 4** Solve: $|2x| + 5 = 7.$

**Solution:** We want the absolute value expression alone on one side of the equation, so begin by subtracting 5 from both sides. Then apply the absolute value property.

$$|2x| + 5 = 7$$
$$|2x| = 2 \qquad \text{Subtract 5 from both sides.}$$
$$2x = 2 \quad \text{or} \quad 2x = -2$$
$$x = 1 \quad \text{or} \quad x = -1$$

The solutions are $-1$ and $1$, or the solution set is $\{-1, 1\}$.

■ Work Practice 4

**Practice 5**

Solve: $|z| = 0.$

**Example 5** Solve: $|y| = 0.$

**Solution:** We are looking for all numbers whose distance from 0 is zero units. The only number is 0. The solution set is $\{0\}$.

■ Work Practice 5

The next two examples illustrate a special case for absolute value equations. This special case occurs when an isolated absolute value is equal to a negative number.

**Practice 6**

Solve: $3|z| + 9 = 7.$

**Example 6** Solve: $2|x| + 25 = 23.$

**Solution:** First, isolate the absolute value.

$$2|x| + 25 = 23$$
$$2|x| = -2 \quad \text{Subtract 25 from both sides.}$$
$$|x| = -1 \quad \text{Divide both sides by 2.}$$

The absolute value of a number is never negative, so this equation has no solution. The solution set is $\{\ \}$ or $\varnothing$.

■ Work Practice 6

**Practice 7**

Solve: $\left|\dfrac{5x + 3}{4}\right| = -8.$

**Example 7** Solve: $\left|\dfrac{3x + 1}{2}\right| = -2.$

**Solution:** Again, the absolute value of any expression is never negative, so no solution exists. The solution set is $\{\ \}$ or $\varnothing$.

■ Work Practice 7

Given two absolute value expressions, we might ask, when are the absolute values of two expressions equal? To see the answer, notice that

$$|2| = |2|, \quad |-2| = |-2|, \quad |-2| = |2|, \quad \text{and} \quad |2| = |-2|$$
$$\underset{\text{same}}{\qquad} \quad \underset{\text{same}}{\qquad} \quad \underset{\text{opposites}}{\qquad} \quad \underset{\text{opposites}}{\qquad}$$

Two absolute value expressions are equal when the expressions inside the absolute value bars are equal to or are opposites of each other.

**Answers**

**4.** $\{-2, 2\}$  **5.** $\{0\}$  **6.** $\{\ \}$ or $\varnothing$
**7.** $\{\ \}$ or $\varnothing$

**Example 8** Solve: $|3x + 2| = |5x - 8|$.

**Solution:** This equation is true if the expressions inside the absolute value bars are equal to or are opposites of each other.

$$3x + 2 = 5x - 8 \quad \text{or} \quad 3x + 2 = -(5x - 8)$$

Next, solve each equation.

$$
\begin{aligned}
3x + 2 &= 5x - 8 &\text{or} \quad 3x + 2 &= -5x + 8 \\
-2x + 2 &= -8 &\text{or} \quad 8x + 2 &= 8 \\
-2x &= -10 &\text{or} \quad 8x &= 6 \\
x &= 5 &\text{or} \quad x &= \frac{3}{4}
\end{aligned}
$$

The solutions are $\frac{3}{4}$ and 5, or the solution set is $\left\{\frac{3}{4}, 5\right\}$.

▰ **Work Practice 8**

**Example 9** Solve: $|x - 3| = |5 - x|$.

**Solution:**
$$
\begin{aligned}
x - 3 &= 5 - x &\text{or} \quad x - 3 &= -(5 - x) \\
2x - 3 &= 5 &\text{or} \quad x - 3 &= -5 + x \\
2x &= 8 &\text{or} \quad x - 3 - x &= -5 + x - x \\
x &= 4 &\text{or} \quad -3 &= -5 \quad \text{False}
\end{aligned}
$$

Recall from Section 2.1 that when an equation simplifies to a false statement, the equation has no solution. Thus, the only solution for the original absolute value equation is 4. The solution set is $\{4\}$.

▰ **Work Practice 9**

✓**Concept Check** True or false? Absolute value equations always have two solutions. Explain your answer.

The following box summarizes the methods shown for solving absolute value equations.

---

**Absolute Value Equations**

$$|X| = a \quad \begin{cases} \text{If } a \text{ is positive, then solve } X = a \text{ or } X = -a. \\ \text{If } a \text{ is 0, solve } X = 0. \\ \text{If } a \text{ is negative, the equation } |X| = a \text{ has no solution.} \end{cases}$$

$|X| = |Y| \quad$ Solve $X = Y$ or $X = -Y$.

---

**Practice 8**
Solve: $|2x + 4| = |3x - 1|$.

**Practice 9**
Solve: $|x - 2| = |8 - x|$.

**Answers**

**8.** $\left\{-\frac{3}{5}, 5\right\}$ 　**9.** $\{5\}$

✓**Concept Check Answer**
false; answers may vary

## Vocabulary, Readiness & Video Check

*Match each absolute value equation with an equivalent statement.*

**1.** $|x - 2| = 5$

**2.** $|x - 2| = 0$

**3.** $|x - 2| = |x + 3|$

**4.** $|x + 3| = 5$

**5.** $|x + 3| = -5$

**A.** $x - 2 = 0$

**B.** $x - 2 = x + 3$ or $x - 2 = -(x + 3)$

**C.** $x - 2 = 5$ or $x - 2 = -5$

**D.** $\varnothing$

**E.** $x + 3 = 5$ or $x + 3 = -5$

*Martin-Gay Interactive Videos*    Watch the section lecture video and answer the following question.

**Objective A** **6.** As explained in Example 3, why is *a* positive in the property "$|X| = a$ is equivalent to $X = a$ or $X = -a$"?

*See Video 2.6*

## 2.6  Exercise Set MyMathLab®

**Objective A** *Solve each absolute value equation. See Examples 1 through 7.*

**1.** $|x| = 7$

**2.** $|y| = 15$

**3.** $|3x| = 12.6$

**4.** $|6n| = 12.6$

**5.** $|2x - 5| = 9$

**6.** $|6 + 2n| = 4$

**7.** $\left|\dfrac{x}{2} - 3\right| = 1$

**8.** $\left|\dfrac{n}{3} + 2\right| = 4$

**9.** $|z| + 4 = 9$

**10.** $|x| + 1 = 3$

**11.** $|3x| + 5 = 14$

**12.** $|2x| - 6 = 4$

**13.** $|2x| = 0$

**14.** $|7z| = 0$

**15.** $|4n + 1| + 10 = 4$

**16.** $|3z - 2| + 8 = 1$

**17.** $|5x - 1| = 0$

**18.** $|3y + 2| = 0$

*Solve. See Examples 8 and 9.*

**19.** $|5x - 7| = |3x + 11|$

**20.** $|9y + 1| = |6y + 4|$

**21.** $|z + 8| = |z - 3|$

**22.** $|2x - 5| = |2x + 5|$

*Solve each absolute value equation. See Examples 1 through 9.*

**23.** $|x| = 4$

**24.** $|x| = 1$

**25.** $|y| = 0$

**26.** $|y| = 8$

**27.** $|z| = -2$

**28.** $|y| = -9$

**29.** $|7 - 3x| = 7$

**30.** $|4m + 5| = 5$

**31.** $|6x| - 1 = 11$

**32.** $|7z| + 1 = 22$

**33.** $|4p| = -8$

**34.** $|5m| = -10$

**▶ 35.** $|x - 3| + 3 = 7$

**36.** $|x + 4| - 4 = 1$

**37.** $\left|\dfrac{z}{4} + 5\right| = -7$

**38.** $\left|\dfrac{c}{5} - 1\right| = -2$

**39.** $|9v - 3| = -8$

**40.** $|1 - 3b| = -7$

**41.** $|8n + 1| = 0$

**42.** $|5x - 2| = 0$

**43.** $|1 + 6c| - 7 = -3$

**44.** $|2 + 3m| - 9 = -7$

**45.** $|5x + 1| = 11$

**46.** $|8 - 6c| = 1$

**47.** $|4x - 2| = |-10|$

**48.** $|3x + 5| = |-4|$

**49.** $|5x + 1| = |4x - 7|$

**50.** $|3 + 6n| = |4n + 11|$

**51.** $|6 + 2x| = -|-7|$

**52.** $|4 - 5y| = -|-3|$

**53.** $|2x - 6| = |10 - 2x|$

**54.** $|4n + 5| = |4n + 3|$

**55.** $\left|\dfrac{2x - 5}{3}\right| = 7$

**56.** $\left|\dfrac{1 + 3n}{4}\right| = 4$

**57.** $2 + |5n| = 17$

**58.** $8 + |4m| = 24$

**59.** $\left|\dfrac{2x - 1}{3}\right| = |-5|$

**60.** $\left|\dfrac{5x + 2}{2}\right| = |-6|$

**▶ 61.** $|2y - 3| = |9 - 4y|$

**62.** $|5z - 1| = |7 - z|$

**63.** $\left|\dfrac{3n + 2}{8}\right| = |-1|$

**64.** $\left|\dfrac{2r - 6}{5}\right| = |-2|$

**65.** $|x + 4| = |7 - x|$

**66.** $|8 - y| = |y + 2|$

**67.** $\left|\dfrac{8c - 7}{3}\right| = -|-5|$

**68.** $\left|\dfrac{5d + 1}{6}\right| = -|-9|$

## Review

*The circle graph shows the types of cheese produced in the United States in 2012. Use this graph to answer Exercises 69 through 72. See Section 2.2.*

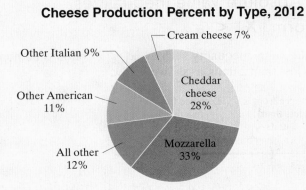

**Cheese Production Percent by Type, 2012**

Cream cheese 7%
Other Italian 9%
Other American 11%
All other 12%
Cheddar cheese 28%
Mozzarella 33%

*Source:* National Agricultural Statistics Service, Department of Agriculture

**69.** In 2012, cheddar cheese made up what percent of U. S. cheese production?

**70.** Which cheese had the highest U.S. production in 2012?

**71.** A circle contains 360°. Find the number of degrees found in the 9% sector for other Italian cheese.

**72.** In 2012, the total production of cheese in the United States was 10,890,000,000 pounds. Find the amount of cream cheese produced during that year.

*List five integer solutions of each inequality. See Sections 1.3 and 1.4.*

**73.** $|x| \le 3$          **74.** $|x| \ge -2$          **75.** $|y| > -10$          **76.** $|y| < 0$

## Concept Extensions

*Without going through a solution procedure, determine the solution of each absolute value equation or inequality.*

**77.** $|x - 7| = -4$                        **78.** $|x - 7| < -4$

*Solve.*

**79.** Write an absolute value equation representing all numbers $x$ whose distance from 0 is 5 units.

**80.** Write an absolute value equation representing all numbers $x$ whose distance from 0 is 2 units.

**81.** Explain why some absolute value equations have two solutions.

**82.** Explain why some absolute value equations have one solution.

**83.** Write an absolute value equation representing all numbers $x$ whose distance from 1 is 5 units.

**84.** Write an absolute value equation representing all numbers $x$ whose distance from 7 is 2 units.

**85.** Describe how solving an absolute value equation such as $|2x - 1| = 3$ is similar to solving an absolute value equation such as $|2x - 1| = |x - 5|$.

**86.** Describe how solving an absolute value equation such as $|2x - 1| = 3$ is different from solving an absolute value equation such as $|2x - 1| = |x - 5|$.

*Write each as an equivalent absolute value equation.*

**87.** $x = 6$ or $x = -6$

**88.** $2x - 1 = 4$ or $2x - 1 = -4$

**89.** $x - 2 = 3x - 4$ or $x - 2 = -(3x - 4)$

**90.** For what value(s) of $c$ will an absolute value equation of the form $|ax + b| = c$ have
    **a.** one solution?
    **b.** no solution?
    **c.** two solutions?

---

## 2.7   Absolute Value Inequalities

### Objectives

**A** Solve Absolute Value Inequalities of the Form $|X| < a$.

**B** Solve Absolute Value Inequalities of the Form $|X| > a$.

### Objective A   Solving Absolute Value Inequalities of the Form $|X| < a$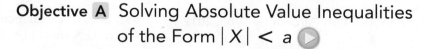

The solution set of an absolute value inequality such as $|x| < 2$ contains all numbers whose distance from 0 is less than 2 units, as shown below.

The solution set is $\{x \mid -2 < x < 2\}$, or $(-2, 2)$ in interval notation.

**Example 1** Solve: $|x| \leq 3$.

**Solution:** The solution set of this inequality contains all numbers whose distance from 0 is less than or equal to 3. Thus, 3, −3, and all numbers between 3 and −3 are in the solution set.

The solution set is $[-3, 3]$.

■ Work Practice 1

In general, we have the following.

---

**Solving Absolute Value Inequalities of the Form $|X| < a$**

If $a$ is a positive number, then $|X| < a$ is equivalent to $-a < X < a$.

---

This property also holds true for the inequality symbol $\leq$.

**Example 2** Solve for $m$: $|m - 6| < 2$.

**Solution:** Replace $X$ with $m - 6$ and $a$ with 2 in the preceding property, and we see that

$|m - 6| < 2$ is equivalent to $-2 < m - 6 < 2$

Solve this compound inequality for $m$ by adding 6 to all three parts.

$$-2 < m - 6 < 2$$
$$-2 + 6 < m - 6 + 6 < 2 + 6 \quad \text{Add 6 to all three parts.}$$
$$4 < m < 8 \quad \text{Simplify.}$$

The solution set is $(4, 8)$, and its graph is shown.

■ Work Practice 2

---

**Helpful Hint**

Before using an absolute value inequality property, isolate the absolute value expression on one side of the inequality.

---

**Example 3** Solve for $x$: $|5x + 1| + 1 \leq 10$.

**Solution:** First, isolate the absolute value expression by subtracting 1 from both sides.

$$|5x + 1| + 1 \leq 10$$
$$|5x + 1| \leq 10 - 1 \quad \text{Subtract 1 from both sides.}$$
$$|5x + 1| \leq 9 \quad \text{Simplify.}$$

*(Continued on next page)*

*(Continued on next page)*

---

**Practice 1**

Solve: $|x| < 3$ and graph the solution set.

**Practice 2**

Solve for $b$: $|b + 1| < 3$. Graph the solution set.

**Practice 3**

Solve for $x$: $|3x - 2| + 5 \leq 9$. Graph the solution set.

**Answers**

**1.** $(-3, 3)$;

**2.** $(-4, 2)$;

**3.** $\left[-\dfrac{2}{3}, 2\right]$;

Since 9 is positive, we apply the absolute value property for $|X| \leq a$.

$$-9 \leq 5x + 1 \leq 9$$

$$-9 - 1 \leq 5x + 1 - 1 \leq 9 - 1 \quad \text{Subtract 1 from all three parts.}$$

$$-10 \leq 5x \leq 8 \quad \text{Simplify.}$$

$$-2 \leq x \leq \frac{8}{5} \quad \text{Divide all three parts by 5.}$$

The solution set is $\left[ -2, \dfrac{8}{5} \right]$, and the graph is shown above.

■ **Work Practice 3**

Examples 4 and 5 illustrate special cases of absolute value inequalities. These special cases occur when an isolated absolute value expression is less than, less than or equal to, greater than, or greater than or equal to a negative number or 0.

**Practice 4**

Solve for $x$: $\left| 3x + \dfrac{5}{8} \right| < -4$.

**Example 4** Solve for $x$: $\left| 2x - \dfrac{1}{10} \right| < -13$.

**Solution:** The absolute value of a number is always nonnegative and can never be less than $-13$. Thus, this absolute value inequality has no solution. The solution set is $\{ \ \}$ or $\varnothing$.

■ **Work Practice 4**

**Practice 5**

Solve for $x$: $\left| \dfrac{3(x - 2)}{5} \right| \leq 0$.

**Example 5** Solve for $x$: $\left| \dfrac{2(x + 1)}{3} \right| \leq 0$.

**Solution:** Recall that "$\leq$" means "less than or equal to." The absolute value of any expression will never be less than 0, but it may be equal to 0. Thus, to solve $\left| \dfrac{2(x + 1)}{3} \right| \leq 0$, we solve $\left| \dfrac{2(x + 1)}{3} \right| = 0$

$$\frac{2(x + 1)}{3} = 0$$

$$3 \left[ \frac{2(x + 1)}{3} \right] = 3(0) \quad \text{Clear the equation of fractions.}$$

$$2x + 2 = 0 \quad \text{Apply the distributive property.}$$

$$2x = -2 \quad \text{Subtract 2 from both sides.}$$

$$x = -1 \quad \text{Divide both sides by 2.}$$

The solution set is $\{-1\}$.

■ **Work Practice 5**

## Objective B Solving Absolute Value Inequalities of the Form $|X| > a$ ▶

Let us now solve an absolute value inequality of the form $|X| > a$, such as $|x| \geq 3$. The solution set contains all numbers whose distance from 0 is 3 or more units. Thus, the graph of the solution set contains 3 and all points to the right of 3 on the number line as well as $-3$ and all points to the left of $-3$ on the number line.

**Answers**

**4.** $\{ \ \}$ or $\varnothing$ **5.** $\{2\}$

This solution set is written as $\{x \mid x \leq -3 \text{ or } x \geq 3\}$. In interval notation, the solution is $(-\infty, -3] \cup [3, \infty)$, since "or" means "union." In general, we have the following.

---

**Solving Absolute Value Inequalities of the Form $|X| > a$**

If $a$ is a positive number, then $|X| > a$ is equivalent to $X < -a \text{ or } X > a$.

---

This property also holds true for the inequality symbol $\geq$.

**Example 6** Solve for $y$: $|y - 3| > 7$.

**Solution:** Since 7 is positive, we apply the property for $|X| > a$.

$|y - 3| > 7$ is equivalent to $y - 3 < -7 \text{ or } y - 3 > 7$

Next, solve the compound inequality.

$$
\begin{array}{ccc}
y - 3 < -7 & or & y - 3 > 7 \\
y - 3 + 3 < -7 + 3 & or & y - 3 + 3 > 7 + 3 \quad \text{Add 3 to both sides.} \\
y < -4 & or & y > 10 \quad \text{Simplify.}
\end{array}
$$

The solution set is $(-\infty, -4) \cup (10, \infty)$, and its graph is shown.

■ **Work Practice 6**

Example 7 illustrates another special case of absolute value inequalities when an isolated absolute value expression is less than, less than or equal to, greater than, or greater than or equal to a negative number or 0.

**Example 7** Solve: $|2x + 9| + 5 > 3$.

**Solution:** First isolate the absolute value expression by subtracting 5 from both sides.

$$
\begin{array}{l}
|2x + 9| + 5 > 3 \\
|2x + 9| + 5 - 5 > 3 - 5 \quad \text{Subtract 5 from both sides.} \\
|2x + 9| > -2 \quad \text{Simplify.}
\end{array}
$$

The absolute value of any number is always nonnegative and thus is always greater than $-2$. This inequality and the original inequality are true for all values of $x$. The solution set is $\{x \mid x \text{ is a real number}\}$ or $(-\infty, \infty)$, and its graph is shown.

■ **Work Practice 7**

**Practice 6**

Solve for $y$: $|y + 4| \geq 6$.

**Practice 7**

Solve: $|4x + 3| + 5 > 3$.
Graph the solution set.

**Answers**

**6.** $(-\infty, -10] \cup [2, \infty)$  **7.** $(-\infty, \infty)$

✓**Concept Check** Without taking any solution steps, how do you know that the absolute value inequality $|3x - 2| > -9$ has a solution? What is its solution?

**Practice 8**

Solve: $\left|\dfrac{x}{2} - 3\right| - 5 > -2$.
Graph the solution set.

**Example 8** Solve: $\left|\dfrac{x}{3} - 1\right| - 7 \geq -5$.

**Solution:** First, isolate the absolute value expression by adding 7 to both sides.

$$\left|\dfrac{x}{3} - 1\right| - 7 \geq -5$$

$$\left|\dfrac{x}{3} - 1\right| - 7 + 7 \geq -5 + 7 \quad \text{Add 7 to both sides.}$$

$$\left|\dfrac{x}{3} - 1\right| \geq 2 \quad \text{Simplify.}$$

Next, write the absolute value inequality as an equivalent compound inequality and solve.

$$\dfrac{x}{3} - 1 \leq -2 \qquad \text{or} \qquad \dfrac{x}{3} - 1 \geq 2$$

$$3\left(\dfrac{x}{3} - 1\right) \leq 3(-2) \quad \text{or} \quad 3\left(\dfrac{x}{3} - 1\right) \geq 3(2) \quad \text{Clear the inequalities of fractions.}$$

$$x - 3 \leq -6 \qquad \text{or} \qquad x - 3 \geq 6 \qquad \text{Apply the distributive property.}$$

$$x \leq -3 \qquad \text{or} \qquad x \geq 9 \qquad \text{Add 3 to both sides.}$$

The solution set is $(-\infty, -3] \cup [9, \infty)$, and its graph is shown.

■ Work Practice 8

The following box summarizes the types of absolute value equations and inequalities.

**Answer**

**8.** $(-\infty, 0) \cup (12, \infty)$

✓**Concept Check Answer**
$(-\infty, \infty)$ since the absolute value is always nonnegative

---

## Solving Absolute Value Equations and Inequalities with $a > 0$

**Algebraic Solution**

$|X| = a$ is equivalent to $X = a$ or $X = -a$.

$|X| < a$ is equivalent to $-a < X < a$.

$|X| > a$ is equivalent to $X < -a$ or $X > a$.

---

## Vocabulary, Readiness & Video Check

*Match each absolute value statement with an equivalent statement.*

**1.** $|2x + 1| = 3$

**2.** $|2x + 1| \leq 3$

**3.** $|2x + 1| < 3$

**4.** $|2x + 1| \geq 3$

**5.** $|2x + 1| > 3$

**A.** $2x + 1 > 3$ or $2x + 1 < -3$

**B.** $2x + 1 \geq 3$ or $2x + 1 \leq -3$

**C.** $-3 < 2x + 1 < 3$

**D.** $2x + 1 = 3$ or $2x + 1 = -3$

**E.** $-3 \leq 2x + 1 \leq 3$

*Martin-Gay Interactive Videos*    *Watch the section lecture video and answer the following questions.*

**Objective A** 6. In ⊟ Example 3, how can you reason that the inequality has no solution even if you don't know the property? ▸

**Objective B** 7. In ⊟ Example 4, why is the union symbol used when the solution is written in interval notation? ▸

*See Video 2.7* 🍎

**2.7  Exercise Set** MyMathLab® ▸

**Objective A** *Solve each inequality. Then graph the solution set and write it in interval notation. See Examples 1 through 4.*

**1.** $|x| \leq 4$

**2.** $|x| < 6$

**3.** $|x - 3| < 2$

**4.** $|y - 7| \leq 5$

**5.** $|x + 3| < 2$

**6.** $|x + 4| < 6$

**7.** $|2x + 7| \leq 13$

**8.** $|5x - 3| \leq 18$

**9.** $|x| + 7 \leq 12$

**10.** $|x| + 6 \leq 7$

▸ **11.** $|3x - 1| < -5$

**12.** $|8x - 3| < -2$

**13.** $|x - 6| - 7 \leq -1$

**14.** $|z + 2| - 7 < -3$

**Objective B** *Solve each inequality. Graph the solution set and write it in interval notation. See Examples 6 through 8.*

▸ **15.** $|x| > 3$

**16.** $|y| \geq 4$

**17.** $|x + 10| \geq 14$

**18.** $|x - 9| \geq 2$

**19.** $|x| + 2 > 6$

**20.** $|x| - 1 > 3$

**21.** $|5x| > -4$

**22.** $|4x - 11| > -1$

**23.** $|6x - 8| + 3 > 7$

**24.** $|10 + 3x| + 1 > 2$

**Objectives** **A** **B** **Mixed Practice** *Solve each inequality. Graph the solution set and write it in interval notation. See Examples 5 and 7.*

**25.** $|x| \le 0$

**26.** $|x| \ge 0$

**27.** $|8x + 3| > 0$

**28.** $|5x - 6| < 0$

*Solve each inequality. Graph the solution set and write it in interval notation. See Examples 1 through 8.*

**29.** $|x| \le 2$

**30.** $|z| < 8$

**31.** $|y| > 1$

**32.** $|x| \ge 10$

**33.** $|x - 3| < 8$

**34.** $|-3 + x| \le 10$

**35.** $|0.6x - 3| > 0.6$

**36.** $|1 + 0.3x| \ge 0.1$

**37.** $5 + |x| \le 2$

**38.** $8 + |x| < 1$

**39.** $|x| > -4$

**40.** $|x| \le -7$

**41.** $|2x - 7| \le 11$

**42.** $|5x + 2| < 8$

**43.** $|x + 5| + 2 \geq 8$

**44.** $|-1 + x| - 6 > 2$

**45.** $|x| > 0$

**46.** $|x| < 0$

**47.** $9 + |x| > 7$

**48.** $5 + |x| \geq 4$

**49.** $6 + |4x - 1| \leq 9$

**50.** $-3 + |5x - 2| \leq 4$

**51.** $\left|\dfrac{2}{3}x + 1\right| > 1$

**52.** $\left|\dfrac{3}{4}x - 1\right| \geq 2$

**53.** $|5x + 3| < -6$

**54.** $|4 + 9x| \geq -6$

**55.** $\left|\dfrac{8x - 3}{4}\right| \leq 0$

**56.** $\left|\dfrac{5x + 6}{2}\right| \leq 0$

**57.** $|1 + 3x| + 4 < 5$

**58.** $|7x - 3| - 1 \leq 10$

**▶59.** $\left|\dfrac{x + 6}{3}\right| > 2$

**60.** $\left|\dfrac{7 + x}{2}\right| \geq 4$

**▶61.** $-15 + |2x - 7| \leq -6$

**62.** $-9 + |3 + 4x| < -4$

**63.** $\left| 2x + \dfrac{3}{4} \right| - 7 \le -2$

**64.** $\left| \dfrac{3}{5} + 4x \right| - 6 < -1$

**Mixed Practice** *(Sections 2.6, 2.7) Solve each equation or inequality for x.*

**65.** $|2x - 3| < 7$

**66.** $|2x - 3| > 7$

**67.** $|2x - 3| = 7$

**68.** $|5 - 6x| = 29$

**69.** $|x - 5| \ge 12$

**70.** $|x + 4| \ge 20$

**71.** $|9 + 4x| = 0$

**72.** $|9 + 4x| \ge 0$

**73.** $|2x + 1| + 4 < 7$

**74.** $8 + |5x - 3| \ge 11$

**75.** $|3x - 5| + 4 = 5$

**76.** $|5x - 3| + 2 = 4$

**77.** $|x + 11| = -1$

**78.** $|4x - 4| = -3$

**79.** $\left| \dfrac{2x - 1}{3} \right| = 6$

**80.** $\left| \dfrac{6 - x}{4} \right| = 5$

**81.** $\left| \dfrac{3x - 5}{6} \right| > 5$

**82.** $\left| \dfrac{4x - 7}{5} \right| < 2$

## Review

*Recall the formula:*

$$\text{Probability of an event} = \dfrac{\text{number of ways that the event can occur}}{\text{number of possible outcomes}}$$

*Find the probability of rolling each number on a single toss of a die. (Recall that a die is a cube with each of its six sides containing 1, 2, 3, 4, 5, or 6 black dots, respectively.) See Section 2.3.*

**83.** $P(\text{rolling a } 2)$

**84.** $P(\text{rolling a } 5)$

**85.** $P(\text{rolling a } 7)$

**86.** $P(\text{rolling a } 0)$

**87.** $P(\text{rolling a 1 or 3})$

**88.** $P(\text{rolling a } 1, 2, 3, 4, 5, \text{ or } 6)$

*Consider the equation $3x - 4y = 12$. For each value of x or y given, find the corresponding value of the other variable that makes the statement true. See Section 2.3.*

**89.** If $x = 2$, find $y$

**90.** If $y = -1$, find $x$

**91.** If $y = -3$, find $x$

**92.** If $x = 4$, find $y$

## Concept Extensions

*Solve.*

**93.** Write an absolute value inequality representing all numbers $x$ whose distance from 0 is less than 7 units.

**94.** Write an absolute value inequality representing all numbers $x$ whose distance from 0 is greater than 4 units.

**95.** Write $-5 \le x \le 5$ as an equivalent inequality containing an absolute value.

**96.** Write $x > 1$ or $x < -1$ as an equivalent inequality containing an absolute value.

**97.** Describe how solving $|x - 3| = 5$ is different from solving $|x - 3| < 5$.

**98.** Describe how solving $|x + 4| = 0$ is similar to solving $|x + 4| \le 0$.

*The expression $|x_T - x|$ is defined to be the absolute error in $x$, where $x_T$ is the true value of a quantity and $x$ is the measured value or value as stored in a computer.*

**99.** If the true value of a quantity is 3.5 and the absolute error must be less than 0.05, find the acceptable measured values.

**100.** If the true value of a quantity is 0.2 and the approximate value stored in a computer is $\dfrac{51}{256}$, find the absolute error.

## Chapter 2 Group Activity

### Nutrition Labels

Since 1994, the Food and Drug Administration (FDA) of the Department of Health and Human Services and the Food Safety and Inspection Service of the U.S. Department of Agriculture (USDA) have required nutrition labels like the one below on most food packaging. The labels were designed to help consumers make healthful food choices by giving standardized nutrition information.

**Nutrition Facts**

Serving Size 36g
Servings Per Container 6

Amount Per Serving

Calories 150    Calories from Fat 54

% Daily Value*

| | % Daily Value* |
|---|---|
| **Total Fat** 6 g | 9% |
| Saturated Fat 2 g | 9% |
| **Cholesterol** 0 mg | 0% |
| **Sodium** 80 mg | 3% |
| **Potassium** 40 mg | 1% |
| **Total Carbohydrate** 24 g | 8% |
| Dietary Fiber 0 g | 0% |
| Sugars 11 g | |
| **Protein** 2 g | |

| | | | |
|---|---|---|---|
| Vitamin A | 25% | Vitamin C | 25% |
| Calcium | 50% | Iron | 25% |
| Vitamin D | 25% | Vitamin E | 25% |
| Thiamin | 25% | Riboflavin | 25% |
| Niacin | 25% | Vitamin B$_6$ | 25% |
| Folate | 25% | Vitamin B$_{12}$ | 25% |
| Biotin | 25% | Pant. Acid | 25% |
| Phosphorus | 30% | Iodine | 25% |
| Magnesium | 25% | Zinc | 25% |
| Copper | 25% | | |

* Percent Daily Values are based on a 2,000 calorie diet. Your daily values may be higher or lower depending on your calorie needs:

One key feature of this labeling is the column of % Daily Value figures. Most of these values are based on a 2000-calorie diet. Critics complain that this diet applies to only a small segment of the population and should be more versatile. However, FDA and USDA officials responded that these are based on 2000 calories as a guideline only to help consumers gauge the relative amount of a nutrient contained by a food product. For instance, a food with 10 grams of saturated fat per serving could be mistaken for a food low in saturated fat. However, in a 2000-calorie diet, 10 grams of saturated fat represents 50% of the allowable daily intake. This percentage signals to consumers that this product is relatively high in saturated fat. Similarly, a food with 125 mg of sodium per serving could be mistaken for a high-sodium food. However, because a person should ingest no more than 2400 mg of sodium per day, a corresponding % Daily Value figure of about 5% conveys to a consumer that a serving of food with 125 mg of sodium is relatively low in sodium.

The % Daily Value figures that depend on the number of calories consumed per day include total fat, saturated fat, carbohydrate, protein, and dietary fiber. The % Daily Value figures for cholesterol, sodium, and potassium do not depend on calories.

For diets that include more or less than 2000 calories, the daily allowable amount of a nutrient that depends on calories can be figured with the following guidelines:

- The daily allowable amount of total fat is based on 30% of calories.
- The daily allowable amount of saturated fat is based on 10% of calories.
- The daily allowable amount of carbohydrate is based on 60% of calories.
- The daily allowable amount of protein is based on 10% of calories.
- The daily allowable amount of dietary fiber is based on 11.5 grams of fiber per 1000 calories.

Additionally, each gram of protein and carbohydrate contains 4 calories. A gram of fat contains 9 calories.

The above information can be used to calculate % Daily Value figures for diets with other calorie levels. For instance, the daily allowable amount of total fat in a 2200-calorie diet is $0.30(2200) = 660$ calories from fat. For the nutrition label shown, one serving contains 6 grams of fat or $6(9) = 54$ calories from fat. The % Daily Value for total fat that one serving of this food provides in a 2200-calorie diet is $54 \div 660 \approx 0.08$ or 8%.

1. Calculate the daily allowable amounts of total fat, saturated fat, carbohydrate, protein, and dietary fiber in 1500-calorie, 1800-calorie, 2500-calorie, and 2800-calorie diets. Summarize your results in a table.

2. Choose five different food products having Nutrition Facts labels. For each product, use the information given on the label to calculate the % Daily Value figures for total fat, saturated fat, carbohydrate, protein, and dietary fiber for (a) a 1500-calorie diet, (b) an 1800-calorie diet, (c) a 2500-calorie diet, and (d) a 2800-calorie diet. Create a chart showing your results for each food product.

3. Use the data given in the food product nutrition labels used in Question 2 to estimate the daily allowable amounts of cholesterol, sodium, and potassium. Recall that the allowable amounts for these nutrients do not depend on calorie intake.

## Chapter 2 Vocabulary Check

*Fill in each blank with one of the words or phrases listed below.*

contradiction          linear inequality in one variable          compound inequality          solution

absolute value        consecutive integers                        identity                            union

formula                 linear equation in one variable            intersection

1. The statement "$x < 5$ *or* $x > 7$" is called a(n) _____.

2. An equation in one variable that has no solution is called a(n) _____.

3. The _____ of two sets is the set of all elements common to both sets.

4. The _____ of two sets is the set of all elements that belong to either of the sets.

5. An equation in one variable that has every number (for which the equation is defined) as a solution is called a(n) _____.

6. The equation $d = rt$ is also called a(n) _____.

7. A number's distance from 0 is called its _____.

8. When a variable in an equation is replaced by a number and the resulting equation is true, then that number is called a(n) _____ of the equation.

9. The integers 17, 18, 19 are examples of _____.

10. The statement $5x - 0.2 < 7$ is an example of a(n) _____.

11. The statement $5x - 0.2 = 7$ is an example of a(n) _____.

**Helpful Hint**

▶ Are you preparing for your test? Don't forget to take the Chapter 2 Test on page 164. Then check your answers at the back of the text and use the Chapter Test Prep Videos to see the fully worked-out solutions to any of the exercises you want to review.

## 2  Chapter Highlights

| Definitions and Concepts | Examples |
|---|---|
| **Section 2.1    Linear Equations in One Variable** | |

| Definitions and Concepts | Examples |
|---|---|
| An **equation** is a statement that two expressions are equal. | $5 = 5 \qquad 7x + 2 = -14 \qquad 3(x - 1)^2 = 9x^2 - 6$ |
| A **linear equation in one variable** is an equation that can be written in the form $ax + b = c$, where $a, b,$ and $c$ are real numbers and $a \neq 0$. | $7x + 2 = -14 \qquad x = -3$ $5(2y - 7) = -2(8y - 1)$ |
| A **solution** of an equation is a value for the variable that makes the equation a true statement. | Determine whether $-1$ is a solution of $$3(x - 1) = 4x - 2$$ $$3(-1 - 1) \overset{?}{=} 4(-1) - 2$$ $$3(-2) \overset{?}{=} -4 - 2$$ $$-6 = -6 \qquad \text{True}$$ Thus, $-1$ is a solution. |
| **Equivalent equations** have the same solution. | $x - 12 = 14$ and $x = 26$ are equivalent equations. |
| The **addition property of equality** guarantees that the same number may be added to (or subtracted from) both sides of an equation and the result is an equivalent equation. | Solve: $\qquad -3x - 2 = 10$ $$-3x - 2 + 2 = 10 + 2 \quad \text{Add 2 to both sides.}$$ $$-3x = 12$$ |
| The **multiplication property of equality** guarantees that both sides of an equation may be multiplied by (or divided by) the same nonzero number and the result is an equivalent equation. | $$\frac{-3x}{-3} = \frac{12}{-3} \qquad \text{Divide both sides by } -3$$ $$x = -4$$ |

*(continued)*

| Definitions and Concepts | Examples |
|---|---|

### Section 2.1    Linear Equations in One Variable (*continued*)

**Solving a Linear Equation in One Variable**

Solve:  $x - \dfrac{x-2}{6} = \dfrac{x-7}{3} + \dfrac{2}{3}$

**Step 1:** Clear the equation of fractions and/or decimals.

**1.**  $6\left(x - \dfrac{x-2}{6}\right) = 6\left(\dfrac{x-7}{3} + \dfrac{2}{3}\right)$   Multiply both sides by 6.

$6x - (x - 2) = 2(x - 7) + 2(2)$   Use the distributive property.

**Step 2:** Remove grouping symbols such as parentheses.

**2.** $6x - x + 2 = 2x - 14 + 4$

**Step 3:** Simplify by combining like terms.

**3.**     $5x + 2 = 2x - 10$

**Step 4:** Write variable terms on one side and numbers on the other side by using the addition property of equality.

**4.** $5x + 2 - 2 = 2x - 10 - 2$   Subtract 2 from both sides.

$5x = 2x - 12$

$5x - 2x = 2x - 12 - 2x$   Subtract 2x from both sides.

$3x = -12$

**Step 5:** Get the variable alone by using the multiplication property of equality.

**5.**     $\dfrac{3x}{3} = \dfrac{-12}{3}$   Divide both sides by 3.

$x = -4$

**Step 6:** Check the proposed solution in the original equation.

**6.**  $-4 - \dfrac{-4-2}{6} \overset{?}{=} \dfrac{-4-7}{3} + \dfrac{2}{3}$   Replace x with −4 in the original equation.

$-4 - \dfrac{-6}{6} \overset{?}{=} \dfrac{-11}{3} + \dfrac{2}{3}$

$-4 - (-1) \overset{?}{=} \dfrac{-9}{3}$

$-3 = -3$   True

### Section 2.2    An Introduction to Problem Solving

**Problem-Solving Strategy**

Colorado is shaped like a rectangle whose length is about 1.3 times its width. If the perimeter of Colorado is 2070 kilometers, find its dimensions.

**1.** UNDERSTAND the problem.

**1.** Read and reread the problem. Guess a solution and check your guess.

Let $x$ = width of Colorado in kilometers. Then $1.3x$ = length of Colorado in kilometers.

**2.** TRANSLATE the problem.

**2.** In words:

| twice the length | + | twice the width | = | perimeter |
|---|---|---|---|---|

Translate: $2(1.3x) + 2x \quad = \quad 2070$

| Definitions and Concepts | Examples |
|---|---|

### Section 2.2　An Introduction to Problem Solving (*continued*)

| | |
|---|---|
| **3.** SOLVE the equation. | **3.** $2.6x + 2x = 2070$ <br> $\qquad 4.6x = 2070$ <br> $\qquad\quad x = 450$ |
| **4.** INTERPRET the results. | **4.** If $x = 450$ kilometers, then $1.3x = 1.3(450) = 585$ kilometers. **Check:** The perimeter of a rectangle whose width is 450 kilometers and length is 585 kilometers is $2(450) + 2(585) = 2070$ kilometers, the required perimeter. **State:** The dimensions of Colorado are 450 kilometers by 585 kilometers. |

### Section 2.3　Formulas and Problem Solving

| | |
|---|---|
| An equation that describes a known relationship among quantities is called a **formula**. | $A = \pi r^2$ ( area of a circle ) <br> $I = Prt$ ( interest $=$ principal $\cdot$ rate $\cdot$ time ) |
| **To solve a formula or equation for a specified variable,** use the steps for solving an equation. Treat the specified variable as the only variable of the equation. | Solve $A = 2HW + 2LW + 2LH$ for $H$. <br><br> $A - 2LW = 2HW + 2LH$　Subtract $2LW$ from both sides. <br> $A - 2LW = H(2W + 2L)$　Use the distributive property. <br> $\dfrac{A - 2LW}{2W + 2L} = \dfrac{H(2W + 2L)}{2W + 2L}$　Divide both sides by $2W + 2L$. <br> $\dfrac{A - 2LW}{2W + 2L} = H$　Simplify. |

### Section 2.4　Linear Inequalities and Problem Solving

| | |
|---|---|
| A **linear inequality in one variable** is an inequality that can be written in the form $ax + b < c$, where $a$, $b$, and $c$ are real numbers and $a \neq 0$. (The inequality symbols $>$, $\leq$, and $\geq$ also apply here.) | $5x - 2 \leq -7, \quad 3y > 1, \quad \dfrac{z}{7} < -9(z - 3)$ |
| The **addition property of inequality** guarantees that the same number may be added to (or subtracted from) both sides of an inequality, and the resulting inequality will have the same solution set. | Solve:　　$x - 9 \leq -16$ <br> $\qquad x - 9 + 9 \leq -16 + 9$　Add 9 to both sides. <br> $\qquad\qquad x \leq -7$ |
| The **multiplication property of inequality** guarantees that both sides of an inequality may be multiplied by (or divided by) the same **positive** number, and the resulting inequality will have the same solution set. We may also multiply (or divide) both sides of an inequality by the same **negative** number and **reverse the direction of the inequality symbol,** and the result will be an inequality with the same solution set. | Solve:　$6x < -66$　Divide both sides by 6. Do not reverse the direction of the inequality symbol. <br> $\qquad \dfrac{6x}{6} < \dfrac{-66}{6}$ <br> $\qquad x < -11$ <br><br> Solve:　$-6x < -66$　Divide both sides by $-6$. Reverse the direction of the inequality symbol. <br> $\qquad \dfrac{-6x}{-6} > \dfrac{-66}{-6}$ <br> $\qquad x > 11$ |

*(continued)*

| Definitions and Concepts | Examples |
|---|---|

### Section 2.4   Linear Inequalities and Problem Solving (*continued*)

**Solving a Linear Inequality in One Variable**

**Step 1:** Clear the equation of fractions and/or decimals.

**Step 2:** Remove grouping symbols such as parentheses.

**Step 3:** Simplify by combining like terms.

**Step 4:** Write variable terms on one side and numbers on the other side using the addition property of inequality.

**Step 5:** Get the variable alone using the multiplication property of inequality.

Solve: $\dfrac{3}{7}(x - 4) \geq x + 2$

**1.** $7\left[\dfrac{3}{7}(x - 4)\right] \geq 7(x + 2)$   Multiply both sides by 7.

$3(x - 4) \geq 7(x + 2)$

**2.** $\qquad 3x - 12 \geq 7x + 14$   Use the distributive property.

**4.** $\qquad -4x - 12 \geq 14$   Subtract $7x$ from both sides.

$\qquad\quad -4x \geq 26$   Add 12 to both sides.

$\qquad \dfrac{-4x}{-4} \leq \dfrac{26}{-4}$   Divide both sides by $-4$. Reverse the direction of the inequality symbol.

$\qquad\quad x \leq -\dfrac{13}{2}$

### Section 2.5   Sets and Compound Inequalities

Two inequalities joined by the words **and** or **or** are called **compound inequalities.**

The solution set of a compound inequality formed by the word **and** is the **intersection** ∩ of the solution sets of the two inequalities.

The solution set of a compound inequality formed by the word **or** is the **union** ∪ of the solution sets of the two inequalities.

$x - 7 \leq 4 \ and \ x \geq -21$

$2x + 7 > x - 3 \ or \ 5x + 2 > -3$

Solve:

$x < 5 \ and \ x < 3$

$\{x \mid x < 5\}$ $(-\infty, 5)$

$\{x \mid x < 3\}$ $(-\infty, 3)$

$\{x \mid x < 3$
$and \ x < 5\}$ $(-\infty, 3)$

Solve:

$x - 2 \geq -3 \quad or \quad 2x \leq -4$

$x \geq -1 \quad or \quad x \leq -2$

$\{x \mid x \geq -1\}$ $[-1, \infty)$

$\{x \mid x \leq -2\}$ $(-\infty, -2]$

$\{x \mid \leq -2$
$or \ x \geq -1\}$ $(-\infty, -2]$
$\cup [-1, \infty)$

| Definitions and Concepts | Examples |
|---|---|

| Section 2.6 | Absolute Value Equations |
|---|---|

| | |
|---|---|
| If $a$ is a positive number, then $\lvert X \rvert = a$ is equivalent to $X = a$ or $X = -a$. | Solve: $\lvert 5y - 1 \rvert - 7 = 4$<br><br>$\qquad\quad \lvert 5y - 1 \rvert = 11$  Add 7 to both sides.<br><br>$5y - 1 = 11$ or $5y - 1 = -11$<br><br>$5y = 12$ or $\qquad 5y = -10$  Add 1 to both sides.<br><br>$y = \dfrac{12}{5}$ or $\qquad 5y = -2$  Divide both sides by 5.<br><br>The solution set is $\left\{ -2, \dfrac{12}{5} \right\}$. |
| If $a$ is negative, then $\lvert X \rvert = a$ has no solution. | Solve: $\left\lvert \dfrac{x}{2} - 7 \right\rvert = -1$<br><br>The solution set is $\{\ \}$, or $\varnothing$. |
| To solve $\lvert X \rvert = \lvert Y \rvert$, solve $X = Y$ or $X = -Y$. | Solve: $\lvert x - 7 \rvert = \lvert 2x + 1 \rvert$<br><br>$x - 7 = 2x + 1$ or $x - 7 = -(2x + 1)$<br><br>$x = 2x + 8$ or $x - 7 = -2x - 1$<br><br>$-x = 8$ or $\qquad x = -2x + 6$<br><br>$x = -8$ or $\qquad 3x = 6$<br><br>$\qquad\qquad\qquad\qquad\qquad x = 2$<br><br>The solution set is $\{-8, 2\}$. |

| Section 2.7 | Absolute Value Inequalities |
|---|---|

| | |
|---|---|
| If $a$ is a positive number, then to solve $\lvert X \rvert < a$, solve $-a < X < a$. | Solve: $\lvert y - 5 \rvert \le 3$<br><br>$-3 \le y - 5 \le 3$<br><br>$-3 + 5 \le y - 5 + 5 \le 3 + 5$  Add 5 to all three parts.<br><br>$2 \le y \le 8$<br><br>The solution set is $[2, 8]$. |
| If $a$ is a positive number, then to solve $\lvert X \rvert > a$, solve $X < -a$ or $X > a$. | Solve: $\left\lvert \dfrac{x}{2} - 3 \right\rvert > 7$<br><br>$\dfrac{x}{2} - 3 < -7$ or $\dfrac{x}{2} - 3 > 7$<br><br>$x - 6 < -14$ or $x - 6 > 14$  Multiply both sides by 2.<br><br>$x < -8$ or $\qquad x > 20$  Add 6 to both sides.<br><br>The solution set is $(-\infty, -8) \cup (20, \infty)$. |

## Chapter 2    Review

**(2.1)** *Solve each linear equation.*

**1.** $4(x - 5) = 2x - 14$

**2.** $x + 7 = -2(x + 8)$

**3.** $3(2y - 1) = -8(6 + y)$

**4.** $-(z + 12) = 5(2z - 1)$

**5.** $n - (8 + 4n) = 2(3n - 4)$

**6.** $4(9v + 2) = 6(1 + 6v) - 10$

**7.** $0.3(x - 2) = 1.2$

**8.** $1.5 = 0.2(c - 0.3)$

**9.** $-4(2 - 3x) = 2(3x - 4) + 6x$

**10.** $6(m - 1) + 3(2 - m) = 0$

**11.** $6 - 3(2g + 4) - 4g = 5(1 - 2g)$

**12.** $20 - 5(p + 1) + 3p = -(2p - 15)$

**13.** $\dfrac{x}{3} - 4 = x - 2$

**14.** $\dfrac{9}{4}y = \dfrac{2}{3}y$

**15.** $\dfrac{3n}{8} - 1 = 3 + \dfrac{n}{6}$

**16.** $\dfrac{z}{6} + 1 = \dfrac{z}{2} + 2$

**17.** $\dfrac{y}{4} - \dfrac{y}{2} = -8$

**18.** $\dfrac{2x}{3} - \dfrac{8}{3} = x$

**19.** $\dfrac{b - 2}{3} = \dfrac{b + 2}{5}$

**20.** $\dfrac{2t - 1}{3} = \dfrac{3t + 2}{15}$

**21.** $\dfrac{2(t + 1)}{3} = \dfrac{2(t - 1)}{3}$

**22.** $\dfrac{3a - 3}{6} = \dfrac{4a + 1}{15} + 2$

**(2.2)** *Solve.*

**23.** Twice the difference of a number and 3 is the same as 1 added to three times the number. Find the number.

**24.** One number is 5 more than another number. If the sum of the numbers is 285, find the numbers.

**25.** Find 40% of 130.

**26.** Find 1.5% of 8.

**27.** In 2012, the median earnings of young adults with a bachelor's degree were $44,800. This represents a 26% increase over the median earnings of young adults with an associate's degree in 2012. Find the median earnings of young adults with associate degrees in 2012. Round to the nearest whole dollar. (*Source: National Center for Educational Statistics*)

**28.** Find four consecutive integers such that twice the first subtracted from the sum of the other three integers is 16.

**29.** Determine whether there are two consecutive odd integers such that 5 times the first exceeds 3 times the second by 54.

△ **30.** The length of a rectangular playing field is 5 meters less than twice its width. If 230 meters of fencing goes around the field, find the dimensions of the field.

**31.** A car rental company charges $19.95 per day for a compact car plus 12 cents per mile for every mile over 100 miles driven per day. If Mr. Woo's bill for 2 days' use is $46.86, find how many miles he drove.

**32.** The cost $C$ of producing $x$ number of scientific calculators is given by $C = 4.50x + 3000$ and the revenue $R$ from selling them is given by $R = 16.50x$. Find the number of calculators that must be sold to break even. (Recall that to break even, revenue = cost.)

**(2.3)** *Solve each equation for the specified variable.*

△ **33.** $V = LWH$ for $W$    △ **34.** $C = 2\pi r$ for $r$    **35.** $5x - 4y = -12$ for $y$    **36.** $5x - 4y = -12$ for $x$

**37.** $y - y_1 = m(x - x_1)$ for $m$    **38.** $y - y_1 = m(x - x_1)$ for $x$    **39.** $E = I(R + r)$ for $r$

**40.** $S = vt + gt^2$ for $g$    **41.** $T = gr + gvt$ for $g$    **42.** $I = Prt + P$ for $P$

**43.** A principal of $3000 is invested in an account paying an annual percentage rate of 3%. Find the amount (to the nearest cent) in the account after 7 years if the amount is compounded
  **a.** semiannually.
  **b.** weekly.

**44.** The high temperature in Slidell, Louisiana, one day was 90° Fahrenheit. Convert this temperature to degrees Celsius.

△ **45.** Angie Applegate has a photograph for which the length is 2 inches longer than the width. If she increases each dimension by 4 inches, the area is increased by 88 square inches. Find the original dimensions.

△ **46.** One-square-foot floor tiles come 24 to a package. Find how many packages are needed to cover a rectangular floor 18 feet by 21 feet.

**(2.4)** *Solve each linear inequality. Write your answers in interval notation.*

**47.** $3(x - 5) > -(x + 3)$

**48.** $-2(x + 7) \geq 3(x + 2)$

**49.** $4x - (5 + 2x) < 3x - 1$

**50.** $3(x - 8) < 7x + 2(5 - x)$

**51.** $24 \geq 6x - 2(3x - 5) + 2x$

**52.** $\dfrac{x}{3} + \dfrac{1}{2} > \dfrac{2}{3}$

**53.** $x + \dfrac{3}{4} < -\dfrac{x}{2} + \dfrac{9}{4}$

**54.** $\dfrac{x - 5}{2} \leq \dfrac{3}{8}(2x + 6)$

*Solve.*

**55.** George Boros can pay his housekeeper $15 per week to do his laundry, or he can have the laundromat do it at a cost of 50 cents per pound for the first 10 pounds and 40 cents for each additional pound. Use an inequality to find the weight at which it is more economical to use the housekeeper than the laundromat.

**56.** In the Olympic gymnastics competition, Nana must average a score of 9.65 to win the silver medal. Seven of the eight judges have reported scores of 9.5, 9.7, 9.9, 9.7, 9.7, 9.6, and 9.5. Use an inequality to find the minimum score that Nana must receive from the last judge to win the silver medal.

**(2.5)** *Solve each inequality. Write your answers in interval notation.*

**57.** $1 \leq 4x - 7 \leq 3$

**58.** $-2 \leq 8 + 5x < -1$

**59.** $-3 < 4(2x - 1) < 12$

**60.** $-6 < x - (3 - 4x) < -3$

**61.** $\dfrac{1}{6} < \dfrac{4x - 3}{3} \leq \dfrac{4}{5}$

**62.** $x \leq 2 \text{ and } x > -5$

**63.** $3x - 5 > 6 \text{ or } -x < -5$

**64.** $-2x \geq 4 \text{ or } 2x \geq 4$

**65.** Ceramic firing temperatures usually range from 500° to 1000° Fahrenheit. Use a compound inequality to convert this range to the Celsius scale. Round to the nearest degree.

**66.** Carol would like to pay cash for a car when she graduates from college and estimates that she can afford a car that costs between $4000 and $8000. She has saved $500 so far and plans to earn the rest of the money by working the next two summers. If Carol plans to save the same amount each summer, use a compound inequality to find the range of money she must save each summer to buy the car.

**(2.6)** *Solve each absolute value equation.*

**67.** $|2x + 9| = 9$

**68.** $|-3x + 4| = 7$

**69.** $|3x - 2| + 6 = 10$

**70.** $5 + |6x + 1| = 5$

**71.** $-5 = |4x - 3|$

**72.** $|5 - 6x| + 8 = 3$

**73.** $\left|\dfrac{3x - 7}{4}\right| = 2$

**74.** $\left|\dfrac{2x - 5}{3}\right| = 1$

**75.** $|6x + 1| = |15 + 4x|$

**76.** $|3x - 7| = |5x - 7|$

**(2.7)** *Solve each absolute value inequality. Graph the solution set and write it in interval notation.*

**77.** $|5x - 1| < 9$

**78.** $|6 + 4x| \geq 10$

**79.** $|3x| - 8 > 1$

**80.** $9 + |5x| < 24$

**81.** $|6x - 5| \leq -1$

**82.** $\left|3x + \dfrac{2}{5}\right| \geq 4$

**83.** $\left|\dfrac{x}{3} + 6\right| - 8 > -5$

**84.** $\left|\dfrac{4(x - 1)}{7}\right| + 10 < 2$

## Mixed Review

*Solve. If an inequality, write your answer in interval notation.*

**85.** $\dfrac{x - 2}{5} + \dfrac{x + 2}{2} = \dfrac{x + 4}{3}$

**86.** $\dfrac{2z - 3}{4} - \dfrac{4 - z}{2} = \dfrac{z + 1}{3}$

**87.** China, the United States, and France are predicted to be the top tourist destinations by 2020. In this year, the United States is predicted to have 9 million more tourists than France, and China is predicted to have 44 million more tourists than France. If the total number of tourists predicted for these three countries is 332 million, find the number predicted for each country in 2020.

△ **88.** $A = \dfrac{h}{2}(B + b)$ for $B$

△ **89.** $V = \dfrac{1}{3}\pi r^2 h$ for $h$

△ **90.** Determine which container holds more ice cream, an 8 inch $\times$ 5 inch $\times$ 3 inch box or a cylinder with radius of 3 inches and height of 6 inches.

**91.** Erasmos Gonzalez left Los Angeles at 11 A.M. and drove nonstop to San Diego, 130 miles away. If he arrived at 1:15 P.M., find his average speed, rounded to the nearest mile per hour.

**92.** $48 + x \geq 5(2x + 4) - 2x$

**93.** $\dfrac{3(x - 2)}{5} > \dfrac{-5(x - 2)}{3}$

**94.** $0 \leq \dfrac{2(3x + 4)}{5} \leq 3$

**95.** $x \leq 2 \ or \ x > -5$

**96.** $-2x \leq 6 \ and \ -2x + 3 < -7$

**97.** $|7x| - 26 = -5$

**98.** $\left|\dfrac{9 - 2x}{5}\right| = -3$

**99.** $|x - 3| = |7 + 2x|$

**100.** $|6x - 5| \geq -1$

**101.** $\left|\dfrac{4x - 3}{5}\right| < 1$

**Answers**

*Solve each equation.*

**1.** $8x + 14 = 5x + 44$

**2.** $9(x + 2) = 5[11 - 2(2 - x) + 3]$

**3.** $3(y - 4) + y = 2(6 + 2y)$

**4.** $7n - 6 + n = 2(4n - 3)$

1. _____

2. _____

**5.** $\dfrac{7w}{4} + 5 = \dfrac{3w}{10} + 1$

**6.** $\dfrac{z + 7}{9} + 1 = \dfrac{2z + 1}{6}$

3. _____

4. _____

**7.** $|6x - 5| - 3 = -2$

**8.** $|8 - 2t| = -6$

5. _____

6. _____

**9.** $|2x - 3| = |4x + 5|$

**10.** $|x - 5| = |x + 2|$

7. _____

8. _____

*Solve each equation for the specified variable.*

9. _____

**11.** $3x - 4y = 8$ for $y$

**12.** $S = gt^2 + gvt$ for $g$

10. _____

11. _____

**13.** $F = \dfrac{9}{5}C + 32$ for $C$

12. _____

13. _____

14. _____

*Solve each inequality. Write your answers in interval notation.*

15. _____

**14.** $3(2x - 7) - 4x > -(x + 6)$

**15.** $\dfrac{3x - 2}{3} - \dfrac{5x + 1}{4} \geq 0$

16. _____

17. _____

**16.** $-3 < 2(x - 3) \leq 4$

**17.** $|3x + 1| > 5$

**18.** $|x - 5| - 4 < -2$

**19.** $x \geq 5 \; and \; x \geq 4$

**20.** $x \geq 5 \; or \; x \geq 4$

**21.** $-1 \leq \dfrac{2x - 5}{3} < 2$

**22.** $6x + 1 > 5x + 4 \; or \; 1 - x > -4$

**23.** Find 12% of 80.

*Solve.*

**24.** In 2012, Ford sold 5,668,000 new vehicles worldwide. This represents a 12.12% increase over the number of new vehicles sold by Ford in 2011. Use this information to find the number of new vehicles sold by Ford in 2011. Round to the nearest thousand. (*Source:* Ford Motor Company)

△ **25.** A circular dog pen has a circumference of 78.5 feet. Approximate $\pi$ by 3.14 and estimate how many hunting dogs could be safely kept in the pen if each dog needs at least 60 square feet of room.

**26.** In 2020, the number of people employed as interpreters or translators in the United States is expected to be 83,000. This represents a 42% increase over the number of people employed as interpreters or translators in 2010. Find the number of interpreters or translators employed in 2010. Round to the nearest thousand. (*Source:* U.S. Bureau of Labor Statistics)

**27.** Find the amount of money in an account after 10 years if a principal of $2500 is invested at 3.5% interest compounded quarterly. (Round to the nearest cent.)

**28.** The three states where international travelers spend the most money are Florida, California, and New York. International travelers spent $4 billion more money in California than New York, and in Florida they spent $1 billion less than twice the amount spent in New York. If total international spending in these three states is $39 billion, find the amount that is spent in each state. (*Source:* Travel Industry Association of America)

**18.** _____

**19.** _____

**20.** _____

**21.** _____

**22.** _____

**23.** _____

**24.** _____

**25.** _____

**26.** _____

**27.** _____

**28.** _____

**Answers**

1. _____

2. a. _____

   b. _____

3. _____

4. _____

5. _____

6. a. _____

   b. _____

   c. _____

7. _____

8. _____

9. _____

10. _____

11. _____

12. _____

13. _____

14. _____

15. _____

16. _____

17. _____

18. _____

19. _____

20. _____

21. _____

22. _____

23. _____

24. _____

**1.** Write the set $\{x \mid x$ is a natural number greater than $100\}$ in roster form.

**2.** List the elements in each set.
   **a.** $\{x \mid x$ is an integer between $-3$ and $5\}$
   **b.** $\{x \mid x$ is a whole number between $3$ and $5\}$

*Write each sentence as an equation.*

**3.** The sum of $x$ and 5 is 20.

**4.** If 19 is subtracted from $x$, the difference is the product of 4 and $x$.

**5.** The quotient of $z$ and 9 amounts to 9 plus $z$.

**6.** Insert $<$, $>$, or $=$ between each pair of numbers to form a true statement.
   **a.** $-3$    $-5$
   **b.** $\dfrac{-12}{-4}$    $3$
   **c.** $0$    $-2$

*Find the opposite, or additive inverse, of each number.*

**7.** $8$

**8.** $0$

**9.** $-\dfrac{1}{5}$

**10.** $-7.3$

*Add.*

**11.** $-3 + (-11)$

**12.** $-20.2 + 7.8$

**13.** $-10 + 15$

**14.** $-\dfrac{1}{2} + \dfrac{7}{2}$

**15.** $-\dfrac{2}{3} + \dfrac{3}{7}$

**16.** Subtract: $1.7 - 8.9$

*Divide.*

**17.** $\dfrac{20}{-4}$

**18.** $\dfrac{30}{0}$

**19.** $\dfrac{0}{-8}$

**20.** $-\dfrac{3}{4} \div \dfrac{9}{4}$

**21.** $\dfrac{-10}{-80}$

**22.** Multiply: $-\dfrac{3}{4}\left(-\dfrac{4}{7}\right)$

**23.** Simplify: $3 + 2 \cdot 30$

**24.** Simplify: $\dfrac{\sqrt{9 + 40} - 2^2}{20 \div 2 \cdot 2}$

**166**

*Simplify by combining like terms.*

**25.** $3x - 5x + 4$

**26.** $-3(4 + y) + \dfrac{1}{3}(6 - 12y)$

**27.** $y + 3y$

**28.** $6 + 2(7x - 1) + 4$

*Use the quotient rule to simplify. Write each answer using positive exponents only.*

**29.** $\dfrac{x^7}{x^4}$

**30.** $\dfrac{a^{-2} \cdot b^{-8}}{b^{-7}}$

**31.** $\dfrac{20x^6}{4x^5}$

*Simplify each expression. Write each answer using positive exponents only.*

**32.** $\dfrac{(-3)^9 z^7 \cdot z^2}{(-3)^{11} z^{-11}}$

**33.** $\left(\dfrac{3x^2 y}{y^{-9} z}\right)^{-2}$

**34.** $(-4a^{-3}b^{-4})(3a^{-2}b)^2$

**35.** $\left(\dfrac{3a^2}{2x^{-1}}\right)^3 \left(\dfrac{x^{-3}}{4a^{-2}}\right)^{-1}$

*Solve.*

**36.** $11.2 = 1.2 - 5x$

**37.** $2x + 5 = 9$

**38.** $2x + 1.5 = -0.2 + 1.6x$

**39.** $6x - 4 = 2 + 6(x - 1)$

**40.** Write the following as algebraic expressions. Then simplify.
   **a.** The sum of three consecutive integers if $x$ is the first consecutive integers.
   **b.** The perimeter of a square with side length $3x + 1$.

**41.** Suppose that a computer store just announced an 8% decrease in the price of a particular computer model. If this computer sells for $2162 after the decrease, find its original price.

**42.** Find two numbers such that the second number is 2 more than three times the first number and the difference of the two numbers is 24.

**43.** Solve $V = lwh$ for $h$.

**44.** Solve $7x - 4y = 10$ for $x$.

**45.** Solve $\dfrac{1}{4}x \le \dfrac{3}{2}$. Graph the solution set.

*Solve.*

**46.** $2(7x - 1) - 5x > -(-7x) + 4$

**47.** $-2x - 5 < -3 \text{ or } 6x < 0$

**48.** $4(x + 1) - 3 < 4x + 1$

**49.** $|p| = 2$

**50.** $|x + 3| = |7 - x|$

25. _____
26. _____
27. _____
28. _____
29. _____
30. _____
31. _____
32. _____
33. _____
34. _____
35. _____
36. _____
37. _____
38. _____
39. _____
40. a. _____
    b. _____
41. _____
42. _____
43. _____
44. _____
45. _____
46. _____
47. _____
48. _____
49. _____
50. _____

# Graphs and Functions

The linear equations we explored in Chapter 2 are statements about a single variable. This chapter examines statements about two variables: linear equations and inequalities in two variables. We focus particularly on graphs of these equations and inequalities, which lead to the notion of relation and to the notion of function, perhaps the single most important and useful concept in all of mathematics.

**London, England was the first city to host the Olympics three times.**

## Which Has More Participants, Summer or Winter Olympic Games?

The Summer Olympic Games, which were first held in 1896, are an international competition held every four years. The Summer Olympic Games are much larger than the Winter Olympic Games because there are many team sports that require multiple participants. London, England, the host of the 2012 Summer Olympic Games, was the first city to host the Olympics three times. In Section 3.1, Exercise 29, and Section 3.6, Exercise 13, we will explore the number of gold medals won by the United States in the 2012 and previous Summer Olympics.

**Gold Medals at 2012 Summer Olympics**

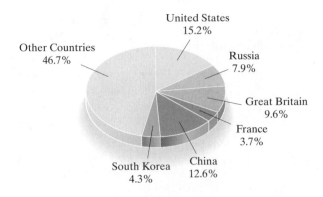

# Graphing Linear Equations

Graphs are widely used today in newspapers, magazines, and all forms of newsletters as a quick way to display data. A few examples of graphs are shown here.

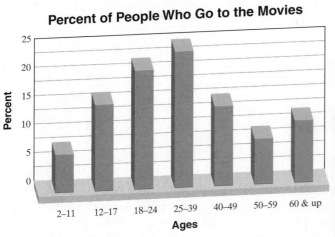

**Percent of People Who Go to the Movies**

*Source:* Motion Picture Association of America

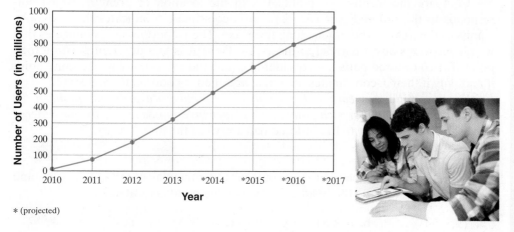

**Predicted Tablet Users Worldwide**

∗ (projected)

To help us understand how to read these graphs, we will review their basis—the rectangular coordinate system.

## Objective A  Plotting Ordered Pairs on a Rectangular Coordinate System

One way to locate points on a plane is by using a **rectangular coordinate system,** which is also called a **Cartesian coordinate system** after its inventor, René Descartes (1596–1650). A rectangular coordinate system consists of two number lines that intersect at right angles at their 0 coordinates. We position these axes on paper such that one number line is horizontal and the other number line is then vertical. The horizontal number line is called the *x*-**axis** (or the axis of the **abscissa**), and the vertical number line is called the *y*-**axis** (or the axis of the **ordinate**). The point of intersection of these axes is named the **origin.**

Notice that the axes divide the plane into four regions. These regions are called **quadrants.** The top-right region is quadrant I. Quadrants II, III, and IV are numbered counterclockwise from the first quadrant as shown on the next page. The *x*-axis and the *y*-axis are not in any quadrant.

Each point in the plane can be located, or **plotted,** by describing its position in terms of distances along each axis from the origin. An **ordered pair,** represented by the notation $(x, y)$, records these distances.

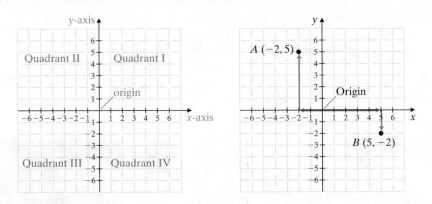

For example, the location of point $A$ in the figure on the right above is described as 2 units to the left of the origin along the $x$-axis and 5 units upward parallel to the $y$-axis. Thus, we identify point $A$ with the ordered pair $(-2, 5)$. Notice that the *order* of these numbers is *critical*. The $x$-value $-2$ is called the **$x$-coordinate** and is associated with the $x$-axis. The $y$-value 5 is called the **$y$-coordinate** and is associated with the $y$-axis.

Compare the location of point $A$ with the location of point $B$, which corresponds to the ordered pair $(5, -2)$. The $x$-coordinate 5 indicates that we move 5 units to the right of the origin along the $x$-axis. The $y$-coordinate $-2$ indicates that we move 2 units down parallel to the $y$-axis. Point $A$ is in a different position than point $B$. Two ordered pairs are considered equal and correspond to the same point if and only if their $x$-coordinates are equal and their $y$-coordinates are equal.

Keep in mind that *each ordered pair corresponds to exactly one point in the coordinate plane and that each point in the plane corresponds to exactly one ordered pair.* Thus, we may refer to the ordered pair $(x, y)$ as the **point** $(x, y)$.

### Practice 1

Plot each ordered pair on a rectangular coordinate system and name the quadrant in which the point is located.

**a.** $(3, -2)$    **b.** $(3.5, 4.5)$

**c.** $(0, 3)$    **d.** $(-4, 1)$

**e.** $(-1, 0)$    **f.** $\left(-2\frac{1}{2}, -3\right)$

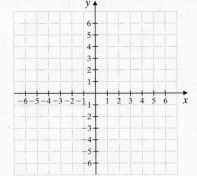

**Example 1**    Plot each ordered pair on a rectangular coordinate system and name the quadrant in which the point is located.

**a.** $(2, -1)$    **b.** $(0, 5)$    **c.** $(-3, 5)$

**d.** $(-2, 0)$    **e.** $\left(-\frac{1}{2}, -4\right)$    **f.** $(1.5, 1.5)$

**Solution:**  The six points are graphed as shown in the figure.

**a.** $(2, -1)$ lies in quadrant IV.

**b.** $(0, 5)$ is not in any quadrant.

**c.** $(-3, 5)$ lies in quadrant II.

**d.** $(-2, 0)$ is not in any quadrant.

**e.** $\left(-\frac{1}{2}, -4\right)$ is in quadrant III.

**f.** $(1.5, 1.5)$ is in quadrant I.

■ **Work Practice 1**

**Answers**

**1.**

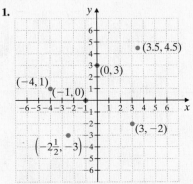

**a.** quadrant IV  **b.** quadrant I

**c.** not in any quadrant  **d.** quadrant II

**e.** not in any quadrant  **f.** quadrant III

Notice that the $y$-coordinate of any point on the $x$-axis is 0. For example, the point with coordinates $(-2, 0)$ lies on the $x$-axis. Also, the $x$-coordinate of any point on the $y$-axis is 0. For example, the point with coordinates $(0, 5)$ lies on the $y$-axis. A point on an axis is called a **quadrantal** point.

✓**Concept Check** Which of the following correctly describes the location of the point $(3, -6)$ in a rectangular coordinate system?

**a.** 3 units to the left of the $y$-axis and 6 units above the $x$-axis

**b.** 3 units above the $x$-axis and 6 units to the left of the $y$-axis

**c.** 3 units to the right of the $y$-axis and 6 units below the $x$-axis

**d.** 3 units below the $x$-axis and 6 units to the right of the $y$-axis

Many types of real-world data occur in pairs. For example, the data pairs below are for Walt Disney World ticket prices in effect at the beginning of the year for the years shown. The graph of paired data, such as the one below, is called a **scatter diagram.** Such diagrams are used to look for patterns and relationships in paired data.

| Paired Data | |
|---|---|
| **Year, $x$** | **Price (in dollars), $y$** |
| 0 | 59.75 |
| 1 | 63 |
| 2 | 67 |
| 3 | 71 |
| 4 | 75 |
| 5 | 79 |
| 6 | 85 |
| 7 | 89 |
| 8 | 95 |

*Note:* The notation 0 ↔ 2005 below the graph means that year 0 corresponds to the year 2005, 1 corresponds to 2006, and so on.

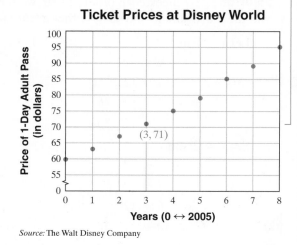

**Ticket Prices at Disney World**

*Source:* The Walt Disney Company

> **Helpful Hint**   Notice, for example, the paired data (3, 71) and its corresponding plotted point, both in blue. Also notice that the symbol ⸕ on the vertical axis is used to note that numbers have been skipped on the axis.

**Practice 2**

Create a scatter diagram for the given paired data.

| Regular Season Completed Passes Made by Drew Brees of the NFL New Orleans Saints | |
|---|---|
| **Year** | **Number of Completed Passes** |
| 2010 | 448 |
| 2011 | 468 |
| 2012 | 422 |
| 2013 | 446 |

*Source:* National Football League

**Completed Passes by Drew Brees**

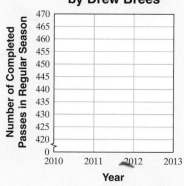

---

**Example 2**  Create a scatter diagram for the given paired data.

| Paired Data | |
|---|---|
| **Year** | **Number of Subscribers (in millions)** |
| 2008 | 40.4 |
| 2009 | 42.6 |
| 2010 | 44.7 |
| 2011 | 46 |
| 2012 | 46.8 |

*Source:* National Cable & Telecommunications Association

**Solution:**  To graph the paired data in the table, we use the first column for the $x$- (or horizontal) axis and the second column for the $y$- (or vertical) axis.

**Digital Video Subscribers**

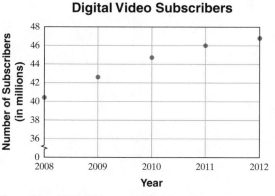

*Source:* National Cable & Telecommunications Association

**Answer**

**2.**  **Completed Passes by Drew Brees**

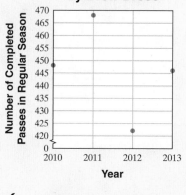

✓**Concept Check Answer**

c

■ **Work Practice 2**

## Objective B Determining Whether an Ordered Pair of Numbers Is a Solution of an Equation

A **solution** of an equation in two variables consists of two numbers that make the equation true. These two numbers can be written as an ordered pair of numbers. Unless we are told otherwise, we will assume that variable values are written as ordered pairs in alphabetical order (that is, $x$ first and then $y$).

**Practice 3**

Determine whether $(0, -6)$, $(1, 4)$, and $(-1, -4)$ are solutions of the equation $2x + y = -6$.

**Example 3** Determine whether $(0, -12)$, $(1, 9)$, and $(2, -6)$ are solutions of the equation $3x - y = 12$.

**Solution:** To check each ordered pair, we replace $x$ with the $x$-coordinate and $y$ with the $y$-coordinate and see whether a true statement results.

Let $x = 0$ and $y = -12$.
$$3x - y = 12$$
$$3(0) - (-12) \stackrel{?}{=} 12$$
$$0 + 12 \stackrel{?}{=} 12$$
$$12 = 12 \quad \text{True}$$

Let $x = 1$ and $y = 9$.
$$3x - y = 12$$
$$3(1) - 9 \stackrel{?}{=} 12$$
$$3 - 9 \stackrel{?}{=} 12$$
$$-6 = 12 \quad \text{False}$$

Let $x = 2$ and $y = -6$.
$$3x - y = 12$$
$$3(2) - (-6) \stackrel{?}{=} 12$$
$$6 + 6 \stackrel{?}{=} 12$$
$$12 = 12 \quad \text{True}$$

We see that $(1, 9)$ is not a solution but both $(0, -12)$ and $(2, -6)$ are solutions.

▶ Work Practice 3

## Objective C Graphing Linear Equations

As we saw in Example 3, some linear equations have more than one ordered pair solution. In fact, the equation $3x - y = 12$ has an infinite number of ordered pair solutions. Since it is impossible to list all solutions, we visualize them by graphing them.

A few more ordered pairs that satisfy $3x - y = 12$ are $(4, 0)$, $(3, -3)$, $(5, 3)$, and $(1, -9)$. These ordered pair solutions, along with the ordered pair solutions from Example 3, are plotted on the following graph. The graph of $3x - y = 12$ is the single line containing these points. Every ordered pair solution of the equation corresponds to a point on this line, and every point on this line corresponds to an ordered pair solution.

| $x$ | $y$ | $3x - y = 12$ |
|---|---|---|
| 5 | 3 | $3 \cdot 5 - 3 = 12$ |
| 4 | 0 | $3 \cdot 4 - 0 = 12$ |
| 3 | -3 | $3 \cdot 3 - (-3) = 12$ |
| 2 | -6 | $3 \cdot 2 - (-6) = 12$ |
| 1 | -9 | $3 \cdot 1 - (-9) = 12$ |
| 0 | -12 | $3 \cdot 0 - (-12) = 12$ |

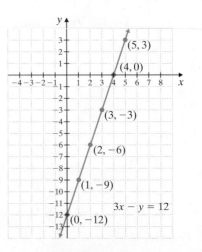

**Answer**

3. yes; no; yes

The equation $3x - y = 12$ is called a *linear equation in two variables*, and *the graph of every linear equation in two variables is a line.*

---

**Linear Equation in Two Variables**

A **linear equation in two variables** is an equation that can be written in the form

$$Ax + By = C$$

where $A$, $B$, and $C$ are real numbers, and $A$ and $B$ are not both 0. This form is called **standard form** and the graph of a linear equation in two variables is a line.

---

Some examples of linear equations in standard form are:

$$3x - y = 12$$
$$-2.1x + 5.6y = 0$$

**Helpful Hint**

Remember that in a linear equation in standard form, all of the variable terms are on one side of the equation and the constant is on the other side.

Recall from geometry that a line is determined by two points. This means that to graph a linear equation in two variables, just two solutions are needed. We will find a third solution, just to check our work. To find ordered pair solutions of linear equations in two variables, we can choose an $x$-value and find its corresponding $y$-value, or we can choose a $y$-value and find its corresponding $x$-value. The number 0 is often a convenient value to choose for $x$ and also for $y$.

**Example 4** Graph: $y = -2x + 3$

**Solution:** This is a linear equation. (In standard form it is $2x + y = 3$.) Find three ordered pair solutions, and plot the ordered pairs. The line through the plotted points is the graph. Since the equation is solved for $y$, let's choose three $x$-values. Let's let $x$ be $0, 2$, and then $-1$ to find our three ordered pair solutions.

| Let $x = 0$ | Let $x = 2$ | Let $x = -1$ |
|---|---|---|
| $y = -2x + 3$ | $y = -2x + 3$ | $y = -2x + 3$ |
| $y = -2 \cdot 0 + 3$ | $y = -2 \cdot 2 + 3$ | $y = -2(-1) + 3$ |
| $y = 3$ Simplify. | $y = -1$ Simplify. | $y = 5$ Simplify. |

The three ordered pairs $(0, 3)$, $(2, -1)$, and $(-1, 5)$ are listed in the table and the graph is shown.

| $x$ | $y$ |
|---|---|
| 0 | 3 |
| 2 | -1 |
| -1 | 5 |

**Work Practice 4**

**Practice 4**

Graph: $y = 4x - 3$

**Answer**

**4.**

**Helpful Hint**

Since the equation $y = -2x + 3$ is solved for $y$, we choose $x$-values for finding points. This way, we simply need to evaluate an expression to find the $y$-value, as shown.

Notice that the graph crosses the $x$-axis at the point $\left(\frac{3}{2}, 0\right)$. This point is called the **$x$-intercept.** (You may sometimes see just the number $\frac{3}{2}$ called the $x$-intercept.) This graph also crosses the $y$-axis at the point $(0, 3)$. This point is called the **$y$-intercept.** (You may also see just the number 3 called the $y$-intercept.)

Since every point on the $x$-axis has a $y$-value of 0, we can find the $x$-intercept of a graph by letting $y = 0$ and solving for $x$. Also, every point on the $y$-axis has an $x$-value of 0. To find the $y$-intercept, we let $x = 0$ and solve for $y$.

### Finding x- and y-Intercepts

To find an $x$-intercept, let $y = 0$ and solve for $x$.
To find a $y$-intercept, let $x = 0$ and solve for $y$.

We will study intercepts further in Section 3.3.

One way to graph a linear equation in two variables is to find $x$- and $y$-intercepts. This provides at most two points, so make sure a third solution point is found to check.

**Practice 5**

Find the intercepts and graph:
$2x + 3y = -6$

**Answer**

**5.**

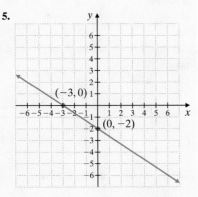

**Example 5**  Find the intercepts and graph: $3x + 4y = -12$

**Solution:** To find the $y$-intercept, we let $x = 0$ and solve for $y$. To find the $x$-intercept, we let $y = 0$ and solve for $x$. Let's let $x = 0, y = 0$, and then let $x = 2$ to find our third point.

Let $x = 0$.
$$3x + 4y = -12$$
$$3 \cdot 0 + 4y = -12$$
$$4y = -12$$
$$y = -3$$

$(0, -3)$

Let $y = 0$.
$$3x + 4y = -12$$
$$3x + 4 \cdot 0 = -12$$
$$3x = -12$$
$$x = -4$$

$(-4, 0)$

Let $x = 2$.
$$3x + 4y = -12$$
$$3 \cdot 2 + 4y = -12$$
$$6 + 4y = -12$$
$$4y = -18$$
$$y = -\frac{18}{4} = -4\frac{1}{2}$$

$\left(2, -4\frac{1}{2}\right)$

The ordered pairs are $(0, -3)$, $(-4, 0)$, and $\left(2, -4\frac{1}{2}\right)$. We plot these points to obtain the graph shown.

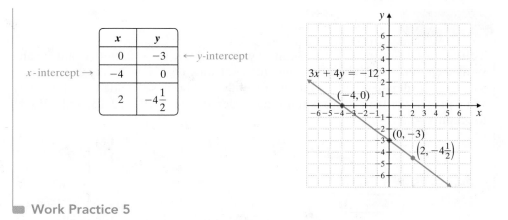

| x | y | |
|---|---|---|
| 0 | $-3$ | ← $y$-intercept |
| $-4$ | 0 | |
| 2 | $-4\frac{1}{2}$ | |

$x$-intercept →

■ Work Practice 5

---

**Example 6** Graph: $y = \frac{1}{3}x$

**Solution:** To graph, we find ordered pair solutions, plot the ordered pairs, and draw a line through the plotted points. We will choose $x$-values and substitute in the equation. To avoid fractions, we choose $x$-values that are multiples of 3. To find the intercepts, we will also let $x = 0$ and solve for $y$, then let $y = 0$ and solve for $x$.

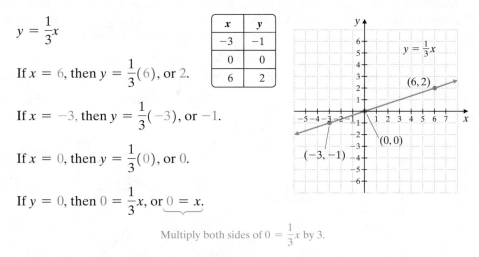

$y = \frac{1}{3}x$

If $x = 6$, then $y = \frac{1}{3}(6)$, or 2.

If $x = -3$, then $y = \frac{1}{3}(-3)$, or $-1$.

If $x = 0$, then $y = \frac{1}{3}(0)$, or 0.

If $y = 0$, then $0 = \frac{1}{3}x$, or $0 = x$.

Multiply both sides of $0 = \frac{1}{3}x$ by 3.

| x | y |
|---|---|
| $-3$ | $-1$ |
| 0 | 0 |
| 6 | 2 |

This graph crosses the $x$-axis at $(0, 0)$ and the $y$-axis at $(0, 0)$. This means that the $x$-intercept is $(0, 0)$ and that the $y$-intercept is $(0, 0)$. This happens when the graph passes through the origin.

■ Work Practice 6

---

## Objective D  Graphing Vertical and Horizontal Lines ▶

The equation $x = c$, where $c$ is a real number constant, is a linear equation in two variables because it can be written in the form $x + 0y = c$. The graph of this equation is a vertical line, as shown in the next example.

---

**Practice 6**

Graph: $y = \frac{1}{4}x$

**Helpful Hint**  Notice that by using multiples of 3 for $x$, we avoid fractions.

**Answer**

6.

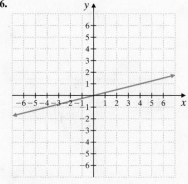

## Practice 7

Graph: $x = -1$

## Example 7    Graph: $x = 2$

**Solution:** The equation $x = 2$ can be written as $x + 0y = 2$. Notice that for any $y$-value chosen, $x$ is 2. No other value for $x$ satisfies $x + 0y = 2$. Any ordered pair whose $x$-coordinate is 2 is a solution to $x + 0y = 2$ because 2 added to 0 times any value of $y$ is $2 + 0$, or 2. We will use the ordered pairs $(2, 3)$, $(2, 0)$, and $(2, -3)$ to graph $x = 2$.

|   | $x$ | $y$ |
|---|-----|-----|
|   | 2 | 3 |
| $x$-intercept → | 2 | 0 |
|   | 2 | -3 |

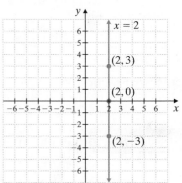

The graph is a vertical line with $x$-intercept $(2, 0)$. It has no $y$-intercept because $x$ is never 0.

■ Work Practice 7

## Practice 8

Graph: $y = 2$

## Example 8    Graph: $y = -3$

**Solution:** The equation $y = -3$ can be written as $0x + y = -3$. For any $x$-value chosen, $y$ is $-3$. If we choose 4, 0, and $-2$ as $x$-values, the ordered pair solutions are $(4, -3)$, $(0, -3)$, and $(-2, -3)$. We will use these ordered pairs to graph $y = -3$.

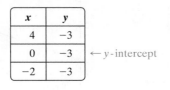

| $x$ | $y$ |
|-----|-----|
| 4 | -3 |
| 0 | -3 | ← $y$-intercept |
| -2 | -3 |

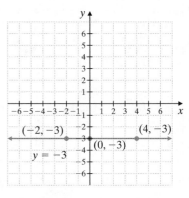

The graph is a horizontal line with $y$-intercept $(0, -3)$. It has no $x$-intercept because $y$ is never 0.

■ Work Practice 8

From Examples 7 and 8, we have the following generalization.

**Answers**

**7.**

**8.**

## Graphing Vertical and Horizontal Lines

The graph of $x = c$, where $c$ is a real number, is a vertical line with $x$-intercept $(c, 0)$.

The graph of $y = c$, where $c$ is a real number, is a horizontal line with $y$-intercept $(0, c)$.

## Calculator Explorations  Graphing

In this section, we begin a study of graphing calculators and graphing software packages for computers.

These graphers use the same point-plotting technique that we introduced in this section. The advantage of this graphing technology is, of course, that graphing calculators and computers can find and plot ordered pair solutions much faster than we can. Note, however, that the features described in these boxes may not be available on all graphing calculators.

The rectangular screen where a portion of the rectangular coordinate system is displayed is called a **window.** We call it a **standard window** for graphing when both the $x$- and $y$-axes display coordinates between $-10$ and $10$. This information is often displayed in the window menu on a graphing calculator as

Xmin = $-10$

Xmax = 10

Xscl = 1      The scale on the $x$-axis is one unit per tick mark.

Ymin = $-10$

Ymax = 10

Yscl = 1      The scale on the $y$-axis is one unit per tick mark.

To use a graphing calculator to graph the equation $y = -5x + 4$, press the $\boxed{Y =}$ key and enter the keystrokes

$\boxed{(-)}$ $\boxed{5}$ $\boxed{X}$ $\boxed{+}$ $\boxed{4}$

↑
Check your owner's manual to make sure the "negative" key is pressed here and not the "subtraction" key.

The top row should now read $Y_1 = -5x + 4$. Next press the $\boxed{GRAPH}$ key, and the display should look like this:

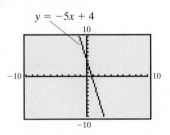

*Use a standard window and graph each linear equation. (Unless otherwise stated, we will use a standard window when graphing.)*

**1.** $y = 6x - 1$

**2.** $y = 3x - 2$

**3.** $y = -3.2x + 7.9$

**4.** $y = -x + 5.85$

**5.** $y = \dfrac{1}{4}x - \dfrac{2}{3}$ (Parentheses may need to be inserted around $\dfrac{1}{4}$.)

**6.** $y = \dfrac{2}{3}x - \dfrac{1}{5}$ (Parentheses may need to be inserted around $\dfrac{2}{3}$.)

## Vocabulary, Readiness & Video Check

*Use the choices below to fill in each blank. Some choices may not be used.*

| | | | |
|---|---|---|---|
| line | vertical | 1 | 3 |
| origin | horizontal | 2 | 4 |

**1.** The intersection of the $x$-axis and $y$-axis is a point called the _____.

**2.** The rectangular coordinate system has _____ quadrants and _____ axes.

**3.** The graph of a single ordered pair of numbers is how many points? _____

4. The graph of $Ax + By = C$, where $A$ and $B$ are not both 0, is a(n) _____.

5. The graph of $y = c$, where $c$ is a real number, is a(n) _____ line.

6. The graph of $x = c$, where $c$ is a real number, is a(n) _____ line.

*Martin-Gay Interactive Videos*    Watch the section lecture video and answer the following questions.

*See Video 3.1*

**Objective A**    7. Several points are plotted in ⊟ Examples 1–3. Using the method of these examples, where do you start to plot a point? How does the 1st coordinate tell you to move? How does the 2nd coordinate tell you to move? Include the role of signs in your answer. ▶

**Objective B**    8. Based on ⊟ Examples 4 and 5, complete the following statement. An ordered pair is a solution of an equation in _____ variables if, when the variables are replaced with their ordered pair values, a _____ statement results. ▶

**Objective C**    9. From ⊟ Example 6 and the lecture before, what is the graph of an equation? If you need only two points to determine a line, why are three ordered pair solutions or points found for a linear equation? ▶

**Objective D**    10. From ⊟ Examples 9 and 10, what can you say about the coefficient of $x$ when the equation of a horizontal line is written in the form $Ax + By = C$? The coefficient of $y$ of a vertical line? ▶

**3.1    Exercise Set** MyMathLab® ▶

**Objective A**    *Determine the coordinates of each point on the graph. See Example 1.*

1. Point $A$

2. Point $B$

3. Point $C$

4. Point $D$

5. Point $E$

6. Point $F$

7. Point $G$

8. Point $H$

*Plot each ordered pair on the given rectangular coordinate system and name the quadrant (or axis) in which the point lies. See Example 1.*

▶9. $(3, 2)$

$(-5, 3)$

$\left(5\frac{1}{2}, -4\right)$

$(0, 3.5)$

$(-2, -4)$

10. $(2, -1)$

$(-3, -1)$

$\left(-2, 6\frac{1}{3}\right)$

$(-2, 4)$

$(-4.2, 0)$

*Without graphing, try to visualize the location of each point. Then give the location of each point by quadrant or x- or y-axis. (Graph the point first, if needed.)*

**11.** $(2, 3)$ | **12.** $(0, 5)$ | **13.** $(-2, 7)$

**14.** $(-3, 0)$ | **15.** $(-1, -4)$ | **16.** $(4, -2)$

**17.** $(0, -100)$ | **18.** $(10, 30)$ | **19.** $(-10, -30)$

**20.** $(0, 0)$ | **21.** $(-87, 0)$ | **22.** $(-42, 17)$

*Given that x is a positive number and y is a positive number, determine the quadrant (or axis) in which each point lies. See Example 1.*

**23.** $(x, -y)$ | **24.** $(-x, y)$ | **25.** $(x, 0)$

**26.** $(0, -y)$ | **27.** $(-x, -y)$ | **28.** $(0, 0)$

*Create a scatter diagram for the given paired data. See Example 2.*

**29.** U.S. Gold Medals in Consecutive Summer Olympics

| Year of Summer Olympics | Number of Gold Medals |
|---|---|
| 1996 | 44 |
| 2000 | 37 |
| 2004 | 36 |
| 2008 | 36 |
| 2012 | 46 |

(*Source:* Int. Olympic Committee)

**30.** Postal Rates for 1-Ounce First Class Letter

| Year | Postal Rate (in cents) |
|---|---|
| 2008 | 42 |
| 2009 | 44 |
| 2010 | 44 |
| 2011 | 44 |
| 2012 | 45 |
| 2013 | 46 |

(*Source:* United States Postal Service)

**U.S. Gold Medals in Summer Olympics**

*Source:* International Olympic Committee

**U.S. Postal Rates for Letters**

*Source:* United State Postal Service

**Objective B** *Determine whether each ordered pair is a solution of the given equation. See Example 3.*

**31.** $y = 3x - 5$; $(0, 5), (-1, -8)$

**32.** $y = -2x + 7$; $(1, 5), (-2, 3)$

**33.** $-6x + 5y = -6$; $(1, 0), \left(2, \dfrac{6}{5}\right)$

**34.** $5x - 3y = 9$; $(0, 3), \left(\dfrac{12}{5}, -1\right)$

**35.** $y = -3$; $(1, -3), (-3, 6)$

**36.** $y = 2$; $(2, 5), (0, 2)$

## Objectives C D Mixed Practice *Graph each linear equation. See Examples 4 through 8.*

**37.** $x - 2y = 4$

**38.** $y - 2x = 4$

**39.** $3x + 2y = 6$

**40.** $2x + 4y = 8$

**41.** $x = 4$

**42.** $y = 5$

**43.** $x - 3y = 6$

**44.** $x - 4y = 8$

**45.** $y = 3x$

**46.** $y = -4x$

**47.** $y = -2$

**48.** $x = -3$

**49.** $4x + 5y = 15$

**50.** $2x + 3y = 9$

**51.** $5y = x - 10$

**52.** $3y = x - 3$

**53.** $x = \dfrac{1}{2}$ **54.** $y = -\dfrac{5}{2}$ **55.** $y = 0.5x$ **56.** $x = 0.5y$

**57.** $y = -4x + 1$ **58.** $y = -3x + 1$ **59.** $2y - 6 = 0$ **60.** $3x + 6 = 0$

**61.** $y = -\dfrac{2}{3}x - 4$ **62.** $y = -\dfrac{3}{2}x + 6$ **63.** $5x - 2y = -10$ **64.** $5x - 3y = -15$

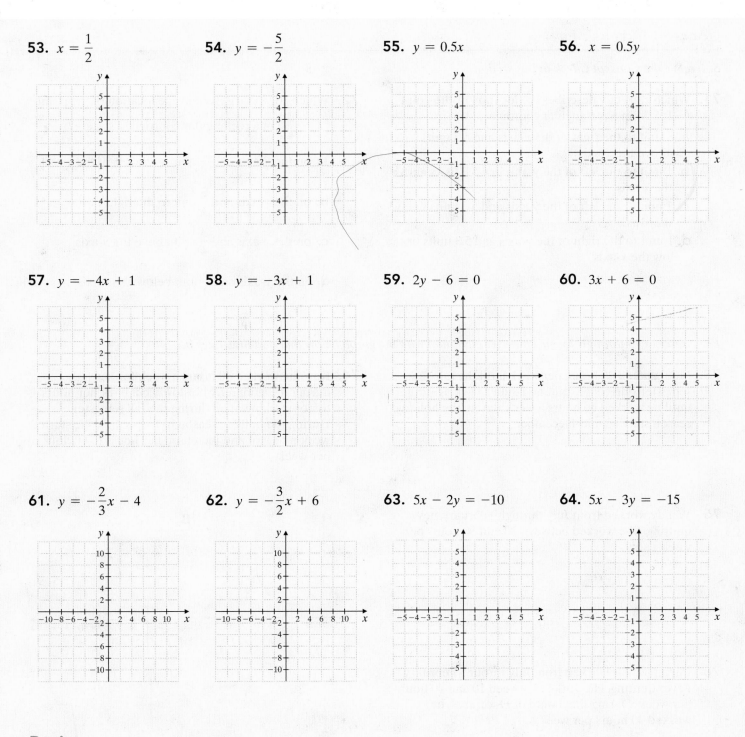

## Review

*Simplify. See Section 1.5.*

**65.** $\dfrac{-6 - 3}{2 - 8}$ **66.** $\dfrac{4 - 5}{-1 - 0}$ **67.** $\dfrac{-8 - (-2)}{-3 - (-2)}$

**68.** $\dfrac{12 - 3}{10 - 9}$ **69.** $\dfrac{0 - 6}{5 - 0}$ **70.** $\dfrac{2 - 2}{3 - 5}$

## Concept Extensions

*Solve. See the Concept Check in this section.*

**71.** Which correctly describes the location of the point $(-1, 5.3)$ in a rectangular coordinate system?

   **a.** 1 unit to the right of the $y$-axis and 5.3 units above the $x$-axis

   **b.** 1 unit to the left of the $y$-axis and 5.3 units above the $x$-axis

   **c.** 1 unit to the left of the $y$-axis and 5.3 units below the $x$-axis

   **d.** 1 unit to the right of the $y$-axis and 5.3 units below the $x$-axis

**72.** Which correctly describes the location of the point $\left(0, -\dfrac{3}{4}\right)$ in a rectangular coordinate system?

   **a.** on the $x$-axis and $\dfrac{3}{4}$ unit to the left of the $y$-axis

   **b.** on the $x$-axis and $\dfrac{3}{4}$ unit to the right of the $y$-axis

   **c.** on the $y$-axis and $\dfrac{3}{4}$ unit above the $x$-axis

   **d.** on the $y$-axis and $\dfrac{3}{4}$ unit below the $x$-axis

*For Exercises 73 through 76, match each description with the graph below that best illustrates it.*

**73.** Moe worked 40 hours per week until the fall semester started. He quit and didn't work again until he worked 60 hours a week during the holiday season starting mid-December.

**74.** Kawana worked 40 hours a week for her father during the summer. She slowly cut back her hours to not working at all during the fall semester. During the holiday season in December, she started working again and increased her hours to 60 hours per week.

**75.** Wendy worked from July through February, never quitting. She worked between 10 and 30 hours per week.

**76.** Bartholomew worked from July through February, never quitting. He worked between 10 and 40 hours per week. During the month of December, he worked 40 hours per week.

*This broken-line graph shows the hourly minimum wage and the years it increased. Use this graph for Exercises 77 through 80.*

**The Federal Hourly Minimum Wage**

*Source:* U.S. Department of Labor

**77.** What was the first year that the minimum hourly wage rose above $5.00?

**78.** What was the first year that the minimum hourly wage rose above $6.00?

**79.** Why do you think that this graph is shaped the way it is?

**80.** The federal hourly minimum wage started in 1938 at $0.25. How much had it increased by 2014?

*Solve.*

△ **81.** The perimeter $y$ of a rectangle whose width is a constant 3 inches and whose length is $x$ inches is given by the equation

$$y = 2x + 6$$

x inches

3 inches

**a.** Draw a graph of this equation in the first quadrant only.

**b.** Read from the graph the perimeter $y$ of a rectangle whose length $x$ is 4 inches.

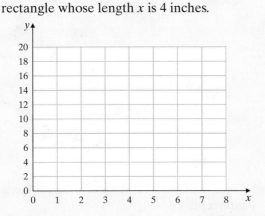

**82.** The distance $y$ traveled in a train moving at a constant speed of 50 miles per hour is given by the equation

$$y = 50x$$

where $x$ is the time in hours traveled.

**a.** Draw a graph of this equation in the first quadrant only.

**b.** Read from the graph the distance $y$ traveled after 6 hours.

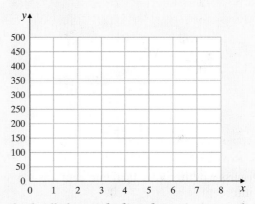

*For income tax purposes, Jason Verges, owner of Copy Services, uses a method called **straight-line depreciation** to show the loss in value of a copy machine he recently purchased. Jason assumes that he can use the machine for 7 years. The following graph shows the value of the machine over the years. Use this graph to answer the following questions.*

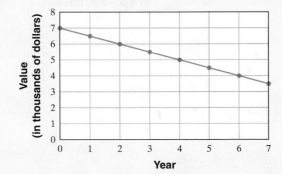

**83.** What was the purchase price of the copy machine?

**84.** What is the depreciated value of the machine in 7 years?

**85.** What loss in value occurred during the first year?

**86.** What loss in value occurred during the second year?

**87.** Why do you think that this method of depreciating is called straight-line depreciation?

**88.** Why is the line tilted downward?

**89.** Broyhill Furniture found that it takes 2 hours to manufacture each table for one of its special dining room sets. Each chair takes 3 hours to manufacture. A total of 1500 hours is available to produce tables and chairs of this style. The linear equation that models this situation is $2x + 3y = 1500$, where $x$ represents the number of tables produced and $y$ the number of chairs produced.

  **a.** Complete the ordered pair solution $(0, \quad)$ of this equation. Describe the manufacturing situation to which this solution corresponds.

  **b.** Complete the ordered pair solution $(\quad, 0)$ for this equation. Describe the manufacturing situation to which this solution corresponds.

  **c.** If 50 tables are produced, find the greatest number of chairs the company can make.

**90.** While manufacturing two different camera models, Kodak found that the basic model costs $55 to produce, whereas the deluxe model costs $75. The weekly budget for those two models is limited to $33,000 in production costs. The linear equation that models this situation is $55x + 75y = 33,000$, where $x$ represents the number of basic models and $y$ the number of deluxe models.

  **a.** Complete the ordered pair solution $(0, \quad)$ of this equation. Describe the manufacturing situation to which this solution corresponds.

  **b.** Complete the ordered pair solution $(\quad, 0)$ of this equation. Describe the manufacturing situation to which this solution corresponds.

  **c.** If 350 deluxe models are produced, find the greatest number of basic models that can be made in one week.

**91.** On the same set of axes, graph $y = 2x$, $y = 2x - 5$, and $y = 2x + 5$. What patterns do you see in these graphs?

**92.** Explain why we generally use three points to graph a line, when only two points are needed.

**93.** Discuss whether a vertical line ever has a $y$-intercept.

**94.** Discuss whether a horizontal line ever has an $x$-intercept.

*Use a grapher to verify the graph of each exercise.*

**95.** Exercise 45

**96.** Exercise 46

**97.** Exercise 57

**98.** Exercise 58

## 3.2 The Slope of a Line ▶

You may have noticed by now that different lines often tilt differently. It is very important in many fields to be able to measure and compare the tilt, or **slope,** of lines. For example, a wheelchair ramp with a slope of $\frac{1}{12}$ means that the ramp rises 1 foot for every 12 horizontal feet. A road with a slope or grade of 8% $\left(\text{or } \frac{8}{100}\right)$ means that the road rises 8 feet for every 100 horizontal feet.

**Objectives**

**A** Find the Slope of a Line Given Two Points on the Line. ▶

**B** Find the Slope of a Line Given the Equation of the Line. ▶

**C** Find the Slopes of Horizontal and Vertical Lines. ▶

**D** Find the Slope of a Line Given the Graph of the Line. ▶

**E** Compare the Slopes of Parallel and Perpendicular Lines. ▶

We measure the slope of a line as a ratio of **vertical change** to **horizontal change.** Slope is usually designated by the letter *m*.

## Objective A Finding Slope from Two Points ▶

Suppose that we want to measure the slope of the following line.

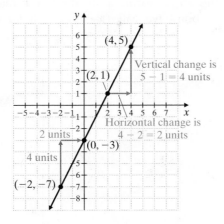

The vertical change between *both* pairs of points on the line is 4 units per horizontal change of 2 units. Then

$$\text{slope } m = \frac{\text{change in } y \text{ (vertical change)}}{\text{change in } x \text{ (horizontal change)}} = \frac{4}{2} = 2$$

Notice that slope is a **rate of change** between points. A slope of 2 or $\frac{2}{1}$ means that between pairs of points on the line, the rate of change is a vertical change of 2 units per horizontal change of 1 unit.

In general, consider the line that passes through the points $(x_1, y_1)$ and $(x_2, y_2)$. (The notation $x_1$ is read "*x*-sub-one.") The vertical change, or **rise,** between these points is the difference in the *y*-coordinates: $y_2 - y_1$. The horizontal change, or **run,** between the points is the difference of the *x*-coordinates: $x_2 - x_1$.

## Slope of a Line

Given a line passing through points $(x_1, y_1)$ and $(x_2, y_2)$ the **slope** $m$ of the line is

$$m = \frac{\text{rise}}{\text{run}} = \frac{y_2 - y_1}{x_2 - x_1}$$

as long as $x_2 \neq x_1$.

$y_2 - y_1$ = vertical change, or rise.

$x_2 - x_1$ = horizontal change, or run.

✔**Concept Check**  In the definition of slope, we state that $x_2 \neq x_1$. Explain why.

**Practice 1**

Find the slope of the line containing the points $(-1, -2)$ and $(2, 5)$. Graph the line.

**Example 1**  Find the slope of the line containing the points $(0, 3)$ and $(2, 5)$. Graph the line.

**Solution:**  We use the slope formula. It does not matter which point we call $(x_1, y_1)$ and which point we call $(x_2, y_2)$. We'll let $(x_1, y_1) = (0, 3)$ and $(x_2, y_2) = (2, 5)$.

$$m = \frac{y_2 - y_1}{x_2 - x_1}$$

$$= \frac{5 - 3}{2 - 0} = \frac{2}{2} = 1$$

$m = 1$

Notice in this example that the slope is *positive* and that the graph of the line containing $(0, 3)$ and $(2, 5)$ moves *upward*—that is, the $y$-values increase—as we go from left to right.

▧ Work Practice 1

**Answer**

**1.** $m = \dfrac{7}{3}$,

✔**Concept Check Answer**

so that the denominator is never 0

### Helpful Hint

The slope of a line is the same no matter which 2 points of a line you choose to calculate slope. The line in Example 1 also contains the point $(-3, 0)$. Below, we calculate the slope of the line using $(0, 3)$ as $(x_1, y_1)$ and $(-3, 0)$ as $(x_2, y_2)$.

$$m = \frac{y_2 - y_1}{x_2 - x_1} = \frac{0 - 3}{-3 - 0} = \frac{-3}{-3} = 1 \quad \text{Same slope as found in Example 1.}$$

**Example 2** Find the slope of the line containing the points $(5, -4)$ and $(-3, 3)$. Graph the line.

**Solution:** We use the slope formula, and let $(x_1, y_1) = (5, -4)$ and $(x_2, y_2) = (-3, 3)$.

$$m = \frac{y_2 - y_1}{x_2 - x_1}$$

$$= \frac{3 - (-4)}{-3 - 5} = \frac{7}{-8} = -\frac{7}{8}$$

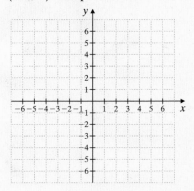

Notice in this example that the slope is negative and that the graph of the line through $(5, -4)$ and $(-3, 3)$ moves downward—that is, the $y$-values decrease—as we go from left to right.

■ **Work Practice 2**

### Helpful Hint

When we are trying to find the slope of a line through two given points, it makes no difference which given point is called $(x_1, y_1)$ and which is called $(x_2, y_2)$. Once an $x$-coordinate is called $x_1$, however, make sure its corresponding $y$-coordinate is called $y_1$.

✓**Concept Check** Find and correct the error in the following calculation of slope of the line containing the points $(12, 2)$ and $(4, 7)$.

$$m = \frac{12 - 4}{2 - 7} = \frac{8}{-5} = -\frac{8}{5}$$

## Objective B Finding Slope from an Equation

As we have seen, the slope of a line is defined by two points on the line. Thus, if we know the equation of a line, we can find its slope.

**Example 3** Find the slope of the line $y = 3x + 2$.

**Solution:** We must find two points on the line defined by $y = 3x + 2$ to find its slope. We will let $x = 0$ and then $x = 1$ to find the required points.

Let $x = 0$.                    Let $x = 1$.

$y = 3x + 2$                    $y = 3x + 2$

$y = 3 \cdot 0 + 2$                $y = 3 \cdot 1 + 2$

$y = 2$                        $y = 5$

Now we use the points $(0, 2)$ and $(1, 5)$ to find the slope. We'll let $(x_1, y_1)$ be $(0, 2)$ and $(x_2, y_2)$ be $(1, 5)$. Then

$$m = \frac{y_2 - y_1}{x_2 - x_1} = \frac{5 - 2}{1 - 0} = \frac{3}{1} = 3$$

■ **Work Practice 3**

---

**Practice 2**

Find the slope of the line containing the points $(1, -1)$ and $(-2, 4)$. Graph the line.

**Practice 3**

Find the slope of the line $y = 2x + 4$.

**Answers**

2. $m = -\dfrac{5}{3}$,

3. $m = 2$

✓**Concept Check Answer**

$$m = \frac{2 - 7}{12 - 4} = \frac{-5}{8} = -\frac{5}{8}$$

Analyzing the results of Example 3, you may notice a striking pattern:

The slope of $y = 3x + 2$ is 3, the same as the coefficient of $x$.
The $y$-coordinate of the $y$-intercept $(0, 2)$, is the same as the constant term.

We have just illustrated, not proved, an amazing pattern with linear equations in two variables. When a linear equation is written in the form $y = mx + b$, $m$ is the slope of the line and $(0, b)$ is its $y$-intercept. The form $y = mx + b$ is appropriately called the *slope-intercept form*.

> ### Slope-Intercept Form
>
> When a linear equation in two variables is written in **slope-intercept form,**
>
> $$\underset{\displaystyle\downarrow}{\text{slope}} \quad \underset{\displaystyle\downarrow}{\text{$y$-intercept is }(0, b)}$$
> $$y = mx + b$$
>
> then $m$ is the slope of the line and $(0, b)$ is the $y$-intercept of the line.

**Practice 4**

Find the slope and the $y$-intercept of the line $2x - 4y = 8$.

**Example 4** Find the slope and the $y$-intercept of the line $3x - 4y = 4$.

**Solution:** We write the equation in slope-intercept form by solving for $y$.

$$3x - 4y = 4$$
$$-4y = -3x + 4 \qquad \text{Subtract } 3x \text{ from both sides.}$$
$$\frac{-4y}{-4} = \frac{-3x}{-4} + \frac{4}{-4} \qquad \text{Divide both sides by } -4.$$
$$y = \frac{3}{4}x - 1 \qquad \text{Simplify.}$$

The coefficient of $x$, $\frac{3}{4}$, is the slope, and $(0, -1)$ is the $y$-intercept.

■ **Work Practice 4**

The graphs of $y = \dfrac{1}{2}x + 1$ and $y = 5x + 1$ are shown below. The graph of $y = \dfrac{1}{2}x + 1$ has a slope of $\dfrac{1}{2}$ and the graph of $y = 5x + 1$ has a slope of 5.

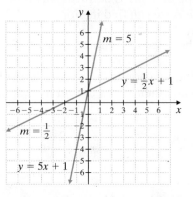

Notice that the line with the slope of 5 is steeper than the line with the slope of $\dfrac{1}{2}$. This is true in general for positive slopes.

> For a line with positive slope $m$, as $m$ increases the line becomes steeper.

**Answer**

4. slope: $\dfrac{1}{2}$; $y$-intercept: $(0, -2)$

## Objective C  Finding Slopes of Horizontal and Vertical Lines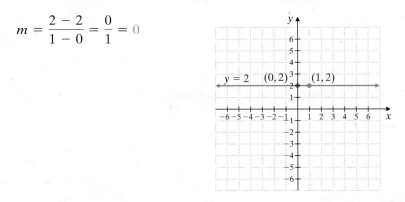

Next we find the slopes of two special types of lines: vertical lines and horizontal lines.

**Example 5**  Find the slope of the line $x = -5$.

**Solution:**   Recall that the graph of $x = -5$ is a vertical line with $x$-intercept $(-5, 0)$. To find the slope, we find two ordered pair solutions of $x = -5$. Of course, solutions of $x = -5$ must have an $x$-value of $-5$. We will let $(x_1, y_1) = (-5, 0)$ and $(x_2, y_2) = (-5, 4)$. Then

$$m = \frac{y_2 - y_1}{x_2 - x_1} = \frac{4 - 0}{-5 - (-5)} = \frac{4}{0}$$

Since $\dfrac{4}{0}$ is undefined, we say that the slope of the vertical line $x = -5$ is undefined.

■ Work Practice 5

**Practice 5**

Find the slope of the line $x = 3$.

**Example 6**  Find the slope of the line $y = 2$.

**Solution:**   Recall that the graph of $y = 2$ is a horizontal line with $y$-intercept $(0, 2)$. To find the slope, we find two points on the line, such as $(0, 2)$ and $(1, 2)$, and use these points to find the slope.

$$m = \frac{2 - 2}{1 - 0} = \frac{0}{1} = 0$$

The slope of the horizontal line $y = 2$ is $0$.

■ Work Practice 6

**Practice 6**

Find the slope of the line $y = -3$.

From the above examples, we have the following generalization.

### Slopes of Vertical and Horizontal Lines

The slope of any vertical line is undefined.
The slope of any horizontal line is 0.

**Answers**
**5.** undefined   **6.** 0

**Helpful Hint**

Slope of 0 and undefined slope are not the same. Vertical lines have undefined slope, whereas horizontal lines have slope of 0.

The following four graphs summarize the overall appearance of lines with positive, negative, zero, and undefined slopes.

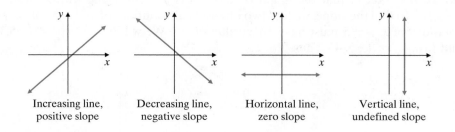

Increasing line,          Decreasing line,          Horizontal line,          Vertical line,
positive slope            negative slope            zero slope                undefined slope

## Objective D Finding the Slope of a Line Given the Graph of the Line

Now that we know the appearance of lines with positive, negative, zero, and undefined slopes, let's practice finding the slope of a line given its graph.

**Practice 7**

Find the slope of the line graphed.

**Example 7**    Find the slope of the line graphed.

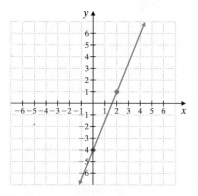

**Solution:**   The two points shown have coordinates $(2, 1)$ and $(0, -4)$. Thus,

$$m = \frac{-4 - 1}{0 - 2} = \frac{-5}{-2} = \frac{5}{2}$$

As a quick check, notice that the line goes upward from left to right, so its slope is positive.

The slope of the line is $\frac{5}{2}$.

▶ Work Practice 7

## Objective E Comparing Slopes of Parallel and Perpendicular Lines

Slopes of lines can help us determine whether lines are parallel. Parallel lines are distinct lines with the same steepness, so it follows that they have the same slope.

**Answer**

**7.** $m = \frac{3}{2}$

## Parallel Lines

Two nonvertical lines are parallel if they have the same slope and different *y*-intercepts.

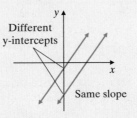

How do the slopes of perpendicular lines compare? (Two lines intersecting at right angles are called **perpendicular lines.**) Suppose that a line has a slope of $\frac{a}{b}$. If the line is rotated 90°, the rise and run are now switched, except that the run is now negative. This means that the new slope is $-\frac{b}{a}$. Notice that

$$\left(\frac{a}{b}\right) \cdot \left(-\frac{b}{a}\right) = -1$$

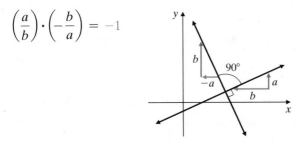

This is how we tell whether two lines are perpendicular.

## Perpendicular Lines

Two nonvertical lines are perpendicular if the product of their slopes is −1.

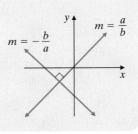

In other words, two nonvertical lines are perpendicular if the slope of one is the negative reciprocal of the slope of the other.

**Example 8**   Determine whether the two lines are parallel, perpendicular, or neither.

$$3x + 7y = 4$$
$$6x + 14y = 7$$

**Solution:**   We find the slope of each line by solving each equation for *y*.

$$3x + 7y = 4 \qquad\qquad 6x + 14y = 7$$
$$7y = -3x + 4 \qquad\qquad 14y = -6x + 7$$
$$\frac{7y}{7} = \frac{-3x}{7} + \frac{4}{7} \qquad\qquad \frac{14y}{14} = \frac{-6x}{14} + \frac{7}{14}$$
$$y = -\frac{3}{7}x + \frac{4}{7} \qquad\qquad y = -\frac{3}{7}x + \frac{1}{2}$$

slope   y-intercept $\left(0, \frac{4}{7}\right)$     slope   y-intercept $\left(0, \frac{1}{2}\right)$

*(Continued on next page)*

**Practice 8**

Determine whether the two lines are parallel, perpendicular, or neither.

$$2x + 5y = 1$$
$$4x + 10y = 3$$

**Answer**

**8.** parallel

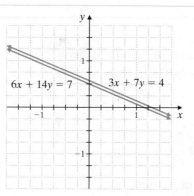

The slopes of both lines are $-\dfrac{3}{7}$. The $y$-intercepts are different. Therefore, the lines are parallel.

■ Work Practice 8

**Practice 9**

Determine whether the two lines are parallel, perpendicular, or neither.

$$x - 4y = 3$$
$$3x + 12y = 7$$

**Example 9** Determine whether the two lines are parallel, perpendicular, or neither.

$$-x + 3y = 2$$
$$2x + 6y = 5$$

**Solution:** When we solve each equation for $y$, we have

$$-x + 3y = 2 \qquad 2x + 6y = 5$$
$$3y = x + 2 \qquad 6y = -2x + 5$$
$$\frac{3y}{3} = \frac{x}{3} + \frac{2}{3} \qquad \frac{6y}{6} = \frac{-2x}{6} + \frac{5}{6}$$
$$y = \frac{1}{3}x + \frac{2}{3} \qquad y = -\frac{1}{3}x + \frac{5}{6}$$

slope   $y$-intercept $\left(0, \dfrac{2}{3}\right)$     slope   $y$-intercept $\left(0, \dfrac{5}{6}\right)$

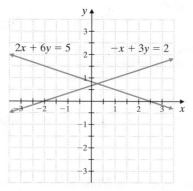

The slopes are not the same and their product is not $-1$ $\left[\left(\dfrac{1}{3}\right) \cdot \left(-\dfrac{1}{3}\right) = -\dfrac{1}{9}\right]$.

Therefore, the lines are neither parallel nor perpendicular.

■ Work Practice 9

**Answer**

9. neither parallel nor perpendicular

✓ **Concept Check Answer**

$y$-intercepts are different

✓ **Concept Check** What is *different* about the equations of two parallel lines in slope-intercept form?

### 🖩 Calculator Explorations   Graphing

It is possible to use a grapher to sketch the graph of more than one equation on the same set of axes. For example, let's graph the equations $y = 2x - 3$ and $y = 2x + 5$ on the same set of axes.

To graph on the same set of axes, press the $\boxed{Y =}$ key and enter the equations on the first two lines.

$Y_1 = 2x - 3$
$Y_2 = 2x + 5$

Then press the $\boxed{\text{GRAPH}}$ key as usual. The screen should look like this:

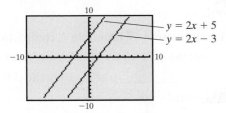

Notice the slopes and $y$-intercepts of the graphs. Since their slopes are the same and they have different $y$-intercepts, we have parallel lines, as shown.

*Graph each pair of equations on the same set of axes. Describe the similarities and differences in their graphs.*

**1.** $y = 3x, y = 3x + 4$

**2.** $y = 5x, y = 5x - 2$

**3.** $y = -\dfrac{2}{3}x + 1, y = -\dfrac{2}{3} + 6$

**4.** $y = -\dfrac{1}{4}x - 3, y = -\dfrac{1}{4}x + 6$

**5.** $y = 4.61x - 1.86, y = 4.61x + 2.11$

**6.** $y = 3.78x + 1.92, y = 3.78x + 8.08$

## Vocabulary, Readiness & Video Check

*Use the choices below to fill in each blank. Some choices may be used more than once and some not at all.*

| | | | | | |
|---|---|---|---|---|---|
| horizontal | the same | $-1$ | $y$-intercepts | $(0, b)$ | slope |
| vertical | different | $m$ | $x$-intercepts | $(b, 0)$ | slope-intercept |

**1.** The measure of the steepness or tilt of a line is called _____.

**2.** The slope of a line through two points is measured by the ratio of _____ change to _____ change.

**3.** If a linear equation is in the form $y = mx + b$, the slope of the line is _____ and the $y$-intercept is

_____.

**4.** The form $y = mx + b$ is called the _____ form.

**5.** The slope of a _____ line is 0.

**6.** The slope of a _____ line is undefined.

**7.** Two nonvertical perpendicular lines have slopes whose product is _____.

**8.** Two nonvertical lines are parallel if they have _____ slope and different _____.

*Decide whether a line with the given slope slants upward or downward from left to right, or is horizontal or vertical.*

**9.** $m = \dfrac{7}{6}$        **10.** $m = -3$        **11.** $m = 0$        **12.** $m$ is undefined

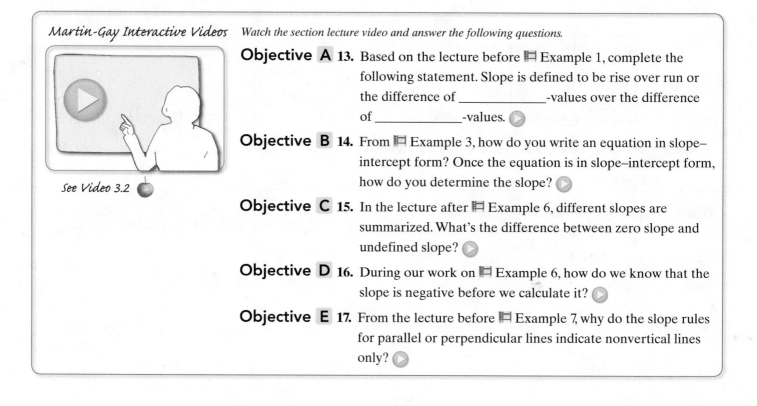

*Martin-Gay Interactive Videos*    *Watch the section lecture video and answer the following questions.*

**Objective A 13.** Based on the lecture before ▣ Example 1, complete the following statement. Slope is defined to be rise over run or the difference of _____-values over the difference of _____-values. ▶

**Objective B 14.** From ▣ Example 3, how do you write an equation in slope–intercept form? Once the equation is in slope–intercept form, how do you determine the slope? ▶

**Objective C 15.** In the lecture after ▣ Example 6, different slopes are summarized. What's the difference between zero slope and undefined slope? ▶

**Objective D 16.** During our work on ▣ Example 6, how do we know that the slope is negative before we calculate it? ▶

**Objective E 17.** From the lecture before ▣ Example 7, why do the slope rules for parallel or perpendicular lines indicate nonvertical lines only? ▶

*See Video 3.2*

**3.2  Exercise Set** MyMathLab® ▶

**Objectives A D  Mixed Practice** *Find the slope of the line containing each pair of points. See Examples 1 and 2.*

**1.** $(3, 2), (8, 11)$        **2.** $(1, 6), (7, 11)$        **3.** $(3, 1), (1, 8)$        **4.** $(2, 9), (6, 4)$

**5.** $(-2, 8), (4, 3)$        **6.** $(3, 7), (-2, 11)$        ▶**7.** $(-2, -6), (4, -4)$        **8.** $(-3, -4), (-1, 6)$

**9.** $(-3, -1), (-12, 11)$        **10.** $(3, -1), (-6, 5)$        **11.** $(-2, 5), (3, 5)$        **12.** $(4, 2), (4, 0)$

*Find the slope of the line graphed. See Example 7.*

**13.**

**14.**

**○15.**

**16.**

*Find each slope. See Examples 1, 2, and 7.*

**17.** Find the pitch, or slope, of the roof shown.

8 ft

12 ft

**18.** Upon takeoff, a Delta Airlines jet climbs to 3 miles as it passes over 25 miles of land below it. Find the slope of its climb.

3 mi

25 mi

**19.** Find the grade, or slope, of the road shown.

15 ft

100 ft

**20.** Driving down Bald Mountain in Wyoming, Bob Dean finds that he descends 1600 feet in elevation by the time he is 2.5 miles (horizontally) away from the high point on the mountain road. Find the slope of his descent. (*Hint:* 1 mile = 5280 feet.)

**Objectives** **B** **C** **Mixed Practice** *Find the slope and the y-intercept of each line. See Examples 3 through 6.*

**21.** $y = -x + 5$

**22.** $y = x + 2$

**○23.** $y = 5x - 2$

**24.** $y = -2x + 6$

**25.** $2x + y = 7$

**26.** $-5x + y = 10$

**○27.** $2x - 3y = 10$

**28.** $-3x - 4y = 6$

**○29.** $x = 4$

**30.** $x = 7.2$

**○31.** $y = -2$

**32.** $y = -3.6$

**33.** $y = \frac{1}{2}x$     **34.** $y = -\frac{1}{4}x$     **35.** $3y + 8 = x$     **36.** $2y - 7 = x$

**37.** $-6x + 5y = 30$     **38.** $4x - 7y = 28$     **39.** $y = 7x$     **40.** $y = \frac{1}{7}x$

**41.** $x + 2 = 0$     **42.** $y - 7 = 0$

*Match each equation with its graph.*

**43.** $y = 2x + 3$     **44.** $y = 2x - 3$     **45.** $y = -2x + 3$     **46.** $y = -2x - 3$

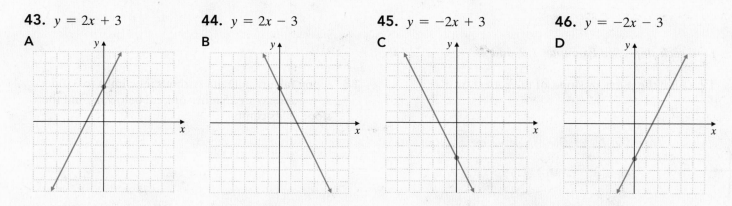

*Two lines are graphed on each set of axes. For each graph, determine whether $l_1$ or $l_2$ has the greater slope.*

**Objective E**  *Determine whether each pair of lines is parallel, perpendicular, or neither. See Examples 8 and 9.*

**53.** $y = 12x + 6$
$y = 12x - 2$

**54.** $y = 5x - 6$
$y = 5x + 2$

▶ **55.** $y = -9x + 3$
$y = \frac{3}{2}x - 7$

**56.** $y = 2x - 12$
$y = \frac{1}{2}x - 6$

**57.** $2x - y = -10$
$2x + 4y = 2$

**58.** $-2x + 3y = 1$
$3x + 2y = 12$

**59.** $y = -3x + 6$
$y = 3x + 5$

**60.** $y = 5x + 8$
$y = \frac{1}{5}x - 8$

**61.** $-4x + 2y = 5$
$2x - y = 7$

**62.** $8x - 11y = 15$
$11x + 8y = 1$

**63.** $7x - y = 1$
$x + 7y = 3$

**64.** $x + 4y = 7$
$-2x - 8y = 0$

**65.** Line 1 goes through $(0, 5)$ and $(10, 0)$.
Line 2 goes through $(0, -10)$ and $(5, 0)$.

**66.** Line 1 goes through $(7, 0)$ and $(0, -21)$.
Line 2 goes through $(21, 0)$ and $(0, 7)$.

**67.** Line 1 goes through $(-3, 8)$ and $(1, -9)$.
Line 2 goes through $(-2, -1)$ and $(-1, 16)$.

**68.** Line 1 goes through $(-4, 10)$ and $(2, -3)$.
Line 2 goes through $(-3, 8)$ and $(2, -5)$.

## Review

*Solve. See Sections 2.6 and 2.7.*

**69.** $|x - 3| = 6$

**70.** $|x + 2| < 4$

**71.** $|2x + 5| > 3$

**72.** $|5x| = 10$

**73.** $|3x - 4| \leq 2$

**74.** $|7x - 2| \geq 5$

## Concept Extensions

*Determine whether each slope calculation is correct or incorrect. If incorrect, then correct the calculation. See the second Concept Check in this section.*

**75.** $(-2, 6)$ and $(7, -14)$
$$m = \frac{-14 - 6}{7 - 2} = \frac{-20}{5} = -4$$

**76.** $(-1, 4)$ and $(-3, 9)$
$$m = \frac{9 - 4}{-3 - 1} = \frac{5}{-4} \text{ or } -\frac{5}{4}$$

**77.** $(-8, -10)$ and $(-11, -5)$
$$m = \frac{-10 - (-5)}{-8 - (-11)} = \frac{-5}{3} \text{ or } -\frac{5}{3}$$

**78.** $(0, -4)$ and $(-6, -6)$
$$m = \frac{0 - (-6)}{-4 - (-6)} = \frac{6}{2} = 3$$

*Solve.*

**79.** Find the slope of a line parallel to the line
$$y = -\frac{7}{2}x - 6.$$

**80.** Find the slope of a line parallel to the line $y = x$.

**81.** Find the slope of a line perpendicular to the line
$$y = -\frac{7}{2}x - 6.$$

**82.** Find the slope of a line perpendicular to the line
$y = x$.

**83.** Find the slope of a line parallel to the line
$5x - 2y = 6.$

**84.** Find the slope of a line parallel to the line
$-3x + 4y = 10.$

**85.** Find the slope of a line perpendicular to the line $5x - 2y = 6$.

**86.** Find the slope of a line perpendicular to the line $-3x + 4y = 10$

*The following graph shows the altitude of a seagull in flight over a time period of 30 seconds. Use this graph to answer Exercises 87 through 90.*

**87.** Find the coordinates of point *B*.

**88.** Find the coordinates of point *C*.

**89.** Find the rate of change of altitude between points *B* and *C*. (Recall that the rate of change between points is the slope between points. This rate of change will be in yards per second.)

**90.** Find the rate of change of altitude (in yards per second) between points *F* and *G*.

*Solve.*

**91.** Professional plumbers suggest that a sewer pipe should be sloped 0.25 inch for every foot. Find the recommended slope for a sewer pipe. (*Source: Rules of Thumb* by Tom Parker, Houghton Mifflin Company)

**92.** A handicap ramp is specified to have a slope no greater than $\frac{1}{12}$. Does a ramp built with a rise of 7 inches every run of 100 inches meet this specification?

**93.** Explain whether two lines, both with positive slopes, can be perpendicular.

**94.** Explain how merely looking at a line can tell us whether its slope is negative, positive, undefined, or zero.

**95.** Each line on the graph has negative slope.

   **a.** Find the slope of each line.

   **b.** Use the results of part (a) to complete the sentence: For lines with negative slopes, the steeper line has the _____ (greater/lesser) slope.

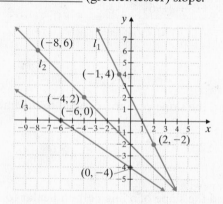

**96.** **a.** On a single screen of a graphing calculator, graph $y = \frac{1}{2}x + 1, y = x + 1,$ and $y = 2x + 1.$ Notice the change in slope for each graph.

   **b.** On a single screen of a graphing calculator, graph $y = -\frac{1}{2}x + 1, y = -x + 1,$ and $y = -2x + 1.$ Notice the change in slope for each graph.

   **c.** Determine whether the following statement is true or false for slope *m* of a given line: As $|m|$ becomes greater, the line becomes steeper.

**The Slope-Intercept Form**

## Objective A Graphing a Line Using Its Slope and y-Intercept

In the last section, we learned that the slope-intercept form of a linear equation is $y = mx + b$. Recall that when an equation is written in this form, the slope of the line is the same as the coefficient $m$ of $x$. Also, the $y$-intercept of the line is $(0, b)$. For example, the slope of the line defined by $y = 2x + 3$ is $2$ and its $y$-intercept is $(0, 3)$.

We may also use the slope-intercept form to graph a linear equation.

**Objectives**

**A** Graph a Line Using Its Slope and y-Intercept.

**B** Use the Slope-Intercept Form to Write an Equation of a Line.

**C** Interpret the Slope-Intercept Form in an Application.

**Example 1** Graph: $y = \dfrac{1}{4}x - 3$

**Solution:** Recall that the slope of the graph of $y = \dfrac{1}{4}x - 3$ is $\dfrac{1}{4}$ and the $y$-intercept is $(0, -3)$. To graph the line, we first plot the $y$-intercept $(0, -3)$. To find another point on the line, we recall that slope is $\dfrac{\text{rise}}{\text{run}} = \dfrac{1}{4}$. We may then plot another point by starting at $(0, -3)$, rising 1 unit up, and then running 4 units to the right. We are now at the point $(4, -2)$. The graph of $y = \dfrac{1}{4}x - 3$ is the line through points $(0, -3)$ and $(4, -2)$, as shown.

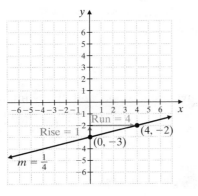

**Practice 1**

Graph: $y = \dfrac{2}{3}x + 1$

■ **Work Practice 1**

**Example 2** Graph: $2x + y = 3$

**Solution:** First, we solve the equation for $y$ to write it in slope-intercept form. In slope-intercept form, the equation is $y = -2x + 3$. Next we plot the $y$-intercept $(0, 3)$. To find another point on the line, we use the slope $-2$, which can be written as $\dfrac{\text{rise}}{\text{run}} = \dfrac{-2}{1}$. We start at $(0, 3)$ and move vertically 2 units down, since the numerator of the slope is $-2$; then we move horizontally 1 unit to the right since the denominator of the slope is 1. We arrive at the point $(1, 1)$. The line through $(1, 1)$ and $(0, 3)$ will have the required slope of $-2$.

**Practice 2**

Graph: $3x + y = -2$

*(Continued on next page)*

**Answers**

**1.**

**2.** See page 200

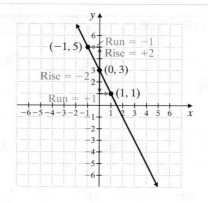

■ Work Practice 2

The slope $-2$ can also be written as $\dfrac{2}{-1}$, so to find another point for Example 2 we could start at $(0, 3)$ and move 2 units up and then 1 unit left. We would stop at the point $(-1, 5)$. The line through $(-1, 5)$ and $(0, 3)$ will have the required slope and will be the same line as shown above through $(1, 1)$ and $(0, 3)$.

## Objective B   Using the Slope-Intercept Form to Write an Equation ▶

Given the slope and $y$-intercept of a line, we may write its equation as well as graph the line.

**Example 3**   Write an equation of the line with $y$-intercept $(0, -7)$ and slope of $\dfrac{2}{3}$.

**Solution:**   We are given the slope and the $y$-intercept. We let $m = \dfrac{2}{3}$ and $b = -7$, and write the equation in slope-intercept form, $y = mx + b$.

$$y = mx + b$$

$$y = \frac{2}{3}x + (-7) \quad \text{Let } m = \frac{2}{3} \text{ and } b = -7.$$

$$y = \frac{2}{3}x - 7 \quad\quad \text{Simplify.}$$

Notice that the graph of this equation will have slope $\dfrac{2}{3}$ and $y$-intercept $(0, -7)$ as desired.

■ Work Practice 3

✓**Concept Check** What is wrong with the following equation of a line with $y$-intercept $(0, 4)$ and slope 2?

$$y = \cancel{4x + 2}$$

## Objective C   Interpreting the Slope-Intercept Form ▶

Recall from Section 3.1 the graph of a one-day pass price for Disney World. Notice that if the points are connected, the resulting graph resembles the graph of a line. Often, businesses depend on equations that "closely fit" lines like this one to model the data and predict future trends. For example, by a method called least squares

### Practice 3

Write an equation of the line with slope $\dfrac{1}{7}$ and $y$-intercept $(0, -5)$.

**Answers**

2.

3. $y = \dfrac{1}{7}x - 5$

✓**Concept Check Answer**

$y$-intercept and slope were switched; should be $y = 2x + 4$

regression, the linear equation $y = 4.383x + 58.44$ approximates the data shown, where $x$ is the number of years since 2005, and $y$ is the ticket price at the beginning of the year. (*Source:* The Walt Disney Company)

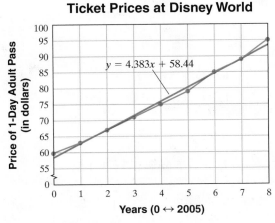

**Ticket Prices at Disney World**

*Source:* The Walt Disney Company

**Helpful Hint**

Don't forget—the notation $0 \leftrightarrow 2005$ means that the number 0 corresponds to the year 2005, 1 corresponds to the year 2006, and so on.

**Example 4**   Predicting Future Prices

The adult one-day pass price $y$ for Disney World is given by

$$y = 4.383x + 58.44$$

where $x$ is the number of years since 2005.

**a.** Use this equation to predict the ticket prices for 2020.
**b.** What does the slope of this equation mean?
**c.** What does the $y$-intercept of this equation mean?

**Solution:**

**a.** To predict the price of a pass in 2020, we need to find $y$ when $x$ is 15. (Since the year 2005 corresponds to $x = 0$, the year 2020 corresponds to the year $2020 - 2005 = 15$).

$$y = 4.383x + 58.44$$
$$= 4.383(15) + 58.44 \quad \text{Let } x = 15$$
$$\approx 124.19$$

We predict that in the year 2020, the price of an adult one-day pass to Disney World will be about $124.19.

**b.** The slope of $y = 4.383x + 58.44$ is $4.383$. We can think of this number as $\dfrac{\text{rise}}{\text{run}}$, or $\dfrac{4.383}{1}$. This means that the ticket price increases on average by $4.383 each year.

**c.** The $y$-intercept of $y = 4.383x + 58.44$ is $58.44$. Notice that it corresponds to the point $(0, 58.44)$ on the graph.

$$\underset{\text{Year} \quad \text{price}}{(0, \; 58.44)}$$

That means that at year $x = 0$, or 2005, the ticket price was about $58.44.

**Work Practice 4**

**Practice 4**

For the period 1980 through 2020, the number of people $y$ age 85 or older living in the United States is given by the equation $y = 110,520x + 2,127,400$, where $x$ is the number of years since 1980. (*Source:* Based on data and estimates from the U.S. Bureau of the Census)

**a.** Estimate the number of people age 85 or older living in the United States in 2010.

**b.** What does the slope of this equation mean?

**c.** What does the $y$-intercept of this equation mean?

**Answers**

**4. a.** 5,443,000  **b.** The number of people age 85 or older in the United States increases at a rate of 110,520 per year.
**c.** At year $x = 0$, or 1980, there were 2,127,400 people age 85 or older in the United States.

## Calculator Explorations   Graphing

You may have noticed by now that to use the $\boxed{Y =}$ key on a grapher to graph an equation, the equation must be solved for $y$.

*Graph each equation by first solving the equation for y.*

**1.** $x = 3.5y$

**2.** $-2.7y = x$

**3.** $5.78x + 2.31y = 10.98$

**4.** $-7.22x + 3.89y = 12.57$

**5.** $y - x = 3.78$

**6.** $3y - 5x = 6x - 4$

**7.** $y - 5.6x = 7.7x + 1.5$

**8.** $y + 2.6x = -3.2$

## Vocabulary, Readiness & Video Check

*Find the slope and the y-intercept of each line.*

**1.** $y = -4x + 12$

**2.** $y = \frac{2}{3}x - \frac{7}{2}$

**3.** $y = 5x$

**4.** $y = -x$

**5.** $y = \frac{1}{2}x + 6$

**6.** $y = -\frac{2}{3}x + 5$

*Martin-Gay Interactive Videos*

*See Video 3.3*

Watch the section lecture video and answer the following questions.

**Objective A 7.** Complete these statements based on ▣ Example 1. To graph a line using its slope and $y$-intercept, first write the equation in _____ form. Graph the one point you now know, the _____. Use the _____ to find a second point. ▶

**Objective B 8.** From ▣ Example 3, given a $y$-intercept point, how do you know which value to use for $b$ in the slope-intercept form? ▶

**Objective C 9.** ▣ Example 4 gives a linear equation that models a real-life application. The equation is rewritten in slope−intercept form. How does this help us better understand how the equation relates to the application? What specific information is discovered? ▶

**3.3** **Exercise Set** MyMathLab®

**Objective A** *Graph each line passing through the given point with the given slope. See Examples 1 and 2.*

**1.** Through $(1, 3)$ with slope $\dfrac{3}{2}$

**2.** Through $(-2, -4)$ with slope $\dfrac{2}{5}$

**3.** Through $(0, 0)$ with slope $5$

**4.** Through $(-5, 2)$ with slope $2$

**5.** Through $(0, 7)$ with slope $-1$

**6.** Through $(3, 0)$ with slope $-3$

*Graph each linear equation using the slope and y-intercept. See Examples 1 and 2.*

**7.** $y = -2x$

**8.** $y = 2x$

**9.** $y = -2x + 3$

**10.** $y = 2x + 6$

**11.** $y = \dfrac{1}{2}x$

**12.** $y = \dfrac{1}{3}x$

**13.** $y = \dfrac{1}{2}x - 4$

**14.** $y = \dfrac{1}{3}x - 2$

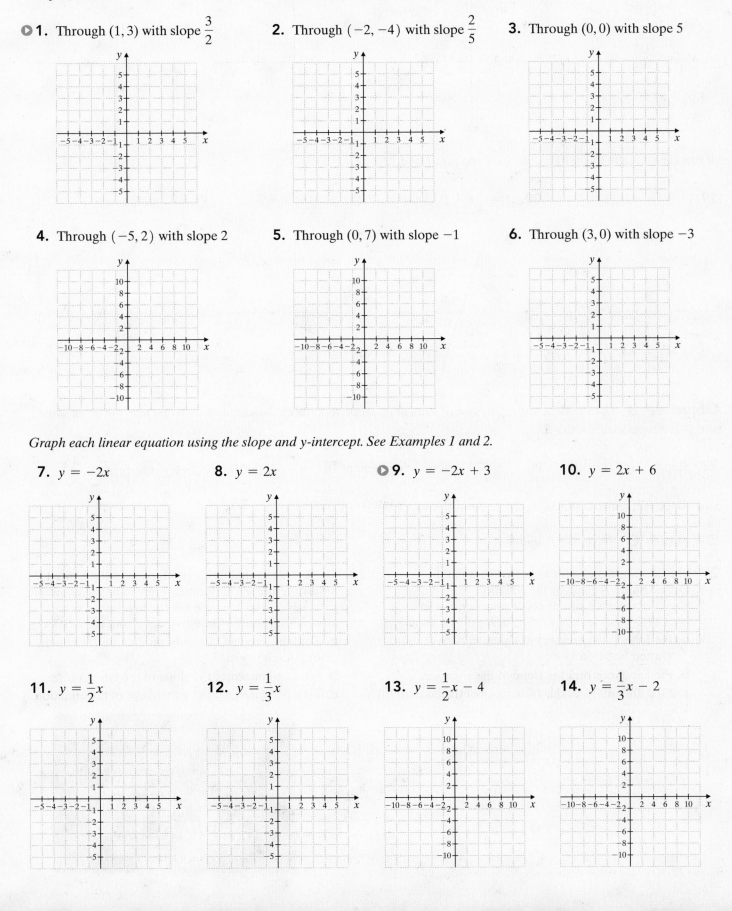

**15.** $7x - 2y = 10$     **16.** $8x - 3y = 9$     **17.** $x + 2y = 8$     **18.** $x - 3y = 3$

*Match each equation with its graph. See Examples 1 and 2.*

**19.** $y = \dfrac{1}{2}x - 1$     **20.** $y = \dfrac{1}{4}x - 1$     **21.** $y = -2x + 3$     **22.** $y = -3x + 3$

A          B          C          D

## Objective B   *Use the slope-intercept form of a linear equation to write the equation of each line with the given slope and y-intercept. See Example 3.*

**23.** Slope $-1$; $y$-intercept $(0, 1)$     **24.** Slope $\dfrac{1}{2}$; $y$-intercept $(0, -6)$     ▸ **25.** Slope $2$; $y$-intercept $\left(0, \dfrac{3}{4}\right)$

**26.** Slope $-3$; $y$-intercept $\left(0, -\dfrac{1}{5}\right)$     **27.** Slope $\dfrac{2}{7}$; $y$-intercept $(0, 0)$     **28.** Slope $-\dfrac{4}{5}$; $y$-intercept $(0, 0)$

## Objective C   *Solve. See Example 4.*

**29.** The life expectancy $y$ for all persons born in the United States is given by the equation $y = 0.19x + 65.8$, where $x$ is the number of years after 1940. (*Source:* U.S. Center for Health Statistics)

   **a.** Find the life expectancy of a person born in the United States in 1990.

   **b.** Find and interpret the slope of the equation.

   **c.** Find and interpret the $y$-intercept of the equation.

**30.** The average annual milk production in the United States (in billions of pounds) is given by the equation $y = 3.2x + 168.2$, where $x$ is the number of years after 2002. (*Source:* Based on data from the USDA)

   **a.** Find the average U.S. milk production (in billions of pounds) in 2012.

   **b.** Find and interpret the slope of the equation.

   **c.** Find and interpret the $y$-intercept of the equation.

**31.** One of the top 10 occupations in terms of job growth over the next few years is expected to be biomedical engineers. The number of people, $y$, in thousands, employed as biomedical engineers in the United States can be estimated by $97x - 100y = -1570$, where $x$ is the number of years after 2010. (*Source:* Based on projections from the U.S. Bureau of Labor Statistics, 2010–2020)
   **a.** Find the slope and $y$-intercept of the linear equation.
   **b.** What does the slope mean in this context?
   **c.** What does the $y$-intercept mean in this context?

**32.** One of the occupations adding the most new jobs in the next few years is registered nurses. The number of new jobs, $y$, in thousands, for registered nurses in the United States can be estimated by $7119x - 100y = -273,740$, where $x$ is the number of years after 2010. (*Source:* Based on projections from the U.S. Bureau of Labor Statistics, 2010–2020)
   **a.** Find the slope and $y$-intercept of the linear equation.
   **b.** What does the slope mean in this context?
   **c.** What does the $y$-intercept mean in this context?

**33.** The number of movies released each year in the United States can be estimated by the linear equation $y = 18.4x + 487.3$, where $x$ is the number of years after 2003. (*Source:* Based on data from the Motion Picture Association of America)
   **a.** Use this equation to estimate the number of movies released in the United States in 2009. (Round to the nearest whole number.)
   **b.** Use this equation to predict in what year the number of movie releases will be more than 700. (*Hint:* Let $y = 700$ and solve for $x$.)
   **c.** Use this equation to estimate the number of movie releases in the present year. How many of these new releases have you and your friends seen?

**34.** The amount of restaurant sales $y$ (in billions of dollars) in the United States can be estimated by the linear equation $y = 27.76x + 213.44$, where $x$ is the number of years after 1990. (*Source:* Based on data from the National Restaurant Association)
   **a.** Use this equation to approximate the amount of restaurant sales in 2008.
   **b.** Use this equation to approximate the year in which the amount of restaurant sales will exceed $900 billion.
   **c.** Use this equation to approximate the amount of restaurant sales in the current year. Do you go out to eat often? Do your friends?

## Review

*Simplify and solve for y. See Section 2.3.*

**35.** $y - 2 = 5(x + 6)$    **36.** $y - 0 = -3[x - (-10)]$    **37.** $y - (-1) = 2(x - 0)$    **38.** $y - 9 = -8[x - (-4)]$

## Concept Extensions

*Determine whether each equation is correct or incorrect. If incorrect, then correct the equation. See the Concept Check in this section.*

**39.** The equation of the line with $y$-intercept $(0, 1.7)$ and slope 3 is $y = 1.7x + 3$.

**40.** The equation of the line with slope $-\dfrac{1}{9}$ and $y$-intercept $(0, -5)$ is $y = -\dfrac{1}{9}x + 5$.

*Solve.*

**41.** In your own words, explain how to find the slope and the $y$-intercept of the graph of an equation given the equation.

**42.** In your own words, explain how to graph an equation using its slope and $y$-intercept.

**43.** Suppose that the revenue of a company has increased at a steady rate of $42,000 per year since 2000. Also, the company's revenue in 2000 was $2,900,000. Write an equation that describes the company's revenue since 2000.

**44.** Suppose that a bird dives off a 500-foot cliff and descends at a rate of 7 feet per second. Write an equation that describes the bird's height at any time $x$.

## Objectives

**A** Use the Point-Slope Form to Write the Equation of a Line.

**B** Write Equations of Vertical and Horizontal Lines.

**C** Write Equations of Parallel and Perpendicular Lines.

**D** Use the Point-Slope Form in Real-World Applications.

## Objective A Using the Point-Slope Form to Write an Equation

When the slope of a line and a point on the line are known, the equation of the line can also be found. To do this, we use the slope formula to write the slope of a line that passes through points $(x, y)$ and $(x_1, y_1)$. We have

$$m = \frac{y - y_1}{x - x_1}$$

We multiply both sides of this equation by $x - x_1$ to obtain

$$y - y_1 = m(x - x_1)$$

This form is called the *point-slope form* of the equation of a line.

---

### Point-Slope Form of the Equation of a Line

The **point-slope form** of the equation of a line is

$$y - y_1 = \overset{\text{slope}}{\underset{\text{point}}{m(x - x_1)}}$$

where $m$ is the slope of the line and $(x_1, y_1)$ is a point on the line.

---

### Practice 1

Write an equation of the line with slope $-2$ and containing the point $(2, -4)$. Write the equation in slope-intercept form, $y = mx + b$.

**Helpful Hint**
Remember, "slope-intercept form" means the equation is "solved for $y$."

**Example 1** Write an equation of the line with slope $-3$ and containing the point $(1, -5)$. Write the equation in slope-intercept form, $y = mx + b$.

**Solution:** Because we know the slope and a point on the line, we use the point-slope form with $m = -3$ and $(x_1, y_1) = (1, -5)$.

$$y - y_1 = m(x - x_1) \quad \text{Point-slope form}$$
$$y - (-5) = -3(x - 1) \quad \text{Let } m = -3 \text{ and } (x_1, y_1) = (1, -5).$$
$$y + 5 = -3x + 3 \quad \text{Use the distributive property.}$$
$$y = -3x - 2 \quad \text{Write in slope-intercept form (solve for } y).$$

In slope-intercept form, the equation is $y = -3x - 2$.

■ Work Practice 1

### Practice 2

Write an equation of the line through points $(3, 0)$ and $(-2, 4)$. Write the equation in standard form, $Ax + By = C$.

**Example 2** Write an equation of the line through points $(4, 0)$ and $(-4, -5)$. Write the equation in standard form $Ax + By = C$.

**Solution:** First we find the slope of the line.

$$m = \frac{-5 - 0}{-4 - 4} = \frac{-5}{-8} = \frac{5}{8}$$

Next we make use of the point-slope form. We replace $(x_1, y_1)$ by either $(4, 0)$ or $(-4, -5)$ in the point-slope equation. We will choose the point $(4, 0)$. The line through $(4, 0)$ with slope $\frac{5}{8}$ is as follows.

### Answers

**1.** $y = -2x$  **2.** $4x + 5y = 12$

$$y - y_1 = m(x - x_1) \quad \text{Point-slope form}$$

$$y - 0 = \frac{5}{8}(x - 4) \quad \text{Let } m = \frac{5}{8} \text{ and } (x_1, y_1) = (4, 0).$$

Let's multiply through by 8 so that the coefficients are integers and are less tedious to work with.

$$8(y - 0) = 8 \cdot \frac{5}{8}(x - 4)$$

$$8y = 5(x - 4) \qquad \text{Simplify.}$$

$$8y = 5x - 20 \qquad \text{Multiply.}$$

$$-5x + 8y = -20 \qquad \text{Write in standard form.}$$

If we multiply both sides of $-5x + 8y = -20$ by $-1$, we have an equivalent equation in standard form. Both $-5x + 8y = -20$ and $5x - 8y = 20$ are acceptable.

■ **Work Practice 2**

**Helpful Hint**    If two points of a line are given, either one may be used with the point-slope form to write an equation of the line.

## Objective B Writing Equations of Vertical and Horizontal Lines ▶

A few special types of linear equations are those whose graphs are vertical and horizontal lines.

**Example 3** Write an equation of the horizontal line containing the point $(2, 3)$.

**Solution:** Recall from Section 3.1 that a horizontal line has an equation of the form $y = b$. Since the line contains the point $(2, 3)$, the equation is $y = 3$.

■ **Work Practice 3**

**Practice 3**

Write an equation of the horizontal line containing the point $(-1, 6)$.

**Example 4** Write an equation of the line containing the point $(2, 3)$ with undefined slope.

**Solution:** Since the line has undefined slope, the line must be vertical. A vertical line has an equation of the form $x = c$, and since the line contains the point $(2, 3)$, the equation is $x = 2$.

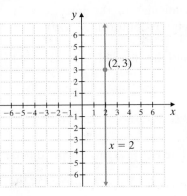

**Practice 4**

Write an equation of the line containing the point $(4, 7)$ with undefined slope.

■ **Work Practice 4**

**Answers**

**3.** $y = 6$  **4.** $x = 4$

## Objective C Writing Equations of Parallel and Perpendicular Lines

Next, we write equations of parallel and perpendicular lines.

**Practice 5**

Write an equation of the line containing the point $(-1, 2)$ and parallel to the line $3x + y = 5$. Write the equation in slope-intercept form $y = mx + b$.

**Example 5** Write an equation of the line containing the point $(4, 4)$ and parallel to the line $2x + y = -6$. Write the equation in slope-intercept form, $y = mx + b$.

**Solution:** Because the line we want to find is *parallel* to the line $2x + y = -6$, the two lines must have equal slopes. So we first find the slope of $2x + y = -6$ by solving the equation for $y$ to write it in the form $y = mx + b$. Here $y = -2x - 6$, so the slope is $-2$.

Now we use the point-slope form to write an equation of the line through $(4, 4)$ with slope $-2$.

$$y - y_1 = m(x - x_1)$$
$$y - 4 = -2(x - 4) \quad \text{Let } m = -2, x_1 = 4, \text{ and } y_1 = 4.$$
$$y - 4 = -2x + 8 \quad \text{Use the distributive property.}$$
$$y = -2x + 12$$

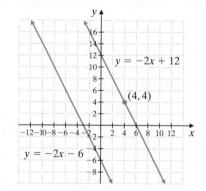

The equation, $y = -2x - 6$, and the new equation, $y = -2x + 12$, have the same slope but different $y$-intercepts so their graphs are parallel. Also, the graph of $y = -2x + 12$ contains the point $(4, 4)$, as desired.

■ Work Practice 5

**Practice 6**

Write an equation of the line containing the point $(3, 4)$ and perpendicular to the line $2x + 4y = 5$. Write the equation in slope-intercept form, $y = mx + b$.

**Example 6** Write an equation of the line containing the point $(-2, 1)$ and perpendicular to the line $3x + 5y = 4$. Write the equation in slope-intercept form, $y = mx + b$.

**Solution:** First we find the slope of $3x + 5y = 4$ by solving the equation for $y$.

$$5y = -3x + 4$$
$$y = -\frac{3}{5}x + \frac{4}{5}$$

The slope of the given line is $-\frac{3}{5}$. A line perpendicular to this line will have a slope that is the negative reciprocal of $-\frac{3}{5}$, or $\frac{5}{3}$. We use the point-slope form to write an equation of a new line through $(-2, 1)$ with slope $\frac{5}{3}$.

**Answers**

5. $y = -3x - 1$   6. $y = 2x - 2$

$$y - y_1 = m(x - x_1)$$

$$y - 1 = \frac{5}{3}[x - (-2)]$$

$$y - 1 = \frac{5}{3}(x + 2) \qquad \text{Simplify.}$$

$$y - 1 = \frac{5}{3}x + \frac{10}{3} \qquad \text{Use the distributive property.}$$

$$y = \frac{5}{3}x + \frac{13}{3} \qquad \text{Add 1 to both sides.}$$

The equation $y = -\frac{3}{5}x + \frac{4}{5}$ and the new equation $y = \frac{5}{3}x + \frac{13}{3}$ have negative reciprocal slopes, so their graphs are perpendicular. Also, the graph of $y = \frac{5}{3}x + \frac{13}{3}$ contains the point $(-2, 1)$, as desired.

■ **Work Practice 6**

## Objective D  Using the Point-Slope Form in Applications ▶

The point-slope form of an equation is very useful for solving real-world problems.

### Example 7  Predicting Sales

Southern Star Realty is an established real estate company that has enjoyed constant growth in sales since 2005. In 2009, the company sold 250 houses, and in 2013, the company sold 330 houses. Use these figures to predict the number of houses this company will sell in 2018.

**Solution:**

1. UNDERSTAND. Read and reread the problem. Then let

    $x$ = the number of years after 2005 and

    $y$ = the number of houses sold in the year corresponding to $x$

The information provided then gives the ordered pairs $(4, 250)$ and $(8, 330)$. To better visualize the sales of Southern Star Realty, we graph the line that passes through the points $(4, 250)$ and $(8, 330)$.

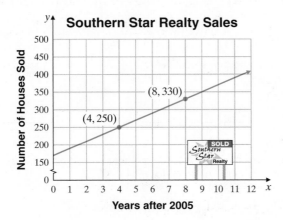

**Southern Star Realty Sales**

(Number of Houses Sold vs. Years after 2005)

(*Continued on next page*)

**Practice 7**

Southwest Regional is an established office product maintenance company that has enjoyed constant growth in new maintenance contracts since 2005. In 2007, the company obtained 15 new contracts, and in 2013, the company obtained 33 new contracts. Use these figures to predict the number of new contracts this company can expect in 2022.

**Answer**

**7.** 60 new contracts

2. **TRANSLATE.** We write the equation of the line that passes through the points $(4, 250)$ and $(8, 330)$. To do so, we first find the slope of the line.

$$m = \frac{330 - 250}{8 - 4} = \frac{80}{4} = 20$$

Then using the point-slope form to write the equation, we have

$$y - y_1 = m(x - x_1)$$
$$y - 250 = 20(x - 4) \quad \text{Let } m = 20 \text{ and } (x_1, y_1) = (4, 250).$$
$$y - 250 = 20x - 80 \quad \text{Multiply.}$$
$$y = 20x + 170 \quad \text{Add 250 to both sides.}$$

3. **SOLVE.** To predict the number of houses sold in 2018, we use $y = 20x + 170$ and complete the ordered pair $(13, \quad)$, since $2018 - 2005 = 13$.

$$y = 20(13) + 170$$
$$y = 430$$

4. **INTERPRET.**

**Check:** Verify that the point $(13, 430)$ is a point on the extended line graphed in Step 1.

**State:** Southern Star Realty should expect to sell 430 houses in 2018.

▬ Work Practice 7

---

### Calculator Explorations  **Graphing**

Many graphing calculators have a TRACE feature. This feature allows you to trace along a graph and see the corresponding $x$- and $y$-coordinates appear on the screen. Use this feature for the following exercises.

*Graph each equation and then use the TRACE feature to complete each ordered pair solution. (Many times the tracer will not show the exact x- or y-value asked for. In each case, trace as closely as you can to the given x- or y-coordinate and approximate the other, unknown coordinate to one decimal place.)*

1. $y = 2.3x + 6.7$;
   $x = 5.1, y = ?$
2. $y = -4.8x + 2.9$;
   $x = -1.8, y = ?$

3. $y = -5.9x - 1.6$;
   $x = ?, y = 7.2$
4. $y = 0.4x - 8.6$;
   $x = ?, y = -4.4$
5. $y = 5.2x - 3.3$;
   $x = 2.3, y = ?$
   $x = ?, y = 36$
6. $y = -6.2x - 8.3$;
   $x = 3.2, y = ?$
   $x = ?, y = 12$

---

## Vocabulary, Readiness & Video Check

*Find the slope of and a point on the line described by each equation.*

1. $y - 4 = -2(x - 1)$

2. $y - 6 = -3(x - 4)$

3. $y - 0 = \frac{1}{4}(x - 2)$

4. $y - 1 = -\frac{2}{3}(x - 0)$

5. $y + 2 = 5(x - 3)$

6. $y - 7 = 4(x + 6)$

*Martin-Gay Interactive Videos*    *Watch the section lecture video and answer the following questions.*

*See Video 3.4*

**Objective A**  **7.** Example 2 discusses how to find an equation of a line given two points. Under what circumstances might the slope-intercept form be chosen over the point-slope form to find an equation?

**Objective B**  **8.** Solve Examples 3 and 4 again, this time using the point $(-1, 3)$ in each exercise.

**Objective C**  **9.** Solve Example 5 again, this time write the equation of the line in slope-intercept form, *parallel* to the given line through the given point.

**Objective D**  **10.** Use the equation found in Example 6 to determine the rate of the rock 2 seconds after it was dropped.

## 3.4  Exercise Set  MyMathLab®

**Objective A**  *Write an equation of each line with the given slope and containing the given point. Write the equation in the slope-intercept form $y = mx + b$. See Example 1.*

**1.** Slope 3; through $(1, 2)$

**2.** Slope 4; through $(5, 1)$

**3.** Slope $-2$; through $(1, -3)$

**4.** Slope $-4$; through $(2, -4)$

**5.** Slope $\frac{1}{2}$; through $(-6, 2)$

**6.** Slope $\frac{2}{3}$; through $(-9, 4)$

**7.** Slope $-\frac{9}{10}$; through $(-3, 0)$

**8.** Slope $-\frac{1}{5}$; through $(4, -6)$

*Write an equation of the line passing through the given points. Write the equation in standard form $Ax + By = C$. See Example 2.*

**9.** $(2, 0)$ and $(4, 6)$

**10.** $(3, 0)$ and $(7, 8)$

**11.** $(-2, 5)$ and $(-6, 13)$

**12.** $(7, -4)$ and $(2, 6)$

**13.** $(-2, -4)$ and $(-4, -3)$

**14.** $(-9, -2)$ and $(-3, 10)$

**15.** $(-3, -8)$ and $(-6, -9)$

**16.** $(8, -3)$ and $(4, -8)$

**17.** $\left(\frac{3}{5}, \frac{4}{10}\right)$ and $\left(-\frac{1}{5}, \frac{7}{10}\right)$

**18.** $\left(\frac{1}{2}, -\frac{1}{4}\right)$ and $\left(\frac{3}{2}, \frac{3}{4}\right)$

**Objective B**  *Write an equation of each line. See Examples 3 and 4.*

**19.** Vertical; through $(2, 6)$

**20.** Slope 0; through $(-2, -4)$

**21.** Horizontal; through $(-3, 1)$

**22.** Vertical; through $(4, 7)$

**23.** Undefined slope; through $(0, 5)$

**24.** Horizontal; through $(0, 5)$

**Objective C** *Write an equation of each line. Write the equation in the form $x = a$, $y = b$, or $y = mx + b$. See Examples 5 and 6.*

**25.** Through $(3, 8)$; parallel to $y = 4x - 2$

**26.** Through $(1, 5)$; parallel to $y = 3x - 4$

**27.** Through $(2, -5)$; perpendicular to $3y = x - 6$

**28.** Through $(-4, 8)$; perpendicular to $2x - 3y = 1$

**29.** Through $(1, 4)$; parallel to $y = 7$

**30.** Through $(-2, 6)$; perpendicular to $y = 7$

**31.** Through $(-2, -3)$; parallel to $3x + 2y = 5$

**32.** Through $(-2, -3)$; perpendicular to $3x + 2y = 5$

**33.** Through $(-1, -5)$; perpendicular to $x = 3$

**34.** Through $(4, -6)$; parallel to $x = -2$

**35.** Through $(-1, 5)$; perpendicular to $x - 4y = 4$

**36.** Through $(2, -3)$; parallel to $x - 5y = 10$

**Objectives A B C Mixed Practice** *Write an equation of each line. Write the equation in standard form unless indicated otherwise. See Examples 1 through 6.*

**37.** Slope 2; through $(-2, 3)$

**38.** Slope 3; through $(-4, 2)$

**39.** Through $(1, 6)$ and $(5, 2)$; use slope-intercept form.

**40.** Through $(2, 9)$ and $(8, 6)$; use slope-intercept form.

**41.** With slope $-\frac{1}{2}$; $y$-intercept $\left(0, \frac{3}{8}\right)$; use slope-intercept form.

**42.** With slope $-4$; $y$-intercept $\left(0, \frac{2}{9}\right)$; use slope-intercept form.

**43.** Through $(-7, -4)$ and $(0, -6)$

**44.** Through $(2, -8)$ and $(-4, -3)$

**45.** Slope $-\frac{4}{3}$; through $(-5, 0)$

**46.** Slope $-\frac{3}{5}$; through $(4, -1)$

**47.** Vertical line; through $(-2, -10)$

**48.** Horizontal line; through $(1, 0)$

**49.** Through $(6, -2)$; parallel to the line $2x + 4y = 9$

**50.** Through $(8, -3)$; parallel to the line $6x + 2y = 5$

**51.** Slope 0; through $(-9, 12)$

**52.** Undefined slope; through $(10, -8)$

**53.** Through $(6, 1)$; parallel to the line $8x - y = 9$

**54.** Through $(3, 5)$; perpendicular to the line $2x - y = 8$

**55.** Through $(5, -6)$; perpendicular to $y = 9$

**56.** Through $(-3, -5)$; parallel to $y = 9$

**57.** Through $(2, -8)$ and $(-6, -5)$; use slope-intercept form.

**58.** Through $(-4, -2)$ and $(-6, 5)$; use slope-intercept form.

**Objective D** *Solve. See Example 7.*

**59.** A rock is dropped from the top of a 400-foot building. After 1 second, the rock is traveling 32 feet per second. After 3 seconds, the rock is traveling 96 feet per second. Let $y$ be the rate of descent and $x$ be the number of seconds since the rock was dropped.

   **a.** Write a linear equation that relates time $x$ to rate $y$. [*Hint:* Use the ordered pairs $(1, 32)$ and $(3, 96)$.]

   **b.** Use this equation to determine the rate of travel of the rock 4 seconds after it was dropped.

**60.** A fruit company recently released a new applesauce. By the end of its first year, profits on this product amounted to $30,000. The anticipated profit for the end of the fourth year is $66,000. The ratio of change in time to change in profit is constant. Let $x$ be years and $y$ be profit.

   **a.** Write a linear equation that relates profit and time. [*Hint:* Use the ordered pairs $(1, 30{,}000)$ and $(4, 66{,}000)$.]

   **b.** Use this equation to predict the company's profit at the end of the seventh year.

   **c.** Predict when the profit should reach $126,000.

**61.** The Whammo Company has learned that by pricing a newly released Frisbee at $6, sales will reach 2000 per day. Raising the price to $8 will cause the sales to fall to 1500 per day. Assume that the ratio of change in price to change in daily sales is constant, and let $x$ be the price of the Frisbee and $y$ be number of sales.

   **a.** Find the linear equation that models the price–sales relationship for this Frisbee. [*Hint:* The line must pass through $(6, 2000)$ and $(8, 1500)$.]

   **b.** Use this equation to predict the daily sales of Frisbees if the price is set at $7.50.

**62.** The Pool Fun Company has learned that, by pricing a newly released Fun Noodle at $3, sales will reach 10,000 Fun Noodles per day during the summer. Raising the price to $5 will cause the sales to fall to 8000 Fun Noodles per day. Let $x$ be price and $y$ be the number sold.

   **a.** Assume that the relationship between sales price and number of Fun Noodles sold is linear and write an equation describing this relationship. [*Hint:* The line must pass through $(3, 10{,}000)$ and $(5, 8000)$.]

   **b.** Use this equation to predict the daily sales of Fun Noodles if the price is $3.50.

**63.** In 2012, the average price of a new home sold in the United States was $292,200. In 2004, the average price of a new home in the United States was $297,000. Let $y$ be the price of a new home in the year $x$, where $x = 0$ represents the year 2004. (*Source:* Based on data from the U.S. Census)

   **a.** Write a linear equation that models the average price of new home in terms of the year $x$. [*Hint:* The line must pass through the points $(0, 297{,}000)$ and $(8, 292{,}200)$.]

   **b.** Use this equation to predict the average price of a new home in 2014.

**64.** The number of McDonald's restaurants worldwide in 2012 was 34,480. In 2007, there were 31,377 McDonald's restaurants worldwide. Let $y$ be the number of McDonald's restaurants in the year $x$, where $x = 0$ represents the year 2007. (*Source:* McDonald's Corporation)

   **a.** Write a linear equation that models the growth in the number of McDonald's restaurants worldwide in terms of the year $x$.

   **b.** Use this information to predict the number of McDonald's restaurants in 2015. (Round to the nearest whole number.)

**65.** The number of theater screens in the United States in 2008 was 38,834. By 2012, the number had grown to 39,918. Let $y$ be the number of theater screens in the United States in year $x$, where $x = 0$ represents the year 2008. (*Source:* Motion Picture Association of America)

   **a.** Write a linear equation that models the number of theater screens in the United States in year $x$.

   **b.** Use this equation to predict the number of theater screens in 2015.

**66.** In 2007, Chrysler Corporation employed 77,778 people worldwide. By 2013, this had shrunk to 70,386. Let $y$ be the number of Chrysler employees worldwide in the year $x$, where $x = 0$ represents the year 2007. (*Source:* Chrysler Corporation)

**a.** Write a linear equation that models the number of Chrysler employees in the year $x$.

**b.** Use this equation to estimate the number of Chrysler employees for the year 2016.

## Review

*Complete each ordered pair for the given equation. See Section 3.1.*

**67.** $y = 7x + 3$; $(4, \quad)$

**68.** $y = 2x - 6$; $(2, \quad)$

**69.** $y = 4.2x$; $(-2, \quad)$

**70.** $y = -1.3x$; $(6, \quad)$

**71.** $y = x^2 + 2x + 1$; $(1, \quad)$

**72.** $y = x^2 - 6x + 4$; $(0, \quad)$

## Concept Extensions

*Find an equation of each line graphed. Write the equation in standard form.*

**73.**          **74.**

**75.**          **76.**

*Answer true or false.*

**77.** A vertical line is always perpendicular to a horizontal line.

**78.** A vertical line is always parallel to another vertical line.

*Solve.*

**79.** Describe how to check to see if the graph of $2x - 4y = 7$ passes through the points $(1.4, -1.05)$ and $(0, -1.75)$. Then follow your directions and check these points.

**80.** Describe how to use the point-slope form to write an equation of a line if the slope and $y$-intercept of a line are given.

*Use a grapher with a TRACE feature to see the results of each exercise.*

**81.** Exercise 25: Graph the equation and verify that it passes through $(3, 8)$ and is parallel to $y = 4x - 2$.

**82.** Exercise 26: Graph the equation and verify that it passes through $(1, 5)$ and is parallel to $y = 3x - 4$.

# Integrated Review

## Linear Equations in Two Variables

*Below is a review of equations of lines.*

### Forms of Linear Equations

| | |
|---|---|
| $Ax + By = C$ | **Standard form** of a linear equation<br>$A$ and $B$ are not both 0. |
| $y = mx + b$ | **Slope-intercept form** of a linear equation<br>The slope is $m$, and the $y$-intercept is $(0, b)$. |
| $y - y_1 = m(x - x_1)$ | **Point-slope form** of a linear equation<br>The slope is $m$, and $(x_1, y_1)$ is a point on the line. |
| $y = c$ | **Horizontal line**<br>The slope is 0, and the $y$-intercept is $(0, c)$. |
| $x = c$ | **Vertical line**<br>The slope is undefined and the $x$-intercept is $(c, 0)$. |

### Parallel and Perpendicular Lines

Nonvertical parallel lines have the same slope.
The product of the slopes of two nonvertical perpendicular lines is −1.

**1.** _____

**2.** _____

**3.** _____

*Graph each linear equation.*

**1.** $y = -2x$

**2.** $3x - 2y = 6$

**3.** $x = -1\dfrac{1}{2}$

**4.** $y = 1.5$

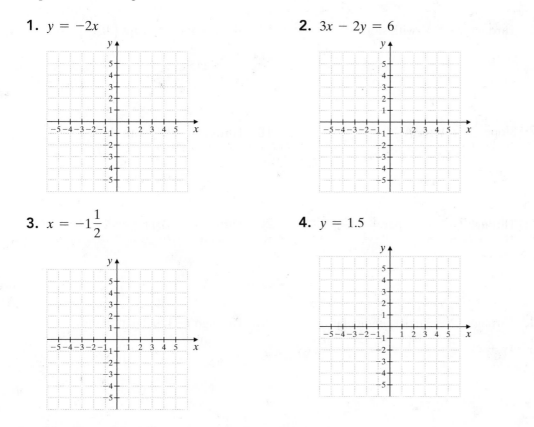

**4.** _____

**5.** _____

*Find the slope of the line containing each pair of points.*

**5.** $(-2, -5), (3, -5)$

**6.** $(5, 2), (0, 5)$

**6.** _____

**7.** _____

**8.** _____

**9.** _____

**10.** _____

**11.** _____

**12.** _____

**13.** _____

**14.** _____

**15.** _____

**16.** _____

**17.** _____

**18.** _____

**19.** _____

**20.** _____

**21.** _____

**22.** _____

**23.** _____

**24.** _____

_Find the slope and y-intercept of each line._

**7.** $y = 3x - 5$

**8.** $5x - 2y = 7$

_Determine whether each pair of lines is parallel, perpendicular, or neither._

**9.** $y = 8x - 6$
$y = 8x + 6$

**10.** $y = \dfrac{2}{3}x + 1$
$2y + 3x = 1$

_Find the equation of each line. Write the equation in the form $x = a$, $y = b$, or $y = mx + b$._

**11.** Through $(1, 6)$ and $(5, 2)$

**12.** Vertical line; through $(-2, -10)$

**13.** Horizontal line; through $(1, 0)$

**14.** Through $(2, -8)$ and $(-6, -5)$

**15.** Through $(-2, 4)$ with slope $-5$

**16.** Slope $-4$; y-intercept $\left(0, \dfrac{1}{3}\right)$

**17.** Slope $\dfrac{1}{2}$; y-intercept $(0, -1)$

**18.** Through $\left(\dfrac{1}{2}, 0\right)$ with slope $3$

**19.** Through $(-1, -5)$; parallel to
$3x - y = 5$

**20.** Through $(0, 4)$; perpendicular to
$4x - 5y = 10$

**21.** Through $(2, -3)$; perpendicular to
$4x + y = \dfrac{2}{3}$

**22.** Through $(-1, 0)$; parallel to
$5x + 2y = 2$

**23.** Undefined slope; through $(-1, 3)$

**24.** $m = 0$; through $(-1, 3)$

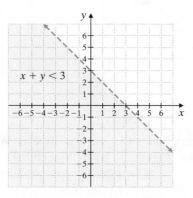

## Objective A Graphing Linear Inequalities

Recall that the graph of a linear equation in two variables is the graph of all ordered pairs that satisfy the equation and that the graph is a line. Here we graph linear inequalities in two variables; that is, we graph all the ordered pairs that satisfy the inequality.

If the equal sign in a linear equation in two variables is replaced with an inequality symbol, the result is a **linear inequality in two variables.**

**Objectives**

A Graph Linear Inequalities.

B Graph the Intersection or Union of Two Linear Inequalities.

**Examples of Linear Inequalities in Two Variables**

$$x + y < 3 \qquad 2x - 4y \geq -3$$
$$4x > 2 \qquad y \leq 5$$

To graph the linear inequality $x + y < 3$, we first graph the related equation $x + y = 3$. The resulting **boundary line** contains all ordered pairs whose coordinates add up to 3. The line separates the plane into two regions called **half-planes.** All points above the boundary line $x + y = 3$ have coordinates that satisfy the inequality $x + y > 3$, and all points below the line have coordinates that satisfy the inequality $x + y < 3$.

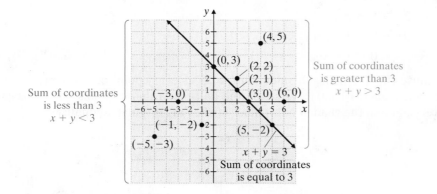

The graph, or **solution region,** for $x + y < 3$, then, is the half-plane below the boundary line and is shown shaded in the figure below. The boundary line is shown dashed since it is not a part of the solution region. The ordered pairs on this line satisfy $x + y = 3$, but not $x + y < 3$.

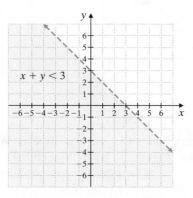

The following steps may be used to graph linear inequalities in two variables.

## Graphing a Linear Inequality in Two Variables

**Step 1:** Graph the boundary line found by replacing the inequality sign with an equal sign. If the inequality sign is $<$ or $>$, graph a dashed line (indicating that points on the line are not solutions of the inequality). If the inequality sign is $\leq$ or $\geq$, graph a solid line (indicating that points on the line are solutions of the inequality).

**Step 2:** Choose a **test point** *not on the boundary line* and substitute the coordinates of this test point into the *original inequality*.

**Step 3:** If a true statement is obtained in Step 2, shade the half-plane that contains the test point. If a false statement is obtained, shade the half-plane that does not contain the test point.

### Practice 1

Graph: $x + 3y > 4$

### Practice 2

Graph: $x \leq 2y$

**Answers**

**1.**

**2.** See page 219.

---

**Example 1**   Graph: $2x - y < 6$

**Solution:**   The boundary line for this inequality is the graph of $2x - y = 6$. We graph a dashed boundary line because the inequality symbol is $<$. Next we choose a test point on either side of the boundary line. The point $(0, 0)$ is not on the boundary line, so we use this point. Replacing $x$ with 0 and $y$ with 0 in the *original inequality* $2x - y < 6$ leads to the following:

$$2x - y < 6$$
$$2(0) - 0 < 6 \quad \text{Let } x = 0 \text{ and } y = 0.$$
$$0 < 6 \quad \text{True}$$

Because $(0, 0)$ satisfies the inequality, so does every point on the same side of the boundary line as $(0, 0)$. We shade the half-plane that contains $(0, 0)$, as shown. Every point in the shaded half-plane satisfies the original inequality.

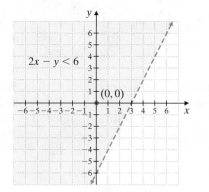

■ Work Practice 1

---

**Example 2**   Graph: $3x \geq y$

**Solution:**   The boundary line is the graph of $3x = y$. We graph a solid boundary line because the inequality symbol is $\geq$. We choose a test point not on the boundary line to determine which half-plane contains points that satisfy the inequality. Let's choose $(0, 1)$ as our test point.

$$3x \geq y$$
$$3(0) \geq 1 \quad \text{Let } x = 0 \text{ and } y = 1.$$
$$0 \geq 1 \quad \text{False}$$

This point does not satisfy the inequality, so the correct half-plane is on the opposite side of the boundary line from $(0, 1)$. The graph of $3x \geq y$ is the boundary line together with the shaded region, as shown.

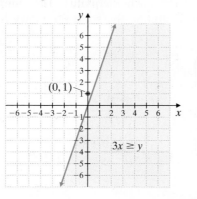

■ Work Practice 2

✓**Concept Check** If a point on the boundary line is included in the solution of an inequality in two variables, should the graph of the boundary line be solid or dashed?

## Objective B Graphing Intersections and Unions ▶

The intersections and the unions of linear inequalities can also be graphed, as shown in the next two examples.

**Example 3** Graph the intersection of $x \geq 1$ and $y \geq 2x - 1$.

**Solution:** First we graph each inequality. The intersection of the two graphs is all points common to both regions, as shown by the *heaviest* shading in the third graph.

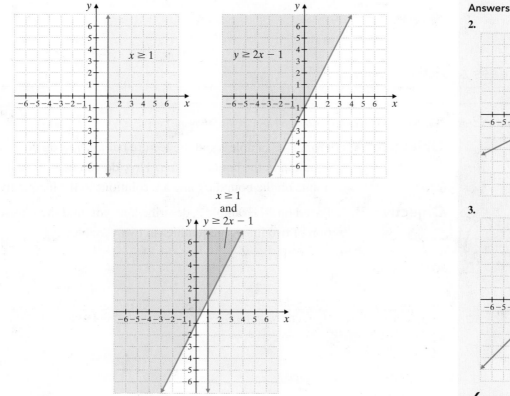

■ Work Practice 3

### Practice 3

Graph the intersection of $x \leq 2$ and $y \geq x + 1$.

**Answers**

2.

3.

✓**Concept Check Answer**
solid

## Practice 4

Graph the union of $x + 2y \le 4$ or $y \ge -1$.

**Answer**

**4.**

**Example 4**    Graph the union of $2x + y \ge -8$ or $y \le -2$.

**Solution:**    First we graph each inequality. The union of the two inequalities is both shaded regions, including the solid boundary lines, as shown in the third graph.

■ Work Practice 4

## Vocabulary, Readiness & Video Check

**Martin-Gay Interactive Videos**    *Watch the section lecture video and answer the following questions.*

**Objective A 1.** From ▤ Example 1 and the lecture before, how do you find the equation of the boundary line? How do you determine if the points on the boundary line are solutions to the inequality? ▶

**Objective B 2.** Based on ▤ Example 2, describe how you find the intersection of two linear inequalities in two variables. ▶

See Video 3.5 🍎

## 3.5 Exercise Set MyMathLab® ▶

**Objectives** **A** *Graph each inequality. See Examples 1 and 2.*

**1.** $x < 2$

**2.** $x > -3$

**3.** $x - y \geq 7$

**4.** $3x + y \leq 1$

▶ **5.** $3x + y > 6$

**6.** $2x + y > 2$

**7.** $y \leq -2x$

**8.** $y \leq 3x$

**9.** $2x + 4y \geq 8$

**10.** $2x + 6y \leq 12$

**11.** $5x + 3y > -15$

**12.** $2x + 5y < -20$

**13.** $y \leq -4.5$

**14.** $y \geq -2\frac{1}{3}$

**15.** $x > 4y$

**16.** $x < -3y$

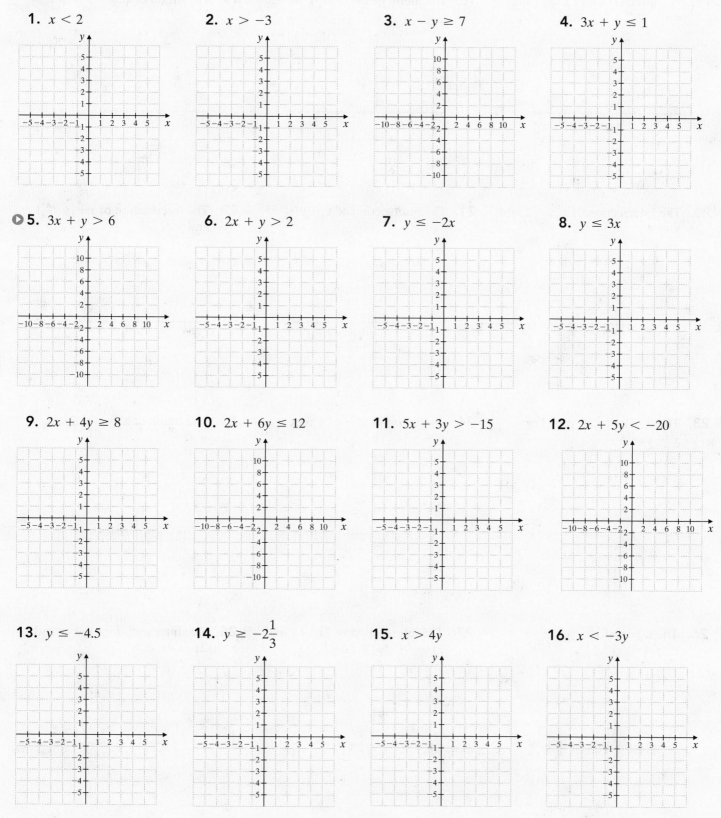

## Objective B  *Graph each union or intersection. See Examples 3 and 4.*

**17.** The intersection of $x \geq 3$ and $y \leq -2$

**18.** The union of $x \geq 3$ or $y \leq -2$

**19.** The union of $x \leq -2$ or $y \geq 4$

**20.** The intersection of $x \leq -2$ and $y \geq 4$

**21.** The intersection of $x - y < 3$ and $x > 4$

**22.** The intersection of $x + y \leq 1$ and $y \leq -1$

**23.** The union of $x + y \leq 3$ or $x - y \geq 5$

**24.** The union of $x - y \leq 3$ or $x + y > -1$

**25.** The union of $x - y \geq 2$ or $y < 5$

**26.** The union of $x - y < 3$ or $x > 4$

**27.** The intersection of $2x > y$ and $3x - 9y > 12$

**28.** The intersection of $y \geq x$ and $2x - 4y \geq 6$

*Match each inequality with its graph.*

**29.** $y \le 2x + 3$       **30.** $y < 2x + 3$       **31.** $y > 2x + 3$       **32.** $y \ge 2x + 3$

A                    B                    C                    D

## Review

*Determine whether the given ordered pair is a solution of both equations. See Section 3.1.*

**33.** $(3, -1); x - y = 4$
$x + 2y = 1$

**34.** $(0, 2); x + 3y = 6$
$4x - y = -2$

**35.** $(-4, 0); 3x + 2y = -12$
$x = 4y$

**36.** $(-5, 2); x + y = -3$
$2x - y = -8$

## Concept Extensions

**37.** Explain when a dashed boundary line should be used in the graph of an inequality.

**38.** Explain why, after the boundary line is sketched, we test a point on either side of this boundary in the original inequality.

*Solve.*

**39.** Chris-Craft manufactures boats out of Fiberglas and wood. Fiberglas hulls require 2 hours of work, whereas wood hulls require 4 hours of work. Employees work at most 40 hours a week. The following inequalities model these restrictions, where $x$ represents the number of Fiberglas hulls produced and $y$ represents the number of wood hulls produced.

$$x \ge 0$$
$$y \ge 0$$
$$2x + 4y \le 40$$

Graph the intersection of these inequalities.

**40.** Rheem Abo-Zahrah decides that she will study at most 20 hours every week and that she must work at least 10 hours every week. Let $x$ represent the hours studying and $y$ represent the hours working. Write two inequalities, in addition to $x \ge 0$ and $y \ge 0$, that model this situation and graph their intersection. (Since hours studied and worked must be 0 or greater than 0, we have the inequalities $x \ge 0$ and $y \ge 0$.)

# Introduction to Functions

## Objective A Defining Relation, Domain, and Range

Equations in two variables, such as $y = 2x + 1$, describe **relations** between $x$-values and $y$-values. For example, if $x = 1$, then this equation describes how to find the $y$-value related to $x = 1$. In words, the equation $y = 2x + 1$ says that twice the $x$-value increased by 1 gives the corresponding $y$-value. The $x$-value of 1 corresponds to the $y$-value of $2(1) + 1 = 3$ for this equation, and we have the ordered pair $(1, 3)$. In other words, for the relationship (or relation) between $x$ and $y$ defined by $y = 2x + 1$, the $x$-value 1 is paired with the $y$-value 3.

There are other ways of describing relations or correspondences between two numbers or, in general, a set of first components (sometimes called the set of *inputs*) and a set of second components (sometimes called the set of *outputs*). For example,

| **First Set: Input** | **Correspondence** | **Second Set: Output** |
|---|---|---|
| People in a certain city | Each person's age, to the nearest year | The set of nonnegative integers |

A few examples of ordered pairs from this relation might be (Ana, 4); (Bob, 36); (Trey, 21); and so on.

Below are just a few other ways of describing relations between two sets and the ordered pairs that they generate.

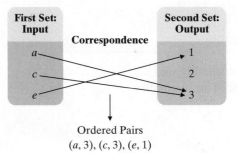

Ordered Pairs
$(a, 3), (c, 3), (e, 1)$

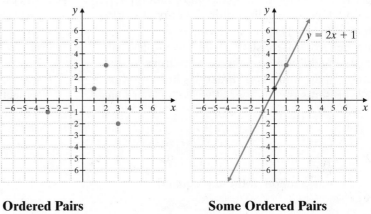

| **Ordered Pairs** | **Some Ordered Pairs** |
|---|---|
| $(-3, -1), (1, 1), (2, 3), (3, -2)$ | $(0, 1), (1, 3)$, and so on |

## Relation, Domain, and Range

A **relation** is a set of ordered pairs.

The **domain** of the relation is the set of all first components of the ordered pairs.

The **range** of the relation is the set of all second components of the ordered pairs.

For example, the domain for our second relation on the previous page is $\{a, c, e\}$ and the range is $\{1, 3\}$. Notice that the range does not include the element 2 of the second set. This is because no element of the first set is assigned to this element. If a relation is defined in terms of $x$- and $y$-values, we will agree that the domain corresponds to $x$-values and that the range corresponds to $y$-values that are paired with $x$-values.

## Helpful Hint

Remember that the range only includes elements that are paired with domain values. For

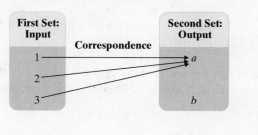

the range is $\{a\}$.

**Examples**    Determine the domain and range of each relation.

**1.** $\{(2, 3), (2, 4), (0, -1), (3, -1)\}$

The domain is the set of all first coordinates of the ordered pairs, $\{2, 0, 3\}$. The range is the set of all second coordinates, $\{3, 4, -1\}$.

**2.**

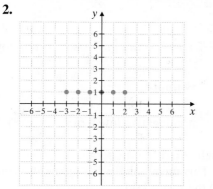

### Helpful Hint

Equivalent domain or range elements that occur more than once need only to be listed once.

The relation is $\{(-3, 1), (-2, 1), (-1, 1), (0, 1), (1, 1), (2, 1)\}$.
The domain is $\{-3, -2, -1, 0, 1, 2\}$.
The range is $\{1\}$.

**3.**

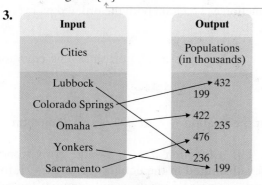

The domain is the set of inputs {Lubbock, Colorado Springs, Omaha, Yonkers, Sacramento}. The range is the numbers in the set of outputs that correspond to elements in the set of inputs {432, 422, 476, 236, 199}.

▪ Work Practice 1–3

## Practice 1–3

Determine the domain and range of each relation.

**1.** $\{(1, 6), (2, 8), (0, 3), (0, -2)\}$

**2.**

**3.**

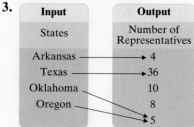

**Answers**

**1.** domain: $\{1, 2, 0\}$; range: $\{6, 8, 3, -2\}$    **2.** domain: $\{1\}$; range: $\{-1, 0, 1, 2, 3\}$    **3.** domain: {Arkansas, Texas, Oklahoma, Oregon}; range: $\{4, 5, 36\}$

**Helpful Hint**   A function is a special type of relation, so all functions are relations. But not all relations are functions.

## Objective B Identifying Functions ▶

Now we consider a special kind of relation called a *function*.

### Function

A **function** is a relation in which each first component in the ordered pairs corresponds to *exactly one* second component.

### Practice 4–6

Determine whether each relation is also a function.

4. $\{(-3, 7), (1, 7), (2, 2)\}$

5.

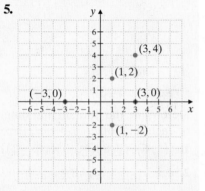

6.

| Input | Correspondence | Output |
|---|---|---|
| People in a certain state | County/ parish that a person lives in | Counties of that state |

**Examples**   Determine whether each relation is also a function.

4. $\{(-2, 5), (2, 7), (-3, 5), (9, 9)\}$

   Although the ordered pairs $(-2, 5)$ and $(-3, 5)$ have the same $y$-value, each $x$-value is assigned to only one $y$-value, so this set of ordered pairs is a function.

5.

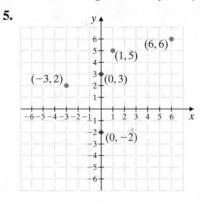

   The $x$-value 0 is assigned to two $y$-values, $-2$ and 3, in this graph, so this relation is not a function.

6.

| Input | Correspondence | Output |
|---|---|---|
| People in a certain city | Each person's age, to the nearest year | The set of nonnegative integers |

This relation is a function because although two different people may have the same age, each person has only one age. This means that each element in the first set is assigned to only one element in the second set.

▦ Work Practice 4–6

✓**Concept Check** Explain why a function can contain both the ordered pairs $(1, 3)$ and $(2, 3)$ but not both $(3, 1)$ and $(3, 2)$.

Recall that an equation such as $y = 2x + 1$ is a relation since this equation defines a set of ordered pair solutions.

### Practice 7

Determine whether the relation $y = 7x - 3$ is also a function.

**Example 7**   Determine whether the relation $y = 2x + 1$ is also a function.

**Solution:** The relation $y = 2x + 1$ is a function if each $x$-value corresponds to just one $y$-value. For each $x$-value substituted in the equation $y = 2x + 1$, the multiplication and addition performed give a single result, so only one $y$-value will be associated with each $x$-value. Thus, $y = 2x + 1$ is a function.

▦ Work Practice 7

**Answers**

4. function   5. not a function
6. function   7. function

✓**Concept Check Answer**

In a function, two different ordered pairs can have the same $y$-value, but not the same $x$-value.

**Example 8** Determine whether the relation $x = y^2$ is also a function.

**Solution:** In $x = y^2$, if $y = 3$, then $x = 9$. Also, if $y = -3$, then $x = 9$. In other words, we have the ordered pairs $(9, 3)$ and $(9, -3)$. Since the $x$-value 9 corresponds to two $y$-values, 3 and $-3$, $x = y^2$ is not a function.

■ Work Practice 8

## Objective C Using the Vertical Line Test

As we have seen, not all relations are functions. Consider the graphs of $y = 2x + 1$ and $x = y^2$ shown next. On the graph of $y = 2x + 1$, notice that each $x$-value corresponds to only one $y$-value. Recall from Example 7 that $y = 2x + 1$ is a function.

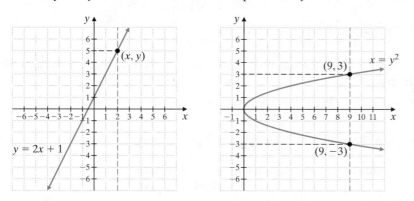

On the graph of $x = y^2$, the $x$-value 9, for example, corresponds to two $y$-values, 3 and $-3$, as shown by the vertical line. Recall from Example 8 that $x = y^2$ is not a function.

Graphs can be used to help determine whether a relation is also a function by the following **vertical line test.**

### Vertical Line Test

If no vertical line can be drawn so that it intersects a graph more than once, the graph is the graph of a function. If such a line can be drawn, the graph is not that of a function.

**Examples** Use the vertical line test to determine which are graphs of functions.

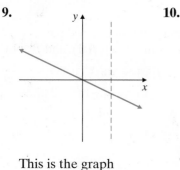

**9.**

This is the graph of a function since no vertical line will intersect this graph more than once.

**10.**

This is the graph of a function.

**11.**

This is not the graph of a function. Note that vertical lines can be drawn that intersect the graph in two points.

*(Continued on next page)*

*(Continued on next page)*

---

**Practice 8**

Determine whether the relation $x = y^2 + 1$ is also a function.

**Practice 9–13**

Use the vertical line test to determine which are graphs of functions.

**9.**

**10.**

**11.**

**12.**

**13.** See page 228

**Answers**

**8.** not a function **9.** function
**10.** function **11.** not a function
**12.** function **13.** not a function

**13.**

**12.**

This is the graph of a function.

**13.**

This is not the graph of a function. A vertical line can be drawn that intersects this line at every point.

■ Work Practice 9–13

✓ **Concept Check** Determine which equations represent functions. Explain your answer.

  **a.** $y = 14$     **b.** $x = -5$     **c.** $x + y = 6$

## Objective D Using Function Notation ▶

Many times letters such as $f$, $g$, and $h$ are used to name functions.

> **Function Notation**
>
> To denote that $y$ is a function of $x$, we can write
>
>   $y = \underbrace{f(x)}_{\text{Function Notation}}$ (Read "$f$ of $x$")
>
> This notation means that **$y$ is a function of $x$** or that $y$ *depends on $x$*. For this reason, $y$ is called the **dependent variable** and $x$ the **independent variable.**

For example, to use function notation with the function $y = 4x + 3$, we write $f(x) = 4x + 3$. The notation $f(1)$ means to replace $x$ with 1 and find the resulting $y$- or function value. Since

  $f(x) = 4x + 3$

then

  $f(1) = 4(1) + 3 = 7$

This means that when $x = 1$, $y$ or $f(x) = 7$. The corresponding ordered pair is $(1, 7)$. Here, the input is 1 and the output is $f(1)$ or 7. Now let's find $f(2), f(0),$ and $f(-1)$.

$f(x) = 4x + 3$     $f(x) = 4x + 3$     $f(x) = 4x + 3$
$f(2) = 4(2) + 3$     $f(0) = 4(0) + 3$     $f(-1) = 4(-1) + 3$
$\quad = 8 + 3$     $\quad = 0 + 3$     $\quad = -4 + 3$
$\quad = 11$     $\quad = 3$     $\quad = -1$

**Ordered Pairs:**

  $(2, 11)$     $(0, 3)$     $(-1, -1)$

> **Helpful Hint**
>
> Note that $f(x)$ is a special symbol in mathematics used to denote a function. The symbol $f(x)$ is read "$f$ of $x$." It does *not* mean $f \cdot x$ ($f$ times $x$).

✓ **Concept Check Answers**

**a, c**

**Examples** Find each function value.

**14.** If $g(x) = 3x - 2$, find $g(1)$.
$$g(1) = 3(1) - 2 = 1$$

**15.** If $g(x) = 3x - 2$, find $g(0)$.
$$g(0) = 3(0) - 2 = -2$$

**16.** If $f(x) = 7x^2 - 3x + 1$, find $f(1)$.
$$f(1) = 7(1)^2 - 3(1) + 1 = 5$$

**17.** If $f(x) = 7x^2 - 3x + 1$, find $f(-2)$.
$$f(-2) = 7(-2)^2 - 3(-2) + 1 = 35$$

■ **Work Practice 14–17**

✓**Concept Check** Suppose $y = f(x)$ and we are told that $f(3) = 9$. Which is not true?

**a.** When $x = 3$, $y = 9$.

**b.** A possible function is $f(x) = x^2$.

**c.** A point on the graph of the function is $(3, 9)$.

**d.** A possible function is $f(x) = 2x + 4$.

If it helps, think of a function, $f$, as a machine that has been programmed with a certain correspondence or rule. An input value (a member of the domain) is then fed into the machine, the machine does the correspondence or rule, and the result is the output (a member of the range).

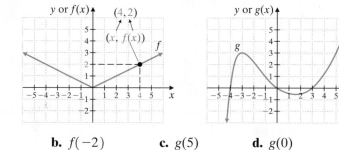

**Example 18** Given the graphs of the functions $f$ and $g$, find each function value by inspecting the graphs.

**a.** $f(4)$      **b.** $f(-2)$      **c.** $g(5)$      **d.** $g(0)$

**e.** Find all $x$-values such that $f(x) = 1$.

**f.** Find all $x$-values such that $g(x) = 0$.

**Solution:**

**a.** To find $f(4)$, find the $y$-value when $x = 4$. We see from the graph that when $x = 4$, $y$ or $f(x) = 2$. Thus, $f(4) = 2$.

**b.** $f(-2) = 1$ from the ordered pair $(-2, 1)$.

**c.** $g(5) = 3$ from the ordered pair $(5, 3)$.

*(Continued on next page)*

---

**Practice 14–17**

Find each function value.

**14.** If $g(x) = 4x + 5$, find $g(0)$.

**15.** If $g(x) = 4x + 5$, find $g(-5)$.

**16.** If $f(x) = 3x^2 - x + 2$, find $f(2)$.

**17.** If $f(x) = 3x^2 - x + 2$, find $f(-1)$.

**Practice 18**

Given the graphs of the functions $f$ and $g$, find each function value by inspecting the graphs.

**a.** $f(1)$      **b.** $f(0)$

**c.** $g(-2)$      **d.** $g(0)$

**e.** Find all $x$-values such that $f(x) = 1$.

**f.** Find all $x$-values such that $g(x) = -2$.

**Answers**

**14.** $g(0) = 5$    **15.** $g(-5) = -15$
**16.** $f(2) = 12$    **17.** $f(-1) = 6$
**18. a.** $-3$   **b.** $-2$   **c.** 3   **d.** 1
**e.** $-1$ and 3   **f.** $-3$

✓**Concept Check Answer**

d

**d.** $g(0) = 0$ from the ordered pair $(0, 0)$.

**e.** To find $x$-values such that $f(x) = 1$, we are looking for any ordered pairs on the graph of $f$ whose $f(x)$ or $y$-value is 1. They are $(2, 1)$ and $(-2, 1)$. Thus $f(2) = 1$ and $f(-2) = 1$. The $x$-values are 2 and $-2$.

**f.** Find ordered pairs on the graph of $g$ whose $g(x)$ or $y$-value is 0. They are $(3, 0)$ $(0, 0)$ and $(-4, 0)$. Thus $g(3) = 0, g(0) = 0$, and $g(-4) = 0$. The $x$-values are $3, 0$ and $-4$.

▄ Work Practice 18

## Objective E Graphing Linear Functions

Recall that the graph of a linear equation in two variables is a line, and a line that is not vertical will always pass the vertical line test. Thus, *all linear equations are functions except those whose graphs are vertical lines.* We call such functions *linear functions.*

> ### Linear Function
>
> A **linear function** is a function that can be written in the form
>
> $$f(x) = mx + b$$

**Practice 19**

Graph the function $f(x) = 3x - 2$.

**Answer**

**19.**

**Example 19**  Graph the function $f(x) = 4x + 2$.

**Solution:** Since $y = f(x)$, we can replace $f(x)$ with $y$ and graph as usual. The graph of $y = 4x + 2$ has slope 4 and $y$-intercept $(0, 2)$. Its graph is shown.

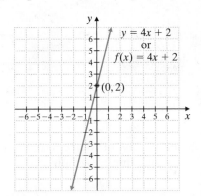

▄ Work Practice 19

## Vocabulary, Readiness & Video Check

*Use the choices below to fill in each blank. Not all choices will be used.*

| | | | |
|---|---|---|---|
| domain | vertical | relation | $(1.7, -2)$ |
| range | horizontal | function | $(-2, 1.7)$ |

**1.** A _____ is a set of ordered pairs.

**2.** The _____ of a relation is the set of all second components of the ordered pairs.

**3.** The _____ of a relation is the set of all first components of the ordered pairs.

**4.** A _____ is a relation in which each first component in the ordered pairs corresponds to *exactly* one second component.

**5.** By the vertical line test, all linear equations are functions except those whose graphs are _____ lines.

**6.** If $f(-2) = 1.7$, the corresponding ordered pair is _____ .

*Martin-Gay Interactive Videos*    *Watch the section lecture video and answer the following questions.*

See Video 3.6

**Objective A**  **7.** Based on the lecture before ⊞ Example 1, why can an equation in two variables define a relation? ▶

**Objective B**  **8.** Based on the lecture before ⊞ Example 2, can a relation in which a second component corresponds to more than one first component be a function? ▶

**Objective C**  **9.** Based on ⊞ Example 3 and the lecture before, explain why the vertical line test works. ▶

**Objective D**  **10.** From ⊞ Examples 5 and 6, what is the connection between function notation to evaluate a function at certain values and ordered pair solutions of the function? ▶

**Objective E**  **11.** From ⊞ Example 7 and the lecture before, do all linear equations in two variables define functions? Explain. ▶

**12.** Based on the lecture before ⊞ Example 7, in what form can a linear function be written? ▶

**3.6**  **Exercise Set** MyMathLab® ▶

**Objectives A B Mixed Practice** *Find the domain and the range of each relation. Also determine whether the relation is a function. See Examples 1 through 6.*

**1.** $\{(-1, 7), (0, 6), (-2, 2), (5, 6)\}$

**2.** $\{(4, 9), (-4, 9), (2, 3), (10, -5)\}$

**3.** $\{(-2, 4), (6, 4), (-2, -3), (-7, -8)\}$

**4.** $\{(6, 6), (5, 6), (5, -2), (7, 6)\}$

**5.** $\{(1, 1), (1, 2), (1, 3), (1, 4)\}$

**6.** $\{(1, 1), (2, 1), (3, 1), (4, 1)\}$

**7.** $\left\{\left(\frac{3}{2}, \frac{1}{2}\right), \left(1\frac{1}{2}, -7\right), \left(0, \frac{4}{5}\right)\right\}$

**8.** $\left\{\left(\frac{1}{2}, \frac{1}{4}\right), \left(0, \frac{7}{8}\right), (0.5, \pi)\right\}$

**9.** $\{(-3, -3), (0, 0), (3, 3)\}$

**10.** $\{(\pi, 0), (0, \pi), (-2, 4), (4, -2)\}$

**11.**

**12.**

**13.**

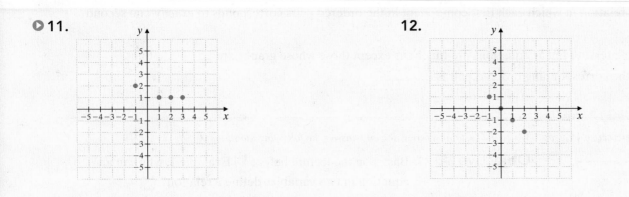

| Input | Output |
|---|---|
| Year of Summer Olympics | Number of Gold Medals Won by U.S. |
| 1996 | 36 |
| 2000 | 37 |
| 2004 | 44 |
| 2008 | 46 |
| 2012 | |

**14.**

| Input | Output |
|---|---|
| Animal | Average Life Span (in years) |
| Polar Bear | 20 |
| Cow | 15 |
| Chimpanzee | 10 |
| Giraffe | 7 |
| Gorilla | |
| Kangaroo | |
| Red Fox | |

**15.**

| Input | Output |
|---|---|
| Degrees Fahrenheit | Degrees Celsius |
| 32 | 0 |
| | 80 |
| 104 | 40 |
| | 37 |
| 212 | 10 |
| 50 | 100 |

**16.**

| Input | Output |
|---|---|
| Words | Number of Letters |
| cat | 3 |
| dog | 7 |
| to | 5 |
| of | 6 |
| given | 2 |
| | 4 |
| | 1 |

**17.**

| Input | Output |
|---|---|
| 2 | 0 |
| −1 | |
| 5 | |
| 100 | |

**18.**

| Input | Output |
|---|---|
| A | 1 |
| B | 2 |
| C | 3 |

*Determine whether each relation is a function. See Examples 4 through 6.*

| | First Set: Input | Correspondence | Second Set: Output |
|---|---|---|---|
| **19.** | Class of algebra students | Final grade average | nonnegative numbers |
| **20.** | People who live in Cincinnati, Ohio | Birth date | days of the year |
| **21.** | blue, green, brown | Eye color | People who live in Cincinnati, Ohio |
| **22.** | Whole numbers from 0 to 4 | Number of children | 50 women in a water aerobics class |

*Determine whether each relation is also a function. See Examples 7 and 8.*

**23.** $y = x + 1$

**24.** $y = x - 1$

**25.** $x = 2y^2$

**26.** $y = x^2$

**27.** $y - x = 7$

**28.** $2x - 3y = 9$

## Objective C  *Use the vertical line test to determine whether each graph is the graph of a function. See Examples 9 through 13.*

**29.**

**30.**

▶ **31.**

**32.**

▶ **33.**

**34.**

**35.**

**36.**

**37.**

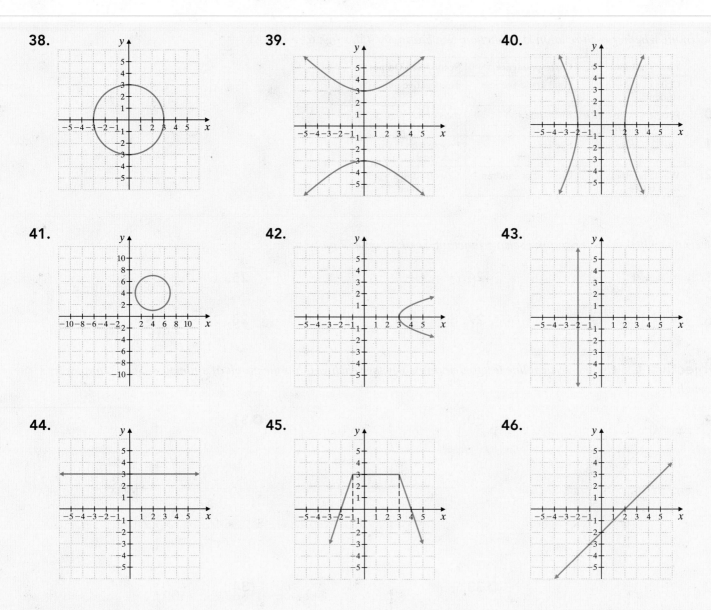

**Objective** **D** *If* $f(x) = 3x + 3$, $g(x) = 4x^2 - 6x + 3$, *and* $h(x) = 5x^2 - 7$, *find each function value. See Examples 14 through 17.*

**47.** $f(4)$          **48.** $f(-1)$          ▶ **49.** $h(-3)$          **50.** $h(0)$

**51.** $g(2)$          **52.** $g(1)$          ▶ **53.** $g(0)$          **54.** $h(-2)$

*For each function, find the indicated values. See Examples 14 through 17.*

**55.** $f(x) = \dfrac{1}{2}x$;

   **a.** $f(0)$

   **b.** $f(2)$

   **c.** $f(-2)$

**56.** $g(x) = -\dfrac{1}{3}x$;

   **a.** $g(0)$

   **b.** $g(-1)$

   **c.** $g(3)$

**57.** $f(x) = -5$;

   **a.** $f(2)$

   **b.** $f(0)$

   **c.** $f(606)$

**58.** $h(x) = 7$;

   **a.** $h(7)$

   **b.** $h(542)$

   **c.** $h\left(-\dfrac{3}{4}\right)$

*Use the graph of the functions below to answer Exercises 59 through 70. See Example 18.*

**59.** If $f(1) = -10$, write the corresponding ordered pair.

**60.** If $f(-5) = -10$, write the corresponding ordered pair.

**61.** If $g(4) = 56$, write the corresponding ordered pair.

**62.** If $g(-2) = 8$, write the corresponding ordered pair.

**63.** Find $f(-1)$.

**64.** Find $f(-2)$.

**65.** Find $g(2)$.

**66.** Find $g(-4)$.

**67.** Find all values of $x$ such that $f(x) = -5$.

**68.** Find all values of $x$ such that $f(x) = -2$.

**69.** Find all positive values of $x$ such that $g(x) = 4$.

**70.** Find all values of $x$ such that $g(x) = 0$.

*The function $A(r) = \pi r^2$ may be used to find the area of a circle if we are given its radius. Use this function to answer Exercises 71 and 72. See Examples 14–17.*

△**71.** Find the area of a circle whose radius is 5 centimeters. (Do not approximate $\pi$.)

△**72.** Find the area of a circular garden whose radius is 8 feet. (Do not approximate $\pi$.)

*The function $V(x) = x^3$ may be used to find the volume of a cube if we are given the length $x$ of a side. Use this function to answer Exercises 73 and 74. See Examples 14–17.*

△**73.** Find the volume of a cube whose side is 14 inches.

△**74.** Find the volume of a die whose side is 1.7 centimeters.

*Forensic scientists use the following functions to find the height of a woman if they are given the height of her femur bone (f) or her tibia bone (t) in centimeters.*

$$H(f) = 2.59f + 47.24$$
$$H(t) = 2.72t + 61.28$$

*Use these functions to answer Exercises 75 and 76. See Examples 14–17.*

**75.** Find the height of a woman whose femur measures 46 centimeters.

**76.** Find the height of a woman whose tibia measures 35 centimeters.

*The dosage in milligrams D of Ivermectin, a heartworm preventive, for a dog who weighs x pounds is given by*

$$D(x) = \frac{136}{25}x$$

*Use this function to answer Exercises 77 and 78. See Examples 14–17.*

**77.** Find the proper dosage for a dog that weighs 30 pounds.

**78.** Find the proper dosage for a dog that weighs 50 pounds.

*Solve. See Examples 14–17.*

**79.** The value of U.S. agricultural exports to China, in billions of dollars, is approximated by the function $E(x) = 1.8x - 1.44$, where $x$ is the number of years since 2000. (*Source:* Based on USDA Economic Research Service data)

**a.** Find and interpret $E(5)$.

**b.** Estimate the value of the agricultural exports to China in 2014.

**80.** The per capita consumption (in kilograms) of all rice in the United States is given by the function $C(x) = 0.11x + 12.8$, where $x$ is the number of years since 2000. (*Source:* Based on data from USA Rice federation)

**a.** Find and interpret $C(6)$.

**b.** Estimate the per capita consumption of rice in the United States in 2015.

**Objective E** *Graph each linear function. See Example 19.*

**81.** $f(x) = 2x + 3$

**82.** $f(x) = 5x - 1$

**83.** $f(x) = -3x$

**84.** $f(x) = -4x$

**85.** $f(x) = -x + 2$

**86.** $f(x) = -x + 1$

## Review

*Solve. See Section 2.4.*

**87.** $2x - 7 \leq 21$

**88.** $-3x + 1 > 0$

**89.** $5(x - 2) \geq 3(x - 1)$

**90.** $-2(x + 1) \leq -x + 10$

**91.** $\dfrac{x}{2} + \dfrac{1}{4} < \dfrac{1}{8}$

**92.** $\dfrac{x}{5} - \dfrac{3}{10} \geq \dfrac{x}{2} - 1$

## Concept Extensions

*Think about the appearance of each graph. Without graphing, determine which equations represent functions. Explain each answer. See the second Concept Check in this section.*

**93.** $x = -1$

**94.** $y = 5$

**95.** $y = 2x$

**96.** $x + y = -5$

*Suppose that $y = f(x)$ and it is true that $f(7) = 50$. For Exercises 97–100, determine whether each statement is true or false. See the third Concept Check in this section.*

**97.** An ordered-pair solution of the function is $(7, 50)$.

**98.** When $x$ is 50, $y$ is 7.

**99.** A possible function is $f(x) = x^2 + 1$.

**100.** A possible function is $f(x) = 10x - 20$.

*Solve.*

**101.** What is the greatest number of $x$-intercepts that a function may have? Explain your answer.

**102.** What is the greatest number of $y$-intercepts that a function may have? Explain your answer.

**103.** In your own words, explain how to find the domain of a function given its graph.

**104.** Explain the vertical line test and how it is used.

*For each function, find the indicated values.*

**105.** $f(x) = x - 12$;
  **a.** $f(12)$
  **b.** $f(a)$
  **c.** $f(-x)$
  **d.** $f(x + h)$

**106.** $f(x) = 2x + 7$
  **a.** $f(2)$
  **b.** $f(a)$
  **c.** $f(-x)$
  **d.** $f(x + h)$

**Finding Domains and Ranges from Graphs and Graphing Piecewise-Defined Functions**

## Objectives

**A** Find the Domain and Range from a Graph.

**B** Graph Piecewise-Defined Functions.

## Objective A Finding the Domain and Range from a Graph

Recall from Section 3.6 that the

> **domain** of a relation is the set of all first components of the ordered pairs of the relation and the
> **range** of a relation is the set of all second components of the ordered pairs of the relation.

In this section, we use the graph of a relation to find its domain and range. Let's use interval notation to write these domains and ranges. Remember, we use a parenthesis to indicate that a number is not part of the domain and we use a bracket to indicate that a number is part of the domain. Of course, as usual, parentheses are placed about infinity symbols indicating that we approach but never reach infinity.

To find the domain of a function (or relation) from its graph, recall that on the rectangular coordinate system, "domain" is the set of first components of the ordered pairs, so this means the *x*-values that are graphed. Similarly, "range" is the set of second components of the ordered pairs, so this means the *y*-values that are graphed.

## Practice 1–4

Find the domain and range of each relation.

**1.**

**2.**

**3.**

**Examples** Find the domain and range of each relation.

**1.**

**2.**

**3.**

**4.**

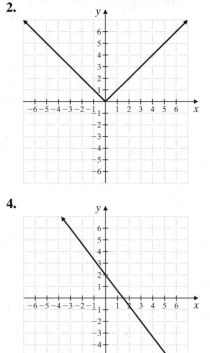

**Answers**

**1.** domain: $[-2, 4]$; range: $[-3, 4]$
**2.** domain: $[0, \infty)$; range: $(-\infty, \infty)$
**3.** domain: $(-\infty, \infty)$; range: $(-\infty, \infty)$

**238**

**Solution:** Notice that the graphs for Examples 1, 2, and 4 are graphs of functions because each passes the vertical line test.

**4.**

**1.**

Range: The *y*-values graphed are from −2 to 4, or [−2, 4].

Domain: The *x*-values graphed are from −3 to 5, or [−3, 5].

**2.**

Range: [0, ∞)

Domain: (−∞, ∞)

**3.**

Range: [−2, 2]

Domain: [−4, 4]

**4.**

Range: (−∞, ∞)

Domain: (−∞, ∞)

■ Work Practice 1–4

# Objective B Graphing Piecewise-Defined Functions ▶

In the last section we graphed linear functions. In this section, we graph functions that are defined by two or more expressions. The expression to use to find the function value depends upon the value of *x*. Before we actually graph these piecewise-defined functions, let's practice finding function values.

**Answer**
**4.** domain: [−2, 2]; range: [−2, 2]

## Practice 5

Evaluate $f(-4), f(3),$ and $f(0)$ for the function

$$f(x) = \begin{cases} 3x + 4 & \text{if } x < 0 \\ -x + 2 & \text{if } x \geq 0 \end{cases}$$

Then write your results in ordered-pair solution form.

**Example 5**  Evaluate $f(2), f(-6),$ and $f(0)$ for the function

$$f(x) = \begin{cases} 2x + 3 & \text{if } x \leq 0 \\ -x - 1 & \text{if } x > 0 \end{cases}$$

Then write your results in ordered-pair form.

**Solution:**  Take a moment and study this function. It is a single function defined by two expressions depending on the value of $x$. From above, if $x \leq 0$, use $f(x) = 2x + 3$. If $x > 0$, use $f(x) = -x - 1$. Thus

| | | |
|---|---|---|
| $f(2) = -(2) - 1$ | $f(-6) = 2(-6) + 3$ | $f(0) = 2(0) + 3$ |
| $\quad = -3$ since $2 > 0$ | $\quad = -9$ since $-6 \leq 0$ | $\quad = 3$ since $0 \leq 0$ |
| $f(2) = -3$ | $f(-6) = -9$ | $f(0) = 3$ |
| Ordered pairs: $(2, -3)$ | $(-6, -9)$ | $(0, 3)$ |

■ **Work Practice 5**

Now, let's graph a piecewise-defined function.

## Practice 6

Graph

$$f(x) = \begin{cases} 3x + 4 & \text{if } x < 0 \\ -x + 2 & \text{if } x \geq 0 \end{cases}$$

**Example 6**  Graph $f(x) = \begin{cases} 2x + 3 & \text{if } x \leq 0 \\ -x - 1 & \text{if } x > 0 \end{cases}$

**Solution:**  Let's graph each piece.

If $x \leq 0$,
$f(x) = 2x + 3$

Values $\leq 0$
| $x$ | $f(x) = 2x + 3$ |
|---|---|
| 0 | 3 Closed circle |
| $-1$ | 1 |
| $-2$ | $-1$ |

If $x > 0$,
$f(x) = -x - 1$

Values $> 0$
| $x$ | $f(x) = -x - 1$ |
|---|---|
| 1 | $-2$ |
| 2 | $-3$ |
| 3 | $-4$ |

The graph of the first part of $f(x)$ listed will look like a ray with a closed-circle endpoint at $(0, 3)$. The graph of the second part of $f(x)$ listed will look like a ray with an open-circle endpoint. To find the exact location of the open-circle endpoint, use $f(x) = -x - 1$ and find $f(0)$. Since $f(0) = -0 - 1 = -1$, we graph the values from the second table and place an open circle at $(0, -1)$.

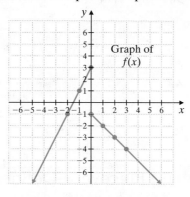

Graph of $f(x)$

Notice that this graph is the graph of a function because it passes the vertical line test. The domain of this function is $(-\infty, \infty)$ and the range is $(-\infty, 3]$.

■ **Work Practice 6**

In the next section, we shall continue our study of graphs of functions.

### Answers

**5.** $f(-4) = -8; f(3) = -1; f(0) = 2;$
$(-4, -8); (3, -1); (0, 2)$

**6.**

# Vocabulary, Readiness & Video Check

*Martin-Gay Interactive Videos*

*Watch the section lecture video and answer the following questions.*

See Video 3.7

**Objective A 1.** Use Examples 1–3 to fill in the blanks. To find the domain of a relation from its graph, we see how far _____ and how far _____ the graph extends. For range, we see how far _____ and how far _____ the graph extends.

**2.** From Example 4, do all linear equations in two variables define functions? Explain.

**Objective B 3.** In Example 1, only one piece of the function is defined for the value $x = -1$. Why do we find $f(-1)$ for $f(x) = x + 3$?

## 3.7 Exercise Set MyMathLab®

**Objective A** *Find the domain and the range of each relation. See Examples 1 through 4.*

9.   10.   ▶11.   12.

13.   14.   15.   16.

## Objective B   *Graph each piecewise-defined function. See Examples 5 and 6.*

17. $f(x) = \begin{cases} 2x & \text{if } x < 0 \\ x + 1 & \text{if } x \ge 0 \end{cases}$

18. $f(x) = \begin{cases} 3x & \text{if } x < 0 \\ x + 2 & \text{if } x \ge 0 \end{cases}$

19. $f(x) = \begin{cases} 4x + 5 & \text{if } x \le 0 \\ \dfrac{1}{4}x + 2 & \text{if } x > 0 \end{cases}$

20. $f(x) = \begin{cases} 5x + 4 & \text{if } x \le 0 \\ \dfrac{1}{3}x - 1 & \text{if } x > 0 \end{cases}$

21. $g(x) = \begin{cases} -x & \text{if } x \le 1 \\ 2x + 1 & \text{if } x > 1 \end{cases}$

22. $g(x) = \begin{cases} 3x - 1 & \text{if } x \le 2 \\ -x & \text{if } x > 2 \end{cases}$

**23.** $f(x) = \begin{cases} 5 & \text{if } x < -2 \\ 3 & \text{if } x \geq -2 \end{cases}$

**24.** $f(x) = \begin{cases} 4 & \text{if } x < -3 \\ -2 & \text{if } x \geq -3 \end{cases}$

**Objectives** A B **Mixed Practice** *Graph each piecewise-defined function. Use the graph to determine the domain and range of the function. See Examples 1 through 6.*

**25.** $f(x) = \begin{cases} -2x & \text{if } x \leq 0 \\ 2x + 1 & \text{if } x > 0 \end{cases}$

**26.** $g(x) = \begin{cases} -3x & \text{if } x \leq 0 \\ 3x + 2 & \text{if } x > 0 \end{cases}$

**27.** $h(x) = \begin{cases} 5x - 5 & \text{if } x < 2 \\ -x + 3 & \text{if } x \geq 2 \end{cases}$

**28.** $f(x) = \begin{cases} 4x - 4 & \text{if } x < 2 \\ -x + 1 & \text{if } x \geq 2 \end{cases}$

**29.** $f(x) = \begin{cases} x + 3 & \text{if } x < -1 \\ -2x + 4 & \text{if } x \geq -1 \end{cases}$

**30.** $h(x) = \begin{cases} x + 2 & \text{if } x < 1 \\ 2x - 1 & \text{if } x \geq 1 \end{cases}$

**31.** $g(x) = \begin{cases} -2 & \text{if } x \leq 0 \\ -4 & \text{if } x \geq 1 \end{cases}$

**32.** $f(x) = \begin{cases} -1 & \text{if } x \leq 0 \\ -3 & \text{if } x \geq 2 \end{cases}$

## Review

*Match each equation with its graph. See Section 3.1.*

**33.** $y = -1$
**A**

**34.** $x = -1$
**B**

**35.** $x = 3$
**C**

**36.** $y = 3$
**D**

## Concept Extensions

**37.** Draw a graph whose domain is $(-\infty, 5]$ and whose range is $[2, \infty)$. Is your graph a function? Discuss why or why not.

**38.** In your own words, describe how to graph a piece-wise-defined function.

**39.** Graph: $f(x) = \begin{cases} -\dfrac{1}{2}x & \text{if } x \le 0 \\ x + 1 & \text{if } 0 < x \le 2 \\ 2x - 1 & \text{if } x > 2 \end{cases}$

**40.** Graph: $f(x) = \begin{cases} \dfrac{1}{3}x & \text{if } x < 0 \\ -x + 2 & \text{if } 0 \le x < 4 \\ 3x - 10 & x \ge 4 \end{cases}$

---

**3.8** **Shifting and Reflecting Graphs of Functions** ▶

### Objectives

**A** Graph Common Equations. ▶

**B** Vertical and Horizontal Shifts. ▶

**C** Reflect Graphs. ▶

In this section, we take common graphs and learn how more complicated graphs are actually formed by shifting and reflecting these common graphs. These shifts and reflections are called transformations, and it is possible to combine transformations. A knowledge of these transformations will help you simplify future graphs.

### Objective A   Graphing Common Equations ▶

Let's begin with the graphs of four common functions.

First, let's graph the linear function $f(x) = x$, or $y = x$. Ordered-pair solutions of this graph consist of ordered pairs whose $x$- and $y$-values are the same.

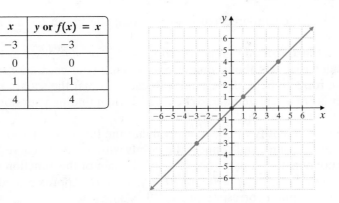

| $x$ | $y$ or $f(x) = x$ |
|----|----|
| $-3$ | $-3$ |
| $0$ | $0$ |
| $1$ | $1$ |
| $4$ | $4$ |

Next, let's graph the nonlinear function $f(x) = x^2$ or $y = x^2$.

This equation is not linear because the $x^2$ term does not allow us to write it in the form $Ax + By = C$. Its graph is not a line. We begin by finding ordered pair solutions. Because this graph is solved for $f(x)$, or $y$, we choose $x$-values and find corresponding $f(x)$, or $y$-values.

If $x = -3$, then $y = (-3)^2$, or 9.

If $x = -2$, then $y = (-2)^2$, or 4.

If $x = -1$, then $y = (-1)^2$, or 1.

If $x = 0$, then $y = 0^2$, or 0.

If $x = 1$, then $y = 1^2$, or 1.

If $x = 2$, then $y = 2^2$, or 4.

If $x = 3$, then $y = 3^2$, or 9.

| $x$ | $f(x)$ or $y$ |
|----|----|
| $-3$ | 9 |
| $-2$ | 4 |
| $-1$ | 1 |
| $0$ | 0 |
| $1$ | 1 |
| $2$ | 4 |
| $3$ | 9 |

Study the table a moment and look for patterns. Notice that the ordered pair solution $(0, 0)$ contains the smallest $y$-value because any other $x$-value squared will give a positive result. This means that the point $(0, 0)$ will be the lowest point on the graph. Also notice that all other $y$-values correspond to two different $x$-values. For example, $3^2 = 9$ and also $(-3)^2 = 9$. This means that the graph will be a mirror image of itself across the $y$-axis. Connect the plotted points with a smooth curve to sketch its graph.

This curve is given a special name, a **parabola**. We will study more about parabolas in later chapters.

Next, let's graph another nonlinear function $f(x) = |x|$ or $y = |x|$.

This is not a linear equation since it cannot be written in the form $Ax + By = C$. Its graph is not a line. Because we do not know the shape of this graph, we find many ordered pair solutions. We will choose $x$-values and substitute to find corresponding $y$-values.

If $x = -3$, then $y = |-3|$, or 3.

If $x = -2$, then $y = |-2|$, or 2.

If $x = -1$, then $y = |-1|$, or 1.

If $x = 0$, then $y = |0|$, or 0.

If $x = 1$, then $y = |1|$, or 1.

If $x = 2$, then $y = |2|$, or 2.

If $x = 3$, then $y = |3|$, or 3.

| $x$ | $f(x)$ or $y$ |
|----|----|
| $-3$ | 3 |
| $-2$ | 2 |
| $-1$ | 1 |
| $0$ | 0 |
| $1$ | 1 |
| $2$ | 2 |
| $3$ | 3 |

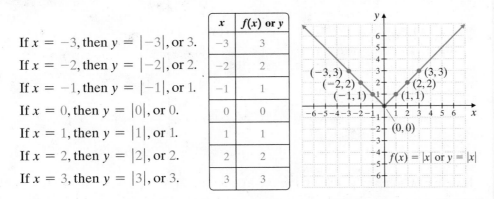

Again, study the table of values for a moment and notice any patterns.

From the plotted ordered pairs, we see that the graph of this absolute value equation is V-shaped.

Finally, a fourth common function, $f(x) = \sqrt{x}$ or $y = \sqrt{x}$. For this graph, you need to recall basic facts about square roots and use your calculator to approximate some square roots to help locate points. Recall also that the square root of a negative number is not a real number, so be careful when finding your domain.

Now let's graph the square root function $f(x) = \sqrt{x}$, or $y = \sqrt{x}$.

To graph, we identify the domain, evaluate the function for several values of $x$, plot the resulting points, and connect the points with a smooth curve. Since $\sqrt{x}$ represents the nonnegative square root of $x$, the domain of this function is the set of all nonnegative numbers, $\{x \mid x \geq 0\}$, or $[0, \infty)$. We have approximated $\sqrt{3}$ below to help us locate the point corresponding to $(3, \sqrt{3})$.

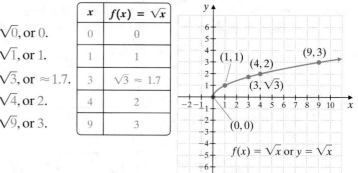

If $x = 0$, then $y = \sqrt{0}$, or $0$.

If $x = 1$, then $y = \sqrt{1}$, or $1$.

If $x = 3$, then $y = \sqrt{3}$, or $\approx 1.7$.

If $x = 4$, then $y = \sqrt{4}$, or $2$.

If $x = 9$, then $y = \sqrt{9}$, or $3$.

| $x$ | $f(x) = \sqrt{x}$ |
|---|---|
| 0 | 0 |
| 1 | 1 |
| 3 | $\sqrt{3} \approx 1.7$ |
| 4 | 2 |
| 9 | 3 |

Notice that the graph of this function passes the vertical line test, as expected.

Following is a summary of our four common graphs. Take a moment and study these graphs. Your success in the rest of this section depends on your knowledge of these graphs.

## Common Graphs

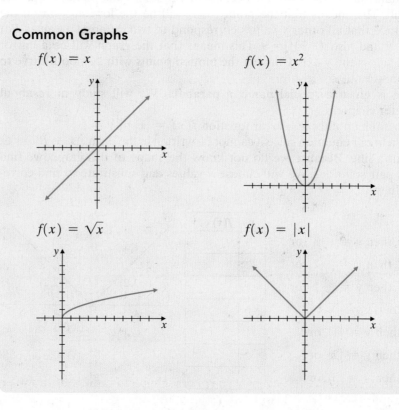

$f(x) = x$

$f(x) = x^2$

$f(x) = \sqrt{x}$

$f(x) = |x|$

## Objective B Vertical and Horizontal Shifting

Your knowledge of the slope-intercept form, $f(x) = mx + b$, will help you understand simple shifting of transformations such as vertical shifts. For example, what is the difference between the graphs of $f(x) = x$ and $g(x) = x + 3$?

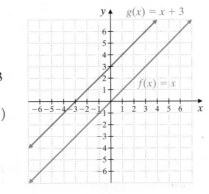

$f(x) = x$
slope, $m = 1$
$y$-intercept is $(0,0)$

$g(x) = x + 3$
slope, $m = 1$
$y$-intercept is $(0,3)$

Notice that the graph of $g(x) = x + 3$ is the same as the graph of $f(x) = x$, but moved upward 3 units. This is an example of a **vertical shift** and is true for graphs in general.

### Vertical Shifts (Upward and Downward)
### Let $k$ Be a Positive Number

| Graph of | Same as | Moved |
|---|---|---|
| $g(x) = f(x) + k$ | $f(x)$ | $k$ units upward |
| $g(x) = f(x) - k$ | $f(x)$ | $k$ units downward |

**Examples** Without plotting points, sketch the graph of each pair of functions on the same set of axes.

**1.** $f(x) = x^2$ and $g(x) = x^2 + 2$   **2.** $f(x) = \sqrt{x}$ and $g(x) = \sqrt{x} - 3$

■ Work Practice 1–2

A horizontal shift to the left or right may be slightly more difficult to understand. Let's graph $g(x) = |x - 2|$ and compare it with $f(x) = |x|$.

**Practice 1–2**

Without plotting points, sketch the graphs of each pair of functions on the same set of axes.

**1.** $f(x) = x^2$ and
$g(x) = x^2 - 1$

**2.** $f(x) = \sqrt{x}$ and
$g(x) = \sqrt{x} + 2$

**Answers**

**1.**

**2.**

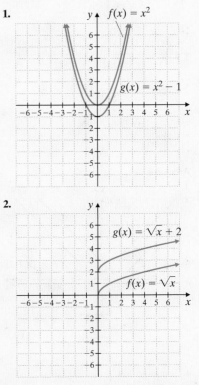

## Practice 3

Without plotting points, sketch the graphs of $f(x) = |x|$ and $g(x) = |x + 2|$ on the same set of axes.

Example 3 Without plotting points, sketch the graphs of $f(x) = |x|$ and $g(x) = |x - 2|$ on the same set of axes.

**Solution:**

| $x$ | $f(x) = |x|$ | $g(x) = |x - 2|$ |
|-----|------|------|
| $-3$ | 3 | 5 |
| $-2$ | 2 | 4 |
| $-1$ | 1 | 3 |
| 0 | 0 | 2 |
| 1 | 1 | 1 |
| 2 | 2 | 0 |
| 3 | 3 | 1 |

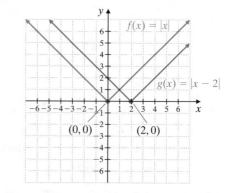

The graph of $g(x) = |x - 2|$ is the same as the graph of $f(x) = |x|$, but moved 2 units to the right. This is an example of a **horizontal shift** and is true for graphs in general. The table is provided to verify the graphs.

■ **Work Practice 3**

### Horizontal Shift (to the Left or Right)
### Let $h$ Be a Positive Number

| Graph of | Same as | Moved |
|----------|---------|-------|
| $g(x) = f(x - h)$ | $f(x)$ | $h$ units to the right |
| $g(x) = f(x + h)$ | $f(x)$ | $h$ units to the left |

## Practice 4

Sketch the graphs of $f(x) = x^2$ and $g(x) = (x + 2)^2 - 1$ on the same set of axes.

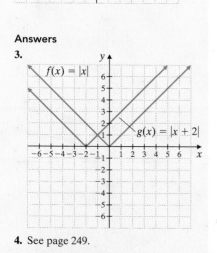

**Helpful Hint**

Notice that $f(x - h)$ corresponds to a shift to the right and $f(x + h)$ corresponds to a shift to the left.

Vertical and horizontal shifts can be combined.

Example 4 Sketch the graphs of $f(x) = x^2$ and $g(x) = (x - 2)^2 + 1$ on the same set of axes.

**Solution:** The graph of $g(x)$ is the same as the graph of $f(x)$ shifted 2 units to the right and 1 unit up.

■ **Work Practice 4**

**Answers**

3.

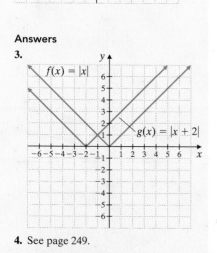

4. See page 249.

## Objective C Reflecting Graphs

Another type of transformation is called a **reflection.** In this section, we will study reflections (mirror images) about the $x$-axis only. For example, take a moment and study these two graphs. The graph of $g(x) = -x^2$ can be found, as usual, by plotting points.

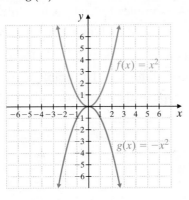

### Reflection About the x-axis

The graph of $g(x) = -f(x)$ is the graph of $f(x)$ reflected about the $x$-axis.

**Example 5** Sketch the graph of $h(x) = -|x - 3| + 2$.

**Solution:** The graph of $h(x) = -|x - 3| + 2$ is the same as the graph of $f(x) = |x|$ reflected about the $x$-axis, then moved three units to the right and two units upward.

■ Work Practice 5

There are other transformations, such as stretching, that won't be covered in this section. For a review of this transformation, see the Appendix.

**Practice 5**

Sketch the graph of $h(x) = -(x - 3)^2 + 2$.

Answers

4.

5.

$h(x) = -(x - 3)^2 + 2$

## Vocabulary, Readiness & Video Check

*Match each equation with its graph.*

**1.** $y = \sqrt{x}$      **2.** $y = x^2$      **3.** $y = x$      **4.** $y = |x|$

**A**      **B**      **C**      **D**

*Martin-Gay Interactive Videos*     *Watch the section lecture video and answer the following questions.*

*See Video 3.8* 🍎

**Objective A** 5. Based on the lecture before ⊞ Example 1, what common function has a graph that is V-shaped? ▶

**Objective B** 6. For ⊞ Examples 1–7, why is it helpful to be familiar with common graphs and their basic shapes? ▶

**Objective C** 7. Based on the lecture before ⊞ Example 8, complete the following statement. The graph of $f(x) = -\sqrt{x} + 6$ has the same shape as the graph of $f(x) = \sqrt{x} + 6$ but is reflected about the _____. ▶

**3.8  Exercise Set** MyMathLab® ▶

**Objectives A B Mixed Practice** *Sketch the graph of each function. See Examples 1 through 4.*

▶ **1.** $f(x) = |x| + 3$

**2.** $f(x) = |x| - 2$

▶ **3.** $f(x) = \sqrt{x} - 2$

**4.** $f(x) = \sqrt{x} + 3$

▶ **5.** $f(x) = |x - 4|$

**6.** $f(x) = |x + 3|$

**7.** $f(x) = \sqrt{x + 2}$

**8.** $f(x) = \sqrt{x - 2}$

▶ **9.** $y = (x - 4)^2$

**10.** $y = (x + 4)^2$

**11.** $f(x) = x^2 + 4$

**12.** $f(x) = x^2 - 4$

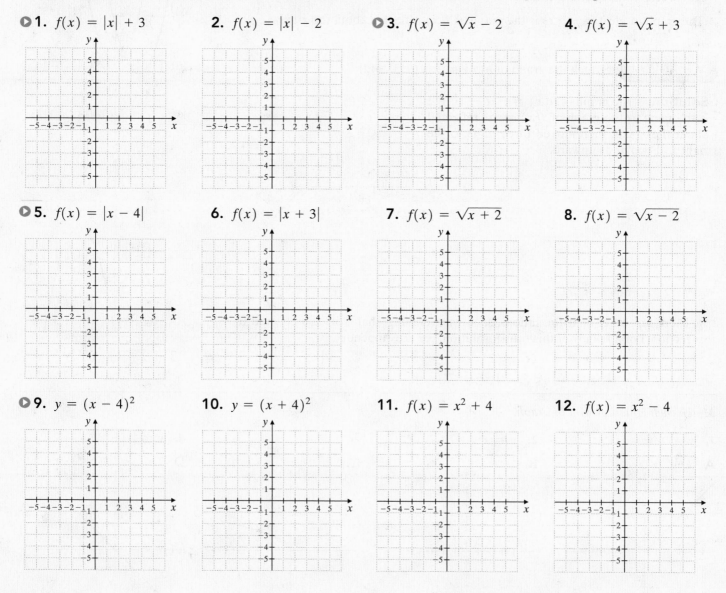

**13.** $f(x) = \sqrt{x-2} + 3$    **14.** $f(x) = \sqrt{x-1} + 3$    **15.** $f(x) = |x-1| + 5$    **16.** $f(x) = |x-3| + 2$

**17.** $f(x) = \sqrt{x+1} + 1$    **18.** $f(x) = \sqrt{x+3} + 2$    **19.** $f(x) = |x+3| - 1$    **20.** $f(x) = |x+1| - 4$

**21.** $g(x) = (x-1)^2 - 1$    **22.** $h(x) = (x+2)^2 + 2$    **23.** $f(x) = (x+3)^2 - 2$    **24.** $f(x) = (x+2)^2 + 4$

**Objectives A B C Mixed Practice** *Sketch the graph of each function. See Examples 1 through 5.*

**25.** $f(x) = -(x-1)^2$    **26.** $g(x) = -(x+2)^2$    **27.** $h(x) = -\sqrt{x} + 3$    **28.** $f(x) = -\sqrt{x+3}$

**29.** $h(x) = -|x + 2| + 3$    **30.** $g(x) = -|x + 1| + 1$    **31.** $f(x) = (x - 3) + 2$    **32.** $f(x) = (x - 1) + 4$

## Review

*Simplify. See Sections 1.5 through 1.7.*

**33.** $-3x^4 \cdot 5x^4$        **34.** $-3x^4 + 5x^4$        **35.** $8(y^7 + y^{11})$        **36.** $y^7 \cdot y^{11}$

## Concept Extensions

**Mixed Practice (*Sections 3.7, 3.8*)** *Write the domain and range of the indicated function in this section.*

**37.** Exercise 13                          **38.** Exercise 14

**39.** Exercise 29                          **40.** Exercise 30

*Without graphing, find the domain of each function.*

**41.** $f(x) = 5\sqrt{x - 20} + 1$        **42.** $g(x) = -3\sqrt{x + 5}$        **43.** $h(x) = 5|x - 20| + 1$

**44.** $f(x) = -3|x + 5.7|$            **45.** $g(x) = 9 - \sqrt{x + 103}$        **46.** $h(x) = \sqrt{x - 17} - 3$

*Sketch the graph of each piecewise-defined function. Write the domain and range of each function.*

**47.** $f(x) = \begin{cases} |x| & \text{if } x \le 0 \\ x^2 & \text{if } x > 0 \end{cases}$

**48.** $f(x) = \begin{cases} x^2 & \text{if } x < 0 \\ \sqrt{x} & \text{if } x \ge 0 \end{cases}$

**49.** $g(x) = \begin{cases} |x - 2| & \text{if } x < 0 \\ -x^2 & \text{if } x \ge 0 \end{cases}$

**50.** $g(x) = \begin{cases} -|x + 1| - 1 & \text{if } x < -2 \\ \sqrt{x + 2} - 4 & \text{if } x \ge -2 \end{cases}$

## Chapter 3 Group Activity

### Linear Modeling

As we saw in Sections 3.3 and 3.4, businesses often depend on equations that "closely fit" data. To *model* the data means to find an equation that describes the relationship between the paired data of two variables, such as time in years and profit. A model that accurately summarizes the relationship between two variables can be used to replace a potentially lengthy listing of the raw data. An accurate model might also be used to predict future trends by answering questions such as "If the trend seen in our company's performance in the last several years continues, what level of profit can we reasonably expect in 3 years?"

There are several ways to find a linear equation that models a set of data. If only two ordered pair data points are involved, an exact equation that contains both points can be found using the methods of Section 3.4. When more than two ordered pair data points are involved, it may be impossible to find a linear equation that contains all of the data points. In this case, the graph of the **best fit equation** should have a majority of the plotted ordered pair data points on the graph or close to it. In statistics, a technique called least squares regression is used to determine an equation that best fits a set of data. Various graphing utilities have built-in capabilities for finding an equation (called a regression equation) that best fits a set of ordered pair data points. Regression capabilities are often found with a graphing utility's statistics features.* A best fit equation can also be estimated using an algebraic method, which is outlined in the Group Activity below. In either case, a useful first step when finding a linear equation that models a set of data is creating a scatter diagram of the ordered pair data points to verify that a linear equation is an appropriate model.

### Group Activity

*The Windows operating system is one of the well-known products of the Microsoft Corporation. This company develops, manufactures, licenses, and supports a wide range of software products for various computing devices. It also provides the MSN network of Internet products and services, and even the Xbox video game system. The table shows Microsoft's total revenues (in billions) for the years 2007–2013. Use the table along with your answers to the questions below to find a linear equation $y = mx + b$ that represents total revenue y (in billions) for Microsoft Corporation, where x represents the number of years after 2007. (Source: Microsoft Corporation)*

| Year | 0 | 1 | 2 | 3 | 4 | 5 | 6 |
|---|---|---|---|---|---|---|---|
| Total Microsoft Revenues (in billions) | 51 | 60 | 58 | 62 | 70 | 74 | 78 |

1. Create a scatter diagram of the paired data given in the table. Does a linear model seem appropriate for the data?

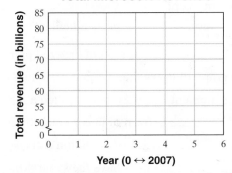

**Total Microsoft Revenue**

2. Use a straightedge to draw on your graph what appears to be the line that "best fits" the data you plotted.

3. Estimate the coordinate of the two points that fit on your best fit line. Use these points to find the equation of the line that passes through both points.

4. Use this equation to find the value of $y$ for $x = 11$. Interpret the meaning of this pair of data.

5. How could this equation be useful to accountants who work at Microsoft?

6. Compare your group's linear equation with other groups' equations. Are they the same or different? Explain why.

7. (Optional) Enter the data from the table into a graphing utility and use the linear regression feature to find a linear equation that models the data. Compare this equation with the one you found in question 3. How are they alike or different?

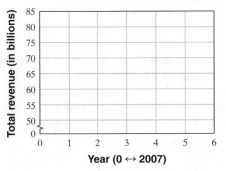

**Total Microsoft Revenue**

8. (Optional) Using corporation annual reports or articles from magazines or newspapers, search for a set of business-related data that could be modeled with a linear equation. Explain how modeling this data could be useful to a business. Then find the best-fit equation for the data.

*To find out more about using a graphing utility to find a regression equation, consult the user's manual for your graphing utility.

## Chapter 3 Vocabulary Check

*Fill in each blank with one of the words or phrases listed below.*

| | | | | |
|---|---|---|---|---|
| relation | line | function | standard | parallel |
| slope-intercept | *x* | *y* | range | domain |
| point-slope | perpendicular | linear inequality | slope | linear function |

1. A _____ is a set of ordered pairs.

2. The graph of every linear equation in two variables is a _____.

3. The statement $-x + 2y > 0$ is called a _____ in two variables.

4. _____ form of linear equation in two variables is $Ax + By = C$.

5. The _____ of a relation is the set of all second components of the ordered pairs of the relation.

6. _____ lines have the same slope and different *y*-intercepts.

7. _____ form of a linear equation in two variables is $y = mx + b$.

8. A _____ is a relation in which each first component in the ordered pairs corresponds to exactly one second component.

9. In the equation $y = 4x - 2$, the coefficient of *x* is the _____ of its corresponding graph.

10. Two lines are _____ if the product of their slopes is $-1$.

11. To find the *x*-intercept of a linear equation, let _____ $= 0$ and solve for the other variable.

12. The _____ of a relation is the set of all first components of the ordered pairs of the relation.

13. A _____ is a function that can be written in the form $f(x) = mx + b$.

14. To find the *y*-intercept of a linear equation, let _____ $= 0$ and solve for the other variable.

15. The equation $y - 8 = -5(x + 1)$ is written in _____ form.

**Helpful Hint**

▶ Are you preparing for your test? Don't forget to take the Chapter 3 Test on page 267. Then check your answers at the back of the text and use the Chapter Test Prep Videos to see the fully worked-out solutions to any of the exercises you want to review.

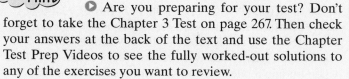

# 3 Chapter Highlights

| Definitions and Concepts | Examples |
|---|---|
| **Section 3.1    Graphing Linear Equations** | |

| | |
|---|---|
| The **rectangular coordinate system,** or **Cartesian coordinate system,** consists of a vertical and a horizontal number line on a plane intersecting at their 0 coordinate. The vertical number line is called the **y-axis,** and the horizontal number line is called **x-axis.** The point of intersection of the axes is called the **origin.** | 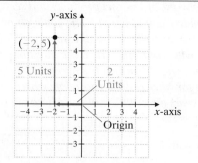 |
| To **plot** or **graph** an ordered pair means to find its corresponding point on a rectangular coordinate system. | Plot or graph the ordered pair $(-2, 5)$. Start at the origin. Move 2 units to the left along the *x*-axis, then 5 units upward parallel to the *y*-axis. |

| Definitions and Concepts | Examples |
|---|---|

### Section 3.1    Graphing Linear Equations (*continued*)

An ordered pair is a **solution** of an equation in two variables if replacing the variables by the corresponding coordinates results in a true statement.

Determine whether $(-2, 3)$ is a solution of $3x + 2y = 0$.

$$3(-2) + 2(3) = 0$$
$$-6 + 6 = 0$$
$$0 = 0 \quad \text{True}$$

$(-2, 3)$ is a solution.

A **linear equation in two variables** is an equation that can be written in the form $Ax + By = C$, where $A$, $B$, and $C$ are real numbers and $A$ and $B$ are not both 0.
The form $Ax + By = C$ is called **standard form.**

$$y = -2x + 5, x = 7,$$
$$y - 3 = 0, 6x - 4y = 10$$

$6x - 4y = 10$ is in standard form.

The **graph of a linear equation** in two variables is a line. To graph a linear equation in two variables, find three ordered pair solutions. Plot the solution points, and draw the line connecting the points.

Graph:    $3x + y = -6$

| x | y |
|---|---|
| 0 | −6 |
| −2 | 0 |
| −3 | 3 |

To find an $x$-intercept, let $y = 0$ and solve for $x$.
To find a $y$-intercept, let $x = 0$ and solve for $y$.

The graph of $x = c$ is a vertical line with $x$-intercept $c$.
The graph of $y = c$ is a horizontal line with $y$-intercept $c$.

### Section 3.2    The Slope of a Line

The **slope** $m$ of the line through $(x_1, y_1)$ and $(x_2, y_2)$ is given by

$$m = \frac{y_2 - y_1}{x_2 - x_1}$$

as long as $x_2 \neq x_1$

Find the slope of the line through $(-1, 7)$ and $(-2, -3)$.

$$m = \frac{y_2 - y_1}{x_2 - x_1} = \frac{-3 - 7}{-2 - (-1)} = \frac{-10}{-1} = 10$$

The **slope-intercept form** of a linear equation is

$$y = mx + b$$

where $m$ is the slope of the line and $(0, b)$ is the $y$-intercept.

Find the slope and $y$-intercept of $-3x + 2y = -8$.

$$2y = 3x - 8$$
$$\frac{2y}{2} = \frac{3x}{2} - \frac{8}{2}$$
$$y = \frac{3}{2}x - 4$$

The slope of the line is $\dfrac{3}{2}$, and the $y$-intercept is $(0, -4)$.

*(Continued)*

| Definitions and Concepts | Examples |
|---|---|

## Section 3.2    The Slope of a Line (continued)

The slope of a horizontal line is 0.
The slope of a vertical line is undefined.
Parallel lines have the same slope but different $y$-intercepts.

The slope of $y = -2$ is 0.
The slope of $x = 5$ is undefined.

If the product of the slopes of two nonvertical lines is $-1$, then the lines are perpendicular.

## Section 3.3    The Slope-Intercept Form

We can use the slope-intercept form to write an equation of a line given its slope and $y$-intercept.

Write an equation of the line with $y$-intercept $(0, -1)$ and slope $\dfrac{2}{3}$.

$$y = mx + b$$
$$y = \frac{2}{3}x - 1$$

## Section 3.4    More Equations of Lines

The **point-slope form** of the equation of a line is

$$y - y_1 = m(x - x_1)$$

where $m$ is the slope of the line and $(x_1, y_1)$ is a point on the line.

Find an equation of the line with slope 2 containing the point $(1, -4)$. Write the equation in standard form: $Ax + By = C$.

$$y - y_1 = m(x - x_1)$$
$$y - (-4) = 2(x - 1)$$
$$y + 4 = 2x - 2$$
$$-2x + y = -6 \qquad \text{Standard form}$$

| Definitions and Concepts | Examples |
|---|---|

### Section 3.5   Graphing Linear Inequalities

If the equal sign in a linear equation in two variables is replaced with an inequality symbol, the result is a **linear inequality in two variables.**

$x \leq -5y, y \geq 2,$

$3x - 2y > 7, x < -5$

**Graphing a Linear Inequality**

Graph:   $2x - 4y > 4$

**Step 1:** Graph the **boundary line** by graphing the related equation. Draw a solid line if the inequality symbol is $\leq$ or $\geq$. Draw a dashed line if the inequality symbol is $<$ or $>$.

1. Graph $2x - 4y = 4$. Draw a dashed line because the inequality symbol is $>$.

**Step 2:** Choose a **test point** not on the line. Substitute its coordinates into the original inequality.

2. Check the test point $(0, 0)$ in the inequality $2x - 4y > 4$.

$2 \cdot 0 - 4 \cdot 0 > 4$    Let $x = 0$ and $y = 0$.

$0 > 4$    False

**Step 3:** If the resulting inequality is true, shade the **half-plane** that contains the test point. If the inequality is not true, shade the half-plane that does not contain the test point.

3. The inequality is false, so shade the half-plane that does not contain $(0, 0)$.

### Section 3.6   Introduction to Functions

A **relation** is a set of ordered pairs. The **domain** of the relation is the set of all first components of the ordered pairs. The **range** of the relation is the set of all second components of the ordered pairs.

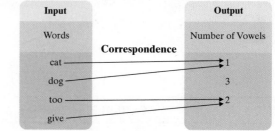

Domain: {cat, dog, too, give}
Range: {1, 2}

A **function** is a relation in which each element of the first set corresponds to exactly one element of the second set.

The previous relation is a function. Each word contains one exact number of vowels.

**Vertical Line Test**

If no vertical line can be drawn so that it intersects a graph more than once, the graph is the graph of a function. If such a line can be drawn, the graph is not that of a function.

Find the domain and the range of the relation. Also determine whether the relation is a function.

Range $(-\infty, 0]$

Domain: $(-\infty, \infty)$

By the vertical line test, this is the graph of a function.

*(Continued)*

| Definitions and Concepts | Examples |
|---|---|

### Section 3.6   Introduction to Functions (*continued*)

The symbol $f(x)$ means **function of $x$** and is called **function notation.**

A **linear function** is a function that can be written in the form

$$f(x) = mx + b$$

**To graph a linear function,** use the slope and $y$-intercept.

If $f(x) = 2x^2 - 5$, find $f(-3)$.

$$f(-3) = 2(-3)^2 - 5 = 2(9) - 5 = 13$$

$$f(x) = -3, g(x) = 5x, h(x) = -\frac{1}{3}x - 7$$

Graph:   $f(x) = -2x$
(or $y = -2x + 0$)

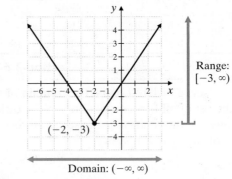

The slope is $\dfrac{2}{-1}$.

The $y$-intercept is $(0, 0)$.

### Section 3.7   Finding Domains and Ranges from Graphs and Graphing Piecewise-Defined Functions

To find the domain of a function (or relation) from its graph, recall that on the rectangular coordinate system, "domain" means the $x$-values that are graphed. Similarly, "range" means the $y$-values that are graphed.

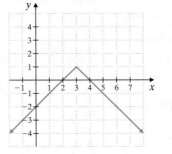

Range:
$[-3, \infty)$

$(-2, -3)$

Domain: $(-\infty, \infty)$

### Section 3.8   Shifting and Reflecting Graphs of Functions

Vertical shifts (upward and downward) let $k$ be a positive number.

| Graph of | Same as | Moved |
|---|---|---|
| $g(x) = f(x) + k$ | $f(x)$ | $k$ units upward |
| $g(x) = f(x) - k$ | $f(x)$ | $k$ units downward |

Horizontal shift (to the left or right) let $h$ be a positive number.

| Graph of | Same as | Moved |
|---|---|---|
| $g(x) = f(x - h)$ | $f(x)$ | $h$ units to the right |
| $g(x) = f(x + h)$ | $f(x)$ | $h$ units to the left |

Reflection about the $x$-axis
The graph of $g(x) = -f(x)$ is the graph of $f(x)$ reflected about the $x$-axis.

The graph of $h(x) = -|x - 3| + 1$ is the same as the graph of $f(x) = |x|$, reflected about the $x$-axis, shifted 3 units right, then 1 unit up.

**(3.1)** *Plot the points and name the quadrant or axis in which each point lies.*

**1.** $A(2, -1), B(-2, 1), C(0, 3), D(-3, -5)$

**2.** $A(-3, 4), B(4, -3), C(-2, 0), D(-4, 1)$

*Create a scatter diagram for the given paired data.*

**3.** U.S. Armed Forces Active Duty Personnel

| Year | Armed Forces (in millions) |
|------|----------------------------|
| 2008 | 1.39 |
| 2009 | 1.45 |
| 2010 | 1.42 |
| 2011 | 1.45 |
| 2012 | 1.43 |
| 2013 | 1.43 |

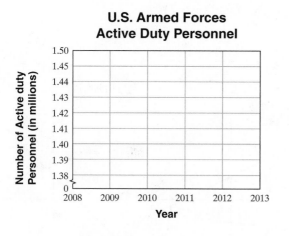

**4.** Per-Person Consumption of Caloric Sweeteners in the United States

| Year | Per Person Consumption (in pounds) |
|------|-------------------------------------|
| 2007 | 137 |
| 2008 | 137 |
| 2009 | 130 |
| 2010 | 131 |
| 2011 | 130 |
| 2012 | 130 |

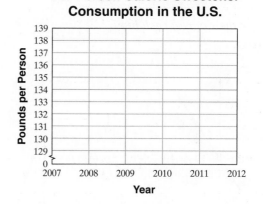

*Determine whether each ordered pair is a solution to the given equation.*

**5.** $7x - 8y = 56; (0, 56), (8, 0)$

**6.** $-2x + 5y = 10; (-5, 0), (1, 1)$

**7.** $x = 13; (13, 5), (13, 13)$

**8.** $y = 2; (7, 2), (2, 7)$

*Graph each linear equation.*

**9.** $3x - y = 3$

**10.** $2x - y = 4$

**11.** $4x + 5y = 20$

**12.** $y = 5$

**13.** $x = -2$

**14.** $y = \frac{1}{3}x$

**(3.2)** *Find the slope of the line through each pair of points.*

**15.** $(2, 8)$ and $(6, -4)$

**16.** $(-3, 9)$ and $(5, 13)$

**17.** $(-7, -4)$ and $(-3, 6)$

**18.** $(7, -2)$ and $(-5, 7)$

*Determine the slope of each line.*

**19.**

**20.**

**21.**

**22.**

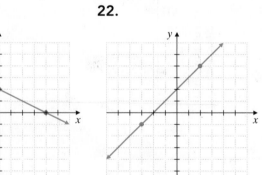

*Two lines are graphed on each set of axes. Determine whether $l_1$ or $l_2$ has the greater slope.*

**23.**    **24.**    **25.**    **26.**

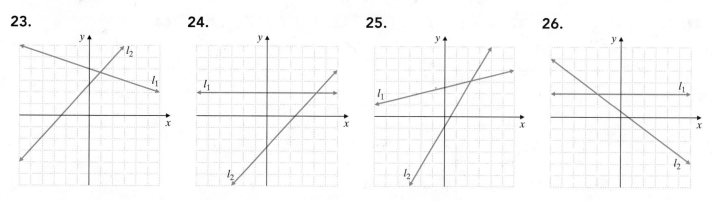

*Find the slope and y-intercept of each line.*

**27.** $y = -3x + \dfrac{1}{2}$

**28.** $y = 2x + 4$

**29.** $6x - 15y = 20$

**30.** $4x + 14y = 21$

*Find the slope of each line.*

**31.** $y - 3 = 0$

**32.** $x = -5$

*Determine whether each pair of lines is parallel, perpendicular, or neither.*

**33.** $y = -2x + 6$
     $y = 2x - 1$

**34.** $-x + 3y = 2$
     $6x - 18y = 3$

**35.** $y = \dfrac{3}{4}x + 1$
     $y = -\dfrac{4}{3}x + 1$

**36.** $x - 2y = 6$
     $4x + y = 8$

**(3.3)** *Graph each line passing through the given point with the given slope.*

**37.** Through $(2, -3)$ with slope $\dfrac{2}{3}$

**38.** Through $(1, -4)$ with slope $\dfrac{1}{2}$

**39.** Through $(0, 1)$ with slope $2$

**40.** Through $(-2, 0)$ with slope $-3$

*Graph each linear equation using the slope and y-intercept.*

**41.** $y = -x + 1$     **42.** $y = 4x - 3$     **43.** $3x - y = 6$     **44.** $y = -5x$

**45.** The cost $C$, in dollars, of renting a minivan for a day is given by the linear equation $C = 0.3x + 42$, where $x$ is number of miles driven.
  **a.** Find the cost of renting the minivan for a day and driving it 150 miles.
  **b.** Find and interpret the slope of this equation.
  **c.** Find and interpret the $y$-intercept of this equation.

**46.** The cost $C$ in dollars, of renting a compact car for a day is given by the linear equation $C = 0.4x + 19$, where $x$ is the number of miles driven.
  **a.** Find the cost of renting the compact car for a day and driving it 325 miles.
  **b.** Find and interpret the slope of this equation.
  **c.** Find and interpret the $y$-intercept of this equation.

**(3.4)** *Write an equation of the line satisfying each set of conditions.*

**47.** Slope undefined; through $(-4, -3)$

**48.** Horizontal; through $(3, -1)$

*Write an equation of the line satisfying each set of conditions. Write the equation in the form $y = mx + b$.*

**49.** Through $(-6, -1)$ and $(-4, -2)$

**50.** Through $(-3, 5)$; slope 3

**51.** Through $(-6, -1)$; perpendicular to $4x + 3y = 5$

**52.** Through $(2, -6)$; parallel to $y = -2x + 3$

**53.** The value of a building bought in 2002 continues to increase as time passes. Seven years after the building was bought, it was worth $210,000; 12 years after it was bought, it was worth $270,000.
  **a.** Assuming that this relationship between the number of years past 2002 and the value of the building is linear, write an equation describing this relationship. [*Hint:* Use ordered pairs of the form (years past 2002, value of the building).]
  **b.** Use this equation to estimate the value of the building in 2020.

**54.** The value of an automobile bought in 2010 continues to decrease as time passes. Two years after the car was bought, it was worth $17,500; four years after it was bought, it was worth $14,300.
  **a.** Assuming that this relationship between the number of years past 2010 and the value of the car is linear, write an equation describing this relationship. [*Hint:* Use ordered pairs of the form (years past 2010, value of the automobile).]
  **b.** Use this equation to estimate the value of the automobile in 2016.

**(3.5)** *Graph each linear inequality.*

**55.** $3x + y > 4$     **56.** $\dfrac{1}{2}x - y < 2$     **57.** $5x - 2y \leq 9$     **58.** $3x - 6y \geq 11$

**59.** $y < 1$     **60.** $x > -2$     **61.** Graph the union of $y > 2x + 3$ or $x \leq -3$.     **62.** Graph the intersection of $2x < 3y + 8$ and $y \geq -2$.

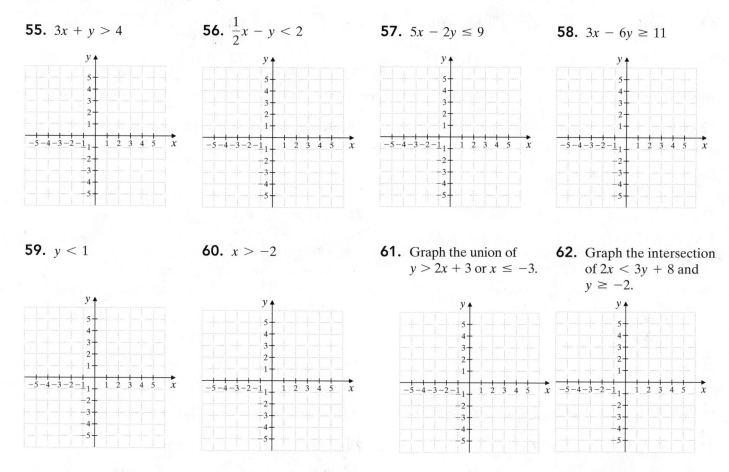

**(3.6)** *Find the domain and range of each relation. Then determine whether the relation is also a function.*

**63.** $\left\{\left(-\dfrac{1}{2}, \dfrac{3}{4}\right), (6, 0.65), (0, -12), (25, 25)\right\}$

**64.** $\left\{\left(\dfrac{3}{4}, -\dfrac{1}{2}\right), (0.65, 6), (-12, 0), (25, 25)\right\}$

**65.**

**66.**

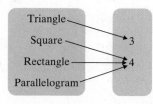

*If* $f(x) = x - 5, g(x) = -3x,$ *and* $h(x) = 2x^2 - 6x + 1,$ *find each function value.*

**67.** $f(2)$                **68.** $g(0)$                **69.** $g(-6)$

**70.** $h(-1)$                **71.** $h(1)$                **72.** $f(5)$

*The function* $J(x) = 2.54x$ *may be used to calculate the weight of an object on Jupiter (J) given its weight on Earth (x).*

**73.** If a person weighs 150 pounds on Earth, find the equivalent weight on Jupiter.

**74.** A 2000-pound probe on Earth weighs how many pounds on Jupiter?

*Graph each linear function.*

**75.** $f(x) = x + 2$

**76.** $f(x) = -\dfrac{1}{2}x + 3$

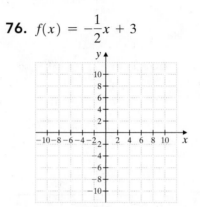

**(3.7)** *Find the domain and range of each relation. For Exercises 77 through 80, also determine whether the relation is a function.*

**77.**        **78.**        **79.**        **80.**

**81.**        **82.**        **83.**        **84.**

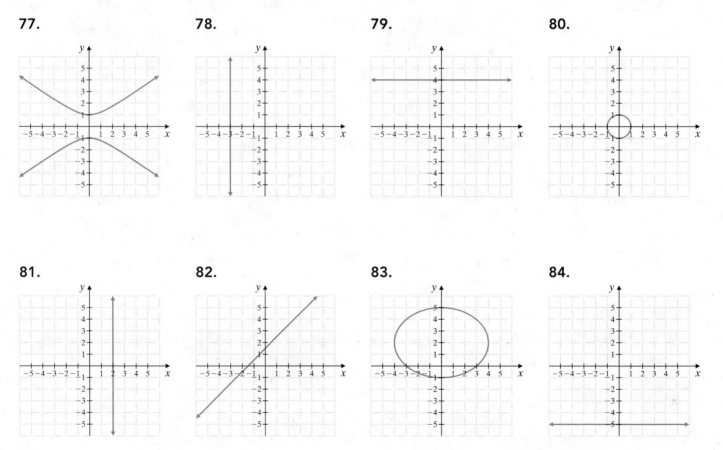

*Graph each function.*

**85.** $f(x) = \begin{cases} -3x & \text{if } x < 0 \\ x - 3 & \text{if } x \geq 0 \end{cases}$

**86.** $g(x) = \begin{cases} -\dfrac{1}{5}x & \text{if } x \leq -1 \\ -4x + 2 & \text{if } x > -1 \end{cases}$

**(3.8)** *Graph each function.*

**87.** $y = \sqrt{x} - 4$    **88.** $f(x) = \sqrt{x - 4}$    **89.** $g(x) = |x - 2| - 2$    **90.** $h(x) = -(x + 3)^2 - 1$

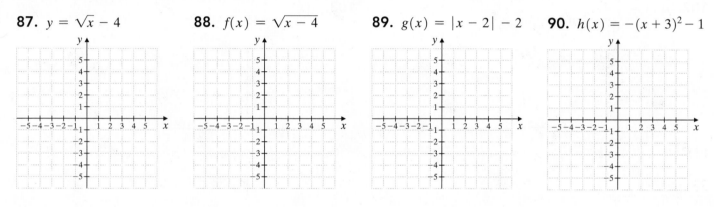

## Mixed Review

*Graph each linear equation or inequality.*

**91.** $3x - 2y = -9$    **92.** $x = -4y$    **93.** $3y \geq x$

*Write an equation of the line satisfying each set of conditions. If possible, write the equation in the form $y = mx + b$.*

**94.** Vertical; through $(-2, -4)$

**95.** Slope 0; through $(2, 5)$

**96.** Slope 2; through $(5, -2)$

**97.** Through $(-5, 3)$ and $(-4, -8)$

**98.** Through $(-4, -2)$; parallel to $y = -\dfrac{3}{2}x + 1$

**99.** Through $(-4, 5)$; perpendicular to $2x - 3y = 6$

*Find the domain and range of each relation.*

**100.**

**101.**

*Graph each piecewise-defined function.*

**102.** $f(x) = \begin{cases} x - 2 & \text{if } x \le 0 \\ -\dfrac{x}{3} & \text{if } x \ge 3 \end{cases}$

**103.** $g(x) = \begin{cases} 4x - 3 & \text{if } x \le 1 \\ 2x & \text{if } x > 1 \end{cases}$

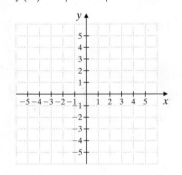

*Graph each function.*

**104.** $f(x) = \sqrt{x - 2}$

**105.** $f(x) = |x + 1| - 3$

**1.** Plot the points, and name the quadrant or axis in which each is located: $A(6, -2)$, $B(4, 0)$, $C(-1, 6)$.

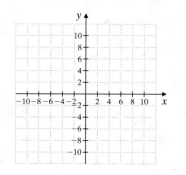

**2.** Create a scatter diagram for the paired data.

| Krispy Kreme Donut Locations | |
|---|---|
| Year | Number of Krispy Kreme Donut Locations |
| 2009 | 523 |
| 2010 | 580 |
| 2011 | 646 |
| 2012 | 694 |
| 2013 | 748 |

**Krispy Kreme Donut Locations**

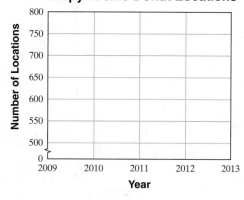

*Graph each linear equation.*

**3.** $-3x + y = -3$

**4.** $2x - 3y = -6$

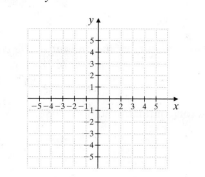

**5.** $4x + 6y = 8$

**6.** $y = -3$

**Answers**

1. _____

2. _____

3. _____

4. _____

5. _____

6. _____

7. _____

8. _____

9. _____

10. _____

11. _____

12. _____

13. _____

14. _____

15. _____

16. _____

17. _____

18. _____

19. _____

20. _____

**7.** Find the slope of the line that passes through $(5, -8)$ and $(-7, 10)$.

**8.** Find the slope and the $y$-intercept of the line $3x + 12y = 8$.

*Match each equation with its graph.*

**9.** $f(x) = 3x + 1$

**10.** $f(x) = 3x - 2$

**11.** $f(x) = 3x + 2$

**12.** $f(x) = 3x - 5$

*Find an equation of the line satisfying each set of conditions. Write the equations in the form $x = a, y = b,$ or $y = mx + b.$ For Exercises 16 and 17, write the equations in standard form, $Ax + By = C.$*

**13.** Horizontal; through $(2, -8)$

**14.** Vertical; through $(-4, -3)$

**15.** Perpendicular to $x = 5$; through $(3, -2)$

**16.** Through $(4, -1)$; slope $-3$

**17.** Through $(0, -2)$; slope $5$

**18.** Through $(4, -2)$ and $(6, -3)$

**19.** Through $(-1, 2)$; perpendicular to $3x - y = 4$

**20.** Parallel to $2y + x = 3$; through $(3, -2)$

**21.** Line $L_1$ has the equation $2x - 5y = 8$. Line $L_2$ passes through the points $(1, 4)$ and $(-1, -1)$. Determine whether these lines are parallel, perpendicular, or neither.

**21.** _____

**22.** For the 2013 Major League Baseball season, the following linear equation describes the relationship between a team's payroll $x$ (in millions of dollars) and the number of games $y$ that team won during the regular season.

$$y = 0.66x + 15.75$$

Round to the nearest whole. (*Sources:* Based on data from Major League Baseball and *Forbes*)

**22. a.** _____

**a.** According to this equation, how many games would have been won during the 2013 season by a team with a payroll of $90 million?

**b.** The Tampa Bay Rays had a payroll of $58 million in 2013. According to this equation, how many games would they have won in 2013?

**b.** _____

**c.** According to this equation, what payroll would have been necessary in 2013 to have won 95 games during the season?

**d.** Find and interpret the slope of the equation.

*Graph each inequality.*

**c.** _____

**23.** $x \le -4$

**24.** $2x - y > 5$

**d.** _____

**25.** Graph the intersection of $2x + 4y < 6$ and $y \le -4$

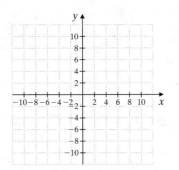

**23.** _____

**24.** _____

*Find the domain and range of each relation. Also determine whether the relation is a function.*

**25.** _____

**26.**

**27.**

**26.** _____

**27.** _____

**28.**

**29.**

**28.** _____

**29.** _____

*Graph each function. For Exercises 30 and 32, state the domain and the range of the function.*

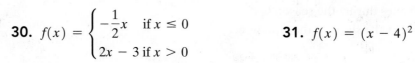

**30.** $f(x) = \begin{cases} -\dfrac{1}{2}x & \text{if } x \le 0 \\ 2x - 3 & \text{if } x > 0 \end{cases}$

**31.** $f(x) = (x - 4)^2$

**30.** _____

**31.** _____

**32.** $g(x) = -|x + 2| - 1$

**33.** $h(x) = \sqrt{x} - 1$

**32.** _____

**33.** _____

*Write each set in roster form.*

**1.** $\{x \mid x$ is a whole number between 1 and 6$\}$

**2.** $\{x \mid x$ is an integer between $-2$ and $3\}$

**3.** $\{x \mid x$ is a natural number greater than than 100$\}$

**4.** Multiply or divide.

    **a.** $\dfrac{-42}{-6}$

    **b.** $\dfrac{0}{14}$

    **c.** $-1(-5)(-2)$

*For Exercises 5 through 7, find the reciprocal, or multiplicative inverse, of each number if one exists.*

**5.** $-9$

**6.** $0$

**7.** $\dfrac{7}{4}$

**8.** Evaluate $2x^2$ for

    **a.** $x = 7$

    **b.** $x = -7$

*Find each absolute value.*

**9.** $|3|$

**10.** $\left| -\dfrac{7}{15} \right|$

**11.** $-|2.7|$

**12.** $-|-102.86|$

**13.** Simplify: $3 - [6(4 - 6) + 2(5 - 9)]$

**14.** Simplify: $-2 + 3[5 - (7 - 10)]$.

*Use the product rule to simplify.*

**15.** $2^2 \cdot 2^5$

**16.** $-7x^9 \cdot \dfrac{3}{14}x^4$

**17.** $y \cdot y^2 \cdot y^4$

**18.** $(-2ab)(-0.3a^7b^{10})(8.1b^2)$

| Answers |
|---|
| 1. |
| 2. |
| 3. |
| 4. a. |
|    b. |
|    c. |
| 5. |
| 6. |
| 7. |
| 8. a. |
|    b. |
| 9. |
| 10. |
| 11. |
| 12. |
| 13. |
| 14. |
| 15. |
| 16. |
| 17. |
| 18. |

**19.** _____

**20.** _____

**21.** _____

**22.** _____

**23.** _____

**24.** _____

**25.** _____

**26.** _____

**27.** _____

**28. a.** _____

**b.** _____

**c.** _____

**29.** _____

**30.** _____

**31.** _____

**32.** _____

**33.** _____

**34.** _____

**35.** _____

**36.** _____

**37.** _____

_Use the power rules to simplify. Write each result using positive exponents only._

**19.** $(5x^2)^3$

**20.** $(3^{-2}y^4)^2$

**21.** $\left(\dfrac{2^{-3}}{y}\right)^{-2}$

**22.** $\left(\dfrac{5a^3}{b^{-2}c^5}\right)^{-3}$

**23.** Solve:  $0.6 = 2 - 3.5c$.

**24.** Solve:  $5(x - 7) = 4x - 35 + x$.

**25.** Find three numbers such that the second number is 3 more than twice the first number, and the third number is four times the first number. The sum of the three numbers is 164.

**26.** Find 3 consecutive odd integers whose sum is 213.

**27.** Marial Callier just received an inheritance of $10,000 and plans to place all the money in a savings account that pays 5% compounded quarterly to help her son go to college in 3 years. How much money will be in the account in 3 years?

**28.** Add.

   **a.** $-4 + (-3)$

   **b.** $\dfrac{1}{2} - \left(-\dfrac{1}{3}\right)$

   **c.** $7 - 20$

_Graph each set on a number line and then write it in interval notation._

**29.** $\{x \mid x \geq 2\}$

**30.** $\left\{x \mid x < -\dfrac{3}{2}\right\}$

**31.** $\{x \mid 0.5 < x \leq 3\}$

_Solve._

**32.** $3x + 10 > \dfrac{5}{2}(x - 1)$

**33.** $x - 7 < 2$ _and_ $2x + 1 < 9$

**34.** $x - 2 < 6$ _or_ $3x + 1 > 1$

**35.** $|x - 3| = |5 - x|$

**36.** $|5x - 2| - 7 = -4$

**37.** $|y - 3| > 7$

**38.** $|-x + 8| - 2 \leq 8$

**39.** Find the intercepts and graph:
$3x + 4y = -12$

**40.** Find the slope and $y$-intercept of $7x + 2y = 10$.

**41.** Find the slope of the line $y = 3x + 2$.

**42.** Find the slope of the line through $(-1, 6)$ and $(0, 9)$.

**43.** Write an equation of the line with $y$-intercept $(0, -7)$ and slope of $\dfrac{2}{3}$.

**44.** Write an equation of the line through $(-2, 5)$ and $(-4, 7)$ in standard form.

**45.** Write an equation of the horizontal line containing the point $(2, 3)$.

**46.** Find an equation of the vertical line through $\left(-2, -\dfrac{3}{4}\right)$.

*Find the following.*

**47.** If $f(x) = 7x^2 - 3x + 1$, find $f(1)$.

**48.** Find the slope of the line defined by $y = -2.3x - 6$.

**49.** If $g(x) = 3x - 2$, find $g(0)$.

**50.** Determine whether the graph below is the graph of a function.

38. _____

39. _____

40. _____

41. _____

42. _____

43. _____

44. _____

45. _____

46. _____

47. _____

48. _____

49. _____

50. _____

# Systems of Equations and Inequalities

**Check Your Progress**

Vocabulary Check

Chapter Highlights

Chapter Review

Chapter Test

Cumulative Review

In this chapter, two or more equations in two or more variables are solved simultaneously. Such a collection of equations is called a *system of equations*. Systems of equations are good mathematical models for many real-world problems because these problems may involve several related patterns. We will study various methods for solving systems of equations and will conclude with a look at systems of inequalities.

## Does Winter Occur Globally During the Same Months?

While North Americans and other citizens of the northern hemisphere, including the English, are enjoying winter and possibly snow, Australians are celebrating the holidays on the beach with a picnic. Why such a difference in temperature? It is because the seasons are reversed between the northern and southern hemispheres. While winter in the northern hemisphere is cold, it is warm in the southern hemisphere. The winter solstice is our shortest day of the year, while below the equator it is the longest day of the year.

Australia and England are examples of countries on opposite sides of the equator. In Exercise 55 of the Chapter Review, we will find the month (between February and July) when the average high temperature in Australia is the same as the average high temperature in England.

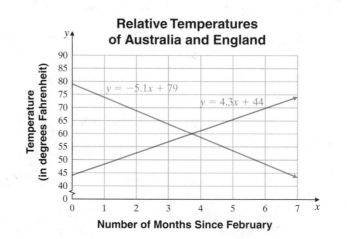

**Relative Temperatures of Australia and England**

$y = -5.1x + 79$

$y = 4.3x + 44$

Temperature (in degrees Fahrenheit)

Number of Months Since February

## 4.1 Solving Systems of Linear Equations in Two Variables

Recall from Chapter 3 that the graph of a linear equation in two variables is a line. Two or more linear equations form a **system of linear equations.** Some examples of systems of linear equations in two variables follow.

$$\begin{cases} x - 2y = -7 \\ 3x + y = 0 \end{cases} \qquad \begin{cases} x = 5 \\ x + \dfrac{y}{2} = 9 \end{cases} \qquad \begin{cases} x - 3 = 2y + 6 \\ y = 1 \end{cases}$$

**Objectives**

**A** Determine Whether an Ordered Pair Is a Solution of a System of Two Linear Equations.

**B** Solve a System of Two Equations by Graphing.

**C** Solve a System Using Substitution.

**D** Solve a System Using Elimination.

### Objective A  Determining Whether an Ordered Pair Is a Solution

Recall that a solution of an equation in two variables is an ordered pair $(x, y)$ that makes the equation true. A **solution of a system** of two equations in two variables is an ordered pair $(x, y)$ that makes both equations true.

---

**Example 1**  Determine whether the ordered pair $(-1, 1)$ is a solution of the system.

$$\begin{cases} -x + y = 2 \\ 2x - y = -3 \end{cases}$$

**Solution:**  We replace $x$ with $-1$ and $y$ with $1$ in each equation.

$$\begin{aligned} -x + y &= 2 \quad &&\text{First equation} \\ -(-1) + (1) &\overset{?}{=} 2 \quad &&\text{Let } x = -1 \text{ and } y = 1. \\ 1 + 1 &\overset{?}{=} 2 \\ 2 &= 2 \quad &&\text{True} \end{aligned}$$

$$\begin{aligned} 2x - y &= -3 \quad &&\text{Second equation} \\ 2(-1) - (1) &\overset{?}{=} -3 \quad &&\text{Let } x = -1 \text{ and } y = 1. \\ -2 - 1 &\overset{?}{=} -3 \\ -3 &= -3 \quad &&\text{True} \end{aligned}$$

Since $(-1, 1)$ makes both equations true, it is a solution.

▶ Work Practice 1

**Practice 1**

Determine whether the ordered pair $(4, 1)$ is a solution of the system.

$$\begin{cases} x - y = 3 \\ 2x - 3y = 5 \end{cases}$$

---

**Example 2**  Determine whether the ordered pair $(-2, 3)$ is a solution of the system.

$$\begin{cases} 5x + 3y = -1 \\ x - y = 1 \end{cases}$$

**Solution:**  We replace $x$ with $-2$ and $y$ with $3$ in each equation.

$$\begin{aligned} 5x + 3y &= -1 \quad &&\text{First equation} \\ 5(-2) + 3(3) &\overset{?}{=} -1 \quad &&\text{Let } x = -2 \text{ and } y = 3. \\ -10 + 9 &\overset{?}{=} -1 \\ -1 &= -1 \quad &&\text{True} \end{aligned}$$

*(Continued on next page)*

**Practice 2**

Determine whether the ordered pair $(-3, 3)$ is a solution of the system.

$$\begin{cases} 3x - y = -12 \\ x - y = 0 \end{cases}$$

**Answers**
1. yes, a solution   2. no, not a solution

$$x - y = 1 \quad \text{Second equation}$$
$$(-2) - (3) \stackrel{?}{=} 1 \quad \text{Let } x = -2 \text{ and } y = 3.$$
$$-5 = 1 \quad \text{False}$$

Since the ordered pair $(-2, 3)$ does not make both equations true, it is not a solution of the system.

◼ Work Practice 2

 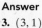
## Objective B Solving a System by Graphing

The graph of each linear equation in a system is a line. Each point on each line corresponds to an ordered pair solution of its equation. If the lines intersect, the point of intersection lies on both lines and corresponds to an ordered pair solution of both equations. In other words, the point of intersection corresponds to an ordered pair solution of the system. Therefore, we can estimate the solutions of a system by graphing the equations on the same rectangular coordinate system and estimating the coordinates of any points of intersection.

**Practice 3**

Solve the system by graphing. If the system has just one solution, estimate the solution.

3. $\begin{cases} x - y = 2 \\ x + 3y = 6 \end{cases}$

**Example 3** Solve the system by graphing.

$$\begin{cases} x + y = 2 \\ 3x - y = -2 \end{cases}$$

**Solution:** First we graph the linear equations on the same rectangular coordinate system. These lines intersect at one point as shown. The coordinates of the point of intersection appear to be $(0, 2)$. We check this estimated solution by replacing $x$ with $0$ and $y$ with $2$ in *both* equations.

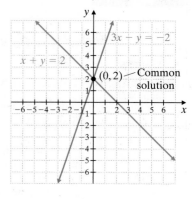

$$x + y = 2 \quad \text{First equation}$$
$$0 + 2 \stackrel{?}{=} 2 \quad \text{Let } x = 0 \text{ and } y = 2.$$
$$2 = 2 \quad \text{True}$$

$$3x - y = -2 \quad \text{Second equation}$$
$$3(0) - 2 \stackrel{?}{=} -2 \quad \text{Let } x = 0 \text{ and } y = 2.$$
$$-2 = -2 \quad \text{True}$$

The ordered pair $(0, 2)$ is the solution of the system. A system that has at least one solution, such as this one, is said to be **consistent.**

◼ Work Practice 3

In Example 3, we have a **consistent system.** To review, a system that has at least one solution is said to be consistent.

Later, we will talk about **dependent equations.** For now, we define an **independent equation** to be an equation in a system of equations that cannot be algebraically derived from any other equation in the system.

Thus, for:

**Answer**

3. $(3, 1)$

Consistent System: the system has at least one solution

One solution

Independent Equations: each equation in the system cannot be algebraically derived from the other

**Example 4**  Solve the system by graphing.

$$\begin{cases} x - 2y = 4 \\ x \quad\quad = 2y \end{cases}$$

**Solution:**  We graph each linear equation.

The lines appear to be parallel. To be sure, let's write each equation in slope-intercept form, $y = mx + b$. To do so, we solve for $y$.

$$x - 2y = 4 \quad\quad \text{First equation}$$

$$-2y = -x + 4 \quad\quad \text{Subtract } x \text{ from both sides.}$$

$$y = \frac{1}{2}x - 2 \quad\quad \text{Divide both sides by } -2.$$

$$x = 2y \quad\quad \text{Second equation}$$

$$\frac{1}{2}x = y \quad\quad \text{Divide both sides by 2.}$$

$$y = \frac{1}{2}x$$

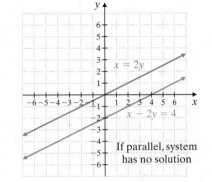

If parallel, system has no solution

The graphs of these equations have the same slope, $\frac{1}{2}$, but different $y$-intercepts, so these lines are parallel. Therefore, the system has no solution since the equations have no common solution (there are no intersection points). A system that has no solution is said to be **inconsistent.**

■ Work Practice 4

In Example 4, we have an **inconsistent system.** To review, a system that has no solution is said to be inconsistent.

Let's now talk about the equations in this system. Each equation in this system cannot be algebraically derived from the other, so each equation is independent of the other.

Thus:

No solution

**Inconsistent System:** the system has no solution

**Independent Equations:** each equation in the system cannot be algebraically derived from the other

**Helpful Hint**

- If a system of equations has *at least one solution,* the system is *consistent.*
- If a system of equations has *no solution,* the system is *inconsistent.*

The pairs of equations in Examples 3 and 4 are called independent because their graphs differ. In Example 5, we see an example of dependent equations.

✓**Concept Check** How can you tell just by looking at the following system that it has no solution?

$$\begin{cases} y = 3x + 5 \\ y = 3x - 7 \end{cases}$$

**Practice 4**

Solve the system by graphing. If the system has just one solution, estimate the solution.

4. $\begin{cases} y = -3x \\ 6x + 2y = 4 \end{cases}$

**Answer**

4. no solution, or $\varnothing$

✓**Concept Check Answer**

answers may vary

## Practice 5

Solve the system by graphing.

$$\begin{cases} -2x + y = 1 \\ 4x - 2y = -2 \end{cases}$$

**Example 5**    Solve the system by graphing.

$$\begin{cases} 2x + 4y = 10 \\ x + 2y = 5 \end{cases}$$

**Solution:**   We graph each linear equation. We see that the graphs of the equations are the same line. To confirm this, notice that if both sides of the second equation are multiplied by 2, the result is the first equation. This means that the equations have identical solutions. Any ordered pair solution of one equation satisfies the other equation also. These equations are said to be **dependent equations.** The solution set of the system is $\{(x, y)\,|\,x + 2y = 5\}$ or, equivalently, $\{(x, y)\,|\,2x + 4y = 10\}$ since the lines describe identical ordered pairs. Written the second way, the solution set is read "the set of all ordered pairs $(x, y)$, such that $2x + 4y = 10$." There is an infinite number of solutions to this system.

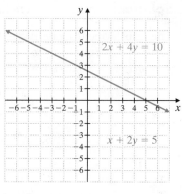

■ **Work Practice 5**

In Example 5, we have a **consistent system** since the system has at least one solution. In fact, the system in Example 5 has an infinite number of solutions.

Let's now define **dependent equations.** We define a **dependent equation** to be an equation in a system of equations that can be algebraically derived from another equation in the system.

Thus:

Consistent System: the system has at least one solution

Infinite number of solutions

Dependent Equations: an equation in a system of equations can be algebraically derived from another

## Answer

**5.** $\{(x, y)\,|\,-2x + y = 1\,|\}$

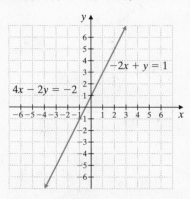

**Helpful Hint**

- If the graphs of two equations *differ*, they are *independent* equations.
- If the graphs of two equations are the *same*, they are *dependent* equations.

✓**Concept Check**  How can you tell just by looking at the following system that it has infinitely many solutions?

$$\begin{cases} x + y = 5 \\ 2x + 2y = 10 \end{cases}$$

✓**Concept Check Answer**

answers may vary

We can summarize the information discovered in Examples 3 through 5 as follows.

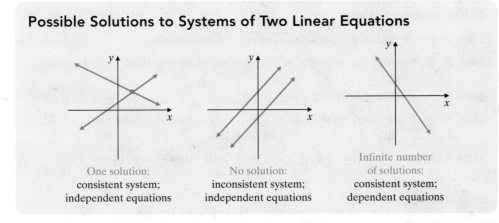

**Possible Solutions to Systems of Two Linear Equations**

One solution:
consistent system;
independent equations

No solution:
inconsistent system;
independent equations

Infinite number
of solutions:
consistent system;
dependent equations

## Objective C  Solving a System Using Substitution

Graphing the equations of a system by hand is often a good method for finding approximate solutions of a system, but it is not a reliable method for finding exact solutions. To find an exact solution, we need to use *algebra*. One *algebraic* method is called the **substitution method.**

**Example 6**  Use the substitution method to solve the system:

$$\begin{cases} 2x + 4y = -6 & \text{First equation} \\ x = 2y - 5 & \text{Second equation} \end{cases}$$

**Solution:**  In the second equation, we are told that $x$ is equal to $2y - 5$. Since they are equal, we can *substitute* $2y - 5$ for $x$ in the first equation. This will give us an equation in one variable, which we can solve for $y$.

$$2x + 4y = -6 \qquad \text{First equation}$$

$$2(2y - 5) + 4y = -6 \qquad \text{Substitute } 2y - 5 \text{ for } x.$$

$$4y - 10 + 4y = -6$$

$$8y = 4$$

$$y = \frac{4}{8} = \frac{1}{2} \quad \text{Solve for } y.$$

The $y$-coordinate of the solution is $\frac{1}{2}$. To find the $x$-coordinate, we replace $y$ with $\frac{1}{2}$ in the second equation,

$$x = 2y - 5$$

$$x = 2y - 5$$

$$x = 2\left(\frac{1}{2}\right) - 5 = 1 - 5 = -4$$

The ordered pair solution is $\left(-4, \frac{1}{2}\right)$. Check to see that $\left(-4, \frac{1}{2}\right)$ satisfies both equations of the system.

■ Work Practice 6

**Practice 6**

Use the substitution method to solve the system:

$$\begin{cases} 6x - 4y = 10 \\ y = 3x - 3 \end{cases}$$

**Answer**

**6.** $\left(\frac{1}{3}, -2\right)$

The substitution method is summarized below. Feel free to use these steps.

> ## Solving a System of Two Equations Using the Substitution Method
>
> **Step 1:** Solve one of the equations for one of its variables.
>
> **Step 2:** Substitute the expression for the variable found in Step 1 into the other equation.
>
> **Step 3:** Find the value of one variable by solving the equation from Step 2.
>
> **Step 4:** Find the value of the other variable by substituting the value found in Step 3 into the equation from Step 1.
>
> **Step 5:** Check the ordered pair solution in *both* original equations.

## Practice 7

Use the substitution method to solve the system:

$$\begin{cases} -\dfrac{x}{2} + \dfrac{y}{4} = \dfrac{1}{2} \\ \dfrac{x}{2} + \dfrac{y}{2} = -\dfrac{1}{8} \end{cases}$$

> **Helpful Hint** If a system of equations contains equations with fractions, the first step is to clear the equations of fractions.

**Answer**

7. $\left( -\dfrac{3}{4}, \dfrac{1}{2} \right)$

---

**Example 7** Use the substitution method to solve the system:

$$\begin{cases} -\dfrac{x}{6} + \dfrac{y}{2} = \dfrac{1}{2} \\ \dfrac{x}{3} - \dfrac{y}{6} = -\dfrac{3}{4} \end{cases}$$

**Solution:** First we multiply each equation by its least common denominator to clear the system of fractions. We multiply the first equation by 6 and the second equation by 12.

$$\begin{cases} 6\left( -\dfrac{x}{6} + \dfrac{y}{2} \right) = 6\left( \dfrac{1}{2} \right) \\ 12\left( \dfrac{x}{3} - \dfrac{y}{6} \right) = 12\left( -\dfrac{3}{4} \right) \end{cases} \text{ simplifies to } \begin{cases} -x + 3y = 3 & \text{First equation} \\ 4x - 2y = -9 & \text{Second equation} \end{cases}$$

We now solve the first equation for $x$ so that we may substitute our findings into the second equation.

$$-x + 3y = 3 \quad \text{First equation}$$
$$3y - 3 = x \quad \text{Solve for } x.$$

Next we replace $x$ with $3y - 3$ in the second equation.

$$4x - 2y = -9 \quad \text{Second equation}$$
$$4(3y - 3) - 2y = -9$$
$$12y - 12 - 2y = -9$$
$$10y = 3$$
$$y = \frac{3}{10} \quad \text{Solve for } y.$$

> **Helpful Hint** To avoid tedious fractions, solve for a variable whose coefficient is 1 or −1, if possible.

The $y$-coordinate is $\dfrac{3}{10}$. To find the $x$-coordinate, we replace $y$ with $\dfrac{3}{10}$ in the equation $x = 3y - 3$. Then

$$x = 3\left( \frac{3}{10} \right) - 3 = \frac{9}{10} - 3 = \frac{9}{10} - \frac{30}{10} = -\frac{21}{10}$$

The ordered pair solution is $\left( -\dfrac{21}{10}, \dfrac{3}{10} \right)$. Check to see that this solution satisfies both original equations.

▪ **Work Practice 7**

## Objective D Solving a System Using Elimination ▶

The **elimination method,** or **addition method,** is a second algebraic technique for solving systems of equations. For this method, we rely on a version of the addition property of equality, which states that "equals added to equals are equal."

If $A = B$ and $C = D$   then   $A + C = B + D$

**Example 8**   Use the elimination method to solve the system:

$$\begin{cases} x - 5y = -12 & \text{First equation} \\ -x + y = \phantom{-}4 & \text{Second equation} \end{cases}$$

**Solution:**   Since the left side of each equation is equal to the right side, we add equal quantities by adding the left sides of the equations and the right sides of the equations. This sum gives us an equation in one variable, $y$, which we can solve for $y$.

$$
\begin{array}{ll}
x - 5y = -12 & \text{First equation} \\
\underline{-x + \phantom{5}y = 4} & \text{Second equation} \\
\phantom{-x}-4y = -8 & \text{Add.} \\
\phantom{-x-4}y = 2 & \text{Solve for } y.
\end{array}
$$

The $y$-coordinate of the solution is 2. To find the corresponding $x$-coordinate, we replace $y$ with 2 in either original equation of the system. Let's use the second equation.

$$
\begin{array}{ll}
-x + y = 4 & \text{Second equation} \\
-x + 2 = 4 & \text{Let } y = 2. \\
-x = 2 & \\
x = -2 &
\end{array}
$$

The ordered pair solution is $(-2, 2)$. Check to see that $(-2, 2)$ satisfies both equations of the system.

▶ **Work Practice 8**

The steps below summarize the elimination method.

**Practice 8**

Use the elimination method to solve the system:

$$\begin{cases} 3x - y = 1 \\ 4x + y = 6 \end{cases}$$

## Solving a System of Two Linear Equations Using the Elimination Method

**Step 1:** Rewrite each equation in standard form, $Ax + By = C$.

**Step 2:** If necessary, multiply one or both equations by some nonzero number so that the coefficient of one variable in one equation is the opposite of the coefficient of that variable in the other equation.

**Step 3:** Add the equations. Your chosen variable should be eliminated.

**Step 4:** Find the value of the remaining variable by solving the equation from Step 3.

**Step 5:** Find the value of the other variable by substituting the value found in Step 4 into either original equation.

**Step 6:** Check the proposed ordered pair solution in *both* original equations.

**Answer**
**8.** $(1, 2)$

## Practice 9

Use the elimination method to solve the system:

$$\begin{cases} \dfrac{x}{3} + 2y = -1 \\ x + 6y = 2 \end{cases}$$

**Example 9**   Use the elimination method to solve the system:

$$\begin{cases} 3x + \dfrac{y}{2} = 2 \\ 6x + y = 5 \end{cases}$$

**Solution:** If we add the two equations, the sum will still be an equation in two variables. Notice, however, that if we multiply both sides of the first equation by $-2$, the coefficients of $x$ in the two equations will be opposites. Then

$$\begin{cases} -2\left(3x + \dfrac{y}{2}\right) = -2(2) \\ 6x + y = 5 \end{cases} \quad \text{simplifies to} \quad \begin{cases} -6x - y = -4 \\ 6x + y = 5 \end{cases}$$

Now we can add the left sides and add the right sides.

$$\begin{array}{r} -6x - y = -4 \\ 6x + y = 5 \\ \hline 0 = 1 \quad \text{False} \end{array}$$

The resulting equation, $0 = 1$, is false for all values of $y$ or $x$. Thus, the system has no solution. The solution set is $\{\ \}$, or $\varnothing$. This system is inconsistent, and the graphs of the equations are parallel lines.

■ Work Practice 9

## Practice 10

Use the elimination method to solve the system:

$$\begin{cases} 2x - 5y = 6 \\ 3x - 4y = 9 \end{cases}$$

**Example 10**   Use the elimination method to solve the system:

$$\begin{cases} 3x - 2y = 10 \\ 4x - 3y = 15 \end{cases}$$

**Solution:** To eliminate $y$, our first step is to multiply both sides of the first equation by 3 and both sides of the second equation by $-2$. Then

$$\begin{cases} 3(3x - 2y) = 3(10) \\ -2(4x - 3y) = -2(15) \end{cases} \quad \text{simplifies to} \quad \begin{cases} 9x - 6y = 30 \\ -8x + 6y = -30 \end{cases}$$

Next we add the left sides and add the right sides.

$$\begin{array}{r} 9x - 6y = 30 \\ -8x + 6y = -30 \\ \hline x \phantom{+ 6y} = 0 \end{array}$$

To find $y$, we let $x = 0$ in either equation of the system

$$\begin{aligned} 3x - 2y &= 10 \quad \text{First equation} \\ 3(0) - 2y &= 10 \quad \text{Let } x = 0. \\ -2y &= 10 \\ y &= -5 \end{aligned}$$

The ordered pair solution is $(0, -5)$. Check to see that $(0, -5)$ satisfies both equations.

■ Work Practice 10

**Answers**

**9.** no solution, or $\varnothing$   **10.** $(3, 0)$

**Example 11**   Use the elimination method to solve the system:

$$\begin{cases} -5x - 3y = 9 \\ 10x + 6y = -18 \end{cases}$$

**Solution:**   To eliminate $x$, our first step is to multiply both sides of the first equation by 2. Then

$$\begin{cases} 2(-5x - 3y) = 2(9) \\ \phantom{2(}10x + 6y = -18 \end{cases} \quad \text{simplifies to} \quad \begin{cases} -10x - 6y = 18 \\ \phantom{-}10x + 6y = -18 \end{cases}$$

Next we add the equations.

$$\begin{aligned} -10x - 6y &= 18 \\ \underline{10x + 6y} &= \underline{-18} \\ 0 &= 0 \end{aligned}$$

The resulting equation, $0 = 0$, is true for all possible values of $y$ or $x$. Notice in the original system that if both sides of the first equation are multiplied by $-2$, the result is the second equation. This means that the two equations are equivalent. They have the same solution set and there is an infinite number of solutions. Thus, the equations of this system are dependent, and the solution set of the system is

$$\{(x, y) \mid -5x - 3y = 9\} \quad \text{or, equivalently,} \quad \{(x, y) \mid 10x + 6y = -18\}$$

■ **Work Practice 11**

**Helpful Hint**

Remember that not all ordered pairs are solutions of the system in Example 11—only the infinite number of ordered pairs that satisfy $-5x - 3y = 9$ or, equivalently, $10x + 6y = -18$.

**Practice 11**

Use the elimination method to solve the system:

$$\begin{cases} 4x - 7y = 10 \\ -8x + 14y = -20 \end{cases}$$

**Answer**

**11.** $\{(x, y) \mid 4x - 7y = 10\}$

---

## Calculator Explorations  **Graphing**

We may use a grapher to approximate solutions of systems of equations by graphing both equations on the same set of axes and approximating any points of intersection. For example, let's approximate the solution of the system

$$\begin{cases} y = -2.6x + 5.6 \\ y = 4.3x - 4.9 \end{cases}$$

We use a standard window and graph the equations on a single screen.

$$y = 4.3x - 4.9$$
$$y = -2.6x + 5.6$$

The two lines intersect. To approximate the point of intersection, we trace to the point of intersection and use an INTERSECT feature of the grapher, a ZOOM IN feature of the grapher, or redefine the window to [0, 3] by [0, 3]. If we redefine the window to [0, 3] by [0, 3], the screen should look like the following:

By tracing along the curves, we can see that the point of intersection has an $x$-value between 1.5 and 1.532. We can continue to zoom and trace or redefine the window until the coordinates of the point of intersection can be determined to the nearest hundredth. The approximate point of intersection is (1.52, 1.64).

*Solve each system of equations. Approximate each solution to two decimal places.*

**1.** $\begin{cases} y = -1.65x + 3.65 \\ y = 4.56x - 9.44 \end{cases}$

**2.** $\begin{cases} y = 7.61x + 3.48 \\ y = -1.26x - 6.43 \end{cases}$

**3.** $\begin{cases} 2.33x - 4.72y = 10.61 \\ 5.86x - 6.22y = -8.89 \end{cases}$

**4.** $\begin{cases} -7.89x - 5.68y = 3.26 \\ -3.65x + 4.98y = 11.77 \end{cases}$

# Vocabulary, Readiness & Video Check

*Match each graph with the solution of the corresponding system.*

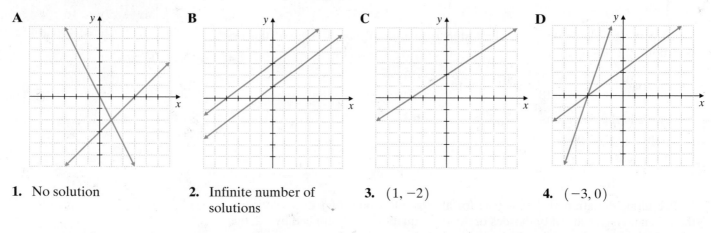

**A**   **B**   **C**   **D**

1. No solution
2. Infinite number of solutions
3. $(1, -2)$
4. $(-3, 0)$

*Martin-Gay Interactive Videos*   *Watch the section lecture video and answer the following questions.*

*See Video 4.1*

**Objective A** 5. In ▣ Example 1, the ordered pair is a solution of the first equation of the system. Why is this not enough to determine if the ordered pair is a solution of the system? ▶

**Objective B** 6. From ▣ Example 2, what potential drawbacks does the graphing method have? ▶

**Objective C** 7. The system in ▣ Example 3 needs one of its equations solved for a variable as a first step. What important point is then emphasized? ▶

**Objective D** 8. Why is Step 2 in Solving a System by Elimination skipped in ▣ Example 5? ▶

## 4.1  Exercise Set  MyMathLab® ▶

**Objective A** *Determine whether the given ordered pair is a solution of the system. See Examples 1 and 2.*

1. $\begin{cases} x - y = 3 \\ 2x - 4y = 8 \end{cases}$ $(2, -1)$

2. $\begin{cases} x - y = -4 \\ 2x + 10y = 4 \end{cases}$ $(-3, 1)$

▶3. $\begin{cases} 2x - 3y = -9 \\ 4x + 2y = -2 \end{cases}$ $(3, 5)$

4. $\begin{cases} 2x - 5y = -2 \\ 3x + 4y = 4 \end{cases}$ $(4, 2)$

5. $\begin{cases} y = -5x \\ x = -2 \end{cases}$ $(-2, 10)$

6. $\begin{cases} y = 6 \\ x = -2y \end{cases}$ $(-12, 6)$

7. $\begin{cases} 3x + 7y = -19 \\ -6x = 5y + 8 \end{cases}$ $\left(\dfrac{2}{3}, -3\right)$

8. $\begin{cases} 4x + 5y = -7 \\ -8x = 3y - 1 \end{cases}$ $\left(\dfrac{3}{4}, -2\right)$

**Objective** **B** *Solve each system by graphing. See Examples 3 through 5.*

9. $\begin{cases} x + y = 1 \\ x - 2y = 4 \end{cases}$

10. $\begin{cases} 2x - y = 8 \\ x + 3y = 11 \end{cases}$

11. $\begin{cases} 2y - 4x = 0 \\ x + 2y = 5 \end{cases}$

12. $\begin{cases} 4x - y = 6 \\ x - y = 0 \end{cases}$

13. $\begin{cases} 3x - y = 4 \\ 6x - 2y = 4 \end{cases}$

14. $\begin{cases} -x + 3y = 6 \\ 3x - 9y = 9 \end{cases}$

15. $\begin{cases} y = -3x \\ 2x - y = -5 \end{cases}$

16. $\begin{cases} y = -2x \\ -3x + y = 10 \end{cases}$

**Objective** **C** *Use the substitution method to solve each system of equations. See Examples 6 and 7.*

17. $\begin{cases} x + y = 10 \\ \phantom{x +} y = 4x \end{cases}$

18. $\begin{cases} 5x + 2y = -17 \\ \phantom{5x +} x = 3y \end{cases}$

19. $\begin{cases} 4x - y = 9 \\ 2x + 3y = -27 \end{cases}$

20. $\begin{cases} 3x - y = 6 \\ -4x + 2y = -8 \end{cases}$

21. $\begin{cases} \dfrac{1}{2}x + \dfrac{3}{4}y = -\dfrac{1}{4} \\ \dfrac{3}{4}x - \dfrac{1}{4}y = 1 \end{cases}$

22. $\begin{cases} \dfrac{2}{5}x + \dfrac{1}{5}y = -1 \\ x + \dfrac{2}{5}y = -\dfrac{8}{5} \end{cases}$

23. $\begin{cases} x = -3y + 4 \\ 3x + 9y = 12 \end{cases}$

24. $\begin{cases} x = 3y - 1 \\ 2x - 6y = -2 \end{cases}$

**Objective** **D** *Use the elimination method to solve each system of equations. See Examples 8 through 11.*

25. $\begin{cases} 2x - 4y = 0 \\ x + 2y = 5 \end{cases}$

26. $\begin{cases} 2x - 3y = 0 \\ 2x + 6y = 3 \end{cases}$

27. $\begin{cases} 5x + 2y = 1 \\ x - 3y = 7 \end{cases}$

28. $\begin{cases} 6x - y = -5 \\ 4x - 2y = 6 \end{cases}$

29. $\begin{cases} 5x - 2y = 27 \\ -3x + 5y = 18 \end{cases}$

30. $\begin{cases} 3x + 4y = 2 \\ 2x + 5y = -1 \end{cases}$

31. $\begin{cases} 3x = 5y + 11 \\ 2x = 6y + 2 \end{cases}$

32. $\begin{cases} 6x = 3y - 3 \\ 4x = -5y - 9 \end{cases}$

33. $\begin{cases} x - 2y = 4 \\ 2x - 4y = 4 \end{cases}$

34. $\begin{cases} -7x + 2y = 5 \\ 14x - 4y = 3 \end{cases}$

35. $\begin{cases} 3x + y = 1 \\ 2y = 2 - 6x \end{cases}$

36. $\begin{cases} y = 2x - 5 \\ 8x - 4y = 20 \end{cases}$

**Objectives C D  Mixed Practice** *Solve each system of equations by substitution or elimination.*

**37.** $\begin{cases} 2x + 5y = 8 \\ 6x + y = 10 \end{cases}$

**38.** $\begin{cases} x - 4y = -5 \\ -3x - 8y = 0 \end{cases}$

**39.** $\begin{cases} 4x + 3y = 5 \\ x - 6y = 8 \end{cases}$

**40.** $\begin{cases} 3x - 2y = 11 \\ x + 10y = 25 \end{cases}$

**41.** $\begin{cases} \dfrac{1}{3}x + y = \dfrac{4}{3} \\ -\dfrac{1}{4}x - \dfrac{1}{2}y = -\dfrac{1}{4} \end{cases}$

**42.** $\begin{cases} \dfrac{3}{4}x - \dfrac{1}{2}y = -\dfrac{1}{2} \\ x + y = -\dfrac{3}{2} \end{cases}$

**43.** $\begin{cases} 4x + 2y = 5 \\ 2x + y = -1 \end{cases}$

**44.** $\begin{cases} 3x + 6y = 15 \\ 2x + 4y = 3 \end{cases}$

**45.** $\begin{cases} 10y - 2x = 1 \\ 5y = 4 - 6x \end{cases}$

**46.** $\begin{cases} 3x + 4y = 0 \\ 7x = 3y \end{cases}$

**47.** $\begin{cases} \dfrac{3}{4}x + \dfrac{5}{2}y = 11 \\ \dfrac{1}{16}x - \dfrac{3}{4}y = -1 \end{cases}$

**48.** $\begin{cases} \dfrac{2}{3}x + \dfrac{1}{4}y = -\dfrac{3}{2} \\ \dfrac{1}{2}x - \dfrac{1}{4}y = -2 \end{cases}$

**49.** $\begin{cases} x = 3y + 2 \\ 5x - 15y = 10 \end{cases}$

**50.** $\begin{cases} x = 7y - 21 \\ 2x - 14y = -42 \end{cases}$

**51.** $\begin{cases} \dfrac{x}{3} + y = \dfrac{4}{3} \\ -x + 2y = 11 \end{cases}$

**52.** $\begin{cases} \dfrac{x}{8} - \dfrac{y}{2} = 1 \\ \dfrac{x}{3} - y = 2 \end{cases}$

**53.** $\begin{cases} 2x = 6 \\ y = 5 - x \end{cases}$

**54.** $\begin{cases} x = 3y + 4 \\ -y = 5 \end{cases}$

**55.** $\begin{cases} \dfrac{x + 5}{2} = \dfrac{6 - 4y}{3} \\ \dfrac{3x}{5} = \dfrac{21 - 7y}{10} \end{cases}$

**56.** $\begin{cases} \dfrac{y}{5} = \dfrac{8 - x}{2} \\ x = \dfrac{2y - 8}{3} \end{cases}$

**57.** $\begin{cases} 4x - 7y = 7 \\ 12x - 21y = 24 \end{cases}$

**58.** $\begin{cases} 2x - 5y = 12 \\ -4x + 10y = 20 \end{cases}$

**59.** $\begin{cases} \dfrac{2}{3}x - \dfrac{3}{4}y = -1 \\ -\dfrac{1}{6}x + \dfrac{3}{8}y = 1 \end{cases}$

**60.** $\begin{cases} \dfrac{1}{2}x - \dfrac{1}{3}y = -3 \\ \dfrac{1}{8}x + \dfrac{1}{6}y = 0 \end{cases}$

**61.** $\begin{cases} 2x - y = -1 \\ y = -2x \end{cases}$

**62.** $\begin{cases} 4y - x = -1 \\ x = -2y \end{cases}$

**63.** $\begin{cases} 0.7x - 0.2y = -1.6 \\ 0.2x - y = -1.4 \end{cases}$

**64.** $\begin{cases} -0.7x + 0.6y = 1.3 \\ 0.5x - 0.3y = -0.8 \end{cases}$

**65.** $\begin{cases} 4x - 1.5y = 10.2 \\ 2x + 7.8y = -25.68 \end{cases}$

**66.** $\begin{cases} x - 3y = -5.3 \\ 6.3x + 6y = 3.96 \end{cases}$

## Review

*Determine whether the given replacement values make each equation true or false. See Section 2.1.*

**67.** $3x - 4y + 2z = 5$;
$x = 1, y = 2,$ and $z = 5$

**68.** $x + 2y - z = 7$;
$x = 2, y = -3,$ and $z = 3$

**69.** $-x - 5y + 3z = 15$;
$x = 0, y = -1,$ and $z = 5$

**70.** $-4x + y - 8z = 4$;
$x = 1, y = 0,$ and $z = -1$

*Add the equations in each system. See this section.*

**71.** $\begin{cases} 3x + 2y - 5z = 10 \\ -3x + 4y + z = 15 \end{cases}$

**72.** $\begin{cases} x + 4y - 5z = 20 \\ 2x - 4y - 2z = -17 \end{cases}$

**73.** $\begin{cases} 10x + 5y + 6z = 14 \\ -9x + 5y - 6z = -12 \end{cases}$

**74.** $\begin{cases} -9x - 8y - z = 31 \\ 9x + 4y - z = 12 \end{cases}$

## Concept Extensions

*Without graphing, determine whether each system in Exercises 75 through 78 has one solution, no solution, or an infinite number of solutions. See the Concept Checks in this section.*

**75.** $\begin{cases} y = 2x - 5 \\ y = 2x + 1 \end{cases}$

**76.** $\begin{cases} y = 3x - \dfrac{1}{2} \\ y = -2x + \dfrac{1}{5} \end{cases}$

**77.** $\begin{cases} x + y = 3 \\ 5x + 5y = 15 \end{cases}$

**78.** $\begin{cases} y = 5x - 2 \\ y = -\dfrac{1}{5}x - 2 \end{cases}$

**79.** Can a system consisting of two linear equations have exactly two solutions? Explain why or why not.

**80.** Suppose the graph of the equations in a system of two equations in two variables consists of a circle and a line. Discuss the possible number of solutions for this system.

*The concept of supply and demand is used often in business. In general, as the unit price of a commodity increases, the demand for that commodity decreases. Also, as a commodity's unit price increases, the manufacturer normally increases the supply. The point where supply is equal to demand is called the equilibrium point. The following shows the graph of a demand equation and the graph of a supply equation for previously rented DVDs. The x-axis represents the number of DVDs in thousands, and the y-axis represents the cost of a DVD. Use this graph to answer Exercises 81 through 84.*

**81.** Find the number of DVDs and the price per DVD when supply equals demand.

**82.** When $x$ is between 3 and 4, is supply greater than demand or is demand greater than supply?

**83.** When $x$ is greater than 6, is supply greater than demand or is demand greater than supply?

**84.** For what $x$-values are the $y$-values corresponding to the supply equation greater than the $y$-values corresponding to the demand equation?

*The revenue equation for a certain brand of toothpaste is $y = 2.5x$, where $x$ is the number of tubes of toothpaste sold and $y$ is the total income for selling $x$ tubes. The cost equation is $y = 0.9x + 3000$, where $x$ is the number of tubes of toothpaste manufactured and $y$ is the cost of producing $x$ tubes. The following set of axes shows the graphs of the cost and revenue equations. Use this graph for Exercises 85 through 90.*

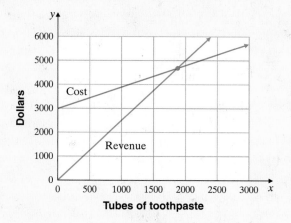

**85.** Find the coordinates of the point of intersection, or break-even point, by solving the system

$$\begin{cases} y = 2.5x \\ y = 0.9x + 3000 \end{cases}$$

**86.** Explain the meaning of the $x$-value of the point of intersection.

**87.** If the company sells 2000 tubes of toothpaste, does the company make money or lose money?

**88.** If the company sells 1000 tubes of toothpaste, does the company make money or lose money?

**89.** For what $x$-values will the company make a profit? (*Hint:* For what $x$-values is the revenue graph "higher" than the cost graph?)

**90.** For what $x$-values will the company lose money? (*Hint:* For what $x$-values is the revenue graph "lower" than the cost graph?)

*Solve.*

**91.** Write a system of two linear equations in $x$ and $y$ that has the ordered pair solution $(2, 5)$.

**92.** Which method would you use to solve the system?

$$\begin{cases} 5x - 2y = 6 \\ 2x + 3y = 5 \end{cases}$$

Explain your choice.

**93.** The amount $y$ of bottled water consumed per person in the United States (in gallons) in the year $x$ can be modeled by the linear equation $y = 0.66x + 21.67$. The amount $y$ of carbonated soft drinks consumed per person in the United States (in gallons) in the year $x$ can be modeled by the linear equation $y = -1.35x + 57.06$. In both models, $x = 0$ represents the year 2002. (*Source:* Based on data from the Economic Research Service, USDA)

  **a.** What does the slope of each equation tell you about the patterns of bottled water and carbonated soft drink consumption in the United States?
  **b.** Solve this system of equations. (Round your final results to the nearest whole numbers.)
  **c.** Explain the meaning of your answer to part (b).

**94.** The amount of U.S. federal government income $y$ (in billions of dollars) for fiscal year $x$, from 2009 to 2013 ($x = 0$ represents 2009), can be modeled by the linear equation $y = 163x + 2033$. The amount of U.S. federal government expenditures $y$ (in billions of dollars) for the same period can be modeled by the linear equation $y = -5x + 3523$. (*Source:* Based on data from Bureau of the Fiscal Service, U.S. Department of the Treasury)

  **a.** What does the slope of each equation tell you about the patterns of the U.S. federal government income and expenditures?
  **b.** Solve this system of equations. (Round your final results to the nearest whole numbers.)
  **c.** Did expenses ever equal income during the period from 2009 through 2013?

*Solve each system. To do so you may want to let* $a = \dfrac{1}{x}$ *(if x is in the denominator) and let* $b = \dfrac{1}{y}$ *(if y is in the denominator.)*

**95.** $\begin{cases} \dfrac{1}{x} + y = 12 \\ \dfrac{3}{x} - y = 4 \end{cases}$

**96.** $\begin{cases} x + \dfrac{2}{y} = 7 \\ 3x + \dfrac{3}{y} = 6 \end{cases}$

**97.** $\begin{cases} \dfrac{1}{x} + \dfrac{1}{y} = 5 \\ \dfrac{1}{x} - \dfrac{1}{y} = 1 \end{cases}$

**98.** $\begin{cases} \dfrac{2}{x} + \dfrac{3}{y} = 5 \\ \dfrac{5}{x} - \dfrac{3}{y} = 2 \end{cases}$

**99.** $\begin{cases} \dfrac{2}{x} + \dfrac{3}{y} = -1 \\ \dfrac{3}{x} - \dfrac{2}{y} = 18 \end{cases}$

**100.** $\begin{cases} \dfrac{3}{x} - \dfrac{2}{y} = -18 \\ \dfrac{2}{x} + \dfrac{3}{y} = 1 \end{cases}$

**101.** $\begin{cases} \dfrac{2}{x} - \dfrac{4}{y} = 5 \\ \dfrac{1}{x} - \dfrac{2}{y} = \dfrac{3}{2} \end{cases}$

**102.** $\begin{cases} \dfrac{5}{x} + \dfrac{7}{y} = 1 \\ -\dfrac{10}{x} - \dfrac{14}{y} = 0 \end{cases}$

## 4.2 Solving Systems of Linear Equations in Three Variables

In this section, we solve systems of linear equations in three variables. We call the equation $3x - y + z = -15$, for example, a **linear equation in three variables** since there are three variables and each variable is raised only to the power 1. A solution of this equation is an **ordered triple (x, y, z)** that makes the equation a true statement.

For example, the ordered triple $(2, 0, -21)$ is a solution of $3x - y + z = -15$ since replacing $x$ with 2, $y$ with 0, and $z$ with $-21$ yields the true statement

$$3(2) - 0 + (-21) = -15$$

The graph of this equation is a plane in three-dimensional space, just as the graph of a linear equation in two variables is a line in two-dimensional space.

Although we will not discuss the techniques for graphing equations in three variables, visualizing the possible patterns of intersecting planes gives us insight into the possible patterns of solutions of a system of three three-variable linear equations. There are four possible patterns.

**Objective**

**A** Solve a System of Three Linear Equations in Three Variables.

1. Three planes have a single point in common. This point represents the single solution of the system. This system is **consistent.**

2. Three planes intersect at no point common to all three. This system has no solution. A few ways that this can occur are shown. This system is **inconsistent.**

3. Three planes intersect at all the points of a single line. The system has infinitely many solutions. This system is **consistent.**

4. Three planes coincide at all points on the plane. The system is **consistent,** and the equations are **dependent.**

## Objective **A** Solving a System of Three Linear Equations in Three Variables

Just as with systems of two equations in two variables, we can use the elimination or substitution method to solve a system of three equations in three variables. To use the elimination method, we eliminate a variable and obtain a system of two equations in two variables. Then we use the methods we learned in the previous section to solve the system of two equations.

**Practice 1**

Solve the system:

$$\begin{cases} 2x - y + 3z = 13 \\ x + y - z = -2 \\ 3x + 2y + 2z = 13 \end{cases}$$

> **Helpful Hint**
> Make sure you add two other equations besides equations (1) and (2) and *also* **eliminate the same variable.** You will see why as you follow this example.

**Example 1** Solve the system:

$$\begin{cases} 3x - y + z = -15 & \text{Equation (1)} \\ x + 2y - z = 1 & \text{Equation (2)} \\ 2x + 3y - 2z = 0 & \text{Equation (3)} \end{cases}$$

**Solution:** Let's add equations (1) and (2) to eliminate $z$.

$$\begin{array}{r} 3x - y + z = -15 \\ \underline{x + 2y - z = 1} \\ 4x + y = -14 \quad \text{Equation (4)} \end{array}$$

Next we add two *other* equations and *eliminate $z$ again.* To do so, we multiply both sides of equation (1) by 2 and add the resulting equation to equation (3). Then

$$\begin{cases} 2(3x - y + z) = 2(-15) \\ 2x + 3y - 2z = 0 \end{cases} \text{ simplifies to } \begin{cases} 6x - 2y + 2z = -30 \\ \underline{2x + 3y - 2z = 0} \\ 8x + y = -30 \end{cases}$$

Equation (5)

We now have two equations (4 and 5) in the same two variables. This means we can solve equations (4) and (5) for $x$ and $y$. To solve by elimination, we multiply both sides of equation (4) by $-1$ and add the resulting equation to equation (5). Then

$$\begin{cases} -1(4x + y) = -1(-14) \\ 8x + y = -30 \end{cases} \text{ simplifies to } \begin{cases} -4x - y = 14 \\ \underline{8x + y = -30} \\ 4x = -16 \quad \text{Add the equations.} \\ x = -4 \quad \text{Solve for } x. \end{cases}$$

**Answer**

**1.** $(1, 1, 4)$

We now replace $x$ with $-4$ in equation (4) or (5).

$$4x + y = -14 \quad \text{Equation (4)}$$
$$4(-4) + y = -14 \quad \text{Let } x = -4.$$
$$y = 2 \quad \text{Solve for } y.$$

Finally, we replace $x$ with $-4$ and $y$ with $2$ in equation (1), (2), or (3).

$$x + 2y - z = 1 \quad \text{Equation (2)}$$
$$-4 + 2(2) - z = 1 \quad \text{Let } x = -4 \text{ and } y = 2.$$
$$-4 + 4 - z = 1$$
$$-z = 1$$
$$z = -1$$

The ordered triple solution is $(-4, 2, -1)$.

**Check:** To check, we let $x = -4$, $y = 2$, and $z = -1$ in *all three* original equations of the system.

**Equation (1)**

$$3x - y + z = -15$$
$$3(-4) - 2 + (-1) \stackrel{?}{=} -15$$
$$-12 - 2 - 1 \stackrel{?}{=} -15$$
$$-15 = -15 \quad \text{True}$$

**Equation (2)**

$$x + 2y - z = 1$$
$$-4 + 2(2) - (-1) \stackrel{?}{=} 1$$
$$-4 + 4 + 1 \stackrel{?}{=} 1$$
$$1 = 1 \quad \text{True}$$

**Equation (3)**

$$2x + 3y - 2z = 0$$
$$2(-4) + 3(2) - 2(-1) \stackrel{?}{=} 0$$
$$-8 + 6 + 2 \stackrel{?}{=} 0$$
$$0 = 0 \quad \text{True}$$

All three statements are true, so the ordered triple solution is $(-4, 2, -1)$.

◼ **Work Practice 1**

---

**Example 2** Solve the system:

$$\begin{cases} 2x - 4y + 8z = 2 & (1) \\ -x - 3y + z = 11 & (2) \\ x - 2y + 4z = 0 & (3) \end{cases}$$

**Solution:** Add equations (2) and (3) to eliminate $x$, and the new equation is

$$-5y + 5z = 11 \quad (4)$$

To eliminate $x$ again, we multiply both sides of equation (2) by 2 and add the resulting equation to equation (1). Then

$$\begin{cases} 2x - 4y + 8z = 2 \\ 2(-x - 3y + z) = 2(11) \end{cases} \quad \text{simplifies to} \quad \begin{cases} 2x - 4y + 8z = 2 \\ \underline{-2x - 6y + 2z = 22} \\ -10y + 10z = 24 \quad (5) \end{cases}$$

Next we solve for $y$ and $z$ using equations (4) and (5). To do so, we multiply both sides of equation (4) by $-2$ and add the resulting equation to equation (5).

$$\begin{cases} -2(-5y + 5z) = -2(11) \\ -10y + 10z = 24 \end{cases} \quad \text{simplifies to} \quad \begin{cases} 10y - 10z = -22 \\ \underline{-10y + 10z = 24} \\ 0 = 2 \quad \text{False} \end{cases}$$

Since the statement is false, this system is inconsistent and has no solution. The solution set is the empty set $\{\ \}$, or $\varnothing$.

◼ **Work Practice 2**

**Practice 2**

Solve the system:

$$\begin{cases} 2x + 4y - 2z = 3 \\ -x + y - z = 6 \\ x + 2y - z = 1 \end{cases}$$

**Answer**

**2.** $\varnothing$

The elimination method is summarized next.

> ## Solving a System of Three Linear Equations by the Elimination Method
>
> **Step 1:** Write each equation in standard form, $Ax + By + Cz = D$.
>
> **Step 2:** Choose a pair of equations and use them to eliminate a variable.
>
> **Step 3:** Choose any other pair of equations and eliminate the *same variable* as in Step 2.
>
> **Step 4:** Two equations in two variables should be obtained from Step 2 and Step 3. Use methods from Section 4.1 to solve this system for both variables.
>
> **Step 5:** To solve for the third variable, substitute the values of the variables found in Step 4 into any of the original equations containing the third variable.
>
> **Step 6:** Check the ordered triple solution in *all three* original equations.

**Helpful Hint** Make sure you read closely and follow Step 3.

✓**Concept Check** In the system

$$\begin{cases} x + y + z = 6 & \text{Equation (1)} \\ 2x - y + z = 3 & \text{Equation (2)} \\ x + 2y + 3z = 14 & \text{Equation (3)} \end{cases}$$

equations (1) and (2) are used to eliminate $y$. Which action could be used to finish solving? Why?

  **a.** Use (1) and (2) to eliminate $z$.

  **b.** Use (2) and (3) to eliminate $y$.

  **c.** Use (1) and (3) to eliminate $x$.

---

## Practice 3

Solve the system:

$$\begin{cases} 3x + 2y \quad\quad = -1 \\ 6x \quad\quad - 2z = 4 \\ \quad\quad y - 3z = 2 \end{cases}$$

**Answer**

**3.** $\left(\frac{1}{3}, -1, -1\right)$

✓**Concept Check Answer**

**b;** answers may vary

---

**Example 3** Solve the system:

$$\begin{cases} 2x + 4y \quad\quad = 1 & (1) \\ 4x \quad\quad - 4z = -1 & (2) \\ \quad\quad y - 4z = -3 & (3) \end{cases}$$

**Solution:** Notice that equation (2) has no term containing the variable $y$. Let's eliminate $y$ using equations (1) and (3). We multiply both sides of equation (3) by $-4$ and add the resulting equation to equation (1). Then

$$\begin{cases} 2x + 4y \quad\quad = 1 \\ -4(y - 4z) = -4(-3) \end{cases} \text{ simplifies to } \begin{cases} 2x + 4y \quad\quad = 1 \\ \underline{\quad -4y + 16z = 12} \\ 2x \quad\quad + 16z = 13 \quad (4) \end{cases}$$

Next we solve for $z$ using equations (4) and (2). We multiply both sides of equation (4) by $-2$ and add the resulting equation to equation (2).

$$\begin{cases} -2(2x + 16z) = -2(13) \\ 4x - 4z = -1 \end{cases} \text{ simplifies to } \begin{cases} -4x - 32z = -26 \\ \underline{\quad 4x - 4z = -1} \\ -36z = -27 \end{cases}$$

$$z = \frac{3}{4}$$

Now we replace $z$ with $\dfrac{3}{4}$ in equation (3) and solve for $y$.

$$y - 4\left(\frac{3}{4}\right) = -3 \quad \text{Let } z = \frac{3}{4} \text{ in equation (3).}$$

$$y - 3 = -3$$

$$y = 0$$

Finally, we replace $y$ with 0 in equation (1) and solve for $x$.

$$2x + 4(0) = 1 \quad \text{Let } y = 0 \text{ in equation (1).}$$

$$2x = 1$$

$$x = \frac{1}{2}$$

The ordered triple solution is $\left(\dfrac{1}{2}, 0, \dfrac{3}{4}\right)$. Check to see that this solution satisfies *all three* equations of the system.

■ Work Practice 3

---

**Example 4** Solve the system:

$$\begin{cases} x - 5y - 2z = 6 & (1) \\ -2x + 10y + 4z = -12 & (2) \\ \dfrac{1}{2}x - \dfrac{5}{2}y - z = 3 & (3) \end{cases}$$

**Solution:** We multiply both sides of equation (3) by 2 to eliminate fractions, and we multiply both sides of equation (2) by $-\dfrac{1}{2}$ so that the coefficient of $x$ is 1. The resulting system is then

$$\begin{cases} x - 5y - 2z = 6 & (1) \\ x - 5y - 2z = 6 & \text{Multiply (2) by } -\dfrac{1}{2}. \\ x - 5y - 2z = 6 & \text{Multiply (3) by 2.} \end{cases}$$

All three resulting equations are identical, and therefore equations (1), (2), and (3) are all equivalent. There are infinitely many solutions of this system. The equations are dependent. The solution set can be written as $\{(x,y,z)\,|\,x - 5y - 2z = 6\}$.

■ Work Practice 4

As mentioned earlier, we can also use the substitution method to solve a system of linear equations in three variables.

**Example 5** Solve the system:

$$\begin{cases} x - 4y - 5z = 35 & (1) \\ x - 3y = 0 & (2) \\ -y + z = -55 & (3) \end{cases}$$

**Solution:** Notice in equations (2) and (3) that a variable is missing. Also notice that both equations contain the variable $y$. Let's use the substitution method by solving equation (2) for $x$ and equation (3) for $z$ and substituting the results in equation (1).

*(Continued on next page)*

---

**Practice 4**

Solve the system:

$$\begin{cases} x - 3y + 4z = 2 \\ -2x + 6y - 8z = -4 \\ \dfrac{1}{2}x - \dfrac{3}{2}y + 2z = 1 \end{cases}$$

**Practice 5**

Solve the system:

$$\begin{cases} 2x + 5y - 3z = 30 & (1) \\ x + y = -3 & (2) \\ 2x - z = 0 & (3) \end{cases}$$

(*Hint:* Equations (2) and (3) each contain the variable $x$ and have a variable missing.)

**Answers**

**4.** $\{(x,y,z)\,|\,x - 3y + 4z = 2\}$

**5.** $(-5, 2, -10)$

$$x - 3y = 0 \qquad (2)$$
$$x = 3y \qquad \text{Solve equation (2) for } x.$$
$$-y + z = -55 \qquad (3)$$
$$z = y - 55 \qquad \text{Solve equation (3) for } z.$$

Now substitute $3y$ for $x$ and $y - 55$ for $z$ in equation (1).

$$x - 4y - 5z = 35 \qquad (1)$$

**Helpful Hint** Do not forget to distribute.

$$3y - 4y - 5(y - 55) = 35 \qquad \text{Let } x = 3y \text{ and } z = y - 55.$$
$$3y - 4y - 5y + 275 = 35 \qquad \text{Use the distributive property and multiply.}$$
$$-6y + 275 = 35 \qquad \text{Combine like terms.}$$
$$-6y = -240 \qquad \text{Subtract 275 from both sides.}$$
$$y = 40 \qquad \text{Solve.}$$

To find $x$, recall that $x = 3y$ and substitute 40 for $y$. Then $x = 3y$ becomes $x = 3 \cdot 40 = 120$. To find $z$, recall that $z = y - 55$ and also substitute 40 for $y$. Then $z = y - 55$ becomes $z = 40 - 55 = -15$. The solution is $(120, 40, -15)$.

■ Work Practice 5

## Vocabulary, Readiness & Video Check

*Solve.*

1. Choose the equation(s) that has $(-1, 3, 1)$ as a solution.
   **a.** $x + y + z = 3$  **b.** $-x + y + z = 5$  **c.** $-x + y + 2z = 0$  **d.** $x + 2y - 3z = 2$

2. Choose the equation(s) that has $(2, 1, -4)$ as a solution.
   **a.** $x + y + z = -1$  **b.** $x - y - z = -3$  **c.** $2x - y + z = -1$  **d.** $-x - 3y - z = -1$

3. Use the result of Exercise 1 to determine whether $(-1, 3, 1)$ is a solution of the system below. Explain your answer.
$$\begin{cases} x + y + z = 3 \\ -x + y + z = 5 \\ x + 2y - 3z = 2 \end{cases}$$

4. Use the result of Exercise 2 to determine whether $(2, 1, -4)$ is a solution of the system below. Explain your answer.
$$\begin{cases} x + y + z = -1 \\ x - y - z = -3 \\ 2x - y + z = -1 \end{cases}$$

*Martin-Gay Interactive Videos*   Watch the section lecture video and answer the following question.

**Objective A** 5. From ⊞ Example 1 and the lecture before, why does Step 3 stress that the same variable be eliminated from two other equations? ▸

*See Video 4.2* 🍎

## 4.2 Exercise Set MyMathLab® ▶

**Objective A** *Solve each system. See Examples 1 through 5.*

**1.** $\begin{cases} x - y + z = -4 \\ 3x + 2y - z = 5 \\ -2x + 3y - z = 15 \end{cases}$

**2.** $\begin{cases} x + y - z = -1 \\ -4x - y + 2z = -7 \\ 2x - 2y - 5z = 7 \end{cases}$

**3.** $\begin{cases} x + y = 3 \\ 2y = 10 \\ 3x + 2y - 3z = 1 \end{cases}$

**4.** $\begin{cases} 5x = 5 \\ 2x + y = 4 \\ 3x + y - 4z = -15 \end{cases}$

**5.** $\begin{cases} 2x + 2y + z = 1 \\ -x + y + 2z = 3 \\ x + 2y + 4z = 0 \end{cases}$

**6.** $\begin{cases} 2x - 3y + z = 5 \\ x + y + z = 0 \\ 4x + 2y + 4z = 4 \end{cases}$

**7.** $\begin{cases} x - 2y + z = -5 \\ -3x + 6y - 3z = 15 \\ 2x - 4y + 2z = -10 \end{cases}$

**8.** $\begin{cases} 3x + y - 2z = 2 \\ -6x - 2y + 4z = -4 \\ 9x + 3y - 6z = 6 \end{cases}$

**9.** $\begin{cases} 4x - y + 2z = 5 \\ 2y + z = 4 \\ 4x + y + 3z = 10 \end{cases}$

**10.** $\begin{cases} 5y - 7z = 14 \\ 2x + y + 4z = 10 \\ 2x + 6y - 3z = 30 \end{cases}$

**11.** $\begin{cases} x + 5z = 0 \\ 5x + y = 0 \\ y - 3z = 0 \end{cases}$

**12.** $\begin{cases} x - 5y = 0 \\ x - z = 0 \\ -x + 5z = 0 \end{cases}$

**13.** $\begin{cases} 6x - 5z = 17 \\ 5x - y + 3z = -1 \\ 2x + y = -41 \end{cases}$

**14.** $\begin{cases} x + 2y = 6 \\ 7x + 3y + z = -33 \\ x - z = 16 \end{cases}$

**15.** $\begin{cases} x + y + z = 8 \\ 2x - y - z = 10 \\ x - 2y - 3z = 22 \end{cases}$

**16.** $\begin{cases} 5x + y + 3z = 1 \\ x - y + 3z = -7 \\ -x + y = 1 \end{cases}$

**17.** $\begin{cases} x + 2y - z = 5 \\ 6x + y + z = 7 \\ 2x + 4y - 2z = 5 \end{cases}$

**18.** $\begin{cases} 4x - y + 3z = 10 \\ x + y - z = 5 \\ 8x - 2y + 6z = 10 \end{cases}$

**▶19.** $\begin{cases} 2x - 3y + z = 2 \\ x - 5y + 5z = 3 \\ 3x + y - 3z = 5 \end{cases}$

**20.** $\begin{cases} 4x + y - z = 8 \\ x - y + 2z = 3 \\ 3x - y + z = 6 \end{cases}$

**21.** $\begin{cases} -2x - 4y + 6z = -8 \\ x + 2y - 3z = 4 \\ 4x + 8y - 12z = 16 \end{cases}$

**22.** $\begin{cases} -6x + 12y + 3z = -6 \\ 2x - 4y - z = 2 \\ -x + 2y + \dfrac{z}{2} = -1 \end{cases}$

**23.** $\begin{cases} 2x + 2y - 3z = 1 \\ y + 2z = -14 \\ 3x - 2y = -1 \end{cases}$

**24.** $\begin{cases} 7x + 4y = 10 \\ x - 4y + 2z = 6 \\ y - 2z = -1 \end{cases}$

**25.** $\begin{cases} x + 2y - z = 5 \\ -3x - 2y - 3z = 11 \\ 4x + 4y + 5z = -18 \end{cases}$

**26.** $\begin{cases} 3x - 3y + z = -1 \\ 3x - y - z = 3 \\ -6x + 4y + 3z = -8 \end{cases}$

**27.** $\begin{cases} \dfrac{3}{4}x - \dfrac{1}{3}y + \dfrac{1}{2}z = 9 \\ \dfrac{1}{6}x + \dfrac{1}{3}y - \dfrac{1}{2}z = 2 \\ \dfrac{1}{2}x - y + \dfrac{1}{2}z = 2 \end{cases}$

**28.** $\begin{cases} \dfrac{1}{3}x - \dfrac{1}{4}y + z = -9 \\ \dfrac{1}{2}x - \dfrac{1}{3}y - \dfrac{1}{4}z = -6 \\ x - \dfrac{1}{2}y - z = -8 \end{cases}$

## Review

*Solve. See Section 2.1.*

**29.** $2(x - 1) - 3x = x - 12$

**30.** $7(2x - 1) + 4 = 11(3x - 2)$

**31.** $-y - 5(y + 5) = 3y - 10$

**32.** $z - 3(z + 7) = 6(2z + 1)$

*Solve. See Section 2.2.*

**33.** The sum of two numbers is 45 and one number is twice the other. Find the numbers.

**34.** The difference between two numbers is 5. Twice the smaller number added to five times the larger number is 53. Find the numbers.

## Concept Extensions

**35.** Write a single linear equation in three variables that has $(-1, 2, -4)$ as a solution. (There are many possibilities.) Explain the process you used to write your equation.

**36.** When solving a system of three equations in three unknowns, explain how to determine that a system has no solution.

**37.** Write a system of linear equations in three variables that has the solution $(-1, 2, -4)$. (There are many possibilities.) Explain the process you used to write your system.

**38.** Write a system of three linear equations in three variables that has $(2, 1, 5)$ as a solution. (There are many possibilities.) Explain the process you used to write your system.

**39.** The fraction $\dfrac{1}{24}$ can be written as the following sum:

$$\frac{1}{24} = \frac{x}{8} + \frac{y}{4} + \frac{z}{3}$$

where the numbers $x, y,$ and $z$ are solutions of

$$\begin{cases} x + y + z = 1 \\ 2x - y + z = 0 \\ -x + 2y + 2z = -1 \end{cases}$$

Solve the system and see that the sum of the fractions is $\dfrac{1}{24}$.

**40.** The fraction $\dfrac{1}{18}$ can be written as the following sum:

$$\frac{1}{18} = \frac{x}{2} + \frac{y}{3} + \frac{z}{9}$$

where the numbers $x, y,$ and $z$ are solutions of

$$\begin{cases} x + 3y + z = -3 \\ -x + y + 2z = -14 \\ 3x + 2y - z = 12 \end{cases}$$

Solve the system and see that the sum of the fractions is $\dfrac{1}{18}$.

*Solving systems involving more than three variables can be accomplished with methods similar to those encountered in this section. For Exercises 41 through 44, apply what you already know to solve each system of equations in four variables.*

**41.**
$$\begin{cases} x + y \quad\;\; - w = 0 \\ \quad\; y + 2z + w = 3 \\ x \qquad - z \qquad = 1 \\ 2x - y \qquad - w = -1 \end{cases}$$

**42.**
$$\begin{cases} 5x + 4y \qquad\qquad = 29 \\ \quad\; y + z - w = -2 \\ 5x \qquad + z \qquad = 23 \\ \quad\; y - z + w = 4 \end{cases}$$

**43.**
$$\begin{cases} x + y + z + w = 5 \\ 2x + y + z + w = 6 \\ x + y + z \qquad = 2 \\ x + y \qquad\quad = 0 \end{cases}$$

**44.**
$$\begin{cases} 2x \qquad - z \qquad = -1 \\ \quad\; y + z + \; w = 9 \\ \quad\; y \qquad - 2w = -6 \\ x + y \qquad\qquad = 3 \end{cases}$$

**45.** Write a system of three linear equations in three variables that are dependent equations.

**46.** What is the solution to the system in Exercise 45?

---

## 4.3   Systems of Linear Equations and Problem Solving ▶

### Objective A   Solving Problems Modeled by Systems of Two Equations ▶

Thus far, we have solved problems by writing one-variable equations and solving for the variable. Some of these problems can be solved, perhaps more easily, by writing a system of equations, as illustrated in this section.

**Objectives**

**A** Solve Problems That Can Be Modeled by Systems of Two Linear Equations. ▶

**B** Solve Problems with Cost and Revenue Functions. ▶

**C** Solve Problems That Can Be Modeled by Systems of Three Linear Equations. ▶

**Example 1**   Predicting Equal Consumption of Red Meat and Poultry

America's consumption of red meat has decreased most years since 2003, while consumption of poultry has increased. The function $y = -1.70x + 121$ approximates the annual pounds of red meat consumed per capita, where $x$ is the number of years since 2003. The function $y = 0.10x + 83$ approximates the annual pounds of poultry consumed per capita, where $x$ is also the number of years since 2003. If these trends continue, determine the year when the annual consumption of red meat and poultry will be equal. (*Source:* Based on data from the USDA: Economic Research Service)

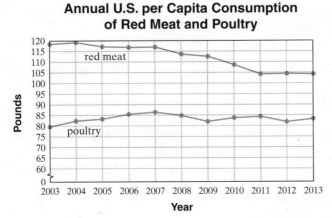

**Annual U.S. per Capita Consumption of Red Meat and Poultry**

*(Continued on next page)*

## Practice 1

Read Example 1. If we use the years 2009, 2011, 2012, and 2013 only to write functions approximating the consumption of red meat and poultry, we have the following:

Red meat: $y = -2.12x + 111$

Poultry: $y = 0.16x + 83$

where $x$ is the number of years since 2009 and $y$ is pounds per year consumed.

**a.** Assuming these trends continue, predict the year when consumption of red meat and poultry will be the same. Round to the nearest year.

**b.** Does your answer differ from the answer to Example 1 Why or why not?

**Solution:**

1. UNDERSTAND. Read and reread the problem and guess a year. Let's guess the year 2023. This year is 20 years since 2003, so $x = 20$. Now let $x = 20$ in each given function.

Red meat: $y = -1.70x + 121 = -1.70(20) + 121 = 87$ pounds

Poultry: $y = 0.10x + 83 = 0.10(20) + 83 = 85$ pounds

Since the projected pounds in 2020 for red meat and poultry are not the same, we guessed incorrectly, but we do have a better understanding of the problem. We know that the year will be later than 2023 because our weight of poultry in pounds is still less than our weight of red meat in pounds.

2. TRANSLATE. We are already given the system of equations.

3. SOLVE. We want to know the year $x$ in which pounds $y$ are the same, so we solve the system:

$$\begin{cases} y = -1.70x + 121 \\ y = 0.10x + 83 \end{cases}$$

Since both equations are solved for $y$, one way to solve is to use the substitution method.

$y = -1.70x + 121$     First equation

$-1.70x + 121 = 0.10x + 83$     Let $y = -1.70x + 121$

$-1.80x = -38$

$x = \dfrac{-38}{-1.80} \approx 21.11$

4. INTERPRET. Since we are only asked to find the year, we need only solve for $x$.

**Check:** To check, see whether $x \approx 21.11$ gives approximately the same number of pounds of red meat and poultry.

Red meat: $y = -1.70x + 121 = -1.70(21.11) + 121 \approx 85.113$ pounds

Poultry: $y = 0.10x + 83 = 0.10(21.11) + 83 \approx 85.111$ pounds

Since we rounded the number of years, the numbers of pounds do differ slightly. They differ only by 0.002, so we can assume we solved correctly.

**State:** The consumption of red meat and poultry will be the same about 21.11 years after 2003, or 2024.11. Thus, in the year 2024, we predict the consumption will be the same.

■ **Work Practice 1**

## Practice 2

A first number is 7 greater than a second number. Twice the first number is 4 more than three times the second. Find the numbers.

**Example 2** Finding Unknown Numbers

A first number is 4 less than a second number. Four times the first number is 6 more than twice the second. Find the numbers.

**Solution:**

1. UNDERSTAND. Read and reread the problem and guess a solution. If one number is 10 and this is 4 less than a second number, the second number is 14. Four times the first number is $4(10)$, or 40. This is not equal to 6 more than twice the second number, which is $2(14) + 6$ or 34. Although we guessed incorrectly, we now have a better understanding of the problem.

Since we are looking for two numbers, we will let

$x =$ first number

$y =$ second number

**Answers**

**1. a.** 2021   **b.** yes; answers may vary
**2.** 17 and 10

**2. TRANSLATE.** Since we have assigned two variables to this problem, we will translate the given facts into two equations. For the first statement we have

In words:   the first number     is     4 less than the second number
                    ↓              ↓                ↓

Translate:          $x$            $=$            $y - 4$

Next we translate the second statement into an equation.

In words:   four times the     is     6 more than twice
            first number               the second number
                  ↓            ↓              ↓

Translate:        $4x$        $=$          $2y + 6$

**3. SOLVE.** Now we solve the system

$$\begin{cases} x = y - 4 \\ 4x = 2y + 6 \end{cases}$$

Since the first equation expresses $x$ in terms of $y$, we will use substitution. We substitute $y - 4$ for $x$ in the second equation and solve for $y$.

$$4x = 2y + 6 \quad \text{Second equation}$$
$$4(y - 4) = 2y + 6 \quad \text{Let } x = y - 4.$$
$$4y - 16 = 2y + 6$$
$$2y = 22$$
$$y = 11$$

Now we replace $y$ with 11 in the equation $x = y - 4$ and solve for $x$. Then $x = y - 4$ becomes $x = 11 - 4 = 7$. The ordered pair solution of the system is $(7, 11)$.

**4. INTERPRET.** Since the solution of the system is $(7, 11)$, the first number we are looking for is 7 and the second number is 11.

**Check:**   Notice that 7 *is* 4 less than 11, and 4 times 7 *is* 6 more than twice 11. The proposed numbers, 7 and 11, are correct.

**State:**   The numbers are 7 and 11.

■ **Work Practice 2**

**Example 3**   Finding the Rate of Speed

Two cars leave Indianapolis, one traveling east and the other west. After 3 hours they are 297 miles apart. If one car is traveling 5 mph faster than the other, what is the speed of each?

**Solution:**

**1. UNDERSTAND.** Read and reread the problem. Let's guess a solution and use the formula $d = r \cdot t$ to check. In this formula, $d$ is distance, $r$ is rate, and $t$ is time. Suppose the faster car is traveling at a rate of 55 mph. This means that the other car is traveling at a rate of 50 mph since we are told that one car is traveling 5 mph faster than the other. To find the distance apart after 3 hours, we will first find the distance traveled by each car. One car's distance is rate · time = $55(3)$ = 165 miles. The other car's distance is rate · time = $50(3)$ = 150 miles. Since one car is traveling east and the other west, their distance apart is the sum of their distances, or 165 miles + 150 miles = 315 miles. Although this distance apart is

*(Continued on next page)*

**Practice 3**

Two trains leave Tulsa, one traveling north and the other south. After 4 hours, they are 376 miles apart. If one train is traveling 10 mph faster than the other, what is the speed of each?

**Answer**
**3.** 42 mph; 52 mph

not the required distance of 297 miles, we now have a better understanding of the problem.

150 + 165 = 315 mi

50(3) = 150 mi          Indianapolis          55(3) = 165 mi

Let's model the problem with a system of equations. We will let

$x$ = speed of one car

$y$ = speed of the other car

We summarize the information on the following chart. Both cars have traveled 3 hours. Since distance = rate · time, their distances are $3x$ and $3y$ miles, respectively.

|  | Rate | · Time | = Distance |
|---|---|---|---|
| **One Car** | $x$ | 3 | $3x$ |
| **Other Car** | $y$ | 3 | $3y$ |

2. TRANSLATE. We can now translate the stated conditions into two equations.

In words:   one car's distance   added to   the other car's distance   is   297

Translate:        $3x$        $+$        $3y$        $=$   297

In words:   one car's speed   is   5 mph faster than the other

Translate:        $x$        $=$        $y + 5$

3. SOLVE. Now we solve the system.

$$\begin{cases} 3x + 3y = 297 \\ x = y + 5 \end{cases}$$

Again, the substitution method is appropriate. We replace $x$ with $y + 5$ in the first equation and solve for $y$.

$$3x + 3y = 297 \quad \text{First equation}$$

$$3(y + 5) + 3y = 297 \quad \text{Let } x = y + 5.$$
$$3y + 15 + 3y = 297$$
$$6y = 282$$
$$y = 47$$

To find $x$, we replace $y$ with 47 in the equation $x = y + 5$. Then $x = 47 + 5 = 52$. The ordered pair solution of the system is $(52, 47)$.

4. INTERPRET. The solution $(52, 47)$ means that the cars are traveling at 52 mph and 47 mph, respectively.

**Check:**   Notice that one car is traveling 5 mph faster than the other. Also, if one car travels 52 mph for 3 hours, the distance is $3(52) = 156$ miles. The other car traveling for 3 hours at 47 mph travels a distance of $3(47) = 141$ miles. The sum of the distances $156 + 141$ is 297 miles, the required distance.

**State:**   The cars are traveling at 52 mph and 47 mph.

**Helpful Hint**
Don't forget to attach units, if appropriate.

■ Work Practice 3

Copyright 2015 Pearson Education, Inc.

## Example 4   Mixing Solutions

Lynn Pike, a pharmacist, needs 70 liters of a 50% alcohol solution. She has available a 30% alcohol solution and an 80% alcohol solution. How many liters of each solution should she mix to obtain 70 liters of a 50% alcohol solution?

**Solution:**

1. UNDERSTAND. Read and reread the problem. Next, guess the solution. Suppose that we need 20 liters of the 30% solution. Then we need $70 - 20 = 50$ liters of the 80% solution. To see if this gives us 70 liters of a 50% alcohol solution, let's find the amount of pure alcohol in each solution.

| number of liters | × | alcohol strength | = | amount of pure alcohol |
|:---:|:---:|:---:|:---:|:---:|
| ↓ | | ↓ | | ↓ |
| 20 liters | × | 0.30 | = | 6 liters |
| 50 liters | × | 0.80 | = | 40 liters |
| 70 liters | × | 0.50 | = | 35 liters |

Since 6 liters + 40 liters = 46 liters and not 35 liters, our guess is incorrect, but we have gained some insight as to how to model and check this problem.

We will let

$x$ = amount of 30% solution, in liters

$y$ = amount of 80% solution, in liters

and use a table to organize the given data.

| | Number of Liters | Alcohol Strength | Amount of Pure Alcohol |
|---|:---:|:---:|:---:|
| **30% Solution** | $x$ | 30% | $0.30x$ |
| **80% Solution** | $y$ | 80% | $0.80y$ |
| **50% Solution Needed** | 70 | 50% | $(0.50)(70)$ |

2. TRANSLATE. We translate the stated conditions into two equations.

| In words: | amount of 30% solution | + | amount of 80% solution | = | 70 |
|:---:|:---:|:---:|:---:|:---:|:---:|
| | ↓ | | ↓ | | ↓ |
| Translate: | $x$ | + | $y$ | = | 70 |

| In words: | amount of pure alcohol in 30% solution | + | amount of pure alcohol in 80% solution | = | amount of pure alcohol in 50% solution |
|:---:|:---:|:---:|:---:|:---:|:---:|
| | ↓ | | ↓ | | ↓ |
| Translate: | $0.30x$ | + | $0.80y$ | = | $(0.50)(70)$ |

(*Continued on next page*)

**Practice 4**

One solution contains 20% acid and a second solution contains 50% acid. How many ounces of each solution should be mixed in order to have 60 ounces of a 30% acid solution?

**Answer**
**4.** 40 oz of 20% solution; 20 oz of 50% solution

**3.** SOLVE. Now we solve the system

$$\begin{cases} x + y = 70 \\ 0.30x + 0.80y = (0.50)(70) \end{cases}$$

To solve this system, we use the elimination method. We multiply both sides of the first equation by $-3$ and both sides of the second equation by 10. Then

$$\begin{cases} -3(x + y) = -3(70) \\ 10(0.30x + 0.80y) = 10(0.50)(70) \end{cases} \text{ simplifies to } \begin{cases} -3x - 3y = -210 \\ \underline{3x + 8y = 350} \\ \phantom{xxxx} 5y = 140 \\ \phantom{xxxxx} y = 28 \end{cases}$$

Now we replace $y$ with 28 in the equation $x + y = 70$ and find that $x + 28 = 70$, or $x = 42$. The ordered pair solution of the system is $(42, 28)$.

**4.** INTERPRET. The solution $(42, 28)$ means 42 liters of the 30% solution and 28 liters of the 80% solution.

**Check:** We check the solution in the same way that we checked our guess.

**State:** The pharmacist needs to mix 42 liters of 30% solution and 28 liters of 80% solution to obtain 70 liters of 50% solution.

■ Work Practice 4

✓**Concept Check** Suppose you mix an amount of 25% acid solution with an amount of 60% acid solution. You then calculate the acid strength of the resulting acid mixture. For which of the following results should you suspect an error in your calculation? Why?

**a.** 14%        **b.** 32%        **c.** 55%

## Objective B  Solving Problems with Cost and Revenue Functions ▶

Recall that businesses are often computing cost and revenue functions or equations to predict sales, to determine whether prices need to be adjusted, and to see whether the company is making or losing money. Recall also that the value at which revenue equals cost is called the break-even point. When revenue is less than cost, the company is losing money; when revenue is greater than cost, the company is making money.

**Example 5**  Finding a Break-Even Point

A manufacturing company recently purchased $3000 worth of new equipment to create new personalized stationery to sell to its customers. The cost of producing a package of personalized stationery is $3.00, and it is sold for $5.50. Find the number of packages that must be sold for the company to break even.

**Practice 5**

A company that manufactures boxes recently purchased $2000 worth of new equipment to make gift boxes to sell to its customers. The cost of producing a package of gift boxes is $1.50 and it is sold for $4.00. Find the number of packages that must be sold for the company to break even.

**Answer**

**5.** 800 packages

✓**Concept Check Answer**

**a;** answers may vary

**Solution:**

1. UNDERSTAND. Read and reread the problem.

   Notice that the cost to the company will include a one-time cost of $3000 for the equipment and then $3.00 per package produced. The revenue will be $5.50 per package sold.

   To model this problem, we will let

   $x$ = number of packages of personalized stationery

   $C(x)$ = total cost for producing $x$ packages of stationery

   $R(x)$ = total revenue for selling $x$ packages of stationery

2. TRANSLATE. The revenue equation is

   | In words: | revenue for selling $x$ packages of stationery | = | price per package | · | number of packages |
   |---|---|---|---|---|---|
   | | ↓ | | ↓ | | ↓ |
   | Translate: | $R(x)$ | = | 5.5 | · | $x$ |

   The cost equation is

   | In words: | cost for producing $x$ packages of stationery | = | cost per package | · | number of packages | · | cost for equipment |
   |---|---|---|---|---|---|---|---|
   | | ↓ | | ↓ | | ↓ | | ↓ |
   | Translate: | $C(x)$ | = | 3 | · | $x$ | + | 3000 |

3. SOLVE. Since the break-even point is when $R(x) = C(x)$, we solve the equation $5.5x = 3x + 3000$.

   $$5.5x = 3x + 3000$$
   $$2.5x = 3000 \qquad \text{Subtract } 3x \text{ from both sides.}$$
   $$x = 1200 \qquad \text{Divide both sides by 2.5.}$$

4. INTERPRET.

**Check:** To see whether the break-even point occurs when 1200 packages are produced and sold, we check to see if revenue equals cost when $x = 1200$. When $x = 1200$,

$$R(x) = 5.5x = 5.5(1200) = 6600$$
$$C(x) = 3x + 3000 = 3(1200) + 3000 = 6600$$

Since $R(1200) = C(1200) = 6600$, the break-even point is 1200.

**State:** The company must sell 1200 packages of stationery to break even. The graph of this system is shown.

■ Work Practice 5

## Objective C Solving Problems Modeled by Systems of Three Equations

To introduce problem solving with systems of three linear equations in three variables, we solve a problem about triangles.

**Practice 6**

The measure of the largest angle of a triangle is 80° more than the measure of the smallest angle, and the measure of the remaining angle is 40° more than the measure of the smallest angle. Find the measure of each angle.

### Example 6 Finding Angle Measures

The measure of the largest angle of a triangle is 80° more than the measure of the smallest angle, and the measure of the remaining angle is 10° more than the measure of the smallest angle. Find the measure of each angle.

**Solution:**

1. UNDERSTAND. Read and reread the problem. Recall that the sum of the measures of the angles of a triangle is 180°. Then guess a solution. If the smallest angle measures 20°, the measure of the largest angle is 80° more, or $20° + 80° = 100°$. The measure of the remaining angle is 10° more than the measure of the smallest angle, or $20° + 10° = 30°$. The sum of these three angles is $20° + 100° + 30° = 150°$, not the required 180°. We now know that the measure of the smallest angle is greater than 20°.

To model this problem, we will let

$x$ = degree measure of the smallest angle

$y$ = degree measure of the largest angle

$z$ = degree measure of the remaining angle

2. TRANSLATE. We translate the given information into three equations.

In words:    the sum of the measures    =    180

Translate:    $x + y + z$    =    180

In words:    the largest angle    is    80 more than the smallest angle

Translate:    $y$    =    $x + 80$

In words:    the remaining angle    is    10 more than the smallest angle

Translate:    $z$    =    $x + 10$

3. SOLVE. We solve the system

$$\begin{cases} x + y + z = 180 \\ y = x + 80 \\ z = x + 10 \end{cases}$$

Since $y$ and $z$ are both expressed in terms of $x$, we will solve using the substitution method. We substitute $y = x + 80$ and $z = x + 10$ in the first equation.

Then

$$x + y + z = 180 \quad \text{First equation}$$

$$x + (x + 80) + (x + 10) = 180 \quad \text{Let } y = x + 80 \text{ and } z = x + 10.$$

$$3x + 90 = 180$$

$$3x = 90$$

$$x = 30$$

Copyright 2015 Pearson Education, Inc.

**Answer**

6. 20°; 60°; 100°

Then $y = x + 80 = 30 + 80 = 110$, and $z = x + 10 = 30 + 10 = 40$. The ordered triple solution is $(30, 110, 40)$.

**4. INTERPRET.**

**Check:** Notice that $30° + 40° + 110° = 180°$. Also, the measure of the largest angle, $110°$, is $80°$ more than the measure of the smallest angle, $30°$. The measure of the remaining angle, $40°$, is $10°$ more than the measure of the smallest angle, $30°$.

**State:** The angles measure $30°$, $110°$, and $40°$.

■ Work Practice 6

## Vocabulary, Readiness & Video Check

*Martin-Gay Interactive Videos*     *Watch the section lecture video and answer the following questions.*

**Objective A** **1.** In ⊞ Example 1 and the lecture before, the problem-solving steps for solving applications are mentioned. What is the difference here from when we've used these steps in the past? ▶

**Objective B** **2.** Based on ⊞ Example 2, explain the meaning of a break-even point. How do you find the break-even point algebraically? ▶

**Objective C** **3.** In ⊞ Example 3, why is the ordered triple not the final stated solution to the application? ▶

*See Video 4.3*

## 4.3  Exercise Set  MyMathLab® ▶

**Objective A** *Solve. See Examples 1 through 4. For Exercises 1 and 2, the solutions have been started for you.*

**1.** One number is two more than a second number. Twice the first is 4 less than 3 times the second. Find the numbers.

**Start the solution:**

**1. UNDERSTAND** the problem. Since we are looking for two numbers, let

$x =$ one number
$y =$ second number

**2. TRANSLATE.** Since we have assigned two variables, we will translate the facts into two equations. (Fill in the blanks.)

First equation:

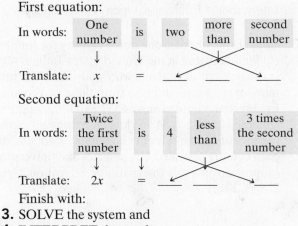

In words: | One number | is | two | more than | second number
Translate: | $x$ | = | ___ ___ ___

Second equation:

In words: | Twice the first number | is | 4 | less than | 3 times the second number
Translate: | $2x$ | = | ___ ___ ___

Finish with:
**3. SOLVE** the system and
**4. INTERPRET** the results.

**2.** Three times one number minus a second is 8, and the sum of the numbers is 12. Find the numbers.

**Start the solution:**

**1. UNDERSTAND** the problem. Since we are looking for two numbers, let

$x =$ one number
$y =$ second number

**2. TRANSLATE.** Since we have assigned two variables, we will translate the facts into two equations. (Fill in the blanks.)

First equation:

In words: | Three times one number | minus | a second number | is | 8
Translate: | $3x$ | ___ | ___ | = | 8

Second equation:

In words: | The sum of the numbers | is | 12
Translate: | $x +$ ___ ___ | 12

Finish with:
**3. SOLVE** the system and
**4. INTERPRET** the results.

**3.** The United States currently has the world's only "large deck" aircraft carriers, which can hold up to 85 fixed-wing aircraft. A Gerald R. Ford class carrier, which is still under production, is longest in length, while a Nimitz class carrier is the second longest. The total length of these two carriers is 2198 feet, while the difference of their lengths is only 14 feet. (*Source:* Naval-technology.com)

  **a.** Find the length of each class of carrier.

  **b.** If a football field has a length of 100 yards, determine the length of a Gerald R. Ford class carrier in terms of the number of football fields. (Round to the nearest tenth.)

**5.** A Delta 727 traveled 560 mph with the wind and 480 mph against the wind. Find the speed of the plane in still air and the speed of the wind.

▶ **7.** Find how many quarts of 4% butterfat milk and 1% butterfat milk should be mixed to yield 60 quarts of 2% butterfat milk.

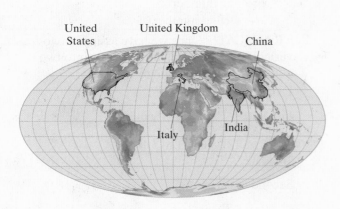

**4.** Baseball has often been called "America's game." For the 2013 season, the team with the highest average attendance was the Los Angeles Dodgers, and the team with the lowest average attendance was the Tampa Bay Rays. The average number of fans who attended a home game of the Dodgers was 8926 more than twice the average number of fans who attended a home game of the Rays. If the total average attendance for both teams was 64,861, find the average attendance for each team for the 2013 Major League Baseball season. (*Source:* ESPN)

**6.** Terry Watkins can row about 10.6 kilometers in 1 hour downstream and 6.8 kilometers upstream in 1 hour. Find how fast he can row in still water, and find the speed of the current.

**8.** A pharmacist needs 500 milliliters of a 20% phenobarbital solution but has only 5% and 25% phenobarbital solutions available. Find how many milliliters of each she should mix to get the desired solution.

**9.** In recent years, the United Kingdom was the most popular host country for U.S. students traveling abroad to study. Italy was the second most popular destination. A total of 64,305 students studied in one of the two countries. If 5015 more U.S. students studied in the United Kingdom than Italy, how many students studied abroad in each country? (*Source:* Institute of International Education, Open Doors, 2013)

**10.** China is the number one home country for foreign students studying in the United States. India is second in the number of students who study here. The total number of foreign students from these two countries for a recent school year was 332,351. If the number of Chinese students is 54,665 less than three times the number of Indian students, how many students from each country attended U.S. colleges and universities? (*Source:* Institute of International Education, Open Doors, 2013)

**11.** Karen Karlin bought some large frames for $15 each and some small frames for $8 each at a closeout sale. If she bought 22 frames for $239, find how many of each type she bought.

**12.** Hilton University Drama Club sold 311 tickets for a play. Student tickets cost 50 cents each; non-student tickets cost $1.50. If total receipts were $385.50, find how many tickets of each type were sold.

**13.** One number is two less than a second number. Twice the first is 4 more than 3 times the second. Find the numbers.

**14.** Twice one number plus a second number is 42, and the first number minus the second number is −6. Find the numbers.

**15.** The international movie box office, in billions of dollars, is in flux. For many years, the EMEA (Europe, Middle East, and Africa) market has been much larger that the Asian Pacific market. This has started to change in recent years, with the Asian Pacific market growing rapidly. From 2008 through 2012, the function $y = 0.37x + 9.8$ can be used to approximate the box office (in billions of dollars) in the EMEA market, while the function $y = 0.9x + 6.6$ approximates the box office in the Asian Pacific market. For both markets, $x$ represents the number of years since 2008. (*Source:* Based on data from Motion Picture Association of America)

   **a.** If these trends continue, predict the year in which the EMEA box office and the Asian Pacific box office will be the same.

   **b.** Use these equations to predict the box office for the EMEA market and the Asian Pacific market for the current year.

**16.** The rate of fatalities per 100 million vehicle-miles has been decreasing on both rural and urban roads. For the years 2008 through 2012, the function $y = -0.072x + 2.05$ can be used to estimate the rate of fatalities per 100 million vehicle-miles on rural roads, while the function $y = -0.014x + 0.77$ can be used to estimate the rate of fatalities per 100 million vehicle-miles on urban roads. For both functions, $x$ is the number of years since 2008. (*Source:* Bureau of Transportation Statistics, U.S. Department of Transportation)

   **a.** If these trends continue, predict the year in which the fatality rate on rural roads equals the fatality rate for urban roads.

   **b.** Use these equations to predict the fatality rate per 100 million vehicle-miles for rural and urban roads for the current year.

**17.** An office supply store in San Diego sells 7 writing tablets and 4 pens for $6.40. Also, 2 tablets and 19 pens cost $5.40. Find the price of each.

**18.** A Candy Barrel shop manager mixes M&M's worth $2.00 per pound with trail mix worth $1.50 per pound. Find how many pounds of each she should use to get 50 pounds of a party mix worth $1.80 per pound.

**19.** An airplane takes 3 hours to travel a distance of 2160 miles with the wind. The return trip takes 4 hours against the wind. Find the speed of the plane in still air and the speed of the wind.

**20.** Two cyclists start at the same point and travel in opposite directions. One travels 4 mph faster than the other. In 4 hours they are 112 miles apart. Find how fast each is traveling.

**21.** The annual U.S. consumption of cheddar cheese has been decreasing in recent years, while the consumption of mozzarella cheese has increased. For the years 2005–2013, the function $y = -0.17x + 11$ approximates the annual U.S. per capita consumption of cheddar cheese in pounds, and the function $y = 0.14x + 10.25$ approximates the annual U.S. per capita consumption of mozzarella cheese in pounds. For both functions, $x$ is the number of years after 2005. (*Source:* Based on data from the U.S. Department of Agriculture)

   **a.** Explain how the given function verifies that the consumption of cheddar cheese is decreasing, while the given function verifies that the consumption of mozzarella cheese is increasing.

   **b.** Based on this information, determine the year in which the pounds of cheddar cheese consumed equaled the pounds of mozzarella cheese consumed.

**22.** Two of the job categories defined by the U.S. Department of Labor are manufacturing technicians and chemical technicians. Manufacturing technician jobs have decreased nearly every year this century. During this same time period, chemical technician jobs have increased. For the years 2012–2022, the function $y = -220x + 68,000$ approximates the number of manufacturing technician jobs forecast for the U.S. labor market, while the function $y = 600x + 63,600$ approximates the forecast for the number of chemical technician jobs. The year 2012 is the year $x = 0$. (*Source:* Based on data from the U.S. Department of Labor)

   **a.** Explain how the decrease in manufacturing technician jobs can be verified by the given function, while the increase in chemical technician jobs can be verified by the given function.

   **b.** Based on this information, determine the year when the number of manufacturing technician jobs and the number of chemical technician jobs are expected to be the same.

△ **23.** The perimeter of a triangle is 93 centimeters. If two sides are equally long and the third side is 9 centimeters longer than the others, find the lengths of the three sides.

**24.** Jack Reinholt, a car salesman, has a choice of two pay arrangements: a weekly salary of $200 plus 5% commission on sales, or a straight 15% commission. Find the amount of weekly sales for which Jack's earnings are the same regardless of the pay arrangement.

**25.** Hertz car rental agency charges $25 daily plus 10 cents per mile. Budget charges $20 daily plus 25 cents per mile. Find the daily mileage for which the Budget charge for the day is twice that of the Hertz charge for the day.

**26.** Carroll Blakemore, a drafting student, bought three templates and a pencil one day for $6.45. Another day, he bought two pads of paper and four pencils for $7.50. If the price of a pad of paper is three times the price of a pencil, find the price of each type of item.

△ **27.** In the figure, line $l$ and line $m$ are parallel lines cut by transversal $t$. Find the values of $x$ and $y$.

△ **28.** Find the values of $x$ and $y$ in the following isosceles triangle.

**Objective B** *Given the cost function $C(x)$ and the revenue function $R(x)$, find the number of units $x$ that must be sold to break even. See Example 5.*

**29.** $C(x) = 30x + 10,000$ $R(x) = 46x$

**30.** $C(x) = 12x + 15,000$ $R(x) = 32x$

**31.** $C(x) = 1.2x + 1500$ $R(x) = 1.7x$

**32.** $C(x) = 0.8x + 900$ $R(x) = 2x$

**33.** $C(x) = 75x + 160,000$ $R(x) = 200x$

**34.** $C(x) = 105x + 70,000$ $R(x) = 245x$

**▶ 35.** The planning department of Abstract Office Supplies has been asked to determine whether the company should introduce a new computer desk next year. The department estimates that $6000 of new manufacturing equipment will need to be purchased and that the cost of constructing each desk will be $200. The department also estimates that the revenue from each desk will be $450.

   **a.** Determine the revenue function $R(x)$ from the sale of $x$ desks.

   **b.** Determine the cost function $C(x)$ for manufacturing $x$ desks.

   **c.** Find the break-even point.

**36.** Baskets, Inc., is planning to introduce a new woven basket. The company estimates that $500 worth of new equipment will be needed to manufacture this new type of basket and that it will cost $15 per basket to manufacture. The company also estimates that the revenue from each basket will be $31.

   **a.** Determine the revenue function $R(x)$ from the sale of $x$ baskets.

   **b.** Determine the cost function $C(x)$ for manufacturing $x$ baskets.

   **c.** Find the break-even point. Round up to the nearest whole basket.

**Objective C** *Solve. See Example 6.*

**37.** Rabbits in a lab are to be kept on a strict daily diet that includes 30 grams of protein, 16 grams of fat, and 24 grams of carbohydrates. The scientist has only three food mixes available with the following grams of nutrients per unit.

| | Protein | Fat | Carbohydrate |
|---|---|---|---|
| Mix A | 4 | 6 | 3 |
| Mix B | 6 | 1 | 2 |
| Mix C | 4 | 1 | 12 |

Find how many units of each mix are needed daily to meet each rabbit's dietary need.

**38.** Gerry Gundersen mixes different solutions with concentrations of 25%, 40%, and 50% to get 200 liters of a 32% solution. If he uses twice as much of the 25% solution as of the 40% solution, find how many liters of each kind he uses.

**△ 39.** The perimeter of a quadrilateral (four-sided polygon) is 29 inches. The longest side is twice as long as the shortest side. The other two sides are equally long and are 2 inches longer than the shortest side. Find the lengths of all four sides.

**△ 40.** The measure of the largest angle of a triangle is 90° more than the measure of the smallest angle, and the measure of the remaining angle is 30° more than the measure of the smallest angle. Find the measure of each angle.

**41.** The sum of three numbers is 40. The first number is five more than the second number. It is also twice the third. Find the numbers.

**42.** The sum of the digits of a three-digit number is 15. The tens-place digit is twice the hundreds-place digit, and the ones-place digit is 1 less than the hundreds-place digit. Find the three-digit number.

**43.** For 2013, the WNBA top scorer was Maya Moore of the Minnesota Lynx. She scored a total of 628 points during the regular season. The number of two-point field goals that Moore made was 55 fewer than three times the number of three-point field goals that she made. The number of free throws (each worth one point) she made was 71 fewer than the number of two-point field goals she made. Find how many free throws, two-point field goals, and three-point field goals Maya Moore made during the 2013 regular season. (*Source:* Women's National Basketball Association)

**44.** During the 2012–2013 regular NBA season, the top-scoring player was Kevin Durant of the Oklahoma City Thunder. Durant scored a total of 2280 points during the regular season. The number of free throws (each worth one point) he made was 123 more than four times the number of three-point field goals he made. The number of two-point field goals that Durant made was 87 less than the number of free throws he made. How many free throws, two-point field goals, and three-point field goals did Kevin Durant make during the 2012–2013 NBA season? (*Source:* National Basketball Association)

△ **45.** Find the values of $x$, $y$, and $z$ in the following triangle.

△ **46.** The sum of the measures of the angles of a quadrilateral is 360°. Find the values of $x$, $y$, and $z$ in the following quadrilateral.

## Review

*Multiply both sides of equation (1) by 2, and add the resulting equation to equation (2). See Section 4.2.*

**47.**  $3x - y + z = 2$  (1)
     $-x + 2y + 3z = 6$  (2)

**48.**  $2x + y + 3z = 7$  (1)
     $-4x + y + 2z = 4$  (2)

*Multiply both sides of equation (1) by −3, and add the resulting equation to equation (2). See Section 4.2.*

**49.**  $x + 2y - z = 0$  (1)
     $3x + y - z = 2$  (2)

**50.** $2x - 3y + 2z = 5$   (1)
     $x - 9y + z = -1$  (2)

## Concept Extensions

**51.** The 2005 amendments to the bankruptcy laws of the United States made it more difficult for citizens to file for bankruptcy. Since this change in the law, the number of bankruptcy petitions has been decreasing. In 2010, the number of bankruptcy petitions was 622,317 less than twice the number of petitions filed in 2013. This is equivalent to a decrease of 485,382 petitions filed from 2010 to 2013. Find how many personal bankruptcy petitions were filed in each year. (*Source:* Based on data from the Administrative Office of the United States Courts)

**52.** In 2012, the median weekly earnings for male registered nurses were $103 more than the median weekly earnings for female registered nurses. The median weekly earnings for female nurses were 0.913 times that of their male counterparts. Also in 2012, the median weekly earnings for female network and computer system administrators were $197 less than the median weekly earnings for male network and computer system administrators. The median weekly earnings for male network and computer system administrators were 1.186 times that of their female counterparts. (*Source:* Based on data from the Bureau of Labor Statistics, U.S. Department of Labor)

**a.** Find the median weekly earnings for female registered nurses in the United States in 2012. (Round to the nearest dollar.)

**b.** Find the median weekly earnings for female network and computer administrators in the United States in 2012. (Round to the nearest dollar.)

**c.** Of the four groups of workers described in the problem, which group makes the greatest weekly earnings? Which group makes the least weekly earnings?

**53.** Find the values of $a$, $b$, and $c$ such that the equation $y = ax^2 + bx + c$ has ordered pair solutions $(1, 6)$, $(-1, -2)$, and $(0, -1)$. To do so, substitute each ordered pair solution into the equation. Each time, the result is an equation in three unknowns: $a$, $b$, and $c$. Then solve the resulting system of three linear equations in three unknowns, $a$, $b$, and $c$.

**54.** Find the values of $a$, $b$, and $c$ such that the equation $y = ax^2 + bx + c$ has ordered pair solutions $(1, 2)$, $(2, 3)$, and $(-1, 6)$. (*Hint:* See Exercise 53.)

**55.** Data $(x, y)$ for the total number (in thousands) of college-bound students who took the ACT assessment in the year $x$ are approximately $(1, 1179)$, $(9, 1495)$ and $(12, 1666)$ where $x = 1$ represents 2001 and $x = 9$ represents 2009. Find the values $a$, $b$, and $c$ such that the equation $y = ax^2 + bx + c$ models these data. According to your model, how many students will take the ACT in 2015? Round to the nearest tenth. (*Source:* ACT, Inc.)

**56.** Monthly normal rainfall data $(x, y)$ for Portland, Oregon, are $(4, 2.47)$, $(7, 0.58)$, $(8, 1.07)$, where $x$ represents time in months (with $x = 1$ representing January) and $y$ represents rainfall in inches. Find the values of $a$, $b$, and $c$ rounded to 2 decimal places such that the equation $y = ax^2 + bx + c$ models this data. According to your model, how much rain should Portland expect during September? (*Source:* National Climatic Data Center)

# Systems of Linear Equations

*The graphs of various systems of equations are shown. Match each graph with the solution of its corresponding system.*

A   B   C   D

**1.** Solution: $(1, 2)$

**2.** Solution: $(-2, 3)$

**3.** No solution

**4.** Infinite number of solutions

*Solve each system by elimination or substitution.*

**5.** $\begin{cases} x + y = 4 \\ y = 3x \end{cases}$

**6.** $\begin{cases} x + y = 1 \\ y = 4x \end{cases}$

**7.** $\begin{cases} x + y = 1 \\ x - 2y = 4 \end{cases}$

**8.** $\begin{cases} 2x - y = 8 \\ x + 3y = 11 \end{cases}$

**9.** $\begin{cases} 2x + 5y = 8 \\ 6x + y = 10 \end{cases}$

**10.** $\begin{cases} \frac{1}{8}x + \frac{1}{2}y = \frac{1}{8} \\ -3x - 8y = 0 \end{cases}$

**11.** $\begin{cases} 4x - 7y = 7 \\ 12x - 21y = 24 \end{cases}$

**12.** $\begin{cases} 2x - 5y = 3 \\ -4x + 10y = -6 \end{cases}$

**13.** $\begin{cases} y = \frac{1}{3}x \\ 5x - 3y = 4 \end{cases}$

**14.** $\begin{cases} y = \frac{1}{4}x \\ 2x - 4y = 3 \end{cases}$

**15.** $\begin{cases} x + y = 2 \\ -3y + z = -7 \\ 2x + y - z = -1 \end{cases}$

**16.** $\begin{cases} y + 2z = -3 \\ x - 2y = 7 \\ 2x - y + z = 5 \end{cases}$

**17.** $\begin{cases} 2x + 4y - 6z = 3 \\ -x + y - z = 6 \\ x + 2y - 3z = 1 \end{cases}$

**18.** $\begin{cases} x - y + 3z = 2 \\ -2x + 2y - 6z = -4 \\ 3x - 3y + 9z = 6 \end{cases}$

**19.** $\begin{cases} x + y - 4z = 5 \\ x - y + 2z = -2 \\ 3x + 2y + 4z = 18 \end{cases}$

**20.** $\begin{cases} 2x - y + 3z = 2 \\ x + y - 6z = 0 \\ 3x + 4y - 3z = 6 \end{cases}$

**21.** A first number is 8 less than a second number. Twice the first number is 11 more than the second number. Find the numbers.

△**22.** The sum of the measures of the angles of a quadrilateral is 360°. The two smallest angles of the quadrilateral shown have the same measure. The third angle measures 30° more than the measure of one of the smallest angles, and the fourth angle measures 50° more than the measure of one of the smallest angles. Find the measure of each angle.

# Solving Systems of Equations Using Matrices ▶

By now, you may have noticed that the solution of a system of equations depends on the coefficients of the equations in the system and not on the variables. In this section, we introduce how to solve a system of equations using a **matrix.**

**Objectives**

**A** Use Matrices to Solve a System of Two Equations. ▶

**B** Use Matrices to Solve a System of Three Equations. ▶

## Objective A Using Matrices to Solve a System of Two Equations ▶

A **matrix** (plural: **matrices**) is a rectangular array of numbers. The following are examples of matrices.

$$\begin{bmatrix} 1 & 0 \\ 0 & 1 \end{bmatrix} \quad \begin{bmatrix} 2 & 1 & 3 & -1 \\ 0 & -1 & 4 & 5 \\ -6 & 2 & 1 & 0 \end{bmatrix} \quad \begin{bmatrix} a & b & c \\ d & e & f \end{bmatrix}$$

The numbers aligned horizontally in a matrix are in the same **row.** The numbers aligned vertically are in the same **column.**

$$\begin{array}{l} \text{Row 1} \rightarrow \\ \text{Row 2} \rightarrow \end{array} \begin{bmatrix} 2 & 1 & 0 \\ -1 & 6 & 2 \end{bmatrix}$$

This matrix has 2 rows and 3 columns. It is called a $2 \times 3$ (read "two by three") matrix.

Column 1  Column 2  Column 3

To see the relationship between systems of equations and matrices, study the example below.

**System of Equations (in Standard Form)**

$$\begin{cases} 2x - 3y = 6 & \text{Equation 1} \\ x + y = 0 & \text{Equation 2} \end{cases}$$

**Corresponding Matrix**

$$\begin{bmatrix} 2 & -3 & | & 6 \\ 1 & 1 & | & 0 \end{bmatrix} \begin{array}{l} \text{Row 1} \\ \text{Row 2} \end{array}$$

**Helpful Hint** Before writing the corresponding matrix associated with a system of equations, make sure that the equations are written in standard form.

Notice that the rows of the matrix correspond to the equations in the system. The coefficients of the variables are placed to the left of a vertical dashed line. The constants are placed to the right. Each of the numbers in the matrix is called an **element.**

The method of solving systems by matrices is to write this matrix as an equivalent matrix from which we can easily identify the solution. Two matrices are equivalent if they represent systems that have the same solution set. The following **row operations** can be performed on matrices, and the result is an equivalent matrix.

### Elementary Row Operations

1. Any two rows in a matrix may be interchanged.
2. The elements of any row may be multiplied (or divided) by the same nonzero number.
3. The elements of any row may be multiplied (or divided) by a nonzero number and added to their corresponding elements in any other row.

**Helpful Hint**

Notice that these *row* operations are the same operations that we can perform on *equations* in a system.

To solve a system of two equations in $x$ and $y$ by matrices, write the corresponding matrix associated with the system. Then use elementary row operations to write equivalent matrices until you have a matrix of the form

$$\begin{bmatrix} 1 & a & b \\ 0 & 1 & c \end{bmatrix},$$

where $a$, $b$, and $c$ are constants. Why? If a matrix associated with a system of equations is in this form, we can easily solve for $x$ and $y$. For example,

**Matrix**                              **System of Equations**

$$\begin{bmatrix} 1 & 2 & -3 \\ 0 & 1 & 5 \end{bmatrix} \text{ corresponds to } \begin{cases} 1x + 2y = -3 \\ 0x + 1y = 5 \end{cases} \text{ or } \begin{cases} x + 2y = -3 \\ y = 5 \end{cases}$$

In the second equation, we have $y = 5$. Substituting this in the first equation, we have $x + 2(5) = -3$ or $x = -13$. The solution of the system is the ordered pair $(-13, 5)$.

## Practice 1

Use matrices to solve the system:

$$\begin{cases} x + 2y = -4 \\ 2x - 3y = 13 \end{cases}$$

---

**Example 1**    Use matrices to solve the system:

$$\begin{cases} x + 3y = 5 \\ 2x - y = -4 \end{cases}$$

**Solution:** The corresponding matrix is $\begin{bmatrix} 1 & 3 & 5 \\ 2 & -1 & -4 \end{bmatrix}$. We use elementary row

operations to write an equivalent matrix that looks like $\begin{bmatrix} 1 & a & b \\ 0 & 1 & c \end{bmatrix}$.

For the matrix given, the element in the first row, first column is already 1, as desired. Next we write an equivalent matrix with a 0 below the 1. To do this, we multiply row 1 by $-2$ and add to row 2. *We will change only row 2.*

$$\begin{bmatrix} 1 & 3 & 5 \\ -2(1) + 2 & -2(3) + (-1) & -2(5) + (-4) \end{bmatrix} \text{ simplifies to}$$

    ↑ ↑         ↑      ↑         ↑    ↑
Row 1 Row 2    Row 1   Row 2    Row 1   Row 2
element element   element   element   element   element    $\begin{bmatrix} 1 & 3 & 5 \\ 0 & -7 & -14 \end{bmatrix}$

Now we change the $-7$ to a 1 by use of an elementary row operation. We divide row 2 by $-7$, then

$$\begin{bmatrix} 1 & 3 & 5 \\ \dfrac{0}{-7} & \dfrac{-7}{-7} & \dfrac{-14}{-7} \end{bmatrix} \text{ simplifies to } \begin{bmatrix} 1 & 3 & 5 \\ 0 & 1 & 2 \end{bmatrix}$$

This last matrix corresponds to the system

$$\begin{cases} x + 3y = 5 \\ y = 2 \end{cases}$$

Thus we know that $y$ is 2. To find $x$, we let $y = 2$ in the first equation, $x + 3y = 5$.

$$x + 3y = 5 \quad \text{First equation}$$
$$x + 3(2) = 5 \quad \text{Let } y = 2.$$
$$x = -1$$

The ordered pair solution is $(-1, 2)$. Check to see that this ordered pair satisfies both original equations.

■ **Work Practice 1**

**Answer**

**1.** $(2, -3)$

**Example 2** Use matrices to solve the system:

$$\begin{cases} 2x - y = 3 \\ 4x - 2y = 5 \end{cases}$$

**Solution:** The corresponding matrix is $\begin{bmatrix} 2 & -1 & | & 3 \\ 4 & -2 & | & 5 \end{bmatrix}$. To get 1 in the row 1, column 1 position, we divide the elements of row 1 by 2.

$$\begin{bmatrix} \dfrac{2}{2} & -\dfrac{1}{2} & | & \dfrac{3}{2} \\ 4 & -2 & | & 5 \end{bmatrix} \text{ simplifies to } \begin{bmatrix} 1 & -\dfrac{1}{2} & | & \dfrac{3}{2} \\ 4 & -2 & | & 5 \end{bmatrix}$$

To get 0 under the 1, we multiply the elements of row 1 by $-4$ and add the new elements to the elements of row 2.

$$\begin{bmatrix} 1 & -\dfrac{1}{2} & | & \dfrac{3}{2} \\ -4(1) + 4 & -4\left(-\dfrac{1}{2}\right) - 2 & | & -4\left(\dfrac{3}{2}\right) + 5 \end{bmatrix} \begin{array}{c} \text{simplifies} \\ \text{to} \end{array} \begin{bmatrix} 1 & -\dfrac{1}{2} & | & \dfrac{3}{2} \\ 0 & 0 & | & -1 \end{bmatrix}$$

The corresponding system is $\begin{cases} x - \dfrac{1}{2}y = -\dfrac{3}{2} \\ 0 = -1 \end{cases}$. The equation $0 = -1$ is false for all $y$ or $x$ values; hence the system is inconsistent and has no solution. The solution set is $\varnothing$ or $\{\ \}$.

■ Work Practice 2

✓**Concept Check** Consider the system

$$\begin{cases} 2x - 3y = 8 \\ x + 5y = -3 \end{cases}$$

What is wrong with its corresponding matrix shown below?

$$\begin{bmatrix} 2 & -3 & | & 8 \\ 0 & 5 & | & -3 \end{bmatrix}$$

## Objective B Using Matrices to Solve a System of Three Equations ▶

To solve a system of three equations in three variables using matrices, we will write the corresponding matrix in the equivalent form

$$\begin{bmatrix} 1 & a & b & | & d \\ 0 & 1 & c & | & e \\ 0 & 0 & 1 & | & f \end{bmatrix}$$

**Example 3** Use matrices to solve the system:

$$\begin{cases} x + 2y + z = 2 \\ -2x - y + 2z = 5 \\ x + 3y - 2z = -8 \end{cases}$$

**Solution:** The corresponding matrix is $\begin{bmatrix} 1 & 2 & 1 & | & 2 \\ -2 & -1 & 2 & | & 5 \\ 1 & 3 & -2 & | & -8 \end{bmatrix}$.

*(Continued on next page)*

(Continued on next page)

**Practice 2**

Use matrices to solve the system:

$$\begin{cases} -3x + y = 0 \\ -6x + 2y = 2 \end{cases}$$

**Practice 3**

Use matrices to solve the system:

$$\begin{cases} x + 3y + z = 5 \\ -3x + y - 3z = 5 \\ x + 2y - 2z = 9 \end{cases}$$

**Answers**

**2.** $\varnothing$ **3.** $(1, 2, -2)$

✓**Concept Check Answer**

matrix should be $\begin{bmatrix} 2 & -3 & | & 8 \\ 1 & 5 & | & -3 \end{bmatrix}$

Our goal is to write an equivalent matrix with 1s along the diagonal (see the numbers in red in the previous matrix) and 0s below the 1s. The element in row 1, column 1 is already 1. Next we get 0s for each element in the rest of column 1. To do this, first we multiply the elements of row 1 by 2 and add the new elements to row 2. Also, we multiply the elements of row 1 by $-1$ and add the new elements to the elements of row 3. *We do not change row 1.* Then

$$\begin{bmatrix} 1 & 2 & 1 & \vdots & 2 \\ 2(1)-2 & 2(2)-1 & 2(1)+2 & \vdots & 2(2)+5 \\ -1(1)+1 & -1(2)+3 & -1(1)-2 & \vdots & -1(2)-8 \end{bmatrix} \text{ simplifies to}$$

$$\begin{bmatrix} 1 & 2 & 1 & \vdots & 2 \\ 0 & 3 & 4 & \vdots & 9 \\ 0 & 1 & -3 & \vdots & -10 \end{bmatrix}$$

We continue down the diagonal and use elementary row operations to get 1 where the element 3 is now. To do this, we interchange rows 2 and 3.

$$\begin{bmatrix} 1 & 2 & 1 & \vdots & 2 \\ 0 & 3 & 4 & \vdots & 9 \\ 0 & 1 & -3 & \vdots & -10 \end{bmatrix} \xrightarrow{\text{is equivalent to}} \begin{bmatrix} 1 & 2 & 1 & \vdots & 2 \\ 0 & 1 & -3 & \vdots & -10 \\ 0 & 3 & 4 & \vdots & 9 \end{bmatrix}$$

Next we want the new row 3, column 2 element to be 0. We multiply the elements of row 2 by $-3$ and add the result to the elements of row 3.

$$\begin{bmatrix} 1 & 2 & 1 & \vdots & 2 \\ 0 & 1 & -3 & \vdots & -10 \\ -3(0)+0 & -3(1)+3 & -3(-3)+4 & \vdots & -3(-10)+9 \end{bmatrix} \text{ simplifies to}$$

$$\begin{bmatrix} 1 & 2 & 1 & \vdots & 2 \\ 0 & 1 & -3 & \vdots & -10 \\ 0 & 0 & 13 & \vdots & 39 \end{bmatrix}$$

Finally, we divide the elements of row 3 by 13 so that the final diagonal element is 1.

$$\begin{bmatrix} 1 & 2 & 1 & \vdots & 2 \\ 0 & 1 & -3 & \vdots & -10 \\ \frac{0}{13} & \frac{0}{13} & \frac{13}{13} & \vdots & \frac{39}{13} \end{bmatrix} \text{ simplifies to } \begin{bmatrix} 1 & 2 & 1 & \vdots & 2 \\ 0 & 1 & -3 & \vdots & -10 \\ 0 & 0 & 1 & \vdots & 3 \end{bmatrix}$$

This matrix corresponds to the system

$$\begin{cases} x + 2y + z = 2 \\ \phantom{x + 2}y - 3z = -10 \\ \phantom{x + 2y + }z = 3 \end{cases}$$

We identify the $z$-coordinate of the solution as 3. Next we replace $z$ with 3 in the second equation and solve for $y$.

$y - 3z = -10$　Second equation

$y - 3(3) = -10$　Let $z = 3$.

$y = -1$

To find $x$, we let $z = 3$ and $y = -1$ in the first equation.

$x + 2y + z = 2$　First equation

$x + 2(-1) + 3 = 2$　Let $z = 3$ and $y = -1$.

$x = 1$

The ordered triple solution is $(1, -1, 3)$. Check to see that it satisfies all three equations in the original system.

■ Work Practice 3

## Vocabulary, Readiness & Video Check

*Use the choices below to fill in each blank.*

column        element        row        matrix

1. A(n) _____ is a rectangular array of numbers.
2. Each of the numbers in a matrix is called a(n) _____.
3. The numbers aligned horizontally in a matrix are in the same _____.
4. The numbers aligned vertically in a matrix are in the same _____.

*Answer true or false for each statement about operations within a matrix forming an equivalent matrix.*

5. Any two columns may be interchanged. _____
6. Any two rows may be interchanged. _____
7. The elements in a row may be added to their corresponding elements in another row. _____
8. The elements of a column may be multiplied by any nonzero number. _____

*Martin-Gay Interactive Videos*    Watch the section lecture video and answer the following questions.

**Objective A**    9. From the lecture before ⊞ Example 1, what elementary row operations can be performed on matrices? Which operation is not performed during Example 1? ▶

**Objective B**    10. In ⊞ Example 2, why do you think the suggestion is made to write neatly when using matrices to solve systems? ▶

*See Video 4.4*

---

## 4.4 Exercise Set MyMathLab® ▶

**Objective A** *Use matrices to solve each system of linear equations. See Example 1.*

1. $\begin{cases} x + y = 1 \\ x - 2y = 4 \end{cases}$

2. $\begin{cases} 2x - y = 8 \\ x + 3y = 11 \end{cases}$

▶3. $\begin{cases} x + 3y = 2 \\ x + 2y = 0 \end{cases}$

4. $\begin{cases} 4x - y = 5 \\ 3x - 3y = 6 \end{cases}$

*Use matrices to solve each system of linear equations. See Example 2.*

5. $\begin{cases} x - 2y = 4 \\ 2x - 4y = 4 \end{cases}$

6. $\begin{cases} -x + 3y = 6 \\ 3x - 9y = 9 \end{cases}$

7. $\begin{cases} 3x - 3y = 9 \\ 2x - 2y = 6 \end{cases}$

8. $\begin{cases} 9x - 3y = 6 \\ -18x + 6y = -12 \end{cases}$

**Objective B** *Use matrices to solve each system of linear equations. See Example 3.*

9. $\begin{cases} x + y = 3 \\ 2y = 10 \\ 3x + 2y - 4z = 12 \end{cases}$

10. $\begin{cases} 5x = 5 \\ 2x + y = 4 \\ 3x + y - 5z = -15 \end{cases}$

11. $\begin{cases} 2y - z = -7 \\ x + 4y + z = -4 \\ 5x - y + 2z = 13 \end{cases}$

12. $\begin{cases} 4y + 3z = -2 \\ 5x - 4y = 1 \\ -5x + 4y + z = -3 \end{cases}$

## Objectives A B Mixed Practice *Solve each system of linear equations using matrices. See Examples 1 through 3.*

13. $\begin{cases} x - 4 = 0 \\ x + y = 1 \end{cases}$

14. $\begin{cases} 3y = 6 \\ x + y = 7 \end{cases}$

15. $\begin{cases} x + y + z = 2 \\ 2x \quad\;\; - z = 5 \\ 3y + z = 2 \end{cases}$

16. $\begin{cases} x + 2y + z = 5 \\ x - y - z = 3 \\ y + z = 2 \end{cases}$

17. $\begin{cases} 5x - 2y = 27 \\ -3x + 5y = 18 \end{cases}$

18. $\begin{cases} 4x - y = 9 \\ 2x + 3y = -27 \end{cases}$

19. $\begin{cases} 4x - 7y = 7 \\ 12x - 21y = 24 \end{cases}$

20. $\begin{cases} 2x - 5y = 12 \\ -4x + 10y = 20 \end{cases}$

21. $\begin{cases} 4x - y + 2z = 5 \\ 2y + z = 4 \\ 4x + y + 3z = 10 \end{cases}$

22. $\begin{cases} 5y - 7z = 14 \\ 2x + y + 4z = 10 \\ 2x + 6y - 3z = 30 \end{cases}$

▶ 23. $\begin{cases} 4x + y + z = 3 \\ -x + y - 2z = -11 \\ x + 2y + 2z = -1 \end{cases}$

24. $\begin{cases} x + y + z = 9 \\ 3x - y + z = -1 \\ -2x + 2y - 3z = -2 \end{cases}$

## Review

*Determine whether each graph is the graph of a function. See Section 3.6.*

25.

26.

27.

28.

## Concept Extensions

*Solve. For Exercises 29 and 30, see the Concept Check in this section.*

29. For the system $\begin{cases} x \quad\;\; + z = 7 \\ y + 2z = -6, \\ 3x - y \quad\;\; = 0 \end{cases}$ which is the correct corresponding matrix?

a. $\begin{bmatrix} 1 & 1 & | & 7 \\ 1 & 2 & | & -6 \\ 3 & -1 & | & 0 \end{bmatrix}$

b. $\begin{bmatrix} 1 & 0 & 1 & | & 7 \\ 1 & 2 & 0 & | & -6 \\ 3 & -1 & 0 & | & 0 \end{bmatrix}$

c. $\begin{bmatrix} 1 & 0 & 1 & | & 7 \\ 0 & 1 & 2 & | & -6 \\ 3 & -1 & 0 & | & 0 \end{bmatrix}$

30. For the system $\begin{cases} x - 6 = 0 \\ 2x - 3y = 1 \end{cases}$, which is the correct corresponding matrix?

a. $\begin{bmatrix} 1 & -6 & | & 0 \\ 2 & -3 & | & 1 \end{bmatrix}$

b. $\begin{bmatrix} 1 & 0 & | & 6 \\ 2 & -3 & | & 1 \end{bmatrix}$

c. $\begin{bmatrix} 1 & 0 & | & -6 \\ 2 & -3 & | & 1 \end{bmatrix}$

**31.** The amount of electricity generated by geothermal sources (in billions of kilowatts) from 2010 to 2013 can be modeled by the linear equation $y - 0.86x = 15.22$, where $x$ represents the number of years since 2010. Similarly, the amount of electricity generated by wind power (in billions of kilowatts) during the same time period can be modeled by the linear equation $y = 21.83x + 94.62$. (*Source:* Based on data from the Energy Information Administration, U.S. Department of Energy)

**a.** The data used to form these two models were incomplete. It is impossible to tell from the data the year in which the electricity generated by geothermal sources was the same as the electricity generated by wind power. Use matrix methods to estimate the year in which this occurred.

**b.** The earliest data for wind power was in 1989, where 2.1 billion kilowatts of electricity was generated. Can this data be determined from the given equation? Why do you think this is?

**c.** According to these models, will the amount of electricity generated by geothermal ever go to zero? Why?

**d.** Can you think of an explanation why the amount of electricity generated by wind power is increasing so much faster than the amount of electricity generated by geothermal power?

**32.** The most popular amusement park in the world (according to annual attendance) is Walt Disney World's Magic Kingdom, whose yearly attendance in thousands can be approximated by the equation $y = 178x + 15,400$, where $x$ is the number of years after 2000. The park that held the distinction of most popular until just recently is Tokyo Disneyland, whose yearly attendance, in thousands, can be approximated by the equation $y = -138x + 16,500$. Find the last year when attendance at Tokyo Disneyland was greater than attendance at Walt Disney World's Magic Kingdom. (*Source:* Themed Entertainment Association)

**33.** For the system $\begin{cases} 2x - 3y = 8 \\ x + 5y = -3 \end{cases}$, explain what is wrong with writing the corresponding matrix as $\begin{bmatrix} 2 & 3 & | & 8 \\ 0 & 5 & | & -3 \end{bmatrix}$.

**34.** For the system $\begin{cases} 5x + 2y = 0 \\ -y = 2 \end{cases}$, explain what is wrong with writing the corresponding matrix as $\begin{bmatrix} 5 & 2 & | & 0 \\ -1 & 0 & | & 2 \end{bmatrix}$.

## 4.5　Systems of Linear Inequalities

### Objective A　Graphing Systems of Linear Inequalities

In Section 3.5, we solved linear inequalities in two variables. Just as two linear equations make a system of linear equations, two linear inequalities make a **system of linear inequalities.** Systems of inequalities are very important in a process called linear programming. Many businesses use linear programming to find the most profitable way to use limited resources such as employees, machines, or buildings.

A **solution of a system of linear inequalities** is an ordered pair that satisfies each inequality in the system. The set of all such ordered pairs is the solution set of the system. Graphing this set gives us a picture of the solution set. We can graph a system of inequalities by graphing each inequality in the system and identifying the region of overlap.

**Objective**

**A** Graph a System of Linear Inequalities.

---

#### Graphing the Solutions of a System of Linear Inequalities

**Step 1:** Graph each inequality in the system on the same set of axes.

**Step 2:** The solutions of the system are the points common to the graphs of all the inequalities in the system.

## Practice 1

Graph the solutions of the system:

$$\begin{cases} 2x \le y \\ x + 4y \ge 4 \end{cases}$$

**Example 1**  Graph the solutions of the system: $\begin{cases} 3x \ge y \\ x + 2y \le 8 \end{cases}$

**Solution:**  We begin by graphing each inequality on the *same* set of axes. The graph of the solutions of the system is the region contained in the graphs of both inequalities. In other words, it is their intersection.

First let's graph $3x \ge y$. The boundary line is the graph of $3x = y$. We sketch a solid boundary line since the inequality $3x \ge y$ means $3x > y$ or $3x = y$. The test point $(1, 0)$ satisfies the inequality, so we shade the half-plane that includes $(1, 0)$.

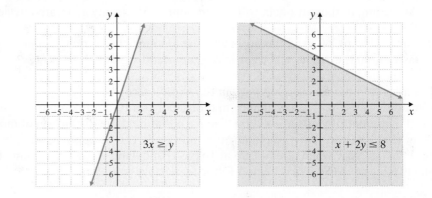

Next we sketch a solid boundary line $x + 2y = 8$ on the same set of axes. The test point $(0, 0)$ satisfies the inequality $x + 2y \le 8$, so we shade the half-plane that includes $(0, 0)$. (For clarity, the graph of $x + 2y \le 8$ is shown here on a separate set of axes.) An ordered pair solution of the system must satisfy both inequalities. These solutions are points that lie in both shaded regions. The solution of the system is the darkest shaded region. This solution includes parts of both boundary lines.

**Answers**

**1.**

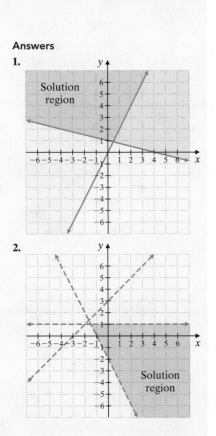

■ **Work Practice 1**

In linear programming, it is sometimes necessary to find the coordinates of the **corner point:** the point at which the two boundary lines intersect. To find the corner point for the system of Example 1, we solve the related linear system

$$\begin{cases} 3x = y \\ x + 2y = 8 \end{cases}$$

using either the substitution or the elimination method. The lines intersect at $\left(\dfrac{8}{7}, \dfrac{24}{7}\right)$, the corner point of the graph.

**Example 2** Graph the solutions of the system:

$$\begin{cases} x - y < 2 \\ x + 2y > -1 \\ \quad y < 2 \end{cases}$$

**Solution:** First we graph all three inequalities on the same set of axes. All boundary lines are dashed lines since the inequality symbols are $<$ and $>$. The solution of the system is the region shown by the shading. In this example, the boundary lines are *not* a part of the solution.

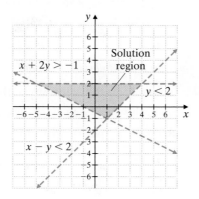

Work Practice 2

✓**Concept Check** Describe the solution of the system of inequalities:

$$\begin{cases} x \le 2 \\ x \ge 2 \end{cases}$$

**Example 3** Graph the solutions of the system: $\begin{cases} -3x + 4y \le 12 \\ \quad x \le 3 \\ \quad x \ge 0 \\ \quad y \ge 0 \end{cases}$

**Solution:** We graph the inequalities on the same set of axes. The intersection of the inequalities is the solution region. It is the only region shaded in this graph and includes the portions of all four boundary lines that border the shaded region.

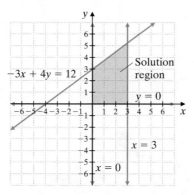

Work Practice 3

**Practice 2**

Graph the solutions of the system:

$$\begin{cases} -x + y < 3 \\ \quad y < 1 \\ 2x + y > -2 \end{cases}$$

**Practice 3**

Graph the solutions of the system:

$$\begin{cases} 2x - 3y \le 6 \\ \quad y \ge 0 \\ \quad y \le 4 \\ \quad x \ge 0 \end{cases}$$

**Answers**

**2.** See page 320.

**3.**

✓**Concept Check Answer**

the line $x = 2$

## Vocabulary, Readiness & Video Check

*Use the choices below to fill in each blank. Not all choices will be used.*

solution      union        system
corner        intersection

**1.** Two or more linear inequalities form a(n) _____ of linear inequalities.

**2.** An ordered pair that satisfies each inequality in a system is a(n) _____ of the system.

**3.** The point where two boundary lines intersect is a(n) _____ point.

**4.** The solution region of a system of inequalities consists of the _____ of the solution regions of the inequalities in the system.

*Martin-Gay Interactive Videos*    *Watch the section lecture video and answer the following question.*

**Objective A 5.** In Example 1, do the solutions of the first inequality of the system limit where we can choose the test point for the second inequality? Why or why not?

*See Video 4.5*

## 4.5 Exercise Set MyMathLab®

**Objective A** *Graph the solutions of each system of linear inequalities. See Examples 1 through 3.*

**1.** $\begin{cases} y \geq x + 1 \\ y \geq 3 - x \end{cases}$

**2.** $\begin{cases} y \geq x - 3 \\ y \geq -1 - x \end{cases}$

**3.** $\begin{cases} y < 3x - 4 \\ y \leq x + 2 \end{cases}$

**4.** $\begin{cases} y \leq 2x + 1 \\ y > x + 2 \end{cases}$

**5.** $\begin{cases} y < -2x - 2 \\ y > x + 4 \end{cases}$

**6.** $\begin{cases} y \leq 2x + 4 \\ y \geq -x - 5 \end{cases}$

**7.** $\begin{cases} y \geq -x + 2 \\ y \leq 2x + 5 \end{cases}$

**8.** $\begin{cases} y \geq x - 5 \\ y \leq -3x + 3 \end{cases}$

9. $\begin{cases} x \geq 3y \\ x + 3y \leq 6 \end{cases}$

10. $\begin{cases} -2x < y \\ x + 2y < 3 \end{cases}$

11. $\begin{cases} x \leq 2 \\ y \geq -3 \end{cases}$

12. $\begin{cases} x \geq -3 \\ y \geq -2 \end{cases}$

13. $\begin{cases} y \geq 1 \\ x < -3 \end{cases}$

14. $\begin{cases} y > 2 \\ x \geq -1 \end{cases}$

15. $\begin{cases} y + 2x \geq 0 \\ 5x - 3y \leq 12 \\ y \leq 2 \end{cases}$

16. $\begin{cases} y + 2x \leq 0 \\ 5x + 3y \geq -2 \\ y \leq 4 \end{cases}$

17. $\begin{cases} 3x - 4y \geq -6 \\ 2x + y \leq 7 \\ y \geq -3 \end{cases}$

18. $\begin{cases} 4x - y \geq -2 \\ 2x + 3y \leq -8 \\ y \geq -5 \end{cases}$

19. $\begin{cases} 2x + y \leq 5 \\ x \leq 3 \\ x \geq 0 \\ y \geq 0 \end{cases}$

20. $\begin{cases} 3x + y \leq 4 \\ x \leq 4 \\ x \geq 0 \\ y \geq 0 \end{cases}$

*Match each system of inequalities to the corresponding graph.*

A

B

C

D

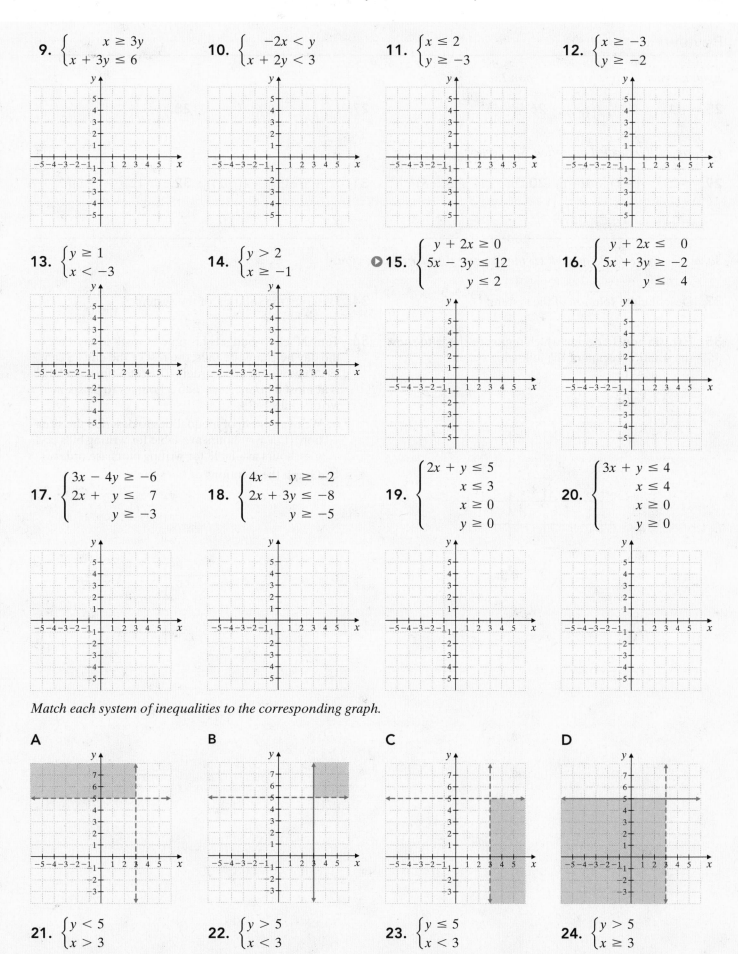

21. $\begin{cases} y < 5 \\ x > 3 \end{cases}$

22. $\begin{cases} y > 5 \\ x < 3 \end{cases}$

23. $\begin{cases} y \leq 5 \\ x < 3 \end{cases}$

24. $\begin{cases} y > 5 \\ x \geq 3 \end{cases}$

## Review

*Evaluate each expression. See Section 1.6.*

**25.** $(-3)^2$     **26.** $(-5)^3$     **27.** $\left(\dfrac{2}{3}\right)^2$     **28.** $\left(\dfrac{3}{4}\right)^3$

*Perform each indicated operation. See Section 1.6.*

**29.** $(-2)^2 - (-3) + 2(-1)$   **30.** $5^2 - 11 + 3(-5)$   **31.** $8^2 + (-13) - 4(-2)$   **32.** $(-12)^2 + (-1)(2) - 6$

## Concept Extensions

*Solve. For Exercises 33 and 34, see the Concept Check in this section.*

**33.** Describe the solution of the system: $\begin{cases} y \le 3 \\ y \ge 3 \end{cases}$

**34.** Describe the solution of the system: $\begin{cases} x \le 5 \\ x \le 3 \end{cases}$

**35.** Explain how to decide which region to shade to show the solution region of the following system.
$$\begin{cases} x \ge 3 \\ y \ge -2 \end{cases}$$

**36.** Tony Noellert budgets his time at work today. Part of the day he can write bills; the rest of the day he can use to write purchase orders. The total time available is at most 8 hours. Less than 3 hours is to be spent writing bills.

**a.** Write a system of inequalities to describe the situation. (Let $x$ = hours available for writing bills and $y$ = hours available for writing purchase orders.)

**b.** Graph the solutions of the system.

## Chapter 4 Group Activity

### Another Mathematical Model

Sometimes mathematical models other than linear models are appropriate for data. Suppose that an equation of the form $y = ax^2 + bx + c$ is an appropriate model for the ordered pairs $(x_1, y_1)$, $(x_2, y_2)$, and $(x_3, y_3)$. Then it is necessary to find the values of $a$, $b$, and $c$ such that the given ordered pairs are solutions of the equation $y = ax^2 + bx + c$. To do so, substitute each ordered pair into the equation. Each time, the result is an equation in three unknowns: $a$, $b$, and $c$. Solving the resulting system of three linear equation in three unknowns will give the required values of $a$, $b$, and $c$.

1. The table gives the supply of electricity (in billions of kilowatts) generated in the United States by wind power in each of the years listed. (*Source:* Based on data from the Energy Information Administration, U.S. Department of Energy)

| **Wind Power–Generated Electricity** | |
|---|---|
| **Year** | **Wind-Generated Electricity (in billions of kilowatts)** |
| 2008 | 55.4 |
| 2010 | 94.6 |
| 2013 | 160.1 |

   **a.** Write the data as ordered pairs of the form $(x, y)$, where $y$ is the amount of electricity generated (in billions of kilowatts) in the year $x$ ($x = 0$ represents 2008).

   **b.** Find the values of $a$, $b$, and $c$ rounded to 2 decimal places (where needed) such that the equation $y = ax^2 + bx + c$ models these data.

   **c.** Verify that the model you found in part (b) gives each of the ordered pair solutions from part (a).

   **d.** According to the model, how much electricity was generated by wind power in 2012?

2. The table gives the percent of American households with broadband Internet access for each of the years listed. (*Source:* Pew Internet and American Life Project)

| **Percent of U.S. Homes with Broadband Internet Access** | |
|---|---|
| **Year** | **Percent of U.S. Households with Broadband Internet Access** |
| 2000 | 3 |
| 2005 | 33 |
| 2013 | 70 |

   **a.** Write the data as ordered pairs of the form $(x, y)$, where $y$ is the percent of U.S. households with broadband Internet access in the year $x$ ($x = 0$ represents 2000).

   **b.** Find the values of $a$, $b$, and $c$ rounded to 2 decimal places (where needed) such that the equation $y = ax^2 + bx + c$ models these data.

   **c.** According to the model, what percent of American homes had broadband Internet access in 2010?

3. **a.** Make up an equation of the form $y = ax^2 + bx + c$.

   **b.** Find three ordered pair solutions of the equation.

   **c.** Without revealing your equation from part (a), exchange lists of ordered pair solutions with another group.

   **d.** Use the method described above to find the values of $a$, $b$, and $c$ such that the equation $y = ax^2 + bx + c$ has the ordered pair solutions you received from the other group.

   **e.** Check with the other group to see if your equation from part (d) is the correct one.

# Chapter 4 Vocabulary Check

*Fill in each blank with one of the words or phrases listed below.*

| matrix | consistent | system of equations | triple | row |
|---|---|---|---|---|
| solution | inconsistent | element | column | |

1. Two or more linear equations in two variables form a(n) _____.

2. A(n) _____ of a system of two equations in two variables is an ordered pair that makes both equations true.

3. A(n) _____ system of equations has at least one solution.

4. A solution of a system of three equations in three variables is an ordered _____ that makes all three equations true.

5. A(n) _____ system of equations has no solution.

6. A(n) _____ is a rectangular array of numbers.

7. Each of the numbers in a matrix is called a(n) _____.

8. The numbers aligned horizontally in a matrix are in the same _____.

9. The numbers aligned vertically in a matrix are in the same _____.

**Helpful Hint**

▶ Are you preparing for your test? Don't forget to take the Chapter 4 Test on page 335. Then check your answers at the back of the text and use the Chapter Test Prep Videos to see the fully worked-out solutions to any of the exercises you want to review.

# 4 Chapter Highlights

| Definitions and Concepts | Examples |
|---|---|
| **Section 4.1   Solving Systems of Linear Equations in Two Variables** | |

| Definitions and Concepts | Examples |
|---|---|
| A **system of linear equations** consists of two or more linear equations. | $\begin{cases} x - 3y = 6 \\ \qquad y = \dfrac{1}{2}x \end{cases}$   $\begin{cases} x + 2y - z = 1 \\ 3x - y + 4z = 0 \\ \qquad 5y + z = 6 \end{cases}$ |
| A **solution** of a system of two equations in two variables is an ordered pair $(x, y)$ that makes both equations true. | Determine whether $(2, -5)$ is a solution of the system. $\begin{cases} x + y = -3 \\ 2x - 3y = 19 \end{cases}$ Replace $x$ with 2 and $y$ with $-5$ in both equations. $\begin{array}{ll} x + y = -3 & 2x - 3y = 19 \\ 2 + (-5) \stackrel{?}{=} -3 & 2(2) - 3(-5) \stackrel{?}{=} 19 \\ \qquad -3 = -3 \;\; \text{True} & 4 + 15 \stackrel{?}{=} 19 \\ & 19 = 19 \;\; \text{True} \end{array}$ $(2, -5)$ is a solution of the system. |
| Geometrically, a solution of a system in two variables is a point of intersection of the graphs of the equations. | Solve by graphing: $\begin{cases} y = 2x - 1 \\ x + 2y = 13 \end{cases}$ |

| Definitions and Concepts | Examples |
|---|---|

### Section 4.1   Solving Systems of Linear Equations in Two Variables (*continued*)

A system of equations with at least one solution is a **consistent system.** A system that has no solution is an **inconsistent system.**

If the graphs of two linear equations are identical, the equations are **dependent.**

If their graphs are different, the equations are **independent.**

One solution:
Independent equations
Consistent system

No solution:
Independent equations
Inconsistent system

Infinite number of solutions:
Dependent equations
Consistent system

**Solving a System of Linear Equations by the Substitution Method**

**Step 1:** Solve one equation for a variable.

**Step 2:** Substitute the expression for the variable into the other equation.

**Step 3:** Solve the equation from Step 2 to find the value of one variable.

**Step 4:** Substitute the value from Step 3 in either original equation to find the value of the other variable.

**Step 5:** Check the solution in both equations.

Solve by substitution:
$$\begin{cases} y = x + 2 \\ 3x - 2y = -5 \end{cases}$$

Since the first equation is solved for $y$, substitute $x + 2$ for $y$ in the second equation.

$$3x - 2y = -5 \quad \text{Second equation}$$
$$3x - 2(x + 2) = -5 \quad \text{Let } y = x + 2.$$
$$3x - 2x - 4 = -5$$
$$x - 4 = -5 \quad \text{Simplify.}$$
$$x = -1 \quad \text{Add 4.}$$

To find $y$, let $x = -1$ in $y = x + 2$, so $y = -1 + 2 = 1$. The solution $(-1, 1)$ checks.

**Solving a System of Linear Equations by the Elimination Method**

**Step 1:** Rewrite each equation in standard form, $Ax + By = C$.

**Step 2:** Multiply one or both equations by a nonzero number so that the coefficients of a variable are opposites.

**Step 3:** Add the equations.

**Step 4:** Find the value of the remaining variable by solving the resulting equation.

**Step 5:** Substitute the value from Step 4 into either original equation to find the value of the other variable.

**Step 6:** Check the solution in both equations.

Solve by elimination:
$$\begin{cases} x - 3y = -3 \\ -2x + y = 6 \end{cases}$$

Multiply both sides of the first equation by 2.

$$\begin{array}{r} 2x - 6y = -6 \\ \underline{-2x + y = 6} \\ -5y = 0 \quad \text{Add.} \\ y = 0 \quad \text{Divide by } -5. \end{array}$$

To find $x$, let $y = 0$ in an original equation.

$$x - 3y = -3$$
$$x - 3 \cdot 0 = -3$$
$$x = -3$$

The solution $(-3, 0)$ checks.

| Definitions and Concepts | Examples |
|---|---|

### Section 4.2   Solving Systems of Linear Equations in Three Variables

A **solution** of an equation in three variables $x$, $y$, and $z$ is an **ordered triple** $(x, y, z)$ that makes the equation a true statement.

Verify that $(-2, 1, 3)$ is a solution of $2x + 3y - 2z = -7$. Replace $x$ with $-2$, $y$ with 1, and $z$ with 3.

$$2(-2) + 3(1) - 2(3) \stackrel{?}{=} -7$$
$$-4 + 3 - 6 \stackrel{?}{=} -7$$
$$-7 = -7 \quad \text{True}$$

$(-2, 1, 3)$ is a solution.

**Solving a System of Three Linear Equations by the Elimination Method**

**Step 1:** Write each equation in standard form, $Ax + By + Cz = D$.

**Step 2:** Choose a pair of equations and use them to eliminate a variable.

**Step 3:** Choose any other pair of equations and eliminate the same variable.

**Step 4:** Solve the system of two equations in two variables from Steps 2 and 3.

**Step 5:** Solve for the third variable by substituting the values of the variables from Step 4 into any of the original equations.

**Step 6:** Check the solution in all three original equations.

Solve:

$$\begin{cases} 2x + y - z = 0 & (1) \\ x - y - 2z = -6 & (2) \\ -3x - 2y + 3z = -22 & (3) \end{cases}$$

**1.** Each equation is written in standard form.

**2.**
$$\begin{array}{rl} 2x + y - z = & 0 \quad (1) \\ \underline{x - y - 2z = -6} & \quad (2) \\ 3x \quad\quad - 3z = -6 & \quad (4) \quad \text{Add.} \end{array}$$

**3.** Eliminate $y$ from equations (1) and (3) also.

$$\begin{array}{rl} 4x + 2y - 2z = & 0 \quad\quad \text{Multiply equation (1) by 2.} \\ \underline{-3x - 2y + 3z = -22} & \quad (3) \\ x \quad\quad + z = -22 & \quad (5) \quad \text{Add.} \end{array}$$

**4.** Solve.

$$\begin{cases} 3x - 3z = -6 & (4) \\ x + z = -22 & (5) \end{cases}$$

$$\begin{array}{rl} x - z = -2 & \quad \text{Divide equation (4) by 3.} \\ \underline{x + z = -22} & \quad (5) \\ 2x \quad\quad = -24 & \\ x = -12 & \end{array}$$

To find $z$, use equation (5).

$$x + z = -22$$
$$-12 + z = -22$$
$$z = -10$$

**5.** To find $y$, use equation (1).

$$2x + y - z = 0$$
$$2(-12) + y - (-10) = 0$$
$$-24 + y + 10 = 0$$
$$y = 14$$

**6.** The solution $(-12, 14, -10)$ checks.

| Definitions and Concepts | Examples |
|---|---|

### Section 4.3   Systems of Linear Equations and Problem Solving

|  | Two numbers have a sum of 11. Twice one number is 3 less than 3 times the other. Find the numbers. |
|---|---|
| **1.** UNDERSTAND the problem. | **1.** Read and reread.<br><br>$x =$ one number<br>$y =$ other number |
| **2.** TRANSLATE. | **2.** In words:<br><br>sum of numbers    is    11<br>$\downarrow$    $\downarrow$    $\downarrow$<br>Translate:  $x + y$    $=$    11<br><br>In words:<br><br>twice one number    is    3 less than 3 times the other number<br>$\downarrow$    $\downarrow$    $\downarrow$<br>Translate:    $2x$    $=$    $3y - 3$ |
| **3.** SOLVE. | **3.** Solve the system: $\begin{cases} x + y = 11 \\ 2x = 3y - 3 \end{cases}$<br>In the first equation, $x = 11 - y$. Substitute into the other equation.<br><br>$$2x = 3y - 3$$<br>$$2(11 - y) = 3y - 3$$<br>$$22 - 2y = 3y - 3$$<br>$$-5y = -25$$<br>$$y = 5$$<br><br>Replace $y$ with 5 in the equation $x = 11 - y$. Then $x = 11 - 5 = 6$. The solution is $(6, 5)$. |
| **4.** INTERPRET. | **4.** *Check:* See that $6 + 5 = 11$ is the required sum and that twice 6 is 3 times 5 less 3. *State:* The numbers are 6 and 5. |

### Section 4.4   Solving Systems of Equations Using Matrices

| A **matrix** is a rectangular array of numbers. | $\begin{bmatrix} -7 & 0 & 3 \\ 1 & 2 & 4 \end{bmatrix}$    $\begin{bmatrix} a & b & c \\ d & e & f \\ g & h & i \end{bmatrix}$ |
|---|---|
| The **matrix** corresponding to a system is composed of the coefficients of the variables and the constants of the system. | The matrix corresponding to the system<br><br>$\begin{cases} x - y = 1 \\ 2x + y = 11 \end{cases}$  is  $\left[\begin{array}{rr|r} 1 & -1 & 1 \\ 2 & 1 & 11 \end{array}\right]$ |

*(continued)*

| Definitions and Concepts | Examples |
|---|---|

### Section 4.4   Solving Systems of Equations Using Matrices (*continued*)

The following **row operations** can be performed on matrices, and the result is an equivalent matrix.

Elementary row operations:

1. Interchange any two rows.

2. Multiply (or divide) the elements of one row by the same nonzero number.

3. Multiply (or divide) the elements of one row by the same nonzero number and add them to their corresponding elements in any other row.

Use matrices to solve: $\begin{cases} x - y = 1 \\ 2x + y = 11 \end{cases}$

The corresponding matrix is

$$\left[\begin{array}{rr|r} 1 & -1 & 1 \\ 2 & 1 & 11 \end{array}\right]$$

Use row operations to write an equivalent matrix with 1s along the diagonal and 0s below each 1 in the diagonal. Multiply row 1 by $-2$ and add to row 2. Change row 2 only.

$$\left[\begin{array}{cc|c} 1 & -1 & 1 \\ -2(1) + 2 & -2(-1) + 1 & -2(1) + 11 \end{array}\right]$$

simplifies to $\left[\begin{array}{rr|r} 1 & -1 & 1 \\ 0 & 3 & 9 \end{array}\right]$

Divide row 2 by 3.

$$\left[\begin{array}{cc|c} 1 & -1 & 1 \\ \dfrac{0}{3} & \dfrac{3}{3} & \dfrac{9}{3} \end{array}\right] \quad \text{simplifies to} \quad \left[\begin{array}{rr|r} 1 & -1 & 1 \\ 0 & 1 & 3 \end{array}\right]$$

This matrix corresponds to the system

$$\begin{cases} x - y = 1 \\ y = 3 \end{cases}$$

Let $y = 3$ in the first equation.

$$x - 3 = 1$$
$$x = 4$$

The ordered pair solution is $(4, 3)$.

### Section 4.5   Systems of Linear Inequalities

A **system of linear inequalities** consists of two or more linear inequalities.

To graph a system of inequalities, graph each inequality in the system. The overlapping region is the solution region of the system.

$$\begin{cases} x - y \geq 3 \\ y \leq -2x \end{cases}$$

Solution region

**(4.1)** *Solve each system of equations in two variables by each method: (a) graphing, (b) substitution, and (c) elimination.*

**1.** $\begin{cases} 3x + 10y = 1 \\ x + 2y = -1 \end{cases}$

**2.** $\begin{cases} y = \dfrac{1}{2}x + \dfrac{2}{3} \\ 4x + 6y = 4 \end{cases}$

**3.** $\begin{cases} 2x - 4y = 22 \\ 5x - 10y = 15 \end{cases}$

**4.** $\begin{cases} 3x - 6y = 12 \\ 2y = x - 4 \end{cases}$

**5.** $\begin{cases} \dfrac{1}{2}x - \dfrac{3}{4}y = -\dfrac{1}{2} \\ \dfrac{1}{8}x + \dfrac{3}{4}y = \dfrac{19}{8} \end{cases}$

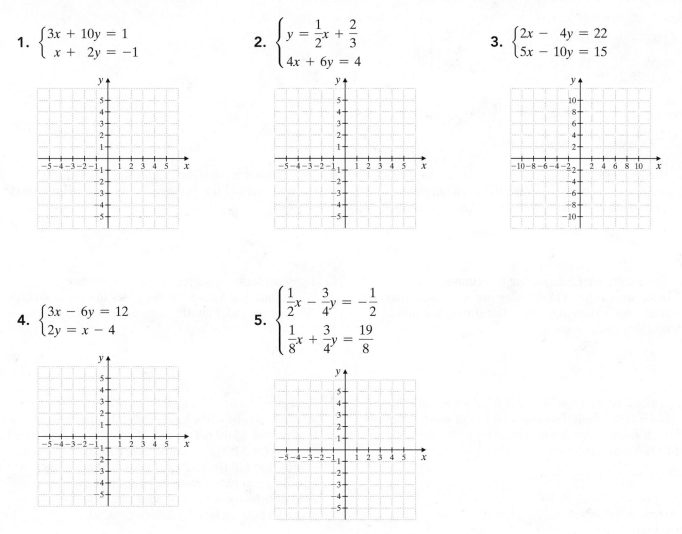

**6.** The revenue equation for a certain style of backpack is $y = 32x$, where $x$ is the number of backpacks sold and $y$ is the income in dollars for selling $x$ backpacks. The cost equation for these units is $y = 15x + 25{,}500$, where $x$ is the number of backpacks manufactured and $y$ is the cost in dollars for manufacturing $x$ backpacks. Find the number of units to be sold for the company to break even. (*Hint:* Solve the system of equations formed by the two given equations.)

**(4.2)** *Solve each system of equations in three variables.*

**7.** $\begin{cases} x + z = 4 \\ 2x - y = 4 \\ x + y - z = 0 \end{cases}$

**8.** $\begin{cases} 2x + 5y = 4 \\ x - 5y + z = -1 \\ 4x - z = 11 \end{cases}$

**9.** $\begin{cases} 4y + 2z = 5 \\ 2x + 8y = 5 \\ 6x + 4z = 1 \end{cases}$

**10.** $\begin{cases} 5x + 7y \phantom{ + z} = 9 \\ \phantom{5x + } 14y - z = 28 \\ 4x \phantom{ + 14y} + 2z = -4 \end{cases}$

**11.** $\begin{cases} 3x - 2y + 2z = 5 \\ -x + 6y + z = 4 \\ 3x + 14y + 7z = 20 \end{cases}$

**12.** $\begin{cases} x + 2y + 3z = 11 \\ \phantom{x + } y + 2z = 3 \\ 2x \phantom{ + 2y} + 2z = 10 \end{cases}$

**13.** $\begin{cases} 7x - 3y + 2z = 0 \\ 4x - 4y - z = 2 \\ 5x + 2y + 3z = 1 \end{cases}$

**14.** $\begin{cases} x - 3y - 5z = -5 \\ 4x - 2y + 3z = 13 \\ 5x + 3y + 4z = 22 \end{cases}$

**(4.3)** *Use systems of equations to solve.*

**15.** The sum of three numbers is 98. The sum of the first and second is two more than the third number, and the second is four times the first. Find the numbers.

**16.** One number is three times a second number, and twice the sum of the numbers is 168. Find the numbers.

**17.** Two cars leave Chicago, one traveling east and the other west. After 4 hours, they are 492 miles apart. If one car is traveling 7 mph faster than the other, find the speed of each.

△**18.** The foundation for a rectangular Hardware Warehouse has a length three times the width and is 296 feet around. Find the dimensions of the building.

**19.** James Callahan has available a 10% alcohol solution and a 60% alcohol solution. Find how many liters of each solution he should mix to make 50 liters of a 40% alcohol solution.

**20.** An employee at See's Candy Store needs a special mixture of candy. She has creme-filled chocolates that sell for $3.00 per pound, chocolate-covered nuts that sell for $2.70 per pound, and chocolate-covered raisins that sell for $2.25 per pound. She wants to have twice as many raisins as nuts in the mixture. Find how many pounds of each she should use to make 45 pounds worth $2.80 per pound.

**21.** Chris Kringler has $2.77 in her coin jar—all in pennies, nickels, and dimes. If she has 53 coins in all and four more nickels than dimes, find how many of each type of coin she has.

**22.** $10,000 and $4000 are invested such that $1250 in interest is earned in one year. If the rate of interest on the larger investment is 2% more than that on the smaller investment, find the rates of interest.

△**23.** The perimeter of an isosceles (two sides equal) triangle is 73 centimeters. If the unequal side is 7 centimeters longer than the two equal sides, find the lengths of the three sides.

**24.** The sum of three numbers is 295. The first number is five more than the second and twice the third. Find the numbers.

**(4.4)** *Use matrices to solve each system.*

**25.** $\begin{cases} 3x + 10y = 1 \\ x + 2y = -1 \end{cases}$

**26.** $\begin{cases} 3x - 6y = 12 \\ 2y = x - 4 \end{cases}$

**27.** $\begin{cases} 3x - 2y = -8 \\ 6x + 5y = 11 \end{cases}$

**28.** $\begin{cases} 6x - 6y = -5 \\ 10x - 2y = 1 \end{cases}$

**29.** $\begin{cases} 3x - 6y = 0 \\ 2x + 4y = 5 \end{cases}$

**30.** $\begin{cases} 5x - 3y = 10 \\ -2x + y = -1 \end{cases}$

**31.** $\begin{cases} 0.2x - 0.3y = -0.7 \\ 0.5x + 0.3y = 1.4 \end{cases}$

**32.** $\begin{cases} 3x + 2y = 8 \\ 3x - y = 5 \end{cases}$

**33.** $\begin{cases} x \quad + z = 4 \\ 2x - y \quad = 0 \\ x + y - z = 0 \end{cases}$

**34.** $\begin{cases} 2x + 5y \quad = 4 \\ x - 5y + z = -1 \\ 4x \quad - z = 11 \end{cases}$

**35.** $\begin{cases} 3x - y \quad = 11 \\ x \quad + 2z = 13 \\ y - z = -7 \end{cases}$

**36.** $\begin{cases} 5x + 7y + 3z = 9 \\ 14y - z = 28 \\ 4x \quad + 2z = -4 \end{cases}$

**37.** $\begin{cases} 7x - 3y + 2z = 0 \\ 4x - 4y - z = 2 \\ 5x + 2y + 3z = 1 \end{cases}$

**38.** $\begin{cases} x + 2y + 3z = 14 \\ y + 2z = 3 \\ 2x \quad - 2z = 10 \end{cases}$

**(4.5)** *Graph the solutions of each system of linear inequalities.*

**39.** $\begin{cases} y \geq 2x - 3 \\ y \leq -2x + 1 \end{cases}$

**40.** $\begin{cases} y \leq -3x - 3 \\ y \leq 2x + 7 \end{cases}$

**41.** $\begin{cases} x + 2y > 0 \\ x - y \leq 6 \end{cases}$

**42.** $\begin{cases} x - 2y \geq 7 \\ x + y \leq -5 \end{cases}$

**43.** $\begin{cases} 3x - 2y \le 4 \\ 2x + y \ge 5 \\ y \le 4 \end{cases}$    **44.** $\begin{cases} 4x - y \le 0 \\ 3x - 2y \ge -5 \\ y \ge -4 \end{cases}$    **45.** $\begin{cases} x + 2y \le 5 \\ x \le 2 \\ x \ge 0 \\ y \ge 0 \end{cases}$    **46.** $\begin{cases} x + 3y \le 7 \\ y \le 5 \\ x \ge 0 \\ y \ge 0 \end{cases}$

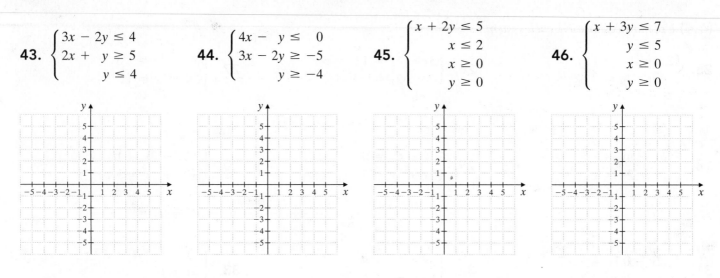

## Mixed Review

*Solve each system.*

**47.** $\begin{cases} y = x - 5 \\ y = -2x + 2 \end{cases}$    **48.** $\begin{cases} \dfrac{2}{5}x + \dfrac{3}{4}y = 1 \\ x + 3y = -2 \end{cases}$    **49.** $\begin{cases} 5x - 2y = 10 \\ x = \dfrac{2}{5}y + 2 \end{cases}$

**50.** $\begin{cases} x - 4y = 4 \\ \dfrac{1}{8}x - \dfrac{1}{2}y = 3 \end{cases}$    **51.** $\begin{cases} x - 3y + 2z = 0 \\ 9y - z = 22 \\ 5x + 3z = 10 \end{cases}$

**52.** One number is five less than three times a second number. If the sum of the numbers is 127, find the numbers.

**53.** The perimeter of a triangle is 126 units. The length of one side is twice the length of the shortest side. The length of the third side is 14 more than the length of the shortest side. Find the lengths of the sides of the triangle.

**54.** Graph the solutions of the system: $\begin{cases} y \le 3x - \dfrac{1}{2} \\ 3x + 4y \ge 6 \end{cases}$

**55.** The temperatures in the northern and southern hemispheres move in opposite directions during the year. As the temperatures rise in the northern hemisphere from February to July, they drop in the southern hemisphere during these same months. The average monthly temperature (in Fahrenheit) for England from February through July can be approximated by the equation $y = 4.3x + 44$, where $x$ is the number of months since February. The average monthly temperature (in Fahrenheit) in Australia from February through July can be approximated by the equation $y = -5.1x + 79$, where $x$ is the number of months since February. Use these equations to determine the month in which the average monthly temperature of England is the same as the average monthly temperature of Australia.

**Answers**

*Solve each system of equations graphically and then solve by the elimination method or the substitution method.*

1. $\begin{cases} 2x - y = -1 \\ 5x + 4y = 17 \end{cases}$

2. $\begin{cases} 7x - 14y = 5 \\ x = 2y \end{cases}$

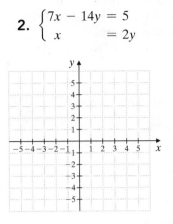

1. _____

2. _____

3. _____

*Solve each system.*

4. _____

3. $\begin{cases} 4x - 7y = 29 \\ 2x + 5y = -11 \end{cases}$

4. $\begin{cases} 15x + 6y = 15 \\ 10x + 4y = 10 \end{cases}$

5. $\begin{cases} 2x - 3y = 4 \\ 3y + 2z = 2 \\ x - z = -5 \end{cases}$

5. _____

6. _____

6. $\begin{cases} 3x - 2y - z = -1 \\ 2x - 2y = 4 \\ 2x - 2z = -12 \end{cases}$

7. $\begin{cases} \dfrac{x}{2} + \dfrac{y}{4} = -\dfrac{3}{4} \\ x + \dfrac{3}{4}y = -4 \end{cases}$

7. _____

*Use matrices to solve each system.*

8. _____

8. $\begin{cases} x - y = -2 \\ 3x - 3y = -6 \end{cases}$

9. $\begin{cases} x + 2y = -1 \\ 2x + 5y = -5 \end{cases}$

10. $\begin{cases} x - y - z = 0 \\ 3x - y - 5z = -2 \\ 2x + 3y = -5 \end{cases}$

9. _____

11. A motel in New Orleans charges $90 per day for double occupancy and $80 per day for single occupancy. If 80 rooms are occupied for a total of $6930, how many rooms of each kind are occupied?

12. The research department of a company that manufactures children's fruit drinks is experimenting with a new flavor. A 17.5% fructose solution is needed, but only 10% and 20% solutions are available. How many gallons of a 10% fructose solution should be mixed with a 20% fructose solution to obtain 20 gallons of a 17.5% fructose solution?

10. _____

11. _____

12. _____

**13.** _____

**14.** _____

**15.** _____

**13.** A company that manufactures boxes recently purchased $2000 worth of new equipment to offer gift boxes to its customers. The cost of producing a package of gift boxes is $1.50 and it is sold for $4.00. Find the number of packages that must be sold for the company to break even.

**14.** The measure of the largest angle of a triangle is three degrees less than 5 times the measure of the smallest angle. The measure of the remaining angle is 1 degree less than twice the measure of the smallest angle. Find the measure of each angle.

_Graph the solutions of the system of linear inequalities._

**15.** $\begin{cases} 2y - x \geq 1 \\ x + y \geq -4 \\ y \leq 2 \end{cases}$

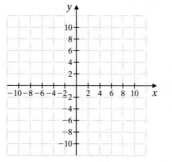

*Determine whether each statement is true or false.*

**1.** $7 \notin \{1, 2, 3\}$

**2.** $6 \in \{1, 2, 3\}$

**3.** $\dfrac{1}{5}$ is an irrational number.

**4.** Every integer is a real number.

*Insert $<$, $>$, or $=$ between each pair of numbers to form a true statement.*

**5.** $-5 \quad 5$

**6.** $-21 \quad -23$

**7.** $-2.5 \quad -2.1$

**8.** $-\dfrac{3}{3} \quad -\dfrac{7.2}{7.2}$

**9.** $\dfrac{2}{3} \quad \dfrac{3}{4}$

**10.** $-\dfrac{1}{11} \quad \dfrac{1}{11}$

*Find each absolute value.*

**11.** $\left| -\dfrac{1}{7} \right|$

**12.** $|-7.8|$

**13.** $-|-8|$

**14.** $-\left| -\dfrac{3}{8} \right|$

**15.** Simplify: $\dfrac{|-2|^3 + 1}{-7 - \sqrt{4}}$

**16.** Subtract.
   **a.** $-7 - (-2)$
   **b.** $14 - 38$

*Simplify and write with positive exponents only.*

**17.** $(3x)^{-1}$

**18.** $9^{-2}$

**19.** $2^{-1} + 3^{-2}$

**20.** $(x^2)^3 \cdot x^{10}$

**21.** Determine whether $-15$ is a solution of $x - 9 = -24$.

**22.** Solve: $8y - 14 = 6y - 14$

**23.** Suppose that a computer store just announced an 8% decrease in the price of a particular computer model. If this computer sells for $2162 after the decrease, find its original price.

**24.** A quadrilateral has four angles whose sum is 360°. In a particular quadrilateral, two angles have the same measure. A third angle is 10° more than the measure of each of the equal angles, and the fourth angle is half the measure of each of the equal angles. Find the measures of the angles.

△**25.** Solve $A = \dfrac{1}{2}(B + b)h$ for $b$.

**26.** Solve:
$2(m - 6) - m = 4(m - 3) - 3m$

**27.** Solve: $4x - 2 < 5x$

**28.** Solve: $5(2x - 1) > -5$

**Answers**

1. _____
2. _____
3. _____
4. _____
5. _____
6. _____
7. _____
8. _____
9. _____
10. _____
11. _____
12. _____
13. _____
14. _____
15. _____
16. a. _____
   b. _____
17. _____
18. _____
19. _____
20. _____
21. _____
22. _____
23. _____
24. _____
25. _____
26. _____
27. _____
28. _____

**337**

**29.** _____

**30.** _____

**31.** _____

**32.** _____

**33.** _____

**34.** _____

**35.** _____

**36. a.** _____

**b.** _____

**c.** _____

**37.** _____

**38. a.** _____

**b.** _____

**39.** _____

**40.** _____

**41.** _____

**42.** _____

**43.** _____

**44.** _____

**45.** _____

**46.** _____

**47.** _____

**48.** _____

**29.** Find the intersection:
$\{2, 4, 6, 8, \} \cap \{3, 4, 5, 6\}$

**30.** Find the union:
$\{2, 4, 6, 8\} \cup \{3, 4, 5, 6\}$

**31.** Solve: $|y| = 0$

**32.** Solve: $|x - 5| = 4$

**33.** Graph: $y = -3$

**34.** Solve: $|2x + 1| > 5$

**35.** Find the slope and the $y$-intercept of the line $3x - 4y = 4$.

**36.** Name the quadrant or axis where each point is located.
**a.** $(-1, -5)$
**b.** $(4, -2)$
**c.** $(0, 2)$

**37.** Write an equation of the line with slope $-3$ containing the point $(1, -5)$. Write the equation in slope-intercept form, $y = mx + b$.

**38.** If $f(x) = 3x^2$, find the following.
**a.** $f(5)$
**b.** $f(-2)$

**39.** Determine whether the relation $y = 2x + 1$ is also a function.

**40.** Find the slope of the line containing $(-2, 6)$ and $(0, 9)$.

**41.** Graph the intersection of $x \geq 1$ and $y \geq 2x - 1$.

**42.** Find an equation of the line through $(-2, 6)$ perpendicular to $f(x) = \dfrac{1}{2}x - \dfrac{1}{3}$. Write the equation in slope-intercept form.

**43.** Use the elimination method to solve the system: $\begin{cases} 3x - 2y = 10 \\ 4x - 3y = 15 \end{cases}$

**44.** Solve the system: $\begin{cases} 5x + y = -2 \\ 4x - 2y = -10 \end{cases}$

**45.** Solve the system:
$\begin{cases} 2x - 4y + 8z = 2 \\ -x - 3y + z = 11 \\ x - 2y + 4z = 0 \end{cases}$

**46.** Solve the system:
$\begin{cases} x - 2y + z = 0 \\ 3x - y - 2z = -15 \\ 2x - 3y + 3z = 7 \end{cases}$

**47.** A first number is 4 less than a second number. Four times the first number is 6 more than twice the second. Find the numbers.

**48.** Solve the system:
$\begin{cases} -6x + 8y = 0 \\ 9x - 12y = 2 \end{cases}$

# Polynomials and Polynomial Functions

Australia

Pacific
Ocean

Aukland

Tasmania

New Zealand

## How is a Tall Tower Constructed "Straight"?

The Sky Tower in Auckland, New Zealand, is the tallest free-standing man-made structure in the southern hemisphere, with a height of 1076 feet. One can travel up the glass-fronted elevators to one of three viewing platforms, look through glass inserts in the floor to the ground below, walk around the narrow pathway outside the building, or even base-jump by harness and wire off the building at a speed of almost 53 miles per hour before halting just above the ground. In addition to being a tourist attraction with restaurants, the Sky Tower is used for telecommunications and broadcasting. See the diagram below to understand how this high tower was built "perfectly straight."

In Section 5.1, Exercise 81, we will use a polynomial function to find the height of an object dropped from the base-jumper platform of the Sky Tower.

Linear equations are important for solving problems. They are not sufficient, however, to solve all problems. Many real-world phenomena are modeled by polynomials. In the first portion of this chapter we will study operations on polynomials. We then look at how polynomials can be used in problem solving.

**Measurements were taken from three existing structures.**

**Lasers on the base pad were shone vertically up to the jump platform.**

**Daily readings of the structure were taken using seven global positioning satellites.**

339

# 5.1 Polynomial Functions and Adding and Subtracting Polynomials

## Objectives

**A** Define Term, Constant, Polynomial, Monomial, Binomial, Trinomial, Degree of a Term, and Degree of a Polynomial.

**B** Combine Like Terms.

**C** Add Polynomials.

**D** Subtract Polynomials.

**E** Evaluate Polynomial Functions.

> **Helpful Hint**
> We will write answers that are polynomials in one variable in descending order.

## Objective A Defining a Polynomial and Related Terms

A **term** is a number or the product of a number and one or more variables raised to powers. The **numerical coefficient,** or simply the **coefficient,** is the numerical factor of a term.

| Term | Numerical Coefficient of Term |
|------|-------------------------------|
| $-1.2x^5$ | $-1.2$ |
| $x^3y$ | $1$ |
| $-z$ | $-1$ |
| $2$ | $2$ |
| $\dfrac{x^9}{7}\left(\text{or } \dfrac{1}{7}x^9\right)$ | $\dfrac{1}{7}$ |

If a term contains only a number, it is called a **constant term,** or simply a **constant.**

A **polynomial** is a finite sum of terms in which all variables are raised to non-negative integer powers and no variables appear in any denominator.

| Polynomials | Not Polynomials | |
|-------------|-----------------|---|
| $4x^5y + 7xz$ | $5x^{-3} + 2x$ | Negative integer exponent |
| $-5x^3 + 2x + \dfrac{2}{3}$ | $\dfrac{6}{x^2} - 5x + 1$ | Variable in denominator |

A polynomial that contains only one variable is called a **polynomial in one variable.** For example, $3x^2 - 2x + 7$ is a **polynomial in x.** This polynomial in $x$ is written in **descending order** since the terms are listed in descending order of the variable's exponents. (The term 7 can be thought of as $7x^0$.) The following examples are polynomials in one variable written in descending order:

$$4x^3 - \frac{7}{8}x^2 + 5 \qquad y^2 - 4.7 \qquad \frac{8a^4}{11} - 7a^2 + \frac{4a}{3}$$

A **monomial** is a polynomial consisting of one term. A **binomial** is a polynomial consisting of two terms. A **trinomial** is a polynomial consisting of three terms.

| Monomials | Binomials | Trinomials |
|-----------|-----------|------------|
| $ax^2$ | $x + y$ | $x^2 + 4xy + y^2$ |
| $-3x$ | $6y^2 - 2.9$ | $-x^4 + 3x^3 + 0.1$ |
| $4$ | $\dfrac{5}{7}z^3 - 2z$ | $8y^2 - 2y - \dfrac{10}{17}$ |

By definition, all monomials, binomials, and trinomials are also polynomials.

Each term of a polynomial has a **degree.**

## Degree of a Term

The **degree of a term** is the sum of the exponents on the *variables* contained in the term.

**Examples**   Determine the degree of each term.

1. $3x^2$       The exponent on $x$ is 2, so the degree of the term is 2.
2. $-2^3x^5$    The exponent on $x$ is 5, so the degree of the term is 5. (Recall that the degree is the sum of the exponents on *only* the *variables*).
3. $y$          The degree of $y$, or $y^1$, is 1.
4. $12x^2yz^3$  The degree is the sum of the exponents on the variables, or $2 + 1 + 3 = 6$.
5. $5.27$       The degree of $5.27$, which can be written as $5.27x^0$, is 0.

■ Work Practice 1–5

From the preceding examples, we can say that the degree of a constant is 0. Also, the term 0 has no degree.

Each polynomial also has a degree.

## Degree of a Polynomial

The **degree of a polynomial** is the greatest degree of any of its terms.

**Examples**   Determine the degree of each polynomial and also indicate whether the polynomial is a monomial, binomial, or trinomial.

|    | Polynomial | Degree | Classification |
|----|------------|--------|----------------|
| 6. | $7x^3 - \dfrac{3}{4}x + 2$ | 3 | Trinomial |
| 7. | $-xyz$ | $1 + 1 + 1 = 3$ | Monomial |
| 8. | $x^4 - 16.5$ | 4 | Binomial |

■ Work Practice 6–8

**Example 9**   Determine the degree of the polynomial

$$3xy + x^2y^2 - 5x^2 - 6.7$$

**Solution:**   The degree of each term is

$$3xy + x^2y^2 - 5x^2 - 6.7$$
$$\downarrow \quad \downarrow \quad \downarrow \quad \downarrow$$
Degree:  2    4    2    0

The greatest degree of any term is 4, so the degree of this polynomial is 4.

■ Work Practice 9

## Objective B  Combining Like Terms ▶

Before we add polynomials, recall from Section 1.5 that terms are considered to be **like terms** if they contain exactly the same variables raised to exactly the same powers.

| Like Terms | Unlike Terms |
|------------|--------------|
| $-5x^2, -x^2$ | $4x^2, 3x$ |
| $7xy^3z, -2xzy^3$ | $12x^2y^3, -2xy^3$ |

To simplify a polynomial, we **combine like terms** by using the distributive property. For example, by the distributive property,

$$5x + 7x = (5 + 7)x = 12x$$

**Practice 1–5**

Determine the degree of each term.

1. $2x^3$   2. $7^2x^4$
3. $x$   4. $15xy^2z^4$
5. $9.1$

**Practice 6–8**

Determine the degree of each polynomial and indicate whether the polynomial is a monomial, binomial, or trinomial.

6. $4x^5 + 7x^3 - 1$
7. $-2xy^2z$
8. $y^3 + 6y$

**Practice 9**

Determine the degree of the polynomial

$$7x^2y - 6x^2yz + 2 - 4y^3.$$

**Answers**
1. 3   2. 4   3. 1   4. 7   5. 0
6. 5; trinomial   7. 4; monomial
8. 3; binomial   9. 4

## Practice 10–11

Simplify each polynomial by combining like terms.

**10.** $10x^3 - 12x^3 - 3x$

**11.** $-6ab + 2a + 12ab - a$

**Examples** Simplify each polynomial by combining like terms.

**10.** $-12x^2 + 7x^2 - 6x = (-12 + 7)x^2 - 6x = -5x^2 - 6x$

**11.** $3xy - 2x + 5xy - x = 3xy + 5xy - 2x - x$

$\qquad = (3 + 5)xy + (-2 - 1)x$

$\qquad = 8xy - 3x$

■ Work Practice 10–11

**Helpful Hint** These two terms are unlike terms. They cannot be combined.

## Objective C  Adding Polynomials ▶

Now we have reviewed the skills we need to add polynomials.

### Adding Polynomials

To add polynomials, combine all like terms.

## Practice 12

Add $14x^4 - 6x^3 + x^2 - 6$ and $x^3 - 5x^2 + 1$.

**Example 12**  Add $11x^3 - 12x^2 + x - 3$ and $x^3 - 10x + 5$.

**Solution:**

$(11x^3 - 12x^2 + x - 3) + (x^3 - 10x + 5)$

$= 11x^3 + x^3 - 12x^2 + x - 10x - 3 + 5$   Group like terms.

$= 12x^3 - 12x^2 - 9x + 2$   Combine like terms.

■ Work Practice 12

Sometimes it is more convenient to add polynomials vertically. To do this, we line up like terms beneath one another and then add like terms.

## Practice 13

Add $10y^3 - y^2 + 4y - 11$ and $y^3 - 4y^2 + 3y$ vertically.

**Example 13**  Add $11x^3 - 12x^2 + x - 3$ and $x^3 - 10x + 5$ vertically.

**Solution:**

$$
\begin{array}{r}
11x^3 - 12x^2 + x - 3 \\
\underline{x^3 \qquad - 10x + 5} \\
12x^3 - 12x^2 - 9x + 2
\end{array}
$$

Line up like terms.

Combine like terms.

This example is the same as Example 12, only here we added vertically.

■ Work Practice 13

## Practice 14

Add:
$(4x^2y - xy^2 + 5) + (-6x^2y - 1)$

**Example 14**  Add:  $(7x^3y - xy^3 + 11) + (6x^3y - 4)$

**Solution:**  To add these polynomials, we remove the parentheses and group like terms.

$(7x^3y - xy^3 + 11) + (6x^3y - 4)$

$= 7x^3y - xy^3 + 11 + 6x^3y - 4$   Remove parentheses.

$= 7x^3y + 6x^3y - xy^3 + 11 - 4$   Group like terms.

$= 13x^3y - xy^3 + 7$   Combine like terms.

■ Work Practice 14

### Answers

**10.** $-2x^3 - 3x$  **11.** $6ab + a$

**12.** $14x^4 - 5x^3 - 4x^2 - 5$

**13.** $11y^3 - 5y^2 + 7y - 11$

**14.** $-2x^2y - xy^2 + 4$

## Objective D Subtracting Polynomials ▶

The definition of subtraction of real numbers can be extended to apply to polynomials. To subtract a number, we add its opposite:

$$a - b = a + (-b)$$

Likewise, to subtract a polynomial we add its opposite. In other words, if $P$ and $Q$ are polynomials, then

$$P - Q = P + (-Q)$$

The polynomial $-Q$ is the **opposite**, or **additive inverse**, of the polynomial $Q$. We can find $-Q$ by changing the sign of each term of $Q$.

✓**Concept Check** Which polynomial is the opposite of $16x^3 - 5x + 7$?

**a.** $-16x^3 - 5x + 7$      **b.** $-16x^3 + 5x - 7$

**c.** $16x^3 + 5x + 7$      **d.** $-16x^3 + 5x + 7$

> ### Subtracting Polynomials
>
> To subtract polynomials, change the signs of the terms of the polynomial being subtracted and then add.

Review the example below.

To subtract, change the signs; then add.

$$(3x^2 + 4x - 7) - (3x^2 - 2x - 5) = (3x^2 + 4x - 7) + (-3x^2 + 2x + 5)$$
$$= 3x^2 + 4x - 7 - 3x^2 + 2x + 5$$
$$= 6x - 2 \qquad \text{Combine like terms.}$$

**Example 15**   Subtract: $(12z^5 - 12z^3 + z) - (-3z^4 + z^3 + 12z)$

**Solution:** First we change the sign of each term of the second polynomial, and then we add the result to the first polynomial.

$$(12z^5 - 12z^3 + z) - (-3z^4 + z^3 + 12z)$$
$$= 12z^5 - 12z^3 + z + 3z^4 - z^3 - 12z \quad \text{Change signs and add.}$$
$$= 12z^5 + 3z^4 - 12z^3 - z^3 + z - 12z \quad \text{Group like terms; write in descending order.}$$
$$= 12z^5 + 3z^4 - 13z^3 - 11z \quad \text{Combine like terms.}$$

■ Work Practice 15

**Practice 15**

Subtract:

$$(7x^4 - 8x^2 + x)$$
$$-(-9x^4 + x^2 - 18)$$

**Example 16**   Subtract $(10x^3 - 7x^2) - (4x^3 - 3x^2 + 2)$ vertically.

**Solution:** To subtract these polynomials, we add the opposite of the second polynomial to the first one.

$$\begin{array}{r} 10x^3 - 7x^2 \\ -(4x^3 - 3x^2 + 2) \end{array} \quad \text{is equivalent to} \quad \begin{array}{r} 10x^3 - 7x^2 \\ -4x^3 + 3x^2 - 2 \\ \hline 6x^3 - 4x^2 - 2 \end{array} \quad \text{Add.}$$

■ Work Practice 16

**Practice 16**

Subtract
$(2y^4 + 4y) - (6y^4 + 7y^3 - 3y)$
vertically.

**Answers**

**15.** $16x^4 - 9x^2 + x + 18$
**16.** $-4y^4 - 7y^3 + 7y$

✓**Concept Check Answer**

**b**

✓**Concept Check**  Why is the following subtraction incorrect?

$$(7z - 5) - (3z - 4)$$
$$= 7z - 5 - 3z - 4$$
$$= 4z - 9$$

### Practice 17

Subtract $3a^2b^3 - 4ab^2 + 6a$ from $7a^2b^3 - ab^2$.

**Example 17**  Subtract $4x^3y^2 - 3x^2y^2 + 2y^2$ from $10x^3y^2 - 7x^2y^2$.

**Solution:**  Notice the order of the expressions, and then write "Subtract $4x^3y^2 - 3x^2y^2 + 2y^2$ from $10x^3y^2 - 7x^2y^2$" as a mathematical expression. (For example, if we subtract 2 from 8, we would write $8 - 2 = 6$.)

$$(10x^3y^2 - 7x^2y^2) - (4x^3y^2 - 3x^2y^2 + 2y^2)$$
$$= 10x^3y^2 - 7x^2y^2 - 4x^3y^2 + 3x^2y^2 - 2y^2 \quad \text{Remove parentheses.}$$
$$= 6x^3y^2 - 4x^2y^2 - 2y^2 \quad \text{Combine like terms.}$$

■ Work Practice 17

## Objective E  Evaluating Polynomial Functions ▶

Recall function notation, first introduced in Section 3.6. At times it is convenient to use function notation to represent polynomials. For example, we may write $P(x)$ to represent the polynomial $3x^4 - 2x^2 - 5$. In symbols, we would write

$$P(x) = 3x^4 - 2x^2 - 5$$

This function is called a **polynomial function** because the expression $3x^4 - 2x^2 - 5$ is a polynomial.

### Practice 18–19

If $P(x) = 5x^4 - 3x^2 + 7$, find each function value.
**18.** $P(2)$     **19.** $P(-1)$

**Helpful Hint**

Recall that the symbol $P(x)$ **does not mean** $P$ times $x$. It is a special symbol used to denote a function.

### Practice 20

The largest natural bridge is in the canyons at the base of Navajo Mountain, Utah. From the base to the top of the arch, it measures 290 feet. Neglecting air resistance, the height of an object dropped off the bridge is given by the polynomial function $P(t) = -16t^2 + 290$ at time $t$ seconds. Find the height of the object at time $t = 0$ second and $t = 2$ seconds.

**Examples**  If $P(x) = 3x^4 - 2x^2 - 5$, find each function value.

**18.** $P(1) = 3(1)^4 - 2(1)^2 - 5 = -4$   Let $x = 1$ in the function $P(x)$.

**19.** $P(-2) = 3(-2)^4 - 2(-2)^2 - 5$   Let $x = -2$ in the function $P(x)$.
$$= 3(16) - 2(4) - 5$$
$$= 35$$

■ Work Practice 18–19

Many real-world phenomena are modeled by polynomial functions. If the polynomial function model is given, we can often find the solution of a problem by evaluating the function at a certain value.

**Example 20**  Finding the Height of an Object

The Siduhe River Bridge, completed in 2009, is the tallest bridge in the world at 1627 feet. It overlooks the Siduhe River in Yesanguanzhen, Hubei, China. An object is dropped from the top of this bridge. Neglecting air resistance, the height of the object at time $t$ seconds is given by the polynomial function $P(t) = -16t^2 + 1627$. Find the height of the object when $t = 1$ second and when $t = 8$ seconds.

### Answers

**17.** $4a^2b^3 + 3ab^2 - 6a$
**18.** $P(2) = 75$  **19.** $P(-1) = 9$
**20.** at 0 sec, height is 290 ft; at 2 sec, height is 226 ft

✓**Concept Check Answer**

with parentheses removed, the expression should be

$$7z - 5 - 3z + 4 = 4z - 1$$

**Solution:** To find the height of the object at 1 second, we find $P(1)$.

$$P(t) = -16t^2 + 1627$$
$$P(1) = -16(1)^2 + 1627$$
$$= 1611$$

When $t = 1$ second, the height of the object is 1611 feet.

To find the height of the object at 8 seconds, we find $P(8)$.

$$P(t) = -16t^2 + 1627$$
$$P(8) = -16(8)^2 + 1627$$
$$= -1024 + 1627$$
$$= 603$$

When $t = 8$ seconds, the height of the object is 603 feet. Notice that as time $t$ increases, the height of the object decreases.

■ Work Practice 20

---

### Calculator Explorations  Graphing

A graphing calculator may be used to check addition and subtraction of polynomials in one variable. For example, to check the polynomial subtraction statement

$$(3x^2 - 6x + 9) - (x^2 - 5x + 6) = 2x^2 - x + 3$$

graph both

$$Y_1 = (3x^2 - 6x + 9) - (x^2 - 5x + 6) \quad \text{Left side of equation}$$

and

$$Y_2 = 2x^2 - x + 3 \quad \text{Right side of equation}$$

on the same screen and see that their graphs coincide. (*Note:* If the graphs do not coincide, we can be sure that a mistake has been made either in combining polynomials or in calculator keystrokes. However, if the graphs appear to coincide, we cannot be sure that our work is correct. This is because it is possible for the graphs to differ so slightly that we do not notice it.)

The graphs of $Y_1$ and $Y_2$ are shown. The graphs appear to coincide, so the subtraction statement

$$(3x^2 - 6x + 9) - (x^2 - 5x + 6) = 2x^2 - x + 3$$

appears to be correct.

*Perform each indicated operation. Then use the procedure described above to check your work.*

**1.** $(2x^2 + 7x + 6) + (x^3 - 6x^2 - 14)$

**2.** $(-14x^3 - x + 2) + (-x^3 + 3x^2 + 4x)$

**3.** $(1.8x^2 - 6.8x - 1.7) - (3.9x^2 - 3.6x)$

**4.** $(-4.8x^2 + 12.5x - 7.8) - (3.1x^2 - 7.8x)$

**5.** $(1.29x - 5.68) + (7.69x^2 - 2.55x + 10.98)$

**6.** $(-0.98x^2 - 1.56x + 5.57) + (4.36x - 3.71)$

## Vocabulary, Readiness & Video Check

*Use the choices below to fill in each blank. Not all choices will be used.*

| | | | | |
|---|---|---|---|---|
| monomial | trinomial | like | degree | coefficient |
| binomial | polynomial | unlike | variables | term |

**1.** The numerical factor of a term is the _____.

**2.** A(n) _____ is a finite sum of terms in which all variables are raised to nonnegative integer powers and no variables appear in any denominator.

**3.** A(n) _____ is a polynomial with 2 terms.

**4.** A(n) _____ is a polynomial with 1 term.

**5.** A(n) _____ is a polynomial with 3 terms.

**6.** The degree of a term is the sum of the exponents on the _____ in the term.

**7.** The _____ of a polynomial is the largest degree of all its terms.

**8.** _____ terms contain the same variables raised to the same powers.

*Add or subtract, if possible.*

**9.** $5x + x$        **10.** $5x - x$        **11.** $y + y$        **12.** $z^2 + z^2$        **13.** $7xy^2 - y^2$        **14.** $x^3 - 9x^3$

*Martin-Gay Interactive Videos*

See Video 5.1

*Watch the section lecture video and answer the following questions.*

**Objective A 15.** In Example 3, why is the degree of each term found when this is not asked for?

**Objective B 16.** How is the polynomial in Example 4 simplified?

**Objective C 17.** From Example 5, how do you add polynomials?

**Objective D 18.** From Examples 6 and 7, how do you change a polynomial subtraction problem into an equivalent addition problem?

**Objective E 19.** From Example 8, finding the value of a polynomial function at a given replacement value is similar to what?

## 5.1 Exercise Set MyMathLab®

**Objective A** *Find the degree of each term. See Examples 1 through 5.*

**1.** $4$          **2.** $7$          **3.** $5x^2$          **4.** $-z^3$          **5.** $-3xy^2$

**6.** $12x^3z$          **7.** $-8^7y^3$          **8.** $-9^{11}y^5$          **9.** $3.78ab^3c^5$          **10.** $9.11r^2st^{12}$

*Find the degree of each polynomial and indicate whether the polynomial is a monomial, binomial, trinomial, or none of these. See Examples 6 through 9.*

**11.** $6x + 0.3$          **12.** $7x - 0.8$

**13.** $3x^2 - 2x + 5$          **14.** $5x^2 - 3x - 2$

**15.** $-3^4xy^2$          **16.** $-7^5abc$

**17.** $x^2y - 4xy^2 + 5x + y^4$

**18.** $-2x^2y - 3y^2 + 4x + y^5$

**Objective** **B** *Simplify each polynomial by combining like terms. See Examples 10 and 11.*

**19.** $5y + y - 6y^2 - y^2$

**20.** $-x + 3x - 4x^2 - 9x^2$

**21.** $4x^2y + 2x - 3x^2y - \dfrac{1}{2} - 7x$

**22.** $-8xy^2 + 4x - x + 2xy^2 - \dfrac{11}{15}$

**23.** $7x^2 - 2xy + 5y^2 - x^2 + xy + 11y^2$

**24.** $-a^2 + 18ab - 2b^2 + 14a^2 - 12ab - b^2$

**Objective** **C** *Add. See Examples 12 through 14.*

**25.** $(9y^2 + y - 8) + (9y^2 - y - 9)$

**26.** $(x^2 + 4x - 7) + (8x^2 + 9x - 7)$

**27.** $(x^2 + xy - y^2)$ and $(2x^2 - 4xy + 7y^2)$

**28.** $(6x^2 + 5x + 7)$ and $(x^2 + 6x - 3)$

**29.** $\quad x^2 - \phantom{0}6x + 3$
$\quad + \phantom{00000}(2x + 5)$
$\overline{\phantom{xxxxxxxxxxxxxxx}}$

**30.** $\quad -2x^2 + \phantom{0}3x - 9$
$\quad + \phantom{00000}(2x - 3)$
$\overline{\phantom{xxxxxxxxxxxxxxx}}$

**31.** $(7x^3y - 4xy + 8) + (5x^3y + 4xy + 8x)$

**32.** $(9xyz + 4x - y) + (-9xyz - 3x + y + 2)$

**33.** $(0.6x^3 + 1.2x^2 - 4.5x + 9.1) + (3.9x^3 - x^2 + 0.7x)$

**34.** $(9.3y^2 - y + 12.8) + (2.6y^2 + 4.4y - 8.9)$

**Objective** **D** *Subtract. See Examples 15 through 17.*

**35.** $(9y^2 - 7y + 5) - (8y^2 - 7y + 2)$

**36.** $(2x^2 + 3x + 12) - (20x^2 - 5x - 7)$

**37.** Subtract $(6x^2 - 3x)$ from $(4x^2 + 2x)$.

**38.** Subtract $(8y^2 + 4x)$ from $(y^2 + x)$.

**39.** $\quad 6y^2 - 6y + 4$
$\quad -(-y^2 + 6y + 7)$
$\overline{\phantom{xxxxxxxxxxxxxxx}}$

**40.** $\quad -4x^3 + 4x^2 - 4x$
$\quad -(2x^3 - 2x^2 + 3x)$
$\overline{\phantom{xxxxxxxxxxxxxxx}}$

**41.** $(9x^3 - 2x^2 + 4x - 7) - (2x^3 - 6x^2 - 4x + 3)$

**42.** $(3x^2 + 6xy + 3y^2) - (8x^2 - 6xy - y^2)$

**43.** Subtract $\left(y^2 + 4yx + \dfrac{1}{7}\right)$ from $\left(-19y^2 + 7yx + \dfrac{1}{7}\right)$.

**44.** Subtract $\left(13x^2 + x^2y - \dfrac{1}{4}\right)$ from $\left(3x^2 - 4x^2y - \dfrac{1}{4}\right)$.

**Objectives** **B** **C** **D** **Mixed Practice** *Perform indicated operations and simplify. See Examples 10 through 17.*

**45.** $(-3x + 8) + (-3x^2 + 3x - 5)$

**46.** $(-5y^2 - 2y + 4) + (3y + 7)$

**47.** $(5y^4 - 7y^2 + x^2 - 3) + (-3y^4 + 2y^2 + 4)$

**48.** $(8x^4 - 14x^2 + x + 6) + (-12x^6 - 21x^4 - 9x^2)$

**49.** $(4x^2 - 6x + 2) - (-x^2 + 3x + 5)$

**50.** $(7x^2 + x + 1) - (6x^2 + x - 1)$

**51.** $(5x^2 + x + 9) - (2x^2 - 9)$

**52.** $(4x - 4) - (-x - 4)$

**53.** $(5x - 11) + (-x - 2)$

**54.** $(3x^2 - 2x) + (-5x^2 - 9x)$

**55.** $(3x^3 - b + 2a - 6) + (-4x^3 + b + 6a - 6)$

**56.** $(9y^3 - a + 7b - 3) + (-2y^3 + a + 6b - 8)$

**57.** $(14ab - 10a^2b + 6b^2) - (18a^2 - 20a^2b - 6b^2)$

**58.** $(13x^2 - 26x^2y^2 + 4) - (19x^2 + x^2y^2 - 11)$

**59.**
$$\begin{array}{r} 3x^2 + 15x + 8 \\ +(2x^2 + \phantom{0}7x + 8) \\ \hline \end{array}$$

**60.**
$$\begin{array}{r} 9x^2 + 9x - 4 \\ +(7x^2 - 3x - 4) \\ \hline \end{array}$$

**61.** $(7x^2 - 5) + (-3x^2 - 2) - (4x^2 - 7)$

**62.** $(9y^2 - 3) + (-4y^2 + 1) - (5y^2 - 2)$

**63.** $(-3 + 4x^2 + 7xy^2) + (2x^3 - x^2 + xy^2)$

**64.** $(-3x^2y + 4) + (-7x^2y - 8y)$

**65.**
$$\begin{array}{r} 3x^2 - 4x + 8 \\ - \phantom{00000}(5x - 7) \\ \hline \end{array}$$

**66.**
$$\begin{array}{r} -3x^2 - 5x + 1 \\ - \phantom{0000}(3x + 12) \\ \hline \end{array}$$

**67.** Subtract $(3x + 7)$ from the sum of $(7x^2 + 4x + 9)$ and $(8x^2 + 7x - 8)$.

**68.** Subtract $(9x + 8)$ from the sum of $(3x^2 - 2x - x^3 + 2)$ and $(5x^2 - 8x - x^3 + 4)$.

**69.** $\left(\dfrac{2}{3}x^2 - \dfrac{1}{6}x + \dfrac{5}{6}\right) - \left(\dfrac{1}{3}x^2 + \dfrac{5}{6}x - \dfrac{1}{6}\right)$

**70.** $\left(\dfrac{3}{16}x^2 + \dfrac{5}{8}x - \dfrac{1}{4}\right) - \left(\dfrac{5}{16}x^2 - \dfrac{3}{8}x + \dfrac{3}{4}\right)$

**Objective E** *If $P(x) = x^2 + x + 1$ and $Q(x) = 5x^2 - 1$, find each function value. See Examples 18 and 19.*

**71.** $P(7)$      **72.** $Q(4)$      ▶ **73.** $Q(-10)$      **74.** $P(-4)$      **75.** $P(0)$      **76.** $Q(0)$

*Solve. See Example 20.*

*The surface area of a rectangular box is given by the polynomial*

$$2HL + 2LW + 2HW$$

*and is measured in square units. In business, surface area is often calculated to help determine cost of materials.*

△ **77.** A rectangular box is to be constructed to hold a new camcorder. The box is to have dimensions 5 inches by 4 inches by 9 inches. Find the surface area of the box.

△ **78.** Suppose it has been determined that a box of dimensions 4 inches by 4 inches by 8.5 inches can be used to contain the camcorder in Exercise 77. Find the surface area of this box and calculate the square inches of material saved by using this box instead of the box in Exercise 77.

**79.** A projectile is fired upward from the ground with an initial velocity of 300 feet per second. Neglecting air resistance, the height of the projectile at any time $t$ can be described by the polynomial function $P(t) = -16t^2 + 300t$. Find the height of the projectile at each given time.

   **a.** $t = 1$ second

   **b.** $t = 2$ seconds

   **c.** $t = 8$ seconds

   **d.** $t = 12$ seconds

   **e.** Explain why the height increases and then decreases as time passes.

   **f.** Approximate (to the nearest second) how long before the object hits the ground.

**80.** A projectile is fired upward from the ground with an initial velocity of 200 feet per second. Neglecting air resistance, the height of the projectile at any time $t$ can be described by the polynomial function $P(t) = -16t^2 + 200t$. Find the height of the projectile at each given time.

   **a.** $t = 1$ second

   **b.** $t = 2$ seconds

   **c.** $t = 5$ seconds

   **d.** $t = 9$ seconds

   **e.** Explain why the height increases and then decreases as time passes.

   **f.** Approximate (to the nearest second) how long before the object hits the ground.

**81.** Two jump masters at the Sky Tower in Auckland, New Zealand, were readying the equipment for the next adventurer when one worker tossed a carabiner to his friend. The carabiner was thrown at a velocity of 5 feet per second, but his coworker missed it, and it went over the edge toward the ground. The height of the carabiner above the ground at any time $t$ can be described by $y = -16t^2 + 5t + 630$. Find the height of the carabiner at each given time.

   **a.** $t = 0$ seconds

   **b.** $t = \dfrac{1}{4}$ second

   **c.** $t = 3$ seconds

   **d.** $t = 5$ seconds

   **e.** Explain why the height increases and then decreases as time passes.

   **f.** Approximate (to the nearest second) how long before the carabiner hits the ground.

carabiner: a link with a safety closure

**82.** Two workers at the Hoover Dam bypass bridge were spending their lunch break tossing a football. One of the workers threw the ball upward with an initial velocity of 10 feet per second, but his partner missed the ball and it went over his head and down toward the river. The height of the football above the Colorado River at any time $t$ can be described by the polynomial function $P(t) = -16t^2 + 10t + 910$. Find the height of the football at each given time.

   **a.** $t = 0$ seconds

   **b.** $t = \dfrac{1}{2}$ second

   **c.** $t = 3$ seconds

   **d.** $t = 5$ seconds

   **e.** Explain why the height increases and then decreases as time passes.

   **f.** Approximate (to the nearest second) how long before the football lands in the river.

**83.** The polynomial function $P(x) = 45x - 100{,}000$ models the relationship between the number of computer briefcases $x$ that a company sells and the profit the company makes, $P(x)$. Find $P(4000)$, the profit from selling 4000 computer briefcases.

**84.** The total revenues (in dollars) for an art supply company to sell $x$ boxes of colored pencils per week over the Internet is given by the polynomial function $R(x) = 11x$. Find the total revenue from selling 1500 boxes of colored pencils.

## Review

*Use the distributive property to multiply. See Section 1.3.*

**85.** $-7(2z - 6y)$

**86.** $5(3x - 2)$

**87.** $5(-3y^2 - 2y + 7)$

**88.** $-2(x^2 - 5x + 6)$

## Concept Extensions

*Solve. See the Concept Checks in this section.*

**89.** Which polynomial(s) is the opposite of $8x - 6$?
  **a.** $-(8x - 6)$
  **b.** $8x + 6$
  **c.** $-8x + 6$
  **d.** $-8x - 6$

**90.** Which polynomial(s) is the opposite of $-y^5 + 10y^3 - 2.3$?
  **a.** $y^5 + 10y^3 + 2.3$
  **b.** $-y^5 - 10y^3 - 2.3$
  **c.** $y^5 + 10y^3 - 2.3$
  **d.** $y^5 - 10y^3 + 2.3$

**91.** Correct the subtraction.

$$(12x - 1.7) - (15x + 6.2) = 12x - 1.7 - 15x + 6.2$$
$$= -3x + 4.5$$

**92.** Correct the addition.

$$(12x - 1.7) + (15x + 6.2) = 12x - 1.7 + 15x + 6.2$$
$$= 27x + 7.9$$

**93.** Write a function, $P(x)$, so that $P(0) = 7$.

**94.** Write a function, $R(x)$, so that $R(1) = 2$.

**95.** In your own words, describe how to find the degree of a term.

**96.** In your own words, describe how to find the degree of a polynomial.

*Solve.*

**97.** The function $f(x) = 0.17x^2 + 6.0x + 67.6$ can be used to approximate the growth of restaurant food-and-drink sales, where $x$ is the number of years since 1970 and $f(x)$ or $y$ is the sales (in billions of dollars.)
  **a.** Approximate the restaurant food-and-drink sales in 2005.
  **b.** Approximate the restaurant food-and-drink sales in 2010.
  **c.** Use this function to estimate the restaurant food-and-drink sales in 2018.
  **d.** From parts (a), (b), and (c), determine whether the restaurant food-and-drink sales is increasing at a steady rate. Explain why or why not.

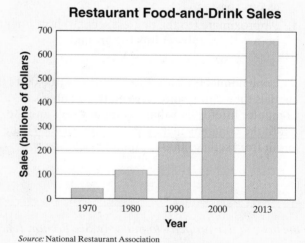

**Restaurant Food-and-Drink Sales**

*Source:* National Restaurant Association

**98.** The function $f(x) = -0.003x^2 + 0.26x + 9.18$ can be used to approximate the total cheese production in the United States from 2005 to 2012, where $x$ is the number of years after 2005 and $y$ is pounds of cheese (in billions). Round answers to the nearest hundredth of a billion. (*Source*: National Agricultural Statistics Service, USDA)

**a.** Approximate the number of pounds of cheese produced in the United States in 2005.

**b.** Approximate the number of pounds of cheese produced in the United States in 2010.

**c.** Use this function to estimate the pounds of cheese produced in the United States in 2015.

**d.** From parts (a), (b), and (c), determine whether the number of pounds of cheese produced in the United States is increasing at a steady rate. Explain why or why not.

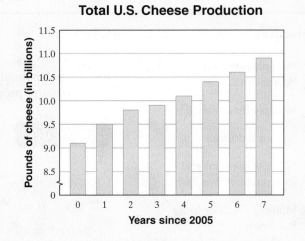

**Total U.S. Cheese Production**

*If $P(x) = 3x + 3$, $Q(x) = 4x^2 - 6x + 3$, and $R(x) = 5x^2 - 7$, find each function.*

**99.** $P(x) + Q(x)$

**100.** $Q(x) - R(x)$

**101.** If $P(x) = 2x - 3$, find $P(a)$, $P(-x)$, and $P(x + h)$.

**102.** If $P(x) = 7x + 5$, find $P(a)$, $P(-x)$, and $P(x + h)$.

*Perform each indicated operation.*

**103.** $(14z^{5x} + 3z^{2x} + z) - (2z^{5x} - 10z^{2x} + 3z)$

**104.** $(8x^{2y} - 7x^y + 3) + (-4x^{2y} + 9x^y - 14)$

*Find each perimeter.*

△ **105.**

$(5xy - y^2)$ units

$(3xy + 4y^2)$ units

$(2x^2 - xy + 7y^2)$ units

△ **106.**

$(x + y)$ units

$(3x^2 - x + 2y)$ units

**Objectives**

**A** Multiply Any Two Polynomials.

**B** Multiply Binomials.

**C** Square Binomials.

**D** Multiply the Sum and Difference of Two Terms.

**E** Multiply Three or More Polynomials.

**F** Evaluate Polynomial Functions.

## Objective A Multiplying Any Two Polynomials

Properties of real numbers and exponents are used continually in the process of multiplying polynomials. To multiply monomials, for example, we apply the commutative and associative properties of real numbers and the product rule for exponents.

**Examples** Multiply.

Group like bases and apply the product rule for exponents.

**1.** $(2x^3)(5x^6) = 2(5)(x^3)(x^6) = 10x^{3+6} = 10x^9$

**2.** $(7y^4z^4)(-xy^{11}z^5) = 7(-1)x(y^4y^{11})(z^4z^5) = -7xy^{4+11}z^{4+5} = -7xy^{15}z^9$

■ Work Practice 1–2

**Helpful Hint**

See Sections 1.6 and 1.7 to review exponential expressions further.

To multiply a monomial by a polynomial other than a monomial, we use an expanded form of the distributive property.

$$a(b + c + d + \cdots + z) = ab + ac + ad + \cdots + az$$

Notice that the monomial $a$ is multiplied by each term of the polynomial.

**Examples** Multiply.

**3.** $2x(5x - 4) = 2x(5x) + 2x(-4)$  Use the distributive property.

$= 10x^2 - 8x$  Multiply.

**4.** $-3x^2(4x^2 - 6x + 1) = -3x^2(4x^2) + (-3x^2)(-6x) + (-3x^2)(1)$

$= -12x^4 + 18x^3 - 3x^2$

**5.** $-xy(7x^2y + 3xy - 11) = -xy(7x^2y) + (-xy)(3xy) + (-xy)(-11)$

$= -7x^3y^2 - 3x^2y^2 + 11xy$

■ Work Practice 3–5

✓**Concept Check** Find the error:

$4x(x - 5) + 2x$

$= 4x(x) + 4x(-5) + 4x(2x)$

$= 4x^2 - 20x + 8x^2$

$= 12x^2 - 20x$

To multiply any two polynomials, we can use the following.

## Multiplying Two Polynomials

To multiply any two polynomials, use the distributive property and multiply each term of one polynomial by each term of the other polynomial. Then combine any like terms.

---

**Practice 1–2**

Multiply.

**1.** $(7y^2)(4y^5)$

**2.** $(-a^2b^3c)(10ab^2c^{12})$

**Practice 3–5**

Multiply.

**3.** $4x(3x - 2)$

**4.** $-2y^3(5y^2 - 2y + 6)$

**5.** $-a^2b(4a^3 - 2ab + b^2)$

**Answers**

**1.** $28y^7$ **2.** $-10a^3b^5c^{13}$ **3.** $12x^2 - 8x$

**4.** $-10y^5 + 4y^4 - 12y^3$

**5.** $-4a^5b + 2a^3b^2 - a^2b^3$

✓**Concept Check Answer**

$4x(x - 5) + 2x$

$= 4x(x) + 4x(-5) + 2x$

$= 4x^2 - 20x + 2x$

$= 4x^2 - 18x$

**Example 6** Multiply: $(x + 3)(2x + 5)$

**Solution:** We multiply each term of $(x + 3)$ by $(2x + 5)$.

$$(x + 3)(2x + 5) = x(2x + 5) + 3(2x + 5) \quad \text{Use the distributive property.}$$
$$= 2x^2 + 5x + 6x + 15 \quad \text{Use the distributive property again.}$$
$$= 2x^2 + 11x + 15 \quad \text{Combine like terms.}$$

■ Work Practice 6

**Practice 6**

Multiply: $(x + 2)(3x + 1)$

**Example 7** Multiply: $(2x - 3)(5x^2 - 6x + 7)$

**Solution:** We multiply each term of $(2x - 3)$ by each term of $(5x^2 - 6x + 7)$.

$$(2x - 3)(5x^2 - 6x + 7) = 2x(5x^2 - 6x + 7) + (-3)(5x^2 - 6x + 7)$$
$$= 10x^3 - 12x^2 + 14x - 15x^2 + 18x - 21$$
$$= 10x^3 - 27x^2 + 32x - 21 \quad \text{Combine like terms.}$$

■ Work Practice 7

**Practice 7**

Multiply:

$$(5x - 1)(2x^2 - x + 4)$$

Sometimes polynomials are easier to multiply vertically, in the same way we multiply real numbers. When multiplying vertically, we line up like terms in the **partial products** vertically. This makes combining like terms easier.

**Example 8** Multiply vertically: $(4x^2 + 7)(x^2 + 2x + 8)$

**Solution:**

$$
\begin{array}{r}
x^2 + 2x + 8 \\
\times \quad 4x^2 + 7 \\
\hline
7x^2 + 14x + 56 \quad 7(x^2 + 2x + 8) \\
4x^4 + 8x^3 + 32x^2 \qquad\qquad 4x^2(x^2 + 2x + 8) \\
\hline
4x^4 + 8x^3 + 39x^2 + 14x + 56 \quad \text{Combine like terms.}
\end{array}
$$

■ Work Practice 8

**Practice 8**

Multiply vertically:

$$(3x^2 + 5)(x^2 - 6x + 1)$$

## Objective B Multiplying Binomials ▶

When multiplying a binomial by a binomial, we can follow a special order for multiplying terms, called the **FOIL** order. The letters of FOIL stand for "First–Outer–Inner–Last." To illustrate this method, let's multiply $(2x - 3)$ by $(3x + 1)$.

| | | |
|---|---|---|
| Multiply the **F**irst terms of each binomial. | $(2x - 3)(3x + 1)$ | F $\quad 2x(3x) = 6x^2$ |
| Multiply the **O**uter terms of each binomial. | $(2x - 3)(3x + 1)$ | O $\quad 2x(1) = 2x$ |
| Multiply the **I**nner terms of each binomial. | $(2x - 3)(3x + 1)$ | I $\quad -3(3x) = -9x$ |
| Multiply the **L**ast terms of each binomial. | $(2x - 3)(3x + 1)$ | L $\quad -3(1) = -3$ |

Combine like terms.

$$6x^2 + 2x - 9x - 3 = 6x^2 - 7x - 3$$

**Answers**

**6.** $3x^2 + 7x + 2$

**7.** $10x^3 - 7x^2 + 21x - 4$

**8.** $3x^4 - 18x^3 + 8x^2 - 30x + 5$

## Practice 9

Use the FOIL order to multiply $(x - 7)(x + 5)$.

## Practice 10–11

Use the FOIL order to multiply.

**10.** $(4x - 3)(x - 6)$

**11.** $(6x^2 + 5y)(2x^2 - y)$

---

**Example 9**  Use the FOIL order to multiply $(x - 1)(x + 2)$.

**Solution:**

$$\begin{array}{cccc} \text{First} & \text{Outer} & \text{Inner} & \text{Last} \\ \downarrow & \downarrow & \downarrow & \downarrow \end{array}$$

$$\begin{aligned} (x - 1)(x + 2) &= x \cdot x + x \cdot 2 + (-1)x + (-1)(2) \\ &= x^2 + 2x - x - 2 \\ &= x^2 + x - 2 \qquad \text{Combine like terms.} \end{aligned}$$

■ Work Practice 9

**Examples**  Use the FOIL order to multiply.

$$\begin{array}{cccc} \text{First} & \text{Outer} & \text{Inner} & \text{Last} \\ \downarrow & \downarrow & \downarrow & \downarrow \end{array}$$

**10.** $\begin{aligned}[t] (2x - 7)(3x - 4) &= 2x(3x) + 2x(-4) + (-7)(3x) + (-7)(-4) \\ &= 6x^2 - 8x - 21x + 28 \\ &= 6x^2 - 29x + 28 \end{aligned}$

$$\begin{array}{cccc} \text{F} & \text{O} & \text{I} & \text{L} \\ \downarrow & \downarrow & \downarrow & \downarrow \end{array}$$

**11.** $\begin{aligned}[t] (3x^2 + y)(5x^2 - 2y) &= 15x^4 - 6x^2y + 5x^2y - 2y^2 \\ &= 15x^4 - x^2y - 2y^2 \end{aligned}$

■ Work Practice 10–11

## Objective C  Squaring Binomials

The **square of a binomial** is a special case of the product of two binomials. By the FOIL order for multiplying two binomials, we have

$$(a + b)^2 = (a + b)(a + b)$$

$$\begin{array}{cccc} \text{F} & \text{O} & \text{I} & \text{L} \\ \downarrow & \downarrow & \downarrow & \downarrow \end{array}$$

$$\begin{aligned} &= a^2 + ab + ba + b^2 \\ &= a^2 + 2ab + b^2 \end{aligned}$$

We can visualize this product geometrically by analyzing areas.

Area of square in the margin: $(a + b)^2$

Sum of areas of smaller rectangles: $a^2 + 2ab + b^2$

Thus, $(a + b)^2 = a^2 + 2ab + b^2$

The same pattern occurs for the square of a difference. In general, we have the following.

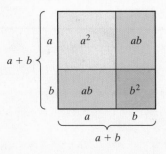

> ### Square of a Binomial
>
> $$(a + b)^2 = a^2 + 2ab + b^2 \qquad (a - b)^2 = a^2 - 2ab + b^2$$

In other words, a binomial squared is the sum of the first term squared, twice the product of both terms, and the second term squared.

**Examples** Multiply.

$$(a + b)^2 = a^2 + 2 \cdot a \cdot b + b^2$$
$$\downarrow \quad \downarrow \qquad \downarrow \quad \downarrow \ \downarrow \ \downarrow \quad \downarrow$$

**12.** $(x + 5)^2 = x^2 + 2 \cdot x \cdot 5 + 5^2 = x^2 + 10x + 25$

**13.** $(x - 9)^2 = x^2 - 2 \cdot x \cdot 9 + 9^2 = x^2 - 18x + 81$

**14.** $(3x + 2z)^2 = (3x)^2 + 2(3x)(2z) + (2z)^2 = 9x^2 + 12xz + 4z^2$

**15.** $(4m^2 - 3n)^2 = (4m^2)^2 - 2(4m^2)(3n) + (3n)^2 = 16m^4 - 24m^2n + 9n^2$

**Work Practice 12–15**

**Helpful Hint**

Note that $(a + b)^2 = a^2 + 2ab + b^2$, *not* $a^2 + b^2$.
   Also, $(a - b)^2 = a^2 - 2ab + b^2$, *not* $a^2 - b^2$.

## Objective D Multiplying the Sum and Difference of Two Terms ▶

Another special product applies to the sum and difference of the same two terms. Multiply $(a + b)(a - b)$ to see a pattern.

$$(a + b)(a - b) = a^2 - ab + ba - b^2 = a^2 - b^2$$

**Product of the Sum and Difference of Two Terms**

$$(a + b)(a - b) = a^2 - b^2$$

In other words, the product of the sum and difference of the same two terms is the difference of the first term squared and the second term squared.

**Examples** Multiply.

$$(a + b) \ (a - b) = a^2 - b^2$$
$$\downarrow \quad \downarrow \ \downarrow \quad \downarrow \qquad \downarrow \quad \downarrow$$

**16.** $(x + 3)(x - 3) = x^2 - 3^2 = x^2 - 9$

**17.** $(4y - 1)(4y + 1) = (4y)^2 - 1^2 = 16y^2 - 1$

**18.** $(x^2 + 2y)(x^2 - 2y) = (x^2)^2 - (2y)^2 = x^4 - 4y^2$

**19.** $\left(3m^2 - \dfrac{1}{2}\right)\left(3m^2 + \dfrac{1}{2}\right) = (3m^2)^2 - \left(\dfrac{1}{2}\right)^2 = 9m^4 - \dfrac{1}{4}$

**Work Practice 16–19**

**Example 20** Multiply $[3 + (2a + b)]^2$.

**Solution:** Think of 3 as the first term and $(2a + b)$ as the second term, and apply the method for squaring a binomial.

$$[a \ + \ b \ ]^2 = a^2 + 2 \cdot a \ \cdot \ b \ + \ b^2$$

$$[3 + \overbrace{(2a + b)}]^2 = 3^2 + 2(3)\overbrace{(2a + b)} + \overbrace{(2a + b)}^2$$
$$= 9 + 6(2a + b) + (2a + b)^2$$
$$= 9 + 12a + 6b + (2a)^2 + 2(2a)(b) + b^2 \quad \text{Square } (2a + b).$$
$$= 9 + 12a + 6b + 4a^2 + 4ab + b^2$$

**Work Practice 20**

**Practice 12–15**

Multiply.
**12.** $(x + 3)^2$
**13.** $(y - 6)^2$
**14.** $(2x + 5y)^2$
**15.** $(6a^2 - 2b)^2$

**Practice 16–19**

Multiply.
**16.** $(x + 4)(x - 4)$
**17.** $(3m - 6)(3m + 6)$
**18.** $(a^2 + 5y)(a^2 - 5y)$
**19.** $\left(4y^2 - \dfrac{1}{3}\right)\left(4y^2 + \dfrac{1}{3}\right)$

**Practice 20**

Multiply:
$[2 - (3x + y)]^2$

**Answers**
**12.** $x^2 + 6x + 9$   **13.** $y^2 - 12y + 36$
**14.** $4x^2 + 20xy + 25y^2$
**15.** $36a^4 - 24a^2b + 4b^2$
**16.** $x^2 - 16$   **17.** $9m^2 - 36$
**18.** $a^4 - 25y^2$   **19.** $16y^4 - \dfrac{1}{9}$
**20.** $4 - 12x - 4y + 9x^2 + 6xy + y^2$

### Practice 21

Multiply:

$[(2x + 3y) - 2][(2x + 3y) + 2]$

**Example 21** Multiply: $[(5x - 2y) - 1][(5x - 2y) + 1]$

**Solution:** We can think of $(5x - 2y)$ as the first term and 1 as the second term. Then we can apply the method for the product of the sum and difference of two terms.

$$
\overset{a}{\overbrace{[(5x - 2y)}} \overset{-\ b}{- 1]} \overset{a}{\overbrace{[(5x - 2y)}} \overset{+\ b}{+ 1]} = \overset{a^2}{(5x - 2y)^2} \overset{-\ b^2}{- 1^2}
$$
$$
= (5x)^2 - 2(5x)(2y) + (2y)^2 - 1 \quad \text{Square } (5x - 2y).
$$
$$
= 25x^2 - 20xy + 4y^2 - 1
$$

■ Work Practice 21

## Objective E  Multiplying Three or More Polynomials ▶

To multiply three or more polynomials, more than one method may be needed.

### Practice 22

Multiply:

$(y - 2)(y + 2)(y^2 - 4)$

**Example 22** Multiply: $(x - 3)(x + 3)(x^2 - 9)$

**Solution:** We multiply the first two binomials, the sum and difference of two terms. Then we multiply the resulting two binomials, the square of a binomial.

$$
(x - 3)(x + 3)(x^2 - 9) = (x^2 - 9)(x^2 - 9) \quad \text{Multiply } (x - 3)(x + 3)
$$
$$
= (x^2 - 9)^2
$$
$$
= x^4 - 18x^2 + 81 \quad \text{Square } (x^2 - 9).
$$

■ Work Practice 22

## Objective F  Evaluating Polynomial Functions ▶

Our work in multiplying polynomials is often useful in evaluating polynomial functions.

### Practice 23

If $f(x) = x^2 - 6x + 1$, find $f(b - 1)$

**Example 23** If $f(x) = x^2 + 5x - 2$, find $f(a + 1)$.

**Solution:** To find $f(a + 1)$, replace $x$ with the expression $a + 1$ in the polynomial function $f(x)$.

$$
f(x) = x^2 + 5x - 2
$$
$$
f(a + 1) = (a + 1)^2 + 5(a + 1) - 2
$$
$$
= a^2 + 2a + 1 + 5a + 5 - 2
$$
$$
= a^2 + 7a + 4
$$

■ Work Practice 23

**Answers**

**21.** $4x^2 + 12xy + 9y^2 - 4$

**22.** $y^4 - 8y^2 + 16$   **23.** $b^2 - 8b + 8$

## Calculator Explorations **Graphing**

In the previous section, we used a graphing calculator to check addition and subtraction of polynomials in one variable. In this section, we use the same method to check multiplication of polynomials in one variable. For example, to see that

$$(x - 2)(x + 1) = x^2 - x - 2$$

graph both $Y_1 = (x - 2)(x + 1)$ and $Y_2 = x^2 - x - 2$ on the same screen and see whether their graphs coincide.

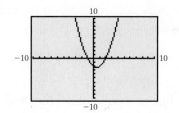

By tracing along both graphs, we see that the graphs of $Y_1$ and $Y_2$ appear to coincide, and thus $(x - 2)(x + 1) = x^2 - x - 2$ appears to be correct.

*Multiply. Then use a graphing calculator to check the results.*

**1.** $(x + 4)(x - 4)$

**2.** $(x + 3)(x + 3)$

**3.** $(3x - 7)^2$

**4.** $(5x - 2)^2$

**5.** $(5x + 1)(x^2 - 3x - 2)$

**6.** $(7x + 4)(2x^2 + 3x - 5)$

## Vocabulary, Readiness & Video Check

*Use the choices to fill in each blank.*

**1.** $(6x^3)\left(\dfrac{1}{2}x^3\right) =$ _____

   **a.** $3x^3$    **b.** $3x^6$    **c.** $10x^6$    **d.** $\dfrac{13}{2}x^6$

**2.** $(x + 7)^2 =$ _____

   **a.** $x^2 + 49$    **b.** $x^2 - 49$    **c.** $x^2 + 14x + 49$    **d.** $x^2 + 7x + 49$

**3.** $(x + 7)(x - 7) =$ _____

   **a.** $x^2 + 49$    **b.** $x^2 - 49$    **c.** $x^2 + 14x - 49$    **d.** $x^2 + 7x - 49$

**4.** The product of $(3x - 1)(4x^2 - 2x + 1)$ is a polynomial of degree _____.

   **a.** 3    **b.** 12    **c.** $12x^3$    **d.** 2

**5.** If $f(x) = x^2 + 1$ then $f(a + 1) =$ _____

   **a.** $(a + 1)^2$    **b.** $a + 1$    **c.** $(a + 1)^2 + (a + 1)$    **d.** $(a + 1)^2 + 1$

**6.** $[x + (2y + 1)]^2 =$ _____

   **a.** $[x + (2y + 1)][x - (2y + 1)]$    **b.** $[x + (2y + 1)][x + (2y + 1)]$    **c.** $[x + (2y + 1)][x + (2y - 1)]$

*Martin-Gay Interactive Videos*

Watch the section lecture video and answer the following questions.

See Video 5.2

**Objective A** **7.** From Examples 1 and 2, what property is key when multiplying polynomials? What exponent rule is often used when applying this property?

**Objective B** **8.** From Examples 3 and 4, for what type of multiplication of polynomials is the FOIL method used?

**Objective C** **9.** Name at least one other method you can use to multiply Example 5.

**Objective D** **10.** From Example 6, why does multiplying the sum and difference of the same two terms give you a binomial answer?

**Objective E** **11.** From Example 9, how do you multiply three or more polynomials?

**Objective F** **12.** From Example 10, can a polynomial function be evaluated only for a numerical value?

## 5.2 Exercise Set MyMathLab®

**Objective A** *Multiply. See Examples 1 through 8.*

**1.** $(-4x^3)(3x^2)$

**2.** $(-6a)(4a)$

**3.** $(8.6a^4b^5c)(10ab^3c^2)$

**4.** $(7.1xy^2z^{11})(10xy^7z)$

**5.** $3x(4x + 7)$

**6.** $5x(6x - 4)$

**7.** $-6xy(4x + y)$

**8.** $-8y(6xy + 4x)$

**9.** $-4ab(xa^2 + ya^2 - 3)$

**10.** $-6b^2z(z^2a + baz - 3b)$

**11.** $(x - 3)(2x + 4)$

**12.** $(y + 5)(3y - 2)$

**13.** $(2x + 3)(x^3 - x + 2)$

**14.** $(a + 2)(3a^2 - a + 5)$

**15.**
$$\begin{array}{r} 3x - 2 \\ \times\ 5x + 1 \end{array}$$

**16.**
$$\begin{array}{r} 2z - 4 \\ \times\ 6z - 2 \end{array}$$

**17.**
$$\begin{array}{r} 3m^2 + 2m - 1 \\ \times\ 5m + 2 \end{array}$$

**18.**
$$\begin{array}{r} 2x^2 - 3x - 4 \\ \times\ 4x + 5 \end{array}$$

**19.** $-6a^2b^2(5a^2b^2 - 6a - 6b)$

**20.** $7x^2y^3(-3ax - 4xy + z)$

**Objective B** *Multiply the binomials. See Examples 9 through 11.*

**21.** $(x - 3)(x + 4)$

**22.** $(c - 3)(c + 1)$

**23.** $(5x - 8y)(2x - y)$

**24.** $(2n - 9m)(n - 7m)$

**25.** $\left(4x + \dfrac{1}{3}\right)\left(4x - \dfrac{1}{2}\right)$

**26.** $\left(4y - \dfrac{1}{3}\right)\left(3y - \dfrac{1}{8}\right)$

**27.** $(5x^2 - 2y^2)(x^2 - 3y^2)$

**28.** $(4x^2 - 5y^2)(x^2 - 2y^2)$

## Objectives C D  *Use special products to multiply. See Examples 12 through 21.*

**29.** $(x + 4)^2$

**30.** $(x - 5)^2$

▶**31.** $(6y - 1)(6y + 1)$

**32.** $(7x - 9)(7x + 9)$

**33.** $(3x - y)^2$

**34.** $(4x + z)^2$

▶**35.** $(7ab + 3c)(7ab - 3c)$

**36.** $(3xy - 2b)(3xy + 2b)$

**37.** $\left(3x + \dfrac{1}{2}\right)\left(3x - \dfrac{1}{2}\right)$

**38.** $\left(2x - \dfrac{1}{3}\right)\left(2x + \dfrac{1}{3}\right)$

**39.** $[3 + (4b + 1)]^2$

**40.** $[5 - (3b - 3)]^2$

▶**41.** $[(2s - 3) - 1][(2s - 3) + 1]$

**42.** $[(2y + 5) + 6][(2y + 5) - 6]$

## Objective E  *Multiply. See Example 22.*

▶**43.** $(x + y)(x - y)(x^2 - y^2)$

**44.** $(z - y)(z + y)(z^2 - y^2)$

**45.** $(x - 2)^4$

**46.** $(x - 1)^4$

**47.** $(x - 5)(x + 5)(x^2 + 25)$

**48.** $(x + 3)(x - 3)(x^2 + 9)$

## Objectives A B C D E  Mixed Practice *Multiply. See Examples 1 through 22.*

**49.** $-8a^2b(3b^2 - 5b + 20)$

**50.** $-9xy^2(3x^2 - 2x + 10)$

▶**51.** $(6x + 1)^2$

**52.** $(4x + 7)^2$

**53.** $(5x^3 + 2y)(5x^3 - 2y)$

**54.** $(3x^4 + 2y)(3x^4 - 2y)$

**55.** $(2x^3 + 5)(5x^2 + 4x + 1)$

**56.** $(3y^3 - 1)(3y^3 - 6y + 1)$

**57.** $(3x^2 + 2x - 1)^2$

**58.** $(4x^2 + 4x - 4)^2$

▶**59.** $(3x - 1)(x + 3)$

**60.** $(5d - 3)(d + 6)$

**61.** $(3x^4 + 1)(3x^2 + 5)$

**62.** $(4x^3 - 5)(5x^2 + 6)$

**63.** $(3x + 1)^2$

**64.** $(4x + 6)^2$

**65.** $(3b - 6y)(3b + 6y)$

**66.** $(2x - 4y)(2x + 4y)$

**67.** $(7x - 3)(7x + 3)$

**68.** $(4x + 1)(4x - 1)$

**69.**  $\begin{array}{r} 3x^2 + 4x - 4 \\ \times \quad\quad 3x + 6 \\ \hline \end{array}$

**70.**  $\begin{array}{r} 6x^2 + 2x - 1 \\ \times \quad\quad 3x - 6 \\ \hline \end{array}$

**71.** $(4x^2 - 2x + 5)(3x + 1)$

**72.** $(5x^2 - x - 2)(2x - 1)$

**73.** $[(xy + 4) - 6]^2$

**74.** $[(2a^2 + 4) - 1]^2$

**75.** $(11a^2 + 1)(2a + 1)$

**76.** $(13x^2 + 1)(3x + 1)$

**77.** $\left(\frac{2}{3}n - 2\right)\left(\frac{1}{2}n - 9\right)$

**78.** $\left(\frac{3}{5}y - 6\right)\left(\frac{1}{3}y - 10\right)$

**79.** $(3x + 1)(3x - 1)(2y + 5x)$

**80.** $(2a + 1)(2a - 1)(6a + 7b)$

**Objective** **F** *If* $f(x) = x^2 - 3x$, *find the following. See Example 23.*

**81.** $f(a)$

**82.** $f(c)$

**○83.** $f(a + h)$

**84.** $f(a + 5)$

**85.** $f(b - 2)$

**86.** $f(a - b)$

## Review

*Simplify. See Section 1.6.*

**87.** $\dfrac{6x^3}{3x}$

**88.** $\dfrac{4x^7}{x^2}$

**89.** $\dfrac{20a^3b^5}{18ab^2}$

**90.** $\dfrac{15x^7y^2}{6xy^2}$

**91.** $\dfrac{8m^4n}{12mn}$

**92.** $\dfrac{6n^6p}{8np}$

## Concept Extensions

*Solve. See the Concept Check in this section.*

**93.** Find the error:  $7y(3z - 2) + 1$
$$= 21yz - 14y + 7y$$
$$= 21yz - 7y$$

**94.** Find the error:  $2x + 3x(12 - x)$
$$= 5x(12 - x)$$
$$= 60x - 5x^2$$

**95.** Explain how to multiply a polynomial by a polynomial.

**96.** Explain why $(3x + 2)^2$ does not equal $9x^2 + 4$.

*Multiply. Assume that variables in exponents represent positive integers.*

**97.** $5x^2y^n(6y^{n+1} - 2)$

**98.** $-3yz^n(2y^3z^{2n} - 1)$

**99.** $(x^a + 5)(x^{2a} - 3)$

**100.** $(x^a + y^{2b})(x^a - y^{2b})$

**101.** Perform each indicated operation.
  **a.** $(3x + 5) + (3x + 7)$
  **b.** $(3x + 5)(3x + 7)$
  **c.** Explain the difference between the two problems.

**102.** Explain when the FOIL method can be used to multiply polynomials.

*Find the area of each shaded region.*

△ **103.**

(3x − 2) in.

(3x − 2) in.

$x$

$x$

△ **104.**

x − 7

2x

x

*Solve.*

△ **105.** Find the area of the circle. Do not approximate $\pi$.

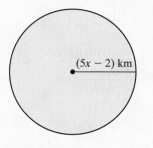

(5x − 2) km

△ **106.** Find the volume of the cylinder. Do not approximate $\pi$.

(y − 3) cm

7y cm

**107.** If $F(x) = x^2 + 3x + 2$, find
  **a.** $F(a + h)$
  **b.** $F(a)$
  **c.** $F(a + h) - F(a)$

**108.** If $g(x) = x^2 + 2x + 1$, find
  **a.** $g(a + h)$
  **b.** $g(a)$
  **c.** $g(a + h) - g(a)$

*If $R(x) = x + 5$, $Q(x) = x^2 - 2$, and $P(x) = 5x$, find the following.*

**109.** $P(x) \cdot R(x)$

**110.** $P(x) \cdot Q(x)$

**111.** $[Q(x)]^2$

**112.** $[R(x)]^2$

**113.** $R(x) \cdot Q(x)$

**114.** $P(x) \cdot R(x) \cdot Q(x)$

## 5.3 Dividing Polynomials and Synthetic Division

### Objectives

**A** Divide a Polynomial by a Monomial.

**B** Divide by a Polynomial.

**C** Use Synthetic Division.

**D** Use the Remainder Theorem to Evaluate Polynomials.

Now that we have added, subtracted, and multiplied polynomials, we will learn how to divide them.

### Objective A Dividing a Polynomial by a Monomial

Recall the following addition fact for fractions with a common denominator:

$$\frac{a}{c} + \frac{b}{c} = \frac{a + b}{c}$$

If $a$, $b$, and $c$ are monomials, we can read this equation from right to left and gain insight into how to divide a polynomial by a monomial.

> **Dividing a Polynomial by a Monomial**
>
> To divide a polynomial by a monomial, divide each term in the polynomial by the monomial.
>
> $$\frac{a + b}{c} = \frac{a}{c} + \frac{b}{c}, \quad c \neq 0$$

**Practice 1**

Divide $16y^3 - 8y^2 + 6y$ by $2y$.

**Example 1** Divide $10x^3 - 5x^2 + 20x$ by $5x$.

**Solution:** We divide each term of $10x^3 - 5x^2 + 20x$ by $5x$ and simplify.

$$\frac{10x^3 - 5x^2 + 20x}{5x} = \frac{10x^3}{5x} - \frac{5x^2}{5x} + \frac{20x}{5x} = 2x^2 - x + 4$$

To check, see that (quotient)(divisor) = dividend, or

$$(2x^2 - x + 4)(5x) = 10x^3 - 5x^2 + 20x$$

■ Work Practice 1

**Practice 2**

Divide:

$$\frac{9a^3b^3 - 6a^2b^2 + a^2b - 4a}{a^2b}$$

**Example 2** Divide: $\dfrac{3x^5y^2 - 15x^3y - x^2y - 6x}{x^2y}$

**Solution:** We divide each term in the numerator by $x^2y$.

$$\frac{3x^5y^2 - 15x^3y - x^2y - 6x}{x^2y} = \frac{3x^5y^2}{x^2y} - \frac{15x^3y}{x^2y} - \frac{x^2y}{x^2y} - \frac{6x}{x^2y}$$

$$= 3x^3y - 15x - 1 - \frac{6}{xy}$$

■ Work Practice 2

### Objective B Dividing by a Polynomial

To divide a polynomial by a polynomial other than a monomial, we use **long division**. Polynomial long division is similar to long division of real numbers. We review long division of real numbers by dividing 7 into 296.

$$\begin{array}{r} 42 \\ 7\overline{)296} \\ -28 \\ \hline 16 \\ -14 \\ \hline 2 \end{array}$$

Divisor: $7\overline{)296}$

$-28$    $4(7) = 28$

$16$    Subtract and bring down the next digit in the dividend.

$-14$    $2(7) = 14$

$2$    Subtract. The remainder is 2.

### Answers

**1.** $8y^2 - 4y + 3$

**2.** $9ab^2 - 6b + 1 - \dfrac{4}{ab}$

The quotient is

$42\dfrac{2}{7}$   $\begin{matrix} \text{remainder} \\ \text{divisor} \end{matrix}$

To check, notice that $42(7) + 2 = 296$, which is the dividend. This same division process can be applied to polynomials, as shown next.

**Example 3**   Divide $2x^2 - x - 10$ by $x + 2$.

**Solution:**   $2x^2 - x - 10$ is the dividend, and $x + 2$ is the divisor.

**Step 1:** Divide $2x^2$ by $x$.

$$\begin{array}{r} 2x \\ x+2\overline{)2x^2 - x - 10} \end{array}$$   $\dfrac{2x^2}{x} = 2x$, so $2x$ is the first term of the quotient.

**Step 2:** Multiply $2x(x + 2)$.

$$\begin{array}{r} 2x \\ x+2\overline{)2x^2 - x - 10} \\ 2x^2 + 4x \end{array}$$   Multiply: $2x(x + 2)$. Like terms are lined up vertically.

**Step 3:** Subtract $(2x^2 + 4x)$ from $(2x^2 - x - 10)$ by changing the signs of $(2x^2 + 4x)$ and adding.

$$\begin{array}{r} 2x \\ x+2\overline{)\;2x^2 - \;x - 10} \\ \underline{\mp 2x^2 \mp 4x} \\ -5x \end{array}$$

**Step 4:** Bring down the next term, $-10$, and start the process over.

$$\begin{array}{r} 2x \\ x+2\overline{)\;2x^2 - \;x - 10} \\ \underline{\mp 2x^2 \mp 4x \quad \downarrow} \\ -5x - 10 \end{array}$$

**Step 5:** Divide $-5x$ by $x$.

$$\begin{array}{r} 2x - 5 \\ x+2\overline{)\;2x^2 - \;x - 10} \\ \underline{\mp 2x^2 \mp 4x \quad \downarrow} \\ -5x - 10 \end{array}$$   $\dfrac{-5x}{x} = -5$ so $-5$ is the second term of the quotient.

**Step 6:** Multiply $-5(x + 2)$.

$$\begin{array}{r} 2x - 5 \\ x+2\overline{)\;2x^2 - \;x - 10} \\ \underline{\mp 2x^2 \mp 4x} \\ -5x - 10 \\ \underline{-5x - 10} \end{array}$$   Multiply: $-5(x + 2)$. Like terms are lined up vertically.

**Step 7:** Subtract by changing the signs of $-5x - 10$ and adding.

$$\begin{array}{r} 2x - 5 \\ x+2\overline{)\;2x^2 - \;x - 10} \\ \underline{\mp 2x^2 \mp 4x} \\ -5x - 10 \\ \underline{\pm 5x \pm 10} \\ 0 \end{array}$$   Subtract.

   Remainder.

*(Continued on next page)*

**Practice 3**

Divide $6x^2 + 11x - 2$ by $x + 2$.

**Answer**

**3.** $6x - 1$

Then $\dfrac{2x^2 - x - 10}{x + 2} = 2x - 5$. There is no remainder.

Check this result by multiplying $2x - 5$ by $x + 2$, the divisor. Their product is $(2x - 5)(x + 2) = 2x^2 - x - 10$, the dividend.

■ Work Practice 3

**Practice 4**

Divide:

$(10x^2 - 17x + 5) \div (5x - 1)$

**Example 4**  Divide:  $(6x^2 - 19x + 12) \div (3x - 5)$

**Solution:**

$$
\begin{array}{r}
2x \\
3x - 5 \overline{)6x^2 - 19x + 12} \\
\underline{6x^2 - 10x} \phantom{+ 12} \\
-9x + 12
\end{array}
$$

Divide: $\dfrac{6x^2}{3x} = 2x$

Multiply: $2x(3x - 5)$

Subtract: $(6x^2 - 19x) - (6x^2 - 10x) = -9x$

Bring down the next term, $+12$.

$$
\begin{array}{r}
2x - 3 \\
3x - 5 \overline{)6x^2 - 19x + 12} \\
\underline{6x^2 - 10x} \phantom{+ 12} \\
-9x + 12 \\
\underline{-9x + 15} \\
-3
\end{array}
$$

Divide: $\dfrac{-9x}{3x} = -3$

Multiply: $-3(3x - 5)$

Subtract: $(-9x + 12) - (-9x + 15) = -3$

**Check:**

$$
\text{divisor} \ \cdot \ \text{quotient} \ + \ \text{remainder}
$$

$$
(3x - 5)(2x - 3) + (-3) = 6x^2 - 19x + 15 - 3
$$
$$
= 6x^2 - 19x + 12 \qquad \text{The dividend}
$$

The division checks, so

> **Helpful Hint**  This fraction is the remainder over the divisor.

$$
\dfrac{6x^2 - 19x + 12}{3x - 5} = 2x - 3 + \dfrac{-3}{3x - 5}
$$

■ Work Practice 4

**Practice 5**

Divide:

$(5x^3 - 4x^2 + 3x - 4) \div (x - 2)$.

**Example 5**  Divide:  $(7x^3 + 16x^2 + 2x - 1) \div (x + 4)$.

**Solution:**

$$
\begin{array}{r}
7x^2 - 12x + 50 \\
x + 4 \overline{)7x^3 + 16x^2 + 2x - 1} \\
\underline{7x^3 + 28x^2} \phantom{+ 2x - 1} \\
-12x^2 + 2x \\
\underline{-12x^2 - 48x} \phantom{- 1} \\
50x - 1 \\
\underline{50x + 200} \\
-201
\end{array}
$$

Divide $\dfrac{7x^3}{x} = 7x^2$.

$7x^2(x + 4)$

Subtract. Bring down $2x$.
$\dfrac{-12x^2}{x} = -12x$, a term of the quotient. $-12x(x + 4)$

Subtract. Bring down $-1$.
$\dfrac{50x}{x} = 50$, a term of the quotient. $50(x + 4)$

Subtract.

Thus, $\dfrac{7x^3 + 16x^2 + 2x - 1}{x + 4} = 7x^2 - 12x + 50 + \dfrac{-201}{x + 4}$ or

$$
7x^2 - 12x + 50 - \dfrac{201}{x + 4}.
$$

■ Work Practice 5

**Answers**

**4.** $2x - 3 + \dfrac{2}{5x - 1}$

**5.** $5x^2 + 6x + 15 + \dfrac{26}{x - 2}$

**Example 6**   Divide $2x^3 + 3x^4 - 8x + 6$ by $x^2 - 1$.

**Solution:**   Before dividing, we write terms in descending order of powers of $x$. Also, we represent any "missing powers" by the product of 0 and the variable raised to the missing power. There is no $x^2$-term in the dividend, so we include $0x^2$ to represent the missing term. Also, there is no $x$-term in the divisor, so we include $0x$ in the divisor.

$$
\begin{array}{r}
3x^2 + 2x + 3 \\
x^2 + 0x - 1 \overline{)\, 3x^4 + 2x^3 + 0x^2 - 8x + 6} \\
\underline{3x^4 \mp 0x^3 \mp 3x^2} \\
2x^3 + 3x^2 - 8x \\
\underline{2x^3 \mp 0x^2 \mp 2x} \\
3x^2 - 6x + 6 \\
\underline{3x^2 \mp 0x \mp 3} \\
-6x + 9
\end{array}
$$

$\dfrac{3x^4}{x^2} = 3x^2$

$3x^2(x^2 + 0x - 1)$
Subtract. Bring down $-8x$.
$2x^3/x^2 = 2x$, a term of the quotient
$2x(x^2 + 0x - 1)$
Subtract. Bring down 6.
$3x^2/x^2 = 3$, a term of the quotient
$3(x^2 + 0x - 1)$
Subtract.

The division process is finished when the degree of the remainder polynomial is less than the degree of the divisor.

Thus,

$$
\frac{2x^3 + 3x^4 - 8x + 6}{x^2 - 1} = 3x^2 + 2x + 3 + \frac{-6x + 9}{x^2 - 1}
$$

▨ **Work Practice 6**

**Example 7**   Divide $27x^3 + 8$ by $3x + 2$.

**Solution:**   We replace the missing terms in the dividend with $0x^2$ and $0x$.

$$
\begin{array}{r}
9x^2 - 6x + 4 \\
3x + 2 \overline{)\, 27x^3 + 0x^2 + 0x + 8} \\
\underline{27x^3 \mp 18x^2} \\
-18x^2 + 0x \\
\underline{\mp 18x^2 \mp 12x} \\
12x + 8 \\
\underline{12x \mp 8} \\
0
\end{array}
$$

$9x^2(3x + 2)$
Subtract. Bring down $0x$.
$-6x(3x + 2)$
Subtract. Bring down 8.
$4(3x + 2)$

Thus, $\dfrac{27x^3 + 8}{3x + 2} = 9x^2 - 6x + 4$.

▨ **Work Practice 7**

**✔Concept Check**   In a division problem, the divisor is $4x^3 - 5$. The division process can be stopped when which of these possible remainder polynomials is reached?

**a.** $2x^4 + x^2 - 3$       **b.** $x^3 - 5^2$       **c.** $4x^2 + 25$

## Objective  C  Using Synthetic Division ▶

When a polynomial is to be divided by a binomial of the form $x - c$, a shortcut process called **synthetic division** may be used. See the next page for a derivation.

**Practice 6**

Divide $4x^2 + 3x^4 - 6x + 1$ by $x^2 + 1$.

**Practice 7**

Divide $64x^3 - 27$ by $4x - 3$.

**Helpful Hint**   The degree of the resulting polynomial (1) is the same as the degree of the divisor (1), so we continue the division process.

**Answers**

**6.** $3x^2 + 1 - \dfrac{6x}{x^2 + 1}$

**7.** $16x^2 + 12x + 9$

**✔Concept Check Answer**

c

On the left is an example of long division, and on the right is the same example showing the coefficients of the variables only.

$$
\begin{array}{r}
2x^2 + 5x + 2 \\
x - 3\overline{\smash{\big)}2x^3 - x^2 - 13x + 1} \\
\underline{2x^3 - 6x^2} \\
5x^2 - 13x \\
\underline{5x^2 - 15x} \\
2x + 1 \\
\underline{2x - 6} \\
7
\end{array}
\qquad
\begin{array}{r}
2 \quad 5 \quad 2 \\
1 - 3\overline{\smash{\big)}2 - 1 - 13 + 1} \\
\underline{2 - 6} \\
5 - 13 \\
\underline{5 - 15} \\
2 + 1 \\
\underline{2 - 6} \\
7
\end{array}
$$

Notice that as long as we keep coefficients of powers of $x$ in the same column, we can perform division of polynomials by performing algebraic operations on the coefficients only. This shorter process of dividing with coefficients only in a special format is called synthetic division. To find $(2x^3 - x^2 - 13x + 1) \div (x - 3)$ by synthetic division, follow the next example.

**Practice 8**

Use synthetic division to divide $3x^3 - 2x^2 + 5x + 4$ by $x - 2$.

**Example 8**  Use synthetic division to divide $2x^3 - x^2 - 13x + 1$ by $x - 3$.

**Solution:**  To use synthetic division, the divisor must be in the form $x - c$. Since we are dividing by $x - 3$, $c$ is 3. We write down 3 and the coefficients of the dividend.

$c$

$$
\begin{array}{c|cccc}
3 & 2 & -1 & -13 & 1 \\
& \downarrow & & & \\
\hline
& 2 & & &
\end{array}
$$

Next, draw a line and bring down the first coefficient of the dividend.

$$
\begin{array}{c|cccc}
3 & 2 & -1 & -13 & 1 \\
& & 6 & & \\
\hline
& 2 & & &
\end{array}
$$

Multiply $3 \cdot 2$ and write down the product, 6.

$$
\begin{array}{c|cccc}
3 & 2 & -1 & -13 & 1 \\
& & 6 & & \\
\hline
& 2 & 5 & &
\end{array}
$$

Add $-1 + 6$. Write down the sum, 5.

$$
\begin{array}{c|cccc}
3 & 2 & -1 & -13 & 1 \\
& & 6 & 15 & \\
\hline
& 2 & 5 & 2 &
\end{array}
$$

$3 \cdot 5 = 15$

$-13 + 15 = 2$

$$
\begin{array}{c|cccc}
3 & 2 & -1 & -13 & 1 \\
& & 6 & 15 & 6 \\
\hline
& 2 & 5 & 2 & 7
\end{array}
$$

$3 \cdot 2 = 6$

$1 + 6 = 7$

The quotient is found in the bottom row. The numbers 2, 5, and 2 are the coefficients of the quotient polynomial, and the number 7 is the remainder. The degree of the quotient polynomial is one less than the degree of the dividend. In our example, the degree of the dividend is 3, so the degree of the quotient polynomial is 2. As we found when we performed the long division, the quotient is

$$2x^2 + 5x + 2, \quad \text{remainder } 7$$

or

$$2x^2 + 5x + 2 + \frac{7}{x - 3}$$

■ **Work Practice 8**

**Answer**

8. $3x^2 + 4x + 13 + \dfrac{30}{x - 2}$

When using synthetic division, if there are missing powers of the variable, insert 0s as coefficients.

**Example 9** Use synthetic division to divide $x^4 - 2x^3 - 11x^2 + 34$ by $x + 2$.

**Solution:** The divisor is $x + 2$, which in the form $x - c$ is $x - (-2)$. Thus, $c$ is $-2$. There is no $x$-term in the dividend, so we insert a coefficient of 0. The dividend coefficients are $1, -2, -11, 0$, and $34$.

$$\begin{array}{r|rrrrr} -2 & 1 & -2 & -11 & 0 & 34 \\ & & -2 & 8 & 6 & -12 \\ \hline & 1 & -4 & -3 & 6 & 22 \end{array}$$

The dividend is a fourth-degree polynomial, so the quotient polynomial is a third-degree polynomial. The quotient is $x^3 - 4x^2 - 3x + 6$ with a remainder of 22. Thus,

$$\frac{x^4 - 2x^3 - 11x^2 + 34}{x + 2} = x^3 - 4x^2 - 3x + 6 + \frac{22}{x + 2}$$

■ Work Practice 9

**Practice 9**

Use synthetic division to divide $x^4 + 3x^3 - 5x + 4$ by $x + 1$.

**Helpful Hint**

Before dividing by long division or by synthetic division, write the dividend in descending order of variable exponents. Any "missing powers" of the variable must be represented by 0 times the variable raised to the missing power.

✓**Concept Check** Which division problems are candidates for the synthetic division process?

**a.** $(3x^2 + 5) \div (x + 4)$ 　　**b.** $(x^3 - x^2 + 2) \div (3x^3 - 2)$
**c.** $(y^4 + y - 3) \div (x^2 + 1)$ 　　**d.** $x^5 \div (x - 5)$

**Example 10** If $P(x) = 2x^3 - 4x^2 + 5$,

**a.** Find $P(2)$ by substitution.
**b.** Use synthetic division to find the remainder when $P(x)$ is divided by $x - 2$.

**Solution:**

**a.** $P(x) = 2x^3 - 4x^2 + 5$
$P(2) = 2(2)^3 - 4(2)^2 + 5$
$\quad = 2(8) - 4(4) + 5 = 16 - 16 + 5 = 5$

Thus, $P(2) = 5$.

**b.** The coefficients of $P(x)$ are $2, -4, 0$, and $5$. The number 0 is a coefficient of the missing power of $x^1$. The divisor is $x - 2$, so $c$ is 2.

$$\begin{array}{r|rrrr} \overset{c}{\phantom{a}} & & & & \\ 2 & 2 & -4 & 0 & 5 \\ & & 4 & 0 & 0 \\ \hline & 2 & 0 & 0 & 5 \end{array} \quad \text{remainder}$$

The remainder when $P(x)$ is divided by $x - 2$ is 5.

■ Work Practice 10

**Practice 10**

If $P(x) = x^3 - 5x - 2$,
**a.** Find $P(2)$ by substitution.
**b.** Use synthetic division to find the remainder when $P(x)$ is divided by $x - 2$.

**Answers**

**9.** $x^3 + 2x^2 - 2x - 3 + \dfrac{7}{x + 1}$

**10. a.** $-4$ **b.** $-4$

✓**Concept Check Answer**
**a** and **d**

## Objective D  Using the Remainder Theorem to Evaluate Polynomials ▶

Notice in the preceding example that $P(2) = 5$ and that the remainder when $P(x)$ is divided by $x - 2$ is 5. This is no accident. This illustrates the **remainder theorem.**

> ### Remainder Theorem
>
> If a polynomial $P(x)$ is divided by $x - c$, then the remainder is $P(c)$.

**Practice 11**

Use the remainder theorem and synthetic division to find $P(3)$ if $P(x) = 2x^5 - 18x^4 + 90x^2 + 59x$.

**Example 11**  Use the remainder theorem and synthetic division to find $P(4)$ if

$$P(x) = 4x^6 - 25x^5 + 35x^4 + 17x^2.$$

**Solution:**  To find $P(4)$ by the remainder theorem, we divide $P(x)$ by $x - 4$. The coefficients of $P(x)$ are $4, -25, 35, 0, 17, 0,$ and $0$. Also, $c$ is 4.

```
c
 ↘
  4 |  4   -25    35    0    17    0    0
    |       16   -36   -4   -16    4   16
    |_____
       4    -9    -1   -4     1    4   16   remainder
                                    ↑_____|
```

Thus, $P(4) = 16$, the remainder.

**Answer**

**11.** 15

■ **Work Practice 11**

## Vocabulary, Readiness & Video Check

*Martin-Gay Interactive Videos*

*Watch the section lecture video and answer the following questions.*

See Video 5.3

**Objective A 1.** In the lecture before ⊞ Example 1, dividing a polynomial by a monomial is compared to adding two fractions. What role does the monomial play in the fraction example? ▶

**Objective B 2.** From ⊞ Example 2, how do you know when to stop your long division? ▶

**Objective C 3.** From ⊞ Example 3, once you've completed the synthetic division, what does the bottom row of numbers mean? What is the degree of the quotient? ▶

**Objective D 4.** From ⊞ Example 4, given a polynomial function $P(x)$, under what circumstances might it be easier/faster to use the remainder theorem to find $P(c)$ rather than substituting the value $c$ for $x$ and then simplifying? ▶

## 5.3 Exercise Set MyMathLab ▶

**Objective A**  *Divide. See Examples 1 and 2.*

**1.** $4a^2 + 8a$ by $2a$

**2.** $6x^4 - 3x^3$ by $3x^2$

**3.** $\dfrac{12a^5b^2 + 16a^4b}{4a^4b}$

**4.** $\dfrac{4x^3y + 12x^2y^2 - 4xy^3}{4xy}$

▶**5.** $\dfrac{4x^2y^2 + 6xy^2 - 4y^2}{2x^2y}$

**6.** $\dfrac{6x^5y + 75x^4y - 24x^3y^2}{3x^4y}$

**Objective B**  *Divide. See Examples 3 through 7.*

**7.** $(x^2 + 3x + 2) \div (x + 2)$

**8.** $(y^2 + 7y + 10) \div (y + 5)$

**9.** $(2x^2 - 6x - 8) \div (x + 1)$

**10.** $(3x^2 + 19x + 20) \div (x + 5)$

**11.** $2x^2 + 3x - 2$ by $2x + 4$

**12.** $6x^2 - 17x - 3$ by $3x - 9$

**13.** $(4x^3 + 7x^2 + 8x + 20) \div (2x + 4)$

**14.** $(8x^3 + 18x^2 + 16x + 24) \div (4x + 8)$

**15.** $(2x^2 + 6x^3 - 18x - 6) \div (3x + 1)$

**16.** $(4x - 15x^2 + 10x^3 - 6) \div (2x - 3)$

▶**17.** $(3x^5 - x^3 + 4x^2 - 12x - 8) \div (x^2 - 2)$

**18.** $(2x^5 - 6x^4 + x^3 - 4x + 3) \div (x^2 - 3)$

**19.** $\left(2x^4 + \dfrac{1}{2}x^3 + x^2 + x\right) \div (x - 2)$

**20.** $\left(x^4 - \dfrac{2}{3}x^3 + x\right) \div (x - 3)$

**Objective C**  *Use synthetic division to divide. See Examples 8 and 9.*

**21.** $\dfrac{x^2 + 3x - 40}{x - 5}$

**22.** $\dfrac{x^2 - 14x + 24}{x - 2}$

**23.** $\dfrac{x^2 + 5x - 6}{x + 6}$

**24.** $\dfrac{x^2 + 12x + 32}{x + 4}$

▶**25.** $\dfrac{x^3 - 7x^2 - 13x + 5}{x - 2}$

**26.** $\dfrac{x^3 + 6x^2 + 4x - 7}{x + 5}$

**27.** $\dfrac{4x^2 - 9}{x - 2}$

**28.** $\dfrac{3x^2 - 4}{x - 1}$

## Objectives A B C Mixed Practice *Divide. See Examples 1–9.*

**29.** $\dfrac{4x^7y^4 + 8xy^2 + 4xy^3}{4xy^3}$

**30.** $\dfrac{15x^3y - 5x^2y + 10xy^2}{5x^2y}$

**31.** $(10x^3 - 5x^2 - 12x + 1) \div (2x - 1)$

**32.** $(20x^3 - 8x^2 + 5x - 5) \div (5x - 2)$

**33.** $(2x^3 - 6x^2 - 4) \div (x - 4)$

**34.** $(3x^3 + 4x - 10) \div (x + 2)$

**35.** $\dfrac{2x^4 - 13x^3 + 16x^2 - 9x + 20}{x - 5}$

**36.** $\dfrac{3x^4 + 5x^3 - x^2 + x - 2}{x + 2}$

**37.** $\dfrac{7x^2 - 4x + 12 + 3x^3}{x + 1}$

**38.** $\dfrac{4x^3 + x^4 - x^2 - 16x - 4}{x - 2}$

**39.** $\dfrac{3x^3 + 2x^2 - 4x + 1}{x - \dfrac{1}{3}}$

**40.** $\dfrac{9y^3 + 9y^2 - y + 2}{y + \dfrac{2}{3}}$

**41.** $\dfrac{x^3 - 1}{x - 1}$

**42.** $\dfrac{y^3 - 8}{y - 2}$

**43.** $(25xy^2 + 75xyz + 125x^2yz) \div (-5x^2y)$

**44.** $(x^6y^6z^2 - x^3y^3z + 7x^3y) \div (-7yz^2)$

**45.** $(9x^5 + 6x^4 - 6x^2 - 4x) \div (3x + 2)$

**46.** $(5x^4 - 5x^2 + 10x^3 - 10x) \div (5x + 10)$

## Objectives C D *For the given polynomial P(x) and the given c, use the remainder theorem to find P(c). See Examples 10 and 11.*

**47.** $P(x) = x^3 + 3x^2 - 7x + 4; 1$

**48.** $P(x) = x^3 + 5x^2 - 4x - 6; 2$

**49.** $P(x) = 3x^3 - 7x^2 - 2x + 5; -3$

**50.** $P(x) = 4x^3 + 5x^2 - 6x - 4; -2$

**▶51.** $P(x) = 4x^4 + x^2 - 2; -1$

**52.** $P(x) = x^4 - 3x^2 - 2x + 5; -2$

**53.** $P(x) = 2x^4 - 3x^2 - 2; \dfrac{1}{3}$

**54.** $P(x) = 4x^4 - 2x^3 + x^2 - x - 4; \dfrac{1}{2}$

**55.** $P(x) = x^5 + x^4 - x^3 + 3; \dfrac{1}{2}$

**56.** $P(x) = x^5 - 2x^3 + 4x^2 - 5x + 6; \dfrac{2}{3}$

## Review

*Solve each inequality. See Section 2.7.*

**57.** $|x + 5| < 4$        **58.** $|x - 1| \le 8$        **59.** $|2x + 7| \ge 9$        **60.** $|4x + 2| > 10$

## Concept Extensions

*Solve. See the Concept Checks in this section.*

*Is the given division problem a candidate for the synthetic division process?*

**61.** $(5x^2 - 3x + 2) \div (x + 2)$

**62.** $(x^4 - 6) \div (x^3 + 3x - 1)$

**63.** $(x^7 - 2) \div (x^5 + 1)$

**64.** $(3x^2 + 7x - 1) \div \left(x - \dfrac{1}{3}\right)$

*Solve.*

**65.** In a long division exercise, if the divisor is $9x^3 - 2x$, then the division process can be stopped when the degree of the remainder is
a. 1    b. 3    c. 9    d. 2

**66.** In a division exercise, if the divisor is $x - 3$, then the division process can be stopped when the degree of the remainder is
a. 1    b. 0    c. 2    d. 3

△ **67.** A board of length $(3x^4 + 6x^2 - 18)$ meters is to be cut into three pieces of the same length. Find the length of each piece.

△ **68.** The perimeter of a regular hexagon is $(12x^5 - 48x^3 + 3)$ miles. Find the length of each side.

△ **69.** If the area of the rectangle is $(15x^2 - 29x - 14)$ square inches, and its length is $(5x + 2)$ inches, find its width.

$(5x + 2)$ in.

△ **70.** If the area of a parallelogram is $(2x^2 - 17x + 35)$ square centimeters and its base is $(2x - 7)$ centimeters, find its height.

$(2x - 7)$ cm

△ **71.** If the area of a parallelogram is $(x^4 - 23x^2 + 9x - 5)$ square centimeters and its base is $(x + 5)$ centimeters, find its height.

Height

$(x + 5)$ centimeters

△ **72.** If the volume of a box is $(x^4 + 6x^3 - 7x^2)$ cubic meters, its height is $x^2$ meters, and its length is $(x + 7)$ meters, find its width.

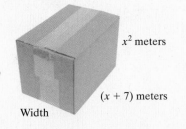

$x^2$ meters

$(x + 7)$ meters

Width

*Divide.*

**73.** $\left(x^4 + \dfrac{2}{3}x^3 + x\right) \div (x - 1)$

**74.** $\left(2x^3 + \dfrac{9}{2}x^2 - 4x - 10\right) \div (x + 2)$

**75.** $\left(3x^4 - x - x^3 + \dfrac{1}{2}\right) \div (2x - 1)$

**76.** $\left(2x^4 + \dfrac{1}{2}x^3 - \dfrac{1}{4}x^2 + x\right) \div (2x + 1)$

**77.** $(5x^4 - 2x^2 + 10x^3 - 4x) \div (5x + 10)$

**78.** $(6x^5 + 3x^4 - 2x^2 - x) \div (2x + 1)$

*For each given $f(x)$ and $g(x)$, find $\dfrac{f(x)}{g(x)}$. Also find any x-values that are not in the domain of $\dfrac{f(x)}{g(x)}$. (Note: Since g(x) is in the denominator, g(x) cannot be 0).*

**79.** $f(x) = 25x^2 - 5x + 30; g(x) = 5x$

**80.** $f(x) = 12x^4 - 9x^3 + 3x - 1; g(x) = 3x$

**81.** $f(x) = 7x^4 - 3x^2 + 2; g(x) = x - 2$

**82.** $f(x) = 2x^3 - 4x^2 + 1; g(x) = x + 3$

*Solve.*

**83.** Try performing the following division without changing the order of the terms. Describe why this makes the process more complicated. Then perform the division again after putting the terms in the dividend in descending order of exponents.

$$\frac{4x^2 - 12x - 12 + 3x^3}{x - 2}$$

**84.** Explain how to check polynomial long division.

**85.** Explain an advantage of using the remainder theorem instead of direct substitution.

**86.** Explain an advantage of using synthetic division instead of long division.

*We say that 2 is a factor of 8 because 2 divides 8 evenly, or with a remainder of 0. In the same manner, the polynomial $x - 2$ is a factor of the polynomial $x^3 - 14x^2 + 24x$ because the remainder is 0 when $x^3 - 14x^2 + 24x$ is divided by $x - 2$. Use this information for Exercises 87 and 88.*

**87.** Use synthetic division to show that $x + 3$ is a factor of $x^3 + 3x^2 + 4x + 12$.

**88.** Use synthetic division to show that $x - 2$ is a factor of $x^3 - 2x^2 - 3x + 6$.

**89.** If a polynomial is divided by $x - 5$, the quotient is $2x^2 + 5x - 6$ and the remainder is 3. Find the original polynomial.

**90.** If a polynomial is divided by $x + 3$, the quotient is $x^2 - x + 10$ and the remainder is $-2$. Find the original polynomial.

**91.** eBay is the leading online auction house. eBay's annual net profit can be modeled by the polynomial function $P(x) = -71x^3 + 393x^2 - 232x + 1884$, where $P(x)$ is net profit in millions of dollars and $x$ is the number of years since 2008. eBay's annual revenue can be modeled by the function $R(x) = 468x^2 - 471x + 8560$, where $R(x)$ is revenue in millions of dollars and $x$ is years since 2008. (*Source:* Based on data from eBay, Inc.)

**a.** Given that

$$\text{net profit margin} = \frac{\text{net profit}}{\text{revenue}},$$

write a function, $m(x)$, that models eBay's net profit margin.

**b.** Use part (a) to predict eBay's profit margin in 2015. Round to the nearest hundredth.

**92.** Kraft Foods is a provider of many of the best-known brands in U.S. supermarkets. Among its well-known brands are Kraft, Oscar Mayer, Maxwell House, and Oreo. Kraft Foods' annual revenues since 2009 can be modeled by the polynomial function $R(x) = -0.25x^3 + 0.93x^2 - 0.16x + 17.28$, where $R(x)$ is revenue in billions of dollars and $x$ is the number of years since 2008. Kraft Foods' net profit can be modeled by the function $P(x) = -0.03x^3 + 0.09x^2 - 0.12x + 1.95$, where $P(x)$ is the net profit in billions of dollars and $x$ is the number of years since 2008. (*Source:* Based on data from Kraft Foods)

**a.** Suppose that a market analyst has found the model $P(x)$ and another analyst at the same firm has found the model $R(x)$. The analysts have been asked by their manager to work together to find a model for Kraft Foods' profit margin. The analysts know that a company's profit margin is the ratio of its profit to its revenue. Describe how these two analysts could collaborate to find a function $m(x)$ that models Kraft Foods' net profit margin based on the work they have done independently.

**b.** Without actually finding $m(x)$, give a general description of what you would expect the answer to be.

## 5.4 The Greatest Common Factor and Factoring by Grouping

**Factoring** is the reverse process of multiplying. It is the process of writing a polynomial as a product

$$\underbrace{6x^2 + 13x - 5}_{} = \underbrace{(3x - 1)(2x + 5)}_{}$$

factoring

multiplying

In the next few sections, we review techniques for factoring polynomials. These techniques are used at the end of this chapter to solve polynomial equations.

**Objectives**

**A** Identify the Greatest Common Factor (GCF).

**B** Factor Out the Greatest Common Factor (GCF) of a Polynomial's Terms.

**C** Factor Polynomials by Grouping.

## Objective A Identifying the GCF

To factor a polynomial, we first factor out the **greatest common factor (GCF)** of its terms, using the distributive property. The GCF of a list of terms or monomials is the product of the GCF of the numerical coefficients and each GCF of the powers of a common variable.

### Finding the GCF of a List of Monomials

**Step 1:** Find the GCF of the numerical coefficients.

**Step 2:** Find the GCF of the variable factors.

**Step 3:** The product of the factors found in Steps 1 and 2 is the GCF of the monomials.

Copyright 2015 Pearson Education, Inc.

**Practice 1**

Find the GCF of $32x^4y^2, 48x^3y$, and $24y^2$.

**Example 1**  Find the GCF of $20x^3y, 10x^2y^2$, and $35x^3$.

**Solution:**  The GCF of the numerical coefficients 20, 10, and 35 is 5, the largest integer that is a factor of each integer. The GCF of the variable factors $x^3$, $x^2$, and $x^3$ is $x^2$ because $x^2$ is the largest factor common to all three powers of $x$. The variable $y$ is not a common factor because it does not appear in all three monomials. The GCF is thus

$$5 \cdot x^2, \quad \text{or} \quad 5x^2$$

To see this in factored form,

$$20x^3y = 2 \cdot 2 \cdot 5 \cdot x^2 \cdot x \cdot y$$
$$10x^2y^2 = 2 \cdot 5 \cdot x^2 \cdot y^2$$
$$35x^3 = 7 \cdot 5 \cdot x^2 \cdot x$$
$$\mathbf{GCF} = 5 \cdot x^2$$

■ Work Practice 1

## Objective B  Factoring Out the GCF of a Polynomial's Terms ▶

A first step in factoring polynomials is to use the distributive property and write the polynomial as a product of the GCF of its monomial terms and a simpler polynomial. This is called **factoring out** the GCF.

**Practice 2**

Factor:  $9x + 3$

**Example 2**  Factor:  $8x + 4$

**Solution:**  The greatest common factor of the terms $8x$ and 4 is 4.

$$8x + 4 = 4 \cdot 2x + 4 \cdot 1 \quad \text{Factor out 4 from each term.}$$
$$= 4(2x + 1) \quad \text{Use the distributive property.}$$

The factored form of $8x + 4$ is $4(2x + 1)$. To check, multiply $4(2x + 1)$ to see that the product is $8x + 4$.

■ Work Practice 2

**Practice 3–5**

Factor each polynomial.
**3.** $20y^2 - 4y^3$
**4.** $6a - 7$
**5.** $6a^4b^2 - 3a^2b^2$

**Examples**  Factor.

**3.** $6x^2 + 3x^3 = 3x^2 \cdot 2 + 3x^2 \cdot x$    The GCF of 6 and 3 is 3 and the GCF of $x^2$ and $x^3$ is $x^2$. Thus, the GCF of the terms is $3x^2$.

$$= 3x^2(2 + x) \quad \text{Use the distributive property.}$$

**4.** $3y + 1$  There is no common factor other than 1.

**5.** $17x^3y^2 - 34x^4y^2 = 17x^3y^2 \cdot 1 - 17x^3y^2 \cdot 2x$    Factor out the greatest common factor, $17x^3y^2$.

$$= 17x^3y^2(1 - 2x) \quad \text{Use the distributive property.}$$

■ Work Practice 3–5

**Answers**

**1.** $8y$  **2.** $3(3x + 1)$  **3.** $4y^2(5 - y)$
**4.** $6a - 7$  **5.** $3a^2b^2(2a^2 - 1)$

Helpful Hint

If the greatest common factor happens to be one of the terms in the polynomial, a factor of 1 will remain for this term when the greatest common factor is factored out. For example, in the polynomial $21x^2 + 7x$, the greatest common factor of $21x^2$ and $7x$ is $7x$, so

$$21x^2 + 7x = 7x(3x) + 7x(1) = 7x(3x + 1)$$

**Helpful Hint** To check that the greatest common factor has been factored out correctly, multiply the factors together and see that their product is the original polynomial.

✓**Concept Check** Which factorization of $12x^2 + 9x - 3$ is correct?

  **a.** $3(4x^2 + 3x + 1)$ $\qquad$ **b.** $3(4x^2 + 3x - 1)$
  **c.** $3(4x^2 + 3x - 3)$ $\qquad$ **d.** $3(4x^2 + 3x)$

**Example 6** Factor: $-3x^3y + 2x^2y - 5xy$

**Practice 6**

Factor: $-2x^2y - 4xy + 10y$

**Solution:** Two possibilities are shown for factoring this polynomial.

First, the common factor $xy$ is factored out.

$$-3x^3y + 2x^2y - 5xy = xy(-3x^2 + 2x - 5)$$

Also, the common factor $-xy$ can be factored out as shown.

$$-3x^3y + 2x^2y - 5xy = (-xy)(3x^2) + (-xy)(-2x) + (-xy)(5)$$
$$= -xy(3x^2 - 2x + 5)$$

Both of these are correct.

 Work Practice 6

**Example 7** Factor: $2(x - 5) + 3a(x - 5)$

**Practice 7**

Factor: $3(x + 7) + 5y(x + 7)$

**Solution:** The greatest common factor is the binomial factor $(x - 5)$.

$$2(x - 5) + 3a(x - 5) = (x - 5)(2 + 3a)$$

 Work Practice 7

**Example 8** Factor: $7x(x^2 + 5y) - (x^2 + 5y)$

**Practice 8**

Factor:
$$6a(2a + 3b) - (2a + 3b)$$

**Solution:**

$$7x(x^2 + 5y) - (x^2 + 5y) = 7x(x^2 + 5y) - 1(x^2 + 5y)$$
$$= (x^2 + 5y)(7x - 1)$$

**Helpful Hint**

Notice that we wrote $-(x^2 + 5y)$ as $-1(x^2 + 5y)$ to aid in factoring.

 Work Practice 8

**Answers**
**6.** $2y(-x^2 - 2x + 5)$ or $-2y(x^2 + 2x - 5)$
**7.** $(x + 7)(3 + 5y)$
**8.** $(2a + 3b)(6a - 1)$

✓**Concept Check Answer**
b

## Objective C Factoring by Grouping ▶

Sometimes it is possible to factor a polynomial by grouping the terms of the polynomial and looking for common factors in each group. This method of factoring is called **factoring by grouping**.

**Practice 9**

Factor:  $xy - 5y + 3x - 15$

**Example 9**   Factor:  $ab - 6a + 2b - 12$

**Solution:**

$$\begin{aligned}
ab - 6a + 2b - 12 &= (ab - 6a) + (2b - 12) && \text{Group pairs of terms.}\\
&= a(b - 6) + 2(b - 6) && \text{Factor each binomial.}\\
&= (b - 6)(a + 2) && \text{Factor out the greatest common}\\
& && \text{factor, } (b - 6).
\end{aligned}$$

To check, multiply $(b - 6)$ and $(a + 2)$ to see that the product is $ab - 6a + 2b - 12$.

■ Work Practice 9

**Helpful Hint**

Notice that the polynomial $a(b - 6) + 2(b - 6)$ is *not* in factored form. It is a *sum*, not a *product*. The factored form is $(b - 6)(a + 2)$.

**Practice 10**

Factor:  $y^3 + 6y^2 + 4y + 24$

**Example 10**   Factor:  $x^3 + 5x^2 + 3x + 15$

**Solution:**

$$\begin{aligned}
x^3 + 5x^2 + 3x + 15 &= (x^3 + 5x^2) + (3x + 15) && \text{Group pairs of terms.}\\
&= x^2(x + 5) + 3(x + 5) && \text{Factor each binomial.}\\
&= (x + 5)(x^2 + 3) && \text{Factor out the common}\\
& && \text{factor, } (x + 5).
\end{aligned}$$

■ Work Practice 10

**Practice 11**

Factor:  $a^2b^2 + a^2 - 3b^2 - 3$

**Example 11**   Factor:  $m^2n^2 + m^2 - 2n^2 - 2$

**Solution:**

$$\begin{aligned}
m^2n^2 + m^2 - 2n^2 - 2 &= (m^2n^2 + m^2) + (-2n^2 - 2) && \text{Group pairs of terms.}\\
&= m^2(n^2 + 1) - 2(n^2 + 1) && \text{Factor each binomial.}\\
&= (n^2 + 1)(m^2 - 2) && \text{Factor out the common}\\
& && \text{factor, } (n^2 + 1).
\end{aligned}$$

■ Work Practice 11

**Practice 12**

Factor:  $ab + 5a - b - 5$

**Example 12**   Factor:  $xy + 2x - y - 2$

**Solution:**

$$\begin{aligned}
xy + 2x - y - 2 &= (xy + 2x) + (-y - 2) && \text{Group pairs of terms.}\\
&= x(y + 2) - 1(y + 2) && \text{Factor each binomial.}\\
&= (y + 2)(x - 1) && \text{Factor out the common factor, } y + 2.
\end{aligned}$$

■ Work Practice 12

**Answers**

9. $(x - 5)(y + 3)$
10. $(y + 6)(y^2 + 4)$
11. $(b^2 + 1)(a^2 - 3)$
12. $(b + 5)(a - 1)$

## Vocabulary, Readiness & Video Check

*Use the choices below to fill in each blank. Some choices will be used more than once and some not at all.*

| least | greatest | sum | product | factoring | $x^3$ | $x^7$ | true | false |

**1.** The reverse process of multiplying is _____.

**2.** The greatest common factor (GCF) of $x^7, x^3, x^5$ is _____.

**3.** In general, the GCF of a list of common variables raised to powers is the expression with the _____ exponent in the list.

**4.** Factoring means writing as a _____.

**5.** True or false: A factored form of $2xy^3 + 10xy$ is $2xy \cdot y^2 + 2xy \cdot 5$. _____

**6.** True or false: A factored form of $x^3 - 6x^2 + x$ is $x(x^2 - 6x)$. _____

**7.** True or false: A factored form of $5x - 5y + x^3 - x^2y$ is $5(x - y) + x^2(x - y)$. _____

**8.** True or false: A factored form of $5x - 5y + x^3 - x^2y$ is $(x - y)(5 + x^2)$. _____

*Martin-Gay Interactive Videos*     Watch the section lecture video and answer the following questions.

**Objective A**   **9.** From Example 1, how do you find the GCF of a list of common variables raised to powers?

**Objective B**   **10.** From Example 2, when factoring out a GCF from a polynomial, what two operations may help? How?

**Objective C**   **11.** In Examples 4 and 5, what are you reminded to do first when factoring a polynomial? How do you know to check to see if factoring by grouping might work?

*See Video 5.4*

## 5.4  Exercise Set MyMathLab®

**Objective A**  *Find the GCF of each list of monomials. See Example 1.*

**1.** $6, 12$

**2.** $9, 27$

**3.** $7x, 14x$

**4.** $8z, 4z$

**5.** $a^8, a^5, a^3$

**6.** $b^9, b^2, b^5$

**7.** $6x^3y, 9x^2y^2, 12x^2y$

**8.** $4xy^2, 16xy^3, 8x^2y^2$

**9.** $10x^3yz^3, 20x^2z^5, 45xz^3$   **10.** $12y^2z^4, 9xy^3z^4, 15x^2y^2z^3$

**Objective B**  *Factor out the greatest common factor. See Examples 2 through 8.*

**11.** $18x - 12$

**12.** $21x + 14$

**13.** $4y^2 - 16xy^3$

**14.** $3z - 21xz^4$

**15.** $6x^5 - 8x^4 + 2x^3$

**16.** $9x + 3x^2 - 6x^3$

**17.** $8a^3b^3 - 4a^2b^2 + 4ab + 16ab^2$

**18.** $12a^3b - 6ab + 18ab^2 - 18a^2b$

**19.** $6(x + 3) + 5a(x + 3)$

**20.** $2(x - 4) + 3y(x - 4)$

**21.** $2x(z + 7) + (z + 7)$

**22.** $9x(y - 2) + (y - 2)$

**23.** $3x(6x^2 + 5) - 2(6x^2 + 5)$

**24.** $4x(2y^2 + 3) - 5(2y^2 + 3)$

**25.** $20a^3 + 5a^2 + 1$

**26.** $12b^4 - 3b^2 + 1$

**27.** $39x^3y^3 - 26x^2y^3$

**28.** $38x^2y^4 - 57xy^4$

## Objective C  *Factor each polynomial by grouping. See Examples 9 through 12.*

**29.** $ab + 3a + 2b + 6$

**30.** $ab + 2a + 5b + 10$

**31.** $ac + 4a - 2c - 8$

**32.** $bc + 8b - 3c - 24$

**33.** $2xy - 3x - 4y + 6$

**34.** $12xy - 18x - 10y + 15$

**35.** $12xy - 8x - 3y + 2$

**36.** $20xy - 15x - 4y + 3$

## Objectives A B C  Mixed Practice  *Factor each polynomial. See Examples 1 through 12.*

**37.** $6x^3 + 9$

**38.** $6x^2 - 8$

**39.** $x^3 + 3x^2$

**40.** $x^4 - 4x^3$

**41.** $8a^3 - 4a$

**42.** $12b^4 + 3b^2$

**43.** $8m^3 + 4m^2 + 1$

**44.** $30n^4 + 15n^2 + 1$

**45.** $-20x^2y + 16xy^3$

**46.** $-18xy^3 + 27x^4y$

**47.** $10a^2b^3 + 5ab^2 - 15ab^3$

**48.** $10ef - 20e^2f^3 + 30e^3f$

**49.** $9abc^2 + 6a^2bc - 6ab + 3bc$

**50.** $4a^2b^2c - 6ab^2c - 4ac + 8a$

**51.** $4x(y - 2) - 3(y - 2)$

**52.** $8y(z + 8) - 3(z + 8)$

**53.** $6xy + 10x + 9y + 15$

**54.** $15xy + 20x + 6y + 8$

**55.** $xy + 3y - 5x - 15$

**56.** $xy + 4y - 3x - 12$

**57.** $6ab - 2a - 9b + 3$

**58.** $16ab - 8a - 6b + 3$

**59.** $12xy + 18x + 2y + 3$

**60.** $20xy + 8x + 5y + 2$

**61.** $2m(n - 8) - (n - 8)$

**62.** $3a(b - 4) - (b - 4)$

**63.** $15x^3y^2 - 18x^2y^2$

**64.** $12x^4y^2 - 16x^3y^3$

**65.** $2x^2 + 3xy + 4x + 6y$

**66.** $3x^2 + 12x + 4xy + 16y$

**67.** $5x^2 + 5xy - 3x - 3y$

**68.** $4x^2 + 2xy - 10x - 5y$

**69.** $x^3 + 3x^2 + 4x + 12$

**70.** $x^3 + 4x^2 + 3x + 12$

**71.** $x^3 - x^2 - 2x + 2$

**72.** $x^3 - 2x^2 - 3x + 6$

## Review

*Find each product by using the FOIL order of multiplying binomials. See Section 5.2.*

**73.** $(x + 2)(x - 5)$

**74.** $(x - 7)(x - 1)$

**75.** $(x + 3)(x + 2)$

**76.** $(x - 4)(x + 2)$

**77.** $(y - 3)(y - 1)$

**78.** $(s + 8)(s + 10)$

## Concept Extensions

*Solve. See the Concept Check in this section.*

**79.** Which factorization of $10x^2 - 2x - 2$ is correct?
  **a.** $2(5x^2 - x + 1)$
  **b.** $2(5x^2 - x)$
  **c.** $2(5x^2 - x - 2)$
  **d.** $2(5x^2 - x - 1)$

**80.** Which factorization of $x^4 + 5x^3 - x^2$ is correct?
  **a.** $-1(x^4 + 5x^3 + x^2)$
  **b.** $x^2(x^2 + 5x^3 - x^2)$
  **c.** $x^2(x^2 + 5x - 1)$
  **d.** $5x^2(x^2 + 5x - 5)$

*Solve.*

△ **81.** The area of the material needed to manufacture a tin can is given by the polynomial $2\pi r^2 + 2\pi rh$, where the radius is $r$ and height is $h$. Factor this expression.

**82.** To estimate the cost of a new product, one expression used by the production department is $4\pi r^2 + \dfrac{4}{3}\pi r^3$. Write an equivalent expression by factoring $4\pi r^2$ from both terms.

**83.** At the end of $T$ years, the amount of money $A$ in a savings account earning simple interest from an initial investment of \$5600 at rate $r$ is given by the formula $A = 5600 + 5600rt$. Write an equivalent equation by factoring the expression $5600 + 5600rt$.

△ **84.** An open-topped box has a square base and a height of 10 inches. If each of the bottom edges of the box has length $x$ inches, find the amount of material needed to construct the box. Write the answer in factored form.

10 in.

**85.** When $3x^2 - 9x + 3$ is factored, the result is $3(x^2 - 3x + 1)$. Explain why it is necessary to include the term 1 in this factored form.

**86.** Construct a trinomial whose greatest common factor is $5x^2y^3$.

**87.** A factored polynomial can be in many forms. For example, a factored form of $xy - 3x - 2y + 6$ is $(x - 2)(y - 3)$. Which of the following is not a factored form of $xy - 3x - 2y + 6$?

**a.** $(2 - x)(3 - y)$  **b.** $(-2 + x)(-3 + y)$
**c.** $(y - 3)(x - 2)$  **d.** $(-x + 2)(-y + 3)$

**88.** Which factorization of $12x^2 + 9x + 3$ is correct?
**a.** $3(4x^2 + 3x + 1)$
**b.** $3(4x^2 + 3x - 1)$
**c.** $3(4x^2 + 3x - 3)$
**d.** $3(4x^2 + 3x)$

**89.** Explain why $9(5 - x) + y(5 - x)$ is not a factored form of $45 - 9x + 5y - xy$.

**90.** Consider the following sequence of algebraic steps:

$$x^3 - 6x^2 + 2x - 10 = (x^3 - 6x^2) + (2x - 10)$$
$$= x^2(x - 6) + 2(x - 5)$$

Explain whether the final result is the factored form of the original polynomial.

**91.** At the end of $T$ years, the amount of money $A$ in a savings account earning simple interest from an initial investment of $P$ dollars at rate $R$ is given by the formula

$$A = P + PRT$$

Write an equivalent equation by factoring the expression $P + PRT$.

**92.** The amount $E$ of voltage in an electrical circuit is given by the formula

$$IR_1 + IR_2 = E$$

Write an equivalent equation by factoring the expression $IR_1 + IR_2$.

*Factor out the greatest common factor. Assume that variables used as exponents represent positive integers.*

**93.** $x^{3n} - 2x^{2n} + 5x^n$   **94.** $3y^n + 3y^{2n} + 5y^{8n}$   **95.** $6x^{8a} - 2x^{5a} - 4x^{3a}$   **96.** $3x^{5a} - 6x^{3a} + 9x^{2a}$

*Solve.*

**97.** An object is thrown upward from the ground with an initial velocity of 64 feet per second. The height $h(t)$ in feet of the object after $t$ seconds is given by the polynomial function

$$h(t) = -16t^2 + 64t$$

**a.** Write an equivalent factored expression for the function $h(t)$ by factoring $-16t^2 + 64t$.
**b.** Find $h(1)$ by using

$$h(t) = -16t^2 + 64t$$

and then by using the factored form of $h(t)$.
**c.** Explain why the values found in part (b) are the same.

**98.** An object is dropped from the gondola of a hot-air balloon at a height of 224 feet. The height $h(t)$ in feet of the object after $t$ seconds is given by the polynomial function

$$h(t) = -16t^2 + 224$$

224 ft

   **a.** Write an equivalent factored expression for the function $h(t)$ by factoring $-16t^2 + 224$.
   **b.** Find $h(2)$ by using

$$h(t) = -16t^2 + 224$$

   and then by using the factored form of the function.
   **c.** Explain why the values found in part (b) are the same.

**99.** The polynomial function $f(x) = -54x^4 + 903x^3 - 3609x^2 + 528x + 91{,}179$ models the number of patents granted to U.S. citizens for the years 2004–2012, where $x$ represents the number of years since 2004 and $f(x)$ is the number of patents granted. Write an equivalent expression for $f(x)$ by factoring the greatest common factor from the terms of $f(x) = -54x^4 + 903x^3 - 3609x^2 + 528x + 91{,}179$. (*Source:* Based on data from the U.S. Patent Office)

**100.** The number of students attending U.S. institutions of higher learning is increasing. The polynomial function $f(x) = -15x^3 + 195x^2 - 90x + 17{,}295$ models the number of students enrolled in American colleges and universities for the years 2004–2013, where $x$ is the years after 2004 and $f(x)$ is the number of college students (in thousands). Write an equivalent expression for $f(x)$ by factoring the greatest common factor from the terms of $f(x) = -15x^3 + 195x^2 - 90x + 17{,}295$. (*Source:* Based on data from the U.S. Department of Education)

---

## 5.5 Factoring Trinomials

### Objective A  Factoring Trinomials of the Form $x^2 + bx + c$

**Objectives**

**A** Factor Trinomials of the Form $x^2 + bx + c$.

**B** Factor Trinomials of the Form $ax^2 + bx + c$.
   **a.** Method 1—Trial and Check
   **b.** Method 2—Grouping

**C** Factor by Substitution.

In the previous section, we used factoring by grouping to factor four-term polynomials. In this section, we present techniques for factoring trinomials. Since $(x - 2)(x + 5) = x^2 + 3x - 10$, we say that $(x - 2)(x + 5)$ is a factored form of $x^2 + 3x - 10$. Taking a close look at how $(x - 2)$ and $(x + 5)$ are multiplied suggests a pattern for factoring trinomials of the form $x^2 + bx + c$.

The pattern for factoring is summarized next.

## Factoring a Trinomial of the Form $x^2 + bx + c$

Find two numbers whose product is $c$ and whose sum is $b$. The factored form of $x^2 + bx + c$ is

$$(x + \text{one number})(x + \text{other number})$$

**Practice 1**

Factor: $x^2 + 8x + 15$

**Example 1** Factor: $x^2 + 10x + 16$

**Solution:** We look for two integers whose product is 16 and whose sum is 10. Since our integers must have a positive product and a positive sum, we look at only positive factors of 16.

| Positive Factors of 16 | Sum of Factors |
|---|---|
| 1, 16 | $1 + 16 = 17$ |
| 4, 4 | $4 + 4 = 8$ |
| 2, 8 | $2 + 8 = 10$ Correct pair |

The correct pair of numbers is 2 and 8 because their product is 16 and their sum is 10. Thus,

$$x^2 + 10x + 16 = (x + 2)(x + 8)$$

**Check:** To check, see that $(x + 2)(x + 8) = x^2 + 10x + 16$.

■ Work Practice 1

**Practice 2**

Factor: $x^2 - 10x + 24$

**Example 2** Factor: $x^2 - 12x + 35$

**Solution:** We need to find two integers whose product is 35 and whose sum is $-12$. Since our integers must have a positive product and a negative sum, we consider only negative factors of 35.

| Negative Factors of 35 | Sum of Factors |
|---|---|
| $-1, -35$ | $-1 + (-35) = -36$ |
| $-5, -7$ | $-5 + (-7) = -12$ Correct pair |

The numbers are $-5$ and $-7$.

$$x^2 - 12x + 35 = [x + (-5)][x + (-7)]$$
$$= (x - 5)(x - 7)$$

**Check:** To check, see that $(x - 5)(x - 7) = x^2 - 12x + 35$.

■ Work Practice 2

**Practice 3**

Factor: $6x^3 + 24x^2 - 30x$

**Example 3** Factor: $5x^3 - 30x^2 - 35x$

**Solution:** First we factor out the greatest common factor, $5x$.

$$5x^3 - 30x^2 - 35x = 5x(x^2 - 6x - 7)$$

Next we factor $x^2 - 6x - 7$ by finding two numbers whose product is $-7$ and whose sum is $-6$. The numbers are 1 and $-7$.

$$5x^3 - 30x^2 - 35x = 5x(x^2 - 6x - 7)$$
$$= 5x(x + 1)(x - 7)$$

■ Work Practice 3

**Answers**

1. $(x + 5)(x + 3)$
2. $(x - 4)(x - 6)$
3. $6x(x - 1)(x + 5)$

## Helpful Hint

If the polynomial to be factored contains a common factor that is factored out, don't forget to include that common factor in the final factored form of the original polynomial.

**Example 4**  Factor:  $2n^2 - 38n + 80$

**Solution:**  The terms of this polynomial have a greatest common factor of 2, which we factor out first.

$$2n^2 - 38n + 80 = 2(n^2 - 19n + 40)$$

Next we factor $n^2 - 19n + 40$ by finding two numbers whose product is 40 and whose sum is $-19$. Both numbers must be negative since their product is positive and their sum is negative. Possibilities are

$$-1 \text{ and } -40, \quad -2 \text{ and } -20, \quad -4 \text{ and } -10, \quad -5 \text{ and } -8$$

None of the pairs has a sum of $-19$, so no further factoring with integers is possible. The factored form of $2n^2 - 38n + 80$ is

$$2n^2 - 38n + 80 = 2(n^2 - 19n + 40)$$

■ Work Practice 4

We call a polynomial such as $n^2 - 19n + 40$, which cannot be factored further, a **prime polynomial.**

## Objective B  Factoring Trinomials of the Form $ax^2 + bx + c$ ▶

Next we factor trinomials of the form $ax^2 + bx + c$, where the coefficient $a$ of $x^2$ is not 1. Don't forget that the first step in factoring any polynomial is to factor out the greatest common factor of its terms.

**Method 1—Factoring $ax^2 + bx + c$ by Trial and Check**

**Example 5**  Factor:  $2x^2 + 11x + 15$

**Solution:**  Factors of $2x^2$ are $2x$ and $x$. Let's try these factors as first terms of the binomials.

$$2x^2 + 11x + 15 = (2x + \quad)(x + \quad)$$

Next we try combinations of factors of 15 until the correct middle term, $11x$, is obtained. We will try only positive factors of 15 since the coefficient of the middle term, 11, is positive. Positive factors of 15 are 1 and 15 and 3 and 5.

Thus, the factored form of $2x^2 + 11x + 15$ is $(2x + 5)(x + 3)$.

■ Work Practice 5

**Practice 4**

Factor:  $3y^2 + 6y + 6$

**Practice 5**

Factor:  $3x^2 + 13x + 4$

**Answers**

**4.** $3(y^2 + 2y + 2)$
**5.** $(3x + 1)(x + 4)$

> ### Factoring a Trinomial of the Form $ax^2 + bx + c$
>
> **Step 1:** Write all pairs of factors of $ax^2$.
>
> **Step 2:** Write all pairs of factors of $c$, the constant term.
>
> **Step 3:** Try various combinations of these factors until the correct middle term $bx$ is found.
>
> **Step 4:** If no combination exists, the polynomial is **prime**.

**Practice 6**

Factor: $5x^2 + 13x - 6$

**Example 6** Factor: $3x^2 - x - 4$

**Solution:** Factors of $3x^2$: $3x \cdot x$

Factors of $-4$: $-1 \cdot 4$, $1 \cdot -4$  $-2 \cdot 2$,  $2 \cdot -2$

Let's try possible combinations of these factors.

$(3x - 1)(x + 4)$  $\qquad$  $(3x + 4)(x - 1)$
$\quad -1x$  $\qquad\qquad\qquad\qquad$  $4x$
$\quad \underline{12x}$  $\qquad\qquad\qquad\qquad$  $\underline{-3x}$
$\quad 11x$  Incorrect middle term  $\qquad$  $1x$  Incorrect middle term

$(3x - 4)(x + 1)$
$\quad -4x$
$\quad \underline{3x}$  Correct middle term
$\quad -1x$

Thus, $3x^2 - x - 4 = (3x - 4)(x + 1)$.

■ **Work Practice 6**

### Helpful Hint

A positive constant in a trinomial tells us to look for two numbers with the same sign. The sign of the coefficient of the middle term tells us whether the signs are both positive or both negative.

both positive → same sign ↗  $\qquad$  both negative ↓  same sign ↓

$2x^2 + 7x + 3 = (2x + 1)(x + 3)$  $\quad$  $2x^2 - 7x + 3 = (2x - 1)(x - 3)$

A negative constant in a trinomial tells us to look for two numbers with opposite signs.

opposite signs ↓  $\qquad\qquad$  opposite signs ↓

$2x^2 - 5x - 3 = (2x + 1)(x - 3)$  $\quad$  $2x^2 + 5x - 3 = (2x - 1)(x + 3)$

**Practice 7**

Factor: $24x^2y^2 - 42xy^2 + 9y^2$

**Answers**

**6.** $(x + 3)(5x - 2)$

**7.** $3y^2(2x - 3)(4x - 1)$

**Example 7** Factor: $12x^3y - 22x^2y + 8xy$

**Solution:** First we factor out the greatest common factor of the terms of this trinomial, $2xy$.

$12x^3y - 22x^2y + 8xy = 2xy(6x^2 - 11x + 4)$

Now we try to factor the trinomial $6x^2 - 11x + 4$.

Factors of $6x^2$: $\quad 2x \cdot 3x, \quad 6x \cdot x$

Let's try $2x$ and $3x$.

$$2xy(6x^2 - 11x + 4) = 2xy(2x + \quad)(3x + \quad)$$

The constant term, 4, is positive and the coefficient of the middle term, $-11$, is negative, so we factor 4 into negative factors only.

Negative factors of 4: $\quad -4(-1), \quad -2(-2)$

Let's try $-4$ and $-1$.

$$2xy(2x - 4)(3x - 1)$$

$$-12x$$
$$-2x$$
$$\overline{-14x} \quad \text{Incorrect middle term}$$

This combination cannot be correct because one of the factors, $(2x - 4)$, has a common factor of 2. This cannot happen if the polynomial $6x^2 - 11x + 4$ has no common factors.

Now let's try $-1$ and $-4$.

$$2xy(2x - 1)(3x - 4)$$

$$-3x$$
$$-8x$$
$$\overline{-11x} \quad \text{Correct middle term}$$

Thus,

$$12x^3y - 22x^2y + 8xy = 2xy(2x - 1)(3x - 4)$$

If this combination had not worked, we would try $-2$ and $-2$ as factors of 4 and then $6x$ and $x$ as factors of $6x^2$.

▨ **Work Practice 7**

> **Helpful Hint** If a trinomial has no common factor (other than 1), then none of its binomial factors will contain a common factor (other than 1).

---

**Example 8** Factor: $16x^2 + 24xy + 9y^2$

**Solution:** No greatest common factor can be factored out of this trinomial.

Factors of $16x^2$: $\quad 16x \cdot x, \quad 8x \cdot 2x, \quad 4x \cdot 4x$
Factors of $9y^2$: $\quad y \cdot 9y, \quad 3y \cdot 3y$

We try possible combinations until the correct factorization is found.

$$16x^2 + 24xy + 9y^2 = (4x + 3y)(4x + 3y) \quad \text{or} \quad (4x + 3y)^2$$

▨ **Work Practice 8**

**Practice 8**

Factor: $4x^2 + 28xy + 49y^2$

---

The trinomial $16x^2 + 24xy + 9y^2$ in Example 8 is an example of a **perfect square trinomial** since its factors are two identical binomials. In the next section, we examine a special method for factoring perfect square trinomials.

## Method 2—Factoring $ax^2 + bx + c$ by Grouping

There is another method we can use when factoring trinomials of the form $ax^2 + bx + c$: Write the trinomial as a four-term polynomial, and then factor by grouping.

**Answer**
**8.** $(2x + 7y)^2$

> ### Factoring a Trinomial of the Form $ax^2 + bx + c$ by Grouping
>
> **Step 1:** Find two numbers whose product is $a \cdot c$ and whose sum is $b$.
>
> **Step 2:** Write the term $bx$ as a sum by using the factors found in Step 1.
>
> **Step 3:** Factor by grouping.

**Practice 9**

Factor: $12x^2 + 11x + 2$

**Example 9**    Factor:  $6x^2 + 13x + 6$

**Solution:**   In this trinomial, $a = 6, b = 13,$ and $c = 6$.

**Step 1:** Find two numbers whose product is $a \cdot c$, or $6 \cdot 6 = 36$, and whose sum is $b$, 13. The two numbers are 4 and 9.

**Step 2:** Write the middle term $13x$ as the sum $4x + 9x$.

$$6x^2 + 13x + 6 = 6x^2 + 4x + 9x + 6$$

**Step 3:** Factor $6x^2 + 4x + 9x + 6$ by grouping.

$$(6x^2 + 4x) + (9x + 6) = 2x(3x + 2) + 3(3x + 2)$$
$$= (3x + 2)(2x + 3)$$

■ Work Practice 9

✓**Concept Check** Name one way that a factorization can be checked.

**Practice 10**

Factor: $14x^2 - x - 3$

**Example 10**    Factor:  $18x^2 - 9x - 2$

**Solution:**   In this trinomial, $a = 18, b = -9,$ and $c = -2$.

**Step 1:** Find two numbers whose product is $a \cdot c$ or $18(-2) = -36$ and whose sum is $b$, $-9$. The two numbers are $-12$ and 3.

**Step 2:** Write the middle term, $-9x$, as the sum $-12x + 3x$.

$$18x^2 - 9x - 2 = 18x^2 - 12x + 3x - 2$$

**Step 3:** Factor by grouping.

$$(18x^2 - 12x) + (3x - 2) = 6x(3x - 2) + 1(3x - 2)$$
$$= (3x - 2)(6x + 1)$$

■ Work Practice 10

## Objective C  Factoring by Substitution ▶

A complicated-looking polynomial may be a simpler trinomial "in disguise." Revealing the simpler trinomial is possible by substitution.

**Practice 11**

Factor: $3(z + 2)^2$
$\quad\quad - 19(z + 2) + 6$

**Example 11**    Factor:  $2(a + 3)^2 - 5(a + 3) - 7$

**Solution:**   The quantity $(a + 3)$ is in two of the terms of this polynomial. If we *substitute* $x$ for $(a + 3)$, the result is the following simpler trinomial.

$$2(a + 3)^2 - 5(a + 3) - 7 \quad \text{Original trinomial}$$
$$= \quad 2(x)^2 \quad - \quad 5(x) \quad - 7 \quad \text{Substitute } x \text{ for } (a + 3).$$

Now we can factor $2x^2 - 5x - 7$.

$$2x^2 - 5x - 7 = (2x - 7)(x + 1)$$

**Answers**

**9.** $(3x + 2)(4x + 1)$

**10.** $(7x + 3)(2x - 1)$

**11.** $(3z + 5)(z - 4)$

✓**Concept Check Answer**

Answers may vary. A sample is: By multiplying the factors to see that the product is the original polynomial.

But the quantity in the original polynomial was $(a + 3)$, not $x$. Thus we need to reverse the substitution and replace $x$ with $(a + 3)$.

$$(2x - 7)(x + 1) \qquad \text{Factored expression}$$
$$= [2(a + 3) - 7][(a + 3) + 1] \qquad \text{Substitute } (a + 3) \text{ for } x.$$
$$= (2a + 6 - 7)(a + 3 + 1) \qquad \text{Remove inside parentheses.}$$
$$= (2a - 1)(a + 4) \qquad \text{Simplify.}$$

Thus, $2(a + 3)^2 - 5(a + 3) - 7 = (2a - 1)(a + 4)$.

■ **Work Practice 11**

---

**Example 12**   Factor:   $5x^4 + 29x^2 - 42$

**Solution:**   Again, substitution may help us factor this polynomial more easily. We will let $y = x^2$, so $y^2 = (x^2)^2$, or $x^4$. Then

$$5x^4 + 29x^2 - 42$$

becomes $\quad \downarrow \qquad \downarrow$

$$5y^2 + 29y - 42$$

which factors as

$$5y^2 + 29y - 42 = (5y - 6)(y + 7)$$

Now we replace $y$ with $x^2$ to get

$$(5x^2 - 6)(x^2 + 7)$$

■ **Work Practice 12**

**Practice 12**

Factor:   $14x^4 + 23x^2 + 3$

**Answer**

**12.** $(2x^2 + 3)(7x^2 + 1)$

---

## Vocabulary, Readiness & Video Check

*Solve each exercise by guessing and checking.*

**1.** Find two numbers whose product is 10 and whose sum is 7.

**2.** Find two numbers whose product is 12 and whose sum is 8.

**3.** Find two numbers whose product is 24 and whose sum is 11.

**4.** Find two numbers whose product is 30 and whose sum is 13.

*Martin-Gay Interactive Videos*

*See Video 5.5*

Watch the section lecture video and answer the following questions.

**Objective A**  **5.** In ▦ Example 2, we know one of the factors of $-24$ is negative. How do we determine which one? ▶

**6.** In ▦ Example 3, once the GCF is factored out how do we know both factors of $+8$ are negative? ▶

**Objective B**  **7.** From ▦ Example 4, explain in your own words how to factor a trinomial with a first-term coefficient $\neq 1$ by trial and error. ▶

**8.** From ▦ Example 6 and the lecture before, why does writing the middle term of your trinomial as the sum of two terms suggest we'd try to factor by grouping? ▶

**Objective C**  **9.** In ▦ Example 8, the goal is to factor the polynomial, so why are we not done once we have substituted, then factored? ▶

## 5.5 Exercise Set MyMathLab® ▶

**Objective A** *Factor each trinomial. See Examples 1 through 4.*

**▶ 1.** $x^2 + 9x + 18$

**2.** $x^2 + 9x + 20$

**3.** $x^2 - 12x + 32$

**4.** $x^2 - 12x + 27$

**5.** $x^2 + 10x - 24$

**6.** $x^2 + 3x - 54$

**▶ 7.** $x^2 - 2x - 24$

**8.** $x^2 - 9x - 36$

**▶ 9.** $3x^2 - 18x + 24$

**10.** $x^2y^2 + 4xy^2 + 3y^2$

**11.** $4x^2z + 28xz + 40z$

**12.** $5x^2 - 45x + 70$

**13.** $2x^2 - 24x - 64$

**14.** $3n^2 - 6n - 51$

**Objective B** *Factor each trinomial. See Examples 5 through 10.*

**15.** $5x^2 + 16x + 3$

**16.** $3x^2 + 8x + 4$

**17.** $2x^2 - 11x + 12$

**18.** $3x^2 - 19x + 20$

**19.** $2x^2 + 25x - 20$

**20.** $6x^2 + 13x + 8$

**21.** $4x^2 - 12x + 9$

**22.** $25x^2 - 30x + 9$

**23.** $12x^2 + 10x - 50$

**24.** $12y^2 - 48y + 45$

**25.** $3y^4 - y^3 - 10y^2$

**26.** $2x^2z + 5xz - 12z$

**27.** $6x^3 + 8x^2 + 24x$

**28.** $18y^3 + 6y^2 + 2y$

**29.** $2x^2 - 5xy - 3y^2$

**30.** $6x^2 + 11xy + 4y^2$

**31.** $28y^2 + 22y + 4$

**32.** $24y^3 - 2y^2 - y$

**33.** $2x^2 + 15x - 27$

**34.** $3x^2 + 14x + 15$

**Objective C** *Use substitution to factor each polynomial completely. See Examples 11 and 12.*

**35.** $x^4 + x^2 - 6$

**36.** $x^4 - x^2 - 20$

**37.** $(5x + 1)^2 + 8(5x + 1) + 7$

**38.** $(3x - 1)^2 + 5(3x - 1) + 6$

**39.** $x^6 - 7x^3 + 12$

**40.** $x^6 - 4x^3 - 12$

**41.** $(a + 5)^2 - 5(a + 5) - 24$

**42.** $(3c + 6)^2 + 12(3c + 6) - 28$

**Objectives A B C** **Mixed Practice** *Factor each polynomial completely. See Examples 1 through 12.*

**43.** $x^2 - 24x - 81$

**44.** $x^2 - 48x - 100$

**45.** $x^2 - 15x - 54$

**46.** $x^2 - 15x + 54$

**47.** $3x^2 - 6x + 3$

**48.** $8x^2 - 8x + 2$

**▶ 49.** $3x^2 - 5x - 2$

**50.** $5x^2 - 14x - 3$

**51.** $8x^2 - 26x + 15$  **52.** $12x^2 - 17x + 6$  ◐ **53.** $18x^4 + 21x^3 + 6x^2$  **54.** $20x^5 + 54x^4 + 10x^3$

**55.** $x^2 + 8xz + 7z^2$  **56.** $a^2 - 2ab - 15b^2$  **57.** $x^2 - x - 12$  **58.** $x^2 + 4x - 5$

**59.** $3a^2 + 12ab + 12b^2$  **60.** $2x^2 + 16xy + 32y^2$  **61.** $x^2 + 4x + 5$  **62.** $x^2 + 5x + 8$

**63.** $2(x + 4)^2 + 3(x + 4) - 5$  **64.** $3(x + 3)^2 + 2(x + 3) - 5$  **65.** $6x^2 - 49x + 30$

**66.** $4x^2 - 39x + 27$  **67.** $x^4 - 5x^2 - 6$  **68.** $x^4 - 5x^2 + 6$

**69.** $6x^3 - x^2 - x$  **70.** $12x^3 + x^2 - x$  **71.** $12a^2 - 29ab + 15b^2$

**72.** $16y^2 + 6yx - 27x^2$  **73.** $9x^2 + 30x + 25$  **74.** $4x^2 + 12x + 9$

**75.** $3x^2y - 11xy + 8y$  **76.** $5xy^2 - 9xy + 4x$  **77.** $2x^2 + 2x - 12$

**78.** $3x^2 + 6x - 45$  ◐ **79.** $(x - 4)^2 + 3(x - 4) - 18$  **80.** $(x - 3)^2 - 2(x - 3) - 8$

**81.** $2x^6 + 3x^3 - 9$  **82.** $3x^6 - 14x^3 + 8$  **83.** $72xy^4 - 24xy^2z + 2xz^2$

**84.** $36xy^2 - 48xyz^2 + 16xz^4$  **85.** $2x^3y + 2x^2y - 12xy$  **86.** $3x^2y^3 + 6x^2y^2 - 45x^2y$

**87.** $x^2 + 6xy + 5y^2$  **88.** $x^2 + 6xy + 8y^2$

## Review

*Multiply. See Section 5.2.*

**89.** $(x - 3)(x + 3)$  **90.** $(x - 4)(x + 4)$  **91.** $(2x + 1)^2$

**92.** $(3x + 5)^2$  **93.** $(x - 2)(x^2 + 2x + 4)$  **94.** $(y + 1)(y^2 - y + 1)$

## Concept Extensions

*Solve.*

**95.** Find all positive and negative integers $b$ such that $x^2 + bx + 6$ is factorable.

**96.** Find all positive and negative integers $b$ such that $x^2 + bx - 10$ is factorable.

△ **97.** The volume $V(x)$ of a box in terms of its height $x$ is given by the function $V(x) = x^3 + 2x^2 - 8x$. Factor this expression for $V(x)$.

△ **98.** Based on your results from Exercise 97, find the length and width of the box if the height is 5 inches and the dimensions of the box are whole numbers.

**99.** Suppose that a movie is being filmed in New York City. An action shot requires an object to be thrown upward with an initial velocity of 80 feet per second off the top of 1 Madison Square Plaza, a height of 576 feet. The height $h(t)$ in feet of the object after $t$ seconds is given by the function $h(t) = -16t^2 + 80t + 576$. (*Source: The World Almanac,* 2001)

**a.** Find the height of the object at $t = 0$ seconds, $t = 2$ seconds, $t = 4$ seconds, and $t = 6$ seconds.

**b.** Explain why the height of the object increases and then decreases as time passes.

**c.** Factor the polynomial $-16t^2 + 80t + 576$.

576 ft

**100.** Suppose that an object is thrown upward with an initial velocity of 64 feet per second off the edge of a 960-foot-cliff. The height $h(t)$ in feet of the object after $t$ seconds is given by the function

$$h(t) = -16t^2 + 64t + 960$$

**a.** Find the height of the object at $t = 0$ seconds, $t = 3$ seconds, $t = 6$ seconds, and $t = 9$ seconds.

**b.** Explain why the height of the object increases and then decreases as time passes.

**c.** Factor the polynomial $-16t^2 + 64t + 960$.

*Factor. Assume that variables used as exponents represent positive integers.*

**101.** $x^{2n} + 10x^n + 16$  **102.** $x^{2n} - 7x^n + 12$  **103.** $x^{2n} - 3x^n - 18$  **104.** $x^{2n} + 7x^n - 18$

**105.** $2x^{2n} + 11x^n + 5$  **106.** $3x^{2n} - 8x^n + 4$  **107.** $4x^{2n} - 12x^n + 9$  **108.** $9x^{2n} + 24x^n + 16$

*Recall that a graphing calculator may be used to check addition, subtraction, and multiplication of polynomials. In the same manner, a graphing calculator may be used to check factoring of polynomials in one variable. For example, to see that*

$$2x^3 - 9x^2 - 5x = x(2x + 1)(x - 5)$$

*graph* $Y_1 = 2x^3 - 9x^2 - 5x$ *and* $Y_2 = x(2x + 1)(x - 5)$. *Then trace along both graphs to see that they coincide. Factor the following and use this method to check your results.*

**109.** $x^4 + 6x^3 + 5x^2$  **110.** $x^3 + 6x^2 + 8x$  **111.** $30x^3 + 9x^2 - 3x$  **112.** $-6x^4 + 10x^3 - 4x^2$

## 5.6 Factoring by Special Products ▶

### Objective A Factoring Perfect Square Trinomials ▶

In the previous section, we considered a variety of ways to factor trinomials of the form $ax^2 + bx + c$. In Example 8, we factored $16x^2 + 24xy + 9y^2$ as

$$16x^2 + 24xy + 9y^2 = (4x + 3y)^2$$

Recall that we called $16x^2 + 24xy + 9y^2$ a perfect square trinomial because its factors are two identical binomials. A trinomial is a perfect square trinomial if it can be written so that its first term is the square of some quantity $a$, its last term is the square of some quantity $b$, and its middle term is twice the product of the quantities $a$ and $b$.

The following special formulas can be used to factor perfect square trinomials.

**Perfect Square Trinomials**

$$a^2 + 2ab + b^2 = (a + b)^2$$
$$a^2 - 2ab + b^2 = (a - b)^2$$

Notice that these equations are the same special products from Section 5.2 for the square of a binomial.

From

$$a^2 + 2ab + b^2 = (a + b)^2$$

we see that

$$16x^2 + 24xy + 9y^2 = (4x)^2 + 2(4x)(3y) + (3y)^2 = (4x + 3y)^2$$

**Example 1** Factor: $m^2 + 10m + 25$

**Solution:** Notice that the first term is a square: $m^2 = (m)^2$, the last term is a square: $25 = 5^2$, and $10m = 2 \cdot 5 \cdot m$.

This is a perfect square trinomial. Thus,

$$m^2 + 10m + 25 = m^2 + 2(m)(5) + 5^2 = (m + 5)^2$$

■ Work Practice 1

**Examples** Factor each trinomial.

**2.** $4x^2 + 4x + 1 = (2x)^2 + 2 \cdot 2x \cdot 1 + 1^2$   See whether it is a perfect square trinomial.
$$= (2x + 1)^2$$   Factor.

**3.** $9x^2 - 12x + 4 = (3x)^2 - 2(3x)(2) + 2^2$   See whether it is a perfect square trinomial.
$$= (3x - 2)^2$$   Factor.

■ Work Practice 2–3

---

**Objectives**

**A** Factor a Perfect Squa[re] Trinomial. ▶

**B** Factor the Difference of Two Squares. ▶

**C** Factor the Sum or Difference of Two Cubes. ▶

**Practice 1**

Factor: $x^2 + 8x + 16$

**Practice 2–3**

Factor.

**2.** $9x^2 + 6x + 1$

**3.** $25x^2 - 20x + 4$

**Answers**

**1.** $(x + 4)^2$   **2.** $(3x + 1)^2$
**3.** $(5x - 2)^2$

$32x^2y + 64xy^2$

**Example 4** Factor: $3a^2x - 12abx + 12b^2x$

**Solution:** The terms of this trinomial have a greatest common factor of $3x$, which we factor out first.

$$3a^2x - 12abx + 12b^2x = 3x(a^2 - 4ab + 4b^2)$$

The polynomial $a^2 - 4ab + 4b^2$ is a perfect square trinomial. Notice that the first term is a square: $a^2 = (a)^2$, the last term is a square: $4b^2 = (2b)^2$, and $4ab = 2(a)(2b)$. The factoring can now be completed as

$$3x(a^2 - 4ab + 4b^2) = 3x(a - 2b)^2$$

■ Work Practice 4

**Helpful Hint**

If you recognize a trinomial as a perfect square trinomial, use the special formulas to factor. However, methods for factoring trinomials in general from Section 5.5 will also result in the correct factored form.

## Objective B Factoring the Difference of Two Squares ▶

We now factor special types of binomials, beginning with the **difference of two squares.** The special product pattern presented in Section 5.2 for the product of a sum and a difference of two terms is used again here. However, the emphasis is now on factoring rather than on multiplying.

**Difference of Two Squares**

$$a^2 - b^2 = (a + b)(a - b)$$

Notice that a binomial is a difference of two squares when it is the difference of the square of some quantity $a$ and the square of some quantity $b$.

**Practice 5–8**

Factor:

**5.** $x^2 - 49$   **6.** $4y^2 - 81$

**7.** $12 - 3a^2$   **8.** $y^2 - \dfrac{1}{25}$

**Examples** Factor.

**5.** $x^2 - 9 = x^2 - 3^2$
$$= (x + 3)(x - 3)$$

**6.** $16y^2 - 9 = (4y)^2 - 3^2$
$$= (4y + 3)(4y - 3)$$

**7.** $50 - 8y^2 = 2(25 - 4y^2)$     Factor out the common factor of 2.
$$= 2[5^2 - (2y)^2]$$
$$= 2(5 + 2y)(5 - 2y)$$

**8.** $x^2 - \dfrac{1}{4} = x^2 - \left(\dfrac{1}{2}\right)^2$
$$= \left(x + \dfrac{1}{2}\right)\left(x - \dfrac{1}{2}\right)$$

**Helpful Hint** The sum of two squares whose greatest common factor is 1 usually cannot be factored by using real numbers.

■ Work Practice 5–8

The binomial $x^2 + 9$ is a **sum of two squares** and cannot be factored by using real numbers. *In general, except for factoring out a greatest common factor, the sum of two squares usually cannot be factored by using real numbers.*

**Answers**

**4.** $4x(x - 4y)^2$   **5.** $(x + 7)(x - 7)$

**6.** $(2y + 9)(2y - 9)$

**7.** $3(2 + a)(2 - a)$

**8.** $\left(y + \dfrac{1}{5}\right)\left(y - \dfrac{1}{5}\right)$

**Example 9** Factor: $p^4 - 16$

**Solution:**

$$p^4 - 16 = (p^2)^2 - 4^2$$
$$= (p^2 + 4)(p^2 - 4)$$

The binomial factor $p^2 + 4$ cannot be factored by using real numbers, but the binomial factor $p^2 - 4$ is a difference of squares.

$$(p^2 + 4)(p^2 - 4) = (p^2 + 4)(p + 2)(p - 2)$$

■ Work Practice 9

Practice 9

Factor: $a^4$ 394

✓**Concept Check** Is $(x - 4)(y^2 - 9)$ completely factored? Why or why not?

**Example 10** Factor: $(x + 3)^2 - 36$

**Solution:**

$$(x + 3)^2 - 36 = (x + 3)^2 - 6^2$$
$$= [(x + 3) + 6][(x + 3) - 6] \quad \text{Factor as the difference of two squares.}$$
$$= [x + 3 + 6][x + 3 - 6] \quad \text{Remove parentheses.}$$
$$= (x + 9)(x - 3) \quad \text{Simplify.}$$

■ Work Practice 10

**Practice 10**

Factor: $(x + 1)^2 - 9$

**Example 11** Factor: $x^2 + 4x + 4 - y^2$

**Solution:** Notice that the first three terms form a perfect square trinomial. Let's try factoring by grouping. To do so, let's group and factor the first three terms.

$$x^2 + 4x + 4 - y^2 = (x^2 + 4x + 4) - y^2 \quad \text{Group the first three terms.}$$
$$= (x + 2)^2 - y^2 \quad \text{Factor the perfect square trinomial.}$$

This is not completely factored yet since we have a *difference,* not a *product.* Since $(x + 2)^2 - y^2$ is a difference of squares, we have

$$(x + 2)^2 - y^2 = [(x + 2) + y][(x + 2) - y]$$
$$= (x + 2 + y)(x + 2 - y)$$

■ Work Practice 11

**Practice 11**

Factor: $a^2 + 2a + 1 - b^2$

## Objective C Factoring the Sum or Difference of Two Cubes ▶

Although the sum of two squares usually cannot be factored, the sum of two cubes, as well as the difference of two cubes, can be factored as follows.

### Sum and Difference of Two Cubes

$$a^3 + b^3 = (a + b)(a^2 - ab + b^2)$$
$$a^3 - b^3 = (a - b)(a^2 + ab + b^2)$$

**Answers**

**9.** $(a^2 + 9)(a + 3)(a - 3)$
**10.** $(x - 2)(x + 4)$
**11.** $(a + 1 + b)(a + 1 - b)$

✓**Concept Check Answer**

no; $(y^2 - 9)$ can be factored

To check the first pattern, let's find the product of $(a + b)$ and $(a^2 - ab + b^2)$.

$$
\begin{aligned}
(a + b)(a^2 - ab + b^2) &= a(a^2 - ab + b^2) + b(a^2 - ab + b^2) \\
&= a^3 - a^2b + ab^2 + a^2b - ab^2 + b^3 \\
&= a^3 + b^3
\end{aligned}
$$

**Practice 12**

Factor: $z^3 + 27$

**Example 12** Factor: $x^3 + 8$

**Solution:** First we write the binomial in the form $a^3 + b^3$. Then we use the formula

$$a^3 + b^3 = (a + b)(a^2 - a \cdot b + b^2), \text{ where } a \text{ is } x \text{ and } b \text{ is } 2.$$

$$x^3 + 8 = x^3 + 2^3 = (x + 2)(x^2 - x \cdot 2 + 2^2)$$

Thus, $x^3 + 8 = (x + 2)(x^2 - 2x + 4)$

■ Work Practice 12

**Practice 13**

Factor: $x^3 + 64y^3$

**Example 13** Factor: $p^3 + 27q^3$

**Solution:**

$$
\begin{aligned}
p^3 + 27q^3 &= p^3 + (3q)^3 \\
&= (p + 3q)[p^2 - (p)(3q) + (3q)^2] \\
&= (p + 3q)(p^2 - 3pq + 9q^2)
\end{aligned}
$$

■ Work Practice 13

**Practice 14**

Factor: $y^3 - 8$

**Example 14** Factor: $y^3 - 64$

**Solution:** This is a difference of cubes since $y^3 - 64 = y^3 - 4^3$.

From $a^3 - b^3 = (a - b)(a^2 + a \cdot b + b^2)$ we have that

$$
\begin{aligned}
y^3 - 4^3 &= (y - 4)(y^2 + y \cdot 4 + 4^2) \\
&= (y - 4)(y^2 + 4y + 16)
\end{aligned}
$$

■ Work Practice 14

> **Helpful Hint**
>
> When factoring sums or differences of cubes, be sure to notice the sign patterns.
>
> Same sign
> $$x^3 + y^3 = (x + y)(x^2 - xy + y^2)$$
> Opposite sign
> Always positive
>
> Same sign
> $$x^3 - y^3 = (x - y)(x^2 + xy + y^2)$$
> Opposite sign

**Answers**

**12.** $(z + 3)(z^2 - 3z + 9)$

**13.** $(x + 4y)(x^2 - 4xy + 16y^2)$

**14.** $(y - 2)(y^2 + 2y + 4)$

**Example 15** Factor: $125q^2 - n^3q^2$

**Solution:** First we factor out a common factor of $q^2$.

$$125q^2 - n^3q^2 = q^2(125 - n^3)$$
$$= q^2(5^3 - n^3)$$

Opposite sign        Positive

$$= q^2(5 - n)[5^2 + (5)(n) + (n^2)]$$
$$= q^2(5 - n)(25 + 5n + n^2)$$

Thus, $125q^2 - n^3q^2 = q^2(5 - n)(25 + 5n + n^2)$. The trinomial $25 + 5n + n^2$ cannot be factored further.

■ Work Practice 15

**Practice 15**

Factor: $27a^2 - b^3a^2$

**Answer**

**15.** $a^2(3 - b)(9 + 3b + b^2)$

## Vocabulary, Readiness & Video Check

*Write each term as a square. For example, $25x^2$ as a square is $(5x)^2$.*

**1.** $81y^2$      **2.** $4z^2$      **3.** $64x^6$      **4.** $49y^6$

*Write each number or term as a cube.*

**5.** 125      **6.** 216      **7.** $8x^3$      **8.** $27y^3$      **9.** $64x^6$      **10.** $x^3y^6$

*Martin-Gay Interactive Videos*

See Video 5.6

*Watch the section lecture video and answer the following questions.*

**Objective A** **11.** From ▣ Example 1, what is the first step to see if you have a perfect square trinomial? How do you then finish determining that you do indeed have a perfect square trinomial? ▶

**Objective B** **12.** In ▣ Example 2, the original binomial is rewritten to write each term as a square. Give two reasons why this is helpful. ▶

**Objective C** **13.** In ▣ Examples 4 and 5, what tips are given to remember how to factor the sum or difference of two cubes rather than memorizing the formulas? ▶

## 5.6 Exercise Set MyMathLab® ▶

**Objective A** *Factor. See Examples 1 through 4.*

**1.** $x^2 + 6x + 9$      **2.** $x^2 - 10x + 25$      ▶**3.** $4x^2 - 12x + 9$      **4.** $25x^2 + 10x + 1$

**5.** $4a^2 + 12a + 9$      **6.** $9a^2 - 30a + 25$      **7.** $3x^2 - 24x + 48$      **8.** $2x^2 + 28x + 98$

**9.** $9y^2x^2 + 12yx^2 + 4x^2$      **10.** $4x^2y^3 - 4xy^3 + y^3$      **11.** $16x^2 - 56xy + 49y^2$      **12.** $81x^2 + 36xy + 4y^2$

**Objective** B  *Factor. See Examples 5 through 11.*

**13.** $x^2 - 25$

**14.** $y^2 - 100$

**15.** $\frac{1}{9} - 4z^2$

**16.** $\frac{1}{16} - y^2$

**17.** $(y + 2)^2 - 49$

**18.** $(x - 1)^2 - z^2$

**19.** $64x^2 - 100$

**20.** $4x^2 - 36$

**21.** $(x + 2y)^2 - 9$

**22.** $(3x + y)^2 - 25$

▶ **23.** $x^2 + 16x + 64 - x^4$

**24.** $x^2 + 20x + 100 - x^4$

**Objective** C  *Factor. See Examples 12 through 15.*

**25.** $x^3 + 27$

**26.** $y^3 + 1$

▶ **27.** $z^3 - 1$

**28.** $x^3 - 8$

**29.** $m^3 + n^3$

**30.** $p^3 + 125q^3$

**31.** $27y^2 - x^3y^2$

**32.** $64q^2 - q^2p^3$

**33.** $8ab^3 + 27a^4$

**34.** $a^3b + 8b^4$

**35.** $250y^3 - 16x^3$

**36.** $54y^3 - 128$

**Objectives** A B C  **Mixed Practice** *Factor completely. See Examples 1 through 15.*

**37.** $x^2 - 12x + 36$

**38.** $x^2 - 18x + 81$

**39.** $18x^2y - 2y$

**40.** $12xy^2 - 108x$

▶ **41.** $9x^2 - 49$

**42.** $25x^2 - 4$

**43.** $x^4 - 1$

**44.** $x^4 - 256$

**45.** $x^6 - y^3$

**46.** $x^3 - y^6$

**47.** $8x^3 + 27y^3$

**48.** $125x^3 + 8y^3$

**49.** $4x^2 + 4x + 1 - z^2$

**50.** $9y^2 + 12y + 4 - x^2$

▶ **51.** $3x^6y^2 + 81y^2$

**52.** $x^2y^9 + x^2y^3$

**53.** $n^3 - \dfrac{1}{27}$

**54.** $p^3 + \dfrac{1}{125}$

**55.** $-16y^2 + 64$

**56.** $-12y^2 + 108$

**57.** $x^2 - 10x + 25 - y^2$

**58.** $x^2 - 18x + 81 - y^2$

**59.** $a^3b^3 + 125$

**60.** $x^3y^3 + 216$

**61.** $\dfrac{x^2}{25} - \dfrac{y^2}{9}$

**62.** $\dfrac{a^2}{4} - \dfrac{b^2}{49}$

**63.** $(x + y)^3 + 125$

**64.** $(r + s)^3 + 27$

## Review

*Solve each equation. See Section 2.1.*

**65.** $x - 5 = 0$

**66.** $x + 7 = 0$

**67.** $3x + 1 = 0$

**68.** $5x - 15 = 0$

**69.** $-2x = 0$

**70.** $3x = 0$

**71.** $-5x + 25 = 0$

**72.** $-4x - 16 = 0$

## Concept Extensions

*Determine whether each polynomial is factored completely or not. See the Concept Check in this section.*

**73.** $5x(x^2 - 4)$

**74.** $x^2y^2(x^3 - y^3)$

**75.** $7y(a^2 + a + 1)$

**76.** $9z(x^2 + 4)$

*Solve.*

△ **77.** A manufacturer of metal washers needs to determine the cross-sectional area of each washer. If the outer radius of the washer is $R$ and the radius of the hole is $r$, express the area of the washer as a polynomial. Factor this polynomial completely.

△ **78.** Express the area of the shaded region as a polynomial. Factor the polynomial completely.

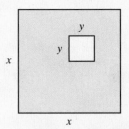

*Express the volume of each solid as a polynomial. To do so, subtract the volume of the "hole" from the volume of the larger solid. Then factor the resulting polynomial.*

△ **79.**    △ **80.**

*Find the value of c that makes each trinomial a perfect square trinomial.*

**81.** $x^2 + 6x + c$    **82.** $y^2 + 10y + c$    **83.** $m^2 - 14m + c$    **84.** $n^2 - 2n + c$

**85.** Factor $x^6 - 1$ completely, using the following methods from this chapter.
 **a.** Factor the expression by treating it as the difference of two squares, $(x^3)^2 - 1^2$.

 **b.** Factor the expression by treating it as the difference of two cubes, $(x^2)^3 - 1^3$.

 **c.** Are the answers to parts (a) and (b) the same? Why or why not?

**86.** Factor $x^{12} - 1$ completely, using the following methods from this chapter:
 **a.** Factor the expression by treating it as the difference of two squares, $(x^3)^4 - 1^4$.

 **b.** Factor the expression by treating it as the difference of two cubes, $(x^4)^3 - 1^3$.

 **c.** Are the answers to parts (a) and (b) the same? Why or why not?

*Factor. Assume that variables used as exponents represent positive integers.*

**87.** $x^{2n} - 25$    **88.** $x^{2n} - 36$    **89.** $36x^{2n} - 49$

**90.** $25x^{2n} - 81$    **91.** $x^{4n} - 16$    **92.** $x^{4n} - 625$

# Operations on Polynomials and Factoring Strategies

**Answers**

## Operations on Polynomials

*Perform each indicated operation.*

**1.** $(-y^2 + 6y - 1) + (3y^2 - 4y - 10)$    **2.** $(5z^4 - 6z^2 + z + 1) - (7z^4 - 2z + 1)$

**3.** Subtract $(x - 5)$ from $(x^2 - 6x + 2)$    **4.** $(2x^2 + 6x - 5) + (5x^2 - 10x)$

**5.** $(5x - 3)^2$    **6.** $(5x^2 - 14x - 3) \div (5x + 1)$

**7.** $(2x^4 - 3x^2 + 5x - 2) \div (x + 2)$    **8.** $(4x - 1)(x^2 - 3x - 2)$

## Factoring Strategies

The key to proficiency in factoring polynomials is to practice until you are comfortable with each technique. A strategy for factoring polynomials completely is given next.

### Factoring a Polynomial

**Step 1:** Are there any common factors? If so, factor out the greatest common factor.

**Step 2:** How many terms are in the polynomial?

    **a.** If there are *two* terms, decide if one of the following formulas may be applied:

      **i.** Difference of two squares: $a^2 - b^2 = (a + b)(a - b)$

      **ii.** Difference of two cubes: $a^3 - b^3 = (a - b)(a^2 + ab + b^2)$

      **iii.** Sum of two cubes: $a^3 + b^3 = (a + b)(a^2 - ab + b^2)$

    **b.** If there are *three* terms, try one of the following:

      **i.** Perfect square trinomial: $a^2 + 2ab + b^2 = (a + b)^2$
                              $a^2 - 2ab + b^2 = (a - b)^2$

      **ii.** If not a perfect square trinomial, factor by using the methods presented in Section 5.5.

    **c.** If there are *four* or more terms, try factoring by grouping.

**Step 3:** See whether any factors in the factored polynomial can be factored further.

*Factor completely.*

**9.** $x^2 - 8x + 16 - y^2$    **10.** $12x^2 - 22x - 20$

**11.** $x^4 - x$    **12.** $(2x + 1)^2 - 3(2x + 1) + 2$

**13.** $14x^2y - 2xy$    **14.** $24ab^2 - 6ab$

**15.** $4x^2 - 16$    **16.** $9x^2 - 81$    **17.** $3x^2 - 8x - 11$

**18.** $5x^2 - 2x - 3$    ▶**19.** $4x^2 + 8x - 12$    **20.** $6x^2 - 6x - 12$

**21.** $4x^2 + 36x + 81$    **22.** $25x^2 + 40x + 16$    **23.** $8x^3 + 125y^3$

1. _____

2. _____

3. _____

4. _____

5. _____

6. _____

7. _____

8. _____

9. _____

10. _____

11. _____

12. _____

13. _____

14. _____

15. _____

16. _____

17. _____

18. _____

19. _____

20. _____

21. _____

22. _____

23. _____

**24.** $27x^3 - 64y^3$

**25.** $64x^2y^3 - 8x^2$

**26.** $27x^5y^4 - 216x^2y$

**27.** $(x + 5)^3 + y^3$

**28.** $(y - 1)^3 + 27x^3$

▶ **29.** $(5a - 3)^2 - 6(5a - 3) + 9$

**30.** $(4r + 1)^2 + 8(4r + 1) + 16$

**31.** $7x^2 - 63x$

**32.** $20x^2 + 23x + 6$

**33.** $ab - 6a + 7b - 42$

**34.** $20x^2 - 220x + 600$

▶ **35.** $x^4 - 1$

**36.** $15x^2 - 20x$

**37.** $10x^2 - 7x - 33$

**38.** $45m^3n^3 - 27m^2n^2$

**39.** $5a^3b^3 - 50a^3b$

**40.** $x^4 + x$

**41.** $16x^2 + 25$

**42.** $20x^3 + 20y^3$

**43.** $10x^3 - 210x^2 + 1100x$

**44.** $9y^2 - 42y + 49$

**45.** $64a^3b^4 - 27a^3b$

**46.** $y^4 - 16$

▶ **47.** $2x^3 - 54$

**48.** $2sr + 10s - r - 5$

▶ **49.** $3y^5 - 5y^4 + 6y - 10$

**50.** $64a^2 + b^2$

**51.** $100z^3 + 100$

**52.** $250x^4 - 16x$

**53.** $4b^2 - 36b + 81$

**54.** $2a^5 - a^4 + 6a - 3$

**55.** $(y - 6)^2 + 3(y - 6) + 2$

**56.** $(c + 2)^2 - 6(c + 2) + 5$

△ **57.** Express the area of the shaded region as a polynomial. Factor the polynomial completely.

**Answer blanks (left margin):**

24.
25.
26.
27.
28.
29.
30.
31.
32.
33.
34.
35.
36.
37.
38.
39.
40.
41.
42.
43.
44.
45.
46.
47.
48.
49.
50.
51.
52.
53.
54.
55.
56.
57.

## 5.7 Solving Equations by Factoring and Solving Problems ▶

### Objective A Solving Polynomial Equations by Factoring ▶

In this section, your efforts to learn factoring will start to pay off. We use factoring to solve polynomial equations.

A **polynomial equation** is the result of setting two polynomials equal to each other. Examples are shown below.

$$3x^3 - 2x^2 = x^2 + 2x - 1 \qquad 2.6x + 7 = -1.3$$
$$-5x^2 - 5 = -9x^2 - 2x + 1$$

A polynomial equation is in **standard form** if one side of the equation is 0, as in the following examples.

$$3x^3 - 3x^2 - 2x + 1 = 0 \qquad 2.6x + 8.3 = 0$$
$$4x^2 + 2x - 6 = 0$$

The degree of a simplified polynomial equation in standard form is the same as the highest degree of any of its terms. A polynomial equation of degree 2 is also called a **quadratic equation.**

A solution of a polynomial equation in one variable is a value of the variable that makes the equation true. The method presented in this section for solving polynomial equations is called the **factoring method.** This method is based on the **zero-factor property.**

---

### Zero-Factor Property

If $a$ and $b$ are real numbers and $a \cdot b = 0$, then $a = 0$ or $b = 0$. This property is true for three or more factors also.

---

In other words, if the product of two or more real numbers is zero, then at least one of the numbers must be zero.

**Example 1** Solve: $(x + 2)(x - 6) = 0$

**Solution:** By the zero-factor property, $(x + 2)(x - 6) = 0$ only if

$$x + 2 = 0 \quad \text{or} \quad x - 6 = 0.$$
$$x + 2 = 0 \quad \text{or} \quad x - 6 = 0 \quad \text{Use the zero-factor property.}$$
$$x = -2 \qquad\qquad x = 6 \quad \text{Solve each linear equation.}$$

To check, let $x = -2$ and then let $x = 6$ in the original equation.

| Let $x = -2$. | Let $x = 6$. |
|---|---|
| $(x + 2)(x - 6) = 0$ | $(x + 2)(x - 6) = 0$ |
| $(-2 + 2)(-2 - 6) = 0$ | $(6 + 2)(6 - 6) = 0$ |
| $(0)(-8) = 0$ | $(8)(0) = 0$ |
| $0 = 0$   True | $0 = 0$   True |

Both $-2$ and 6 check, so the solution set is $\{-2, 6\}$.

■ Work Practice 1

### Objectives

**A** Solve Polynomial Equations by Factoring. ▶

**B** Solve Problems That Can Be Modeled by Polynomial Equations. ▶

**C** Find the x-Intercepts of a Polynomial Function. ▶

**Practice 1**

Solve: $(x - 3)(x + 5) = 0$

**Answer**
**1.** $\{-5, 3\}$

**Practice 2**

Solve: $3x^2 + 5x - 2 = 0$

**Example 2** Solve: $2x^2 + 9x - 5 = 0$

**Solution:** To use the zero-factor property, one side of the equation must be 0, and the other side must be in factored form.

$$2x^2 + 9x - 5 = 0$$
$$(2x - 1)(x + 5) = 0 \qquad \text{Factor.}$$
$$2x - 1 = 0 \quad \text{or} \quad x + 5 = 0 \qquad \text{Set each factor equal to 0.}$$
$$2x = 1 \qquad\qquad x = -5 \qquad \text{Solve each linear equation.}$$
$$x = \frac{1}{2}$$

To check, let $x = \frac{1}{2}$ in the original equation; then let $x = -5$ in the original equation. The solution set is $\left\{ -5, \frac{1}{2} \right\}$.

■ Work Practice 2

---

### Solving a Polynomial Equation by Factoring

**Step 1:** Write the equation in standard form so that one side of the equation is 0.

**Step 2:** Factor the polynomial completely.

**Step 3:** Set each factor containing a variable equal to 0.

**Step 4:** Solve the resulting equations.

**Step 5:** Check each solution in the original equation.

---

Since it is not always possible to factor a polynomial, not all polynomial equations can be solved by factoring. Other methods of solving polynomial equations are presented in Chapter 8.

**Practice 3**

Solve: $x(5x - 7) = -2$

**Example 3** Solve: $x(2x - 7) = 4$

**Solution:** We first write the equation in standard form; then we factor.

$$x(2x - 7) = 4$$
$$2x^2 - 7x = 4 \qquad \text{Multiply.}$$
$$2x^2 - 7x - 4 = 0 \qquad \text{Write in standard form.}$$
$$(2x + 1)(x - 4) = 0 \qquad \text{Factor.}$$
$$2x + 1 = 0 \quad \text{or} \quad x - 4 = 0 \qquad \text{Set each factor equal to 0.}$$
$$2x = -1 \qquad\qquad x = 4 \qquad \text{Solve.}$$
$$x = -\frac{1}{2}$$

Check both solutions in the original equation. The solution set is $\left\{ -\frac{1}{2}, 4 \right\}$.

■ Work Practice 3

**Helpful Hint**

To apply the zero-factor property, one side of the equation must be 0, and the other side of the equation must be factored. To solve the equation $x(2x - 7) = 4$, for example, you may *not* set each factor equal to 4.

**Answers**

**2.** $\left\{ -2, \frac{1}{3} \right\}$ **3.** $\left\{ \frac{2}{5}, 1 \right\}$

**Example 4** Solve: $3(x^2 + 4) + 5 = -6(x^2 + 2x) + 13$

**Solution:** We rewrite the equation so that one side is 0.

$$3(x^2 + 4) + 5 = -6(x^2 + 2x) + 13$$
$$3x^2 + 12 + 5 = -6x^2 - 12x + 13 \qquad \text{Use the distributive property.}$$
$$9x^2 + 12x + 4 = 0 \qquad \begin{array}{l}\text{Rewrite the equation in standard} \\ \text{form so that one side is 0.}\end{array}$$
$$(3x + 2)(3x + 2) = 0 \qquad \text{Factor.}$$
$$3x + 2 = 0 \quad \text{or} \quad 3x + 2 = 0 \qquad \text{Set each factor equal to 0.}$$
$$3x = -2 \qquad\qquad 3x = -2 \qquad \text{Solve each equation.}$$
$$x = -\frac{2}{3} \qquad\qquad x = -\frac{2}{3}$$

Check by substituting $-\frac{2}{3}$ into the original equation. The solution set is $\left\{-\frac{2}{3}\right\}$.

■ **Work Practice 4**

If the equation contains fractions, we clear the equation of fractions as a first step.

**Example 5** Solve: $2x^2 = \frac{17}{3}x + 1$

**Solution:**

$$2x^2 = \frac{17}{3}x + 1$$
$$3(2x^2) = 3\left(\frac{17}{3}x + 1\right) \qquad \text{Clear the equation of fractions.}$$
$$6x^2 = 17x + 3 \qquad \text{Use the distributive property.}$$
$$6x^2 - 17x - 3 = 0 \qquad \begin{array}{l}\text{Rewrite the equation in standard form} \\ \text{so that one side is 0.}\end{array}$$
$$(6x + 1)(x - 3) = 0 \qquad \text{Factor.}$$
$$6x + 1 = 0 \quad \text{or} \quad x - 3 = 0 \qquad \text{Set each factor equal to 0.}$$
$$6x = -1 \qquad\qquad x = 3 \qquad \text{Solve each equation.}$$
$$x = -\frac{1}{6}$$

Check by substituting into the original equation. The solution set is $\left\{-\frac{1}{6}, 3\right\}$.

■ **Work Practice 5**

**Example 6** Solve: $x^3 = 4x$

**Solution:**

$$x^3 = 4x$$
$$x^3 - 4x = 0 \qquad \begin{array}{l}\text{Rewrite the equation in standard form} \\ \text{so that one side is 0.}\end{array}$$
$$x(x^2 - 4) = 0 \qquad \text{Factor out the greatest common factor.}$$
$$x(x + 2)(x - 2) = 0 \qquad \text{Factor the difference of squares.}$$
$$x = 0 \quad \text{or} \quad x + 2 = 0 \quad \text{or} \quad x - 2 = 0 \qquad \text{Set each factor equal to 0.}$$
$$x = -2 \qquad\qquad x = 2 \qquad \text{Solve each equation.}$$

Check by substituting into the original equation. The solution set is $\{-2, 0, 2\}$.

■ **Work Practice 6**

Notice that the *third*-degree equation of Example 6 yielded *three* solutions.

**Practice 4**

Solve:

$$2(x^2 + 5) + 10$$
$$= -2(x^2 + 10x) - 5$$

**Practice 5**

Solve: $2x^2 + \frac{5}{2}x = 3$

**Practice 6**

Solve: $x^3 = x^2 + 6x$

**Answers**

**4.** $\left\{-\frac{5}{2}\right\}$  **5.** $\left\{-2, \frac{3}{4}\right\}$  **6.** $\{-2, 0, 3\}$

## Practice 7

Solve: $x^3 + 7x^2 = 4x + 28$

**Example 7** Solve: $x^3 - x = -5x^2 + 5$.

**Solution:** First, write the equation so that one side is 0.

$$x^3 - x + 5x^2 - 5 = 0$$
$$(x^3 - x) + (5x^2 - 5) = 0 \qquad \text{Factor by grouping.}$$
$$x(x^2 - 1) + 5(x^2 - 1) = 0$$
$$(x^2 - 1)(x + 5) = 0$$
$$(x + 1)(x - 1)(x + 5) = 0 \qquad \text{Factor the difference of squares.}$$
$$x + 1 = 0 \quad \text{or} \quad x - 1 = 0 \quad \text{or} \quad x + 5 = 0 \qquad \text{Set each factor equal to 0.}$$
$$x = -1 \quad \text{or} \qquad x = 1 \quad \text{or} \qquad x = -5 \quad \text{Solve each equation.}$$

The solution set is $\{-5, -1, 1\}$. Check in the original equation.

■ Work Practice 7

✓ **Concept Check** Which solution strategies are incorrect? Why?

**a.** Solve $(y - 2)(y + 2) = 4$ by setting each factor equal to 4.

**b.** Solve $(x + 1)(x + 3) = 0$ by setting each factor equal to 0.

**c.** Solve $z^2 + 5z + 6 = 0$ by factoring $z^2 + 5z + 6$ and setting each factor equal to 0.

**d.** Solve $x^2 + 6x + 8 = 10$ by factoring $x^2 + 6x + 8$ and setting each factor equal to 0.

## Objective B Solving Problems Modeled by Polynomial Equations ▶

Some problems may be modeled by polynomial equations. To solve these problems, we use the same problem-solving steps that were introduced in Section 2.2. When solving these problems, keep in mind that a solution of an equation that models a problem is not always a solution to the problem. For example, a person's weight or the length of a side of a geometric figure is always a positive number. Discard solutions that do not make sense as solutions of the problem.

## Practice 8

A model rocket is launched from the ground. Its height $h$ in feet at time $t$ seconds is approximated by the equation

$$h = -16t^2 + 112t$$

Find how long it takes the rocket to return to the ground.

**Example 8** Finding the Return Time of a Rocket

An Alpha III model rocket is launched from the ground with an A8-3 engine. Without a parachute the height $h$ in feet of the rocket at time $t$ seconds is approximated by the equation

$$h = -16t^2 + 144t$$

Find how long it takes the rocket to return to the ground.

**Solution:**

**1. UNDERSTAND.** Read and reread the problem. The equation $h = -16t^2 + 144t$ models the height of the rocket. Familiarize yourself with this equation by finding a few values.

When $t = 1$ second, the height of the rocket is

$$h = -16(1)^2 + 144(1) = 128 \text{ feet}$$

When $t = 2$ seconds, the height of the rocket is

$$h = -16(2)^2 + 144(2) = 224 \text{ feet}$$

**Answers**

**7.** $\{-7, -2, 2\}$   **8.** 7 sec

✓ **Concept Check Answer**

**a** and **d**; the zero-factor property works only if one side of the equation is 0

**2. TRANSLATE.** To find how long it takes the rocket to return to the ground, we want to know what value of $t$ makes the height $h$ equal to 0. That is, we want to solve for $t$ when $h = 0$.

$$-16t^2 + 144t = 0$$

**3. SOLVE** the quadratic equation by factoring.

$$-16t^2 + 144t = 0$$
$$-16t(t - 9) = 0$$
$$-16t = 0 \quad \text{or} \quad t - 9 = 0$$
$$t = 0 \qquad\qquad t = 9$$

**4. INTERPRET.** The height $h$ is 0 feet at time 0 seconds (when the rocket is launched) and at time 9 seconds.

**Check:** See that the height of the rocket at 9 seconds equals 0.

$$h = -16(9)^2 + 144(9) = -1296 + 1296 = 0$$

**State:** The rocket returns to the ground 9 seconds after it is launched.

■ Work Practice 8

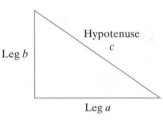

Some of the exercises at the end of this section make use of the **Pythagorean theorem.** Before we review this theorem, recall that a **right triangle** is a triangle that contains a 90° angle, or right angle. The **hypotenuse** of a right triangle is the side opposite the right angle and is the longest side of the triangle. The **legs** of a right triangle are the other sides of the triangle.

---

**Pythagorean Theorem**

In a right triangle, the sum of the squares of the lengths of the two legs is equal to the square of the length of the hypotenuse.

$$(\text{leg})^2 + (\text{leg})^2 = (\text{hypotenuse})^2 \quad \text{or} \quad a^2 + b^2 = c^2$$

---

△ **Example 9** Using the Pythagorean Theorem

While framing an addition to an existing home, Kim Menzies, a carpenter, used the Pythagorean theorem to determine whether a wall was "square"—that is, whether the wall formed a right angle with the floor. He used a triangle whose sides are three consecutive integers. Find a right triangle whose sides are three consecutive integers.

(*Continued on next page*)

**Practice 9**

Find a right triangle whose sides are three consecutive even integers.

**Answer**

**9.** 6, 8, and 10 units

**Solution:**

1. **UNDERSTAND.** Read and reread the problem. Let $x$, $x + 1$, and $x + 2$ be three consecutive integers. Since these integers represent lengths of the sides of a right triangle, we have

$x$ = one leg,

$x + 1$ = other leg, and

$x + 2$ = hypotenuse (longest side)

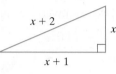

2. **TRANSLATE.** By the Pythagorean theorem, we have

In words:   $(\text{leg})^2 + (\text{leg})^2 = (\text{hypotenuse})^2$

$\qquad\qquad\quad \downarrow \qquad\quad \downarrow \qquad\qquad \downarrow$

Translate:   $(x)^2 + (x + 1)^2 = (x + 2)^2$

3. **SOLVE** the equation.

$$x^2 + (x + 1)^2 = (x + 2)^2$$
$$x^2 + x^2 + 2x + 1 = x^2 + 4x + 4 \qquad \text{Multiply.}$$
$$2x^2 + 2x + 1 = x^2 + 4x + 4$$
$$x^2 - 2x - 3 = 0 \qquad \text{Write in standard form.}$$
$$(x - 3)(x + 1) = 0$$
$$x - 3 = 0 \quad \text{or} \quad x + 1 = 0$$
$$x = 3 \qquad\qquad x = -1$$

4. **INTERPRET.** Discard $x = -1$ since length cannot be negative. If $x = 3$, then $x + 1 = 4$ and $x + 2 = 5$.

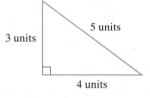

**Check:**   To check, see that $(\text{leg})^2 + (\text{leg})^2 = (\text{hypotenuse})^2$.

$$3^2 + 4^2 = 5^2$$
$$9 + 16 = 25 \quad \text{True}$$

**State:**   The lengths of the sides of the right triangle are 3, 4, and 5 units. Kim used this information by marking off lengths of 3 and 4 feet on the floor and framing, respectively. If the diagonal length between these marks was 5 feet, the wall was square. If not, adjustments were made.

◼ **Work Practice 9**

## Objective C Finding the $x$-Intercepts of a Polynomial Function ▶

Recall that to find the $x$-intercepts of the graph of a function, let $f(x) = 0$, or $y = 0$, and solve for $x$. This fact gives us a visual interpretation of the results of this section.

From Example 1, we know that the solutions of the equation $(x + 2)(x - 6) = 0$ are $-2$ and 6. These solutions give us important information about the related polynomial function $p(x) = (x + 2)(x - 6)$. We know that when $x$ is $-2$ or when $x$ is 6, the value of $p(x)$ is 0.

$$p(x) = (x + 2)(x - 6)$$
$$p(-2) = (-2 + 2)(-2 - 6) = (0)(-8) = 0$$
$$p(6) = (6 + 2)(6 - 6) = (8)(0) = 0$$

Thus, we know that $(-2, 0)$ and $(6, 0)$ are the $x$-intercepts of the graph of $p(x)$.

We also know that the graph of $p(x)$ does not cross the $x$-axis at any other point. For this reason, and the fact that $p(x) = (x + 2)(x - 6) = x^2 - 4x - 12$ has degree 2, we conclude that the graph of $p$ must look something like one of these two graphs:

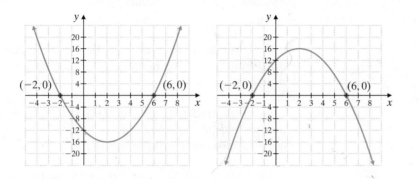

In a later chapter, we explore these graphs more fully. For the moment, know that the solutions of a polynomial equation are the $x$-intercepts of the graph of the related function and that the $x$-intercepts of the graph of a polynomial function are the solutions of the related polynomial equation. These values are also called **roots,** or **zeros,** of the polynomial function.

**Example 10** Match Each Function with Its Graph

$$f(x) = (x - 3)(x + 2) \qquad g(x) = x(x + 2)(x - 2)$$
$$h(x) = (x - 2)(x + 2)(x - 1)$$

**A**                **B**                **C**

**Solution:** The graph of the function $f(x) = (x - 3)(x + 2)$ has two $x$-intercepts, $(3, 0)$ and $(-2, 0)$, because the equation $0 = (x - 3)(x + 2)$ has two solutions, 3 and $-2$.

The graph of $f(x)$ is graph B.

The graph of the function $g(x) = x(x + 2)(x - 2)$ has three $x$-intercepts, $(0, 0)$, $(-2, 0)$, and $(2, 0)$, because the equation $0 = x(x + 2)(x - 2)$ has three solutions, $0$, $-2$, and 2.

The graph of $g(x)$ is graph C.

The graph of the function $h(x) = (x - 2)(x + 2)(x - 1)$ has three $x$-intercepts, $(-2, 0)$, $(1, 0)$, and $(2, 0)$, because the equation $0 = (x - 2)(x + 2)(x - 1)$ has three solutions, $-2$, 1, and 2.

The graph of $h(x)$ is graph A.

■ **Work Practice 10**

**Practice 10**

Match each function with its graph.

$$f(x) = (x + 1)(x - 3)$$
$$g(x) = (x - 5)(x - 1)(x + 1)$$
$$h(x) = x(x - 1)(x + 1)$$

**a.**

**b.**

**c.**

**Answers**

**10. a.** $g(x) = (x - 5)(x - 1)(x + 1)$

**b.** $f(x) = (x + 1)(x - 3)$

**c.** $h(x) = x(x - 1)(x + 1)$

## Calculator Explorations   Graphing

We can use a graphing calculator to approximate real number solutions of any quadratic equation in standard form, whether the associated polynomial is factorable or not. For example, let's solve the quadratic equation $x^2 - 2x - 4 = 0$. The solutions of this equation will be the $x$-intercepts of the graph of the function $f(x) = x^2 - 2x - 4$. (Recall that to find $x$-intercepts, we let $f(x) = 0$, or $y = 0$.) When we use a standard window, the graph of this function looks like this.

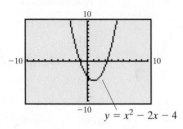

$y = x^2 - 2x - 4$

The graph appears to have one $x$-intercept between $-2$ and $-1$ and one between 3 and 4. To find the $x$-intercept between 3 and 4 to the nearest hundredth, we can use a Zero feature, a Zoom feature, which magnifies a portion of the graph around the cursor, or we can redefine our window. If we redefine our window to

| | |
|---|---|
| Xmin = 2 | Ymin = $-1$ |
| Xmax = 5 | Ymax = 1 |
| Xscl = 1 | Yscl = 1 |

the resulting screen is

By using the Trace feature, we can now see that one of the intercepts is between 3.21 and 3.25. To approximate to the nearest hundredth, Zoom again or redefine the window to

| | |
|---|---|
| Xmin = 3.2 | Ymin = $-0.1$ |
| Xmax = 3.3 | Ymax = 0.1 |
| Xscl = 1 | Yscl = 1 |

If we use the Trace feature again, we see that, to the nearest hundredth, the $x$-intercept is 3.24. By repeating this process, we can approximate the other $x$-intercept to be $-1.24$.

To check, find $f(3.24)$ and $f(-1.24)$. Both of these values should be close to 0. (They will not be exactly 0 since we approximated these solutions.)

$$f(3.24) = 0.0176 \quad \text{and} \quad f(-1.24) = 0.0176$$

*Solve each of these quadratic equations by graphing a related function and approximating the x-intercepts to the nearest thousandth.*

**1.** $x^2 + 3x - 2 = 0$

**2.** $5x^2 - 7x + 1 = 0$

**3.** $2.3x^2 - 4.4x - 5.6 = 0$

**4.** $0.2x^2 + 6.2x + 2.1 = 0$

**5.** $0.09x^2 - 0.13x - 0.08 = 0$

**6.** $x^2 + 0.08x - 0.01 = 0$

## Vocabulary, Readiness & Video Check

*Solve each equation for the variable.*

**1.** $(x - 3)(x + 5) = 0$

**2.** $(y + 5)(y + 3) = 0$

**3.** $(z - 3)(z + 7) = 0$

**4.** $(c - 2)(c - 4) = 0$

**5.** $x(x - 9) = 0$

**6.** $w(w + 7) = 0$

*Martin-Gay Interactive Videos*    *Watch the section lecture video and answer the following questions.*

**Objective** **A** **7.** As shown in ▥ Examples 1–3, what two things have to be true in order to use the zero factor property? ▶

**Objective** **B** **8.** In ▥ Example 5, why aren't both solutions to the equation accepted? ▶

**Objective** **C** **9.** From ▥ Example 6, how can finding the $x$-intercepts of the graph of a quadratic equation in two variables lead to solving a quadratic equation by factoring? ▶

*See Video 5.7* ▶

**5.7** **Exercise Set** MyMathLab® ▶

**Objective** **A** *Solve each equation. See Example 1.*

▶**1.** $(x + 3)(3x - 4) = 0$

**2.** $(5x + 1)(x - 2) = 0$

**3.** $3(2x - 5)(4x + 3) = 0$

**4.** $8(3x - 4)(2x - 7) = 0$

*Solve each equation. See Examples 2 through 5.*

**5.** $x^2 + 11x + 24 = 0$

**6.** $y^2 - 10y + 24 = 0$

**7.** $12x^2 + 5x - 2 = 0$

**8.** $3y^2 - y - 14 = 0$

**9.** $z^2 + 9 = 10z$

**10.** $n^2 + n = 72$

▶**11.** $x(5x + 2) = 3$

**12.** $n(2n - 3) = 2$

**13.** $x^2 - 6x = x(8 + x)$

**14.** $n(3 + n) = n^2 + 4n$

**15.** $\dfrac{z^2}{6} - \dfrac{z}{2} - 3 = 0$

**16.** $\dfrac{c^2}{20} - \dfrac{c}{4} + \dfrac{1}{5} = 0$

**17.** $\dfrac{x^2}{2} + \dfrac{x}{20} = \dfrac{1}{10}$

**18.** $\dfrac{y^2}{30} = \dfrac{y}{15} + \dfrac{1}{2}$

**19.** $\dfrac{4t^2}{5} = \dfrac{t}{5} + \dfrac{3}{10}$

**20.** $\dfrac{5x^2}{6} - \dfrac{7x}{2} + \dfrac{2}{3} = 0$

*Solve each equation. See Examples 6 and 7.*

**21.** $(x + 2)(x - 7)(3x - 8) = 0$

**22.** $(4x + 9)(x - 4)(x + 1) = 0$

**23.** $y^3 = 9y$

**24.** $n^3 = 16n$

▶**25.** $x^3 - x = 2x^2 - 2$

**26.** $m^3 = m^2 + 12m$

*Solve each equation. This section of exercises contains some linear equations. See Examples 1 through 7.*

**27.** $(2x + 7)(x - 10) = 0$

**28.** $(x + 4)(5x - 1) = 0$

**29.** $3x(x - 5) = 0$

**30.** $4x(2x + 3) = 0$

**31.** $x^2 - 2x - 15 = 0$

**32.** $x^2 + 6x - 7 = 0$

**33.** $12x^2 + 2x - 2 = 0$

**34.** $8x^2 + 13x + 5 = 0$

**35.** $w^2 - 5w = 36$

**36.** $x^2 + 32 = 12x$

**37.** $25x^2 - 40x + 16 = 0$

**38.** $9n^2 + 30n + 25 = 0$

**39.** $2r^3 + 6r^2 = 20r$

**40.** $-2t^3 = 108t - 30t^2$

**41.** $z(5z - 4)(z + 3) = 0$

**42.** $2r(r - 3)(5r + 4) = 0$

▶ **43.** $2z(z + 6) = 2z^2 + 12z - 8$

**44.** $3c^2 - 8c + 2 = c(3c - 8)$

**45.** $(x - 1)(x + 4) = 24$

**46.** $(2x - 1)(x + 2) = -3$

**47.** $\dfrac{x^2}{4} - \dfrac{5}{2}x + 6 = 0$

**48.** $\dfrac{x^2}{18} + \dfrac{x}{2} + 1 = 0$

**49.** $y^2 + \dfrac{1}{4} = -y$

**50.** $\dfrac{x^2}{10} + \dfrac{5}{2} = x$

**51.** $y^3 + 4y^2 = 9y + 36$

**52.** $x^3 + 5x^2 = x + 5$

**53.** $2x^3 = 50x$

**54.** $m^5 = 36m^3$

**55.** $x^2 + (x + 1)^2 = 61$

**56.** $y^2 + (y + 2)^2 = 34$

**57.** $m^2(3m - 2) = m$

**58.** $x^2(5x + 3) = 26x$

**59.** $3x^2 = -x$

**60.** $y^2 = -5y$

**61.** $x(x - 3) = x^2 + 5x + 7$

**62.** $z^2 - 4z + 10 = z(z - 5)$

**63.** $3(t - 8) + 2t = 7 + t$

**64.** $7c - 2(3c + 1) = 5(4 - 2c)$

**65.** $-3(x - 4) + x = 5(3 - x)$

**66.** $-4(a + 1) - 3a = -7(2a - 3)$

## Objective B  *Solve. See Examples 8 and 9.*

**67.** One number exceeds another by five, and their product is 66. Find the numbers.

**68.** If the sum of two numbers is 4 and their product is $\dfrac{15}{4}$, find the numbers.

△ **69.** An electrician needs to run a cable from the top of a 60-foot tower to a transmitter box located 45 feet away from the base of the tower. Find how long he should cut the cable.

**Tower Rewiring Plan**

△ **70.** A stereo-system installer needs to run speaker wire above the ceiling along the two diagonals of a rectangular room whose dimensions are 40 feet by 75 feet. Find how much speaker wire she needs.

**71.** If the cost, $C(x)$, for manufacturing $x$ units of a certain product is given by $C(x) = x^2 - 15x + 50$, find the number of units manufactured at a cost of $9500.

△ **72.** Determine whether any three consecutive integers represent the lengths of the sides of a right triangle.

▶ **73.** The shorter leg of a right triangle is 3 centimeters less than the other leg. Find the length of the two legs if the hypotenuse is 15 centimeters.

△ **74.** The longer leg of a right triangle is 4 feet longer than the other leg. Find the length of the two legs if the hypotenuse is 20 feet.

**75.** Marie Mulroney has a rectangular board 12 inches by 16 inches around which she wants to put a uniform border of shells. If she has enough shells for a border whose area is 128 square inches, determine the width of the border.

△ **76.** A gardener has a rose garden that measures 30 feet by 20 feet. He wants to put a uniform border of pine bark around the outside of the garden. Find how wide the border should be if he has enough pine bark to cover 336 square feet.

**77.** While hovering near the top of Ribbon Falls in Yosemite National Park at 1600 feet, a helicopter pilot accidentally drops his sunglasses. The height $h(t)$ of the sunglasses after $t$ seconds is given by the polynomial function

$$h(t) = -16t^2 + 1600$$

When will the sunglasses hit the ground?

**78.** After $t$ seconds, the height $h(t)$ of a model rocket launched from the ground into the air is given by the function

$$h(t) = -16t^2 + 80t$$

Find how long it takes the rocket to reach a height of 96 feet.

△ **79.** The floor of a shed has an area of 90 square feet. The floor is in the shape of a rectangle whose length is 3 feet less than twice the width. Find the length and the width of the floor of the shed.

△ **80.** A vegetable garden with an area of 200 square feet is to be fertilized. If the length of the garden is 1 foot less than three times the width, find the dimensions of the garden.

**81.** The function $W(x) = 0.5x^2$ gives the number of servings of wedding cake that can be obtained from a two-layer $x$-inch-square wedding cake tier. What size square wedding cake tier is needed to serve 50 people? (*Source:* Based on data from the *Wilton 2000 Yearbook of Cake Decorating*)

**82.** Use the function in Exercise 81 to determine what size wedding cake tier is needed to serve 200 people.

## Objective C  *Match each polynomial function with its graph (A–F). See Example 10.*

**83.** $f(x) = (x - 2)(x + 5)$

**84.** $g(x) = (x + 1)(x - 6)$

**85.** $h(x) = x(x + 3)(x - 3)$

**86.** $F(x) = (x + 1)(x - 2)(x + 5)$

**87.** $G(x) = 2x^2 + 9x + 4$

**88.** $H(x) = 2x^2 - 7x - 4$

A

B

C

D

E

F

## Review

*Write the x- and y-intercepts for each graph. See Section 3.1.*

**89.**    **90.**    **91.**    **92.**

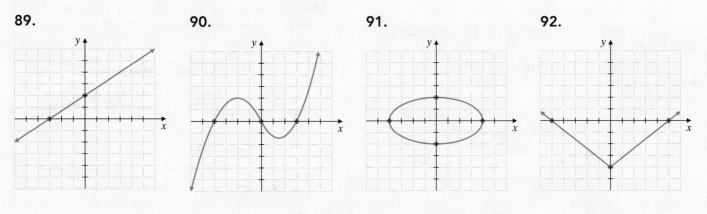

## Concept Extensions

*Each exercise contains an error. Find and correct the error. See the Concept Check in the section.*

**93.**  $(x - 5)(x + 2) = 0$
  $x - 5 = 0$   or   $x + 2 = 0$
      $x = -5$   or       $x = -2$

**94.**  $(4x - 5)(x + 7) = 0$
  $4x - 5 = 0$   or   $x + 7 = 0$
      $x = \dfrac{4}{5}$   or       $x = -7$

**95.**  $y(y - 5) = -6$
  $y = -6$   or   $y - 5 = -6$
  $y = -6$   or       $y = -1$

**96.**  $3x^2 - 19x = 14$
       $-16x = 14$
        $x = -\dfrac{14}{16}$
        $x = -\dfrac{7}{8}$

*Solve.*

**97.**  $(x^2 + x - 6)(3x^2 - 14x - 5) = 0$

**98.**  $(x^2 - 9)(x^2 + 8x + 16) = 0$

**99.** Explain how solving $2(x - 3)(x - 1) = 0$ differs from solving $2x(x - 3)(x - 1) = 0$.

**100.** Explain why the zero-factor property works for more than two numbers whose product is 0.

**101.** Is the following step correct? Why or why not?
  $x(x - 3) = 5$
  $x = 5$   or   $x - 3 = 5$

**102.** Are the following steps correct? Why or why not?
  $x^2 + x = 6$
  $x(x + 1) = 6$
  $x = 6$   or   $x + 1 = 6$

*Write a quadratic equation that has the given numbers as solutions.*

**103.** $5, 3$

**104.** $6, 7$

## Chapter 5 Group Activity

### Finding the Largest Area

*This activity may be completed by working in groups or individually.*

A picture framer has a piece of wood that measures 1 inch wide by 50 inches long. She would like to make a picture frame with the largest possible interior area. Complete the following activity to help her determine the dimensions of the frame that she should use to achieve her goal.

1. Use the situation shown in the figure to write an equation in $x$ and $y$ for the *outer* perimeter of the frame. (Remember that the outer perimeter will equal 50 inches.)

2. Use your equation from Question 1 to help you find the value of $y$ for each value of $x$ given in the table. Complete the $y$ column of the table. (*Note:* The first two columns of the table give possible combinations for the outer dimensions of the frame.)

3. How is the interior width of the frame related to the exterior width of the frame? How is the interior height of the frame related to the exterior height of the frame? Use these relationships to complete the two columns of the table labeled "Interior Width" and "Interior Height."

4. Complete the last column of the table labeled "Interior Area" by using the columns of dimensions for the interior width and height.

5. From the table, what appears to be the largest interior area of the frame? Which exterior dimensions of the frame provide this area?

6. Use the patterns in the table to write an algebraic expression in terms of $x$ for the interior width of the frame.

7. Use the patterns in the table to write an algebraic expression in terms of $y$ for the interior height of the frame.

8. Use the perimeter equation from Question 1 to rewrite the algebraic expression for the interior height of the frame in terms of $x$.

| Frame's Interior Dimensions | | | | |
|---|---|---|---|---|
| $x$ | $y$ | Interior Width (in.) | Interior Height (in.) | Interior Area (sq in.) |
| 2.0 | | | | |
| 2.5 | | | | |
| 3.0 | | | | |
| 3.5 | | | | |
| 4.0 | | | | |
| 4.5 | | | | |
| 5.0 | | | | |
| 5.5 | | | | |
| 6.0 | | | | |
| 6.5 | | | | |
| 7.0 | | | | |
| 7.5 | | | | |
| 8.0 | | | | |
| 8.5 | | | | |
| 9.0 | | | | |
| 9.5 | | | | |
| 10.0 | | | | |
| 10.5 | | | | |
| 11.0 | | | | |
| 11.5 | | | | |
| 12.0 | | | | |
| 12.5 | | | | |
| 13.0 | | | | |
| 13.5 | | | | |
| 14.0 | | | | |
| 14.5 | | | | |
| 15.0 | | | | |

9. Find a function $A$ that gives the interior area of the frame in terms of its exterior width $x$. (*Hint:* Study the patterns in the table. How could the expressions from Questions 6 and 8 be used to write this function?)

10. Graph the function $A$. Locate and label the point from the table that represents the maximum interior area. Describe the location of the point in relation to the rest of the graph.

## Chapter 5 Vocabulary Check

*Fill in each blank with one of the words or phrases listed below.*

| | | | | | |
|---|---|---|---|---|---|
| quadratic equation | synthetic division | polynomial | FOIL | monomial | 0 |
| binomial | trinomial | degree of a polynomial | degree of a term | factoring | |

1. A _____ is a finite sum of terms in which all variables are raised to nonnegative integer powers and no variables appear in any denominator.

2. _____ is the process of writing a polynomial as a product.

3. The _____ is the sum of the exponents on the variables contained in the term.

4. A _____ is a polynomial with one term.

5. A _____ is a polynomial with three terms.

6. A polynomial equation of degree 2 is also called a _____.

7. The _____ is the largest degree of all of its terms.

8. A _____ is a polynomial with two terms.

9. If $a$ and $b$ are real numbers and $a \cdot b =$ _____, then $a = 0$ or $b = 0$.

10. The _____ order may be used when multiplying two binomials.

11. A shortcut method called _____ may be used to divide a polynomial by a binomial of the form $x - c$.

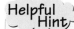

**Helpful Hint**

▶ Are you preparing for your test? Don't forget to take the Chapter 5 Test on page 423. Then check your answers at the back of the text and use the Chapter Test Prep Videos to see the fully worked-out solutions to any of the exercises you want to review.

## 5 Chapter Highlights

| Definitions and Concepts | Examples |
|---|---|
| **Section 5.1    Polynomial Functions and Adding and Subtracting Polynomials** | |
| A **polynomial** is a finite sum of terms in which all variables are raised to nonnegative integer powers and no variables appear in any denominator. | $1.3x^2$  Monomial <br><br> $-\dfrac{1}{3}y + 5$  Binomial <br><br> $6z^2 - 5z + 7$  Trinomial |
| To **add polynomials,** combine all like terms. | Add: <br><br> $(3y^2x - 2yx + 11) + (-5y^2x - 7)$ <br> $= -2y^2x - 2yx + 4$ |
| To **subtract polynomials,** change the signs of the terms of the polynomial being subtracted; then add. | Subtract: <br><br> $(-2z^3 - z + 1) - (3z^3 + z - 6)$ <br> $= -2z^3 - z + 1 - 3z^3 - z + 6$ <br> $= -5z^3 - 2z + 7$ |
| A function $P$ is a **polynomial function** if $P(x)$ is a polynomial. | For the polynomial function <br><br> $P(x) = -x^2 + 6x - 12$ <br><br> find $P(-2)$. <br><br> $P(-2) = -(-2)^2 + 6(-2) - 12 = -28$ |

| Definitions and Concepts | Examples |
|---|---|

### Section 5.2    Multiplying Polynomials

**Multiplying Two Polynomials**

Use the distributive property and multiply each term of one polynomial by each term of the other polynomial; then combine like terms.

Multiply.

$$(x^2 - 2x)(3x^2 - 5x + 1)$$
$$= 3x^4 - 5x^3 + x^2 - 6x^3 + 10x^2 - 2x$$
$$= 3x^4 - 11x^3 + 11x^2 - 2x$$

**Special Products**

$$(a + b)^2 = a^2 + 2ab + b^2$$
$$(a - b)^2 = a^2 - 2ab + b^2$$
$$(a + b)(a - b) = a^2 - b^2$$

The **FOIL order** may be used when multiplying two binomials.

Multiply.

$$(3m + 2n)^2 = 9m^2 + 12mn + 4n^2$$
$$(z^2 - 5)^2 = z^4 - 10z^2 + 25$$
$$(7y + 1)(7y - 1) = 49y^2 - 1$$

Multiply.

$$(x^2 + 5)(2x^2 - 9)$$

$$\begin{array}{cccc} F & O & I & L \\ \downarrow & \downarrow & \downarrow & \downarrow \end{array}$$

$$= x^2(2x^2) + x^2(-9) + 5(2x^2) + 5(-9)$$
$$= 2x^4 - 9x^2 + 10x^2 - 45$$
$$= 2x^4 + x^2 - 45$$

### Section 5.3    Dividing Polynomials and Synthetic Division

**Dividing a Polynomial by a Monomial**

Divide each term in the polynomial by the monomial.

$$\frac{12a^5b^3 - 6a^2b^2 + ab}{6a^2b^2}$$

$$= \frac{12a^5b^3}{6a^2b^2} - \frac{6a^2b^2}{6a^2b^2} + \frac{ab}{6a^2b^2}$$

$$= 2a^3b - 1 + \frac{1}{6ab}$$

**Dividing a Polynomial by a Polynomial Other Than a Monomial**

Use **long division.**

Divide $2x^3 - x^2 - 8x - 1$ by $x - 2$.

$$\begin{array}{r} 2x^2 + 3x - 2 \\ x - 2 \overline{)\ 2x^3 - \ x^2 - 8x - 1} \\ \underline{2x^3 - 4x^2} \\ 3x^2 - 8x \\ \underline{3x^2 - 6x} \\ -2x - 1 \\ \underline{-2x + 4} \\ -5 \end{array}$$

The quotient is $2x^2 + 3x - 2 - \dfrac{5}{x - 2}$.

A shortcut method called **synthetic division** may be used to divide a polynomial by a binomial of the form $x - c$.

Use synthetic division to divide $2x^3 - x^2 - 8x - 1$ by $x - 2$.

$$\begin{array}{r|rrrr} 2 & 2 & -1 & -8 & -1 \\ & \downarrow & 4 & 6 & -4 \\ \hline & 2 & 3 & -2 & -5 \end{array}$$

The quotient is $2x^2 + 3x - 2 - \dfrac{5}{x - 2}$.

| Definitions and Concepts | Examples |
|---|---|

### Section 5.4   The Greatest Common Factor and Factoring by Grouping

The **greatest common factor** of the terms of a polynomial is the product of the greatest common factor of the numerical coefficients and the greatest common factor of the variable factors.

Factor:   $14xy^3 - 2xy^2 = 2 \cdot 7 \cdot x \cdot y^3 - 2 \cdot x \cdot y^2$
The greatest common factor is $2 \cdot x \cdot y^2$, or $2xy^2$.

$$14xy^3 - 2xy^2 = 2xy^2(7y - 1)$$

**Factoring a Polynomial by Grouping**

Group the terms so that each group has a common factor. Factor out these common factors. Then see if the new groups have a common factor.

Factor:   $x^4y - 5x^3 + 2xy - 10$
$$= x^3(xy - 5) + 2(xy - 5)$$
$$= (xy - 5)(x^3 + 2)$$

### Section 5.5   Factoring Trinomials

**Factoring $ax^2 + bx + c$**

**Step 1:** Write all pairs of factors of $ax^2$.

**Step 2:** Write all pairs of factors of $c$.

**Step 3:** Try combinations of these factors until the middle term $bx$ is found.

Factor: $28x^2 - 27x - 10$
Factors of $28x^2$: $28x$ and $x$, $2x$ and $14x$, $4x$ and $7x$.
Factors of $-10$: $-2$ and $5$, $2$ and $-5$, $-10$ and $1$, $10$ and $-1$.
$$28x^2 - 27x - 10 = (7x + 2)(4x - 5)$$

### Section 5.6   Factoring by Special Products

**Perfect Square Trinomial**
$$a^2 + 2ab + b^2 = (a + b)^2$$
$$a^2 - 2ab + b^2 = (a - b)^2$$

Factor.
$$25x^2 + 30x + 9 = (5x + 3)^2$$
$$49z^2 - 28z + 4 = (7z - 2)^2$$

**Difference of Two Squares**
$$a^2 - b^2 = (a + b)(a - b)$$

Factor.
$$36x^2 - y^2 = (6x + y)(6x - y)$$

**Sum and Difference of Two Cubes**
$$a^3 + b^3 = (a + b)(a^2 - ab + b^2)$$
$$a^3 - b^3 = (a - b)(a^2 + ab + b^2)$$

Factor.
$$8y^3 + 1 = (2y + 1)(4y^2 - 2y + 1)$$
$$27p^3 - 64q^3 = (3p - 4q)(9p^2 + 12pq + 16q^2)$$

### Section 5.7   Solving Equations by Factoring and Solving Problems

**Solving a Polynomial Equation by Factoring**

**Step 1:** Write the equation so that one side is 0.

**Step 2:** Factor the polynomial completely.

**Step 3:** Set each factor equal to 0.

**Step 4:** Solve the resulting equations.

**Step 5:** Check each solution.

Solve:          $2x^3 - 5x^2 = 3x$
$$2x^3 - 5x^2 - 3x = 0$$
$$x(2x + 1)(x - 3) = 0$$
$$x = 0 \quad \text{or} \quad 2x + 1 = 0 \quad \text{or} \quad x - 3 = 0$$
$$x = 0 \qquad\qquad x = -\frac{1}{2} \qquad\qquad x = 3$$

**(5.1)** *Find the degree of each polynomial.*

**1.** $x^2y - 3xy^3z + 5x + 7y$

**2.** $3x + 2$

*Simplify by combining like terms.*

**3.** $4x + 8x - 6x^2 - 6x^2y$

**4.** $-8xy^3 + 4xy^3 - 3x^3y$

*Add or subtract as indicated.*

**5.** $(3x + 7y) + (4x^2 - 3x + 7) + (y - 1)$

**6.** $(4x^2 - 6xy + 9y^2) - (8x^2 - 6xy - y^2)$

**7.** $(3x^2 - 4b + 28) + (9x^2 - 30) - (4x^2 - 6b + 20)$

**8.** Add $(9xy + 4x^2 + 18)$ and $(7xy - 4x^3 - 9x)$.

**9.** Subtract $(3x - 5)$ from $2x$.

**10.** Subtract $(x - 7)$ from the sum of $(3x^2y - 7xy - 4)$ and $(9x^2y + x)$.

**11.**
$$\begin{array}{r} x^2 - 5x + 7 \\ -\quad\ (x + 4) \\ \hline \end{array}$$

**12.**
$$\begin{array}{r} x^3\ \ + 2xy^2 - y \\ +\ (x - 4xy^2\quad\ - 7) \\ \hline \end{array}$$

*If $P(x) = 9x^2 - 7x + 8$, find each function value.*

**13.** $P(6)$

**14.** $P(0)$

**15.** $P(-2)$

**16.** $P(-3)$

*If $P(x) = 2x - 1$ and $Q(x) = x^2 + 2x - 5$, find each function value.*

**17.** $P(x) + Q(x)$

**18.** $2 \cdot P(x) - Q(x)$

△ **19.** Find the perimeter of the rectangle.

$(x^2y + 5)$ cm

$(2x^2y - 6x + 1)$ cm

**20.** Find the perimeter of the triangle.

$(3x + 11)$ in.    $(3x + 11)$ in.

$3x$ in.

**(5.2)** *Multiply.*

**21.** $-6x(4x^2 - 6x + 1)$

**22.** $-4ab^2(3ab^3 + 7ab + 1)$

**23.** $(x - 4)(2x + 9)$

**24.** $(-3xa + 4b)^2$

**25.** $(9x^2 + 4x + 1)(4x - 3)$

**26.** $(2x - 1)(x^2 + 2x - 5)$

**27.** $(5x - 9y)(3x + 9y)$ **28.** $\left(x - \dfrac{1}{3}\right)\left(x + \dfrac{2}{3}\right)$ **29.** $(x^2 + 9x + 1)^2$

**30.** $(m^2 + m - 2)^2$

*Use special products to multiply.*

**31.** $(3x - y)^2$ **32.** $(4x + 9)^2$ **33.** $(x + 3y)(x - 3y)$

**34.** $[4 + (3a - b)][4 - (3a - b)]$  △ **35.** Find the area of the rectangle. **36.** Find the area of the square.

$(3y - 7z)$ units

$(3y + 7z)$ units

$(11y + 9)$ units

**(5.3)** *Divide.*

**37.** $(4xy + 2x^2 - 9) \div (4xy)$ **38.** $12xb^2 + 16xb^4$ by $4xb^3$

**39.** $(3x^4 - 25x^2 - 20) \div (x - 3)$ **40.** $(-x^2 + 2x^4 + 5x - 12) \div (x - 3)$

**41.** $(2x^4 - x^3 + 2x^2 - 3x + 1) \div \left(x - \dfrac{1}{2}\right)$ **42.** $(x^3 + 3x^2 - 2x + 2) \div \left(x - \dfrac{1}{2}\right)$

**43.** $(3x^4 + 5x^3 + 7x^2 + 3x - 2) \div (x^2 + x + 2)$ **44.** $(9x^4 - 6x^3 + 3x^2 - 12x - 30) \div (3x^2 - 2x - 5)$

*Use synthetic division to find each quotient.*

**45.** $(3x^3 + 12x - 4) \div (x - 2)$ **46.** $(4x^3 + 2x^2 - 4x - 2) \div \left(x + \dfrac{3}{2}\right)$.

**47.** $(x^5 - 1) \div (x + 1)$ **48.** $(x^3 - 81) \div (x - 3)$

**49.** $(x^3 - x^2 + 3x^4 - 2) \div (x - 4)$ **50.** $(3x^4 - 2x^2 + 10) \div (x + 2)$

**(5.4)** *Factor out the greatest common factor.*

**51.** $16x^3 - 24x^2$

**52.** $36y - 24y^2$

**53.** $6ab^2 + 8ab - 4a^2b^2$

**54.** $14a^2b^2 - 21ab^2 + 7ab$

**55.** $6a(a + 3b) - 5(a + 3b)$

**56.** $4x(x - 2y) - 5(x - 2y)$

*Factor by grouping.*

**57.** $xy - 6y + 3x - 18$

**58.** $ab - 8b + 4a - 32$

**59.** $pq - 3p - 5q + 15$

**60.** $x^3 - x^2 - 2x + 2$

*For Exercises 61 and 62, smaller squares are cut from larger rectangles. Write the area of each shaded region as a factored polynomial.*

**61.**

**62.**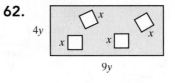

**(5.5)** *Factor each polynomial completely.*

**63.** $x^2 - 14x - 72$

**64.** $x^2 + 16x - 80$

**65.** $2x^2 - 18x + 28$

**66.** $3x^2 + 33x + 54$

**67.** $2x^3 - 7x^2 - 9x$

**68.** $3x^2 + 2x - 16$

**69.** $6x^2 + 17x + 10$

**70.** $15x^2 - 91x + 6$

**71.** $4x^2 + 2x - 12$

**72.** $9x^2 - 12x - 12$

**73.** $y^2(x + 6)^2 - 2y(x + 6)^2 - 3(x + 6)^2$

**74.** $(x + 5)^2 + 6(x + 5) + 8$

**75.** $x^4 - 6x^2 - 16$

**76.** $x^4 + 8x^2 - 20$

**(5.6)** *Factor each polynomial completely.*

**77.** $x^2 - 100$

**78.** $x^2 - 81$

**79.** $2x^2 - 32$

**80.** $6x^2 - 54$

**81.** $81 - x^4$

**82.** $16 - y^4$

**83.** $(y + 2)^2 - 25$

**84.** $(x - 3)^2 - 16$

**85.** $x^3 + 216$

**86.** $y^3 + 512$

**87.** $8 - 27y^3$

**88.** $1 - 64y^3$

**89.** $6x^4y + 48xy$

**90.** $2x^5 + 16x^2y^3$

**91.** $x^2 - 2x + 1 - y^2$

**92.** $x^2 - 6x + 9 - 4y^2$

**93.** $4x^2 + 12x + 9$

**94.** $16a^2 - 40ab + 25b^2$

**(5.7)** *Solve each polynomial equation for the variable.*

**95.** $(3x - 1)(x + 7) = 0$

**96.** $3(x + 5)(8x - 3) = 0$

**97.** $5x(x - 4)(2x - 9) = 0$

**98.** $6(x + 3)(x - 4)(5x + 1) = 0$

**99.** $2x^2 = 12x$

**100.** $4x^3 - 36x = 0$

**101.** $(1 - x)(3x + 2) = -4x$

**102.** $2x(x - 12) = -40$

**103.** $3x^2 + 2x = 12 - 7x$

**104.** $2x^2 + 3x = 35$

**105.** $x^3 - 18x = 3x^2$

**106.** $19x^2 - 42x = -x^3$

**107.** $12x = 6x^3 + 6x^2$

**108.** $8x^3 + 10x^2 = 3x$

**109.** The sum of a number and twice its square is 105. Find the number.

△ **110.** The volume of the cylindrical shell is $\pi R^2 h - \pi r^2 h$ cubic units. Write this volume as a factored expression.

△ **111.** The length of a rectangular piece of carpet is 2 meters less than 5 times its width. Find the dimensions of the carpet if its area is 16 square meters.

**112.** A scene from an adventure film calls for a stunt dummy to be dropped from above the second-story platform of the Eiffel Tower, a distance of 400 feet. Its height $h$ in feet at the time $t$ seconds is given by $h = -16t^2 + 400$. Determine how long before the stunt dummy reaches the ground.

400 ft

## Mixed Review

*Perform the indicated operation.*

**113.** $(x + 5)(3x^2 - 2x + 1)$

**114.** $(3x^2 + 4x - 1.2) - (5x^2 - x + 5.7)$

**115.** $(7x^3 - 16x^2 - 13x - 4) \div (x - 3)$

**116.** $\left(7ab - \dfrac{1}{2}\right)^2$

*If $P(x) = -x^2 + x - 4$, find*

**117.** $P(5)$

**118.** $P(-2)$

*Factor each polynomial completely.*

**119.** $12y^5 - 6y^4$

**120.** $x^2y + 4x^2 - 3y - 12$

**121.** $6x^2 - 34x - 12$

**122.** $y^2(4x + 3)^2 - 19y(4x + 3)^2 - 20(4x + 3)^2$

**123.** $4z^7 - 49z^5$

**124.** $5x^4 + 4x^2 - 9$

*Solve each equation.*

**125.** $8x^2 = 24x$

**126.** $x(x - 11) = 26$

*Perform each indicated operation.*

**1.** $(4x^3 - 3x - 4) - (9x^3 + 8x + 5)$

**2.** $-3xy(4x + y)$

**3.** $(3x + 4)(4x - 7)$

**4.** $(5a - 2b)(5a + 2b)$

**5.** $(6m + n)^2$

**6.** $(2x - 1)(x^2 - 6x + 4)$

**7.** $(4x^2y + 9x + z) \div (3xz)$

**8.** $(4x^5 - 2x^4 + 4x^2 - 6x + 3) \div (2x - 1)$

**9.** Use synthetic division to divide $4x^4 - 3x^3 + 2x^2 - x - 1$ by $x + 3$.

*Factor each polynomial completely.*

**10.** $16x^3y - 12x^2y^4$

**11.** $x^2 - 13x - 30$

**12.** $4y^2 + 20y + 25$

**13.** $6x^2 - 15x - 9$

**14.** $4x^2 - 25$

**15.** $x^3 + 64$

**16.** $3x^2y - 27y^3$

**17.** $16y^3 - 2$

**18.** $x^2y - 9y - 3x^2 + 27$

**Answers**

1. _____

2. _____

3. _____

4. _____

5. _____

6. _____

7. _____

8. _____

9. _____

10. _____

11. _____

12. _____

13. _____

14. _____

15. _____

16. _____

17. _____

18. _____

19. _____

20. _____

21. _____

22. _____

23. a. _____

   b. _____

   c. _____

   d. _____

*Solve each equation for the variable.*

**19.** $3n(7n - 20) = 96$

**20.** $(x + 2)(x - 2) = 5(x + 4)$

**21.** $2x^3 + 5x^2 = 8x + 20$

△ **22.** Write the area of the shaded region as a factored polynomial.

**23.** A pebble is hurled upward from the top of the 880-foot-tall Canada Trust Tower with an initial velocity of 96 feet per second. Neglecting air resistance, the height $h(t)$ in feet of the pebble after $t$ seconds is given by the polynomial function

$$h(t) = -16t^2 + 96t + 880$$

  **a.** Find the height of the pebble when $t = 1$.

  **b.** Find the height of the pebble when $t = 5.1$.

  **c.** Factor the polynomial $-16t^2 + 96t + 880$.

  **d.** When will the pebble hit the ground?

*Simplify.*

**1.** $-2(x + 3) + 7$

**2.** $6z - 9z - \dfrac{1}{2} + z - \dfrac{3}{8}$

**3.** $7x + 3 - 5(x - 4)$

**4.** $7 + 2(9x - 1) - (1.7x + 5.8)$

**5.** Use scientific notation to simplify:
$\dfrac{2000 \times 0.000021}{700}$

**6.** Simplify each expression. Write the answers with positive exponents only.

   **a.** $(4a^{-1}b^0)^{-3}$     **b.** $\left(\dfrac{a^{-6}}{a^{-8}}\right)^{-2}$

   **c.** $\left(\dfrac{2}{3}\right)^{-3}$     **d.** $\dfrac{3^{-2}a^{-2}b^{12}}{a^4b^{-5}}$

**7.** Solve: $3(x - 3) = 5x - 9$

**8.** Solve: $0.3y + 2.4 = 0.1y + 4$

**9.** Solve $3y - 2x = 7$ for $y$.

**10.** A gallon of latex paint can cover 400 square feet. How many gallon containers of paint should be bought to paint two coats on each wall of a rectangular room whose dimensions are 14 feet by 18 feet? (Assume 8-foot ceilings.)

*Solve. Write the solution set in interval notation.*

**11.** $-(x - 3) + 2 \le 3(2x - 5) + x$

**12.** $x + 2 < \dfrac{1}{4}(x - 7)$

**13.** $-1 \le \dfrac{2}{3}x + 5 < 2$

**14.** Solve: $-\dfrac{1}{3} < \dfrac{3x + 1}{6} \le \dfrac{1}{3}$

**Answers**

1. _____

2. _____

3. _____

4. _____

5. _____

6. a. _____

   b. _____

   c. _____

   d. _____

7. _____

8. _____

9. _____

10. _____

11. _____

12. _____

13. _____

14. _____

**15.** _____

**16.** _____

**17.** _____

**18.** _____

**19.** _____

**20.** _____

**21.** _____

**22.** _____

**23.** _____

**24.** _____

**25.** _____

**26.** _____

**27.** _____

**28. a.** _____

**b.** _____

**29.** _____

**30.** _____

**31.** _____

**32.** _____

**15.** Solve: $|3x + 2| = |5x - 8|$

**16.** Solve: $|5x - 1| + 9 > 5$

**17.** Find the slope of the line $x = -5$.

**18.** Find the slope of the line, $f(x) = -2x - 3$.

**19.** Graph: $y = \dfrac{1}{4}x - 3$

**20.** Graph: $y = 3x$

**21.** Write an equation of the line containing the point $(-2, 1)$ and perpendicular to the line $3x + 5y = 4$. Write the equation in slope-intercept form, $y = mx + b$.

**22.** Find an equation of the vertical line containing the point $(-3, 2)$.

**23.** Determine the domain and range of the relation: $\{(2, 3), (2, 4), (0, -1), (3, -1)\}$

**24.** Find an equation of the line containing the point $(-2, 3)$ and slope of 0.

_Find each function value._

**25.** If $g(x) = 3x - 2$, find $g(0)$.

**26.** If $f(x) = 3x^2 + 2x + 3$, find $f(-3)$.

**27.** If $f(x) = 7x^2 - 3x + 1$, find $f(1)$.

**28.** Multiply.
   **a.** $(3y^6)(4.7y^2)$   **b.** $(6a^3b^2)(-a^2bc^4)$

**29.** Graph: $2x - y < 6$

**30.** One solution contains 20% acid and a second solution contains 60% acid. How many ounces of each solution should be mixed in order to have 50 ounces of a 30% acid solution?

**31.** Use the substitution method to solve the system: $\begin{cases} 2x + 4y = -6 \\ x = 2y - 5 \end{cases}$

**32.** Solve the system: $\begin{cases} 4y = 8 \\ x + y = 7 \end{cases}$

**33.** Solve the system: $\begin{cases} 3x - y + z = -15 \\ x + 2y - z = 1 \\ 2x + 3y - 2z = 0 \end{cases}$

**34.** Solve the system: $\begin{cases} x + y + z = 0 \\ 2x - 3y + z = 5 \\ 2x + y + 2z = 2 \end{cases}$

**35.** Use matrices to solve the system: $\begin{cases} x + 3y = 5 \\ 2x - y = -4 \end{cases}$

**36.** Solve the system: $\begin{cases} x + y - \dfrac{3}{2}z = \dfrac{1}{2} \\ -y - 2z = 14 \\ x - \dfrac{2}{3}y = -\dfrac{1}{3} \end{cases}$

**37.** Graph the solutions of the system:
$\begin{cases} x - y < 2 \\ x + 2y > -1 \\ y < 2 \end{cases}$

**38.** Write each number in scientific notation.
  **a.** 8,250,000
  **b.** 0.0000346

**39.** Determine the degree of the polynomial $3xy + x^2y^2 - 5x^2 - 6.7$.

**40.** Subtract $(5x^2 + 3x)$ from $(3x^2 - 2x)$.

*Multiply.*

**41.** $2x(5x - 4)$

**42.** $(2x - 1)(5x^2 - 6x + 2)$

**43.** $-xy(7x^2y + 3xy - 11)$

**44.** $\left(7x - \dfrac{1}{2}\right)^2$

**45.** Divide $10x^3 - 5x^2 + 20x$ by $5x$.

*Factor.*

**46.** $12x^3y - 3xy^3$

**47.** $x^2 - 12x + 35$

**48.** $5a^2 + 14a - 3$

**49.** Solve: $2x^2 + 9x - 5 = 0$

**50.** Solve: $3x^2 - 10x - 8 = 0$

33. _____

34. _____

35. _____

36. _____

37. _____

38. a. _____

b. _____

39. _____

40. _____

41. _____

42. _____

43. _____

44. _____

45. _____

46. _____

47. _____

48. _____

49. _____

50. _____

# 6

# Rational Expressions

## Check Your Progress

Dividing one polynomial by another in Section 5.3, we found quotients of polynomials. When the remainder part of the quotient was not 0, the remainder was a fraction, such as $\dfrac{x}{x^2 + 1}$. This fraction is not a polynomial, since it cannot be written as the sum of whole number powers. Instead, it is called a *rational expression*. In this chapter, we study these algebraic forms, the operations that can be performed on them, and the *rational functions* they generate.

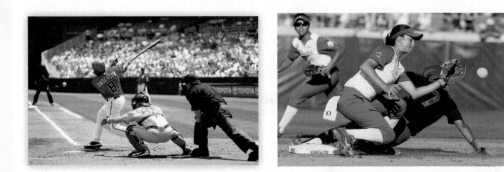

## Statistics and Sports Go Together

Statistics and sports seem to go together. While many people watch professional sports merely for entertainment, there are multitudes more that live and breathe the statistics. In baseball or softball, for example, the batting average of each player is usually reported each time he or she comes to the plate. The number of home runs, the number of steals, and the number of strikeouts are all studied by baseball (softball) enthusiasts, and those are only the batting statistics. There is also a complete set of statistics for the pitching staff; the ERA or earned run average, the number of strikeouts, the number of walks, and the complete innings pitched matter. In Section 6.3, Exercise 62, and in the Chapter Test, Exercise 27, we will explore how a major league baseball pitcher's earned run average is computed.

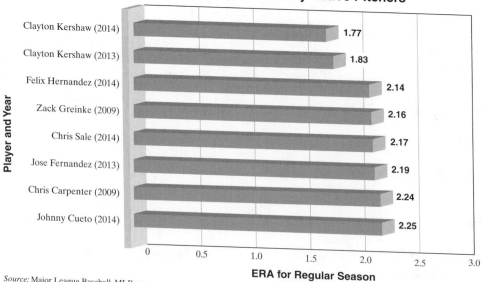

**Major League Baseball Top Eight ERA's by Active Pitchers**

| Player and Year | ERA |
|---|---|
| Clayton Kershaw (2014) | 1.77 |
| Clayton Kershaw (2013) | 1.83 |
| Felix Hernandez (2014) | 2.14 |
| Zack Greinke (2009) | 2.16 |
| Chris Sale (2014) | 2.17 |
| Jose Fernandez (2013) | 2.19 |
| Chris Carpenter (2009) | 2.24 |
| Johnny Cueto (2014) | 2.25 |

ERA for Regular Season

*Source:* Major League Baseball, MLB.com

# 6.1 Rational Functions and Multiplying and Dividing Rational Expressions

Recall that a *rational number,* or *fraction,* is a number that can be written as the quotient $\frac{p}{q}$ of two integers $p$ and $q$ as long as $q$ is not 0. A **rational expression** is an expression that can be written as the quotient $\frac{P}{Q}$ of two polynomials $P$ and $Q$ as long as $Q$ is not 0.

**Objectives**

**A** Find the Domain of a Rational Function.

**B** Simplify Rational Expressions.

**C** Multiply Rational Expressions.

**D** Divide Rational Expressions.

**E** Use Rational Functions in Real-World Applications.

### Examples of Rational Expressions

$$\frac{8x^3 + 7x^2 + 20}{2} \qquad \frac{5x^2 - 3}{x - 1} \qquad \frac{7x - 2}{x^2 - 2x - 15}$$

Rational expressions are sometimes used to describe functions. For example, we call the function $f(x) = \frac{x^2 + 2}{x - 3}$ a **rational function** since $\frac{x^2 + 2}{x - 3}$ is a rational expression.

## Objective A  Finding Domains of Rational Functions

As with fractions, a rational expression is **undefined** if the denominator is 0. If a variable in a rational expression is replaced with a number that makes the denominator 0, we say that the rational expression is **undefined** for this value of the variable. For example, the rational expression $\frac{x^2 + 2}{x - 3}$ is undefined when $x$ is 3, because replacing $x$ with 3 results in a denominator of 0. For this reason, we must exclude 3 from the domain of the function $f(x) = \frac{x^2 + 2}{x - 3}$.

The domain of $f$ is then

$$\{x \,|\, x \text{ is a real number and } x \neq 3\}$$

"The set of all $x$ such that $x$ is a real number and $x$ is not equal to 3."
In this section, we will use this set builder notation to write domains.

Unless told otherwise, we assume that the domain of a function described by an equation is the set of all real numbers for which the equation is defined.

---

**Example 1**  Find the domain of each rational function.

**a.** $f(x) = \dfrac{8x^3 + 7x^2 + 20}{2}$    **b.** $g(x) = \dfrac{5x^2 - 3}{x - 1}$

**c.** $f(x) = \dfrac{7x - 2}{x^2 - 2x - 15}$

**Solution:**  The domain of each function will contain all real numbers except those values that make the denominator 0.

**a.** No matter what the value of $x$, the denominator of $f(x) = \dfrac{8x^3 + 7x^2 + 20}{2}$ is never 0, so the domain of $f$ is $\{x \,|\, x \text{ is a real number}\}$.

**b.** To find the values of $x$ that make the denominator of $g(x)$ equal to 0, we solve the equation "denominator = 0":

$$x - 1 = 0, \quad \text{or} \quad x = 1$$

The domain must exclude 1 since the rational expression is undefined when $x$ is 1. The domain of $g$ is $\{x \,|\, x \text{ is a real number and } x \neq 1\}$.

*(Continued on next page)*

---

**Practice 1**

Find the domain of each rational function.

**a.** $f(x) = \dfrac{x^2 + 1}{x - 6}$

**b.** $g(x) = \dfrac{5x + 4}{x^2 - 3x - 10}$

**c.** $h(x) = \dfrac{x^2 - 9}{4}$

**Answers**

**1. a.** $\{x \,|\, x \text{ is a real number}, x \neq 6\}$
**b.** $\{x \,|\, x \text{ is a real number}, x \neq 5, x \neq -2\}$
**c.** $\{x \,|\, x \text{ is a real number}\}$

**429**

**c.** We find the domain by setting the denominator equal to 0.

$$x^2 - 2x - 15 = 0 \quad \text{Set the denominator equal to 0 and solve.}$$
$$(x - 5)(x + 3) = 0$$
$$x - 5 = 0 \quad \text{or} \quad x + 3 = 0$$
$$x = 5 \quad \text{or} \quad x = -3$$

If $x$ is replaced with $5$ or with $-3$, the rational expression is undefined.

The domain of $f$ is $\{x \mid x \text{ is a real number and } x \neq 5, x \neq -3\}$.

■ Work Practice 1

## Objective B  Simplifying Rational Expressions ▶

Recall that a fraction is in lowest terms or simplest form if the numerator and denominator have no common factors other than 1 (or $-1$). For example, $\dfrac{3}{13}$ is in lowest terms since 3 and 13 have no common factors other than 1 (or $-1$).

To **simplify** a rational expression, or to write it in lowest terms, we use a method similar to simplifying a fraction.

Recall that to simplify a fraction, we essentially "remove factors of 1." Our ability to do this comes from these facts:

- If $c \neq 0$, then $\dfrac{c}{c} = 1$. For example, $\dfrac{7}{7} = 1$ and $\dfrac{-8.65}{-8.65} = 1$.

- $n \cdot 1 = n$. For example, $-5 \cdot 1 = -5$, $126.8 \cdot 1 = 126.8$, and $\dfrac{a}{b} \cdot 1 = \dfrac{a}{b}, b \neq 0$.

In other words, we have the following:

$$\frac{a \cdot c}{b \cdot c} = \underbrace{\frac{a}{b} \cdot \frac{c}{c}}_{\text{Since } \frac{a}{b} \cdot 1 = \frac{a}{b}} = \frac{a}{b}$$

Let's practice simplifying a fraction by simplifying $\dfrac{15}{65}$.

$$\frac{15}{65} = \frac{3 \cdot 5}{13 \cdot 5} = \frac{3}{13} \cdot \frac{5}{5} = \frac{3}{13} \cdot 1 = \frac{3}{13}$$

Let's use the same technique and simplify the rational expression $\dfrac{(x + 2)^2}{x^2 - 4}$.

$$\frac{(x + 2)^2}{x^2 - 4} = \frac{(x + 2)(x + 2)}{(x - 2)(x + 2)}$$
$$= \frac{(x + 2)}{(x - 2)} \cdot \frac{x + 2}{x + 2}$$
$$= \frac{x + 2}{x - 2} \cdot 1$$
$$= \frac{x + 2}{x - 2}$$

This means that the rational expression $\dfrac{(x + 2)^2}{x^2 - 4}$ has the same value as the rational expression $\dfrac{x + 2}{x - 2}$ for all values of $x$ except 2 and $-2$. (Remember that when $x$ is 2, the denominators of both rational expressions are 0, and that when $x$ is $-2$, the original rational expression has a denominator of 0.)

As we simplify rational expressions, we will assume that the simplified rational expression is equivalent to the original rational expression for all real numbers except those for which either denominator is 0.

Just as for numerical fractions, we can use a shortcut notation. Remember that as long as exact factors in both the numerator and denominator are divided out, we are "removing a factor of 1." We can use the following notation:

$$\frac{(x+2)^2}{x^2-4} = \frac{(x+2)\,(x+2)}{(x-2)\,(x+2)}$$  A factor of 1 is identified by the shading.

$$= \frac{x+2}{x-2}$$  "Remove" the factor of 1.

In general, the following steps may be used to simplify rational expressions or to write rational expressions in lowest terms.

---

## Simplifying or Writing a Rational Expression in Lowest Terms

**Step 1:** Completely factor the numerator and denominator of the rational expression.

**Step 2:** Divide out factors common to the numerator and denominator. (This is the same as "removing a factor of 1.")

---

For now, we assume that variables in a rational expression do not represent values that make the denominator 0.

**Examples** Simplify each rational expression.

**2.** $\dfrac{2x^2}{10x^3 - 2x^2} = \dfrac{2x^2 \cdot 1}{2x^2\,(5x-1)}$  Factor the numerator and denominator.

$$= 1 \cdot \frac{1}{5x-1}$$  Since $\frac{2x^2}{2x^2} = 1$.

$$= \frac{1}{5x-1}$$  Simplest form.

**3.** $\dfrac{9x^2 + 13x + 4}{8x^2 + x - 7} = \dfrac{(9x+4)\,(x+1)}{(8x-7)\,(x+1)}$  Factor the numerator and denominator.

$$= \frac{9x+4}{8x-7} \cdot 1$$  Since $\frac{x+1}{x+1} = 1$.

$$= \frac{9x+4}{8x-7}$$  Simplest form.

■ Work Practices 2–3

**Examples** Simplify each rational expression.

**4.** $\dfrac{2+x}{x+2} = \dfrac{x+2}{x+2} = 1$  By the commutative property of addition, $2 + x = x + 2$.

**5.** $\dfrac{2-x}{x-2}$

The terms in the numerator of $\dfrac{2-x}{x-2}$ differ by sign from the terms of the denominator, so the polynomials are opposites of each other and the expression simplifies to $-1$. To see this, we factor out $-1$ from the numerator or the denominator.

*(Continued on next page)*

## Practice 2–3

Simplify each rational expression.

**2.** $\dfrac{3y^3}{6y^4 - 3y^3}$

**3.** $\dfrac{5x^2 + 13x + 6}{6x^2 + 7x - 10}$

## Practice 4–6

Simplify each rational expression.

**4.** $\dfrac{5+x}{x+5}$   **5.** $\dfrac{5-x}{x-5}$

**6.** $\dfrac{3 - 3x^2}{x^2 + x - 2}$

**Answers**

**2.** $\dfrac{1}{2y-1}$   **3.** $\dfrac{5x+3}{6x-5}$   **4.** 1   **5.** $-1$

**6.** $-\dfrac{3(x+1)}{x+2}$

When the numerator and the denominator of a rational expression are opposites of each other, the expression simplifies to $-1$.

If $-1$ is factored from the *numerator*, then

$$\left\{ \frac{2 - x}{x - 2} = \frac{-1(-2 + x)}{x - 2} = \frac{-1(x - 2)}{x - 2} = \frac{-1}{1} = -1 \right.$$

If $-1$ is factored from the *denominator*, the result is the same.

$$\left\{ \frac{2 - x}{x - 2} = \frac{2 - x}{-1(-x + 2)} = \frac{2 - x}{-1(2 - x)} = \frac{1}{-1} = -1 \right.$$

**6.** $\dfrac{18 - 2x^2}{x^2 - 2x - 3} = \dfrac{2(9 - x^2)}{(x + 1)(x - 3)}$    Factor.

$$= \frac{2(3 + x)(3 - x)}{(x + 1)(x - 3)}$$    Factor completely.

Notice the opposites $3 - x$ and $x - 3$. We write $3 - x$ as $-1(x - 3)$ and simplify.

$$\frac{2(3 + x)(3 - x)}{(x + 1)(x - 3)} = \frac{2(3 + x) \cdot -1(x - 3)}{(x + 1)(x - 3)} = -\frac{2(3 + x)}{x + 1}$$

■ Work Practice 4–6

### Helpful Hint

Recall that for a fraction

$$\frac{a}{-b} = \frac{-a}{b} = -\frac{a}{b}$$

For example,

$$\frac{-(x + 1)}{(x + 2)} = \frac{(x + 1)}{-(x + 2)} = -\frac{x + 1}{x + 2}$$

✓ **Concept Check** Which of the following expressions are equivalent to $\dfrac{x}{8 - x}$?

**a.** $\dfrac{-x}{x - 8}$    **b.** $\dfrac{-x}{8 - x}$    **c.** $\dfrac{x}{x - 8}$    **d.** $\dfrac{-x}{-8 + x}$

**Practice 7–8**

Simplify each rational expression.

**7.** $\dfrac{x^3 + 27}{x + 3}$

**8.** $\dfrac{3x^2 + 6}{x^3 - 3x^2 + 2x - 6}$

**Examples** Simplify each rational expression.

**7.** $\dfrac{x^3 + 8}{x + 2} = \dfrac{(x + 2)(x^2 - 2x + 4)}{x + 2}$    Factor the sum of the two cubes.

$$= x^2 - 2x + 4$$    Simplest form.

**8.** $\dfrac{2y^2 + 2}{y^3 - 5y^2 + y - 5} = \dfrac{2(y^2 + 1)}{(y^3 - 5y^2) + (y - 5)}$    Factor the numerator.

$$= \frac{2(y^2 + 1)}{y^2(y - 5) + 1(y - 5)}$$    Factor the denominator by grouping.

$$= \frac{2(y^2 + 1)}{(y - 5)(y^2 + 1)}$$

$$= \frac{2}{y - 5}$$    Simplest form.

■ Work Practice 7–8

**Answers**

**7.** $x^2 - 3x + 9$  **8.** $\dfrac{3}{x - 3}$

✓ **Concept Check Answer**

**a** and **d**

✓**Concept Check** Does $\dfrac{n}{n+2}$ simplify to $\dfrac{1}{2}$? Why or why not?

## Objective C  Multiplying Rational Expressions 🔵

Arithmetic operations on rational expressions are performed in the same way as they are on rational numbers. To multiply rational expressions, we multiply numerators and multiply denominators.

> **Multiplying Rational Expressions**
>
> The rule for multiplying rational expressions is
>
> $$\frac{P}{Q} \cdot \frac{R}{S} = \frac{PR}{QS} \quad \text{as long as } Q \neq 0 \text{ and } S \neq 0.$$
>
> To multiply rational expressions, you may use these steps:
>
> **Step 1:** Completely factor each numerator and denominator.
>
> **Step 2:** Use the rule above and multiply the numerators and the denominators.
>
> **Step 3:** Simplify the product.

When we multiply rational expressions, notice that we factor each numerator and denominator first. This helps when we check to see whether the product is in simplest form.

**Examples**  Multiply.

**9.** $\dfrac{3n+1}{2n} \cdot \dfrac{2n-4}{3n^2-2n-1} = \dfrac{3n+1}{2n} \cdot \dfrac{2(n-2)}{(3n+1)(n-1)}$  Factor.

$\qquad = \dfrac{(3n+1) \cdot 2(n-2)}{2n(3n+1)(n-1)}$  Multiply.

$\qquad = \dfrac{n-2}{n(n-1)}$  Simplest form.

**10.** $\dfrac{x^3-1}{-3x+3} \cdot \dfrac{15x^2}{x^2+x+1} = \dfrac{(x-1)(x^2+x+1)}{-3(x-1)} \cdot \dfrac{15x^2}{x^2+x+1}$  Factor.

$\qquad = \dfrac{(x-1)(x^2+x+1) \cdot 3 \cdot 5x^2}{-1 \cdot 3(x-1)(x^2+x+1)}$  Factor.

$\qquad = \dfrac{5x^2}{-1} = -5x^2$  Simplest form.

■ **Work Practice 9–10**

## Objective D  Dividing Rational Expressions 🔵

Recall that two numbers are reciprocals of each other if their product is 1. Similarly, if $\dfrac{P}{Q}$ is a rational expression and $P \neq 0$, then $\dfrac{Q}{P}$ is its **reciprocal**, since

$$\frac{P}{Q} \cdot \frac{Q}{P} = \frac{P \cdot Q}{Q \cdot P} = 1$$

**Practice 9–10**

Multiply.

**9.** $\dfrac{2x-3}{5x} \cdot \dfrac{5x+5}{2x^2-x-3}$

**10.** $\dfrac{x^3+27}{-2x-6} \cdot \dfrac{4x^3}{x^2-3x+9}$

**Answers**

**9.** $\dfrac{1}{x}$   **10.** $-2x^3$

✓**Concept Check Answer**

no; answers may vary

The following are examples of expressions and their reciprocals.

| Expression | Reciprocal |
|---|---|
| $\dfrac{3}{x}$ | $\dfrac{x}{3}$ |
| $\dfrac{2 + x^2}{4x - 3}$ | $\dfrac{4x - 3}{2 + x^2}$ |
| $x^3$ | $\dfrac{1}{x^3}$ |
| $0$ | no reciprocal |

## Dividing Rational Expressions

The rule for dividing rational expressions is

$$\frac{P}{Q} \div \frac{R}{S} = \frac{P}{Q} \cdot \frac{S}{R} = \frac{PS}{QR} \quad \text{as long as } Q \neq 0, S \neq 0, \text{ and } R \neq 0.$$

To divide by a rational expression, use the rule above and multiply by its reciprocal. Then simplify if possible.

Notice that division of rational expressions is the same as for rational numbers.

**Practice 11–12**

Divide.

**11.** $\dfrac{12y^3}{5y^2 - 5} \div \dfrac{6}{1 - y}$

**12.** $\dfrac{8z^2 + 14z + 3}{20z^2 + z - 1}$
$\div \dfrac{2z^2 + 7z + 6}{35z^2 + 3z - 2}$

**Examples**   Divide.

**11.** $\dfrac{8m^2}{3m^2 - 12} \div \dfrac{40}{2 - m} = \dfrac{8m^2}{3m^2 - 12} \cdot \dfrac{2 - m}{40}$    Multiply by the reciprocal of the divisor.

$= \dfrac{8m^2(2 - m)}{3(m + 2)(m - 2) \cdot 40}$    Factor and multiply.

$= \dfrac{8 m^2 \cdot -1(m - 2)}{3(m + 2)(m - 2) \cdot 8 \cdot 5}$    Write $(2 - m)$ as $-1(m - 2)$.

$= -\dfrac{m^2}{15(m + 2)}$    Simplify.

**12.** $\dfrac{18y^2 + 9y - 2}{24y^2 - 10y + 1} \div \dfrac{3y^2 + 17y + 10}{8y^2 + 18y - 5}$

$= \dfrac{18y^2 + 9y - 2}{24y^2 - 10y + 1} \cdot \dfrac{8y^2 + 18y - 5}{3y^2 + 17y + 10}$    Multiply by the reciprocal.

$= \dfrac{(6y - 1)(3y + 2) \cdot (4y - 1)(2y + 5)}{(6y - 1)(4y - 1) \cdot (3y + 2)(y + 5)}$    Factor.

$= \dfrac{2y + 5}{y + 5}$    Simplest form.

■ Work Practice 11–12

### Helpful Hint

When dividing rational expressions, do not divide out common factors until the division problem is rewritten as a multiplication problem.

**Answers**

**11.** $-\dfrac{2y^3}{5(y + 1)}$   **12.** $\dfrac{7z + 2}{z + 2}$

**Example 13** Perform each indicated operation.

$$\frac{x^2 - 25}{(x+5)^2} \cdot \frac{3x + 15}{4x} \div \frac{x^2 - 3x - 10}{x}$$

**Solution:**

$$\frac{x^2 - 25}{(x+5)^2} \cdot \frac{3x + 15}{4x} \div \frac{x^2 - 3x - 10}{x}$$

$$= \frac{x^2 - 25}{(x+5)^2} \cdot \frac{3x + 15}{4x} \cdot \frac{x}{x^2 - 3x - 10} \qquad \text{To divide, multiply by the reciprocal.}$$

$$= \frac{(x+5)\,(x-5)}{(x+5)\,(x+5)} \cdot \frac{3\,(x+5)}{4\,x} \cdot \frac{x}{(x-5)\,(x+2)}$$

$$= \frac{3}{4(x+2)}$$

**Work Practice 13**

**Practice 13**

Perform each indicated operation.

$$\frac{(x+3)^2}{x^2 - 9} \cdot \frac{2x - 6}{5x}$$
$$\div \frac{x^2 + 7x + 12}{x}$$

## Objective E Applications with Rational Functions

Rational functions are often used to model real-life situations. Don't forget to be aware of the domains of these functions. See the Calculator Explorations on the following page for further domain exercises.

**Example 14** Finding Unit Cost

For the ICL Production Company, the rational function $C(x) = \dfrac{2.6x + 10,000}{x}$ describes the company's cost per disc of pressing $x$ DVDs. Find the cost per disc for pressing:

**a.** 100 DVDs
**b.** 1000 DVDs

**Practice 14**

A company's cost per book for printing $x$ particular books is given by the rational function $C(x) = \dfrac{0.8x + 5000}{x}$.

Find the cost per book for printing:

**a.** 100 books
**b.** 1000 books

**Solution:**

**a.** $C(100) = \dfrac{2.6(100) + 10,000}{100} = \dfrac{10,260}{100} = 102.6$

The cost per disc for pressing 100 DVDs is \$102.60.

**b.** $C(1000) = \dfrac{2.6(1000) + 10,000}{1000} = \dfrac{12,600}{1000} = 12.6$

The cost per disc for pressing 1000 DVDs is \$12.60. Notice that as more DVDs are produced, the cost per disc decreases.

**Work Practice 14**

## Calculator Explorations  Graphing

Recall that since the rational expression $\dfrac{7x - 2}{(x - 2)(x + 5)}$ is not defined when $x = 2$ or when $x = -5$, we say that the domain of the rational function $f(x) = \dfrac{7x - 2}{(x - 2)(x + 5)}$ is all real numbers except 2 and $-5$. This domain can be written as $\{x \mid x \text{ is a real number and } x \neq 2, x \neq -5\}$. This means that the graph of $f$ should not cross the vertical lines $x = 2$ and $x = -5$.

The graph of $f$ in *connected* mode is shown below. In connected mode the grapher tries to connect all dots of the graph so that the result is a smooth curve. This is what has happened in the graph. Notice that the graph appears to contain vertical lines at $x = 2$ and at $x = -5$. We know that this cannot happen because the function is not defined at $x = 2$ and at $x = -5$. We also know that this cannot happen because the graph of this function would not pass the vertical line test.

If we graph $f$ in *dot* mode, the graph appears as below. In dot mode the grapher will not connect dots with a smooth curve. Notice that the vertical lines have disappeared, and we have a better picture of the graph. It actually appears more like the hand-drawn graph at the top of the next column. By using a TABLE feature or a CALCULATE VALUE feature or by tracing, we can see that the function is not defined at $x = 2$ and at $x = -5$.

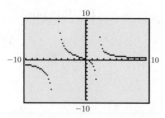

*Note:* Some calculator manufacturers now offer downloadable operating systems that eliminate the need to use dot mode to graph rational functions.

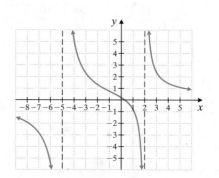

*Find the domain of each rational function. Then graph each rational function and use the graph to confirm the domain.*

**1.** $f(x) = \dfrac{5x}{x - 6}$

**2.** $f(x) = \dfrac{x}{x + 4}$

**3.** $f(x) = \dfrac{x + 1}{x^2 - 4}$

**4.** $g(x) = \dfrac{5x}{x^2 - 9}$

**5.** $h(x) = \dfrac{x^2}{2x^2 + 7x - 4}$

**6.** $f(x) = \dfrac{3x + 2}{4x^2 - 19x - 5}$

**7.** $g(x) = \dfrac{x^2 + x + 1}{5}$

**8.** $h(x) = \dfrac{x^2 + 25}{2}$

## Vocabulary, Readiness & Video Check

*Use the choices below to fill in each blank. Not all choices will be used.*

| | | | | | | |
|---|---|---|---|---|---|---|
| 1 | true | rational | simplified | $\dfrac{-a}{-b}$ | $\dfrac{-a}{b}$ | $\dfrac{a}{-b}$ |
| $-1$ | false | domain | 0 | | | |

**1.** A _____ expression is an expression that can be written as the quotient $\dfrac{P}{Q}$ of two polynomials $P$ and $Q$ as long as $Q \neq 0$.

**2.** A rational expression is undefined if the denominator is _____.

**3.** The _____ of the rational function $f(x) = \dfrac{2}{x}$ is $\{x \mid x$ is a real number and $x \neq 0\}$.

**4.** A rational expression is _____ if the numerator and denominator have no common factors other than 1 or $-1$.

**5.** The expression $\dfrac{x^2 + 2}{2 + x^2}$ simplifies to _____.

**6.** The expression $\dfrac{y - z}{z - y}$ simplifies to _____.

**7.** For a rational expression, $-\dfrac{a}{b} =$ _____ $=$ _____.

**8.** True or false: $\dfrac{a - 6}{a + 2} = \dfrac{-(a - 6)}{-(a + 2)} = \dfrac{-a + 6}{-a - 2}$. _____

*Multiply.*

**9.** $\dfrac{x}{5} \cdot \dfrac{y}{2}$

**10.** $\dfrac{y}{6} \cdot \dfrac{z}{5}$

**11.** $\dfrac{2}{x} \cdot \dfrac{y}{3}$

**12.** $\dfrac{a}{5} \cdot \dfrac{7}{b}$

**13.** $\dfrac{m}{6} \cdot \dfrac{m}{6}$

**14.** $\dfrac{9}{x} \cdot \dfrac{8}{x}$

*Martin-Gay Interactive Videos*

See Video 6.1

*Watch the section lecture video and answer the following questions.*

**Objective A** **15.** From ⊞ Examples 1 and 2, why can't the denominators of rational expressions be zero? How does this relate to the domain of a rational function? ▶

**Objective B** **16.** What conclusion does ⊞ Example 4 lead us to? ▶

**Objective C** **17.** Does a person need to be comfortable with factoring polynomials in order to be successful with multiplying rational expressions? Explain, referencing ⊞ Example 6 in your answer. ▶

**Objective D** **18.** Based on ⊞ Example 7, complete the following statements. Division of rational expressions is similar to division of _____. Therefore, to divide rational expressions, it's the first rational expression times the _____ of the second rational expression. ▶

**Objective E** **19.** In ⊞ Example 8d, we find the domain of the function. Do you think the domain should be restricted in any way given the context of this application? Explain. ▶

## 6.1 Exercise Set MyMathLab® ▶

### Objective A  *Find the domain of each rational expression. See Example 1.*

**1.** $f(x) = \dfrac{5x - 7}{4}$

**2.** $g(x) = \dfrac{4 - 3x}{2}$

**3.** $s(t) = \dfrac{t^2 + 1}{2t}$

**4.** $v(t) = -\dfrac{5t + t^2}{3t}$

**5.** $f(x) = \dfrac{3x}{7 - x}$

**6.** $f(x) = \dfrac{-4x}{-2 + x}$

**7.** $f(x) = \dfrac{x}{3x - 1}$

**8.** $g(x) = \dfrac{-2}{2x + 5}$

**9.** $R(x) = \dfrac{3 + 2x}{x^3 + x^2 - 2x}$

**10.** $h(x) = \dfrac{5 - 3x}{2x^2 - 14x + 20}$

**11.** $C(x) = \dfrac{x + 3}{x^2 - 4}$

**12.** $R(x) = \dfrac{5}{x^2 - 7x}$

### Objective B  *Simplify each rational expression. See Examples 2 through 8.*

**13.** $\dfrac{8x - 16x^2}{8x}$

**14.** $\dfrac{3x - 6x^2}{3x}$

**15.** $\dfrac{x^2 - 9}{3 + x}$

**16.** $\dfrac{x^2 - 25}{5 + x}$

**17.** $\dfrac{9y - 18}{7y - 14}$

**18.** $\dfrac{6y - 18}{2y - 6}$

**19.** $\dfrac{x^2 + 6x - 40}{x + 10}$

**20.** $\dfrac{x^2 - 8x + 16}{x - 4}$

**21.** $\dfrac{x - 9}{9 - x}$

**22.** $\dfrac{x - 4}{4 - x}$

**23.** $\dfrac{x^2 - 49}{7 - x}$

**24.** $\dfrac{x^2 - y^2}{y - x}$

**25.** $\dfrac{2x^2 - 7x - 4}{x^2 - 5x + 4}$

**26.** $\dfrac{3x^2 - 11x + 10}{x^2 - 7x + 10}$

**27.** $\dfrac{x^3 - 125}{2x - 10}$

**28.** $\dfrac{4x + 4}{x^3 + 1}$

**29.** $\dfrac{3x^2 - 5x - 2}{6x^3 + 2x^2 + 3x + 1}$

**30.** $\dfrac{2x^2 - x - 3}{2x^3 - 3x^2 + 2x - 3}$

**31.** $\dfrac{9x^2 - 15x + 25}{27x^3 + 125}$

**32.** $\dfrac{8x^3 - 27}{4x^2 + 6x + 9}$

**Objective C** *Multiply and simplify. See Examples 9 and 10.*

**33.** $\dfrac{2x - 4}{15} \cdot \dfrac{6}{2 - x}$

**34.** $\dfrac{10 - 2x}{7} \cdot \dfrac{14}{5x - 25}$

**35.** $\dfrac{18a - 12a^2}{4a^2 + 4a + 1} \cdot \dfrac{4a^2 + 8a + 3}{4a^2 - 9}$

**36.** $\dfrac{a - 5}{a^2 + a} \cdot \dfrac{1 - a^2}{10 - 2a}$

**▶ 37.** $\dfrac{9x + 9}{4x + 8} \cdot \dfrac{2x + 4}{3x^2 - 3}$

**38.** $\dfrac{2x^2 - 2}{10x + 30} \cdot \dfrac{12x + 36}{3x - 3}$

**39.** $\dfrac{2x^3 - 16}{6x^2 + 6x - 36} \cdot \dfrac{9x + 18}{3x^2 + 6x + 12}$

**40.** $\dfrac{x^2 - 3x + 9}{5x^2 - 20x - 105} \cdot \dfrac{x^2 - 49}{x^3 + 27}$

**41.** $\dfrac{a^3 + a^2b + a + b}{5a^3 + 5a} \cdot \dfrac{6a^2}{2a^2 - 2b^2}$

**42.** $\dfrac{4a^2 - 8a}{ab - 2b + 3a - 6} \cdot \dfrac{8b + 24}{3a + 6}$

**43.** $\dfrac{x^2 - 6x - 16}{2x^2 - 128} \cdot \dfrac{x^2 + 16x + 64}{3x^2 + 30x + 48}$

**44.** $\dfrac{2x^2 + 12x - 32}{x^2 + 16x + 64} \cdot \dfrac{x^2 + 10x + 16}{x^2 - 3x - 10}$

**45.** $\dfrac{15x^2 - x - 2}{10x^2 - 19x + 6} \cdot \dfrac{8x^2 - 18x + 9}{9x^2 + 6x + 1}$

**46.** $\dfrac{21x^2 + 17x + 2}{16x^2 - 8x + 1} \cdot \dfrac{20x^2 + 3x - 2}{15x^2 + 16x + 4}$

**Objective D** *Divide and simplify. See Examples 11 and 12.*

**47.** $\dfrac{2x}{5} \div \dfrac{6x + 12}{5x + 10}$

**48.** $\dfrac{7}{3x} \div \dfrac{14 - 7x}{18 - 9x}$

**49.** $\dfrac{a + b}{ab} \div \dfrac{a^2 - b^2}{4a^3b}$

**50.** $\dfrac{6a^2b^2}{a^2 - 4} \div \dfrac{3ab^2}{a - 2}$

**51.** $\dfrac{x^2 - 6x + 9}{x^2 - x - 6} \div \dfrac{x^2 - 9}{4}$

**52.** $\dfrac{x^2 - 4}{3x + 6} \div \dfrac{2x^2 - 8x + 8}{x^2 + 4x + 4}$

**53.** $\dfrac{3x^2 + x - 14}{6x^2 + 23x - 4} \div \dfrac{4x^2 - 5x - 6}{4x^2 + 19x + 12}$

**54.** $\dfrac{5x^2 - 13x - 6}{9x^2 + 17x - 2} \div \dfrac{8x^2 - 23x - 3}{8x^2 + 17x + 2}$

**55.** $\dfrac{x^2 - 6x - 16}{2x^2 - 128} \div \dfrac{x^2 + 10x + 16}{x^2 + 16x + 64}$

**56.** $\dfrac{a^2 - a - 6}{a^2 - 81} \div \dfrac{a^2 - 7a - 18}{4a + 36}$

**▶ 57.** $\dfrac{3x - x^2}{x^3 - 27} \div \dfrac{x}{x^2 + 3x + 9}$

**58.** $\dfrac{x^2 - 3x}{x^3 - 27} \div \dfrac{2x}{2x^2 + 6x + 18}$

**59.** $\dfrac{8b + 24}{3a + 6} \div \dfrac{ab - 2b + 3a - 6}{a^2 - 4a + 4}$

**60.** $\dfrac{2a^2 - 2b^2}{a^3 + a^2b + a + b} \div \dfrac{6a^2}{a^3 + a}$

## Objectives B C D Mixed Practice *Perform each indicated operation. See Examples 2 through 13.*

**61.** $\dfrac{x^2 - 9}{4} \cdot \dfrac{x^2 - x - 6}{x^2 - 6x + 9}$

**62.** $\dfrac{x^2 - 4}{9} \cdot \dfrac{x^2 - 6x + 9}{x^2 - 5x + 6}$

**63.** $\dfrac{2x^2 - 4x - 30}{5x^2 - 40x - 75} \div \dfrac{x^2 - 8x + 15}{x^2 - 6x + 9}$

**64.** $\dfrac{4a + 36}{a^2 - 7a - 18} \div \dfrac{a^2 - a - 6}{a^2 - 81}$

**65.** Simplify: $\dfrac{r^3 + s^3}{r + s}$

**66.** Simplify: $\dfrac{m^3 - n^3}{m - n}$

**67.** $\dfrac{4}{x} \div \dfrac{3xy}{x^2} \cdot \dfrac{6x^2}{x^4}$

**68.** $\dfrac{4}{x} \cdot \dfrac{3xy}{x^2} \div \dfrac{6x^2}{x^4}$

**69.** $\dfrac{5x^4 + 3x^2 - 2}{x - 1} \cdot \dfrac{x + 1}{x^4 - 1}$

**70.** $\dfrac{3x^4 - 10x^2 - 8}{x - 2} \cdot \dfrac{3x + 6}{15x^2 + 10}$

**71.** $\dfrac{3x^2 - 5x - 2}{y^2 + y - 2} \cdot \dfrac{y^2 + 4y - 5}{12x^2 + 7x + 1} \div \dfrac{5x^2 - 9x - 2}{8x^2 - 2x - 1}$

**72.** $\dfrac{x^2 + x - 2}{3y^2 - 5y - 2} \cdot \dfrac{12y^2 + y - 1}{x^2 + 4x - 5} \div \dfrac{8y^2 - 6y + 1}{5y^2 - 9y - 2}$

## Objective E *Find each function value. See Example 14.*

**73.** If $f(x) = \dfrac{x + 8}{2x - 1}$, find $f(2), f(0)$, and $f(-1)$.

**74.** If $f(x) = \dfrac{x - 2}{-5 + x}$, find $f(-5), f(0)$, and $f(10)$.

**75.** $g(x) = \dfrac{x^2 + 8}{x^3 - 25x}$; $g(3), g(-2), g(1)$

**76.** $s(t) = \dfrac{t^3 + 1}{t^2 + 1}$; $s(-1), s(1), s(2)$

**77.** The total revenue from the sale of a popular book is approximated by the rational function $R(x) = \dfrac{1000x^2}{x^2 + 4}$, where $x$ is the number of years since publication and $R(x)$ is the total revenue in millions of dollars.

a. Find the total revenue at the end of the first year.

b. Find the total revenue at the end of the second year.

c. Find the revenue during the second year only.

d. Find the domain of function $R$.

**78.** The function $f(x) = \dfrac{100{,}000x}{100 - x}$ models the cost in dollars for removing $x$ percent of the pollutants from a bayou in which a nearby company dumped creosol.

a. Find the cost of removing 20% of the pollutants from the bayou. [*Hint:* Find $f(20)$.]

b. Find the cost of removing 60% of the pollutants and then 80% of the pollutants.

c. Find $f(90)$, then $f(95)$, and then $f(99)$. What happens to the cost as $x$ approaches 100%?

d. Find the domain of function $f$.

## Review

*Perform each indicated operation. See Section 1.4.*

**79.** $\dfrac{4}{5} + \dfrac{3}{5}$

**80.** $\dfrac{4}{10} - \dfrac{7}{10}$

**81.** $\dfrac{5}{28} - \dfrac{2}{21}$

**82.** $\dfrac{5}{13} + \dfrac{2}{7}$

**83.** $\dfrac{3}{8} + \dfrac{1}{2} - \dfrac{3}{16}$

**84.** $\dfrac{2}{9} - \dfrac{1}{6} + \dfrac{2}{3}$

## Concept Extensions

*Solve. For Exercises 85 and 86, see the first Concept Check in this section; for Exercises 87 and 88, see the second Concept Check.*

**85.** Which of the expressions are equivalent to $\dfrac{x}{5 - x}$?

    **a.** $\dfrac{-x}{5 - x}$      **b.** $\dfrac{-x}{-5 + x}$

    **c.** $\dfrac{x}{x - 5}$      **d.** $\dfrac{-x}{x - 5}$

**86.** Which of the expressions are equivalent to $\dfrac{-2 + x}{x}$?

    **a.** $\dfrac{2 - x}{-x}$      **b.** $-\dfrac{2 - x}{x}$

    **c.** $\dfrac{x - 2}{x}$      **d.** $\dfrac{x - 2}{-x}$

**87.** Does $\dfrac{x}{x + 5}$ simplify to $\dfrac{1}{5}$? Why or why not?

**88.** Does $\dfrac{x + 7}{x}$ simplify to 7? Why or why not?

△ **89.** Find the area of the rectangle.

$\dfrac{5x}{x^2 - 4}$ m

$\dfrac{x + 2}{x}$ m

△ **90.** Find the area of the triangle.

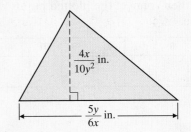

$\dfrac{4x}{10y^2}$ in.

$\dfrac{5y}{6x}$ in.

△ **91.** A parallelogram has an area of $\dfrac{x^2 + x - 2}{x^3}$ square feet and a height of $\dfrac{x^2}{x - 1}$ feet. Express the length of its base as a rational expression in $x$. (*Hint:* Since $A = b \cdot h$, then $b = \dfrac{A}{h}$ or $b = A \div h$.)

**92.** A lottery prize of $\dfrac{15x^3}{y^2}$ dollars is to be divided among $5x$ people. Express the amount of money each person is to receive as a rational expression in $x$ and $y$.

**93.** In your own words explain how to simplify a rational expression.

**94.** In your own words, explain the difference between multiplying rational expressions and dividing rational expressions.

**95.** Decide whether each rational expression equals 1, $-1$, or neither.

   **a.** $\dfrac{x+5}{5+x}$

   **b.** $\dfrac{x-5}{5-x}$

   **c.** $\dfrac{x+5}{x-5}$

   **d.** $\dfrac{-x-5}{x+5}$

   **e.** $\dfrac{x-5}{-x+5}$

   **f.** $\dfrac{-5+x}{x-5}$

**96.** Find the polynomial in the second numerator such that the following statement is true.

$$\frac{x^2-4}{x^2-7x+10} \cdot \frac{?}{2x^2+11x+14} = 1$$

**97.** In our definition of division for

$$\frac{P}{Q} \div \frac{R}{S}$$

   we stated that $Q \neq 0$, $S \neq 0$, and $R \neq 0$. Explain why $R$ cannot equal 0.

**98.** In your own words, explain how to find the domain of a rational function.

*Simplify. Assume that no denominator is 0.*

**99.** $\dfrac{p^x-4}{4-p^x}$

**100.** $\dfrac{3+q^n}{q^n+3}$

**101.** $\dfrac{x^n+4}{x^{2n}-16}$

**102.** $\dfrac{x^{2k}-9}{3+x^k}$

*Solve.*

**103.** Graph a portion of the function $f(x) = \dfrac{20x}{100-x}$.
To do so, complete the given table, plot the points, and then connect the plotted points with a smooth curve.

| $x$ | 0 | 10 | 30 | 50 | 70 | 90 | 95 | 99 |
|---|---|---|---|---|---|---|---|---|
| $y$ or $f(x)$ | | | | | | | | |

**104.** The domain of the function $f(x) = \dfrac{1}{x}$ is all real numbers except 0. This means that the graph of this function will be in two pieces: one piece corresponding to $x$ values less than 0 and one piece corresponding to $x$ values greater than 0. Graph the function by completing the following tables, separately plotting the points, and connecting each set of plotted points with a smooth curve.

| $x$ | $\dfrac{1}{4}$ | $\dfrac{1}{2}$ | 1 | 2 | 4 |
|---|---|---|---|---|---|
| $y$ or $f(x)$ | | | | | |

| $x$ | $-4$ | $-2$ | $-1$ | $-\dfrac{1}{2}$ | $-\dfrac{1}{4}$ |
|---|---|---|---|---|---|
| $y$ or $f(x)$ | | | | | |

## 6.2 Adding and Subtracting Rational Expressions ▶

## Objective A Adding or Subtracting Rational Expressions with the Same Denominator ▶

We add or subtract rational expressions just as we add or subtract fractions.

> **Adding or Subtracting Rational Expressions with Common Denominators**
>
> If $\dfrac{P}{Q}$ and $\dfrac{R}{Q}$ are rational expressions, then
>
> $$\frac{P}{Q} + \frac{R}{Q} = \frac{P + R}{Q} \quad \text{and} \quad \frac{P}{Q} - \frac{R}{Q} = \frac{P - R}{Q}$$
>
> To add or subtract rational expressions with common denominators, add or subtract the numerators and write the sum or difference over the common denominator.

**Examples** Add.

**1.** $\dfrac{5}{7z^2} + \dfrac{x}{7z^2} = \dfrac{5 + x}{7z^2}$     Add the numerators and write the result over the common denominator.

**2.** $\dfrac{x}{4} + \dfrac{5x}{4} = \dfrac{x + 5x}{4} = \dfrac{6x}{4} = \dfrac{3x}{2}$

■ Work Practice 1–2

**Examples** Subtract.

**3.** $\dfrac{x^2}{x + 7} - \dfrac{49}{x + 7} = \dfrac{x^2 - 49}{x + 7}$     Subtract the numerators and write the result over the common denominator.

$$= \frac{(x + 7)(x - 7)}{x + 7}$$  Factor the numerator.

$$= x - 7$$  Simplify.

**4.** $\dfrac{x}{3y^2} - \dfrac{x + 1}{3y^2} = \dfrac{x - (x + 1)}{3y^2}$  Subtract the numerators.

$$= \frac{x - x - 1}{3y^2}$$  Use the distributive property.

$$= -\frac{1}{3y^2}$$  Simplify.

> **Helpful Hint**
> Be sure to insert parentheses here so that the entire second numerator is subtracted.

■ Work Practice 3–4

**✓Concept Check** Find and correct the error.

$$\frac{3 + 2y}{y^2 - 1} - \frac{y + 3}{y^2 - 1}$$

$$= \frac{3 + 2y - y + 3}{y^2 - 1}$$

$$= \frac{y + 6}{y^2 - 1}$$

---

### Objectives

**A** Add or Subtract Rational Expressions with the Same Denominator. ▶

**B** Find the Least Common Denominator (LCD) of Two or More Rational Expressions. ▶

**C** Add or Subtract Rational Expressions with Different Denominators. ▶

**Practice 1–2**
Add.

**1.** $\dfrac{9}{11x^4} + \dfrac{y}{11x^4}$    **2.** $\dfrac{x}{6} + \dfrac{7x}{6}$

**Practice 3–4**
Subtract.

**3.** $\dfrac{x^2}{x + 3} - \dfrac{9}{x + 3}$

**4.** $\dfrac{a}{5b^3} - \dfrac{a + 2}{5b^3}$

**Answers**

**1.** $\dfrac{9 + y}{11x^4}$   **2.** $\dfrac{4x}{3}$   **3.** $x - 3$   **4.** $-\dfrac{2}{5b^3}$

**✓Concept Check Answer**

$$\frac{3 + 2y}{y^2 - 1} - \frac{y + 3}{y^2 - 1} = \frac{3 + 2y - y - 3}{y^2 - 1}$$

$$= \frac{y}{y^2 - 1}$$

## Objective B Finding the LCD of Rational Expressions ▶

To add or subtract rational expressions with unlike, or different, denominators, we first write the rational expressions as equivalent rational expressions with a common denominator.

The **least common denominator (LCD)** is usually the easiest common denominator to work with. The LCD of a list of rational expressions is a polynomial of least degree whose factors include all the factors of the denominators in the list.

The following steps can be used to find the LCD.

---

**Finding the Least Common Denominator (LCD)**

**Step 1:** Factor each denominator completely.

**Step 2:** The LCD is the product of all unique factors each raised to a power equal to the greatest number of times that the factor appears in any one factored denominator.

---

### Practice 5

Find the LCD of the rational expressions in each list.

**a.** $\dfrac{7}{20a^2b^3}, \dfrac{9}{15ab^4}$

**b.** $\dfrac{6x}{x-2}, \dfrac{5}{x+2}$

**c.** $\dfrac{x+4}{x^2-36}, \dfrac{x}{x^2+12x+36}, \dfrac{x^3}{3x^2+19x+6}$

**d.** $\dfrac{6}{x^2-1}, \dfrac{7}{2-2x}$

**Example 5** Find the LCD of the rational expressions in each list.

**a.** $\dfrac{2}{3x^5y^2}, \dfrac{3z}{5xy^3}$

**b.** $\dfrac{7}{z+1}, \dfrac{z}{z-1}$

**c.** $\dfrac{m-1}{m^2-25}, \dfrac{2m}{2m^2-9m-5}, \dfrac{7}{m^2-10m+25}$

**d.** $\dfrac{x}{x^2-4}, \dfrac{11}{6-3x}$

**Solution:**

**a.** First we factor each denominator.

$$3x^5y^2 = 3 \cdot x^5 \cdot y^2$$
$$5xy^3 = 5 \cdot x \cdot y^3$$
$$\text{LCD} = 3 \cdot 5 \cdot x^5 \cdot y^3 = 15x^5y^3$$

> **Helpful Hint** The greatest power of $x$ is 5, so we have a factor of $x^5$. The greatest power of $y$ is 3, so we have a factor of $y^3$.

**b.** The denominators $z+1$ and $z-1$ do not factor further.

$$(z+1) = (z+1)$$
$$(z-1) = (z-1)$$
$$\text{LCD} = (z+1)(z-1)$$

**c.** We first factor each denominator.

$$m^2-25 = (m+5)(m-5)$$
$$2m^2-9m-5 = (2m+1)(m-5)$$
$$m^2-10m+25 = (m-5)(m-5)$$
$$\text{LCD} = (m+5)(2m+1)(m-5)^2$$

**d.** We factor each denominator.

$$x^2-4 = (x+2)(x-2)$$
$$6-3x = 3(2-x) = 3(-1)(x-2)$$
$$\text{LCD} = 3(-1)(x+2)(x-2)$$
$$= -3(x+2)(x-2)$$

> **Helpful Hint** $(x-2)$ and $(2-x)$ are opposite factors. Notice that a $-1$ was factored from $(2-x)$ so that the factors are identical.

■ Work Practice 5

> **Helpful Hint**
>
> If opposite factors occur, do not use both in the LCD. Instead, factor $-1$ from one of the opposite factors so that the factors are then identical.

## Objective C Adding or Subtracting Rational Expressions with Different Denominators ▶

To add or subtract rational expressions with different denominators, we write each rational expression as an equivalent rational expression with the LCD as the denominator. To do this, we use the multiplication property of 1 and multiply each rational expression by a form of 1 so that each denominator becomes the LCD.

> **Adding or Subtracting Rational Expressions with Different Denominators**
>
> **Step 1:** Find the LCD of the rational expressions.
>
> **Step 2:** Write each rational expression as an equivalent rational expression whose denominator is the LCD found in Step 1.
>
> **Step 3:** Add or subtract numerators, and write the result over the common denominator.
>
> **Step 4:** Simplify the resulting rational expression.

**Example 6** Add: $\dfrac{2}{x^2} + \dfrac{5}{3x^3}$

**Solution:** The LCD is $3x^3$. To write the first rational expression as an equivalent rational expression with denominator $3x^3$, we multiply by 1 in the form of $\dfrac{3x}{3x}$.

$$\frac{2}{x^2} + \frac{5}{3x^3} = \frac{2 \cdot 3x}{x^2 \cdot 3x} + \frac{5}{3x^3} \quad \text{The second expression already has a denominator of } 3x^3.$$

$$= \frac{6x}{3x^3} + \frac{5}{3x^3}$$

$$= \frac{6x + 5}{3x^3} \quad \text{Add the numerators.}$$

■ Work Practice 6

**Practice 6**

Add: $\dfrac{7}{a^3} + \dfrac{9}{2a^4}$

**Example 7** Add: $\dfrac{3}{x + 2} + \dfrac{2x}{x - 2}$

**Solution:** The LCD is the product of the two denominators: $(x + 2)(x - 2)$.

$$\frac{3}{x + 2} + \frac{2x}{x - 2} = \frac{3 \cdot (x - 2)}{(x + 2) \cdot (x - 2)} + \frac{2x \cdot (x + 2)}{(x - 2) \cdot (x + 2)} \quad \begin{array}{l}\text{Write equivalent}\\\text{rational expressions.}\end{array}$$

$$= \frac{3x - 6}{(x + 2)(x - 2)} + \frac{2x^2 + 4x}{(x + 2)(x - 2)} \quad \begin{array}{l}\text{Multiply in the}\\\text{numerators.}\end{array}$$

$$= \frac{3x - 6 + 2x^2 + 4x}{(x + 2)(x - 2)} \quad \begin{array}{l}\text{Add the}\\\text{numerators.}\end{array}$$

$$= \frac{2x^2 + 7x - 6}{(x + 2)(x - 2)} \quad \text{Simplify.}$$

■ Work Practice 7

**Practice 7**

Add: $\dfrac{1}{x + 5} + \dfrac{6x}{x - 5}$

**Answers**

**6.** $\dfrac{14a + 9}{2a^4}$  **7.** $\dfrac{6x^2 + 31x - 5}{(x + 5)(x - 5)}$

## Practice 8

Subtract: $\dfrac{3m - 26}{m - 6} - \dfrac{8}{6 - m}$

**Example 8**　Subtract: $\dfrac{2x - 6}{x - 1} - \dfrac{4}{1 - x}$

**Solution:**　The LCD is either $x - 1$ or $1 - x$. To get a common denominator of $x - 1$, we factor $-1$ from the denominator of the second rational expression.

$$\frac{2x - 6}{x - 1} - \frac{4}{1 - x} = \frac{2x - 6}{x - 1} - \frac{4}{-1(x - 1)} \qquad \text{Write } 1 - x \text{ as } -1(x - 1).$$

$$= \frac{2x - 6}{x - 1} - \frac{-1 \cdot 4}{x - 1} \qquad \text{Write } \frac{4}{-1(x - 1)} \text{ as } \frac{-1 \cdot 4}{x - 1}.$$

$$= \frac{2x - 6 - (-4)}{x - 1}$$

$$= \frac{2x - 6 + 4}{x - 1} \qquad \text{Simplify.}$$

$$= \frac{2x - 2}{x - 1}$$

$$= \frac{2(x - 1)}{x - 1}$$

$$= 2$$

■ **Work Practice 8**

## Practice 9

Subtract:

$$\frac{2x}{x^2 - 9} - \frac{3}{x^2 - 4x + 3}$$

**Example 9**　Subtract: $\dfrac{5k}{k^2 - 4} - \dfrac{2}{k^2 + k - 2}$

**Solution:**

$$\frac{5k}{k^2 - 4} - \frac{2}{k^2 + k - 2} = \frac{5k}{(k + 2)(k - 2)} - \frac{2}{(k + 2)(k - 1)} \quad \begin{array}{l}\text{Factor each}\\\text{denominator to}\\\text{find the LCD.}\end{array}$$

The LCD is $(k + 2)(k - 2)(k - 1)$. We write equivalent rational expressions with the LCD as the denominator.

$$\frac{5k}{(k + 2)(k - 2)} - \frac{2}{(k + 2)(k - 1)}$$

$$= \frac{5k \cdot (k - 1)}{(k + 2)(k - 2) \cdot (k - 1)} - \frac{2 \cdot (k - 2)}{(k + 2)(k - 1) \cdot (k - 2)}$$

$$= \frac{5k^2 - 5k}{(k + 2)(k - 2)(k - 1)} - \frac{2k - 4}{(k + 2)(k - 2)(k - 1)} \quad \begin{array}{l}\text{Multiply in the}\\\text{numerators.}\end{array}$$

$$= \frac{5k^2 - 5k - 2k + 4}{(k + 2)(k - 2)(k - 1)} \quad \begin{array}{l}\text{Subtract the}\\\text{numerators.}\end{array}$$

$$= \frac{5k^2 - 7k + 4}{(k + 2)(k - 2)(k - 1)} \quad \text{Simplify.}$$

The numerator is a prime polynomial, so the expression cannot be simplified further.

■ **Work Practice 9**

> **Helpful Hint**
> **Very important:** Because we are subtracting, notice the sign change on 4.

**Answers**

**8.** 3　**9.** $\dfrac{2x^2 - 5x - 9}{(x + 3)(x - 3)(x - 1)}$

**Example 10** Add: $\dfrac{2x - 1}{2x^2 - 9x - 5} + \dfrac{x + 3}{6x^2 - x - 2}$

**Solution:**

$\dfrac{2x - 1}{2x^2 - 9x - 5} + \dfrac{x + 3}{6x^2 - x - 2} = \dfrac{2x - 1}{(2x + 1)(x - 5)} + \dfrac{x + 3}{(2x + 1)(3x - 2)}$

Factor the denominators.

The LCD is $(2x + 1)(x - 5)(3x - 2)$.

$= \dfrac{(2x - 1) \cdot (3x - 2)}{(2x + 1)(x - 5) \cdot (3x - 2)} + \dfrac{(x + 3) \cdot (x - 5)}{(2x + 1)(3x - 2) \cdot (x - 5)}$

$= \dfrac{6x^2 - 7x + 2}{(2x + 1)(x - 5)(3x - 2)} + \dfrac{x^2 - 2x - 15}{(2x + 1)(x - 5)(3x - 2)}$    Multiply in the numerators.

$= \dfrac{6x^2 - 7x + 2 + x^2 - 2x - 15}{(2x + 1)(x - 5)(3x - 2)}$    Add the numerators.

$= \dfrac{7x^2 - 9x - 13}{(2x + 1)(x - 5)(3x - 2)}$    Simplify.

The numerator is a prime polynomial, so the expression cannot be simplified further.

■ Work Practice 10

**Example 11** Perform each indicated operation:

$\dfrac{7}{x - 1} + \dfrac{10x}{x^2 - 1} - \dfrac{5}{x + 1}$

**Solution:**

$\dfrac{7}{x - 1} + \dfrac{10x}{x^2 - 1} - \dfrac{5}{x + 1} = \dfrac{7}{x - 1} + \dfrac{10x}{(x - 1)(x + 1)} - \dfrac{5}{x + 1}$

The LCD is $(x - 1)(x + 1)$.

$= \dfrac{7 \cdot (x + 1)}{(x - 1) \cdot (x + 1)} + \dfrac{10x}{(x - 1) \cdot (x + 1)} - \dfrac{5 \cdot (x - 1)}{(x + 1) \cdot (x - 1)}$

$= \dfrac{7x + 7}{(x - 1)(x + 1)} + \dfrac{10x}{(x - 1)(x + 1)} - \dfrac{5x - 5}{(x + 1)(x - 1)}$    Multiply in the numerators.

$= \dfrac{7x + 7 + 10x - 5x + 5}{(x - 1)(x + 1)}$    Add and subtract the numerators.

$= \dfrac{12x + 12}{(x - 1)(x + 1)}$    Simplify.

$= \dfrac{12 (x + 1)}{(x - 1) (x + 1)}$    Factor the numerator.

$= \dfrac{12}{x - 1}$    Simplify.

■ Work Practice 11

**Practice 10**

Add:

$\dfrac{x + 1}{x^2 + x - 12} + \dfrac{2x - 1}{x^2 + 6x + 8}$

**Practice 11**

Perform each indicated operation.

$\dfrac{6}{x - 5} + \dfrac{x - 35}{x^2 - 5x} - \dfrac{2}{x}$

**Answers**

**10.** $\dfrac{3x^2 - 4x + 5}{(x - 3)(x + 4)(x + 2)}$   **11.** $\dfrac{5}{x}$

---

📱 **Calculator Explorations** **Graphing**

A grapher can be used to support the results of operations on rational expressions. For example, to verify the result of Example 7, graph

on the same set of axes. The graphs should be the same. Use a TABLE feature or a TRACE feature to see that this is true.

$$Y_1 = \dfrac{3}{x + 2} + \dfrac{2x}{x - 2} \quad \text{and} \quad Y_2 = \dfrac{2x^2 + 7x - 6}{(x + 2)(x - 2)}$$

## Vocabulary, Readiness & Video Check

*Name the operation(s) below that make each statement true.*

    **a.** Addition        **b.** Subtraction        **c.** Multiplication        **d.** Division

**1.** The denominators must be the same before performing the operation. _____

**2.** To perform this operation, you multiply the first rational expression by the reciprocal of the second rational expression. _____

**3.** Numerator times numerator all over denominator times denominator. _____

**4.** These operations are commutative (order doesn't matter.) _____

*For the rational expressions $\dfrac{5}{y}$ and $\dfrac{7}{y}$, perform each operation mentally.*

**5.** Addition        **6.** Subtraction        **7.** Multiplication        **8.** Division

*Martin-Gay Interactive Videos*

*Watch the section lecture video and answer the following questions.*

**Objective A**  **9.** In Example 1, why are we told to be especially careful with subtraction when the second numerator has more than one term?

**Objective B**  **10.** In Example 2, $(a - b)$ appears as a factor three times within the two factored denominators. Why does the LCD only contain two factors of $(a - b)$?

*See Video 6.2*

**Objective C**  **11.** Based on Example 3, complete the following statements. To write an equivalent rational expression, you multiply the _____ of the expression by the exact same thing as the denominator. This is the same as multiplying the original rational expression by _____, which doesn't change the _____ of the original expression.

## 6.2 Exercise Set MyMathLab®

**Objective A** *Be careful when subtracting! For example,* $\dfrac{8}{x + 1} - \dfrac{x + 5}{x + 1} = \dfrac{8 - (x + 5)}{x + 1} = \dfrac{3 - x}{x + 1}$ *or* $\dfrac{-x + 3}{x + 1}$.
*Use this example to help you perform the subtractions. See Examples 3 and 4.*

**1.** $\dfrac{5}{2x} - \dfrac{x + 1}{2x} =$ _____

**2.** $\dfrac{9}{5x} - \dfrac{6 - x}{5x} =$ _____

**3.** $\dfrac{y + 11}{y - 2} - \dfrac{y - 5}{y - 2} =$ _____

**4.** $\dfrac{z - 1}{z + 6} - \dfrac{z + 4}{z + 6} =$ _____

*Add or subtract as indicated. Simplify each answer. See Examples 1 through 4.*

**5.** $\dfrac{2}{xz^2} - \dfrac{5}{xz^2}$

**6.** $\dfrac{4}{x^2y} + \dfrac{2}{x^2y}$

**7.** $\dfrac{2}{x-2} + \dfrac{x}{x-2}$

**8.** $\dfrac{x}{5-x} + \dfrac{7}{5-x}$

**9.** $\dfrac{x^2}{x+2} - \dfrac{4}{x+2}$

**10.** $\dfrac{x^2}{x+6} - \dfrac{36}{x+6}$

**11.** $\dfrac{2x-6}{x^2+x-6} + \dfrac{3-3x}{x^2+x-6}$

**12.** $\dfrac{5x+2}{x^2+2x-8} + \dfrac{2-4x}{x^2+2x-8}$

▶ **13.** $\dfrac{x-5}{2x} - \dfrac{x+5}{2x}$

**14.** $\dfrac{x+4}{4x} - \dfrac{x-4}{4x}$

**Objective** **B** *Find the LCD of the rational expressions in each list. See Example 5.*

**15.** $\dfrac{2}{7}, \dfrac{3}{5x}$

**16.** $\dfrac{4}{5y}, \dfrac{3}{4y^2}$

**17.** $\dfrac{3}{x}, \dfrac{2}{x+1}$

**18.** $\dfrac{5}{2x}, \dfrac{7}{2+x}$

**19.** $\dfrac{12}{x+7}, \dfrac{8}{x-7}$

**20.** $\dfrac{1}{2x-1}, \dfrac{8}{2x+1}$

**21.** $\dfrac{5}{3x+6}, \dfrac{2x}{2x-4}$

**22.** $\dfrac{2}{3a+9}, \dfrac{5}{5a-15}$

▶ **23.** $\dfrac{2a}{a^2-b^2}, \dfrac{1}{a^2-2ab+b^2}$

**24.** $\dfrac{2a}{a^2+8a+16}, \dfrac{7a}{a^2+a-12}$

**25.** $\dfrac{x}{x^2-9}, \dfrac{5}{x}, \dfrac{7}{12-4x}$

**26.** $\dfrac{9}{x^2-25}, \dfrac{1}{50-10x}, \dfrac{6}{x}$

**Objective** **C** *Add or subtract as indicated. Simplify each answer. See Examples 6 and 7.*

▶ **27.** $\dfrac{4}{3x} + \dfrac{3}{2x}$

**28.** $\dfrac{10}{7x} - \dfrac{5}{2x}$

**29.** $\dfrac{5}{2y^2} - \dfrac{2}{7y}$

**30.** $\dfrac{4}{11x^4} - \dfrac{1}{4x^2}$

▶ **31.** $\dfrac{x-3}{x+4} - \dfrac{x+2}{x-4}$

**32.** $\dfrac{x-1}{x-5} - \dfrac{x+2}{x+5}$

**33.** $\dfrac{1}{x-5} + \dfrac{2x-19}{(x-5)(x+4)}$

**34.** $\dfrac{4x-2}{(x-5)(x+4)} - \dfrac{2}{x+4}$

*Perform the indicated operation. If possible, simplify your answer. See Example 8.*

**35.** $\dfrac{1}{a-b} + \dfrac{1}{b-a}$

**36.** $\dfrac{1}{a-3} - \dfrac{1}{3-a}$

**37.** $\dfrac{x+1}{1-x} - \dfrac{1}{x-1}$

**38.** $\dfrac{5}{1-x} - \dfrac{1}{x-1}$

**39.** $\dfrac{5}{x-2} + \dfrac{x+4}{2-x}$

**40.** $\dfrac{3}{5-x} + \dfrac{x+2}{x-5}$

*Perform each indicated operation. If possible, simplify your answer. See Examples 6 through 10.*

▶ **41.** $\dfrac{y + 1}{y^2 - 6y + 8} - \dfrac{3}{y^2 - 16}$

**42.** $\dfrac{x + 2}{x^2 - 36} - \dfrac{x}{x^2 + 9x + 18}$

**43.** $\dfrac{x + 4}{3x^2 + 11x + 6} + \dfrac{x}{2x^2 + x - 15}$

**44.** $\dfrac{x + 3}{5x^2 + 12x + 4} + \dfrac{6}{x^2 - x - 6}$

**45.** $\dfrac{x}{x^2 + 4x + 3} - \dfrac{7}{x^2 - x - 2}$

**46.** $\dfrac{a}{a^2 + 10a + 25} - \dfrac{4}{a^2 + 6a + 5}$

**47.** $\dfrac{x}{2x^2 + x - 15} - \dfrac{x + 4}{3x^2 + 11x + 6}$

**48.** $\dfrac{x}{x^2 - x - 6} - \dfrac{x + 3}{5x^2 + 12x + 4}$

**49.** $\dfrac{2}{a^2 + 2a + 1} + \dfrac{3}{a^2 - 1}$

**50.** $\dfrac{9x + 2}{3x^2 - 2x - 8} + \dfrac{7}{3x^2 + x - 4}$

**Objectives** <kbd>A</kbd> <kbd>B</kbd> <kbd>C</kbd> **Mixed Practice** *Add or subtract as indicated. If possible, simplify your answer.*
*See Examples 1 through 10.*

**51.** $\dfrac{4}{3x^2y^3} + \dfrac{5}{3x^2y^3}$

**52.** $\dfrac{7}{2xy^4} + \dfrac{1}{2xy^4}$

**53.** $\dfrac{13x - 5}{2x} - \dfrac{13x + 5}{2x}$

**54.** $\dfrac{17x + 4}{4x} - \dfrac{17x - 4}{4x}$

**55.** $\dfrac{3}{2x + 10} + \dfrac{8}{3x + 15}$

**56.** $\dfrac{10}{3x - 3} + \dfrac{1}{7x - 7}$

**57.** $\dfrac{-2}{x^2 - 3x} - \dfrac{1}{x^3 - 3x^2}$

**58.** $\dfrac{-3}{2a + 8} - \dfrac{8}{a^2 + 4a}$

**59.** $\dfrac{ab}{a^2 - b^2} + \dfrac{b}{a + b}$

**60.** $\dfrac{x}{25 - x^2} + \dfrac{2}{3x - 15}$

**61.** $\dfrac{5}{x^2 - 4} - \dfrac{3}{x^2 + 4x + 4}$

**62.** $\dfrac{3z}{z^2 - 9} - \dfrac{2}{3 - z}$

**63.** $\dfrac{3x}{2x^2 - 11x + 5} + \dfrac{7}{x^2 - 2x - 15}$

**64.** $\dfrac{2x}{3x^2 - 13x + 4} + \dfrac{5}{x^2 - 2x - 8}$

## Objective C  *Perform each indicated operation. Simplify each answer. See Example 11.*

**65.** $\dfrac{2}{x + 1} - \dfrac{3x}{3x + 3} + \dfrac{1}{2x + 2}$

**66.** $\dfrac{5}{3x - 6} - \dfrac{x}{x - 2} + \dfrac{3 + 2x}{5x - 10}$

**67.** $\dfrac{3}{x + 3} + \dfrac{5}{x^2 + 6x + 9} - \dfrac{x}{x^2 - 9}$

**68.** $\dfrac{x + 2}{x^2 - 2x - 3} + \dfrac{x}{x - 3} - \dfrac{x}{x + 1}$

**69.** $\dfrac{x}{x^2 - 9} + \dfrac{3}{x^2 - 6x + 9} - \dfrac{1}{x + 3}$

**70.** $\dfrac{3}{x^2 - 9} - \dfrac{x}{x^2 - 6x + 9} + \dfrac{1}{x + 3}$

**71.** $\left(\dfrac{1}{x} + \dfrac{2}{3}\right) - \left(\dfrac{1}{x} - \dfrac{2}{3}\right)$

**72.** $\left(\dfrac{1}{2} + \dfrac{2}{x}\right) - \left(\dfrac{1}{2} - \dfrac{1}{x}\right)$

## Review

*Use the distributive property to multiply each expression. See Sections 1.3 and 1.4.*

**73.** $12\left(\dfrac{2}{3} + \dfrac{1}{6}\right)$

**74.** $14\left(\dfrac{1}{7} + \dfrac{3}{14}\right)$

**75.** $x^2\left(\dfrac{4}{x^2} + 1\right)$

**76.** $5y^2\left(\dfrac{1}{y^2} - \dfrac{1}{5}\right)$

## Concept Extensions

*Find and correct each error. See the Concept Check in this section.*

**77.**

$$\dfrac{2x - 3}{x^2 + 1} - \dfrac{x - 6}{x^2 + 1} = \dfrac{2x - 3 - x - 6}{x^2 + 1}$$
$$= \dfrac{x - 9}{x^2 + 1}$$

**78.**
$$\dfrac{7}{x + 7} - \dfrac{x + 3}{x + 7} = \dfrac{7 - x - 3}{(x + 7)^2}$$
$$= \dfrac{-x + 4}{(x + 7)^2}$$

*Solve.*

△ **79.** Find the perimeter and the area of the square.

$\dfrac{x}{x + 5}$ ft

△ **80.** Find the perimeter of the quadrilateral.

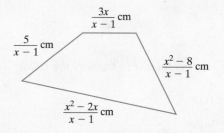

$\dfrac{3x}{x - 1}$ cm

$\dfrac{5}{x - 1}$ cm

$\dfrac{x^2 - 8}{x - 1}$ cm

$\dfrac{x^2 - 2x}{x - 1}$ cm

**81.** When is the LCD of two rational expressions equal to the product of their denominators? ( *Hint:* What is the LCD of $\dfrac{1}{x}$ and $\dfrac{7}{x + 5}$? )

**82.** When is the LCD of two rational expressions with different denominators equal to one of the denominators? ( *Hint:* What is the LCD of $\dfrac{3x}{x + 2}$ and $\dfrac{7x + 1}{(x + 2)^3}$? )

**83.** In your own words, explain how to add rational expressions with different denominators.

**84.** In your own words, explain how to multiply rational expressions.

**85.** In your own words, explain how to divide rational expressions.

**86.** In your own words, explain how to subtract rational expressions with different denominators.

## Mixed Practice (Sections 6.1, 6.2) *Perform the indicated operation. If possible, simplify your answer.*

**87.** $\left(\dfrac{2}{3} - \dfrac{1}{x}\right) \cdot \left(\dfrac{3}{x} + \dfrac{1}{2}\right)$

**88.** $\left(\dfrac{2}{3} - \dfrac{1}{x}\right) \div \left(\dfrac{3}{x} + \dfrac{1}{2}\right)$

**89.** $\left(\dfrac{2a}{3}\right)^2 \div \left(\dfrac{a^2}{a+1} - \dfrac{1}{a+1}\right)$

**90.** $\left(\dfrac{x+2}{2x} - \dfrac{x-2}{2x}\right) \cdot \left(\dfrac{5x}{4}\right)^2$

**91.** $\left(\dfrac{2x}{3}\right)^2 \div \left(\dfrac{x}{3}\right)^2$

**92.** $\left(\dfrac{2x}{3}\right)^2 \cdot \left(\dfrac{3}{x}\right)^2$

**93.** $\left(\dfrac{x}{x+1} - \dfrac{x}{x-1}\right) \div \dfrac{x}{2x+2}$

**94.** $\dfrac{x}{2x+2} \div \left(\dfrac{x}{x+1} + \dfrac{x}{x-1}\right)$

**95.** $\dfrac{4}{x} \cdot \left(\dfrac{2}{x+2} - \dfrac{2}{x-2}\right)$

**96.** $\dfrac{1}{x+1} \cdot \left(\dfrac{5}{x} + \dfrac{2}{x-3}\right)$

*Perform each indicated operation. (Hint: First write each expression with positive exponents.)*

**97.** $x^{-1} + (2x)^{-1}$

**98.** $y^{-1} + (4y)^{-1}$

**99.** $4x^{-2} - 3x^{-1}$

**100.** $(4x)^{-2} - (3x)^{-1}$

*Use a graphing calculator to support the results of each exercise.*

**101.** Exercise 7

**102.** Exercise 8

## 6.3 Simplifying Complex Fractions

A rational expression whose numerator, denominator, or both contain one or more rational expressions is called a **complex rational expression** or a **complex fraction.** Examples are

$$\dfrac{\dfrac{1}{a}}{\dfrac{b}{2}} \qquad \dfrac{\dfrac{x}{2y^2}}{\dfrac{6x-2}{9y}} \qquad \dfrac{x+\dfrac{1}{y}}{y+1}$$

The parts of a complex fraction are

$$\left.\dfrac{\dfrac{x}{y+2}}{7+\dfrac{1}{y}}\right\}$$

← Numerator of complex fraction

← Main fraction bar

← Denominator of complex fraction

Our goal in this section is to simplify complex fractions. A complex fraction is simplified when it is in the form $\dfrac{P}{Q}$, where $P$ and $Q$ are polynomials that have no common factors. Two methods of simplifying complex fractions are introduced.

**Objectives**

**A** Simplify Complex Fractions by Simplifying the Numerator and Denominator and Then Dividing. ▶

**B** Simplify Complex Fractions by Multiplying by the Least Common Denominator (LCD). ▶

**C** Simplify Expressions with Negative Exponents. ▶

## Objective A  Method 1: Simplifying a Complex Fraction by Simplifying the Numerator and Denominator and Then Dividing ▶

In the first method we study, we simplify complex fractions by simplifying and dividing.

---

### Simplifying a Complex Fraction: Method 1

**Step 1:** Simplify the numerator and the denominator of the complex fraction so that each is a single fraction.

**Step 2:** Perform the indicated division by multiplying the numerator of the complex fraction by the reciprocal of the denominator of the complex fraction.

**Step 3:** Simplify if possible.

---

**Example 1**  Simplify: $\dfrac{\dfrac{2x}{27y^2}}{\dfrac{6x^2}{9}}$

**Solution:** The numerator of the complex fraction is already a single fraction, and so is the denominator. Perform the indicated division by multiplying the numerator, $\dfrac{2x}{27y^2}$, by the reciprocal of the denominator, $\dfrac{6x^2}{9}$. Then simplify.

*(Continued on next page)*

**Practice 1**

Use Method 1 to simplify:

$$\dfrac{\dfrac{3a}{25b^2}}{\dfrac{9a^2}{5b}}$$

**Answer**

**1.** $\dfrac{1}{15ab}$

453

$$\frac{\dfrac{2x}{27y^2}}{\dfrac{6x^2}{9}} = \frac{2x}{27y^2} \div \frac{6x^2}{9}$$

$$= \frac{2x}{27y^2} \cdot \frac{9}{6x^2} \qquad \text{Multiply by the reciprocal of } \frac{6x^2}{9}.$$

$$= \frac{2x \cdot 9}{27y^2 \cdot 6x^2}$$

$$= \frac{1}{9xy^2}$$

■ Work Practice 1

## Practice 2

Use Method 1 to simplify:

$$\frac{\dfrac{6x}{x-5}}{\dfrac{12}{x+5}}$$

**Helpful Hint**    Both the numerator and denominator are single fractions, so we perform the indicated division.

**Example 2**  Simplify: $\dfrac{\dfrac{5x}{x+2}}{\dfrac{10}{x-2}}$

**Solution:**

$$\frac{\left\{\dfrac{5x}{x+2}\right.}{\left\{\dfrac{10}{x-2}\right.} = \frac{5x}{x+2} \div \frac{10}{x-2} = \frac{5x}{x+2} \cdot \frac{x-2}{10} \quad \text{Multiply by the reciprocal of } \dfrac{10}{x-2}.$$

$$= \frac{5\,x(x-2)}{2\cdot 5\,(x+2)}$$

$$= \frac{x(x-2)}{2(x+2)} \qquad \text{Simplify.}$$

■ Work Practice 2

✓ **Concept Check**  Which of the following are equivalent to $\dfrac{\dfrac{1}{x}}{\dfrac{3}{y}}$?

**a.** $\dfrac{1}{x} \div \dfrac{3}{y}$      **b.** $\dfrac{1}{x} \cdot \dfrac{y}{3}$      **c.** $\dfrac{1}{x} \div \dfrac{y}{3}$

## Practice 3

Use Method 1 to simplify:

$$\frac{\dfrac{x}{y^2} - \dfrac{1}{y}}{\dfrac{y}{x^2} - \dfrac{1}{x}}$$

**Example 3**  Simplify: $\dfrac{\dfrac{x}{y^2} + \dfrac{1}{y}}{\dfrac{y}{x^2} + \dfrac{1}{x}}$

**Solution:**  First we simplify the numerator and the denominator of the complex fraction separately so that each is a single fraction.

$$\frac{\dfrac{x}{y^2} + \dfrac{1}{y}}{\dfrac{y}{x^2} + \dfrac{1}{x}} = \frac{\dfrac{x}{y^2} + \dfrac{1 \cdot y}{y \cdot y}}{\dfrac{y}{x^2} + \dfrac{1 \cdot x}{x \cdot x}} \qquad \begin{array}{l}\text{The LCD is } y^2. \\[2ex] \text{The LCD is } x^2.\end{array}$$

$$= \frac{\dfrac{x+y}{y^2}}{\dfrac{y+x}{x^2}} \qquad \begin{array}{l}\text{Add.} \\[2ex] \text{Add.}\end{array}$$

$$= \frac{x + y}{y^2} \cdot \frac{x^2}{y + x} \qquad \text{Multiply by the reciprocal of } \frac{y + x}{x^2}.$$

$$= \frac{x^2 (x + y)}{y^2 (y + x)}$$

$$= \frac{x^2}{y^2} \qquad \text{Simplify.}$$

■ Work Practice 3

## Objective B Method 2: Simplifying a Complex Fraction by Multiplying the Numerator and Denominator by the LCD ▶

With this method, we multiply the numerator and the denominator of the complex fraction by the least common denominator (LCD) of all fractions in the complex fraction.

---

**Simplifying a Complex Fraction: Method 2**

**Step 1:** Multiply the numerator and the denominator of the complex fraction by the LCD of all the fractions in both the numerator and the denominator.

**Step 2:** Simplify.

---

**Example 4** Simplify: $\dfrac{\dfrac{5x}{x + 2}}{\dfrac{10}{x - 2}}$

**Solution:** Notice we are reworking Example 2 using Method 2. The least common denominator of $\dfrac{5x}{x + 2}$ and $\dfrac{10}{x - 2}$ is $(x + 2)(x - 2)$. We multiply both the numerator, $\dfrac{5x}{x + 2}$, and the denominator, $\dfrac{10}{x - 2}$, by this LCD.

$$\frac{\dfrac{5x}{x + 2}}{\dfrac{10}{x - 2}} = \frac{\left(\dfrac{5x}{x + 2}\right) \cdot (x + 2)(x - 2)}{\left(\dfrac{10}{x - 2}\right) \cdot (x + 2)(x - 2)} \qquad \begin{array}{l}\text{Multiply the numerator and denominator by} \\ \text{the LCD.}\end{array}$$

$$= \frac{5x \cdot (x - 2)}{2 \cdot 5 \cdot (x + 2)} \qquad \text{Simplify.}$$

$$= \frac{x(x - 2)}{2(x + 2)} \qquad \text{Simplify.}$$

■ Work Practice 4

**Practice 4**

Use Method 2 to simplify:

$$\frac{\dfrac{6x}{x - 5}}{\dfrac{12}{x + 5}}$$

Examples 2 and 4 are the same and simplify to the same rational expression. Regardless of what method you choose to use, the simplification is the same.

**Answer**

4. $\dfrac{x(x + 5)}{2(x - 5)}$

## Practice 5

Use Method 2 to simplify:

$$\dfrac{\dfrac{x}{y^2} - \dfrac{1}{y}}{\dfrac{y}{x^2} - \dfrac{1}{x}}$$

**Example 5** Simplify: $\dfrac{\dfrac{x}{y^2} + \dfrac{1}{y}}{\dfrac{y}{x^2} + \dfrac{1}{x}}$

**Solution:** The least common denominator of $\dfrac{x}{y^2}, \dfrac{1}{y}, \dfrac{y}{x^2},$ and $\dfrac{1}{x}$ is $x^2y^2$.

$$\dfrac{\dfrac{x}{y^2} + \dfrac{1}{y}}{\dfrac{y}{x^2} + \dfrac{1}{x}} = \dfrac{\left(\dfrac{x}{y^2} + \dfrac{1}{y}\right) \cdot x^2 y^2}{\left(\dfrac{y}{x^2} + \dfrac{1}{x}\right) \cdot x^2 y^2}$$    Multiply the numerator and denominator by the LCD.

$$= \dfrac{\dfrac{x}{y^2} \cdot x^2 y^2 + \dfrac{1}{y} \cdot x^2 y^2}{\dfrac{y}{x^2} \cdot x^2 y^2 + \dfrac{1}{x} \cdot x^2 y^2}$$    Use the distributive property.

$$= \dfrac{x^3 + x^2 y}{y^3 + xy^2}$$    Simplify.

$$= \dfrac{x^2(x + y)}{y^2(y + x)}$$    Factor.

$$= \dfrac{x^2}{y^2}$$    Simplify.

■ Work Practice 5

**Helpful Hint**

Just as for Examples 2 and 4, Examples 3 and 5 are the same and they simplify to the same rational expression. Note that regardless of which method you use, the result is the same.

## Objective C Simplifying Expressions with Negative Exponents

Some expressions containing negative exponents can be written as complex fractions. To simplify these expressions, we first write them as equivalent expressions with positive exponents.

## Practice 6

Simplify: $\dfrac{2x^{-1} + 3y^{-1}}{x^{-1} - 2y^{-1}}$

**Example 6** Simplify: $\dfrac{x^{-1} + 2xy^{-1}}{x^{-2} - x^{-2}y^{-1}}$

**Solution:** This fraction does not appear to be a complex fraction. However, if we write it by using only positive exponents, we see that it is a complex fraction.

$$\dfrac{x^{-1} + 2xy^{-1}}{x^{-2} - x^{-2}y^{-1}} = \dfrac{\dfrac{1}{x} + \dfrac{2x}{y}}{\dfrac{1}{x^2} - \dfrac{1}{x^2 y}}$$

The LCD of $\dfrac{1}{x}, \dfrac{2x}{y}, \dfrac{1}{x^2},$ and $\dfrac{1}{x^2 y}$ is $x^2 y$. We multiply both the numerator and denominator by $x^2 y$.

## Answers

5. $-\dfrac{x^2}{y^2}$   6. $\dfrac{2y + 3x}{y - 2x}$

$$\frac{\dfrac{1}{x} + \dfrac{2x}{y}}{\dfrac{1}{x^2} - \dfrac{1}{x^2 y}} = \frac{\left(\dfrac{1}{x} + \dfrac{2x}{y}\right) \cdot x^2 y}{\left(\dfrac{1}{x^2} - \dfrac{1}{x^2 y}\right) \cdot x^2 y}$$

$$= \frac{\dfrac{1}{x} \cdot x^2 y + \dfrac{2x}{y} \cdot x^2 y}{\dfrac{1}{x^2} \cdot x^2 y - \dfrac{1}{x^2 y} \cdot x^2 y} \qquad \text{Use the distributive property.}$$

$$= \frac{xy + 2x^3}{y - 1} \qquad \text{Simplify.}$$

$$\text{or } \frac{x(y + 2x^2)}{y - 1}$$

■ **Work Practice 6**

**Example 7** Simplify: $\dfrac{(2x)^{-1} + 1}{2x^{-1} - 1}$

**Solution:**

$$\frac{(2x)^{-1} + 1}{2x^{-1} - 1} = \frac{\dfrac{1}{2x} + 1}{\dfrac{2}{x} - 1} \qquad \text{Write using positive exponents.}$$

$$= \frac{\left(\dfrac{1}{2x} + 1\right) \cdot 2x}{\left(\dfrac{2}{x} - 1\right) \cdot 2x} \qquad \text{The LCD of } \dfrac{1}{2x} \text{ and } \dfrac{2}{x} \text{ is } 2x.$$

$$= \frac{\dfrac{1}{2x} \cdot 2x + 1 \cdot 2x}{\dfrac{2}{x} \cdot 2x - 1 \cdot 2x} \qquad \text{Use distributive property.}$$

$$= \frac{1 + 2x}{4 - 2x} \qquad \text{Simplify.}$$

■ **Work Practice 7**

**Practice 7**

Simplify: $\dfrac{5 - 3x^{-1}}{2 + (3x)^{-1}}$

**Helpful Hint** Don't forget that $(2x)^{-1} = \dfrac{1}{2x}$ but $2x^{-1} = 2 \cdot \dfrac{1}{x} = \dfrac{2}{x}$.

**Answer**

7. $\dfrac{15x - 9}{6x + 1}$ or $\dfrac{3(5x - 3)}{6x + 1}$

## Vocabulary, Readiness & Video Check

*Complete the steps by writing the simplified complex fraction.*

**1.** $\dfrac{\dfrac{7}{x}}{\dfrac{1}{x} + \dfrac{z}{x}} = \dfrac{x\left(\dfrac{7}{x}\right)}{x\left(\dfrac{1}{x}\right) + x\left(\dfrac{z}{x}\right)} = $ _____

**2.** $\dfrac{\dfrac{x}{4}}{\dfrac{x^2}{2} + \dfrac{1}{4}} = \dfrac{4\left(\dfrac{x}{4}\right)}{4\left(\dfrac{x^2}{2}\right) + 4\left(\dfrac{1}{4}\right)} = $ _____

*Write each with positive exponents.*

**3.** $x^{-2} =$ _____

**4.** $y^{-3} =$ _____

**5.** $2x^{-1} =$ _____

**6.** $(2x)^{-1} =$ _____

**7.** $(9y)^{-1} =$ _____

**8.** $9y^{-2} =$ _____

*Martin-Gay Interactive Videos*　　*Watch the section lecture video and answer the following questions.*

**Objective A 9.** From ⊞ Example 2, before you can rewrite the complex fraction as division, describe how it must appear. ▶

**Objective B 10.** How does finding an LCD in Method 2, as in ⊞ Example 3, differ from finding an LCD in Method 1? In your answer, mention the purpose of the LCD in each method. ▶

**Objective C 11.** Based on ⊞ Example 4, what connection is there between negative exponents and complex fractions? ▶

*See Video 6.3* 🍎

## 6.3 Exercise Set MyMathLab® ▶

**Objectives A B Mixed Practice** *Simplify each complex fraction. See Examples 1 through 5.*

**▶1.** $\dfrac{\dfrac{10}{3x}}{\dfrac{5}{6x}}$

**2.** $\dfrac{\dfrac{15}{2x}}{\dfrac{5}{6x}}$

**3.** $\dfrac{1 + \dfrac{2}{5}}{2 + \dfrac{3}{5}}$

**4.** $\dfrac{2 + \dfrac{1}{7}}{3 - \dfrac{4}{7}}$

**5.** $\dfrac{\dfrac{4}{x-1}}{\dfrac{x}{x-1}}$

**6.** $\dfrac{\dfrac{x}{x+2}}{\dfrac{2}{x+2}}$

**7.** $\dfrac{1 - \dfrac{2}{x}}{x + \dfrac{4}{9x}}$

**8.** $\dfrac{5 - \dfrac{3}{x}}{x + \dfrac{2}{3x}}$

**▶9.** $\dfrac{\dfrac{4x^2 - y^2}{xy}}{\dfrac{2}{y} - \dfrac{1}{x}}$

**10.** $\dfrac{\dfrac{x^2 - 9y^2}{xy}}{\dfrac{1}{y} - \dfrac{3}{x}}$

**11.** $\dfrac{\dfrac{x+1}{3}}{\dfrac{2x-1}{6}}$

**12.** $\dfrac{\dfrac{x+3}{12}}{\dfrac{4x-5}{15}}$

**13.** $\dfrac{\dfrac{2}{x} + \dfrac{3}{x^2}}{\dfrac{4}{x^2} - \dfrac{9}{x}}$

**14.** $\dfrac{\dfrac{2}{x^2} + \dfrac{1}{x}}{\dfrac{4}{x^2} - \dfrac{1}{x}}$

**15.** $\dfrac{\dfrac{1}{x} + \dfrac{2}{x^2}}{x + \dfrac{8}{x^2}}$

**16.** $\dfrac{\dfrac{1}{y} + \dfrac{3}{y^2}}{y + \dfrac{27}{y^2}}$

**17.** $\dfrac{\dfrac{4}{5-x} + \dfrac{5}{x-5}}{\dfrac{2}{x} + \dfrac{3}{x-5}}$

**18.** $\dfrac{\dfrac{3}{x-4} - \dfrac{2}{4-x}}{\dfrac{2}{x-4} - \dfrac{2}{x}}$

**▶19.** $\dfrac{\dfrac{x+2}{x} - \dfrac{2}{x-1}}{\dfrac{x+1}{x} + \dfrac{x+1}{x-1}}$

**20.** $\dfrac{\dfrac{5}{a+2} - \dfrac{1}{a-2}}{\dfrac{3}{2+a} + \dfrac{6}{2-a}}$

**21.** $\dfrac{\dfrac{2}{x} + 3}{\dfrac{4}{x^2} - 9}$

**22.** $\dfrac{2 + \dfrac{1}{x}}{4x - \dfrac{1}{x}}$

**23.** $\dfrac{1 - \dfrac{1}{y}}{\dfrac{1}{y^2} - 1}$

**24.** $\dfrac{1 - \dfrac{2}{x}}{x - \dfrac{4}{x}}$

**25.** $\dfrac{\dfrac{-2x}{x^2 - xy}}{\dfrac{y}{x^2}}$

**26.** $\dfrac{\dfrac{7y}{x^2 + xy}}{\dfrac{y^2}{x^2}}$

**27.** $\dfrac{\dfrac{2}{x} + \dfrac{1}{x^2}}{\dfrac{y}{x^2} + 1}$

**28.** $\dfrac{\dfrac{5}{x^2} - \dfrac{2}{x}}{\dfrac{1}{x} + 2}$

**29.** $\dfrac{\dfrac{x}{9} - \dfrac{1}{x}}{1 + \dfrac{3}{x}}$

**30.** $\dfrac{\dfrac{x}{4} - \dfrac{4}{x}}{1 - \dfrac{4}{x}}$

**31.** $\dfrac{\dfrac{x-1}{x^2 - 4}}{1 + \dfrac{1}{x-2}}$

**32.** $\dfrac{\dfrac{x-2}{x^2 - 9}}{1 + \dfrac{1}{x-3}}$

**33.** $\dfrac{\dfrac{2}{x+5} + \dfrac{4}{x+3}}{\dfrac{3x+13}{x^2 + 8x + 15}}$

**34.** $\dfrac{\dfrac{2}{x+2} + \dfrac{6}{x+7}}{\dfrac{4x+13}{x^2 + 9x + 14}}$

**Objective C**   *Simplify. See Examples 6 and 7.*

**35.** $\dfrac{x^{-1}}{x^{-2} + y^{-2}}$

**36.** $\dfrac{a^{-3} + b^{-1}}{a^{-2}}$

▶ **37.** $\dfrac{2a^{-1} + 3b^{-2}}{a^{-1} - b^{-1}}$

**38.** $\dfrac{x^{-1} + y^{-1}}{3x^{-2} + 5y^{-2}}$

**39.** $\dfrac{1}{x - x^{-1}}$

**40.** $\dfrac{x^{-2}}{x + 3x^{-1}}$

**41.** $\dfrac{a^{-1} + 1}{a^{-1} - 1}$

**42.** $\dfrac{a^{-1} - 4}{4 + a^{-1}}$

**43.** $\dfrac{3x^{-1} + (2y)^{-1}}{x^{-2}}$

**44.** $\dfrac{5x^{-2} - 3y^{-1}}{x^{-1} + y^{-1}}$

**45.** $\dfrac{2a^{-1} + (2a)^{-1}}{a^{-1} + 2a^{-2}}$

**46.** $\dfrac{a^{-1} + 2a^{-2}}{2a^{-1} + (2a)^{-1}}$

**47.** $\dfrac{5x^{-1} + 2y^{-1}}{x^{-2}y^{-2}}$

**48.** $\dfrac{x^{-2}y^{-2}}{5x^{-1} + 2y^{-1}}$

**49.** $\dfrac{5x^{-1} - 2y^{-1}}{25x^{-2} - 4y^{-2}}$

**50.** $\dfrac{3x^{-1} + 3y^{-1}}{4x^{-2} - 9y^{-2}}$

## Review

*Solve each equation for x. See Sections 2.1 and 5.7.*

**51.** $7x + 2 = x - 3$

**52.** $4 - 2x = 17 - 5x$

**53.** $x^2 = 4x - 4$

**54.** $5x^2 + 10x = 15$

**55.** $\dfrac{x}{3} - 5 = 13$

**56.** $\dfrac{2x}{9} + 1 = \dfrac{7}{9}$

## Concept Extensions

*Solve. For Exercises 57 through 60, see the Concept Check in the section.*

**57.** Which of the following are equivalent to $\dfrac{\frac{x+1}{9}}{\frac{y-2}{5}}$?

 **a.** $\dfrac{x+1}{9} \div \dfrac{y-2}{5}$

 **b.** $\dfrac{x+1}{9} \cdot \dfrac{y-2}{5}$

 **c.** $\dfrac{x+1}{9} \cdot \dfrac{5}{y-2}$

**58.** Which of the following are equivalent to $\dfrac{\frac{a}{7}}{\frac{b}{13}}$?

 **a.** $\dfrac{a}{7} \cdot \dfrac{b}{13}$

 **b.** $\dfrac{a}{7} \div \dfrac{b}{13}$

 **c.** $\dfrac{a}{7} \div \dfrac{13}{b}$

 **d.** $\dfrac{a}{7} \cdot \dfrac{13}{b}$

**59.** Which of the following are equivalent to $\dfrac{\frac{1}{x}}{\frac{3}{y}}$?

 **a.** $\dfrac{1}{x} \div \dfrac{3}{y}$     **b.** $\dfrac{1}{x} \cdot \dfrac{y}{3}$     **c.** $\dfrac{1}{x} \div \dfrac{y}{3}$

**60.** Which of the following are equivalent to $\dfrac{\frac{5}{2}}{a}$?

 **a.** $\dfrac{5}{1} \div \dfrac{2}{a}$     **b.** $\dfrac{1}{5} \div \dfrac{2}{a}$     **c.** $\dfrac{5}{1} \cdot \dfrac{2}{a}$

**61.** When the source of a sound is traveling toward a listener, the pitch that the listener hears due to the Doppler effect is given by the complex rational expression $\dfrac{a}{1 - \dfrac{s}{770}}$, where $a$ is the actual pitch of the sound and $s$ is the speed of the sound source. Simplify this expression.

**62.** In baseball, the earned run average (ERA) statistic gives the average number of earned runs scored on a pitcher per game. It is computed with the following expression: $\dfrac{E}{\dfrac{I}{9}}$, where $E$ is the number of earned runs scored on a pitcher and $I$ is the total number of innings pitched by the pitcher. Simplify this expression.

**63.** In your own words, explain one method for simplifying a complex fraction.

**64.** Explain your favorite method for simplifying a complex fraction and why.

*Simplify.*

**65.** $\dfrac{\dfrac{2}{y^2} - \dfrac{5}{xy} - \dfrac{3}{x^2}}{\dfrac{2}{y^2} + \dfrac{7}{xy} + \dfrac{3}{x^2}}$

**66.** $\dfrac{\dfrac{2}{x^2} - \dfrac{1}{xy} - \dfrac{1}{y^2}}{\dfrac{1}{x^2} - \dfrac{3}{xy} + \dfrac{2}{y^2}}$

**67.** $\dfrac{1}{1 + (1 + x)^{-1}}$

**68.** $\dfrac{(x + 2)^{-1} + (x - 2)^{-1}}{(x^2 - 4)^{-1}}$

**69.** $\dfrac{x}{1 - \dfrac{1}{1 + \dfrac{1}{x}}}$

**70.** $\dfrac{x}{1 - \dfrac{1}{1 - \dfrac{1}{x}}}$

*In the study of calculus, the difference quotient $\dfrac{f(a + h) - f(a)}{h}$ is often found and simplified. Find and simplify this quotient for each function $f(x)$ by following steps **a** through **d**.*

**a.** *Find $f(a + h)$.*

**b.** *Find $f(a)$.*

**c.** *Use steps **a** and **b** to find $\dfrac{f(a + h) - f(a)}{h}$*

**d.** *Simplify the result of step **c**.*

**71.** $f(x) = \dfrac{1}{x}$

**72.** $f(x) = \dfrac{5}{x}$

**73.** $f(x) = \dfrac{3}{x + 1}$

**74.** $f(x) = \dfrac{2}{x^2}$

**Solving Equations Containing Rational Expressions** ▶

## Objective A Solving Equations Containing Rational Expressions ▶

In this section, we solve rational equations. A *rational equation* is an equation containing at least one rational expression. Before beginning this section, make sure that you understand the difference between an *equation* and an *expression*. An **equation** contains an equal sign and an **expression** does not.

| **Equation** | **Expression** |
|:---:|:---:|
| $\dfrac{x}{2} + \dfrac{x}{6} \underset{\uparrow}{=} \dfrac{2}{3}$ | $\dfrac{x}{2} + \dfrac{x}{6}$ |
| equal sign | |

> **Helpful Hint** The method described is for equations only. It may *not* be used for performing operations on expressions.

### Solving an Equation Containing Rational Expressions

To solve an *equation* containing rational expressions, first clear the equation of fractions by multiplying both sides of the equation by the LCD of all rational expressions. Then solve as usual.

✔ **Concept Check** True or false? Clearing fractions is valid when solving an equation and when simplifying rational expressions. Explain.

**Practice 1**

Solve: $\dfrac{5x}{6} + \dfrac{1}{2} = \dfrac{x}{3}$

**Example 1** Solve: $\dfrac{4x}{5} + \dfrac{3}{2} = \dfrac{3x}{10}$

**Solution:** The LCD of $\dfrac{4x}{5}, \dfrac{3}{2},$ and $\dfrac{3x}{10}$ is 10. We multiply both sides of the equation by 10.

$$\dfrac{4x}{5} + \dfrac{3}{2} = \dfrac{3x}{10}$$

$$10\left(\dfrac{4x}{5} + \dfrac{3}{2}\right) = 10\left(\dfrac{3x}{10}\right) \quad \text{Multiply both sides by the LCD.}$$

$$10 \cdot \dfrac{4x}{5} + 10 \cdot \dfrac{3}{2} = 10 \cdot \dfrac{3x}{10} \quad \text{Use the distributive property.}$$

$$8x + 15 = 3x \quad \text{Simplify.}$$

$$15 = -5x \quad \text{Subtract } 8x \text{ from both sides.}$$

$$-3 = x \quad \text{Solve.}$$

We verify this solution by replacing $x$ with $-3$ in the original equation.

**Check:** $\dfrac{4x}{5} + \dfrac{3}{2} = \dfrac{3x}{10}$

$$\dfrac{4(-3)}{5} + \dfrac{3}{2} \overset{?}{=} \dfrac{3(-3)}{10}$$

$$\dfrac{-12}{5} + \dfrac{3}{2} \overset{?}{=} \dfrac{-9}{10}$$

$$-\dfrac{24}{10} + \dfrac{15}{10} \overset{?}{=} -\dfrac{9}{10}$$

$$-\dfrac{9}{10} = -\dfrac{9}{10} \quad \text{True}$$

The solution set is $\{-3\}$.

**Answer**

1. $\{-1\}$

✔ **Concept Check Answer**

false; answers may vary

▇ Work Practice 1

The important difference about the equations in this section is that the denominator of a rational expression may contain a variable. Recall that a rational expression is undefined for values of the variable that make the denominator 0. If a proposed solution makes a denominator 0, then it must be rejected as a solution of the original equation. Such proposed solutions are called **extraneous solutions.**

The following steps may be used to solve equations containing rational expressions.

---

### To Solve an Equation Containing Rational Expressions

**Step 1:** Multiply both sides of the equation by the LCD of all rational expressions in the equation.

**Step 2:** Simplify both sides.

**Step 3:** Determine whether the equation is linear, quadratic, or higher degree and solve accordingly.

**Step 4:** Check the solution in the original equation.

---

**Example 2** Solve: $\dfrac{3}{x} - \dfrac{x+21}{3x} = \dfrac{5}{3}$

**Solution:** The LCD of the denominators $x$, $3x$, and $3$ is $3x$. We multiply both sides by $3x$. (We are allowed to do this as long as $x \neq 0$.)

$$\frac{3}{x} - \frac{x+21}{3x} = \frac{5}{3}$$

$$3x\left(\frac{3}{x} - \frac{x+21}{3x}\right) = 3x\left(\frac{5}{3}\right) \qquad \text{Multiply both sides by the LCD.}$$

$$3x \cdot \frac{3}{x} - 3x \cdot \frac{x+21}{3x} = 3x \cdot \frac{5}{3} \qquad \text{Use the distributive property.}$$

$$9 - (x+21) = 5x \qquad \text{Simplify.}$$

$$9 - x - 21 = 5x$$

$$-12 = 6x$$

$$-2 = x \qquad \text{Solve.}$$

The proposed solution is $-2$.

**Check:** We check the proposed solution in the original equation.

$$\frac{3}{x} - \frac{x+21}{3x} = \frac{5}{3}$$

$$\frac{3}{-2} - \frac{-2+21}{3(-2)} \overset{?}{=} \frac{5}{3}$$

$$-\frac{9}{6} + \frac{19}{6} \overset{?}{=} \frac{5}{3}$$

$$\frac{10}{6} = \frac{5}{3} \qquad \text{True}$$

The solution set is $\{-2\}$.

■ Work Practice 2

**Practice 2**

Solve: $\dfrac{5}{x} - \dfrac{3x+6}{2x} = \dfrac{7}{2}$

Let's talk more about multiplying both sides of an equation by the LCD of the rational expressions in the equation. In Example 3 that follows, the LCD is $x - 2$, so we will first multiply both sides of the equation by $x - 2$. Recall that the multiplication property for equations allows us to multiply both sides of an equation by

**Answer**

2. $\left\{\dfrac{2}{5}\right\}$

any *nonzero* number. In other words, for Example 3 below, we may multiply both sides of the equation by $x - 2$ *as long as* $x - 2 \neq 0$ or as long as $x \neq 2$. Keep this in mind when solving these equations.

**Practice 3**

Solve: $\dfrac{x + 5}{x - 3} = \dfrac{2(x + 1)}{x - 3}$

**Example 3**  Solve: $\dfrac{x + 6}{x - 2} = \dfrac{2(x + 2)}{x - 2}$

**Solution:** First multiply both sides of the equation by the LCD, $x - 2$. (Remember, we can only do this if $x \neq 2$, so that we are not multiplying by 0.)

$$\frac{x + 6}{x - 2} = \frac{2(x + 2)}{x - 2}$$

$$(x - 2) \cdot \frac{x + 6}{x - 2} = (x - 2) \cdot \frac{2(x + 2)}{x - 2} \qquad \text{Multiply both sides by } x - 2.$$

$$x + 6 = 2(x + 2) \qquad \text{Simplify.}$$

$$x + 6 = 2x + 4 \qquad \text{Use the distributive property.}$$

$$2 = x \qquad \text{Solve.}$$

From above, we assumed that $x \neq 2$, so this equation has no solution. This will also show as we attempt to check this proposed solution.

**Check:**   The proposed solution is 2. Notice that 2 makes the denominators 0 in the original equation. This can also be seen in a check. Check the proposed solution 2 in the original equation.

$$\frac{x + 6}{x - 2} = \frac{2(x + 2)}{x - 2}$$

$$\frac{2 + 6}{6 - 2} = \frac{2(2 + 2)}{2 - 2}$$

$$\frac{8}{0} = \frac{2(4)}{0}$$

The denominators are 0, so 2 is not a solution of the original equation. The solution set is $\varnothing$ or $\{\ \}$.

■ **Work Practice 3**

**Practice 4**

Solve: $\dfrac{3x}{3x - 1} + \dfrac{1}{x} = \dfrac{1}{3x - 1}$

**Example 4**  Solve: $\dfrac{2x}{2x - 1} + \dfrac{1}{x} = \dfrac{1}{2x - 1}$

**Solution:**   The LCD is $x(2x - 1)$. Multiply both sides by $x(2x - 1)$. By the distributive property, this is the same as multiplying each term by $x(2x - 1)$.

$$x(2x - 1) \cdot \frac{2x}{2x - 1} + x(2x - 1) \cdot \frac{1}{x} = x(2x - 1) \cdot \frac{1}{2x - 1}$$

$$x(2x) + (2x - 1) = x$$

$$2x^2 + 2x - 1 - x = 0$$

$$2x^2 + x - 1 = 0$$

$$(x + 1)(2x - 1) = 0$$

$$x + 1 = 0 \quad \text{or} \quad 2x - 1 = 0$$

$$x = -1 \qquad\qquad x = \frac{1}{2}$$

The number $\dfrac{1}{2}$ makes the denominator $2x - 1$ equal 0, so it is not a solution. The solution set is $\{-1\}$.

■ **Work Practice 4**

**Answers**

**3.** $\varnothing$   **4.** $\{-1\}$

**Example 5** Solve: $\dfrac{2x}{x-3} + \dfrac{6-2x}{x^2-9} = \dfrac{x}{x+3}$

**Solution:** We factor the second denominator to find that $(x+3)(x-3)$ is the LCD. We multiply both sides of the equation by $(x+3)(x-3)$. By the distributive property, this is the same as multiplying each term by $(x+3)(x-3)$.

$$\dfrac{2x}{x-3} + \dfrac{6-2x}{x^2-9} = \dfrac{x}{x+3}$$

$$(x+3)(x-3) \cdot \dfrac{2x}{x-3} + (x+3)(x-3) \cdot \dfrac{6-2x}{(x+3)(x-3)}$$

$$= (x+3)(x-3)\left(\dfrac{x}{x+3}\right)$$

$2x(x+3) + (6-2x) = x(x-3)$    Simplify.

$2x^2 + 6x + 6 - 2x = x^2 - 3x$    Use the distributive property.

Next we solve this quadratic equation by the factoring method. To do so, we first write the equation so that one side is 0.

$$x^2 + 7x + 6 = 0$$

$(x+6)(x+1) = 0$    Factor.

$x = -6$   or   $x = -1$    Set each factor equal to 0 and solve.

Neither $-6$ nor $-1$ makes any denominator 0. Check to see that the solution set is $\{-6, -1\}$.

■ **Work Practice 5**

**Example 6** Solve: $\dfrac{z}{2z^2+3z-2} - \dfrac{1}{2z} = \dfrac{3}{z^2+2z}$

**Solution:** Factor the denominators to find that the LCD is $2z(z+2)(2z-1)$. Multiply both sides by the LCD. Remember, by using the distributive property, this is the same as multiplying each term by $2z(z+2)(2z-1)$.

$$\dfrac{z}{2z^2+3z-2} - \dfrac{1}{2z} = \dfrac{3}{z^2+2z}$$

$$\dfrac{z}{(2z-1)(z+2)} - \dfrac{1}{2z} = \dfrac{3}{z(z+2)}$$

$$2z(z+2)(2z-1) \cdot \dfrac{z}{(2z-1)(z+2)} - 2z(z+2)(2z-1) \cdot \dfrac{1}{2z}$$

$$= 2z(z+2)(2z-1) \cdot \dfrac{3}{z(z+2)}$$    Apply the distributive property.

$2z(z) - (z+2)(2z-1) = 3 \cdot 2(2z-1)$    Simplify.

$2z^2 - (2z^2 + 3z - 2) = 12z - 6$

$2z^2 - 2z^2 - 3z + 2 = 12z - 6$

$-3z + 2 = 12z - 6$

$-15z = -8$

$z = \dfrac{8}{15}$    Solve.

The proposed solution $\dfrac{8}{15}$ does not make any denominator 0; the solution set is $\left\{\dfrac{8}{15}\right\}$.

■ **Work Practice 6**

**Practice 5**

Solve:
$$\dfrac{2x}{x-4} + \dfrac{10-5x}{x^2-16} = \dfrac{x}{x+4}$$

**Practice 6**

Solve:
$$\dfrac{3z}{3z^2+7z-6} - \dfrac{1}{z} = \dfrac{4}{3z^2-2z}$$

**Answers**

**5.** $\{-5, -2\}$   **6.** $\left\{-\dfrac{6}{11}\right\}$

A graph can be helpful in visualizing solutions of equations. For example, to visualize the solution of the equation $\dfrac{3}{x} - \dfrac{x+21}{3x} = \dfrac{5}{3}$ in Example 2, the graph of the related rational function $f(x) = \dfrac{3}{x} - \dfrac{x+21}{3x}$ is shown. A solution of the equation is an $x$-value that corresponds to a $y$-value of $\dfrac{5}{3}$.

Notice that an $x$-value of $-2$ corresponds to a $y$-value of $\dfrac{5}{3}$. The solution of the equation is indeed $-2$ as shown in Example 2.

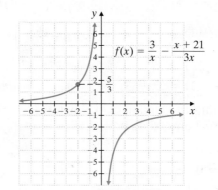

## Vocabulary, Readiness & Video Check

*Determine whether each is an equation or an expression. Do not solve or simplify.*

**1.** $\dfrac{x}{2} = \dfrac{3x}{5} + \dfrac{x}{6}$

**2.** $\dfrac{3x}{5} + \dfrac{x}{6}$

**3.** $\dfrac{x}{x-1} + \dfrac{2x}{x+1}$

**4.** $\dfrac{x}{x-1} + \dfrac{2x}{x+1} = 5$

**5.** $\dfrac{y+7}{2} = \dfrac{y+1}{6} + \dfrac{1}{y}$

**6.** $\dfrac{y+1}{6} + \dfrac{1}{y}$

*Choose the least common denominator (LCD) for the rational expressions in each equation. Do not solve.*

**7.** $\dfrac{x}{7} - \dfrac{x}{2} = \dfrac{1}{2}$; LCD = _____

    **a.** 7    **b.** 2    **c.** 14    **d.** 28

**8.** $\dfrac{9}{x+1} + \dfrac{5}{(x+1)^2} = \dfrac{x}{x+1}$; LCD = _____

    **a.** $x+1$    **b.** $(x+1)^2$    **c.** $(x+1)^3$

**9.** $\dfrac{7}{x-4} = \dfrac{x}{x^2-16} + \dfrac{1}{x+4}$; LCD = _____

    **a.** $(x+4)(x-4)$    **b.** $x-4$    **c.** $x+4$    **d.** $(x^2-16)(x-4)(x+4)$

**10.** $3 = \dfrac{1}{x-5} - \dfrac{2}{x^2-5x}$; LCD = _____

    **a.** $x-5$    **b.** $3(x-5)$    **c.** $3x(x-5)$    **d.** $x(x-5)$

*Martin-Gay Interactive Videos*    Watch the section lecture video and answer the following questions.

See Video 6.4

**Objective A** **11.** From Examples 2 and 3, why is it important to determine whether you have a linear or a quadratic equation before you finish solving the equation?

**12.** From Examples 2 and 3, what extra check do you make for a proposed solution before you see that it satisfies the original equation?

## 6.4 Exercise Set MyMathLab®

**Objective A**  *Solve each equation. See Examples 1 and 2.*

**1.** $\dfrac{x}{2} - \dfrac{x}{3} = 12$

**2.** $x = \dfrac{x}{2} - 4$

**3.** $\dfrac{x}{3} = \dfrac{1}{6} + \dfrac{x}{4}$

**4.** $\dfrac{x}{2} = \dfrac{21}{10} - \dfrac{x}{5}$

**5.** $\dfrac{2}{x} + \dfrac{1}{2} = \dfrac{5}{x}$

**6.** $\dfrac{5}{3x} + 1 = \dfrac{7}{6}$

**7.** $\dfrac{x^2 + 1}{x} = \dfrac{5}{x}$

**8.** $\dfrac{x^2 - 14}{2x} = -\dfrac{5}{2x}$

*Solve each equation. See Examples 3 through 6.*

**9.** $\dfrac{x + 5}{x + 3} = \dfrac{2}{x + 3}$

**10.** $\dfrac{x - 7}{x - 1} = \dfrac{11}{x - 1}$

**11.** $\dfrac{5}{x - 2} - \dfrac{2}{x + 4} = \dfrac{-4}{x^2 + 2x - 8}$

**12.** $\dfrac{1}{x - 1} + \dfrac{1}{x + 1} = \dfrac{2}{x^2 - 1}$

**13.** $\dfrac{1}{x - 1} = \dfrac{2}{x + 1}$

**14.** $\dfrac{6}{x + 3} = \dfrac{4}{x - 3}$

**15.** $\dfrac{x^2 - 23}{2x^2 - 5x - 3} + \dfrac{2}{x - 3} = \dfrac{-1}{2x + 1}$

**16.** $\dfrac{4x^2 - 24x}{3x^2 - x - 2} + \dfrac{3}{3x + 2} = \dfrac{-4}{x - 1}$

**17.** $\dfrac{1}{x - 4} - \dfrac{3x}{x^2 - 16} = \dfrac{2}{x + 4}$

**18.** $\dfrac{3}{2x + 3} - \dfrac{1}{2x - 3} = \dfrac{4}{4x^2 - 9}$

**19.** $\dfrac{1}{x - 4} = \dfrac{8}{x^2 - 16}$

**20.** $\dfrac{2}{x^2 - 4} = \dfrac{1}{2x - 4}$

**21.** $\dfrac{1}{x - 2} - \dfrac{2}{x^2 - 2x} = 1$

**22.** $\dfrac{12}{3x^2 + 12x} = 1 - \dfrac{1}{x + 4}$

## Mixed Practice *Solve each equation. See Examples 1 through 6.*

**23.** $\dfrac{5}{x} = \dfrac{20}{12}$

**24.** $\dfrac{2}{x} = \dfrac{10}{5}$

**25.** $1 - \dfrac{4}{a} = 5$

**26.** $7 + \dfrac{6}{a} = 5$

**27.** $\dfrac{x^2 + 5}{x} - 1 = \dfrac{5(x + 1)}{x}$

**28.** $\dfrac{x^2 + 6}{x} + 5 = \dfrac{2(x + 3)}{x}$

**29.** $\dfrac{1}{2x} - \dfrac{1}{x + 1} = \dfrac{1}{3x^2 + 3x}$

**30.** $\dfrac{2}{x - 5} + \dfrac{1}{2x} = \dfrac{5}{3x^2 - 15x}$

**31.** $\dfrac{1}{x} - \dfrac{x}{25} = 0$

**32.** $\dfrac{x}{4} + \dfrac{5}{x} = 3$

**33.** $5 - \dfrac{2}{2y - 5} = \dfrac{3}{2y - 5}$

**34.** $1 - \dfrac{5}{y + 7} = \dfrac{4}{y + 7}$

**35.** $\dfrac{x - 1}{x + 2} = \dfrac{2}{3}$

**36.** $\dfrac{6x + 7}{2x + 9} = \dfrac{5}{3}$

**37.** $\dfrac{x + 3}{x + 2} = \dfrac{1}{x + 2}$

**▶ 38.** $\dfrac{2x + 1}{4 - x} = \dfrac{9}{4 - x}$

**39.** $\dfrac{1}{a - 3} + \dfrac{2}{a + 3} = \dfrac{1}{a^2 - 9}$

**40.** $\dfrac{12}{9 - a^2} + \dfrac{3}{3 + a} = \dfrac{2}{3 - a}$

**41.** $\dfrac{64}{x^2 - 16} + 1 = \dfrac{2x}{x - 4}$

**42.** $2 + \dfrac{3}{x} = \dfrac{2x}{x + 3}$

**43.** $\dfrac{-15}{4y + 1} + 4 = y$

**44.** $\dfrac{36}{x^2 - 9} + 1 = \dfrac{2x}{x + 3}$

**45.** $\dfrac{28}{x^2 - 9} + \dfrac{2x}{x - 3} + \dfrac{6}{x + 3} = 0$

**▶ 46.** $\dfrac{x^2 - 20}{x^2 - 7x + 12} = \dfrac{3}{x - 3} + \dfrac{5}{x - 4}$

**47.** $\dfrac{x + 2}{x^2 + 7x + 10} = \dfrac{1}{3x + 6} - \dfrac{1}{x + 5}$

**48.** $\dfrac{3}{2x - 5} + \dfrac{2}{2x + 3} = 0$

## Review

*Write each sentence as an equation and solve. See Section 2.2.*

**49.** Four more than 3 times a number is 19. Find the number.

**50.** The sum of two consecutive integers is 147. Find the integers.

**△ 51.** The length of a rectangle is 5 inches more than the width. Its perimeter is 50 inches. Find the length and width.

**52.** Five times the sum of a number and 7 is 60. Find the number.

*The following graph is from a recent survey of state and federal prisons. Use this graph to answer Exercises 53 through 58. See Section 2.2.*

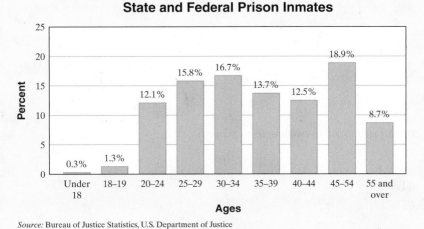

**State and Federal Prison Inmates**

Source: Bureau of Justice Statistics, U.S. Department of Justice

**53.** What percent of state and federal prison inmates are age 20 to 24?

**54.** What percent of state and federal prison inmates are age 30 to 34?

**55.** What age category shows the highest percent of prison inmates?

**56.** What percent of state and federal prison inmates are age 18 to 19?

**57.** At the end of 2012, there were 134,534 inmates under the jurisdiction of state and federal correction authorities in the state of California. Approximately how many 25- to 29-year-old inmates would you expect to have been held in California at the end of 2012? Round to the nearest whole.

**58.** Use the data from Exercise 57 to answer the following:

**a.** Approximate the number of 35- to 39-year-old inmates you might expect to have been held in California at the end of 2012. Round to the nearest whole.

**b.** Is your answer to part (a) greater than or less than your answer to Exercise 57? Is this reasonable? Why or why not?

## Concept Extensions

**59.** In your own words, explain the differences between equations and expressions.

**60.** In your own words, explain why it is necessary to check solutions to equations containing rational expressions.

**61.** The average cost of producing $x$ game disks for a computer is given by the function $C(x) = 3.3 + \dfrac{5400}{x}$. Find the number of game disks that must be produced for the average cost to be $5.10.

**62.** The average cost of producing $x$ electric pencil sharpeners is given by the function $C(x) = 20 + \dfrac{4000}{x}$. Find the number of electric pencil sharpeners that must be produced for the average cost to be $25.

*Solve each equation. Begin by writing each equation with positive exponents only.*

**63.** $x^{-2} - 19x^{-1} + 48 = 0$

**64.** $x^{-2} - 5x^{-1} - 36 = 0$

**65.** $p^{-2} + 4p^{-1} - 5 = 0$

**66.** $6p^{-2} - 5p^{-1} + 1 = 0$

*Solve each equation. Round solutions to two decimal places.*

**67.** $\dfrac{1.4}{x - 2.6} = \dfrac{-3.5}{x + 7.1}$

**68.** $\dfrac{-8.5}{x + 1.9} = \dfrac{5.7}{x - 3.6}$

**69.** $\dfrac{10.6}{y} - 14.7 = \dfrac{9.92}{3.2} + 7.6$

**70.** $\dfrac{12.2}{x} + 17.3 = \dfrac{9.6}{x} - 14.7$

*Solve each equation by substitution. For example, to solve Exercise 71, first let $u = x - 1$. After substituting, we have $u^2 + 3u + 2 = 0$. Solve for $u$ and then substitute back to solve for $x$.*

**71.** $(x - 1)^2 + 3(x - 1) + 2 = 0$

**72.** $(4 - x)^2 - 5(4 - x) + 6 = 0$

**73.** $\left(\dfrac{3}{x - 1}\right)^2 + 2\left(\dfrac{3}{x - 1}\right) + 1 = 0$

**74.** $\left(\dfrac{5}{2 + x}\right)^2 + \left(\dfrac{5}{2 + x}\right) - 20 = 0$

*Use a graphing calculator to verify the solution of each given exercise.*

**75.** Exercise 23        **76.** Exercise 24        **77.** Exercise 35        **78.** Exercise 36

# Expressions and Equations Containing Rational Expressions

It is very important that you understand the difference between an expression and an equation containing rational expressions. An equation contains an equal sign; an expression does not.

**Expression to Be Simplified**

$$\frac{x}{2} + \frac{x}{6}$$

Write both rational expressions with the LCD, 6, as the denominator.

$$\frac{x}{2} + \frac{x}{6} = \frac{x \cdot 3}{2 \cdot 3} + \frac{x}{6}$$

$$= \frac{3x}{6} + \frac{x}{6}$$

$$= \frac{4x}{6} = \frac{2x}{3}$$

**Equation to Be Solved**

$$\frac{x}{2} + \frac{x}{6} = \frac{2}{3}$$

Multiply both sides by the LCD, 6.

$$6\left(\frac{1}{2} + \frac{x}{6}\right) = 6\left(\frac{2}{3}\right)$$

$$3 + x = 4$$

$$x = 1$$

Check to see that the solution set is {1}.

**Helpful Hint**

Remember: Equations can be cleared of fractions; expressions cannot.

*Perform each indicated operation and either simplify the expression, or solve the equation for the variable.*

**1.** $\dfrac{x}{2} = \dfrac{1}{8} + \dfrac{x}{4}$

**2.** $\dfrac{x}{4} = \dfrac{3}{2} + \dfrac{x}{10}$

**3.** $\dfrac{1}{8} + \dfrac{x}{4}$

**4.** $\dfrac{3}{2} + \dfrac{x}{10}$

**5.** $\dfrac{4}{x+2} - \dfrac{2}{x-1}$

**6.** $\dfrac{5}{x-2} - \dfrac{10}{x+4}$

**7.** $\dfrac{4}{x+2} = \dfrac{2}{x-1}$

**8.** $\dfrac{5}{x-2} = \dfrac{10}{x+4}$

**9.** $\dfrac{2}{x^2-4} = \dfrac{1}{x+2} - \dfrac{3}{x-2}$

**10.** $\dfrac{3}{x^2-25} = \dfrac{1}{x+5} + \dfrac{2}{x-5}$

**11.** $\dfrac{5}{x^2-3x} + \dfrac{4}{2x-6}$

**12.** $\dfrac{5}{x^2-3x} \div \dfrac{4}{2x-6}$

**13.** $\dfrac{x-1}{x+1} + \dfrac{x+7}{x-1} = \dfrac{4}{x^2-1}$

**14.** $\left(1 - \dfrac{y}{x}\right) \div \left(1 - \dfrac{x}{y}\right)$

**15.** $\dfrac{a^2-9}{a-6} \cdot \dfrac{a^2-5a-6}{a^2-a-6}$

**16.** $\dfrac{2}{a-6} + \dfrac{3a}{a^2-5a-6} - \dfrac{a}{5a+5}$

**17.** $\dfrac{2x+3}{3x-2} = \dfrac{4x+1}{6x+1}$

**18.** $\dfrac{5x-3}{2x} = \dfrac{10x+3}{4x+1}$

**19.** $\dfrac{a}{9a^2-1} + \dfrac{2}{6a-2}$

**20.** $\dfrac{3}{4a-8} - \dfrac{a+2}{a^2-2a}$

1. _____
2. _____
3. _____
4. _____
5. _____
6. _____
7. _____
8. _____
9. _____
10. _____
11. _____
12. _____
13. _____
14. _____
15. _____
16. _____
17. _____
18. _____
19. _____
20. _____

**21.** _____

**22.** _____

**23.** _____

**24.** _____

**25.** _____

**26.** _____

**27. a.** _____

**b.** _____

**c.** _____

**28. a.** _____

**b.** _____

**c.** _____

**29.** _____

**30.** _____

**31.** _____

**32.** _____

**33.** _____

**21.** $-\dfrac{3}{x^2} - \dfrac{1}{x} + 2 = 0$

**22.** $\dfrac{x}{2x+6} + \dfrac{5}{x^2-9}$

**23.** $\dfrac{x-8}{x^2-x-2} + \dfrac{2}{x-2}$

**24.** $\dfrac{x-8}{x^2-x-2} + \dfrac{2}{x-2} = \dfrac{3}{x+1}$

**25.** $\dfrac{3}{a} - 5 = \dfrac{7}{a} - 1$

**26.** $\dfrac{7}{3z-9} + \dfrac{5}{z}$

_Use_ $\dfrac{x}{5} - \dfrac{x}{4} = \dfrac{1}{10}$ _and_ $\dfrac{x}{5} - \dfrac{x}{4} + \dfrac{1}{10}$ _for Exercises 27 and 28._

**27. a.** Which one above is an expression?
  **b.** Describe the first step to simplify this expression.
  **c.** Simplify the expression.

**28. a.** Which one above is an equation?
  **b.** Describe the first step to solve this equation.
  **c.** Solve the equation.

_For each exercise, choose the correct statement.* Each figure represents a real number and no denominators are 0._

**29. a.** $\dfrac{\triangle + \square}{\triangle} = \square$

  **b.** $\dfrac{\triangle + \square}{\triangle} = 1 + \dfrac{\square}{\triangle}$

  **c.** $\dfrac{\triangle + \square}{\triangle} = \dfrac{\square}{\triangle}$

  **d.** $\dfrac{\triangle + \square}{\triangle} = 1 + \square$

  **e.** $\dfrac{\triangle + \square}{\triangle - \square} = -1$

**30. a.** $\dfrac{\triangle}{\square} + \dfrac{\square}{\triangle} = \dfrac{\triangle + \square}{\square + \triangle} = 1$

  **b.** $\dfrac{\triangle}{\square} + \dfrac{\square}{\triangle} = \dfrac{\triangle + \square}{\triangle\square}$

  **c.** $\dfrac{\triangle}{\square} + \dfrac{\square}{\triangle} = \triangle\triangle + \square\square$

  **d.** $\dfrac{\triangle}{\square} + \dfrac{\square}{\triangle} = \dfrac{\triangle\triangle + \square\square}{\square\triangle}$

  **e.** $\dfrac{\triangle}{\square} + \dfrac{\square}{\triangle} = \dfrac{\triangle\square}{\square\triangle} = 1$

**31. a.** $\dfrac{\triangle}{\square} \cdot \dfrac{\bigcirc}{\square} = \dfrac{\triangle\bigcirc}{\square}$

  **b.** $\dfrac{\triangle}{\square} \cdot \dfrac{\bigcirc}{\square} = \triangle\bigcirc$

  **c.** $\dfrac{\triangle}{\square} \cdot \dfrac{\bigcirc}{\square} = \dfrac{\triangle + \bigcirc}{\square + \square}$

  **d.** $\dfrac{\triangle}{\square} \cdot \dfrac{\bigcirc}{\square} = \dfrac{\triangle\bigcirc}{\square\square}$

**32. a.** $\dfrac{\triangle}{\square} \div \dfrac{\bigcirc}{\triangle} = \dfrac{\triangle\triangle}{\square\bigcirc}$

  **b.** $\dfrac{\triangle}{\square} \div \dfrac{\bigcirc}{\triangle} = \dfrac{\bigcirc\square}{\triangle\triangle}$

  **c.** $\dfrac{\triangle}{\square} \div \dfrac{\bigcirc}{\triangle} = \dfrac{\bigcirc}{\square}$

  **d.** $\dfrac{\triangle}{\square} \div \dfrac{\bigcirc}{\triangle} = \dfrac{\triangle + \triangle}{\square + \bigcirc}$

**33. a.** $\dfrac{\dfrac{\triangle + \square}{\bigcirc}}{\dfrac{\triangle}{\bigcirc}} = \square$

  **b.** $\dfrac{\dfrac{\triangle + \square}{\bigcirc}}{\dfrac{\triangle}{\bigcirc}} = \dfrac{\triangle\triangle + \triangle\square}{\bigcirc\bigcirc}$

  **c.** $\dfrac{\dfrac{\triangle + \square}{\bigcirc}}{\dfrac{\triangle}{\bigcirc}} = 1 + \square$

  **d.** $\dfrac{\dfrac{\triangle + \square}{\bigcirc}}{\dfrac{\triangle}{\bigcirc}} = \dfrac{\triangle + \square}{\triangle}$

*My thanks to Kelly Champange for permission to use her exercises for 29 through 33.

## Objective A Solving Rational Equations for a Specified Variable ▶

In Section 2.3, we solved equations for a specified variable. In this section, we continue practicing this skill by solving equations containing rational expressions for a specified variable. The steps given in Section 2.3 for solving equations for a specified variable are repeated here.

**Objectives**

**A** Solve an Equation Containing Rational Expressions for a Specified Variable. ▶

**B** Solve Number Problems by Writing Equations Containing Rational Expressions. ▶

**C** Solve Problems Modeled by Proportions. ▶

**D** Solve Problems About Work. ▶

**E** Solve Problems About Distance, Rate, and Time. ▶

> **Solving an Equation for a Specified Variable**
>
> **Step 1:** Clear the equation of fractions or rational expressions by multiplying each side of the equation by the least common denominator (LCD) of all denominators in the equation.
>
> **Step 2:** Use the distributive property to remove grouping symbols such as parentheses.
>
> **Step 3:** Combine like terms on each side of the equation.
>
> **Step 4:** Use the addition property of equality to rewrite the equation as an equivalent equation with terms containing the specified variable on one side and all other terms on the other side.
>
> **Step 5:** Use the distributive property and the multiplication property of equality to get the specified variable alone.

**Example 1** Solve $\dfrac{1}{x} + \dfrac{1}{y} = \dfrac{1}{z}$ for $x$.

**Solution:** To clear this equation of fractions, we multiply both sides of the equation by $xyz$, the LCD of $\dfrac{1}{x}, \dfrac{1}{y},$ and $\dfrac{1}{z}$.

$$\frac{1}{x} + \frac{1}{y} = \frac{1}{z}$$

$$xyz\left(\frac{1}{x} + \frac{1}{y}\right) = xyz\left(\frac{1}{z}\right) \qquad \text{Multiply both sides by } xyz.$$

$$xyz\left(\frac{1}{x}\right) + xyz\left(\frac{1}{y}\right) = xyz\left(\frac{1}{z}\right) \qquad \text{Use the distributive property.}$$

$$yz + xz = xy \qquad \text{Simplify.}$$

Notice the two terms that contain the specified variable, $x$.

Next, we subtract $xz$ from both sides so that all terms containing the specified variable $x$ are on one side of the equation and all other terms are on the other side.

$$yz = xy - xz$$

Now we use the distributive property to factor $x$ from $xy - xz$ and then the multiplication property of equality to solve for $x$.

$$yz = x(y - z)$$

$$\frac{yz}{y - z} = x \quad \text{or} \quad x = \frac{yz}{y - z} \qquad \text{Divide both sides by } y - z.$$

■ Work Practice 1

**Practice 1**

Solve $\dfrac{1}{x} + \dfrac{1}{y} = \dfrac{1}{z}$ for $y$.

**Answer**

**1.** $y = \dfrac{xz}{x - z}$

Problem solving sometimes involves modeling a described situation with an equation containing rational expressions. In Examples 2 through 5, we practice solving such problems and use the problem-solving steps first introduced in Section 2.2.

## Objective B Solving Number Problems Modeled by Rational Equations ▶

Let's review our problem-solving steps by solving a direct translation example.

**Example 2**   Finding an Unknown Number

If a certain number is subtracted from the numerator and added to the denominator of $\frac{9}{19}$, the new fraction is equivalent to $\frac{1}{3}$. Find the number.

**Solution:**

1. **UNDERSTAND** the problem. Read and reread the problem and try guessing the solution. For example, if the unknown number is 3, we have

$$\frac{9-3}{19+3} = \frac{6}{22} = \frac{3}{11} \neq \frac{1}{3}$$

Thus, $\frac{9-3}{19+3} \neq \frac{1}{3}$ and 3 is not the correct number. Remember that the purpose of this step is not to guess the correct solution but to gain an understanding of the problem posed.

   We will let $n$ = the number to be subtracted from the numerator and added to the denominator.

2. **TRANSLATE** the problem.

| In words: | when the number is subtracted from the numerator and added to the denominator of the fraction $\frac{9}{19}$ | this is equivalent to | $\frac{1}{3}$ |
|---|---|---|---|
| | ↓ | ↓ | ↓ |
| Translate: | $\frac{9-n}{19+n}$ | $=$ | $\frac{1}{3}$ |

3. **SOLVE** the equation for $n$.

$$\frac{9-n}{19+n} = \frac{1}{3}$$

To solve for $n$, we begin by multiplying both sides by the LCD, $3(19+n)$.

$$3(19+n) \cdot \frac{9-n}{19+n} = 3(19+n) \cdot \frac{1}{3} \qquad \text{Multiply both sides by the LCD.}$$
$$3(9-n) = 19+n \qquad \text{Simplify.}$$
$$27 - 3n = 19 + n$$
$$8 = 4n$$
$$2 = n \qquad \text{Solve.}$$

4. **INTERPRET** the results.

**Check:**   If we subtract 2 from the numerator and add 2 to the denominator of $\frac{9}{19}$, we have $\frac{9-2}{19+2} = \frac{7}{21} = \frac{1}{3}$, and the problem checks.

**State:**   The unknown number is 2.

■ **Work Practice 2**

---

**Practice 2**

Find the number that, when added to the numerator and subtracted from the denominator of $\frac{1}{20}$, results in a fraction equivalent to $\frac{2}{5}$.

**Answer**

**2.** 5

## Objective C Solving Problems Modeled by Proportions

A **ratio** is the quotient of two numbers or two quantities. Since rational expressions are quotients of quantities, rational expressions are ratios also. A **proportion** is a mathematical statement that two ratios are equal.

Let's review two methods for solving a proportion such as $\dfrac{x-3}{10} = \dfrac{7}{15}$. We can multiply both sides of the equation by the LCD, 30.

Multiply both sides by the LCD, 30.

$$30 \cdot \frac{x-3}{10} = 30 \cdot \frac{7}{15}$$

$$3(x-3) = 2 \cdot 7$$

$$3x - 9 = 14$$

$$3x = 23$$

$$x = \frac{23}{3}$$

We can also solve a proportion by setting cross products equal. Here, we are using the fact that if $\dfrac{a}{b} = \dfrac{c}{d}$, then $ad = bc$.

$$\frac{x-3}{10} = \frac{7}{15}$$

$15(x-3) = 10 \cdot 7$     Set cross products equal.

$15x - 45 = 70$     Use the distributive property.

$$15x = 115$$

$$x = \frac{115}{15} \quad \text{or} \quad \frac{23}{3}$$

A ratio of two different quantities is called a **rate.** For example $\dfrac{3 \text{ miles}}{2 \text{ hours}}$ or 1.5 miles/hour is a rate. The proportions we write to solve problems will sometimes include rates. When this happens, make sure that the rates contain units written in the same order.

**Example 3** In the United States, 11 out of every 25 homes are heated by electricity. At this rate, how many homes in a community of 36,000 homes would you predict are heated by electricity? (*Source: Census Survey*)

**Solution:**

1. UNDERSTAND. Read and reread the problem. Try to estimate a reasonable solution. For example, since 11 is less than $\dfrac{1}{2}$ of 25, we might reason that the solution would be less than $\dfrac{1}{2}$ of 36,000, or 18,000.

   Let's let $x$ = number of homes in the community heated by electricity.

2. TRANSLATE.

   homes heated by electricity $\rightarrow$   $\dfrac{11}{25} = \dfrac{x}{36,000}$   $\leftarrow$ homes heated by electricity
   total homes   $\rightarrow$                  $\leftarrow$ total homes

3. SOLVE. To solve this proportion, we will set cross products equal.

   $$\frac{11}{25} = \frac{x}{36,000}$$

   $$11 \cdot 36,000 = 25x$$

   $$\frac{396,000}{25} = x$$

   $$15,840 = x$$

(*Continued on next page*)

**Practice 3**

In the United States, 1 out of 40 homes is heated by wood. At this rate, how many homes in a community of 36,000 homes are heated by wood? (*Source: Census Survey*)

**Answer**

3. 900 homes

4. **INTERPRET.**

**Check:**   To check, replace $x$ with 15,840 in the proportion and see that a true statement results. Notice that our answer is reasonable since it is less than 18,000 as we stated in Step 1. UNDERSTAND on the previous page.

**State:**   We predict that 15,840 homes are heated by electricity.

▪ Work Practice 3

## Objective D  Solving Problems About Work ▶

The following work example leads to an equation containing rational expressions.

**Practice 4**

Greg Guillot can paint a room alone in 3 hours. His brother Phillip can do the same job alone in 5 hours. How long would it take them to paint the room if they work together?

> **Example 4**   Calculating Work Hours

Melissa Scarlatti can clean the house in 4 hours, whereas her husband, Zack, can do the same job in 5 hours. They have agreed to clean together so that they can finish in time to watch a movie on TV that starts in 2 hours. How long will it take them to clean the house together? Can they finish before the movie starts?

**Solution:**

1.  Read and reread the problem. The key idea here is the relationship between the *time* (in hours) it takes to complete the job and the *part of the job* completed in 1 unit of time (1 hour). For example, if the *time* it takes Melissa to complete the job is 4 hours, the *part of the job* she can complete in 1 hour is $\frac{1}{4}$. Similarly, Zack can complete $\frac{1}{5}$ of the job in 1 hour.

    We will let $t =$ the *time* in hours it takes Melissa and Zack to clean the house together. Then $\frac{1}{t}$ represents the *part of the job* they complete in 1 hour. We summarize the given information on a chart.

|               | Hours to Complete the Job | Part of Job Completed in 1 Hour |
|---------------|:-------------------------:|:-------------------------------:|
| **Melissa Alone** | 4 | $\frac{1}{4}$ |
| **Zack Alone**    | 5 | $\frac{1}{5}$ |
| **Together**      | $t$ | $\frac{1}{t}$ |

2.  **TRANSLATE.**

| In words: | part of job Melissa can complete in 1 hour | added to | part of job Zack can complete in 1 hour | is equal to | part of job they can complete together in 1 hour |
|:---------:|:---:|:---:|:---:|:---:|:---:|
|  | ↓ | ↓ | ↓ | ↓ | ↓ |
| Translate: | $\frac{1}{4}$ | $+$ | $\frac{1}{5}$ | $=$ | $\frac{1}{t}$ |

**Answer**

4. $1\frac{7}{8}$ hr

**3. SOLVE:**

$$\frac{1}{4} + \frac{1}{5} = \frac{1}{t}$$

$$20t\left(\frac{1}{4} + \frac{1}{5}\right) = 20t\left(\frac{1}{t}\right) \quad \text{Multiply both sides by the LCD, } 20t.$$

$$5t + 4t = 20$$

$$9t = 20$$

$$t = \frac{20}{9} \quad \text{or} \quad 2\frac{2}{9} \quad \text{Solve.}$$

**4. INTERPRET.**

**Check:** The proposed solution is $2\frac{2}{9}$. That is, Melissa and Zack would take $2\frac{2}{9}$ hours to clean the house together. This proposed solution is reasonable since $2\frac{2}{9}$ hours is more than half of Melissa's time and less than half of Zack's time. Check this solution in the originally stated problem.

**State:** Melissa and Zack can clean the house together in $2\frac{2}{9}$ hours. They cannot complete the job before the movie starts.

■ **Work Practice 4**

## Objective E  Solving Problems About Distance, Rate, and Time ▶

Before we solve Example 5 on the next page, let's review what we learned in Chapter 2 about the formula

$$d = r \cdot t, \quad \text{or} \quad \text{distance} = \text{rate} \cdot \text{time}$$

For example, if we travel at a rate or speed of 60 mph for a time of 3 hours, the distance we travel is

$$d = 60 \text{ mph} \cdot 3 \text{ hr}$$
$$= 180 \text{ mi}$$

The formula $d = r \cdot t$ is solved for distance, $d$. We can also solve this formula for rate $r$ or for time $t$.

Solve $d = r \cdot t$ for $r$.

$$d = r \cdot t$$
$$\frac{d}{t} = \frac{r \cdot t}{t}$$
$$\frac{d}{t} = r$$

Solve $d = r \cdot t$ for $t$.

$$d = r \cdot t$$
$$\frac{d}{r} = \frac{r \cdot t}{r}$$
$$\frac{d}{r} = t$$

All three forms of the distance formula are useful, as we shall see.

## Practice 5

A fisherman traveling on the Pearl River takes $\frac{2}{3}$ times as long to travel 60 miles downstream in his boat than to return. If the boat's speed is 25 mph in still water, find the speed of the current.

**Answer**

5. 5 mph

---

**Example 5**   Finding the Speed of a Current

Steve Deitmer takes $1\frac{1}{2}$ times as long to go 72 miles upstream in his boat as he does to return. If the boat cruises at 30 mph in still water, what is the speed of the current?

### Solution:

1. **UNDERSTAND.** Read and reread the problem. Guess a solution. Suppose that the current is 4 mph. The speed of the boat upstream is slowed down by the current: $30 - 4$, or 26 mph, and the speed of the boat downstream is speeded up by the current: $30 + 4$, or 34 mph. Next let's find out how long it takes to travel 72 miles upstream and 72 miles downstream. To do so, we use the formula $d = r \cdot t$, or $\frac{d}{r} = t$.

| Upstream | Downstream |
|---|---|
| $\dfrac{d}{r} = t$ | $\dfrac{d}{r} = t$ |
| $\dfrac{72}{26} = t$ | $\dfrac{72}{34} = t$ |
| $2\dfrac{10}{13} = t$ | $2\dfrac{2}{17} = t$ |

Since the time upstream $\left(2\dfrac{10}{13}\text{ hours}\right)$ is not $1\dfrac{1}{2}$ times the time downstream $\left(2\dfrac{2}{17}\text{ hours}\right)$, our guess is not correct. We do, however, have a better understanding of the problem.

   We will let

$$x = \text{the speed of the current}$$
$$30 + x = \text{the speed of the boat downstream}$$
$$30 - x = \text{the speed of the boat upstream}$$

This information is summarized in the following chart, where we use the formula $\frac{d}{r} = t$.

|  | Distance | Rate | Time $\left(\dfrac{d}{r}\right)$ |
|---|---|---|---|
| **Upstream** | 72 | $30 - x$ | $\dfrac{72}{30 - x}$ |
| **Downstream** | 72 | $30 + x$ | $\dfrac{72}{30 + x}$ |

**2. TRANSLATE.** Since the time spent traveling upstream is $1\frac{1}{2}$ times the time spent traveling downstream, we have

| In words: | time upstream | is | $1\frac{1}{2}$ | times | time downstream |
|---|---|---|---|---|---|
| | ↓ | ↓ | ↓ | ↓ | ↓ |
| Translate: | $\dfrac{72}{30-x}$ | $=$ | $\dfrac{3}{2}$ | $\cdot$ | $\dfrac{72}{30+x}$ |

**3. SOLVE.**
$$\frac{72}{30-x} = \frac{3}{2} \cdot \frac{72}{30+x}$$

First we multiply both sides by the LCD, $2(30+x)(30-x)$.

$$2(30+x)(30-x) \cdot \frac{72}{30-x} = 2(30+x)(30-x)\left(\frac{3}{2} \cdot \frac{72}{30+x}\right)$$

$$72 \cdot 2(30+x) = 3 \cdot 72 \cdot (30-x) \qquad \text{Simplify.}$$

$$2(30+x) = 3(30-x) \qquad \text{Divide both sides by 72.}$$

$$60 + 2x = 90 - 3x \qquad \text{Use the distributive property.}$$

$$5x = 30$$

$$x = 6 \qquad \text{Solve.}$$

**4. INTERPRET.**

**Check:** Check the proposed solution of 6 mph in the originally stated problem.

**State:** The current's speed is 6 mph.

■ **Work Practice 5**

## Vocabulary, Readiness & Video Check

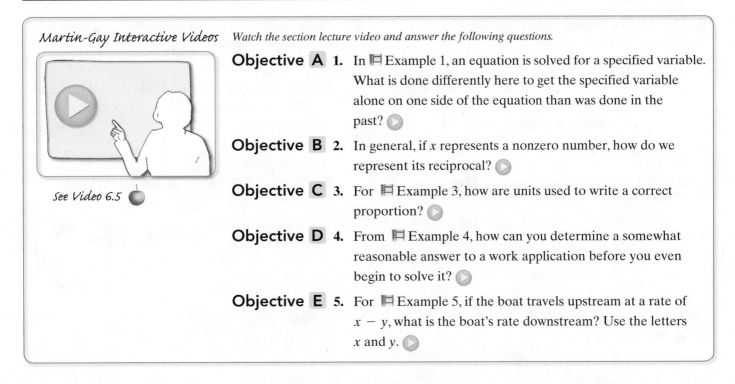

*Martin-Gay Interactive Videos*

*See Video 6.5* 🍎

*Watch the section lecture video and answer the following questions.*

**Objective A** 1. In ▤ Example 1, an equation is solved for a specified variable. What is done differently here to get the specified variable alone on one side of the equation than was done in the past? ▶

**Objective B** 2. In general, if $x$ represents a nonzero number, how do we represent its reciprocal? ▶

**Objective C** 3. For ▤ Example 3, how are units used to write a correct proportion? ▶

**Objective D** 4. From ▤ Example 4, how can you determine a somewhat reasonable answer to a work application before you even begin to solve it? ▶

**Objective E** 5. For ▤ Example 5, if the boat travels upstream at a rate of $x - y$, what is the boat's rate downstream? Use the letters $x$ and $y$. ▶

## 6.5 Exercise Set MyMathLab®

**Objective A**  *Solve each equation for the specified variable. See Example 1.*

**1.** $F = \frac{9}{5}C + 32$ for $C$

△ **2.** $V = \frac{1}{3}\pi r^2 h$ for $h$

**3.** $Q = \frac{A - I}{L}$ for $I$

**4.** $P = 1 - \frac{C}{S}$ for $S$

▶ **5.** $\frac{1}{R} = \frac{1}{R_1} + \frac{1}{R_2}$ for $R$

**6.** $\frac{1}{R} = \frac{1}{R_1} + \frac{1}{R_2}$ for $R_1$

**7.** $S = \frac{n(a + L)}{2}$ for $n$

**8.** $S = \frac{n(a + L)}{2}$ for $a$

△ **9.** $A = \frac{h(a + b)}{2}$ for $b$

△ **10.** $A = \frac{h(a + b)}{2}$ for $h$

**11.** $\frac{P_1 V_1}{T_1} = \frac{P_2 V_2}{T_2}$ for $T_2$

**12.** $H = \frac{kA(T_1 - T_2)}{L}$ for $T_2$

**13.** $f = \frac{f_1 f_2}{f_1 + f_2}$ for $f_2$

**14.** $I = \frac{E}{R + r}$ for $r$

**15.** $\lambda = \frac{2L}{n}$ for $L$

**16.** $S = \frac{a_1 - a_n r}{1 - r}$ for $a_1$

**17.** $\frac{\theta}{\omega} = \frac{2L}{c}$ for $c$

**18.** $F = \frac{-GMm}{r^2}$ for $M$

**Objective B**  *Solve. For Exercises 19 and 20, the solutions have been started for you. See Example 2.*

▶ **19.** The sum of a number and 5 times its reciprocal is 6. Find the number(s).

**Start the solution:**

1. UNDERSTAND the problem. Reread it as many times as needed. Let's let

   $x$ = a number. Then

   $\frac{1}{x}$ = its reciprocal.

2. TRANSLATE into an equation. (Fill in the blanks below.)

| a number | plus | 5 | times | its reciprocal | is | 6 |
|---|---|---|---|---|---|---|
| ↓ | ↓ | ↓ | ↓ | ↓ | ↓ | ↓ |
| _____ | + | 5 | · | _____ | = | 6 |

Finish with:

3. SOLVE and
4. INTERPRET

**20.** The quotient of a number and 9 times its reciprocal is 1. Find the number(s).

**Start the solution:**

1. UNDERSTAND the problem. Reread it as many times as needed. Let's let

   $x$ = a number. Then

   $\frac{1}{x}$ = its reciprocal.

2. TRANSLATE into an equation. (Fill in the blanks below.)

| a number | divided by | 9 | times | its reciprocal | is | 1 |
|---|---|---|---|---|---|---|
| ↓ | ↓ | ↓ | ↓ | ↓ | ↓ | ↓ |
| _____ | ÷ | 9 | · | _____ | = | 1 |

Finish with:

3. SOLVE and
4. INTERPRET

**21.** If a number is added to the numerator of $\frac{12}{41}$ and twice the number is added to the denominator of $\frac{12}{41}$, the resulting fraction is equivalent to $\frac{1}{3}$. Find the number.

**22.** If a number is subtracted from the numerator of $\frac{13}{8}$ and added to the denominator of $\frac{13}{8}$, the resulting fraction is equivalent to $\frac{2}{5}$. Find the number.

## Objective C  *Solve. See Example 3.*

**23.** An Arabian camel can drink 15 gallons of water in 10 minutes. At this rate, how much water can the camel drink in 3 minutes? (*Source:* Grolier, Inc.)

**24.** An Arabian camel can travel 20 miles in 8 hours, carrying a 300-pound load on its back. At this rate, how far can the camel travel in 10 hours? (*Source:* Grolier, Inc.)

**25.** In 2013, 12.5 out of every 100 Coast Guard personnel were women. If there were 41,362 total Coast Guard personnel on active duty, estimate the number of women. Round to the nearest whole. (*Source:* U.S. Coast Guard)

**26.** In 2013, 161 out of every 200 Air Force personnel were men. If there were 322,900 total Air Force personnel in 2013, estimate the number of men. Round to the nearest whole. (*Source:* Defense Manpower Research)

## Objective D  *Solve. See Example 4.*

**27.** An experienced roofer can roof a house in 26 hours. A beginning roofer needs 39 hours to complete the same job. Find how long it takes for the two to do the job together.

**28.** Alan Cantrell can word process a research paper in 6 hours. With Steve Isaac's help, the paper can be processed in 4 hours. Find how long it takes Steve to word process the paper alone.

**29.** Three postal workers can sort a stack of mail in 20 minutes, 30 minutes, and 60 minutes, respectively. Find how long it takes them to sort the mail if all three work together.

**30.** A new printing press can print newspapers twice as fast as the old one can. The old one can print the afternoon edition in 4 hours. Find how long it takes to print the afternoon edition if both presses are operating.

## Objective E  *Solve. See Example 5.*

**31.** Mattie Evans drove 150 miles in the same amount of time that it took a turbopropeller plane to travel 600 miles. The speed of the plane was 150 mph faster than the speed of the car. Find the speed of the plane.

**32.** An F-100 plane and a Toyota truck leave the same town at sunrise and head for a town 450 miles away. The speed of the plane is three times the speed of truck, and the plane arrives 6 hours ahead of the truck. Find the speed of the truck.

**33.** The speed of Lazy River's current is 5 mph. If a boat travels 20 miles downstream in the same time that it takes to travel 10 miles upstream, find the speed of the boat in still water.

**34.** The speed of a boat in still water is 24 mph. If the boat travels 54 miles upstream in the same time that it takes to travel 90 miles downstream, find the speed of the current.

## Objectives B C D E  Mixed Practice *Solve. See Examples 2 through 5.*

**35.** The sum of the reciprocals of two consecutive integers is $-\frac{15}{56}$. Find the two integers.

**36.** The sum of the reciprocals of two consecutive odd integers is $\frac{20}{99}$. Find the two integers.

**37.** One hose can fill a goldfish pond in 45 minutes, and if a second hose is used, the two hoses can fill the same pond in 20 minutes. Find how long it takes the second hose alone to fill the pond.

**38.** If Sarah Clark can do a job in 5 hours and Dick Belli and Sarah working together can do the same job in 2 hours, find how long it takes Dick to do the job alone.

**39.** Two trains going in opposite directions leave at the same time. One train travels 15 mph faster than the other. In 6 hours the trains are 630 miles apart. Find the speed of each.

**40.** Justin Wilson recorded the fastest lap in the 2013 Indianapolis 500. His speed for this lap was 39.51 mph faster than the average speed for the race. If Wilson traveled 2.5 miles in the time that a driver at the average speed covered 2.06 miles, find the speed of Wilson, rounded to the nearest tenth. (*Source:* Indianapolis 500)

**41.** A giant tortoise can travel 0.17 miles in 1 hour. At this rate, how long would it take the tortoise to travel 1 mile? Round to the nearest tenth of an hour. (*Source: The World Almanac*)

**42.** A black mamba snake can travel 88 feet in 3 seconds. At this rate, how long does it take to travel 300 feet (the length of a football field)? Round to the nearest tenth of a second. (*Source: The World Almanac,* 2002).

**43.** Moo Dairy has three machines to fill half-gallon milk cartons. The machines can fill the daily quota in 5 hours, 6 hours, and 7.5 hours, respectively. Find how long it takes to fill the daily quota if all three machines are running.

**44.** The inlet pipe of an oil tank can fill the tank in 1 hour 30 minutes. The outlet pipe can empty the tank in 1 hour. Find how long it takes to empty a full tank if both pipes are open.

**45.** A plane flies 465 miles with the wind and 345 miles against the wind in the same length of time. If the speed of the wind is 20 mph, find the speed of the plane in still air.

**46.** Two rockets are launched. The first travels at 9000 mph. Fifteen minutes later the second is launched at 10,000 mph. Find the distance at which both rockets are the same distance from Earth.

**47.** Two joggers, one averaging 8 mph and one averaging 6 mph, start from a designated initial point. The slower jogger arrives at the end of the run a half-hour after the other jogger. Find the distance of the run.

**48.** A semi truck travels 300 miles through the flatland in the same amount of time that it travels 180 miles through the Great Smoky Mountains. The rate of the truck is 20 miles per hour slower in the mountains than in the flatland. Find both the flatland rate and mountain rate.

**49.** The denominator of a fraction is 1 more than the numerator. If both the numerator and the denominator are decreased by 3, the resulting fraction is equivalent to $\frac{4}{5}$. Find the fraction.

**50.** The numerator of a fraction is 4 less than the denominator. If both the numerator and the denominator are increased by 2, the resulting fraction is equivalent to $\frac{2}{3}$. Find the fraction.

**51.** In 2 minutes, a conveyor belt can move 300 pounds of recyclable aluminum from the delivery truck to a storage area. A smaller belt can move the same quantity of cans the same distance in 6 minutes. If both belts are used, find how long it takes to move the cans to the storage area.

**52.** Gary Marcus and Tony Alva work at Lombardo's Pipe and Concrete. Mr. Lombardo is preparing an estimate for a customer. He knows that Gary can lay a slab of concrete in 6 hours. Tony can lay the same size slab in 4 hours. If both work on the job and the cost of labor is $45.00 per hour, determine what the labor estimate should be.

**53.** The world record for the largest white bass caught is held by Ronald Sprouse of Virginia. The bass weighed 6 pounds 13 ounces. If Ronald rows to his favorite fishing spot 9 miles downstream in the same amount of time that he rows 3 miles upstream and if the current is 6 mph, find how long it takes him to cover the 12 miles.

**54.** An amateur cyclist training for a road race rode the first 20-mile portion of his workout at a constant rate. For the 16-mile cooldown portion of his workout, he reduced his speed by 2 miles per hour. Each portion of the workout took equal time. Find the cyclist's rate during the first portion and his rate during the cooldown portion.

**55.** Smith Engineering is in the process of reviewing the salaries of their surveyors. During this review, the company found that an experienced surveyor can survey a roadbed in 4 hours. An apprentice surveyor needs 5 hours to survey the same stretch of road. If the two work together, find how long it takes them to complete the job.

**56.** Mr. Dodson can paint his house by himself in four days. His son will need an additional day to complete the job if he works by himself. If they work together, find how long it takes to paint the house.

**57.** An experienced bricklayer can construct a small wall in 3 hours. An apprentice can complete the job in 6 hours. Find how long it takes if they work together.

**58.** A marketing manager travels 1080 miles in a corporate jet and then an additional 240 miles by car. If the car ride takes 1 hour longer, and if the rate of the jet is 6 times the rate of the car, find the time the manager travels by jet and find the time she travels by car.

## Review

*Solve each equation. See Section 2.1.*

**59.** $\dfrac{x}{5} = \dfrac{x+2}{3}$

**60.** $\dfrac{x}{4} = \dfrac{x+3}{6}$

**61.** $\dfrac{x-3}{2} = \dfrac{x-5}{6}$

**62.** $\dfrac{x-6}{4} = \dfrac{x-2}{5}$

## Concept Extensions

*Calculating body-mass index (BMI) is a way to gauge whether a person should lose weight. Doctors recommend that body-mass index values fall between 19 and 25. The formula for body-mass index B is $B = \dfrac{705w}{h^2}$, where w is weight in pounds and h is height in inches. Use this formula to answer Exercises 63 and 64.*

**63.** A patient is 5 ft 8 in. tall. What should his or her weight be to have a body-mass index of 25? Round to the nearest whole pound.

**64.** A doctor recorded a body-mass index of 47 on a patient's chart. Later, a nurse notices that the doctor recorded the patient's weight as 240 pounds but neglected to record the patient's height. Explain how the nurse can use the information from the chart to find the patient's height. Then find the height.

*In physics, when the source of a sound is traveling toward an observer, the relationship between the actual pitch a of the sound and the pitch h that the observer hears due to the Doppler effect is described by the formula $h = \dfrac{a}{1 - \dfrac{s}{770}}$, where s is the speed of the sound source in miles per hour. Use this formula to answer Exercises 65 and 66.*

**65.** An emergency vehicle has a single-tone siren with the pitch of the musical note E. As it approaches an observer standing by the road, the vehicle is traveling 50 mph. Is the pitch that the observer hears due to the Doppler effect lower or higher than the actual pitch? To which musical note is the pitch that the observer hears closest?

**66.** Suppose an emergency van has a single-tone siren with the pitch of the musical note G. If the van is traveling at 80 mph approaching a standing observer, calculate the pitch the observer hears and the musical note closest to that pitch.

| Pitch of an Octave of Musical Notes in Hertz (Hz) | |
|---|---|
| Note | Pitch |
| Middle C | 261.63 |
| D | 293.66 |
| E | 329.63 |
| F | 349.23 |
| G | 392.00 |
| A | 440.00 |
| B | 493.88 |

*Note:* Greater numbers indicate higher pitches (acoustically).
(*Source:* American Standards Association)

*In electronics, the relationship among the resistances $R_1$ and $R_2$ of two resistors wired in a parallel circuit and their combined resistance R is described by the formula $\dfrac{1}{R} = \dfrac{1}{R_1} + \dfrac{1}{R_2}$. Use this formula to solve Exercises 67 through 69.*

**67.** If the combined resistance is 2 ohms and one of the two resistances is 3 ohms, find the other resistance.

**68.** Find the combined resistance of two resistors of 12 ohms each when they are wired in a parallel circuit.

**69.** The relationship among resistance of two resistors wired in a parallel circuit and their combined resistance may be extended to three resistors of resistances $R_1$, $R_2$, and $R_3$. Write an equation you think may describe the relationship, and use it to find the combined resistance if $R_1$ is 5, $R_2$ is 6, and $R_3$ is 2.

**70.** For the formula $\dfrac{1}{x} = \dfrac{1}{y} + \dfrac{1}{z} - \dfrac{1}{w}$, find x if $y = 2$, $z = 7$, and $w = 6$.

**Variation and Problem Solving** ▶

## Objective A Solving Problems Involving Direct Variation ▶

A very familiar example of **direct variation** is the relationship of the circumference $C$ of a circle to its radius $r$. The formula $C = 2\pi r$ expresses that the circumference is always $2\pi$ times the radius. In other words, $C$ is always a constant multiple $(2\pi)$ of $r$. Because it is, we say that $C$ *varies directly as* $r$, that $C$ *varies directly with* $r$, or that $C$ *is directly proportional to* $r$.

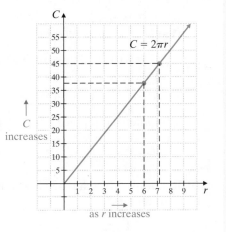

$$C = \underset{\underset{\text{constant}}{\uparrow}}{2\pi} r$$

### Direct Variation

**$y$ varies directly as $x$,** or **$y$ is directly proportional to $x$,** if there is a nonzero constant $k$ such that

$$y = kx$$

The number $k$ is called the **constant of variation** or the **constant of proportionality**.

In the above definition, the relationship described between $x$ and $y$ is a linear one. In other words, the graph of $y = kx$ is a line. The slope of the line is $k$, and the line passes through the origin.

For example, the graph of the direct variation equation $C = 2\pi r$ is shown. The horizontal axis represents the radius $r$, and the vertical axis is the circumference $C$. From the graph we can read that when the radius is 6 units, the circumference is approximately 38 units. Also, when the circumference is 45 units, the radius is between 7 and 8 units. Notice that as the radius increases, the circumference increases.

**Example 1** Suppose that $y$ varies directly as $x$. If $y$ is 5 when $x$ is 30, find the constant of variation and the direct variation equation.

**Solution:** Since $y$ varies directly as $x$, we write $y = kx$. If $y = 5$ when $x = 30$, we have that

$$y = kx$$

$$5 = k(30) \qquad \text{Replace } y \text{ with 5 and } x \text{ with 30.}$$

$$\frac{1}{6} = k \qquad \text{Solve for } k.$$

The constant of variation is $\frac{1}{6}$.

After finding the constant of variation $k$, the direct variation equation can be written as $y = \frac{1}{6}x$.

▬ **Work Practice 1**

Objectives

**A** Solve Problems Involving Direct Variation. ▶

**B** Solve Problems Involving Inverse Variation. ▶

**C** Solve Problems Involving Joint Variation. ▶

**D** Solve Problems Involving Combined Variation. ▶

**Practice 1**

Suppose that $y$ varies directly as $x$. If $y$ is 24 when $x$ is 8, find the constant of variation and the direct variation equation.

**Answer**

**1.** $k = 3; y = 3x$

**485**

## Practice 2

Use Hooke's law as stated in Example 2. If a 56-pound weight attached to a spring stretches the spring 8 inches, find the distance that an 85-pound weight attached to the spring stretches the spring.

**Example 2** Using Direct Variation and Hooke's Law

Hooke's law states that the distance a spring stretches is directly proportional to the weight attached to the spring. If a 40-pound weight attached to the spring stretches the spring 5 inches, find the distance that a 65-pound weight attached to the spring stretches the spring.

**Solution:**

1. UNDERSTAND. Read and reread the problem. Notice that we are given that the distance a spring stretches is *directly proportional* to the weight attached. We let

   $d$ = the distance stretched.

   $w$ = the weight attached

   The constant of variation is represented by $k$.

2. TRANSLATE. Because $d$ is directly proportional to $w$, we write

   $$d = kw$$

3. SOLVE. When a weight of 40 pounds is attached, the spring stretches 5 inches. That is, when $w = 40, d = 5$.

   $5 = k(40)$     Replace $d$ with 5 and $w$ with 40.

   $\dfrac{1}{8} = k$     Solve for $k$.

   Now when we replace $k$ with $\dfrac{1}{8}$ in the equation $d = kw$, we have

   $$d = \frac{1}{8}w$$

   To find the stretch when a weight of 65 pounds is attached, we replace $w$ with 65 to find $d$.

   $$d = \frac{1}{8}(65)$$

   $$= \frac{65}{8} = 8\frac{1}{8} \text{ or } 8.125$$

4. INTERPRET.

   **Check:** Check the proposed solution of 8.125 inches in the original problem.

   **State:** The spring stretches 8.125 inches when a 65-pound weight is attached.

■ Work Practice 2

## Objective B Solving Problems Involving Inverse Variation ▶

When $y$ is proportional to the *reciprocal* of another variable $x$, we say that $y$ *varies inversely as* $x$, or that $y$ *is inversely proportional to* $x$. An example of the **inverse variation** relationship is the relationship between the pressure that a gas exerts and

**Answer**

2. $12\dfrac{1}{7}$ in.

the volume of its container. As the volume of a container decreases, the pressure of the gas it contains increases.

> ### Inverse Variation
>
> **y varies inversely as x,** or **y is inversely proportional to x,** if there is a nonzero constant $k$ such that
>
> $$y = \frac{k}{x}$$
>
> The number $k$ is called the **constant of variation** or the **constant of proportionality**.

Notice that $y = \dfrac{k}{x}$ is an equation containing a rational expression. Its graph for $k > 0$ and $x > 0$ is shown. From the graph, we can see that as $x$ increases, $y$ decreases.

$$y = \frac{k}{x}, k > 0, x > 0$$

y decreases

as $x$ increases

**Example 3**  Suppose that $u$ varies inversely as $w$. If $u$ is 3 when $w$ is 5, find the constant of variation and the inverse variation equation.

**Solution:**  Since $u$ varies inversely as $w$, we have $u = \dfrac{k}{w}$. We let $u = 3$ and $w = 5$, and we solve for $k$.

$$u = \frac{k}{w}$$

$$3 = \frac{k}{5} \quad \text{Let } u = 3 \text{ and } w = 5.$$

$$15 = k \quad \text{Multiply both sides by 5.}$$

The constant of variation $k$ is 15. This gives the inverse variation equation

$$u = \frac{15}{w}$$

■ **Work Practice 3**

**Example 4**  Using Inverse Variation and Boyle's Law

Boyle's law says that if the temperature stays the same, the pressure $P$ of a gas is inversely proportional to the volume $V$. If a cylinder in a steam engine has a pressure of 960 kilopascals when the volume is 1.4 cubic meters, find the pressure when the volume increases to 2.5 cubic meters.

*(Continued on next page)*

### Practice 3

Suppose that $y$ varies inversely as $x$. If $y$ is 6 when $x$ is 3, find the constant of variation and the inverse variation equation.

### Practice 4

The speed $r$ at which one needs to drive in order to travel a constant distance is inversely proportional to the time $t$. A fixed distance can be driven in 5 hours at a rate of 24 mph. Find the rate needed to drive the same distance in 4 hours.

**Answers**

**3.** $k = 18; y = \dfrac{18}{x}$  **4.** 30 mph

**Solution:**

1. **UNDERSTAND.** Read and reread the problem. Notice that we are given that the pressure of a gas is *inversely proportional* to the volume. We will let $P =$ the pressure and $V =$ the volume. The constant of variation is represented by $k$.

2. **TRANSLATE.** Because $P$ is inversely proportional to $V$, we write

$$P = \frac{k}{V}.$$

When $P = 960$ kilopascals, the volume $V = 1.4$ cubic meters. We use this information to find $k$.

$$960 = \frac{k}{1.4} \quad \text{Let } P = 960 \text{ and } V = 1.4.$$

$$1344 = k \quad \text{Multiply both sides by 1.4.}$$

Thus, the value of $k$ is 1344. Replacing $k$ with 1344 in the variation equation, we have

$$P = \frac{1344}{V}$$

Next we find $P$ when $V$ is 2.5 cubic meters.

3. **SOLVE.**

$$P = \frac{1344}{2.5} \quad \text{Let } V = 2.5$$

$$= 537.6$$

4. **INTERPRET.** *Check* the proposed solution in the original problem.

**State:**  When the volume is 2.5 cubic meters, the pressure is 537.6 kilopascals.

■ Work Practice 4

## Objective C Solving Problems Involving Joint Variation ▶

Sometimes the ratio of a variable to the product of many other variables is constant. For example, the ratio of distance traveled to the product of speed and time traveled is constantly 1:

$$\frac{d}{rt} = 1 \quad \text{or} \quad d = rt$$

Such a relationship is called **joint variation.**

> ### Joint Variation
>
> If the ratio of a variable $y$ to the product of two or more variables is constant, then $y$ **varies jointly as,** or **is jointly proportional to,** the other variables. If
>
> $$y = kxz$$
>
> then the number $k$ is the **constant of variation** or the **constant of proportionality.**

✓**Concept Check** Which type of variation is represented by the equation $xy = 8$? Explain.

  **a.** Direct variation

  **b.** Inverse variation

  **c.** Joint variation

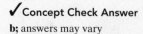
✓**Concept Check Answer**

**b;** answers may vary

**Example 5** The lateral surface area of a cylinder varies jointly as its radius and height. Express surface area $S$ in terms of radius $r$ and height $h$.

**Solution:** Because the surface area varies jointly as the radius $r$ and the height $h$, we equate $S$ to a constant multiple of $r$ and $h$:

$$S = krh$$

*Note:* From actual values of $S, r$, and $h$, it can be determined that the constant $k$ is $2\pi$, and we then have the formula $S = 2\pi rh$.

■ Work Practice 5

Practice 5

The area of a triangle varies jointly as its base and height. Express the area $A$ in terms of base $b$ and height $h$.

## Objective D  Solving Problems Involving Combined Variation ▶

Some examples of variation involve combinations of direct, inverse, and joint variation. We will call these variations **combined variation.**

**Example 6** Suppose that $y$ varies directly as the square of $x$. If $y$ is 24 when $x$ is 2, find the constant of variation and the variation equation.

**Solution:** Since $y$ varies directly as the square of $x$, we have

$$y = kx^2$$

Now let $y = 24$ and $x = 2$ and solve for $k$.

$$y = kx^2$$
$$24 = k \cdot 2^2$$
$$24 = 4k$$
$$6 = k$$

The constant of variation is 6, so the variation equation is

$$y = 6x^2$$

■ Work Practice 6

Practice 6

Suppose that $y$ varies inversely as the square of $x$. If $y$ is 24 when $x$ is 2, find the constant of variation and the variation equation.

**Example 7** Using Combined Variation

The maximum weight that a circular column can support is directly proportional to the fourth power of its diameter and inversely proportional to the square of its height. A 2-meter-wide column that is 8 meters in height can support 1 ton. Find the weight that a 1-meter-wide column that is 4 meters in height can support.

**Solution:**

1. UNDERSTAND. Read and reread the problem. Let $w$ = weight, $d$ = diameter, $h$ = height, and $k$ = the constant of variation.

*(Continued on next page)*

Practice 7

The maximum weight that a rectangular beam can support varies jointly as its width and the square of its height and inversely as its length. If a beam $\frac{1}{3}$ foot wide, 1 foot high, and 10 feet long can support 3 tons, find how much weight a similar beam can support if it is 1 foot wide, $\frac{1}{3}$ foot high, and 9 feet long.

**Answers**

**5.** $A = kbh$  **6.** $k = 96; y = \dfrac{96}{x^2}$

**7.** $1\frac{1}{9}$ tons

**2. TRANSLATE.** Since $w$ is directly proportional to $d^4$ and inversely proportional to $h^2$, we have

$$w = \frac{kd^4}{h^2}$$

**3. SOLVE.** To find $k$, we are given that a 2-meter-wide column that is 8 meters in height can support 1 ton. That is, $w = 1$ when $d = 2$ and $h = 8$, or

$$1 = \frac{k \cdot 2^4}{8^2} \quad \text{Let } w = 1, d = 2, \text{ and } h = 8.$$

$$1 = \frac{k \cdot 16}{64}$$

$$4 = k \quad \text{Solve for } k.$$

Now we replace $k$ with 4 in the equation $w = \frac{kd^4}{h^2}$:

$$w = \frac{4d^4}{h^2}$$

To find weight, $w$, for a 1-meter-wide column that is 4 meters in height, we let $d = 1$ and $h = 4$.

$$w = \frac{4 \cdot 1^4}{4^2}$$

$$w = \frac{4}{16} = \frac{1}{4}$$

**4. INTERPRET.** *Check* the proposed solution in the original problem.

**State:**  The 1-meter-wide column that is 4 meters in height can hold $\frac{1}{4}$ ton of weight.

■ Work Practice 7

## Vocabulary, Readiness & Video Check

*State whether each equation represents direct, inverse, or joint variation.*

**1.** $y = 5x$

**2.** $y = \dfrac{700}{x}$

**3.** $y = 5xz$

**4.** $y = \dfrac{1}{2}abc$

**5.** $y = \dfrac{9.1}{x}$

**6.** $y = 2.3x$

**7.** $y = \dfrac{2}{3}x$

**8.** $y = 3.1st$

*Martin-Gay Interactive Videos*

*See Video 6.6*

*Watch the section lecture video and answer the following questions.*

**Objective A**  **9.** Based on the lecture before ⊞ Example 1, what kind of equation is a direct variation equation? What does $k$, the constant of variation, represent in this equation? ▶

**Objective B**  **10.** In ⊞ Example 3, why is it not necessary to replace the given values of $x$ and $y$ in the inverse variation equation in order to find $k$? ▶

**Objective C**  **11.** Based on ⊞ Example 5 and the lecture before, what is the variation equation for "$y$ varies jointly as the square of $a$ and the fifth power of $b$"? ▶

**Objective D**  **12.** From ⊞ Example 6, what kind of variation does a combined variation application involve? ▶

## 6.6 Exercise Set MyMathLab®

**Objective A** *If y varies directly as x, find the constant of variation and the direct variation equation for each situation. See Example 1.*

**1.** $y = 4$ when $x = 20$

**2.** $y = 6$ when $x = 30$

**3.** $y = 6$ when $x = 4$

**4.** $y = 12$ when $x = 8$

**5.** $y = 7$ when $x = \frac{1}{2}$

**6.** $y = 11$ when $x = \frac{1}{3}$

**7.** $y = 0.2$ when $x = 0.8$

**8.** $y = 0.4$ when $x = 2.5$

**Objectives A D Mixed Practice** *Solve. See Examples 2 and 7.*

**9.** The weight of a synthetic ball varies directly with the cube of its radius. A ball with a radius of 2 inches weighs 1.20 pounds. Find the weight of a ball of the same material with a 3-inch radius.

**10.** At sea, the distance to the horizon is directly proportional to the square root of the elevation of the observer. If a person who is 36 feet above the water can see 7.4 miles, find how far a person 64 feet above the water can see. Round to the nearest tenth of a mile.

**11.** The amount $P$ of pollution varies directly with the population $N$ of people. Kansas City had a population of 464,000 in 2012 and produced 170,900 tons of pollution. Find how many tons of pollution we should expect St. Louis produced if we know its population was 318,000. Round to the nearest whole ton. (*Source:* Kansas City Government Web site)

**12.** Charles's law states that if the pressure $P$ stays the same, the volume $V$ of a gas is directly proportional to its temperature $T$. If a balloon is filled with 20 cubic meters of a gas at a temperature of 300 K, find the new volume if the temperature rises 360 K while the pressure stays the same.

**Objective B** *If y varies inversely as x, find the constant of variation and the inverse variation equation for each situation. See Example 3.*

**13.** $y = 6$ when $x = 5$

**14.** $y = 20$ when $x = 9$

**15.** $y = 100$ when $x = 7$

**16.** $y = 63$ when $x = 3$

**17.** $y = \frac{1}{8}$ when $x = 16$

**18.** $y = \frac{1}{10}$ when $x = 40$

**19.** $y = 0.2$ when $x = 0.7$

**20.** $y = 0.6$ when $x = 0.3$

**Objectives B D Mixed Practice** *Solve. See Examples 4 and 7.*

**21.** Pairs of markings a set distance apart are made on highways so that police can detect drivers exceeding the speed limit. Over a fixed distance, the speed $R$ varies inversely with the time $T$. In one particular pair of markings, $R$ is 45 mph when $T$ is 6 seconds. Find the speed of a car that travels the given distance in 5 seconds.

**22.** The weight of an object on or above the surface of Earth varies inversely as the square of the distance between the object and Earth's center. If a person weighs 160 pounds on Earth's surface, find the individual's weight if he moves 200 miles above Earth. Round to the nearest whole pound. (Assume that Earth's radius is 4000 miles.)

**23.** If the voltage $V$ in an electric circuit is held constant, the current $I$ is inversely proportional to the resistance $R$. If the current is 40 amperes when the resistance is 270 ohms, find the current when the resistance is 150 ohms.

**24.** Because it is more efficient to produce larger numbers of items, the cost of producing a certain computer DVD is inversely proportional to the number produced. If 4000 can be produced at a cost of $1.20 each, find the cost per DVD when 6000 are produced.

**25.** The intensity $I$ of light varies inversely as the square of the distance $d$ from the light source. If the distance from the light source is doubled (see the figure), determine what happens to the intensity of light at the new location.

△ **26.** The maximum weight that a circular column can hold is inversely proportional to the square of its height. If an 8-foot column can hold 2 tons, find how much weight a 10-foot column can hold.

**Objectives C D Mixed Practice** *Write each statement as an equation. Use k as the constant of variation. See Example 5.*

**27.** $x$ varies jointly as $y$ and $z$.

**28.** $P$ varies jointly as $R$ and the square of $S$.

**29.** $r$ varies jointly as $s$ and the cube of $t$.

**30.** $a$ varies jointly as $b$ and $c$.

*For each statement, find the constant of variation and the variation equation. See Examples 5 and 6.*

**31.** $y$ varies directly as the cube of $x$; $y = 9$ when $x = 3$

**32.** $y$ varies directly as the cube of $x$; $y = 32$ when $x = 4$

**33.** $y$ varies directly as the square root of $x$; $y = 0.4$ when $x = 4$

**34.** $y$ varies directly as the square root of $x$; $y = 2.1$ when $x = 9$

**35.** $y$ varies inversely as the square of $x$; $y = 0.052$ when $x = 5$

**36.** $y$ varies inversely as the square of $x$; $y = 0.011$ when $x = 10$

▶ **37.** $y$ varies jointly as $x$ and the cube of $z$; $y = 120$ when $x = 5$ and $z = 2$

**38.** $y$ varies jointly as $x$ and the square of $z$; $y = 360$ when $x = 4$ and $z = 3$

*Solve. See Example 7.*

**39.** The maximum weight that a rectangular beam can support varies jointly as its width and the square of its height and inversely as its length. If a beam $\frac{1}{2}$ foot wide, $\frac{1}{3}$ foot high, and 10 feet long can support 12 tons, find how much a similar beam can support if the beam is $\frac{2}{3}$ foot wide, $\frac{1}{2}$ foot high, and 16 feet long.

**40.** The number of cars manufactured on an assembly line at a General Motors plant varies jointly as the number of workers and the time they work. If 200 workers can produce 60 cars in 2 hours, find how many cars 240 workers should be able to make in 3 hours.

**41.** The volume of a cone varies jointly as its height and the square of its radius. If the volume of a cone is $32\pi$ cubic inches when the radius is 4 inches and the height is 6 inches, find the volume of a cone when the radius is 3 inches and the height is 5 inches.

**42.** When a wind blows perpendicularly against a flat surface, its force is jointly proportional to the surface area and the speed of the wind. A sail whose surface area is 12 square feet experiences a 20-pound force when the wind speed is 10 miles per hour. Find the force on an 8-square-foot sail if the wind speed is 12 miles per hour.

**43.** The intensity of light (in foot-candles) varies inversely as the square of $x$, the distance in feet from the light source. The intensity of light 2 feet from the source is 80 foot-candles. How far away is the source if the intensity of light is 5 foot-candles?

**44.** The horsepower that can be safely transmitted to a shaft varies jointly as the shaft's angular speed of rotation (in revolutions per minute) and the cube of its diameter. A 2-inch shaft making 120 revolutions per minute safely transmits 40 horsepower. Find how much horsepower can be safely transmitted by a 3-inch shaft making 80 revolutions per minute.

**Objectives A B C D Mixed Practice** *Write an equation to describe each variation. Use k for the constant of variation. See Examples 1 through 7.*

**45.** $y$ varies directly as $x$

**46.** $p$ varies directly as $q$

**47.** $a$ varies inversely as $b$

**48.** $y$ varies inversely as $x$

**49.** $y$ varies jointly as $x$ and $z$

**50.** $y$ varies jointly as $q$, $r$, and $t$

**51.** $y$ varies inversely as $x^3$

**52.** $y$ varies inversely as $a^4$

**53.** $y$ varies directly as $x$ and inversely as $p^2$

**54.** $y$ varies directly as $a^5$ and inversely as $b$

## Review

*Find the exact circumference C and area A of each circle. Use the formulas $C = 2\pi r$ and $A = \pi r^2$ where r is the radius of the circle. See also Section 2.3.*

**55.**    **56.**    **57.**    **58.**

*Find the slope of the line containing each pair of points. See Section 3.2.*

**59.** $(3, 6), (-2, 6)$    **60.** $(-5, -2), (0, 7)$    **61.** $(4, -1), (5, -2)$    **62.** $(2, 1), (2, -3)$

## Concept Extensions

*Solve. See the Concept Check in this section. Choose the type of variation that each equation represents.*
**a.** *Direct variation*   **b.** *Inverse variation*   **c.** *Joint variation*

**63.** $y = \dfrac{2}{3}x$    **64.** $y = \dfrac{0.6}{x}$    **65.** $y = 9ab$    **66.** $xy = \dfrac{2}{11}$

*Solve.*

△ **67.** The volume of a cylinder varies jointly as the height and the square of the radius. If the height is halved and the radius is doubled, determine what happens to the volume.

**68.** The horsepower to drive a boat varies directly as the cube of the speed of the boat. If the speed of the boat is to double, determine the corresponding increase in horsepower required.

**69.** Suppose that $y$ varies directly as $x^2$. If $x$ is doubled, what is the effect on $y$?

**70.** Suppose that $y$ varies directly as $x$. If $x$ is doubled, what is the effect on $y$?

## Chapter 6 Group Activity

### Fastest-Growing Occupations

According to the U.S. Bureau of Labor Statistics projections, the careers listed in the following table will be among the top 10 fastest-growing jobs in the next decade, according to the percent increase in the number of jobs.

| Employment (in thousands) | | | |
|---|---|---|---|
| Occupation | 2012 | 2022 | % Change |
| Personal care aides | 1191 | 1771 | |
| Interpreters and translators | 64 | 93 | |
| Diagnostic medical sonographers | 110 | 153 | |
| Skincare specialists | 44 | 61 | |
| Stonemasons and tile and marble setters | 1285 | 1710 | |
| Occupational therapy assistants | 39 | 51 | |
| Home health aides | 875 | 1299 | |
| Insulation workers | 52 | 77 | |
| Genetic counselors | 2 | 3 | |
| Physical therapist assistants | 121 | 170 | |

What do all these fast-growing occupations have in common? They all require some knowledge of math! For some careers, such as diagnostic medical sonographers, stonemasons, and insulation workers, the ways math is used on the job may be obvious. For other occupations, the use of math may not be quite as apparent. However, tasks common to many jobs, like filling in a time sheet, figuring a bill, writing up an expense or mileage report, planning a budget, ordering supplies, and even making a work schedule, all require math.

### Group Activity

1. Find the percent change in the number of jobs available from 2012 to 2022 for each occupation on the list. If necessary, round to the nearest tenth of a percent.

2. Rank these top 10 occupations according to percent growth, from greatest to least.

3. Which occupation will be the fastest growing during this period?

4. How many occupations have greater than or equal to 48% more positions in 2022 than in 2012?

5. Which of the listed occupations will be the slowest growing during this period?

## Chapter 6 Vocabulary Check

*Fill in each blank with one of the words or phrases listed below.*

| | | | | |
|---|---|---|---|---|
| rational expression | equation | complex fraction | opposites | directly |
| least common denominator | expression | inversely | jointly | |

1. A rational expression whose numerator, denominator, or both contain one or more rational expressions is called a(n) _____.

2. In the equation $y = kx$, $y$ varies _____ as $x$.

3. In the equation $y = \dfrac{k}{x}$, $y$ varies _____ as $x$.

4. The _____ of a list of rational expressions is a polynomial of least degree whose factors include the factors of all denominators in the list.

5. In the equation $y = kxz$, $y$ varies _____ as $x$ and $z$.

6. The expressions $(x - 5)$ and $(5 - x)$ are called _____.

7. A(n) _____ is an expression that can be written as the quotient of $\dfrac{P}{Q}$ of two polynomials $P$ and $Q$ as long as $Q$ is not 0.

8. Which is an expression and which is an equation? An example

    of an _____ is $\dfrac{2}{x} + \dfrac{2}{x^2} = 7$ and an

    example of an _____ is $\dfrac{2}{x} + \dfrac{5}{x^2}$.

**Helpful Hint**

> ▶ Are you preparing for your test? Don't forget to take the Chapter 6 Test on page 503. Then check your answers at the back of the text and use the Chapter Test Prep Videos to see the fully worked-out solutions to any of the exercises you want to review.

# 6  Chapter Highlights

| Definitions and Concepts | Examples |
|---|---|
| **Section 6.1    Rational Functions and Multiplying and Dividing Rational Expressions** | |

| Definitions and Concepts | Examples |
|---|---|
| A **rational expression** is the quotient $\dfrac{P}{Q}$ of two polynomials $P$ and $Q$, as long as $Q$ is not 0. | $\dfrac{2x - 6}{7}, \dfrac{t^2 - 3t + 5}{t - 1}$ |
| **Simplifying a Rational Expression**<br><br>**Step 1:** Completely factor the numerator and the denominator.<br><br>**Step 2:** Divide out common factors. | Simplify.<br><br>$\dfrac{2x^2 + 9x - 5}{x^2 - 25} = \dfrac{(2x - 1)(x + 5)}{(x - 5)(x + 5)}$<br><br>$\qquad\qquad = \dfrac{2x - 1}{x - 5}$ |
| **Multiplying Rational Expressions**<br><br>**Step 1:** Completely factor numerators and denominators.<br><br>**Step 2:** Multiply the numerators and multiply the denominators.<br><br>**Step 3:** Simplify the product. | Multiply: $\dfrac{x^3 + 8}{12x - 18} \cdot \dfrac{14x^2 - 21x}{x^2 + 2x}$<br><br>$= \dfrac{(x + 2)(x^2 - 2x + 4)}{6(2x - 3)} \cdot \dfrac{7x(2x - 3)}{x(x + 2)}$<br><br>$= \dfrac{7(x^2 - 2x + 4)}{6}$ |

*(Continued)*

| Definitions and Concepts | Examples |
|---|---|

### Section 6.1   Rational Functions and Multiplying and Dividing Rational Expressions (*continued*)

**Dividing Rational Expressions**

Multiply the first rational expression by the reciprocal of the second rational expression.

Divide: $\dfrac{x^2 + 6x + 9}{5xy - 5y} \div \dfrac{x + 3}{10y}$

$= \dfrac{(x + 3)(x + 3)}{5y(x - 1)} \cdot \dfrac{2 \cdot 5y}{x + 3}$

$= \dfrac{2(x + 3)}{x - 1}$

A **rational function** is a function described by a rational expression.

$f(x) = \dfrac{2x - 6}{7}, h(t) = \dfrac{t^2 - 3t + 5}{t - 1}$

### Section 6.2   Adding and Subtracting Rational Expressions

**Adding or Subtracting Rational Expressions**

**Step 1:** Find the LCD.

**Step 2:** Write each rational expression as an equivalent rational expression whose denominator is the LCD.

**Step 3:** Add or subtract numerators and write the sum or difference over the common denominator.

**Step 4:** Simplify the result.

Subtract: $\dfrac{3}{x + 2} - \dfrac{x + 1}{x - 3}$

$= \dfrac{3 \cdot (x - 3)}{(x + 2) \cdot (x - 3)} - \dfrac{(x + 1) \cdot (x + 2)}{(x - 3) \cdot (x + 2)}$

$= \dfrac{3(x - 3) - (x + 1)(x + 2)}{(x + 2)(x - 3)}$

$= \dfrac{3x - 9 - (x^2 + 3x + 2)}{(x + 2)(x - 3)}$

$= \dfrac{3x - 9 - x^2 - 3x - 2}{(x + 2)(x - 3)}$

$= \dfrac{-x^2 - 11}{(x + 2)(x - 3)}$

### Section 6.3   Simplifying Complex Fractions

Simplify: $\dfrac{\dfrac{x + 2}{x}}{x - \dfrac{4}{x}}$

**Method 1:** Simplify the numerator and the denominator so that each is a single fraction. Then perform the indicated division and simplify if possible.

Method 1: $\dfrac{\dfrac{x + 2}{x}}{\dfrac{x \cdot x}{1 \cdot x} - \dfrac{4}{x}} = \dfrac{\dfrac{x + 2}{x}}{\dfrac{x^2 - 4}{x}}$

$= \dfrac{x + 2}{x} \cdot \dfrac{x}{(x + 2)(x - 2)} = \dfrac{1}{x - 2}$

**Method 2:** Multiply the numerator and the denominator of the complex fraction by the LCD of all the fractions in both the numerator and the denominator. Then simplify if possible.

Method 2: $\dfrac{\left(\dfrac{x + 2}{x}\right) \cdot x}{\left(x - \dfrac{4}{x}\right) \cdot x} = \dfrac{x + 2}{x \cdot x - \dfrac{4}{x} \cdot x}$

$= \dfrac{x + 2}{x^2 - 4} = \dfrac{x + 2}{(x + 2)(x - 2)} = \dfrac{1}{x - 2}$

| Definitions and Concepts | Examples |
|---|---|

### Section 6.4   Solving Equations Containing Rational Expressions

**Solving an Equation Containing Rational Expressions**

Multiply both sides of the equation by the LCD of all rational expressions. Then use the distributive property and simplify. Solve the resulting equation and then check each proposed solution to see whether it makes any denominator 0. Discard any solutions that do.

Solve:     $x - \dfrac{3}{x} = \dfrac{1}{2}$

$2x\left(x - \dfrac{3}{x}\right) = 2x\left(\dfrac{1}{2}\right)$    The LCD is $2x$.

$2x \cdot x - 2x\left(\dfrac{3}{x}\right) = 2x\left(\dfrac{1}{2}\right)$    Distribute.

$2x^2 - 6 = x$

$2x^2 - x - 6 = 0$    Subtract $x$ from both sides.

$(2x + 3)(x - 2) = 0$    Factor.

$x = -\dfrac{3}{2}$   or   $x = 2$

Both $-\dfrac{3}{2}$ and 2 check. The solution set is $\left\{2, -\dfrac{3}{2}\right\}$.

### Section 6.5   Rational Equations and Problem Solving

**Solving an Equation for a Specified Variable**
Treat the specified variable as the only variable of the equation and solve as usual.

Solve for $x$.

$A = \dfrac{2x + 3y}{5}$

$5A = 2x + 3y$    Multiply both sides by 5.

$5A - 3y = 2x$    Subtract $3y$ from both sides.

$\dfrac{5A - 3y}{2} = x$    Divide both sides by 2.

**Solving a Problem that Involves a Rational Equation**

Jeanee and David Dillon volunteer every year to clean a strip of Lake Ponchartrain beach. Jeanee can clean all the trash in this area of beach in 6 hours; David takes 5 hours. Find how long it will take them to clean the area of beach together.

**1.** UNDERSTAND.

**1.** Read and reread the problem. Let $x =$ time in hours that it takes Jeanee and David to clean the beach together.

| | Hours to Complete | Part Completed in 1 Hour |
|---|---|---|
| **Jeanee Alone** | 6 | $\dfrac{1}{6}$ |
| **David Alone** | 5 | $\dfrac{1}{5}$ |
| **Together** | $x$ | $\dfrac{1}{x}$ |

**2.** TRANSLATE.

**2.** In words:

| part Jeanee can complete in 1 hour | | part David can complete in 1 hour | | part they can complete together in 1 hour |
|---|---|---|---|---|
| ↓ | + | ↓ | = | ↓ |

Translate:

$\dfrac{1}{6}$   $+$   $\dfrac{1}{5}$   $=$   $\dfrac{1}{x}$

*(Continued)*

| Definitions and Concepts | Examples |
|---|---|
| **Section 6.5    Rational Equations and Problem Solving (*continued*)** | |
| **3.** SOLVE. | **3.** $\dfrac{1}{6} + \dfrac{1}{5} = \dfrac{1}{x}$ <br><br> $5x + 6x = 30$     Multiply both sides by 30x. <br> $11x = 30$ <br> $x = \dfrac{30}{11}$   or   $2\dfrac{8}{11}$ |
| **4.** INTERPRET. | **4.** *Check* and then *state*. Together, they can clean the beach in $2\dfrac{8}{11}$ hours. |

| | |
|---|---|
| **Section 6.6    Variation and Problem Solving** | |
| *y* **varies directly** as *x*, or *y* is **directly proportional** to *x*, if there is a nonzero constant *k* such that <br><br> $y = kx$ | The circumference of a circle *C* varies directly as its radius *r*. <br><br> $C = \underset{k}{\underbrace{2\pi}}r$ |
| *y* **varies inversely** as *x*, or *y* is **inversely proportional** to *x*, if there is a nonzero constant *k* such that <br><br> $y = \dfrac{k}{x}$ | Pressure *P* varies inversely with volume *V*. <br><br> $P = \dfrac{k}{V}$ |
| *y* **varies jointly** as *x* and *z*, or *y* is **jointly proportional** to *x* and *z*, if there is a nonzero constant *k* such that <br><br> $y = kxz$ | The lateral surface area *S* of a cylinder varies jointly as its radius *r* and height *h*. <br><br> $S = \underset{k}{\underbrace{2\pi}}rh$ |

# Chapter 6  Review

**(6.1)** *Find the domain for each rational function.*

**1.** $f(x) = \dfrac{3 - 5x}{7}$

**2.** $g(x) = \dfrac{2x + 4}{11}$

**3.** $F(x) = \dfrac{-3x^2}{x - 5}$

**4.** $h(x) = \dfrac{4x}{3x - 12}$

**5.** $f(x) = \dfrac{x^3 + 2}{x^2 + 8x}$

**6.** $G(x) = \dfrac{20}{3x^2 - 48}$

*Write each rational expression in lowest terms.*

**7.** $\dfrac{x - 12}{12 - x}$

**8.** $\dfrac{5x - 15}{25x - 75}$

**9.** $\dfrac{2x}{2x^2 - 2x}$

**10.** $\dfrac{x + 7}{x^2 - 49}$

**11.** $\dfrac{2x^2 + 4x - 30}{x^2 + x - 20}$

**12.** The average cost of manufacturing $x$ bookcases is given by the rational function.

$$C(x) = \frac{35x + 4200}{x}$$

**a.** Find the average cost per bookcase of manufacturing 50 bookcases.

**b.** Find the average cost per bookcase of manufacturing 100 bookcases.

**c.** As the number of bookcases increases, does the average cost per bookcase increase or decrease? (See parts (a) and (b).)

*Perform each indicated operation. Write your answers in lowest terms.*

**13.** $\dfrac{4 - x}{5} \cdot \dfrac{15}{2x - 8}$

**14.** $\dfrac{x^2 - 6x + 9}{2x^2 - 18} \cdot \dfrac{4x + 12}{5x - 15}$

**15.** $\dfrac{a - 4b}{a^2 + ab} \cdot \dfrac{b^2 - a^2}{8b - 2a}$

**16.** $\dfrac{x^2 - x - 12}{2x^2 - 32} \cdot \dfrac{x^2 + 8x + 16}{3x^2 + 21x + 36}$

**17.** $\dfrac{4x + 8y}{3} \div \dfrac{5x + 10y}{9}$

**18.** $\dfrac{x^2 - 25}{3} \div \dfrac{x^2 - 10x + 25}{x^2 - x - 20}$

**19.** $\dfrac{a - 4b}{a^2 + ab} \div \dfrac{20b - 5a}{b^2 - a^2}$

**20.** $\dfrac{3x + 3}{x - 1} \div \dfrac{x^2 - 6x - 7}{x^2 - 1}$

**21.** $\dfrac{2x - x^2}{x^3 - 8} \div \dfrac{x^2}{x^2 + 2x + 4}$

**22.** $\dfrac{5x - 15}{3 - x} \cdot \dfrac{x + 2}{10x + 20} \cdot \dfrac{x^2 - 9}{x^2 - x - 6}$

**(6.2)** *Find the LCD of the rational expressions in each list.*

**23.** $\dfrac{5}{4x^2y^5}, \dfrac{3}{10x^2y^4}, \dfrac{x}{6y^4}$

**24.** $\dfrac{5}{2x}, \dfrac{7}{x - 2}$

**25.** $\dfrac{3}{5x}, \dfrac{2}{x - 5}$

**26.** $\dfrac{1}{5x^3}, \dfrac{4}{x^2 + 3x - 28}, \dfrac{11}{10x^2 - 30x}$

*Perform each indicated operation. Write your answers in lowest terms.*

**27.** $\dfrac{4}{x - 4} + \dfrac{x}{x - 4}$

**28.** $\dfrac{4}{3x^2} + \dfrac{2}{3x^2}$

**29.** $\dfrac{1}{x - 2} - \dfrac{1}{4 - 2x}$

**30.** $\dfrac{1}{10 - x} + \dfrac{x - 1}{x - 10}$

**31.** $\dfrac{x}{9 - x^2} - \dfrac{2}{5x - 15}$

**32.** $2x + 1 - \dfrac{1}{x - 3}$

**33.** $\dfrac{2}{a^2 - 2a + 1} + \dfrac{3}{a^2 - 1}$

**34.** $\dfrac{x}{9x^2 + 12x + 16} - \dfrac{3x + 4}{27x^3 - 64}$

*Perform each indicated operation. Write your answers in lowest terms.*

**35.** $\dfrac{2}{x-1} - \dfrac{3x}{3x-3} + \dfrac{1}{2x-2}$

△ **36.** Find the perimeter of the heptagon (a polygon with seven sides).

**(6.3)** *Simplify each complex fraction.*

**37.** $\dfrac{1 - \dfrac{3x}{4}}{2 + \dfrac{x}{4}}$

**38.** $\dfrac{\dfrac{x^2}{15}}{\dfrac{x+1}{5x}}$

**39.** $\dfrac{2 - \dfrac{3}{2x}}{x - \dfrac{2}{5x}}$

**40.** $\dfrac{1 + \dfrac{x}{y}}{\dfrac{x^2}{y^2} - 1}$

**41.** $\dfrac{\dfrac{5}{x} + \dfrac{1}{xy}}{\dfrac{3}{x^2}}$

**42.** $\dfrac{\dfrac{x}{3} - \dfrac{3}{x}}{1 + \dfrac{3}{x}}$

**43.** $\dfrac{\dfrac{1}{x-1} + 1}{\dfrac{1}{x+1} - 1}$

**44.** $\dfrac{\dfrac{x-3}{x+3} + \dfrac{x+3}{x-3}}{\dfrac{x-3}{x+3} - \dfrac{x+3}{x-3}}$

**(6.4)** *Solve each equation.*

**45.** $\dfrac{3}{x} + \dfrac{1}{3} = \dfrac{5}{x}$

**46.** $\dfrac{2x+3}{5x-9} = \dfrac{3}{2}$

**47.** $\dfrac{1}{x-2} - \dfrac{3x}{x^2-4} = \dfrac{2}{x+2}$

**48.** $\dfrac{7}{x} - \dfrac{x}{7} = 0$

*Solve each equation or perform each indicated operation. Simplify.*

**49.** $\dfrac{5}{x^2-7x} + \dfrac{4}{2x-14}$

**50.** $\dfrac{4}{3-x} - \dfrac{7}{2x-6} + \dfrac{5}{x}$

**51.** $3 - \dfrac{5}{x} - \dfrac{2}{x^2} = 0$

**52.** $2 + \dfrac{15}{x^2} = \dfrac{13}{x}$

**(6.5)** *Solve each equation for the specified variable.*

△ **53.** $A = \dfrac{h(a+b)}{2}$ for $a$

**54.** $\dfrac{1}{R} = \dfrac{1}{R_1} + \dfrac{1}{R_2}$ for $R_2$

**55.** $I = \dfrac{E}{R+r}$ for $R$

**56.** $A = P + Prt$ for $r$

**57.** $\dfrac{1}{x} = \dfrac{1}{y} - \dfrac{1}{z}$ for $x$

**58.** $H = \dfrac{kA(T_1 - T_2)}{L}$ for $A$

*Solve.*

**59.** The sum of a number and twice its reciprocal is 3. Find the number(s).

**60.** If a number is added to the numerator of $\frac{3}{7}$, and twice that number is added to the denominator of $\frac{3}{7}$, the result is equivalent to $\frac{10}{21}$. Find the number.

**61.** Three boys can paint a fence in 4 hours, 5 hours, and 6 hours, respectively. Find how long it will take all three boys to paint the fence.

**62.** If Sue Katz can type a certain number of mailing labels in 6 hours and Tom Neilson and Sue working together can type the same number of mailing labels in 4 hours, find how long it takes Tom alone to type the mailing labels.

**63.** The speed of a Ranger boat in still water is 32 mph. If the boat travels 72 miles upstream in the same time that it takes to travel 120 miles downstream, find the speed of the current.

**64.** The speed of a jogger is 3 mph faster than the speed of a walker. If the jogger travels 14 miles in the same amount of time that the walker travels 8 miles, find the speed of the walker.

**(6.6)** *Solve each variation problem.*

**65.** $A$ is directly proportional to $B$. If $A = 6$ when $B = 14$, find $A$ when $B = 21$.

**66.** According to Boyle's law, the pressure exerted by a gas is inversely proportional to the volume, as long as the temperature stays the same. If a gas exerts a pressure of 1250 kilopascals when the volume is 2 cubic meters, find the volume when the pressure is 800 kilopascals.

## Mixed Review

*For expressions, perform the indicated operation and/or simplify. For equations, solve the equation for the unknown variable.*

**67.** $\dfrac{22x + 8}{11x + 4}$

**68.** $\dfrac{xy - 3x + 2y - 6}{x^2 + 4x + 4}$

**69.** $\dfrac{2}{5x} \div \dfrac{4 - 18x}{6 - 27x}$

**70.** $\dfrac{7x + 28}{2x + 4} \div \dfrac{x^2 + 2x - 8}{x^2 - 2x - 8}$

**71.** $\dfrac{5a^2 - 20}{a^3 + 2a^2 + a + 2} \div \dfrac{7a}{a^3 + a}$

**72.** $\dfrac{4a + 8}{5a^2 - 20} \cdot \dfrac{3a^2 - 6a}{a + 3} \div \dfrac{2a^2}{5a + 15}$

**73.** $\dfrac{x - 2}{x + 1} - \dfrac{x - 3}{x - 1}$

**74.** $\dfrac{7}{2x} + \dfrac{5}{6x}$

**75.** $\dfrac{2}{x^2 - 16} - \dfrac{3x}{x^2 + 8x + 16} + \dfrac{3}{x + 4}$

**76.** $\dfrac{2x + 1}{x^2 + x - 6} + \dfrac{2 - x}{x^2 + x - 6}$

**77.** $\dfrac{\dfrac{1}{x} - \dfrac{2}{3x}}{\dfrac{5}{2x} - \dfrac{1}{3}}$

**78.** $\dfrac{2}{1 - \dfrac{2}{x}}$

**79.** $\dfrac{\dfrac{x^2 + 5x - 6}{4x + 3}}{\dfrac{(x + 6)^2}{8x + 6}}$

**80.** $\dfrac{\dfrac{3}{x - 1} - \dfrac{2}{1 - x}}{\dfrac{2}{x - 1} - \dfrac{2}{x}}$

**81.** $4 + \dfrac{8}{x} = 8$

**82.** $\dfrac{x - 2}{x^2 - 7x + 10} = \dfrac{1}{5x - 10} - \dfrac{1}{x - 5}$

**83.** The denominator of a fraction is 2 more than the numerator. If the numerator is decreased by 3 and the denominator is increased by 5, the resulting fraction is equivalent to $\dfrac{2}{3}$. Find the fraction.

**84.** The sum of the reciprocals of two consecutive even integers is $-\dfrac{9}{40}$. Find the two integers.

**85.** The inlet pipe of a water tank can fill the tank in 2 hours and 30 minutes. The outlet pipe can empty the tank in 2 hours. Find how long it takes to empty a full tank if both pipes are open.

**86.** Timmy Garnica drove 210 miles in the same amount of time that it took a DC-10 jet to travel 1715 miles. The speed of the jet was 430 mph faster than the speed of the car. Find the speed of the jet.

**87.** Two Amtrak trains traveling on parallel tracks leave Tucson at the same time. In 6 hours the faster train is 382 miles from Tucson and the trains are 112 miles apart. Find how fast each train is traveling.

△ **88.** The surface area of a sphere varies directly as the square of its radius. If the surface area is $36\pi$ square inches when the radius is 3 inches, find the surface area when the radius is 4 inches.

**89.** $C$ is directly proportional to $D$. If $C = 12$ when $D = 8$, find $C$ when $D = 24$.

**90.** $C$ is inversely proportional to $D$. If $C = 12$ when $D = 8$, find $C$ when $D = 24$.

Step-by-step test solutions are found on the Chapter Test Prep Videos. Where available: **MyMathLab®** or **You Tube** Test

**Chapter 6**

*Find the domain of each rational function.*

**1.** $f(x) = \dfrac{5x^2}{1 - x}$

**2.** $g(x) = \dfrac{9x^2 - 9}{x^2 + 4x + 3}$

*Write each rational expression in lowest terms.*

**3.** $\dfrac{7x - 21}{24 - 8x}$

**4.** $\dfrac{x^2 - 4x}{x^2 + 5x - 36}$

**5.** $\dfrac{x^3 - 8}{x - 2}$

*Perform the indicated operation. If possible, simplify your answer.*

**6.** $\dfrac{2x^3 + 16}{6x^2 + 12x} \cdot \dfrac{5}{x^2 - 2x + 4}$

**7.** $\dfrac{5}{4x^3} + \dfrac{7}{4x^3}$

**8.** $\dfrac{3x^2 - 12}{x^2 + 2x - 8} \div \dfrac{6x + 18}{x + 4}$

**9.** $\dfrac{4x - 12}{2x - 9} \div \dfrac{3 - x}{4x^2 - 81} \cdot \dfrac{x + 3}{5x + 15}$

**10.** $\dfrac{3 + 2x}{10 - x} + \dfrac{13 + x}{x - 10}$

**11.** $\dfrac{2x^2 + 7}{2x^4 - 18x^2} - \dfrac{6x + 7}{2x^4 - 18x^2}$

**12.** $\dfrac{3}{x^2 - x - 6} + \dfrac{2}{x^2 - 5x + 6}$

**13.** $\dfrac{5}{x - 7} - \dfrac{2x}{3x - 21} + \dfrac{x}{2x - 14}$

**14.** $\dfrac{3x}{5} \cdot \left( \dfrac{5}{x} - \dfrac{5}{2x} \right)$

*Simplify each complex fraction.*

**15.** $\dfrac{\dfrac{5}{x} - \dfrac{7}{3x}}{\dfrac{9}{8x} - \dfrac{1}{x}}$

**16.** $\dfrac{\dfrac{x^2 - 5x + 6}{x + 3}}{\dfrac{x^2 - 4x + 4}{x^2 - 9}}$

*Solve each equation for x.*

**17.** $\dfrac{5x + 3}{3x - 7} = \dfrac{19}{7}$

**18.** $\dfrac{x}{x - 4} = 3 - \dfrac{4}{x - 4}$

**Answers**

1. _____

2. _____

3. _____

4. _____

5. _____

6. _____

7. _____

8. _____

9. _____

10. _____

11. _____

12. _____

13. _____

14. _____

15. _____

16. _____

17. _____

18. _____

19. _____

20. _____

21. _____

22. _____

23. _____

24. _____

25. _____

26. _____

27. _____

**19.** $\dfrac{3}{x+2} - \dfrac{1}{5x} = \dfrac{2}{5x^2+10x}$

**20.** $\dfrac{x^2+8}{x} - 1 = \dfrac{2(x+4)}{x}$

**21.** Solve for $x$: $\dfrac{x+b}{a} = \dfrac{4x-7a}{b}$

**22.** The product of one more than a number and twice the reciprocal of the number is $\dfrac{12}{5}$. Find the number.

**23.** If Jan can weed the garden in 2 hours and her husband can weed it in 1 hour and 30 minutes, find how long it takes them to weed the garden together.

**24.** Suppose that $W$ is inversely proportional to $V$. If $W = 20$ when $V = 12$, find $W$ when $V = 15$.

**25.** Suppose that $Q$ is jointly proportional to $R$ and the square of $S$. If $Q = 24$ when $R = 3$ and $S = 4$, find $Q$ when $R = 2$ and $S = 3$.

**26.** When an anvil is dropped into a gorge, the speed with which it strikes the ground is directly proportional to the square root of the distance it falls. An anvil that falls 400 feet hits the ground at a speed of 160 feet per second. Find the height of a cliff over the gorge if a dropped anvil hits the ground at a speed of 128 feet per second.

**27.** In baseball, a pitcher's earned run average statistic $A$ is computed as $A = \dfrac{E}{\dfrac{I}{9}}$, where

$E$ is the number of earned runs scored on the pitcher and $I$ is the total number of innings he or she pitched. During the 2014 Major League Baseball season, pitcher Clayton Kershaw of the Los Angeles Dodgers led the major leagues with an earned run average of 1.77. Kershaw pitched a total of 198.1 innings during the season. How many earned runs were scored on Clayton Kershaw? Round to the nearest whole. (*Source:* Major League Baseball)

**1.** Determine whether the statement is true or false.
$3 \in \{x \mid x \text{ is a natural number}\}$

**2.** Translate each phrase to an algebraic expression. Use the variable $x$ to represent each unknown number.
   **a.** One third subtracted from a number.
   **b.** Six less than five times a number.
   **c.** Three more than eight times a number.
   **d.** The quotient of seven and the difference of two and a number.

*Evaluate each expression.*

**3.** $-7^0$

**4.** $(-5xy^7)(-0.2xy^{-5})$

**5.** $(2x + 5)^0$

**6.** $\dfrac{(3a^{-2}b)^{-2}}{(2ab^{-3})^{-3}}$

**7.** Solve: $2x \geq 0 \text{ and } 4x - 1 \leq -9$

**8.** Solve: $\left|\dfrac{2x - 1}{3}\right| + 6 = 3$

**9.** Solve: $|2x| + 5 = 7$

**10.** Solve: $\left|\dfrac{3(x - 1)}{4}\right| \geq 2$

**11.** Determine whether $(0, -12), (1, 9),$ and $(2, -6)$ are solutions of the equation $3x - y = 12$.

**12.** If $f(x) = -x^2 + 3x - 2$, find
   **a.** $f(0)$    **b.** $f(-3)$
   **c.** $f\left(\dfrac{1}{3}\right)$

**13.** Determine whether the two lines are parallel, perpendicular, or neither.
$$3x + 7y = 4$$
$$6x + 14y = 7$$

**14.** Solve: $\dfrac{x}{7} + \dfrac{x}{5} = \dfrac{12}{5}$

**15.** Write an equation of the line through points $(4, 0)$ and $(-4, -5)$. Write the equation in standard form $Ax + By = C$.

**16.** Find an equation of the line with slope $\dfrac{1}{2}$ containing the point $(-1, 3)$.
Use function notation to write the equation.

*Find the domain and range of each relation.*

**17.**

**18.**

| Answers |
|---|
| **1.** _____ |
| **2. a.** _____ |
| **b.** _____ |
| **c.** _____ |
| **d.** _____ |
| **3.** _____ |
| **4.** _____ |
| **5.** _____ |
| **6.** _____ |
| **7.** _____ |
| **8.** _____ |
| **9.** _____ |
| **10.** _____ |
| **11.** _____ |
| **12. a.** _____ |
| **b.** _____ |
| **c.** _____ |
| **13.** _____ |
| **14.** _____ |
| **15.** _____ |
| **16.** _____ |
| **17.** _____ |
| **18.** _____ |

19. _____

20. _____

21. _____

22. _____

23. _____

24. _____

25. _____

26. _____

27. _____

28. _____

29. _____

30. a. _____

   b. _____

   c. _____

31. _____

32. a. _____

   b. _____

**19.**

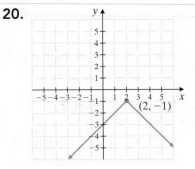

**20.**

**21.** Use the substitution method to solve the system:

$$\begin{cases} -\dfrac{x}{6} + \dfrac{y}{2} = \dfrac{1}{2} \\ \dfrac{x}{3} - \dfrac{y}{6} = -\dfrac{3}{4} \end{cases}$$

**22.** Use the substitution method to solve the system.

$$\begin{cases} -2x + 3y = 6 \\ 3x - y = 5 \end{cases}$$

**23.** Solve the system:

$$\begin{cases} 2x + 4y = 1 \\ 4x - 4z = -1 \\ y - 4z = -3 \end{cases}$$

**24.** Solve the system:

$$\begin{cases} 2x - 2y + 4z = 6 \\ -4x - y + z = -8 \\ 3x - y + z = 6 \end{cases}$$

**25.** Two cars leave Indianapolis, one traveling east and the other west. After 3 hours they are 297 miles apart. If one car is traveling 5 mph faster than the other, what is the speed of each?

**26.** Kernersville office supply sold three reams of paper and two boxes of manila folders for $21.90. Also, five reams of paper and one box of manila folders cost $24.25. Find the price of a ream of paper and a box of manila folders.

**27.** Use matrices to solve the system:

$$\begin{cases} x + 2y + z = 2 \\ -2x - y + 2z = 5 \\ x + 3y - 2z = -8 \end{cases}$$

**28.** Use matrices to solve the system.

$$\begin{cases} x + y + z = 9 \\ 2x - 2y + 3z = 2 \\ -3x + y - z = 1 \end{cases}$$

**29.** Simplify the expression. Write the answer using positive exponents only.

$$\left( \frac{3x^2 y}{y^{-9} z} \right)^{-2}$$

**30.** Simplify the following. Write answers with positive exponents.

  **a.** $2^{-2} + 3^{-1}$     **b.** $-6a^0$

  **c.** $\dfrac{x^{-5}}{x^{-2}}$

**31.** Graph the solutions of the system:

$$\begin{cases} 3x \ge y \\ x + 2y \le 8 \end{cases}$$

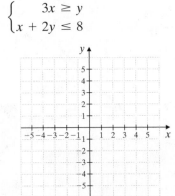

**32.** Simplify each. Assume that $a$ and $b$ are integers and that $x$ and $y$ are not 0.

  **a.** $3x^{4a} \left( 4x^{-a} \right)^2$

  **b.** $\dfrac{\left( y^{4b} \right)^3}{y^{2b-3}}$

*Perform indicated operations, if necessary. Then simplify by combining like terms.*

**33.** $-12x^2 + 7x^2 - 6x$

**34.** Subtract $(2x - 7)$ from $2x^2 + 8x - 3$

**35.** $3xy - 2x + 5xy - x$

**36.** $7 + 2(x - 7y) + \dfrac{1}{2} - \dfrac{1}{4}(8x - 16y)$

**37.** Multiply vertically:
$(4x^2 + 7)(x^2 + 2x + 8)$

**38.** Multiply: $[4 + (3x - y)]^2$

**39.** Divide: $2x^2 - x - 10$ by $x + 2$.

**40.** Factor: $xy + 2x - 5y - 10$

**41.** Factor: $-3x^3y + 2x^2y - 5xy$

**42.** Factor: $6x^2 - x - 35$

**43.** Factor: $x^2 + 10x + 16$

**44.** Factor: $4x^2 - 4x + 1 - 9y^2$

**45.** Factor: $p^4 - 16$

**46.** For the graph of $f(x)$, answer the following:
  **a.** Find the domain and range.
  **b.** List the $x$- and $y$-intercepts.
  **c.** Find the coordinates of the point with the greatest $y$-value.
  **d.** Find the coordinates of the point with the least $y$-value.
  **e.** List the $x$-values whose $y$-values are equal to 0.
  **f.** List the $x$-values whose $y$-values are less than 0.
  **g.** Find the solutions of $f(x) = 0$.

**47.** Solve: $(x + 2)(x - 6) = 0$

**48.** Solve: $2x(3x + 1)(x - 3) = 0$

**49.** Simplify: $\dfrac{\dfrac{5x}{x + 2}}{\dfrac{10}{x - 2}}$

**50.** Solve: $\dfrac{3x - 4}{2x} = -\dfrac{8}{x}$

33. _____

34. _____

35. _____

36. _____

37. _____

38. _____

39. _____

40. _____

41. _____

42. _____

43. _____

44. _____

45. _____

46. a. _____

    b. _____

    c. _____

    d. _____

    e. _____

    f. _____

    g. _____

47. _____

48. _____

49. _____

50. _____

# 7 Rational Exponents, Radicals, and Complex Numbers

In this chapter, radical notation is reviewed, and then rational exponents are introduced. As the name implies, rational exponents are exponents that are rational numbers. We present an interpretation of rational exponents that is consistent with the meaning and rules already established for integer exponents, and we present two forms of notation for roots: radical and exponent. We conclude this chapter with complex numbers, a natural extension of the real number system.

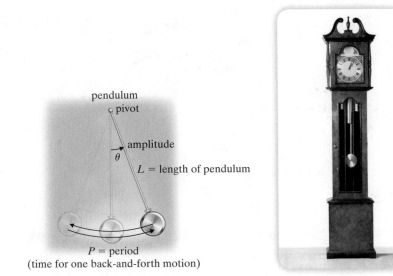

pendulum
pivot
amplitude
$\theta$
$L$ = length of pendulum

$P$ = period
(time for one back-and-forth motion)

## How Do Pendulums Keep Time?

A pendulum is a weight suspended from a pivot. The weight is called a bob, and we call the time for one back-and-forth swing of a pendulum a period. Interestingly enough, the time for one back-and-forth swing of a pendulum depends only on the length of the pendulum. Galileo discovered this around the early 1600s, and it was this regular period (time) that made pendulums useful for timekeeping.

A floor clock, also known as a grandfather clock, is a tall pendulum-driven clock. These early clocks were 6–8 feet tall and were the world's most accurate method of timekeeping until the 1930s. (Source: Wikipedia)

For small angles of motion, or amplitudes, the formula $P = 2\pi\sqrt{\dfrac{L}{32}}$ approximates the time, in seconds, for a back-and-forth swing (period). In this formula, $P$ is the period (the time in seconds for one full back-and-forth swing) and $L$ is the length, in feet, of a pendulum. Below is the graph of this formula, and in Sections 7.3 and 7.6, we compute the periods of pendulums of different lengths.

**Length vs. Period**
**(time of a back-and-forth swing)**
**of a Pendulum**

Period (time in seconds for one back-and-forth swing)

$P = 2\pi\sqrt{\dfrac{L}{32}}$

**Length of Pendulum (in feet)**

Note: This graph has the appearance of a square root function, as it should because of $\sqrt{L}$ in the formula.

## 7.1 Radical Expressions and Radical Functions ▶

### Objective A Finding Square Roots ▶

Recall from Section 1.4 that to find a *square root* of a number $a$, we find a number that was squared to get $a$.

> **Square Root**
>
> The number $b$ is a **square root** of $a$ if $b^2 = a$.

**Examples**   Find the real square roots of each number.

1. 25   Since $5^2 = 25$ and $(-5)^2 = 25$, the square roots of 25 are 5 and $-5$.
2. 49   Since $7^2 = 49$ and $(-7)^2 = 49$, the square roots of 49 are 7 and $-7$.
3. $-4$   There is no real number whose square is $-4$. The number $-4$ has no real number square root.

▥ Work Practice 1–3

Recall that we denote the *nonnegative*, or *principal*, *square root* with the **radical sign:**

$$\sqrt{25} = 5$$

We denote the *negative square root* with the **negative radical sign:**

$$-\sqrt{25} = -5$$

An expression containing a radical sign is called a **radical expression.** An expression within, or "under," a radical sign is called a **radicand.**

radical expression:   $\overset{\text{radical sign}}{\sqrt{\underset{\text{radicand}}{a}}}$

> **Principal and Negative Square Roots**
>
> The **principal square root** of a nonnegative number $a$ is its nonnegative square root. The principal square root is written as $\sqrt{a}$. The **negative square root** of $a$ is written as $-\sqrt{a}$.

**Examples**   Find each square root. Assume that all variables represent non-negative real numbers.

▶ 4. $\sqrt{36} = 6$ because $6^2 = 36$.
▶ 5. $\sqrt{0} = 0$ because $0^2 = 0$.
▶ 6. $\sqrt{\dfrac{4}{49}} = \dfrac{2}{7}$ because $\left(\dfrac{2}{7}\right)^2 = \dfrac{4}{49}$.
▶ 7. $\sqrt{0.25} = 0.5$ because $(0.5)^2 = 0.25$.
▶ 8. $\sqrt{x^6} = x^3$ because $(x^3)^2 = x^6$.
▶ 9. $\sqrt{9x^{12}} = 3x^6$ because $(3x^6)^2 = 9x^{12}$.

*(Continued on next page)*

### Objectives

**A** Find Square Roots. ▶

**B** Approximate Roots Using a Calculator. ▶

**C** Find Cube Roots. ▶

**D** Find $n$th Roots. ▶

**E** Find $\sqrt[n]{a^n}$ when $a$ Is Any Real Number. ▶

**F** Find Function Values and Graph Square and Cube Root Functions. ▶

### Practice 1–3

Find the real square roots of each number.

1. 36   2. 81   3. $-16$

### Practice 4–11

Find each square root. Assume that all variables represent nonnegative real numbers.

4. $\sqrt{25}$      5. $\sqrt{0}$

6. $\sqrt{\dfrac{9}{25}}$      7. $\sqrt{0.36}$

8. $\sqrt{x^{10}}$      9. $\sqrt{36x^6}$

10. $-\sqrt{25}$      11. $\sqrt{-25}$

**Answers**

1. 6, $-6$   2. 9, $-9$
3. no real number square root
4. 5   5. 0   6. $\dfrac{3}{5}$   7. 0.6   8. $x^5$
9. $6x^3$   10. $-5$   11. not a real number

**509**

▶ **10.** $-\sqrt{81} = -9$. The negative in front of the radical indicates the negative square root of 81.

▶ **11.** $\sqrt{-81}$ is not a real number. (There is no real number whose square is $-81$.)

■ Work Practice 4–11

See Example 3 for a discussion of the square root of a negative number. For example, can we simplify $\sqrt{-4}$? That is, can we find a real number whose square is $-4$? No, there is no real number whose square is $-4$, and we say that $\sqrt{-4}$ is not a real number. In general:

*The square root of a negative number is not a real number.*

**Helpful Hint**

- Remember: $\sqrt{0} = 0$.
- Don't forget that the square root of a negative number is not a real number. For example,

$$\sqrt{-9} \text{ is not a real number}$$

because there is no real number that when multiplied by itself would give a product of $-9$. In Section 7.7, we will see what kind of a number $\sqrt{-9}$ is.

## Objective B Approximating Roots ▶

Recall that numbers such as 1, 4, 9, and 25 are called **perfect squares,** since $1 = 1^2, 4 = 2^2, 9 = 3^2$, and $25 = 5^2$. Square roots of perfect square radicands simplify to rational numbers. What happens when we try to simplify a root such as $\sqrt{3}$? Since 3 is not a perfect square, $\sqrt{3}$ is not a rational number. It is called an **irrational number,** and we can find a decimal **approximation** of it. To find decimal approximations, we can use a calculator. For example, an approximation for $\sqrt{3}$ is

$$\sqrt{3} \approx 1.732$$
$$\uparrow$$
approximation symbol

To see if the approximation is reasonable, notice that since

$$1 < 3 < 4, \quad \text{then}$$
$$\sqrt{1} < \sqrt{3} < \sqrt{4}, \quad \text{or}$$
$$1 < \sqrt{3} < 2.$$

We found $\sqrt{3} \approx 1.732$, a number between 1 and 2, so our result is reasonable.

**Practice 12**

Use a calculator to approximate $\sqrt{30}$. Round the approximation to three decimal places and check to see that your approximation is reasonable.

**Example 12** Use a calculator to approximate $\sqrt{20}$. Round the approximation to three decimal places and check to see that your approximation is reasonable.

**Solution:**

$$\sqrt{20} \approx 4.472$$

Is this reasonable? Since $16 < 20 < 25$, then $\sqrt{16} < \sqrt{20} < \sqrt{25}$, or $4 < \sqrt{20} < 5$. The approximation is between 4 and 5 and is thus reasonable.

■ Work Practice 12

## Objective C Finding Cube Roots ▶

Finding roots can be extended to other roots such as cube roots. For example, since $2^3 = 8$, we call 2 the *cube root* of 8. In symbols, we write

$$\sqrt[3]{8} = 2$$

**Answer**

**12.** 5.477

## Cube Root

The **cube root** of a real number $a$ is written as $\sqrt[3]{a}$, and

$$\sqrt[3]{a} = b \quad \text{only if } b^3 = a$$

From this definition, we have

$\sqrt[3]{64} = 4$ since $4^3 = 64$

$\sqrt[3]{-27} = -3$ since $(-3)^3 = -27$

$\sqrt[3]{x^3} = x$ since $x^3 = x^3$

Notice that, unlike with square roots, *it is possible to have a negative radicand when finding a cube root.* This is so because the *cube* of a negative number is a negative number. Therefore, the *cube root* of a negative number is a negative number.

**Examples**  Find each cube root.

**13.** $\sqrt[3]{1} = 1$ because $1^3 = 1$.

**14.** $\sqrt[3]{-64} = -4$ because $(-4)^3 = -64$.

**15.** $\sqrt[3]{\dfrac{8}{125}} = \dfrac{2}{5}$ because $\left(\dfrac{2}{5}\right)^3 = \dfrac{8}{125}$.

**16.** $\sqrt[3]{x^6} = x^2$ because $(x^2)^3 = x^6$.

**17.** $\sqrt[3]{-8x^9} = -2x^3$ because $(-2x^3)^3 = -8x^9$.

▢ Work Practice 13–17

**Practice 13–17**

Find each cube root.

**13.** $\sqrt[3]{0}$ **14.** $\sqrt[3]{-8}$

**15.** $\sqrt[3]{\dfrac{1}{64}}$ **16.** $\sqrt[3]{x^9}$

**17.** $\sqrt[3]{-64x^6}$

## Objective D  Finding *n*th Roots

Just as we can raise a real number to powers other than 2 or 3, we can find roots other than square roots and cube roots. In fact, we can find the **nth root** of a number, where $n$ is any natural number. In symbols, the *n*th root of $a$ is written as $\sqrt[n]{a}$, where $n$ is called the **index.** The index 2 is usually omitted for square roots.

**Helpful Hint**

If the index is even, such as in $\sqrt{\phantom{x}}$, $\sqrt[4]{\phantom{x}}$, $\sqrt[6]{\phantom{x}}$, and so on, the radicand must be nonnegative for the root to be a real number. For example,

$\sqrt[4]{16} = 2$, but $\sqrt[4]{-16}$ is not a real number,

$\sqrt[6]{64} = 2$, but $\sqrt[6]{-64}$ is not a real number,

If the index is odd, such as in $\sqrt[3]{\phantom{x}}$, $\sqrt[5]{\phantom{x}}$, and so on, the radicand may be any real number. For example,

$\sqrt[3]{64} = 4$ and $\sqrt[3]{-64} = -4$,

$\sqrt[5]{32} = 2$ and $\sqrt[5]{-32} = -2$

✓**Concept Check**  Which one is not a real number?

**a.** $\sqrt[3]{-15}$ **b.** $\sqrt[4]{-15}$ **c.** $\sqrt[5]{-15}$ **d.** $\sqrt{(-15)^2}$

**Answers**

**13.** 0 **14.** $-2$ **15.** $\dfrac{1}{4}$ **16.** $x^3$ **17.** $-4x^2$

✓**Concept Check Answer**

**b**

## Practice 18–22

Find each root.

**18.** $\sqrt[4]{16}$    **19.** $\sqrt[5]{-32}$

**20.** $-\sqrt{36}$    **21.** $\sqrt[4]{-16}$

**22.** $\sqrt[3]{8x^6}$

> **Examples**    Find each root.
>
> **18.** $\sqrt[4]{81} = 3$ because $3^4 = 81$ and 3 is positive.
> **19.** $\sqrt[5]{-243} = -3$ because $(-3)^5 = -243$.
> **20.** $-\sqrt{25} = -5$ because $-5$ is the opposite of $\sqrt{25}$.
> **21.** $\sqrt[4]{-81}$ is not a real number. There is no real number that, when raised to the fourth power, is $-81$.
> **22.** $\sqrt[3]{64x^3} = 4x$ because $(4x)^3 = 64x^3$.
>
> ▮ **Work Practice 18–22**

## Objective E  Finding $\sqrt[n]{a^n}$ when $a$ Is Any Real Number ▶

Recall that the notation $\sqrt{a^2}$ indicates the positive square root of $a^2$ only. For example,

$$\sqrt{(-5)^2} = \sqrt{25} = 5$$

When variables are present in the radicand and it is *unclear whether the variable represents a positive number or a negative number,* absolute value bars are sometimes needed to ensure that the result is a positive number. For example,

$$\sqrt{x^2} = |x|$$

This ensures that the result is positive. This same situation may occur when the index is any *even* positive integer. When the index is any *odd* positive integer, absolute value bars are not necessary.

> **Finding $\sqrt[n]{a^n}$**
>
> If $n$ is an *even* positive integer, then $\sqrt[n]{a^n} = |a|$.
> If $n$ is an *odd* positive integer, then $\sqrt[n]{a^n} = a$.

## Practice 23–29

Simplify. Assume that the variables represent any real number.

**23.** $\sqrt{(-5)^2}$    **24.** $\sqrt{x^6}$

**25.** $\sqrt[4]{(x+6)^4}$    **26.** $\sqrt[3]{(-3)^3}$

**27.** $\sqrt[5]{(7x-1)^5}$    **28.** $\sqrt{36x^2}$

**29.** $\sqrt{x^2 + 6x + 9}$

> **Examples**    Simplify. Assume that the variables represent any real number.
>
> **23.** $\sqrt{(-3)^2} = |-3| = 3$     When the index is even, the absolute value bars ensure that the result is not negative.
> **24.** $\sqrt{x^2} = |x|$
> **25.** $\sqrt[4]{(x-2)^4} = |x-2|$
> **26.** $\sqrt[3]{(-5)^3} = -5$     Absolute value bars are not needed when the index is odd.
> **27.** $\sqrt[5]{(2x-7)^5} = 2x - 7$
> **28.** $\sqrt{25x^2} = 5|x|$
> **29.** $\sqrt{x^2 + 2x + 1} = \sqrt{(x+1)^2} = |x+1|$
>
> ▮ **Work Practice 23–29**

## Objective F  Find Function Values and Graph Radical Functions ▶

In general, **radical functions** are functions of the form

$$f(x) = \sqrt[n]{x}.$$

Recall that the domain of a function in $x$ is the set of all possible replacement values of $x$. This means that if $n$ is even, the domain is the set of all nonnegative numbers, or $\{x \mid x \geq 0\}$ or $[0, \infty)$. If $n$ is odd, the domain is the set of all real numbers, or $(-\infty, \infty)$. Keep this in mind as we find function values. In Section 3.8, we graphed square root functions and discussed their domains.

**Answers**

**18.** 2  **19.** $-2$  **20.** $-6$
**21.** not a real number  **22.** $2x^2$  **23.** 5
**24.** $|x^3|$  **25.** $|x+6|$  **26.** $-3$
**27.** $7x - 1$  **28.** $6|x|$  **29.** $|x + 3|$

**Examples** If $f(x) = \sqrt{x - 4}$ and $g(x) = \sqrt[3]{x + 2}$, find each function value.

**30.** $f(8) = \sqrt{8 - 4} = \sqrt{4} = 2$

**31.** $f(6) = \sqrt{6 - 4} = \sqrt{2}$

**32.** $g(-1) = \sqrt[3]{-1 + 2} = \sqrt[3]{1} = 1$

**33.** $g(1) = \sqrt[3]{1 + 2} = \sqrt[3]{3}$

▰ Work Practice 30–33

Notice that for the function $f(x) = \sqrt{x - 4}$, the domain includes all real numbers that make the radicand $\geq 0$. To see what numbers these are, solve $x - 4 \geq 0$ and find that $x \geq 4$. The domain is $\{x \mid x \geq 4\}$ or $[4, \infty)$.

The domain of the cube root function $g(x) = \sqrt[3]{x + 2}$ is the set of real numbers or $(-\infty, \infty)$.

Recall from Section 3.8 that the graph of $f(x) = \sqrt{x}$ is

Let's now graph the function $f(x) = \sqrt{x - 2}$.

**Example 34** Identify the domain and then graph the function $f(x) = \sqrt{x - 2}$.

**Solution:** To graph, we identify the domain, evaluate the function for several values of $x$, plot the resulting points, and connect the points with a smooth curve. The domain of $f(x) = \sqrt{x + 2}$ includes all real numbers that make the radicand $\geq 0$. By solving $x - 2 \geq 0$, we find that $x \geq 2$. The domain is $\{x \mid x \geq 2\}$ or $[2, \infty)$.

| $x$ | $f(x) = \sqrt{x - 2}$ |
|---|---|
| 2 | 0 |
| 3 | 1 |
| 5 | $\sqrt{3} \approx 1.7$ |
| 6 | 2 |
| 11 | 3 |

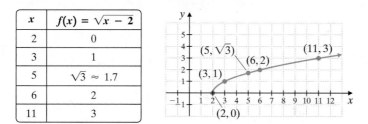

Notice that the graph of this function passes the vertical line test, as expected.

▰ Work Practice 34

The equation $f(x) = \sqrt[3]{x}$ also describes a function. Here, $x$ may be any real number, so the domain of this function is the set of all real numbers, or $(-\infty, \infty)$. A few function values are given next.

$f(0) = \sqrt[3]{0} = 0$

$f(1) = \sqrt[3]{1} = 1$

$f(-1) = \sqrt[3]{-1} = -1$

$f(6) = \sqrt[3]{6}$  ⎫ Here, there is no rational number
$f(-6) = \sqrt[3]{-6}$  ⎬ whose cube is 6. Thus, the radicals
  ⎭ do not simplify to rational numbers.

$f(8) = \sqrt[3]{8} = 2$

$f(-8) = \sqrt[3]{-8} = -2$

**Practice 30–33**

If $f(x) = \sqrt{x + 2}$ and $g(x) = \sqrt[3]{x - 1}$, find each function value.

**30.** $f(7)$     **31.** $g(9)$

**32.** $f(0)$     **33.** $g(10)$

**Practice 34**

Identify the function and then graph the function $h(x) = \sqrt{x + 2}$.

**Answers**

**30.** 3  **31.** 2  **32.** $\sqrt{2}$  **33.** $\sqrt[3]{9}$

**34.**

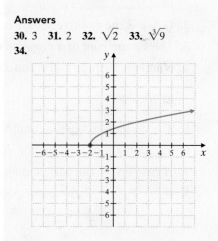

## Practice 35

Identify the domain and then graph the cube root function $h(x) = \sqrt[3]{x} + 2$.

**Answer**

**35.**

**Example 35**    Identify the domain and then graph the function $f(x) = \sqrt[3]{x}$.

**Solution:**    To graph, we identify the domain, plot points, and connect the points with a smooth curve. The domain of this function is the set of all real numbers, or $(-\infty, \infty)$. The table comes from the function values obtained earlier. (We have approximated $\sqrt[3]{6}$ and $\sqrt[3]{-6}$ for graphing purposes.)

| $x$ | $f(x) = \sqrt[3]{x}$ |
|-----|------------------------|
| 0 | 0 |
| 1 | 1 |
| $-1$ | $-1$ |
| 6 | $\sqrt[3]{6} \approx 1.8$ |
| $-6$ | $\sqrt[3]{-6} \approx -1.8$ |
| 8 | 2 |
| $-8$ | $-2$ |

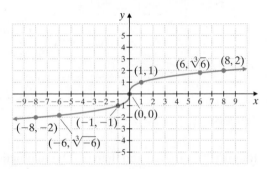

The graph of this function passes the vertical line test, as expected.

🔳 **Work Practice 35**

## Vocabulary, Readiness & Video Check

*Use the choices below to fill in each blank. Not all choices will be used.*

| is | cubes | $-\sqrt{a}$ | radical sign | index |
|----|-------|-------------|--------------|-------|
| is not | squares | $\sqrt{-a}$ | radicand | |

1. In the expression $\sqrt[n]{a}$, the $n$ is called the _____, the $\sqrt{\phantom{x}}$ is called the _____, and $a$ is called the _____.

2. If $\sqrt{a}$ is the positive square root of $a$, $a \neq 0$, then _____ is the negative square root of $a$.

3. The square root of a negative number _____ a real number.

4. Numbers such as 1, 4, 9, and 25 are called perfect _____.

*Fill in the blank.*

5. The domain of the function $f(x) = \sqrt{x}$ is _____.

6. The domain of the function $f(x) = \sqrt[3]{x}$ is _____.

7. If $f(16) = 4$, the corresponding ordered pair is _____.

8. If $g(-8) = -2$, the corresponding ordered pair is _____.

*Choose the correct letter or letters. No pencil is needed; just think your way through these.*

**9.** Which radical is not a real number?

   **a.** $\sqrt{3}$     **b.** $-\sqrt{11}$     **c.** $\sqrt[3]{-10}$     **d.** $\sqrt{-10}$

**10.** Which radical(s) simplify to 3?

   **a.** $\sqrt{9}$     **b.** $\sqrt{-9}$     **c.** $\sqrt[3]{27}$     **d.** $\sqrt[3]{-27}$

**11.** Which radical(s) simplify to $-3$?

   **a.** $\sqrt{9}$     **b.** $\sqrt{-9}$     **c.** $\sqrt[3]{27}$     **d.** $\sqrt[3]{-27}$

**12.** Which radical does not simplify to a whole number?

   **a.** $\sqrt{64}$     **b.** $\sqrt[3]{64}$     **c.** $\sqrt{8}$     **d.** $\sqrt[3]{8}$

---

*Martin-Gay Interactive Videos*    *Watch the section lecture video and answer the following questions.*

See Video 7.1

**Objective** **A** **13.** From Examples 5 and 6, when simplifying radicals containing variables with exponents, describe a shortcut you can use.

**Objective** **B** **14.** From Example 9, how can you determine a reasonable approximation for a non-perfect square root without using a calculator?

**Objective** **C** **15.** From Example 11, what is an important difference between the square root and the cube root of a negative number?

**Objective** **D** **16.** From Example 12, what conclusion is made about the even root of a negative number?

**Objective** **E** **17.** From the lecture before Example 17, why do you think no absolute value bars are used when $n$ is odd?

**Objective** **F** **18.** In Example 19, the domain is found by looking at the graph. How can the domain be found by looking at the function?

---

## 7.1 Exercise Set MyMathLab®

**Objective** **A**   *Find the real square roots of each number. See Examples 1 through 3.*

**1.** 4                           **2.** 9                       **3.** $-25$

**4.** $-49$                     **5.** 100                   **6.** 64

*Find each square root. Assume that all variables represent nonnegative real numbers. See Examples 4 through 11.*

**7.** $\sqrt{100}$         **8.** $\sqrt{400}$         **9.** $\sqrt{\dfrac{1}{4}}$         **10.** $\sqrt{\dfrac{9}{25}}$

**11.** $\sqrt{0.0001}$     **12.** $\sqrt{0.04}$     **13.** $-\sqrt{36}$     **14.** $-\sqrt{9}$

**15.** $\sqrt{x^{10}}$         **16.** $\sqrt{x^{16}}$         **17.** $\sqrt{16y^6}$     **18.** $\sqrt{64y^{20}}$

**Objective** **B**   *Use a calculator to approximate each square root to three decimal places. Check to see that each approximation is reasonable. See Example 12.*

**19.** $\sqrt{7}$     **20.** $\sqrt{11}$     **21.** $\sqrt{38}$     **22.** $\sqrt{56}$     **23.** $\sqrt{200}$     **24.** $\sqrt{300}$

**Objective C** *Find each cube root. See Examples 13 through 17.*

**25.** $\sqrt[3]{64}$

**26.** $\sqrt[3]{27}$

▶ **27.** $\sqrt[3]{\dfrac{1}{8}}$

**28.** $\sqrt[3]{\dfrac{27}{64}}$

**29.** $\sqrt[3]{-1}$

**30.** $\sqrt{-125}$

**31.** $\sqrt[3]{x^{12}}$

**32.** $\sqrt[3]{x^{15}}$

▶ **33.** $\sqrt[3]{-27x^9}$

**34.** $\sqrt[3]{-64x^6}$

**Objective D** *Find each root. Assume that all variables represent nonnegative real numbers. See Examples 18 through 22.*

**35.** $-\sqrt[4]{16}$

**36.** $\sqrt[5]{-243}$

▶ **37.** $\sqrt[4]{-16}$

**38.** $\sqrt{-16}$

▶ **39.** $\sqrt[5]{-32}$

**40.** $\sqrt[5]{-1}$

**41.** $\sqrt[5]{x^{20}}$

**42.** $\sqrt[4]{x^{20}}$

▶ **43.** $\sqrt[6]{64x^{12}}$

**44.** $\sqrt[5]{-32x^{15}}$

**45.** $\sqrt{81x^4}$

**46.** $\sqrt[4]{81x^4}$

**47.** $\sqrt[4]{256x^8}$

**48.** $\sqrt{256x^8}$

**Objective E** *Simplify. Assume that the variables represent any real number. See Examples 23 through 29.*

▶ **49.** $\sqrt{(-8)^2}$

**50.** $\sqrt{(-7)^2}$

▶ **51.** $\sqrt[3]{(-8)^3}$

**52.** $\sqrt[5]{(-7)^5}$

**53.** $\sqrt{4x^2}$

**54.** $\sqrt[4]{16x^4}$

**55.** $\sqrt[3]{x^3}$

**56.** $\sqrt[5]{x^5}$

▶ **57.** $\sqrt{(x-5)^2}$

**58.** $\sqrt{(y-6)^2}$

**59.** $\sqrt{x^2 + 4x + 4}$
   (*Hint:* Factor the polynomial first.)

**60.** $\sqrt{x^2 - 8x + 16}$
   (*Hint:* Factor the polynomial first.)

**Objectives A B C D Mixed Practice** *Simplify each radical. Assume that all variables represent positive real numbers. See Examples 1 through 22.*

**61.** $-\sqrt{121}$

**62.** $-\sqrt[3]{125}$

**63.** $\sqrt[3]{8x^3}$

**64.** $\sqrt{16x^8}$

**65.** $\sqrt{y^{12}}$

**66.** $\sqrt[3]{y^{12}}$

**67.** $\sqrt{25a^2b^{20}}$

**68.** $\sqrt{9x^4y^6}$

**69.** $\sqrt[3]{-27x^{12}y^9}$

**70.** $\sqrt[3]{-8a^{21}y^6}$

**71.** $\sqrt[4]{a^{16}b^4}$

**72.** $\sqrt[4]{x^8y^{12}}$

▶ **73.** $\sqrt[5]{-32x^{10}y^5}$

**74.** $\sqrt[5]{-243z^{15}}$

**75.** $\sqrt{\dfrac{25}{49}}$

**76.** $\sqrt{\dfrac{4}{81}}$

**77.** $\sqrt{\dfrac{x^2}{4y^2}}$

**78.** $\sqrt{\dfrac{y^{10}}{9x^6}}$

**79.** $-\sqrt[3]{\dfrac{z^{21}}{27x^3}}$

**80.** $-\sqrt[3]{\dfrac{64a^3}{b^9}}$

**81.** $\sqrt[4]{\dfrac{x^4}{16}}$

**82.** $\sqrt[4]{\dfrac{y^4}{81x^4}}$

**Objective** F *If* $f(x) = \sqrt{2x + 3}$ *and* $g(x) = \sqrt[3]{x - 8}$, *find each function value. See Examples 30 through 33.*

**83.** $f(0)$

**84.** $g(0)$

**85.** $g(7)$

**86.** $f(-1)$

**87.** $g(-19)$

**88.** $f(3)$

**89.** $f(2)$

**90.** $g(1)$

*Identify the domain and then graph each function. See Example 34.*

**91.** $f(x) = \sqrt{x} + 2$
**92.** $f(x) = \sqrt{x} - 2$
**93.** $f(x) = \sqrt{x - 3}$
**94.** $f(x) = \sqrt{x + 1}$

*Identify the domain and then graph each function. See Example 35.*

**95.** $f(x) = \sqrt[3]{x} + 1$
**96.** $f(x) = \sqrt[3]{x} - 2$
**97.** $g(x) = \sqrt[3]{x - 1}$
**98.** $g(x) = \sqrt[3]{x + 1}$

## Review

*Simplify each exponential expression. See Sections 1.6 and 1.7.*

**99.** $(-2x^3y^2)^5$

**100.** $(4y^6z^7)^3$

**101.** $(-3x^2y^3z^5)(20x^5y^7)$

**102.** $(-14a^5bc^2)(2abc^4)$

**103.** $\dfrac{7x^{-1}y}{14(x^5y^2)^{-2}}$

**104.** $\dfrac{(2a^{-1}b^2)^3}{(8a^2b)^{-2}}$

*Which of the following are not real numbers? See the Concept Check in this section.*

**105.** $\sqrt{-17}$

**106.** $\sqrt[3]{-17}$

**107.** $\sqrt[10]{-17}$

**108.** $\sqrt[15]{-17}$

## Concept Extensions

**109.** Explain why $\sqrt{-64}$ is not a real number.

**110.** Explain why $\sqrt[3]{-64}$ is a real number.

*For Exercises 111 through 114, do not use a calculator.*

**111.** $\sqrt{160}$ is closest to
   **a.** 10   **b.** 13   **c.** 20   **d.** 40

**112.** $\sqrt{1000}$ is closest to
   **a.** 10   **b.** 30   **c.** 100   **d.** 500

△ **113.** The perimeter of the triangle is closest to
   **a.** 12   **b.** 18   **c.** 66   **d.** 132

**114.** The length of the bent wire is closest to
   **a.** 5   **b.** $\sqrt{28}$   **c.** 7   **d.** 14

*The Mosteller formula for calculating adult body surface area is* $B = \sqrt{\dfrac{hw}{3131}}$, *where B is an individual's body surface area in square meters, h is the individual's height in inches, and w is the individual's weight in pounds. Use this information to answer Exercises 115 and 116. Round answers to 2 decimal places.*

△ **115.** Find the body surface area of an individual who is 66 inches tall and who weighs 135 pounds.

△ **116.** Find the body surface area of an individual who is 74 inches tall and who weighs 225 pounds.

*Solve.*

**117.** Escape velocity is the minimum speed that an object must reach to escape a planet's pull of gravity. Escape velocity $v$ is given by the equation
$$v = \sqrt{\frac{2Gm}{r}},$$ where $m$ is the mass of the planet, $r$ is its radius, and $G$ is the universal gravitational constant, which has a value of $G = 6.67 \times 10^{-11}$ m³/kg·s². The mass of Earth is $5.97 \times 10^{24}$ kg, and its radius is $6.37 \times 10^{6}$ m. Use this information to find the escape velocity for Earth in meters per second. Round to the nearest whole number. (*Source:* National Space Science Data Center)

**118.** Use the formula from Exercise 117 to determine the escape velocity from the moon. The mass of the moon is $7.35 \times 10^{22}$ kg, and its radius is $1.74 \times 10^{6}$ m. Round to the nearest whole. (*Source:* National Space Science Data Center)

**119.** Suppose a classmate tells you that $\sqrt{13} \approx 5.7$. Without a calculator, how can you convince your classmate that he or she must have made an error?

**120.** Suppose a classmate tells you that $\sqrt[3]{10} \approx 3.2$. Without a calculator, how can you convince your friend that he or she must have made an error?

## Objective A Understanding $a^{1/n}$

So far in this text, we have not defined expressions with rational exponents such as $3^{1/2}$, $x^{2/3}$, and $-9^{-1/4}$. We will define these expressions so that the rules for exponents shall apply to these rational exponents as well.

Suppose that $x = 5^{1/3}$. Then

$$x^3 = (5^{1/3})^3 = 5^{1/3 \cdot 3} = 5^1 \text{ or } 5$$

$\underset{\text{for exponents}}{\underset{\text{using rules} \uparrow}{\quad}}$

Since $x^3 = 5$, then $x$ is the number whose cube is 5, or $x = \sqrt[3]{5}$. Notice that we also know that $x = 5^{1/3}$. This means that

$$5^{1/3} = \sqrt[3]{5}$$

### Definition of $a^{1/n}$

If $n$ is a positive integer greater than 1 and $\sqrt[n]{a}$ is a real number, then

$$a^{1/n} = \sqrt[n]{a}$$

Notice that the denominator of the rational exponent corresponds to the index of the radical.

**Examples** Use radical notation to rewrite each expression. Simplify if possible.

1. $4^{1/2} = \sqrt{4} = 2$
2. $64^{1/3} = \sqrt[3]{64} = 4$
3. $x^{1/4} = \sqrt[4]{x}$
4. $-9^{1/2} = -\sqrt{9} = -3$
5. $(81x^8)^{1/4} = \sqrt[4]{81x^8} = 3x^2$
6. $5y^{1/3} = 5\sqrt[3]{y}$

■ Work Practice 1–6

## Objective B Understanding $a^{m/n}$

As we expand our use of exponents to include $\dfrac{m}{n}$, we define their meaning so that rules for exponents still hold true. For example, by properties of exponents,

$$8^{2/3} = (8^{1/3})^2 = (\sqrt[3]{8})^2 \quad \text{or} \quad 8^{2/3} = (8^2)^{1/3} = \sqrt[3]{8^2}$$

### Definition of $a^{m/n}$

If $m$ and $n$ are positive integers greater than 1 with $\dfrac{m}{n}$ in simplest form, then

$$a^{m/n} = \sqrt[n]{a^m} = (\sqrt[n]{a})^m$$

as long as $\sqrt[n]{a}$ is a real number.

### Objectives

A Understand the Meaning of $a^{1/n}$.

B Understand the Meaning of $a^{m/n}$.

C Understand the Meaning of $a^{-m/n}$.

D Use Rules for Exponents to Simplify Expressions That Contain Rational Exponents.

E Use Rational Exponents to Simplify Radical Expressions.

### Practice 1–6

Use radical notation to rewrite each expression. Simplify if possible.

1. $25^{1/2}$  2. $27^{1/3}$
3. $x^{1/5}$  4. $-25^{1/2}$
5. $(-27y^6)^{1/3}$  6. $7x^{1/5}$

**Answers**
1. 5  2. 3  3. $\sqrt[5]{x}$  4. $-5$
5. $-3y^2$  6. $7\sqrt[5]{x}$

Notice that the denominator $n$ of the rational exponent corresponds to the index of the radical. The numerator $m$ of the rational exponent indicates that the base is to be raised to the $m$th power. This means that

$$8^{2/3} = \sqrt[3]{8^2} = \sqrt[3]{64} = 4 \quad \text{or} \quad 8^{2/3} = \left(\sqrt[3]{8}\right)^2 = 2^2 = 4$$

**Helpful Hint**

Most of the time, $\left(\sqrt[n]{a}\right)^m$ will be easier to calculate than $\sqrt[n]{a^m}$.

**Practice 7–11**

Use radical notation to rewrite each expression. Simplify if possible.

**7.** $9^{3/2}$      **8.** $-256^{3/4}$

**9.** $(-32)^{2/5}$    **10.** $\left(\dfrac{1}{4}\right)^{3/2}$

**11.** $(2x + 1)^{2/7}$

**Examples**   Use radical notation to rewrite each expression. Simplify if possible.

**7.** $4^{3/2} = \left(\sqrt{4}\right)^3 = 2^3 = 8$

**8.** $-16^{3/4} = -\left(\sqrt[4]{16}\right)^3 = -(2)^3 = -8$

**9.** $(-27)^{2/3} = \left(\sqrt[3]{-27}\right)^2 = (-3)^2 = 9$

**10.** $\left(\dfrac{1}{9}\right)^{3/2} = \left(\sqrt{\dfrac{1}{9}}\right)^3 = \left(\dfrac{1}{3}\right)^3 = \dfrac{1}{27}$

**11.** $(4x - 1)^{3/5} = \sqrt[5]{(4x - 1)^3}$

■ Work Practice 7–11

**Helpful Hint**

The *denominator* of a rational exponent is the index of the corresponding radical. For example, $x^{1/5} = \sqrt[5]{x}$, and $z^{2/3} = \sqrt[3]{z^2}$ or $z^{2/3} = \left(\sqrt[3]{z}\right)^2$.

## Objective C  Understanding $a^{-m/n}$

The rational exponents we have given meaning to exclude negative rational numbers. To complete the set of definitions, we define $a^{-m/n}$.

> **Definition of $a^{-m/n}$**
>
> $$a^{-m/n} = \dfrac{1}{a^{m/n}}$$
>
> as long as $a^{m/n}$ is a nonzero real number.

**Practice 12–13**

Write each expression with a positive exponent. Then simplify.

**12.** $27^{-2/3}$    **13.** $-256^{-3/4}$

**Examples**   Write each expression with a positive exponent. Then simplify.

**12.** $16^{-3/4} = \dfrac{1}{16^{3/4}} = \dfrac{1}{\left(\sqrt[4]{16}\right)^3} = \dfrac{1}{2^3} = \dfrac{1}{8}$

**13.** $(-27)^{-2/3} = \dfrac{1}{(-27)^{2/3}} = \dfrac{1}{\left(\sqrt[3]{-27}\right)^2} = \dfrac{1}{(-3)^2} = \dfrac{1}{9}$

■ Work Practice 12–13

**Answers**

**7.** 27   **8.** $-64$   **9.** 4   **10.** $\dfrac{1}{8}$

**11.** $\sqrt[7]{(2x + 1)^2}$   **12.** $\dfrac{1}{9}$   **13.** $-\dfrac{1}{64}$

## Helpful Hint

If an expression contains a negative rational exponent, you may want to first write the expression with a positive exponent, then interpret the rational exponent. Notice that the sign of the base is not affected by the sign of its exponent. For example,

$$9^{-3/2} = \frac{1}{9^{3/2}} = \frac{1}{(\sqrt{9})^3} = \frac{1}{27}$$

Also,

$$(-27)^{-1/3} = \frac{1}{(-27)^{1/3}} = -\frac{1}{3}$$

✓**Concept Check** Which one is correct?

**a.** $-8^{2/3} = \dfrac{1}{4}$   **b.** $8^{-2/3} = -\dfrac{1}{4}$   **c.** $8^{-2/3} = -4$   **d.** $-8^{-2/3} = -\dfrac{1}{4}$

## Objective D  Using Rules for Exponents ▶

It can be shown that the properties of integer exponents hold for rational exponents. By using these properties and definitions, we can now simplify expressions that contain rational exponents. These rules are repeated here for review.

### Summary of Exponent Rules

If $m$ and $n$ are rational numbers, and $a$, $b$, and $c$ are numbers for which the expressions below exist, then

| | |
|---|---|
| Product rule for exponents: | $a^m \cdot a^n = a^{m+n}$ |
| Power rule for exponents: | $(a^m)^n = a^{m \cdot n}$ |
| Power rules for products and quotients: | $(ab)^n = a^n b^n$  and |
| | $\left(\dfrac{a}{c}\right)^n = \dfrac{a^n}{c^n},\ c \neq 0$ |
| Quotient rule for exponents: | $\dfrac{a^m}{a^n} = a^{m-n},\ a \neq 0$ |
| Zero exponent: | $a^0 = 1,\ a \neq 0$ |
| Negative exponent: | $a^{-n} = \dfrac{1}{a^n},\ a \neq 0$ |

**Examples**  Use the properties of exponents to simplify.

**14.** $x^{1/2}x^{1/3} = x^{1/2+1/3} = x^{3/6+2/6} = x^{5/6}$    Use the product rule.

**15.** $\dfrac{7^{1/3}}{7^{4/3}} = 7^{1/3-4/3} = 7^{-3/3} = 7^{-1} = \dfrac{1}{7}$    Use the quotient rule.

**16.** $y^{-4/7} \cdot y^{6/7} = y^{-4/7+6/7} = y^{2/7}$    Use the product rule.

**17.** $(5^{3/8})^4 = 5^{3/8 \cdot 4} = 5^{12/8} = 5^{3/2}$    Use the power rule.

(*Continued on next page*)

### Practice 14–18

Use the properties of exponents to simplify.

**14.** $x^{1/3}x^{1/4}$    **15.** $\dfrac{9^{2/5}}{9^{12/5}}$

**16.** $y^{-3/10} \cdot y^{6/10}$  **17.** $(11^{2/9})^3$

**18.** $\dfrac{(3x^{2/3})^3}{x^2}$

**Answers**

**14.** $x^{7/12}$  **15.** $\dfrac{1}{81}$  **16.** $y^{3/10}$  **17.** $11^{2/3}$

**18.** 27

✓**Concept Check Answer**

d

**18.** $\dfrac{(2x^{2/5})^5}{x^2} = \dfrac{2^5(x^{2/5})^5}{x^2}$   Use the power rule.

$\quad = \dfrac{32x^2}{x^2}$   Simplify.

$\quad = 32x^{2-2}$   Use the quotient rule.

$\quad = 32x^0$   Simplify.

$\quad = 32 \cdot 1 \text{ or } 32$   Substitute 1 for $x^0$.

■ Work Practice 14–18

## Objective E  Using Rational Exponents to Simplify Radical Expressions

We can simplify some radical expressions by first writing the expression with rational exponents. Then we can use the properties of exponents to simplify, and finally convert the expression back to radical notation.

**Practice 19–21**

Use rational exponents to simplify. Assume that all variables represent positive real numbers.

**19.** $\sqrt[10]{y^5}$   **20.** $\sqrt[4]{9}$

**21.** $\sqrt[9]{a^6b^3}$

**Examples**  Use rational exponents to simplify. Assume that all variables represent positive real numbers.

**19.** $\sqrt[8]{x^4} = x^{4/8}$   Write with rational exponents.

$\quad = x^{1/2}$   Simplify the exponent.

$\quad = \sqrt{x}$   Write with radical notation.

**20.** $\sqrt[6]{25} = 25^{1/6}$   Write with rational exponents.

$\quad = (5^2)^{1/6}$   Write 25 as $5^2$.

$\quad = 5^{2/6}$   Use the power rule.

$\quad = 5^{1/3}$   Simplify the exponent.

$\quad = \sqrt[3]{5}$   Write with radical notation.

**21.** $\sqrt[6]{r^2s^4} = (r^2s^4)^{1/6}$   Write with rational exponents.

$\quad = r^{2/6}s^{4/6}$   Use the power rule.

$\quad = r^{1/3}s^{2/3}$   Simplify the exponents.

$\quad = (rs^2)^{1/3}$   Use $a^nb^n = (ab)^n$.

$\quad = \sqrt[3]{rs^2}$   Write with radical notation.

■ Work Practice 19–21

**Practice 22–24**

Use rational exponents to write as a single radical.

**22.** $\sqrt{y} \cdot \sqrt[3]{y}$   **23.** $\dfrac{\sqrt[3]{x}}{\sqrt[4]{x}}$

**24.** $\sqrt{5} \cdot \sqrt[3]{2}$

**Examples**  Use rational exponents to write as a single radical expression.

**22.** $\sqrt{x} \cdot \sqrt[4]{x} = x^{1/2} \cdot x^{1/4} = x^{1/2+1/4}$

$\quad = x^{3/4} = \sqrt[4]{x^3}$

**23.** $\dfrac{\sqrt{x}}{\sqrt[3]{x}} = \dfrac{x^{1/2}}{x^{1/3}} = x^{1/2-1/3} = x^{3/6-2/6}$

$\quad = x^{1/6} = \sqrt[6]{x}$

**24.** $\sqrt[3]{3} \cdot \sqrt{2} = 3^{1/3} \cdot 2^{1/2}$   Write with rational exponents.

$\quad = 3^{2/6} \cdot 2^{3/6}$   Write the exponents so that they have the same denominator.

$\quad = (3^2 \cdot 2^3)^{1/6}$   Use $a^nb^n = (ab)^n$.

$\quad = \sqrt[6]{3^2 \cdot 2^3}$   Write with radical notation.

$\quad = \sqrt[6]{72}$   Multiply $3^2 \cdot 2^3$.

■ Work Practice 22–24

**Answers**

**19.** $\sqrt{y}$   **20.** $\sqrt{3}$   **21.** $\sqrt[3]{a^2b}$

**22.** $\sqrt[6]{y^5}$   **23.** $\sqrt[12]{x}$   **24.** $\sqrt[6]{500}$

## Vocabulary, Readiness & Video Check

*Answer each true or false.*

**1.** $9^{-1/2}$ is a positive number. _____

**2.** $9^{-1/2}$ is an integer. _____

**3.** $\dfrac{1}{a^{-m/n}} = a^{m/n}$ (where $a^{m/n}$ is a nonzero real number). _____

*Fill in the blank with the correct choice.*

**4.** To simplify $x^{2/3} \cdot x^{1/5}$, _____ the exponents.

   **a.** add     **b.** subtract     **c.** multiply     **d.** divide

**5.** To simplify $(x^{2/3})^{1/5}$, _____ the exponents.

   **a.** add     **b.** subtract     **c.** multiply     **d.** divide

**6.** To simplify $\dfrac{x^{2/3}}{x^{1/5}}$, _____ the exponents.

   **a.** add     **b.** subtract     **c.** multiply     **d.** divide

*Choose the correct letter for each exercise. Letters will be used more than once. No pencil is needed; just think about the meaning of each expression.*

A = 2, B = −2, C = not a real number

**7.** $4^{1/2}$ ___      **8.** $-4^{1/2}$ ___      **9.** $(-4)^{1/2}$ ___      **10.** $8^{1/3}$ ___      **11.** $-8^{1/3}$ ___      **12.** $(-8)^{1/3}$ ___

*Martin-Gay Interactive Videos*

See Video 7.2

*Watch the section lecture video and answer the following questions.*

**Objective A** **13.** From looking at Example 2, what is $-(3x)^{1/5}$ in radical notation?

**Objective B** **14.** From Examples 3 and 4, in a fractional exponent, what do the numerator and denominator each represent in radical form?

**Objective C** **15.** Based on Example 5, complete the following statements. A negative fractional exponent will move a base from the numerator to the _____ with the fractional exponent becoming _____.

**Objective D** **16.** Based on Examples 7–8, complete the following statements. Assume you have an expression with fractional exponents. If applying the product rule of exponents, you _____ the exponents. If applying the quotient rule of exponents, you _____ the exponents. If applying the power rule of exponents, you _____ the exponents.

**Objective E** **17.** From Example 10, describe a way to simplify a radical of a variable raised to a power if the index and the power have a common factor.

## 7.2 Exercise Set MyMathLab®

**Objective A** *Use radical notation to rewrite each expression. Simplify if possible. See Examples 1 through 6.*

**1.** $49^{1/2}$

**2.** $64^{1/3}$

**3.** $27^{1/3}$

**4.** $8^{1/3}$

**5.** $\left(\dfrac{1}{16}\right)^{1/4}$

**6.** $\left(\dfrac{1}{64}\right)^{1/2}$

**7.** $169^{1/2}$

**8.** $81^{1/4}$

**9.** $2m^{1/3}$

**10.** $(2m)^{1/3}$

**11.** $(9x^4)^{1/2}$

**12.** $(16x^8)^{1/2}$

**13.** $(-27)^{1/3}$

**14.** $-64^{1/2}$

**15.** $-16^{1/4}$

**16.** $(-32)^{1/5}$

**Objective B** *Use radical notation to rewrite each expression. Simplify if possible. See Examples 7 through 11.*

**17.** $16^{3/4}$

**18.** $4^{5/2}$

**19.** $(-64)^{2/3}$

**20.** $(-8)^{4/3}$

**21.** $(-16)^{3/4}$

**22.** $(-9)^{3/2}$

**23.** $(2x)^{3/5}$

**24.** $2x^{3/5}$

**25.** $(7x + 2)^{2/3}$

**26.** $(x - 4)^{3/4}$

**27.** $\left(\dfrac{16}{9}\right)^{3/2}$

**28.** $\left(\dfrac{49}{25}\right)^{3/2}$

**Objective C** *Write with positive exponents. Simplify if possible. See Examples 12 and 13.*

**29.** $8^{-4/3}$

**30.** $64^{-2/3}$

**31.** $(-64)^{-2/3}$

**32.** $(-8)^{-4/3}$

**33.** $(-4)^{-3/2}$

**34.** $(-16)^{-5/4}$

**35.** $x^{-1/4}$

**36.** $y^{-1/6}$

**37.** $\dfrac{1}{a^{-2/3}}$

**38.** $\dfrac{1}{n^{-8/9}}$

**39.** $\dfrac{5}{7x^{-3/4}}$

**40.** $\dfrac{2}{3y^{-5/7}}$

**Objective D** *Use the properties of exponents to simplify each expression. Write with positive exponents. See Examples 14 through 18.*

**41.** $a^{2/3}a^{5/3}$

**42.** $b^{9/5}b^{8/5}$

**43.** $x^{-2/5} \cdot x^{7/5}$

**44.** $y^{4/3} \cdot y^{-1/3}$

**45.** $3^{1/4} \cdot 3^{3/8}$

**46.** $5^{1/2} \cdot 5^{1/6}$

**47.** $\dfrac{y^{1/3}}{y^{1/6}}$

**48.** $\dfrac{x^{3/4}}{x^{1/8}}$

**49.** $(4u^2)^{3/2}$

**50.** $(32^{1/5}x^{2/3})^3$

**51.** $\dfrac{b^{1/2}b^{3/4}}{-b^{1/4}}$

**52.** $\dfrac{a^{1/4}a^{-1/2}}{a^{2/3}}$

**53.** $\dfrac{(x^3)^{1/2}}{x^{7/2}}$

**54.** $\dfrac{y^{11/3}}{(y^5)^{1/3}}$

**55.** $\dfrac{(3x^{1/4})^3}{x^{1/12}}$

**56.** $\dfrac{(2x^{1/5})^4}{x^{3/10}}$

**57.** $\dfrac{(y^3z)^{1/6}}{y^{-1/2}z^{1/3}}$

**58.** $\dfrac{(m^2n)^{1/4}}{m^{-1/2}n^{5/8}}$

**59.** $\dfrac{(x^3y^2)^{1/4}}{(x^{-5}y^{-1})^{-1/2}}$

**60.** $\dfrac{(a^{-2}b^3)^{1/8}}{(a^{-3}b)^{-1/4}}$

**Objective E** *Use rational exponents to simplify each radical. Assume that all variables represent positive real numbers. See Examples 19 through 21.*

**▶ 61.** $\sqrt[6]{x^3}$  **62.** $\sqrt[9]{a^3}$  **63.** $\sqrt[6]{4}$  **64.** $\sqrt[4]{36}$  **▶ 65.** $\sqrt[4]{16x^2}$  **66.** $\sqrt[8]{4y^2}$

**67.** $\sqrt[4]{(x+3)^2}$  **68.** $\sqrt[8]{(y+1)^4}$  **69.** $\sqrt[8]{x^4y^4}$  **70.** $\sqrt[9]{y^6z^3}$  **71.** $\sqrt[12]{a^8b^4}$  **72.** $\sqrt[10]{a^5b^5}$

*Use rational expressions to write as a single radical expression. See Examples 22 through 24.*

**73.** $\sqrt[3]{y} \cdot \sqrt[5]{y^2}$  **74.** $\sqrt[3]{y^2} \cdot \sqrt[6]{y}$  **75.** $\dfrac{\sqrt[3]{b^2}}{\sqrt[4]{b}}$  **76.** $\dfrac{\sqrt[4]{a}}{\sqrt[5]{a}}$  **77.** $\sqrt[3]{x} \cdot \sqrt[4]{x} \cdot \sqrt[8]{x^3}$

**78.** $\sqrt[6]{y} \cdot \sqrt[3]{y} \cdot \sqrt[5]{y^2}$  **79.** $\dfrac{\sqrt[3]{a^2}}{\sqrt[6]{a}}$  **80.** $\dfrac{\sqrt[5]{b^2}}{\sqrt[10]{b^3}}$  **81.** $\sqrt{3} \cdot \sqrt[3]{4}$  **82.** $\sqrt[3]{5} \cdot \sqrt{2}$

**83.** $\sqrt[5]{7} \cdot \sqrt[3]{y}$  **84.** $\sqrt[4]{5} \cdot \sqrt[3]{x}$  **85.** $\sqrt{5r} \cdot \sqrt[3]{s}$  **86.** $\sqrt[3]{b} \cdot \sqrt[4]{4a}$

## Review

*Write each integer as a product of two integers such that one of the factors is a perfect square. For example, write 18 as $9 \cdot 2$, because 9 is a perfect square. See Sections 1.4 and 7.1.*

**87.** 75  **88.** 20  **89.** 48  **90.** 45

*Write each integer as a product of two integers such that one of the factors is a perfect cube. For example, write 24 as $8 \cdot 3$, because 8 is a perfect cube. See Sections 1.4 and 7.1.*

**91.** 16  **92.** 56  **93.** 54  **94.** 80

## Concept Extensions

*Basal metabolic rate (BMR) is the number of calories per day a person needs to maintain life. A person's basal metabolic rate $B(w)$ in calories per day can be estimated with the function $B(w) = 70w^{3/4}$, where w is the person's weight in kilograms. Use this information to answer Exercises 95 and 96.*

**95.** Estimate the BMR for a person who weighs 60 kilograms. Round to the nearest calorie. (*Note:* 60 kilograms is approximately 132 pounds.)

**96.** Estimate the BMR for a person who weighs 90 kilograms. Round to the nearest calorie. (*Note:* 90 kilograms is approximately 198 pounds.)

*The number of tablet computer users worldwide from 2010 to 2017 can be predicted by $f(x) = 76x^{13/10}$, where $f(x)$ is the number of tablet computer users worldwide in millions, x years after 2010. (Source: Forrester Research: World Tablet Adoption Forecast 2012–2017) Use this information to answer Exercises 97 and 98.*

**97.** Use this model to estimate the number of tablet computer users worldwide in 2015. Round to the nearest tenth of a million.

**98.** Predict the number of tablet computer users worldwide in 2020. Round to the nearest tenth of a million.

**99.** Explain how writing $x^{-7}$ with positive exponents is similar to writing $x^{-1/4}$ with positive exponents.

**100.** Explain how writing $2x^{-5}$ with positive exponents is similar to writing $2x^{-3/4}$ with positive exponents.

*Fill in each box with the correct expression.*

**101.** $\Box \cdot a^{2/3} = a^{3/3}$, or $a$

**102.** $\Box \cdot x^{1/8} = x^{4/8}$, or $x^{1/2}$

**103.** $\dfrac{\Box}{x^{-2/5}} = x^{3/5}$

**104.** $\dfrac{\Box}{y^{-3/4}} = y^{4/4}$, or $y$

*Use a calculator to write a four-decimal-place approximation of each number.*

**105.** $8^{1/4}$

**106.** $18^{3/5}$

*Solve.*

**107.** In physics, the speed of a wave traveling over a stretched string with tension $t$ and density $u$ is given by the expression $\dfrac{\sqrt{t}}{\sqrt{u}}$. Write this expression with rational exponents.

**108.** In electronics, the angular frequency of oscillations in a certain type of circuit is given by the expression $(LC)^{-1/2}$. Use radical notation to write this expression.

---

## 7.3  Simplifying Radical Expressions

### Objectives

**A** Use the Product Rule for Radicals.

**B** Use the Quotient Rule for Radicals.

**C** Simplify Radicals.

**D** Use the Distance and Midpoint Formulas.

### Objective A  Using the Product Rule

It is possible to simplify some radicals that do not evaluate to rational numbers. To do so, we use a product rule and a quotient rule for radicals. To discover the product rule, notice the following pattern:

$$\sqrt{9} \cdot \sqrt{4} = 3 \cdot 2 = 6$$
$$\sqrt{9 \cdot 4} = \sqrt{36} = 6$$

Since both expressions simplify to 6, it is true that

$$\sqrt{9} \cdot \sqrt{4} = \sqrt{9 \cdot 4}$$

This pattern suggests the following product rule for radicals.

## Product Rule for Radicals

If $\sqrt[n]{a}$ and $\sqrt[n]{b}$ are real numbers, then

$$\sqrt[n]{a} \cdot \sqrt[n]{b} = \sqrt[n]{ab}$$

Notice that the product rule is the relationship $a^{1/n} \cdot b^{1/n} = (ab)^{1/n}$ stated in radical notation.

**Examples** Use the product rule to multiply.

**1.** $\sqrt{3} \cdot \sqrt{5} = \sqrt{3 \cdot 5} = \sqrt{15}$

**2.** $\sqrt{21} \cdot \sqrt{x} = \sqrt{21x}$

**3.** $\sqrt[3]{4} \cdot \sqrt[3]{2} = \sqrt[3]{4 \cdot 2} = \sqrt[3]{8} = 2$

**4.** $\sqrt[4]{5} \cdot \sqrt[4]{2x^3} = \sqrt[4]{5 \cdot 2x^3} = \sqrt[4]{10x^3}$

**5.** $\sqrt{\dfrac{2}{a}} \cdot \sqrt{\dfrac{b}{3}} = \sqrt{\dfrac{2}{a} \cdot \dfrac{b}{3}} = \sqrt{\dfrac{2b}{3a}}$

■ Work Practice 1–5

**Practice 1–5**

Use the product rule to multiply.

**1.** $\sqrt{2} \cdot \sqrt{13}$     **2.** $\sqrt{17} \cdot \sqrt{y}$

**3.** $\sqrt[3]{2} \cdot \sqrt[3]{32}$     **4.** $\sqrt[4]{6} \cdot \sqrt[4]{3x^2}$

**5.** $\sqrt{\dfrac{3}{x}} \cdot \sqrt{\dfrac{y}{2}}$

## Objective B  Using the Quotient Rule ▶

To discover the quotient rule for radicals, notice the following pattern:

$$\sqrt{\frac{4}{9}} = \frac{2}{3}$$

$$\frac{\sqrt{4}}{\sqrt{9}} = \frac{2}{3}$$

Since both expressions simplify to $\dfrac{2}{3}$, it is true that

$$\sqrt{\frac{4}{9}} = \frac{\sqrt{4}}{\sqrt{9}}$$

This pattern suggests the following quotient rule for radicals.

## Quotient Rule for Radicals

If $\sqrt[n]{a}$ and $\sqrt[n]{b}$ are real numbers and $\sqrt[n]{b}$ is not zero, then

$$\sqrt[n]{\frac{a}{b}} = \frac{\sqrt[n]{a}}{\sqrt[n]{b}}$$

Notice that the quotient rule is the relationship $\left(\dfrac{a}{b}\right)^{1/n} = \dfrac{a^{1/n}}{b^{1/n}}$ stated in radical notation. We can use the quotient rule to simplify radical expressions by reading the rule from left to right or to divide radicals by reading the rule from right to left.

For example.

$$\sqrt{\frac{x}{16}} = \frac{\sqrt{x}}{\sqrt{16}} = \frac{\sqrt{x}}{4} \qquad \text{Using } \sqrt[n]{\frac{a}{b}} = \frac{\sqrt[n]{a}}{\sqrt[n]{b}}$$

$$\frac{\sqrt{50}}{\sqrt{2}} = \sqrt{\frac{50}{2}} = \sqrt{25} = 5 \qquad \text{Using } \frac{\sqrt[n]{a}}{\sqrt[n]{b}} = \sqrt[n]{\frac{a}{b}}$$

**Answers**

**1.** $\sqrt{26}$     **2.** $\sqrt{17y}$     **3.** 4

**4.** $\sqrt[4]{18x^2}$     **5.** $\sqrt{\dfrac{3y}{2x}}$

Note: *For the remainder of this chapter, we will assume that variables represent positive real numbers. If this is so, we need not insert absolute value bars when we simplify even roots.*

## Practice 6–9

Use the quotient rule to simplify. Assume that all variables represent positive real numbers.

**6.** $\sqrt{\dfrac{9}{25}}$    **7.** $\sqrt{\dfrac{y}{36}}$

**8.** $\sqrt[3]{\dfrac{27}{64}}$    **9.** $\sqrt[5]{\dfrac{7}{32x^5}}$

**Examples**    Use the quotient rule to simplify.

**6.** $\sqrt{\dfrac{25}{49}} = \dfrac{\sqrt{25}}{\sqrt{49}} = \dfrac{5}{7}$

**7.** $\sqrt{\dfrac{x}{9}} = \dfrac{\sqrt{x}}{\sqrt{9}} = \dfrac{\sqrt{x}}{3}$

**8.** $\sqrt[3]{\dfrac{8}{27}} = \dfrac{\sqrt[3]{8}}{\sqrt[3]{27}} = \dfrac{2}{3}$

**9.** $\sqrt[4]{\dfrac{3}{16y^4}} = \dfrac{\sqrt[4]{3}}{\sqrt[4]{16y^4}} = \dfrac{\sqrt[4]{3}}{2y}$

🔳 Work Practice 6–9

## Objective C  Simplifying Radicals

Both the product and quotient rules can be used to simplify a radical. If the product rule is read from right to left, we have that $\sqrt[n]{ab} = \sqrt[n]{a} \cdot \sqrt[n]{b}$. We use this to simplify the following radicals.

## Practice 10

Simplify:   $\sqrt{18}$

**Example 10**    Simplify:   $\sqrt{50}$

**Solution:**   We factor 50 such that one factor is the largest perfect square that divides 50. The largest perfect square factor of 50 is 25, so we write 50 as $25 \cdot 2$ and use the product rule for radicals to simplify.

$$\sqrt{50} = \sqrt{25 \cdot 2} = \sqrt{25} \cdot \sqrt{2} = 5\sqrt{2}$$

⌞— the largest perfect square factor of 50

**Helpful Hint**    Don't forget that, for example, $5\sqrt{2}$ means $5 \cdot \sqrt{2}$.

🔳 Work Practice 10

## Practice 11–13

Simplify.

**11.** $\sqrt[3]{40}$    **12.** $\sqrt{14}$

**13.** $\sqrt[4]{162}$

**Examples**    Simplify.

**11.** $\sqrt[3]{24} = \sqrt[3]{8 \cdot 3} = \sqrt[3]{8} \cdot \sqrt[3]{3} = 2\sqrt[3]{3}$

⌞— the largest perfect cube factor of 24

**12.** $\sqrt{26}$ The largest perfect square factor of 26 is 1, so $\sqrt{26}$ cannot be simplified further.

**13.** $\sqrt[4]{32} = \sqrt[4]{16 \cdot 2} = \sqrt[4]{16} \cdot \sqrt[4]{2} = 2\sqrt[4]{2}$

⌞— the largest 4th power factor of 32

🔳 Work Practice 11–13

After simplifying a radical such as a square root, always check the radicand to see that it contains no other perfect square factors. It may, if the largest perfect square factor of the radicand was not originally recognized. For example,

$$\sqrt{200} = \sqrt{4 \cdot 50} = \sqrt{4} \cdot \sqrt{50} = 2\sqrt{50}$$

Notice that the radicand 50 still contains the perfect square factor 25. This is because 4 is not the largest perfect square factor of 200. We continue as follows:

$$2\sqrt{50} = 2\sqrt{25 \cdot 2} = 2 \cdot \sqrt{25} \cdot \sqrt{2} = 2 \cdot 5 \cdot \sqrt{2} = 10\sqrt{2}$$

The radical is now simplified since 2 contains no perfect square factors (other than 1).

## Answers

**6.** $\dfrac{3}{5}$   **7.** $\dfrac{\sqrt{y}}{6}$   **8.** $\dfrac{3}{4}$   **9.** $\dfrac{\sqrt[5]{7}}{2x}$

**10.** $3\sqrt{2}$   **11.** $2\sqrt[3]{5}$   **12.** $\sqrt{14}$

**13.** $3\sqrt[4]{2}$

**Helpful Hint**

To recognize the largest perfect power factors of a radicand, it will help if you are familiar with some perfect powers. A few are listed below.

Perfect Squares          1, 4, 9, 16, 25, 36, 49, 64, 81, 100, 121, 144
$\quad\quad\quad\quad\quad\quad\quad\quad\quad 1^2 \ 2^2 \ 3^2 \ 4^2 \ 5^2 \ 6^2 \ 7^2 \ 8^2 \ 9^2 \ 10^2 \ 11^2 \ 12^2$

Perfect Cubes            1, 8, 27, 64, 125
$\quad\quad\quad\quad\quad\quad\quad\quad\quad 1^3 \ 2^3 \ 3^3 \ 4^3 \ 5^3$

Perfect 4th powers       1, 16, 81, 256
$\quad\quad\quad\quad\quad\quad\quad\quad\quad 1^4 \ 2^4 \ 3^4 \ 4^4$

**Helpful Hint**

We say that a radical of the form $\sqrt[n]{a}$ is simplified when the radicand $a$ contains no factors that are perfect $n$th powers (other than 1 or $-1$).

**Examples**   Simplify. Assume that all variables represent positive real numbers.

**14.** $\sqrt{25x^3} = \sqrt{25 \cdot x^2 \cdot x}$      Find the largest perfect square factor.

$\quad\quad\quad\quad = \sqrt{25 \cdot x^2} \cdot \sqrt{x}$      Use the product rule.

$\quad\quad\quad\quad = 5x\sqrt{x}$      Simplify.

**15.** $\sqrt[3]{54x^6y^8} = \sqrt[3]{27 \cdot 2 \cdot x^6 \cdot y^6 \cdot y^2}$      Factor the radicand and identify perfect cube factors.

$\quad\quad\quad\quad = \sqrt[3]{27 \cdot x^6 \cdot y^6 \cdot 2y^2}$

$\quad\quad\quad\quad = \sqrt[3]{27 \cdot x^6 \cdot y^6} \cdot \sqrt[3]{2y^2}$      Use the product rule.

$\quad\quad\quad\quad = 3x^2y^2\sqrt[3]{2y^2}$      Simplify.

**16.** $\sqrt[4]{81z^{11}} = \sqrt[4]{81 \cdot z^8 \cdot z^3}$      Factor the radicand and identify perfect 4th power factors.

$\quad\quad\quad\quad = \sqrt[4]{81 \cdot z^8} \cdot \sqrt[4]{z^3}$      Use the product rule.

$\quad\quad\quad\quad = 3z^2\sqrt[4]{z^3}$      Simplify.

▬ Work Practice 14–16

**Examples**   Use the quotient rule to divide. Then simplify if possible. Assume that all variables represent positive real numbers.

**17.** $\dfrac{\sqrt{20}}{\sqrt{5}} = \sqrt{\dfrac{20}{5}}$      Use the quotient rule.

$\quad\quad\quad = \sqrt{4}$      Simplify.

$\quad\quad\quad = 2$      Simplify.

**18.** $\dfrac{\sqrt{50x}}{2\sqrt{2}} = \dfrac{1}{2} \cdot \sqrt{\dfrac{50x}{2}}$      Use the quotient rule.

$\quad\quad\quad = \dfrac{1}{2} \cdot \sqrt{25x}$      Simplify.

$\quad\quad\quad = \dfrac{1}{2} \cdot \sqrt{25} \cdot \sqrt{x}$      Factor $25x$.

$\quad\quad\quad = \dfrac{1}{2} \cdot 5 \cdot \sqrt{x}$      Simplify.

$\quad\quad\quad = \dfrac{5}{2}\sqrt{x}$

*(Continued on next page)*

**Practice 14–16**

Simplify. Assume that all variables represent positive real numbers.

**14.** $\sqrt{49a^5}$     **15.** $\sqrt[3]{24x^9y^7}$

**16.** $\sqrt[4]{16z^9}$

**Practice 17–20**

Use the quotient rule to divide. Then simplify if possible. Assume that all variables represent positive real numbers.

**17.** $\dfrac{\sqrt{75}}{\sqrt{3}}$     **18.** $\dfrac{\sqrt{80y}}{3\sqrt{5}}$

**19.** $\dfrac{5\sqrt[3]{162x^8}}{\sqrt[3]{3x^2}}$     **20.** $\dfrac{3\sqrt[4]{243x^9y^6}}{\sqrt[4]{x^{-3}y}}$

**Answers**

**14.** $7a^2\sqrt{a}$   **15.** $2x^3y^2\sqrt[3]{3y}$   **16.** $2z^2\sqrt[4]{z}$

**17.** 5   **18.** $\dfrac{4}{3}\sqrt{y}$   **19.** $15x^2\sqrt[3]{2}$

**20.** $9x^3y\sqrt[4]{3y}$

**19.** $\dfrac{7\sqrt[3]{48y^4}}{\sqrt[3]{2y}} = 7\sqrt[3]{\dfrac{48y^4}{2y}} = 7\sqrt[3]{24y^3} = 7\sqrt[3]{8\cdot y^3\cdot 3}$

$= 7\sqrt[3]{8\cdot y^3}\cdot\sqrt[3]{3} = 7\cdot 2y\sqrt[3]{3} = 14y\sqrt[3]{3}$

**20.** $\dfrac{2\sqrt[4]{32a^8b^6}}{\sqrt[4]{a^{-1}b^2}} = 2\sqrt[4]{\dfrac{32a^8b^6}{a^{-1}b^2}} = 2\sqrt[4]{32a^9b^4} = 2\sqrt[4]{16\cdot a^8\cdot b^4\cdot 2\cdot a}$

$= 2\sqrt[4]{16\cdot a^8\cdot b^4}\cdot\sqrt[4]{2\cdot a} = 2\cdot 2a^2b\cdot\sqrt[4]{2a} = 4a^2b\sqrt[4]{2a}$

■ Work Practice 17–20

✓**Concept Check** Find and correct the error:

$$\dfrac{\sqrt[3]{27}}{\sqrt{9}} = \sqrt[3]{\dfrac{27}{9}} = \sqrt[3]{3}$$

## Objective D Using the Distance and Midpoint Formulas

Now that we know how to simplify radicals, we can derive and use the distance formula. The midpoint formula is often confused with the distance formula, so to clarify both, we will also review the midpoint formula.

The Cartesian coordinate system helps us visualize the distance between points. To find the distance between two points, we use the distance formula, which is derived from the Pythagorean theorem.

To find the distance $d$ between two points $(x_1, y_1)$ and $(x_2, y_2)$, draw vertical and horizontal lines so that a right triangle is formed, as shown. Notice that the length of leg $a$ is $x_2 - x_1$ and that the length of leg $b$ is $y_2 - y_1$. Thus, the Pythagorean theorem tells us that

$$d^2 = a^2 + b^2$$

or

$$d^2 = (x_2 - x_1)^2 + (y_2 - y_1)^2$$

or

$$d = \sqrt{(x_2 - x_1)^2 + (y_2 - y_1)^2}$$

This formula gives us the distance between any two points on the real plane.

### Distance Formula

The distance $d$ between two points $(x_1, y_1)$ and $(x_2, y_2)$ is given by

$$d = \sqrt{(x_2 - x_1)^2 + (y_2 - y_1)^2}$$

**Example 21**  Find the distance between $(2, -5)$ and $(1, -4)$. Give the exact distance and a three-decimal-place approximation.

**Solution:**  To use the distance formula, it makes no difference which point we call $(x_1, y_1)$ and which point we call $(x_2, y_2)$. We will let $(x_1, y_1) = (2, -5)$ and $(x_2, y_2) = (1, -4)$.

$$
\begin{aligned}
d &= \sqrt{(x_2 - x_1)^2 + (y_2 - y_1)^2} \\
  &= \sqrt{(1 - 2)^2 + [-4 - (-5)]^2} \\
  &= \sqrt{(-1)^2 + (1)^2} \\
  &= \sqrt{1 + 1} \\
  &= \sqrt{2} \approx 1.414
\end{aligned}
$$

The distance between the two points is exactly $\sqrt{2}$ units, or approximately 1.414 units.

📖 **Work Practice 21**

**Practice 21**

Find the distance between $(-1, 3)$ and $(-2, 6)$. Give the exact distance and a three-decimal-place approximation.

The **midpoint** of a line segment is the **point** located exactly halfway between the two endpoints of the line segment. On the following graph, the point $M$ is the midpoint of line segment $PQ$. Thus, the distance between $M$ and $P$ equals the distance between $M$ and $Q$. *Note:* We usually need no knowledge of roots to calculate the midpoint of a line segment. We review midpoint here only because it is often confused with the distance between two points.

The $x$-coordinate of $M$ is at half the distance between the $x$-coordinates of $P$ and $Q$, and the $y$-coordinate of $M$ is at half the distance between the $y$-coordinates of $P$ and $Q$. That is, the $x$-coordinate of $M$ is the average of the $x$-coordinates of $P$ and $Q$; the $y$-coordinate of $M$ is the average of the $y$-coordinates of $P$ and $Q$.

**Midpoint Formula**

The midpoint of the line segment whose endpoints are $(x_1, y_1)$ and $(x_2, y_2)$ is the point with coordinates

$$
\left( \frac{x_1 + x_2}{2}, \frac{y_1 + y_2}{2} \right)
$$

**Answer**
**21.** $\sqrt{10} \approx 3.162$

**Practice 22**

Find the midpoint of the line segment that joins points $P(-2, 5)$ and $Q(4, -6)$.

**Example 22** Find the midpoint of the line segment that joins points $P(-3, 3)$ and $Q(1, 0)$.

**Solution:** To use the midpoint formula, it makes no difference which point we call $(x_1, y_1)$ and which point we call $(x_2, y_2)$. We will let $(x_1, y_1) = (-3, 3)$ and $(x_2, y_2) = (1, 0)$.

$$\text{midpoint} = \left( \frac{x_1 + x_2}{2}, \frac{y_1 + y_2}{2} \right)$$

$$= \left( \frac{-3 + 1}{2}, \frac{3 + 0}{2} \right)$$

$$= \left( \frac{-2}{2}, \frac{3}{2} \right)$$

$$= \left( -1, \frac{3}{2} \right)$$

The midpoint of the segment is $\left( -1, \dfrac{3}{2} \right)$.

▶ **Work Practice 22**

**Helpful Hint**

The distance between two points is a distance. The midpoint of a line segment is the point halfway between the endpoints of the segment.

distance—measured in units

midpoint—it is a point

**Answer**

**22.** $\left( 1, -\dfrac{1}{2} \right)$

## Vocabulary, Readiness & Video Check

*Use the choices below to fill in each blank. Some choices may be used more than once.*

distance          midpoint          point

**1.** The _____ of a line segment is a _____ exactly halfway between the two endpoints of the line segment.

**2.** The _____ formula is $d = \sqrt{(x_2 - x_1)^2 + (y_2 - y_1)^2}$.

**3.** The _____ formula is $\left( \dfrac{x_1 + x_2}{2}, \dfrac{y_1 + y_2}{2} \right)$.

*Answer true or false. Assume all radicals represent nonzero real numbers.*

**4.** $\sqrt[n]{a} \cdot \sqrt[n]{b} = \sqrt[n]{ab}$, _____

**5.** $\sqrt[3]{7} \cdot \sqrt[3]{11} = \sqrt[3]{18}$, _____

**6.** $\sqrt[3]{7} \cdot \sqrt{11} = \sqrt{77}$, _____

**7.** $\sqrt{x^7 y^8} = \sqrt{x^7} \cdot \sqrt{y^8}$, _____

**8.** $\dfrac{\sqrt[n]{a}}{\sqrt[n]{b}} = \sqrt[n]{\dfrac{a}{b}}$, _____

**9.** $\dfrac{\sqrt[3]{12}}{\sqrt[3]{4}} = \sqrt[3]{8}$, _____

**10.** $\dfrac{\sqrt[n]{x^7}}{\sqrt[n]{x}} = \sqrt[n]{x^6}$, _____

*Martin-Gay Interactive Videos*          *Watch the section lecture video and answer the following questions.*

See Video 7.3

**Objective A** 11. From ▣ Example 1 and the lecture before, in order to apply the product rule for radicals, what must be true about the indexes of the radicals being multiplied? ▶

**Objective B** 12. From ▣ Examples 2–6, when might you apply the quotient rule (in either direction) in order to simplify a fractional radical expression? ▶

**Objective C** 13. From ▣ Example 8, we know that an even power of a variable is a perfect square factor of the variable, leaving no factor in the radicand once simplified. Therefore, what must be true about the power of any variable left in the radicand of a simplified square root? Explain. ▶

**Objective D** 14. From ▣ Example 10, the formula uses the coordinates of two points similar to the slope formula. What caution should you take when replacing values in the formula? ▶

15. Based on ▣ Example 11, complete the following statement. The $x$-value of the midpoint is the _____ of the $x$-values of the endpoints and the $y$-value of the midpoint is the _____ of the $y$-values of the endpoints. ▶

## 7.3 Exercise Set MyMathLab® ▶

**Objective A** *Use the product rule to multiply. Assume that all variables represent positive real numbers.*
*See Examples 1 through 5.*

1. $\sqrt{7} \cdot \sqrt{2}$

2. $\sqrt{11} \cdot \sqrt{10}$

3. $\sqrt[4]{8} \cdot \sqrt[4]{2}$

4. $\sqrt[4]{27} \cdot \sqrt[4]{3}$

5. $\sqrt[3]{4} \cdot \sqrt[3]{9}$

6. $\sqrt[3]{10} \cdot \sqrt[3]{5}$

▶ 7. $\sqrt{2} \cdot \sqrt{3x}$

8. $\sqrt{3y} \cdot \sqrt{5x}$

9. $\sqrt{\dfrac{7}{x}} \cdot \sqrt{\dfrac{2}{y}}$

10. $\sqrt{\dfrac{6}{m}} \cdot \sqrt{\dfrac{n}{5}}$

11. $\sqrt[4]{4x^3} \cdot \sqrt[4]{5}$

12. $\sqrt[4]{ab^2} \cdot \sqrt[4]{27ab}$

**Objective B** *Use the quotient rule to simplify. Assume that all variables represent positive real numbers. See Examples 6 through 9.*

▶ 13. $\sqrt{\dfrac{6}{49}}$

14. $\sqrt{\dfrac{10}{81}}$

15. $\sqrt{\dfrac{2}{49}}$

16. $\sqrt{\dfrac{5}{121}}$

▶ 17. $\sqrt[4]{\dfrac{x^3}{16}}$

18. $\sqrt[4]{\dfrac{y}{81x^4}}$

19. $\sqrt[3]{\dfrac{4}{27}}$

20. $\sqrt[3]{\dfrac{3}{64}}$

21. $\sqrt[4]{\dfrac{8}{x^8}}$

22. $\sqrt[4]{\dfrac{a^3}{81}}$

23. $\sqrt[3]{\dfrac{2x}{27y^{12}}}$

24. $\sqrt[3]{\dfrac{3y}{8x^6}}$

**25.** $\sqrt{\dfrac{x^2y}{169}}$    **26.** $\sqrt{\dfrac{y^2z}{225}}$    ▶**27.** $\sqrt{\dfrac{5x^2}{4y^2}}$    **28.** $\sqrt{\dfrac{y^{10}}{9x^6}}$    **29.** $-\sqrt[3]{\dfrac{z^7}{125x^3}}$    **30.** $-\sqrt[3]{\dfrac{1000a}{b^9}}$

## Objective C  *Simplify. Assume that all variables represent positive real numbers. See Examples 10 through 16.*

▶**31.** $\sqrt{32}$    **32.** $\sqrt{27}$    **33.** $\sqrt[3]{192}$    **34.** $\sqrt[3]{108}$    **35.** $5\sqrt{75}$    **36.** $3\sqrt{8}$

**37.** $\sqrt{24}$    **38.** $\sqrt{20}$    **39.** $\sqrt{100x^5}$    **40.** $\sqrt{64y^9}$    **41.** $\sqrt[3]{16y^7}$    **42.** $\sqrt[3]{128y^{10}}$

**43.** $\sqrt[4]{a^8b^7}$    **44.** $\sqrt[5]{32z^{12}}$    **45.** $\sqrt{y^5}$    **46.** $\sqrt[3]{y^5}$    ▶**47.** $\sqrt{25a^2b^3}$    **48.** $\sqrt{9x^5y^7}$

▶**49.** $\sqrt[5]{-32x^{10}y}$    **50.** $\sqrt[5]{-243z^9}$    **51.** $\sqrt[3]{50x^{14}}$    **52.** $\sqrt[3]{40y^{10}}$    **53.** $-\sqrt{32a^8b^7}$    **54.** $-\sqrt{20ab^6}$

**55.** $\sqrt{9x^7y^9}$    **56.** $\sqrt{12r^9s^{12}}$    **57.** $\sqrt[3]{125r^9s^{12}}$    **58.** $\sqrt[3]{8a^6b^9}$    **59.** $\sqrt[4]{32x^{12}y^5}$    **60.** $\sqrt[4]{162x^7y^{20}}$

*Use the quotient rule to divide. Then simplify if possible. Assume that all variables represent positive real numbers. See Examples 17 through 20.*

▶**61.** $\dfrac{\sqrt{14}}{\sqrt{7}}$    **62.** $\dfrac{\sqrt{45}}{\sqrt{9}}$    **63.** $\dfrac{\sqrt[3]{24}}{\sqrt[3]{3}}$    **64.** $\dfrac{\sqrt[3]{10}}{\sqrt[3]{2}}$

**65.** $\dfrac{5\sqrt[4]{48}}{\sqrt[4]{3}}$    **66.** $\dfrac{7\sqrt[4]{162}}{\sqrt[4]{2}}$    ▶**67.** $\dfrac{\sqrt{x^5y^3}}{\sqrt{xy}}$    **68.** $\dfrac{\sqrt{a^7b^6}}{\sqrt{a^3b^2}}$

**69.** $\dfrac{8\sqrt[3]{54m^7}}{\sqrt[3]{2m}}$    **70.** $\dfrac{\sqrt[3]{128x^3}}{-3\sqrt[3]{2x}}$    **71.** $\dfrac{3\sqrt{100x^2}}{2\sqrt{2x^{-1}}}$    **72.** $\dfrac{\sqrt{270y^2}}{5\sqrt{3y^{-4}}}$

**73.** $\dfrac{\sqrt[4]{96a^{10}b^3}}{\sqrt[4]{3a^2b^3}}$    **74.** $\dfrac{\sqrt[4]{160x^{10}y^5}}{\sqrt[4]{2x^2y^2}}$    **75.** $\dfrac{\sqrt[5]{64x^{10}y^3}}{\sqrt[5]{2x^3y^{-7}}}$    **76.** $\dfrac{\sqrt[5]{192x^6y^{12}}}{\sqrt[5]{2x^{-1}y^{-3}}}$

**Objective D** *Find the distance between each pair of points. Give the exact distance and a three-decimal-place approximation. See Example 21.*

**77.** $(5, 1)$ and $(8, 5)$

**78.** $(2, 3)$ and $(14, 8)$

▶ **79.** $(-3, 2)$ and $(1, -3)$

**80.** $(3, -2)$ and $(-4, 1)$

**81.** $(-9, 4)$ and $(-8, 1)$

**82.** $(-5, -2)$ and $(-6, -6)$

**83.** $(0, -\sqrt{2})$ and $(\sqrt{3}, 0)$

**84.** $(-\sqrt{5}, 0)$ and $(0, \sqrt{7})$

**85.** $(1.7, -3.6)$ and $(-8.6, 5.7)$

**86.** $(9.6, 2.5)$ and $(-1.9, -3.7)$

*Find the midpoint of each line segment whose endpoints are given. See Example 22.*

**87.** $(6, -8); (2, 4)$

**88.** $(3, 9); (7, 11)$

▶ **89.** $(-2, -1); (-8, 6)$

**90.** $(-3, -4); (6, -8)$

**91.** $(6, 3); (-1, -3)$

**92.** $(-2, 5); (2, 6)$

**93.** $\left(\dfrac{1}{2}, \dfrac{3}{8}\right); \left(-\dfrac{3}{2}, \dfrac{5}{8}\right)$

**94.** $\left(-\dfrac{2}{5}, \dfrac{7}{15}\right); \left(-\dfrac{2}{5}, -\dfrac{4}{15}\right)$

**95.** $(\sqrt{2}, 3\sqrt{5}); (\sqrt{2}, -2\sqrt{5})$

**96.** $(\sqrt{8}, -\sqrt{12}); (3\sqrt{2}, 7\sqrt{3})$

**97.** $(4.6, -3.5); (7.8, -9.8)$

**98.** $(-4.6, 2.1); (-6.7, 1.9)$

## Review

*Perform each indicated operation. See Sections 1.5 and 5.2.*

**99.** $6x + 8x$

**100.** $(6x)(8x)$

**101.** $(2x + 3)(x - 5)$

**102.** $(2x + 3) + (x - 5)$

**103.** $9y^2 - 8y^2$

**104.** $(9y^2)(-8y^2)$

**105.** $(x - 4)^2$

**106.** $(2x + 1)^2$

## Concept Extensions

*Find and correct the error. See the Concept Check in this section.*

**107.** $\dfrac{\sqrt[3]{64}}{\sqrt{64}} = \sqrt[3]{\dfrac{64}{64}} = \sqrt[3]{1} = 1$

**108.** $\dfrac{\sqrt[4]{16}}{\sqrt{4}} = \sqrt[4]{\dfrac{16}{4}} = \sqrt[4]{4}$

*Solve.*

**109.** The formula for the radius $r$ of a sphere with surface area $A$ is given by $r = \sqrt{\dfrac{A}{4\pi}}$. Calculate the radius of a standard zorb whose outside surface area is 32.17 sq m. Round to the nearest tenth. (A zorb is a large inflated ball within a ball in which a person, strapped inside, may choose to roll down a hill. *Source:* Zorb, Ltd.)

**110.** Before Mount Vesuvius, a volcano in Italy, erupted violently in 79 A.D., its height was 4190 feet. Vesuvius was roughly cone shaped, and its base had a radius of approximately 25,200 feet. Use the formula $A = \pi r \sqrt{r^2 + h^2}$ for the lateral surface area $A$ of a cone with radius $r$ and height $h$ to approximate the surface area of this volcano before it erupted. Round to the nearest square foot. (*Source:* Global Volcanism Network)

**111.** The time it takes a pendulum to swing back and forth is dependent on the length of the pendulum. The formula for this is $P = 2\pi \sqrt{\dfrac{L}{32}}$, where $P$ is the time in seconds it takes the pendulum to swing back and forth, $L$ is the length in feet of the pendulum, and 32 represents the force due to gravity.

a. Approximate to one decimal place the time of the pendulum's swing if the length of the pendulum is 4 feet.

b. Approximate to one decimal place the time of the pendulum's swing if the length of the pendulum is 1.5 feet.

c. Explain how the formula can be used to determine the length of the pendulum if the time of the swing is known.

### 7.4 Adding, Subtracting, and Multiplying Radical Expressions

**Objectives**

**A** Add or Subtract Radical Expressions.

**B** Multiply Radical Expressions.

### Objective A  Adding or Subtracting Radical Expressions

We have learned that the sum or difference of like terms can be simplified. To simplify these sums or differences, we use the distributive property. For example,

$$2x + 3x = (2 + 3)x = 5x$$

The distributive property can also be used to add *like radicals*.

## Like Radicals

Radicals with the same index and the same radicand are **like radicals.** The example below shows how to use the distributive property to simplify an expression containing like radicals.

$$2\sqrt{7} + 3\sqrt{7} = (2 + 3)\sqrt{7} = 5\sqrt{7}$$

Like radicals

**Helpful Hint**

The expression

$$5\sqrt{7} - 3\sqrt{6}$$

does not contain like radicals and cannot be simplified further.

**Examples**   Add or subtract as indicated.

1. $4\sqrt{11} + 8\sqrt{11} = (4 + 8)\sqrt{11} = 12\sqrt{11}$
2. $5\sqrt[3]{3x} - 7\sqrt[3]{3x} = (5 - 7)\sqrt[3]{3x} = -2\sqrt[3]{3x}$
3. $2\sqrt{7} + 2\sqrt[3]{7}$   This expression cannot be simplified since $2\sqrt{7}$ and $2\sqrt[3]{7}$ do not contain like radicals.

▶ Work Practice 1–3

✔**Concept Check** True or false:

$$\sqrt{a} + \sqrt{b} = \sqrt{a + b}$$

Explain your answer.

When adding or subtracting radicals, always check first to see whether any radicals can be simplified.

**Examples**   Add or subtract as indicated. Assume that all variables represent positive real numbers.

4.
$$
\begin{aligned}
\sqrt{20} + 2\sqrt{45} &= \sqrt{4 \cdot 5} + 2\sqrt{9 \cdot 5} & \text{Factor 20 and 45.}\\
&= \sqrt{4} \cdot \sqrt{5} + 2 \cdot \sqrt{9} \cdot \sqrt{5} & \text{Use the product rule.}\\
&= 2 \cdot \sqrt{5} + 2 \cdot 3 \cdot \sqrt{5} & \text{Simplify } \sqrt{4} \text{ and } \sqrt{9}.\\
&= 2\sqrt{5} + 6\sqrt{5} & \text{Add like radicals.}\\
&= 8\sqrt{5}
\end{aligned}
$$

5.
$$
\begin{aligned}
\sqrt[3]{54} - 5\sqrt[3]{16} + \sqrt[3]{2} &\\
&= \sqrt[3]{27} \cdot \sqrt[3]{2} - 5 \cdot \sqrt[3]{8} \cdot \sqrt[3]{2} + \sqrt[3]{2} & \text{Factor and use the product rule.}\\
&= 3 \cdot \sqrt[3]{2} - 5 \cdot 2 \cdot \sqrt[3]{2} + \sqrt[3]{2} & \text{Simplify } \sqrt[3]{27} \text{ and } \sqrt[3]{8}.\\
&= 3\sqrt[3]{2} - 10\sqrt[3]{2} + \sqrt[3]{2} & \text{Write } 5 \cdot 2 \text{ as } 10.\\
&= -6\sqrt[3]{2} & \text{Combine like radicals.}
\end{aligned}
$$

6.
$$
\begin{aligned}
\sqrt{27x} - 2\sqrt{9x} + \sqrt{72x} &\\
&= \sqrt{9} \cdot \sqrt{3x} - 2 \cdot \sqrt{9} \cdot \sqrt{x} + \sqrt{36} \cdot \sqrt{2x} & \text{Factor and use the product rule.}\\
&= 3 \cdot \sqrt{3x} - 2 \cdot 3 \cdot \sqrt{x} + 6 \cdot \sqrt{2x} & \text{Simplify } \sqrt{9} \text{ and } \sqrt{36}.\\
&= 3\sqrt{3x} - 6\sqrt{x} + 6\sqrt{2x} & \text{Write } 2 \cdot 3 \text{ as } 6.
\end{aligned}
$$

*(Continued on next page)*

---

**Practice 1–3**

Add or subtract as indicated.
1. $5\sqrt{15} + 2\sqrt{15}$
2. $9\sqrt[3]{2y} - 15\sqrt[3]{2y}$
3. $6\sqrt{10} - 3\sqrt[3]{10}$

**Practice 4–8**

Add or subtract as indicated. Assume that all variables represent positive real numbers.
4. $\sqrt{50} + 5\sqrt{18}$
5. $\sqrt[3]{24} - 4\sqrt[3]{192} + \sqrt[3]{3}$
6. $\sqrt{20x} - 6\sqrt{16x} + \sqrt{32x}$
7. $\sqrt[4]{32} + \sqrt{32}$
8. $\sqrt[3]{8y^5} + \sqrt[3]{27y^5}$

**Helpful Hint**   None of these terms contains like radicals. We can simplify no further.

**Answers**
1. $7\sqrt{15}$  2. $-6\sqrt[3]{2y}$  3. $6\sqrt{10} - 3\sqrt[3]{10}$
4. $20\sqrt{2}$  5. $-13\sqrt[3]{3}$
6. $2\sqrt{5x} - 24\sqrt{x} + 4\sqrt{2x}$
7. $2\sqrt[4]{2} + 4\sqrt{2}$  8. $5y\sqrt[3]{y^2}$

✔**Concept Check Answer**
false; answers may vary

**7.** $\sqrt[3]{98} + \sqrt{98} = \sqrt[3]{98} + \sqrt{49} \cdot \sqrt{2}$    Factor and use the product rule.

$= \sqrt[3]{98} + 7\sqrt{2}$    No further simplification is possible.

**8.** $\sqrt[3]{48y^4} + \sqrt[3]{6y^4} = \sqrt[3]{8y^3} \cdot \sqrt[3]{6y} + \sqrt[3]{y^3} \cdot \sqrt[3]{6y}$    Factor and use the product rule.

$= 2y\sqrt[3]{6y} + y\sqrt[3]{6y}$    Simplify $\sqrt[3]{8y^3}$ and $\sqrt[3]{y^3}$.

$= 3y\sqrt[3]{6y}$    Combine like radicals.

■ Work Practice 4–8

## Practice 9–10

Add or subtract as indicated. Assume that all variables represent positive real numbers.

**9.** $\dfrac{\sqrt{75}}{9} - \dfrac{\sqrt{3}}{2}$

**10.** $\sqrt[3]{\dfrac{5x}{27}} + 4\sqrt[3]{5x}$

**Examples**    Add or subtract as indicated. Assume that all variables represent positive real numbers.

**9.** $\dfrac{\sqrt{45}}{4} - \dfrac{\sqrt{5}}{3} = \dfrac{3\sqrt{5}}{4} - \dfrac{\sqrt{5}}{3}$    To subtract, notice that the LCD is 12.

$= \dfrac{3\sqrt{5} \cdot 3}{4 \cdot 3} - \dfrac{\sqrt{5} \cdot 4}{3 \cdot 4}$    Write each expression as an equivalent expression with a denominator of 12.

$= \dfrac{9\sqrt{5}}{12} - \dfrac{4\sqrt{5}}{12}$    Multiply factors in the numerators and the denominators.

$= \dfrac{5\sqrt{5}}{12}$    Subtract.

**10.** $\sqrt[3]{\dfrac{7x}{8}} + 2\sqrt[3]{7x} = \dfrac{\sqrt[3]{7x}}{\sqrt[3]{8}} + 2\sqrt[3]{7x}$    Use the quotient rule for radicals.

$= \dfrac{\sqrt[3]{7x}}{2} + 2\sqrt[3]{7x}$    Simplify.

$= \dfrac{\sqrt[3]{7x}}{2} + \dfrac{2\sqrt[3]{7x} \cdot 2}{2}$    Write each expression as an equivalent expression with a denominator of 2.

$= \dfrac{\sqrt[3]{7x}}{2} + \dfrac{4\sqrt[3]{7x}}{2}$

$= \dfrac{5\sqrt[3]{7x}}{2}$    Add.

■ Work Practice 9–10

## Objective B Multiplying Radical Expressions ▶

We can multiply radical expressions by using many of the same properties used to multiply polynomial expressions. For instance, to multiply $\sqrt{2}(\sqrt{6} - 3\sqrt{2})$, we use the distributive property and multiply $\sqrt{2}$ by each term inside the parentheses.

$\sqrt{2}(\sqrt{6} - 3\sqrt{2}) = \sqrt{2}(\sqrt{6}) - \sqrt{2}(3\sqrt{2})$    Use the distributive property.

$= \sqrt{2 \cdot 6} - 3\sqrt{2 \cdot 2}$

$= \sqrt{2 \cdot 2 \cdot 3} - 3 \cdot 2$    Use the product rule for radicals.

$= 2\sqrt{3} - 6$

**Answers**

**9.** $\dfrac{\sqrt{3}}{18}$    **10.** $\dfrac{13\sqrt[3]{5x}}{3}$

**Example 11**  Multiply:  $\sqrt{3}(5 + \sqrt{30})$

**Solution:**

$$\sqrt{3}(5 + \sqrt{30}) = \sqrt{3}(5) + \sqrt{3}(\sqrt{30})$$
$$= 5\sqrt{3} + \sqrt{3 \cdot 30}$$
$$= 5\sqrt{3} + \sqrt{3 \cdot 3 \cdot 10}$$
$$= 5\sqrt{3} + 3\sqrt{10}$$

■ **Work Practice 11**

**Practice 11**

Multiply:  $\sqrt{2}(6 + \sqrt{10})$

**Examples**  Multiply. Assume that all variables represent positive real numbers.

**12.** $(\sqrt{5} - \sqrt{6})(\sqrt{7} + 1) = \overset{\text{First}}{\sqrt{5} \cdot \sqrt{7}} + \overset{\text{Outer}}{\sqrt{5} \cdot 1} - \overset{\text{Inner}}{\sqrt{6} \cdot \sqrt{7}} - \overset{\text{Last}}{\sqrt{6} \cdot 1}$

                                      Using the FOIL order.

               $= \sqrt{35} + \sqrt{5} - \sqrt{42} - \sqrt{6}$    Simplify.

**13.** $(\sqrt{2x} + 5)(\sqrt{2x} - 5) = (\sqrt{2x})^2 - 5^2$    Multiply the sum and difference of
                                        two terms: $(a + b)(a - b) = a^2 - b^2$

               $= 2x - 25$

**14.** $(\sqrt{3} - 1)^2 = (\sqrt{3})^2 - 2 \cdot \sqrt{3} \cdot 1 + 1^2$    Square the binomial:
                                   $(a - b)^2 = a^2 - 2ab + b^2$

              $= 3 - 2\sqrt{3} + 1$

              $= 4 - 2\sqrt{3}$

**15.** $(\sqrt{x - 3} + 5)^2 = (\sqrt{x - 3})^2 + 2 \cdot \sqrt{x - 3} \cdot 5 + 5^2$    Square the binomial:
          $a$       $b$       $a^2$   $+ 2 \cdot$   $a$   $\cdot b + b^2$   $(a + b)^2 = a^2 + 2ab + b^2$

              $= x - 3 + 10\sqrt{x - 3} + 25$    Simplify.

              $= x + 22 + 10\sqrt{x - 3}$    Combine like terms.

■ **Work Practice 12–15**

**Practice 12–15**

Multiply. Assume that all variables represent positive real numbers.

**12.** $(\sqrt{3} - \sqrt{5})(\sqrt{2} + 7)$
**13.** $(\sqrt{5y} + 2)(\sqrt{5y} - 2)$
**14.** $(\sqrt{3} - 7)^2$
**15.** $(\sqrt{x + 1} + 2)^2$

**Answers**
**11.** $6\sqrt{2} + 2\sqrt{5}$
**12.** $\sqrt{6} + 7\sqrt{3} - \sqrt{10} - 7\sqrt{5}$
**13.** $5y - 4$  **14.** $52 - 14\sqrt{3}$
**15.** $x + 5 + 4\sqrt{x + 1}$

## Vocabulary, Readiness & Video Check

Complete the table with "Like" or "Unlike."

| | Terms | Like or Unlike Radical Terms? |
|---|---|---|
| **1.** | $\sqrt{7}, \sqrt[3]{7}$ | |
| **2.** | $\sqrt[3]{x^2y}, \sqrt[3]{yx^2}$ | |
| **3.** | $\sqrt[3]{abc}, \sqrt[3]{cba}$ | |
| **4.** | $2x\sqrt{5}, 2x\sqrt{10}$ | |

Simplify. Assume that all variables represent positive real numbers.

**5.** $2\sqrt{3} + 4\sqrt{3} =$ _____

**6.** $5\sqrt{7} + 3\sqrt{7} =$ _____

**7.** $8\sqrt{x} - \sqrt{x} =$ _____

**8.** $3\sqrt{y} - \sqrt{y} =$ _____

**9.** $7\sqrt[3]{x} + \sqrt[3]{x} =$ _____

**10.** $8\sqrt[3]{z} + \sqrt[3]{z} =$ _____

Add or subtract if possible.

**11.** $\sqrt{11} + \sqrt[3]{11} =$ _____

**12.** $9\sqrt{13} - \sqrt[4]{13} =$ _____

**13.** $8\sqrt[3]{2x} + 3\sqrt[3]{2x} - \sqrt[3]{2x} =$ _____

**14.** $8\sqrt[3]{2x} + 3\sqrt[3]{2x^2} - \sqrt[3]{2x} =$ _____

*Martin-Gay Interactive Videos*     *Watch the section lecture video and answer the following questions.*

**Objective A** **15.** From ▭ Examples 1 and 2, why should you always check to see if all terms in your expression are simplified before attempting to add or subtract radicals? ▶

**Objective B** **16.** In ▭ Example 4, what are you told to remember about the square root of a positive number times the square root of the same positive number? ▶

*See Video 7.4* 🍒

## 7.4 Exercise Set MyMathLab® ▶

**Objective A** *Add or subtract as indicated. Assume that all variables represent positive real numbers. See Examples 1 through 10.*

**1.** $\sqrt{8} - \sqrt{32}$

**2.** $\sqrt{27} - \sqrt{75}$

**3.** $2\sqrt{2x^3} + 4x\sqrt{8x}$

**4.** $3\sqrt{45x^3} + x\sqrt{5x}$

▶**5.** $2\sqrt{50} - 3\sqrt{125} + \sqrt{98}$

**6.** $4\sqrt{32} - \sqrt{18} + 2\sqrt{128}$

**7.** $\sqrt[3]{16x} - \sqrt[3]{54x}$

**8.** $2\sqrt[3]{3a^4} - 3a\sqrt[3]{81a}$

**9.** $\sqrt{9b^3} - \sqrt{25b^3} + \sqrt{49b^3}$

**10.** $\sqrt{4x^7} + 9x^2\sqrt{x^3} - 5x\sqrt{x^5}$

**11.** $\dfrac{5\sqrt{2}}{3} + \dfrac{2\sqrt{2}}{5}$

**12.** $\dfrac{\sqrt{3}}{2} + \dfrac{4\sqrt{3}}{3}$

▶**13.** $\sqrt[3]{\dfrac{11}{8}} - \dfrac{\sqrt[3]{11}}{6}$

**14.** $\dfrac{2\sqrt[3]{4}}{7} - \dfrac{\sqrt[3]{4}}{14}$

**15.** $\dfrac{\sqrt{20x}}{9} + \sqrt{\dfrac{5x}{9}}$

**16.** $\dfrac{3x\sqrt{7}}{5} + \sqrt{\dfrac{7x^2}{100}}$

**17.** $7\sqrt{9} - 7 + \sqrt{3}$

**18.** $\sqrt{16} - 5\sqrt{10} + 7$

**19.** $2 + 3\sqrt{y^2} - 6\sqrt{y^2} + 5$

**20.** $3\sqrt{7} - \sqrt[3]{x} + 4\sqrt{7} - 3\sqrt[3]{x}$

**21.** $3\sqrt{108} - 2\sqrt{18} - 3\sqrt{48}$

**22.** $-\sqrt{75} + \sqrt{12} - 3\sqrt{3}$

**23.** $-5\sqrt[3]{625} + \sqrt[3]{40}$

**24.** $-2\sqrt[3]{108} - \sqrt[3]{32}$

**25.** $3\sqrt{a^5b^7} - ab\sqrt{25a^3b^5} + a^2\sqrt{16ab^7}$

**26.** $\sqrt{4x^7y^5} + 9x^2\sqrt{x^3y^5} - 5xy\sqrt{x^5y^3}$

**27.** $5y\sqrt{8y} + 2\sqrt{50y^3}$

**28.** $3\sqrt{8x^2y^3} - 2x\sqrt{32y^3}$

**29.** $\sqrt[3]{54xy^3} - 5\sqrt[3]{2xy^3} + y\sqrt[3]{128x}$

**30.** $2\sqrt[3]{24x^3y^4} + 4x\sqrt[3]{81y^4}$

**31.** $6\sqrt[3]{11} + 8\sqrt{11} - 12\sqrt{11}$

**32.** $3\sqrt[3]{5} + 4\sqrt{5} - 2\sqrt{5}$

**33.** $-2\sqrt[4]{x^7} + 3\sqrt[4]{16x^7}$

**34.** $6\sqrt[3]{24x^3} - 2\sqrt[3]{81x^3} - x\sqrt[3]{3}$

**35.** $\dfrac{4\sqrt{3}}{3} - \dfrac{\sqrt{12}}{3}$

**36.** $\dfrac{\sqrt{45}}{10} + \dfrac{7\sqrt{5}}{10}$

**37.** $\dfrac{\sqrt[3]{8x^4}}{7} + \dfrac{3x\sqrt[3]{x}}{7}$

**38.** $\dfrac{\sqrt[4]{48}}{5x} - \dfrac{2\sqrt[4]{3}}{10x}$

**39.** $\sqrt{\dfrac{28}{x^2}} + \sqrt{\dfrac{7}{4x^2}}$

**40.** $\dfrac{\sqrt{99}}{5x} - \sqrt{\dfrac{44}{x^2}}$

**41.** $\sqrt[3]{\dfrac{16}{27}} - \dfrac{\sqrt[3]{54}}{6}$

**42.** $\dfrac{\sqrt[3]{3}}{10} + \sqrt[3]{\dfrac{24}{125}}$

**43.** $-\dfrac{\sqrt[3]{2x^4}}{9} + \sqrt[3]{\dfrac{250x^4}{27}}$

**44.** $\dfrac{\sqrt[3]{y^5}}{8} + \dfrac{5y\sqrt[3]{y^2}}{4}$

△ **45.** Find the perimeter of the trapezoid.

△ **46.** Find the perimeter of the triangle.

**Objective B**  *Multiply. Then simplify if possible. Assume that all variables represent positive real numbers. See Examples 11 through 15.*

**▶ 47.** $\sqrt{7}(\sqrt{5} + \sqrt{3})$

**48.** $\sqrt{2}(\sqrt{15} - \sqrt{35})$

**49.** $(\sqrt{5} - \sqrt{2})^2$

**50.** $(3x - \sqrt{2})(3x - \sqrt{2})$

**51.** $\sqrt{3x}(\sqrt{3} - \sqrt{x})$

**52.** $\sqrt{5y}(\sqrt{y} + \sqrt{5})$

**53.** $(2\sqrt{x} - 5)(3\sqrt{x} + 1)$

**54.** $(8\sqrt{y} + z)(4\sqrt{y} - 1)$

**55.** $(\sqrt[3]{a} - 4)(\sqrt[3]{a} + 5)$

**56.** $(\sqrt[3]{a} + 2)(\sqrt[3]{a} + 7)$

**57.** $6(\sqrt{2} - 2)$

**58.** $\sqrt{5}(6 - \sqrt{5})$

**59.** $\sqrt{2}(\sqrt{2} + x\sqrt{6})$

**60.** $\sqrt{3}(\sqrt{3} - 2\sqrt{5x})$

**▶ 61.** $(2\sqrt{7} + 3\sqrt{5})(\sqrt{7} - 2\sqrt{5})$

**62.** $(\sqrt{x} - y)(\sqrt{x} + y)$

**63.** $(\sqrt{6} - 4\sqrt{2})(3\sqrt{6} + 1)$

**64.** $(3\sqrt{x} + 2)(\sqrt{3x} - 2)$

**65.** $(\sqrt{3} + x)^2$

**66.** $(\sqrt{y} - 3x)^2$

**67.** $(\sqrt{5x} - 3\sqrt{2})(\sqrt{5x} - 3\sqrt{3})$

**68.** $(5\sqrt{3x} - \sqrt{y})(4\sqrt{x} + 1)$

**69.** $(\sqrt[3]{4} + 2)(\sqrt[3]{2} - 1)$

**70.** $(\sqrt[3]{3} + \sqrt[3]{2})(\sqrt[3]{9} - \sqrt[3]{4})$

**71.** $(\sqrt[3]{x} + 1)(\sqrt[3]{x} - 4\sqrt{x} + 7)$

**72.** $(\sqrt[3]{3x} + 3)(\sqrt[3]{2x} - 3x - 1)$

**73.** $(\sqrt{x-1} + 5)^2$

**74.** $(\sqrt{3x+1} + 2)^2$

**75.** $(\sqrt{2x+5} - 1)^2$

**76.** $(\sqrt{x-6} - 7)^2$

## Review

*Factor each numerator and denominator. Then simplify if possible. See Section 6.1.*

**77.** $\dfrac{2x - 14}{2}$

**78.** $\dfrac{8x - 24y}{4}$

**79.** $\dfrac{7x - 7y}{x^2 - y^2}$

**80.** $\dfrac{x^3 - 8}{4x - 8}$

**81.** $\dfrac{6a^2b - 9ab}{3ab}$

**82.** $\dfrac{14r - 28r^2s^2}{7rs}$

**83.** $\dfrac{-4 + 2\sqrt{3}}{6}$

**84.** $\dfrac{-5 + 10\sqrt{7}}{5}$

## Concept Extensions

△ **85.** Find the perimeter and area of the rectangle.

△ **86.** Find the perimeter and area of the trapezoid. (*Hint:* The area of a trapezoid is the product of half the height $6\sqrt{3}$ meters and the sum of the bases $2\sqrt{63}$ and $7\sqrt{7}$ meters.)

**87. a.** Add: $\sqrt{3} + \sqrt{3}$
  **b.** Multiply: $\sqrt{3} \cdot \sqrt{3}$
  **c.** Describe the differences in parts (a) and (b).

**88. a.** Add: $2\sqrt{5} + \sqrt{5}$
  **b.** Multiply: $2\sqrt{5} \cdot \sqrt{5}$
  **c.** Describe the differences in parts (a) and (b).

**89.** Multiply: $(\sqrt{2} + \sqrt{3} - 1)^2$

**90.** Multiply: $(\sqrt{5} - \sqrt{2} + 1)^2$

## 7.5 Rationalizing Numerators and Denominators of Radical Expressions

### Objective A Rationalizing Denominators

Often in mathematics it is helpful to write a radical expression such as $\dfrac{\sqrt{3}}{\sqrt{2}}$ either without a radical in the denominator or without a radical in the numerator. The process of writing this expression as an equivalent expression but without a radical in the denominator is called **rationalizing the denominator**. To rationalize the denominator of $\dfrac{\sqrt{3}}{\sqrt{2}}$, we multiply the numerator and the denominator by $\sqrt{2}$. Recall that this is the same as multiplying by $\dfrac{\sqrt{2}}{\sqrt{2}}$, which simplifies to 1.

$$\frac{\sqrt{3}}{\sqrt{2}} = \frac{\sqrt{3} \cdot \sqrt{2}}{\sqrt{2} \cdot \sqrt{2}} = \frac{\sqrt{6}}{\sqrt{4}} = \frac{\sqrt{6}}{2}$$

**Objectives**

A Rationalize Denominators.

B Rationalize Denominators Having Two Terms.

C Rationalize Numerators.

**Example 1** Rationalize the denominator of $\dfrac{2}{\sqrt{5}}$.

**Solution:** To rationalize the denominator, we multiply the numerator and denominator by a factor that makes the radicand in the denominator a perfect square.

$$\frac{2}{\sqrt{5}} = \frac{2 \cdot \sqrt{5}}{\sqrt{5} \cdot \sqrt{5}} = \frac{2\sqrt{5}}{5} \qquad \text{The denominator is now rationalized.}$$

◼ Work Practice 1

**Practice 1**

Rationalize the denominator of $\dfrac{7}{\sqrt{2}}$.

**Example 2** Rationalize the denominator of $\dfrac{2\sqrt{16}}{\sqrt{9x}}$.

**Solution:** First we simplify the radicals; then we rationalize the denominator.

$$\frac{2\sqrt{16}}{\sqrt{9x}} = \frac{2(4)}{\sqrt{9} \cdot \sqrt{x}} = \frac{8}{3\sqrt{x}}$$

To rationalize the denominator, we multiply the numerator and the denominator by $\sqrt{x}$.

$$\frac{8}{3\sqrt{x}} = \frac{8 \cdot \sqrt{x}}{3\sqrt{x} \cdot \sqrt{x}} = \frac{8\sqrt{x}}{3x}$$

◼ Work Practice 2

**Practice 2**

Rationalize the denominator of $\dfrac{2\sqrt{9}}{\sqrt{16y}}$.

**Example 3** Rationalize the denominator of $\sqrt[3]{\dfrac{1}{2}}$.

**Solution:** $\sqrt[3]{\dfrac{1}{2}} = \dfrac{\sqrt[3]{1}}{\sqrt[3]{2}} = \dfrac{1}{\sqrt[3]{2}}$

Now we rationalize the denominator. Since $\sqrt[3]{2}$ is a cube root, we want to multiply by a value that will make the radicand 2 a perfect cube. If we multiply by $\sqrt[3]{2^2}$, we get $\sqrt[3]{2^3} = 2$. Thus,

$$\frac{1 \cdot \sqrt[3]{2^2}}{\sqrt[3]{2} \cdot \sqrt[3]{2^2}} = \frac{\sqrt[3]{4}}{\sqrt[3]{2^3}} = \frac{\sqrt[3]{4}}{2} \qquad \text{Multiply numerator and denominator by } \sqrt[3]{2^2} \text{ and then simplify.}$$

◼ Work Practice 3

**Practice 3**

Rationalize the denominator of $\sqrt[3]{\dfrac{2}{25}}$.

**Answers**

1. $\dfrac{7\sqrt{2}}{2}$  2. $\dfrac{3\sqrt{y}}{2y}$  3. $\dfrac{\sqrt[3]{10}}{5}$

✔**Concept Check** Determine by which number both the numerator and denominator should be multiplied to rationalize the denominator of the radical expression.

**a.** $\dfrac{1}{\sqrt[3]{7}}$    **b.** $\dfrac{1}{\sqrt[4]{8}}$

## Practice 4

Rationalize the denominator of $\sqrt{\dfrac{5m}{11n}}$. Assume that all variables represent positive real numbers.

**Example 4** Rationalize the denominator of $\sqrt{\dfrac{7x}{3y}}$. Assume that all variables represent positive real numbers.

**Solution:**

$$\sqrt{\dfrac{7x}{3y}} = \dfrac{\sqrt{7x}}{\sqrt{3y}}$$ 
  Use the quotient rule. No radical may be simplified further.

$$= \dfrac{\sqrt{7x} \cdot \sqrt{3y}}{\sqrt{3y} \cdot \sqrt{3y}}$$ 
  Multiply numerator and denominator by $\sqrt{3y}$ so that the radicand in the denominator is a perfect square.

$$= \dfrac{\sqrt{21xy}}{3y}$$ 
  Use the product rule in the numerator and denominator. Remember that $\sqrt{3y} \cdot \sqrt{3y} = 3y$.

■ **Work Practice 4**

## Practice 5

Rationalize the denominator of $\dfrac{\sqrt[5]{a^2}}{\sqrt[5]{32b^{12}}}$. Assume that all variables represent positive real numbers.

**Example 5** Rationalize the denominator of $\dfrac{\sqrt[4]{x}}{\sqrt[4]{81y^5}}$. Assume that all variables represent positive real numbers.

**Solution:** First we simplify each radical if possible.

$$\dfrac{\sqrt[4]{x}}{\sqrt[4]{81y^5}} = \dfrac{\sqrt[4]{x}}{\sqrt[4]{81y^4} \cdot \sqrt[4]{y}}$$ 
  Use the product rule in the denominator.

$$= \dfrac{\sqrt[4]{x}}{3y\sqrt[4]{y}}$$ 
  Write $\sqrt[4]{81y^4}$ as $3y$.

$$= \dfrac{\sqrt[4]{x} \cdot \sqrt[4]{y^3}}{3y\sqrt[4]{y} \cdot \sqrt[4]{y^3}}$$ 
  Multiply numerator and denominator by $\sqrt[4]{y^3}$ so that the radicand in the denominator is a perfect 4th power.

$$= \dfrac{\sqrt[4]{xy^3}}{3y\sqrt[4]{y^4}}$$ 
  Use the product rule in the numerator and denominator.

$$= \dfrac{\sqrt[4]{xy^3}}{3y^2}$$ 
  In the denominator, $\sqrt[4]{y^4} = y$ and $3y \cdot y = 3y^2$.

■ **Work Practice 5**

## Objective B Rationalizing Denominators Having Two Terms ▶

Remember the product of the sum and difference of two terms?

$$(a + b)(a - b) = a^2 - b^2$$

These two expressions are called **conjugates** of each other.

**Answers**

**4.** $\dfrac{\sqrt{55mn}}{11n}$  **5.** $\dfrac{\sqrt[5]{a^2b^3}}{2b^3}$

✔**Concept Check Answers**

**a.** $\sqrt[3]{7^2}$ or $\sqrt[3]{49}$  **b.** $\sqrt[4]{2}$

To rationalize a denominator that is a sum or difference of two terms, we use conjugates. To see how and why this works, let's rationalize the denominator of the expression $\dfrac{5}{\sqrt{3}-2}$. To do so, we multiply both the numerator and the denominator by $\sqrt{3}+2$, the *conjugate* of the denominator $\sqrt{3}-2$, and see what happens.

$$\frac{5}{\sqrt{3}-2}=\frac{5(\sqrt{3}+2)}{(\sqrt{3}-2)(\sqrt{3}+2)}$$

$$=\frac{5(\sqrt{3}+2)}{(\sqrt{3})^2-2^2} \quad \text{Multiply the sum and difference of two terms: } (a+b)(a-b)=a^2-b^2.$$

$$=\frac{5(\sqrt{3}+2)}{3-4}$$

$$=\frac{5(\sqrt{3}+2)}{-1}$$

$$=-5(\sqrt{3}+2) \text{ or } -5\sqrt{3}-10$$

Notice in the denominator that the product of $(\sqrt{3}-2)$ and its conjugate, $(\sqrt{3}+2)$, is $-1$. In general, the product of an expression and its conjugate will contain no radical terms. This is why, when rationalizing a denominator or a numerator containing two terms, we multiply by its conjugate. Examples of conjugates are

$$\sqrt{a}-\sqrt{b} \quad \text{and} \quad \sqrt{a}+\sqrt{b}$$
$$x+\sqrt{y} \quad \text{and} \quad x-\sqrt{y}$$

**Example 6** Rationalize the denominator of $\dfrac{2}{3\sqrt{2}+4}$.

**Solution:** We multiply the numerator and the denominator by the conjugate of $3\sqrt{2}+4$.

$$\frac{2}{3\sqrt{2}+4}=\frac{2(3\sqrt{2}-4)}{(3\sqrt{2}+4)(3\sqrt{2}-4)}$$

$$=\frac{2(3\sqrt{2}-4)}{(3\sqrt{2})^2-4^2} \quad \text{Multiply the sum and difference of two terms: } (a+b)(a-b)=a^2-b^2.$$

$$=\frac{2(3\sqrt{2}-4)}{18-16} \quad \text{Write } (3\sqrt{2})^2 \text{ as } 9\cdot2 \text{ or } 18 \text{ and } 4^2 \text{ as } 16.$$

$$=\frac{2(3\sqrt{2}-4)}{2}=3\sqrt{2}-4$$

■ **Work Practice 6**

As we saw in Example 6, it is often helpful to leave a numerator in factored form to help determine whether the expression can be simplified.

**Practice 6**

Rationalize the denominator of $\dfrac{3}{2\sqrt{5}+1}$.

**Practice 7**

Rationalize the denominator of $\dfrac{\sqrt{5}+3}{\sqrt{3}-\sqrt{2}}$.

**Example 7** Rationalize the denominator of $\dfrac{\sqrt{6}+2}{\sqrt{5}-\sqrt{3}}$.

**Solution:** We multiply the numerator and the denominator by the conjugate of $\sqrt{5}-\sqrt{3}$.

$$\frac{\sqrt{6}+2}{\sqrt{5}-\sqrt{3}} = \frac{(\sqrt{6}+2)(\sqrt{5}+\sqrt{3})}{(\sqrt{5}-\sqrt{3})(\sqrt{5}+\sqrt{3})}$$

$$= \frac{\sqrt{6}\sqrt{5}+\sqrt{6}\sqrt{3}+2\sqrt{5}+2\sqrt{3}}{(\sqrt{5})^2-(\sqrt{3})^2}$$

$$= \frac{\sqrt{30}+\sqrt{18}+2\sqrt{5}+2\sqrt{3}}{5-3}$$

$$= \frac{\sqrt{30}+3\sqrt{2}+2\sqrt{5}+2\sqrt{3}}{2}$$

■ Work Practice 7

**Practice 8**

Rationalize the denominator of $\dfrac{3}{2-\sqrt{x}}$. Assume that all variables represent positive real numbers.

**Example 8** Rationalize the denominator of $\dfrac{2\sqrt{m}}{3\sqrt{x}+\sqrt{m}}$. Assume that all variables represent positive real numbers.

**Solution:** We multiply by the conjugate of $3\sqrt{x}+\sqrt{m}$ to eliminate the radicals from the denominator.

$$\frac{2\sqrt{m}}{3\sqrt{x}+\sqrt{m}} = \frac{2\sqrt{m}(3\sqrt{x}-\sqrt{m})}{(3\sqrt{x}+\sqrt{m})(3\sqrt{x}-\sqrt{m})} = \frac{6\sqrt{mx}-2m}{(3\sqrt{x})^2-(\sqrt{m})^2}$$

$$= \frac{6\sqrt{mx}-2m}{9x-m}$$

■ Work Practice 8

## Objective C Rationalizing Numerators

As mentioned earlier, it is also often helpful to write an expression such as $\dfrac{\sqrt{3}}{\sqrt{2}}$ as an equivalent expression without a radical in the numerator. This process is called **rationalizing the numerator.** To rationalize the numerator of $\dfrac{\sqrt{3}}{\sqrt{2}}$, we multiply the numerator and the denominator by $\sqrt{3}$.

$$\frac{\sqrt{3}}{\sqrt{2}} = \frac{\sqrt{3}\cdot\sqrt{3}}{\sqrt{2}\cdot\sqrt{3}} = \frac{\sqrt{9}}{\sqrt{6}} = \frac{3}{\sqrt{6}}$$

**Practice 9**

Rationalize the numerator of $\dfrac{\sqrt{18}}{\sqrt{75}}$.

**Example 9** Rationalize the numerator of $\dfrac{\sqrt{7}}{\sqrt{45}}$.

**Solution:** First we simplify $\sqrt{45}$.

$$\frac{\sqrt{7}}{\sqrt{45}} = \frac{\sqrt{7}}{\sqrt{9\cdot5}} = \frac{\sqrt{7}}{3\sqrt{5}}$$

Next we rationalize the numerator by multiplying the numerator and the denominator by $\sqrt{7}$.

$$\frac{\sqrt{7}}{3\sqrt{5}} = \frac{\sqrt{7}\cdot\sqrt{7}}{3\sqrt{5}\cdot\sqrt{7}} = \frac{7}{3\sqrt{5}\cdot7} = \frac{7}{3\sqrt{35}}$$

■ Work Practice 9

**Answers**

**7.** $\sqrt{15}+\sqrt{10}+3\sqrt{3}+3\sqrt{2}$

**8.** $\dfrac{6+3\sqrt{x}}{4-x}$  **9.** $\dfrac{6}{5\sqrt{6}}$

**Example 10** Rationalize the numerator of $\dfrac{\sqrt[3]{2x^2}}{\sqrt[3]{5y}}$.

**Solution:**

$$\dfrac{\sqrt[3]{2x^2}}{\sqrt[3]{5y}} = \dfrac{\sqrt[3]{2x^2} \cdot \sqrt[3]{2^2 x}}{\sqrt[3]{5y} \cdot \sqrt[3]{2^2 x}}$$    Multiply the numerator and denominator by $\sqrt[3]{2^2 x}$ so that the radicand in the numerator is a perfect cube.

$$= \dfrac{\sqrt[3]{2^3 x^3}}{\sqrt[3]{5y \cdot 2^2 x}}$$    Use the product rule in the numerator and denominator.

$$= \dfrac{2x}{\sqrt[3]{20xy}}$$    Simplify.

■ **Work Practice 10**

**Practice 10**

Rationalize the numerator of $\dfrac{\sqrt[3]{3a}}{\sqrt[3]{7b}}$.

Just as for denominators, to rationalize a numerator that is a sum or difference of two terms, we use conjugates.

**Example 11** Rationalize the numerator of $\dfrac{\sqrt{x} + 2}{5}$. Assume that all variables represent positive real numbers.

**Solution:** We multiply the numerator and the denominator by the conjugate of $\sqrt{x} + 2$, the numerator.

$$\dfrac{\sqrt{x} + 2}{5} = \dfrac{(\sqrt{x} + 2)(\sqrt{x} - 2)}{5(\sqrt{x} - 2)}$$    Multiply by $\sqrt{x} - 2$, the conjugate of $\sqrt{x} + 2$.

$$= \dfrac{(\sqrt{x})^2 - 2^2}{5(\sqrt{x} - 2)}$$    $(a + b)(a - b) = a^2 - b^2$.

$$= \dfrac{x - 4}{5(\sqrt{x} - 2)}$$

■ **Work Practice 11**

**Practice 11**

Rationalize the numerator of $\dfrac{\sqrt{x} + 5}{3}$. Assume that all variables represent positive real numbers.

**Answers**

**10.** $\dfrac{3a}{\sqrt[3]{63a^2 b}}$    **11.** $\dfrac{x - 25}{3(\sqrt{x} - 5)}$

## Vocabulary, Readiness & Video Check

*Use the choices below to fill in each blank. Not all choices will be used.*

     rationalizing the numerator        conjugate        $\dfrac{\sqrt{3}}{\sqrt{3}}$

     rationalizing the denominator        $\dfrac{5}{5}$

**1.** The _____ of $a + b$ is $a - b$.

**2.** The process of writing an equivalent expression, but without a radical in the denominator, is called _____.

**3.** The process of writing an equivalent expression, but without a radical in the numerator, is called _____.

**4.** To rationalize the denominator of $\dfrac{5}{\sqrt{3}}$, we multiply by _____.

*Find the conjugate of each expression.*

**5.** $\sqrt{2} + x$        **6.** $\sqrt{3} + y$        **7.** $5 - \sqrt{a}$        **8.** $6 - \sqrt{b}$

**9.** $-7\sqrt{5} + 8\sqrt{x}$        **10.** $-9\sqrt{2} - 6\sqrt{y}$

**Martin-Gay Interactive Videos**    *Watch the section lecture video and answer the following questions.*

**Objective A** **11.** From ⊞ Examples 1–3, what is the goal of rationalizing a denominator?

**Objective B** **12.** From ⊞ Example 4, why will multiplying a denominator by its conjugate always rationalize the denominator?

**Objective C** **13.** From ⊞ Example 5, is the process of rationalizing a numerator any different from rationalizing a denominator?

*See Video 7.5*

## 7.5 Exercise Set MyMathLab®

**Objective A** *Rationalize each denominator. Assume that all variables represent positive real numbers. See Examples 1 through 5.*

**1.** $\dfrac{\sqrt{2}}{\sqrt{7}}$

**2.** $\dfrac{\sqrt{5}}{\sqrt{2}}$

**3.** $\sqrt{\dfrac{1}{5}}$

**4.** $\sqrt{\dfrac{1}{2}}$

**5.** $\dfrac{4}{\sqrt[3]{3}}$

**6.** $\dfrac{6}{\sqrt[3]{9}}$

**7.** $\dfrac{3}{\sqrt{8x}}$

**8.** $\dfrac{5}{\sqrt{27a}}$

**9.** $\dfrac{3}{\sqrt[3]{4x^2}}$

**10.** $\dfrac{5}{\sqrt[3]{3y}}$

**11.** $\dfrac{9}{\sqrt{3a}}$

**12.** $\dfrac{x}{\sqrt{5}}$

**13.** $\dfrac{3}{\sqrt[3]{2}}$

**14.** $\dfrac{5}{\sqrt[3]{9}}$

**15.** $\dfrac{2\sqrt{3}}{\sqrt{7}}$

**16.** $\dfrac{-5\sqrt{2}}{\sqrt{11}}$

**17.** $\sqrt{\dfrac{2x}{5y}}$

**18.** $\sqrt{\dfrac{13a}{2b}}$

**19.** $\sqrt[3]{\dfrac{3}{5}}$

**20.** $\sqrt[3]{\dfrac{7}{10}}$

**21.** $\sqrt{\dfrac{3x}{50}}$

**22.** $\sqrt{\dfrac{11y}{45}}$

**23.** $\dfrac{1}{\sqrt{12z}}$

**24.** $\dfrac{1}{\sqrt{32x}}$

**25.** $\dfrac{\sqrt[3]{2y^2}}{\sqrt[3]{9x^2}}$

**26.** $\dfrac{\sqrt[3]{3x}}{\sqrt[3]{4y^4}}$

**27.** $\sqrt[4]{\dfrac{16}{9x^7}}$

**28.** $\sqrt[5]{\dfrac{32}{m^6n^{13}}}$

**29.** $\dfrac{5a}{\sqrt[5]{8a^9b^{11}}}$

**30.** $\dfrac{9y}{\sqrt[4]{4y^9}}$

**Objective B** *Rationalize each denominator. Assume that all variables represent positive real numbers. See Examples 6 through 8.*

**31.** $\dfrac{6}{2-\sqrt{7}}$

**32.** $\dfrac{3}{\sqrt{7}-4}$

**33.** $\dfrac{-7}{\sqrt{x}-3}$

**34.** $\dfrac{-8}{\sqrt{y}+4}$

**35.** $\dfrac{\sqrt{2}-\sqrt{3}}{\sqrt{2}+\sqrt{3}}$

**36.** $\dfrac{\sqrt{3}+\sqrt{4}}{\sqrt{2}+\sqrt{3}}$

**37.** $\dfrac{\sqrt{a} + 1}{2\sqrt{a} - \sqrt{b}}$

**38.** $\dfrac{2\sqrt{a} - 3}{2\sqrt{a} - \sqrt{b}}$

**39.** $\dfrac{8}{1 + \sqrt{10}}$

**40.** $\dfrac{-3}{\sqrt{6} - 2}$

**41.** $\dfrac{\sqrt{x}}{\sqrt{x} + \sqrt{y}}$

**42.** $\dfrac{2\sqrt{a}}{2\sqrt{x} - \sqrt{y}}$

**43.** $\dfrac{2\sqrt{3} + \sqrt{6}}{4\sqrt{3} - \sqrt{6}}$

**44.** $\dfrac{4\sqrt{5} + \sqrt{2}}{2\sqrt{5} - \sqrt{2}}$

**Objective C** *Rationalize each numerator. Assume that all variables represent positive real numbers. See Examples 9 and 10.*

**45.** $\sqrt{\dfrac{5}{3}}$

**46.** $\sqrt{\dfrac{3}{2}}$

**47.** $\sqrt{\dfrac{18}{5}}$

**48.** $\sqrt{\dfrac{12}{7}}$

**49.** $\dfrac{\sqrt{4x}}{7}$

**50.** $\dfrac{\sqrt{3x^5}}{6}$

**51.** $\dfrac{\sqrt[3]{5y^2}}{\sqrt[3]{4x}}$

**52.** $\dfrac{\sqrt[3]{4x}}{\sqrt[3]{z^4}}$

**53.** $\sqrt{\dfrac{2}{5}}$

**54.** $\sqrt{\dfrac{3}{7}}$

**55.** $\dfrac{\sqrt{2x}}{11}$

**56.** $\dfrac{\sqrt{y}}{7}$

**57.** $\sqrt[3]{\dfrac{7}{8}}$

**58.** $\sqrt[3]{\dfrac{25}{2}}$

**59.** $\dfrac{\sqrt[3]{3x^5}}{10}$

**60.** $\sqrt[3]{\dfrac{9y}{7}}$

**61.** $\sqrt{\dfrac{18x^4y^6}{3z}}$

**62.** $\sqrt{\dfrac{8x^5y}{2z}}$

**63.** When rationalizing the denominator of $\dfrac{\sqrt{5}}{\sqrt{7}}$, explain why both the numerator and the denominator must be multiplied by $\sqrt{7}$.

**64.** When rationalizing the numerator of $\dfrac{\sqrt{5}}{\sqrt{7}}$, explain why both the numerator and the denominator must be multiplied by $\sqrt{5}$.

*Rationalize each numerator. Assume that all variables represent positive real numbers. See Example 11.*

**65.** $\dfrac{2 - \sqrt{11}}{6}$

**66.** $\dfrac{\sqrt{15} + 1}{2}$

**67.** $\dfrac{2 - \sqrt{7}}{-5}$

**68.** $\dfrac{\sqrt{5} + 2}{\sqrt{2}}$

**69.** $\dfrac{\sqrt{x} + 3}{\sqrt{x}}$

**70.** $\dfrac{5 + \sqrt{2}}{\sqrt{2x}}$

**71.** $\dfrac{\sqrt{x} + 1}{\sqrt{x} - 1}$

**72.** $\dfrac{\sqrt{x} + \sqrt{y}}{\sqrt{x} - \sqrt{y}}$

## Review

*Solve each equation. See Sections 2.1 and 5.7.*

**73.** $2x - 7 = 3(x - 4)$

**74.** $9x - 4 = 7(x - 2)$

**75.** $(x - 6)(2x + 1) = 0$

**76.** $(y + 2)(5y + 4) = 0$

**77.** $x^2 - 8x = -12$

**78.** $x^3 = x$

## Concept Extensions

△ **79.** The formula of the radius $r$ of a sphere with surface area $A$ is

$$r = \sqrt{\frac{A}{4\pi}}$$

Rationalize the denominator of the radical expression in this formula.

△ **80.** The formula for the radius $r$ of a cone with height 7 centimeters and volume $V$ is

$$r = \sqrt{\frac{3V}{7\pi}}$$

Rationalize the numerator of the radical expression in this formula.

**81.** Explain why rationalizing the denominator does not change the value of the original expression.

**82.** Explain why rationalizing the numerator does not change the value of the original expression.

**83.** Given $\dfrac{\sqrt{5y^3}}{\sqrt{12x^3}}$, rationalize the denominator by following parts (a) and (b).

    **a.** Multiply the numerator and denominator by $\sqrt{12x^3}$.

    **b.** Multiply the numerator and denominator by $\sqrt{3x}$.

    **c.** What can you conclude from parts (a) and (b)?

**84.** Given $\dfrac{\sqrt[3]{5y}}{\sqrt[3]{4}}$, rationalize the denominator by following parts (a) and (b).

    **a.** Multiply the numerator and denominator by $\sqrt[3]{16}$.

    **b.** Multiply the numerator and denominator by $\sqrt[3]{2}$.

    **c.** What can you conclude from parts (a) and (b)?

*Determine the smallest number both the numerator and denominator should be multiplied by to rationalize the denominator of the radical expression. See the Concept Check in this section.*

**85.** $\dfrac{9}{\sqrt[3]{5}}$

**86.** $\dfrac{5}{\sqrt{27}}$

# Radicals and Rational Exponents

*Throughout this review, assume that all variables represent positive real numbers. Find each root.*

**1.** $\sqrt{81}$

**2.** $\sqrt[3]{-8}$

**3.** $\sqrt[4]{\dfrac{1}{16}}$

**4.** $\sqrt{x^6}$

**5.** $\sqrt[3]{y^9}$

**6.** $\sqrt{4y^{10}}$

**7.** $\sqrt[5]{-32y^5}$

**8.** $\sqrt[4]{81b^{12}}$

*Use radical notation to rewrite each expression. Simplify if possible.*

**9.** $36^{1/2}$

**10.** $(3y)^{1/4}$

**11.** $64^{-2/3}$

**12.** $(x + 1)^{3/5}$

*Use the properties of exponents to simplify each expression. Write with positive exponents.*

**13.** $y^{-1/6} \cdot y^{7/6}$

**14.** $\dfrac{(2x^{1/3})^4}{x^{5/6}}$

**15.** $\dfrac{x^{1/4}x^{3/4}}{x^{-1/4}}$

**16.** $4^{1/3} \cdot 4^{2/5}$

*Use rational exponents to simplify each radical.*

**17.** $\sqrt[3]{8x^6}$

**18.** $\sqrt[12]{a^9b^6}$

*Use rational exponents to write each as a single radical expression.*

**19.** $\sqrt[4]{x} \cdot \sqrt{x}$

**20.** $\sqrt{5} \cdot \sqrt[3]{2}$

*Simplify.*

**21.** $\sqrt{40}$

**22.** $\sqrt[4]{16x^7y^{10}}$

**23.** $\sqrt[3]{54x^4}$

**24.** $\sqrt[5]{-64b^{10}}$

1. _____

2. _____

3. _____

4. _____

5. _____

6. _____

7. _____

8. _____

9. _____

10. _____

11. _____

12. _____

13. _____

14. _____

15. _____

16. _____

17. _____

18. _____

19. _____

20. _____

21. _____

22. _____

23. _____

24. _____

| | |
|---|---|
| 25. _____ | *Multiply or divide. Then simplify if possible.* |

**25.** $\sqrt{5} \cdot \sqrt{x}$          **26.** $\sqrt[3]{8x} \cdot \sqrt[3]{8x^2}$

26. _____

27. _____

**27.** $\dfrac{\sqrt{98y^6}}{\sqrt{2y}}$          **28.** $\dfrac{\sqrt[4]{48a^9b^3}}{\sqrt[4]{ab^3}}$

28. _____

29. _____

*Perform each indicated operation.*

30. _____

**29.** $\sqrt{20} - \sqrt{75} + 5\sqrt{7}$          **30.** $\sqrt[3]{54y^4} - y\sqrt[3]{16y}$

31. _____

32. _____

**31.** $\sqrt{3}(\sqrt{5} - \sqrt{2})$          **32.** $(\sqrt{7} + \sqrt{3})^2$

33. _____

34. _____

**33.** $(2x - \sqrt{5})(2x + \sqrt{5})$          **34.** $(\sqrt{x+1} - 1)^2$

35. _____

*Rationalize each denominator.*

36. _____

**35.** $\sqrt{\dfrac{7}{3}}$          **36.** $\dfrac{5}{\sqrt[3]{2x^2}}$          **37.** $\dfrac{\sqrt{3} - \sqrt{7}}{2\sqrt{3} + \sqrt{7}}$

37. _____

38. _____

*Rationalize each numerator.*

39. _____

**38.** $\sqrt{\dfrac{7}{3}}$          **39.** $\sqrt[3]{\dfrac{9y}{11}}$          **40.** $\dfrac{\sqrt{x} - 2}{\sqrt{x}}$

40. _____

## 7.6 Radical Equations and Problem Solving ▶

### Objective A Solving Equations That Contain Radical Expressions ▶

In this section, we present techniques to solve equations containing radical expressions such as

$$\sqrt{2x - 3} = 9$$

We use the power rule to help us solve these radical equations.

This property *does not* say that raising both sides of an equation to a power yields an equivalent equation. A solution of the new equation *may* or *may not* be a solution of the original equation. Thus, *each solution of the new equation must be checked* to make sure it is a solution of the original equation. Recall that a proposed solution that is not a solution of the original equation is called an extraneous solution.

**Example 1** Solve: $\sqrt{2x - 3} = 9$

**Solution:** We use the power rule to square both sides of the equation to eliminate the radical.

$$\sqrt{2x - 3} = 9$$
$$(\sqrt{2x - 3})^2 = 9^2$$
$$2x - 3 = 81$$
$$2x = 84$$
$$x = 42$$

Now we check the solution in the original equation.

**Check:**
$$\sqrt{2x - 3} = 9$$
$$\sqrt{2(42) - 3} \stackrel{?}{=} 9 \quad \text{Let } x = 42.$$
$$\sqrt{84 - 3} \stackrel{?}{=} 9$$
$$\sqrt{81} \stackrel{?}{=} 9$$
$$9 = 9 \quad \text{True}$$

The solution checks, so we conclude that the solution set is $\{42\}$.

■ Work Practice 1

To solve a radical equation, first isolate a radical on one side of the equation.

**Objectives**

**A** Solve Equations That Contain Radical Expressions. ▶

**B** Use the Pythagorean Theorem to Model Problems. ▶

**Practice 1**

Solve: $\sqrt{3x - 2} = 5$

**Answer**
1. $\{9\}$

The following steps may be used to solve a radical equation.

> ### Solving a Radical Equation
>
> **Step 1:** Isolate one radical on one side of the equation.
> **Step 2:** Raise each side of the equation to a power equal to the index of the radical and simplify.
> **Step 3:** If the equation still contains a radical term, repeat Steps 1 and 2. If not, solve the equation.
> **Step 4:** Check all proposed solutions in the original equation.

**Practice 2**

Solve: $\sqrt{9x - 2} - 2x = 0$

**Example 2** Solve: $\sqrt{-10x - 1} + 3x = 0$

**Solution:** First we isolate the radical on one side of the equation. To do this, we subtract $3x$ from both sides.

$$\sqrt{-10x - 1} + 3x = 0$$
$$\sqrt{-10x - 1} + 3x - 3x = 0 - 3x$$
$$\sqrt{-10x - 1} = -3x$$

Next we use the power rule to eliminate the radical.

$$(\sqrt{-10x - 1})^2 = (-3x)^2$$
$$-10x - 1 = 9x^2$$

Since this is a quadratic equation, we can set the equation equal to 0 and try to solve by factoring.

$$9x^2 + 10x + 1 = 0$$
$$(9x + 1)(x + 1) = 0 \qquad \text{Factor.}$$
$$9x + 1 = 0 \quad \text{or} \quad x + 1 = 0 \qquad \text{Set each factor equal to 0.}$$
$$x = -\frac{1}{9} \qquad\qquad x = -1$$

**Check:**  Let $x = -\frac{1}{9}$.

$$\sqrt{-10x - 1} + 3x = 0$$
$$\sqrt{-10\left(-\frac{1}{9}\right) - 1} + 3\left(-\frac{1}{9}\right) \overset{?}{=} 0$$
$$\sqrt{\frac{10}{9} - \frac{9}{9}} - \frac{3}{9} \overset{?}{=} 0$$
$$\sqrt{\frac{1}{9}} - \frac{1}{3} \overset{?}{=} 0$$
$$\frac{1}{3} - \frac{1}{3} = 0 \quad \text{True}$$

Let $x = -1$.

$$\sqrt{-10x - 1} + 3x = 0$$
$$\sqrt{-10(-1) - 1} + 3(-1) \overset{?}{=} 0$$
$$\sqrt{10 - 1} - 3 \overset{?}{=} 0$$
$$\sqrt{9} - 3 \overset{?}{=} 0$$
$$3 - 3 = 0 \quad \text{True}$$

Both solutions check. The solution set is $\left\{-\frac{1}{9}, -1\right\}$.

■ Work Practice 2

**Practice 3**

Solve: $\sqrt[3]{x - 5} + 2 = 1$

**Example 3** Solve: $\sqrt[3]{x + 1} + 5 = 3$

**Solution:** First we isolate the radical by subtracting 5 from both sides of the equation.

$$\sqrt[3]{x + 1} + 5 = 3$$
$$\sqrt[3]{x + 1} = -2$$

**Answers**

**2.** $\left\{\frac{1}{4}, 2\right\}$  **3.** $\{4\}$

Next we raise both sides of the equation to the third power to eliminate the radical.

$$(\sqrt[3]{x + 1})^3 = (-2)^3$$
$$x + 1 = -8$$
$$x = -9$$

The solution checks in the original equation, so the solution set is $\{-9\}$.

■ Work Practice 3

---

**Example 4**   Solve:   $\sqrt{4 - x} = x - 2$

**Solution:**

$$\sqrt{4 - x} = x - 2$$
$$(\sqrt{4 - x})^2 = (x - 2)^2$$
$$4 - x = x^2 - 4x + 4 \quad \text{Write the quadratic equation in standard form.}$$
$$x^2 - 3x = 0$$
$$x(x - 3) = 0 \qquad \text{Factor.}$$
$$x = 0 \quad \text{or} \quad x - 3 = 0 \qquad \text{Set each factor equal to 0.}$$
$$x = 3$$

**Check:**

$$\sqrt{4 - x} = x - 2 \qquad\qquad \sqrt{4 - x} = x - 2$$
$$\sqrt{4 - 0} \stackrel{?}{=} 0 - 2 \quad \text{Let } x = 0. \qquad \sqrt{4 - 3} \stackrel{?}{=} 3 - 2 \quad \text{Let } x = 3.$$
$$2 = -2 \quad \text{False} \qquad\qquad 1 = 1 \qquad \text{True}$$

The proposed solution 3 checks, but 0 does not. Since 0 is an extraneous solution, the solution set is $\{3\}$.

■ Work Practice 4

---

**Practice 4**

Solve:   $\sqrt{9 + x} = x + 3$

---

**Helpful Hint**

In Example 4, notice that $(x - 2)^2 = x^2 - 4x + 4$. Make sure binomials are squared correctly.

---

✓**Concept Check**   How can you immediately tell that the equation $\sqrt{2y + 3} = -4$ has no real solution?

---

**Example 5**   Solve:   $\sqrt{2x + 5} + \sqrt{2x} = 3$

**Solution:**   We get one radical alone by subtracting $\sqrt{2x}$ from both sides.

$$\sqrt{2x + 5} + \sqrt{2x} = 3$$
$$\sqrt{2x + 5} = 3 - \sqrt{2x}$$

Now we use the power rule to begin eliminating the radicals. First we square both sides.

$$(\sqrt{2x + 5})^2 = (3 - \sqrt{2x})^2$$
$$2x + 5 = 9 - 6\sqrt{2x} + 2x \quad \text{Multiply:} \ (3 - \sqrt{2x})(3 - \sqrt{2x})$$

*(Continued on next page)*

---

**Practice 5**

Solve:   $\sqrt{3x + 1} + \sqrt{3x} = 2$

---

**Answers**

4. $\{0\}$   5. $\left\{\dfrac{3}{16}\right\}$

✓**Concept Check Answer**

answers may vary

There is still a radical in the equation, so we get the radical alone again. Then we square both sides.

$$2x + 5 = 9 - 6\sqrt{2x} + 2x$$
$$6\sqrt{2x} = 4 \qquad \text{Get the radical alone.}$$
$$(6\sqrt{2x})^2 = 4^2 \qquad \text{Square both sides of the equation to eliminate the radical.}$$
$$36(2x) = 16$$
$$72x = 16 \qquad \text{Multiply.}$$
$$x = \frac{16}{72} \qquad \text{Solve.}$$
$$x = \frac{2}{9} \qquad \text{Simplify.}$$

The proposed solution $\frac{2}{9}$ checks in the original equation. The solution set is $\left\{\frac{2}{9}\right\}$.

■ Work Practice 5

**Helpful Hint**

Make sure expressions are squared correctly. In Example 5, we squared $(3 - \sqrt{2x})$ as

$$(3 - \sqrt{2x})^2 = (3 - \sqrt{2x})(3 - \sqrt{2x})$$
$$= 3 \cdot 3 - 3\sqrt{2x} - 3\sqrt{2x} + \sqrt{2x} \cdot \sqrt{2x}$$
$$= 9 - 6\sqrt{2x} + 2x$$

✔ **Concept Check** What is wrong with the following solution?

$$\sqrt{2x + 5} + \sqrt{4 - x} = 8$$
$$(\sqrt{2x + 5} + \sqrt{4 - x})^2 = 8^2$$
$$(2x + 5) + (4 - x) = 64$$
$$x + 9 = 64$$
$$x = 55$$

## Objective B  Using the Pythagorean Theorem ▶

Recall that the Pythagorean theorem states that in a right triangle, the length of the hypotenuse squared equals the sum of the lengths of each of the legs squared.

**Pythagorean Theorem**

If $a$ and $b$ are the lengths of the legs of a right triangle and $c$ is the length of the hypotenuse, then $a^2 + b^2 = c^2$.

Hypotenuse

$c$

$a$

Legs

$b$

✔ **Concept Check Answer**

From the second line of the solution to the third line of the solution, the left side of the equation is squared incorrectly.

**Example 6** Find the length of the unknown leg of the right triangle.

**Practice 6**

Find the length of the unknown leg of the right triangle.

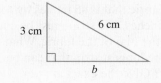

**Solution:** In the formula $a^2 + b^2 = c^2$, $c$ is the hypotenuse. Here, $c = 10$, the length of the hypotenuse, and $a = 4$. We solve for $b$. Then $a^2 + b^2 = c^2$ becomes

$$4^2 + b^2 = 10^2$$
$$16 + b^2 = 100$$
$$b^2 = 84 \quad \text{Subtract 16 from both sides.}$$

Recall from Section 7.1 our definition of square root that if $b^2 = a$, then $b$ is a square root of $a$. Since $b$ is a length and thus is positive, we have that

$$b = \sqrt{84} = \sqrt{4 \cdot 21} = 2\sqrt{21}$$

The unknown leg of the triangle is exactly $2\sqrt{21}$ meters long. Using a calculator, this is approximately 9.2 meters.

■ Work Practice 6

**Example 7** Calculating Distance Across a Quadrangle

**Practice 7**

Find the length of a sidewalk that is the diagonal of a square whose side length is 60 feet.

The current route of the Great Court Run at Trinity College runs along the perimeter of the grassy rectangle. If the dimensions of the grassy rectangle are 70 meters by 80 meters, how far would a student travel by walking across the grass from corner to corner?

70 m

80 m

**Solution:** In the formula $a^2 + b^2 = c^2$, $a$ and $b$ represent the legs of the triangle, and $c$ represents the hypotenuse. Here, $a = 80$ and $b = 70$, and we wish to solve for $c$, the length of the hypotenuse.

Then $a^2 + b^2 = c^2$ becomes
$$80^2 + 70^2 = c^2$$
$$6400 + 4900 = c^2$$
$$11{,}300 = c^2$$

Recall from Section 7.1 our definition of square root that if $c^2 = a$, then $c$ is a square root of $a$. Since $c$ is a length, and thus is positive, we have that

$$c = \sqrt{11{,}300} = \sqrt{100 \cdot 113} = 10\sqrt{113}$$

The distance across the grassy area of the quad is therefore $10\sqrt{113}$ meters. Using a calculator, this is approximately 106.3 meters.

■ Work Practice 7

**Answers**

**6.** $3\sqrt{3}$ cm $\approx 5.2$ cm

**7.** $60\sqrt{2}$ ft $\approx 84.9$ ft

## Practice 8

A furniture upholsterer wishes to cut a strip from a piece of fabric that is 45 inches by 45 inches. The strip must be cut on the bias of the fabric. What is the longest strip that can be cut? Give the exact answer and a two-decimal-place approximation.

45 in.

45 in.

**Example 8** Calculating Placement of a Wire

A 50-foot supporting wire is to be attached to a 75-foot antenna. Because of surrounding buildings, sidewalks, and roadways, the wire must be anchored exactly 20 feet from the base of the antenna.

**a.** How high from the base of the antenna must the wire be attached?

**b.** Local regulations require that a supporting wire be attached at a height no less than $\frac{3}{5}$ of the total height the antenna. From part (a), have local regulations been met?

### Solution:

**1. UNDERSTAND.** Read and reread the problem. From the diagram we notice that a right triangle is formed with hypotenuse 50 feet and one leg 20 feet. We let $x$ = the height from the base of the antenna to the attached wire.

**2. TRANSLATE.** We'll use the Pythagorean theorem.

$$a^2 + b^2 = c^2$$
$$20^2 + x^2 = 50^2 \quad a = 20, c = 50$$

**3. SOLVE.**

$$20^2 + x^2 = 50^2$$
$$400 + x^2 = 2500$$
$$x^2 = 2100 \quad \text{Subtract 400 from both sides.}$$
$$x = \sqrt{2100}$$
$$= 10\sqrt{21}$$

**4. INTERPRET.** *Check* the work and *state* the solution.

**a.** The wire is attached exactly $10\sqrt{21}$ feet from the base of the pole, or approximately 45.8 feet.

**b.** The supporting wire must be attached at a height no less than $\frac{3}{5}$ of the total height of the antenna. This height is $\frac{3}{5}$ (75 feet), or 45 feet.

Since we know from part (a) that the wire is to be attached at a height of approximately 45.8 feet, local regulations have been met.

■ **Work Practice 8**

**Answer**

**8.** $45\sqrt{2}$ in. $\approx 63.64$ in.

## Calculator Explorations  Graphing

We can use a graphing calculator to solve radical equations. For example, to use a graphing calculator to approximate the solutions of the equation solved in Example 4, we graph the following:

$$Y_1 = \sqrt{4 - x} \quad \text{and} \quad Y_2 = x - 2$$

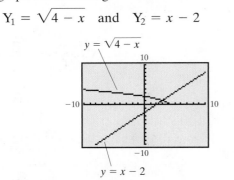

The $x$-value of the point of intersection is the solution. Use the INTERSECT feature or the ZOOM and TRACE features of your graphing calculator to see that the solution is 3.

*Use a graphing calculator to solve each radical equation. Round all solutions to the nearest hundredth.*

1. $\sqrt{x + 7} = x$
2. $\sqrt{3x + 5} = 2x$
3. $\sqrt{2x + 1} = \sqrt{2x + 2}$
4. $\sqrt{10x - 1} = \sqrt{-10x + 10} - 1$
5. $1.2x = \sqrt{3.1x + 5}$
6. $\sqrt{1.9x^2 - 2.2} = -0.8x + 3$

## Vocabulary, Readiness & Video Check

*Use the choices below to fill in each blank. Not all choices will be used.*

hypotenuse            right            $x^2 + 25$            $16 - 8\sqrt{7x} + 7x$

extraneous solution   legs             $x^2 - 10x + 25$     $16 + 7x$

1. A proposed solution that is not a solution of the original equation is called a(n) _____.
2. The Pythagorean theorem states that $a^2 + b^2 = c^2$ where $a$ and $b$ are the lengths of the _____ of a(n) _____ triangle and $c$ is the length of the _____.
3. The square of $x - 5$, or $(x - 5)^2 =$ _____.
4. The square of $4 - \sqrt{7x}$, or $(4 - \sqrt{7x})^2 =$ _____.

*Martin-Gay Interactive Videos*  *Watch the section lecture video and answer the following questions.*

**Objective A** 5. From ⊞ Examples 1–4, why must you be careful to check your proposed solution(s) in the original equation? ▶

**Objective B** 6. From ⊞ Example 5, when solving problems using the Pythagorean theorem, what two things must you remember? ▶

7. What important reminder is given as the final answer to ⊞ Example 5 is being found? ▶

*See Video 7.6* 🍎

## 7.6 Exercise Set MyMathLab® ▶

**Objective A** *Solve. See Examples 1 and 2.*

**1.** $\sqrt{2x} = 4$

**2.** $\sqrt{3x} = 3$

▶**3.** $\sqrt{x-3} = 2$

**4.** $\sqrt{x+1} = 5$

**5.** $\sqrt{2x} = -4$

**6.** $\sqrt{5x} = -5$

**7.** $\sqrt{4x-3} - 5 = 0$

**8.** $\sqrt{x-3} - 1 = 0$

**9.** $\sqrt{2x-3} - 2 = 1$

**10.** $\sqrt{3x+3} - 4 = 8$

*Solve. See Example 3.*

**11.** $\sqrt[3]{6x} = -3$

**12.** $\sqrt[3]{4x} = -2$

**13.** $\sqrt[3]{x-2} - 3 = 0$

**14.** $\sqrt[3]{2x-6} - 4 = 0$

*Solve. See Examples 4 and 5.*

**15.** $\sqrt{13-x} = x - 1$

**16.** $\sqrt{2x-3} = 3 - x$

▶**17.** $x - \sqrt{4-3x} = -8$

**18.** $2x + \sqrt{x+1} = 8$

**19.** $\sqrt{y+5} = 2 + \sqrt{y-2}$

**20.** $\sqrt{x+3} + \sqrt{x-5} = 3$

**21.** $\sqrt{x-3} + \sqrt{x+2} = 5$

**22.** $\sqrt{2x-4} - \sqrt{3x+4} = -2$

*Solve. See Examples 1 through 5.*

**23.** $\sqrt{3x-2} = 5$

**24.** $\sqrt{5x-4} = 9$

**25.** $-\sqrt{2x} + 4 = -6$

**26.** $-\sqrt{3x} + 2 = -10$

**27.** $\sqrt{3x+1} + 2 = 0$

**28.** $\sqrt{3x+1} - 2 = 0$

**29.** $\sqrt[4]{4x+1} - 2 = 0$

**30.** $\sqrt[4]{2x-9} - 3 = 0$

**31.** $\sqrt{3x+4} = 5$

**32.** $\sqrt{3x+9} = 12$

**33.** $\sqrt[3]{6x-3} - 3 = 0$

**34.** $\sqrt[3]{3x} + 4 = 7$

**35.** $\sqrt[3]{2x - 3} - 2 = -5$    **36.** $\sqrt[3]{x - 4} - 5 = -7$    **37.** $\sqrt{x + 4} = \sqrt{2x - 5}$    **38.** $\sqrt{3y + 6} = \sqrt{7y - 6}$

**39.** $x - \sqrt{1 - x} = -5$    **40.** $x - \sqrt{x - 2} = 4$    **41.** $\sqrt[3]{-6x - 1} = \sqrt[3]{-2x - 5}$

**42.** $x + \sqrt{x + 5} = 7$    **43.** $\sqrt{5x - 1} - \sqrt{x + 2} = 3$    **44.** $\sqrt{2x - 1} - 4 = -\sqrt{x - 4}$

**45.** $\sqrt{2x - 1} = \sqrt{1 - 2x}$    **46.** $\sqrt{7x - 4} = \sqrt{4 - 7x}$    **47.** $\sqrt{3x + 4} - 1 = \sqrt{2x + 1}$

**48.** $\sqrt{x - 2} + 3 = \sqrt{4x + 1}$    **49.** $\sqrt{y + 3} - \sqrt{y - 3} = 1$    **50.** $\sqrt{x + 1} - \sqrt{x - 1} = 2$

**Objective B**  *Find the length of the unknown side of each triangle. See Example 6.*

**51.**  6 ft, 3 ft

**52.**  7 in., 8 in.

**53.**  3 m, 7 m

**54.**  4 cm, 7 cm

*Find the length of the unknown side of each triangle. Give the exact length and a one-decimal-place approximation. See Example 6.*

**55.**  9 m, $11\sqrt{5}$ m

**56.**  $5\sqrt{3}$ cm, 10 cm

**57.**  7 mm, 7.2 mm

**58.**  2.7 in., 2.3 in.

*Solve. Give exact answers and two-decimal-place approximations where appropriate. For Exercises 59 and 60, the solutions have been started for you. See Examples 7 and 8.*

**59.** A wire is needed to support a vertical pole 15 feet tall. The cable will be anchored to a stake 8 feet from the base of the pole. How much cable is needed?

**Start the solution:**

1. UNDERSTAND the problem. Reread it as many times as needed. Notice that a right triangle is formed with legs of length 8 ft and 15 ft. Since we are looking for how much cable is needed, let

   $x =$ amount of cable needed

2. TRANSLATE into an equation. We use the Pythagorean theorem. (Fill in the blanks below.)

   $$a^2 \quad + \quad b^2 \quad = \quad c^2$$
   $$\downarrow \qquad\qquad \downarrow$$
   $$\underline{\phantom{xx}}^2 \quad + \quad \underline{\phantom{xx}}^2 \quad = \quad \underline{x^2}$$

   Finish with:

3. SOLVE and 4. INTERPRET

**60.** The tallest structure in the United States is a TV tower in Blanchard, North Dakota. Its height is 2063 feet. A 2382-foot length of wire is to be used as a guy wire attached to the top of the tower. Approximate to the nearest foot how far from the base of the tower the guy wire must be anchored. (*Source:* U.S. Geological Survey)

**Start the solution:**

1. UNDERSTAND the problem. Reread it as many times as needed. Notice that a right triangle is formed with hypotenuse 2382 ft and one leg 2063 ft. Since we are looking for how far from the base of the tower the guy wire is anchored, let

   $x =$ distance from base of tower to where guy wire is anchored.

2. TRANSLATE into an equation. We use the Pythagorean theorem. (Fill in the blanks below.)

   $$a^2 \quad + \quad b^2 \quad = \quad c^2$$
   $$\downarrow \qquad\quad \downarrow \qquad\quad \downarrow$$
   $$\underline{\phantom{xx}}^2 \quad + \quad \underline{x}^2 \quad = \quad \underline{\phantom{xx}}^2$$

   Finish with:

3. SOLVE and 4. INTERPRET

**61.** A spotlight is mounted on the eaves of a house, 12 feet above the ground. A flower bed runs between the house and the sidewalk, so the closest the ladder can be placed to the house is 5 feet. How long a ladder is needed so that an electrician can reach the place where the light is mounted?

**62.** A wire is to be attached to support a telephone pole. Because of surrounding buildings, sidewalks, and roadway, the wire must be anchored exactly 15 feet from the base of the pole. Telephone company workers have only 30 feet of cable, and 2 feet of that must be used to attach the cable to the pole and to the stake on the ground. How high from the base of the pole can the wire be attached?

△ **63.** The radius of the moon is 1080 miles. Use the formula for the radius $r$ of a sphere given its surface area $A$.

$$r = \sqrt{\frac{A}{4\pi}}$$

to find the surface area of the moon. Round to the nearest square mile. (*Source:* National Space Science Data Center)

**64.** Police departments find it very useful to be able to approximate driving speeds in skidding accidents. If the road surface is wet concrete, the function $S(x) = \sqrt{10.5x}$ is used, where $S(x)$ is the speed of the car in miles per hour and $x$ is the distance skidded in feet. Find how fast a car was moving if it skidded 280 feet on wet concrete.

*The formula $v = \sqrt{2gh}$ relates the velocity $v$, in feet per second, of an object after it falls $h$ feet accelerated by gravity $g$, in feet per second squared. If $g$ is approximately 32 feet per second squared, find how far an object has fallen for each given velocity.*

**65.** Velocity of 80 feet per second

**66.** Velocity of 40 feet per second

**67.** Harvard Yard is known as the heart of Harvard University. It is a great rectangular lawn area in the midst of Cambridge, Massachusetts. This "old yard" has dimensions of 750 feet by 250 feet. How far does a student walk to travel across the diagonal of the yard?

**68.** Two tractors are pulling a tree stump from a field. If two forces $A$ and $B$ pull at right angles (90°) to each other, the resulting force $R$ is given by the formula $R = \sqrt{A^2 + B^2}$. If tractor $A$ is exerting 600 pounds of force and the resulting force is 850 pounds, find how much force tractor $B$ is exerting.

600 lb

*In psychology, it has been suggested that the number $S$ of nonsense syllables that a person can repeat consecutively depends on his or her IQ score $I$ according to the equation $S = 2\sqrt{I} - 9$.*

**69.** Use this relationship to estimate the IQ of a person who can repeat 11 nonsense syllables consecutively.

**70.** Use this relationship to estimate the IQ of a person who can repeat 15 nonsense syllables consecutively.

*The **period** of a pendulum is the time it takes for the pendulum to make one full back-and-forth swing. The period of a pendulum depends on the length of the pendulum. The formula for the period P, in seconds, is $P = 2\pi\sqrt{\dfrac{L}{32}}$, where L is the length of the pendulum in feet. Use this formula for Exercises 71 through 76.*

**71.** Find the period of a pendulum whose length is 2 feet. Give the exact answer and a two-decimal-place approximation.

2 feet

**72.** Klockit sells a 43-inch lyre pendulum. Find the period of this pendulum. Round your answer to 2 decimal places. (*Hint:* First convert inches to feet.)

**73.** Find the length of a pendulum whose period is 4 seconds. Round your answer to 2 decimal places.

**74.** Find the length of a pendulum whose period is 3 seconds. Round your answer to 3 decimal places.

**75.** Study the relationship between period and pendulum length in Exercises 71 through 74 and make a conjecture about this relationship.

**76.** Galileo experimented with pendulums. He supposedly made conjectures about pendulums of equal length with different bob weights. Try this experiment. Make two pendulums 3 feet long. Attach a heavy weight (lead) to one and a light weight (a cork) to the other. Pull both pendulums back the same angle measure and release. Make a conjecture from your observations.

*If the three lengths of the sides of a triangle are known, Heron's formula can be used to find its area. If a, b, and c are the three lengths of the sides, Heron's formula for area is*

$$A = \sqrt{s(s-a)(s-b)(s-c)}$$

*where s is half the perimeter of the triangle, or $s = \dfrac{1}{2}(a + b + c)$. Use this formula to find the area of each triangle. Give an exact answer and then a two-decimal-place approximation.*

△ **77.**

6 mi     10 mi

14 mi

△ **78.**

2 cm     3 cm

3 cm

**79.** Describe when Heron's formula might be useful.

**80.** In your own words, explain why you think s in Heron's formula is called the *semiperimeter*.

*The maximum distance $D(h)$ in kilometers that a person can see from a height h kilometers above the ground is given by the function $D(h) = 111.7\sqrt{h}$. Use this function for Exercises 81 and 82. Round your answers to two decimal places.*

**81.** Find the height that would allow a person to see 80 kilometers.

**82.** Find the height that would allow a person to see 40 kilometers.

## Review

*Simplify. See Section 6.3.*

**83.** $\dfrac{\dfrac{x}{6}}{\dfrac{2x}{3} + \dfrac{1}{2}}$

**84.** $\dfrac{\dfrac{1}{y} + \dfrac{4}{5}}{\dfrac{-3}{20}}$

**85.** $\dfrac{\dfrac{z}{5} + \dfrac{1}{10}}{\dfrac{z}{20} - \dfrac{z}{5}}$

**86.** $\dfrac{\dfrac{1}{y} + \dfrac{1}{x}}{\dfrac{1}{y} - \dfrac{1}{x}}$

## Concept Extensions

**87.** Consider the equations $\sqrt{2x} = 4$ and $\sqrt[3]{2x} = 4$.
  **a.** Explain the difference in solving these equations.
  **b.** Explain the similarity in solving these equations.

**88.** Explain why proposed solutions of radical equations must be checked.

**89.** Find and correct the error in the following solution. See the second Concept Check in this section.

$$\sqrt{5x - 1} + 4 = 7$$
$$(\sqrt{5x - 1} + 4)^2 = 7^2$$
$$5x - 1 + 16 = 49$$
$$5x = 34$$
$$x = \frac{34}{5}$$

**90.** Solve: $\sqrt{\sqrt{x + 3} + \sqrt{x}} = \sqrt{3}$

**91.** The cost $C(x)$ in dollars per day to operate a small delivery service is given by $C(x) = 80\sqrt[3]{x} + 500$, where $x$ is the number of deliveries per day. In July, the manager decides that it is necessary to keep delivery costs below $1620.00. Find the greatest number of deliveries this company can make per day and still keep overhead below $1620.00.

## Objectives

**A** Write Square Roots of Negative Numbers in the Form $bi$. ⏵

**B** Add or Subtract Complex Numbers. ⏵

**C** Multiply Complex Numbers. ⏵

**D** Divide Complex Numbers. ⏵

**E** Raise $i$ to Powers. ⏵

## Objective A Writing Numbers in the Form $bi$ ⏵

Our work with radical expressions has excluded expressions such as $\sqrt{-16}$ because $\sqrt{-16}$ is not a real number; there is no real number whose square is $-16$. In this section, we discuss a number system that includes roots of negative numbers. This number system is the **complex number system,** and it includes the set of real numbers as a subset. The complex number system allows us to solve equations such as $x^2 + 1 = 0$ that have no real number solutions. The set of complex numbers includes the *imaginary unit*.

### Imaginary Unit

The **imaginary unit,** written $i$, is the number whose square is $-1$. That is,

$$i^2 = -1 \quad \text{and} \quad i = \sqrt{-1}$$

To write the square root of a negative number in terms of $i$, we use the property that if $a$ is a positive number, then

$$\sqrt{-a} = \sqrt{-1} \cdot \sqrt{a}$$
$$= i \cdot \sqrt{a}$$

Using $i$, we can write $\sqrt{-16}$ as

$$\sqrt{-16} = \sqrt{-1 \cdot 16} = \sqrt{-1} \cdot \sqrt{16} = i \cdot 4 \text{ or } 4i$$

**Practice 1–3**

Write using $i$ notation.
1. $\sqrt{-25}$
2. $\sqrt{-17}$
3. $-\sqrt{-50}$

**Examples** Write using $i$ notation.

1. $\sqrt{-36} = \sqrt{-1 \cdot 36} = \sqrt{-1} \cdot \sqrt{36} = i \cdot 6 \text{ or } 6i$
2. $\sqrt{-5} = \sqrt{-1(5)} = \sqrt{-1} \cdot \sqrt{5} = i\sqrt{5}$

> **Helpful Hint** Since $\sqrt{5}i$ can easily be confused with $\sqrt{5i}$, we write $\sqrt{5}i$ as $i\sqrt{5}$.

3. $-\sqrt{-20} = -\sqrt{-1 \cdot 20} = -\sqrt{-1} \cdot \sqrt{4 \cdot 5} = -i \cdot 2\sqrt{5} = -2i\sqrt{5}$

▪ Work Practice 1–3

The product rule for radicals does not necessarily hold true for imaginary numbers. *To multiply square roots of negative numbers, first we write each number in terms of the imaginary unit i.* For example, to multiply $\sqrt{-4}$ and $\sqrt{-9}$, we first write each number in the form $bi$:

$$\sqrt{-4} \cdot \sqrt{-9} = 2i(3i) = 6i^2 = 6(-1) = -6 \quad \text{Correct.}$$

Make sure you notice that the product rule does not work for this example. In other words, $\sqrt{-4} \cdot \sqrt{-9} = \sqrt{(-4)(-9)} = \sqrt{36} = 6$ is incorrect!

**Practice 4–7**

Multiply or divide as indicated.
4. $\sqrt{-3} \cdot \sqrt{-7}$
5. $\sqrt{-25} \cdot \sqrt{-1}$
6. $\sqrt{27} \cdot \sqrt{-3}$
7. $\dfrac{\sqrt{-8}}{\sqrt{2}}$

**Examples** Multiply or divide as indicated.

4. $\sqrt{-3} \cdot \sqrt{-5} = i\sqrt{3}(i\sqrt{5}) = i^2\sqrt{15} = -1\sqrt{15} = -\sqrt{15}$
5. $\sqrt{-36} \cdot \sqrt{-1} = 6i(i) = 6i^2 = 6(-1) = -6$
6. $\sqrt{8} \cdot \sqrt{-2} = 2\sqrt{2}(i\sqrt{2}) = 2i(\sqrt{2}\sqrt{2}) = 2i(2) = 4i$
7. $\dfrac{\sqrt{-125}}{\sqrt{5}} = \dfrac{i\sqrt{125}}{\sqrt{5}} = i\sqrt{25} = 5i$

▪ Work Practice 4–7

**Answers**

1. $5i$  2. $i\sqrt{17}$  3. $-5i\sqrt{2}$
4. $-\sqrt{21}$  5. $-5$  6. $9i$  7. $2i$

Now that we have practiced working with the imaginary unit, we define *complex numbers*.

### Complex Numbers

A **complex number** is a number that can be written in the form $a + bi$, where $a$ and $b$ are real numbers.

Notice that the set of real numbers is a subset of the complex numbers since any real number can be written in the form of a complex number. For example,

$$16 = 16 + 0i$$

In general, a complex number $a + bi$ is a real number if $b = 0$. Also, a complex number is called an **imaginary number** if $a = 0$. For example,

$$3i = 0 + 3i \quad \text{and} \quad i\sqrt{7} = 0 + i\sqrt{7}$$

are imaginary numbers.

The following diagram shows the relationship between complex numbers and their subsets.

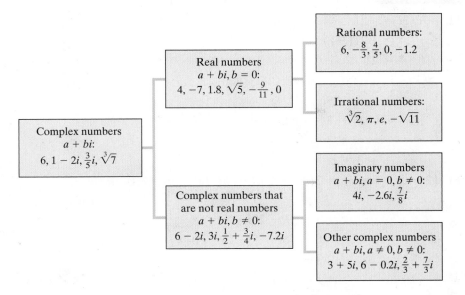

✔**Concept Check** True or false? Every complex number is also a real number.

## Objective B Adding or Subtracting Complex Numbers ▶

Two complex numbers $a + bi$ and $c + di$ are equal if and only if $a = c$ and $b = d$. Complex numbers can be added or subtracted by adding or subtracting their real parts and then adding or subtracting their imaginary parts.

### Sum or Difference of Complex Numbers

If $a + bi$ and $c + di$ are complex numbers, then their sum is

$$(a + bi) + (c + di) = (a + c) + (b + d)i$$

Their difference is

$$(a + bi) - (c + di) = a + bi - c - di = (a - c) + (b - d)i$$

## Practice 8–10

Add or subtract as indicated.

**8.** $(5 + 2i) + (4 - 3i)$

**9.** $6i - (2 - i)$

**10.** $(-2 - 4i) - (-3)$

**Examples**    Add or subtract as indicated.

**8.** $(2 + 3i) + (-3 + 2i) = (2 - 3) + (3 + 2)i = -1 + 5i$

**9.** $5i - (1 - i) = 5i - 1 + i$
$$= -1 + (5 + 1)i$$
$$= -1 + 6i$$

**10.** $(-3 - 7i) - (-6) = -3 - 7i + 6$
$$= (-3 + 6) - 7i$$
$$= 3 - 7i$$

▶ **Work Practice 8–10**

## Objective C  Multiplying Complex Numbers ▶

To multiply two complex numbers of the form $a + bi$, we multiply as though they were binomials. Then we use the relationship $i^2 = -1$ to simplify.

## Practice 11–15

Multiply.

**11.** $-5i \cdot 3i$

**12.** $-2i(6 - 2i)$

**13.** $(3 - 4i)(6 + i)$

**14.** $(1 - 2i)^2$

**15.** $(6 + 5i)(6 - 5i)$

**Examples**    Multiply.

**11.** $-7i \cdot 3i = -21i^2$
$$= -21(-1) \quad \text{Replace } i^2 \text{ with } -1.$$
$$= 21$$

**12.** $3i(2 - i) = 3i \cdot 2 - 3i \cdot i \quad \text{Use the distributive property.}$
$$= 6i - 3i^2 \qquad \text{Multiply.}$$
$$= 6i - 3(-1) \quad \text{Replace } i^2 \text{ with } -1.$$
$$= 6i + 3$$
$$= 3 + 6i$$

**13.** $(2 - 5i)(4 + i) = 2(4) + 2(i) - 5i(4) - 5i(i) \quad \text{Use the FOIL order.}$
$$\qquad\qquad\qquad\quad \text{F} \qquad \text{O} \qquad \text{I} \qquad \text{L} \quad \text{(First, Outer, Inner, Last)}$$
$$= 8 + 2i - 20i - 5i^2$$
$$= 8 - 18i - 5(-1) \quad i^2 = -1$$
$$= 8 - 18i + 5$$
$$= 13 - 18i$$

**14.** $(2 - i)^2 = (2 - i)(2 - i)$
$$= 2(2) - 2(i) - 2(i) + i^2$$
$$= 4 - 4i + (-1) \qquad\qquad i^2 = -1$$
$$= 3 - 4i$$

**15.** $(7 + 3i)(7 - 3i) = 7(7) - 7(3i) + 3i(7) - 3i(3i)$
$$= 49 - 21i + 21i - 9i^2$$
$$= 49 - 9(-1) \qquad\qquad\qquad i^2 = -1$$
$$= 49 + 9$$
$$= 58$$

▶ **Work Practice 11–15**

Notice that if you add, subtract, or multiply two complex numbers, the result is a complex number.

## Objective D  Dividing Complex Numbers ▶

From Example 15, notice that the product of $7 + 3i$ and $7 - 3i$ is a real number. These two complex numbers are called *complex conjugates* of one another. In general, we have the following definition.

**Answers**

**8.** $9 - i$   **9.** $-2 + 7i$   **10.** $1 - 4i$
**11.** 15   **12.** $-4 - 12i$   **13.** $22 - 21i$
**14.** $-3 - 4i$   **15.** 61

## Complex Conjugates

The complex numbers $(a + bi)$ and $(a - bi)$ are called **complex conjugates** of each other, and

$$(a + bi)(a - bi) = a^2 + b^2$$

To see that the product of a complex number $a + bi$ and its conjugate $a - bi$ is the real number $a^2 + b^2$, we multiply:

$$(a + bi)(a - bi) = a^2 - abi + abi - b^2i^2$$
$$= a^2 - b^2(-1)$$
$$= a^2 + b^2$$

We use complex conjugates to divide by a complex number.

**Example 16**  Divide and write in the form $a + bi$: $\dfrac{2 + i}{1 - i}$

**Solution:**  We multiply the numerator and the denominator by the complex conjugate of $1 - i$ to get a real number in the denominator.

$$\frac{2 + i}{1 - i} = \frac{(2 + i)(1 + i)}{(1 - i)(1 + i)}$$

$$= \frac{2(1) + 2(i) + 1(i) + i^2}{1^2 - i^2}$$

$$= \frac{2 + 3i - 1}{1 + 1}$$

$$= \frac{1 + 3i}{2} = \frac{1}{2} + \frac{3}{2}i$$

■ Work Practice 16

**Practice 16**

Divide and write in the form

$a + bi$: $\dfrac{3 + i}{2 - 3i}$

**Example 17**  Divide and write in the form $a + bi$: $\dfrac{7}{3i}$

**Solution:**   We multiply the numerator and the denominator by the conjugate of $3i$. Note that $3i = 0 + 3i$, so its conjugate is $0 - 3i$ or $-3i$.

$$\frac{7}{3i} = \frac{7(-3i)}{(3i)(-3i)} = \frac{-21i}{-9i^2} = \frac{-21i}{-9(-1)} = \frac{-21i}{9} = \frac{-7i}{3} = -\frac{7}{3}i$$

■ Work Practice 17

**Practice 17**

Divide and write in the form

$a + bi$: $\dfrac{6}{5i}$

## Objective E  Finding Powers of $i$ ▶

We can use the fact that $i^2 = -1$ to simplify $i^3$ and $i^4$.

$$i^3 = i^2 \cdot i = (-1)i = -i$$
$$i^4 = i^2 \cdot i^2 = (-1) \cdot (-1) = 1$$

We continue this process and use the fact that $i^4 = 1$ and $i^2 = -1$ to simplify $i^5$ and $i^6$.

$$i^5 = i^4 \cdot i = 1 \cdot i = i$$
$$i^6 = i^4 \cdot i^2 = 1 \cdot (-1) = -1$$

**Helpful Hint**

$-\dfrac{7}{3}i$ is also $0 - \dfrac{7}{3}i$.

When writing in $a + bi$ form, if $a$ is 0, we will write $a + bi$ as $bi$ and if $b = 0$, we will write $a + bi$ as $a$.

**Answers**

16. $\dfrac{3}{13} + \dfrac{11}{13}i$  17. $-\dfrac{6}{5}i$

If we continue finding powers of $i$, we generate the following pattern. Notice that the values $i$, $-1$, $-i$, and 1 repeat as $i$ is raised to higher and higher powers.

$$i^1 = i \qquad\qquad i^5 = i \qquad\qquad i^9 = i$$
$$i^2 = -1 \qquad\quad i^6 = -1 \qquad\quad i^{10} = -1$$
$$i^3 = -i \qquad\quad i^7 = -i \qquad\quad i^{11} = -i$$
$$i^4 = 1 \qquad\qquad i^8 = 1 \qquad\qquad i^{12} = 1$$

This pattern allows us to find other powers of $i$. To do so, we will use the fact that $i^4 = 1$ and rewrite a power of $i$ in terms of $i^4$. For example,

$$i^{22} = i^{20} \cdot i^2 = (i^4)^5 \cdot i^2 = 1^5 \cdot (-1) = 1 \cdot (-1) = -1$$

**Practice 18–21**

Find the powers of $i$.
18. $i^{11}$      19. $i^{40}$
20. $i^{50}$      21. $i^{-10}$

**Answers**
18. $-i$   19. 1   20. $-1$   21. $-1$

**Examples**   Find each power of $i$.

18. $i^7 = i^4 \cdot i^3 = 1(-i) = -i$

19. $i^{20} = (i^4)^5 = 1^5 = 1$

20. $i^{46} = i^{44} \cdot i^2 = (i^4)^{11} \cdot i^2 = 1^{11}(-1) = -1$

21. $i^{-12} = \dfrac{1}{i^{12}} = \dfrac{1}{(i^4)^3} = \dfrac{1}{(1)^3} = \dfrac{1}{1} = 1$

▣ Work Practice 18–21

## Vocabulary, Readiness & Video Check

*Use the choices below to fill in each blank. Not all choices will be used.*

$-1$     $\sqrt{-1}$     real     1     $\sqrt{1}$     complex     imaginary     imaginary unit

1. A(n) _____ number is one that can be written in the form $a + bi$ where $a$ and $b$ are real numbers.

2. In the complex number system, $i$ denotes the _____.

3. $i^2 =$ _____

4. $i =$ _____

5. A complex number, $a + bi$, is a(n) _____ number if $b = 0$.

6. A complex number, $a + bi$, is a(n) _____ number if $a = 0$ and $b \neq 0$.

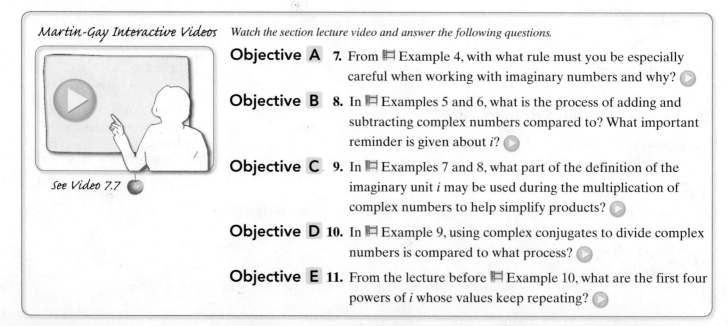

*Martin-Gay Interactive Videos*

*Watch the section lecture video and answer the following questions.*

See Video 7.7 ●

**Objective A**   7. From ▤ Example 4, with what rule must you be especially careful when working with imaginary numbers and why? ▶

**Objective B**   8. In ▤ Examples 5 and 6, what is the process of adding and subtracting complex numbers compared to? What important reminder is given about $i$? ▶

**Objective C**   9. In ▤ Examples 7 and 8, what part of the definition of the imaginary unit $i$ may be used during the multiplication of complex numbers to help simplify products? ▶

**Objective D**   10. In ▤ Example 9, using complex conjugates to divide complex numbers is compared to what process? ▶

**Objective E**   11. From the lecture before ▤ Example 10, what are the first four powers of $i$ whose values keep repeating? ▶

## 7.7 Exercise Set MyMathLab® ▶

**Objective A** *Simplify and write using i notation, if possible. See Examples 1 through 3.*

▶**1.** $\sqrt{-81}$

**2.** $\sqrt{-49}$

▶**3.** $\sqrt{-7}$

**4.** $\sqrt{-3}$

**5.** $-\sqrt{16}$

**6.** $-\sqrt{4}$

**7.** $\sqrt{-64}$

**8.** $\sqrt{-100}$

▶**9.** $\sqrt{-24}$

**10.** $\sqrt{-32}$

**11.** $-\sqrt{-36}$

**12.** $-\sqrt{-121}$

**13.** $8\sqrt{-63}$

**14.** $4\sqrt{-20}$

**15.** $-\sqrt{54}$

**16.** $-\sqrt{63}$

*Multiply or divide as indicated. See Examples 4 through 7.*

**17.** $\sqrt{-2} \cdot \sqrt{-7}$

**18.** $\sqrt{-11} \cdot \sqrt{-3}$

▶**19.** $\sqrt{-5} \cdot \sqrt{-10}$

**20.** $\sqrt{-2} \cdot \sqrt{-6}$

**21.** $\sqrt{16} \cdot \sqrt{-1}$

**22.** $\sqrt{3} \cdot \sqrt{-27}$

**23.** $\dfrac{\sqrt{-9}}{\sqrt{3}}$

**24.** $\dfrac{\sqrt{49}}{\sqrt{-10}}$

**25.** $\dfrac{\sqrt{-80}}{\sqrt{-10}}$

**26.** $\dfrac{\sqrt{-40}}{\sqrt{-8}}$

**Objective B** *Add or subtract as indicated. Write your answers in the form a + bi. See Examples 8 through 10.*

▶**27.** $(4 - 7i) + (2 + 3i)$

**28.** $(2 - 4i) - (2 - i)$

**29.** $(6 + 5i) - (8 - i)$

**30.** $(8 - 3i) + (-8 + 3i)$

▶**31.** $6 - (8 + 4i)$

**32.** $(9 - 4i) - 9$

**33.** $(6 - 3i) - (4 - 2i)$

**34.** $(-2 - 4i) - (6 - 8i)$

**35.** $(5 - 6i) - 4i$

**36.** $(6 - 2i) + 7i$

**37.** $(2 + 4i) + (6 - 5i)$

**38.** $(5 - 3i) + (7 - 8i)$

**Objective C** *Multiply. Write your answers in the form a + bi. See Examples 11 through 15.*

**39.** $6i \cdot 2i$

**40.** $5i \cdot 7i$

**41.** $-9i \cdot 7i$

**42.** $-6i \cdot 4i$

▶**43.** $-10i \cdot -4i$

**44.** $-2i \cdot -11i$

**45.** $6i(2 - 3i)$

**46.** $5i(4 - 7i)$

**47.** $-3i(-1 + 9i)$

**48.** $-5i(-2 + i)$

**49.** $(4 + i)(5 + 2i)$

**50.** $(3 + i)(2 + 4i)$

**51.** $(\sqrt{3} + 2i)(\sqrt{3} - 2i)$

**52.** $(\sqrt{5} - 5i)(\sqrt{5} + 5i)$

**53.** $(4 - 2i)^2$

**54.** $(6 - 3i)^2$

**55.** $(6 - 2i)(3 + i)$

**56.** $(2 - 4i)(2 - i)$

**57.** $(1 - i)(1 + i)$

**58.** $(6 + 2i)(6 - 2i)$

**59.** $(9 + 8i)^2$

**60.** $(4 + 7i)^2$

**61.** $(1 - i)^2$

**62.** $(2 - 2i)^2$

## Objective D  *Divide. Write your answers in the form a + bi. See Examples 16 and 17.*

**63.** $\dfrac{4}{i}$

**64.** $\dfrac{5}{6i}$

**65.** $\dfrac{7}{4 + 3i}$

**66.** $\dfrac{9}{1 - 2i}$

**67.** $\dfrac{6i}{1 - 2i}$

**68.** $\dfrac{3i}{5 + i}$

**69.** $\dfrac{3 + 5i}{1 + i}$

**70.** $\dfrac{6 + 2i}{4 - 3i}$

**71.** $\dfrac{4 - 5i}{2i}$

**72.** $\dfrac{6 + 8i}{3i}$

**73.** $\dfrac{16 + 15i}{-3i}$

**74.** $\dfrac{2 - 3i}{-7i}$

**75.** $\dfrac{2}{3 + i}$

**76.** $\dfrac{5}{3 - 2i}$

**77.** $\dfrac{2 - 3i}{2 + i}$

**78.** $\dfrac{6 + 5i}{6 - 5i}$

## Objective E  *Find each power of i. See Examples 18 through 21.*

**79.** $i^8$

**80.** $i^{10}$

**81.** $i^{21}$

**82.** $i^{15}$

**83.** $i^{11}$

**84.** $i^{40}$

**85.** $i^{-6}$

**86.** $i^{-9}$

**87.** $(2i)^6$

**88.** $(5i)^4$

**89.** $(-3i)^5$

**90.** $(-2i)^7$

## Review

*Thirty people were recently polled about the average monthly balance in their checking account. The results of this poll are shown in the bar graph. Use this graph to answer Exercises 91 through 96. See Section 3.1.*

**Average Monthly Checking Account Balances**

Number of People / Balances (in dollars)
0 to 100, 101 to 200, 201 to 300, 301 to 400, 401 to 500, Over 500

**91.** How many people polled reported an average checking balance of $201 to $300?

**92.** How many people polled reported an average checking balance of $0 to $100?

**93.** How many people polled reported an average checking balance of $200 or less?

**94.** How many people polled reported an average checking balance of $301 or more?

**95.** What percent of people polled reported an average checking balance of $201 to $300?

**96.** What percent of people polled reported an average checking balance of 0 to $100?

## Concept Extensions

*Write each expression in the form a + bi.*

**97.** $i^3 + i^4$     **98.** $i^8 - i^7$     **99.** $i^6 + i^8$     **100.** $i^4 + i^{12}$     **101.** $2 + \sqrt{-9}$     **102.** $7 - \sqrt{-25}$

**103.** $5 - \sqrt{-16}$     **104.** $\dfrac{6 + \sqrt{-18}}{3}$     **105.** $\dfrac{4 - \sqrt{-8}}{2}$     **106.** $\dfrac{5 - \sqrt{-75}}{10}$

**107.** Describe how to find the conjugate of a complex number.

**108.** Explain why the product of a complex number and its complex conjugate is a real number.

*Simplify.*

**109.** $(8 - \sqrt{-3}) - (2 + \sqrt{-12})$

**110.** $(8 - \sqrt{-4}) - (2 + \sqrt{-16})$

**111.** Determine whether $2i$ is a solution of $x^2 + 4 = 0$.

**112.** Determine whether $-1 + i$ is a solution of $x^2 + 2x = -2$.

### Chapter 7 Group Activity

#### Heron of Alexandria

Heron (also Hero) was a Greek mathematician and engineer. He lived and worked in Alexandria, Egypt, around 75 A.D. During his prolific work life, Heron developed a rotary steam engine called an aeolipile and a surveying tool called a dioptra as well as a wind organ and a fire engine. As an engineer, he must have had the need to approximate square roots because he described an iterative method for doing so in his work *Metrica*. Heron's method for approximating a square root can be summarized as follows:

Suppose that $x$ is not a perfect square and $a^2$ is the nearest perfect square to $x$. For a rough estimate of the value of $\sqrt{x}$, find the value of $y_1 = \dfrac{1}{2}\left(a + \dfrac{x}{a}\right)$.

This estimate can be improved by calculating a second estimate using the first estimate $y_1$ in place of $a$:

$$y_2 = \dfrac{1}{2}\left(y_1 + \dfrac{x}{y_1}\right).$$

Repeating this process several times will give more and more accurate estimates of $\sqrt{x}$.

#### Critical Thinking

**1. a.** Which perfect square is closest to 80?

**b.** Use Heron's method for approximating square roots to calculate the first estimate of the square root of 80.

**c.** Use the first estimate of the square root of 80 to find a more refined second estimate.

**d.** Use a calculator to find the actual value of the square root of 80. List all digits shown on your calculator's display.

**e.** Compare the actual value from part (d) to the values of the first and second estimates. What do you notice?

**f.** How many iterations of this process are necessary to get an estimate that differs no more than one digit from the actual value recorded in part (d)?

**2.** Repeat Question 1 for finding an estimate of the square root of 30.

**3.** Repeat Question 1 for finding an estimate of the square root of 4572.

**4.** Why would this iterative method have been important to people of Heron's era? Would you say that this method is as important today? Why or why not?

## Chapter 7 Vocabulary Check

*Fill in each blank with one of the words or phrases listed below.*

| | | | | | |
|---|---|---|---|---|---|
| index | rationalizing | conjugate | principal square root | cube root | midpoint |
| complex number | like radicals | radicand | imaginary unit | distance | |

1. The _____ of $\sqrt{3} + 2$ is $\sqrt{3} - 2$.

2. The _____ of a nonnegative number $a$ is written as $\sqrt{a}$.

3. The process of writing a radical expression as an equivalent expression but without a radical in the denominator is called _____ the denominator.

4. The _____, written $i$, is the number whose square is $-1$.

5. The _____ of a number is written as $\sqrt[3]{a}$.

6. In the notation $\sqrt[n]{a}$, $n$ is called the _____ and $a$ is called the _____.

7. Radicals with the same index and the same radicand are called _____.

8. A(n) _____ is a number that can be written in the form $a + bi$, where $a$ and $b$ are real numbers.

9. The _____ formula is $d = \sqrt{(x_2 - x_1)^2 + (y_2 - y_1)^2}$.

10. The _____ formula is $\left( \dfrac{x_1 + x_2}{2}, \dfrac{y_1 + y_2}{2} \right)$.

**Helpful Hint**

▶ Are you preparing for your test? Don't forget to take the Chapter 7 Test on page 582. Then check your answers at the back of the text and use the Chapter Test Prep Videos to see the fully worked-out solutions to any of the exercises you want to review.

## 7 Chapter Highlights

| Definitions and Concepts | Examples |
|---|---|
| **Section 7.1    Radical Expressions and Radical Functions** | |

| Definitions and Concepts | Examples |
|---|---|
| The **positive**, or **principal**, **square root** of a nonnegative number $a$ is written as $\sqrt{a}$. <br><br> $\sqrt{a} = b$ only if $b^2 = a$ and $b \geq 0$ | $\sqrt{36} = 6$    $\sqrt{\dfrac{9}{100}} = \dfrac{3}{10}$ |
| The **negative square root** of $a$ is written as $-\sqrt{a}$. | $-\sqrt{36} = -6$    $-\sqrt{0.04} = -0.2$ |
| The **cube root** of a real number $a$ is written as $\sqrt[3]{a}$. <br><br> $\sqrt[3]{a} = b$ only if $b^3 = a$ | $\sqrt[3]{27} = 3$    $\sqrt[3]{-\dfrac{1}{8}} = -\dfrac{1}{2}$ <br><br> $\sqrt[3]{y^6} = y^2$    $\sqrt[3]{64x^9} = 4x^3$ |
| If $n$ is an even positive integer, then $\sqrt[n]{a^n} = \lvert a \rvert$. | $\sqrt{(-3)^2} = \lvert -3 \rvert = 3$ |
| If $n$ is an odd positive integer, then $\sqrt[n]{a^n} = a$. | $\sqrt{(-7)^3} = -7$ |
| A **radical function** in $x$ is a function defined by an expression containing a root of $x$. | If $f(x) = \sqrt{x} + 2$, <br><br> $f(1) = \sqrt{1} + 2 = 1 + 2 = 3$ <br> $f(3) = \sqrt{3} + 2 \approx 3.73$ |

| Definitions and Concepts | Examples |
|---|---|

### Section 7.2    Rational Exponents

$a^{1/n} = \sqrt[n]{a}$ if $\sqrt[n]{a}$ is a real number.

If $m$ and $n$ are positive integers greater than 1 with $\dfrac{m}{n}$ in lowest terms and $\sqrt[n]{a}$ is a real number, then

$$a^{m/n} = (a^{1/n})^m = (\sqrt[n]{a})^m$$

$a^{-m/n} = \dfrac{1}{a^{m/n}}$ as long as $a^{m/n}$ is a nonzero number.

**Exponent rules** are true for rational exponents.

$81^{1/2} = \sqrt{81} = 9$

$(-8x^3)^{1/3} = \sqrt[3]{-8x^3} = -2x$

$4^{5/2} = (\sqrt{4})^5 = 2^5 = 32$

$27^{2/3} = (\sqrt[3]{27})^2 = 3^2 = 9$

$16^{-3/4} = \dfrac{1}{16^{3/4}} = \dfrac{1}{(\sqrt[4]{16})^3} = \dfrac{1}{2^3} = \dfrac{1}{8}$

$x^{2/3} \cdot x^{-5/6} = x^{2/3-5/6} = x^{-1/6} = \dfrac{1}{x^{1/6}}$

$(8^{14})^{1/7} = 8^2 = 64$

$\dfrac{a^{4/5}}{a^{-2/5}} = a^{4/5-(-2/5)} = a^{6/5}$

### Section 7.3    Simplifying Radical Expressions

**Product and Quotient Rules**

If $\sqrt[n]{a}$ and $\sqrt[n]{b}$ are real numbers,

$$\sqrt[n]{a} \cdot \sqrt[n]{b} = \sqrt[n]{a \cdot b}$$

$$\dfrac{\sqrt[n]{a}}{\sqrt[n]{b}} = \sqrt[n]{\dfrac{a}{b}}, \text{ provided } \sqrt[n]{b} \neq 0$$

A radical of the form $\sqrt[n]{a}$ is **simplified** when $a$ contains no factors that are perfect $n$th powers.

Multiply or divide as indicated:

$\sqrt{11} \cdot \sqrt{3} = \sqrt{33}$

$\dfrac{\sqrt[3]{40x}}{\sqrt[3]{5x}} = \sqrt[3]{8} = 2$

$\sqrt{40} = \sqrt{4 \cdot 10} = 2\sqrt{10}$

$\sqrt{36x^5} = \sqrt{36x^4 \cdot x} = 6x^2\sqrt{x}$

$\sqrt[3]{24x^7y^3} = \sqrt[3]{8x^6y^3 \cdot 3x} = 2x^2y\sqrt[3]{3x}$

**Distance Formula**

The distance $d$ between two points $(x_1, y_1)$ and $(x_2, y_2)$ is given by

$$d = \sqrt{(x_2 - x_1)^2 + (y_2 - y_1)^2}$$

Find the distance between points $(-1, 6)$ and $(-2, -4)$.
Let $(x_1, y_1) = (-1, 6)$ and $(x_2, y_2) = (-2, -4)$.

$$\begin{aligned} d &= \sqrt{(x_2 - x_1)^2 + (y_2 - y_1)^2} \\ &= \sqrt{(-2 - (-1))^2 + (-4 - 6)^2} \\ &= \sqrt{1 + 100} = \sqrt{101} \end{aligned}$$

**Midpoint Formula**

The midpoint of the line segment whose endpoints are $(x_1, y_1)$ and $(x_2, y_2)$ is the point with coordinates

$$\left(\dfrac{x_1 + x_2}{2}, \dfrac{y_1 + y_2}{2}\right)$$

Find the midpoint of the line segment whose endpoints are $(-1, 6)$ and $(-2, -4)$.

$$\left(\dfrac{-1 + (-2)}{2}, \dfrac{6 + (-4)}{2}\right)$$

The midpoint is $\left(-\dfrac{3}{2}, 1\right)$.

### Section 7.4    Adding, Subtracting, and Multiplying Radical Expressions

Radicals with the same index and the same radicand are **like radicals.**

The distributive property can be used to add like radicals.

$5\sqrt{6} + 2\sqrt{6} = (5 + 2)\sqrt{6} = 7\sqrt{6}$

$-\sqrt[3]{3x} - 10\sqrt[3]{3x} + 3\sqrt[3]{10x}$

$= (-1 - 10)\sqrt[3]{3x} + 3\sqrt[3]{10x}$

$= -11\sqrt[3]{3x} + 3\sqrt[3]{10x}$

*(continued)*

| Definitions and Concepts | Examples |
|---|---|

### Section 7.4   Adding, Subtracting, and Multiplying Radical Expressions (*continued*)

Radical expressions are multiplied by using many of the same properties used to multiply polynomials.

Multiply:

$$(\sqrt{5} - \sqrt{2x})(\sqrt{2} + \sqrt{2x})$$
$$= \sqrt{10} + \sqrt{10x} - \sqrt{4x} - 2x$$
$$= \sqrt{10} + \sqrt{10x} - 2\sqrt{x} - 2x$$

$$(2\sqrt{3} - \sqrt{8x})(2\sqrt{3} + \sqrt{8x})$$
$$= 4(3) - 8x = 12 - 8x$$

### Section 7.5   Rationalizing Numerators and Denominators of Radical Expressions

The **conjugate** of $a + b$ is $a - b$.
The process of writing the denominator of a radical expression without a radical is called **rationalizing the denominator.**

The conjugate of $\sqrt{7} + \sqrt{3}$ is $\sqrt{7} - \sqrt{3}$.
Rationalize each denominator:

$$\frac{\sqrt{5}}{\sqrt{3}} = \frac{\sqrt{5} \cdot \sqrt{3}}{\sqrt{3} \cdot \sqrt{3}} = \frac{\sqrt{15}}{3}$$

$$\frac{6}{\sqrt{7} + \sqrt{3}} = \frac{6(\sqrt{7} - \sqrt{3})}{(\sqrt{7} + \sqrt{3})(\sqrt{7} - \sqrt{3})}$$
$$= \frac{6(\sqrt{7} - \sqrt{3})}{7 - 3}$$
$$= \frac{6(\sqrt{7} - \sqrt{3})}{4} = \frac{3(\sqrt{7} - \sqrt{3})}{2}$$

The process of writing the numerator of a radical expression without a radical is called **rationalizing the numerator.**

Rationalize each numerator:

$$\frac{\sqrt[3]{9}}{\sqrt[3]{5}} = \frac{\sqrt[3]{9} \cdot \sqrt[3]{3}}{\sqrt[3]{5} \cdot \sqrt[3]{3}} = \frac{\sqrt[3]{27}}{\sqrt[3]{15}} = \frac{3}{\sqrt[3]{15}}$$

$$\frac{\sqrt{9} + \sqrt{3x}}{12} = \frac{(\sqrt{9} + \sqrt{3x})(\sqrt{9} - \sqrt{3x})}{12(\sqrt{9} - \sqrt{3x})}$$
$$= \frac{9 - 3x}{12(\sqrt{9} - \sqrt{3x})}$$
$$= \frac{3(3 - x)}{3 \cdot 4(3 - \sqrt{3x})} = \frac{3 - x}{4(3 - \sqrt{3x})}$$

### Section 7.6   Radical Equations and Problem Solving

**Solving A Radical Equation**

**Step 1:** Write the equation so that one radical is by itself on one side of the equation.

**Step 2:** Raise each side of the equation to a power equal to the index of the radical and simplify.

**Step 3:** If the equation still contains a radical, repeat Steps 1 and 2. If not, solve the equation.

**Step 4:** Check all proposed solutions in the original equation.

Solve:  $x = \sqrt{4x + 9} + 3$

**1.**    $x - 3 = \sqrt{4x + 9}$

**2.**    $(x - 3)^2 = (\sqrt{4x + 9})^2$
$$x^2 - 6x + 9 = 4x + 9$$

**3.**    $x^2 - 10x = 0$
$$x(x - 10) = 0$$
$$x = 0 \quad \text{or} \quad x = 10$$

**4.** The proposed solution 10 checks, but 0 does not. The solution set is {10}.

| Definitions and Concepts | Examples |
|---|---|

## Section 7.7   Complex Numbers

A **complex number** is a number that can be written in the form $a + bi$, where $a$ and $b$ are real numbers.

$$i^2 = -1 \quad \text{and} \quad i = \sqrt{-1}$$

Simplify:   $\sqrt{-9}$

$$\sqrt{-9} = \sqrt{-1 \cdot 9} = \sqrt{-1} \cdot \sqrt{9} = i \cdot 3, \text{ or } 3i$$

| **Complex Numbers** | **Written in Form $a + bi$** |
|---|---|
| 12 | $12 + 0i$ |
| $-5i$ | $0 + (-5)i$ |
| $-2 - 3i$ | $-2 + (-3)i$ |

Multiply.

$$\sqrt{-3} \cdot \sqrt{-7} = i\sqrt{3} \cdot i\sqrt{7}$$
$$= i^2\sqrt{21}$$
$$= -\sqrt{21}$$

**To add or subtract complex numbers,** add or subtract their real parts and add or subtract their imaginary parts.

Perform each indicated operation.

$$(-3 + 2i) - (7 - 4i) = -3 + 2i - 7 + 4i$$
$$= -10 + 6i$$

**To multiply complex numbers,** multiply as though they were binomials.

$$(-7 - 2i)(6 + i) = -42 - 7i - 12i - 2i^2$$
$$= -42 - 19i - 2(-1)$$
$$= -42 - 19i + 2$$
$$= -40 - 19i$$

The complex numbers $(a + bi)$ and $(a - bi)$ are called **complex conjugates.**

The complex conjugate of

$$(3 + 6i) \text{ is } (3 - 6i).$$

The product of a complex number and its complex conjugate is a real number.

$$(3 - 6i)(3 + 6i) = 9 - 36i^2$$
$$= 9 - 36(-1) = 9 + 36 = 45$$

**To divide complex numbers,** multiply the numerator and the denominator by the conjugate of the denominator.

Divide:

$$\frac{4}{2 - i} = \frac{4(2 + i)}{(2 - i)(2 + i)}$$
$$= \frac{4(2 + i)}{4 - i^2}$$
$$= \frac{4(2 + i)}{4 - (-1)}$$
$$= \frac{8 + 4i}{5} = \frac{8}{5} + \frac{4}{5}i$$

**(7.1)** *Find each root. Assume that all variables represent positive real numbers.*

**1.** $\sqrt{81}$

**2.** $\sqrt[4]{81}$

**3.** $\sqrt[3]{-8}$

**4.** $\sqrt[4]{-16}$

**5.** $-\sqrt{\dfrac{1}{49}}$

**6.** $\sqrt{x^{64}}$

**7.** $-\sqrt{36}$

**8.** $\sqrt[3]{64}$

**9.** $\sqrt[3]{-a^6 b^9}$

**10.** $\sqrt{16a^4 b^{12}}$

**11.** $\sqrt[5]{32a^5 b^{10}}$

**12.** $\sqrt[5]{-32x^{15} y^{20}}$

**13.** $\sqrt{\dfrac{x^{12}}{36y^2}}$

**14.** $\sqrt[3]{\dfrac{27y^3}{z^{12}}}$

*Simplify. Use absolute value bars when necessary.*

**15.** $\sqrt{x^2}$

**16.** $\sqrt[4]{(x^2 - 4)^4}$

**17.** $\sqrt[3]{(-27)^3}$

**18.** $\sqrt[5]{(-5)^5}$

**19.** $-\sqrt[5]{x^5}$

**20.** $-\sqrt[3]{x^3}$

**21.** $\sqrt[4]{16(2y + z)^4}$

**22.** $\sqrt{25(x - y)^2}$

**23.** $\sqrt[5]{y^5}$

**24.** $\sqrt[6]{x^6}$

**25.** If $f(x) = \sqrt{x} + 3$, find $f(0)$ and $f(9)$.

**26.** If $g(x) = \sqrt[3]{x} - 3$, find $g(11)$ and $g(20)$.

**(7.2)** *Evaluate.*

**27.** $\left(\dfrac{1}{81}\right)^{1/4}$

**28.** $\left(-\dfrac{1}{27}\right)^{1/3}$

**29.** $(-27)^{-1/3}$

**30.** $(-64)^{-1/3}$

**31.** $-9^{3/2}$

**32.** $64^{-1/3}$

**33.** $(-25)^{5/2}$

**34.** $\left(\dfrac{25}{49}\right)^{-3/2}$

**35.** $\left(\dfrac{8}{27}\right)^{-2/3}$

**36.** $\left(-\dfrac{1}{36}\right)^{-1/4}$

*Write with rational exponents.*

**37.** $\sqrt[3]{x^2}$

**38.** $\sqrt[5]{5x^2 y^3}$

*Write using radical notation.*

**39.** $y^{4/5}$

**40.** $5(xy^2 z^5)^{1/3}$

**41.** $(x + 2)^{-1/3}$

**42.** $(x + 2y)^{-1/2}$

*Simplify each expression. Assume that all variables represent positive real numbers. Write with only positive exponents.*

**43.** $a^{1/3} a^{4/3} a^{1/2}$

**44.** $\dfrac{b^{1/3}}{b^{4/3}}$

**45.** $(a^{1/2} a^{-2})^3$

**46.** $(x^{-3} y^6)^{1/3}$

**47.** $\left(\dfrac{b^{3/4}}{a^{-1/2}}\right)^8$

**48.** $\dfrac{x^{1/4} x^{-1/2}}{x^{2/3}}$

**49.** $\left(\dfrac{49c^{5/3}}{a^{-1/4} b^{5/6}}\right)^{-1}$

**50.** $a^{-1/4}(a^{5/4} - a^{9/4})$

📱 *Use a calculator and write a three-decimal-place approximation of each number.*

**51.** $\sqrt{20}$

**52.** $\sqrt[3]{-39}$

**53.** $\sqrt[4]{726}$

**54.** $56^{1/3}$

**55.** $-78^{3/4}$

**56.** $105^{-2/3}$

*Use rational exponents to write each as a single radical.*

**57.** $\sqrt[3]{2} \cdot \sqrt{7}$

**58.** $\sqrt[3]{3} \cdot \sqrt[4]{x}$

**(7.3)** *Perform each indicated operation and then simplify if possible. Assume that all variables represent positive real numbers.*

**59.** $\sqrt{3} \cdot \sqrt{8}$

**60.** $\sqrt[3]{7y} \cdot \sqrt[3]{x^2z}$

**61.** $\dfrac{\sqrt{44x^3}}{\sqrt{11x}}$

**62.** $\dfrac{\sqrt[4]{a^6b^{13}}}{\sqrt[4]{a^2b}}$

*Simplify.*

**63.** $\sqrt{60}$

**64.** $-\sqrt{75}$

**65.** $\sqrt[3]{162}$

**66.** $\sqrt[3]{-32}$

**67.** $\sqrt{36x^7}$

**68.** $\sqrt[3]{24a^5b^7}$

**69.** $\sqrt{\dfrac{p^{17}}{121}}$

**70.** $\sqrt[3]{\dfrac{y^5}{27x^6}}$

**71.** $\sqrt[4]{\dfrac{xy^6}{81}}$

**72.** $\sqrt{\dfrac{2x^3}{49y^4}}$

△ *The formula for the radius r of a circle of area A is* $r = \sqrt{\dfrac{A}{\pi}}$. *Use this for Exercises 73 and 74.*

**73.** Find the exact radius of a circle whose area is 25 square meters.

📱 **74.** Approximate to two decimal places the radius of a circle whose area is 104 square inches.

*Find the distance between each pair of points. Give the exact value and a three-decimal-place approximation.*

**75.** $(-6, 3)$ and $(8, 4)$

**76.** $(-4, -6)$ and $(-1, 5)$

**77.** $(-1, 5)$ and $(2, -3)$

**78.** $(-\sqrt{2}, 0)$ and $(0, -4\sqrt{6})$

**79.** $(-\sqrt{5}, -\sqrt{11})$ and $(-\sqrt{5}, -3\sqrt{11})$

📱 **80.** $(7.4, -8.6)$ and $(-1.2, 5.6)$

*Find the midpoint of each line segment whose endpoints are given.*

**81.** $(2, 6)$; $(-12, 4)$

**82.** $(-6, -5)$; $(-9, 7)$

**83.** $(4, -6)$; $(-15, 2)$

**84.** $\left(0, -\dfrac{3}{8}\right)$; $\left(\dfrac{1}{10}, 0\right)$

**85.** $\left(\dfrac{3}{4}, -\dfrac{1}{7}\right)$; $\left(-\dfrac{1}{4}, -\dfrac{3}{7}\right)$

**86.** $(\sqrt{3}, -2\sqrt{6})$ and $(\sqrt{3}, -4\sqrt{6})$

**(7.4)** *Perform each indicated operation. Assume that all variables represent positive real numbers.*

**87.** $\sqrt{20} + \sqrt{45} - 7\sqrt{5}$

**88.** $x\sqrt{75x} - \sqrt{27x^3}$

**89.** $\sqrt[3]{128} + \sqrt[3]{250}$

**90.** $3\sqrt[4]{32a^5} - a\sqrt[4]{162a}$

**91.** $\dfrac{5}{\sqrt{4}} + \dfrac{\sqrt{3}}{3}$

**92.** $\sqrt{\dfrac{8}{x^2}} - \sqrt{\dfrac{50}{16x^2}}$

**93.** $2\sqrt{50} - 3\sqrt{125} + \sqrt{98}$

**94.** $2a\sqrt[4]{32b^5} - 3b\sqrt[4]{162a^4b} + \sqrt[4]{2a^4b^5}$

*Multiply and then simplify if possible. Assume that all variables represent positive real numbers.*

**95.** $\sqrt{3}(\sqrt{27} - \sqrt{3})$

**96.** $(\sqrt{x} - 3)^2$

**97.** $(\sqrt{5} - 5)(2\sqrt{5} + 2)$

**98.** $(2\sqrt{x} - 3\sqrt{y})(2\sqrt{x} + 3\sqrt{y})$

**99.** $(\sqrt{a} + 3)(\sqrt{a} - 3)$

**100.** $(\sqrt[3]{a} + 2)^2$

**101.** $(\sqrt[3]{5x} + 9)(\sqrt[3]{5x} - 9)$

**102.** $(\sqrt[3]{a} + 4)(\sqrt[3]{a^2} - 4\sqrt[3]{a} + 16)$

**(7.5)** *Rationalize each denominator. Assume that all variables represent positive real numbers.*

**103.** $\dfrac{3}{\sqrt{7}}$

**104.** $\sqrt{\dfrac{x}{12}}$

**105.** $\dfrac{5}{\sqrt[3]{4}}$

**106.** $\sqrt{\dfrac{24x^5}{3y}}$

**107.** $\sqrt[3]{\dfrac{15x^6y^7}{z^2}}$

**108.** $\sqrt[4]{\dfrac{81}{8x^{10}}}$

**109.** $\dfrac{3}{\sqrt{y} - 2}$

**110.** $\dfrac{\sqrt{2} - \sqrt{3}}{\sqrt{2} + \sqrt{3}}$

*Rationalize each numerator. Assume that all variables represent positive real numbers.*

**111.** $\dfrac{\sqrt{11}}{3}$

**112.** $\sqrt{\dfrac{18}{y}}$

**113.** $\dfrac{\sqrt[3]{9}}{7}$

**114.** $\sqrt{\dfrac{24x^5}{3y^2}}$

**115.** $\sqrt[3]{\dfrac{xy^2}{10z}}$

**116.** $\dfrac{\sqrt{x} + 5}{-3}$

**(7.6)** *Solve each equation.*

**117.** $\sqrt{y - 7} = 5$

**118.** $\sqrt{2x} + 10 = 4$

**119.** $\sqrt[3]{2x - 6} = 4$

**120.** $\sqrt{x + 6} = \sqrt{x + 2}$

**121.** $2x - 5\sqrt{x} = 3$

**122.** $\sqrt{x + 9} = 2 + \sqrt{x - 7}$

*Find each unknown length.*

△ **123.**

3 cm

3 cm

△ **124.**

7 ft

$8\sqrt{3}$ ft

△ **125.** Craig and Daniel Cantwell want to determine the distance $x$ across a pond on their property. They are able to measure the distances shown on the following diagram. Find how wide the lake is at the crossing point indicated by the triangle to the nearest tenth of a foot.

△ **126.** Andrea Roberts, a pipefitter, needs to connect two underground pipelines that are offset by 3 feet, as pictured in the diagram. Neglecting the joints needed to join the pipes, find the length of the shortest possible connecting pipe rounded to the nearest hundredth of a foot.

**(7.7)** *Perform each indicated operation and simplify. Write the results in the form a + bi.*

**127.** $\sqrt{-8}$

**128.** $-\sqrt{-6}$

**129.** $\sqrt{-4} + \sqrt{-16}$

**130.** $\sqrt{-2} \cdot \sqrt{-5}$

**131.** $(12 - 6i) + (3 + 2i)$

**132.** $(-8 - 7i) - (5 - 4i)$

**133.** $(2i)^6$

**134.** $(3i)^4$

**135.** $-3i(6 - 4i)$

**136.** $(3 + 2i)(1 + i)$

**137.** $(2 - 3i)^2$

**138.** $(\sqrt{6} - 9i)(\sqrt{6} + 9i)$

**139.** $\dfrac{2 + 3i}{2i}$

**140.** $\dfrac{1 + i}{-3i}$

## Mixed Review

*Simplify. Use absolute value bars when necessary.*

**141.** $\sqrt[3]{x^3}$

**142.** $\sqrt{(x + 2)^2}$

*Simplify. Assume that all variables represent positive real numbers. If necessary, write answers with positive exponents only.*

**143.** $-\sqrt{100}$

**144.** $\sqrt[3]{-x^{12}y^3}$

**145.** $\sqrt[4]{\dfrac{y^{20}}{16x^{12}}}$

**146.** $9^{1/2}$

**147.** $64^{-1/2}$

**148.** $\left(\dfrac{27}{64}\right)^{-2/3}$

**149.** $\dfrac{(x^{2/3}x^{-3})^3}{x^{-1/2}}$

**150.** $\sqrt{200x^9}$

**151.** $\sqrt{\dfrac{3n^3}{121m^{10}}}$

**152.** $3\sqrt{20} - 7x\sqrt[3]{40} + 3\sqrt[3]{5x^3}$

**153.** $(2\sqrt{x} - 5)^2$

**154.** Find the distance between $(-3, 5)$ and $(-8, 9)$.

**155.** Find the midpoint of the line segment joining $(-3, 8)$ and $(11, 24)$.

*Rationalize each denominator.*

**156.** $\dfrac{7}{\sqrt{13}}$

**157.** $\dfrac{2}{\sqrt{x} + 3}$

*Solve.*

**158.** $\sqrt{x} + 2 = x$

**159.** $\sqrt{2x - 1} + 2 = x$

# Chapter 7 Test

Step-by-step test solutions are found on the Chapter Test Prep Videos. Where available: MyMathLab® or You Tube™

**Answers**

*Raise to the power or find the root. Assume that all variables represent positive real numbers. Write with only positive exponents.*

**1.** $\sqrt{216}$

**2.** $-\sqrt[4]{x^{64}}$

**3.** $\left(\dfrac{1}{125}\right)^{1/3}$

**4.** $\left(\dfrac{1}{125}\right)^{-1/3}$

**5.** $\left(\dfrac{8x^3}{27}\right)^{2/3}$

**6.** $\sqrt[3]{-a^{18}b^9}$

**7.** $\left(\dfrac{64c^{4/3}}{a^{-2/3}b^{5/6}}\right)^{1/2}$

**8.** $a^{-2/3}(a^{5/4} - a^3)$

*Find each root. Use absolute value bars when necessary.*

**9.** $\sqrt[4]{(4xy)^4}$

**10.** $\sqrt[3]{(-27)^3}$

*Rationalize each denominator. Assume that all variables represent positive real numbers.*

**11.** $\sqrt{\dfrac{9}{y}}$

**12.** $\dfrac{4 - \sqrt{x}}{4 + 2\sqrt{x}}$

**13.** $\sqrt[3]{\dfrac{8}{9x}}$

**14.** Rationalize the numerator of $\dfrac{\sqrt{6} + x}{8}$ and simplify.

*Perform each indicated operation. Assume that all variables represent positive real numbers.*

**15.** $\sqrt{125x^3} - 3\sqrt{20x^3}$

**16.** $\sqrt{3}(\sqrt{16} - \sqrt{2})$

**17.** $(\sqrt{x} + 1)^2$

**18.** $(\sqrt{2} - 4)(\sqrt{3} + 1)$

**19.** $(\sqrt{5} + 5)(\sqrt{5} - 5)$

Answers

1. _____

2. _____

3. _____

4. _____

5. _____

6. _____

7. _____

8. _____

9. _____

10. _____

11. _____

12. _____

13. _____

14. _____

15. _____

16. _____

17. _____

18. _____

19. _____

Use a calculator to approximate each number to three decimal places.

**20.** $\sqrt{561}$

**21.** $386^{-2/3}$

Solve.

**22.** $x = \sqrt{x - 2} + 2$   **23.** $\sqrt{x^2 - 7} + 3 = 0$   **24.** $\sqrt{x + 5} = \sqrt{2x - 1}$

Perform each indicated operation and simplify. Write the results in the form $a + bi$.

**25.** $\sqrt{-2}$

**26.** $-\sqrt{-8}$

**27.** $(12 - 6i) - (12 - 3i)$

**28.** $(6 - 2i)(6 + 2i)$

**29.** $(4 + 3i)^2$

**30.** $\dfrac{1 + 4i}{1 - i}$

△ **31.** Find $x$.

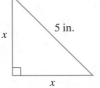

5 in.

**32.** If $g(x) = \sqrt{x + 2}$, find $g(0)$ and $g(23)$.

**33.** Find the distance between the points $(-6, 3)$ and $(-8, -7)$.

**34.** Find the distance between the points $(-2\sqrt{5}, \sqrt{10})$ and $(-\sqrt{5}, 4\sqrt{10})$.

**35.** Find the midpoint of the line segment whose endpoints are $(-2, -5)$ and $(-6, 12)$.

**36.** Find the midpoint of the line segment whose endpoints are $\left(-\dfrac{2}{3}, -\dfrac{1}{5}\right)$ and $\left(-\dfrac{1}{3}, \dfrac{4}{5}\right)$.

Solve.

**37.** The function $V = \sqrt{2.5r}$ can be used to estimate the maximum safe velocity, $V$, in miles per hour, at which a car can travel if it is driven along a curved road with a *radius of curvature, r,* in feet. To the nearest whole number, find the maximum safe speed if a cloverleaf exit on an expressway has a radius of curvature of 300 feet.

**38.** Use the formula from Exercise 37 to find the radius of curvature if the safe velocity is 30 miles per hour.

**20.** _____

**21.** _____

**22.** _____

**23.** _____

**24.** _____

**25.** _____

**26.** _____

**27.** _____

**28.** _____

**29.** _____

**30.** _____

**31.** _____

**32.** _____

**33.** _____

**34.** _____

**35.** _____

**36.** _____

**37.** _____

**38.** _____

## Answers

1. _____

2. _____

3. _____

4. _____

5. _____

6. _____

7. _____

8. _____

9. _____

10. _____

11. _____

12. _____

13. _____

14. a. _____

b. _____

c. _____

15. _____

16. _____

*Simplify. Assume that a and b are integers and that x and y are not 0.*

1. $x^{-b}(2x^b)^2$

2. $\left(-\dfrac{1}{3}x^a\right)\left(-\dfrac{2}{5}x^a\right)$

3. $\dfrac{(y^{3a})^2}{y^{a-6}}$

4. $\dfrac{2.8y^{7a}}{-2y^{-2a}}$

5. Solve:
$$|3x + 2| = |5x - 8|$$

6. Solve:
$$|3x - 2| + 5 = 5$$

7. Write an equation of the line containing the point $(4, 4)$ and parallel to the line $2x + y = -6$. Write the equation in slope-intercept form, $y = mx + b$.

8. Solve: $\dfrac{a - 1}{2} + a = 2 - \dfrac{2a + 7}{8}$

9. Use the elimination method to solve the system: $\begin{cases} x - 5y = -12 \\ -x + y = 4 \end{cases}$

10. Find the slope of $y = -3$.

11. Solve the system:
$$\begin{cases} x - 5y - 2z = 6 \\ -2x + 10y + 4z = -12 \\ \dfrac{1}{2}x - \dfrac{5}{2}y - z = 3 \end{cases}$$

12. Use the substitution method to solve the system.
$$\begin{cases} \dfrac{x}{6} - \dfrac{y}{2} = 1 \\ \dfrac{x}{3} - \dfrac{y}{4} = 2 \end{cases}$$

13. Use matrices to solve the system:
$$\begin{cases} 2x - y = 3 \\ 4x - 2y = 5 \end{cases}$$

14. Simplify each expression.
   a. $2(x - 3) + (5x + 3)$
   b. $4(3x + 2) - 3(5x - 1)$
   c. $7x + 2(x - 7) - 3x$

15. Add:
$$(7x^3y - xy^3 + 11) + (6x^3y - 4)$$

16. Use scientific notation to simplify and write the answer in scientific notation.
$$\dfrac{0.0000035 \times 4000}{0.28}$$

If $P(x) = 3x^4 - 2x^2 - 5$, find the following.

**17.** $P(1)$      **18.** $P(0)$      **19.** $P(-2)$      **20.** $P\left(\dfrac{1}{2}\right)$

*Multiply.*

**21.** $(x + 5)^2$

**22.** $(y - 2)(3y + 4)$

**23.** $(4m^2 - 3n)^2$

**24.** $(3y - 1)(2y^2 + 3y - 1)$

**25.** Use synthetic division to divide $2x^3 - x^2 - 13x + 1$ by $x - 3$.

**26.** Use synthetic division to divide $4y^3 - 12y^2 - y + 12$ by $y - 3$.

*Factor.*

**27.** $ab - 6a + 2b - 12$

**28.** Factor: $x^3 - x^2 + 4x - 4$

**29.** $2n^2 - 38n + 80$

**30.** $2x^3 - 16$

**31.** $16x^2 + 24xy + 9y^2$

**32.** $x^2 + 2x + 1 - y^2$

**33.** $50 - 8y^2$

**34.** $x^4 - 1$

17. _____

18. _____

19. _____

20. _____

21. _____

22. _____

23. _____

24. _____

25. _____

26. _____

27. _____

28. _____

29. _____

30. _____

31. _____

32. _____

33. _____

34. _____

**35.** _____

**36.** _____

**37.** _____

**38. a.** _____

   **b.** _____

**39.** _____

**40.** _____

**41.** _____

**42.** _____

**43.** _____

**44. a.** _____

   **b.** _____

   **c.** _____

**45.** _____

**46.** _____

**47.** _____

**48.** _____

**49.** _____

**50.** _____

**35.** Solve: $x(2x - 7) = 4$

**36.** Divide $x^3 - 2x^2 + 3x - 6$ by $x - 2$.

**37.** Solve: $x^3 = 4x$

**38.** Simplify each complex fraction.

**a.** $\dfrac{\dfrac{y - 2}{16}}{\dfrac{2y + 3}{12}}$     **b.** $\dfrac{\dfrac{x}{16} - \dfrac{1}{x}}{1 - \dfrac{4}{x}}$

_Simplify each rational expression._

**39.** $\dfrac{2 + x}{x + 2}$

**40.** $\dfrac{a^3 - 8}{2 - a}$

**41.** $\dfrac{2x^2}{10x^3 - 2x^2}$

**42.** $\dfrac{3a^2 - 3}{a^3 + 5a^2 - a - 5}$

**43.** Add: $\dfrac{5}{7z^2} + \dfrac{x}{7z^2}$

**44.** Perform the indicated operations.

**a.** $\dfrac{3}{xy^2} - \dfrac{2}{3x^2y}$     **b.** $\dfrac{5x}{x + 3} - \dfrac{2x}{x - 3}$

**c.** $\dfrac{x}{x - 2} - \dfrac{5}{2 - x}$

**45.** Solve: $\dfrac{3}{x} - \dfrac{x + 21}{3x} = \dfrac{5}{3}$

**46.** Solve: $\dfrac{28}{9 - a^2} = \dfrac{2a}{a - 3} + \dfrac{6}{a + 3}$

_Write each expression with a positive exponent, and then simplify._

**47.** $16^{-3/4}$

**48.** $81^{-3/4}$

**49.** $(-27)^{-2/3}$

**50.** $(-125)^{-2/3}$

# Quadratic Equations and Functions

8

## What Are Supertall and Megatall Buildings?

The term "supertall" refers to a building over 300 meters. According to the Council on Tall Buildings and Urban Habitat (CTBUH), this term is no longer adequate to rank the tallest buildings in the world. By 2020, it is predicted that 8 buildings will have a height over 600 meters tall. We now use the term "megatall" to describe these buildings over 600 meters. By 2020, it is predicted that the tallest building in the world, the Kingdom Tower, will have a height of 1000 meters (about 3280 feet).

In Section 8.1, Exercises 75–78, we will explore the heights of some tall buildings. (*Source*: Council on Tall Buildings and Urban Habitat)

World's Megatall Buildings in 2020 (Projected)
(Including the Tallest Building in the U.S.)

Average Height of Tallest 20 Buildings in World

An important part of algebra is learning to model and solve problems. Often, the model of a problem is a quadratic equation or a function containing a second-degree polynomial. In this chapter, we continue the work from Chapter 5, solving polynomial equations in one variable by factoring. Two other methods of solving quadratic equations are analyzed in this chapter, along with methods of solving nonlinear inequalities in one variable and the graphs of quadratic functions.

587

# Solving Quadratic Equations by Completing the Square

## Objectives

**A** Use the Square Root Property to Solve Quadratic Equations.

**B** Write Perfect Square Trinomials.

**C** Solve Quadratic Equations by Completing the Square.

**D** Use Quadratic Equations to Solve Problems.

## Objective A  Using the Square Root Property

In Chapter 5, we solved quadratic equations by factoring. Recall that a **quadratic,** or **second-degree, equation** is an equation that can be written in the form $ax^2 + bx + c = 0$, where $a, b,$ and $c$ are real numbers and $a$ is not 0. To solve a quadratic equation such as $x^2 = 9$ by factoring, we use the zero-factor property. To use the zero-factor property, the equation must first be written in the standard form $ax^2 + bx + c = 0$.

$$x^2 = 9$$
$$x^2 - 9 = 0 \qquad \text{Subtract 9 from both sides to write in standard form.}$$
$$(x + 3)(x - 3) = 0 \qquad \text{Factor.}$$
$$x + 3 = 0 \quad \text{or} \quad x - 3 = 0 \quad \text{Set each factor equal to 0.}$$
$$x = -3 \qquad\qquad x = 3 \quad \text{Solve.}$$

The solution set is $\{-3, 3\}$, the positive and negative square roots of 9.

Not all quadratic equations can be solved by factoring, so we need to explore other methods. Notice that the solutions of the equation $x^2 = 9$ are two numbers whose square is 9:

$$3^2 = 9 \qquad \text{and} \qquad (-3)^2 = 9$$

Thus, we can solve the equation $x^2 = 9$ by taking the square root of both sides. Be sure to include both $\sqrt{9}$ and $-\sqrt{9}$ as solutions since both $\sqrt{9}$ and $-\sqrt{9}$ are numbers whose square is 9.

$$x^2 = 9$$
$$x = \pm\sqrt{9} \quad \text{The notation } \pm\sqrt{9} \text{ (read as "plus or minus } \sqrt{9}\text{") indicates the pair of numbers } +\sqrt{9} \text{ and } -\sqrt{9}.$$
$$x = \pm 3$$

This illustrates the square root property.

### Helpful Hint

The notation $\pm 3$, for example, is read as "plus or minus 3." It is a shorthand notation for the pair of numbers $+3$ and $-3$.

### Square Root Property

If $b$ is a real number and if $a^2 = b$, then $a = \pm\sqrt{b}$.

**Practice 1**

Use the square root property to solve $x^2 = 45$.

**Example 1** Use the square root property to solve $x^2 = 50$.

**Solution:**

$$x^2 = 50$$
$$x = \pm\sqrt{50} \quad \text{Use the square root property.}$$
$$x = \pm 5\sqrt{2} \quad \text{Simplify the radical.}$$

**Answer**

**1.** $\{3\sqrt{5}, -3\sqrt{5}\}$

**Check:**    Let $x = 5\sqrt{2}$.                           Let $x = -5\sqrt{2}$.

$$x^2 = 50 \qquad\qquad\qquad\qquad x^2 = 50$$
$$(5\sqrt{2})^2 \overset{?}{=} 50 \qquad\qquad (-5\sqrt{2})^2 \overset{?}{=} 50$$
$$25 \cdot 2 \overset{?}{=} 50 \qquad\qquad\qquad 25 \cdot 2 \overset{?}{=} 50$$
$$50 = 50 \quad \text{True} \qquad\qquad 50 = 50 \quad \text{True}$$

The solution set is $\{5\sqrt{2}, -5\sqrt{2}\}$.

◼ Work Practice 1

**Example 2**  Use the square root property to solve $2x^2 = 14$.

**Solution:**  First we get the squared variable alone on one side of the equation.

$$2x^2 = 14$$
$$x^2 = 7 \qquad \text{Divide both sides by 2.}$$
$$x = \pm\sqrt{7} \quad \text{Use the square root property.}$$

**Check:**     Let $x = \sqrt{7}$.                              Let $x = -\sqrt{7}$.

$$2x^2 = 14 \qquad\qquad\qquad\qquad 2x^2 = 14$$
$$2(\sqrt{7})^2 \overset{?}{=} 14 \qquad\qquad 2(-\sqrt{7})^2 \overset{?}{=} 14$$
$$2 \cdot 7 \overset{?}{=} 14 \qquad\qquad\qquad 2 \cdot 7 \overset{?}{=} 14$$
$$14 = 14 \quad \text{True} \qquad\qquad 14 = 14 \quad \text{True}$$

The solution set is $\{\sqrt{7}, -\sqrt{7}\}$.

◼ Work Practice 2

**Example 3**  Use the square root property to solve $(x + 1)^2 = 12$.

**Solution:**

$$(x + 1)^2 = 12$$
$$x + 1 = \pm\sqrt{12} \qquad \text{Use the square root property.}$$
$$x + 1 = \pm 2\sqrt{3} \qquad \text{Simplify the radical.}$$
$$x = -1 \pm 2\sqrt{3} \quad \text{Subtract 1 from both sides.}$$

**Check:**  Below is a check for $-1 + 2\sqrt{3}$. The check for $-1 - 2\sqrt{3}$ is almost the same and is left for you to do on your own.

$$(x + 1)^2 = 12$$
$$(-1 + 2\sqrt{3} + 1)^2 \overset{?}{=} 12$$
$$(2\sqrt{3})^2 \overset{?}{=} 12$$
$$4 \cdot 3 \overset{?}{=} 12$$
$$12 = 12 \quad \text{True}$$

The solution set is $\{-1 + 2\sqrt{3}, -1 - 2\sqrt{3}\}$.

◼ Work Practice 3

**Practice 2**

Use the square root property to solve $5x^2 = 55$.

**Practice 3**

Use the square root property to solve $(x + 2)^2 = 18$.

**Helpful Hint**    Don't forget that $-1 \pm 2\sqrt{3}$, for example, means $-1 + 2\sqrt{3}$ and $-1 - 2\sqrt{3}$. In other words, the equation in Example 3 has two solutions.

**Answers**
2. $\{\sqrt{11}, -\sqrt{11}\}$
3. $\{-2 + 3\sqrt{2}, -2 - 3\sqrt{2}\}$

## Practice 4

Use the square root property to solve $(3x - 1)^2 = -4$.

**Example 4** Use the square root property to solve $(2x - 5)^2 = -16$.

**Solution:**

$$(2x - 5)^2 = -16$$
$$2x - 5 = \pm\sqrt{-16} \quad \text{Use the square root property.}$$
$$2x - 5 = \pm 4i \quad \text{Simplify the radical.}$$
$$2x = 5 \pm 4i \quad \text{Add 5 to both sides.}$$
$$x = \frac{5 \pm 4i}{2} \quad \text{Divide both sides by 2.}$$

Check each proposed solution in the original equation to see that the solution set is $\left\{ \dfrac{5 + 4i}{2}, \dfrac{5 - 4i}{2} \right\}$.

■ Work Practice 4

✓**Concept Check** How do you know just by looking that $(x - 2)^2 = -81$ has complex solutions?

## Objective B  Writing Perfect Square Trinomials ▶

Notice from Examples 3 and 4 that, if we write a quadratic equation so that one side is the square of a binomial, we can solve by using the square root property. To write the square of a binomial, we must have a perfect square trinomial. Recall that a perfect square trinomial is a trinomial that can be factored into two identical binomial factors, that is, as a binomial squared.

| **Perfect Square Trinomials** | **Factored Form** |
|---|---|
| $x^2 + 8x + 16$ | $(x + 4)^2$ |
| $x^2 - 6x + 9$ | $(x - 3)^2$ |
| $x^2 + 3x + \dfrac{9}{4}$ | $\left( x + \dfrac{3}{2} \right)^2$ |

Notice that for each perfect square trinomial, **the constant term of the trinomial is the square of half the coefficient of the $x$-term.** For example,

$$x^2 + 8x + 16 \qquad\qquad x^2 - 6x + 9$$

$$\frac{1}{2}(8) = 4 \text{ and } 4^2 = 16 \qquad \frac{1}{2}(-6) = -3 \text{ and } (-3)^2 = 9$$

## Practice 5

Add the proper constant to $x^2 + 12x$ so that the result is a perfect square trinomial. Then factor.

**Example 5** Add the proper constant to $x^2 + 6x$ so that the result is a perfect square trinomial. Then factor.

**Solution:** We add the square of half the coefficient of $x$.

$$x^2 + 6x + 9 \quad = \quad (x + 3)^2 \quad \text{In factored form}$$

$$\frac{1}{2}(6) = 3 \text{ and } 3^2 = 9$$

■ Work Practice 5

**Answers**

**4.** $\left\{ \dfrac{1 - 2i}{3}, \dfrac{1 + 2i}{3} \right\}$

**5.** $x^2 + 12x + 36 = (x + 6)^2$

✓**Concept Check Answer**

answers may vary

**Example 6** Add the proper constant to $x^2 - 3x$ so that the result is a perfect square trinomial. Then factor.

**Solution:** We add the square of half the coefficient of $x$.

$$x^2 - 3x + \frac{9}{4} = \left(x - \frac{3}{2}\right)^2 \quad \text{In factored form}$$

$$\frac{1}{2}(-3) = -\frac{3}{2} \text{ and } \left(-\frac{3}{2}\right)^2 = \frac{9}{4}$$

■ Work Practice 6

## Objective C  Solving by Completing the Square

The process of writing a quadratic equation so that one side is a perfect square trinomial is called **completing the square.** We will use this process in the next examples.

**Example 7** Solve $p^2 + 2p = 4$ by completing the square.

**Solution:** First we add the square of half the coefficient of $p$ to both sides so that the resulting trinomial will be a perfect square trinomial. The coefficient of $p$ is 2.

$$\frac{1}{2}(2) = 1 \quad \text{and} \quad 1^2 = 1$$

Now we add 1 to both sides of the original equation.

$$p^2 + 2p = 4$$
$$p^2 + 2p + 1 = 4 + 1 \quad \text{Add 1 to both sides.}$$
$$(p + 1)^2 = 5 \quad \text{Factor the trinomial; simplify the right side.}$$

We may now use the square root property and solve for $p$.

$$p + 1 = \pm \sqrt{5} \quad \text{Use the square root property.}$$
$$p = -1 \pm \sqrt{5} \quad \text{Subtract 1 from both sides.}$$

Don't forget that there are two solutions: $-1 + \sqrt{5}$ and $-1 - \sqrt{5}$. The solution set is $\{-1 + \sqrt{5}, -1 - \sqrt{5}\}$.

■ Work Practice 7

**Example 8** Solve $m^2 - 7m - 1 = 0$ by completing the square.

**Solution:** First we add 1 to both sides of the equation so that the left side has no constant term. We can then add the constant term on both sides that will make the left side a perfect square trinomial.

$$m^2 - 7m - 1 = 0$$
$$m^2 - 7m = 1$$

Now we find the constant term that makes the left side a perfect square trinomial by squaring half the coefficient of $m$. We add this constant to both sides of the equation.

$$\frac{1}{2}(-7) = -\frac{7}{2} \quad \text{and} \quad \left(-\frac{7}{2}\right)^2 = \frac{49}{4}$$

*(Continued on next page)*

**Practice 6**

Add the proper constant to $y^2 - 5y$ so that the result is a perfect square trinomial. Then factor.

**Practice 7**

Solve $x^2 + 8x = 1$ by completing the square.

**Practice 8**

Solve $y^2 - 5y + 2 = 0$ by completing the square.

**Answers**

**6.** $y^2 - 5y + \frac{25}{4} = \left(y - \frac{5}{2}\right)^2$

**7.** $\{-4 - \sqrt{17}, -4 + \sqrt{17}\}$

**8.** $\left\{\frac{5 - \sqrt{17}}{2}, \frac{5 + \sqrt{17}}{2}\right\}$

$$m^2 - 7m + \frac{49}{4} = 1 + \frac{49}{4} \qquad \text{Add } \frac{49}{4} \text{ to both sides of the equation.}$$

$$\left(m - \frac{7}{2}\right)^2 = \frac{53}{4} \qquad \begin{array}{l}\text{Factor the perfect square trinomial and}\\ \text{simplify the right side.}\end{array}$$

$$m - \frac{7}{2} = \pm\sqrt{\frac{53}{4}} \qquad \text{Use the square root property.}$$

$$m = \frac{7}{2} \pm \frac{\sqrt{53}}{2} \qquad \text{Add } \frac{7}{2} \text{ to both sides and simplify } \sqrt{\frac{53}{4}}.$$

$$m = \frac{7 \pm \sqrt{53}}{2} \qquad \text{Simplify.}$$

The solution set is $\left\{\dfrac{7 + \sqrt{53}}{2}, \dfrac{7 - \sqrt{53}}{2}\right\}$.

■ Work Practice 8

The following steps may be used to solve a quadratic equation such as $ax^2 + bx + c = 0$ by completing the square. This method may be used whether or not the polynomial $ax^2 + bx + c$ is factorable.

### Solving a Quadratic Equation in x by Completing the Square

**Step 1:** If the coefficient of $x^2$ is 1, go to Step 2. Otherwise, divide both sides of the equation by the coefficient of $x^2$.

**Step 2:** Get all variable terms alone on one side of the equation.

**Step 3:** Complete the square for the resulting binomial by adding the square of half of the coefficient of $x$ to both sides of the equation.

**Step 4:** Factor the resulting perfect square trinomial and write it as the square of a binomial.

**Step 5:** Use the square root property to solve for $x$.

**Practice 9**

Solve $2x^2 - 2x + 7 = 0$ by completing the square.

**Example 9**  Solve $4x^2 - 24x + 41 = 0$ by completing the square.

**Solution:**  First we divide both sides of the equation by 4 so that the coefficient of $x^2$ is 1.

$$4x^2 - 24x + 41 = 0$$

**Step 1:** $x^2 - 6x + \dfrac{41}{4} = 0 \qquad$ Divide both sides of the equation by 4.

**Step 2:** $\quad x^2 - 6x = -\dfrac{41}{4} \qquad$ Subtract $\dfrac{41}{4}$ from both sides.

Since $\dfrac{1}{2}(-6) = -3$ and $(-3)^2 = 9$, we add 9 to both sides of the equation.

**Step 3:** $\quad x^2 - 6x + 9 = -\dfrac{41}{4} + 9 \qquad$ Add 9 to both sides.

**Step 4:** $\quad (x - 3)^2 = -\dfrac{41}{4} + \dfrac{36}{4} \qquad$ Factor the perfect square trinomial.

$$(x - 3)^2 = -\dfrac{5}{4}$$

**Answer**

9. $\left\{\dfrac{1 + i\sqrt{13}}{2}, \dfrac{1 - i\sqrt{13}}{2}\right\}$

**Step 5:** $x - 3 = \pm\sqrt{-\dfrac{5}{4}}$  Use the square root property.

$\qquad x - 3 = \pm\dfrac{i\sqrt{5}}{2}$  Simplify the radical.

$\qquad\quad x = 3 \pm \dfrac{i\sqrt{5}}{2}$  Add 3 to both sides.

$\qquad\qquad = \dfrac{6}{2} \pm \dfrac{i\sqrt{5}}{2}$  Find a common denominator.

$\qquad\qquad = \dfrac{6 \pm i\sqrt{5}}{2}$  Simplify.

The solution set is $\left\{\dfrac{6 + i\sqrt{5}}{2}, \dfrac{6 - i\sqrt{5}}{2}\right\}$.

■ Work Practice 9

## Objective D Solving Problems Modeled by Quadratic Equations ▶

Recall the **simple interest** formula $I = Prt$, where $I$ is the interest earned, $P$ is the principal, $r$ is the rate of interest, and $t$ is time. If \$100 is invested at a simple interest rate of 5% annually, at the end of 3 years the total interest $I$ earned is

$I = P \cdot r \cdot t$

or

$I = 100 \cdot 0.05 \cdot 3 = \$15$

and the new principal is

$\$100 + \$15 = \$115$

Most of the time, the interest computed on money borrowed or money deposited is **compound interest.** Unlike simple interest, compound interest is computed on original principal *and* on interest already earned. To see the difference between simple interest and compound interest, suppose that \$100 is invested at a rate of 5% compounded annually. To find the total amount of money at the end of 3 years, we calculate as follows:

$I = P \cdot r \cdot t$

First year:     Interest $= \$100 \cdot 0.05 \cdot 1 = \$5.00$
                New principal $= \$100.00 + \$5.00 = \$105.00$

Second year:    Interest $= \$105.00 \cdot 0.05 \cdot 1 = \$5.25$
                New principal $= 105.00 + \$5.25 = \$110.25$

Third year:     Interest $= \$110.25 \cdot 0.05 \cdot 1 \approx \$5.51$
                New principal $= \$110.25 + \$5.51 = \$115.76$

At the end of the third year, the total compound interest earned is \$15.76, whereas the total simple interest earned is \$15.

It is tedious to calculate compound interest as we did above, so we use a compound interest formula. The formula for calculating the total amount of money when interest is compounded annually is

$A = P(1 + r)^{t}$

where $P$ is the original investment, $r$ is the interest rate per compounding period, and $t$ is the number of periods. For example, the amount of money $A$ at the end of 3 years if \$100 is invested at 5% compounded annually is

$A = \$100(1 + 0.05)^{3} \approx 100(1.1576) = \$115.76$

as we previously calculated.

## Practice 10

Use the formula from Example 10 to find the interest rate $r$ if $1600 compounded annually grows to $1764 in 2 years.

### Example 10    Finding an Interest Rate

Find the interest rate $r$ if $2000 compounded annually grows to $2420 in 2 years.

**Solution:**

1. **UNDERSTAND** the problem. For this example, make sure that you understand the formula for compounding interest annually.

2. **TRANSLATE.** We substitute the given values into the formula.

$$A = P(1 + r)^t$$
$$2420 = 2000(1 + r)^2 \qquad \text{Let } A = 2420, P = 2000, \text{ and } t = 2.$$

3. **SOLVE.** We now solve the equation for $r$.

$$2420 = 2000(1 + r)^2$$

$$\frac{2420}{2000} = (1 + r)^2 \qquad \text{Divide both sides by 2000.}$$

$$\frac{121}{100} = (1 + r)^2 \qquad \text{Simplify the fraction.}$$

$$\pm\sqrt{\frac{121}{100}} = 1 + r \qquad \text{Use the square root property.}$$

$$\pm\frac{11}{10} = 1 + r \qquad \text{Simplify.}$$

$$-1 \pm \frac{11}{10} = r$$

$$-\frac{10}{10} \pm \frac{11}{10} = r$$

$$\frac{1}{10} = r \quad \text{or} \quad -\frac{21}{10} = r$$

4. **INTERPRET.** The rate cannot be negative, so we reject $-\dfrac{21}{10}$.

**Check:** $\dfrac{1}{10} = 0.10 = 10\%$ per year. If we invest $2000 at 10% compounded annually, in 2 years the amount in the account would be $2000(1 + 0.10)^2 = 2420$ dollars, the desired amount.

**State:**  The interest rate is 10% compounded annually.

■ Work Practice 10

## Vocabulary, Readiness & Video Check

*Use the choices below to fill in each blank. Not all choices will be used.*

| binomial | $\sqrt{b}$ | $\pm\sqrt{b}$ | $b^2$ | 9 | 25 | completing the square |
| quadratic | $-\sqrt{b}$ | $\dfrac{b}{2}$ | $\left(\dfrac{b}{2}\right)^2$ | 3 | 5 | |

1. By the square root property, if $b$ is a real number, and $a^2 = b$, then $a =$ _____.

2. A _____ equation can be written in the form $ax^2 + bx + c = 0, a \neq 0$.

3. The process of writing a quadratic equation so that one side is a perfect square trinomial is called _____.

4. A perfect square trinomial is one that can be factored as a _____ squared.

5. To solve $x^2 + 6x = 10$ by completing the square, add _____ to both sides.

6. To solve $x^2 + bx = c$ by completing the square, add _____ to both sides.

---

*Martin-Gay Interactive Videos*    *Watch the section lecture video and answer the following questions.*

*See Video 8.1*

**Objective A**  7. From ⊞ Examples 2 and 3, explain a step you can perform so that you may easily apply the square root property to $2x^2 = 16$. Explain why you perform this step. ▶

**Objective B**  8. Use ⊞ Example 5 to explain why $r^2 - r + 1$ isn't a perfect square trinomial. ▶

**Objective C**  9. In ⊞ Example 7, why is the equation first divided through by 3? ▶

**Objective D**  10. In ⊞ Example 8, why is the negative solution not considered? ▶

---

**8.1  Exercise Set** MyMathLab® ▶

**Objective A** *Use the square root property to solve each equation. See Examples 1 through 4.*

▶**1.** $x^2 = 16$          **2.** $x^2 = 49$          **3.** $x^2 - 7 = 0$          **4.** $x^2 - 11 = 0$

**5.** $x^2 = 18$          **6.** $y^2 = 20$          **7.** $3z^2 - 30 = 0$          **8.** $2x^2 = 4$

**9.** $(x + 5)^2 = 9$          **10.** $(y - 3)^2 = 4$          ▶**11.** $(z - 6)^2 = 18$          **12.** $(y + 4)^2 = 27$

**13.** $(2x - 3)^2 = 8$          **14.** $(4x + 9)^2 = 6$          **15.** $x^2 + 9 = 0$          **16.** $x^2 + 4 = 0$

▶**17.** $x^2 - 6 = 0$          **18.** $y^2 - 10 = 0$          **19.** $2z^2 + 16 = 0$          **20.** $3p^2 + 36 = 0$

**21.** $(3x - 1)^2 = -16$          **22.** $(4y + 2)^2 = -25$          **23.** $(z + 7)^2 = 5$

**24.** $(x + 10)^2 = 11$          **25.** $(x + 3)^2 + 8 = 0$          **26.** $(y - 4)^2 + 18 = 0$

**Objective B** *Fill in the blank with the number needed to make the expression a perfect square trinomial. See Examples 5 and 6.*

**27.** $m^2 + 2m +$ _____

**28.** $m^2 - 2m +$ _____

**29.** $y^2 - 14y +$ _____

**30.** $z^2 + z +$ _____

*Add the proper constant to each binomial so that the resulting trinomial is a perfect square trinomial. Then factor the trinomial. See Examples 5 and 6.*

**▶ 31.** $x^2 + 16x$

**32.** $y^2 + 2y$

**33.** $z^2 - 12z$

**34.** $x^2 - 8x$

**35.** $p^2 + 9p$

**36.** $n^2 + 5n$

**▶ 37.** $r^2 - r$

**38.** $p^2 - 7p$

**Objective C** *Solve each equation by completing the square. See Examples 7 through 9.*

**39.** $x^2 + 8x = -15$

**40.** $y^2 + 6y = -8$

**▶ 41.** $x^2 + 6x + 2 = 0$

**42.** $x^2 - 2x - 2 = 0$

**43.** $x^2 + x - 1 = 0$

**44.** $x^2 + 3x - 2 = 0$

**45.** $x^2 + 2x - 5 = 0$

**46.** $x^2 - 6x + 3 = 0$

**47.** $y^2 + y - 7 = 0$

**48.** $x^2 - 7x - 1 = 0$

**49.** $x^2 + 8x + 1 = 0$

**50.** $x^2 - 10x + 2 = 0$

**51.** $3p^2 - 12p + 2 = 0$

**52.** $2x^2 + 14x - 1 = 0$

**53.** $2x^2 + 7x = 4$

**54.** $3x^2 - 4x = 4$

**▶ 55.** $3y^2 + 6y - 4 = 0$

**56.** $2y^2 + 12y + 3 = 0$

**57.** $y^2 + 2y + 2 = 0$

**58.** $x^2 + 4x + 6 = 0$

**59.** $2a^2 + 8a = -12$

**60.** $3x^2 + 12x = -14$

**61.** $2x^2 - x + 6 = 0$

**62.** $4x^2 - 2x + 5 = 0$

**63.** $x^2 + 10x + 28 = 0$

**64.** $y^2 + 8y + 18 = 0$

**65.** $z^2 + 3z - 4 = 0$

**66.** $y^2 + y - 2 = 0$

**67.** $2x^2 - 4x + 3 = 0$

**68.** $9x^2 - 36x = -40$

**69.** $3x^2 + 3x = 5$

**70.** $5y^2 - 15y = 1$

**Objective D** *Use the formula* $A = P(1 + r)^t$ *to solve Exercises 71 through 74. See Example 10.*

**71.** Find the rate $r$ at which \$3000 grows to \$4320 in 2 years.

**72.** Find the rate $r$ at which \$800 grows to \$882 in 2 years.

**73.** Find the rate $r$ at which \$810 grows to approximately \$1000 in 2 years.

**74.** Find the rate $r$ at which \$2000 grows to \$2880 in 2 years.

*Neglecting air resistance, the distance $s(t)$ in feet traveled by a freely falling object is given by the function $s(t) = 16t^2$, where t is time in seconds. For Exercises 75–78, use this formula to find the time it takes for an object to fall to the ground from the top of each building. Round answers to two decimal places. (Source: Council on Tall Buildings and Urban Habitat)*

**75.** The Petronas Towers are located in Kuala Lumpur, Malaysia. Each tower is 1483 feet tall.

**76.** The Burj Khalifa, is located in Dubai, United Arab Emirates. It is estimated to be 2717 feet tall.

**77.** The Kingdom Tower in Jeddah, Saudi Arabia has a projected completion date of 2019. It is estimated to be 3281 feet tall.

**78.** One World Trade Center in New York City was completed in 2014. It has a height of 1776 feet.

*Use the formula $s(t) = 16t^2$ to solve. See Exercises 75–78.*

**79.** The Rogun Dam in Tajikistan (part of the former USSR that borders Afghanistan) is the tallest dam in the world at 1100 feet. How long would it take an object to fall from the top to the base of the dam? (*Source:* U.S. Committee on Large Dams of the International Commission on Large Dams)

**80.** The Hoover Dam, located on the Colorado River on the border of Nevada and Arizona near Las Vegas, is 725 feet tall. How long would it take an object to fall from the top to the base of the dam? (*Source:* U.S. Committee on Large Dams of the International Commission on Large Dams)

*Solve.*

△ **81.** The area of a square room is 225 square feet. Find the dimensions of the room.

△ **82.** The area of a circle is $36\pi$ square inches. Find the radius of the circle.

△ **83.** An isosceles right triangle has legs of equal length. If the hypotenuse is 20 centimeters long, find the length of each leg.

**84.** The top of a square coffee table has a diagonal that measures 30 inches. Find the length of each side of the top of the coffee table.

## Review

*Simplify each expression. See Section 6.1.*

**85.** $\dfrac{6 + 4\sqrt{5}}{2}$

**86.** $\dfrac{10 - 20\sqrt{3}}{2}$

**87.** $\dfrac{3 - 9\sqrt{2}}{6}$

**88.** $\dfrac{12 - 8\sqrt{7}}{16}$

*Evaluate $\sqrt{b^2 - 4ac}$ for each set of values. See Section 7.3.*

**89.** $a = 2, b = 4, c = -1$

**90.** $a = 1, b = 6, c = 2$

**91.** $a = 3, b = -1, c = -2$

**92.** $a = 1, b = -3, c = -1$

## Concept Extensions

*For Exercises 93 through 98, without solving, determine whether the solutions of each equation are real numbers or complex, but not real numbers. See the Concept Check in this section.*

**93.** $(x + 1)^2 = -1$

**94.** $(y - 5)^2 = -9$

**95.** $3z^2 = 10$

**96.** $4x^2 = 17$

**97.** $(2y - 5)^2 + 7 = 3$

**98.** $(3m + 2)^2 + 4 = 1$

**99.** In your own words, explain how to calculate the number that will complete the square on an expression such as $x^2 - 5x$.

**100.** In your own words, what is the difference between simple interest and compound interest?

**101.** If you are depositing money in an account that pays 4%, would you prefer the interest to be simple or compound? Explain your answer.

**102.** If you are borrowing money at a rate of 10%, would you prefer the interest to be simple or compound? Explain your answer.

*Find two possible missing terms so that each is a perfect square trinomial.*

**103.** $x^2 + \square + 16$

**104.** $y^2 + \square + 9$

*A common equation used in business is a demand equation. It expresses the relationship between the unit price of some commodity and the quantity demanded. For Exercises 105 and 106, p represents the unit price and x represents the quantity demanded in thousands.*

**105.** A manufacturing company has found that the demand equation for a certain type of scissors is given by the equation $p = -x^2 + 47$. Find the demand for the scissors if the price is $11 per pair.

**106.** Acme, Inc., sells desk lamps and has found that the demand equation for a certain style of desk lamp is given by the equation $p = -x^2 + 15$. Find the demand for the desk lamp if the price is $7 per lamp.

## 8.2 Solving Quadratic Equations by Using the Quadratic Formula

### Objectives

**A** Solve Quadratic Equations by Using the Quadratic Formula.

**B** Determine the Number and Type of Solutions of a Quadratic Equation by Using the Discriminant.

**C** Solve Problems Modeled by Quadratic Equations.

### Objective A Solving Equations by Using the Quadratic Formula

Any quadratic equation can be solved by completing the square. Since the same sequence of steps is repeated each time we complete the square, let's complete the square for a general quadratic equation, $ax^2 + bx + c = 0$. By doing so, we will find a pattern for the solutions of a quadratic equation known as the **quadratic formula.**

Recall that to complete the square for an equation such as $ax^2 + bx + c = 0$, $a \neq 0$, we first divide both sides by the coefficient of $x^2$.

$$ax^2 + bx + c = 0$$

$$x^2 + \frac{b}{a}x + \frac{c}{a} = 0 \quad \text{Divide both sides by } a, \text{ the coefficient of } x^2.$$

$$x^2 + \frac{b}{a}x = -\frac{c}{a} \quad \text{Subtract the constant } \frac{c}{a} \text{ from both sides.}$$

Next we find the square of half $\frac{b}{a}$, the coefficient of $x$.

$$\frac{1}{2}\left(\frac{b}{a}\right) = \left(\frac{b}{2a}\right) \quad \text{and} \quad \left(\frac{b}{2a}\right)^2 = \frac{b^2}{4a^2}$$

Now we add this result to both sides of the equation.

$$x^2 + \frac{b}{a}x + \frac{b^2}{4a^2} = -\frac{c}{a} + \frac{b^2}{4a^2} \quad \text{Add } \frac{b^2}{4a^2} \text{ to both sides.}$$

$$x^2 + \frac{b}{a}x + \frac{b^2}{4a^2} = \frac{-c \cdot 4a}{a \cdot 4a} + \frac{b^2}{4a^2} \quad \text{Find a common denominator on the right side.}$$

$$x^2 + \frac{b}{a}x + \frac{b^2}{4a^2} = \frac{b^2 - 4ac}{4a^2} \quad \text{Simplify the right side.}$$

$$\left(x + \frac{b}{2a}\right)^2 = \frac{b^2 - 4ac}{4a^2} \quad \text{Factor the perfect square trinomial on the left side.}$$

$$x + \frac{b}{2a} = \pm\sqrt{\frac{b^2 - 4ac}{4a^2}} \quad \text{Use the square root property.}$$

$$x + \frac{b}{2a} = \pm\frac{\sqrt{b^2 - 4ac}}{2a} \quad \text{Simplify the radical.}$$

$$x = -\frac{b}{2a} \pm \frac{\sqrt{b^2 - 4ac}}{2a} \quad \text{Subtract } \frac{b}{2a} \text{ from both sides.}$$

$$x = \frac{-b \pm \sqrt{b^2 - 4ac}}{2a} \quad \text{Simplify.}$$

The resulting equation identifies the solutions of the general quadratic equation in standard form and is called the quadratic formula. It can be used to solve any equation written in standard form $ax^2 + bx + c = 0$ as long as $a$ is not 0.

### Quadratic Formula

A quadratic equation written in the form $ax^2 + bx + c = 0$, $a \neq 0$, has the solutions

$$x = \frac{-b \pm \sqrt{b^2 - 4ac}}{2a}$$

**Example 1**   Solve:   $3x^2 + 16x + 5 = 0$

**Solution:**   This equation is in standard form with $a = 3, b = 16$, and $c = 5$. We substitute these values into the quadratic formula.

$$x = \frac{-b \pm \sqrt{b^2 - 4ac}}{2a} \qquad \text{Quadratic formula}$$

$$= \frac{-16 \pm \sqrt{16^2 - 4(3)(5)}}{2(3)} \qquad \text{Let } a = 3, b = 16, \text{ and } c = 5.$$

$$= \frac{-16 \pm \sqrt{256 - 60}}{6}$$

$$= \frac{-16 \pm \sqrt{196}}{6} = \frac{-16 \pm 14}{6}$$

$$x = \frac{-16 + 14}{6} = -\frac{1}{3} \quad \text{or} \quad x = \frac{-16 - 14}{6} = -\frac{30}{6} = -5$$

The solution set is $\left\{ -\dfrac{1}{3}, -5 \right\}$.

■ **Work Practice 1**

**Helpful Hint**

To replace $a, b$, and $c$ correctly in the quadratic formula, write the quadratic equation in standard form, $ax^2 + bx + c = 0$.

**Example 2**   Solve:   $2x^2 - 4x = 3$

**Solution:**   First we write the equation in standard form by subtracting 3 from both sides.

$$2x^2 - 4x - 3 = 0$$

Now $a = 2, b = -4$, and $c = -3$. We substitute these values into the quadratic formula.

$$x = \frac{-b \pm \sqrt{b^2 - 4ac}}{2a}$$

$$= \frac{-(-4) \pm \sqrt{(-4)^2 - 4(2)(-3)}}{2(2)}$$

$$= \frac{4 \pm \sqrt{16 + 24}}{4}$$

$$= \frac{4 \pm \sqrt{40}}{4}$$

$$= \frac{4 \pm 2\sqrt{10}}{4}$$

$$= \frac{2(2 \pm \sqrt{10})}{2 \cdot 2}$$

$$= \frac{2 \pm \sqrt{10}}{2}$$

The solution set is $\left\{ \dfrac{2 + \sqrt{10}}{2}, \dfrac{2 - \sqrt{10}}{2} \right\}$.

■ **Work Practice 2**

**Practice 1**

Solve:   $2x^2 + 9x + 10 = 0$

**Practice 2**

Solve:   $2x^2 - 6x = 1$

**Answers**

1. $\left\{ -\dfrac{5}{2}, -2 \right\}$

2. $\left\{ \dfrac{3 + \sqrt{11}}{2}, \dfrac{3 - \sqrt{11}}{2} \right\}$

✓**Concept Check** For the quadratic equation $x^2 = 7$, which substitution is correct?

  **a.** $a = 1, b = 0,$ and $c = -7$      **b.** $a = 1, b = 0,$ and $c = 7$

  **c.** $a = 0, b = 0,$ and $c = 7$      **d.** $a = 1, b = 1,$ and $c = -7$

**Helpful Hint**

To simplify the expression $\dfrac{4 \pm 2\sqrt{10}}{4}$ in Example 2, note that we factored 2 out of both terms of the numerator *before* simplifying.

$$\frac{4 \pm 2\sqrt{10}}{4} = \frac{2(2 \pm \sqrt{10})}{2 \cdot 2} = \frac{2 \pm \sqrt{10}}{2}$$

**Practice 3**

Solve: $\dfrac{1}{6}x^2 - \dfrac{1}{3}x - 1 = 0$

**Example 3**    Solve: $\dfrac{1}{4}m^2 - m + \dfrac{1}{2} = 0$

**Solution:** We could use the quadratic formula with $a = \dfrac{1}{4}, b = -1,$ and $c = \dfrac{1}{2}$.

Instead, let's find a simpler, equivalent, standard-form equation whose coefficients are not fractions.

First we multiply both sides of the equation by 4 to clear the fractions.

$$4\left(\frac{1}{4}m^2 - m + \frac{1}{2}\right) = 4 \cdot 0$$

$$m^2 - 4m + 2 = 0 \qquad \text{Simplify.}$$

Now we can substitute $a = 1, b = -4,$ and $c = 2$ into the quadratic formula and simplify.

$$m = \frac{-(-4) \pm \sqrt{(-4)^2 - 4(1)(2)}}{2(1)}$$

$$= \frac{4 \pm \sqrt{16 - 8}}{2}$$

$$= \frac{4 \pm \sqrt{8}}{2} = \frac{4 \pm 2\sqrt{2}}{2} = \frac{2(2 \pm \sqrt{2})}{2} = 2 \pm \sqrt{2}$$

The solution set is $\{2 + \sqrt{2}, 2 - \sqrt{2}\}$.

▥ **Work Practice 3**

**Practice 4**

Solve: $x = -4x^2 - 4$

**Example 4**    Solve: $p = -3p^2 - 3$

**Solution:** The equation in standard form is $3p^2 + p + 3 = 0$. Thus, $a = 3, b = 1,$ and $c = 3$ in the quadratic formula.

$$p = \frac{-1 \pm \sqrt{1^2 - 4(3)(3)}}{2(3)} = \frac{-1 \pm \sqrt{1 - 36}}{6}$$

$$= \frac{-1 \pm \sqrt{-35}}{6} = \frac{-1 \pm i\sqrt{35}}{6}$$

The solution set is $\left\{\dfrac{-1 + i\sqrt{35}}{6}, \dfrac{-1 - i\sqrt{35}}{6}\right\}$.

▥ **Work Practice 4**

**Answers**

**3.** $\{1 + \sqrt{7}, 1 - \sqrt{7}\}$

**4.** $\left\{\dfrac{-1 - 3i\sqrt{7}}{8}, \dfrac{-1 + 3i\sqrt{7}}{8}\right\}$

✓**Concept Check Answer**

**a**

✓**Concept Check** What is the first step in solving $-3x^2 = 5x - 4$ using the quadratic formula?

## Objective B Using the Discriminant ▶

In the quadratic formula $x = \dfrac{-b \pm \sqrt{b^2 - 4ac}}{2a}$, the radicand $b^2 - 4ac$ is called the **discriminant** because when we know its value, we can **discriminate** among the possible number and type of solutions of a quadratic equation. Possible values of the discriminant and their meanings are summarized next.

---

### Discriminant

The following table relates the discriminant $b^2 - 4ac$ of a quadratic equation of the form $ax^2 + bx + c = 0$ with the number and type of solutions of the equation.

| $b^2 - 4ac$ | Number and Type of Solutions |
| --- | --- |
| Positive | Two real solutions |
| Zero | One real solution |
| Negative | Two complex but not real solutions |

---

**Example 5**   Use the discriminant to determine the number and type of solutions of $x^2 + 2x + 1 = 0$.

**Solution:** In $x^2 + 2x + 1 = 0$, $a = 1$, $b = 2$, and $c = 1$. Thus,

$$b^2 - 4ac = 2^2 - 4(1)(1) = 0$$

Since $b^2 - 4ac = 0$, this quadratic equation has one real solution.

■ Work Practice 5

**Practice 5**

Use the discriminant to determine the number and type of solutions of $x^2 + 4x + 4 = 0$.

---

**Example 6**   Use the discriminant to determine the number and type of solutions of $3x^2 + 2 = 0$.

**Solution:** In this equation, $a = 3$, $b = 0$, and $c = 2$. Then

$$b^2 - 4ac = 0^2 - 4(3)(2) = -24$$

Since $b^2 - 4ac$ is negative, this quadratic equation has two complex but not real solutions.

■ Work Practice 6

**Practice 6**

Use the discriminant to determine the number and type of solutions of $5x^2 + 7 = 0$.

---

**Practice 7**

Use the discriminant to determine the number and type of solutions of $3x^2 - 2x - 2 = 0$.

**Example 7**   Use the discriminant to determine the number and type of solutions of $2x^2 - 7x - 4 = 0$.

**Solution:** In this equation, $a = 2$, $b = -7$, and $c = -4$. Then

$$b^2 - 4ac = (-7)^2 - 4(2)(-4) = 81$$

Since $b^2 - 4ac$ is positive, this quadratic equation has two real solutions.

■ Work Practice 7

**Answers**

**5.** one real solution
**6.** two complex but not real solutions
**7.** two real solutions

✓**Concept Check Answer**
Write the equation in standard form.

The discriminant helps us determine the number and type of solutions of a quadratic equation, $ax^2 + bx + c = 0$. Recall from Chapter 5 that the solutions of this equation are the same as the $x$-intercepts of its related graph $f(x) = ax^2 + bx + c$. This means that the discriminant of $ax^2 + bx + c = 0$ also tells us the number of $x$-intercepts for the graph of $f(x) = ax^2 + bx + c$, or, equivalently, $y = ax^2 + bx + c$.

### Graph of $f(x) = ax^2 + bx + c$ or $y = ax^2 + bx + c$

| $b^2 - 4ac > 0$, $f(x)$ has two $x$-intercepts | $b^2 - 4ac = 0$, $f(x)$ has one $x$-intercept | $b^2 - 4ac < 0$, $f(x)$ has no $x$-intercepts |

## Objective C Solving Problems Modeled by Quadratic Equations ▶

The quadratic formula is useful in solving problems that are modeled by quadratic equations.

Practice 8

Given the diagram below, approximate to the nearest foot how many feet of walking distance a person saves by cutting across the lawn instead of walking on the sidewalk.

20 ft

$x + 8$

$x$

### Example 8    Calculating Distance Saved

At a local university, students often leave the sidewalk and cut across the lawn to save walking distance. Given the diagram below of a favorite place to cut across the lawn, approximate to the nearest foot how many feet of walking distance a student saves by cutting across the lawn instead of walking on the sidewalk.

50 ft

$x$        $x + 20$

**Solution:**

1. UNDERSTAND. Read and reread the problem. You may want to review the Pythagorean theorem.

2. TRANSLATE. By the Pythagorean theorem, we have

   In words: $(\text{leg})^2 + (\text{leg})^2 = (\text{hypotenuse})^2$

   Translate: $x^2 + (x + 20)^2 = 50^2$

3. SOLVE. Use the quadratic formula to solve.

   $x^2 + x^2 + 40x + 400 = 2500$   Square $(x + 20)$ and 50.

   $2x^2 + 40x - 2100 = 0$   Write the equation in standard form.

   We can use the quadratic formula right now with $a = 2$, $b = 40$, and $c = -2100$. Instead, just as in Example 3, you may want to find a simpler, equivalent equation by dividing both sides of the equation by 2.

   $x^2 + 20x - 1050 = 0$   Divide by 2.

**Answer**

**8.** 8 ft

Here, $a = 1$, $b = 20$, and $c = -1050$. By the quadratic formula,

$$x = \frac{-20 \pm \sqrt{20^2 - 4(1)(-1050)}}{2 \cdot 1}$$

$$= \frac{-20 \pm \sqrt{400 + 4200}}{2} = \frac{-20 \pm \sqrt{4600}}{2}$$

$$= \frac{-20 \pm \sqrt{100 \cdot 46}}{2} = \frac{-20 \pm 10\sqrt{46}}{2}$$

$$= -10 \pm 5\sqrt{46} \qquad \text{Simplify.}$$

**Check:**

**4. INTERPRET.** We check our calculations from the quadratic formula. The length of a side of a triangle can't be negative, so we reject $-10 - 5\sqrt{46}$. Since $-10 + 5\sqrt{46} \approx 24$ feet, the walking distance along the sidewalk is

$$x + (x + 20) \approx 24 + (24 + 20) = 68 \text{ feet.}$$

**State:** A person saves about $68 - 50$ or 18 feet of walking distance by cutting across the lawn.

 **Work Practice 8**

**Example 9** Calculating Landing Time

An object is thrown upward from the top of a 200-foot cliff with a velocity of 12 feet per second. The height above ground $h$ in feet of the object after $t$ seconds is

$$h = -16t^2 + 12t + 200$$

How long after the object is thrown will it strike the ground? Round to the nearest tenth of a second.

200 ft

**Solution:**

**1. UNDERSTAND.** Read and reread the problem.

**2. TRANSLATE.** Since we want to know when the object strikes the ground, we want to know when the height $h = 0$, or

$$0 = -16t^2 + 12t + 200$$

**3. SOLVE.** First we divide both sides of the equation by $-4$.

$$0 = 4t^2 - 3t - 50 \qquad \text{Divide both sides by } -4.$$

Here, $a = 4$, $b = -3$, and $c = -50$. By the quadratic formula,

$$t = \frac{-(-3) \pm \sqrt{(-3)^2 - 4(4)(-50)}}{2 \cdot 4}$$

$$= \frac{3 \pm \sqrt{9 + 800}}{8}$$

$$= \frac{3 \pm \sqrt{809}}{8}$$

(*Continued on next page*)

**Practice 9**

How long after the object in Example 9 is thrown will it be 100 feet from the ground? Round to the nearest tenth of a second.

**Answer**
**9.** 2.9 sec

**Check:**

4. INTERPRET. We check our calculations from the quadratic formula. Since the time won't be negative, we reject the proposed solution $\dfrac{3 - \sqrt{809}}{8}$.

**State:** The time it takes for the object to strike the ground is exactly $\dfrac{3 + \sqrt{809}}{8}$ seconds $\approx 3.9$ seconds.

■ **Work Practice 9**

---

## Calculator Explorations  Graphing

In Section 5.7, we showed how we can use a grapher to approximate real number solutions of a quadratic equation written in standard form. We can also use a grapher to solve a quadratic equation when it is not written in standard form. For example, to solve $(x + 1)^2 = 12$, the quadratic equation in Example 3 of Section 8.1, we graph the following on the same set of axes. We use   Xmin $= -10$, Xmax $= 10$, Ymin $= -13$,   and Ymax $= 13$.

$$Y_1 = (x + 1)^2 \quad \text{and} \quad Y_2 = 12$$

We use the INTERSECT feature or the ZOOM and TRACE features to locate the points of intersection of the graphs. The $x$-values of these points are the solutions of $(x + 1)^2 = 12$. The solutions, rounded to two decimal points, are 2.46 and $-4.46$.

Check to see that these numbers are approximations of the exact solutions, $-1 \pm 2\sqrt{3}$.

*Use a grapher to solve each quadratic equation. Round all solutions to the nearest hundredth.*

1. $x(x - 5) = 8$
2. $x(x + 2) = 5$
3. $x^2 + 0.5x = 0.3x + 1$
4. $x^2 - 2.6x = -2.2x + 3$
5. Use a grapher to solve $(2x - 5)^2 = -16$, (Example 4, Section 8.1) using the window

    Xmin $= -20$
    Xmax $= 20$
    Xscl $= 1$
    Ymin $= -20$
    Ymax $= 20$
    Yscl $= 1$

    Explain the results. Compare your results with the solution found in Example 4 of Section 8.1.
6. What are the advantages and disadvantages of using a grapher to solve quadratic equations?

---

## Vocabulary, Readiness & Video Check

*Fill in each blank.*

1. The quadratic formula is _____.
2. For $2x^2 + x + 1 = 0$, if $a = 2$, then $b = $ _____ and $c = $ _____.
3. For $5x^2 - 5x - 7 = 0$, if $a = 5$, then $b = $ _____ and $c = $ _____.
4. For $7x^2 - 4 = 0$, if $a = 7$, then $b = $ _____ and $c = $ _____.
5. For $x^2 + 9 = 0$, if $c = 9$, then $a = $ _____ and $b = $ _____.
6. The correct simplified form of $\dfrac{5 \pm 10\sqrt{2}}{5}$ is _____.

   **a.** $1 \pm 10\sqrt{2}$    **b.** $2\sqrt{2}$    **c.** $1 \pm 2\sqrt{2}$    **d.** $\pm 5\sqrt{2}$

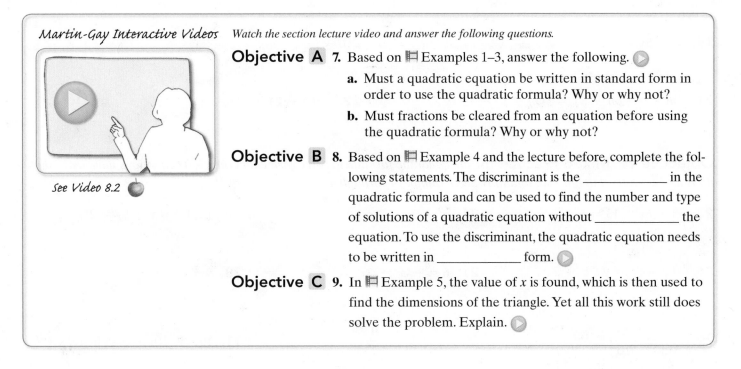

*Martin-Gay Interactive Videos*    *Watch the section lecture video and answer the following questions.*

See Video 8.2

**Objective A** **7.** Based on ▤ Examples 1–3, answer the following.

    **a.** Must a quadratic equation be written in standard form in order to use the quadratic formula? Why or why not?

    **b.** Must fractions be cleared from an equation before using the quadratic formula? Why or why not?

**Objective B** **8.** Based on ▤ Example 4 and the lecture before, complete the following statements. The discriminant is the _____ in the quadratic formula and can be used to find the number and type of solutions of a quadratic equation without _____ the equation. To use the discriminant, the quadratic equation needs to be written in _____ form.

**Objective C** **9.** In ▤ Example 5, the value of $x$ is found, which is then used to find the dimensions of the triangle. Yet all this work still does solve the problem. Explain.

## 8.2 Exercise Set MyMathLab®

**Objective A**  *Use the quadratic formula to solve each equation. These equations have real number solutions only. See Examples 1 through 3.*

**1.** $m^2 + 5m - 6 = 0$

**2.** $p^2 + 11p - 12 = 0$

**3.** $2y = 5y^2 - 3$

**4.** $5x^2 - 3 = 14x$

**5.** $x^2 - 6x + 9 = 0$

**6.** $y^2 + 10y + 25 = 0$

**7.** $x^2 + 7x + 4 = 0$

**8.** $y^2 + 5y + 3 = 0$

**9.** $8m^2 - 2m = 7$

**10.** $11n^2 - 9n = 1$

**11.** $3m^2 - 7m = 3$

**12.** $x^2 - 13 = 5x$

**13.** $\frac{1}{2}x^2 - x - 1 = 0$

**14.** $\frac{1}{6}x^2 + x + \frac{1}{3} = 0$

**15.** $\frac{2}{5}y^2 + \frac{1}{5}y = \frac{3}{5}$

**16.** $\frac{1}{8}x^2 + x = \frac{5}{2}$

**17.** $\frac{1}{3}y^2 - y - \frac{1}{6} = 0$

**18.** $\frac{1}{2}y^2 = y + \frac{1}{2}$

**19.** $x^2 + 5x = -2$

**20.** $y^2 - 8 = 4y$

**21.** $(m + 2)(2m - 6) = 5(m - 1) - 12$

**22.** $7p(p - 2) + 2(p + 4) = 3$

**Objective** **A** **Mixed Practice** *Use the quadratic formula to solve each equation. These equations have real solutions and complex, but not real, solutions. See Examples 1 through 4.*

**23.** $x^2 + 6x + 13 = 0$

**24.** $x^2 + 2x + 2 = 0$

▶ **25.** $(x + 5)(x - 1) = 2$

**26.** $x(x + 6) = 2$

**27.** $6 = -4x^2 + 3x$

**28.** $9x^2 + x + 2 = 0$

**29.** $\dfrac{x^2}{3} - x = \dfrac{5}{3}$

**30.** $\dfrac{x^2}{2} - 3 = -\dfrac{9}{2}x$

**31.** $10y^2 + 10y + 3 = 0$

**32.** $3y^2 + 6y + 5 = 0$

**33.** $x(6x + 2) = 3$

**34.** $x(7x + 1) = 2$

▶ **35.** $\dfrac{2}{5}y^2 + \dfrac{1}{5}y + \dfrac{3}{5} = 0$

**36.** $\dfrac{1}{8}x^2 + x + \dfrac{5}{2} = 0$

**37.** $\dfrac{1}{2}y^2 = y - \dfrac{1}{2}$

**38.** $\dfrac{2}{3}x^2 - \dfrac{20}{3}x = -\dfrac{100}{6}$

**39.** $(n - 2)^2 = 2n$

**40.** $\left(p - \dfrac{1}{2}\right)^2 = \dfrac{p}{2}$

**Objective** **B** *Use the discriminant to determine the number and types of solutions of each equation. See Examples 5 through 7.*

**41.** $x^2 - 5 = 0$

**42.** $x^2 - 7 = 0$

**43.** $4x^2 + 12x = -9$

**44.** $9x^2 + 1 = 6x$

**45.** $3x = -2x^2 + 7$

**46.** $3x^2 = 5 - 7x$

▶ **47.** $6 = 4x - 5x^2$

**48.** $5 - 4x + 12x^2 = 0$

**49.** $9x - 2x^2 + 5 = 0$

**50.** $8x = 3 - 9x^2$

## Objective C  *Solve. See Examples 8 and 9.*

**51.** Nancy, Thelma, and John Varner live on a corner lot. Often, neighborhood children cut across their lot to save walking distance. Given the diagram below, approximate to the nearest foot how many feet of walking distance children save by cutting across their property instead of walking around the lot.

**52.** Given the diagram below, approximate to the nearest foot how many feet of walking distance a person saves by cutting across the lawn instead of walking on the sidewalk.

**53.** The hypotenuse of an isosceles right triangle is 2 centimeters longer than either of its legs. Find the exact length of each side. (*Hint:* An isosceles right triangle is a right triangle whose legs are the same length.)

**54.** The hypotenuse of an isosceles right triangle is one meter longer than either of its legs. Find the length of each side. (See the hint given in Exercise 53.)

**55.** Bailey's rectangular dog pen for her Irish setter must have an area of 400 square feet. Also, the length must be 10 feet longer than the width. Find the dimensions of the pen.

**56.** An entry in the Peach Festival Poster Contest must be rectangular and have an area of 1200 square inches. Furthermore, its length must be 20 inches longer than its width. Find the dimensions each entry must have.

**57.** A holding pen for cattle must be square and have a diagonal length of 100 meters.

   **a.** Find the length of a side of the pen.

   **b.** Find the area of the pen.

**58.** A rectangle is three times longer than it is wide. It has a diagonal of length 50 centimeters.

   **a.** Find the dimensions of the rectangle.

   **b.** Find the perimeter of the rectangle.

**59.** The heaviest reported door in the world is the 708.6-ton radiation shield door in the National Institute for Fusion Science at Toki, Japan. If the height of the door is 1.1 feet longer than its width, and its front area (neglecting depth) is 1439.9 square feet, find its width and height. [Interesting note: The door is 6.6 feet thick.] (*Source: Guiness World Records*)

**60.** Christi and Robbie Wegmann are constructing a rectangular stained glass window whose length is 7.3 inches longer than its width. If the area of the window is 569.9 square inches, find its width and length.

**61.** The base of a triangle is four more than twice its height. If the area of the triangle is 42 square centimeters, find its base and height.

**62.** If a point $B$ divides a line segment such that the smaller portion is to the larger portion as the larger is to the whole, the whole is the length of the *golden ratio*.

The golden ratio was thought by the Greeks to be the most pleasing to the eye, and many of their buildings contained numerous examples of the golden ratio. The value of the golden ratio is the positive solution of the following equation.

$$\frac{x-1}{1} = \frac{1}{x}$$

Find this value.

*The Wollomombi Falls in Australia have a height of 1100 feet. A pebble is thrown upward from the top of the falls with an initial velocity of 20 feet per second. The height of the pebble h in feet after t seconds is given by the equation* $h = -16t^2 + 20t + 1100$. *Use this equation for Exercises 63 and 64.*

**63.** How long after the pebble is thrown will it hit the ground? Round to the nearest tenth of a second.

**64.** How long after the pebble is thrown will it be 550 feet from the ground? Round to the nearest tenth of a second.

*A ball is thrown downward from the top of a 180-foot building with an initial velocity of 20 feet per second. The height of the ball h in feet after t seconds is given by the equation $h = -16t^2 - 20t + 180$. Use this equation to answer Exercises 65 and 66.*

180 ft

50 ft

**65.** How long after the ball is thrown will it strike the ground? Round the result to the nearest tenth of a second.

**66.** How long after the ball is thrown will it be 50 feet from the ground? Round the result to the nearest tenth of a second.

## Review

*Solve each equation. See Sections 6.4 and 7.6.*

**67.** $\sqrt{5x - 2} = 3$

**68.** $\sqrt{y + 2} + 7 = 12$

**69.** $\dfrac{1}{x} + \dfrac{2}{5} = \dfrac{7}{x}$

**70.** $\dfrac{10}{z} = \dfrac{5}{z} - \dfrac{1}{3}$

*Factor. See Section 5.5 and 5.6.*

**71.** $x^4 + x^2 - 20$

**72.** $2y^4 + 11y^2 - 6$

**73.** $z^4 - 13z^2 + 36$

**74.** $x^4 - 1$

## Concept Extensions

*For each quadratic equation in Exercises 75 and 76, choose the correct substitution for a, b, and c in the standard form $ax^2 + bx + c = 0$. See the first Concept Check in this section.*

**75.** $x^2 = -10$
a. $a = 1, b = 0, c = -10$
b. $a = 1, b = 0, c = 10$
c. $a = 0, b = 1, c = -10$
d. $a = 1, b = 1, c = 10$

**76.** $x^2 + 5 = -x$
a. $a = 1, b = 5, c = -1$
b. $a = 1, b = -1, c = 5$
c. $a = 1, b = 5, c = 1$
d. $a = 1, b = 1, c = 5$

**77.** Solve Exercise 1 by factoring. Explain the result.

**78.** Solve Exercise 2 by factoring. Explain the result.

*Use the quadratic formula and a calculator to approximate each solution to the nearest tenth.*

**79.** $2x^2 - 6x + 3 = 0$

**80.** $3.6x^2 + 1.8x - 4.3 = 0$

*The graph shows the daily low temperatures for one week in New Orleans, Louisiana. Use this graph to answer Exercises 81 through 84.*

**81.** Which day of the week shows the greatest decrease in the low temperature?

**82.** Which day of the week shows the greatest increase in the low temperature?

**83.** Which day of the week had the lowest temperature?

**84.** Use the graph to estimate the low temperature on Thursday.

*Notice that the shape of the temperature graph for Exercises 81 through 84 is similar to a parabola (see Section 3.8). In fact, this graph can be approximated by the quadratic function $f(x) = 3x^2 - 18x + 57$, where $f(x)$ is the temperature in degrees Fahrenheit and $x$ is the number of days from Sunday. Use this function to answer Exercises 85 and 86.*

**85.** Use the given quadratic function to approximate the low temperature on Thursday. Does your answer agree with the graph?

**86.** Use the given function and the quadratic formula to find when the low temperature was 35°F. [*Hint:* Let $f(x) = 35$ and solve for $x$.] Round your answer to one decimal place and interpret your result. Does your answer agree with the graph above?

*Solve.*

**87.** The amount of electricity generated by wind power in the United States can be modeled by the quadratic function $f(x) = 0.45x^2 + 18.71x + 55.42$, where $f(x)$ is the amount of electricity in billions of kilowatts and $x$ is the number of years after 2008. (*Source:* Based on data from the Energy Information Administration, U.S. Department of Energy)

  **a.** Find the amount of electricity generated by wind power in 2010.

  **b.** If the trend described by this model continues, find the year after 2008 when the amount of electricity generated by wind power tops 200 billion kilowatts.

**88.** The function $f(x) = 0.0054x^2 - 1.46x + 95$ can be used to estimate the expected number of years of life remaining for a person of age $x$ years, when $30 \leq x \leq 100$. (*Source:* Based on data from the U.S. Census Bureau)

  **a.** Estimate the expected remaining years of life for a person of age 50.

  **b.** Estimate the expected remaining years of life for a person of 30.

  **c.** Why do you think that this function is valid only for ages 30 through 100?

**89.** The number of tablet computer users worldwide (in millions) can be projected by the function $f(x) = 4x^2 + 107x - 15$, where $x$ is the number of years since 2010. (*Source:* Forrester Research: World Tablet Adoption Forecast 2012–2017)

  **a.** Find the projected number of tablet computer users in 2012.

  **b.** According to this model, find the projected number of tablet computer users in 2015.

  **c.** If the trend represented by this model continues, in what year will the projected number of tablet computer users reach 1000 million?

**90.** The relationship between body weight and the Recommended Dietary Allowance (RDA) for vitamin A in children up to age 10 is modeled by the quadratic equation $y = 0.149x^2 - 4.475x + 406.478$, where $y$ is the RDA for vitamin A in micrograms for a child whose weight is $x$ pounds. (*Source:* Based on data from the Food and Nutrition Board, National Academy of Sciences—Institute of Medicine, 1989)

  **a.** Determine the vitamin A requirements of a child who weighs 35 pounds.

  **b.** What is the weight of a child whose RDA of vitamin A is 600 micrograms? Round your answer to the nearest pound.

**91.** Use a grapher to solve Exercise 79.

**92.** Use a grapher to solve Exercise 80.

## 8.3 Solving Equations by Using Quadratic Methods ▶

### Objective A Solving Equations That Are Quadratic in Form ▶

In this section, we discuss various types of equations that can be solved in part by using the methods for solving quadratic equations.

Once each equation is simplified, you may want to use these steps when deciding what method to use to solve the quadratic equation.

**Objectives**

**A** Solve Various Equations That Are Quadratic in Form. ▶

**B** Solve Problems That Lead to Quadratic Equations. ▶

---

**Solving a Quadratic Equation**

**Step 1:** If the equation is in the form $(ax + b)^2 = c$, use the square root property and solve. If not, go to Step 2.

**Step 2:** Write the equation in standard form by setting it equal to 0: $ax^2 + bx + c = 0$.

**Step 3:** Try to solve the equation by the factoring method. If not possible, go to Step 4.

**Step 4:** Solve the equation by the quadratic formula.

---

The first example is a radical equation that becomes a quadratic equation once we square both sides.

**Example 1** Solve: $x - \sqrt{x} - 6 = 0$

**Solution:** Recall that to solve a radical equation, we first get the radical alone on one side of the equation. Then we square both sides.

$$x - 6 = \sqrt{x} \quad \text{Add } \sqrt{x} \text{ to both sides.}$$
$$x^2 - 12x + 36 = x \quad \text{Square both sides.}$$
$$x^2 - 13x + 36 = 0 \quad \text{Set the equation equal to 0.}$$
$$(x - 9)(x - 4) = 0 \quad \text{Factor.}$$
$$x - 9 = 0 \quad \text{or} \quad x - 4 = 0 \quad \text{Set each factor equal to 0.}$$
$$x = 9 \qquad\qquad x = 4 \quad \text{Solve.}$$

**Check:** Let $x = 9$.

$$x - \sqrt{x} - 6 = 0$$
$$9 - \sqrt{9} - 6 \stackrel{?}{=} 0$$
$$9 - 3 - 6 \stackrel{?}{=} 0$$
$$0 = 0 \quad \text{True}$$

Let $x = 4$.

$$x - \sqrt{x} - 6 = 0$$
$$4 - \sqrt{4} - 6 \stackrel{?}{=} 0$$
$$4 - 2 - 6 \stackrel{?}{=} 0$$
$$-4 = 0 \quad \text{False}$$

The solution set is $\{9\}$.

■ Work Practice 1

**Practice 1**

Solve: $x - \sqrt{x - 1} - 3 = 0$

**Answer**

**1.** $\{5\}$

---

**Practice 2**

Solve:

$$\frac{2x}{x-1} - \frac{x+2}{x} = \frac{5}{x(x-1)}$$

**Example 2**  Solve: $\dfrac{3x}{x-2} - \dfrac{x+1}{x} = \dfrac{6}{x(x-2)}$

**Solution:** In this equation, $x$ cannot be either 2 or 0 because these values cause denominators to equal zero. To solve for $x$, we first multiply both sides of the equation by $x(x-2)$ to clear the fractions. By the distributive property, this means that we multiply each term by $x(x-2)$.

$$x(x-2)\left(\frac{3x}{x-2}\right) - x(x-2)\left(\frac{x+1}{x}\right) = x(x-2)\left[\frac{6}{x(x-2)}\right]$$

$$3x^2 - (x-2)(x+1) = 6 \quad \text{Simplify.}$$
$$3x^2 - (x^2 - x - 2) = 6 \quad \text{Multiply.}$$
$$3x^2 - x^2 + x + 2 = 6$$
$$2x^2 + x - 4 = 0 \quad \text{Simplify.}$$

This equation cannot be factored using integers, so we solve by the quadratic formula.

$$x = \frac{-1 \pm \sqrt{1^2 - 4(2)(-4)}}{2\cdot 2} \quad \text{Let } a = 2, b = 1, \text{ and } c = -4, \text{ in the quadratic formula.}$$

$$= \frac{-1 \pm \sqrt{1 + 32}}{4} \quad \text{Simplify.}$$

$$= \frac{-1 \pm \sqrt{33}}{4}$$

Neither proposed solution will make the denominators 0.

The solution set is $\left\{\dfrac{-1 + \sqrt{33}}{4}, \dfrac{-1 - \sqrt{33}}{4}\right\}$.

■ Work Practice 2

**Practice 3**

Solve: $x^4 + 5x^2 - 36 = 0$

**Example 3**  Solve: $p^4 - 3p^2 - 4 = 0$

**Solution:**  First we factor the trinomial.

$$p^4 - 3p^2 - 4 = 0$$
$$(p^2 - 4)(p^2 + 1) = 0 \quad \text{Factor.}$$
$$(p - 2)(p + 2)(p^2 + 1) = 0 \quad \text{Factor further.}$$
$$p - 2 = 0 \quad \text{or} \quad p + 2 = 0 \quad \text{or} \quad p^2 + 1 = 0 \quad \text{Set each factor equal}$$
$$p = 2 \qquad\qquad p = -2 \qquad\qquad p^2 = -1 \quad \text{to 0 and solve.}$$
$$p = \pm\sqrt{-1} = \pm i$$

The solution set is $\{2, -2, i, -i\}$.

■ Work Practice 3

**Helpful Hint**

Example 3 can be solved using substitution also. Think of $p^4 - 3p^2 - 4 = 0$ as

$$(p^2)^2 - 3p^2 - 4 = 0 \quad \text{Then let } x = p^2, \text{ and solve and substitute back.}$$
$$\qquad\qquad\qquad\qquad \text{The solution set will be the same.}$$
$$x^2 - 3x - 4 = 0$$

**Answers**

**2.** $\left\{\dfrac{1 + \sqrt{13}}{2}, \dfrac{1 - \sqrt{13}}{2}\right\}$

**3.** $\{2, -2, 3i, -3i\}$

## ✓Concept Check

a. True or false? The maximum number of solutions that a quadratic equation can have is 2.

b. True or false? The maximum number of solutions that an equation in quadratic form can have is 2.

**Example 4**    Solve:  $(x - 3)^2 - 3(x - 3) - 4 = 0$

**Solution:**  Notice that the quantity $(x - 3)$ is repeated in this equation. Sometimes it is helpful to substitute a variable (in this case other than $x$) for the repeated quantity. We will let $u = x - 3$. Then

$$(x - 3)^2 - 3(x - 3) - 4 = 0$$

becomes

$$u^2 - 3u - 4 = 0 \quad \text{Let } x - 3 = u.$$
$$(u - 4)(u + 1) = 0 \quad \text{Factor.}$$

To solve, we use the zero-factor property.

$$u - 4 = 0 \quad \text{or} \quad u + 1 = 0 \quad \text{Set each factor equal to 0.}$$
$$u = 4 \qquad\qquad u = -1 \quad \text{Solve.}$$

To find values of $x$, we substitute back. That is, we substitute $x - 3$ for $u$.

$$x - 3 = 4 \quad \text{or} \quad x - 3 = -1$$
$$x = 7 \qquad\qquad x = 2$$

Both 2 and 7 check. The solution is $\{2, 7\}$.

■ Work Practice 4

**Practice 4**

Solve:
$(x + 4)^2 - (x + 4) - 6 = 0$

> **Helpful Hint**    When using substitution, don't forget to substitute back to the original variable.

**Example 5**    Solve:  $x^{2/3} - 5x^{1/3} + 6 = 0$

**Solution:**    The key to solving this equation is recognizing that $x^{2/3} = (x^{1/3})^2$. We replace $x^{1/3}$ with $m$ so that

$$(x^{1/3})^2 - 5x^{1/3} + 6 = 0$$

becomes

$$m^2 - 5m + 6 = 0$$

Now we solve by factoring.

$$m^2 - 5m + 6 = 0$$
$$(m - 3)(m - 2) = 0 \qquad \text{Factor.}$$
$$m - 3 = 0 \quad \text{or} \quad m - 2 = 0 \quad \text{Set each factor equal to 0.}$$
$$m = 3 \qquad\qquad m = 2$$

Since $m = x^{1/3}$, we have

$$x^{1/3} = 3 \qquad \text{or} \quad x^{1/3} = 2$$
$$x = 3^3 = 27 \quad \text{or} \qquad x = 2^3 = 8$$

Both 8 and 27 check. The solution set is $\{8, 27\}$.

■ Work Practice 5

**Practice 5**

Solve:  $x^{2/3} - 7x^{1/3} + 10 = 0$

**Answers**
4. $\{-1, -6\}$    5. $\{8, 125\}$

**✓ Concept Check Answers**
a. true    b. false

## Objective B Solving Problems That Lead to Quadratic Equations ▶

The next example is a work problem. This problem is modeled by a rational equation that simplifies to a quadratic equation.

### Practice 6

Together, Karen and Doug Lewis can clean a strip of beach in 5 hours. Alone, Karen can clean the strip of beach 1 hour faster than Doug. Find the time that each person can clean the strip of beach alone. Give an exact answer and a one-decimal-place approximation.

### Example 6 Finding Work Time

Together, an experienced word processor and an apprentice word processor can create a document in 6 hours. Alone, the experienced word processor can process the document 2 hours faster than the apprentice word processor can. Find the time in which each person can create the document alone.

**Solution:**

1. UNDERSTAND. Read and reread the problem. The key idea here is the relationship between the *time* (hours) it takes to complete the job and the *part of the job* completed in one unit of time (hour). For example, because they can complete the job together in 6 hours, the *part of the job* they can complete in 1 hour is $\frac{1}{6}$. We let

> $x$ = the *time* in hours it takes the apprentice word processor to complete the job alone, and
>
> $x - 2$ = the *time* in hours it takes the experienced word processor to complete the job alone

We can summarize in a chart the information discussed.

| | Total Hours to Complete Job | Part of Job Completed in 1 Hour |
|---|---|---|
| **Apprentice Word Processor** | $x$ | $\frac{1}{x}$ |
| **Experienced Word Processor** | $x - 2$ | $\frac{1}{x - 2}$ |
| **Together** | 6 | $\frac{1}{6}$ |

2. TRANSLATE.

| In words: | part of job completed by apprentice word processor in 1 hour | added to | part of job completed by experienced word processor in 1 hour | is equal to | part of job completed together in 1 hour |
|---|---|---|---|---|---|
| | ↓ | ↓ | ↓ | ↓ | ↓ |
| Translate: | $\frac{1}{x}$ | $+$ | $\frac{1}{x - 2}$ | $=$ | $\frac{1}{6}$ |

**Answer**

6. Doug: $\dfrac{11 + \sqrt{101}}{2} \approx 10.5$ hr;

Karen: $\dfrac{9 + \sqrt{101}}{2} \approx 9.5$ hr

**3. SOLVE.**

$$\frac{1}{x} + \frac{1}{x-2} = \frac{1}{6}$$

$$6x(x-2)\left(\frac{1}{x} + \frac{1}{x-2}\right) = 6x(x-2)\cdot\frac{1}{6}$$ Multiply both sides by the LCD $6x(x-2)$.

$$6x(x-2)\cdot\frac{1}{x} + 6x(x-2)\cdot\frac{1}{x-2} = 6x(x-2)\cdot\frac{1}{6}$$ Use the distributive property.

$$6(x-2) + 6x = x(x-2)$$

$$6x - 12 + 6x = x^2 - 2x$$

$$0 = x^2 - 14x + 12$$

Now we can substitute $a = 1, b = -14$, and $c = 12$ into the quadratic formula and simplify.

$$x = \frac{-(-14) \pm \sqrt{(-14)^2 - 4(1)(12)}}{2\cdot 1} = \frac{14 \pm \sqrt{148}}{2}$$

Using a calculator or a square root table, we see that $\sqrt{148} \approx 12.2$ rounded to one decimal place. Thus,

$$x \approx \frac{14 \pm 12.2}{2}$$

$$x \approx \frac{14 + 12.2}{2} = 13.1 \quad \text{or} \quad x \approx \frac{14 - 12.2}{2} = 0.9$$

**4. INTERPRET.**

**Check:** If the apprentice word processor completes the job alone in 0.9 hours, the experienced word processor completes the job alone in $x - 2 = 0.9 - 2 = -1.1$ hours. Since this is not possible, we reject the solution of 0.9. The approximate solution thus is 13.1 hours.

**State:** The apprentice word processor can complete the job alone in approximately 13.1 hours, and the experienced word processor can complete the job alone in approximately

$$x - 2 = 13.1 - 2 = 11.1 \text{ hours}$$

■ **Work Practice 6**

---

**Example 7**  Finding Driving Speeds

Beach and Fargo are about 400 miles apart. A salesperson travels from Fargo to Beach one day at a certain speed. She returns to Fargo the next day and drives 10 mph faster. Her total travel time was $14\frac{2}{3}$ hours. Find her speed to Beach and the return speed to Fargo.

Beach   *x* mph   ————— 400 mi —————   *x* + 10 mph   Fargo

*(Continued on next page)*

**Practice 7**

A family drives 500 miles to the beach for a vacation. The return trip was made at a speed that was 10 miles per hour faster. The total traveling time was $18\frac{1}{3}$ hours. Find the speed to the beach and the return speed.

**Answer**

**7.** 50 mph to the beach; 60 mph returning

**Solution:**

1. UNDERSTAND. Read and reread the problem. Let

$$x = \text{the speed to Beach, so}$$

$$x + 10 = \text{the return speed to Fargo}$$

Then organize the given information in a table.

Since $d = rt$, then $t = \dfrac{d}{r}$. The time column was completed using $\dfrac{d}{r}$.

| | Distance | = | Rate | · | Time | |
|---|---|---|---|---|---|---|
| **To Beach** | 400 | | $x$ | | $\dfrac{400}{x}$ | ← distance / ← rate |
| **Return to Fargo** | 400 | | $x + 10$ | | $\dfrac{400}{x + 10}$ | ← distance / ← rate |

2. TRANSLATE.

In words:   time to Beach   +   return time to Fargo   =   $14\dfrac{2}{3}$ hours

Translate:   $\dfrac{400}{x} + \dfrac{400}{x + 10} = \dfrac{44}{3}$

3. SOLVE.

$$\frac{400}{x} + \frac{400}{x + 10} = \frac{44}{3}$$

This next step is optional. Notice that all three numerators in our equation are divisible by 4. To keep the numbers in our equation as simple as possible, we will take a step and divide through by 4.

$$\frac{100}{x} + \frac{100}{x + 10} = \frac{11}{3} \qquad \text{Divide both sides by 4.}$$

$$3x(x + 10)\left(\frac{100}{x} + \frac{100}{x + 10}\right) = 3x(x + 10) \cdot \frac{11}{3} \qquad \text{Multiply both sides by the LCD, } 3x(x + 10).$$

$$3x(x + 10)\frac{100}{x} + 3x(x + 10)\frac{100}{x + 10} = 3x(x + 10) \cdot \frac{11}{3} \qquad \text{Use the distributive property.}$$

$$3(x + 10)100 + 3x(100) = x(x + 10)11$$

$$300x + 3000 + 300x = 11x^2 + 110x$$

$$0 = 11x^2 - 490x - 3000 \qquad \text{Set equation equal to 0.}$$

$$0 = (11x + 60)(x - 50) \qquad \text{Factor.}$$

$$11x + 60 = 0 \qquad \text{or} \qquad x - 50 = 0 \qquad \text{Set each factor equal to 0.}$$

$$x = -\frac{60}{11} = -5\frac{5}{11} \qquad\qquad x = 50$$

4. INTERPRET.

**Check:** The speed is not negative, so it's not $-5\dfrac{5}{11}$. The number 50 does check.

**State:** The speed to Beach was 50 miles per hour and the return speed to Fargo was 60 miles per hour.

▣ Work Practice 7

## Vocabulary, Readiness & Video Check

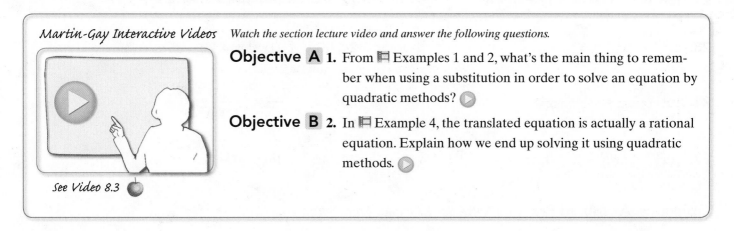

*Martin-Gay Interactive Videos*    *Watch the section lecture video and answer the following questions.*

**Objective A** **1.** From ▤ Examples 1 and 2, what's the main thing to remember when using a substitution in order to solve an equation by quadratic methods? ▶

**Objective B** **2.** In ▤ Example 4, the translated equation is actually a rational equation. Explain how we end up solving it using quadratic methods. ▶

*See Video 8.3* 🍎

## 8.3 **Exercise Set** MyMathLab® ▶

**Objective A** *Solve. See Example 1.*

**1.** $2x = \sqrt{10 + 3x}$

**2.** $3x = \sqrt{8x + 1}$

**3.** $x - 2\sqrt{x} = 8$

**4.** $x - \sqrt{2x} = 4$

**5.** $\sqrt{9x} = x + 2$

**6.** $\sqrt{16x} = x + 3$

*Solve. See Example 2.*

▶**7.** $\dfrac{2}{x} + \dfrac{3}{x - 1} = 1$

**8.** $\dfrac{6}{x^2} = \dfrac{3}{x + 1}$

**9.** $\dfrac{3}{x} + \dfrac{4}{x + 2} = 2$

**10.** $\dfrac{5}{x - 2} + \dfrac{4}{x + 2} = 1$

**11.** $\dfrac{7}{x^2 - 5x + 6} = \dfrac{2x}{x - 3} - \dfrac{x}{x - 2}$

**12.** $\dfrac{11}{2x^2 + x - 15} = \dfrac{5}{2x - 5} - \dfrac{x}{x + 3}$

*Solve. See Example 3.*

**13.** $p^4 - 16 = 0$

**14.** $z^4 = 81$

**15.** $z^4 - 5z^2 - 36 = 0$

**16.** $x^4 + 2x^2 - 3 = 0$

**17.** $4x^4 + 11x^2 = 3$

**18.** $9x^4 + 5x^2 - 4 = 0$

*Solve. See Examples 4 and 5.*

▶**19.** $x^{2/3} - 3x^{1/3} - 10 = 0$

**20.** $x^{2/3} + 2x^{1/3} + 1 = 0$

**21.** $(5n + 1)^2 + 2(5n + 1) - 3 = 0$

**22.** $(m - 6)^2 + 5(m - 6) + 4 = 0$

**23.** $2x^{2/3} - 5x^{1/3} = 3$

**24.** $3x^{2/3} + 11x^{1/3} = 4$

**25.** $1 + \dfrac{2}{3t - 2} = \dfrac{8}{(3t - 2)^2}$

**26.** $2 - \dfrac{7}{x + 6} = \dfrac{15}{(x + 6)^2}$

**27.** $20x^{2/3} - 6x^{1/3} - 2 = 0$

**28.** $4x^{2/3} + 16x^{1/3} = -15$

## Objective A Mixed Practice *Solve. See Examples 1 through 5.*

**29.** $a^4 - 5a^2 + 6 = 0$

**30.** $x^4 - 12x^2 + 11 = 0$

**31.** $\dfrac{2x}{x - 2} + \dfrac{x}{x + 3} = \dfrac{-5}{x + 3}$

**32.** $\dfrac{5}{x - 3} + \dfrac{x}{x + 3} = \dfrac{19}{x^2 - 9}$

**▶ 33.** $(p + 2)^2 = 9(p + 2) - 20$

**34.** $2(4m - 3)^2 - 9(4m - 3) = 5$

**35.** $2x = \sqrt{11x + 3}$

**36.** $4x = \sqrt{2x + 3}$

**37.** $x^{2/3} - 8x^{1/3} + 15 = 0$

**38.** $x^{2/3} - 2x^{1/3} - 8 = 0$

**39.** $x - \sqrt{19 - 2x} - 2 = 0$

**40.** $x - \sqrt{17 - 4x} - 3 = 0$

**41.** $2x^{2/3} + 3x^{1/3} - 2 = 0$

**42.** $6x^{2/3} - 25x^{1/3} - 25 = 0$

**43.** $(t + 3)^2 - 2(t + 3) - 8 = 0$

**44.** $(2n - 3)^2 - 7(2n - 3) + 12 = 0$

**45.** $x - \sqrt{x} = 2$

**46.** $x - \sqrt{3x} = 6$

**47.** $\dfrac{x}{x - 1} + \dfrac{1}{x + 1} = \dfrac{2}{x^2 - 1}$

**48.** $\dfrac{x}{x - 5} + \dfrac{5}{x + 5} = \dfrac{-1}{x^2 - 25}$

**49.** $p^4 - p^2 - 20 = 0$

**50.** $x^4 - 10x^2 + 9 = 0$

**51.** $1 = \dfrac{4}{x - 7} + \dfrac{5}{(x - 7)^2}$

**52.** $3 + \dfrac{1}{2p + 4} = \dfrac{10}{(2p + 4)^2}$

**53.** $27y^4 + 15y^2 = 2$

**54.** $8z^4 + 14z^2 = -5$

## Objective B   *Solve. For Exercises 59 and 60, the solutions have been started for you. See Examples 6 and 7.*

**55.** A jogger ran 3 miles, decreased her speed by 1 mile per hour, and then ran another 4 miles. If her total time jogging was $1\frac{3}{5}$ hours, find her speed for each part of her run.

**56.** Mark Keaton's workout consists of jogging for 3 miles and then riding his bike for 5 miles at a speed 4 miles per hour faster than he jogs. If his total workout time is 1 hour, find his jogging speed and his biking speed.

**57.** A Chinese restaurant in Mandeville, Louisiana, has a large goldfish pond around the restaurant. Suppose that an inlet pipe and a hose together can fill the pond in 8 hours. The inlet pipe alone can complete the job in 1 hour less time than the hose alone. Find the time that the hose can complete the job alone and the time that the inlet pipe can complete the job alone. Round each to the nearest tenth of an hour.

**58.** A water tank on a farm in Flatonia, Texas, can be filled with a large inlet pipe and a small inlet pipe in 3 hours. The large inlet pipe alone can fill the tank in 2 hours less time than the small inlet pipe alone. Find the time to the nearest tenth of an hour each pipe can fill the tank alone.

**59.** Roma Sherry drove 330 miles from her home town to Tucson. During her return trip, she was able to increase her speed by 11 miles per hour. If her return trip took 1 hour less time, find her original speed and her speed returning home.

**Start the solution:**

**1.** UNDERSTAND the problem. Reread it as many times as needed. Let

$$x = \text{original speed}$$
$$x + 11 = \text{return-trip speed}$$

Organize the information in a table.

| | distance | = | rate | · | time |
|---|---|---|---|---|---|
| **To Tucson** | 330 | | $x$ | | $\dfrac{330}{x}$ |
| **Return trip** | 330 | | ____ | | $\dfrac{330}{\rule{1cm}{0.4pt}}$ |

**2.** TRANSLATE into an equation. (Fill in the blanks below.)

| Time to Tucson | equals | Return trip time | plus | 1 hour |
|---|---|---|---|---|
| ↓ | ↓ | ↓ | ↓ | ↓ |
| ____ | = | ____ | + | 1 |

Finish with:

**3.** SOLVE and **4.** INTERPRET

**60.** A salesperson drove to Portland, a distance of 300 miles. During the last 80 miles of his trip, heavy rainfall forced him to decrease his speed by 15 miles per hour. If his total driving time was 6 hours, find his original speed and his speed during the rainfall.

**Start the solution:**

**1.** UNDERSTAND the problem. Reread it as many times as needed. Let

$$x = \text{original speed}$$
$$x - 15 = \text{rainfall speed}$$

Organize the information in a table.

| | distance | = | rate | · | time |
|---|---|---|---|---|---|
| **First part of trip** | 300 − 80, or 220 | | $x$ | | $\dfrac{220}{x}$ |
| **Heavy rainfall part of trip** | 80 | | $x - 15$ | | $\dfrac{80}{\rule{1cm}{0.4pt}}$ |

**2.** TRANSLATE into an equation. (Fill in the blanks below.)

| Time during first part of trip | plus | Time during heavy rainfall | equals | 6 hr |
|---|---|---|---|---|
| ↓ | ↓ | ↓ | ↓ | ↓ |
| ____ | + | ____ | = | 6 |

Finish with:

**3.** SOLVE and **4.** INTERPRET

**61.** Bill Shaughnessy and his son Billy can clean the house together in 4 hours. When the son works alone, it takes him an hour longer to clean than it takes his dad alone. Find how long to the nearest tenth of an hour it takes the son to clean alone.

**62.** Together, Noodles and Freckles eat a 50-pound bag of dog food in 30 days. Noodles by herself eats a 50-pound bag in 2 weeks less time than Freckles does by himself. How many days to the nearest whole day would a 50-pound bag of dog food last Freckles?

**63.** The product of a number and 4 less than the number is 96. Find the number.

**64.** A whole number increased by its square is two more than twice itself. Find the number.

△ **65.** Suppose that we want to make an open box from a square sheet of cardboard by cutting out squares from each corner as shown and then folding along the dotted lines. If the box is to have a volume of 300 cubic inches, find the original dimensions of the sheet of cardboard.

**a.** The ? in the drawing to the right will be the length (and also the width) of the box as shown in the drawing to the left. Represent this length in terms of $x$.

**b.** Use the formula for volume of a box, $V = l \cdot w \cdot h$, to write an equation in $x$.

**c.** Solve the equation for $x$ and give the dimensions of the sheet of cardboard. Check your solution.

△ **66.** Suppose that we want to make an open box from a square sheet of cardboard by cutting out squares from each corner as shown and then folding along the dotted lines. If the box is to have a volume of 128 cubic inches, find the original dimensions of the sheet of cardboard. (*Hint:* Use Exercise 65 parts (a), (b), and (c) to help you.)

△ **67.** A sprinkler that sprays water in a circular pattern is to be used to water a square garden. If the area of the garden is 920 square feet, find the smallest whole number *radius* that the sprinkler can be adjusted to so that the entire garden is watered.

△ **68.** Suppose that a square field has an area of 6270 square feet. See Exercise 67 and find the required sprinkler radius.

## Review

*Solve each inequality. See Section 2.4.*

**69.** $\frac{5x}{3} + 2 \leq 7$

**70.** $\frac{2x}{3} + \frac{1}{6} \geq 2$

**71.** $\frac{y-1}{15} > -\frac{2}{5}$

**72.** $\frac{z-2}{12} < \frac{1}{4}$

## Concept Extensions

*Solve.*

**73.** $y^3 + 9y - y^2 - 9 = 0$

**74.** $x^3 + x - 3x^2 - 3 = 0$

**75.** $x^{-2} - x^{-1} - 6 = 0$

**76.** $y^{-2} - 8y^{-1} + 7 = 0$

**77.** $2x^3 = -54$

**78.** $y^3 - 216 = 0$

**79.** Write a polynomial equation that has three solutions: $2, 5,$ and $-7$.

**80.** Write a polynomial equation that has three solutions: $0, 2i,$ and $-2i$.

**81.** At the FIS World Cup Slalom event in Flachau, Austria, on January 14, 2014, Mikaela Shiffrin of the United States beat Frida Hansdotter of Sweden. During the first run, Shiffrin posted a time that was 1.09 seconds faster than Hansdotter's. On the second run, however, Hansdotter beat Shiffrin's time by 0.26 second. The total of the two runs for Shiffrin was 105.83 seconds, and the first run for Hansdotter was 52.81 seconds.

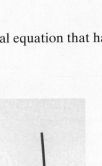

   **a.** Find the time for Shiffrin's first run.

   **b.** Find the time for Shiffrin's second run.

   **c.** Find the final time for Hansdotter's two runs.

   **d.** Find the time for Hansdotter's second run.

**82.** Use a grapher to solve Exercise 29. Compare the solution with the solution from Exercise 29. Explain any differences.

## Summary on Solving Quadratic Equations

**Answers**

1. _____

2. _____

3. _____

4. _____

5. _____

6. _____

7. _____

8. _____

9. _____

10. _____

11. _____

12. _____

13. _____

14. _____

15. _____

16. _____

17. _____

18. _____

*Use the square root property to solve each equation.*

**1.** $x^2 - 10 = 0$

**2.** $3x^2 + 24 = 0$

**3.** $(x - 1)^2 = 8$

**4.** $(2x + 5)^2 = 12$

*Solve each equation by completing the square.*

**5.** $x^2 + 2x - 12 = 0$

**6.** $x^2 - 12x + 11 = 0$

**7.** $4x^2 + 12x = 8$

**8.** $16y^2 + 16y = 1$

*Use the quadratic formula to solve each equation.*

**9.** $2x^2 - 4x + 1 = 0$

**10.** $\frac{1}{2}x^2 + 3x + 2 = 0$

**11.** $x^2 + 4x = -7$

**12.** $5x^2 + 6x = -3$

*Solve each equation. Use the method of your choice.*

**13.** $x^2 + 3x + 6 = 0$

**14.** $2x^2 + 18 = 0$

**15.** $x^2 + 17x = 0$

**16.** $4x^2 - 2x - 3 = 0$

**17.** $(x - 2)^2 = 27$

**18.** $\frac{1}{2}x^2 - 2x + \frac{1}{2} = 0$

**19.** $3x^2 + 2x = 8$

**20.** $2x^2 = -5x - 1$

**21.** $x(x - 2) = 5$

**22.** $x^2 - 31 = 0$

**23.** $4x^2 - 48 = 0$

**24.** $5x^2 + 55 = 0$

**25.** $x(x + 5) = 66$

**26.** $5x^2 + 6x - 2 = 0$

**27.** $2x^2 + 3x = 1$

**28.** $x - \sqrt{13 - 3x} - 3 = 0$

**29.** $\dfrac{5x}{x - 2} - \dfrac{x + 1}{x} = \dfrac{3}{x(x - 2)}$

△ **30.** The diagonal of a square room measures 20 feet. Find the exact length of a side of the room. Then approximate the length to the nearest tenth of a foot.

20 ft
x

**31.** Jack and Lucy Hoag together can prepare a crawfish boil for a large party in 4 hours. Lucy alone can complete the job in 2 hours less time than Jack alone. Find the time in which each person can prepare the crawfish boil alone. Round each time to the nearest tenth of an hour.

**32.** Kraig Blackwelder exercises at Total Body Gym. On the treadmill, he runs 5 miles, then increases his speed by 1 mile per hour and runs an additional 2 miles. If his total time on the treadmill is $1\dfrac{1}{3}$ hours, find his speed during each part of his run.

**19.** _____

**20.** _____

**21.** _____

**22.** _____

**23.** _____

**24.** _____

**25.** _____

**26.** _____

**27.** _____

**28.** _____

**29.** _____

**30.** _____

**31.** _____

**32.** _____

**Nonlinear Inequalities in One Variable** ▶

## Objectives

**A** Solve Polynomial Inequalities of Degree 2 or Greater. ▶

**B** Solve Inequalities That Contain Rational Expressions with Variables in the Denominator. ▶

## Objective A Solving Polynomial Inequalities ▶

Just as we can solve linear inequalities in one variable, we can also solve quadratic and higher-degree inequalities in one variable. Let's begin with quadratic inequalities. A **quadratic inequality** is an inequality that can be written so that one side is a quadratic expression and the other side is 0. Here are examples of quadratic inequalities in one variable. Each is written in **standard form.**

$$x^2 - 10x + 7 \le 0 \qquad 3x^2 + 2x - 6 > 0$$
$$2x^2 + 9x - 2 < 0 \qquad x^2 - 3x + 11 \ge 0$$

A solution of a quadratic inequality in one variable is a value of the variable that makes the inequality a true statement.

The value of an expression such as $x^2 - 3x - 10$ will sometimes be positive, sometimes negative, and sometimes 0, depending on the value substituted for $x$. To solve the inequality $x^2 - 3x - 10 < 0$, we look for all values of $x$ that make the expression $x^2 - 3x - 10$ less than 0, or negative. To understand how we find these values, we'll study the graph of the quadratic function $y = x^2 - 3x - 10$.

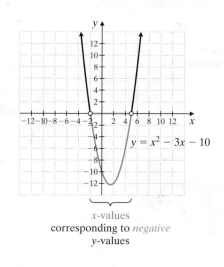

$x$-values
corresponding to *negative*
$y$-values

Notice that the $x$-values for which $y$ or $x^2 - 3x - 10$ is positive are separated from the $x$-values for which $y$ or $x^2 - 3x - 10$ is negative by the values for which $y$ or $x^2 - 3x - 10$ is 0, the $x$-intercepts. Thus, the solution set of $x^2 - 3x - 10 < 0$ consists of all real numbers from $-2$ to 5 or, in interval notation, $(-2, 5)$.

It is not necessary to graph $y = x^2 - 3x - 10$ to solve the related inequality $x^2 - 3x - 10 < 0$. Instead, we can draw a number line representing the $x$-axis and keep the following in mind: *A region on the number line for which the value of $x^2 - 3x - 10$ is positive is separated from a region on the number line for which the value of $x^2 - 3x - 10$ is negative by a value for which the expression is 0.*

Let's find these values for which the expression is 0 by solving the related equation, $x^2 - 3x - 10 = 0$.

$$x^2 - 3x - 10 = 0$$
$$(x - 5)(x + 2) = 0 \qquad \text{Factor.}$$
$$x - 5 = 0 \quad \text{or} \quad x + 2 = 0 \qquad \text{Set each factor equal to 0.}$$
$$x = 5 \qquad\qquad x = -2 \qquad \text{Solve.}$$

These two numbers $-2$ and $5$ divide the number line into three regions. We will call the regions $A$, $B$, and $C$.

These regions are important because if the value of $x^2 - 3x - 10$ is negative when a number from a region is substituted for $x$, then $x^2 - 3x - 10$ is negative when any number in that region is substituted for $x$. Similarly, if the value of $x^2 - 3x - 10$ is positive when a number from a region is substituted for $x$, then $x^2 - 3x - 10$ is positive when any number in that region is substituted for $x$.

To see whether the inequality $x^2 - 3x - 10 < 0$ is true or false in each region, we choose a test point from each region and substitute its value for $x$ in the inequality $x^2 - 3x - 10 < 0$. If the resulting inequality is true, the region containing the test point is a solution region.

| Region | Test Point Value | $(x - 5)(x + 2) < 0$ | Result |
|--------|------------------|----------------------|--------|
| $A$ | $-3$ | $(-8)(-1) < 0$ | False |
| $B$ | $0$ | $(-5)(2) < 0$ | True |
| $C$ | $6$ | $(1)(8) < 0$ | False |

The values in region $B$ satisfy the inequality. The numbers $-2$ and $5$ are not included in the solution set since the inequality symbol is $<$. The solution set is $(-2, 5)$, and its graph is shown.

---

**Example 1** Solve: $(x + 3)(x - 3) > 0$

**Solution:** First we solve the related equation, $(x + 3)(x - 3) = 0$.

$$(x + 3)(x - 3) = 0$$
$$x + 3 = 0 \quad \text{or} \quad x - 3 = 0$$
$$x = -3 \qquad \qquad x = 3$$

The two numbers $-3$ and $3$ separate the number line into three regions, $A$, $B$, and $C$.

Now we substitute the value of a test point from each region. If the test value satisfies the inequality, every value in the region containing the test value is a solution.

| Region | Test Point Value | $(x + 3)(x - 3) > 0$ | Result |
|--------|------------------|----------------------|--------|
| $A$ | $-4$ | $(-1)(-7) > 0$ | True |
| $B$ | $0$ | $(3)(-3) > 0$ | False |
| $C$ | $4$ | $(7)(1) > 0$ | True |

The points in regions $A$ and $C$ satisfy the inequality. The numbers $-3$ and $3$ are not included in the solution since the inequality symbol is $>$. The solution set is $(-\infty, -3) \cup (3, \infty)$, and its graph is shown.

■ **Work Practice 1**

**Practice 1**

Solve: $(x - 2)(x + 4) > 0$

**Answer**
**1.** $(-\infty, -4) \cup (2, \infty)$

The steps below may be used to solve a polynomial inequality of degree 2 or greater.

> ## Solving a Polynomial Inequality of Degree 2 or Greater
>
> **Step 1:** Write the inequality in standard form and solve the related equation.
>
> **Step 2:** Separate the number line into regions using the solutions from Step 1.
>
> **Step 3:** For each region, choose a test point and determine whether its value satisfies the *original inequality*.
>
> **Step 4:** The solution set includes the regions whose test point value is a solution. If the inequality symbol is $\leq$ or $\geq$, the values from Step 1 are solutions; if $<$ or $>$, they are not.

**✓Concept Check** When choosing a test point in Step 4, why would the solutions from Step 1 not make good choices for test points?

**Practice 2**

Solve: $x^2 - 6x \leq 0$

**Example 2** Solve: $x^2 - 4x \leq 0$

**Solution:** First we solve the related equation, $x^2 - 4x = 0$.

$$x^2 - 4x = 0$$
$$x(x - 4) = 0$$
$$x = 0 \quad \text{or} \quad x = 4$$

The numbers 0 and 4 separate the number line into three regions, $A$, $B$, and $C$.

We check a test value in each region in the original inequality. Values in region $B$ satisfy the inequality. The numbers 0 and 4 are included in the solution since the inequality symbol is $\leq$. The solution set is $[0, 4]$, and its graph is shown.

■ **Work Practice 2**

**Practice 3**

Solve:
$(x - 2)(x + 1)(x + 5) \leq 0$

**Example 3** Solve: $(x + 2)(x - 1)(x - 5) \leq 0$

**Solution:** First we solve $(x + 2)(x - 1)(x - 5) = 0$. By inspection, we see that the solutions are $-2, 1$, and $5$. They separate the number line into four regions, $A, B, C$, and $D$. Next we check test points from each region.

| Region | Test Point Value | $(x + 2)(x - 1)(x - 5) \leq 0$ | Result |
|--------|------------------|-------------------------------|--------|
| $A$ | $-3$ | $(-1)(-4)(-8) \leq 0$ | True |
| $B$ | $0$ | $(2)(-1)(-5) \leq 0$ | False |
| $C$ | $2$ | $(4)(1)(-3) \leq 0$ | True |
| $D$ | $6$ | $(8)(5)(1) \leq 0$ | False |

The solution set is $(-\infty, -2] \cup [1, 5]$, and its graph is shown. We include the numbers $-2, 1$, and $5$ because the inequality symbol is $\leq$.

**Answers**

**2.** $[0, 6]$   **3.** $(-\infty, -5] \cup [-1, 2]$

**✓Concept Check Answer**

The solutions found in Step 1 have a value of 0 in the original inequality.

■ **Work Practice 3**

## Objective B Solving Rational Inequalities

Inequalities containing rational expressions with variables in the denominator are solved by using a similar procedure. Notice as we solve an example that unlike quadratic inequalities, we must also consider values for which the rational inequality is undefined. Why? As usual, these values may not be solution values for the inequality.

**Example 4** Solve: $\dfrac{x + 2}{x - 3} \le 0$

**Solution:** First we find all values that make the denominator equal to 0. To do this, we solve $x - 3 = 0$, or $x = 3$.

Next, we solve the related equation, $\dfrac{x + 2}{x - 3} = 0$.

$$\dfrac{x + 2}{x - 3} = 0 \qquad \text{Multiply both sides by the LCD, } x - 3.$$

$$x + 2 = 0$$

$$x = -2$$

Now we place these numbers on a number line and proceed as before, checking test point values in the original inequality.

| **Choose $-3$ from region $A$** | **Choose $0$ from region $B$** | **Choose $4$ from region $C$** |
|---|---|---|
| $\dfrac{x + 2}{x - 3} \le 0$ | $\dfrac{x + 2}{x - 3} \le 0$ | $\dfrac{x + 2}{x - 3} \le 0$ |
| $\dfrac{-3 + 2}{-3 - 3} \le 0$ | $\dfrac{0 + 2}{0 - 3} \le 0$ | $\dfrac{4 + 2}{4 - 3} \le 0$ |
| $\dfrac{-1}{-6} \le 0$ | $-\dfrac{2}{3} \le 0$  True | $6 \le 0$  False |
| $\dfrac{1}{6} \le 0$  False | | |

The solution set is $[-2, 3)$. This interval includes $-2$ because $-2$ satisfies the original inequality. This interval does not include $3$ because $3$ would make the denominator $0$.

■ **Work Practice 4**

The steps below may be used to solve a rational inequality with variables in the denominator.

---

### Solving a Rational Inequality

**Step 1:** Solve for values that make any denominator $0$.

**Step 2:** Solve the related equation.

**Step 3:** Separate the number line into regions using the solutions from Steps 1 and 2.

**Step 4:** For each region, choose a test point and determine whether its value satisfies the *original inequality*.

**Step 5:** The solution set includes the regions whose test point value is a solution. Check whether to include values from Step 2. Be sure *not* to include values that make any denominator $0$.

---

**Practice 4**

Solve: $\dfrac{x - 3}{x + 5} \le 0$

**Answer**

**4.** $(-5, 3]$

## Practice 5

Solve: $\dfrac{3}{x-2} < 2$

**Answer**

5. $(-\infty, 2) \cup \left(\dfrac{7}{2}, \infty\right)$

**Example 5**  Solve: $\dfrac{5}{x+1} < -2$

**Solution:**  First we find values for $x$ that make the denominator equal to 0.

$$x + 1 = 0$$
$$x = -1$$

Next we solve $\dfrac{5}{x+1} = -2$.

$$(x+1) \cdot \dfrac{5}{x+1} = (x+1) \cdot -2 \quad \text{Multiply both sides by the LCD, } x+1.$$
$$5 = -2x - 2 \qquad \text{Simplify.}$$
$$7 = -2x$$
$$-\dfrac{7}{2} = x$$

We use these two solutions to divide a number line into three regions and choose test points. Only a test point value from region $B$ satisfies the *original inequality*.

The solution set is $\left(-\dfrac{7}{2}, -1\right)$, and its graph is shown.

**Answer**

5. $(-\infty, 2) \cup \left(\dfrac{7}{2}, \infty\right)$

■ Work Practice 5

---

## Vocabulary, Readiness & Video Check

*Write the graphed solution set in interval notation.*

1. _____
   $-7$   $3$

2. _____
   $-1$   $5$

3. _____
   $0$

4. _____
   $-8$

5. _____
   $-12$   $-10$

6. _____
   $-3$   $4$

---

*See Video 8.4*

Watch the section lecture video and answer the following questions.

**Objective A**  **7.** From ▤ Examples 1–3, how does solving a related equation help you solve a polynomial inequality? Are the solutions to the related equation ever solutions to the inequality? ▶

**Objective B**  **8.** In ▤ Example 4, one of the values that separates the number line into regions is 4. The inequality is $\geq$, so why isn't 4 included in the solution set? ▶

## 8.4 Exercise Set MyMathLab ▶

**Objective A** *Solve. Write the solution set in interval notation. See Examples 1 through 3.*

**1.** $(x + 1)(x + 5) > 0$

**2.** $(x + 1)(x + 5) \le 0$

▶ **3.** $(x - 3)(x + 4) \le 0$

**4.** $(x + 4)(x - 1) > 0$

**5.** $x^2 + 8x + 15 \ge 0$

**6.** $x^2 - 7x + 10 \le 0$

**7.** $3x^2 + 16x < -5$

**8.** $2x^2 - 5x < 7$

**9.** $(x - 6)(x - 4)(x - 2) > 0$

**10.** $(x - 6)(x - 4)(x - 2) \le 0$

**11.** $x(x - 1)(x + 4) \le 0$

**12.** $x(x - 6)(x + 2) > 0$

**13.** $(x^2 - 9)(x^2 - 4) > 0$

**14.** $(x^2 - 16)(x^2 - 1) \le 0$

**Objective B** *Solve. Write the solution set in interval notation. See Examples 4 and 5.*

**15.** $\dfrac{x + 7}{x - 2} < 0$

**16.** $\dfrac{x - 5}{x - 6} > 0$

**17.** $\dfrac{5}{x + 1} > 0$

**18.** $\dfrac{3}{y - 5} < 0$

▶ **19.** $\dfrac{x + 1}{x - 4} \ge 0$

**20.** $\dfrac{x + 1}{x - 4} \le 0$

**21.** $\dfrac{3}{x - 2} < 4$

**22.** $\dfrac{-2}{y + 3} > 2$

**23.** $\dfrac{x^2 + 6}{5x} \ge 1$

**24.** $\dfrac{y^2 + 15}{8y} \le 1$

**25.** $\dfrac{x + 2}{x - 3} < 1$

**26.** $\dfrac{x - 1}{x + 4} > 2$

**Objectives A B Mixed Practice** *Solve each inequality. Write the solution set in interval notation. See Examples 1 through 5.*

**27.** $(2x - 3)(4x + 5) \le 0$

**28.** $(6x + 7)(7x - 12) > 0$

▶ **29.** $x^2 > x$

**30.** $x^2 < 25$

**31.** $\dfrac{x}{x - 10} < 0$

**32.** $\dfrac{x + 10}{x - 10} > 0$

**33.** $(2x - 8)(x + 4)(x - 6) \le 0$

**34.** $(3x - 12)(x + 5)(2x - 3) \ge 0$

**35.** $6x^2 - 5x \ge 6$

**36.** $12x^2 + 11x \leq 15$

**37.** $\dfrac{x - 5}{x + 4} \geq 0$

**38.** $\dfrac{x - 3}{x + 2} \leq 0$

**39.** $\dfrac{-1}{x - 1} > -1$

**40.** $\dfrac{4}{y + 2} < -2$

**41.** $4x^3 + 16x^2 - 9x - 36 > 0$

**42.** $x^3 + 2x^2 - 4x - 8 < 0$

▶ **43.** $x^4 - 26x^2 + 25 \geq 0$

**44.** $16x^4 - 40x^2 + 9 \leq 0$

**45.** $\dfrac{x(x + 6)}{(x - 7)(x + 1)} \geq 0$

**46.** $\dfrac{(x - 2)(x + 2)}{(x + 1)(x - 4)} \leq 0$

**47.** $\dfrac{x}{x + 4} \leq 2$

**48.** $\dfrac{4x}{x - 3} \geq 5$

**49.** $(2x - 7)(3x + 5) > 0$

**50.** $(4x - 9)(2x + 5) < 0$

**51.** $\dfrac{z}{z - 5} \geq 2z$

**52.** $\dfrac{p}{p + 4} \leq 3p$

**53.** $\dfrac{(x + 1)^2}{5x} > 0$

**54.** $\dfrac{(2x - 3)^2}{x} < 0$

## Review

*Fill in each table so that each ordered pair is a solution of the given function. See Section 3.6.*

**55.** $f(x) = x^2$

| x | y |
|---|---|
| 0 | |
| 1 | |
| −1 | |
| 2 | |
| −2 | |

**56.** $f(x) = 2x^2$

| x | y |
|---|---|
| 0 | |
| 1 | |
| −1 | |
| 2 | |
| −2 | |

**57.** $f(x) = -x^2$

| x | y |
|---|---|
| 0 | |
| 1 | |
| −1 | |
| 2 | |
| −2 | |

**58.** $f(x) = -3x^2$

| x | y |
|---|---|
| 0 | |
| 1 | |
| −1 | |
| 2 | |
| −2 | |

## Concept Extensions

**59.** Explain why $\dfrac{x + 2}{x - 3} > 0$ and $(x + 2)(x - 3) > 0$ have the same solution sets.

**60.** Explain why $\dfrac{x + 2}{x - 3} \geq 0$ and $(x + 2)(x - 3) \geq 0$ do not have the same solution sets.

*Find all numbers that satisfy each statement.*

**61.** A number minus its reciprocal is less than zero. Find the numbers.

**62.** Twice a number, added to its reciprocal is nonnegative. Find the numbers.

*Solve.*

**63.** The total profit $P(x)$ for a company producing $x$ thousand units is given by the function $P(x) = -2x^2 + 26x - 44$. Find the values of $x$ for which the company makes a profit. [*Hint:* The company makes a profit when $P(x) > 0$.]

**64.** A projectile is fired straight up from the ground with an initial velocity of 80 feet per second. Its height $s(t)$ in feet at any time $t$ in seconds is given by the function $s(t) = -16t^2 + 80t$. Find the interval of time for which the height of the projectile is greater than 96 feet.

*Solve each inequality, then use a graphing calculator to check.*

**65.** $x^2 - x - 56 > 0$

**66.** $x^2 - 4x - 5 < 0$

---

## 8.5 Quadratic Functions and Their Graphs

### Objective A Graphing $f(x) = x^2 + k$

We first graphed the quadratic function $f(x) = x^2$ in Section 3.8. In that section, we discovered that the graph of a quadratic function is a parabola opening upward or downward. Now, as we continue our study, we will discover more details about quadratic functions and their graphs.

First, let's recall the definition of a *quadratic function*.

**Quadratic Function**

A **quadratic function** is a function that can be written in the form $f(x) = ax^2 + bx + c$, where $a, b,$ and $c$ are real numbers and $a \neq 0$.

Notice that equations of the form $y = ax^2 + bx + c$, where $a \neq 0$, also define quadratic functions since $y$ is a function of $x$ or $y = f(x)$.

Recall that if $a > 0$, the parabola opens upward, and if $a < 0$, the parabola opens downward. Also, the vertex of a parabola is the lowest point if the parabola opens upward and the highest point if the parabola opens downward. The axis of symmetry is the vertical line that passes through the vertex.

**Objectives**

**A** Graph Quadratic Functions of the Form $f(x) = x^2 + k$.

**B** Graph Quadratic Functions of the Form $f(x) = (x - h)^2$.

**C** Graph Quadratic Functions of the Form $f(x) = (x - h)^2 + k$.

**D** Graph Quadratic Functions of the Form $f(x) = ax^2$.

**E** Graph Quadratic Functions of the Form $f(x) = a(x - h)^2 + k$.

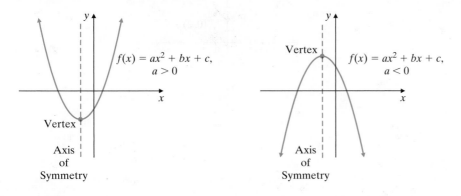

## Practice 1

Graph $f(x) = x^2$ and $g(x) = x^2 + 4$ on the same set of axes.

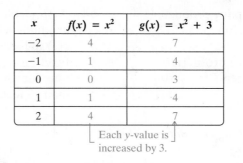

## Practice 2–3

Graph each function.

**2.** $F(x) = x^2 + 1$

**3.** $g(x) = x^2 - 2$

**Answers**

**1.**

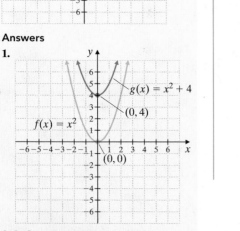

2–3. See answers on page 635.

**Example 1**   Graph $f(x) = x^2$ and $g(x) = x^2 + 3$ on the same set of axes.

**Solution:**   First we construct a table of values for $f$ and plot the points. Notice that for each $x$-value, the corresponding value of $g(x)$ must be 3 more than the corresponding value of $f(x)$, since $f(x) = x^2$ and $g(x) = x^2 + 3$. In other words, the graph of $g(x) = x^2 + 3$ is the same as the graph of $f(x) = x^2$ shifted upward 3 units. The axis of symmetry for both graphs is the $y$-axis.

| $x$ | $f(x) = x^2$ | $g(x) = x^2 + 3$ |
|-----|-----|-----|
| $-2$ | 4 | 7 |
| $-1$ | 1 | 4 |
| 0 | 0 | 3 |
| 1 | 1 | 4 |
| 2 | 4 | 7 |

Each $y$-value is increased by 3.

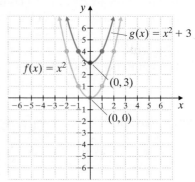

■ Work Practice 1

In general, we have the following properties.

### Graphing the Parabola Defined by $f(x) = x^2 + k$

If $k$ is positive, the graph of $f(x) = x^2 + k$ is the graph of $y = x^2$ shifted upward $k$ units.

If $k$ is negative, the graph of $f(x) = x^2 + k$ is the graph of $y = x^2$ shifted downward $|k|$ units.

The vertex is $(0, k)$, and the axis of symmetry is the $y$-axis.

**Examples**   Graph each function.

**2.** $F(x) = x^2 + 2$

The graph of $F(x) = x^2 + 2$ is obtained by shifting the graph of $y = x^2$ upward 2 units.

**3.** $g(x) = x^2 - 3$

The graph of $g(x) = x^2 - 3$ is obtained by shifting the graph of $y = x^2$ downward 3 units.

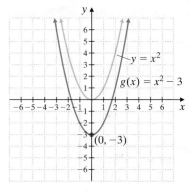

■ Work Practice 2–3

## Objective B Graphing $f(x) = (x - h)^2$ ▶

Now we will graph functions of the form $f(x) = (x - h)^2$.

**Example 4** Graph $f(x) = x^2$ and $g(x) = (x - 2)^2$ on the same set of axes.

**Solution:** By plotting points, we see that for each $x$-value, the corresponding value of $g(x)$ is the same as the value of $f(x)$ when the $x$-value is increased by 2. Thus, the graph of $g(x) = (x - 2)^2$ is the graph of $f(x) = x^2$ shifted to the right 2 units. The axis of symmetry for the graph of $g(x) = (x - 2)^2$ is also shifted 2 units to the right and is the line $x = 2$.

| $x$ | $f(x) = x^2$ | $x$ | $g(x) = (x - 2)^2$ |
|-----|-----|-----|-----|
| $-2$ | 4 | 0 | 4 |
| $-1$ | 1 | 1 | 1 |
| 0 | 0 | 2 | 0 |
| 1 | 1 | 3 | 1 |
| 2 | 4 | 4 | 4 |

Each $x$-value increased by 2 corresponds to same $y$-value.

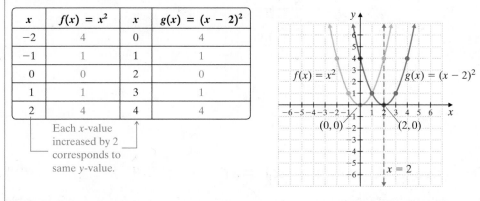

■ Work Practice 4

In general, we have the following properties.

> ### Graphing the Parabola Defined by $f(x) = (x - h)^2$
>
> If $h$ is positive, the graph of $f(x) = (x - h)^2$ is the graph of $y = x^2$ shifted to the right $h$ units.
>
> If $h$ is negative, the graph of $f(x) = (x - h)^2$ is the graph of $y = x^2$ shifted to the left $|h|$ units.
>
> The vertex is $(h, 0)$, and the axis of symmetry is the vertical line $x = h$.

**Practice 4**

Graph $f(x) = x^2$ and $g(x) = (x - 1)^2$ on the same set of axes.

**Answers**

**2.**

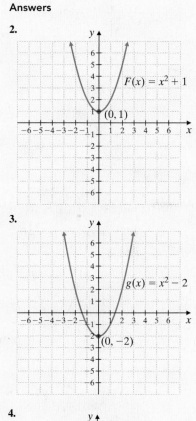

$F(x) = x^2 + 1$
$(0, 1)$

**3.**

$g(x) = x^2 - 2$
$(0, -2)$

**4.**

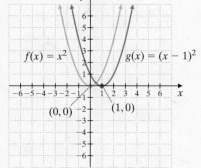

$f(x) = x^2$    $g(x) = (x - 1)^2$
$(0, 0)$    $(1, 0)$

## Practice 5–6

Graph each function.

**5.** $G(x) = (x - 4)^2$

**6.** $F(x) = (x + 2)^2$

**Answers**

**5.**

**6.**

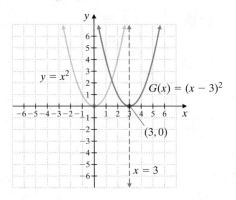

**Examples**   Graph each function.

**5.** $G(x) = (x - 3)^2$

The graph of $G(x) = (x - 3)^2$ is obtained by shifting the graph of $y = x^2$ to the right 3 units.

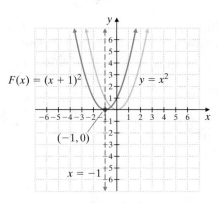

**6.** $F(x) = (x + 1)^2$

The equation $F(x) = (x + 1)^2$ can be written as $F(x) = [x - (-1)]^2$. The graph of $F(x) = [x - (-1)]^2$ is obtained by shifting the graph of $y = x^2$ to the left 1 unit.

**Work Practice 5–6**

## Objective C  Graphing $f(x) = (x - h)^2 + k$ ▶

As we will see in graphing functions of the form $f(x) = (x - h)^2 + k$, it is possible to combine vertical and horizontal shifts.

### Graphing the Parabola Defined by $f(x) = (x - h)^2 + k$

The parabola has the same shape as $y = x^2$.

The vertex is $(h, k)$, and the axis of symmetry is the vertical line $x = h$.

**Example 7** Graph: $F(x) = (x - 3)^2 + 1$

**Solution:** The graph of $F(x) = (x - 3)^2 + 1$ is the graph of $y = x^2$ shifted 3 units to the right and 1 unit up. The vertex is then $(3, 1)$, and the axis of symmetry is $x = 3$. A few ordered pair solutions are plotted to aid in graphing.

| $x$ | $F(x) = (x - 3)^2 + 1$ |
|-----|------------------------|
| 1 | 5 |
| 2 | 2 |
| 4 | 2 |
| 5 | 5 |

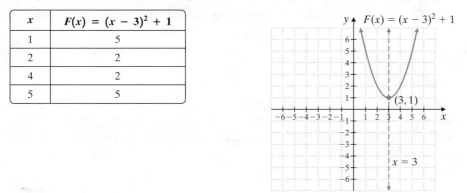

■ Work Practice 7

## Objective D Graphing $f(x) = ax^2$ ▶

Next, we discover the change in the shape of the graph when the coefficient of $x^2$ is not 1.

**Example 8** Graph $f(x) = x^2$, $g(x) = 3x^2$, and $h(x) = \frac{1}{2}x^2$ on the same set of axes.

**Solution:** Comparing the table of values, we see that for each $x$-value, the corresponding value of $g(x)$ is triple the corresponding value of $f(x)$. Similarly, the value of $h(x)$ is half the value of $f(x)$.

| $x$ | $f(x) = x^2$ |
|-----|--------------|
| $-2$ | 4 |
| $-1$ | 1 |
| 0 | 0 |
| 1 | 1 |
| 2 | 4 |

| $x$ | $g(x) = 3x^2$ |
|-----|---------------|
| $-2$ | 12 |
| $-1$ | 3 |
| 0 | 0 |
| 1 | 3 |
| 2 | 12 |

| $x$ | $h(x) = \frac{1}{2}x^2$ |
|-----|-------------------------|
| $-2$ | 2 |
| $-1$ | $\frac{1}{2}$ |
| 0 | 0 |
| 1 | $\frac{1}{2}$ |
| 2 | 2 |

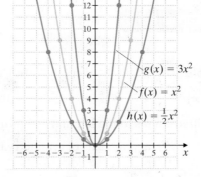

The result is that the graph of $g(x) = 3x^2$ is narrower than the graph of $f(x) = x^2$ and the graph of $h(x) = \frac{1}{2}x^2$ is wider. The vertex for each graph is $(0, 0)$, and the axis of symmetry is the $y$-axis.

■ Work Practice 8

**Practice 7**

Graph: $F(x) = (x - 2)^2 + 3$

**Practice 8**

Graph $f(x) = x^2$, $g(x) = 2x^2$, and $h(x) = \frac{1}{3}x^2$ on the same set of axes.

**Answers**

7.

8.

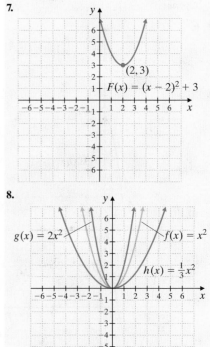

## Practice 9

Graph: $f(x) = -3x^2$

## Practice 10

Graph: $f(x) = 2(x + 3)^2 - 4$. Find the vertex and axis of symmetry.

**9.**

**10.**

$f(x) = 2(x + 3)^2 - 4$

$(-3, -4)$

$x = -3$

### Graphing the Parabola Defined by $f(x) = ax^2$

If $a$ is positive, the parabola opens upward, and if $a$ is negative, the parabola opens downward.

If $|a| > 1$, the graph of the parabola is narrower than the graph of $y = x^2$.

If $|a| < 1$, the graph of the parabola is wider than the graph of $y = x^2$.

**Example 9**  Graph: $f(x) = -2x^2$

**Solution:**  Because $a = -2$, a negative value, this parabola opens downward. Since $|-2| = 2$ and $2 > 1$, the parabola is narrower than the graph of $y = x^2$. The vertex is $(0, 0)$, and the axis of symmetry is the $y$-axis. We verify this by plotting a few points.

| $x$ | $f(x) = -2x^2$ |
|-----|----------------|
| $-2$ | $-8$ |
| $-1$ | $-2$ |
| $0$ | $0$ |
| $1$ | $-2$ |
| $2$ | $-8$ |

$f(x) = -2x^2$

■ Work Practice 9

## Objective E Graphing $f(x) = a(x - h)^2 + k$

Now we will see the shape of the graph of a quadratic function of the form $f(x) = a(x - h)^2 + k$.

**Example 10**  Graph: $g(x) = \dfrac{1}{2}(x + 2)^2 + 5$. Find the vertex and the axis of symmetry.

**Solution:**  The function $g(x) = \dfrac{1}{2}(x + 2)^2 + 5$ may be written as $g(x) = \dfrac{1}{2}[x - (-2)]^2 + 5$. Thus, this graph is the same as the graph of $y = x^2$ shifted 2 units to the left and 5 units upward and widened because $a$ is $\dfrac{1}{2}$. The vertex is $(-2, 5)$, and the axis of symmetry is $x = -2$. We plot a few points to verify.

| $x$ | $g(x) = \dfrac{1}{2}(x + 2)^2 + 5$ |
|-----|------------------------------------|
| $-4$ | $7$ |
| $-3$ | $5\dfrac{1}{2}$ |
| $-2$ | $5$ |
| $-1$ | $5\dfrac{1}{2}$ |
| $0$ | $7$ |

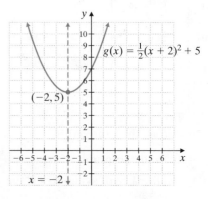

$g(x) = \dfrac{1}{2}(x + 2)^2 + 5$

$(-2, 5)$

$x = -2$

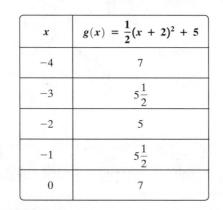

■ Work Practice 10

In general, the following holds.

## Graphing a Quadratic Function

The graph of a quadratic function written in the form $f(x) = a(x - h)^2 + k$ is a parabola with vertex $(h, k)$.

If $a > 0$, the parabola opens upward.
If $a < 0$, the parabola opens downward. The axis of symmetry is the line whose equation is $x = h$.

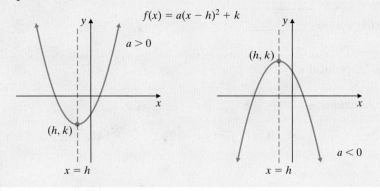

✓**Concept Check** Which description of the graph of $f(x) = -0.35(x + 3)^2 - 4$ is correct?

**a.** The graph opens downward and has its vertex at $(-3, 4)$.

**b.** The graph opens upward and has its vertex at $(-3, 4)$.

**c.** The graph opens downward and has its vertex at $(-3, -4)$.

**d.** The graph is narrower than the graph of $y = x^2$.

✓**Concept Check Answer**

c

---

🖩 **Calculator Explorations   Graphing**

*Use a graphing calculator to graph the first function of each pair. Then use its graph to predict the graph of the second function. Check your prediction by graphing both on the same set of axes. See this section and Section 3.8.*

**1.** $F(x) = \sqrt{x}$;
$G(x) = \sqrt{x} + 1$

**2.** $g(x) = x^3$;
$H(x) = x^3 - 2$

**3.** $H(x) = |x|$;
$f(x) = |x - 5|$

**4.** $h(x) = x^3 + 2; g(x) = (x - 3)^3 + 2$

**5.** $f(x) = |x + 4|; F(x) = |x + 4| + 3$

**6.** $G(x) = \sqrt{x} - 2; g(x) = \sqrt{x - 4} - 2$

## Vocabulary, Readiness & Video Check

*Use the choices below to fill in each blank. Some choices will be used more than once.*

upward     highest     parabola     downward     lowest     quadratic

**1.** A _____ function is one that can be written in the form $f(x) = ax^2 + bx + c, a \neq 0$.

**2.** The graph of a quadratic function is a _____ opening _____ or _____.

**3.** If $a > 0$, the graph of the quadratic function opens _____.

**4.** If $a < 0$, the graph of the quadratic function opens _____.

**5.** The vertex of a parabola is the _____ point if $a > 0$.

**6.** The vertex of a parabola is the _____ point if $a < 0$.

*State the vertex of the graph of each quadratic function.*

**7.** $f(x) = x^2$

**8.** $f(x) = -5x^2$

**9.** $g(x) = (x - 2)^2$

**10.** $g(x) = (x + 5)^2$

**11.** $f(x) = 2x^2 + 3$

**12.** $h(x) = x^2 - 1$

**13.** $g(x) = (x + 1)^2 + 5$

**14.** $h(x) = (x - 10)^2 - 7$

*Martin-Gay Interactive Videos*

*See Video 8.5*

Watch the section lecture video and answer the following questions.

**Objective A** **15.** From ▥ Examples 1 and 2 and the lecture before, how do graphs of the form $f(x) = x^2 + k$ differ from $y = x^2$? Consider the location of the vertex $(0, k)$ on these graphs of the form $f(x) = x^2 + k$. By what other name do we call this point on a graph? ▶

**Objective B** **16.** From ▥ Example 3 and the lecture before, how do graphs of the form $f(x) = (x - h)^2$ differ from $y = x^2$? Consider the location of the vertex $(h, 0)$ on these graphs of the form $f(x) = (x - h)^2$. By what other name do we call this point on a graph? ▶

**Objective C** **17.** From ▥ Example 4 and the lecture before, what general information does the equation $f(x) = (x - h)^2 + k$ tell us about its graph? ▶

**Objective D** **18.** From the lecture before ▥ Example 5, besides the direction a parabola opens, what other graphing information can the value of $a$ tell us? ▶

**Objective E** **19.** In ▥ Examples 6 and 7, what four properties of the graph did we learn from the equations that help us locate and draw the general shape of each parabola? ▶

## 8.5 Exercise Set MyMathLab®

**Objectives** **A** **B** **Mixed Practice** *Graph each quadratic function. Label the vertex and sketch and label the axis of symmetry. See Examples 1 through 6.*

**1.** $f(x) = x^2 - 1$

**2.** $h(x) = x^2 + 3$

**3.** $h(x) = x^2 + 5$

**4.** $h(x) = x^2 - 4$

**5.** $f(x) = (x - 5)^2$

**6.** $g(x) = (x + 5)^2$

**7.** $h(x) = (x + 2)^2$

**8.** $H(x) = (x - 1)^2$

**9.** $g(x) = x^2 + 7$

**10.** $f(x) = x^2 - 2$

**11.** $G(x) = (x + 3)^2$

**12.** $f(x) = (x - 6)^2$

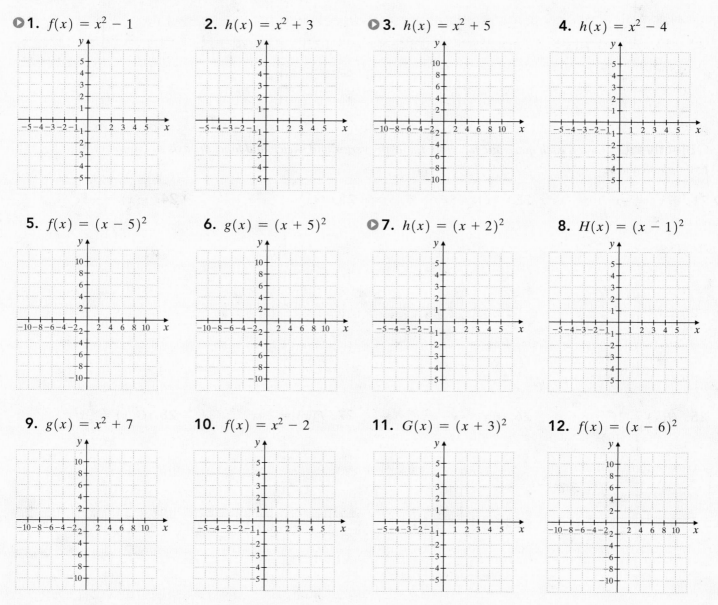

**Objective** **C** *Graph each quadratic function. Label the vertex and sketch and label the axis of symmetry. See Example 7.*

**13.** $f(x) = (x - 2)^2 + 5$

**14.** $g(x) = (x - 6)^2 + 1$

**15.** $h(x) = (x + 1)^2 + 4$

**16.** $G(x) = (x + 3)^2 + 3$

**17.** $g(x) = (x + 2)^2 - 5$    **18.** $h(x) = (x + 4)^2 - 6$    **19.** $h(x) = (x - 3)^2 + 2$    **20.** $F(x) = (x - 2)^2 - 3$

## Objective D    Graph each quadratic function. Label the vertex and sketch and label the axis of symmetry. See Examples 8 and 9.

**21.** $g(x) = -x^2$    **22.** $f(x) = 5x^2$    **23.** $h(x) = \frac{1}{3}x^2$    **24.** $g(x) = -3x^2$

**25.** $H(x) = 2x^2$    **26.** $f(x) = -\frac{1}{4}x^2$    **27.** $F(x) = -4x^2$    **28.** $G(x) = \frac{1}{5}x^2$

## Objective E    Graph each quadratic function. Label the vertex and sketch and label the axis of symmetry. See Example 10.

**29.** $f(x) = 10(x + 4)^2 - 6$    **30.** $g(x) = 4(x - 4)^2 + 2$    **31.** $h(x) = -3(x + 3)^2 + 1$

**32.** $f(x) = -(x - 2)^2 - 6$

**33.** $H(x) = \frac{1}{2}(x - 6)^2 - 3$

**34.** $G(x) = \frac{1}{5}(x + 4)^2 + 3$

**35.** $f(x) = -(x - 1)^2$

**36.** $f(x) = 2(x + 3)^2$

**37.** $F(x) = \left(x + \frac{1}{2}\right)^2 - 2$

**38.** $H(x) = \left(x + \frac{1}{4}\right)^2 - 3$

**39.** $F(x) = -x^2 + 2$

**40.** $G(x) = 3x^2 + 1$

**Objectives** **A** **B** **C** **D** **E** **Mixed Practice** *Sketch the graph of each quadratic function. Label the vertex and sketch and label the axis of symmetry. See Examples 1 through 10.*

**41.** $f(x) = -(x - 2)^2$

**42.** $g(x) = -(x + 6)^2$

**43.** $F(x) = -x^2 + 4$

**44.** $H(x) = -x^2 + 10$

**45.** $F(x) = 2x^2 - 5$

**46.** $g(x) = \frac{1}{2}x^2 - 2$

**47.** $h(x) = (x - 6)^2 + 4$

**48.** $f(x) = (x - 5)^2 + 2$

**49.** $F(x) = \left(x + \frac{1}{2}\right)^2 - 2$

**50.** $H(x) = \left(x + \frac{1}{2}\right)^2 - 3$

**51.** $F(x) = \frac{3}{2}(x + 7)^2 + 1$

**52.** $g(x) = -\frac{3}{2}(x - 1)^2 - 5$

**53.** $f(x) = \frac{1}{4}x^2 - 9$

**54.** $H(x) = \frac{3}{4}x^2 - 2$

**55.** $G(x) = 5\left(x + \frac{1}{2}\right)^2$

**56.** $F(x) = 3\left(x - \frac{3}{2}\right)^2$

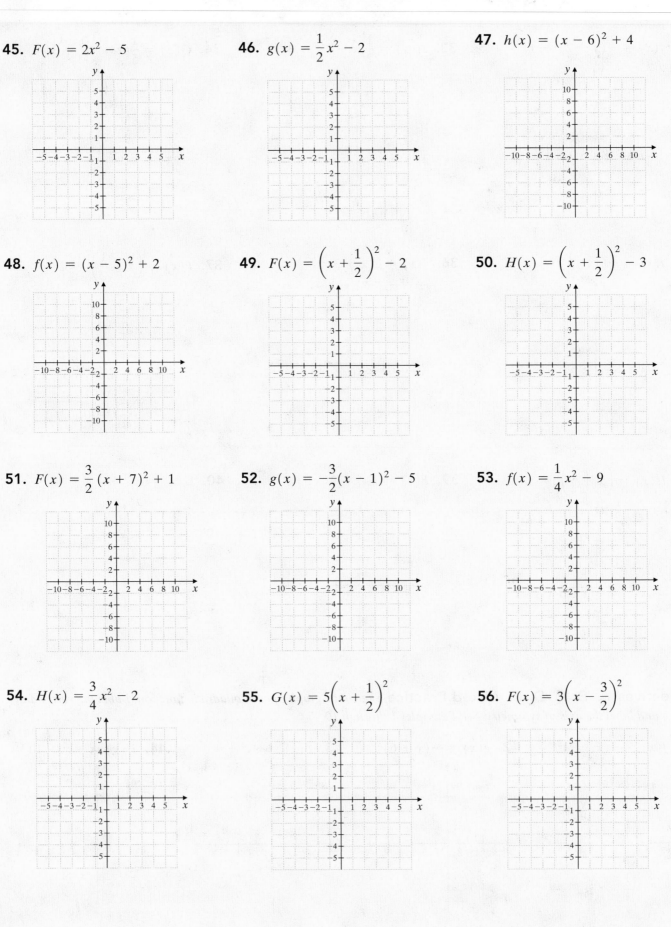

**57.** $h(x) = -(x - 1)^2 - 1$

**58.** $f(x) = -3(x + 2)^2 + 2$

**59.** $f(x) = 2(x - 1)^2 + 3$

**60.** $h(x) = 8(x + 1)^2 + 9$

**61.** $f(x) = -2(x - 4)^2 + 5$

**62.** $G(x) = -4(x + 9)^2 - 1$

## Review

*Add the proper constant to each binomial so that the resulting trinomial is a perfect square trinomial. See Section 8.1.*

**63.** $x^2 + 8x$

**64.** $y^2 + 4y$

**65.** $z^2 - 16z$

**66.** $x^2 - 10x$

**67.** $y^2 + y$

**68.** $z^2 - 3z$

## Concept Extensions

*Write the equation of the parabola that has the same shape as $f(x) = 5x^2$ but with the given vertex. Call each function $g(x)$.*

**69.** $(2, 3)$

**70.** $(1, 6)$

**71.** $(-3, 6)$

**72.** $(4, -1)$

*Recall from Section 3.8 that the shifting properties covered in this section apply to the graphs of all functions. Given the accompanying graph of $y = f(x)$, graph each function.*

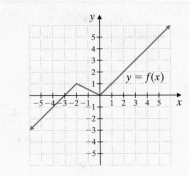

**73.** $y = f(x) + 1$

**74.** $y = f(x) - 2$

**75.** $y = f(x - 3)$

**76.** $y = f(x + 3)$

**77.** $y = f(x + 2) + 2$

**78.** $y = f(x - 1) + 1$

*Solve. See the Concept Check in this section.*

**79.** Which description of $f(x) = -213(x - 0.1)^2 + 3.6$ is correct?

| Graph Opens | Vertex |
|---|---|
| **a.** upward | $(0.1, 3.6)$ |
| **b.** upward | $(-213, 3.6)$ |
| **c.** downward | $(0.1, 3.6)$ |
| **d.** downward | $(-0.1, 3.6)$ |

**80.** Which description of $f(x) = 5\left(x + \dfrac{1}{2}\right)^2 + \dfrac{1}{2}$ is correct?

| Graph Opens | Vertex |
|---|---|
| **a.** upward | $\left(\dfrac{1}{2}, \dfrac{1}{2}\right)$ |
| **b.** upward | $\left(-\dfrac{1}{2}, \dfrac{1}{2}\right)$ |
| **c.** downward | $\left(\dfrac{1}{2}, -\dfrac{1}{2}\right)$ |
| **d.** downward | $\left(-\dfrac{1}{2}, -\dfrac{1}{2}\right)$ |

## 8.6 Further Graphing of Quadratic Functions

### Objective A Writing Quadratic Functions in the Form $y = a + (x - h)^2 + k$

We know that the graph of a quadratic function is a parabola. If a quadratic function is written in the form

$$f(x) = a(x - h)^2 + k$$

we can easily find the vertex $(h, k)$ and graph the parabola. To write a quadratic function in this form, we need to complete the square. (See Section 8.1 for a review of completing the square.)

**Objectives**

**A** Write Quadratic Functions in the Form $y = a + (x - h)^2 + k$.

**B** Derive a Formula for Finding the Vertex of a Parabola.

**C** Find the Minimum or Maximum Value of a Quadratic Function.

**Example 1**  Graph: $f(x) = x^2 - 4x - 12$. Find the vertex and any intercepts.

**Solution:** The graph of this quadratic function is a parabola. To find the vertex of the parabola, we complete the square on the binomial $x^2 - 4x$. To simplify our work, we let $f(x) = y$.

$$y = x^2 - 4x - 12 \quad \text{Let } f(x) = y.$$
$$y + 12 = x^2 - 4x \qquad \text{Add 12 to both sides to get the } x\text{-variable terms alone.}$$

Now we add the square of half of $-4$ to both sides.

$$\frac{1}{2}(-4) = -2 \quad \text{and} \quad (-2)^2 = 4$$

$$y + 12 + 4 = x^2 - 4x + 4 \qquad \text{Add 4 to both sides.}$$
$$y + 16 = (x - 2)^2 \qquad \text{Factor the trinomial.}$$
$$y = (x - 2)^2 - 16 \quad \text{Subtract 16 from both sides.}$$
$$f(x) = (x - 2)^2 - 16 \quad \text{Replace } y \text{ with } f(x).$$

From this equation, we can see that the vertex of the parabola is $(2, -16)$, a point in quadrant IV, and the axis of symmetry is the line $x = 2$.

Notice that $a = 1$. Since $a > 0$, the parabola opens upward. This parabola opening upward with vertex $(2, -16)$ will have two $x$-intercepts.

To find the $x$-intercepts, we let $f(x)$ or $y = 0$.

$$0 = x^2 - 4x - 12$$
$$0 = (x - 6)(x + 2)$$
$$0 = x - 6 \quad \text{or} \quad 0 = x + 2$$
$$6 = x \qquad \qquad -2 = x$$

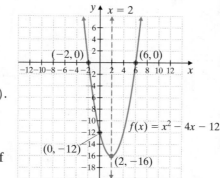

The two $x$-intercepts are $(6, 0)$ and $(-2, 0)$.
To find the $y$-intercept, we let $x = 0$.

$$f(0) = 0^2 - 4 \cdot 0 - 12 = -12$$

The $y$-intercept is $(0, -12)$. The sketch of $f(x) = x^2 - 4x - 12$ is shown.

■ **Work Practice 1**

**Practice 1**

Graph: $f(x) = x^2 - 4x - 5$. Find the vertex and any intercepts.

**Answer**

**1.** vertex: $(2, -9)$; $x$-intercepts: $(-1, 0)$, $(5, 0)$; $y$-intercept: $(0, -5)$

## Practice 2

Graph: $f(x) = 2x^2 + 2x + 5$. Find the vertex and any intercepts.

 **Example 2** Graph: $f(x) = 3x^2 + 3x + 1$. Find the vertex and any intercepts.

**Solution:** We replace $f(x)$ with $y$ and complete the square on $x$ to write the equation in the form $y = a(x - h)^2 + k$.

$$y = 3x^2 + 3x + 1 \quad \text{Replace } f(x) \text{ with } y.$$

$$y - 1 = 3x^2 + 3x \quad \text{Get the } x\text{-variable terms alone.}$$

Next we factor 3 from the terms $3x^2 + 3x$ so that the coefficient of $x^2$ is 1.

$$y - 1 = 3(x^2 + x) \quad \text{Factor out 3.}$$

The coefficient of $x$ is 1. Then $\frac{1}{2}(1) = \frac{1}{2}$ and $\left(\frac{1}{2}\right)^2 = \frac{1}{4}$. Since we are adding $\frac{1}{4}$ inside the parentheses, we are really adding $3\left(\frac{1}{4}\right)$, so we *must* add $3\left(\frac{1}{4}\right)$ to the left side.

$$y - 1 + 3\left(\frac{1}{4}\right) = 3\left(x^2 + x + \frac{1}{4}\right)$$

$$y - \frac{1}{4} = 3\left(x + \frac{1}{2}\right)^2 \quad \text{Simplify the left side and factor the right side.}$$

$$y = 3\left(x + \frac{1}{2}\right)^2 + \frac{1}{4} \quad \text{Add } \frac{1}{4} \text{ to both sides.}$$

$$f(x) = 3\left(x + \frac{1}{2}\right)^2 + \frac{1}{4} \quad \text{Replace } y \text{ with } f(x).$$

Then $a = 3$, $h = -\frac{1}{2}$, and $k = \frac{1}{4}$. This means that the parabola opens upward with vertex $\left(-\frac{1}{2}, \frac{1}{4}\right)$ and that the axis of symmetry is the line $x = -\frac{1}{2}$. This parabola has no $x$-intercepts since the vertex is in the second quadrant and it opens upward.

To find the $y$-intercept, we let $x = 0$. Then

$$f(0) = 3(0)^2 + 3(0) + 1 = 1$$

We use the vertex, axis of symmetry, and $y$-intercept to graph the parabola.

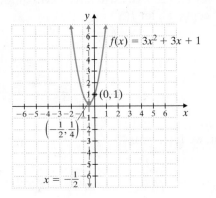

**Work Practice 2**

## Answer

**2.** vertex: $\left(-\frac{1}{2}, \frac{9}{2}\right)$; $y$-intercept: $(0, 5)$

Parabola Opens Upward
Vertex in quadrant I or II: no $x$-intercepts
Vertex in quadrant III or IV: 2 $x$-intercepts

Parabola Opens Downward
Vertex in quadrant I or II: 2 $x$-intercepts
Vertex in quadrant III or IV: no $x$-intercepts

**Practice 3**

Graph: $f(x) = -x^2 - 2x + 8$.
Find the vertex and any intercepts.

**Example 3** Graph: $f(x) = -x^2 - 2x + 3$. Find the vertex and any intercepts.

**Solution:** We write $f(x)$ in the form $a(x - h)^2 + k$ by completing the square. First we replace $f(x)$ with $y$.

$$f(x) = -x^2 - 2x + 3$$
$$y = -x^2 - 2x + 3$$
$$y - 3 = -x^2 - 2x \qquad \text{Subtract 3 from both sides to get the } x\text{-variable terms alone.}$$
$$y - 3 = -1(x^2 + 2x) \qquad \text{Factor } -1 \text{ from the terms } -x^2 - 2x.$$

The coefficient of $x$ is 2. Then $\frac{1}{2}(2) = 1$ and $1^2 = 1$. We add 1 to the right side inside the parentheses and add $-1(1)$ to the left side.

$$y - 3 - 1(1) = -1(x^2 + 2x + 1)$$
$$y - 4 = -1(x + 1)^2 \qquad \text{Simplify the left side and factor the right side.}$$
$$y = -1(x + 1)^2 + 4 \qquad \text{Add 4 to both sides.}$$
$$f(x) = -1(x + 1)^2 + 4 \qquad \text{Replace } y \text{ with } f(x).$$

Since $a = -1$, the parabola opens downward with vertex $(-1, 4)$ and axis of symmetry $x = -1$.
To find the $x$-intercepts, we let $y$ or $f(x) = 0$ and solve for $x$.

$$f(x) = -x^2 - 2x + 3$$
$$0 = -x^2 - 2x + 3 \qquad \text{Let } f(x) = 0.$$

(*Continued on next page*)

This can be written as $f(x) = -1[x - (-1)]^2 + 4$. Notice that the vertex is $(-1, 4)$.

**Answer**

**3.** vertex: $(-1, 9)$; $x$-intercepts: $(-4, 0)$, $(2, 0)$; $y$-intercept: $(0, 8)$

Now we divide both sides by $-1$ so that the coefficient of $x^2$ is 1. (If you prefer, you may factor $-1$ from the trinomial on the right side.)

$$\frac{0}{-1} = \frac{-x^2}{-1} - \frac{2x}{-1} + \frac{3}{-1} \qquad \text{Divide both sides by } -1.$$

$$0 = x^2 + 2x - 3 \qquad \text{Simplify.}$$

$$0 = (x + 3)(x - 1) \qquad \text{Factor.}$$

$$x + 3 = 0 \quad \text{or} \quad x - 1 = 0 \qquad \text{Set each factor equal to 0.}$$

$$x = -3 \qquad\qquad x = 1 \qquad \text{Solve.}$$

The $x$-intercepts are $(-3, 0)$ and $(1, 0)$.

To find the $y$-intercept, we let $x = 0$ and solve for $y$. Then

$$f(0) = -0^2 - 2(0) + 3 = 3$$

Thus, $(0, 3)$ is the $y$-intercept. We use these points to graph the parabola.

■ **Work Practice 3**

## Objective B  Deriving a Formula for Finding the Vertex ▶

As you have seen in previous examples, it can sometimes be tedious to find the vertex of a parabola by completing the square. There is a formula for finding the vertex of a parabola. Now that we have practiced completing the square, we will show that the $x$-coordinate of the vertex of the graph of $f(x)$ or $y = ax^2 + bx + c$ can be found by the formula $x = \dfrac{-b}{2a}$. To do so, we complete the square on $x$ and write the equation in the form $y = (x - h)^2 + k$.

First we get the $x$-variable terms alone by subtracting $c$ from both sides.

$$y = ax^2 + bx + c$$

$$y - c = ax^2 + bx$$

$$y - c = a\left(x^2 + \frac{b}{a}x\right) \qquad \text{Factor } a \text{ from the terms } ax^2 + bx.$$

Now we add the square of half of $\dfrac{b}{a}$, or $\left(\dfrac{b}{2a}\right)^2 = \dfrac{b^2}{4a^2}$, to the right side inside the parentheses. Because of the factor $a$, what we really added is $a\left(\dfrac{b^2}{4a^2}\right)$ and this must be added to the left side as well.

$$y - c + a\left(\frac{b^2}{4a^2}\right) = a\left(x^2 + \frac{b}{a}x + \frac{b^2}{4a^2}\right)$$

$$y - c + \frac{b^2}{4a} = a\left(x + \frac{b}{2a}\right)^2$$

$$y = a\left(x + \frac{b}{2a}\right)^2 + c - \frac{b^2}{4a} \qquad \begin{array}{l}\text{Simplify the left side and factor the right} \\ \text{side. Add } c \text{ to both sides and subtract } \frac{b^2}{4a} \\ \text{from both sides.}\end{array}$$

Compare this form with $f(x)$ or $y = a(x - h)^2 + k$ and see that $h$ is $\dfrac{-b}{2a}$, which means that the $x$-coordinate of the vertex of the graph of $f(x) = ax^2 + bx + c$ is $\dfrac{-b}{2a}$.

Let's use the vertex formula below to find the vertex of the parabola we graphed in Example 1.

> **Vertex Formula**
>
> The graph of $f(x) = ax^2 + bx + c$, when $a \neq 0$, is a parabola with vertex
> $$\left( \frac{-b}{2a}, f\left( \frac{-b}{2a} \right) \right)$$

**Example 4**  Find the vertex of the graph of $f(x) = x^2 - 4x - 12$.

**Solution:**  In the quadratic function $f(x) = x^2 - 4x - 12$, notice that $a = 1$, $b = -4$, and $c = -12$.

$$\frac{-b}{2a} = \frac{-(-4)}{2(1)} = 2$$

The $x$-value of the vertex is 2. To find the corresponding $f(x)$ or $y$-value, find $f(2)$. Then

$$f(2) = 2^2 - 4(2) - 12 = 4 - 8 - 12 = -16$$

The vertex is $(2, -16)$. These results agree with our findings in Example 1.

🔊 **Work Practice 4**

**Practice 4**

Find the vertex of the graph of $f(x) = x^2 - 4x - 5$. Compare your result with the result of Practice 1.

## Objective C  Finding Minimum and Maximum Values ▶

The quadratic function whose graph is a parabola that opens upward has a minimum value, and the quadratic function whose graph is a parabola that opens downward has a maximum value. The $f(x)$- or $y$-value of the vertex is the minimum or maximum value of the function.

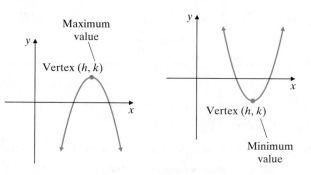

✔**Concept Check** Without making any calculations, tell whether the graph of $f(x) = 7 - x - 0.3x^2$ has a maximum value or a minimum value. Explain your reasoning.

**Answer**

**4.** $(2, -9)$

✔**Concept Check Answer**

$f(x)$ has a maximum value since it opens downward.

## Practice 5

An object is thrown upward from the top of a 100-foot cliff. Its height in feet above ground after $t$ seconds is given by the function $f(t) = -16t^2 + 10t + 100$. Find the maximum height of the object and the number of seconds it took for the object to reach its maximum height.

### Answer

**5.** maximum height: $101\frac{9}{16}$ ft in $\frac{5}{16}$ sec

## Example 5  Finding Maximum Height

A rock is thrown upward from the ground. Its height in feet above ground after $t$ seconds is given by the function $f(t) = -16t^2 + 20t$. Find the maximum height of the rock and the number of seconds it took for the rock to reach its maximum height.

### Solution:

1. **UNDERSTAND.** The maximum height of the rock is the largest value of $f(t)$. Since the function $f(t) = -16t^2 + 20t$ is a quadratic function, its graph is a parabola. It opens downward since $-16 < 0$. Thus, the maximum value of $f(t)$ is the $f(t)$- or $y$-value of the vertex of its graph.

2. **TRANSLATE.** To find the vertex $(h, k)$, we notice that for $f(t) = -16t^2 + 20t$, $a = -16, b = 20$, and $c = 0$. We will use these values and the vertex formula

$$\left(\frac{-b}{2a}, f\left(\frac{-b}{2a}\right)\right)$$

3. **SOLVE.** $h = \dfrac{-b}{2a} = \dfrac{-20}{-32} = \dfrac{5}{8}$

$$f\left(\frac{5}{8}\right) = -16\left(\frac{5}{8}\right)^2 + 20\left(\frac{5}{8}\right) = -16\left(\frac{25}{64}\right) + \frac{25}{2} = -\frac{25}{4} + \frac{50}{4} = \frac{25}{4}$$

4. **INTERPRET.** The graph of $f(t)$ is a parabola opening downward with vertex $\left(\frac{5}{8}, \frac{25}{4}\right)$. This means that the rock's maximum height is $\frac{25}{4}$ feet, or $6\frac{1}{4}$ feet, which was reached in $\frac{5}{8}$ second.

■ **Work Practice 5**

## Vocabulary, Readiness & Video Check

*Fill in each blank.*

1. If a quadratic function is in the form $f(x) = a(x - h)^2 + k$, the vertex of its graph is _____.

2. The graph of $f(x) = ax^2 + bx + c, a \neq 0$ is a parabola whose vertex has $x$-value of _____.

*Martin-Gay Interactive Videos*

See Video 8.6

Watch the section lecture video and answer the following questions.

**Objective A  3.** From ▤ Example 1, how does writing a quadratic function in the form $f(x) = a(x - h)^2 + k$ help us graph the function? What procedure can we use to write a quadratic function in this form? ▶

**Objective B  4.** From ▤ Example 2, how can locating the vertex and knowing whether parabola opens upward or downward potentially help save unnecessary work? Explain. ▶

**Objective C  5.** From ▤ Example 4, when an application involving a quadratic function asks for the maximum or minimum, what part of a parabola should we find? ▶

**8.6** **Exercise Set** MyMathLab®

### Objective A   *Fill in each blank. See Examples 1 through 3.*

|      | Parabola Opens | Vertex Location | Number of $x$-intercept(s) | Number of $y$-intercept(s) |
|------|----------------|-----------------|------------------------------|------------------------------|
| 1.   | up             | Q I             |                              |                              |
| 2.   | up             | Q III           |                              |                              |
| 3.   | down           | Q II            |                              |                              |
| 4.   | down           | Q IV            |                              |                              |
| 5.   | up             | $x$-axis        |                              |                              |
| 6.   | down           | $x$-axis        |                              |                              |
| 7.   |                | Q III           | 0                            |                              |
| 8.   |                | Q I             | 2                            |                              |
| 9.   |                | Q IV            | 2                            |                              |
| 10.  |                | Q II            | 0                            |                              |

### Objectives A B   Mixed Practice   *Find the vertex of the graph of each quadratic function by completing the square or using the vertex formula. See Examples 1 through 4.*

**11.** $f(x) = x^2 + 8x + 7$     **12.** $f(x) = x^2 + 6x + 5$     **13.** $f(x) = -x^2 + 10x + 5$   **14.** $f(x) = -x^2 - 8x + 2$

**15.** $f(x) = 5x^2 - 10x + 3$   **16.** $f(x) = -3x^2 + 6x + 4$   **17.** $f(x) = -x^2 + x + 1$     **18.** $f(x) = x^2 - 9x + 8$

*Match each function with its graph. See Examples 1 through 4.*

**19.** $f(x) = x^2 - 4x + 3$     **20.** $f(x) = x^2 + 2x - 3$     **21.** $f(x) = x^2 - 2x - 3$     **22.** $f(x) = x^2 + 4x + 3$

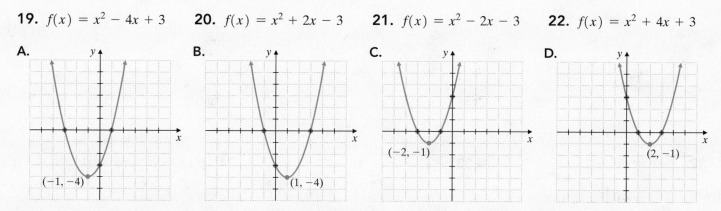

A.          B.          C.          D.

$(-1, -4)$          $(1, -4)$          $(-2, -1)$          $(2, -1)$

*Find the vertex of the graph of each quadratic function. Determine whether the graph opens upward or downward, find any intercepts, and graph the function. See Examples 1 through 4.*

**23.** $f(x) = x^2 + 4x - 5$          **24.** $f(x) = x^2 + 2x - 3$          ▶**25.** $f(x) = -x^2 + 2x - 1$

**26.** $f(x) = -x^2 + 4x - 4$

**27.** $f(x) = x^2 - 4$

**28.** $f(x) = x^2 - 1$

**29.** $f(x) = 4x^2 + 4x - 3$

**30.** $f(x) = 2x^2 - x - 3$

**31.** $f(x) = \dfrac{1}{2}x^2 + 4x + \dfrac{15}{2}$

**32.** $f(x) = \dfrac{1}{5}x^2 + 2x + \dfrac{9}{5}$

**33.** $f(x) = x^2 - 4x + 5$

**34.** $f(x) = x^2 - 6x + 11$

**35.** $f(x) = 2x^2 + 4x + 5$    **36.** $f(x) = 3x^2 + 12x + 16$    **37.** $f(x) = -2x^2 + 12x$    **38.** $f(x) = -4x^2 + 8x$

## Objective C  *Solve. See Example 5.*

**39.** If a projectile is fired straight upward from the ground with an initial speed of 96 feet per second, then its height $h$ in feet after $t$ seconds is given by the function $h(t) = -16t^2 + 96t$. Find the maximum height of the projectile.

**40.** If Rheam Gaspar throws a ball upward with an initial speed of 32 feet per second, then its height $h$ in feet after $t$ seconds is given by the function $h(t) = -16t^2 + 32t$. Find the maximum height of the ball.

**41.** The cost $C$ in dollars of manufacturing $x$ bicycles at Holladay's Production Plant is given by the function $C(x) = 2x^2 - 800x + 92,000$.

  **a.** Find the number of bicycles that must be manufactured to minimize the cost.

  **b.** Find the minimum cost.

**42.** The Utah Ski Club sells calendars to raise money. The profit $P$, in cents, from selling $x$ calendars is given by the function $P(x) = 360x - x^2$.

  **a.** Find how many calendars must be sold to maximize profit.

  **b.** Find the maximum profit.

**43.** Find two numbers whose sum is 60 and whose product is as large as possible. [*Hint:* Let $x$ and $60 - x$ be the two positive numbers. Their product can be described by the function $f(x) = x(60 - x)$.]

**44.** Find two numbers whose sum is 11 and whose product is as large as possible. (Use the hint for Exercise 43.)

**45.** Find two numbers whose difference is 10 and whose product is as small as possible. (Use the hint for Exercise 43.)

**46.** Find two numbers whose difference is 8 and whose product is as small as possible.

**47.** The length and width of a rectangle must have a sum of 40. Find the dimensions of the rectangle that will have the maximum area. (Use the hint for Exercise 43.)

**48.** The length and width of a rectangle must have a sum of 50. Find the dimensions of the rectangle that will have maximum area.

## Review

*Find the vertex of the graph of each function. See Section 8.5.*

**49.** $f(x) = x^2 + 2$    **50.** $f(x) = (x - 3)^2$    **51.** $g(x) = (x + 2)^2$    **52.** $h(x) = x^2 - 3$

**53.** $f(x) = (x + 5)^2 + 2$    **54.** $f(x) = 2(x - 3)^2 + 2$    **55.** $f(x) = 3(x - 4)^2 + 1$    **56.** $f(x) = (x + 1)^2 + 4$

## Concept Extensions

*Without calculating, tell whether each graph has a minimum value or a maximum value. See the Concept Check in the section.*

**57.** $f(x) = 2x^2 - 5$      **58.** $g(x) = -7x^2 + x + 1$      **59.** $F(x) = 3 - \frac{1}{2}x^2$      **60.** $G(x) = 3 - \frac{1}{2}x + 0.8x^2$

*Find the vertex of the graph of each quadratic function. Determine whether the graph opens upward or downward, find the y-intercept, approximate the x-intercepts to one decimal place, and graph the function.*

**61.** $f(x) = x^2 + 10x + 15$          **62.** $f(x) = 2x^2 + 4x - 1$

*Use a graphing calculator to verify the graph of each exercise.*

**63.** Exercise 31.          **64.** Exercise 32.

*Find the maximum or minimum value of each function. Approximate to two decimal places.*

**65.** $f(x) = 2.3x^2 - 6.1x + 3.2$          **66.** $f(x) = 7.6x^2 + 9.8x - 2.1$

**67.** The percent of U.S. households with broadband Internet access can be modeled by the function $f(x) = -0.11x^2 + 6.53x + 3$, where $f(x)$ is the projected percent of American households with broadband Internet access and $x$ is the number of years since 2000. (*Source:* Based on data from Pew Internet and American Life Project)

    **a.** Will this function have a maximum or a minimum? How can you tell?

    **b.** According to this model, in what year will the percent of American homes with broadband Internet access be at its maximum or minimum? Round to the nearest whole year.

    **c.** What is the predicted maximum/minimum percent of homes with broadband Internet access? (Use your rounded answer from part b.)

**68.** Methane is a gas produced by landfills, natural gas systems, and coal mining that contributes to the greenhouse effect and global warming. Projected methane emissions in the United States can be modeled by the quadratic function

$$f(x) = -0.072x^2 + 1.93x + 173.9$$

where $f(x)$ is the amount of methane produced in millions of metric tons and $x$ is the number of years after 2000. (*Source:* Based on data from the U.S. Environmental Protection Agency, 2000–2020)

    **a.** According to this model, what will U.S. emissions of methane be in 2018?

    **b.** Will this function have a maximum or a minimum? How can you tell?

    **c.** In what year will methane emissions in the United States be at their maximum or minimum? Round to the nearest whole year.

    **d.** What is the level of methane emissions for that year? (Use your rounded answer from part c.)

## Chapter 8 Group Activity

### Recognizing Linear and Quadratic Models

*This activity may be completed by working in groups or individually.*

We have seen in this and previous chapters that data can be modeled by both linear models and quadratic models. However, when we are given a set of data to model, how can we tell if a linear or quadratic model is appropriate? The best answer requires looking at a scatter diagram of the data. If the plotted data points fall roughly on a line, a linear model is usually the better choice. If the plotted data points seem to fall on a definite curve or if a maximum or minimum point is apparent, a quadratic model is usually the better choice.

One of the sets of data shown in the tables is best modeled by a linear function and one is best modeled by a quadratic function. In each case, the variable $x$ represents the number of years after 2008.

| North American Total Hybrid Car Sales | | | | | |
|---|---|---|---|---|---|
| Year | 2009 | 2010 | 2011 | 2012 | 2013 |
| $x$ | 1 | 2 | 3 | 4 | 5 |
| Number of Hybrids Sold, $y$ | 290,271 | 274,210 | 268,752 | 434,498 | 495,685 |
| *(Source: The Christian Science Monitor)* | | | | | |

| Total Number of Target Stores | | | | | |
|---|---|---|---|---|---|
| Year | 2009 | 2010 | 2011 | 2012 | 2013 |
| $x$ | 1 | 2 | 3 | 4 | 5 |
| Number of Stores, $y$ | 1740 | 1750 | 1763 | 1778 | 1790 |
| *(Source: Target, Inc.)* | | | | | |

**1.** Make a scatter diagram for each set of data. Which type of model should be used for each set of data?

**North American Hybrid Car Sales**

**Number of Target Stores**

**2.** For the set of data that you have determined to be linear, find a linear function that fits the data points. Explain the method that you used.

**3.** For the set of data that you have determined to be quadratic, identify the point on your scatter diagram that appears to be the vertex of the parabola. Use the coordinates of this vertex in the quadratic model $f(x) = a(x - h)^2 + k$.

**4.** Solve for the remaining unknown constant in the quadratic model by substituting the coordinates for another data point into the function. Write the final form of the quadratic model for this data set.

**5.** Use your models to estimate the number of hybrid cars sold and the number of Target stores in 2014.

**6.** (Optional) For each set of data, enter the data from the table into a graphing calculator and use either the linear regression feature or the quadratic regression feature to find an appropriate function that models the data.* Compare these functions with the ones you found in Exercise 2. How are they alike or different?

*To find out more about using your graphing calculator to find a regression equation, consult your user's manual.

## Chapter 8 Vocabulary Check

*Fill in each blank with one of the words or phrases listed below.*

| | | |
|---|---|---|
| quadratic formula | quadratic | discriminant |
| completing the square | quadratic inequality | |

$(h, k)$ $\qquad$ $(0, k)$ $\qquad$ $(h, 0)$ $\qquad$ $\dfrac{-b}{2a}$ $\qquad$ $\pm \sqrt{b}$

1. The _____ helps us find the number and type of solutions of a quadratic equation.

2. If $a^2 = b$, then $a = $ _____.

3. The graph of $f(x) = ax^2 + bx + c$, where $a$ is not 0, is a parabola whose vertex has an $x$-value of _____.

4. A(n) _____ is an inequality that can be written so that one side is a quadratic expression and the other side is 0.

5. The process of writing a quadratic equation so that one side is a perfect square trinomial is called _____.

6. The graph of $f(x) = x^2 + k$ has vertex _____.

7. The graph of $f(x) = (x - h)^2$ has vertex _____.

8. The graph of $f(x) = (x - h)^2 + k$ has vertex _____.

9. The formula $x = \dfrac{-b \pm \sqrt{b^2 - 4ac}}{2a}$ is called the _____.

10. A(n) _____ equation is one that can be written in the form $ax^2 + bx + c = 0$, where $a$, $b$, and $c$ are real numbers and $a$ is not 0.

> **Helpful Hint**
>
> ▶ Are you preparing for your test? Don't forget to take the Chapter 8 Test on page 665. Then check your answers at the back of the text and use the Chapter Test Prep Videos to see the fully worked-out solutions to any of the exercises you want to review.

## 8 Chapter Highlights

| Definitions and Concepts | Examples |
|---|---|
| **Section 8.1   Solving Quadratic Equations by Completing the Square** | |

| | |
|---|---|
| **Square Root Property** | Solve: $(x + 3)^2 = 14$ |
| If $b$ is a real number and if $a^2 = b$, then $a = \pm \sqrt{b}$. | $\qquad x + 3 = \pm \sqrt{14}$ |
| | $\qquad\qquad x = -3 \pm \sqrt{14}$ |
| **Solving a Quadratic Equation in $x$ by Completing the Square** | Solve: $3x^2 - 12x - 18 = 0$ |
| **Step 1:** If the coefficient of $x^2$ is not 1, divide both sides of the equation by the coefficient of $x^2$. | 1. $x^2 - 4x - 6 = 0$ |
| **Step 2:** Get the variable terms alone. | 2. $\qquad x^2 - 4x = 6$ |
| **Step 3:** Complete the square by adding the square of half of the coefficient of $x$ to both sides. | 3. $\dfrac{1}{2}(-4) = -2$ and $(-2)^2 = 4$ |
| **Step 4:** Write the resulting trinomial as the square of a binomial. | $\qquad x^2 - 4x + 4 = 6 + 4$ |
| | 4. $(x - 2)^2 = 10$ |
| **Step 5:** Use the square root property. | 5. $x - 2 = \pm \sqrt{10}$ |
| | $\qquad\quad x = 2 \pm \sqrt{10}$ |

| Definitions and Concepts | Examples |
|---|---|

### Section 8.2    Solving Quadratic Equations by Using the Quadratic Formula

**Quadratic Formula**

A quadratic equation written in the form $ax^2 + bx + c = 0$ has solutions

$$x = \frac{-b \pm \sqrt{b^2 - 4ac}}{2a}$$

Solve:  $x^2 - x - 3 = 0$

$$a = 1, b = -1, c = -3$$

$$x = \frac{-(-1) \pm \sqrt{(-1)^2 - 4(1)(-3)}}{2 \cdot 1}$$

$$x = \frac{1 \pm \sqrt{13}}{2}$$

### Section 8.3    Solving Equations by Using Quadratic Methods

Substitution is often helpful in solving an equation that contains a repeated variable expression.

Solve:  $(2x + 1)^2 - 5(2x + 1) + 6 = 0$

Let $m = 2x + 1$, then the equation is

$$m^2 - 5m + 6 = 0 \qquad \text{Let } m = 2x + 1$$

$$(m - 3)(m - 2) = 0$$

$$m = 3 \quad \text{or} \quad m = 2$$

$$2x + 1 = 3 \qquad 2x + 1 = 2 \quad \text{Substitute back.}$$

$$x = 1 \qquad x = \frac{1}{2}$$

### Section 8.4    Nonlinear Inequalities in One Variable

**Solving a Polynomial Inequality**

**Step 1:** Write the inequality in standard form and solve the related equation.

**Step 2:** Use solutions from Step 1 to separate the number line into regions.

**Step 3:** Use a test point to determine whether values in each region satisfy the original inequality.

**Step 4:** Write the solution set as the union of regions whose test point values are solutions.

Solve:  $x^2 \geq 6x$

**1.** $x^2 - 6x \geq 0$

$$x^2 - 6x = 0$$
$$x(x - 6) = 0$$
$$x = 0 \quad \text{or} \quad x = 6$$

**2.**

**3.**

| Region | Test Point Value | $x^2 \geq 6x$ | Result |
|---|---|---|---|
| A | $-2$ | $(-2)^2 \geq 6(-2)$ | True |
| B | $1$ | $1^2 \geq 6(1)$ | False |
| C | $7$ | $7^2 \geq 6(7)$ | True |

**4.**

The solution set is $(-\infty, 0] \cup [6, \infty)$.

**Solving a Rational Inequality**

**Step 1:** Solve for values that make all denominators 0.

**Step 2:** Solve the related equation.

Solve:  $\dfrac{6}{x - 1} < -2$

**1.** $x - 1 = 0$    Set the denominator equal to 0.

$$x = 1$$

**2.** $\dfrac{6}{x - 1} = -2$

$$6 = -2(x - 1) \quad \text{Multiply by } (x - 1).$$
$$6 = -2x + 2$$
$$4 = -2x$$
$$-2 = x$$

*(continued)*

| Definitions and Concepts | Examples |
| --- | --- |

### Section 8.4   Nonlinear Inequalities in One Variable (*continued*)

**Step 3:** Use solutions from Steps 1 and 2 to separate the number line into regions.

**3.**

$$A \qquad B \qquad C$$
$$-2 \qquad 1$$

**Step 4:** Use a test point to determine whether values in each region satisfy the original inequality.

**4.** Only a test value from region *B* satisfies the original inequality.

**Step 5:** Write the solution set as the union of regions whose test point values are solutions.

**5.**

$$-2 \qquad 1$$

The solution set is $(-2, 1)$.

### Section 8.5   Quadratic Functions and Their Graphs

**Graphing a Quadratic function**

The graph of a quadratic function written in the form $f(x) = a(x - h)^2 + k$ is a parabola with vertex $(h, k)$.

If $a > 0$, the parabola opens upward.

If $a < 0$, the parabola opens downward.

The axis of symmetry is the line whose equation is $x = h$.

$$f(x) = a(x - h)^2 + k$$

$a > 0$

$(h, k)$

$x = h$

$a < 0$

$(h, k)$

$x = h$

Graph:  $g(x) = 3(x - 1)^2 + 4$

The graph is a parabola with vertex $(1, 4)$ and axis of symmetry $x = 1$. Since $a = 3$ is positive, the graph opens upward.

$g(x) = 3(x - 1)^2 + 4$

$(1, 4)$

$x = 1$

### Section 8.6   Further Graphing of Quadratic Functions

The graph of $f(x) = ax^2 + bx + c, a \neq 0$, is a parabola with vertex

$$\left( \frac{-b}{2a}, f\left( \frac{-b}{2a} \right) \right)$$

Graph: $f(x) = x^2 - 2x - 8$. Find the vertex and *x*- and *y*-intercepts.

$$\frac{-b}{2a} = \frac{-(-2)}{2 \cdot 1} = 1$$

$$f(1) = 1^2 - 2(1) - 8 = -9$$

The vertex is $(1, -9)$.

$$0 = x^2 - 2x - 8$$

$$0 = (x - 4)(x + 2)$$

$$x = 4 \quad \text{or} \quad x = -2$$

The *x*-intercepts are $(4, 0)$ and $(-2, 0)$.

$$f(0) = 0^2 - 2 \cdot 0 - 8 = -8$$

The *y*-intercept is $(0, -8)$.

$(-2, 0)$    $(4, 0)$

$(0, -8)$

$(1, -9)$

**(8.1)** *Solve by factoring.*

**1.** $x^2 - 15x + 14 = 0$

**2.** $7a^2 = 29a + 30$

*Use the square root property to solve each equation.*

**3.** $4m^2 = 196$

**4.** $(5x - 2)^2 = 2$

*Solve by completing the square.*

**5.** $z^2 + 3z + 1 = 0$

**6.** $(2x + 1)^2 = x$

**7.** If $P$ dollars are invested, the formula $A = P(1 + r)^2$ gives the amount $A$ in an account paying interest rate $r$ compounded annually after 2 years. Find the interest rate $r$ such that \$2500 increases to \$2717 in 2 years. Round the result to the nearest hundredth of a percent.

**8.** Two ships leave a port at the same time and travel at the same speed. One ship is traveling due north and the other due east. In a few hours, the ships are 150 miles apart. How many miles has each ship traveled? Give the exact answer and a one-decimal-place approximation.

**(8.2)** *If the discriminant of a quadratic equation has the given value, determine the number and type of solutions of the equation.*

**9.** $-8$

**10.** $48$

**11.** $100$

**12.** $0$

*Use the quadratic formula to solve each equation.*

**13.** $x^2 - 16x + 64 = 0$

**14.** $x^2 + 5x = 0$

**15.** $2x^2 + 3x = 5$

**16.** $6x^2 + 7 = 5x$

**17.** $9x^2 + 4 = 2x$

**18.** $(2x - 3)^2 = x$

**19.** Cadets graduating from military school usually toss their hats high into the air at the end of the ceremony. One cadet threw his hat so that its distance $d(t)$ in feet above the ground $t$ seconds after it was thrown was $d(t) = -16t^2 + 30t + 6$.

   **a.** Find the distance above the ground of the hat 1 second after it was thrown.

   **b.** Find the time it took that hat to hit the ground. Give the exact time and a one-decimal-place approximation.

△ **20.** The hypotenuse of an isosceles right triangle is 6 centimeters longer than either of the legs. Find the length of the legs. (*Hint:* Don't forget that an isosceles triangle has two sides of equal length.)

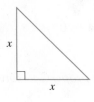

**(8.3)** *Solve each equation.*

**21.** $x^3 = 27$

**22.** $y^3 = -64$

**23.** $\dfrac{5}{x} + \dfrac{6}{x-2} = 3$

**24.** $x^4 - 21x^2 - 100 = 0$

**25.** $5(x+3)^2 - 19(x+3) = 4$

**26.** $x^{2/3} - 6x^{1/3} + 5 = 0$

**27.** $a^6 - a^2 = a^4 - 1$

**28.** $y^{-2} + y^{-1} = 20$

**29.** Two postal workers, Jerome Grant and Tim Bozik, can sort a stack of mail in 5 hours. Working alone, Tim can sort the mail in 1 hour less time than Jerome can. Find the time in which each postal worker can sort the mail alone. Round the result to one decimal place.

**30.** A negative number decreased by its reciprocal is $-\dfrac{24}{5}$. Find the number.

**(8.4)** *Solve each inequality for x. Write each solution set in interval notation.*

**31.** $2x^2 - 50 \le 0$

**32.** $\dfrac{1}{4}x^2 < \dfrac{1}{16}$

**33.** $(x^2 - 16)(x^2 - 1) > 0$

**34.** $(x^2 - 4)(y^2 - 25) \le 0$

**35.** $\dfrac{x-5}{x-6} < 0$

**36.** $\dfrac{(4x+3)(x-5)}{x(x+6)} > 0$

**37.** $(x+5)(x-6)(x+2) \le 0$

**38.** $x^3 + 3x^2 - 25x - 75 > 0$

**39.** $\dfrac{x^2+4}{3x} \le 1$

**40.** $\dfrac{(5x+6)(x-3)}{x(6x-5)} < 0$

**(8.5)** *Graph each function. Label the vertex and the axis of symmetry of each graph.*

**41.** $f(x) = x^2 - 4$

**42.** $g(x) = x^2 + 7$

**43.** $H(x) = 2x^2$

**44.** $h(x) = -\dfrac{1}{3}x^2$

**45.** $F(x) = (x - 1)^2$

**46.** $G(x) = (x + 5)^2$

**47.** $f(x) = (x - 4)^2 - 2$

**48.** $f(x) = -3(x - 1)^2 + 1$

**(8.6)** *Graph each function. Find the vertex and any intercepts of each graph.*

**49.** $f(x) = x^2 + 10x + 25$

**50.** $f(x) = -x^2 + 6x - 9$

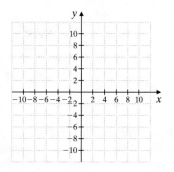

**51.** $f(x) = 4x^2 - 1$

**52.** $f(x) = -5x^2 + 5$

**53.** Find the vertex of the graph of
$f(x) = -3x^2 - 5x + 4$. Determine whether the
graph opens upward or downward, find the
y-intercept, approximate the x-intercepts to one
decimal place, and graph the function.

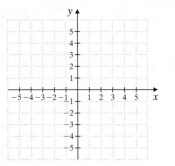

**54.** The function $h(t) = -16t^2 + 120t + 300$ gives the
height in feet of a projectile fired from the top of a
building at $t$ seconds.

**a.** When will the object reach a height of 350 feet?
Round your answer to one decimal place.

**b.** Explain why part (a) has two answers.

**55.** Find two numbers whose sum is 420 and whose
product is as large as possible.

**56.** Find two numbers whose difference is 100 and whose
product is as small as possible.

## Mixed Review

*Solve each equation.*

**57.** $x^2 - x - 30 = 0$

**58.** $10x^2 = 3x + 4$

**59.** $9y^2 = 36$

**60.** $(9n + 1)^2 = 9$

**61.** $x^2 + x + 7 = 0$

**62.** $(3x - 4)^2 = 10x$

**63.** $x^2 + 11 = 0$

**64.** $x^2 + 7 = 0$

**65.** $(5x - 2)^2 - x = 0$

**66.** $\dfrac{7}{8} = \dfrac{8}{x^2}$

**67.** $x^{2/3} - 6x^{1/3} = -8$

**68.** $(2x - 3)(4x + 5) \geq 0$

**69.** $\dfrac{x(x + 5)}{4x - 3} \geq 0$

**70.** $\dfrac{3}{x - 2} > 2$

**71.** The busiest airport in the world is the Hartsfield
International Airport in Atlanta, Georgia. The total
amount of passenger traffic through Atlanta during
the period 2008 through 2013 can be modeled by the
equation $y = 592x^2 - 1002x + 89,401$, where $y$ is
the number of passengers enplaned and deplaned,
in thousands, and $x$ is the number of years after
2008. (*Source:* Based on data from Airports Council
International)

**a.** Estimate the passenger traffic at Atlanta's
Hartsfield International Airport in 2015.

**b.** According to this model, in what year after
2008 will the passenger traffic at Atlanta's
Hartsfield International Airport surpass
150,000 passengers?

**72.** The Cliffs of Moher, at their peak, rise 702 feet
straight up from the Atlantic Ocean. Visitors are
often portrayed lying on the ground, looking over the
edge, straight down into the cold water. If a visitor's
sunglasses were to drop off while he or she was
looking over the edge, how long would it take for
the sunglasses to land in the water? (Use the formula
$s(t) = 16t^2$ from Section 8.1 and round to 2 decimal
places.)

*Solve each equation.*

**1.** $5x^2 - 2x = 7$

**2.** $(x + 1)^2 = 10$

**3.** $m^2 - m + 8 = 0$

**4.** $u^2 - 6u + 2 = 0$

**5.** $7x^2 + 8x + 1 = 0$

**6.** $y^2 - 3y = 5$

**7.** $\dfrac{4}{x + 2} + \dfrac{2x}{x - 2} = \dfrac{6}{x^2 - 4}$

**8.** $x^4 - 8x^2 - 9 = 0$

**9.** $x^6 + 1 = x^4 + x^2$

**10.** $(x + 1)^2 - 15(x + 1) + 56 = 0$

*Solve by completing the square.*

**11.** $x^2 - 6x = -2$

**12.** $2a^2 + 5 = 4a$

*Solve each inequality. Write each solution set in interval notation.*

**13.** $2x^2 - 7x > 15$

**14.** $(x^2 - 16)(x^2 - 25) \geq 0$

**15.** $\dfrac{5}{x + 3} < 1$

**16.** $\dfrac{7x - 14}{x^2 - 9} \leq 0$

*Graph each function. Label the vertex for each graph.*

**17.** $f(x) = 3x^2$

**18.** $G(x) = -2(x - 1)^2 + 5$

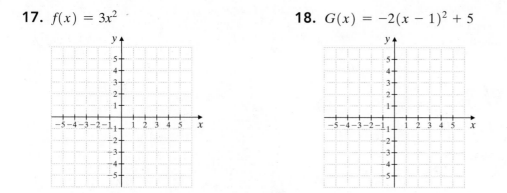

**Answers**

1. _____

2. _____

3. _____

4. _____

5. _____

6. _____

7. _____

8. _____

9. _____

10. _____

11. _____

12. _____

13. _____

14. _____

15. _____

16. _____

17. _____

18. _____

**19.** _____

**20.** _____

**21.** _____

**22.** _____

**23. a.** _____

**b.** _____

*Graph each function. Find and label the vertex, y-intercept, and x-intercepts (if any) for each graph.*

**19.** $h(x) = x^2 - 4x + 4$

**20.** $F(x) = 2x^2 - 8x + 9$

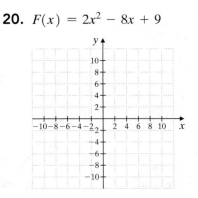

△ **21.** Given the diagram shown, approximate to the nearest foot how many feet of walking distance a person saves by cutting across the lawn instead of walking on the sidewalk.

**22.** Dave and Sandy Hartranft can paint a room together in 4 hours. Working alone, Dave can paint the room in 2 hours less time than Sandy can. Find how long it takes Sandy to paint the room alone. Give the exact answer and a two-decimal-place approximation.

**23.** A stone is thrown upward from a bridge. The stone's height $s(t)$ in feet, above the water $t$ seconds after the stone is thrown is given by the function $s(t) = -16t^2 + 32t + 256$.

    **a.** Find the maximum height of the stone.

    **b.** Find the time it takes the stone to hit the water. Round to the nearest hundredth of a second.

**1.** Write 730,000 in scientific notation.

**2.** Write $2.068 \times 10^{-3}$ in standard form.

**3.** Solve: $2|x| + 25 = 23$

**4.** Solve: $|3x - 2| = -5$

**5.** Determine whether the relation $x = y^2$ is also a function.

**6.** Find the vertex and any intercepts of $f(x) = x^2 + x - 12$.

**7.** Use the elimination method to solve the system: $\begin{cases} 3x + \dfrac{y}{2} = 2 \\ 6x + y = 5 \end{cases}$

**8.** Use the elimination method to solve the system.
$\begin{cases} -6x + y = 5 \\ 4x - 2y = 6 \end{cases}$

**9.** Subtract vertically:
$(10x^3 - 7x^2) - (4x^3 - 3x^2 + 2)$

**10.** Simplify: $\dfrac{x^2 - 4x + 4}{2 - x}$

*Multiply.*

**11.** $(2x - 7)(3x - 4)$

**12.** $(4a - 3)(7a - 2)$

**13.** $(3x^2 + y)(5x^2 - 2y)$

**14.** $(2a + b)(3a - 5b)$

**15.** Divide $2x^3 + 3x^4 - 8x + 6$ by $x^2 - 1$.

**16.** Simplify. Use positive exponents to write each answer.
  **a.** $(a^{-2}bc^3)^{-3}$

  **b.** $\left(\dfrac{a^{-4}b^2}{c^3}\right)^{-2}$

  **c.** $\left(\dfrac{3a^8b^2}{12a^5b^5}\right)^{-2}$

**17.** Factor: $6x^2 + 3x^3$

**18.** Factor: $9x^3 + 27x^2 - 15x$

**Answers**

1. _____

2. _____

3. _____

4. _____

5. _____

6. _____

7. _____

8. _____

9. _____

10. _____

11. _____

12. _____

13. _____

14. _____

15. _____

16. a. _____

  b. _____

  c. _____

17. _____

18. _____

**19.** Factor: $2(a + 3)^2 - 5(a + 3) - 7$

**20.** Factor: $2x(3y - 2) - 5(3y - 2)$

**21.** Factor: $3a^2x - 12abx + 12b^2x$

**22.** Factor: $2xy + 6x - y - 3$

**23.** Solve: $2x^2 = \dfrac{17}{3}x + 1$

**24.** Solve: $2(a^2 + 2) - 8 = -2a(a - 2) - 5$

**25.** Multiply: $\dfrac{3n + 1}{2n} \cdot \dfrac{2n - 4}{3n^2 - 2n - 1}$

**26.** Multiply: $(\sqrt{3} - 4)(2\sqrt{3} + 2)$

**27.** Multiply: $\dfrac{x^3 - 1}{-3x + 3} \cdot \dfrac{15x^2}{x^2 + x + 1}$

**28.** Subtract: $\dfrac{3}{x^2 - 4} - \dfrac{1 - x}{x^2 - 4}$

**29.** Add: $\dfrac{2}{x^2} + \dfrac{5}{3x^3}$

**30.** Subtract: $\dfrac{a + 1}{a^2 - 6a + 8} - \dfrac{3}{16 - a^2}$

**31.** Simplify: $\dfrac{\dfrac{x}{y^2} + \dfrac{1}{y}}{\dfrac{y}{x^2} + \dfrac{1}{x}}$

**32.** Simplify: $\dfrac{(2a)^{-1} + b^{-1}}{a^{-1} + (2b)^{-1}}$

**33.** Solve: $\dfrac{2x}{x - 3} + \dfrac{6 - 2x}{x^2 - 9} = \dfrac{x}{x + 3}$

**34.** Solve: $\dfrac{x + 3}{x^2 + 5x + 6} = \dfrac{3}{2x + 4} - \dfrac{1}{x + 3}$

**35.** Solve $\dfrac{1}{x} + \dfrac{1}{y} = \dfrac{1}{z}$ for $x$.

**36.** Divide $x^3 - 3x^2 - 10x + 24$ by $x + 3$.

19. _____

20. _____

21. _____

22. _____

23. _____

24. _____

25. _____

26. _____

27. _____

28. _____

29. _____

30. _____

31. _____

32. _____

33. _____

34. _____

35. _____

36. _____

△ **37.** The lateral surface area of a cylinder varies jointly as its radius and height. Express surface area $S$ in terms of radius $r$ and height $h$.

**38.** Suppose that $y$ varies inversely as $x$. If $y$ is 8 when $x$ is 24, find the constant of variation and the variation equation.

*Find each cube root.*

**39.** $\sqrt[3]{-64}$

**40.** $\sqrt[3]{24a^7b^9}$

**41.** $\sqrt[3]{\dfrac{8}{125}}$

**42.** Rationalize the denominator: $\dfrac{2\sqrt{16}}{\sqrt{9x}}$

*Multiply.*

**43.** $\sqrt[3]{4} \cdot \sqrt[3]{2}$

**44.** $(\sqrt{5} - x)^2$

**45.** $\sqrt{\dfrac{2}{a}} \cdot \sqrt{\dfrac{b}{3}}$

**46.** $(\sqrt{a} + b)(\sqrt{a} - b)$

**47.** Simplify: $\sqrt[3]{54} - 5\sqrt[3]{16} + \sqrt[3]{2}$

**48.** Solve: $\sqrt{x-2} = \sqrt{4x+1} - 3$

**49.** Add: $\sqrt[3]{\dfrac{7x}{8}} + 2\sqrt[3]{7x}$

**50.** Simplify: $\sqrt[4]{48x^9}$

**51.** Solve $p^2 + 2p = 4$ by completing the square.

**52.** Use the square root property to solve $(y - 1)^2 = 24$.

**53.** Use the quadratic formula to solve $2x^2 - 4x = 3$.

**54.** Use the quadratic formula to solve $m^2 = 4m + 8$.

**55.** An object is thrown upward from the top of a 200-foot cliff with a velocity of 12 feet per second. The height above ground $h$ in feet of the object after $t$ seconds is $h = -16t^2 + 12t + 200$. How long after the object is thrown will it strike the ground? Round to the nearest tenth of a second.

**56.** Mr. Briley can roof his house in 24 hours. His son can roof the same house in 40 hours. If they work together, how long will it take to roof the house?

37. _____

38. _____

39. _____

40. _____

41. _____

42. _____

43. _____

44. _____

45. _____

46. _____

47. _____

48. _____

49. _____

50. _____

51. _____

52. _____

53. _____

54. _____

55. _____

56. _____

# 9 Exponential and Logarithmic Functions

In this chapter, we discuss two closely related types of functions: exponential and logarithmic functions. These functions are vital to applications in economics, finance, engineering, the sciences, education, and other fields. Models of tumor growth and learning curves are two examples of the uses of exponential and logarithmic functions.

## Are LED Light Bulbs the Light Bulbs of Our Immediate Future?

An LED light bulb is a light-emitting diode (LED) product that has a predicted future in lighting fixtures. LED light bulbs have a life span much greater than current rivals in the market. The graph on the bottom of this page shows a comparison of some current light bulbs available. The other graph on this page shows a predicted exponential growth of LED products. There is also ongoing research in the production of organic LEDs (OLED).

Since this chapter is about exponential and logarithmic functions, it should be noted that the light output of a compact fluorescent lamp (or light) (CFL) decays exponentially. Also, the response of the human eye to light is logarithmic.

**Global LED Lighting Market Growth**

$y = 2.62(1.688)^x$

*predicted

**Approximate Watts Used by Different Globes to Deliver the Same Light as a 100 Watt Incandescent Bulb**

# The Algebra of Functions

## Objective A  Adding, Subtracting, Multiplying, and Dividing Functions

As we have seen in earlier chapters, it is possible to add, subtract, multiply, and divide functions. Although we have not stated them as such, the sums, differences, products, and quotients of functions are themselves functions. For example, if $f(x) = 3x$ and $g(x) = x + 1$, their product, $f(x) \cdot g(x) = 3x(x + 1) = 3x^2 + 3x$, is a new function. We can use the notation $(f \cdot g)(x)$ to denote this new function. Using the sum, difference, product, and quotient of functions to generate new functions is called the **algebra of functions.**

---

**Objectives**

**A** Add, Subtract, Multiply, and Divide Functions.

**B** Construct Composite Functions.

---

### Algebra of Functions

Let $f$ and $g$ be functions. New functions from $f$ and $g$ are defined as follows:

| | |
|---|---|
| Sum | $(f + g)(x) = f(x) + g(x)$ |
| Difference | $(f - g)(x) = f(x) - g(x)$ |
| Product | $(f \cdot g)(x) = f(x) \cdot g(x)$ |
| Quotient | $\left(\dfrac{f}{g}\right)(x) = \dfrac{f(x)}{g(x)}, \; g(x) \neq 0$ |

---

**Example 1**  If $f(x) = x - 1$ and $g(x) = 2x - 3$, find the following.

**a.** $(f + g)(x)$    **b.** $(f - g)(x)$    **c.** $(f \cdot g)(x)$    **d.** $\left(\dfrac{f}{g}\right)(x)$

**Solution:**  Use the algebra of functions and replace $f(x)$ by $x - 1$ and $g(x)$ by $2x - 3$. Then simplify.

**a.** $(f + g)(x) = f(x) + g(x)$
$$= (x - 1) + (2x - 3)$$
$$= 3x - 4$$

**b.** $(f - g)(x) = f(x) - g(x)$
$$= (x - 1) - (2x - 3)$$
$$= x - 1 - 2x + 3$$
$$= -x + 2$$

**c.** $(f \cdot g)(x) = f(x) \cdot g(x)$
$$= (x - 1)(2x - 3)$$
$$= 2x^2 - 5x + 3$$

**d.** $\left(\dfrac{f}{g}\right)(x) = \dfrac{f(x)}{g(x)} = \dfrac{x - 1}{2x - 3}$, where $x \neq \dfrac{3}{2}$

▬ Work Practice 1

---

**Practice 1**

If $f(x) = x + 3$ and $g(x) = 3x - 1$, find

**a.** $(f + g)(x)$
**b.** $(f - g)(x)$
**c.** $(f \cdot g)(x)$
**d.** $\left(\dfrac{f}{g}\right)(x)$

---

There is an interesting but not surprising relationship between the graphs of functions and the graphs of their sum, difference, product, and quotient. For example, the graph of $(f + g)$ can be found by adding the graph of $f$ to the graph of $g$. We add two graphs by adding corresponding $y$-values. (*Note:* These graphs on the top of the next page are not from Example 1.)

**Answers**
**1. a.** $4x + 2$   **b.** $-2x + 4$
**c.** $3x^2 + 8x - 3$
**d.** $\dfrac{x + 3}{3x - 1}$, where $x \neq \dfrac{1}{3}$

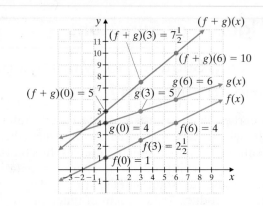

## Objective B   Constructing Composite Functions ▶

Another way to combine functions is called **function composition.** To understand this new way of combining functions, study the diagrams below. The left diagram (in the margin) shows an illustration by thermometers, and the right diagram (below) is the same illustration but by tables. In both illustrations, we show degrees Celsius $f(x)$ as a function of degrees Fahrenheit $x$, then Kelvins $g(x)$ as a function of degrees Celsius $x$. (The Kelvin scale is a temperature scale devised by Lord Kelvin in 1848.) The first function we will call $f$, and the second function we will call $g$.

**Thermometer Illustration**

**Table Illustration**

| $x$ = **Degrees Fahrenheit** (**Input**) | −13 | 32 | 68 | 212 |
|---|---|---|---|---|
| $f(x)$ = **Degrees Celsius** (**Output**) | −25 | 0 | 20 | 100 |

| $x$ = **Degrees Celsius** (**Input**) | −25 | 0 | 20 | 100 |
|---|---|---|---|---|
| $g(x)$ = **Kelvins** (**Output**) | 248.15 | 273.15 | 293.15 | 373.15 |

Suppose that we want a function that shows a direct conversion from degrees Fahrenheit to Kelvins. In other words, suppose that a function is needed that shows Kelvins as a function of degrees Fahrenheit. This can easily be done because the output of the first function $f(x)$ is the same as the input of the second function. If we use $g(f(x))$ to represent this, then we get the diagram to the left.

| $x$ = **Degrees Fahrenheit** (**Input**) | −13 | 32 | 68 | 212 |
|---|---|---|---|---|
| $f(g(x))$ = **Kelvins** (**Output**) | 248.15 | 273.15 | 293.15 | 373.15 |

For example $g(f(-13)) = 248.15$, and so on.

Since the output of the first function is used as the input of the second function, we write the new function as $g(f(x))$. The new function is formed from the composition of the other two functions. The mathematical symbol for this composition is $(g \circ f)(x)$. Thus, $(g \circ f)(x) = g(f(x))$.

It is possible to find an equation for the composition of the two functions $f$ and $g$. In other words, we can find a function that converts degrees Fahrenheit directly to Kelvins. The function $f(x) = \dfrac{5}{9}(x - 32)$ converts degrees Fahrenheit to degrees Celsius, and the function $g(x) = x + 273.15$ converts degrees Celsius to Kelvins. Thus,

$$(g \circ f)(x) = g(f(x)) = g\left(\frac{5}{9}(x - 32)\right) = \frac{5}{9}(x - 32) + 273.15$$

In general, the notation $g(f(x))$ means "$g$ composed with $f$" and can be written as $(g \circ f)(x)$. Also $f(g(x))$, or $(f \circ g)(x)$, means "$f$ composed with $g$."

---

### Composition of Functions

The composition of functions $f$ and $g$ is

$$(f \circ g)(x) = f(g(x))$$

---

**Helpful Hint**

$(f \circ g)(x)$ does not mean the same as $(f \cdot g)(x)$.

$$(f \circ g)(x) = f(g(x)) \text{ while } (f \cdot g)(x) = f(x) \cdot g(x)$$

↑                 ↑

Composition of functions      Multiplication of functions

---

**Example 2**   If $f(x) = x^2$ and $g(x) = x + 3$, find each composition.

**a.** $(f \circ g)(2)$ and $(g \circ f)(2)$
**b.** $(f \circ g)(x)$ and $(g \circ f)(x)$

**Solution:**

**a.** $(f \circ g)(2) = f(g(2))$
$\quad\quad\quad\quad\quad = f(5)$       Since $g(x) = x + 3$, then $g(2) = 2 + 3 = 5$.
$\quad\quad\quad\quad\quad = 5^2 = 25$
$\quad (g \circ f)(2) = g(f(2))$
$\quad\quad\quad\quad\quad = g(4)$       Since $f(x) = x^2$, then $f(2) = 2^2 = 4$.
$\quad\quad\quad\quad\quad = 4 + 3 = 7$

**b.** $(f \circ g)(x) = f(g(x))$
$\quad\quad\quad\quad\quad = f(x + 3)$       Replace $g(x)$ with $x + 3$.
$\quad\quad\quad\quad\quad = (x + 3)^2$       $f(x + 3) = (x + 3)^2$
$\quad\quad\quad\quad\quad = x^2 + 6x + 9$    Square $(x + 3)$.
$\quad (g \circ f)(x) = g(f(x))$
$\quad\quad\quad\quad\quad = g(x^2)$       Replace $f(x)$ with $x^2$.
$\quad\quad\quad\quad\quad = x^2 + 3$       $g(x^2) = x^2 + 3$

■ **Work Practice 2**

---

**Example 3**   If $f(x) = |x|$ and $g(x) = x - 2$, find each composition.

**a.** $(f \circ g)(x)$            **b.** $(g \circ f)(x)$

**Solution:**

**a.** $(f \circ g)(x) = f(g(x)) = f(x - 2) = |x - 2|$
**b.** $(g \circ f)(x) = g(f(x)) = g(|x|) = |x| - 2$

■ **Work Practice 3**

---

**Practice 2**

If $f(x) = x^2$ and $g(x) = 2x + 1$, find each composition.

**a.** $(f \circ g)(3)$ and $(g \circ f)(3)$
**b.** $(f \circ g)(x)$ and $(g \circ f)(x)$

**Practice 3**

If $f(x) = \sqrt{x}$ and $g(x) = x + 1$, find each composition.

**a.** $(f \circ g)(x)$     **b.** $(g \circ f)(x)$

**Answers**

**2. a.** $49; 19$   **b.** $4x^2 + 4x + 1; 2x^2 + 1$
**3. a.** $\sqrt{x + 1}$   **b.** $\sqrt{x} + 1$

Copyright 2015 Pearson Education, Inc.

## Helpful Hint

In Examples 2 and 3, notice that $(g \circ f)(x) \neq (f \circ g)(x)$. In general, $(g \circ f)(x)$ *may* or *may not* equal $(f \circ g)(x)$.

## Practice 4

If $f(x) = 2x, g(x) = x + 5$, and $h(x) = |x|$, write each function as a composition of $f, g,$ or $h$.

**a.** $F(x) = |x + 5|$

**b.** $G(x) = 2x + 5$

---

**Example 4** If $f(x) = 5x, g(x) = x - 2$, and $h(x) = \sqrt{x}$, write each function as a composition with $f, g,$ or $h$.

**a.** $F(x) = \sqrt{x - 2}$

**b.** $G(x) = 5x - 2$

**Solution:**

**a.** Notice the order in which the function $F$ operates on an input value $x$. First, 2 is subtracted from $x$, and then the square root of that result is taken. This means that $F(x) = (h \circ g)(x)$. To check, we find $(h \circ g)(x)$.

$$(h \circ g)(x) = h(g(x)) = h(x - 2) = \sqrt{x - 2}$$

**b.** Notice the order in which the function $G$ operates on an input value $x$. First, $x$ is multiplied by 5, and then 2 is subtracted from the result. This means that $G(x) = (g \circ f)(x)$. To check, we find $(g \circ f)(x)$.

$$(g \circ f)(x) = g(f(x)) = g(5x) = 5x - 2$$

■ **Work Practice 4**

## Answers

**4. a.** $(h \circ g)(x)$  **b.** $(g \circ f)(x)$

---

## Calculator Explorations  Graphing

If $f(x) = \dfrac{1}{2}x + 2$ and $g(x) = \dfrac{1}{3}x^2 + 4$, then

$$(f + g)(x) = f(x) + g(x)$$
$$= \left(\frac{1}{2}x + 2\right) + \left(\frac{1}{3}x^2 + 4\right)$$
$$= \frac{1}{3}x^2 + \frac{1}{2}x + 6$$

$y = \frac{1}{3}x^2 + \frac{1}{2}x + 6$

$y = \frac{1}{3}x^2 + 4$

$y = \frac{1}{2}x + 2$

To visualize this addition of functions with a grapher, graph

$Y_1 = \dfrac{1}{2}x + 2$, $Y_2 = \dfrac{1}{3}x^2 + 4$, and $Y_3 = \dfrac{1}{3}x^2 + \dfrac{1}{2}x + 6$

Use a TABLE feature to verify that for a given $x$ value, $Y_1 + Y_2 = Y_3$. For example, verify that when $x = 0$, $Y_1 = 2, Y_2 = 4,$ and $Y_3 = 2 + 4 = 6$.

## Vocabulary, Readiness & Video Check

*Match each function (Exercises 1–6) with its definition (Choices A–F).*

**1.** $(f \circ g)(x)$

**2.** $(f \cdot g)(x)$

**3.** $(f - g)(x)$

**4.** $(g \circ f)(x)$

**5.** $\left(\dfrac{f}{g}\right)(x)$

**6.** $(f + g)(x)$

**A.** $g(f(x))$

**B.** $f(x) + g(x)$

**C.** $f(g(x))$

**D.** $\dfrac{f(x)}{g(x)}, g(x) \neq 0$

**E.** $f(x) \cdot g(x)$

**F.** $f(x) - g(x)$

*Martin-Gay Interactive Videos*　Watch the section lecture video and answer the following questions.

**Objective A** **7.** From ▥ Example 1 and the lecture before, we know that $(f + g)(x) = f(x) + g(x)$. Use this fact to explain two ways you can find $(f + g)(2)$. ▸

**Objective B** **8.** From ▥ Example 3, given two functions $f(x)$ and $g(x)$, can $f(g(x))$ ever equal $g(f(x))$? ▸

*See Video 9.1*

---

**9.1　Exercise Set** MyMathLab® ▸

**Objective A** *For the functions f and g, find* ***a.*** $(f + g)(x)$, ***b.*** $(f - g)(x)$, ***c.*** $(f \cdot g)(x)$, *and* ***d.*** $\left(\dfrac{f}{g}\right)(x)$. *See Example 1.*

**1.** $f(x) = x - 7; g(x) = 2x + 1$　　**2.** $f(x) = x + 4; g(x) = 5x - 2$　▸**3.** $f(x) = x^2 + 1; g(x) = 5x$

**4.** $f(x) = x^2 - 2; g(x) = 3x$　　**5.** $f(x) = \sqrt{x}; g(x) = x + 5$　　**6.** $f(x) = \sqrt[3]{x}; g(x) = x - 3$

**7.** $f(x) = -3x; g(x) = 5x^2$　　**8.** $f(x) = 4x^3; g(x) = -6x$

**Objective B** *If* $f(x) = x^2 - 6x + 2$, $g(x) = -2x$, *and* $h(x) = \sqrt{x}$, *find each composition. See Example 2.*

**9.** $(f \circ g)(2)$　　　**10.** $(h \circ f)(-2)$　　　▸**11.** $(g \circ f)(-1)$

**12.** $(f \circ h)(1)$　　　**13.** $(g \circ h)(0)$　　　**14.** $(h \circ g)(0)$

*Find* $(f \circ g)(x)$ *and* $(g \circ f)(x)$. *See Examples 2 and 3.*

▸**15.** $f(x) = x^2 + 1; g(x) = 5x$　　　　**16.** $f(x) = x - 3; g(x) = x^2$

**17.** $f(x) = 2x - 3; g(x) = x + 7$　　　　**18.** $f(x) = x + 10; g(x) = 3x + 1$

**19.** $f(x) = x^3 + x - 2; g(x) = -2x$

**20.** $f(x) = -4x; g(x) = x^3 + x^2 - 6$

**21.** $f(x) = |x|; g(x) = 10x - 3$

**22.** $f(x) = |x|; g(x) = 14x - 8$

**23.** $f(x) = \sqrt{x}; g(x) = -5x + 2$

**24.** $f(x) = 7x - 1; g(x) = \sqrt[3]{x}$

*If $f(x) = 3x, g(x) = \sqrt{x}$, and $h(x) = x^2 + 2$, write each function as a composition with f, g, or h. See Example 4.*

▶ **25.** $H(x) = \sqrt{x^2 + 2}$

**26.** $G(x) = \sqrt{3x}$

**27.** $F(x) = 9x^2 + 2$

**28.** $H(x) = 3x^2 + 6$

**29.** $G(x) = 3\sqrt{x}$

**30.** $F(x) = x + 2$

*Find f(x) and g(x) so that the given function $h(x) = (f \circ g)(x)$.*

**31.** $h(x) = (x + 2)^2$

**32.** $h(x) = |x - 1|$

**33.** $h(x) = \sqrt{x + 5} + 2$

**34.** $h(x) = (3x + 4)^2 + 3$

**35.** $h(x) = \dfrac{1}{2x - 3}$

**36.** $h(x) = \dfrac{1}{x + 10}$

## Review

*Solve each equation for y. See Section 2.3.*

**37.** $x = y + 2$

**38.** $x = y - 5$

**39.** $x = 3y$

**40.** $x = -6y$

**41.** $x = -2y - 7$

**42.** $x = 4y + 7$

## Concept Extensions

*Solve.*

**43.** Businesspeople are concerned with cost functions, revenue functions, and profit functions. Recall that the profit $P(x)$ obtained from selling $x$ units of a product is equal to the revenue $R(x)$ from selling the $x$ units minus the cost $C(x)$ of manufacturing the $x$ units. Write an equation expressing this relationship among $C(x), R(x)$, and $P(x)$.

**44.** Suppose the revenue $R(x)$ for $x$ units of a product can be described by $R(x) = 25x$ and the cost $C(x)$ can be described by $C(x) = 50 + x^2 + 4x$. Find the profit $P(x)$ for $x$ units.

**45.** If you are given $f(x)$ and $g(x)$, explain in your own words how to find $(f \circ g)(x)$, and then how to find $(g \circ f)(x)$.

**46.** Given $f(x)$ and $g(x)$, describe in your own words the difference between $(f \circ g)(x)$ and $(f \cdot g)(x)$.

In the next sections, we begin a study of two new functions: exponential and logarithmic functions. As we learn more about these functions, we will discover that they share a special relation to each other; they are inverses of each other.

Before we study these functions, we need to learn about inverses. We begin by defining one-to-one functions.

## Objective A Determining Whether a Function Is One-to-One

Study the following table.

| Degrees Fahrenheit (Input) | −31 | −13 | 32 | 68 | 149 | 212 |
|---|---|---|---|---|---|---|
| Degrees Celsius (Output) | −35 | −25 | 0 | 20 | 65 | 100 |

Recall that since each degrees Fahrenheit (input) corresponds to exactly one degrees Celsius (output), this table of inputs and outputs does describe a function. Also notice that each output corresponds to a different input. This type of function is given a special name—a *one-to-one function*.

Does the set $f = \{(0,1), (2,2), (-3,5), (7,6)\}$ describe a one-to-one function? It is a function since each $x$-value corresponds to a unique $y$-value. For this particular function $f$, each $y$-value corresponds to a unique $x$-value. Thus, this function is also a one-to-one function.

### One-to-One Function

For a **one-to-one function**, each $x$-value (input) corresponds to only one $y$-value (output) and each $y$-value (output) corresponds to only one $x$-value (input).

**Examples** Determine whether each function described is one-to-one.

1. $f = \{(6,2), (5,4), (-1,0), (7,3)\}$

   The function $f$ is one-to-one since each $y$-value corresponds to only one $x$-value.

2. $g = \{(3,9), (-4,2), (-3,9), (0,0)\}$

   The function $g$ is not one-to-one because the $y$-value 9 in $(3,9)$ and $(-3,9)$ corresponds to two different $x$-values.

3. $h = \{(1,1), (2,2), (10,10), (-5,-5)\}$

   The function $h$ is one-to-one since each $y$-value corresponds to only one $x$-value.

4.

| Mineral (Input) | Talc | Gypsum | Diamond | Topaz | Stibnite |
|---|---|---|---|---|---|
| Hardness on the Mohs Scale (Output) | 1 | 2 | 10 | 8 | 2 |

   This table does not describe a one-to-one function since the output 2 corresponds to two different inputs, gypsum and stibnite.

(*Continued on next page*)

## Objectives

**A** Determine Whether a Function Is a One-to-One Function.

**B** Use the Horizontal Line Test to Decide Whether a Function Is a One-to-One Function.

**C** Find the Inverse of a Function.

**D** Find an Equation of the Inverse of a Function.

**E** Graph Functions and Their Inverses.

**F** Determine Whether Two Functions Are Inverses of Each Other.

## Practice 1–5

Determine whether each function described is one-to-one.

1. $f = \{(7,3), (-1,1), (5,0), (4,-2)\}$

2. $g = \{(-3,2), (6,3), (2,14), (-6,2)\}$

3. $h = \{(0,0), (1,2), (3,4), (5,6)\}$

4.

| State (Input) | Colorado | Mississippi | Nevada | New Mexico | Utah |
|---|---|---|---|---|---|
| Number of Colleges and Universities (Output) | 16 | 7 | 4 | 10 | 7 |

5.

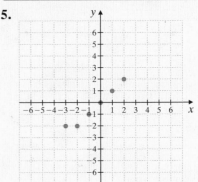

### Answers

1. one-to-one  2. not one-to-one
3. one-to-one  4. not one-to-one
5. not one-to-one

**5.**

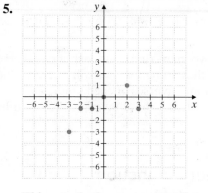

This graph does not describe a one-to-one function since the $y$-value $-1$ corresponds to three different $x$-values, $-2$, $-1$ and 3, as shown to the right.

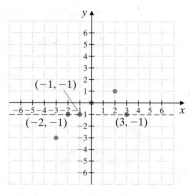

■ Work Practice 1–5

## Objective B Using the Horizontal Line Test ▶

Recall that we recognize the graph of a function when it passes the vertical line test. Since every $x$-value of the function corresponds to exactly one $y$-value, each vertical line intersects the function's graph at most once. The graph shown next, for instance, is the graph of a function.

Is this function a *one-to-one* function? The answer is no. To see why not, notice that the $y$-value of the ordered pair $(-3, 3)$, for example, is the same as the $y$-value of the ordered pair $(3, 3)$. This function is therefore not one-to-one.

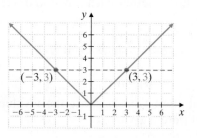

To test whether a graph is the graph of a one-to-one function, we can apply the vertical line test to see whether it is a function, and then apply a similar **horizontal line test** to see whether it is a one-to-one function.

### Horizontal Line Test

If every horizontal line intersects the graph of a function at most once, then the function is a one-to-one function.

**Example 6** Use the vertical and horizontal line tests to determine whether each graph is the graph of a one-to-one function.

**a.**

**b.**

**c.**

**d.**

**e.**

**Solution:** Graphs **a, b, c,** and **d** all pass the vertical line test, so only these graphs are graphs of functions. But, of these, only **b** and **c** pass the horizontal line test, so only **b** and **c** are graphs of one-to-one functions.

■ Work Practice 6

**Helpful Hint**

All linear equations are one-to-one functions except those whose graphs are horizontal or vertical lines. A vertical line does not pass the vertical line test and hence is not the graph of a function. A horizontal line is the graph of a function but does not pass the horizontal line test and hence is not the graph of a one-to-one function.

not a function          function, but not one-to-one

## Objective C Finding the Inverse of a Function ▶

One-to-one functions are special in that their graphs pass the vertical and horizontal line tests. They are special, too, in another sense: We can find the **inverse function** for any one-to-one function by switching the coordinates of the ordered pairs of the

**Practice 6**

Use the vertical and horizontal line tests to determine whether each graph is the graph of a one-to-one function.

**a.**

**b.**

**c.**

**d.**

**e.**

**Answers**

**6. a.** not a function   **b.** not a function
**c.** one-to-one   **d.** one-to-one
**e.** not one-to-one

f

Input:
degrees
Fahrenheit

Output:
degrees
Celsius

212 → 100

68 → 20

32 → 0

−13 → −25

$f^{-1}$

function, or the inputs and the outputs. For example, the inverse of the one-to-one function

| Degrees Fahrenheit (Input) | −31 | −13 | 32 | 68 | 149 | 212 |
| --- | --- | --- | --- | --- | --- | --- |
| Degrees Celsius (Output) | −35 | −25 | 0 | 20 | 65 | 100 |

is the function

| Degrees Celsius (Input) | −35 | −25 | 0 | 20 | 65 | 100 |
| --- | --- | --- | --- | --- | --- | --- |
| Degrees Fahrenheit (Output) | −31 | −13 | 32 | 68 | 149 | 212 |

Notice that the ordered pair $(-31, -35)$ of the function, for example, becomes the ordered pair $(-35, -31)$ of its inverse.

Also, the inverse of the one-to-one function $f = \{(2, -3), (5, 10), (9, 1)\}$ is $\{(-3, 2), (10, 5), (1, 9)\}$. For a function $f$, we use the notation $f^{-1}$, read "$f$ inverse," to denote its inverse function. Notice that since the coordinates of each ordered pair have been switched, the domain (set of inputs) of $f$ is the range (set of outputs) of $f^{-1}$, and the range of $f$ is the domain of $f^{-1}$.

### Inverse Function

The inverse of a one-to-one function $f$ is the one-to-one function $f^{-1}$ that consists of the set of all ordered pairs $(y, x)$ where $(x, y)$ belongs to $f$.

### Helpful Hint

If a function is not one-to-one, it does not have an inverse function.

**Practice 7**

Find the inverse of the one-to-one function:

$f = \{(2, -4), (-1, 13),$
$\qquad (0, 0), (-7, -8)\}$

**Example 7**  Find the inverse of the one-to-one function:

$$f = \{(0, 1), (-2, 7), (3, -6), (4, 4)\}$$

**Solution:**

$$f^{-1} = \{(1, 0), (7, -2), (-6, 3), (4, 4)\}$$

———————————————— Switch coordinates of each ordered pair.

■ Work Practice 7

✓**Concept Check**  Suppose that $f$ is a one-to-one function and that $f(1) = 5$.
   **a.** Write the corresponding ordered pair.
   **b.** Write one point that we know must belong to the inverse function $f^{-1}$.

## Objective D  Finding an Equation of the Inverse of a Function ▶

If a one-to-one function $f$ is defined as a set of ordered pairs, we can find $f^{-1}$ by interchanging the $x$- and $y$-coordinates of the ordered pairs. If a one-to-one function $f$ is given in the form of an equation, we can find an equation of $f^{-1}$ by using a similar procedure.

**Answer**

7. $f^{-1} = \{(-4, 2), (13, -1),$
$\qquad (0, 0), (-8, -7)\}$

✓**Concept Check Answers**

**a.** $(1, 5)$

**b.** $(5, 1)$

### Finding an Equation of the Inverse of a One-to-One Function f

**Step 1:** Replace $f(x)$ with $y$.

**Step 2:** Interchange $x$ and $y$.

**Step 3:** Solve the equation for $y$.

**Step 4:** Replace $y$ with the notation $f^{-1}(x)$.

**Helpful Hint**

The symbol $f^{-1}$ is the single symbol used to denote the inverse of the function $f$.

It is read as "$f$ inverse." This symbol *does not mean* $\dfrac{1}{f}$.

---

**Example 8**  Find the equation of the inverse of $f(x) = x + 3$.

**Solution:**  $f(x) = x + 3$

**Step 1:**          $y = x + 3$    Replace $f(x)$ with $y$.

**Step 2:**          $x = y + 3$    Interchange $x$ and $y$.

**Step 3:**  $x - 3 = y$       Solve for $y$.

**Step 4:**  $f^{-1}(x) = x - 3$    Replace $y$ with $f^{-1}(x)$.

The inverse of $f(x) = x + 3$ is $f^{-1}(x) = x - 3$. Notice that, for example,

$$f(1) = 1 + 3 = 4 \quad \text{and} \quad f^{-1}(4) = 4 - 3 = 1$$

Ordered pair: $(1, 4)$        Ordered pair: $(4, 1)$

The coordinates are switched, as expected.

**■ Work Practice 8**

---

**Practice 8**

Find the equation of the inverse of $f(x) = x - 6$.

---

**Example 9**  Find the equation of the inverse of $f(x) = 3x - 5$. Graph $f$ and $f^{-1}$ on the same set of axes.

**Solution:**  $f(x) = 3x - 5$

**Step 1:**          $y = 3x - 5$          Replace $f(x)$ with $y$.

**Step 2:**          $x = 3y - 5$          Interchange $x$ and $y$.

**Step 3:**  $x + 5 = 3y$          Solve for $y$.

$$\dfrac{x + 5}{3} = y \quad \text{or} \quad y = \dfrac{x + 5}{3}$$

**Step 4:**  $f^{-1}(x) = \dfrac{x + 5}{3}$      Replace $y$ with $f^{-1}(x)$.

Now we graph $f$ and $f^{-1}$ on the same set of axes. Both $f(x) = 3x - 5$ and $f^{-1}(x) = \dfrac{x + 5}{3}$ are linear functions, so each graph is a line.

| $f(x) = 3x - 5$ | |
|---|---|
| $x$ | $y = f(x)$ |
| 1 | $-2$ |
| 0 | $-5$ |
| $\dfrac{5}{3}$ | 0 |

| $f^{-1}(x) = \dfrac{x + 5}{3}$ | |
|---|---|
| $x$ | $y = f^{-1}(x)$ |
| $-2$ | 1 |
| $-5$ | 0 |
| 0 | $\dfrac{5}{3}$ |

*(Continued on next page)*

---

**Practice 9**

Find the equation of the inverse of $f(x) = 2x + 3$. Graph $f$ and $f^{-1}$ on the same set of axes.

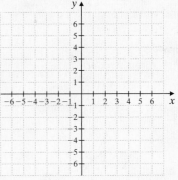

**Answers**

**8.** $f^{-1}(x) = x + 6$

**9.** $f^{-1}(x) = \dfrac{x - 3}{2}$

## Practice 10

Graph the inverse of each function.

**a.**

**b.**

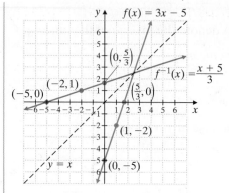

■ Work Practice 9

## Objective E Graphing Inverse Functions

Notice that the graphs of $f$ and $f^{-1}$ in Example 9 are mirror images of each other, and the "mirror" is the dashed line $y = x$. This is true for every function and its inverse. For this reason, we say that the *graphs of f and $f^{-1}$ are symmetric about the line $y = x$.*

To see why this happens, study the graph of a few ordered pairs and their switched coordinates.

**Answers**

**10. a.**

**b.**

### Example 10    Graph the inverse of each function.

**Solution:** The function is graphed in blue and the inverse is graphed in red.

**a.**

**b.**

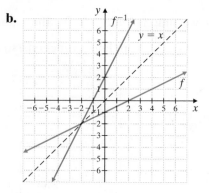

■ Work Practice 10

## Objective F Determining Whether Functions Are Inverses of Each Other ▶

Notice in the table of values in Example 9 that $f(0) = -5$ and $f^{-1}(-5) = 0$, as expected. Also, for example, $f(1) = -2$ and $f^{-1}(-2) = 1$. In words, we say that for some input $x$, the function $f^{-1}$ takes the output of $x$, called $f(x)$, back to $x$.

$$x \rightarrow f(x) \quad \text{and} \quad f^{-1}(f(x)) \rightarrow x$$
$$\downarrow \quad \downarrow \quad\quad\quad \downarrow \quad\quad \downarrow$$
$$f(0) = -5 \quad \text{and} \quad f^{-1}(-5) = 0$$
$$f(1) = -2 \quad \text{and} \quad f^{-1}(-2) = 1$$

In general,

If $f$ is a one-to-one function, then the inverse of $f$ is the function $f^{-1}$ such that

$$(f^{-1} \circ f)(x) = x \quad \text{and} \quad (f \circ f^{-1})(x) = x$$

**Example 11** Show that if $f(x) = 3x + 2$, then $f^{-1}(x) = \dfrac{x - 2}{3}$.

**Solution:** See that $(f^{-1} \circ f)(x) = x$ and $(f \circ f^{-1})(x) = x$.

$$(f^{-1} \circ f)(x) = f^{-1}(f(x))$$
$$= f^{-1}(3x + 2) \quad \text{Replace } f(x) \text{ with } 3x + 2.$$
$$= \frac{3x + 2 - 2}{3}$$
$$= \frac{3x}{3}$$
$$= x$$
$$(f \circ f^{-1})(x) = f(f^{-1}(x))$$
$$= f\left(\frac{x - 2}{3}\right) \quad \text{Replace } f^{-1}(x) \text{ with } \frac{x - 2}{3}.$$
$$= 3\left(\frac{x - 2}{3}\right) + 2$$
$$= x - 2 + 2$$
$$= x$$

■ Work Practice 11

**Practice 11**

Show that if $f(x) = 4x - 1$, then $f^{-1}(x) = \dfrac{x + 1}{4}$.

**Answer**

**11.** $f(f^{-1}(x)) = 4\left(\dfrac{x + 1}{4}\right) - 1$
$$= x + 1 - 1 = x$$
$$f^{-1}(f(x)) = \frac{(4x - 1) + 1}{4}$$
$$= \frac{4x}{4} = x$$

---

🖩 **Calculator Explorations  Graphing**

A grapher can be used to visualize functions and their inverses. Recall that the graph of a function $f$ and its inverse $f^{-1}$ are mirror images of each other across the line $y = x$. To see this for the function $f(x) = 3x + 2$, use a square window (see Appendix C) and graph

the given function: $Y_1 = 3x + 2$

its inverse: $Y_2 = \dfrac{x - 2}{3}$

and the line $Y_3 = x$

Exercises will follow in Exercise Set 9.2.

## Vocabulary, Readiness & Video Check

*Use the choices below to fill in each blank. Some choices will not be used, and some will be used more than once.*

| | | | | | |
|---|---|---|---|---|---|
| vertical | $(3, 7)$ | $(11, 2)$ | $y = x$ | $x$ | true |
| horizontal | $(7, 3)$ | $(2, 11)$ | $\dfrac{1}{f}$ | the inverse of $f$ | false |

**1.** If $f(2) = 11$, the corresponding ordered pair is _____.

**2.** If $(7, 3)$ is an ordered pair solution of $f(x)$, and $f(x)$ has an inverse, then an ordered pair solution of $f^{-1}(x)$ is _____.

**3.** The symbol $f^{-1}$ means _____.

**4.** True or false: The function notation $f^{-1}(x)$ means $\dfrac{1}{f(x)}$. _____

**5.** To tell whether a graph is the graph of a function, use the _____ line test.

**6.** To tell whether the graph of a function is also a one-to-one function, use the _____ line test.

**7.** The graphs of $f$ and $f^{-1}$ are symmetric about the line _____.

**8.** Two functions are inverses of each other if $(f \circ f^{-1})(x) = $ _____ and $(f^{-1} \circ f)(x) = $ _____.

---

### Martin-Gay Interactive Videos

*See Video 9.2*

*Watch the section lecture video and answer the following questions.*

**Objective A   9.** From ▣ Example 1 and the definition before, what makes a one-to-one function different from other types of functions? ⊙

**Objective B   10.** From ▣ Examples 2 and 3, if a graph passes the horizontal line test but not the vertical line test, is it a one-to-one function? Explain. ⊙

**Objective C   11.** From ▣ Example 4 and the lecture before, if you find the inverse of a one-to-one function, is this inverse function also a one-to-one function? How do you know? ⊙

**Objective D   12.** From ▣ Examples 5 and 6, explain why the interchanging of $x$ and $y$ when finding an inverse equation makes sense given the definition of an inverse function. ⊙

**Objective E   13.** From ▣ Example 7, if you have the equation or graph of a one-to-one function, how can you graph its inverse without finding the inverse's equation? ⊙

**Objective F   14.** Based on ▣ Example 8 and the lecture before, what's wrong with the following statement? "If $f$ is a one-to-one function, you can prove that $f$ and $f^{-1}$ are inverses of each other by showing that $f(f^{-1}(x)) = f^{-1}(f(x))$." ⊙

## 9.2 Exercise Set MyMathLab®

**Objectives A C Mixed Practice** *Determine whether each function is a one-to-one function. If it is one-to-one, list the inverse function by switching coordinates, or inputs and outputs. See Examples 1 through 5 and 7.*

**1.** $g = \{(0, 3), (3, 7), (6, 7), (-2, -2)\}$

**2.** $g = \{(8, 6), (9, 6), (3, 4), (-4, 4)\}$

**3.** $h = \{(10, 10)\}$

**4.** $r = \{(1, 2), (3, 4), (5, 6), (6, 7)\}$

**5.** $f = \{(11, 12), (4, 3), (3, 4), (6, 6)\}$

**6.** $f = \{(-1, -1), (1, 1), (0, 2), (2, 0)\}$

**7.**

| Month of 2014 (Input) | May | June | July | August | September | October |
|---|---|---|---|---|---|---|
| Unemployment Rate in Percent (Output) | 6.3 | 6.1 | 6.2 | 6.1 | 5.9 | 5.8 |

(*Source:* U.S. Bureau of Labor Statistics)

**8.**

| State (Input) | Maine | South Dakota | New Hampshire | Idaho | Michigan |
|---|---|---|---|---|---|
| Number of Two-Year Colleges (Output) | 8 | 6 | 8 | 3 | 30 |

(*Source:* U.S. University of Texas at Austin)

**9.**

| State (Input) | California | Alaska | Indiana | Louisiana | New Mexico | Ohio |
|---|---|---|---|---|---|---|
| Rank in Population (Output) | 1 | 47 | 16 | 25 | 36 | 7 |

(*Source:* U.S. Bureau of the Census)

**10.**

| Shape (Input) | Triangle | Pentagon | Quadrilateral | Hexagon | Decagon |
|---|---|---|---|---|---|
| Number of Sides (Output) | 3 | 5 | 4 | 6 | 10 |

*Given the one-to-one function* $f(x) = x^3 + 2$, *find the following. (Hint: You do not need to find the equation for* $f^{-1}$.)

**11. a.** $f(1)$
 **b.** $f^{-1}(3)$

**12. a.** $f(0)$
 **b.** $f^{-1}(2)$

**13. a.** $f(-1)$
 **b.** $f^{-1}(1)$

**14. a.** $f(-2)$
 **b.** $f^{-1}(-6)$

## Objective B  *Determine whether the graph of each function is the graph of a one-to-one function. See Example 6.*

**15.**  **16.**  **17.**  **18.**

**19.**  **20.**  **21.**  **22.**

## Objectives D E  **Mixed Practice** *Each of the following functions is one-to-one. Find the inverse of each function and graph the function and its inverse on the same set of axes. See Examples 8 through 10.*

**23.** $f(x) = x + 4$  **24.** $f(x) = x - 5$  **25.** $f(x) = 2x - 3$  **26.** $f(x) = 4x + 9$

**27.** $f(x) = \frac{1}{2}x - 1$  **28.** $f(x) = -\frac{1}{2}x + 2$  **29.** $f(x) = x^3$  **30.** $f(x) = x^3 - 1$

*Find the inverse of each one-to-one function. See Examples 8 and 9.*

**31.** $f(x) = 5x + 2$

**32.** $f(x) = 6x - 1$

**33.** $f(x) = \dfrac{x - 2}{5}$

**34.** $f(x) = \dfrac{4x - 3}{2}$

**35.** $f(x) = \sqrt[3]{x}$

**36.** $f(x) = \sqrt[3]{x + 1}$

▶ **37.** $f(x) = \dfrac{5}{3x + 1}$

**38.** $f(x) = \dfrac{7}{2x + 4}$

**39.** $f(x) = (x + 2)^3$

**40.** $f(x) = (x - 5)^3$

*Graph the inverse of each function on the same set of axes. See Example 10.*

**41.**

**42.**

**43.**

**44.**

**45.**

**46.**

## Objective F   *Solve. See Example 11.*

▶ **47.** If $f(x) = 2x + 1$, show that $f^{-1}(x) = \dfrac{x - 1}{2}$.

**48.** If $f(x) = 3x - 10$, show that $f^{-1}(x) = \dfrac{x + 10}{3}$.

**49.** If $f(x) = x^3 + 6$, show that $f^{-1}(x) = \sqrt[3]{x - 6}$.

**50.** If $f(x) = x^3 - 5$, show that $f^{-1}(x) = \sqrt[3]{x + 5}$.

## Review

*Evaluate each of the following. See Section 7.2.*

**51.** $25^{1/2}$

**52.** $49^{1/2}$

**53.** $16^{3/4}$

**54.** $27^{2/3}$

**55.** $9^{-3/2}$

**56.** $81^{-3/4}$

*If $f(x) = 3^x$, find the following. In Exercises 59 and 60, give an exact answer and a two-decimal-place approximation. See Sections 1.6, 3.6, and 7.2.*

**57.** $f(2)$

**58.** $f(0)$

**59.** $f\left(\dfrac{1}{2}\right)$

**60.** $f\left(\dfrac{2}{3}\right)$

## Concept Extensions

*Solve. See the Concept Check in this section.*

**61.** Suppose that $f$ is a one-to-one function and that
$f(2) = 9$.

   **a.** Write the corresponding ordered pair.

   **b.** Name one ordered pair that we know is a solution
   of the inverse of $f$, or $f^{-1}$.

**62.** Suppose that $F$ is a one-to-one function and that
$F\left(\dfrac{1}{2}\right) = -0.7$.

   **a.** Write the corresponding ordered pair.

   **b.** Name one ordered pair that we know is a solution
   of the inverse of $F$, or $F^{-1}$.

*For Exercises 63 and 64.*

   **a.** *Write the ordered pairs for f whose points are highlighted. (Include the points whose coordinates are given.)*

   **b.** *Write the corresponding ordered pairs for the inverse of f, f⁻¹.*

   **c.** *Graph the ordered pairs for f⁻¹ found in part (b).*

   **d.** *Graph f⁻¹ by drawing a smooth curve through the plotted points.*

**63. a.**

**c. d.**

**64. a.**

**b. c.**

**65.** If you are given the graph of a function, describe
how you can tell from the graph whether the
function has an inverse.

**66.** Describe the appearance of the graphs of a function
and its inverse.

*Find the inverse of each one-to-one function. Then graph the function and its inverse in a square window.*

**67.** $f(x) = 3x + 1$    **68.** $f(x) = -2x - 6$    **69.** $f(x) = \sqrt[3]{x + 3}$    **70.** $f(x) = x^3 - 3$

**Exponential Functions**

In earlier chapters, we gave meaning to exponential expressions such as $2^x$, where $x$ is a rational number. Recall the following examples.

$$2^3 = 2 \cdot 2 \cdot 2 \qquad \text{Three factors; each factor is 2}$$
$$2^{3/2} = (2^{1/2})^3 = \sqrt{2} \cdot \sqrt{2} \cdot \sqrt{2} \qquad \text{Three factors; each factor is } \sqrt{2}$$

When $x$ is an irrational number (for example, $\sqrt{3}$), what meaning can we give to $2^{\sqrt{3}}$?

It is beyond the scope of this book to give precise meaning to $2^x$ if $x$ is irrational. We can confirm your intuition and say that $2^{\sqrt{3}}$ is a real number, and since $1 < \sqrt{3} < 2$, then $2^1 < 2^{\sqrt{3}} < 2^2$. We can also use a calculator and approximate $2^{\sqrt{3}}$: $2^{\sqrt{3}} \approx 3.321997$. In fact, as long as the base $b$ is positive, $b^x$ is a real number for all real numbers $x$. Finally, the rules of exponents apply whether $x$ is rational or irrational, as long as $b$ is positive.

In this section, we are interested in functions of the form $f(x) = b^x$, or $y = b^x$, where $b > 0$. A function of this form is called an *exponential function*.

**Objectives**

**A** Graph Exponential Functions.

**B** Solve Equations of the Form $b^x = b^y$.

**C** Solve Problems Modeled by Exponential Equations.

---

**Exponential Function**

A function of the form

$$f(x) = b^x$$

is called an **exponential function**, where $b > 0$, $b$ is not 1, and $x$ is a real number.

---

## Objective A  Graphing Exponential Functions

Now let's practice graphing exponential functions.

**Example 1** Graph the exponential functions $f(x) = 2^x$ and $g(x) = 3^x$ on the same set of axes.

**Solution:** To graph these functions, we find some ordered pair solutions, plot the points, and connect them with a smooth curve. Remember throughout that $y = f(x)$.

| $f(x) = 2^x$ | $x$ | 0 | 1 | 2 | 3 | $-1$ | $-2$ |
|---|---|---|---|---|---|---|---|
| | $f(x)$ | 1 | 2 | 4 | 8 | $\frac{1}{2}$ | $\frac{1}{4}$ |

| $g(x) = 3^x$ | $x$ | 0 | 1 | 2 | 3 | $-1$ | $-2$ |
|---|---|---|---|---|---|---|---|
| | $g(x)$ | 1 | 3 | 9 | 27 | $\frac{1}{3}$ | $\frac{1}{9}$ |

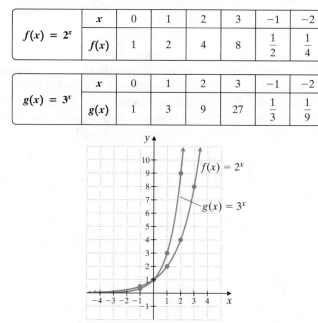

**Work Practice 1**

**Practice 1**

Graph the exponential function $f(x) = 6^x$.

**Answer**

**1.**

A number of things should be noted about the two graphs of exponential functions in Example 1. First, the graphs show that $f(x) = 2^x$ and $g(x) = 3^x$ are one-to-one functions since each graph passes the vertical and horizontal line tests. The $y$-intercept of each graph is $(0, 1)$, but neither graph has an $x$-intercept. From the graph, we can also see that the domain of each function is all real numbers and that the range is $(0, \infty)$. We can also see that as $x$-values are increasing, $y$-values are increasing also.

## Practice 2

Graph the exponential function

$$f(x) = \left(\frac{1}{6}\right)^x.$$

**Example 2**  Graph the exponential functions $f(x) = \left(\frac{1}{2}\right)^x$ and $g(x) = \left(\frac{1}{3}\right)^x$ on the same set of axes.

**Solution:**  As before, we find some ordered pair solutions, plot the points, and connect them with a smooth curve.

| $f(x) = \left(\frac{1}{2}\right)^x$ | $x$ | 0 | 1 | 2 | 3 | −1 | −2 |
|---|---|---|---|---|---|---|---|
| | $y$ | 1 | $\frac{1}{2}$ | $\frac{1}{4}$ | $\frac{1}{8}$ | 2 | 4 |

| $g(x) = \left(\frac{1}{3}\right)^x$ | $x$ | 0 | 1 | 2 | 3 | −1 | −2 |
|---|---|---|---|---|---|---|---|
| | $y$ | 1 | $\frac{1}{3}$ | $\frac{1}{9}$ | $\frac{1}{27}$ | 3 | 9 |

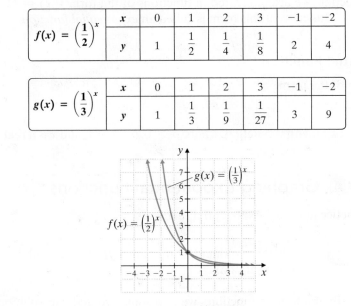

**Work Practice 2**

Each function in Example 2 again is a one-to-one function. The $y$-intercept of both is $(0, 1)$. The domain is the set of all real numbers, and the range is $(0, \infty)$.

Notice the difference between the graphs of Example 1 and the graphs of Example 2. An exponential function is always increasing if the base is greater than 1. When the base is between 0 and 1, the graph is always decreasing. The following figures summarize these characteristics of exponential functions.

**Answer**

**2.**

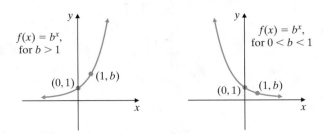

**Example 3** Graph the exponential function $f(x) = 3^{x+2}$.

**Solution:** As before, we find and plot a few ordered pair solutions. Then we connect the points with a smooth curve.

| $f(x) = 3^{x+2}$ | $x$ | 0 | −1 | −2 | −3 | −4 |
|---|---|---|---|---|---|---|
| | $y$ | 9 | 3 | 1 | $\frac{1}{3}$ | $\frac{1}{9}$ |

$f(x) = 3^{x+2}$

■ Work Practice 3

✓**Concept Check** Which functions are exponential functions?

**a.** $f(x) = x^3$  **b.** $g(x) = \left(\frac{2}{3}\right)^x$  **c.** $h(x) = 5^{x-2}$  **d.** $w(x) = (2x)^2$

# Objective B  Solving Equations of the Form $b^x = b^y$ ▶

We have seen that an exponential function $y = b^x$ is a one-to-one function. Another way of stating this fact is a property that we can use to solve exponential equations.

> **Uniqueness of $b^x$**
>
> Let $b > 0$ and $b \neq 1$. Then $b^x = b^y$ is equivalent to $x = y$.

Thus, one way to solve an exponential equation depends on whether it's possible to write each side of the equation with the same base; that is, $b^x = b^y$. We solve by this method first.

**Example 4** Solve: $2^x = 16$

**Solution:** We write 16 as a power of 2 so that each side of the equation has the same base. Then we use the uniqueness of $b^x$ to solve.

$2^x = 16$
$2^x = 2^4$

Since the bases are the same and are nonnegative, by the uniqueness of $b^x$ we then have that the exponents are equal. Thus,

$x = 4$

To check, we replace $x$ with 4 in the original equation. The solution set is {4}.

■ Work Practice 4

**Example 5** Solve: $25^x = 125$

**Solution:** Since both 25 and 125 are powers of 5, we can use the uniqueness of $b^x$.

$25^x = 125$
$(5^2)^x = 5^3$   Write 25 and 125 as powers of 5.
$5^{2x} = 5^3$
$2x = 3$   Use the uniqueness of $b^x$.
$x = \dfrac{3}{2}$   Divide both sides by 2.

*(Continued on next page)*

## Practice 3

Graph the exponential function $f(x) = 2^{x-1}$.

## Practice 4

Solve: $5^x = 125$

## Practice 5

Solve: $4^x = 8$

**Answers**

**3.**

$f(x) = 2^{x-1}$

**4.** {3}  **5.** $\left\{\dfrac{3}{2}\right\}$

✓**Concept Check Answer**
**b** and **c**

To check, we replace $x$ with $\frac{3}{2}$ in the original equation.

The solution set is $\left\{\frac{3}{2}\right\}$.

■ Work Practice 5

**Practice 6**

Solve:  $9^{x-1} = 27^x$

**Example 6**   Solve: $4^{x+3} = 8^x$

**Solution:**   We write both 4 and 8 as powers of 2, and then use the uniqueness of $b^x$.

$$4^{x+3} = 8^x$$
$$(2^2)^{x+3} = (2^3)^x$$
$$2^{2x+6} = 2^{3x}$$
$$2x + 6 = 3x \qquad \text{Use the uniqueness of } b^x.$$
$$6 = x \qquad \text{Subtract } 2x \text{ from both sides.}$$

Check to see that the solution set is $\{6\}$.

■ Work Practice 6

There is one major problem with the preceding technique. Often the two sides of an equation, $4 = 3^x$ for example, cannot easily be written as powers of a common base. We explore how to solve such an equation with the help of *logarithms* later.

## Objective C  Solving Problems Modeled by Exponential Equations

The bar graph below shows the increase in the number of cellular phone users. Notice that the graph of the exponential function $y = 246.98(1.05)^x$ approximates the heights of the bars. This is just one example of how the world abounds with patterns that can be modeled by exponential functions. To make these applications realistic, we use numbers that warrant a calculator.

**Cellular Phone Subscriptions**

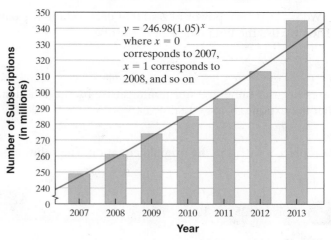

Source: ICT database of ITU/United Nations

Another application of an exponential function has to do with interest rates on loans. The exponential function defined by $A = P\left(1 + \dfrac{r}{n}\right)^{nt}$ models the pattern relating the dollars $A$ accrued (or owed) after $P$ dollars are invested (or loaned) at an annual rate of interest $r$ compounded $n$ times each year for $t$ years. This function is known as the *compound interest formula*.

**Answer**

6. $\{-2\}$

**Example 7**  Using the Compound Interest Formula

Find the amount owed at the end of 5 years if $1600 is loaned at a rate of 9% compounded monthly.

**Solution:** Use the formula $A = P\left(1 + \dfrac{r}{n}\right)^{nt}$, with the following values:

$P = \$1600$ (the amount of the loan)

$r = 9\% = 0.09$ (the annual rate of interest)

$n = 12$ (the number of times interest is compounded each year)

$t = 5$ (the duration of the loan, in years)

$$A = P\left(1 + \frac{r}{n}\right)^{nt} \qquad \text{Compound interest formula}$$

$$= 1600\left(1 + \frac{0.09}{12}\right)^{12(5)} \qquad \text{Substitute known values.}$$

$$= 1600(1.0075)^{60}$$

To approximate $A$, use the $\boxed{y^x}$ or $\boxed{\wedge}$ key on your calculator.

$\boxed{2505.0896}$

Thus, the amount $A$ owed is approximately $2505.09.

■ **Work Practice 7**

**Practice 7**

a. As a result of the Chernobyl nuclear accident, radioactive debris was carried through the atmosphere. One immediate concern was the impact that the debris had on the milk supply. The percent $y$ of radioactive material in raw milk $t$ days after the accident is estimated by $y = 100(2.7)^{-0.1t}$. Estimate the expected percent of radioactive material in the milk after 30 days.

b. Find the amount owed at the end of 6 years if $23,000 is loaned at a rate of 12% compounded quarterly (4 times a year). Round your answer to the nearest cent.

**Answers**

**7. a.** approximately 5.08%

**b.** $46,754.26

---

📟 **Calculator Explorations**  **Graphing**

We can use a graphing calculator and its TRACE feature to solve Practice 7a graphically.

To estimate the percent of radioactive material in the milk after 30 days, enter $Y_1 = 100(2.7)^{-0.1x}$. The graph does not appear on a standard viewing window, so we need to determine an appropriate viewing window. Because it doesn't make sense to look at radioactivity *before* the Chernobyl nuclear accident, we use Xmin = 0. We are interested in finding the percent of radioactive material in the milk when $x = 30$, so we choose Xmax = 35 to leave enough space to see the graph at $x = 30$. Because the values of $y$ are percents, it seems appropriate that $0 \le y \le 100$. (We also use Xscl = 1 and Yscl = 10.) Now we graph the function.

We can use the TRACE feature to obtain an approximation of the expected percent of radioactive material in the milk when $x = 30$. (A TABLE feature may also be used to approximate the percent.) To obtain a better approximation, let's use the ZOOM feature several times to zoom in near $x = 30$.

X=30.001689  Y=5.0796742

The percent of radioactive material in the milk 30 days after the Chernobyl accident was 5.08%, accurate to two decimal places.

*Use a graphing calculator to find each percent. Round your solutions to two decimal places.*

1. Estimate the percent of radioactive material in the milk 2 days after the Chernobyl nuclear accident.

2. Estimate the percent of radioactive material in the milk 10 days after the Chernobyl nuclear accident.

3. Estimate the percent of radioactive material in the milk 15 days after the Chernobyl nuclear accident.

4. Estimate the percent of radioactive material in the milk 25 days after the Chernobyl nuclear accident.

## Vocabulary, Readiness & Video Check

*Use the choices to fill in each blank.*

1. A function such as $f(x) = 2^x$ is a(n) _____ function.
      **a.** linear      **b.** quadratic     **c.** exponential

2. If $7^x = 7^y$, then _____.
      **a.** $x = 7^y$    **b.** $x = y$       **c.** $y = 7^x$       **d.** $7 = 7^y$

*Answer the questions about the graph of $y = 2^x$, shown to the right.*

3. Is this a function? _____
4. Is this a one-to-one function? _____
5. Is there an $x$-intercept? _____ If so, name the coordinates. _____
6. Is there a $y$-intercept? _____ If so, name the coordinates. _____
7. The domain of this function, in interval notation, is _____.
8. The range of this function, in interval notation, is _____.

**Martin-Gay Interactive Videos**    *Watch the section lecture video and answer the following questions.*

**Objective A**   **9.** From the lecture before ▢ Example 1, what's the main difference between a polynomial function and an exponential function? ▶

**Objective B**   **10.** From ▢ Examples 2 and 3, you can only apply the uniqueness of $b^x$ to solve an exponential equation if you're able to do what? ▶

**Objective B**   **11.** For ▢ Example 4, write the equation and find how much uranium will remain after 101 days. Round your answer to the nearest tenth. ▶

*See Video 9.3* 🍎

---

**9.3**   **Exercise Set** MyMathLab® ▶

**Objective A**   *Graph each exponential function. See Examples 1 through 3.*

1. $y = 5^x$

2. $y = 4^x$

3. $y = 1 + 2^x$

4. $y = 3^x - 1$

**5.** $y = \left(\dfrac{1}{4}\right)^x$      **6.** $y = \left(\dfrac{1}{5}\right)^x$      **7.** $y = \left(\dfrac{1}{2}\right)^x - 2$      **8.** $y = \left(\dfrac{1}{3}\right)^x + 2$

**9.** $y = -2^x$      **10.** $y = -3^x$      **11.** $y = 3^x - 2$      **12.** $y = 2^x - 3$

**13.** $y = -\left(\dfrac{1}{4}\right)^x$      **14.** $y = -\left(\dfrac{1}{5}\right)^x$      **15.** $y = \left(\dfrac{1}{3}\right)^x + 1$      **16.** $y = \left(\dfrac{1}{2}\right)^x + 2$

**17.** $f(x) = 2^{x-2}$      **18.** $g(x) = 2^{x+1}$      **19.** $F(x) = 5^{x+1}$      **20.** $G(x) = 3^{x-2}$

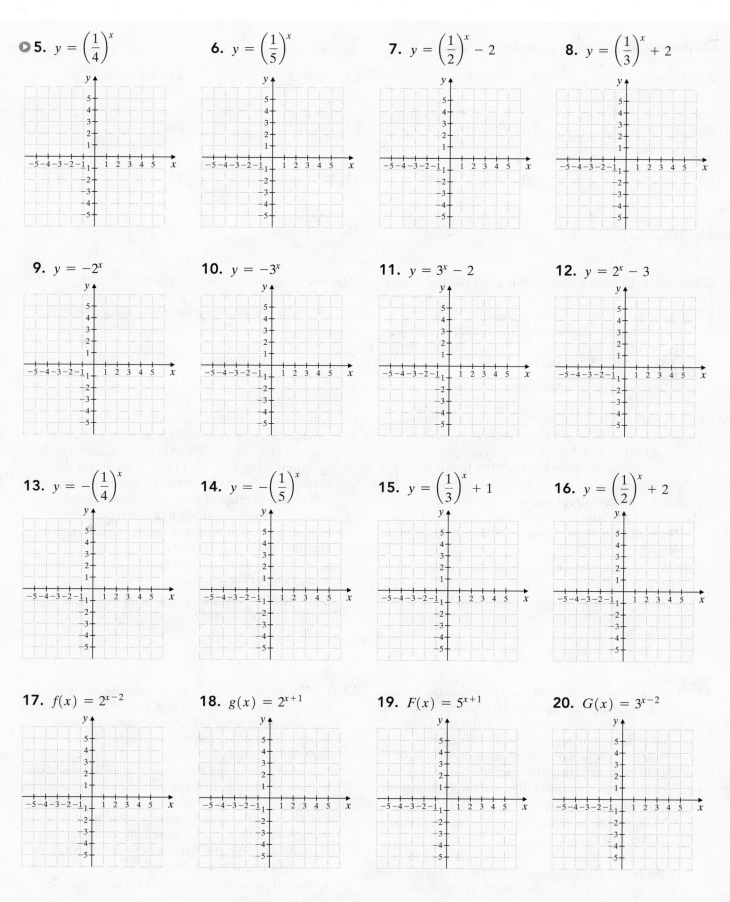

## Objective B  *Solve. See Examples 4 through 6.*

**21.** $3^x = 27$

**22.** $6^x = 36$

**23.** $16^x = 8$

**24.** $64^x = 16$

**25.** $32^{2x-3} = 2$

**26.** $9^{2x+1} = 81$

**27.** $\dfrac{1}{4} = 2^{3x}$

**28.** $\dfrac{1}{27} = 3^{2x}$

**29.** $9^x = 27$

**30.** $32^x = 4$

**31.** $27^{x+1} = 9$

**32.** $125^{x-2} = 25$

**33.** $81^{x-1} = 27^{2x}$

**34.** $4^{3x-7} = 32^{2x}$

**35.** $\left(\dfrac{1}{8}\right)^x = 16^{1-x}$

**36.** $\left(\dfrac{1}{9}\right)^x = 27^{2-x}$

## Objective C  *Solve. Unless otherwise indicated, round results to one decimal place. See Example 7.*

**37.** One type of uranium has a radioactive decay rate of 0.4% per day. If 30 pounds of this uranium is available today, how much will still remain after 50 days? Use $y = 30(0.996)^x$, and let $x$ be 50.

**38.** The nuclear waste from an atomic energy plant decays at a rate of 3% each century. If 150 pounds of nuclear waste is disposed of, how much of it will still remain after 10 centuries? Use $y = 150(0.97)^x$, and let $x$ be 10.

**39.** Cheese production in the United States is currently growing at a rate of 3.7% per year. The equation $y = 1158.6(1.037)^x$ models the cheese production in the United States from 1970 to 2012. In this equation, $y$ is the amount of cheese produced, in thousands of metric tons, and $x$ represents the number of years after 1970. Round answers to the nearest tenth of a thousand metric tons. (*Source:* National Agricultural Statistics Service)

   **a.** Estimate the total cheese production in the United States in 2012.

   **b.** Assuming this equation continues to be valid in the future, use the equation to predict the total amount of cheese produced in the United States in 2020.

**40.** Americans are living longer. The population of Americans age 85 and older is projected to keep growing at a rate of 3.2%. Answer the following questions using $y = 1.09(1.032)^t$, where $y$ is the population, in millions, of people age 85 and older and $t$ is the number of years after 1960. Round answers to the nearest tenth of a million people. (*Source:* Administration on Aging, Department of Health and Human Services)

   **a.** According to the model, how many Americans, age 85 and older, were alive in 2010?

   **b.** If this trend continues, how many Americans age 85 and older would be expected to be alive in 2040?

**41.** The equation $y = 259,184(1.02)^x$ models the number of American college students who studied abroad each year from 2007 through 2012. In the equation, $y$ is the number of American students studying abroad and $x$ represents the number of years after 2007. Round answers to the nearest whole. (*Source:* Based on data from the Institute of International Education, Open Doors)

   **a.** Estimate the number of American students studying abroad in 2012.

   **b.** Assuming this equation continues to be valid in the future, use this equation to predict the number of American college students studying abroad in 2020.

**42.** Carbon dioxide ($CO_2$) is a greenhouse gas that contributes to global warming. Partially due to the combustion of fossil fuels, the amount of $CO_2$ in Earth's atmosphere has been increasing by 0.5% annually over the past decade. In 2005, the concentration of $CO_2$ in the atmosphere was 379.8 parts per million by volume. To make the following predictions, use $y = 379.8(1.005)^t$, where $y$ is the concentration of $CO_2$ in parts per million and $t$ is the number of years after 2005. (*Source:* Based on data from the National Oceanic and Atmospheric Administration, U.S. Department of Commerce)

   **a.** Predict the concentration of $CO_2$ in the atmosphere in the year 2015.

   **b.** Predict the concentration of $CO_2$ in the atmosphere in the year 2030.

*The equation $y = 246.98(1.05)^x$ gives the number of cellular phone subscriptions y (in millions) in the United States for the years 2007 through 2013. In this equation, $x = 0$ corresponds to 2007, $x = 1$ corresponds to 2008, and so on. Use this model to solve Exercises 43 and 44. Round answers to the nearest tenth of a million.*

**43.** Predict the number of cellular subscriptions in the year 2017.

**44.** Predict the number of cellular subscriptions in 2020.

*Solve. Use $A = P\left(1 + \dfrac{r}{n}\right)^{nt}$. Round answers to two decimal places. See Example 7.*

**45.** Find the amount a college student owes at the end of 3 years if $6000 is loaned to her at a rate of 8% compounded monthly.

**46.** Find the amount owed at the end of 5 years if $3000 is loaned at a rate of 10% compounded quarterly.

**47.** Find the total amount a college student has in a savings account if $2000 was invested and earned 6% compounded semiannually for 12 years.

**48.** Find the amount accrued if $500 is invested and earns 7% compounded monthly for 4 years.

## Review

*Solve each equation. See Section 2.1.*

**49.** $5x - 2 = 18$

**50.** $3x - 7 = 11$

**51.** $3x - 4 = 3(x + 1)$

**52.** $2 - 6x = 6(1 - x)$

## Concept Extensions

*Is the given function an exponential function? See the Concept Check in this section.*

**53.** $f(x) = 1.5x^2$

**54.** $g(x) = 3^x$

**55.** $h(x) = \left(\dfrac{1}{2}x\right)^2$

**56.** $F(x) = 0.4^{x+1}$

*Match each exponential function with its graph.*

**57.** $f(x) = \left(\dfrac{1}{2}\right)^x$

**58.** $f(x) = 2^x$

**59.** $f(x) = \left(\dfrac{1}{4}\right)^x$

**60.** $f(x) = 3^x$

**61.** Explain why the graph of an exponential function $y = b^x$ contains the point $(1, b)$.

**62.** Explain why an exponential function $y = b^x$ has a $y$-intercept of $(0, 1)$.

*Use a graphing calculator to solve. Estimate your results to two decimal places.*

 **63.** Verify the results of Exercise 37.

**64.** Verify the results of Exercise 38.

 **65.** From Exercise 37, estimate the number of pounds of uranium that will remain after 100 days.

 **66.** From Exercise 37, estimate the number of pounds of uranium that will remain after 120 days.

## 9.4  Exponential Growth and Decay Functions

### Objectives

**A** Model Exponential Growth.

**B** Model Exponential Decay.

Now that we can graph exponential functions, let's learn about exponential growth and exponential decay.

A quantity that grows or decays by the same percent at regular time periods is said to have **exponential growth** or **exponential decay**. There are many real-life examples of exponential growth and decay, such as population, bacteria, viruses, and radioactive substances, just to name a few.

Recall the graphs of exponential functions.

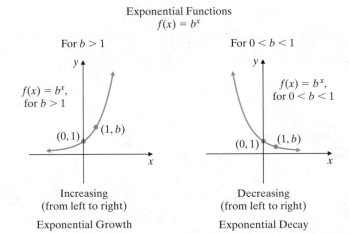

Exponential Functions
$f(x) = b^x$

For $b > 1$

$f(x) = b^x$, for $b > 1$

$(0, 1)$  $(1, b)$

Increasing
(from left to right)

Exponential Growth

For $0 < b < 1$

$f(x) = b^x$, for $0 < b < 1$

$(0, 1)$  $(1, b)$

Decreasing
(from left to right)

Exponential Decay

## Objective A  Modeling Exponential Growth

We begin with exponential growth, as described below.

**Exponential Growth**

initial amount

number of time intervals

$$y = C(1 + r)^x$$

$(1 + r)$ is growth factor; $r$ is growth rate (often a percent)

**Example 1** In 2005, let's suppose a town named La Combe had a population of 15,500 and was consistently increasing by 10% per year. If this yearly increase continues, predict the city's population in 2025. (Round to the nearest whole.)

**Solution:** Let's begin to understand by calculating the city's population each year:

| Time Interval | $x = 1$ | $x = 2$ | 3 | 4 | 5 | and so on ... |
|---|---|---|---|---|---|---|
| Year | 1996 | 1997 | 1998 | 1999 | 2000 | |
| Population | 17,050 | 18,755 | 20,631 | 22,694 | 24,963 | |

$15,500 + 0.10(15,500)$    $17,050 + 0.10(17,050)$

This is an example of exponential growth, so let's use our formula with

$$C = 15,500; r = 0.10; x = 2015 - 1995 = 20$$

Then,

$$y = C(1 + r)^x$$
$$= 15,500(1 + 0.10)^{20}$$
$$= 15,500(1.1)^{20}$$
$$\approx 104,276$$

In 2025, we predict the population of La Combe to be 104,276.

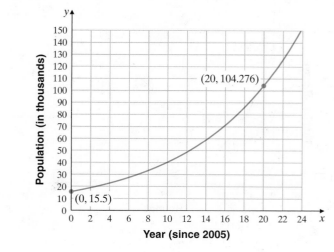

**Work Practice 1**

*Note:* The exponential growth formula, $y = C(1 + r)^x$, should remind you of the compound interest formula from the previous section, $A = P(1 + \frac{r}{n})^{nt}$. In fact, if the number of compoundings per year, $n$, is 1, the interest formula becomes $A = P(1 + r)^t$, which is the exponential growth formula written with different variables.

## Objective B  Modeling Exponential Decay

Now let's study exponential decay.

**Exponential Decay**

initial amount — number of time intervals

$$y = C(1 - r)^x$$

$(1 - r)$ is decay factor; $r$ is decay rate (often a percent)

**Practice 1**

In 2010, the town of La Combe (from Example 1) had a population of 25,000 and started consistently increasing by 12% per year. If this yearly increase continues, predict the city's population in 2025. Round to the nearest whole.

**Answer**
**1.** 136,839

## Practice 2

A tournament with 800 persons is played so that after each round the number of players decreases by 30%. Find the number of players after round 9. Round your answer to the nearest whole.

**Example 2**   A large golf country club holds a singles tournament each year. At the start of the tournament for a particular year there are 512 players. After each round, half the players are eliminated. How many players remain after 6 rounds?

**Solution:**   This is an example of exponential decay.

Let's begin to understand by calculating the number of players after a few rounds.

| Round (same as interval) | 1 | 2 | 3 | 4 | and so on … |
|---|---|---|---|---|---|
| Players (at end of round) | 256 | 128 | 64 | 32 | |

$$512 - 0.50(512) \qquad 256 - 0.50(256)$$

Here, $C = 512$;  $r = \dfrac{1}{2}$ or $50\% = 0.50$;  $x = 6$

Thus,

$$y = 512(1 - 0.50)^6$$
$$= 512(0.50)^6$$
$$= 8$$

After 6 rounds, there are 8 players remaining.

**Work Practice 2**

The **half-life** of a substance is the amount of time it takes for half of the substance to decay.

## Practice 3

Use the information from Example 3 and calculate how much of a 500-gram sample of DDT will remain after 51 years. Round to the nearest tenth.

**Example 3**   A form of DDT pesticide (banned in 1972) has a half-life of approximately 15 years. If a storage unit had 400 pounds of DDT, find how much DDT is remaining after 72 years. Round to the nearest tenth.

**Solution:**   Here, we need to be careful because each time interval is 15 years, the half-life.

| Time Interval | 1 | 2 | 3 | 4 | 5 | and so on … |
|---|---|---|---|---|---|---|
| Years Passed | 15 | $2 \cdot 15 = 30$ | 45 | 60 | 75 | |
| Pounds of DDT | 200 | 100 | 50 | 25 | 12.5 | |

From the table we see that after 72 years, between 4 and 5 intervals, there should be between 12.5 and 25 pounds of DDT remaining.

**Answers**

**2.** 32   **3.** 47.4 g

Let's calculate $x$, the number of time intervals.

$$x = \frac{72 \text{ (years)}}{15 \text{ (half-life)}} = 4.8$$

Now, using our exponential decay formula and the definition of half-life, for each time interval $x$, the decay rate $r$ is $\frac{1}{2}$ or 50% or 0.50.

$$y = 400(1 - 0.50)^{4.8} \text{—time intervals for 72 years}$$

original amount     decay rate

$$y = 400(0.50)^{4.8}$$
$$y \approx 14.4$$

In 72 years, 14.4 pounds of DDT remain.

■ Work Practice 3

## Vocabulary, Readiness & Video Check

*Martin-Gay Interactive Videos*     *Watch the section lecture video and answer the following questions.*

**Objective A 1.** ▤ Example 1 reviews exponential growth. Explain how you find the growth rate and the correct number of time intervals. ▶

**Objective B 2.** Explain how you know that ▤ Example 2 has to do with exponential decay and not exponential growth. ▶

**3.** For ▤ Example 3, which has to do with half-life, explain how to calculate the number of time intervals. Also, what is the decay rate for half-life and why? ▶

*See Video 9.4*

**9.4   Exercise Set** MyMathLab® ▶

**Objective A** *Practice using the exponential growth formula by completing the table below. Round final amounts to the nearest whole. See Example 1.*

| | Original Amount | Growth Rate per Year | Number of Years, $x$ | Final Amount After $x$ Years of Growth |
|---|---|---|---|---|
| ▶ **1.** | 305 | 5% | 8 | |
| **2.** | 402 | 7% | 5 | |
| **3.** | 2000 | 11% | 41 | |
| **4.** | 1000 | 47% | 19 | |
| **5.** | 17 | 29% | 28 | |
| **6.** | 29 | 61% | 12 | |

**Objective B** *Practice using the exponential decay formula by completing the table below. Round final amounts to the nearest whole. See Example 2.*

| | Original Amount | Decay Rate per Year | Number of Years, $x$ | Final Amount After $x$ Years of Decay |
|---|---|---|---|---|
| **7.** | 305 | 5% | 8 | |
| **8.** | 402 | 7% | 5 | |
| **9.** | 10,000 | 12% | 15 | |
| **10.** | 15,000 | 16% | 11 | |
| **11.** | 207,000 | 32% | 25 | |
| **12.** | 325,000 | 29% | 31 | |

**Objectives A B Mixed Practice** *Solve. Unless noted otherwise, round answers to the nearest whole. See Examples 1 and 2.*

**13.** Suppose a city with population 500,000 has been growing at a rate of 3% per year. If this rate continues, find the population of this city in 12 years.

**14.** Suppose a city with population 320,000 has been growing at a rate of 4% per year. If this rate continues, find the population of this city in 20 years.

**15.** The number of employees for a certain company has been decreasing each year by 5%. If the company currently has 640 employees and this rate continues, find the number of employees in 10 years.

**16.** The number of students attending summer school at a local community college has been decreasing each year by 7%. If 984 students currently attend summer school and this rate continues, find the number of students attending summer school in 5 years.

**17.** National Park Service personnel are trying to increase the size of the bison population of Theodore Roosevelt National Park. If 260 bison currently live in the park and if the population's rate of growth is 2.5% annually, find how many bison there should be in 10 years.

**18.** The size of the rat population of a wharf area grows at a rate of 8% monthly. If there are 200 rats in January, find how many rats should be expected by next January.

**19.** A rare isotope of a nuclear material is very unstable, decaying at a rate of 15% each second. Find how much isotope remains 10 seconds after 5 grams of the isotope is created.

**20.** An accidental spill of 75 grams of radioactive material in a local stream has led to the presence of radioactive debris decaying at a rate of 4% each day. Find how much debris still remains after 14 days.

*Practice using the exponential decay formula with half-lives by completing the table below. The first row has been completed for you. See Example 3.*

| | Original Amount | Half-Life (in years) | Number of Years | Time Intervals, $x\left(\dfrac{\text{Years}}{\text{Half-Life}}\right)$ Rounded to Tenths If Needed | Final Amount After $x$ Time Intervals (rounded to tenths) | Is Your Final Amount Reasonable? |
|---|---|---|---|---|---|---|
| | 60 | 8 | 10 | $\dfrac{10}{8} = 1.25$ | 25.2 | yes |
| **21.** a. | 40 | 7 | 14 | | | |
| b. | 40 | 7 | 11 | | | |
| **22.** a. | 200 | 12 | 36 | | | |
| b. | 200 | 12 | 40 | | | |
| **23.** | 21 | 152 | 500 | | | |
| **24.** | 35 | 119 | 500 | | | |

*Solve. Round answers to the nearest tenth.*

**25.** A form of nickel has a half-life of 96 years. How much of a 30-gram sample is left after 250 years?

**26.** A form of uranium has a half-life of 72 years. How much of a 100-gram sample is left after 500 years?

## Review

*By inspection, find the value for x that makes each statement true. See Section 9.3.*

**27.** $2^x = 8$

**28.** $3^x = 9$

**29.** $5^x = \dfrac{1}{5}$

**30.** $4^x = 1$

## Concept Extensions

**31.** An item is on sale for 40% off its original price. If it is then marked down an additional 60%, does this mean the item is free? Discuss why or why not.

**32.** Uranium U-232 has a half-life of 72 years. What eventually happens to a 10 gram sample? Does it ever completely decay and disappear? Discuss why or why not.

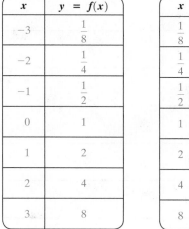

## Objectives

**A** Write Exponential Equations with Logarithmic Notation and Write Logarithmic Equations with Exponential Notation.

**B** Solve Logarithmic Equations by Using Exponential Notation.

**C** Identify and Graph Logarithmic Functions.

### Objective A Using Logarithmic Notation

Since the exponential function $f(x) = 2^x$ is a one-to-one function, it has an inverse. We can create a table of values for $f^{-1}$ by switching the coordinates in the accompanying table of values for $f(x) = 2^x$.

| $x$ | $y = f(x)$ |
|---|---|
| $-3$ | $\frac{1}{8}$ |
| $-2$ | $\frac{1}{4}$ |
| $-1$ | $\frac{1}{2}$ |
| $0$ | $1$ |
| $1$ | $2$ |
| $2$ | $4$ |
| $3$ | $8$ |

| $x$ | $y = f^{-1}(x)$ |
|---|---|
| $\frac{1}{8}$ | $-3$ |
| $\frac{1}{4}$ | $-2$ |
| $\frac{1}{2}$ | $-1$ |
| $1$ | $0$ |
| $2$ | $1$ |
| $4$ | $2$ |
| $8$ | $3$ |

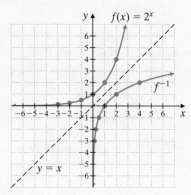

The graphs of $f$ and its inverse are shown in the margin. Notice that the graphs of $f$ and $f^{-1}$ are symmetric about the line $y = x$, as expected.

Now we would like to be able to write an equation for $f^{-1}$. To do so, we follow the steps for finding the equation of an inverse.

$$f(x) = 2^x$$

**Step 1:** Replace $f(x)$ by $y$.  $\quad y = 2^x$

**Step 2:** Interchange $x$ and $y$.  $\quad x = 2^y$

**Step 3:** Solve for $y$.

At this point, we are stuck. To solve this equation for $y$, a new notation, **logarithmic notation,** is needed.

The symbol $\log_b x$ means "the power to which $b$ is raised to produce a result of $x$." In other words,

$$\log_b x = y \quad \text{means} \quad b^y = x$$

We say that $\log_b x$ is "the logarithm of $x$ to the base $b$" or "the log of $x$ to the base $b$."

---

**Logarithmic Definition**

If $b > 0$, and $b \neq 1$, then

$$y = \log_b x \quad \text{means} \quad x = b^y$$

for every $x > 0$ and every real number $y$.

---

Before returning to the function $x = 2^y$ and solving it for $y$ in terms of $x$, let's practice using the new notation $\log_b x$.

It is important to be able to write exponential equations with logarithmic notation, and vice versa. The following table shows examples of both forms.

| Logarithmic Equation | Corresponding Exponential Equation |
|---|---|
| $\log_3 9 = 2$ | $3^2 = 9$ |
| $\log_6 1 = 0$ | $6^0 = 1$ |
| $\log_2 8 = 3$ | $2^3 = 8$ |
| $\log_4 \dfrac{1}{16} = -2$ | $4^{-2} = \dfrac{1}{16}$ |
| $\log_8 2 = \dfrac{1}{3}$ | $8^{1/3} = 2$ |

**Helpful Hint**    Notice that a *logarithm* is an *exponent*. In other words, $\log_3 9$ is the *power* that we raise 3 to in order to get 9.

**Examples**   Write as an exponential equation.

**1.** $\log_5 25 = 2$ means $5^2 = 25$.

**2.** $\log_6 \dfrac{1}{6} = -1$ means $6^{-1} = \dfrac{1}{6}$.

**3.** $\log_2 \sqrt{2} = \dfrac{1}{2}$ means $2^{1/2} = \sqrt{2}$.

■ Work Practice 1–3

**Examples**   Write as a logarithmic equation.

**4.** $9^3 = 729$ means $\log_9 729 = 3$.

**5.** $6^{-2} = \dfrac{1}{36}$ means $\log_6 \dfrac{1}{36} = -2$.

**6.** $5^{1/3} = \sqrt[3]{5}$ means $\log_5 \sqrt[3]{5} = \dfrac{1}{3}$.

■ Work Practice 4–6

**Example 7**   Find the value of each logarithmic expression.

**a.** $\log_4 16$    **b.** $\log_{10} \dfrac{1}{10}$    **c.** $\log_9 3$

**Solution:**

**a.** $\log_4 16 = 2$ because $4^2 = 16$.

**b.** $\log_{10} \dfrac{1}{10} = -1$ because $10^{-1} = \dfrac{1}{10}$.

**c.** $\log_9 3 = \dfrac{1}{2}$ because $9^{1/2} = \sqrt{9} = 3$.

■ Work Practice 7

## Objective B  Solving Logarithmic Equations ▶

The ability to interchange the logarithmic and exponential forms of a statement is often the key to solving logarithmic equations.

**Example 8**   Solve:  $\log_5 x = 3$

**Solution:**  $\log_5 x = 3$
$5^3 = x$   Write as an exponential equation.
$125 = x$

The solution set is $\{125\}$.

■ Work Practice 8

---

**Practice 1–3**

Write as an exponential equation.

**1.** $\log_7 49 = 2$

**2.** $\log_8 \dfrac{1}{8} = -1$

**3.** $\log_3 \sqrt{3} = \dfrac{1}{2}$

**Practice 4–6**

Write as a logarithmic equation.

**4.** $3^4 = 81$    **5.** $2^{-3} = \dfrac{1}{8}$

**6.** $7^{1/3} = \sqrt[3]{7}$

**Practice 7**

Find the value of each logarithmic expression.

**a.** $\log_5 125$   **b.** $\log_7 \dfrac{1}{49}$

**c.** $\log_{100} 10$

**Practice 8**

Solve:  $\log_2 x = 4$

**Answers**

**1.** $7^2 = 49$  **2.** $8^{-1} = \dfrac{1}{8}$  **3.** $3^{1/2} = \sqrt{3}$

**4.** $\log_3 81 = 4$  **5.** $\log_2 \dfrac{1}{8} = -3$

**6.** $\log_7 \sqrt[3]{7} = \dfrac{1}{3}$  **7. a.** 3  **b.** $-2$  **c.** $\dfrac{1}{2}$

**8.** $\{16\}$

## Practice 9

Solve: $\log_x 9 = 2$

**Example 9**    Solve: $\log_x 25 = 2$

**Solution:**
$$\log_x 25 = 2$$
$$x^2 = 25 \quad \text{Write as an exponential equation.}$$
$$x = 5$$

Even though $(-5)^2 = 25$, the base $b$ of a logarithm must be positive. The solution set is $\{5\}$.

■ Work Practice 9

## Practice 10

Solve: $\log_2 1 = x$

**Example 10**    Solve: $\log_3 1 = x$

**Solution:**
$$\log_3 1 = x$$
$$3^x = 1 \quad \text{Write as an exponential equation.}$$
$$3^x = 3^0 \quad \text{Write 1 as } 3^0.$$
$$x = 0 \quad \text{Use the uniqueness of } b^x.$$

The solution set is $\{0\}$.

■ Work Practice 10

In Example 10, we illustrated an important property of logarithms. That is, $\log_b 1$ is always 0. This property as well as two important others are given below.

---

### Properties of Logarithms

If $b$ is a real number, $b > 0$ and $b \neq 1$, then

1. $\log_b 1 = 0$
2. $\log_b b^x = x$
3. $b^{\log_b x} = x$

---

To see that $\log_b b^x = x$, we change the logarithmic form to exponential form. Then, $\log_b b^x = x$ means $b^x = b^x$. In exponential form, the statement is true, so in logarithmic form, the statement is also true.

## Practice 11

Simplify.
**a.** $\log_6 6^3$    **b.** $\log_{11} 11^{-4}$
**c.** $7^{\log_7 13}$    **d.** $3^{\log_3 10}$

**Example 11**    Simplify.

**a.** $\log_3 3^2$    **b.** $\log_7 7^{-1}$
**c.** $5^{\log_5 3}$    **d.** $2^{\log_2 6}$

**Solution:**

**a.** From property 2, $\log_3 3^2 = 2$.
**b.** From property 2, $\log_7 7^{-1} = -1$.
**c.** From property 3, $5^{\log_5 3} = 3$.
**d.** From property 3, $2^{\log_2 6} = 6$.

■ Work Practice 11

## Objective C  Graphing Logarithmic Functions

Let us now return to the function $f(x) = 2^x$ and write an equation for its inverse, $f^{-1}$. Recall our earlier work.

$$f(x) = 2^x$$

**Step 1:** Replace $f(x)$ by $y$.    $y = 2^x$
**Step 2:** Interchange $x$ and $y$.    $x = 2^y$

**Answers**
9. $\{3\}$  10. $\{0\}$  11. a. 3  b. $-4$
c. 13  d. 10

Having gained proficiency with the notation $\log_b x$, we can now complete the steps for writing the inverse equation.

**Step 3:** Solve for $y$. $\qquad\qquad\qquad y = \log_2 x$

**Step 4:** Replace $y$ with $f^{-1}(x)$. $\quad f^{-1}(x) = \log_2 x$

Thus, $f^{-1}(x) = \log_2 x$ defines a function that is the inverse function of the function $f(x) = 2^x$. The function $f^{-1}(x)$ or $y = \log_2 x$ is called a *logarithmic function*.

> ### Logarithmic Function
>
> If $x$ is a positive real number, $b$ is a constant positive real number, and $b$ is not 1, then a **logarithmic function** is a function that can be defined by
>
> $$f(x) = \log_b x$$
>
> The domain of $f$ is the set of positive real numbers, and the range of $f$ is the set of real numbers.

✓**Concept Check** Let $f(x) = \log_3 x$ and $g(x) = 3^x$. These two functions are inverses of each other. Since $(2, 9)$ is an ordered pair solution of $g(x)$, what ordered pair do we know to be a solution of $f(x)$? Explain why.

We can explore logarithmic functions by graphing them.

**Example 12** Graph the logarithmic function $y = \log_2 x$.

**Solution:** First we write the equation with exponential notation as $2^y = x$. Then we find some ordered pair solutions that satisfy this equation. Finally, we plot the points and connect them with a smooth curve. The domain of this function is $(0, \infty)$, and the range is all real numbers.

Since $x = 2^y$ is solved for $x$, we choose $y$-values and compute corresponding $x$-values.

If $y = 0, x = 2^0 = 1.$
If $y = 1, x = 2^1 = 2.$
If $y = 2, x = 2^2 = 4$
If $y = -1, x = 2^{-1} = \dfrac{1}{2}.$

| $x = 2^y$ | $y$ |
|-----------|-----|
| 1 | 0 |
| 2 | 1 |
| 4 | 2 |
| $\dfrac{1}{2}$ | $-1$ |

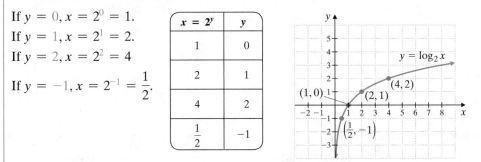

■ **Work Practice 12**

**Example 13** Graph the logarithmic function $f(x) = \log_{1/3} x$.

**Solution:** We can replace $f(x)$ with $y$, and write the result with exponential notation.

$$f(x) = \log_{1/3} x$$
$$y = \log_{1/3} x \qquad \text{Replace } f(x) \text{ with } y.$$
$$\left(\frac{1}{3}\right)^y = x \qquad \text{Write in exponential form.}$$

(*Continued on next page*)

Now we can find ordered pair solutions that satisfy $\left(\dfrac{1}{3}\right)^y = x$, plot these points, and connect them with a smooth curve.

If $y = 0$, $x = \left(\dfrac{1}{3}\right)^0 = 1.$

If $y = 1$, $x = \left(\dfrac{1}{3}\right)^1 = \dfrac{1}{3}.$

If $y = -1$, $x = \left(\dfrac{1}{3}\right)^{-1} = 3.$

If $y = -2$, $x = \left(\dfrac{1}{3}\right)^{-2} = 9.$

| $x = \left(\dfrac{1}{3}\right)^y$ | $y$ |
|---|---|
| 1 | 0 |
| $\dfrac{1}{3}$ | 1 |
| 3 | -1 |
| 9 | -2 |

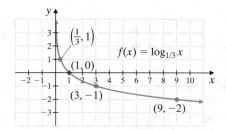

The domain of this function is $(0, \infty)$, and the range is the set of all real numbers.

■ Work Practice 13

The following figures summarize characteristics of logarithmic functions.

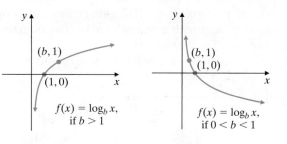

# Vocabulary, Readiness & Video Check

*Use the choices to fill in each blank.*

1. A function such as $y = \log_2 x$ is a(n) _____ function.
   **a.** linear    **b.** logarithmic    **c.** quadratic    **d.** exponential

2. If $y = \log_2 x$, then _____.
   **a.** $x = y$    **b.** $2^x = y$    **c.** $2^y = x$    **d.** $2y = x$

*Answer the questions about the graph of $y = \log_2 x$, shown to the left.*

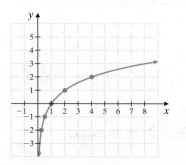

3. Is this a one-to-one function? _____

4. Is there an $x$-intercept? _____ If so, name the coordinates. _____

5. Is there a $y$-intercept? _____ If so, name the coordinates. _____

6. The domain of this function, in interval notation, is _____.

7. The range of this function, in interval notation, is _____.

*Martin-Gay Interactive Videos*    Watch the section lecture video and answer the following questions.

See Video 9.5

**Objective A**  **8.** Notice from the definition of a logarithm and from ⊞ Examples 1–4 that a logarithmic statement equals the power in the exponent statement, such as $b^y = x$. What conclusion can you make about logarithms and exponents? ◉

**Objective B**  **9.** From ⊞ Examples 8 and 9, how do you solve these types of logarithmic equations? ◉

**Objective C**  **10.** In ⊞ Example 12, why is it easier to choose values for $y$ when finding ordered pairs for the graph? ◉

## 9.5  Exercise Set MyMathLab® ◉

**Objective A**  *Write each as an exponential equation. See Examples 1 through 3.*

**◉ 1.** $\log_6 36 = 2$

**2.** $\log_2 32 = 5$

**3.** $\log_3 \dfrac{1}{27} = -3$

**4.** $\log_5 \dfrac{1}{25} = -2$

**5.** $\log_{10} 1000 = 3$

**6.** $\log_{10} 10 = 1$

**7.** $\log_e x = 4$

**8.** $\log_e y = 7$

**9.** $\log_e \dfrac{1}{e^2} = -2$

**10.** $\log_e \dfrac{1}{e} = -1$

**◉ 11.** $\log_7 \sqrt{7} = \dfrac{1}{2}$

**12.** $\log_{11} \sqrt[4]{11} = \dfrac{1}{4}$

**13.** $\log_{0.7} 0.343 = 3$

**14.** $\log_{1.2} 1.44 = 2$

**15.** $\log_3 \dfrac{1}{81} = -4$

**16.** $\log_{1/4} 16 = -2$

*Write each as a logarithmic equation. See Examples 4 through 6.*

**◉ 17.** $2^4 = 16$

**18.** $5^3 = 125$

**19.** $10^2 = 100$

**20.** $10^4 = 10,000$

**21.** $e^3 = x$

**22.** $e^5 = y$

**◉ 23.** $10^{-1} = \dfrac{1}{10}$

**24.** $10^{-2} = \dfrac{1}{100}$

**25.** $4^{-2} = \dfrac{1}{16}$

**26.** $3^{-4} = \dfrac{1}{81}$

**27.** $5^{1/2} = \sqrt{5}$

**28.** $4^{1/3} = \sqrt[3]{4}$

*Find the value of each logarithmic expression. See Example 7.*

**◉ 29.** $\log_2 8$

**30.** $\log_3 9$

**31.** $\log_2 \dfrac{1}{4}$

**32.** $\log_2 \dfrac{1}{32}$

**33.** $\log_{25} 5$

**34.** $\log_8 \dfrac{1}{2}$

**◉ 35.** $\log_{1/2} 2$

**36.** $\log_{2/3} \dfrac{4}{9}$

**◉ 37.** $\log_6 1$

**38.** $\log_9 9$

**39.** $\log_{10} 100$

**40.** $\log_{10} \dfrac{1}{10}$

**41.** $\log_3 81$     **42.** $\log_2 16$     **43.** $\log_4 \dfrac{1}{64}$     **44.** $\log_3 \dfrac{1}{9}$

## Objective B   *Solve. See Examples 8 through 10.*

**45.** $\log_3 9 = x$     **46.** $\log_2 8 = x$     **47.** $\log_3 x = 4$     **48.** $\log_2 x = 3$

**49.** $\log_x 49 = 2$     **50.** $\log_x 8 = 3$     **51.** $\log_2 \dfrac{1}{8} = x$     **52.** $\log_3 \dfrac{1}{81} = x$

**53.** $\log_3 \dfrac{1}{27} = x$     **54.** $\log_5 \dfrac{1}{125} = x$     ▶ **55.** $\log_8 x = \dfrac{1}{3}$     **56.** $\log_9 x = \dfrac{1}{2}$

**57.** $\log_4 16 = x$     **58.** $\log_2 16 = x$     **59.** $\log_{3/4} x = 3$     **60.** $\log_{2/3} x = 2$

▶ **61.** $\log_x 100 = 2$     **62.** $\log_x 27 = 3$     **63.** $\log_2 2^4 = x$     **64.** $\log_6 6^{-2} = x$

**65.** $3^{\log_3 5} = x$     **66.** $5^{\log_5 7} = x$     **67.** $\log_x \dfrac{1}{7} = \dfrac{1}{2}$     **68.** $\log_x 2 = -\dfrac{1}{3}$

*Simplify. See Example 11.*

**69.** $\log_5 5^3$     **70.** $\log_6 6^2$     ▶ **71.** $2^{\log_2 3}$     **72.** $7^{\log_7 4}$

▶ **73.** $\log_9 9$     **74.** $\log_2 2$     **75.** $\log_8 (8)^{-1}$     **76.** $\log_{11} (11)^{-1}$

## Objective C   *Graph each logarithmic function. See Examples 12 and 13.*

▶ **77.** $y = \log_3 x$     **78.** $y = \log_2 x$     **79.** $f(x) = \log_{1/4} x$     **80.** $f(x) = \log_{1/2} x$

**81.** $f(x) = \log_5 x$     **82.** $f(x) = \log_6 x$     **83.** $f(x) = \log_{1/6} x$     **84.** $f(x) = \log_{1/5} x$

## Review

*Simplify each rational expression. See Section 6.1.*

**85.** $\dfrac{x+3}{3+x}$

**86.** $\dfrac{x-5}{5-x}$

**87.** $\dfrac{x^2-8x+16}{2x-8}$

**88.** $\dfrac{x^2-3x-10}{2+x}$

## Concept Extensions

*Solve. See the Concept Check in this section.*

**89.** Let $f(x)=\log_5 x$. Then $g(x)=5^x$ is the inverse of $f(x)$. The ordered pair $(2,25)$ is a solution of the function $g(x)$.

    **a.** Write this solution using function notation.

    **b.** Write an ordered pair that we know to be a solution of $f(x)$.

    **c.** Use the answer to part (b) and write the solution using function notation.

**90.** Let $f(x)=\log_{0.3} x$. Then $g(x)=0.3^x$ is the inverse of $f(x)$. The ordered pair $(3,0.027)$ is a solution of the function $g(x)$.

    **a.** Write this solution using function notation.

    **b.** Write an ordered pair that we know to be a solution of $f(x)$.

    **c.** Use the answer to part (b) and write the solution using function notation.

**91.** Explain why negative numbers are not included as logarithmic bases.

**92.** Explain why 1 is not included as a logarithmic base.

*For Exercises 93–96, graph each function and its inverse on the same set of axes.*

**93.** $y=4^x;\ y=\log_4 x$

**94.** $y=3^x;\ y=\log_3 x$

**95.** $y=\left(\dfrac{1}{3}\right)^x;\ y=\log_{1/3} x$

**96.** $y=\left(\dfrac{1}{2}\right)^x;\ y=\log_{1/2} x$

**97.** Explain why the graph of the function $y=\log_b x$ contains the point $(1,0)$ no matter what $b$ is.

**98.** $\log_3 10$ is between which two integers? Explain your answer.

*Solve. Round Exercises 99 and 100 to four decimal places.*

**99.** The formula $\log_{10}(1-k)=\dfrac{-0.3}{H}$ models the relationship between the half-life $H$ of a radioactive material and its rate of decay $k$. Find the rate of decay of the iodine isotope I-131 if its half-life is 8 days.

**100.** The formula $\text{pH}=-\log_{10}(\text{H}^+)$ gives the pH for a liquid, where $\text{H}^+$ stands for the concentration of hydronium ions in moles per liter. Find the concentration of hydronium ions of lemonade, whose pH is 2.3.

**Properties of Logarithms** ⓘ

## Objectives

**A** Use the Product Property of Logarithms. ⓘ

**B** Use the Quotient Property of Logarithms. ⓘ

**C** Use the Power Property of Logarithms. ⓘ

**D** Use the Properties of Logarithms Together. ⓘ

In the previous section, we explored some basic properties of logarithms. We now introduce and study additional properties. Because a logarithm is an exponent, logarithmic properties are just restatements of exponential properties.

## Objective A  Using the Product Property ⓘ

The first of these properties is called the **product property of logarithms** because it deals with the logarithm of a product.

> ### Product Property of Logarithms
>
> If $x$, $y$, and $b$ are positive real numbers and $b \neq 1$, then
>
> $$\log_b xy = \log_b x + \log_b y$$

To prove this, we let $\log_b x = M$ and $\log_b y = N$. Now we write each logarithm with exponential notation.

$$\log_b x = M \quad \text{is equivalent to} \quad b^M = x$$
$$\log_b y = N \quad \text{is equivalent to} \quad b^N = y$$

When we multiply the left sides and the right sides of the exponential equations, we have that

$$xy = (b^M)(b^N) = b^{M+N}$$

If we write the equation $xy = b^{M+N}$ in equivalent logarithmic form, we have

$$\log_b xy = M + N$$

But since $M = \log_b x$ and $N = \log_b y$, we can write

$$\log_b xy = \log_b x + \log_b y \quad \text{Let } M = \log_b x \text{ and } N = \log_b y.$$

In other words, the logarithm of a product is the sum of the logarithms of the factors. This property is sometimes used to simplify logarithmic expressions.

### Practice 1

Write as a single logarithm:
$\log_2 7 + \log_2 5$

**Example 1**  Write as a single logarithm:   $\log_{11} 10 + \log_{11} 3$

**Solution:**  $\log_{11} 10 + \log_{11} 3 = \log_{11}(10 \cdot 3)$   Use the product property.
$$= \log_{11} 30$$

■ Work Practice 1

### Practice 2

Write as a single logarithm:
$\log_3 x + \log_3(x - 9)$

**Example 2**  Write as a single logarithm:   $\log_2(x + 2) + \log_2 x$

**Solution:**  $\log_2(x + 2) + \log_2 x = \log_2[(x + 2) \cdot x] = \log_2(x^2 + 2x)$

■ Work Practice 2

## Objective B  Using the Quotient Property ⓘ

The second property is the **quotient property of logarithms**.

### Answers

**1.** $\log_2 35$   **2.** $\log_3(x^2 - 9x)$

## Quotient Property of Logarithms

If $x$, $y$, and $b$ are positive real numbers and $b \neq 1$, then

$$\log_b \frac{x}{y} = \log_b x - \log_b y$$

The proof of the quotient property of logarithms is similar to the proof of the product property. Notice that the quotient property says that the logarithm of a quotient is the difference of the logarithms of the dividend and divisor.

✓**Concept Check** Which of the following is the correct way to rewrite $\log_5 \frac{7}{2}$?

**a.** $\log_5 7 - \log_5 2$    **b.** $\log_5(7 - 2)$    **c.** $\dfrac{\log_5 7}{\log_5 2}$    **d.** $\log_5 14$

**Example 3**    Write as a single logarithm:   $\log_{10} 27 - \log_{10} 3$

**Solution:**   $\log_{10} 27 - \log_{10} 3 = \log_{10} \dfrac{27}{3}$   Use the quotient property.

$$= \log_{10} 9$$

■ Work Practice 3

**Example 4**    Write as a single logarithm:   $\log_3(x^2 + 5) - \log_3(x^2 + 1)$

**Solution:**   $\log_3(x^2 + 5) - \log_3(x^2 + 1) = \log_3 \dfrac{x^2 + 5}{x^2 + 1}$   Use the quotient property.

■ Work Practice 4

## Objective C  Using the Power Property ▶

The third and final property we introduce is the **power property of logarithms.**

## Power Property of Logarithms

If $x$ and $b$ are positive real numbers, $b \neq 1$, and $r$ is a real number, then

$$\log_b x^r = r \log_b x$$

**Examples**    Use the power property to rewrite each expression.

**5.** $\log_5 x^3 = 3 \log_5 x$

**6.** $\log_4 \sqrt{2} = \log_4 2^{1/2} = \dfrac{1}{2} \log_4 2$

■ Work Practice 5–6

## Objective D  Using More Than One Property ▶

Many times we must use more than one property of logarithms to simplify logarithmic expressions.

---

**Practice 3**

Write as a single logarithm:
$\log_7 40 - \log_7 8$

**Practice 4**

Write as a single logarithm:
$\log_3(x^3 + 4) - \log_3(x^2 + 2)$

**Practice 5–6**

Use the power property to rewrite each expression.
**5.** $\log_3 x^5$    **6.** $\log_7 \sqrt[3]{4}$

**Answers**

**3.** $\log_7 5$   **4.** $\log_3 \dfrac{x^3 + 4}{x^2 + 2}$   **5.** $5 \log_3 x$

**6.** $\dfrac{1}{3} \log_7 4$

✓**Concept Check Answer**

a

## Practice 7–8

Write as a single logarithm.

**7.** $3\log_4 2 + 2\log_4 5$

**8.** $5\log_2(2x - 1) - \log_2 x$

---

**Examples**   Write as a single logarithm.

**7.** $2\log_5 3 + 3\log_5 2 = \log_5 3^2 + \log_5 2^3$    Use the power property.

$\qquad\qquad\qquad = \log_5 9 + \log_5 8$

$\qquad\qquad\qquad = \log_5(9 \cdot 8)$    Use the product property.

$\qquad\qquad\qquad = \log_5 72$

**8.** $3\log_9 x - \log_9(x + 1) = \log_9 x^3 - \log_9(x + 1)$    Use the power property.

$\qquad\qquad\qquad\qquad\quad = \log_9 \dfrac{x^3}{x + 1}$    Use the quotient property.

 **Work Practice 7–8**

---

## Practice 9–10

Write each expression as sums or differences of logarithms.

**9.** $\log_7 \dfrac{6 \cdot 2}{5}$

**10.** $\log_3 \dfrac{x^4}{y^3}$

---

**Examples**   Write each expression as sums or differences of logarithms.

**9.** $\log_3 \dfrac{5 \cdot 7}{4} = \log_3(5 \cdot 7) - \log_3 4$    Use the quotient property.

$\qquad\qquad = \log_3 5 + \log_3 7 - \log_3 4$    Use the product property.

**10.** $\log_2 \dfrac{x^5}{y^2} = \log_2(x^5) - \log_2(y^2)$    Use the quotient property.

$\qquad\qquad = 5\log_2 x - 2\log_2 y$    Use the power property.

**Work Practice 9–10**

---

**Helpful Hint**

Notice that we are not able to simplify further a logarithmic expression such as $\log_5(2x - 1)$. None of the basic properties gives a way to write the logarithm of a difference (or sum) in some equivalent form.

---

✓**Concept Check**   What is wrong with the following?

$\log_{10}(x^2 + 5) = \log_{10} x^2 + \log_{10} 5$

$\qquad\qquad\quad = 2\log_{10} x + \log_{10} 5$

Use a numerical example to demonstrate that the result is incorrect.

---

## Practice 11–13

If $\log_b 4 = 0.86$ and $\log_b 7 = 1.21$, use the properties of logarithms to evaluate each expression.

**11.** $\log_b 28$    **12.** $\log_b 49$

**13.** $\log_b \sqrt[3]{4}$

---

**Examples**   If $\log_b 2 = 0.43$ and $\log_b 3 = 0.68$, use the properties of logarithms to evaluate each expression.

**11.** $\log_b 6 = \log_b(2 \cdot 3)$    Write 6 as $2 \cdot 3$.

$\qquad\quad = \log_b 2 + \log_b 3$    Use the product property.

$\qquad\quad = 0.43 + 0.68$    Substitute given values.

$\qquad\quad = 1.11$    Simplify.

**12.** $\log_b 9 = \log_b 3^2$    Write 9 as $3^2$.

$\qquad\quad = 2\log_b 3$    Use the power property.

$\qquad\quad = 2(0.68)$    Substitute the given value.

$\qquad\quad = 1.36$    Simplify.

**13.** $\log_b \sqrt{2} = \log_b 2^{1/2}$    Write $\sqrt{2}$ as $2^{1/2}$.

$\qquad\quad = \dfrac{1}{2}\log_b 2$    Use the power property.

$\qquad\quad = \dfrac{1}{2}(0.43)$    Substitute the given value.

$\qquad\quad = 0.215$    Simplify.

**Work Practice 11–13**

---

### Answers

**7.** $\log_4 200$    **8.** $\log_2 \dfrac{(2x - 1)^5}{x}$

**9.** $\log_7 6 + \log_7 2 - \log_7 5$

**10.** $4\log_3 x - 3\log_3 y$    **11.** 2.07

**12.** 2.42    **13.** $0.28\overline{6}$

✓**Concept Check Answer**

The properties do not give any way to simplify the logarithm of a sum; answers may vary.

# Vocabulary, Readiness & Video Check

*Select the correct choice.*

**1.** $\log_b 12 + \log_b 3 = \log_b$ _____

   **a.** 36      **b.** 15      **c.** 4      **d.** 9

**2.** $\log_b 12 - \log_b 3 = \log_b$ _____

   **a.** 36      **b.** 15      **c.** 4      **d.** 9

**3.** $7 \log_b 2 =$ _____

   **a.** $\log_b 14$    **b.** $\log_b 2^7$    **c.** $\log_b 7^2$    **d.** $(\log_b 2)^7$

**4.** $\log_b 1 =$ _____

   **a.** $b$      **b.** 1      **c.** 0      **d.** no answer

**5.** $b^{\log_b x} =$ _____

   **a.** $x$      **b.** $b$      **c.** 1      **d.** 0

**6.** $\log_5 5^2 =$ _____

   **a.** 25      **b.** 2      **c.** $5^{5^2}$      **d.** 32

*Martin-Gay Interactive Videos*    Watch the section lecture video and answer the following questions.

*See Video 9.6*

**Objective A**   **7.** Can the product property of logarithms be used again on the bottom line of ⊞ Example 2 to write $\log_{10}(10x^2 + 20)$ as a sum of logarithms, $\log_{10} 10x^2 + \log_{10} 20$? Explain. ▶

**Objective B**   **8.** From ⊞ Example 3 and the lecture before, what must be true about bases before you can apply the quotient property of logarithms? ▶

**Objective C**   **9.** Based on ⊞ Example 5, explain why $\log_2 \dfrac{1}{x} = -\log_2 x$. ▶

**Objective D**   **10.** From the lecture before ⊞ Example 6, where do the logarithmic properties come from? ▶

## 9.6   Exercise Set   MyMathLab® ▶

**Objective A**   *Write each sum as a single logarithm. Assume that variables represent positive numbers. See Examples 1 and 2.*

▶ **1.** $\log_5 2 + \log_5 7$          **2.** $\log_3 8 + \log_3 4$          **3.** $\log_4 9 + \log_4 x$

**4.** $\log_2 x + \log_2 y$          **5.** $\log_6 x + \log_6 (x + 1)$          **6.** $\log_5 y^3 + \log_5 (y - 7)$

▶ **7.** $\log_{10} 5 + \log_{10} 2 + \log_{10}(x^2 + 2)$      **8.** $\log_6 3 + \log_6(x + 4) + \log_6 5$

**Objective B** *Write each difference as a single logarithm. Assume that variables represent positive numbers. See Examples 3 and 4.*

**9.** $\log_5 12 - \log_5 4$ **10.** $\log_7 20 - \log_7 4$ **▶11.** $\log_3 8 - \log_3 2$

**12.** $\log_5 12 - \log_5 3$ **13.** $\log_2 x - \log_2 y$ **14.** $\log_3 12 - \log_3 z$

**15.** $\log_2 (x^2 + 6) - \log_2 (x^2 + 1)$ **16.** $\log_7 (x + 9) - \log_7 (x^2 + 10)$

**Objective C** *Use the power property to rewrite each expression. See Example 5 and 6.*

**▶17.** $\log_3 x^2$ **18.** $\log_2 x^5$ **19.** $\log_4 5^{-1}$

**20.** $\log_6 7^{-2}$ **▶21.** $\log_5 \sqrt{y}$ **22.** $\log_5 \sqrt[3]{x}$

**Objective D** *Write each as a single logarithm. Assume that variables represent positive numbers. See Examples 7 and 8.*

**23.** $\log_2 5 + \log_2 x^3$ **24.** $\log_5 2 + \log_5 y^2$ **25.** $3 \log_4 2 + \log_4 6$ **26.** $2 \log_3 5 + \log_3 2$

**▶27.** $3 \log_5 x + 6 \log_5 z$ **28.** $2 \log_7 y + 6 \log_7 z$ **▶29.** $\log_4 2 + \log_4 10 - \log_4 5$

**30.** $\log_6 21 + \log_6 2 - \log_6 7$ **31.** $\log_7 6 + \log_7 3 - \log_7 4$ **32.** $\log_8 5 + \log_8 15 - \log_8 20$

**33.** $\log_{10} x - \log_{10} (x + 1) + \log_{10} (x^2 - 2)$ **34.** $\log_9 (4x) - \log_9 (x - 3) + \log_9 (x^3 + 1)$

**35.** $3 \log_2 x + \frac{1}{2} \log_2 x - 2 \log_2 (x + 1)$ **36.** $2 \log_5 x + \frac{1}{3} \log_5 x - 3 \log_5 (x + 5)$

**37.** $2 \log_8 x - \frac{2}{3} \log_8 x + 4 \log_8 x$ **38.** $5 \log_6 x - \frac{3}{4} \log_6 x + 3 \log_6 x$

*Write each expression as a sum or difference of logarithms. Assume that variables represent positive numbers. See Examples 9 and 10.*

**39.** $\log_3 \dfrac{4y}{5}$        **40.** $\log_7 \dfrac{5x}{4}$        **41.** $\log_4 \dfrac{2}{9z}$

**42.** $\log_9 \dfrac{7}{8y}$        ▶ **43.** $\log_2 \dfrac{x^3}{y}$        **44.** $\log_5 \dfrac{x}{y^4}$

**45.** $\log_b \sqrt{7x}$        **46.** $\log_b \sqrt{\dfrac{3}{y}}$        **47.** $\log_6 x^4 y^5$

**48.** $\log_2 y^3 z$        **49.** $\log_5 x^3(x+1)$        **50.** $\log_3 x^2(x-9)$

**51.** $\log_6 \dfrac{x^2}{x+3}$                **52.** $\log_3 \dfrac{(x+5)^2}{x}$

*If $\log_b 3 = 0.5$ and $\log_b 5 = 0.7$, evaluate each expression. See Examples 11 through 13.*

**53.** $\log_b \dfrac{5}{3}$        **54.** $\log_b 25$        **55.** $\log_b 15$

**56.** $\log_b \dfrac{3}{5}$        **57.** $\log_b \sqrt{5}$        **58.** $\log_b \sqrt[4]{3}$

*If $\log_b 2 = 0.43$ and $\log_b 3 = 0.68$, evaluate each expression. See Examples 11 through 13.*

**59.** $\log_b 8$    **60.** $\log_b 81$    **61.** $\log_b \dfrac{3}{9}$    **62.** $\log_b \dfrac{4}{32}$    **63.** $\log_b \sqrt{\dfrac{2}{3}}$    **64.** $\log_b \sqrt{\dfrac{3}{2}}$

## Review

*Graph each function on the same set of axes. See Sections 9.3 and 9.5.*

**65.** $y = 10^x$               **66.** $y = \log_{10} x$

*Evaluate each expression. See Section 9.5.*

**67.** $\log_{10} 100$

**68.** $\log_{10} \dfrac{1}{10}$

**69.** $\log_7 7^2$

**70.** $\log_7 \sqrt{7}$

## Concept Extensions

*Solve. See the Concept Checks in this section.*

**71.** Which of the following is the correct way to rewrite $\log_3 \dfrac{14}{11}$?

    **a.** $\dfrac{\log_3 14}{\log_3 11}$

    **b.** $\log_3 14 - \log_3 11$

    **c.** $\log_3 (14 - 11)$

    **d.** $\log_3 154$

**72.** Which of the following is the correct way to rewrite $\log_9 \dfrac{21}{3}$?

    **a.** $\log_9 7$

    **b.** $\log_9 (21 - 3)$

    **c.** $\dfrac{\log_9 21}{\log_9 3}$

    **d.** $\log_9 21 - \log_9 3$

*Determine whether each statement is true or false.*

**73.** $\log_2 x^3 = 3 \log_2 x$

**74.** $\log_3 (x + y) = \log_3 x + \log_3 y$

**75.** $\dfrac{\log_7 10}{\log_7 5} = \log_7 2$

**76.** $\log_7 \dfrac{14}{8} = \log_7 14 - \log_7 8$

**77.** $\dfrac{\log_7 x}{\log_7 y} = \log_7 x - \log_7 y$

**78.** $(\log_3 6) \cdot (\log_3 4) = \log_3 24$

**79.** It is true that $\log_b 8 = \log_b (8 \cdot 1) = \log_b 8 + \log_b 1$. Explain how $\log_b 8$ can equal $\log_b 8 + \log_b 1$.

**80.** It is true that $\log_b 7 = \log_b \dfrac{7}{1} = \log_b 7 - \log_b 1$. Explain how $\log_b 7$ can equal $\log_b 7 - \log_b 1$.

# Functions and Properties of Logarithms

**Answers**

If $f(x) = x - 6$ and $g(x) = x^2 + 1$, find each value.

**1.** $(f + g)(x)$      **2.** $(f - g)(x)$      **3.** $(f \cdot g)(x)$      **4.** $\left(\dfrac{f}{g}\right)(x)$

If $f(x) = \sqrt{x}$ and $g(x) = 3x - 1$, find each value.

**5.** $(f \circ g)(x)$                 **6.** $(g \circ f)(x)$

*Determine whether each is a one-to-one function. If it is, find its inverse.*

**7.** $f = \{(-2, 6), (4, 8), (2, -6), (3, 3)\}$     **8.** $g = \{(4, 2), (-1, 3), (5, 3), (7, 1)\}$

*Determine whether the graph of each function is the graph of a one-to-one function.*

**9.**

**10.**

**11.**

*Each function listed is one-to-one. Find the inverse of each function.*

**12.** $f(x) = 3x$                    **13.** $f(x) = x + 4$

**14.** $f(x) = 5x - 1$            **15.** $f(x) = 3x + 2$

*Graph each function.*

**16.** $y = \left(\dfrac{1}{2}\right)^x$                 **17.** $y = 2^x + 1$

1. _____

2. _____

3. _____

4. _____

5. _____

6. _____

7. _____

8. _____

9. _____

10. _____

11. _____

12. _____

13. _____

14. _____

15. _____

16. _____

17. _____

18. _____

19. _____

20. _____

21. _____

22. _____

23. _____

24. _____

25. _____

26. _____

27. _____

28. _____

29. _____

30. _____

31. _____

32. _____

33. _____

34. _____

35. _____

36. _____

37. _____

38. _____

**18.** $y = \log_3 x$

**19.** $y = \log_{1/3} x$

*Solve.*

**20.** $2^x = 8$

**21.** $9 = 3^{x-5}$

**22.** $4^{x-1} = 8^{x+2}$

**23.** $25^x = 125^{x-1}$

**24.** $\log_4 16 = x$

**25.** $\log_{49} 7 = x$

**26.** $\log_2 x = 5$

**27.** $\log_x 64 = 3$

**28.** $\log_x \dfrac{1}{125} = -3$

**29.** $\log_3 x = -2$

*Write each as a single logarithm.*

**30.** $\log_2 x + \log_2 14$

**31.** $x \log_2 5 + \log_2 8$

**32.** $3 \log_5 x - 5 \log_5 y$

**33.** $9 \log_5 x + 3 \log_5 y$

**34.** $\log_2 x + \log_2(x - 3) - \log_2(x^2 + 4)$

**35.** $\log_3 y - \log_3(y + 2) + \log_3 (y^3 + 11)$

*Write each expression as a sum or difference of logarithms.*

**36.** $\log_7 \dfrac{9x^2}{y}$

**37.** $\log_6 \dfrac{5y}{z^2}$

**38.** An unusually wet spring has caused the size of the Cape Cod mosquito population to increase by 8% each day. If an estimated 200,000 mosquitoes are on Cape Cod on May 12, find how many mosquitoes will inhabit the Cape on May 25. Round to the nearest thousand.

## 9.7 Common Logarithms, Natural Logarithms, and Change of Base

In this section, we look closely at two particular logarithmic bases. These two logarithmic bases are used so frequently that logarithms to their bases are given special names. **Common logarithms** are logarithms to base 10. **Natural logarithms** are logarithms to base $e$, which we define in this section. The work in this section is based on the use of a calculator that has both the "common log" $\boxed{\text{LOG}}$ and the "natural log" $\boxed{\text{LN}}$ keys.

### Objective A Approximating Common Logarithms

Logarithms to base 10—**common logarithms**—are used frequently because our number system is a base 10 decimal system. The notation $\log x$ means the same as $\log_{10} x$.

> **Common Logarithm**
>
> $\log x$    means    $\log_{10} x$

**Example 1**   Use a calculator to approximate $\log 7$ to four decimal places.

**Solution:**   Press the following sequence of keys:

$\boxed{7}\ \boxed{\text{LOG}}$   or   $\boxed{\text{LOG}}\ \boxed{7}\ \boxed{\text{ENTER}}$

To four decimal places,

$\log 7 \approx 0.8451$

■ Work Practice 1

### Objective B Evaluating Common Logarithms of Powers of 10

To evaluate the common log of a power of 10, a calculator is not needed. According to the property of logarithms,

$\log_b b^x = x$

It follows that if $b$ is replaced with 10, we have

$\log 10^x = x$

> **Helpful Hint**
> Remember that the base of this logarithm is understood to be 10.

**Examples**   Find the exact value of each logarithm.

**2.** $\log 10 = \log 10^1 = 1$

**3.** $\log \dfrac{1}{10} = \log 10^{-1} = -1$

**4.** $\log 100{,}000 = \log 10^5 = 5$

**5.** $\log \sqrt[4]{10} = \log 10^{1/4} = \dfrac{1}{4}$

■ Work Practice 2–5

### Objectives

**A** Identify Common Logarithms and Approximate Them with a Calculator.

**B** Evaluate Common Logarithms of Powers of 10.

**C** Identify Natural Logarithms and Approximate Them with a Calculator.

**D** Evaluate Natural Logarithms of Powers of $e$.

**E** Use the Change of Base Formula.

**Practice 1**

Use a calculator to approximate $\log 21$ to four decimal places.

**Practice 2–5**

Find the exact value of each logarithm.

**2.** $\log 1000$

**3.** $\log \dfrac{1}{100}$

**4.** $\log 10{,}000$

**5.** $\log \sqrt[3]{10}$

**Answers**

**1.** 1.3222   **2.** 3   **3.** −2   **4.** 4   **5.** $\dfrac{1}{3}$

**721**

## Practice 6

Solve: $\log x = 2.9$. Give the exact solution and then approximate the solution to four decimal places.

**Helpful Hint**

The understood base is 10.

As we will soon see, equations containing common logs are useful models of many natural phenomena.

**Example 6**  Solve: $\log x = 1.2$. Give the exact solution and then approximate the solution to four decimal places.

**Solution:** Remember that the base of a common log is understood to be 10.

$$\log x = 1.2$$
$$10^{1.2} = x \quad \text{Write with exponential notation.}$$

The exact solution is $10^{1.2}$ or the solution set is $\{10^{1.2}\}$. To four decimal places, $x \approx 15.8489$.

■ Work Practice 6

### Objective C Approximating Natural Logarithms ▶

Natural logarithms are also frequently used, especially to describe natural events; hence the label "natural logarithm." **Natural logarithms** are logarithms to the base $e$, which is a constant approximately equal to 2.7183. The number $e$ is an irrational number, as is $\pi$. The notation $\log_e x$ is usually abbreviated to $\ln x$. (The abbreviation ln is read "el en.")

> **Natural Logarithm**
>
> $\ln x$ means $\log_e x$

## Practice 7

Use a calculator to approximate ln 11 to four decimal places.

**Example 7** Use a calculator to approximate ln 8 to four decimal places.

**Solution:** Press the following sequence of keys:

$\boxed{8}$ $\boxed{\ln}$   or   $\boxed{\ln}$ $\boxed{8}$ $\boxed{\text{ENTER}}$

To four decimal places,
$$\ln 8 \approx 2.0794$$

■ Work Practice 7

### Objective D Evaluating Natural Logarithms of Powers of $e$ ▶

As a result of the property $\log_b b^x = x$, we know that $\log_e e^x = x$, or $\ln e^x = x$.

## Practice 8–9

Find the exact value of each natural logarithm.
**8.** $\ln e^9$
**9.** $\ln \sqrt[3]{e}$

**Examples** Find the exact value of each natural logarithm.

**8.** $\ln e^3 = 3$

**9.** $\ln \sqrt[7]{e} = \ln e^{1/7} = \dfrac{1}{7}$

■ Work Practice 8–9

**Answers**

**6.** $\{10^{2.9}\}$; $\{794.3282\}$   **7.** 2.3979

**8.** 9   **9.** $\dfrac{1}{3}$

**Example 10** Solve: $\ln 3x = 5$. Give the exact solution and then approximate the solution to four decimal places.

**Solution:** Remember that the base of a natural logarithm is understood to be $e$.

> **Helpful Hint**
> The understood base is $e$.

$$\ln 3x = 5$$
$$e^5 = 3x \quad \text{Write with exponential notation.}$$
$$\frac{e^5}{3} = x \quad \text{Solve for } x.$$

The exact solution is $\dfrac{e^5}{3}$ or the solution set is $\left\{\dfrac{e^5}{3}\right\}$. To four decimal places, $x \approx 49.4711$.

■ **Work Practice 10**

**Practice 10**

Solve: $\ln 7x = 10$. Give the exact solution and then approximate the solution to four decimal places.

Recall from Section 9.3 the formula $A = P\left(1 + \dfrac{r}{n}\right)^{nt}$ for compound interest, where $n$ represents the number of compoundings per year. When interest is compounded continuously, we use the formula $A = Pe^{rt}$, where $r$ is the annual interest rate and interest is compounded continuously for $t$ years.

**Example 11** Finding the Amount Owed on a Loan

Find the amount owed at the end of 5 years if $1600 is loaned at a rate of 9% compounded continuously.

**Solution:** We use the formula $A = Pe^{rt}$ and the following values of the variables.

$$P = \$1600 \quad \text{(the amount of the loan)}$$
$$r = 9\% = 0.09 \quad \text{(the rate of interest)}$$
$$t = 5 \quad \text{(the 5-year duration of the loan)}$$
$$A = Pe^{rt}$$
$$= 1600e^{0.09(5)} \quad \text{Substitute known values.}$$
$$= 1600e^{0.45}$$

Now we can use a calculator to approximate the solution.

$$A \approx 2509.30$$

The total amount of money owed is approximately $2509.30.

■ **Work Practice 11**

**Practice 11**

Find the amount owed at the end of 3 years if $1200 is loaned at a rate of 8% compounded continuously.

## Objective E  Using the Change of Base Formula ▶

Calculators are handy tools for approximating natural and common logarithms. Unfortunately, most calculators cannot be used to approximate logarithms to bases other than $e$ or 10—at least not directly. In such cases, we use the **change of base formula.**

> **Change of Base**
>
> If $a, b,$ and $c$ are positive real numbers and neither $b$ nor $c$ is 1, then
>
> $$\log_b a = \frac{\log_c a}{\log_c b}$$

**Answers**

**10.** $\left\{\dfrac{e^{10}}{7}\right\}$; $\{3146.6380\}$

**11.** $1525.50

## Practice 12

Approximate $\log_7 5$ to four decimal places.

**Answer**

**12.** 0.8271

✓ **Concept Check Answer**

$$f(x) = \frac{\log x}{\log 5}$$

**Example 12**  Approximate $\log_5 3$ to four decimal places.

**Solution:**  We use the change of base property to write $\log_5 3$ as a quotient of logarithms to base 10.

$$\log_5 3 = \frac{\log 3}{\log 5} \qquad \text{Use the change of base property.}$$

$$\approx \frac{0.4771213}{0.69897} \qquad \text{Approximate the logarithms by calculator.}$$

$$\approx 0.6826063 \qquad \text{Simplify by calculator.}$$

To four decimal places, $\log_5 3 \approx 0.6826$.

■ **Work Practice 12**

✓ **Concept Check**  If a graphing calculator cannot directly evaluate logarithms to base 5, describe how you could use the graphing calculator to graph the function $f(x) = \log_5 x$.

# Vocabulary, Readiness & Video Check

*Use the choices to fill in each blank.*

**1.** The base of log 7 is _____.
 **a.** $e$    **b.** 7    **c.** 10    **d.** no answer

**2.** The base of ln 7 is _____.
 **a.** $e$    **b.** 7    **c.** 10    **d.** no answer

**3.** $\log_{10} 10^7 =$ _____.
 **a.** $e$    **b.** 7    **c.** 10    **d.** no answer

**4.** $\log_7 1 =$ _____.
 **a.** $e$    **b.** 7    **c.** 10    **d.** 0

**5.** $\log_e e^5 =$ _____.
 **a.** $e$    **b.** 5    **c.** 0    **d.** 1

**6.** Study exercise 5 to the left. Then answer:
 $\ln e^5 =$ _____.
 **a.** $e$    **b.** 5    **c.** 0    **d.** 1

**7.** $\log_2 7 =$ _____ (There may be more than one answer.)

 **a.** $\dfrac{\log 7}{\log 2}$    **b.** $\dfrac{\ln 7}{\ln 2}$    **c.** $\dfrac{\log 2}{\log 7}$    **d.** $\log \dfrac{7}{2}$

*Martin-Gay Interactive Videos*

*See Video 9.7*

Watch the section lecture video and answer the following questions.

**Objective A**  **8.** From ⊟ Example 1 and the lecture before, what is the understood base of a common logarithm? ▶

**Objective B**  **9.** From ⊟ Example 2, why can you find exact values of common logarithms of powers of 10? ▶

**Objective C**  **10.** From ⊟ Example 4 and the lecture before, what is the understood base of a natural logarithm? ▶

**Objective D**  **11.** In ⊟ Examples 5 and 6, consider how the expression is rewritten and the resulting answer. What logarithm property is actually used here? ▶

**Objective E**  **12.** From ⊟ Example 8, what two equivalent fractions will give you the exact value of $\log_6 4$? ▶

## 9.7  Exercise Set MyMathLab® ▶

**Objectives** A C **Mixed Practice** *Use a calculator to approximate each logarithm to four decimal places. See Examples 1 and 7.*

▶ **1.** log 8

**2.** log 6

**3.** log 2.31

**4.** log 4.86

**5.** ln 2

**6.** ln 3

**7.** ln 0.0716

**8.** ln 0.0032

**9.** log 12.6

**10.** log 25.9

▶ **11.** ln 5

**12.** ln 7

**13.** log 41.5

**14.** ln 41.5

**Objectives** B D **Mixed Practice** *Find the exact value of each logarithm. See Examples 2 through 5, 8, and 9.*

▶ **15.** log 100

**16.** log 10,000

**17.** $\log \dfrac{1}{1000}$

**18.** $\log \dfrac{1}{10}$

▶ **19.** $\ln e^2$

**20.** $\ln e^4$

**21.** $\ln \sqrt[4]{e}$

**22.** $\ln \sqrt[5]{e}$

**23.** $\log 10^3$

**24.** $\ln e^5$

**25.** $\ln e^{3.1}$

**26.** $\log 10^7$

**27.** log 0.0001

**28.** log 0.001

▶ **29.** $\ln \sqrt{e}$

**30.** $\log \sqrt{10}$

*Solve each equation. Give the exact solution and a four-decimal-place approximation. See Examples 6 and 10.*

**31.** $\log x = 1.3$

**32.** $\log x = 2.1$

**33.** $\ln x = 1.4$

**34.** $\ln x = 2.1$

**35.** $\log x = 2.3$

**36.** $\log x = 3.1$

**37.** $\ln x = -2.3$

**38.** $\ln x = -3.7$

▶ **39.** $\log 2x = 1.1$

**40.** $\log 3x = 1.3$

**41.** $\ln 4x = 0.18$

**42.** $\ln 3x = 0.76$

**43.** $\ln(3x - 4) = 2.3$

**44.** $\ln(2x + 5) = 3.4$

**45.** $\log(2x + 1) = -0.5$

**46.** $\log(3x - 2) = -0.8$

*Use the formula $A = Pe^{rt}$ to solve. See Example 11.*

▶ **47.** How much money does Dana Jones have after 12 years if she invests $1400 at 8% interest compounded continuously?

**48.** Determine the size of an account in which $3500 earns 6% interest compounded continuously for 1 year.

**49.** How much money does Barbara Mack owe at the end of 4 years if 6% interest is compounded continuously on her $2000 debt?

**50.** Find the amount of money for which a $2500 certificate of deposit is redeemable if it has been earning 10% interest compounded continuously for 3 years.

## Objective E  *Approximate each logarithm to four decimal places. See Example 12.*

**51.** $\log_2 3$    **52.** $\log_3 2$    **53.** $\log_8 6$    **54.** $\log_6 8$    ▶**55.** $\log_4 9$

**56.** $\log_9 4$    **57.** $\log_3 \dfrac{1}{6}$    **58.** $\log_6 \dfrac{2}{3}$    **59.** $\log_{1/2} 5$    **60.** $\log_{1/3} 2$

## Review

*Solve for x. See Sections 2.1, 2.3, and 5.7*

**61.** $6x - 3(2 - 5x) = 6$

**62.** $2x + 3 = 5 - 2(3x - 1)$

**63.** $2x + 3y = 6x$

**64.** $4x - 8y = 10x$

**65.** $x^2 + 7x = -6$

**66.** $x^2 + 4x = 12$

## Concept Extensions

**67.** Use a calculator to try to approximate log 0. Describe what happens and explain why.

**68.** Use a calculator to try to approximate ln 0. Describe what happens and explain why.

*Graph each function by finding ordered pair solutions, plotting the solutions, and then drawing a smooth curve through the plotted points.*

**69.** $f(x) = e^x$    **70.** $f(x) = e^{2x}$    **71.** $f(x) = \ln x$    **72.** $f(x) = \log x$

*Solve.*

**73.** Without using a calculator, explain which of log 50 or ln 50 must be larger.

**74.** Without using a calculator, explain which of log $50^{-1}$ or ln $50^{-1}$ must be larger.

*The Richter scale measures the intensity, or magnitude, of an earthquake. The formula for the magnitude R of an earthquake is $R = \log\left(\dfrac{a}{T}\right) + B$, where a is the amplitude in micrometers of the vertical motion of the ground at the recording station, T is the number of seconds between successive seismic waves, and B is an adjustment factor that takes into account the weakening of the seismic wave as the distance increases from the epicenter of the earthquake.*

*Use the Richter scale formula to find the magnitude R of the earthquake that fits the description given. Round answers to one decimal place.*

**75.** Amplitude $a$ is 200 micrometers, time $T$ between waves is 1.6 seconds, and $B$ is 2.1.

**76.** Amplitude $a$ is 150 micrometers, time $T$ between waves is 3.6 seconds, and $B$ is 1.9.

**77.** Amplitude $a$ is 400 micrometers, time $T$ between waves is 2.6 seconds, and $B$ is 3.1.

**78.** Amplitude $a$ is 450 micrometers, time $T$ between waves is 4.2 seconds, and $B$ is 2.7.

# 9.8 Exponential and Logarithmic Equations and Problem Solving

## Objective A Solving Exponential Equations

**Objectives**

**A** Solve Exponential Equations.

**B** Solve Logarithmic Equations.

**C** Solve Problems That Can Be Modeled by Exponential and Logarithmic Equations.

In Section 9.3, we solved exponential equations such as $2^x = 16$ by writing both sides in terms of the same base. Here, we write 16 as a power of 2 and using the uniqueness of $b^x$.

$$2^x = 16$$
$$2^x = 2^4 \quad \text{Write 16 as } 2^4.$$
$$x = 4 \quad \text{Use the uniqueness of } b^x.$$

How do we solve an exponential equation when the bases cannot easily be written the same? For example, how do we solve an equation such as $3^x = 7$? We use the fact that $f(x) = \log_b x$ is a one-to-one function. Another way of stating this fact is as a property of equality.

> **Logarithm Property of Equality**
>
> Let $a$, $b$, and $c$ be real numbers such that $\log_b a$ and $\log_b c$ are real numbers and $b$ is not 1. Then
>
> $$\log_b a = \log_b c \quad \text{is equivalent to} \quad a = c$$

**Example 1** Solve: $3^x = 7$. Give the exact answer and a four-decimal-place approximation.

**Solution:** We use the logarithm property of equality and take the logarithm of both sides. For this example, we use the common logarithm.

$$3^x = 7$$
$$\log 3^x = \log 7 \quad \text{Take the common log of both sides.}$$
$$x \log 3 = \log 7 \quad \text{Use the power property of logarithms.}$$
$$x = \frac{\log 7}{\log 3} \quad \text{Divide both sides by log 3.}$$

The exact solution is $\dfrac{\log 7}{\log 3}$. When we approximate to four decimal places, we have

$$\frac{\log 7}{\log 3} \approx \frac{0.845098}{0.477121} \approx 1.7712$$

The solution set is $\left\{ \dfrac{\log 7}{\log 3} \right\}$, or approximately $\{1.7712\}$.

■ Work Practice 1

**Practice 1**

Solve: $2^x = 5$. Give the exact answer and a four-decimal-place approximation.

## Objective B Solving Logarithmic Equations

By applying the appropriate properties of logarithms, we can solve a broad variety of logarithmic equations.

**Answer**

1. $\left\{ \dfrac{\log 5}{\log 2} \right\}$; $\{2.3219\}$

## Practice 2

Solve: $\log_3(x + 5) = 2$

**Example 2** Solve: $\log_4(x - 2) = 2$

**Solution:** Notice that $x - 2$ must be positive, so $x$ must be greater than 2. With this in mind, we first write the equation with exponential notation.

$$\log_4(x - 2) = 2$$
$$4^2 = x - 2$$
$$16 = x - 2$$
$$18 = x \qquad \text{Add 2 to both sides.}$$

To check, we replace $x$ with 18 in the original equation.

$$\log_4(x - 2) = 2$$
$$\log_4(18 - 2) \stackrel{?}{=} 2 \qquad \text{Let } x = 18.$$
$$\log_4 16 \stackrel{?}{=} 2$$
$$4^2 = 16 \qquad \text{True}$$

The solution set is $\{18\}$.

■ **Work Practice 2**

## Practice 3

Solve: $\log_6 x + \log_6(x + 1) = 1$

**Example 3** Solve: $\log_2 x + \log_2(x - 1) = 1$

**Solution:** Notice that $x - 1$ must be positive, so $x$ must be greater than 1. We use the product property on the left side of the equation.

$$\log_2 x + \log_2(x - 1) = 1$$
$$\log_2[x(x - 1)] = 1 \qquad \text{Use the product property.}$$
$$\log_2(x^2 - x) = 1$$

Next we write the equation with exponential notation and solve for $x$.

$$2^1 = x^2 - x$$
$$0 = x^2 - x - 2 \qquad \text{Subtract 2 from both sides.}$$
$$0 = (x - 2)(x + 1) \qquad \text{Factor.}$$
$$0 = x - 2 \quad \text{or} \quad 0 = x + 1 \qquad \text{Set each factor equal to 0.}$$
$$2 = x \qquad\qquad -1 = x$$

Recall that $-1$ cannot be a solution because $x$ must be greater than 1. If we forgot this, we would still reject $-1$ after checking. To see this, we replace $x$ with $-1$ in the original equation.

$$\log_2 x + \log_2(x - 1) = 1$$
$$\log_2(-1) + \log_2(-1 - 1) \stackrel{?}{=} 1 \qquad \text{Let } x = -1.$$

Because the logarithm of a negative number is undefined, $-1$ is rejected. Check to see that the solution set is $\{2\}$.

■ **Work Practice 3**

**Answers**

**2.** $\{4\}$    **3.** $\{2\}$

**Example 4** Solve: $\log(x + 2) - \log x = 2$

**Solution:** We use the quotient property of logarithms on the left side of the equation.

$$\log(x + 2) - \log x = 2$$

$$\log\frac{x + 2}{x} = 2 \qquad \text{Use the quotient property.}$$

$$10^2 = \frac{x + 2}{x} \qquad \text{Write using exponential notation.}$$

$$100 = \frac{x + 2}{x}$$

$$100x = x + 2 \qquad \text{Multiply both sides by } x.$$

$$99x = 2 \qquad \text{Subtract } x \text{ from both sides.}$$

$$x = \frac{2}{99} \qquad \text{Divide both sides by 99.}$$

Check to see that the solution set is $\left\{\dfrac{2}{99}\right\}$.

■ Work Practice 4

**Practice 4**

Solve:
$\log(x + 1) - \log x = 1$

## Objective C Solving Problems Modeled by Exponential and Logarithmic Equations ▶

Logarithmic and exponential functions are used in a variety of scientific, technical, and business settings. A few examples follow.

**Example 5** Estimating Population Size

The population size $y$ of a community of lemmings varies according to the relationship $y = y_0 e^{0.15t}$. In this formula, $t$ is time in months, and $y_0$ is the initial population at time 0. Estimate the population after 6 months if there were originally 5000 lemmings.

**Solution:** We substitute 5000 for $y_0$ and 6 for $t$.

$$y = y_0 e^{0.15t}$$

$$= 5000e^{0.15(6)} \qquad \text{Let } t = 6 \text{ and } y_0 = 5000.$$

$$= 5000e^{0.9} \qquad \text{Multiply.}$$

Using a calculator, we find that $y \approx 12{,}298.016$. In 6 months the population will be approximately 12,300 lemmings.

■ Work Practice 5

**Practice 5**

Use the equation in Example 5 to estimate the lemming population in 8 months.

**Example 6** Doubling an Investment

How long does it take an investment of $2000 to double if it is invested at 5% interest compounded quarterly? The necessary formula is $A = P\left(1 + \dfrac{r}{n}\right)^{nt}$, where $A$ is the accrued amount, $P$ is the principal invested, $r$ is the annual rate of interest, $n$ is the number of compounding periods per year, and $t$ is the number of years.

*(Continued on next page)*

**Practice 6**

How long does it take an investment of $1000 to double if it is invested at 6% interest compounded quarterly?

**Answers**

4. $\left\{\dfrac{1}{9}\right\}$ 5. approximately 16,600

lemmings 6. $11\dfrac{3}{4}$ yr

**Solution:** We are given that $P = \$2000$ and $r = 5\% = 0.05$. Compounding quarterly means 4 times a year, so $n = 4$. The investment is to double, so $A$ must be $4000. We substitute these values and solve for $t$.

$$A = P\left(1 + \frac{r}{n}\right)^{nt}$$

$$4000 = 2000\left(1 + \frac{0.05}{4}\right)^{4t} \qquad \text{Substitute known values.}$$

$$4000 = 2000(1.0125)^{4t} \qquad \text{Simplify } 1 + \frac{0.05}{4}.$$

$$2 = (1.0125)^{4t} \qquad \text{Divide both sides by 2000.}$$

$$\log 2 = \log 1.0125^{4t} \qquad \text{Take the logarithm of both sides.}$$

$$\log 2 = 4t(\log 1.0125) \qquad \text{Use the power property.}$$

$$\frac{\log 2}{4 \log 1.0125} = t \qquad \text{Divide both sides by } 4 \log 1.0125.$$

$$13.949408 \approx t \qquad \text{Approximate by calculator.}$$

It takes approximately 14 years for the money to double in value.

■ Work Practice 6

---

## Calculator Explorations  Graphing

Use a grapher to find how long it takes an investment of $1500 to triple if it is invested at 8% interest compounded monthly. First, let $P = \$1500$, $r = 0.08$, and $n = 12$ (for monthly compounding) in the formula

$$A = P\left(1 + \frac{r}{n}\right)^{nt}$$

Notice that when the investment has tripled, the accrued amount $A$ is $4500. Thus,

$$4500 = 1500\left(1 + \frac{0.08}{12}\right)^{12t}$$

Determine an appropriate viewing window and enter and graph the equations

$$Y_1 = 1500\left(1 + \frac{0.08}{12}\right)^{12x}$$

and

$$Y_2 = 4500$$

The point of intersection of the two curves is the solution. The $x$-coordinate tells how long it takes for the investment to triple.

Use a TRACE feature or an INTERSECT feature to approximate the coordinates of the point of intersection of the two curves. It takes approximately 13.78 years, or 13 years and 10 months, for the investment to triple in value to $4500.

*Use this graphical solution method to solve each problem. Round each answer to the nearest hundredth. Because of rounding and the timing of interest earned, the actual answers may be slightly different from the answers found using this method.*

1. Find how long it takes an investment of $5000 to grow to $6000 if it is invested at 5% interest compounded quarterly.

2. Find how long it takes an investment of $1000 to double if it is invested at 4.5% interest compounded daily. (Use 365 days in a year.)

3. Find how long it takes an investment of $10,000 to quadruple if it is invested at 6% interest compounded monthly.

4. Find how long it takes $500 to grow to $800 if it is invested at 4% interest compounded semiannually.

## Vocabulary, Readiness & Video Check

*Martin-Gay Interactive Videos*   Watch the section lecture video and answer the following questions.

**Objective A  1.** From the lecture before ⊞ Example 1, explain why
ln (4x − 2) = ln 3 is equivalent to 4x − 2 = 3. ▶

**Objective B  2.** Why is the possible solution of −8 rejected in
⊞ Example 3? ▶

**Objective C  3.** For ⊞ Example 4, write the equation and find the number of
years it takes $1000 to double at 7% interest compounded
monthly. Explain the similarity to the answer to Example 4.
Round your answer to the nearest tenth. ▶

*See Video 9.8*

## 9.8  Exercise Set  MyMathLab® ▶

**Objective A**  *Solve each equation. Give the exact solution and a four-decimal-place approximation. See Example 1.*

**1.** $3^x = 6$

**2.** $4^x = 7$

**3.** $9^x = 5$

**4.** $3^x = 11$

**5.** $3^{2x} = 3.8$

**6.** $5^{3x} = 5.6$

**7.** $e^{6x} = 5$

**8.** $e^{2x} = 8$

▶**9.** $2^{x-3} = 5$

**10.** $8^{x-2} = 12$

**11.** $4^{x+7} = 3$

**12.** $6^{x+3} = 2$

**13.** $7^{3x-4} = 11$

**14.** $5^{2x-6} = 12$

**Objective B**  *Solve each equation. See Examples 2 through 4.*

▶**15.** $\log_2(x + 5) = 4$

**16.** $\log_2(x - 5) = 3$

**17.** $\log_4 2 + \log_4 x = 0$

**18.** $\log_3 5 + \log_3 x = 1$

**19.** $\log_2 6 - \log_2 x = 3$

**20.** $\log_4 10 - \log_4 x = 2$

**21.** $\log_2(x^2 + x) = 1$

**22.** $\log_6(x^2 - x) = 1$

▶**23.** $\log_4 x + \log_4(x + 6) = 2$

**24.** $\log_3 x + \log_3(x + 6) = 3$

**25.** $\log_5(x + 3) - \log_5 x = 2$

**26.** $\log_6(x + 2) - \log_6 x = 2$

**27.** $\log_4(x^2 - 3x) = 1$

**28.** $\log_8(x^2 - 2x) = 1$

**29.** $\log_2 x + \log_2(3x + 1) = 1$

**30.** $\log_3 x + \log_3(x - 8) = 2$

## Objective C   *Solve. See Example 5.*

**31.** The size of the wolf population at Isle Royale National Park increases according to the formula $y = y_0 e^{0.043t}$. In this formula, $t$ is time in years and $y_0$ is the initial population at time 0. If the size of the current population is 83 wolves, find how many there should be in 5 years. Round to the nearest whole number.

**32.** The number of victims of a flu epidemic is increasing according to the formula $y = y_0 e^{0.075t}$. In this formula, $t$ is time in weeks and $y_0$ is the infected population at time 0. If 20,000 people are currently infected, how many might be infected in 3 weeks? Round to the nearest whole number.

**33.** The population of Brazil is increasing according to the formula $y = y_0 e^{0.009t}$. In this formula, $t$ is time in years and $y_0$ is the initial population at time 0. If the size of the population in 2009 was 191 million, use the formula to predict the population of Brazil in year 2020. Round to the nearest tenth of a million. (*Source:* World Bank)

**34.** The population of the Faeroe Islands is decreasing according to the formula $y = y_0 e^{-0.001t}$. In this formula, $t$ is the time in years and $y_0$ is the initial population at time 0. If the size of the population in 2009 was 48,778, use the formula to predict the population of the Faeroe Islands in the year 2025. Round to the nearest whole number. (*Source:* World Bank and ArcticStat)

**35.** The population of the Cook Islands is decreasing according to the formula $y = y_0 e^{-0.0277t}$. In this formula, $t$ is time in years and $y_0$ is the initial population at time 0. If the size of the population in 2009 was 11,870, use the formula to predict the population of Cook Islands in the year 2025. Round to the nearest whole number. (*Source:* The World Almanac)

**36.** The population of Saint Barthelemy is decreasing according to the formula $y = y_0 e^{-0.0034t}$. In this formula, $t$ is time in years and $y_0$ is the initial population at time 0. If the size of the population in 2009 was 7448, use the formula to predict the population of Saint Barthelemy in the year 2025. Round to the nearest whole number. (*Source:* The World Almanac)

*Use the formula $A = P\left(1 + \dfrac{r}{n}\right)^{nt}$ to solve these compound interest problems. Round to the nearest tenth. See Example 6.*

▶ **37.** How long does it take for $600 to double if it is invested at 7% interest compounded monthly?

**38.** How long does it take for $600 to double if it is invested at 12% interest compounded monthly?

**39.** How long does it take for a $1200 investment to earn $200 interest if it is invested at 9% interest compounded quarterly?

**40.** How long does it take for a $1500 investment to earn $200 interest if it is invested at 10% interest compounded semiannually?

**41.** How long does it take for $1000 to double if it is invested at 8% interest compounded semiannually?

**42.** How long does it take for $1000 to double if it is invested at 8% interest compounded monthly?

*The formula $w = 0.00185h^{2.67}$ is used to estimate the normal weight $w$ in pounds of a boy $h$ inches tall. Use this formula to solve Exercises 43 and 44. Round to the nearest tenth.*

**43.** Find the expected height of a boy who weighs 85 pounds.

**44.** Find the expected height of a boy who weighs 140 pounds.

*The formula $P = 14.7e^{-0.21x}$ gives the average atmospheric pressure P, in pounds per square inch, at an altitude x, in miles above sea level. Use this formula to solve Exercises 45 through 48. Round to the nearest tenth.*

**45.** Find the average atmospheric pressure of Denver, which is 1 mile above sea level.

**46.** Find the average atmospheric pressure of Pikes Peak, which is 2.7 miles above sea level.

**47.** Find the elevation of a Delta jet if the atmospheric pressure outside the jet is 7.5 pounds per square inch.

**48.** Find the elevation of a remote Himalayan peak if the atmospheric pressure atop the peak is 6.5 pounds per square inch.

*Psychologists call the graph of the formula $t = \dfrac{1}{c} \ln\left(\dfrac{A}{A - N}\right)$ the learning curve since the formula relates time t passed, in weeks, to a measure N of learning achieved, to a measure A of maximum learning possible, and to a measure c of an individual's learning style. Use this formula to answer Exercises 49 through 52. Round to the nearest whole number.*

**49.** Norman Weidner is learning to type. If he wants to type at a rate of 50 words per minute ($N = 50$) and his expected maximum rate is 75 words per minute ($A = 75$), how many weeks should it take him to achieve his goal? Assume that c is 0.09.

**50.** An experiment of teaching chimpanzees sign language shows that a typical chimp can master a maximum of 65 signs. How many weeks should it take a chimpanzee to master 30 signs if c is 0.03?

**51.** Janine Jenkins is working on her dictation skills. She wants to take dictation at a rate of 150 words per minute and believes that the maximum rate she can hope for is 210 words per minute. How many weeks should it take her to achieve the 150-word level if c is 0.07?

**52.** A psychologist is measuring human capability to memorize nonsense syllables. How many weeks should it take a subject to learn 15 nonsense syllables if the maximum possible to learn is 24 syllables and c is 0.17?

## Review

*If $x = -2$, $y = 0$, and $z = 3$, find the value of each expression. See Section 1.5.*

**53.** $\dfrac{x^2 - y + 2z}{3x}$

**54.** $\dfrac{x^3 - 2y + z}{2z}$

**55.** $\dfrac{3z - 4x + y}{x + 2z}$

**56.** $\dfrac{4y - 3x + z}{2x + y}$

## Concept Extensions

*The formula $y = y_0 e^{kt}$ gives the population size y of a population that experiences a relative growth rate k (k is positive if growth is increasing and k is negative if growth is decreasing). In this formula, t is time in years and $y_0$ is the initial population at time 0. Use this formula to solve Exercises 57 and 58. Round answers to the nearest year.*

**57.** In 2009, the population of Michigan was approximately 9,970,000 and decreasing according to the formula $y = y_0 e^{-0.003t}$. Assume that the population continues to decrease according to the given formula and predict how many years after which the population of Michigan will be 9,500,000. (*Hint:* Let $y_0 = 9,970,000$; $y = 9,500,000$, and solve for t.)

**58.** In 2009, the population of Illinois was approximately 12,910,000 and increasing according to the formula $y = y_0 e^{0.005t}$. Assume that the population continues to increase according to the given formula and predict how many years after which the population of Illinois will be 13,500,000. (See the Hint for Exercise 57.) (Source: U.S. Census Bureau and Federal Reserve Bank of Chicago)

*Solve.*

**59.** When solving a logarithmic equation, explain why you must check possible solutions in the original equation.

**60.** Solve $5^x = 9$ by taking the common logarithm of both sides of the equation. Next, solve this equation by taking the natural logarithm of both sides. Compare your solutions. Are they the same? Why or why not?

*Use a graphing calculator to solve. Round your answers to two decimal places.*

**61.** $e^{0.3x} = 8$

**62.** $10^{0.5x} = 7$

## Chapter 9 Group Activity

### Sound Intensity

The decibel (dB) measures sound intensity, or the relative loudness or strength of a sound. One decibel is the smallest difference in sound levels that is detectable by humans. The decibel is a logarithmic unit. This means that for approximately every 3-decibel increase in sound intensity, the relative loudness of the sound is doubled. For example, a 35 dB sound is twice as loud as a 32 dB sound.

In the modern world, noise pollution has increasingly become a concern. Sustained exposure to high sound intensities can lead to hearing loss. Regular exposure to 90 dB sounds can eventually lead to loss of hearing. Sounds of 130 dB and more can cause permanent loss of hearing instantaneously.

The relative loudness of a sound $D$ in decibels is given by the equation

$$D = 10 \log_{10} \frac{I}{10^{-16}}$$

where $I$ is the intensity of a sound given in watts per square centimeter. Some sound intensities of common noises are listed in the table in order of increasing sound intensity.

### Group Activity

1. Work together to create a table of the relative loudness (in decibels) of the sounds listed in the table.

2. Research the loudness of other common noises. Add these sounds and their decibel levels to your table. Be sure to list the sounds in order of increasing sound intensity.

| Some Sound Intensities of Common Noises | |
| --- | --- |
| Noise | Intensity (watts/cm$^2$) |
| Whispering | $10^{-15}$ |
| Rustling leaves | $10^{-14.2}$ |
| Normal conversation | $10^{-13}$ |
| Background noise in a quiet residence | $10^{-12.2}$ |
| Normal office noise or a quiet stream | $10^{-11}$ |
| Air conditioning | $10^{-10}$ |
| Freight train at 50 feet | $10^{-8.5}$ |
| Vacuum cleaner | $10^{-8}$ |
| Nearby thunder | $10^{-7}$ |
| Air hammer | $10^{-6.5}$ |
| Jet plane at takeoff | $10^{-6}$ |
| Threshold of pain | $10^{-4}$ |

## Chapter 9 Vocabulary Check

*Fill in each blank with one of the words or phrases listed below. Some words or phrases may be used more than once.*

| inverse | common | composition | symmetric | exponential |
|---------|--------|-------------|-----------|-------------|
| vertical | logarithmic | natural | half-life | horizontal |

1. For a one-to-one function, we can find its _____ function by switching the coordinates of the ordered pairs of the function.

2. The _____ of functions $f$ and $g$ is $(f \circ g)(x) = f(g(x))$.

3. A function of the form $f(x) = b^x$ is called a(n) _____ function if $b > 0$, $b$ is not 1, and $x$ is a real number.

4. The graphs of $f$ and $f^{-1}$ are _____ about the line $y = x$.

5. _____ logarithms are logarithms to base $e$.

6. _____ logarithms are logarithms to base 10.

7. To see whether a graph is the graph of a one-to-one function, apply the _____ line test to see whether it is a function and then apply the _____ line test to see whether it is a one-to-one function.

8. A(n) _____ function is a function that can be defined by $f(x) = \log_b x$ where $x$ is a positive real number, $b$ is a constant positive real number, and $b$ is not 1.

9. _____ is the amount of time it takes for half of the amount of a substance to decay.

10. A quantity that grows or decays by the same percent at regular time periods is said to have _____ growth or decay.

**Helpful Hint**

▶ Are you preparing for your test? Don't forget to take the Chapter 9 Test on page 745. Then check your answers at the back of the text and use the Chapter Test Prep Videos to see the fully worked-out solutions to any of the exercises you want to review.

# 9 Chapter Highlights

| Definitions and Concepts | Examples |
|---|---|

### Section 9.1   The Algebra of Functions

**Algebra of Functions**

Let $f$ and $g$ be functions.

| | |
|---|---|
| Sum | $(f + g)(x) = f(x) + g(x)$ |
| Difference | $(f - g)(x) = f(x) - g(x)$ |
| Product | $(f \cdot g)(x) = f(x) \cdot g(x)$ |
| Quotient | $\left(\dfrac{f}{g}\right)(x) = \dfrac{f(x)}{g(x)}, g(x) \neq 0$ |

If $f(x) = 7x$ and $g(x) = x^2 + 1$,

$$(f + g)(x) = f(x) + g(x) = 7x + x^2 + 1$$
$$(f - g)(x) = f(x) - g(x) = 7x - (x^2 + 1)$$
$$= 7x - x^2 - 1$$
$$(f \cdot g)(x) = f(x) \cdot g(x) = 7x(x^2 + 1)$$
$$= 7x^3 + 7x^2$$
$$\left(\frac{f}{g}\right)(x) = \frac{f(x)}{g(x)} = \frac{7x}{x^2 + 1}$$

**Composite Functions**

The notation $(f \circ g)(x)$ means "$f$ composed with $g$."

$$(f \circ g)(x) = f(g(x))$$
$$(g \circ f)(x) = g(f(x))$$

If $f(x) = x^2 + 1$ and $g(x) = x - 5$, find $(f \circ g)(x)$.

$$(f \circ g)(x) = f(g(x))$$
$$= f(x - 5)$$
$$= (x - 5)^2 + 1$$
$$= x^2 - 10x + 26$$

| Definitions and Concepts | Examples |
|---|---|

## Section 9.2  Inverse Functions

### One-to-One Function

If $f$ is a function, then $f$ is a **one-to-one function** only if each $y$-value (output) corresponds to only one $x$-value (input).

### Horizontal Line Test

If every horizontal line intersects the graph of a function at most once, then the function is a one-to-one function.

Determine whether each graph is a one-to-one function.

Graphs A and C pass the vertical line test, so only these are graphs of functions. Of graphs A and C, only graph A passes the horizontal line test, so only graph A is the graph of a one-to-one function.

The **inverse** of a one-to-one function $f$ is the one-to-one function $f^{-1}$ that is the set of all ordered pairs $(b, a)$ such that $(a, b)$ belongs to $f$.

### Finding the Inverse of a One-to-One Function f

**Step 1:**  Replace $f(x)$ with $y$.

**Step 2:**  Interchange $x$ and $y$.

**Step 3:**  Solve for the equation for $y$.

**Step 4:**  Replace $y$ with the notation $f^{-1}(x)$.

Find the inverse of $f(x) = 2x + 7$.

$$y = 2x + 7 \quad \text{Replace } f(x) \text{ with } y.$$
$$x = 2y + 7 \quad \text{Interchange } x \text{ and } y.$$
$$2y = x - 7 \quad \text{Solve for } y.$$
$$y = \frac{x - 7}{2}$$
$$f^{-1}(x) = \frac{x - 7}{2} \quad \text{Replace } y \text{ with } f^{-1}(x).$$

The inverse of $f(x) = 2x + 7$ is $f^{-1}(x) = \dfrac{x - 7}{2}$.

## Section 9.3  Exponential Functions

### Exponential Function

A function of the form $f(x) = b^x$ is an **exponential function,** where $b > 0$, $b \neq 1$, and $x$ is a real number.

Graph the exponential function $y = 4^x$.

| x | y |
|---|---|
| $-2$ | $\dfrac{1}{16}$ |
| $-1$ | $\dfrac{1}{4}$ |
| $0$ | $1$ |
| $1$ | $4$ |
| $2$ | $16$ |

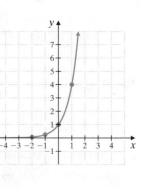

| Definitions and Concepts | Examples |
|---|---|

### Section 9.3    Exponential Functions (*continued*)

**Uniqueness of $b^x$**

If $b > 0$ and $b \neq 1$, then $b^x = b^y$ is equivalent to $x = y$.

Solve:    $2^{x+5} = 8$

$\quad\quad 2^{x+5} = 2^3$    Write 8 as $2^3$.

$\quad\quad x + 5 = 3$    Use the uniqueness of $b^x$.

$\quad\quad\quad\quad x = -2$    Subtract 5 from both sides.

### Section 9.4    Exponential Growth and Decay Functions

A quantity that grows or decays by the same percent at regular time periods is said to have **exponential growth** or **exponential decay**.

**Exponential Growth**

initial amount — number of time intervals

$$y = C(1 + r)^x$$

$(1 + r)$ is growth factor; $r$ is growth rate (often a percent)

**Exponential Decay**

initial amount — number of time intervals

$$y = C(1 - r)^x$$

$(1 - r)$ is decay factor; $r$ is decay rate (often a percent)

A city has a current population of 37,000 that has been increasing at a rate of 3% per year. At this rate, find the city's population in 20 years.

$$y = C(1 + r)^x$$
$$y = 37{,}000(1 + 0.03)^{20}$$
$$y \approx 66{,}826.12$$

In 20 years, the predicted population of the city is 66,826.

A city has a current population of 37,000 that has been decreasing at a rate of 3% per year. At this rate, find the city's population in 20 years.

$$y = C(1 - r)^x$$
$$y = 37{,}000(1 - 0.03)^{20}$$
$$y \approx 20{,}120.39$$

In 20 years, predicted population of the city is 20,120.

### Section 9.5    Logarithmic Functions

**Logarithmic Definition**

If $b > 0$ and $b \neq 1$, then

$\quad y = \log_b x \quad$ means $\quad x = b^y$

for any positive number $x$ and real number $y$.

**Properties of Logarithms**

If $b$ is a real number, $b > 0$ and $b \neq 1$, then

$\quad \log_b 1 = 0, \quad \log_b b^x = x, \quad$ and $\quad b^{\log_b x} = x$

**Logarithmic function**

If $b > 0$ and $b \neq 1$, then a **logarithmic function** is a function that can be defined as

$\quad f(x) = \log_b x$

The domain of $f$ is the set of positive real numbers, and the range of $f$ is the set of real numbers.

| Logarithmic Form | Corresponding Exponential Statement |
|---|---|
| $\log_5 25 = 2$ | $5^2 = 25$ |
| $\log_9 3 = \dfrac{1}{2}$ | $9^{1/2} = 3$ |

$\log_5 1 = 0, \quad \log_7 7^2 = 2, \quad$ and $\quad 3^{\log_3 6} = 6$

Graph:    $y = \log_3 x$

Write $y = \log_3 x$ as $3^y = x$. Plot the ordered pair solutions listed in the table, and connect them with a smooth curve.

| $x$ | $y$ |
|---|---|
| 3 | 1 |
| 1 | 0 |
| $\dfrac{1}{3}$ | $-1$ |
| $\dfrac{1}{9}$ | $-2$ |

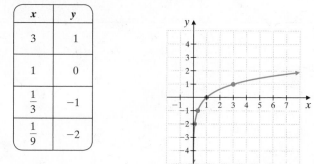

| Definitions and Concepts | Examples |
|---|---|

## Section 9.6    Properties of Logarithms

Let $x, y$, and $b$ be positive numbers, $b \neq 1$, and $r$ be a real number.

**Product Property**

$$\log_b xy = \log_b x + \log_b y$$

**Quotient Property**

$$\log_b \frac{x}{y} = \log_b x - \log_b y$$

**Power Property**

$$\log_b x^r = r \log_b x$$

Write as a single logarithm:

$$2\log_5 6 + \log_5 x - \log_5 (y + 2)$$

$$= \log_5 6^2 + \log_5 x - \log_5 (y + 2) \quad \text{Power property}$$

$$= \log_5 36 \cdot x - \log_5 (y + 2) \quad \text{Product property}$$

$$= \log_5 \frac{36x}{y + 2} \quad \text{Quotient property}$$

## Section 9.7    Common Logarithms, Natural Logarithms, and Change of Base

**Common Logarithms**

$$\log x \quad \text{means} \quad \log_{10} x$$

**Natural Logarithms**

$$\ln x \quad \text{means} \quad \log_e x$$

**Continuously Compounded Interest Formula**

$$A = Pe^{rt}$$

where $r$ is the annual interest rate for $P$ dollars invested for $t$ years.

**Change of Base Formula**

If $a, b$, and $c$ are positive real numbers and neither $b$ nor $c$ is 1, then

$$\log_b a = \frac{\log_c a}{\log_c b}$$

$$\log 5 = \log_{10} 5 \approx 0.6990$$

$$\ln 7 = \log_e 7 \approx 1.9459$$

Find the amount in an account at the end of 3 years if $1000 is invested at an interest rate of 4% compounded continuously.
Here, $t = 3$ years, $P = \$1000$, and $r = 0.04$.

$$A = Pe^{rt}$$

$$= \$1000e^{0.04(3)}$$

$$\approx \$1127.50$$

## Section 9.8    Exponential and Logarithmic Equations and Problem Solving

**Logarithm Property of Equality**

Let $\log_b a$ and $\log_b c$ be real numbers and $b \neq 1$. Then

$$\log_b a = \log_b c \quad \text{is equivalent to} \quad a = c$$

Solve: $2^x = 5$

$$\log 2^x = \log 5 \quad \text{Log property of equality}$$

$$x \log 2 = \log 5 \quad \text{Power property}$$

$$x = \frac{\log 5}{\log 2} \quad \text{Divide both sides by log 2.}$$

$$x \approx 2.3219 \quad \text{Use a calculator.}$$

## Chapter 9 Review

**(9.1)** *If $f(x) = x - 5$ and $g(x) = 2x + 1$, find the following.*

**1.** $(f + g)(x)$

**2.** $(f - g)(x)$

**3.** $(f \cdot g)(x)$

**4.** $\left(\dfrac{g}{f}\right)(x)$

*If $f(x) = x^2 - 2$, $g(x) = x + 1$, and $h(x) = x^3 - x^2$, find each composition.*

**5.** $(f \circ g)(x)$

**6.** $(g \circ f)(x)$

**7.** $(h \circ g)(2)$

**8.** $(f \circ f)(x)$

**9.** $(f \circ g)(-1)$

**10.** $(h \circ h)(2)$

**(9.2)** *Determine whether each function is a one-to-one function. If it is one-to-one, list the elements of its inverse.*

**11.** $h = \{(-9, 14), (6, 8), (-11, 12), (15, 15)\}$

**12.** $f = \{(-5, 5), (0, 4), (13, 5), (11, -6)\}$

**13.**

| U.S. Region (Input) | Northeast | Midwest | South | West |
|---|---|---|---|---|
| Rank in Housing Starts for 2013 (Output) | 4 | 3 | 1 | 2 |

**14.**

| Shape (Input) | Square | Triangle | Parallelogram | Rectangle |
|---|---|---|---|---|
| Number of Sides (Output) | 4 | 3 | 4 | 4 |

*Determine whether each function is a one-to-one function.*

**15.**     **16.**     **17.**     **18.**

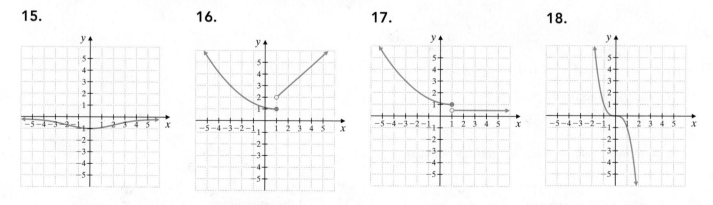

*Find an equation defining the inverse function of each one-to-one function.*

**19.** $f(x) = 6x + 11$     **20.** $f(x) = 12x$     **21.** $f(x) = 3x - 5$     **22.** $f(x) = 2x + 1$

*Graph each one-to-one function and its inverse on the same set of axes.*

**23.** $f(x) = -2x + 3$                          **24.** $f(x) = 5x - 5$

**(9.3)** *Solve each equation.*

**25.** $4^x = 64$     **26.** $2^{3x} = \dfrac{1}{16}$     **27.** $9^{x+1} = 243$     **28.** $25^{x-1} = 125$

*Graph each exponential function.*

**29.** $y = 3^x$     **30.** $y = \left(\dfrac{1}{3}\right)^x$     **31.** $y = 2^{x-4}$     **32.** $y = 2^x + 4$

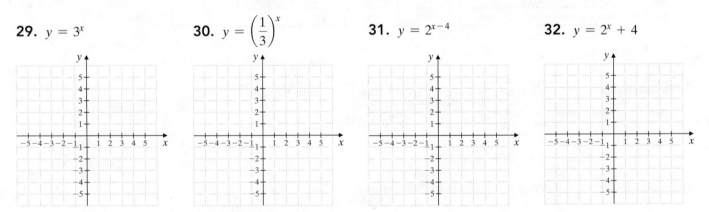

*Use the formula $A = P\left(1 + \dfrac{r}{n}\right)^{nt}$ to solve Exercises 33 and 34. In this formula,*

   $A =$ amount accrued (or owed)

   $P =$ principal invested (or loaned)

   $r =$ rate of interest

   $n =$ number of compounding periods per year

   $t =$ time in years

**33.** A total of $800 is invested in a 7% certificate of deposit for which interest is compounded quarterly. Find the value that this certificate will have at the end of 5 years.

**34.** Find the amount accrued if $1600 is invested at 9% interest compounded semiannually for 7 years.

**(9.4)** *Solve. Round each answer to the nearest whole.*

**35.** The city of Henderson, Nevada, has been growing at a rate of 4.8% per year since the year 2000. If the population of Henderson was 179,087 in 2000 and this rate continues, predict the city's population in 2020.

**36.** The city of Raleigh, North Carolina, has been growing at a rate of 3.9% per year since the year 2000. If the population of Raleigh was 287,370 in 2000 and this rate continues, predict the city's population in 2021.

**37.** A summer camp tournament starts with 1024 players. After each round, half the players are eliminated. How many players remain after 7 rounds?

**38.** The bear population in a certain national park is decreasing by 11% each year. If this rate continues, and there is currently an estimated bear population of 1280, find the bear population in 6 years.

**(9.5)** *Write each exponential equation with logarithmic notation.*

**39.** $49 = 7^2$

**40.** $2^{-4} = \dfrac{1}{16}$

*Write each logarithmic equation with exponential notation.*

**41.** $\log_{1/2} 16 = -4$

**42.** $\log_{0.4} 0.064 = 3$

*Solve.*

**43.** $\log_4 x = -3$

**44.** $\log_3 x = 2$

**45.** $\log_3 1 = x$

**46.** $\log_x 64 = 2$

**47.** $\log_4 4^5 = x$

**48.** $\log_7 7^{-2} = x$

**49.** $5^{\log_5 4} = x$

**50.** $2^{\log_2 9} = x$

**51.** $\log_3 (2x + 5) = 2$

**52.** $\log_8 (x^2 + 7x) = 1$

*Graph each pair of equations on the same set of axes.*

**53.** $y = 2^x;\ y = \log_2 x$

**54.** $y = \left(\dfrac{1}{2}\right)^x;\ y = \log_{1/2} x$

**(9.6)** *Write each expression as a single logarithm.*

**55.** $\log_3 8 + \log_3 4$

**56.** $\log_2 6 + \log_2 3$

**57.** $\log_7 15 - \log_7 20$

**58.** $\log_e 18 - \log_e 12$      **59.** $\log_{11} 8 + \log_{11} 3 - \log_{11} 6$      **60.** $\log_5 14 + \log_5 3 - \log_5 21$

**61.** $2\log_5 x - 2\log_5 (x + 1) + \log_5 x$      **62.** $4\log_3 x - \log_3 x + \log_3 (x + 2)$

*Use properties of logarithms to write each expression as a sum or difference of logarithms.*

**63.** $\log_3 \dfrac{x^3}{x + 2}$      **64.** $\log_4 \dfrac{x + 5}{x^2}$

**65.** $\log_2 \dfrac{3x^2 y}{z}$      **66.** $\log_7 \dfrac{yz^3}{x}$

*If $\log_b 2 = 0.36$ and $\log_b 5 = 0.83$, evaluate each expression.*

**67.** $\log_b 50$      **68.** $\log_b \dfrac{4}{5}$

**(9.7)** *Use a calculator to approximate each logarithm to four decimal places.*

**69.** $\log 3.6$      **70.** $\log 0.15$      **71.** $\ln 1.25$      **72.** $\ln 4.63$

*Find the exact value of each logarithm.*

**73.** $\log 1000$      **74.** $\log \dfrac{1}{10}$      **75.** $\ln \dfrac{1}{e}$      **76.** $\ln e^4$

*Solve each equation. Give the exact solution and a four-decimal approximation where necessary.*

**77.** $\log 2x = 2$      **78.** $\ln(3x) = 1.6$      **79.** $\ln(2x - 3) = -1$      **80.** $\ln(3x + 1) = 2$

*Approximate each logarithm to four decimal places.*

**81.** $\log_5 1.6$      **82.** $\log_3 4$

*Use the formula $A = Pe^{rt}$ to solve Exercises 83 and 84, in which interest is compounded continuously. In this formula,*

     $A =$ amount accrued (or owed)
     $P =$ principal invested (or loaned)
     $r =$ rate of interest
     $t =$ time in years

**83.** Bank of New York offers a 5-year 6% continuously compounded investment option. Find the amount accrued if $1450 is invested.

**84.** Find the amount to which a $940 investment grows if it is invested at 11% interest compounded continuously for 3 years.

**(9.8)** *Solve each exponential equation. Give the exact solution and a four-decimal-place approximation.*

**85.** $7^x = 20$

**86.** $3^{2x} = 7$

**87.** $3^{2x+1} = 6$

**88.** $8^{4x-2} = 3$

*Solve each equation.*

**89.** $\log_5 2 + \log_5 x = 2$

**90.** $\log 5x - \log(x + 1) = 4$

**91.** $\log_2 x + \log_2 2x - 3 = 1$

**92.** $\log_3(x^2 - 8x) = 2$

*Use the formula $y = y_0 e^{kt}$ to solve Exercises 93 through 96. In this formula,*

$\quad y$ = size of population

$\quad y_0$ = initial count of population

$\quad k$ = rate of growth, expressed as a decimal

$\quad t$ = time

*Round each answer to the nearest tenth.*

**93.** Use this formula to calculate the rate of growth of the population of California condors. In 1987, the population of California condors was only 87 birds. They were all brought in from the wild and an intensive breeding program was instituted. By 2013, there were 416 California condors. *(Source:* National Parks Conservation Association)

**94.** France is experiencing an annual growth rate of 0.5%. In 2012, the population of France was approximately 65,300,000. How long will it take for the population to reach 67,000,000? Round to the nearest tenth. *(Source:* Population Reference Bureau)

**95.** In 2013, the population of Australia was approximately 23,400,000. How long will it take Australia to double its population if its growth rate is 1.5% annually? *(Source:* Population Reference Bureau)

**96.** Monaco's population is increasing at a rate of 0.6% per year. How long will it take for its 2013 population of 30,429 to double in size? *(Source:* Population Reference Bureau)

*Use the compound interest equation $A = P\left(1 + \dfrac{r}{n}\right)^{nt}$ to solve Exercises 97 and 98. (See the directions for Exercises 33 and 34 for an explanation of this formula.) Round answers to the nearest tenth.*

**97.** How long does it take for a $5000 investment to grow to $10,000 if it is invested at 8% interest compounded quarterly?

**98.** An investment of $6000 has grown to $10,000 while the money was invested at 6% interest compounded monthly. How long was it invested?

## Mixed Review

*Solve each equation. Give exact answers.*

**99.** $3^x = \dfrac{1}{9}$

**100.** $5^{2x} = 125$

**101.** $8^{3x-2} = 4$

**102.** $9^{x-2} = 27$

**103.** $\log_4 64 = x$

**104.** $\log_x 81 = 4$

**105.** $\log_4 (x^2 - 3x) = 1$

**106.** $\log_2 (3x - 1) = 4$

**107.** $\ln x = -1.2$

**108.** $\log_3 x + \log_3 10 = 2$

**109.** $\ln 3x - \ln (x - 3) = 2$

**110.** $\log_6 x - \log_6 (4x + 7) = 1$

*If $f(x) = x$ and $g(x) = 2x - 3$, find the following.*

**1.** $(f \cdot g)(x)$

**2.** $(f - g)(x)$

*If $f(x) = x, g(x) = x - 7$, and $h(x) = x^2 - 6x + 5$, find each composition.*

**3.** $(f \circ h)(0)$

**4.** $(g \circ f)(x)$

**5.** $(g \circ h)(x)$

**Answers**

**1.** _____

*Graph the one-to-one function and its inverse on the same set of axes.*

**6.** $f(x) = 7x - 14$

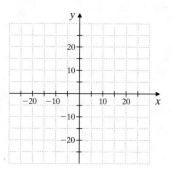

**2.** _____

**3.** _____

**4.** _____

**5.** _____

*Determine whether each graph is the graph of a one-to-one function.*

**7.**

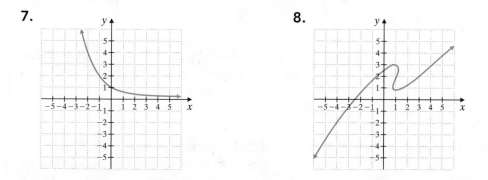

**8.**

**6.** _____

**7.** _____

**8.** _____

*Determine whether each function is one-to-one. If it is one-to-one, find an equation or a set of ordered pairs that defines the inverse function of the given function.*

**9.** _____

**9.** $f(x) = 6 - 2x$

**10.** $f = \{(0, 0), (2, 3), (-1, 5)\}$

**10.** _____

**11.**

| Word (Input) | Dog | Cat | House | Desk | Circle |
|---|---|---|---|---|---|
| First Letter of Word (Output) | d | c | h | d | c |

**11.** _____

*Use the properties of logarithms to write each expression as a single logarithm.*

**12.** $\log_3 6 + \log_3 4$

**13.** $\log_5 x + 3\log_5 x - \log_5(x+1)$

**14.** Write the expression $\log_6 \dfrac{2x}{y^3}$ as the sum or difference of logarithms.

**15.** If $\log_b 3 = 0.79$ and $\log_b 5 = 1.16$, find the value of $\log_b \dfrac{3}{25}$.

**16.** Approximate $\log_7 8$ to four decimal places.

**17.** Solve $8^{x-1} = \dfrac{1}{64}$ for $x$. Give the exact solution.

**18.** Solve $3^{2x+5} = 4$ for $x$. Give the exact solution and a four-decimal-place approximation.

*Solve each logarithmic equation. Give the exact solution.*

**19.** $\log_3 x = -2$

**20.** $\ln \sqrt{e} = x$

**21.** $\log_8(3x - 2) = 2$

**22.** $\log_5 x + \log_5 3 = 2$

**23.** $\log_4(x+1) - \log_4(x-2) = 3$

**24.** Solve $\ln(3x + 7) = 1.31$. Round the solution to four decimal places.

**25.** Graph $f(x) = \left(\dfrac{1}{2}\right)^x + 1$.

**26.** Graph the functions $y = 3^x$ and $y = \log_3 x$ on the same set of axes.

12. _____

13. _____

14. _____

15. _____

16. _____

17. _____

18. _____

19. _____

20. _____

21. _____

22. _____

23. _____

24. _____

25. _____

26. _____

*Use the formula $A = P\left(1 + \dfrac{r}{n}\right)^{nt}$ or $A = Pe^{rt}$ to solve Exercises 27 through 29.*

**27.** Find the amount in an account in which $4000 is invested for 3 years at 9% interest compounded monthly.

**28.** How long will it take $2000 to grow to $3000 if the money is invested at 7% interest compounded semiannually? Round to the nearest whole year.

**29.** Suppose you have $3000 to invest. Which investment, rounded to the nearest dollar, yields the greater return over 10 years: 6.5% compounded semiannually or 6% compounded monthly? How much more is yielded by the better investment?

*Solve. Round answers to the nearest whole.*

**30.** Suppose a city with population of 150,000 has been decreasing at a rate of 2% per year. If this rate continues, predict the population of the city in 20 years.

**31.** The prairie dog population of the Grand Forks area now stands at 57,000 animals. If the population is growing at a rate of 2.6% annually, how many prairie dogs will there be in that area 5 years from now?

**32.** In an attempt to save an endangered species of wood duck, naturalists would like to increase the wood duck population from 400 to 1000 ducks. If the annual population growth rate is 6.2%, how long will it take the naturalists to reach their goal? Round to the nearest whole year.

*The reliability of a new model of CD player can be described by the exponential function $R(t) = 2.7^{-(1/3)t}$, where the reliability R is the probability (as a decimal) that the CD player is still working t years after it is manufactured. Round answers to the nearest hundredth. Then write your answers as percents.*

**33.** What is the probability that the CD player will still work half a year after it is manufactured?

**34.** What is the probability that the CD player will still work 2 years after it is manufactured?

27. _____

28. _____

29. _____

30. _____

31. _____

32. _____

33. _____

34. _____

Copyright 2015 Pearson Education, Inc.

**Answers**

1. _____

2. _____

3. _____

4. _____

5. _____

6. _____

7. _____

8. _____

9. _____

10. _____

11. _____

12. _____

13. _____

14. _____

15. _____

16. _____

17. _____

18. _____

19. _____

20. _____

21. _____

22. _____

*Find each root.*

**1.** $\sqrt[3]{27}$

**2.** $\sqrt[3]{64x^7y^2}$

**3.** $\sqrt[4]{16}$

**4.** $\sqrt[4]{162a^4b^5}$

△ **5.** The measure of the largest angle of a triangle is 80° more than the measure of the smallest angle, and the measure of the remaining angle is 10° more than the measure of the smallest angle. Find the measure of each angle.

**6.** Line $l$ and line $m$ are parallel lines cut by transversal $t$. Find the values of $x$ and $y$.

**7.** Factor: $7x(x^2 + 5y) - (x^2 + 5y)$

**8.** Solve: $\dfrac{1}{3}(x - 2) = \dfrac{1}{4}(x + 1)$

**9.** Subtract: $\dfrac{5k}{k^2 - 4} - \dfrac{2}{k^2 + k - 2}$

**10.** Perform the indicated operation and simplify if possible.

$$\dfrac{5}{x - 2} + \dfrac{3}{x^2 + 4x + 4} - \dfrac{6}{x + 2}$$

*If $f(x) = \sqrt{x - 4}$ and $g(x) = \sqrt[3]{x + 2}$, find each function value.*

**11.** $f(8)$

**12.** $f(28)$

**13.** $g(-1)$

**14.** $g(0)$

*Use the properties of exponents to simplify. Write all results using positive exponents.*

**15.** $x^{1/2}x^{1/3}$

**16.** $(a^{-2}b^3c^{-4})^{-2}$

**17.** $\dfrac{(2x^{2/5})^5}{x^2}$

**18.** $\left(\dfrac{3^{-2}}{x}\right)^{-3}$

*Use the quotient rule to simplify. Assume that all variables represent positive real numbers.*

**19.** $\sqrt{\dfrac{x}{9}}$

**20.** $\dfrac{\sqrt{108a^2}}{3\sqrt{3}}$

**21.** $\sqrt[4]{\dfrac{3}{16y^4}}$

**22.** $\dfrac{3\sqrt[3]{81a^5b^{10}}}{\sqrt[3]{3b^4}}$

*Multiply.*

**23.** $(\sqrt{2x} + 5)(\sqrt{2x} - 5)$  **24.** $\left(\frac{1}{2}x + 3\right)\left(\frac{1}{2}x - 3\right)$  **25.** $(\sqrt{3} - 1)^2$

**26.** Multiply.
a. $a^{1/4}(a^{3/4} - a^8)$
b. $(x^{1/2} - 3)(x^{1/2} + 5)$

**27.** Rationalize the numerator of $\dfrac{\sqrt{7}}{\sqrt{45}}$.

**28.** Rationalize the denominator. $\sqrt[3]{\dfrac{27}{m^4 n^8}}$

**29.** Solve: $\sqrt{4 - x} = x - 2$

**30.** Find an equation of the line through $(-2, 6)$ and perpendicular to $f(x) = -3x + 4$. Write the equation using function notation.

**31.** Add: $(2 + 3i) + (-3 + 2i)$

**32.** Find the following powers of $i$.
a. $i^8$  b. $i^{21}$
c. $i^{42}$  d. $i^{-13}$

**33.** Solve $4x^2 - 24x + 41 = 0$ by completing the square.

**34.** Solve $4x^2 + 8x - 1 = 0$ by completing the square.

**35.** Solve: $\dfrac{1}{4}m^2 - m + \dfrac{1}{2} = 0$

**36.** Solve by using the quadratic formula. $\left(x - \dfrac{1}{2}\right)^2 = \dfrac{x}{2}$

**37.** Solve: $x^{2/3} - 5x^{1/3} + 6 = 0$

**38.** Solve:
$$\dfrac{1}{a + 5} = \dfrac{1}{3a + 6} - \dfrac{a + 2}{a^2 + 7a + 10}$$

**39.** Solve: $\dfrac{5}{x + 1} < -2$

**40.** Find the length of the unknown side of the triangle.

**41.** Graph $f(x) = 3x^2 + 3x + 1$. Find the vertex and any intercepts.

23. _____

24. _____

25. _____

26. a. _____

b. _____

27. _____

28. _____

29. _____

30. _____

31. _____

32. a. _____

b. _____

c. _____

d. _____

33. _____

34. _____

35. _____

36. _____

37. _____

38. _____

39. _____

40. _____

41. _____

**42.** _____

**43. a.** _____

**b.** _____

**44. a.** _____

**b.** _____

**c.** _____

**45.** _____

**46.** _____

**47.** _____

**48.** _____

**49.** _____

**50. a.** _____

**b.** _____

**51.** _____

**52.** _____

**53.** _____

**54.** _____

**42.** Divide $x^3 - 8$ by $x - 2$.

**43.** If $f(x) = |x|$ and $g(x) = x - 2$, find
   **a.** $(f \circ g)(x)$
   **b.** $(g \circ f)(x)$

**44.** Simplify each complex fraction.

   **a.** $\dfrac{\dfrac{a}{5}}{\dfrac{a - 1}{10}}$

   **b.** $\dfrac{\dfrac{3}{2 + a} + \dfrac{6}{2 - a}}{\dfrac{5}{a + 2} - \dfrac{1}{a - 2}}$

   **c.** $\dfrac{x^{-1} + y^{-1}}{xy}$

**45.** Find an equation of the inverse of $f(x) = 3x - 5$.

**46.** Use synthetic division to divide:
   $(8x^2 - 12x - 7) \div (x - 2)$

**47.** Solve: $4^{x+3} = 8^x$

**48.** Suppose that $y$ varies directly as $x$. If $y = \dfrac{1}{2}$ when $x = 12$, find the constant of variation and the variation equation.

**49.** Solve: $\log_x 25 = 2$

**50.** Add or subtract as indicated.

   **a.** $\dfrac{\sqrt{20}}{3} + \dfrac{\sqrt{5}}{4}$

   **b.** $\sqrt[3]{\dfrac{24x}{27}} - \dfrac{\sqrt[3]{3x}}{2}$

**51.** Find the amount owed at the end of 5 years if $1600 is loaned at a rate of 9% compounded continuously.

**52.** Write as a single logarithm:
   $7 \log_3 x + 9 \log_3 y$

**53.** Solve: $\log(x + 2) - \log x = 2$

**54.** Solve: $\log_7(8x - 6) = 2$

# Conic Sections

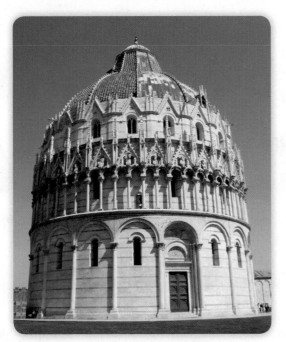

**The circular Baptistery in Pisa, Italy**

## What Are the Conic Sections, and How Are They Used?

Applications of conic sections—the parabola, the circle, the ellipse, and the hyperbola are too numerous to mention. For example, a parabolic orbit is often seen in comets. Many telescopes contain a parabolic dish because of the reflective properties of parabolas. The orbits of planets are in the shape of an ellipse. For this chapter opener, we study a building whose base is in the shape of a circle.

Located in the Piazza Dei Miracoli, in Italy, are the Duomo di Pisa, the great cathedral, and the Baptistery building. The circular Baptistery is the largest such building in Italy, and it is even taller than the Leaning Tower of Pisa, which is also in the Piazza.

In Section 10.1, Exercise 100, we will explore the dimensions of this interesting building.

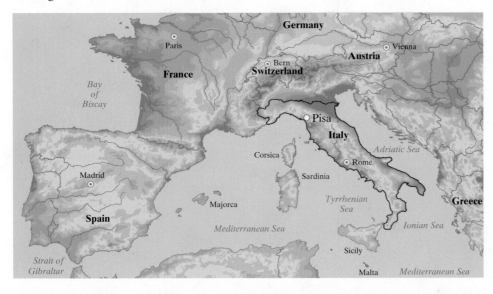

In Chapter 8, we analyzed some of the important connections between a parabola and its equation. Parabolas are interesting in their own right but are more interesting still because they are part of a collection of curves known as conic sections. This chapter is devoted to quadratic equations in two variables and their conic section graphs: the parabola, circle, ellipse, and hyperbola.

## Objectives

**A** Graph Parabolas of the Forms $y = a(x - h)^2 + k$ and $x = a(y - k)^2 + h$.

**B** Graph Circles of the Form $(x - h)^2 + (y - k)^2 = r^2$.

**C** Find the Center and the Radius of a Circle, Given Its Equation.

**D** Write the Equation of a Circle, Given Its Center and Radius.

**Conic sections** are called such because each conic section is the intersection of a right circular cone and a plane. The circle, parabola, ellipse, and hyperbola are the conic sections.

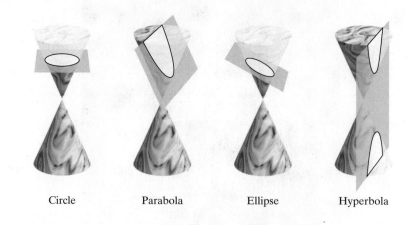

Circle      Parabola      Ellipse      Hyperbola

## Objective A  Graphing Parabolas

Thus far, we have seen that $f(x)$ or $y = a(x - h)^2 + k$ is the equation of a parabola that opens upward if $a > 0$ or downward if $a < 0$. Parabolas can also open left or right or even on a slant. Equations of these parabolas are not functions of $x$, of course, since a parabola opening any way other than upward or downward fails the vertical line test. In this section, we introduce parabolas that open to the left and to the right. Parabolas opening on a slant will not be developed in this book.

Just as $y = a(x - h)^2 + k$ is the equation of a parabola that opens upward or downward, $x = a(y - k)^2 + h$ is the equation of a parabola that opens to the right or to the left. The parabola opens to the right if $a > 0$ and to the left if $a < 0$. The parabola has vertex $(h, k)$, and its axis of symmetry is the line $y = k$.

### Parabolas

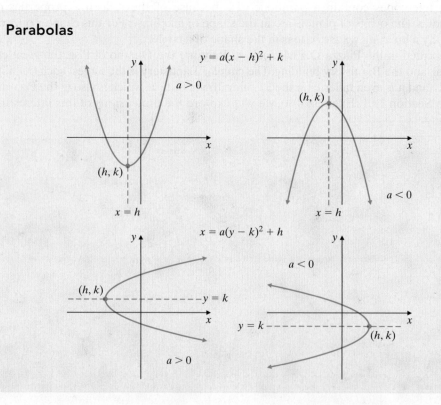

The forms $y = a(x - h)^2 + k$ and $x = a(y - k)^2 + h$ are called **standard forms.**

✓**Concept Check** Does the graph of the parabola given by the equation $x = -3y^2$ open to the left, to the right, upward, or downward?

**Example 1**   Graph:   $x = 2y^2$

**Solution:**   Written in standard form, the equation $x = 2y^2$ is $x = 2(y - 0)^2 + 0$ with $a = 2$, $h = 0$, and $k = 0$. Its graph is a parabola with vertex $(0, 0)$, and its axis of symmetry is the line $y = 0$. Since $a > 0$, this parabola opens to the right. We use a table to obtain a few more ordered pair solutions to help us graph $x = 2y^2$.

| $x$ | $y$ |
|---|---|
| 8 | −2 |
| 2 | −1 |
| 0 | 0 |
| 2 | 1 |
| 8 | 2 |

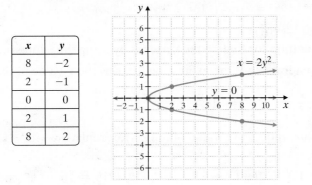

🔳 **Work Practice 1**

**Example 2**   Graph:   $x = -3(y - 1)^2 + 2$

**Solution:**   The equation $x = -3(y - 1)^2 + 2$ is in the form $x = a(y - k)^2 + h$ with $a = -3$, $k = 1$, and $h = 2$. Since $a < 0$, the parabola opens to the left. The vertex $(h, k)$ is $(2, 1)$, and the axis of symmetry is the horizontal line $y = 1$. When $y = 0$, the $x$-value is $-1$, so the $x$-intercept is $(-1, 0)$.

Again, we use a table to obtain a few ordered pair solutions and then graph the parabola.

| $x$ | $y$ |
|---|---|
| 2 | 1 |
| −1 | 0 |
| −1 | 2 |

🔳 **Work Practice 2**

## Practice 3

Graph: $y = -x^2 - 4x + 12$

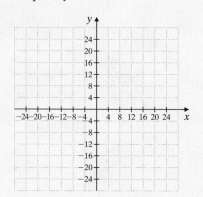

## Practice 4

Graph: $x = 3y^2 + 12y + 13$

**Answers**

**3.**

**4.**

**Example 3**     Graph: $y = -x^2 - 2x + 15$

**Solution:** Notice that this equation is not written in standard form, $y = a(x - h)^2 + k$. There are two methods that we can use to find the vertex. The first method is completing the square.

$$y - 15 = -x^2 - 2x \qquad \text{Subtract 15 from both sides.}$$
$$y - 15 = -1(x^2 + 2x) \qquad \text{Factor } -1 \text{ from the terms } -x^2 - 2x.$$

The coefficient of $x$ is 2, so we find the square of half of 2.

$$\frac{1}{2}(2) = 1 \quad \text{and} \quad 1^2 = 1$$

$$y - 15 - 1(1) = -1(x^2 + 2x + 1) \qquad \text{Add } -1(1) \text{ to both sides.}$$
$$y - 16 = -1(x + 1)^2 \qquad \text{Simplify the left side, and factor the right side.}$$
$$y = -(x + 1)^2 + 16 \qquad \text{Add 16 to both sides.}$$

The vertex is $(-1, 16)$.

The second method for finding the vertex is by using the expression $\dfrac{-b}{2a}$. Since the equation is quadratic in $x$, the expression gives us the $x$-value of the vertex.

$$x = \frac{-(-2)}{2(-1)} = \frac{2}{-2} = -1$$

To find the corresponding $y$-value of the vertex, replace $x$ with $-1$ in the original equation.

$$y = -(-1)^2 - 2(-1) + 15 = -1 + 2 + 15 = 16$$

Again, we see that the vertex is $(-1, 16)$, and the axis of symmetry is the vertical line $x = -1$. The $y$-intercept is $(0, 15)$. Now we can use a few more ordered pair solutions to graph the parabola.

| $x$ | $y$ |
|-----|-----|
| $-5$ | $0$ |
| $-3$ | $12$ |
| $-2$ | $15$ |
| $-1$ | $16$ |
| $0$ | $15$ |
| $1$ | $12$ |
| $3$ | $0$ |

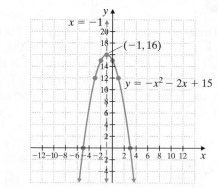

**Work Practice 3**

**Example 4**     Graph: $x = 2y^2 + 4y + 5$

**Solution:**   We notice that this equation is quadratic in $y$ so its graph is a parabola that opens to the left or the right. We can complete the square on $y$ or we can use the expression $\dfrac{-b}{2a}$ to find the vertex.

Since the equation is quadratic in $y$, the expression gives us the $y$-value of the vertex.

$$y = \frac{-4}{2 \cdot 2} = \frac{-4}{4} = -1$$

$$x = 2(-1)^2 + 4(-1) + 5 = 2 \cdot 1 - 4 + 5 = 3$$

The vertex is $(3, -1)$, and the axis of symmetry is the line $y = -1$. The parabola opens to the right since $a > 0$. The $x$-intercept is $(5, 0)$.

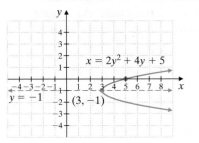

**■ Work Practice 4**

## Objective B Graphing Circles ▶

Another conic section is the **circle**. A circle is the set of all points in a plane that are the same distance from a fixed point called the **center**. The distance is called the **radius** of the circle. To find a standard equation for a circle, let $(h, k)$ represent the center of the circle, and let $(x, y)$ represent any point on the circle. The distance between $(h, k)$ and $(x, y)$ is defined to be the radius, $r$ units. We can find this distance $r$ by using the distance formula. (For a review of the distance formula, see Section 7.3.)

$r = \sqrt{(x - h)^2 + (y - k)^2}$    The distance formula.

$r^2 = (x - h)^2 + (y - k)^2$        Square both sides.

> ### Circle
>
> The graph of $(x - h)^2 + (y - k)^2 = r^2$ is a circle with center $(h, k)$ and radius $r$.
>
>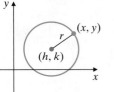

The form $(x - h)^2 + (y - k)^2 = r^2$ is called **standard form.**

If an equation can be written in the standard form

$(x - h)^2 + (y - k)^2 = r^2$

then its graph is a circle, which we can draw by graphing the center $(h, k)$ and using the radius $r$.

## Practice 5

Graph: $x^2 + y^2 = 36$

## Practice 6

Graph: $x^2 + (y + 2)^2 = 6$

**Answers**

**5.**

**6.**

### Helpful Hint

Notice that the radius is the *distance* from the center of the circle to any point of the circle. Also notice that the *midpoint* of a diameter of a circle is the center of the circle.

---

**Example 5**   Graph: $x^2 + y^2 = 4$

**Solution:**   The equation can be written in standard form as

$$(x - 0)^2 + (y - 0)^2 = 2^2$$

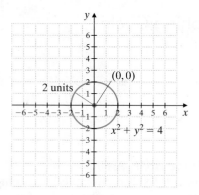

The center of the circle is $(0, 0)$, and the radius is 2. The graph of the circle is shown above.

■ **Work Practice 5**

### Helpful Hint

Notice the difference between the equation of a circle and the equation of a parabola. The equation of a circle contains both $x^2$- and $y^2$-terms on the same side of the equation with equal coefficients. The equation of a parabola has either an $x^2$-term or a $y^2$-term but not both.

---

**Example 6**   Graph: $(x + 1)^2 + y^2 = 8$

**Solution:**   The equation can be written as $(x - (-1))^2 + (y - 0)^2 = 8$ with $h = -1$, $k = 0$, and $r = \sqrt{8}$. The center is $(-1, 0)$, and the radius is $\sqrt{8} = 2\sqrt{2} \approx 2.8$. We use the decimal approximation to approximate the radius when graphing.

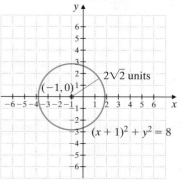

■ **Work Practice 6**

✓**Concept Check** In the graph of the equation $(x - 3)^2 + (y - 2)^2 = 5$, what is the distance between the center of the circle and any point on the circle?

## Objective C Finding the Center and the Radius of a Circle ▶

To find the center and the radius of a circle from its equation, we write the equation in standard form. To write the equation of a circle in standard form, we complete the square on both $x$ and $y$.

**Example 7** Graph: $x^2 + y^2 + 4x - 8y = 16$

**Solution:** Since this equation contains $x^2$- and $y^2$-terms on the same side of the equation with equal coefficients, its graph is a circle. To write the equation in standard form, we group the terms involving $x$ and the terms involving $y$, and then complete the square on each variable.

$$(x^2 + 4x) + (y^2 - 8y) = 16$$

Now, $\frac{1}{2}(4) = 2$ and $2^2 = 4$. Also, $\frac{1}{2}(-8) = -4$ and $(-4)^2 = 16$. We add 4 and then 16 to both sides.

$$(x^2 + 4x + 4) + (y^2 - 8y + 16) = 16 + 4 + 16$$
$$(x + 2)^2 + (y - 4)^2 = 36 \qquad \text{Factor.}$$

This circle has the center $(-2, 4)$ and radius 6, as shown.

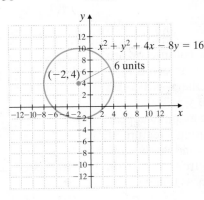

**Work Practice 7**

## Objective D Writing Equations of Circles ▶

Since a circle is determined entirely by its center and radius, this information is all we need to write an equation of a circle.

**Example 8** Write an equation of the circle with center $(-7, 3)$ and radius 10.

**Solution:** Using the given values $h = -7$, $k = 3$, and $r = 10$, we write the equation

$$(x - h)^2 + (y - k)^2 = r^2$$

or

$$(x - (-7))^2 + (y - 3)^2 = 10^2 \qquad \text{Substitute the given values.}$$

or

$$(x + 7)^2 + (y - 3)^2 = 100$$

**Work Practice 8**

---

**Practice 7**

Graph: $x^2 + y^2 - 2x + 6y = 6$

**Practice 8**

Write an equation of the circle with the center $(2, -5)$ and radius 7.

**Answers**

7.

8. $(x - 2)^2 + (y + 5)^2 = 49$

✓**Concept Check Answer**

$\sqrt{5}$ units

## 🖩 Calculator Explorations **Graphing**

To graph an equation such as $x^2 + y^2 = 25$ with a graphing calculator, we first solve the equation for $y$.

$$x^2 + y^2 = 25$$
$$y^2 = 25 - x^2$$
$$y = \pm\sqrt{25 - x^2}$$

The graph of $y = \sqrt{25 - x^2}$ will be the top half of the circle, and the graph of $y = -\sqrt{25 - x^2}$ will be the bottom half of the circle.

To graph, we press $\boxed{Y =}$ and enter $Y_1 = \sqrt{25 - x^2}$ and $Y_2 = -\sqrt{25 - x^2}$. We insert parentheses about $25 - x^2$ so that $\sqrt{25 - x^2}$ and not $\sqrt{25} - x^2$ is graphed.

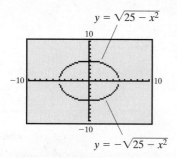

The graph does not appear to be a circle because we are currently using a standard window and the screen is rectangular. This causes the tick marks on the $x$-axis to be farther apart than the tick marks on the $y$-axis and thus creates the distorted circle. If we want the graph to appear circular, we define a square window by using a feature of the graphing calculator or redefine the window to show the $x$-axis from $-15$ to $15$ and the $y$-axis from $-10$ to $10$. Using a square window, the graph appears as follows:

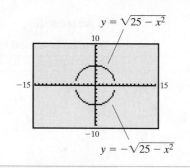

*Use a graphing calculator to graph each circle.*

**1.** $x^2 + y^2 = 55$

**2.** $x^2 + y^2 = 20$

**3.** $7x^2 + 7y^2 - 89 = 0$

**4.** $3x^2 + 3y^2 - 35 = 0$

## Vocabulary, Readiness & Video Check

*Use the choices below to fill in each blank. Some choices may be used more than once.*

radius          center          vertex

diameter          circle          conic sections

1. The circle, parabola, ellipse, and hyperbola are called the _____.
2. For a parabola that opens upward, the lowest point is the _____.
3. A _____ is the set of all points in a plane that are the same distance from a fixed point. The fixed point is called the _____.
4. The midpoint of a diameter of a circle is the _____.
5. The distance from the center of a circle to any point of the circle is called the _____.
6. Twice a circle's radius is its _____.

*Martin-Gay Interactive Videos*     Watch the section lecture video and answer the following questions.

See Video 10.1

**Objective A**   7. Based on ▤ Example 1 and the lecture before, would you say that parabolas of the form $x = a(y - k)^2 + h$ are functions? Why or why not? ▶

**Objective B**   8. Based on the lecture before ▤ Example 2, what would be the standard form of a circle with its center at the origin? Simplify your answer. ▶

**Objective C**   9. From ▤ Example 3, why do we need to complete the square twice when writing this equation of a circle in standard form? ▶

**Objective D**  10. From ▤ Example 4, if you know the center and radius of a circle, how can you write that circle's equation? ▶

## 10.1  Exercise Set MyMathLab® ▶

**Objective A**  *The graph of each equation is a parabola. Determine whether the parabola opens upward, downward, to the left, or to the right. See Examples 1 through 4.*

1. $y = x^2 - 7x + 5$

2. $y = -x^2 + 16$

3. $x = -y^2 - y + 2$

4. $x = 3y^2 + 2y - 5$

5. $y = -x^2 + 2x + 1$

6. $x = -y^2 + 2y - 6$

*The graph of each equation is a parabola. Find the vertex of the parabola and then graph it. See Examples 1 through 4.*

**7.** $x = 3y^2$

**8.** $x = 5y^2$

**9.** $x = -2y^2$

**10.** $x = -4y^2$

**11.** $y = -4x^2$

**12.** $y = -2x^2$

**13.** $x = (y - 2)^2 + 3$

**14.** $x = (y - 4)^2 - 1$

**15.** $y = -3(x - 1)^2 + 5$

**16.** $y = -4(x - 2)^2 + 2$

**17.** $x = y^2 + 6y + 8$

**18.** $x = y^2 - 6y + 6$

**19.** $y = x^2 + 10x + 20$

**20.** $y = x^2 + 4x - 5$

**21.** $x = -2y^2 + 4y + 6$

**22.** $x = 3y^2 + 6y + 7$

**Objectives** **B** **C** **Mixed Practice** *The graph of each equation is a circle. Find the center and the radius, and then graph the circle. See Examples 5 through 7.*

**23.** $x^2 + y^2 = 9$

**24.** $x^2 + y^2 = 25$

**25.** $x^2 + (y - 2)^2 = 1$

**26.** $(x - 3)^2 + y^2 = 9$

**27.** $(x - 5)^2 + (y + 2)^2 = 1$

**28.** $(x + 3)^2 + (y + 3)^2 = 4$

**29.** $x^2 + y^2 + 6y = 0$

**30.** $x^2 + 10x + y^2 = 0$

**31.** $x^2 + y^2 + 2x - 4y = 4$

**32.** $x^2 + y^2 + 6x - 4y = 3$

**33.** $(x + 2)^2 + (y - 3)^2 = 7$

**34.** $(x + 1)^2 + (y - 2)^2 = 5$

**35.** $x^2 + y^2 - 4x - 8y - 2 = 0$

**36.** $x^2 + y^2 - 2x - 6y - 5 = 0$

*Hint: For Exercises 37 through 42, first divide the equation through by the coefficient of $x^2$ (or $y^2$).*

**37.** $3x^2 + 3y^2 = 75$

**38.** $2x^2 + 2y^2 = 18$

**39.** $6(x - 4)^2 + 6(y - 1)^2 = 24$

**40.** $7(x - 1)^2 + 7(y - 3)^2 = 63$

**41.** $4(x + 1)^2 + 4(y - 3)^2 = 12$

**42.** $5(x - 2)^2 + 5(y + 1) = 50$

## Objective D   *Write an equation of the circle with the given center and radius. See Example 8.*

**43.** $(2, 3); 6$

**44.** $(-7, 6); 2$

**45.** $(0, 0); \sqrt{3}$

**46.** $(0, -6); \sqrt{2}$

**47.** $(-5, 4); 3\sqrt{5}$

**48.** The origin; $4\sqrt{7}$

## Objectives A B C Mixed Practice *Sketch the graph of each equation. If the graph is a parabola, find its vertex. If the graph is a circle, find its center and radius.*

**49.** $x = y^2 - 3$

**50.** $x = y^2 + 2$

**51.** $y = (x - 2)^2 - 2$

**52.** $y = (x + 3)^2 + 3$

**53.** $x^2 + y^2 = 1$

**54.** $x^2 + y^2 = 49$

**55.** $x = (y + 3)^2 - 1$

**56.** $x = (y - 1)^2 + 4$

**57.** $(x - 2)^2 + (y - 2)^2 = 16$

**58.** $(x + 3)^2 + (y - 1)^2 = 9$

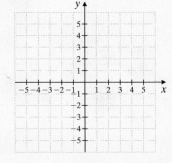

**59.** $x = -(y - 1)^2$

**60.** $x = -2(y + 5)^2$

**61.** $(x - 4)^2 + y^2 = 7$

**62.** $x^2 + (y + 5)^2 = 5$

**63.** $y = 5(x + 5)^2 + 3$

**64.** $y = 3(x - 4)^2 + 2$

**65.** $\dfrac{x^2}{8} + \dfrac{y^2}{8} = 2$

**66.** $2x^2 + 2y^2 = \dfrac{1}{2}$

**67.** $y = x^2 + 7x + 6$

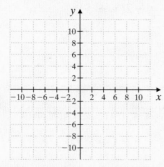

**68.** $y = x^2 - 2x - 15$

**69.** $x^2 + y^2 + 2x + 12y - 12 = 0$

**70.** $x^2 + y^2 + 6x + 10y - 2 = 0$

**71.** $x = y^2 + 8y - 4$

**72.** $x = y^2 + 6y + 2$

**73.** $x^2 - 10y + y^2 + 4 = 0$

**74.** $x^2 + y^2 - 8y + 5 = 0$

**75.** $x = -3y^2 + 30y$

**76.** $x = -2y^2 - 4y$

**77.** $5x^2 + 5y^2 = 25$

**78.** $\dfrac{x^2}{3} + \dfrac{y^2}{3} = 2$

**79.** $y = 5x^2 - 20x + 16$

**80.** $y = 4x^2 - 40x + 105$

# Review

*Graph each equation. See Sections 3.1 and 3.3.*

**81.** $y = 2x + 5$

**82.** $y = -3x + 3$

**83.** $y = 3$

**84.** $x = -2$

*Rationalize each denominator and simplify if possible. See Section 7.5.*

**85.** $\dfrac{1}{\sqrt{3}}$

**86.** $\dfrac{\sqrt{5}}{\sqrt{8}}$

**87.** $\dfrac{4\sqrt{7}}{\sqrt{6}}$

**88.** $\dfrac{10}{\sqrt{5}}$

## Concept Extensions

*For Exercises 89 and 90, explain the error in each statement.*

**89.** The graph of $x = 5(y + 5)^2 + 1$ is a parabola with vertex $(-5, 1)$ and opening to the right.

**90.** The graph of $x^2 + (y + 3)^2 = 10$ is a circle with center $(0, -3)$ and radius 5.

**91. The Sarsen Circle:** The first image that comes to mind when one thinks of Stonehenge is the very large sandstone blocks with sandstone lintels across the top. The Sarsen Circle of Stonehenge is the outer circle of the sandstone blocks, each of which weighs up to 50 tons. There were originally 30 of these monolithic blocks, but only 17 remain upright to this day. The "altar stone" lies at the center of this circle, which has a diameter of 33 meters.

**a.** What is the radius of the Sarsen Circle?

**b.** What is the circumference of the Sarsen Circle? Round your result to 2 decimal places.

**c.** Since there were originally 30 Sarsen stones located on the circumference, how far apart would the centers of the stones have been? Round to the nearest tenth of a meter.

**d.** Using the axes in the drawing, what are the coordinates of the center of the circle?

**e.** Use parts **a** and **d** to write an equation of the Sarsen Circle.

**92.** There is heavy competition to be the tallest observation wheel in the world. The largest observation wheel at the moment is the "High Roller" in Las Vegas, although New York, Dubai, and China all have larger wheels in the planning stage. From the High Roller, you can see all of Las Vegas. Each of its 28-passenger cabins can accommodate 40 people and completes a full rotation every 30 minutes. The High Roller has a diameter of 520 feet and a height of 550 feet. (*Source:* NBCnews.com)

**a.** What is the radius of this observation wheel?

**b.** How close is the wheel to the ground?

**c.** How high is the center of the wheel from the ground?

**d.** Using the axes in the drawing, what are the coordinates of the center of the wheel?

**e.** Use parts **a** and **d** to write an equation of the wheel.

**93.** In 1893, Pittsburgh bridge builder George Ferris designed and built a gigantic revolving steel wheel whose height was 264 feet and diameter was 250 feet. This Ferris wheel opened at the 1893 exposition in Chicago. It had 36 wooden cars, each capable of holding 60 passengers. (*Source: The Handy Science Answer Book*)

   **a.** What was the radius of this Ferris wheel?
   **b.** How close is the wheel to the ground?
   **c.** How high is the center of the wheel from the ground?
   **d.** Using the axes in the drawing, what are the coordinates of the center of the wheel?
   **e.** Use parts **a** and **d** to write an equation of the wheel.

**94.** The Cosmo Clock 21 is a large Ferris wheel currently operating in Yokohama City, Japan. It has a 60-armed wheel, its diameter is 100 meters, and it has a height of 105 meters. (*Source: The Handy Science Answer Book*)

   **a.** What is the radius of this Ferris wheel?
   **b.** How close is the wheel to the ground?
   **c.** How high is the center of the wheel from the ground?
   **d.** Using the axes in the drawing, what are the coordinates of the center of the wheel?
   **e.** Use parts **a** and **d** to write an equation of the wheel.

**95.** If you are given a list of equations of circles and parabolas and none are in standard form, explain how you would determine which is an equation of a circle and which is an equation of a parabola. Explain also how you would distinguish the upward or downward parabolas from the left-opening or right-opening parabolas.

△ **96.** Determine whether the triangle with vertices $(2, 6)$, $(0, -2)$, and $(5, 1)$ is an isosceles triangle.

*Solve.*

**97.** Two surveyors need to find the distance across a lake. They place a reference pole at point $A$ in the diagram. Point $B$ is 3 meters east and 1 meter north of the reference point $A$. Point $C$ is 19 meters east and 13 meters north of point $A$. Find the distance across the lake, from $B$ to $C$.

**98.** A bridge constructed over a bayou has a supporting arch in the shape of a parabola. Find an equation of the parabolic arch if the length of the road over the arch is 100 meters and the maximum height of the arch is 40 meters. (*Hint:* Use the lower horizontal line in the figure as the $x$-axis.).

**99.** Cindy Brown, an architect, is drawing plans on grid paper for a circular pool with a fountain in the middle. The paper is marked off in centimeters, and each centimeter represents 1 foot. On the paper, the diameter of the "pool" is 20 centimeters, and "fountain" is the point $(0, 0)$.

  **a.** Sketch the architect's drawing. Be sure to label the axes.

  **b.** Write an equation that describes the circular pool.

  **c.** Cindy plans to place a circle of lights around the fountain such that each light is 5 feet from the fountain. Write an equation for the circle of lights and sketch the circle on your drawing.

**100.** The circular Baptistery building in Pisa is the largest such building in Italy, with a height of 180 feet and a circumference of approximately 352 feet. (*Source: towerofpisa.net*)

  **a.** What is the diameter of the base floor (circular imprint) of the Baptistery? Round to two decimal places.

  **b.** What is the radius of the circular footprint of the Baptistery? Round to two decimal places.

  **c.** Using the axes in the drawing, what are the coordinates of the center of the circular imprint?

  **d.** Write an equation for the circular footprint of the Baptistry.

180 ft

circular footprint

Circumference of base (circular imprint): 352 ft

## 10.2    The Ellipse and the Hyperbola

### Objectives

**A** Define and Graph Ellipses.

**B** Define and Graph Hyperbolas.

### Objective A  Graphing Ellipses

An **ellipse** can be thought of as the set of points in a plane such that the sum of the distances of each of those points from two fixed points is constant. Each of the two fixed points is called a **focus**. The plural of focus is **foci.** The point midway between the foci is called the **center.**

    An ellipse may be drawn by hand by using two tacks, a piece of string, and a pencil. Secure the two tacks into a piece of cardboard, for example, and tie each end of the string to a tack. Use your pencil to pull the string tight and draw the ellipse. The two tacks are the foci of the drawn ellipse.

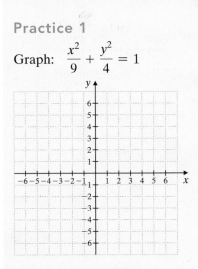

## Practice 1

Graph: $\dfrac{x^2}{9} + \dfrac{y^2}{4} = 1$

## Ellipse with Center (0, 0)

The graph of an equation of the form $\dfrac{x^2}{a^2} + \dfrac{y^2}{b^2} = 1$ is an ellipse with center $(0,0)$.

The $x$-intercepts are $(a, 0)$ and $(-a, 0)$, and the $y$-intercepts are $(0, b)$ and $(0, -b)$.

The **standard form** of the equation of an ellipse with center $(0, 0)$ is $\dfrac{x^2}{a^2} + \dfrac{y^2}{b^2} = 1$.

## Practice 2

Graph: $4x^2 + 36y^2 = 144$

**Example 1** Graph: $\dfrac{x^2}{9} + \dfrac{y^2}{16} = 1$

**Solution:** The equation is of the form $\dfrac{x^2}{a^2} + \dfrac{y^2}{b^2} = 1$ with $a = 3$ and $b = 4$, so its graph is an ellipse with center $(0, 0)$, $x$-intercepts $(3, 0)$ and $(-3, 0)$, and $y$-intercepts $(0, 4)$ and $(0, -4)$.

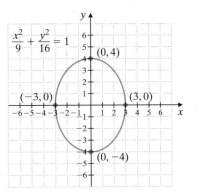

■ Work Practice 1

**Example 2** Graph: $4x^2 + 16y^2 = 64$

**Solution:** Although this equation contains a sum of squared terms in $x$ and $y$ on the same side of an equation, this is not the equation of a circle since the coefficients of $x^2$ and $y^2$ are not the same. When this happens, the graph is an ellipse. Since the standard form of the equation of an ellipse has 1 on one side, we divide both sides of this equation by 64 to get it in standard form.

$$4x^2 + 16y^2 = 64$$

$$\frac{4x^2}{64} + \frac{16y^2}{64} = \frac{64}{64} \quad \text{Divide both sides by 64.}$$

$$\frac{x^2}{16} + \frac{y^2}{4} = 1 \quad \text{Simplify.}$$

*(Continued on next page)*

**Answers**

1.

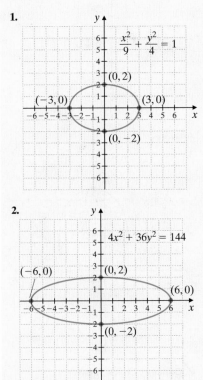

2.

We now recognize the equation of an ellipse with center $(0, 0)$, $x$-intercepts $(4, 0)$ and $(-4, 0)$, and $y$-intercepts $(0, 2)$ and $(0, -2)$.

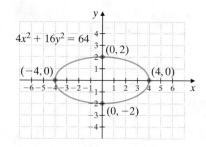

**Work Practice 2**

The center of an ellipse is not always $(0, 0)$, as shown in the next example.

> ### Ellipse with Center (*h, k*)
>
> The standard form of the equation of an ellipse with center $(h, k)$ is
>
> $$\frac{(x - h)^2}{a^2} + \frac{(y - k)^2}{b^2} = 1$$

**Practice 3**

Graph:

$$\frac{(x - 1)^2}{9} + \frac{(y - 3)^2}{16} = 1$$

**Answer**

**3.**

$$\frac{(x - 1)^2}{9} + \frac{(y - 3)^2}{16} = 1$$

**Example 3**  Graph:  $\dfrac{(x + 3)^2}{25} + \dfrac{(y - 2)^2}{36} = 1$

**Solution:**  This ellipse has center $(-3, 2)$. Notice that $a = 5$ and $b = 6$. To find four points on the graph of the ellipse, we first graph the center, $(-3, 2)$. Since $a = 5$, we count 5 units right and then 5 units left of the point with coordinates $(-3, 2)$. Next, since $b = 6$, we start at $(-3, 2)$ and count 6 units up and then 6 units down to find two more points on the ellipse.

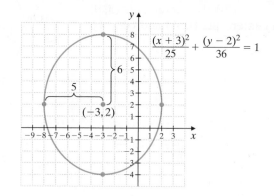

**Work Practice 3**

**✓Concept Check**  In the graph of the equation $\dfrac{x^2}{64} + \dfrac{y^2}{36} = 1$, which distance is longer: the distance between the $x$-intercepts or the distance between the $y$-intercepts? How much longer? Explain.

## Objective B  Graphing Hyperbolas ▶

The final conic section is the **hyperbola.** A hyperbola is the set of points in a plane such that for each point in the set, the absolute value of the difference of the

**✓Concept Check Answer**

$x$-intercepts, by 4 units

distances from two fixed points is constant. Each of the two fixed points is called a **focus.** The point midway between the foci is called the **center.**

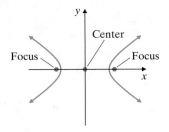

Using the distance formula, we can show that the graph of $\dfrac{x^2}{a^2} - \dfrac{y^2}{b^2} = 1$ is a hyperbola with center $(0, 0)$ and $x$-intercepts $(a, 0)$ and $(-a, 0)$. Also, the graph of $\dfrac{y^2}{b^2} - \dfrac{x^2}{a^2} = 1$ is a hyperbola with center $(0, 0)$ and $y$-intercepts $(0, b)$ and $(0, -b)$.

## Hyperbola with Center (0, 0)

The graph of an equation of the form $\dfrac{x^2}{a^2} - \dfrac{y^2}{b^2} = 1$ is a hyperbola with center $(0, 0)$ and $x$-intercepts $(a, 0)$ and $(-a, 0)$.

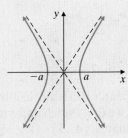

The graph of an equation of the form $\dfrac{y^2}{b^2} - \dfrac{x^2}{a^2} = 1$ is a hyperbola with center $(0, 0)$ and $y$-intercepts $(0, b)$ and $(0, -b)$.

The equations $\dfrac{x^2}{a^2} - \dfrac{y^2}{b^2} = 1$ and $\dfrac{y^2}{b^2} - \dfrac{x^2}{a^2} = 1$ are the **standard forms** for the equation of a hyperbola.

## Helpful Hint

Notice the difference between the equation of an ellipse and a hyperbola. The equation of an ellipse contains $x^2$- and $y^2$-terms on the same side of the equation with same-sign coefficients. For a hyperbola, the coefficients on the same side of the equation have different signs.

Graphing a hyperbola such as $\dfrac{y^2}{b^2} - \dfrac{x^2}{a^2} = 1$ is made easier by recognizing one of its important characteristics. Examining the figure below, notice how the sides of the branches of the hyperbola extend indefinitely and seem to approach, but not intersect, the dashed lines in the figure. These dashed lines are called the **asymptotes** of the hyperbola.

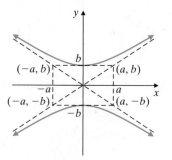

To sketch these lines, or asymptotes, draw a rectangle with vertices $(a, b)$, $(-a, b)$, $(a, -b)$, and $(-a, -b)$. The asymptotes of the hyperbola are the extended diagonals of this rectangle.

**Practice 4**

Graph: $\dfrac{x^2}{9} - \dfrac{y^2}{4} = 1$

**Example 4**   Graph: $\dfrac{x^2}{16} - \dfrac{y^2}{25} = 1$

**Solution:**   This equation has the form $\dfrac{x^2}{a^2} - \dfrac{y^2}{b^2} = 1$, with $a = 4$ and $b = 5$. Thus, its graph is a hyperbola with center $(0, 0)$ and $x$-intercepts of $(4, 0)$ and $(-4, 0)$. To aid in graphing the hyperbola, we first sketch its asymptotes. The extended diagonals of the rectangle with coordinates $(4, 5)$, $(4, -5)$, $(-4, 5)$, and $(-4, -5)$ are the asymptotes of the hyperbola. Then we use the asymptotes to aid in graphing the hyperbola.

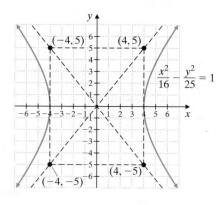

**Work Practice 4**

**Answer**

**4.**

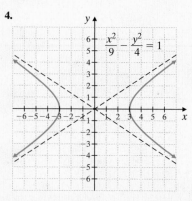

**Example 5** Graph: $4y^2 - 9x^2 = 36$

**Solution:** Since this is a difference of squared terms in $x$ and $y$ on the same side of the equation, its graph is a hyperbola, as opposed to an ellipse or a circle. The standard form of the equation of a hyperbola has a 1 on one side, so we divide both sides of the equation by 36 to get it in standard form.

$$4y^2 - 9x^2 = 36$$

$$\frac{4y^2}{36} - \frac{9x^2}{36} = \frac{36}{36} \quad \text{Divide both sides by 36.}$$

$$\frac{y^2}{9} - \frac{x^2}{4} = 1 \quad \text{Simplify.}$$

The equation is of the form $\dfrac{y^2}{b^2} - \dfrac{x^2}{a^2} = 1$ with $a = 2$ and $b = 3$, so the hyperbola is centered at $(0, 0)$ with $y$-intercepts $(0, 3)$ and $(0, -3)$.

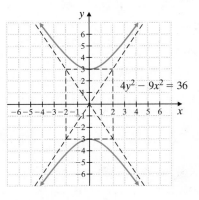

**Work Practice 5**

Although this is beyond the scope of this text, the standard forms of the equations of hyperbolas with center $(h, k)$ are given below. The Concept Extensions section in this Exercise Set (10.2) contains some hyperbolas of this form.

### Hyperbola with Center ($h$, $k$)

Standard forms of the equations of hyperbolas with centers $(h, k)$ are:

$$\frac{(x - h)^2}{a^2} - \frac{(y - k)^2}{b^2} = 1 \qquad \frac{(y - k)^2}{b^2} - \frac{(x - h)^2}{a^2} = 1$$

**Practice 5**

Graph: $9y^2 - 16x^2 = 144$

**Answer**

**5.**

## Calculator Explorations Graphing

To find the graph of an ellipse using a graphing calculator, use the same procedure as for graphing a circle. For example, to graph $x^2 + 3y^2 = 22$, first solve for $y$.

$$3y^2 = 22 - x^2$$

$$y^2 = \frac{22 - x^2}{3}$$

$$y = \pm\sqrt{\frac{22 - x^2}{3}}$$

Next press the $\boxed{Y=}$ key and enter $Y_1 = \sqrt{\frac{22 - x^2}{3}}$ and $Y_2 = -\sqrt{\frac{22 - x^2}{3}}$. (Insert two sets of parentheses in the radicand as in $\sqrt{((22 - x^2)/3)}$ so that the desired graph is obtained.) The graph appears as follows:

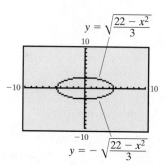

$$y = \sqrt{\frac{22 - x^2}{3}}$$

$$y = -\sqrt{\frac{22 - x^2}{3}}$$

Note: Use a square window for a better view of the shape of each ellipse.

*Use a graphing calculator to graph each ellipse.*

**1.** $10x^2 + y^2 = 32$

**2.** $20x^2 + 5y^2 = 100$

**3.** $7.3x^2 + 15.5y^2 = 95.2$

**4.** $18.8x^2 + 36.1y^2 = 205.8$

## Vocabulary, Readiness & Video Check

*Use the choices below to fill in each blank. Some choices will be used more than once and some not at all.*

| | | | | | |
|---|---|---|---|---|---|
| ellipse | $(0,0)$ | $x$ | $(a,0)$ and $(-a,0)$ | $(0,a)$ and $(0,-a)$ | focus |
| hyperbola | center | $y$ | $(b,0)$ and $(-b,0)$ | $(0,b)$ and $(0,-b)$ | |

1. A(n) _____ is the set of points in a plane such that the absolute value of the differences of their distances from two fixed points is constant.

2. A(n) _____ is the set of points in a plane such that the sum of their distances from two fixed points is constant.

*For Exercises 1 and 2 above,*

3. The two fixed points are each called a(n) _____ .

4. The point midway between the foci is called the _____ .

5. The graph of $\dfrac{x^2}{a^2} - \dfrac{y^2}{b^2} = 1$ is a(n) _____ with center _____ and _____ -intercepts of _____ .

6. The graph of $\dfrac{x^2}{a^2} + \dfrac{y^2}{b^2} = 1$ is a(n) _____ with center _____ and $y$-intercepts of _____ .

*Martin-Gay Interactive Videos*     Watch the section lecture video and answer the following questions.

**Objective A** 7. From ⊞ Example 1, what information do the values of $a$ and $b$ give us about the graph of an ellipse? Answer this same question for ⊞ Example 2. ⊙

**Objective B** 8. From ⊞ Example 3, we know the points $(a,b)$, $(a,-b)$, $(-a,b)$, and $(-a,-b)$ are not part of the graph. Explain the role of these points. ⊙

*See Video 10.2* 🍎

**10.2 Exercise Set** MyMathLab® ⊙

**Objective A** *Graph each ellipse. See Examples 1 and 2.*

1. $\dfrac{x^2}{4} + \dfrac{y^2}{25} = 1$

2. $\dfrac{x^2}{16} + \dfrac{y^2}{9} = 1$

3. $\dfrac{x^2}{9} + y^2 = 1$

4. $x^2 + \dfrac{y^2}{4} = 1$

**5.** $9x^2 + 4y^2 = 36$     **6.** $x^2 + 4y^2 = 16$     **7.** $4x^2 + 25y^2 = 100$     **8.** $36x^2 + y^2 = 36$

*Graph each ellipse. See Example 3.*

**9.** $\dfrac{(x+1)^2}{36} + \dfrac{(y-2)^2}{49} = 1$

**10.** $\dfrac{(x-3)^2}{9} + \dfrac{(y+3)^2}{16} = 1$

**11.** $\dfrac{(x-1)^2}{4} + \dfrac{(y-1)^2}{25} = 1$

**12.** $\dfrac{(x+3)^2}{16} + \dfrac{(y+2)^2}{4} = 1$

## Objective B   *Graph each hyperbola. See Examples 4 and 5.*

**13.** $\dfrac{x^2}{4} - \dfrac{y^2}{9} = 1$     **14.** $\dfrac{x^2}{36} - \dfrac{y^2}{36} = 1$     **15.** $\dfrac{y^2}{25} - \dfrac{x^2}{16} = 1$     **16.** $\dfrac{y^2}{25} - \dfrac{x^2}{49} = 1$

**17.** $x^2 - 4y^2 = 16$    **18.** $4x^2 - y^2 = 36$    **19.** $16y^2 - x^2 = 16$    **20.** $4y^2 - 25x^2 = 100$

**Objectives A B Mixed Practice** *Identify the graph of each equation as an ellipse or a hyperbola. See Examples 1 through 5.*

**21.** $\dfrac{x^2}{16} + \dfrac{y^2}{4} = 1$    **22.** $\dfrac{x^2}{16} - \dfrac{y^2}{4} = 1$    **23.** $x^2 - 5y^2 = 3$

**24.** $-x^2 + 5y^2 = 3$    **25.** $-\dfrac{y^2}{25} + \dfrac{x^2}{36} = 1$    **26.** $\dfrac{y^2}{25} + \dfrac{x^2}{36} = 1$

*Graph each equation. See Examples 1 through 5.*

**27.** $\dfrac{y^2}{36} = 1 - x^2$    **28.** $\dfrac{x^2}{36} = 1 - y^2$    **29.** $4(x - 1)^2 + 9(y + 2)^2 = 36$

**30.** $25(x + 3)^2 + 4(y - 3)^2 = 100$    **31.** $8x^2 + 2y^2 = 32$    **32.** $3x^2 + 12y^2 = 48$

**33.** $25x^2 - y^2 = 25$

**34.** $x^2 - 9y^2 = 9$

**Mixed Practice—Sections 10.1, 10.2** *Identify whether each equation, when graphed, will be a parabola, circle, ellipse, or hyperbola. Sketch the graph of each equation.*

*If a parabola, label the vertex.*
*If a circle, label the center note the radius.*
*If an ellipse, label the center.*
*If a hyperbola, label the x- or y-intercepts.*

**35.** $(x - 7)^2 + (y - 2)^2 = 4$

**36.** $x^2 + y^2 = 16$

**37.** $y = x^2 + 12x + 36$

**38.** $y = x^2 + 4$

**39.** $\dfrac{y^2}{9} - \dfrac{x^2}{9} = 1$

**40.** $\dfrac{x^2}{16} - \dfrac{y^2}{4} = 1$

**41.** $\dfrac{x^2}{16} + \dfrac{y^2}{4} = 1$

**42.** $\dfrac{x^2}{4} + \dfrac{y^2}{9} = 1$

**43.** $x = y^2 + 4y - 1$

**44.** $x = -y^2 + 6y$          **45.** $9x^2 - 4y^2 = 36$          **46.** $y^2 = x^2 + 16$

**47.** $\dfrac{(x-1)^2}{49} + \dfrac{(y+2)^2}{25} = 1$          **48.** $\dfrac{(x+2)^2}{36} + \dfrac{(y-3)^2}{16} = 1$          **49.** $\left(x + \dfrac{1}{2}\right)^2 + \left(y - \dfrac{1}{2}\right)^2 = 1$

**50.** $\left(x - \dfrac{3}{2}\right)^2 + \left(y + \dfrac{3}{2}\right)^2 = 9$          **51.** $y = -2x^2 + 4x - 3$          **52.** $y = -\dfrac{1}{2}x^2 + 2x - 4$

## Review

*Perform each indicated operation. See Sections 5.1 and 5.2.*

**53.** $(2x^3)(-4x^2)$          **54.** $2x^3 - 4x^3$          **55.** $-5x^2 + x^2$          **56.** $(-5x^2)(x^2)$

## Concept Extensions

*The graph of each equation is an ellipse. Determine which distance is longer, the distance between the x-intercepts or the distance between the y-intercepts. How much longer? See the Concept Check in this section.*

**57.** $\dfrac{x^2}{16} + \dfrac{y^2}{25} = 1$          **58.** $\dfrac{x^2}{100} + \dfrac{y^2}{49} = 1$          **59.** $4x^2 + y^2 = 16$          **60.** $x^2 + 4y^2 = 36$

*Solve.*

**61.** If you are given a list of equations of circles, parabolas, ellipses, and hyperbolas, explain how you could distinguish the different conic sections from their equations.

**62.** We know that $x^2 + y^2 = 25$ is the equation of a circle. Rewrite the equation so that the right side is equal to 1. Which type of conic section does this equation form resemble? In fact, the circle is a special case of this type of conic section. Describe the conditions under which this type of conic section is a circle.

*The orbits of stars, planets, comets, asteroids, and satellites all have the shape of one of the conic sections. Astronomers use a measure called eccentricity to describe the shape and elongation of an orbital path. For the circle and ellipse, eccentricity e is calculated with the formula $e = \dfrac{c}{d}$, where $c^2 = |a^2 - b^2|$ and d is the larger value of a or b. For a hyperbola, eccentricity e is calculated with the formula $e = \dfrac{c}{d}$, where $c^2 = a^2 + b^2$ and the value of d is equal to a if the hyperbola has x-intercepts or equal to b if the hyperbola has y-intercepts. Use equations A–H to answer Exercises 63 through 72.*

**A.** $\dfrac{x^2}{36} - \dfrac{y^2}{13} = 1$     **B.** $\dfrac{x^2}{4} + \dfrac{y^2}{4} = 1$     **C.** $\dfrac{x^2}{25} + \dfrac{y^2}{16} = 1$     **D.** $\dfrac{y^2}{25} - \dfrac{x^2}{39} = 1$

**E.** $\dfrac{x^2}{17} + \dfrac{y^2}{81} = 1$     **F.** $\dfrac{x^2}{36} + \dfrac{y^2}{36} = 1$     **G.** $\dfrac{x^2}{16} - \dfrac{y^2}{65} = 1$     **H.** $\dfrac{x^2}{144} + \dfrac{y^2}{140} = 1$

**63.** Identify the type of conic section represented by each of the equations A–H.

**64.** For each of the equations A–H, identify the values of $a^2$ and $b^2$.

**65.** For each of the equations A–H, calculate the value of $c^2$ and $c$.

**66.** For each of the equations A–H, find the value of $d$.

**67.** For each of the equations A–H, calculate the eccentricity $e$.

**68.** What do you notice about the values of $e$ for the equations you identified as ellipses?

**69.** What do you notice about the values of $e$ for the equations you identified as circles?

**70.** What do you notice about the values of $e$ for the equations you identified as hyperbolas?

**71.** The eccentricity of a parabola is exactly 1. Use this information and the observations you made in Exercises 68, 69, and 70 to describe a way that could be used to identify the type of conic section based on its eccentricity value.

**72.** Graph each of the conic sections given in equations A–H. What do you notice about the shape of the ellipses for increasing values of eccentricity? Which is the most elliptical? Which is the least elliptical, that is, the most circular?

**A.**

**B.**

**C.**

**D.**

**E.**

**F.**

**G.**

**H.**

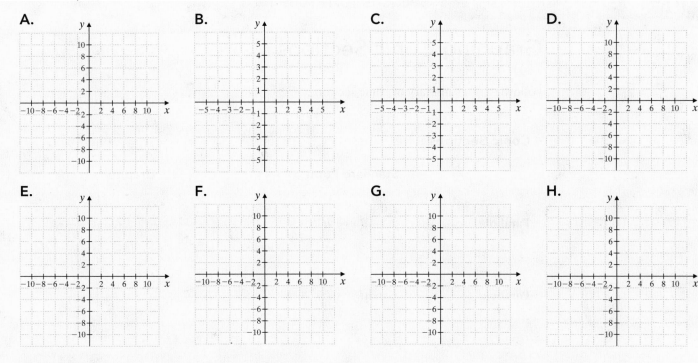

*For Exercises 73 through 78, see the example below.*

**Example**

Sketch the graph of $\dfrac{(x - 2)^2}{25} - \dfrac{(y - 1)^2}{9} = 1$.

*Solution*

This hyperbola has center $(2, 1)$. Notice that $a = 5$ and $b = 3$.

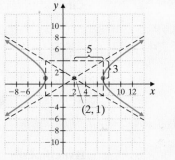

*Sketch the graph of each equation.*

**73.** $\dfrac{(x - 1)^2}{4} - \dfrac{(y + 1)^2}{25} = 1$

**74.** $\dfrac{(x + 2)^2}{9} - \dfrac{(y - 1)^2}{4} = 1$

**75.** $\dfrac{y^2}{16} - \dfrac{(x + 3)^2}{9} = 1$

**76.** $\dfrac{(y + 4)^2}{4} - \dfrac{x^2}{25} = 1$

**77.** $\dfrac{(x + 5)^2}{16} - \dfrac{(y + 2)^2}{25} = 1$

**78.** $\dfrac{(x - 3)^2}{9} - \dfrac{(y - 2)^2}{4} = 1$

## Graphing Conic Sections

**Answers**

1. _____

2. _____

3. _____

Following is a summary of conic sections.

**Conic Sections**

| | Standard Form | Graph |
|---|---|---|
| **Parabola** | $y = a(x - h)^2 + k$ |  |
| **Parabola** | $x = a(y - k)^2 + h$ | |
| **Circle** | $(x - h)^2 + (y - k)^2 = r^2$ | |
| **Ellipse** | $\dfrac{x^2}{a^2} + \dfrac{y^2}{b^2} = 1$ | |
| **Hyperbola** | $\dfrac{x^2}{a^2} - \dfrac{y^2}{b^2} = 1$ | |
| **Hyperbola** | $\dfrac{y^2}{b^2} - \dfrac{x^2}{a^2} = 1$ | |

*Identify whether each equation, when graphed, will be a parabola, circle, ellipse, or hyperbola. Then graph each equation.*

**1.** $(x - 7)^2 + (y - 2)^2 = 4$    **2.** $y = x^2 + 4$    **3.** $y = x^2 + 12x + 36$

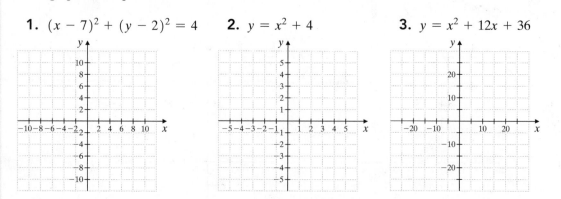

**4.** $\dfrac{x^2}{4} + \dfrac{y^2}{9} = 1$

**5.** $\dfrac{y^2}{9} - \dfrac{x^2}{9} = 1$

**6.** $\dfrac{x^2}{16} - \dfrac{y^2}{4} = 1$

**7.** $2x^2 + 8y^2 = 32$

**8.** $x^2 + y^2 = 16$

**9.** $x = y^2 + 4y - 1$

**10.** $x = -y^2 + 6y$

**11.** $9x^2 - 4y^2 = 36$

**12.** $18x^2 + 8y^2 = 72$

**13.** $\dfrac{(x-1)^2}{49} + \dfrac{(y+2)^2}{25} = 1$

**14.** $y^2 = x^2 + 16$

**15.** $\left(x + \dfrac{1}{2}\right)^2 + \left(y - \dfrac{1}{2}\right)^2 = 1$

4. _____

5. _____

6. _____

7. _____

8. _____

9. _____

10. _____

11. _____

12. _____

13. _____

14. _____

15. _____

**Solving Nonlinear Systems of Equations**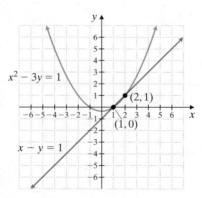

## Objectives

**A** Solve a Nonlinear System by Substitution.

**B** Solve a Nonlinear System by Elimination.

In Section 4.1, we used graphing, substitution, and elimination methods to find solutions of systems of linear equations in two variables. We now apply these same methods to nonlinear systems of equations in two variables. A **nonlinear system of equations** is a system of equations at least one of which is not linear. Since we will be graphing the equations in each system, we are interested in real number solutions only.

### Objective A Solving Nonlinear Systems by Substitution

First we solve nonlinear systems by the substitution method.

**Practice 1**

Solve the system:

$$\begin{cases} x^2 - 2y = 5 \\ x + y = -1 \end{cases}$$

**Example 1** Solve the system:

$$\begin{cases} x^2 - 3y = 1 \\ x - y = 1 \end{cases}$$

**Solution:** We can solve this system by substitution if we solve one equation for one of the variables. Solving the first equation for $x$ is not the best choice since doing so introduces a radical. Also, solving for $y$ in the first equation introduces a fraction. Thus, we solve the second equation for $y$.

$$x - y = 1 \quad \text{Second equation}$$
$$x - 1 = y \quad \text{Solve for } y.$$

Now we replace $y$ with $x - 1$ in the first equation and then solve for $x$.

$$x^2 - 3y = 1 \quad \text{First equation}$$
$$x^2 - 3(x - 1) = 1 \quad \text{Replace } y \text{ with } x - 1.$$
$$x^2 - 3x + 3 = 1$$
$$x^2 - 3x + 2 = 0$$
$$(x - 2)(x - 1) = 0$$
$$x = 2 \quad \text{or} \quad x = 1$$

Now we let $x = 2$ and then $x = 1$ in the equation $y = x - 1$ to find corresponding $y$-values.

Let $x = 2$.

$$y = x - 1$$
$$y = 2 - 1 = 1$$

Let $x = 1$.

$$y = x - 1$$
$$y = 1 - 1 = 0$$

When we check $(2, 1)$ and $(1, 0)$ in the equations, we find that both ordered pairs satisfy both equations. Thus, the solution set for the system is $\{(2, 1), (1, 0)\}$. The graph of each equation in the system is shown.

Work Practice 1

**Answer**

**1.** $\{(-3, 2), (1, -2)\}$

**Example 2** Solve the system:

$$\begin{cases} y = \sqrt{x} \\ x^2 + y^2 = 6 \end{cases}$$

**Solution:** This system is ideal for the substitution method since $y$ is expressed in terms of $x$ in the first equation. Notice that if $y = \sqrt{x}$, then both $x$ and $y$ must be nonnegative if they are real numbers. Let's substitute $\sqrt{x}$ for $y$ in the second equation and solve for $x$.

$$x^2 + y^2 = 6$$
$$x^2 + (\sqrt{x})^2 = 6 \quad \text{Let } y = \sqrt{x}.$$
$$x^2 + x = 6$$
$$x^2 + x - 6 = 0$$
$$(x + 3)(x - 2) = 0$$
$$x = -3 \quad \text{or} \quad x = 2$$

The solution $-3$ is discarded because we have noted that $x$ must be nonnegative. To see this, we let $x = -3$ and $x = 2$ in the first equation to find the corresponding $y$-values.

Let $x = -3$.

$$y = \sqrt{x}$$
$$y = \sqrt{-3} \quad \text{Not a real number}$$

Let $x = 2$.

$$y = \sqrt{x}$$
$$y = \sqrt{2}$$

Since we are interested only in real number solutions, the only solution is $(2, \sqrt{2})$. The solution set is $\{(2, \sqrt{2})\}$. Check to see that this solution satisfies both equations. The graph of each equation in this system is shown.

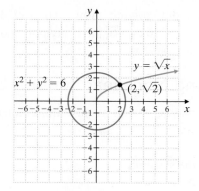

**Work Practice 2**

**Example 3** Solve the system:

$$\begin{cases} x^2 + y^2 = 4 \\ x + y = 3 \end{cases}$$

**Solution:** We use the substitution method and solve the second equation for $x$

$$x + y = 3 \quad \text{Second equation}$$
$$x = 3 - y$$

*(Continued on next page)*

*(Continued on next page)*

**Practice 2**
Solve the system:

$$\begin{cases} y = \sqrt{x} \\ x^2 + y^2 = 30 \end{cases}$$

**Practice 3**
Solve the system:

$$\begin{cases} x^2 + y^2 = 1 \\ x + y = 4 \end{cases}$$

**Answers**
**2.** $\{(5, \sqrt{5})\}$  **3.** no solutions

Now we let $x = 3 - y$ in the first equation.

$$x^2 + y^2 = 4 \quad \text{First equation}$$

$$(3 - y)^2 + y^2 = 4 \quad \text{Let } x = 3 - y.$$
$$9 - 6y + y^2 + y^2 = 4$$
$$2y^2 - 6y + 5 = 0$$

By the quadratic formula, where $a = 2$, $b = -6$, and $c = 5$, we have

$$y = \frac{6 \pm \sqrt{(-6)^2 - 4 \cdot 2 \cdot 5}}{2 \cdot 2} = \frac{6 \pm \sqrt{-4}}{4}$$

Since $\sqrt{-4}$ is not a real number, there is no solution. Graphically, the circle and the line do not intersect, as shown.

■ Work Practice 3

✓**Concept Check** Without solving, how can you tell that the graphs of $x^2 + y^2 = 9$ and $x^2 + y^2 = 16$ do not have any points of intersection?

## Objective B  Solving Nonlinear Systems by Elimination ▶

Some nonlinear systems may be solved by the elimination method.

**Example 4**  Solve the system:

$$\begin{cases} x^2 + 2y^2 = 10 \\ x^2 - y^2 = 1 \end{cases}$$

**Solution:**  We will use the elimination, or addition, method to solve this system. To eliminate $x^2$ when we add the two equations, we multiply both sides of the second equation by $-1$. Then

$$\begin{array}{r} x^2 + 2y^2 = 10 \\ (-1)(x^2 - y^2) = -1 \cdot 1 \end{array} \quad \begin{array}{c} \text{is} \\ \text{equivalent} \\ \text{to} \end{array} \quad \begin{cases} x^2 + 2y^2 = 10 \\ -x^2 + y^2 = -1 \end{cases}$$

$$3y^2 = 9 \quad \text{Add.}$$
$$y^2 = 3$$
$$y = \pm\sqrt{3} \quad \text{Divide both sides by 3.}$$

---

**Practice 4**

Solve the equation:

$$\begin{cases} x^2 + 3y^2 = 21 \\ x^2 - y^2 = 1 \end{cases}$$

**Answer**

**4.** $\{(\sqrt{6}, \sqrt{5}), (\sqrt{6}, -\sqrt{5}),$
$(-\sqrt{6}, \sqrt{5}), (-\sqrt{6}, -\sqrt{5})\}$

✓**Concept Check Answer**

$x^2 + y^2 = 9$ is a circle inside the circle $x^2 + y^2 = 16$; therefore, they do not have any points of intersection.

To find the corresponding $x$-values, we let $y = \sqrt{3}$ and $y = -\sqrt{3}$ in either original equation. We choose the second equation.

$$\text{Let } y = \sqrt{3}.$$
$$x^2 - y^2 = 1$$
$$x^2 - (\sqrt{3})^2 = 1$$
$$x^2 - 3 = 1$$
$$x^2 = 4$$
$$x = \pm\sqrt{4} = \pm 2$$

$$\text{Let } y = -\sqrt{3}.$$
$$x^2 - y^2 = 1$$
$$x^2 - (-\sqrt{3})^2 = 1$$
$$x^2 - 3 = 1$$
$$x^2 = 4$$
$$x = \pm\sqrt{4} = \pm 2$$

The solution set is $\{(2, \sqrt{3}), (-2, \sqrt{3}), (2, -\sqrt{3}), (-2, -\sqrt{3})\}$. Check all four ordered pairs in both equations of the system. The graph of each equation in this system is shown.

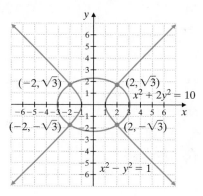

■ Work Practice 4

## Vocabulary, Readiness & Video Check

*Martin-Gay Interactive Videos*     *Watch the section lecture video and answer the following questions.*

**Objective A**  **1.** In ▤ Example 1, why do we choose not to solve either equation for $y$? ▶

**Objective B**  **2.** In ▤ Example 2, what important reminder is made as the second equation is multiplied by a number to get opposite coefficients of $x$? ▶

*See Video 10.3* 🍎

## 10.3 Exercise Set MyMathLab® ▶

**Objectives A B Mixed Practice** *Solve each nonlinear system of equations. See Examples 1 through 4.*

1. $\begin{cases} x^2 + y^2 = 25 \\ 4x + 3y = 0 \end{cases}$

2. $\begin{cases} x^2 + y^2 = 25 \\ 3x + 4y = 0 \end{cases}$

3. $\begin{cases} x^2 + 4y^2 = 10 \\ y = x \end{cases}$

4. $\begin{cases} 4x^2 + y^2 = 10 \\ y = x \end{cases}$

▶5. $\begin{cases} y^2 = 4 - x \\ x - 2y = 4 \end{cases}$

6. $\begin{cases} x^2 + y^2 = 4 \\ x + y = -2 \end{cases}$

7. $\begin{cases} x^2 + y^2 = 9 \\ 16x^2 - 4y^2 = 64 \end{cases}$

8. $\begin{cases} 4x^2 + 3y^2 = 35 \\ 5x^2 + 2y^2 = 42 \end{cases}$

9. $\begin{cases} x^2 + 2y^2 = 2 \\ x - y = 2 \end{cases}$

10. $\begin{cases} x^2 + 2y^2 = 2 \\ x^2 - 2y^2 = 6 \end{cases}$

11. $\begin{cases} y = x^2 - 3 \\ 4x - y = 6 \end{cases}$

12. $\begin{cases} y = x + 1 \\ x^2 - y^2 = 1 \end{cases}$

13. $\begin{cases} y = x^2 \\ 3x + y = 10 \end{cases}$

14. $\begin{cases} 6x - y = 5 \\ xy = 1 \end{cases}$

15. $\begin{cases} y = 2x^2 + 1 \\ x + y = -1 \end{cases}$

16. $\begin{cases} x^2 + y^2 = 9 \\ x + y = 5 \end{cases}$

17. $\begin{cases} y = x^2 - 4 \\ y = x^2 - 4x \end{cases}$

18. $\begin{cases} x = y^2 - 3 \\ x = y^2 - 3y \end{cases}$

▶19. $\begin{cases} 2x^2 + 3y^2 = 14 \\ -x^2 + y^2 = 3 \end{cases}$

20. $\begin{cases} 4x^2 - 2y^2 = 2 \\ -x^2 + y^2 = 2 \end{cases}$

21. $\begin{cases} x^2 + y^2 = 1 \\ x^2 + (y + 3)^2 = 4 \end{cases}$

22. $\begin{cases} x^2 + 2y^2 = 4 \\ x^2 - y^2 = 4 \end{cases}$

23. $\begin{cases} y = x^2 + 2 \\ y = -x^2 + 4 \end{cases}$

24. $\begin{cases} x = -y^2 - 3 \\ x = y^2 - 5 \end{cases}$

25. $\begin{cases} 3x^2 + y^2 = 9 \\ 3x^2 - y^2 = 9 \end{cases}$

26. $\begin{cases} x^2 + y^2 = 25 \\ x = y^2 - 5 \end{cases}$

27. $\begin{cases} x^2 + 3y^2 = 6 \\ x^2 - 3y^2 = 10 \end{cases}$

28. $\begin{cases} x^2 + y^2 = 1 \\ y = x^2 - 9 \end{cases}$

29. $\begin{cases} x^2 + y^2 = 36 \\ y = \dfrac{1}{6}x^2 - 6 \end{cases}$

30. $\begin{cases} x^2 + y^2 = 16 \\ y = -\dfrac{1}{4}x^2 + 4 \end{cases}$

31. $\begin{cases} y = \sqrt{x} \\ x^2 + y^2 = 12 \end{cases}$

32. $\begin{cases} y = \sqrt{x} \\ x^2 + y^2 = 20 \end{cases}$

## Review

*Graph each inequality in two variables. See Section 3.5.*

33. $x > -3$

34. $y \le 1$

35. $y < 2x - 1$

36. $3x - y \le 4$

*Find the perimeter of each geometric figure. See Section 5.1.*

△ **37.**

$x$ inches

$(2x - 5)$ inches

$(5x - 20)$ inches

△ **38.**

$(3x + 2)$ centimeters

△ **39.**

$(x^2 + 3x + 1)$ meters

$x^2$ meters

△ **40.**

$2x^2$ feet

$4x$ feet

$(3x^2 + 1)$ feet

$(3x^2 + 7)$ feet

## Concept Extensions

*For the exercises below, see the Concept Check in this section.*

**41.** Without solving, how can you tell that the graphs of $x^2 + y^2 = 1$ and $x^2 + y^2 = 4$ do not have any points of intersection?

**42.** Without solving, how can you tell that the graphs of $y = 2x + 3$ and $y = 2x + 7$ do not have any points of intersection?

**43.** How many real solutions are possible for a system of equations whose graphs are a circle and a parabola?

**44.** How many real solutions are possible for a system of equations whose graphs are an ellipse and a line?

*Solve.*

**45.** The sum of the squares of two numbers is 130. The difference of the squares of the two numbers is 32. Find the two numbers.

**46.** The sum of the squares of two numbers is 20. Their product is 8. Find the two numbers.

△ **47.** During the development stage of a new rectangular keypad for a security system, it was decided that the area of the rectangle should be 285 square centimeters and the perimeter should be 68 centimeters. Find the dimensions of the keypad.

△ **48.** A rectangular holding pen for cattle is to be designed so that its perimeter is 92 feet and its area is 525 feet. Find the dimensions of the holding pen.

*Recall that in business, a demand function expresses the quantity of a commodity demanded as a function of the commodity's unit price. A supply function expresses the quantity of a commodity supplied as a function of the commodity's unit price. When the quantity produced and supplied is equal to the quantity demanded, then we have what is called* **market equilibrium.** *Use this information for Exercises 49 and 50.*

**49.** The demand function for a certain compact disc is given by the function $p(x) = -0.01x^2 - 0.2x + 9$ and the corresponding supply function is given by $p(x) = 0.01x^2 - 0.1x + 3$, where $p(x)$ is in dollars and $x$ is in thousands of units. Find the equilibrium quantity and the corresponding price by solving the system consisting of the two given equations.

**50.** The demand function for a certain style of picture frame is given by the function $p(x) = -2x^2 + 90$ and the corresponding supply function is given by $p(x) = 9x + 34$, where $p(x)$ is in dollars and $x$ is in thousands of units. Find the equilibrium quantity and the corresponding price by solving the system consisting of the two given equations.

*Use a grapher to verify the results of each exercise.*

**51.** Exercise 3    **52.** Exercise 4    **53.** Exercise 23    **54.** Exercise 24

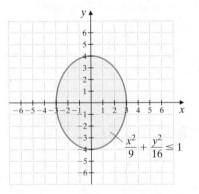
## Objectives

**A** Graph a Nonlinear Inequality.

**B** Graph a System of Nonlinear Inequalities.

## Objective **A** Graphing Nonlinear Inequalities

We can graph a nonlinear inequality in two variables such as $\frac{x^2}{9} + \frac{y^2}{16} \leq 1$ in a way similar to the way we graphed a linear inequality in two variables in Section 3.5. First, we graph the related equation $\frac{x^2}{9} + \frac{y^2}{16} = 1$. The graph of the equation is our boundary. Then, using test points, we determine and shade the region whose points satisfy the inequality.

### Practice 1

Graph: $\frac{x^2}{25} + \frac{y^2}{4} \leq 1$

**Example 1** Graph: $\frac{x^2}{9} + \frac{y^2}{16} \leq 1$

**Solution:** First we graph the equation $\frac{x^2}{9} + \frac{y^2}{16} = 1$. We sketch a solid curve because of the inequality symbol $\leq$. It means that the graph of $\frac{x^2}{9} + \frac{y^2}{16} \leq 1$ includes the graph of $\frac{x^2}{9} + \frac{y^2}{16} = 1$. The graph is an ellipse, and it divides the plane into two regions, the "inside" and the "outside" of the ellipse. Recall from Section 3.5 that to determine which region contains the solutions, we select a test point in either region and determine whether the coordinates of the point satisfy the inequality. We choose $(0, 0)$ as the test point.

$$\frac{x^2}{9} + \frac{y^2}{16} \leq 1$$

$$\frac{0^2}{9} + \frac{0^2}{16} \leq 1 \quad \text{Let } x = 0 \text{ and } y = 0.$$

$$0 \leq 1 \quad \text{True}$$

Since this statement is true, the solution set is the region containing $(0, 0)$. The graph of the solution set includes the points on and inside the ellipse, as shaded in the figure.

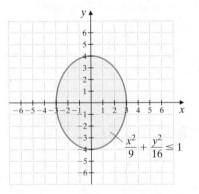

**Work Practice 1**

## Answer

**1.**

### Example 2 — Graph: $4y^2 > x^2 + 16$

**Solution:** The related equation is $4y^2 = x^2 + 16$, or $\dfrac{y^2}{4} - \dfrac{x^2}{16} = 1$, which is a hyperbola. We graph the hyperbola as a dashed curve because of the inequality symbol $>$. It means that the graph of $4y^2 > x^2 + 16$ does *not* include the graph of $4y^2 = x^2 + 16$. The hyperbola divides the plane into three regions. We select a test point in each region—not on a boundary line—to determine whether that region contains solutions of the inequality.

| Test Region A with $(0, 4)$ | Test Region B with $(0, 0)$ | Test Region C with $(0, -4)$ |
|---|---|---|
| $4y^2 > x^2 + 16$ | $4y^2 > x^2 + 16$ | $4y^2 > x^2 + 16$ |
| $4(4)^2 > 0^2 + 16$ | $4(0)^2 > 0^2 + 16$ | $4(-4)^2 > 0^2 + 16$ |
| $64 > 16$   True | $0 > 16$   False | $64 > 16$   True |

The graph of the solution set includes the shaded regions $A$ and $C$ only, not the boundary.

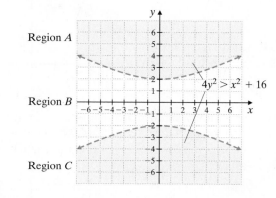

Region A

Region B

$4y^2 > x^2 + 16$

Region C

■ Work Practice 2

## Objective B  Graphing Systems of Nonlinear Inequalities

In Section 4.5, we graphed systems of linear inequalities. Recall that the graph of a system of inequalities is the intersection of the graphs of the inequalities.

### Example 3 — Graph the system:

$$\begin{cases} x \le 1 - 2y \\ y \le x^2 \end{cases}$$

**Solution:** We graph each inequality on the same set of axes. The intersection is the darkest shaded region along with its boundary lines. The coordinates of the points of intersection can be found by solving the related system

$$\begin{cases} x = 1 - 2y \\ y = x^2 \end{cases}$$

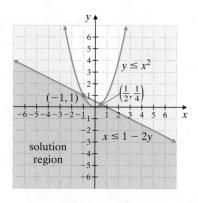

$y \le x^2$

$\left(\frac{1}{2}, \frac{1}{4}\right)$

$(-1, 1)$

$x \le 1 - 2y$

solution region

■ Work Practice 3

---

**Practice 2**

Graph:  $9x^2 > 4y^2 + 144$

**Practice 3**

Graph the system:

$$\begin{cases} y \ge x^2 \\ y \le -4x + 2 \end{cases}$$

**Answers**

2.

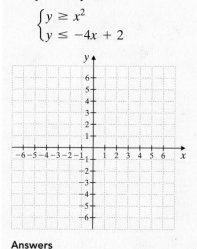

$9x^2 > 4y^2 + 144$

3.

solution region

$y \ge x^2$

$y \le -4x + 2$

## Practice 4

Graph the system:

$$\begin{cases} x^2 + y^2 < 9 \\ \dfrac{x^2}{9} - \dfrac{y^2}{4} < 1 \\ \qquad y > x - 2 \end{cases}$$

**Example 4**    Graph the system:

$$\begin{cases} x^2 + y^2 < 25 \\ \dfrac{x^2}{9} - \dfrac{y^2}{25} < 1 \\ \qquad y < x + 3 \end{cases}$$

**Solution:**    We graph each inequality. The graph of $x^2 + y^2 < 25$ contains points "inside" the circle that has center $(0, 0)$ and radius 5. The graph of $\dfrac{x^2}{9} - \dfrac{y^2}{25} < 1$ is the region between the two branches of the hyperbola with $x$-intercepts $(-3, 0)$ and $(3, 0)$ and center $(0, 0)$. The graph of $y < x + 3$ is the region "below" the line with slope 1 and $y$-intercept $(0, 3)$. The graph of the solution set of the system is the intersection of all the graphs. This intersection region is shown as the shaded region on the fourth graph. The boundary of this region is not part of the solution.

■ **Work Practice 4**

# Vocabulary, Readiness & Video Check

*Martin-Gay Interactive Videos*    Watch the section lecture video and answer the following questions.

**Objective A**   **1.** From ▦ Example 1, explain the similarities between graphing linear inequalities and graphing nonlinear inequalities. ▶

**Objective B**   **2.** From ▦ Example 2, describe one possible illustration of graphs of two circle inequalities in which the system has no solution—that is, the graph of the inequalities in the system do not overlap. ▶

*See Video 10.4*

## 10.4 Exercise Set MyMathLab®

**Objective A**   *Graph each inequality. See Examples 1 and 2.*

**1.** $y < x^2$

**2.** $y < -x^2$

**3.** $x^2 + y^2 \geq 16$

**4.** $x^2 + y^2 < 36$

**5.** $\dfrac{x^2}{4} - y^2 < 1$

**6.** $x^2 - \dfrac{y^2}{9} \geq 1$

**7.** $y > (x - 1)^2 - 3$

**8.** $y > (x + 3)^2 + 2$

**9.** $x^2 + y^2 \leq 9$

**10.** $x^2 + y^2 > 4$

**11.** $y > -x^2 + 5$

**12.** $y < -x^2 + 5$

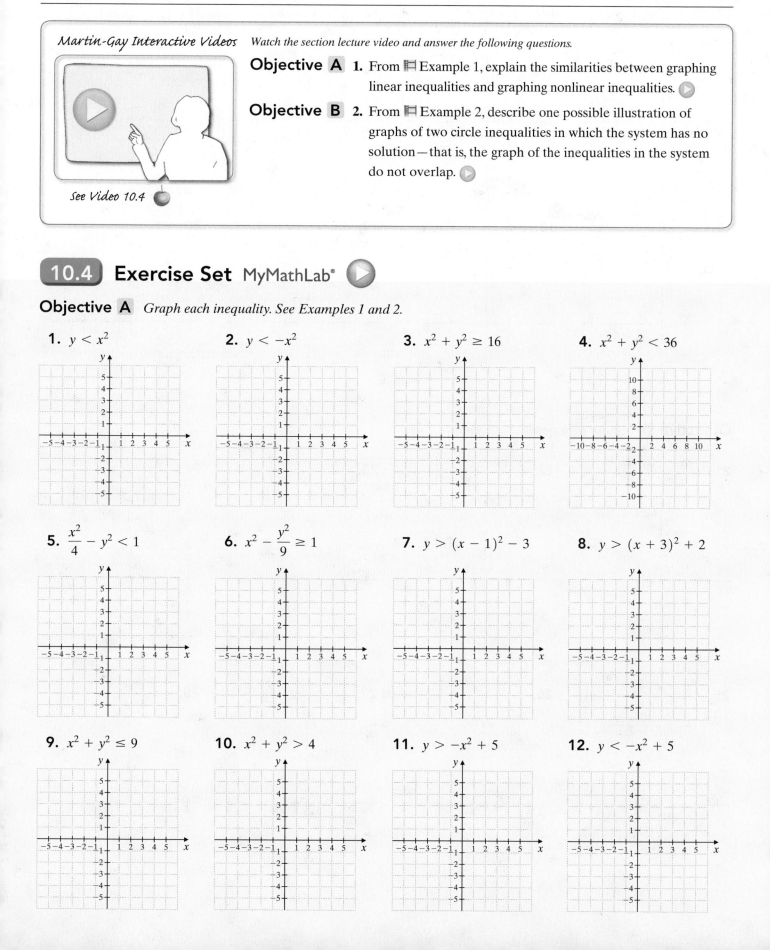

**13.** $\dfrac{x^2}{4} + \dfrac{y^2}{9} \leq 1$ **14.** $\dfrac{x^2}{25} + \dfrac{y^2}{4} \geq 1$ **15.** $\dfrac{y^2}{4} - x^2 \leq 1$ **16.** $\dfrac{y^2}{16} - \dfrac{x^2}{9} > 1$

**17.** $y < (x - 2)^2 + 1$ **18.** $y > (x - 2)^2 + 1$ **19.** $y \leq x^2 + x - 2$ **20.** $y > x^2 + x - 2$

**Objective B** *Graph each system. See Examples 3 and 4.*

**21.** $\begin{cases} 4x + 3y \geq 12 \\ x^2 + y^2 < 16 \end{cases}$ **22.** $\begin{cases} 3x - 4y \leq 12 \\ x^2 + y^2 < 16 \end{cases}$ **23.** $\begin{cases} x^2 + y^2 \leq 9 \\ x^2 + y^2 \geq 1 \end{cases}$ **24.** $\begin{cases} x^2 + y^2 \geq 9 \\ x^2 + y^2 \geq 16 \end{cases}$

**25.** $\begin{cases} y > x^2 \\ y \geq 2x + 1 \end{cases}$ **26.** $\begin{cases} y \leq -x^2 + 3 \\ y \leq 2x - 1 \end{cases}$ **27.** $\begin{cases} x^2 + y^2 > 9 \\ y > x^2 \end{cases}$ **28.** $\begin{cases} x^2 + y^2 \leq 9 \\ y < x^2 \end{cases}$

29. $\begin{cases} \dfrac{x^2}{4} + \dfrac{y^2}{9} \geq 1 \\ x^2 + y^2 \geq 4 \end{cases}$

30. $\begin{cases} x^2 + (y-2)^2 \geq 9 \\ \dfrac{x^2}{4} + \dfrac{y^2}{25} < 1 \end{cases}$

31. $\begin{cases} x^2 - y^2 \geq 1 \\ y \geq 0 \end{cases}$

32. $\begin{cases} x^2 - y^2 \geq 1 \\ x \geq 0 \end{cases}$

33. $\begin{cases} x + y \geq 1 \\ 2x + 3y < 1 \\ x > -3 \end{cases}$

34. $\begin{cases} x - y < -1 \\ 4x - 3y > 0 \\ y > 0 \end{cases}$

35. $\begin{cases} x^2 - y^2 < 1 \\ \dfrac{x^2}{16} + y^2 \leq 1 \\ x \geq -2 \end{cases}$

36. $\begin{cases} x^2 - y^2 \geq 1 \\ \dfrac{x^2}{16} + \dfrac{y^2}{4} \leq 1 \\ y \geq 1 \end{cases}$

## Review

*Determine whether each graph is the graph of a function. See Section 3.6*

**37.**     **38.**     **39.**     **40.**

*Find each function value if $f(x) = 3x^2 - 2$. See Section 3.6.*

**41.** $f(-1)$     **42.** $f(-3)$     **43.** $f(a)$     **44.** $f(b)$

## Concept Extensions

**45.** Discuss how graphing a linear inequality such as $x + y < 9$ is similar to graphing a nonlinear inequality such as $x^2 + y^2 < 9$.

**46.** Discuss how graphing a linear inequality such as $x + y < 9$ is different from graphing a nonlinear inequality such as $x^2 + y^2 < 9$.

**47.** Graph the system:

$$\begin{cases} y \le x^2 \\ y \ge x + 2 \\ x \ge 0 \\ y \ge 0 \end{cases}$$

**48.** Graph the system:

$$\begin{cases} x \ge 0 \\ y \ge 0 \\ y \ge x^2 + 1 \\ y \le 4 - x \end{cases}$$

---

### Chapter 10 Group Activity

**Modeling Conic Sections**

**Materials**

- two thumbtacks (or nails)
- graph paper
- cardboard
- tape
- string
- pencil
- ruler

**Figure 1**

**Figure 2**

**Figure 3**

*This activity may be completed by working in groups or individually.*

**1.** Draw an $x$-axis and a $y$-axis on the graph paper as shown in Figure 1.

**2.** Place the graph paper on the cardboard and use tape to attach.

**3.** Locate two points on the $x$-axis each about $1\frac{1}{2}$ inches from the origin and on opposite sides of the origin (see Figure 1). Insert thumbtacks (or nails) at each of these locations.

**4.** Fasten a 9-inch piece of string to the thumbtacks as shown in Figure 2. Use your pencil to draw and keep the string taut while you carefully move the pencil in a path all around the thumbtacks.

**5.** Using the grid of the graph paper as a guide, find an approximate equation of the ellipse you drew.

**6.** Experiment by moving the tacks closer together or farther apart and drawing new ellipses. What do you observe?

**7.** Write a paragraph explaining why the figure drawn by the pencil is an ellipse. How might you use the same materials to draw a circle?

**8.** (Optional) Choose one of the ellipses you drew with the string and pencil. Use a ruler to draw any six tangent lines to the ellipse. (A line is tangent to the ellipse if it intersects, or just touches, the ellipse at only one point. See Figure 3.) Extend the tangent lines to yield six points of intersection among the tangents. Use a straightedge to draw lines connecting each pair of opposite points of intersection. What do you observe? Repeat with a different ellipse. Can you make a conjecture about the relationship among the lines that connect opposite points of intersection?

## Chapter 10 Vocabulary Check

*Fill in each blank with one of the words or phrases listed below.*

| circle | ellipse | hyperbola |
|---|---|---|
| conic sections | vertex | diameter |
| center | radius | nonlinear system of equations |

1. A(n) _____ is the set of all points in a plane that are the same distance from a fixed point, called the _____ .

2. A(n) _____ is a system of equations at least one of which is not linear.

3. A(n) _____ is the set of points in a plane such that the sum of the distances of those points from two fixed points is a constant.

4. In a circle, the distance from the center to a point of the circle is called its _____ .

5. A(n) _____ is the set of points in a plane such that the absolute value of the difference of the distance from two fixed points is constant.

6. The circle, parabola, ellipse, and hyperbola are called the _____ .

7. For a parabola that opens upward, the lowest point is the _____ .

8. Twice a circle's radius is its _____ .

**Helpful Hint**

▶ Are you preparing for your test? Don't forget to take the Chapter 10 Test on page 805. Then check your answers at the back of the text and use the Chapter Test Prep Videos to see the fully worked-out solutions to any of the exercises you want to review.

## 10 Chapter Highlights

| Definitions and Concepts | Examples |
|---|---|

### Section 10.1　The Parabola and the Circle

**Parabolas**

$$y = a(x - h)^2 + k$$

$$x = a(y - k)^2 + h$$

Graph:　$x = 3y^2 - 12y + 13$

$$x - 13 = 3(y^2 - 4y)$$

$$x - 13 + 3(4) = 3(y^2 - 4y + 4)$$

$$x = 3(y - 2)^2 + 1$$

Since $a = 3$, this parabola opens to the right with vertex $(1, 2)$. Its axis of symmetry is $y = 2$. The $x$-intercept is $(13, 0)$.

*(continued)*

| Definitions and Concepts | Examples |
|---|---|

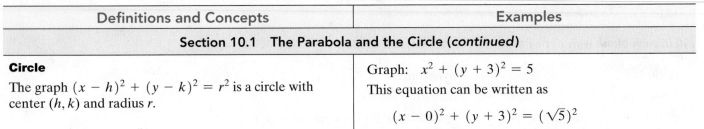

### Section 10.1   The Parabola and the Circle (*continued*)

**Circle**

The graph $(x - h)^2 + (y - k)^2 = r^2$ is a circle with center $(h, k)$ and radius $r$.

Graph:  $x^2 + (y + 3)^2 = 5$

This equation can be written as

$$(x - 0)^2 + (y + 3)^2 = (\sqrt{5})^2$$

with $h = 0$, $k = -3$, and $r = \sqrt{5}$. The center of this circle is $(0, -3)$, and the radius is $\sqrt{5}$.

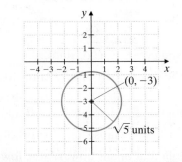

### Section 10.2   The Ellipse and the Hyperbola

**Ellipse with center (0, 0)**

The graph of an equation of the form $\dfrac{x^2}{a^2} + \dfrac{y^2}{b^2} = 1$ is an ellipse with center $(0, 0)$. The $x$-intercepts are $(a, 0)$ and $(-a, 0)$, and the $y$-intercepts are $(0, b)$ and $(0, -b)$.

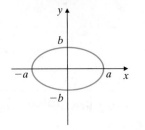

Graph:  $4x^2 + 9y^2 = 36$

$$\dfrac{x^2}{9} + \dfrac{y^2}{4} = 1 \quad \text{Divide both sides by 36.}$$

$$\dfrac{x^2}{3^2} + \dfrac{y^2}{2^2} = 1$$

The ellipse has center $(0, 0)$, $x$-intercepts $(3, 0)$ and $(-3, 0)$, and $y$-intercepts $(0, 2)$ and $(0, -2)$.

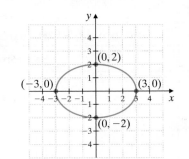

| Definitions and Concepts | Examples |
|---|---|

### Section 10.2 The Ellipse and the Hyperbola (*continued*)

**Hyperbola with center (0, 0)**

The graph of an equation of the form $\dfrac{x^2}{a^2} - \dfrac{y^2}{b^2} = 1$ is a hyperbola with center $(0, 0)$ and $x$-intercepts $(a, 0)$ and $(-a, 0)$.

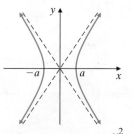

The graph of an equation of the form $\dfrac{y^2}{b^2} - \dfrac{x^2}{a^2} = 1$ is a hyperbola with center $(0, 0)$ and $y$-intercepts $(0, b)$ and $(0, -b)$.

Graph: $\dfrac{x^2}{9} - \dfrac{y^2}{4} = 1$. Here $a = 3$ and $b = 2$.

### Section 10.3 Solving Nonlinear Systems of Equations

A **nonlinear system of equations** is a system of equations at least one of which is not linear. Both the substitution method and the elimination method may be used to solve a nonlinear system of equations.

Solve the nonlinear system: $\begin{cases} y = x + 2 \\ 2x^2 + y^2 = 3 \end{cases}$

Substitute $x + 2$ for $y$ in the second equation:

$$2x^2 + y^2 = 3$$
$$2x^2 + (x + 2)^2 = 3$$
$$2x^2 + x^2 + 4x + 4 = 3$$
$$3x^2 + 4x + 1 = 0$$
$$(3x + 1)(x + 1) = 0$$
$$x = -\frac{1}{3}, \quad \text{or} \quad x = -1$$

If $x = -\dfrac{1}{3}, y = x + 2 = -\dfrac{1}{3} + 2 = \dfrac{5}{3}$.

If $x = -1, y = x + 2 = -1 + 2 = 1$.

The solution set is $\left\{ \left( -\dfrac{1}{3}, \dfrac{5}{3} \right), (-1, 1) \right\}$

| Definitions and Concepts | Examples |
|---|---|

### Section 10.4   Nonlinear Inequalities and Systems of Inequalities

The **graph of a system of inequalities** is the intersection of the graphs of the inequalities.

Graph the system: $\begin{cases} x \geq y^2 \\ x + y \leq 4 \end{cases}$

The graph of the system is the darkest shaded region along with its boundary lines.

## Chapter 10   Review

**(10.1)** *Write an equation of the circle with the given center and radius or diameter. For Exercises 3 and 4, begin by writing diameters as radii.*

**1.** Center $(-4, 4)$, radius $3$

**2.** Center $(-7, -9)$, radius $\sqrt{11}$

**3.** Center $(5, 0)$, diameter $10$

**4.** Center $(0, 0)$, diameter $7$

*Graph each equation. If the graph is a circle, find its center and radius. If the graph is a parabola, find its vertex.*

**5.** $x^2 + y^2 = 4$

**6.** $x = 2(y - 5)^2 + 4$

**7.** $x = -(y + 2)^2 + 3$

**8.** $(x - 1)^2 + (y - 2)^2 = 4$

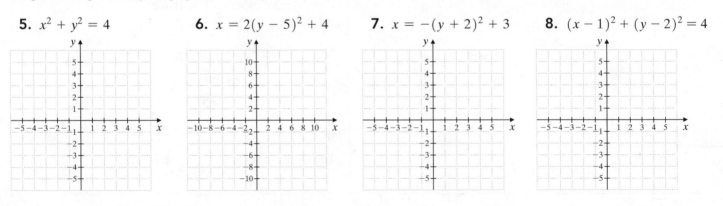

**9.** $y = -x^2 + 4x + 10$

**10.** $x = -y^2 - 4y + 6$

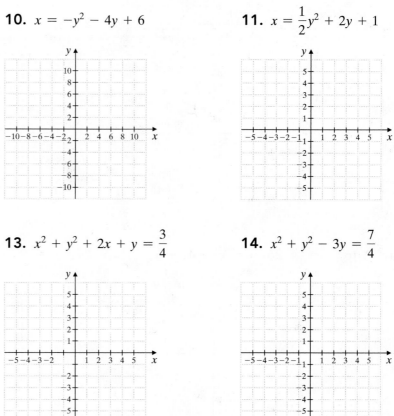

**11.** $x = \dfrac{1}{2}y^2 + 2y + 1$

**12.** $y = -3x^2 + \dfrac{1}{2}x + 4$

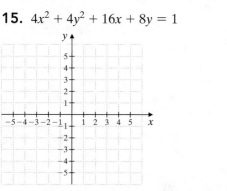

**13.** $x^2 + y^2 + 2x + y = \dfrac{3}{4}$

**14.** $x^2 + y^2 - 3y = \dfrac{7}{4}$

**15.** $4x^2 + 4y^2 + 16x + 8y = 1$

**16.** $3x^2 + 3y^2 + 18x - 12y = -12$

**(10.1, 10.2)** *Graph each equation.*

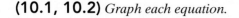

**17.** $x^2 - \dfrac{y^2}{4} = 1$

**18.** $x^2 + \dfrac{y^2}{4} = 1$

**19.** $4y^2 + 9x^2 = 36$

**20.** $-5x^2 + 25y^2 = 125$

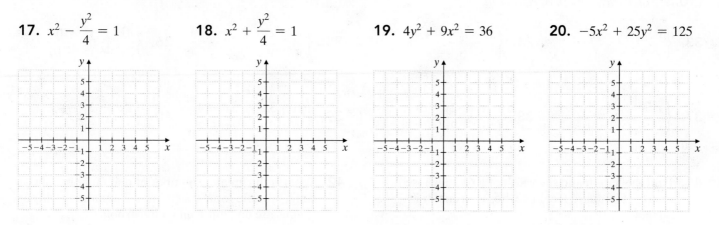

**21.** $x^2 - y^2 = 1$

**22.** $\dfrac{(x+3)^2}{9} + \dfrac{(y-4)^2}{25} = 1$

**23.** $y = x^2 + 9$

**24.** $36y^2 - 49x^2 = 1764$

**25.** $x = 4y^2 - 16$

**26.** $y = x^2 + 4x + 6$

**27.** $y^2 + 2(x-1)^2 = 8$

**28.** $x - 4y = y^2$

**29.** $x^2 - 4 = y^2$

**30.** $x^2 = 4 - y^2$

**31.** $36y^2 = 576 + 16x^2$

**32.** $3(x-7)^2 + 3(y+4)^2 = 1$

**(10.3)** *Solve each system of equations.*

**33.** $\begin{cases} y = 2x - 4 \\ y^2 = 4x \end{cases}$

**34.** $\begin{cases} x^2 + y^2 = 4 \\ x - y = 4 \end{cases}$

**35.** $\begin{cases} y = x + 2 \\ y = x^2 \end{cases}$

**36.** $\begin{cases} 4x - y^2 = 0 \\ 2x^2 + y^2 = 16 \end{cases}$

**37.** $\begin{cases} x^2 + 4y^2 = 16 \\ x^2 + y^2 = 4 \end{cases}$

**38.** $\begin{cases} x^2 + 2y = 9 \\ 5x - 2y = 5 \end{cases}$

**39.** $\begin{cases} y = 3x^2 + 5x - 4 \\ y = 3x^2 - x + 2 \end{cases}$

**40.** $\begin{cases} x^2 - 3y^2 = 1 \\ 4x^2 + 5y^2 = 21 \end{cases}$

△ **41.** Find the length and the width of a room whose area is 150 square feet and whose perimeter is 50 feet.

**42.** What is the greatest number of real number solutions possible for a system of two equations whose graphs are an ellipse and a hyperbola?

**(10.4)** *Graph each inequality or system of inequalities.*

**43.** $y \le -x^2 + 3$

**44.** $x < y^2 - 1$

**45.** $x^2 + y^2 < 9$

**46.** $\dfrac{x^2}{4} + \dfrac{y^2}{9} \ge 1$

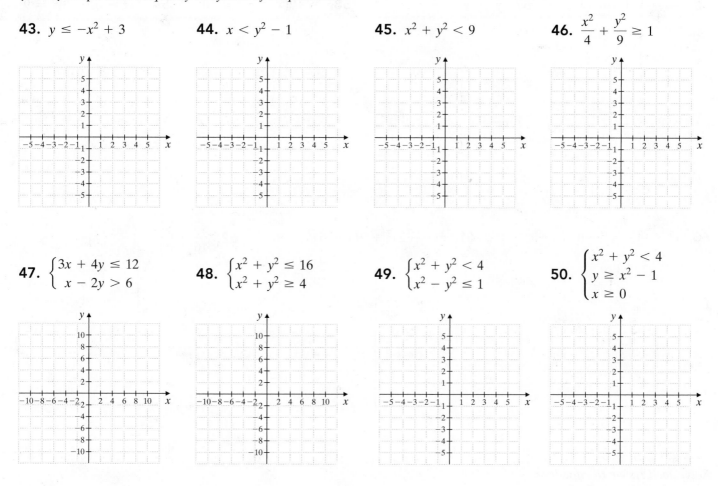

**47.** $\begin{cases} 3x + 4y \le 12 \\ x - 2y > 6 \end{cases}$

**48.** $\begin{cases} x^2 + y^2 \le 16 \\ x^2 + y^2 \ge 4 \end{cases}$

**49.** $\begin{cases} x^2 + y^2 < 4 \\ x^2 - y^2 \le 1 \end{cases}$

**50.** $\begin{cases} x^2 + y^2 < 4 \\ y \ge x^2 - 1 \\ x \ge 0 \end{cases}$

## Mixed Review

**51.** Write an equation of the circle with center $(-7, 8)$ and radius 5.

*Graph each equation.*

**52.** $y = x^2 + 6x + 9$

**53.** $x = y^2 + 6y + 9$

**54.** $\dfrac{y^2}{4} - \dfrac{x^2}{16} = 1$

**55.** $\dfrac{y^2}{4} + \dfrac{x^2}{16} = 1$

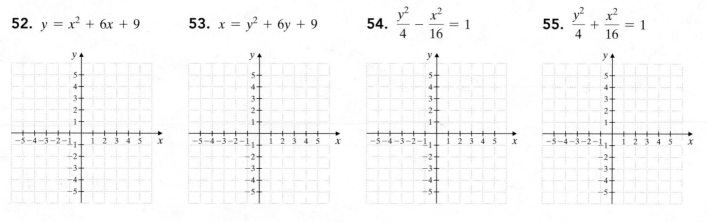

**56.** $\dfrac{(x-2)^2}{4} + (y-1)^2 = 1$    **57.** $y^2 = x^2 + 6$    **58.** $y^2 + (x-2)^2 = 10$    **59.** $3x^2 + 6x + 3y^2 = 9$

**60.** $x^2 + y^2 - 8y = 0$    **61.** $6(x-2)^2 + 9(y+5)^2 = 36$    **62.** $\dfrac{x^2}{16} - \dfrac{y^2}{25} = 1$

*Solve each system of equations.*

**63.** $\begin{cases} y = x^2 - 5x + 1 \\ y = -x + 6 \end{cases}$

**64.** $\begin{cases} x^2 + y^2 = 10 \\ 9x^2 + y^2 = 18 \end{cases}$

*Graph each inequality or system of inequalities.*

**65.** $x^2 - y^2 < 1$

**66.** $\begin{cases} y > x^2 \\ x + y \geq 3 \end{cases}$

*Graph each equation.*

**Answers**

**1.** $x^2 + y^2 = 36$

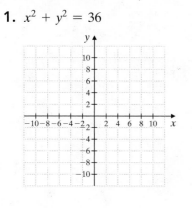

**2.** $x^2 - y^2 = 36$

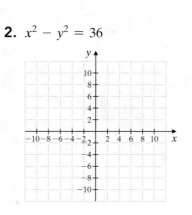

**3.** $16x^2 + 9y^2 = 144$

**4.** $y = x^2 - 8x + 16$

**5.** $x^2 + y^2 + 6x = 16$

**6.** $x = y^2 + 8y - 3$

1. _____

2. _____

3. _____

4. _____

5. _____

6. _____

**7.** _____

**7.** $\dfrac{(x-4)^2}{16} + \dfrac{(y-3)^2}{9} = 1$

**8.** _____

**8.** $y^2 - x^2 = 1$

**9.** _____

*Solve each system.*

**10.** _____

**9.** $\begin{cases} x^2 + y^2 = 169 \\ 5x + 12y = 0 \end{cases}$

**10.** $\begin{cases} x^2 + y^2 = 26 \\ x^2 - 2y^2 = 23 \end{cases}$

**11.** $\begin{cases} y = x^2 - 5x + 6 \\ y = 2x \end{cases}$

**12.** $\begin{cases} x^2 + 4y^2 = 5 \\ y = x \end{cases}$

**11.** _____

*Graph each system.*

**12.** _____

**13.** $\begin{cases} 2x + 5y \ge 10 \\ y \ge x^2 + 1 \end{cases}$

**14.** $\begin{cases} \dfrac{x^2}{4} + y^2 \le 1 \\ x + y > 1 \end{cases}$

**13.** _____

**14.** _____

**15.** $\begin{cases} x^2 + y^2 > 1 \\ \dfrac{x^2}{4} - y^2 \ge 1 \end{cases}$

**16.** $\begin{cases} x^2 + y^2 \ge 4 \\ x^2 + y^2 < 16 \\ y \ge 0 \end{cases}$

**15.** _____

**16.** _____

**17.** Which graph best resembles the graph of $x = a(y - k)^2 + h$ if $a > 0, h < 0$, and $k > 0$?

**17.** _____

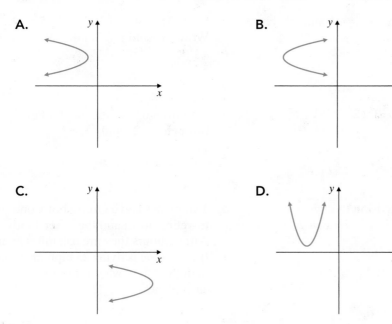

A.

B.

C.

D.

**18.** A bridge has an arch in the shape of half an ellipse. If the equation of the ellipse, measured in feet, is $100x^2 + 225y^2 = 22{,}500$, find the height of the arch from the road and the width of the arch.

**18.** _____

Copyright 2015 Pearson Education, Inc.

**Answers**

1. _____

2. _____

3. _____

4. _____

5. _____

6. _____

7. _____

8. _____

9. _____

10. _____

11. _____

12. _____

13. _____

14. _____

15. _____

16. _____

17. _____

18. _____

**1.** Add $11x^3 - 12x^2 + x - 3$ and $x^3 - 10x + 5$.

**2.** Solve $3x + 4 > 1$ and $2x - 5 \le 9$. Write the solution in interval notation.

**3.** Multiply: $(x + 3)(2x + 5)$

**4.** Find the slope of the line that goes through $(3, 2)$ and $(1, -4)$.

**5.** Use synthetic division to divide $x^4 - 2x^3 - 11x^2 + 34$ by $x + 2$.

**6.** Two planes leave Greensboro, one traveling north and the other south. After 2 hours they are 650 miles apart. If one plane is flying 25 mph faster than the other, what is the speed of each plane?

*Factor completely.*

**7.** $2(x - 5) + 3a(x - 5)$

**8.** $3y^2 + 14y + 15$

**9.** $x^2 - 12x + 35$

**10.** $20a^5 + 54a^4 + 10a^3$

**11.** $(x + 3)^2 - 36$

**12.** $(y - 3)^2 - 2(y - 3) - 8$

**13.** Solve: $2x^2 + 9x - 5 = 0$

**14.** Solve: $\dfrac{2}{x + 3} = \dfrac{1}{x^2 - 9} - \dfrac{1}{x - 3}$.

*Simplify each rational expression.*

**15.** $\dfrac{2 + x}{x + 2}$

**16.** $\dfrac{8x^3 - 1}{2x - 1}$

**17.** $\dfrac{2 - x}{x - 2}$

**18.** $\dfrac{4x^9 - 25x^5y^2}{2x^7 + 5x^5y}$

*Perform the indicated operation. Simplify if possible.*

**19.** $\dfrac{x^2}{x+7} - \dfrac{49}{x+7}$

**20.** $\dfrac{2}{3a-15} - \dfrac{a}{25-a^2}$

**21.** $\dfrac{x}{3y^2} - \dfrac{x+1}{3y^2}$

**22.** $\dfrac{x}{3x^2-13x-10} + \dfrac{9}{9x^2-4}$

**23.** Simplify: $\dfrac{\dfrac{5x}{x+2}}{\dfrac{10}{x-2}}$

**24.** Suppose that $y$ varies inversely as $x$. If $y = 3$ when $x = \dfrac{2}{3}$, find the constant of variation and the variation equation.

**25.** Solve: $\dfrac{2x}{x-3} + \dfrac{6-2x}{x^2-9} = \dfrac{x}{x+3}$

**26.** Multiply and simplify if possible.
 **a.** $\sqrt{5}(2+\sqrt{15})$
 **b.** $(\sqrt{3}-\sqrt{5})(\sqrt{7}-1)$
 **c.** $(2\sqrt{5}-1)^2$
 **d.** $(3\sqrt{2}+5)(3\sqrt{2}-5)$

**27.** If a certain number is subtracted from the numerator and added to the denominator of $\dfrac{9}{19}$, the new fraction is equivalent to $\dfrac{1}{3}$. Find the number.

**28.** Rationalize the denominator. $\dfrac{-2}{\sqrt{3}+3}$

*Find each square root. Assume that all variables represent positive real numbers.*

**29.** $\sqrt{0}$

**30.** $\sqrt{\dfrac{x^7}{49}}$

**31.** $\sqrt{0.25}$

**32.** $\sqrt{90a^5b^2}$

*Use rational exponents to simplify. Assume that all variables represent positive real numbers.*

**33.** $\sqrt[8]{x^4}$

**34.** $\sqrt[4]{x^2y^4}$

**35.** $\sqrt[6]{r^2s^4}$

**36.** Solve: $\sqrt{2x-3} = x-3$

*Simplify. Assume that all variables represent positive real numbers.*

**37.** $\sqrt[3]{24}$

**38.** $\sqrt{144y^2}$

**39.** $\sqrt[4]{32}$

**40.** $-\sqrt[3]{-27x^3}$

19. _____

20. _____

21. _____

22. _____

23. _____

24. _____

25. _____

26. a. _____

 b. _____

 c. _____

 d. _____

27. _____

28. _____

29. _____

30. _____

31. _____

32. _____

33. _____

34. _____

35. _____

36. _____

37. _____

38. _____

39. _____

40. _____

**41.** _____

**42. a.** _____

  **b.** _____

  **c.** _____

**43.** _____

**44.** _____

**45.** _____

**46.** _____

**47.** _____

**48.** _____

**49.** _____

**50. a.** _____

  **b.** _____

  **c.** _____

**51.** _____

**52.** _____

**53.** _____

**54. a.** _____

  **b.** _____

  **c.** _____

  **d.** _____

**55.** _____

**56.** _____

**57.** _____

**41.** Rationalize the denominator of $\dfrac{2}{\sqrt{5}}$.

**42.** Use the quotient rule to simplify.

  **a.** $\dfrac{\sqrt{32}}{\sqrt{4}}$    **b.** $\dfrac{\sqrt[5]{64x^9y^2}}{\sqrt[5]{2x^2y^{-8}}}$

  **c.** $\dfrac{\sqrt[3]{240y^2}}{5\sqrt[3]{3y^{-4}}}$

**43.** Solve:  $\sqrt{-10x-1}+3x=0$

**44.** Solve:  $x^3+2x^2-4x>8$

**45.** Multiply:  $(2-5i)(4+i)$

**46.** Find the inverse of $f(x)=\dfrac{x+1}{2}$.

**47.** Solve $p^2+2p=4$ by completing the square.

**48.** Find the maximum value of $f(x)=-x^2-6x+4$.

**49.** Solve:  $2x^2-4x=3$

**50.** Solve each equation for $x$.
  **a.** $64^x=4$
  **b.** $125^{x-3}=25$
  **c.** $\dfrac{1}{81}=3^{2x}$

**51.** Find the distance between $(2,-5)$ and $(1,-4)$. Give the exact distance and a three-decimal-place approximation.

**52.** Solve:  $\dfrac{4}{x-2}-\dfrac{x}{x+2}=\dfrac{16}{x^2-4}$

**53.** Graph:  $\dfrac{x^2}{9}+\dfrac{y^2}{16}=1$

**54.** If $f(x)=x^2-3x+2$ and $g(x)=-3x+5$, find
  **a.** $(f\circ g)(x)$
  **b.** $(f\circ g)(-2)$
  **c.** $(g\circ f)(x)$
  **d.** $(g\circ f)(5)$

_Solve each system._

**55.** $\begin{cases} x^2+y^2=4 \\ x+y=3 \end{cases}$

**56.** Solve the system $\begin{cases} x^2+y^2=36 \\ y=x+6 \end{cases}$

**57.** $\begin{cases} x^2+2y^2=10 \\ x^2-y^2=1 \end{cases}$

# Fractions

A quotient of two numbers such as $\frac{2}{9}$ is called a **fraction.** The parts of a fraction are:

Fraction bar → $\dfrac{2}{9}$  ← Numerator
← Denominator

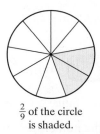

$\frac{2}{9}$ of the circle
is shaded.

A fraction may be used to refer to part of a whole. For example, $\frac{2}{9}$ of the circle in the figure is shaded. The denominator 9 tells us how many equal parts the whole circle is divided into and the numerator 2 tells us how many equal parts are shaded.

In this section, we will use numerators that are **whole numbers** and denominators that are nonzero whole numbers. The whole numbers consist of 0 and the natural numbers.

Whole numbers: 0, 1, 2, 3, 4, 5, and so on

## Objectives

**A** Write Equivalent Fractions.

**B** Write Fractions in Simplest Form.

**C** Multiply and Divide Fractions.

**D** Add and Subtract Fractions.

**E** Perform Operations on Mixed Numbers.

## Objective A Writing Equivalent Fractions

More than one fraction can be used to name the same part of a whole. Such fractions are called **equivalent fractions.**

$\frac{1}{3}$ $=$ $\frac{2}{6}$ $=$ $\frac{4}{12}$

Equivalent fractions

### Equivalent Fractions

Fractions that represent the same portion of a whole are called **equivalent fractions.**

To write equivalent fractions, we use the **fundamental principle of fractions.** This principle guarantees that, if we multiply both the numerator and the denominator by the same nonzero number, the result is an equivalent fraction. For example, if we multiply the numerator and denominator of $\frac{1}{3}$ by the same number, 2, the result is the equivalent fraction $\frac{2}{6}$.

$$\frac{1 \cdot 2}{3 \cdot 2} = \frac{2}{6}$$

## Fundamental Principle of Fractions

If $a$, $b$, and $c$ are numbers, then

$$\frac{a}{b} = \frac{a \cdot c}{b \cdot c} \quad \text{or} \quad \frac{a \cdot c}{b \cdot c} = \frac{a}{b}$$

as long as $b$ and $c$ are not 0.

**Practice 1**

Write $\frac{1}{4}$ as an equivalent fraction with a denominator of 20.

**Example 1**  Write $\frac{2}{5}$ as an equivalent fraction with a denominator of 15.

**Solution:**  Since $5 \cdot 3 = 15$, we use the fundamental principle of fractions and multiply the numerator and denominator of $\frac{2}{5}$ by 3.

$$\frac{2}{5} = \frac{2 \cdot 3}{5 \cdot 3} = \frac{6}{15}$$

Thus, $\frac{2}{5}$ is equivalent to $\frac{6}{15}$. They both represent the same part of a whole.

■ Work Practice 1

## Objective B  Simplifying Fractions

A fraction is said to be **simplified** or in **lowest terms** when the numerator and the denominator have no factors in common other than 1. For example, the fraction $\frac{5}{11}$ is in lowest terms since 5 and 11 have no common factors other than 1.

One way to simplify fractions is to write both the numerator and the denominator as products of primes and then apply the fundamental principle of fractions.

**Practice 2**

Simplify: $\frac{20}{35}$

**Example 2**  Simplify: $\frac{42}{49}$

**Solution:**  We write the numerator and the denominator as products of primes. Then we apply the fundamental principle of fractions to the common factor 7.

$$\frac{42}{49} = \frac{2 \cdot 3 \cdot 7}{7 \cdot 7} = \frac{2 \cdot 3}{7} = \frac{6}{7}$$

■ Work Practice 2

✔**Concept Check**  Explain the errors in the following steps.

**a.** $\frac{15}{55} = \frac{15}{55} = \frac{1}{5}$    **b.** $\frac{6}{7} = \frac{5 + 1}{5 + 2} = \frac{1}{2}$

**Practice 3–4**

Simplify each fraction.

**3.** $\frac{7}{20}$    **4.** $\frac{12}{40}$

**Examples**  Simplify each fraction.

**3.** $\frac{11}{27} = \frac{11}{3 \cdot 3 \cdot 3}$  There are no common factors other than 1, so $\frac{11}{27}$ is already simplified.

**4.** $\frac{88}{20} = \frac{2 \cdot 2 \cdot 2 \cdot 11}{2 \cdot 2 \cdot 5} = \frac{22}{5}$

■ Work Practice 3–4

**Answers**

**1.** $\frac{5}{20}$  **2.** $\frac{4}{7}$  **3.** $\frac{7}{20}$  **4.** $\frac{3}{10}$

✔**Concept Check Answer**

answers may vary

A **proper fraction** is a fraction whose numerator is less than its denominator. The fraction $\frac{22}{5}$ from Example 4 is called an improper fraction. An **improper fraction** is a fraction whose numerator is greater than or equal to its denominator.

The improper fraction $\frac{22}{5}$ may be written as the mixed number $4\frac{2}{5}$. Notice that a **mixed number** has a whole number part and a fraction part. We review operations on mixed numbers in Objective E in this section.

We may simplify some fractions by recalling that the fraction bar means division.

$$\frac{6}{6} = 6 \div 6 = 1 \quad \text{and} \quad \frac{3}{1} = 3 \div 1 = 3$$

**Examples** Simplify by dividing the numerator by the denominator.

**5.** $\frac{3}{3} = 1$   Since $3 \div 3 = 1$.

**6.** $\frac{4}{2} = 2$   Since $4 \div 2 = 2$.

**7.** $\frac{7}{7} = 1$   Since $7 \div 7 = 1$.

**8.** $\frac{8}{1} = 8$   Since $8 \div 1 = 8$.

▨ Work Practice 5–8

In general, if the numerator and the denominator are the same nonzero number, the fraction is equivalent to 1. Also, if the denominator of a fraction is 1, the fraction is equivalent to the numerator.

> If $a$ is any number other than 0, then $\frac{a}{a} = 1$.
>
> Also, if $a$ is any number, $\frac{a}{1} = a$.

## Objective C  Multiplying and Dividing Fractions ▶

To multiply two fractions, we multiply numerator times numerator to obtain the numerator of the product. Then we multiply denominator times denominator to obtain the denominator of the product.

> **Multiplying Fractions**
>
> $$\frac{a}{b} \cdot \frac{c}{d} = \frac{a \cdot c}{b \cdot d} \quad \text{if } b \neq 0 \text{ and } d \neq 0$$

**Example 9** Multiply: $\frac{2}{15} \cdot \frac{5}{13}$. Simplify the product if possible.

**Solution:** $\frac{2}{15} \cdot \frac{5}{13} = \frac{2 \cdot 5}{15 \cdot 13}$   Multiply numerators. Multiply denominators.

To simplify the product, we divide the numerator and the denominator by any common factors.

$$\frac{2}{15} \cdot \frac{5}{13} = \frac{2 \cdot 5}{3 \cdot 5 \cdot 13} = \frac{2}{39}$$

▨ Work Practice 9

**Practice 5–8**

Simplify by dividing the numerator by the denominator.

**5.** $\frac{4}{4}$    **6.** $\frac{9}{3}$

**7.** $\frac{10}{10}$    **8.** $\frac{5}{1}$

**Practice 9**

Multiply: $\frac{3}{10} \cdot \frac{2}{5}$. Simplify the product if possible.

**Answers**

**5.** 1  **6.** 3  **7.** 1  **8.** 5  **9.** $\frac{3}{25}$

Before we divide fractions, we first define **reciprocals.** Two numbers are reciprocals of each other if their product is 1.

The reciprocal of $\frac{2}{3}$ is $\frac{3}{2}$ because $\frac{2}{3} \cdot \frac{3}{2} = \frac{6}{6} = 1$.

The reciprocal of 5 is $\frac{1}{5}$ because $5 \cdot \frac{1}{5} = \frac{5}{1} \cdot \frac{1}{5} = \frac{5}{5} = 1$.

To divide fractions, we multiply the first fraction by the reciprocal of the second fraction. For example,

$$\frac{1}{2} \div \frac{5}{7} = \frac{1}{2} \cdot \frac{7}{5} = \frac{1 \cdot 7}{2 \cdot 5} = \frac{7}{10}$$

**Helpful Hint**

To divide, multiply by the reciprocal.

### Dividing Fractions

$$\frac{a}{b} \div \frac{c}{d} = \frac{a}{b} \cdot \frac{d}{c}, \quad \text{if } b \neq 0, d \neq 0, \text{ and } c \neq 0$$

**Practice 10–12**

Divide and simplify.

**10.** $\frac{2}{9} \div \frac{3}{4}$ **11.** $\frac{8}{11} \div 24$

**12.** $\frac{5}{4} \div \frac{5}{8}$

**Examples** Divide and simplify.

**10.** $\frac{4}{5} \div \frac{5}{16} = \frac{4}{5} \cdot \frac{16}{5} = \frac{4 \cdot 16}{5 \cdot 5} = \frac{64}{25}$

**11.** $\frac{7}{10} \div 14 = \frac{7}{10} \div \frac{14}{1} = \frac{7}{10} \cdot \frac{1}{14} = \frac{7 \cdot 1}{2 \cdot 5 \cdot 2 \cdot 7} = \frac{1}{20}$

**12.** $\frac{3}{8} \div \frac{3}{10} = \frac{3}{8} \cdot \frac{10}{3} = \frac{3 \cdot 2 \cdot 5}{2 \cdot 2 \cdot 2 \cdot 3} = \frac{5}{4}$

■ Work Practice 10–12

## Objective D Adding and Subtracting Fractions ▶

To add or subtract fractions with the same denominator, we combine numerators and place the sum or difference over the common denominator.

### Adding and Subtracting Fractions with the Same Denominator

$$\frac{a}{b} + \frac{c}{b} = \frac{a + c}{b}, \quad \text{if } b \neq 0$$

$$\frac{a}{b} - \frac{c}{b} = \frac{a - c}{b}, \quad \text{if } b \neq 0$$

**Answers**

**10.** $\frac{8}{27}$ **11.** $\frac{1}{33}$ **12.** 2

**Examples** Add or subtract as indicated. Then simplify if possible.

**13.** $\dfrac{2}{7} + \dfrac{4}{7} = \dfrac{2+4}{7} = \dfrac{6}{7}$

**14.** $\dfrac{3}{10} + \dfrac{2}{10} = \dfrac{3+2}{10} = \dfrac{5}{10} = \dfrac{5}{2\cdot 5} = \dfrac{1}{2}$

**15.** $\dfrac{9}{7} - \dfrac{2}{7} = \dfrac{9-2}{7} = \dfrac{7}{7} = 1$

**16.** $\dfrac{5}{3} - \dfrac{1}{3} = \dfrac{5-1}{3} = \dfrac{4}{3}$

■ Work Practice 13–16

To add or subtract with different denominators, we first write the fractions as **equivalent fractions** with the same denominator. We use the smallest or **least common denominator,** or **LCD.** The LCD is the same as the least common multiple of the denominators.

**Example 17** Add: $\dfrac{2}{5} + \dfrac{1}{4}$

**Solution:** We first must find the least common denominator before the fractions can be added. The least common multiple for the denominators 5 and 4 is 20. This is the LCD we will use.

We write both fractions as equivalent fractions with denominators of 20. Since

$$\frac{2}{5} = \frac{2\cdot 4}{5\cdot 4} = \frac{8}{20} \quad \text{and} \quad \frac{1}{4} = \frac{1\cdot 5}{4\cdot 5} = \frac{5}{20}$$

then

$$\frac{2}{5} + \frac{1}{4} = \frac{8}{20} + \frac{5}{20} = \frac{13}{20}$$

■ Work Practice 17

**Example 18** Subtract and simplify: $\dfrac{19}{6} - \dfrac{23}{12}$

**Solution:** The LCD is 12. We write both fractions as equivalent fractions with denominators of 12.

$$\frac{19}{6} - \frac{23}{12} = \frac{19\cdot 2}{6\cdot 2} - \frac{23}{12}$$
$$= \frac{38}{12} - \frac{23}{12}$$
$$= \frac{15}{12}$$
$$= \frac{3\cdot 5}{2\cdot 2\cdot 3}$$
$$= \frac{5}{4}$$

■ Work Practice 18

**Practice 13–16**

Add or subtract as indicated. Then simplify if possible.

**13.** $\dfrac{2}{11} + \dfrac{5}{11}$   **14.** $\dfrac{1}{8} + \dfrac{3}{8}$

**15.** $\dfrac{13}{10} - \dfrac{3}{10}$   **16.** $\dfrac{7}{6} - \dfrac{2}{6}$

**Practice 17**

Add: $\dfrac{3}{8} + \dfrac{1}{20}$

**Practice 18**

Subtract and simplify: $\dfrac{8}{15} - \dfrac{1}{3}$

**Answers**

**13.** $\dfrac{7}{11}$ **14.** $\dfrac{1}{2}$ **15.** 1 **16.** $\dfrac{5}{6}$

**17.** $\dfrac{17}{40}$ **18.** $\dfrac{1}{5}$

## Objective E  Performing Operations on Mixed Numbers

To perform operations on mixed numbers, first write each mixed number as an improper fraction. To recall how this is done, let's write $3\frac{1}{5}$ as an improper fraction.

$$3\frac{1}{5} = 3 + \frac{1}{5} = \frac{15}{5} + \frac{1}{5} = \frac{16}{5}$$

Because of the steps above, notice we can use a shortcut process for writing a mixed number as an improper fraction.

$$3\frac{1}{5} = \frac{5 \cdot 3 + 1}{5} = \frac{16}{5}$$

**Practice 19**

Multiply: $5\frac{1}{6} \cdot 4\frac{2}{5}$

> **Example 19**  Divide: $2\frac{1}{8} \div 1\frac{2}{3}$
>
> **Solution:**  First write each mixed number as an improper fraction.
>
> $$2\frac{1}{8} = \frac{8 \cdot 2 + 1}{8} = \frac{17}{8}; \quad 1\frac{2}{3} = \frac{3 \cdot 1 + 2}{3} = \frac{5}{3}$$
>
> Now divide as usual.
>
> $$2\frac{1}{8} \div 1\frac{2}{3} = \frac{17}{8} \div \frac{5}{3} = \frac{17}{8} \cdot \frac{3}{5} = \frac{51}{40} \quad \text{or} \quad 1\frac{11}{40}$$
>
> ■ Work Practice 19

As a general rule, if the original exercise contains mixed numbers, write the result as a mixed number, if possible.

**Practice 20**

Subtract: $7\frac{3}{8} - 6\frac{1}{4}$

> **Example 20**  Add: $2\frac{1}{8} + 1\frac{2}{3}$
>
> **Solution:**
>
> $$2\frac{1}{8} + 1\frac{2}{3} = \frac{17}{8} + \frac{5}{3} = \frac{51}{24} + \frac{40}{24} = \frac{91}{24} \quad \text{or} \quad 3\frac{19}{24}$$
>
> ■ Work Practice 20

**Practice 21**

Add: $76\frac{1}{9} + 35\frac{3}{4}$

When adding or subtracting larger mixed numbers, you might want to use the following method.

> **Example 21**  Subtract: $50\frac{1}{6} - 38\frac{1}{3}$
>
> **Solution:**
>
> $$
> \begin{array}{r}
> 50\frac{1}{6} = \quad 50\frac{1}{6} = \quad 49\frac{7}{6} \\
> -38\frac{1}{3} = -38\frac{2}{6} = -38\frac{2}{6} \\
> \hline
> 11\frac{5}{6}
> \end{array}
> $$
>
> $50\frac{1}{6} = 49 + 1 + \frac{1}{6} = 49\frac{7}{6}$
>
> ■ Work Practice 21

**Answers**

**19.** $22\frac{11}{15}$  **20.** $1\frac{1}{8}$  **21.** $111\frac{31}{36}$

## A Exercise Set MyMathLab®

**Objective A** *Write each fraction as an equivalent fraction with the given denominator. See Example 1.*

1. $\frac{7}{10}$ with a denominator of 30

2. $\frac{2}{3}$ with a denominator of 9

3. $\frac{2}{9}$ with a denominator of 18

4. $\frac{8}{7}$ with a denominator of 56

5. $\frac{4}{5}$ with a denominator of 20

6. $\frac{4}{5}$ with a denominator of 25

**Objective B** *Simplify each fraction. See Examples 2 through 8.*

7. $\frac{2}{4}$

8. $\frac{3}{6}$

9. $\frac{10}{15}$

10. $\frac{15}{20}$

11. $\frac{3}{7}$

12. $\frac{5}{9}$

13. $\frac{20}{20}$

14. $\frac{24}{24}$

15. $\frac{35}{7}$

16. $\frac{42}{6}$

17. $\frac{18}{30}$

18. $\frac{42}{45}$

19. $\frac{24}{30}$

20. $\frac{8}{40}$

21. $\frac{66}{48}$

22. $\frac{64}{24}$

23. $\frac{120}{244}$

24. $\frac{360}{700}$

25. $\frac{192}{264}$

26. $\frac{455}{525}$

**Objectives C E** **Mixed Practice** *Multiply or divide as indicated. See Examples 9 through 12 and 19.*

27. $\frac{1}{2} \cdot \frac{3}{4}$

28. $\frac{10}{6} \cdot \frac{3}{5}$

29. $\frac{2}{3} \cdot \frac{3}{4}$

30. $\frac{7}{8} \cdot \frac{3}{21}$

31. $5\frac{1}{9} \cdot 3\frac{2}{3}$

32. $2\frac{3}{4} \cdot 1\frac{7}{8}$

33. $7\frac{2}{5} \div \frac{1}{5}$

34. $9\frac{5}{6} \div \frac{1}{6}$

35. $\frac{1}{2} \div \frac{7}{12}$

36. $\frac{7}{12} \div \frac{1}{2}$

37. $\frac{3}{4} \div \frac{1}{20}$

38. $\frac{3}{5} \div \frac{9}{10}$

39. $\frac{7}{10} \cdot \frac{5}{21}$

40. $\frac{3}{35} \cdot \frac{10}{63}$

41. $\frac{9}{20} \div 12$

42. $\frac{25}{36} \div 10$

43. $4\frac{2}{11} \cdot 2\frac{1}{2}$

44. $6\frac{6}{7} \cdot 3\frac{1}{2}$

45. $8\frac{3}{5} \div 2\frac{9}{10}$

46. $1\frac{7}{8} \div 3\frac{8}{9}$

**Objectives D E** **Mixed Practice** *Add or subtract as indicated. See Examples 13 through 18, 20, and 21.*

47. $\frac{4}{5} + \frac{1}{5}$

48. $\frac{6}{7} + \frac{1}{7}$

49. $\frac{4}{5} - \frac{1}{5}$

50. $\frac{6}{7} - \frac{1}{7}$

51. $\frac{23}{105} + \frac{4}{105}$

52. $\frac{13}{132} + \frac{35}{132}$

53. $\frac{17}{21} - \frac{10}{21}$

54. $\frac{18}{35} - \frac{11}{35}$

55. $9\frac{7}{8} + 2\frac{3}{8}$

56. $8\frac{1}{8} - 6\frac{3}{8}$

**57.** $5\frac{2}{5} - 3\frac{4}{5}$   **58.** $7\frac{3}{4} + 2\frac{1}{4}$   **59.** $\frac{2}{3} + \frac{3}{7}$   **60.** $\frac{3}{4} + \frac{1}{6}$   **61.** $\frac{10}{3} - \frac{5}{21}$

**62.** $\frac{11}{7} - \frac{3}{35}$   **63.** $\frac{10}{21} + \frac{5}{21}$   **64.** $\frac{11}{35} + \frac{3}{35}$   **65.** $\frac{5}{22} - \frac{5}{33}$   **66.** $\frac{7}{10} - \frac{8}{15}$

**67.** $8\frac{11}{12} - 1\frac{5}{6}$   **68.** $4\frac{7}{8} - 2\frac{3}{16}$   **69.** $17\frac{2}{5} + 30\frac{2}{3}$   **70.** $26\frac{11}{20} + 40\frac{7}{10}$

**71.** $\frac{12}{5} - 1$   **72.** $2 - \frac{3}{8}$   **73.** $\frac{2}{3} - \frac{5}{9} + \frac{5}{6}$   **74.** $\frac{8}{11} - \frac{1}{4} + \frac{1}{2}$

**75.** In your own words, describe how to add or subtract fractions.

**76.** In your own words, describe how to divide fractions.

## Concept Extensions

*Each circle below represents a whole, or 1. Determine the unknown part of the circle.*

**77.**    **78.**   **79.**    **80.**

**81.** During the 2012 Summer Olympic Games, Sandra Perković of Croatia took the gold medal in the women's discus throw with a distance of approximately $226\frac{3}{4}$ feet. However, the Olympic record for the women's discus throw was set in 1988 by Martina Hellmann of East Germany with a distance of $237\frac{1}{6}$ feet. How much longer was the Olympic record discus throw than the gold medal throw in 2012? (*Source:* wikipedia)

**82.** Approximately $\frac{43}{50}$ of all Americans use the Internet. What fraction of Americans do not use the Internet? (*Source:* PEW)

# Solving Systems of Equations Using Determinants

## Objectives

**A** Define and Evaluate a 2 × 2 Determinant. ▷

**B** Use Cramer's Rule to Solve a System of Two Linear Equations in Two Variables. ▷

**C** Define and Evaluate a 3 × 3 Determinant. ▷

**D** Use Cramer's Rule to Solve a System of Three Linear Equations in Three Variables. ▷

We have solved systems of two linear equations in two variables in four different ways: graphically, by substitution, by elimination, and by matrices. Now we analyze another method, called **Cramer's rule.**

## Objective A Evaluating 2 × 2 Determinants ▷

Recall that a matrix is a rectangular array of numbers. If a matrix has the same number of rows and columns, it is called a **square matrix.** Examples of square matrices are

$$\begin{bmatrix} 1 & 6 \\ 5 & 2 \end{bmatrix} \qquad \begin{bmatrix} 2 & 4 & 1 \\ 0 & 5 & 2 \\ 3 & 6 & 9 \end{bmatrix}$$

A **determinant** is a real number associated with a square matrix. The determinant of a square matrix is denoted by placing vertical bars about the array of numbers. Thus,

The determinant of the square matrix $\begin{bmatrix} 1 & 6 \\ 5 & 2 \end{bmatrix}$ is $\begin{vmatrix} 1 & 6 \\ 5 & 2 \end{vmatrix}$.

The determinant of the square matrix $\begin{bmatrix} 2 & 4 & 1 \\ 0 & 5 & 2 \\ 3 & 6 & 9 \end{bmatrix}$ is $\begin{vmatrix} 2 & 4 & 1 \\ 0 & 5 & 2 \\ 3 & 6 & 9 \end{vmatrix}$.

We define the determinant of a 2 × 2 matrix first. (Recall that 2 × 2 is read "two by two." It means that the matrix has 2 rows and 2 columns.)

### Determinant of a 2 × 2 Matrix

$$\begin{vmatrix} a & b \\ c & d \end{vmatrix} = ad - bc$$

**Example 1** Evaluate each determinant.

**a.** $\begin{vmatrix} -1 & 2 \\ 3 & -4 \end{vmatrix}$ **b.** $\begin{vmatrix} 2 & 0 \\ 7 & -5 \end{vmatrix}$

**Solution:** First we identify the values of $a$, $b$, $c$, and $d$. Then we perform the evaluation.

**a.** Here $a = -1, b = 2, c = 3$, and $d = -4$.

$$\begin{vmatrix} -1 & 2 \\ 3 & -4 \end{vmatrix} = ad - bc = (-1)(-4) - (2)(3) = -2$$

**b.** In this example, $a = 2, b = 0, c = 7$, and $d = -5$.

$$\begin{vmatrix} 2 & 0 \\ 7 & -5 \end{vmatrix} = ad - bc = 2(-5) - (0)(7) = -10$$

■ Work Practice 1

**Practice 1**

Evaluate each determinant.

**a.** $\begin{vmatrix} -3 & 6 \\ 2 & 1 \end{vmatrix}$

**b.** $\begin{vmatrix} 4 & 5 \\ 0 & -5 \end{vmatrix}$

**Answers**

**1. a.** $-15$ **b.** $-20$

## Objective B  Using Cramer's Rule to Solve a System of Two Linear Equations

To develop Cramer's rule, we solve the system $\begin{cases} ax + by = h \\ cx + dy = k \end{cases}$ using elimination.

First, we eliminate $y$ by multiplying both sides of the first equation by $d$ and both sides of the second equation by $-b$ so that the coefficients of $y$ are opposites. The result is that

$$\begin{cases} d(ax + by) = d \cdot h \\ -b(cx + dy) = -b \cdot k \end{cases} \quad \text{simplifies to} \quad \begin{cases} adx + bdy = hd \\ -bcx - bdy = -bk \end{cases}$$

We now add the two equations and solve for $x$.

$$\begin{array}{r} adx + bdy = hd \\ \underline{-bcx - bdy = -bk} \\ adx - bcx = hd - bk \quad \text{Add the equations.} \\ (ad - bc)x = hd - bk \\ x = \dfrac{hd - bk}{ad - bc} \quad \text{Solve for } x. \end{array}$$

When we replace $x$ with $\dfrac{hd - bk}{ad - bc}$ in the equation $ax + by = h$ and solve for $y$, we find that $y = \dfrac{ak - hc}{ad - bc}$.

Notice that the numerator of the value of $x$ is the determinant of

$$\begin{vmatrix} h & b \\ k & d \end{vmatrix} = hd - bk$$

Also, the numerator of the value of $y$ is the determinant of

$$\begin{vmatrix} a & h \\ c & k \end{vmatrix} = ak - hc$$

Finally, the denominators of the values of $x$ and $y$ are the same and are the determinant of

$$\begin{vmatrix} a & b \\ c & d \end{vmatrix} = ad - bc$$

This means that the values of $x$ and $y$ can be written in determinant notation:

$$x = \dfrac{\begin{vmatrix} h & b \\ k & d \end{vmatrix}}{\begin{vmatrix} a & b \\ c & d \end{vmatrix}} \quad \text{and} \quad y = \dfrac{\begin{vmatrix} a & h \\ c & k \end{vmatrix}}{\begin{vmatrix} a & b \\ c & d \end{vmatrix}}$$

For convenience, we label the determinants $D$, $D_x$, and $D_y$.

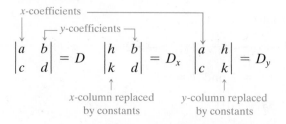

$$\begin{vmatrix} a & b \\ c & d \end{vmatrix} = D \qquad \begin{vmatrix} h & b \\ k & d \end{vmatrix} = D_x \qquad \begin{vmatrix} a & h \\ c & k \end{vmatrix} = D_y$$

$x$-column replaced by constants    $y$-column replaced by constants

These determinant formulas for the coordinates of the solution of a system are known as **Cramer's rule.**

## Cramer's Rule for Two Linear Equations in Two Variables

The solution of the system $\begin{cases} ax + by = h \\ cx + dy = k \end{cases}$ is given by

$$x = \frac{\begin{vmatrix} h & b \\ k & d \end{vmatrix}}{\begin{vmatrix} a & b \\ c & d \end{vmatrix}} = \frac{D_x}{D} \qquad y = \frac{\begin{vmatrix} a & h \\ c & k \end{vmatrix}}{\begin{vmatrix} a & b \\ c & d \end{vmatrix}} = \frac{D_y}{D}$$

as long as $D = ad - bc$ is not 0.

When $D = 0$, the system is either inconsistent or the equations are dependent. When this happens, we need to use another method to see which is the case.

**Example 2**   Use Cramer's rule to solve the system:

$$\begin{cases} 3x + 4y = -7 \\ x - 2y = -9 \end{cases}$$

**Solution:**   First we find $D$, $D_x$, and $D_y$.

$$\begin{array}{ccc} a & b & h \\ \downarrow & \downarrow & \downarrow \end{array}$$
$$\begin{cases} 3x + 4y = -7 \\ x - 2y = -9 \end{cases}$$
$$\begin{array}{ccc} \uparrow & \uparrow & \uparrow \\ c & d & k \end{array}$$

$$D = \begin{vmatrix} a & b \\ c & d \end{vmatrix} = \begin{vmatrix} 3 & 4 \\ 1 & -2 \end{vmatrix} = 3(-2) - 4(1) = -10$$

$$D_x = \begin{vmatrix} h & b \\ k & d \end{vmatrix} = \begin{vmatrix} -7 & 4 \\ -9 & -2 \end{vmatrix} = (-7)(-2) - 4(-9) = 50$$

$$D_y = \begin{vmatrix} a & h \\ c & k \end{vmatrix} = \begin{vmatrix} 3 & -7 \\ 1 & -9 \end{vmatrix} = 3(-9) - (-7)(1) = -20$$

Then $x = \dfrac{D_x}{D} = \dfrac{50}{-10} = -5$ and $y = \dfrac{D_y}{D} = \dfrac{-20}{-10} = 2$.

The ordered pair solution is $(-5, 2)$.
As always, check the solution in both original equations.

■ **Work Practice 2**

**Example 3**   Use Cramer's rule to solve the system:

$$\begin{cases} 5x + y = 5 \\ -7x - 2y = -7 \end{cases}$$

**Solution:**   First we find $D$, $D_x$, and $D_y$.

$$D = \begin{vmatrix} 5 & 1 \\ -7 & -2 \end{vmatrix} = 5(-2) - (-7)(1) = -3$$

$$D_x = \begin{vmatrix} 5 & 1 \\ -7 & -2 \end{vmatrix} = 5(-2) - (-7)(1) = -3$$

$$D_y = \begin{vmatrix} 5 & 5 \\ -7 & -7 \end{vmatrix} = 5(-7) - 5(-7) = 0$$

(Continued on next page)

**Practice 2**

Use Cramer's rule to solve the system.

$$\begin{cases} x - y = -4 \\ 2x + 3y = 2 \end{cases}$$

**Practice 3**

Use Cramer's rule to solve the system.

$$\begin{cases} 4x + y = 3 \\ 2x - 3y = -9 \end{cases}$$

**Answers**
**2.** $(-2, 2)$   **3.** $(0, 3)$

Then

$$x = \frac{D_x}{D} = \frac{-3}{-3} = 1 \text{ and } y = \frac{D_y}{D} = \frac{0}{-3} = 0$$

The ordered pair solution is $(1, 0)$.

■ Work Practice 3

## Objective C  Evaluating 3 × 3 Determinants ▶

A 3 × 3 determinant can be used to solve a system of three equations in three variables. The determinant of a 3 × 3 matrix, however, is considerably more complex than a 2 × 2 one.

### Determinant of a 3 × 3 Matrix

$$\begin{vmatrix} a_1 & b_1 & c_1 \\ a_2 & b_2 & c_2 \\ a_3 & b_3 & c_3 \end{vmatrix} = a_1 \cdot \begin{vmatrix} b_2 & c_2 \\ b_3 & c_3 \end{vmatrix} - a_2 \cdot \begin{vmatrix} b_1 & c_1 \\ b_3 & c_3 \end{vmatrix} + a_3 \cdot \begin{vmatrix} b_1 & c_1 \\ b_2 & c_2 \end{vmatrix}$$

Notice that the determinant of a 3 × 3 matrix is related to the determinants of three 2 × 2 matrices. Each determinant of these 2 × 2 matrices is called a **minor,** and every element of a 3 × 3 matrix has a minor associated with it. For example, the minor of $c_2$ is the determinant of the 2 × 2 matrix found by deleting the row and column containing $c_2$.

$$\begin{array}{ccc} a_1 & b_1 & c_1 \\ a_2 & b_2 & c_2 \\ a_3 & b_3 & c_3 \end{array} \quad \text{The minor of } c_2 \text{ is} \quad \begin{vmatrix} a_1 & b_1 \\ a_3 & b_3 \end{vmatrix}$$

Also, the minor of element $a_1$ is the determinant of the 2 × 2 matrix that has no row or column containing $a_1$.

$$\begin{array}{ccc} a_1 & b_1 & c_1 \\ a_2 & b_2 & c_2 \\ a_3 & b_3 & c_3 \end{array} \quad \text{The minor of } a_1 \text{ is} \quad \begin{vmatrix} b_2 & c_2 \\ b_3 & c_3 \end{vmatrix}$$

So the determinant of a 3 × 3 matrix can be written as

$$a_1 \cdot (\text{minor of } a_1) - a_2 \cdot (\text{minor of } a_2) + a_3 \cdot (\text{minor of } a_3)$$

Finding the determinant by using minors of elements in the first column is called **expanding** by the minors of the first column. *The value of a determinant can be found by expanding by the minors of any row or column.* The following **array of signs** is helpful in determining whether to add or subtract the product of an element and its minor.

$$\begin{array}{ccc} + & - & + \\ - & + & - \\ + & - & + \end{array}$$

If an element is in a position marked $+$, we add. If marked $-$, we subtract.

✓ **Concept Check** Suppose you are interested in finding the determinant of a 4 × 4 matrix. Study the pattern shown in the array of signs for a 3 × 3 matrix. Use the pattern to expand the array of signs for use with a 4 × 4 matrix.

### Practice 4

Evaluate by expanding by the minors of the given row or column.

a. First column   b. Third row

$$\begin{vmatrix} 2 & 0 & 1 \\ -1 & 3 & 2 \\ 5 & 1 & 4 \end{vmatrix}$$

**Answers**

4. a. 4  b. 4

✓ **Concept Check Answer**

$$\begin{array}{cccc} + & - & + & - \\ - & + & - & + \\ + & - & + & - \\ - & + & - & + \end{array}$$

**Example 4**   Evaluate by expanding by the minors of the given row or column.

$$\begin{vmatrix} 0 & 5 & 1 \\ 1 & 3 & -1 \\ -2 & 2 & 4 \end{vmatrix}$$

a. First column         b. Second row

## Solution:

**a.** The elements of the first column are $0, 1$, and $-2$. The first column of the array of signs is $+, -, +$.

$$\begin{vmatrix} 0 & 5 & 1 \\ 1 & 3 & -1 \\ -2 & 2 & 4 \end{vmatrix} = 0 \cdot \begin{vmatrix} 3 & -1 \\ 2 & 4 \end{vmatrix} - 1 \cdot \begin{vmatrix} 5 & 1 \\ 2 & 4 \end{vmatrix} + (-2) \cdot \begin{vmatrix} 5 & 1 \\ 3 & -1 \end{vmatrix}$$

$$= 0(12 - (-2)) - 1(20 - 2) + (-2)(-5 - 3)$$

$$= 0 - 18 + 16 = -2$$

**b.** The elements of the second row are $1, 3$, and $-1$. This time, the signs begin with $-$ and again alternate.

$$\begin{vmatrix} 0 & 5 & 1 \\ 1 & 3 & -1 \\ -2 & 2 & 4 \end{vmatrix} = -1 \cdot \begin{vmatrix} 5 & 1 \\ 2 & 4 \end{vmatrix} + 3 \cdot \begin{vmatrix} 0 & 1 \\ -2 & 4 \end{vmatrix} - (-1) \cdot \begin{vmatrix} 0 & 5 \\ -2 & 2 \end{vmatrix}$$

$$= -1(20 - 2) + 3(0 - (-2)) - (-1)(0 - (-10))$$

$$= -18 + 6 + 10 = -2$$

Notice that the determinant of the $3 \times 3$ matrix is the same regardless of the row or column you select to expand by.

■ Work Practice 4

✓**Concept Check** Why would expanding by minors of the second row be a good

choice for the determinant $\begin{vmatrix} 3 & 4 & -2 \\ 5 & 0 & 0 \\ 6 & -3 & 7 \end{vmatrix}$?

## Objective D  Using Cramer's Rule to Solve a System of Three Linear Equations ▶

A system of three equations in three variables may be solved with Cramer's rule also. Using the elimination process to solve a system with unknown constants as coefficients leads to the following.

### Cramer's Rule for Three Equations in Three Variables

The solution of the system $\begin{cases} a_1x + b_1y + c_1z = k_1 \\ a_2x + b_2y + c_2z = k_2 \\ a_3x + b_3y + c_3z = k_3 \end{cases}$ is given by

$$x = \frac{D_x}{D} \qquad y = \frac{D_y}{D} \qquad \text{and} \qquad z = \frac{D_z}{D}$$

where

$$D = \begin{vmatrix} a_1 & b_1 & c_1 \\ a_2 & b_2 & c_2 \\ a_3 & b_3 & c_3 \end{vmatrix} \qquad D_x = \begin{vmatrix} k_1 & b_1 & c_1 \\ k_2 & b_2 & c_2 \\ k_3 & b_3 & c_3 \end{vmatrix}$$

$$D_y = \begin{vmatrix} a_1 & k_1 & c_1 \\ a_2 & k_2 & c_2 \\ a_3 & k_3 & c_3 \end{vmatrix} \qquad D_z = \begin{vmatrix} a_1 & b_1 & k_1 \\ a_2 & b_2 & k_2 \\ a_3 & b_3 & k_3 \end{vmatrix}$$

as long as $D$ is not 0.

✓**Concept Check Answer**
Two elements of the second row are 0, which makes calculations easier.

## Practice 5

Use Cramer's rule to solve the system:

$$\begin{cases} x + 2y - z = 3 \\ 2x - 3y + z = -9 \\ -x + y - 2z = 0 \end{cases}$$

**Example 5** Use Cramer's rule to solve the system:

$$\begin{cases} x - 2y + z = 4 \\ 3x + y - 2z = 3 \\ 5x + 5y + 3z = -8 \end{cases}$$

**Solution:** First we find $D$, $D_x$, $D_y$, and $D_z$. Beginning with $D$, we expand by the minors of the first column.

$$D = \begin{vmatrix} 1 & -2 & 1 \\ 3 & 1 & -2 \\ 5 & 5 & 3 \end{vmatrix} = 1 \cdot \begin{vmatrix} 1 & -2 \\ 5 & 3 \end{vmatrix} - 3 \cdot \begin{vmatrix} -2 & 1 \\ 5 & 3 \end{vmatrix} + 5 \cdot \begin{vmatrix} -2 & 1 \\ 1 & -2 \end{vmatrix}$$

$$= 1(3 - (-10)) - 3(-6 - 5) + 5(4 - 1)$$

$$= 13 + 33 + 15 = 61$$

$$D_x = \begin{vmatrix} 4 & -2 & 1 \\ 3 & 1 & -2 \\ -8 & 5 & 3 \end{vmatrix} = 4 \cdot \begin{vmatrix} 1 & -2 \\ 5 & 3 \end{vmatrix} - 3 \cdot \begin{vmatrix} -2 & 1 \\ 5 & 3 \end{vmatrix} + (-8) \cdot \begin{vmatrix} -2 & 1 \\ 1 & -2 \end{vmatrix}$$

$$= 4(3 - (-10)) - 3(-6 - 5) + (-8)(4 - 1)$$

$$= 52 + 33 - 24 = 61$$

$$D_y = \begin{vmatrix} 1 & 4 & 1 \\ 3 & 3 & -2 \\ 5 & -8 & 3 \end{vmatrix} = 1 \cdot \begin{vmatrix} 3 & -2 \\ -8 & 3 \end{vmatrix} - 3 \cdot \begin{vmatrix} 4 & 1 \\ -8 & 3 \end{vmatrix} + 5 \cdot \begin{vmatrix} 4 & 1 \\ 3 & -2 \end{vmatrix}$$

$$= 1(9 - 16) - 3(12 - (-8)) + 5(-8 - 3)$$

$$= -7 - 60 - 55 = -122$$

$$D_z = \begin{vmatrix} 1 & -2 & 4 \\ 3 & 1 & 3 \\ 5 & 5 & -8 \end{vmatrix} = 1 \cdot \begin{vmatrix} 1 & 3 \\ 5 & -8 \end{vmatrix} - 3 \cdot \begin{vmatrix} -2 & 4 \\ 5 & -8 \end{vmatrix} + 5 \cdot \begin{vmatrix} -2 & 4 \\ 1 & 3 \end{vmatrix}$$

$$= 1(-8 - 15) - 3(16 - 20) + 5(-6 - 4)$$

$$= -23 + 12 - 50 = -61$$

From these determinants, we calculate the solution:

$$x = \frac{D_x}{D} = \frac{61}{61} = 1 \quad y = \frac{D_y}{D} = \frac{-122}{61} = -2 \quad z = \frac{D_z}{D} = \frac{-61}{61} = -1$$

The ordered triple solution is $(1, -2, -1)$. Check this solution by verifying that it satisfies each equation of the system.

■ Work Practice 5

**Answer**

**5.** $(-1, 3, 2)$

## B Exercise Set MyMathLab®

**Objective A** *Evaluate each determinant. See Example 1.*

**1.** $\begin{vmatrix} 7 & 2 \\ 0 & 8 \end{vmatrix}$

**2.** $\begin{vmatrix} 6 & 0 \\ 1 & 2 \end{vmatrix}$

**3.** $\begin{vmatrix} -4 & 2 \\ 0 & 8 \end{vmatrix}$

**4.** $\begin{vmatrix} 5 & 0 \\ 3 & -5 \end{vmatrix}$

**5.** $\begin{vmatrix} -2 & 0 \\ 3 & -10 \end{vmatrix}$

**6.** $\begin{vmatrix} -1 & 4 \\ 0 & -18 \end{vmatrix}$

**7.** $\begin{vmatrix} 3 & 5 \\ -1 & 7 \end{vmatrix}$

**8.** $\begin{vmatrix} -5 & 1 \\ 1 & -4 \end{vmatrix}$

**9.** $\begin{vmatrix} 9 & -2 \\ 4 & -3 \end{vmatrix}$

**10.** $\begin{vmatrix} 4 & -1 \\ 9 & 8 \end{vmatrix}$

**11.** $\begin{vmatrix} -2 & 9 \\ 4 & -18 \end{vmatrix}$

**12.** $\begin{vmatrix} -40 & 8 \\ 70 & -14 \end{vmatrix}$

**13.** $\begin{vmatrix} \frac{3}{4} & \frac{5}{2} \\ -\frac{1}{6} & \frac{7}{3} \end{vmatrix}$

**14.** $\begin{vmatrix} \frac{5}{7} & \frac{1}{3} \\ \frac{6}{7} & \frac{2}{3} \end{vmatrix}$

**Objective B** *Use Cramer's rule, if possible, to solve each system of linear equations. See Examples 2 and 3.*

**15.** $\begin{cases} 2y - 4 = 0 \\ x + 2y = 5 \end{cases}$

**16.** $\begin{cases} 4x - y = 5 \\ 3x - 3 = 0 \end{cases}$

**17.** $\begin{cases} 3x + y = 1 \\ 2y = 2 - 6x \end{cases}$

**18.** $\begin{cases} y = 2x - 5 \\ 8x - 4y = 20 \end{cases}$

**19.** $\begin{cases} 5x - 2y = 27 \\ -3x + 5y = 18 \end{cases}$

**20.** $\begin{cases} 4x - y = 9 \\ 2x + 3y = -27 \end{cases}$

**21.** $\begin{cases} 2x - 5y = 4 \\ x + 2y = -7 \end{cases}$

**22.** $\begin{cases} 3x - y = 2 \\ -5x + 2y = 0 \end{cases}$

**23.** $\begin{cases} \frac{2}{3}x - \frac{3}{4}y = -1 \\ -\frac{1}{6}x + \frac{3}{4}y = \frac{5}{2} \end{cases}$

**24.** $\begin{cases} \frac{1}{2}x - \frac{1}{3}y = -3 \\ \frac{1}{8}x + \frac{1}{6}y = 0 \end{cases}$

**Objective C** *Evaluate. See Example 4.*

**25.** $\begin{vmatrix} 2 & 1 & 0 \\ 0 & 5 & -3 \\ 4 & 0 & 2 \end{vmatrix}$

**26.** $\begin{vmatrix} -6 & 4 & 2 \\ 1 & 0 & 5 \\ 0 & 3 & 1 \end{vmatrix}$

**27.** $\begin{vmatrix} 4 & -6 & 0 \\ -2 & 3 & 0 \\ 4 & -6 & 1 \end{vmatrix}$

**28.** $\begin{vmatrix} 5 & 2 & 1 \\ 3 & -6 & 0 \\ -2 & 8 & 0 \end{vmatrix}$

**29.** $\begin{vmatrix} 1 & 0 & 4 \\ 1 & -1 & 2 \\ 3 & 2 & 1 \end{vmatrix}$

**30.** $\begin{vmatrix} 0 & 1 & 2 \\ 3 & -1 & 2 \\ 3 & 2 & -2 \end{vmatrix}$

**31.** $\begin{vmatrix} 3 & 6 & -3 \\ -1 & -2 & 3 \\ 4 & -1 & 6 \end{vmatrix}$

**32.** $\begin{vmatrix} 2 & -2 & 1 \\ 4 & 1 & 3 \\ 3 & 1 & 2 \end{vmatrix}$

## Objective D   *Use Cramer's rule, if possible, to solve each system of linear equations. See Example 5.*

**33.** $\begin{cases} 3x \quad\;\; + z = -1 \\ -x - 3y + z = \;\; 7 \\ \quad\;\;\; 3y + z = \;\; 5 \end{cases}$

**34.** $\begin{cases} \quad\; 4y - 3z = -2 \\ 8x - 4y \quad\;\;\; = \;\; 4 \\ -8x + 4y + \;\; z = -2 \end{cases}$

**35.** $\begin{cases} x + \;\; y + \;\; z = \;\; 8 \\ 2x - \;\; y - \;\; z = 10 \\ x - 2y + 3z = 22 \end{cases}$

**36.** $\begin{cases} 5x + y + 3z = \;\; 1 \\ \;\; x - y - 3z = -7 \\ -x + y \quad\quad\;\; = \;\; 1 \end{cases}$

**37.** $\begin{cases} 2x + 2y + \;\; z = 1 \\ -x + \;\; y + 2z = 3 \\ \;\; x + 2y + 4z = 0 \end{cases}$

**38.** $\begin{cases} 2x - 3y + \;\; z = 5 \\ \;\; x + \;\; y + \;\; z = 0 \\ 4x + 2y + 4z = 4 \end{cases}$

**39.** $\begin{cases} x - 2y + \;\; z = -5 \\ \quad\quad\; 3y + 2z = \;\; 4 \\ 3x - \;\; y \quad\quad\;\; = -2 \end{cases}$

**40.** $\begin{cases} 4x + 5y \quad\quad\; = 10 \\ \quad\quad\; 3y + 2z = -6 \\ \;\; x + \;\; y + \;\; z = \;\; 3 \end{cases}$

## Concept Extensions

*Find the value of x that will make each a true statement.*

**41.** $\begin{vmatrix} 1 & x \\ 2 & 7 \end{vmatrix} = -3$

**42.** $\begin{vmatrix} 6 & 1 \\ -2 & x \end{vmatrix} = 26$

**43.** If all the elements in a single row of a determinant are zero, what is the value of the determinant? Explain your answer.

**44.** If all the elements in a single column of a determinant are 0, what is the value of the determinant? Explain your answer.

# An Introduction to Using a Graphing Utility

## Appendix C

**Objectives**

**A** View Window and Interpret Window Settings.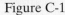

**B** Graph Equations and Use Square Viewing Windows.

## Objective A  Viewing Window and Interpreting Window Settings

In this appendix, we will use the term **graphing utility** to mean a graphing calculator or a computer software graphing package. All graphing utilities graph equations by plotting points on a screen. While plotting several points can be slow and sometimes tedious for us, a graphing utility can quickly and accurately plot hundreds of points. How does a graphing utility show plotted points? A computer or calculator screen is made up of a grid of small rectangular areas called **pixels.** If a pixel contains a point to be plotted, the pixel is turned "on"; otherwise, the pixel remains "off." The graph of an equation is then a collection of pixels turned "on." The graph of $y = 3x + 1$ from a graphing calculator is shown in Figure C-1. Notice the irregular shape of the line caused by the rectangular pixels.

Figure C-1

The portion of the coordinate plane shown on the screen in Figure C-1 is called the **viewing window** or the **viewing rectangle.** Notice the $x$-axis and the $y$-axis on the graph. While tick marks are shown on the axes, they are not labeled. This means that from this screen alone, we do not know how many units each tick mark represents. To see what each tick mark represents and the minimum and maximum values on the axes, check the *window setting* of the graphing utility. It defines the viewing window. The window of the graph of $y = 3x + 1$ shown in Figure C-1 has the following setting (Figure C-2):

| | |
|---|---|
| $\text{Xmin} = -10$ | The minimum $x$-value is $-10$. |
| $\text{Xmax} = 10$ | The maximum $x$-value is 10. |
| $\text{Xscl} = 1$ | The $x$-axis scale is 1 unit per tick mark. |
| $\text{Ymin} = -10$ | The minimum $y$-value is $-10$. |
| $\text{Ymax} = 10$ | The maximum $y$-value is 10. |
| $\text{Yscl} = 1$ | The $y$-axis scale is 1 unit per tick mark. |

Figure C-2

By knowing the scale, we can find the minimum and the maximum values on the axes simply by counting tick marks. For example, if both the Xscl ($x$-axis scale) and the Yscl are 1 unit per tick mark on the graph in Figure C-3, we can count the tick marks and find that the minimum $x$-value is $-10$ and the maximum $x$-value is 10. Also, the minimum $y$-value is $-10$ and the maximum $y$-value is 10. If the Xscl ($x$-axis scale) changes to 2 units per tick mark (shown in Figure C-4), by counting tick marks, we see that the minimum $x$-value is now $-20$ and the maximum $x$-value is now 20.

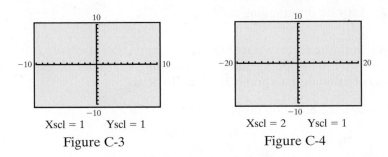

Figure C-3          Figure C-4

Sometimes window settings are given in the form [Xmin, Xmax] by [Ymin, Ymax]. For example, the window setting shown in Figure C-4 can be written as [−20, 20] by [−10 10].

Figure C-5

It is also true that if we know the Xmin and the Xmax values, we can calculate the Xscl by the displayed axes. For example, the Xscl of the graph in Figure C-5 must be 2 units per tick mark for the maximum and minimum $x$-values to be as shown. Also, the Yscl of that graph must be 1 unit per tick mark for the maximum and minimum $y$-values to be as shown.

We will call the viewing window in Figure C-3 a *standard* viewing window or rectangle. Although a standard viewing window is sufficient for much of this text, special care must be taken to ensure that all key features of a graph are shown. Figures C-6, C-7, and C-8 show the graph of $y = x^2 + 11x - 1$ on three different viewing windows. Note that certain viewing windows for this equation are misleading.

Figure C-6      Figure C-7      Figure C-8

How do we ensure that all distinguishing features of the graph of an equation are shown? It helps to know about the equation that is being graphed. For example, the equation $y = x^2 + 11x - 1$ is not a linear equation, and its graph is not a line. This equation is a quadratic equation, and therefore its graph is a parabola. By knowing this information, we know that the graph shown in Figure C-6, although correct, is misleading. Of the three viewing rectangles shown, the graph in Figure C-8 is best because it shows more of the distinguishing features of the parabola. Properties of equations needed for graphing are studied in this text.

## Objective B   Graphing Equations and Square Viewing Windows ▶

In general, the following steps may be used to graph an equation on a standard viewing window.

> **To Graph an Equation in x and y with a Graphing Utility on a Standard Viewing Window**
>
> **Step 1:** Solve the equation for $y$.
>
> **Step 2:** Use your graphing utility and enter the equation in the form
> $Y = expression\ involving\ x$
>
> **Step 3:** Activate the graphing utility.

Special care must be taken when entering the *expression involving x* in Step 2. You must be sure that the graphing utility you are using interprets the expression as you want it to. For example, let's graph $3y = 4x$. To do so,

**Step 1:** Solve the equation for $y$.

$$3y = 4x \qquad \frac{3y}{3} = \frac{4x}{3} \qquad y = \frac{4}{3}x$$

**Step 2:** Using your graphing utility, enter the expression $\frac{4}{3}x$ after the $Y =$ prompt.

In order for your graphing utility to correctly interpret the expression, you may need to enter $(4/3)x$ or $(4 \div 3)x$.

**Step 3:** Activate the graphing utility. The graph should appear as in Figure C-9.

Figure C-9

Distinguishing features of the graph of a line include showing all the intercepts of the line. For example, the window of the graph of the line in Figure C-10 does not show both intercepts of the line, but the window of the graph of the same line in Figure C-11 does show both intercepts.

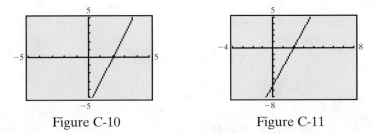

Figure C-10                    Figure C-11

On a standard viewing window, the tick marks on the *y*-axis are closer than the tick marks on the *x*-axis. This happens because the viewing window is a rectangle, and so 10 equally spaced tick marks on the positive *y*-axis will be closer together than 10 equally spaced tick marks on the positive *x*-axis. This causes the appearance of graphs to be distorted.

For example, notice the different appearances of the same line graphed using different viewing windows. The line in Figure C-12 is distorted because the tick marks along the *x*-axis are farther apart than the tick marks along the *y*-axis. The graph of the same line in Figure C-13 is not distorted because the viewing rectangle has been selected so that there is equal spacing between tick marks on both axes.

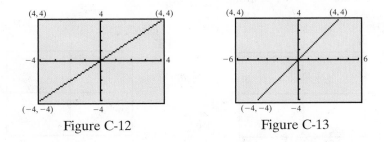

Figure C-12                    Figure C-13

We say that the line in Figure C-13 is graphed on a *square* setting. Some graphing utilities have a built-in program that, if activated, will automatically provide a square setting. A square setting is especially helpful when we are graphing perpendicular lines, circles, or when a true geometric perspective is desired. Some examples of square screens are shown in Figures C-14 and C-15.

Other features of a graphing utility such as Trace, Zoom, Intersect, and Table are discussed in appropriate Calculator Explorations in this text.

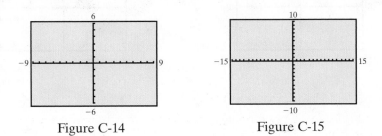

Figure C-14                    Figure C-15

## C   Exercise Set MyMathLab®  ▶

**Objective A** *In Exercises 1–4, determine whether all ordered pairs listed will lie within a standard viewing rectangle.*

**1.** $(-9, 0), (5, 8), (1, -8)$

**2.** $(4, 7), (0, 0), (-8, 9)$

**3.** $(-11, 0), (2, 2), (7, -5)$

**4.** $(3, 5), (-3, -5), (15, 0)$

*In Exercises 5–10, choose an Xmin, Xmax, Ymin, and Ymax so that all ordered pairs listed will lie within the viewing rectangle.*

**5.** $(-90, 0), (55, 80), (0, -80)$

**6.** $(4, 70), (20, 20), (-18, 90)$

**7.** $(-11, 0), (2, 2), (7, -5)$

**8.** $(3, 5), (-3, -5), (15, 0)$

**9.** $(200, 200), (50, -50), (70, -50)$

**10.** $(40, 800), (-30, 500), (15, 0)$

*Write the window setting for each viewing window shown. Use the following format:*

Xmin =          Ymin =
Xmax =          Ymax =
Xscl =          Yscl =

**11.**    **12.**    **13.**

**14.**    **15.**    **16.**

**17.**
Xscl = 1, Yscl = 3

**18.**
Xscl = 10, Yscl = 2

**19.**
Xscl = 5, Yscl = 10

**20.**
Xscl = 100, Yscl = 200

**Objective B** *Graph each linear equation in two variables, using the two different window settings given. Determine which setting shows all intercepts of the line.*

**21.** $y = 2x + 12$
Setting A:  $[-10, 10]$ by $[-10, 10]$
Setting B:  $[-10, 10]$ by $[-10, 15]$

**22.** $y = -3x + 25$
Setting A:  $[-5, 5]$ by $[-30, 10]$
Setting B:  $[-10, 10]$ by $[-10, 30]$

**23.** $y = -x - 41$
Setting A:  $[-50, 10]$ by $[-10, 10]$
Setting B:  $[-50, 10]$ by $[-50, 15]$

**24.** $y = 6x - 18$
Setting A:  $[-10, 10]$ by $[-20, 10]$
Setting B:  $[-10, 10]$ by $[-10, 10]$

**25.** $y = \dfrac{1}{2}x - 15$
Setting A:  $[-10, 10]$ by $[-20, 10]$
Setting B:  $[-10, 35]$ by $[-20, 15]$

**26.** $y = -\dfrac{2}{3}x - \dfrac{29}{3}$
Setting A:  $[-10, 10]$ by $[-10, 10]$
Setting B:  $[-15, 5]$ by $[-15, 5]$

*The graph of each equation is a line. Use a graphing utility and a standard viewing window to graph each equation.*

**27.** $3x = 5y$       **28.** $7y = -3x$       **29.** $9x - 5y = 30$       **30.** $4x + 6y = 20$

**31.** $y = -7$       **32.** $y = 2$       **33.** $x + 10y = -5$       **34.** $x - 5y = 9$

*Graph the following equations using the square setting given. Some keystrokes that may be helpful are given.*

**35.** $y = \sqrt{x}$   $[-12, 12]$ by $[-8, 8]$
Suggested keystrokes:   $\sqrt{x}$

**36.** $y = \sqrt{2x}$   $[-12, 12]$ by $[-8, 8]$
Suggested keystrokes:   $\sqrt{(2x)}$

**37.** $y = x^2 + 2x + 1$   $[-15, 15]$ by $[-10, 10]$
Suggested keystrokes:   $x \wedge 2 + 2x + 1$

**38.** $y = x^2 - 5$   $[-15, 15]$ by $[-10, 10]$
Suggested keystrokes:   $x \wedge 2 - 5$

**39.** $y = |x|$   $[-9, 9]$ by $[-6, 6]$
Suggested keystrokes:   ABS $(x)$

**40.** $y = |x - 2|$   $[-9, 9]$ by $[-6, 6]$
Suggested keystrokes:   ABS $(x - 2)$

*Graph the line on a single set of axes. Use a standard viewing window; then, if necessary, change the viewing window so that all intercepts of the line show.*

**41.** $x + 2y = 30$

**42.** $1.5x - 3.7y = 40.3$

# Stretching and Compressing Graphs of Absolute Value Functions

In Section 3.8, we learned to shift and reflect graphs of common functions: $f(x) = x$, $f(x) = x^2$, $f(x) = |x|$, and $f(x) = \sqrt{x}$. Since other common functions are studied throughout this text, in this appendix we concentrate on the absolute value function.

Recall that the graph of $h(x) = -|x - 1| + 2$, for example, is the same as the graph of $f(x) = |x|$ reflected about the $x$-axis, moved 1 unit to the right and 2 units upward. In other words,

$$h(x) = -|x - 1| + 2$$

opens   $(1, 2)$ location of vertex of
downward     V-shape

Let's now study the graphs of a few other absolute value functions.

**Example 1**   Graph $h(x) = 2|x|$, and $g(x) = \dfrac{1}{2}|x|$

**Solution:**   Let's find and plot ordered-pair solutions for the functions.

| $x$ | $h(x)$ | $g(x)$ |
|-----|--------|--------|
| $-2$ | $4$ | $1$ |
| $-1$ | $2$ | $\dfrac{1}{2}$ |
| $0$ | $0$ | $0$ |
| $1$ | $2$ | $\dfrac{1}{2}$ |
| $2$ | $4$ | $1$ |

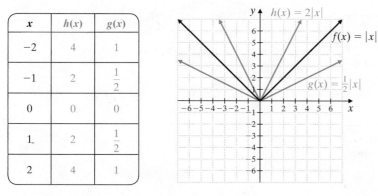

Notice that the graph of $h(x) = 2|x|$ is narrower than the graph of $f(x) = |x|$ and the graph of $g(x) = \dfrac{1}{2}|x|$ is wider than the graph of $f(x) = |x|$.

■ Work Practice 1

## Practice 1

Graph $h(x) = 4|x|$ and $g(x) = \dfrac{1}{5}|x|$ on the same set of axes.

**Answer**

1.

In general, for the absolute function, we have the following:

### The Graph of the Absolute Value Function

The graph of $f(x) = a|x - h| + k$

- Has vertex $(h, k)$ and is V-shaped.
- Opens upward if $a > 0$ and downward if $a < 0$.
- If $|a| < 1$, the graph is wider than the graph of $y = |x|$.
- If $|a| > 1$, the graph is narrower than a graph of $y = |x|$.

## Practice 2

Graph $f(x) = -\dfrac{1}{2}|x + 1| + 3$

**Example 2**    Graph $f(x) = -\dfrac{1}{3}|x + 2| + 4$

**Solution:** Let's write this function in the form $f(x) = a|x - h| + k$. For our function, we have $f(x) = -\dfrac{1}{3}|x - (-2)| + 4$. Thus:

- vertex is $(-2, 4)$
- since $a < 0$, V-shape opens down
- since $|a| = \left|-\dfrac{1}{3}\right| = \dfrac{1}{3} < 1$, the graph is wider than $y = |x|$

We will also find and plot ordered-pair solutions.

If $x = -5, f(-5) = -\dfrac{1}{3}|-5 + 2| + 4$, or $3$

If $x = 1, f(1) = -\dfrac{1}{3}|1 + 2| + 4$, or $3$

If $x = 3, f(3) = -\dfrac{1}{3}|3 + 2| + 4$, or $\dfrac{7}{3}$, or $2\dfrac{1}{3}$

| $x$ | $f(x)$ |
|-----|--------|
| $-5$ | $3$ |
| $1$ | $3$ |
| $3$ | $2\dfrac{1}{3}$ |

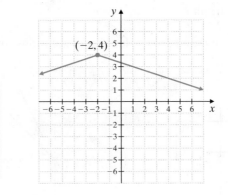

Work Practice 2

**Answer**

2.

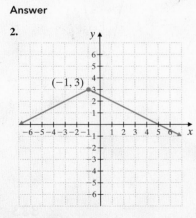

**D** **Exercise Set** MyMathLab®

*Sketch the graph of each function. Label the vertex of the V-shape. See Examples 1 and 2.*

**1.** $f(x) = 3|x|$

**2.** $f(x) = 5|x|$

**3.** $f(x) = \frac{1}{4}|x|$

**4.** $f(x) = \frac{1}{3}|x|$

**5.** $g(x) = 2|x| + 3$

**6.** $g(x) = 3|x| + 2$

**7.** $h(x) = -\frac{1}{2}|x|$

**8.** $h(x) = -\frac{1}{3}|x|$

**9.** $f(x) = 4|x - 1|$

**10.** $f(x) = 3|x - 2|$

**11.** $g(x) = -\frac{1}{3}|x| - 2$

**12.** $g(x) = -\frac{1}{2}|x| - 3$

**13.** $f(x) = -2|x - 3| + 4$

**14.** $f(x) = -3|x - 1| + 5$

**15.** $f(x) = \frac{2}{3}|x + 2| - 5$

**16.** $f(x) = \frac{3}{4}|x + 1| - 4$

# Review of Angles, Lines, and Special Triangles

The word **geometry** is formed from the Greek words, **geo,** meaning earth, and **metron,** meaning measure. Geometry literally means to measure the earth.

This appendix contains a review of some basic geometric ideas. It will be assumed that fundamental ideas of geometry such as point, line, ray, and angle are known. In this appendix, the notation $\angle 1$ is read "angle 1" and the notation $m\angle 1$ is read "the measure of angle 1."

We first review types of angles.

---

### Angles

An angle whose measure is greater than 0° but less than 90° is called an **acute angle.**

A **right angle** is an angle whose measure is 90°. A right angle can be indicated by a square drawn at the vertex of the angle, as shown below.

An angle whose measure is greater than 90° but less than 180° is called an **obtuse angle.**

An angle whose measure is 180° is called a **straight angle.**

Two angles are said to be **complementary** if the sum of their measures is 90°. Each angle is called the **complement** of the other.

Two angles are said to be **supplementary** if the sum of their measures is 180°. Each angle is called the **supplement** of the other.

---

**Example 1**    If an angle measures 28°, find its complement.

**Solution:**   Two angles are complementary if the sum of their measures is 90°. The complement of a 28° angle is an angle whose measure is $90° - 28° = 62°$. To check, notice that $28° + 62° = 90°$.

**Plane** is an undefined term that we will describe. A plane can be thought of as a flat surface with infinite length and width, but no thickness. A plane is two dimensional.

The arrows in the following diagram indicate that a plane extends indefinitely and has no boundaries.

Figures that lie on a plane are called **plane figures.** Lines that lie in the same plane are called **coplanar.**

## Lines

Two lines are **parallel** if they lie in the same plane but never meet. **Intersecting lines** meet or cross in one point.

Two lines that form right angles when they intersect are said to be **perpendicular.**

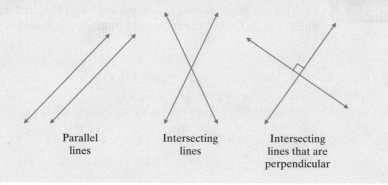

|  Parallel lines | Intersecting lines | Intersecting lines that are perpendicular |

Two intersecting lines form **vertical angles.** Angles 1 and 3 are vertical angles. Also angles 2 and 4 are vertical angles. It can be shown that **vertical angles have equal measures.**

$$m\angle 1 = m\angle 3$$
$$m\angle 2 = m\angle 4$$

**Adjacent angles** have the same vertex and share a side. Angles 1 and 2 are adjacent angles. Other pairs of adjacent angles are angles 2 and 3, angles 3 and 4, and angles 4 and 1.

A **transversal** is a line that intersects two or more lines in the same plane. Line $l$ is a transversal that intersects lines $m$ and $n$. The eight angles formed are numbered and certain pairs of these angles are given special names.

**Corresponding angles:** $\angle 1$ and $\angle 5$, $\angle 3$ and $\angle 7$, $\angle 2$ and $\angle 6$, and $\angle 4$ and $\angle 8$.

**Exterior angles:** $\angle 1$, $\angle 2$, $\angle 7$, and $\angle 8$.

**Interior angles:** $\angle 3$, $\angle 4$, $\angle 5$, and $\angle 6$.

**Alternate interior angles:** $\angle 3$ and $\angle 6$, $\angle 4$ and $\angle 5$.

These angles and parallel lines are related in the following manner.

---

### Parallel Lines Cut by a Transversal

**1.** If two parallel lines are cut by a transversal, then
    **a. corresponding angles are equal** and
    **b. alternate interior angles are equal.**

**2.** If corresponding angles formed by two lines and a transversal are equal, then the lines are parallel.

**3.** If alternate interior angles formed by two lines and a transversal are equal, then the lines are parallel.

---

**Example 2**    Given that lines $m$ and $n$ are parallel and that the measure of angle 1 is $100°$, find the measures of angles 2, 3, and 4.

**Solution:**

| | |
|---|---|
| $m\angle 2 = 100°$ | since angles 1 and 2 are vertical angles |
| $m\angle 4 = 100°$ | since angles 1 and 4 are alternate interior angles |
| $m\angle 3 = 180° - 100° = 80°$ | since angles 4 and 3 are supplementary angles |

A **polygon** is the union of three or more coplanar line segments that intersect each other only at each endpoint, with each endpoint shared by exactly two segments.

A **triangle** is a polygon with three sides. The sum of the measures of the three angles of a triangle is $180°$. In the following figure, $m\angle 1 + m\angle 2 + m\angle 3 = 180°$.

**Example 3**    Find the measure of the third angle of the triangle shown.

**Solution:**   The sum of the measures of the angles of a triangle is 180°. Since one angle measures 45° and the other angle measures 95°, the third angle measures $180° − 45° − 95° = 40°$.

Two triangles are **congruent** if they have the same size and the same shape. In congruent triangles, the measures of corresponding angles are equal and the lengths of corresponding sides are equal. The following triangles are congruent.

Corresponding angles are equal: $m\angle 1 = m\angle 4$, $m\angle 2 = m\angle 5$, and $m\angle 3 = m\angle 6$. Also, lengths of corresponding sides are equal: $a = x$, $b = y$, and $c = z$.

Any one of the following may be used to determine whether two triangles are congruent.

## Congruent Triangles

**1.** If the measures of two angles of a triangle equal the measures of two angles of another triangle and the lengths of the sides between each pair of angles are equal, the triangles are congruent.

$$m\angle 1 = m\angle 3$$
$$m\angle 2 = m\angle 4$$
$$\text{and}$$
$$a = x$$

**2.** If the lengths of the three sides of a triangle equal the lengths of corresponding sides of another triangle, the triangles are congruent.

$$a = x$$
$$b = y$$
$$\text{and}$$
$$c = z$$

**3.** If the lengths of two sides of a triangle equal the lengths of corresponding sides of another triangle, and the measures of the angles between each pair of sides are equal, the triangles are congruent.

$$a = x$$
$$b = y$$
$$\text{and}$$
$$m\angle 1 = m\angle 2$$

Two triangles are **similar** if they have the same shape but not necessarily the same size. In similar triangles, the measures of corresponding angles are equal and corresponding sides are in proportion. The following triangles are similar. (All similar triangles drawn in this appendix will be oriented the same.)

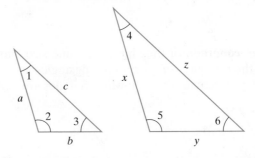

Corresponding angles are equal: $m\angle 1 = m\angle 4, m\angle 2 = m\angle 5,$ and $m\angle 3 = m\angle 6.$
Also, corresponding sides are proportional: $\dfrac{a}{x} = \dfrac{b}{y} = \dfrac{c}{z}.$

Any one of the following may be used to determine whether two triangles are similar.

### Similar Triangles

1. If the measures of two angles of a triangle equal the measures of two angles of another triangle, the triangles are similar.

$$m\angle 1 = m\angle 2$$
$$\text{and}$$
$$m\angle 3 = m\angle 4$$

2. If three sides of one triangle are proportional to three sides of another triangle, the triangles are similar.

$$\frac{a}{x} = \frac{b}{y} = \frac{c}{z}$$

3. If two sides of a triangle are proportional to two sides of another triangle and the measures of the included angles are equal, the triangles are similar.

$$m\angle 1 = m\angle 2$$
$$\text{and}$$
$$\frac{a}{x} = \frac{b}{y}$$

**Example 4**   Given that the following triangles are similar, find the missing length $x$.

**Solution:** Since the triangles are similar, corresponding sides are in proportion. Thus, $\dfrac{2}{3} = \dfrac{10}{x}$. To solve this equation for $x$, we cross multiply.

$$\dfrac{2}{3} = \dfrac{10}{x}$$

$$2x = 30$$

$$x = 15$$

The missing length is 15 units.

A **right triangle** contains a right angle. The side opposite the right angle is called the **hypotenuse,** and the other two sides are called the **legs.** The **Pythagorean theorem** gives a formula that relates the lengths of the three sides of a right triangle.

### The Pythagorean Theorem

If $a$ and $b$ are the lengths of the legs of a right triangle and $c$ is the length of the hypotenuse, then $a^2 + b^2 = c^2$.

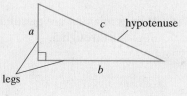

**Example 5** Find the length of the hypotenuse of a right triangle whose legs have lengths of 3 centimeters and 4 centimeters.

**Solution:** Because we have a right triangle, we use the Pythagorean theorem. The legs are 3 centimeters and 4 centimeters, so let $a = 3$ and $b = 4$ in the formula.

$$a^2 + b^2 = c^2$$

$$3^2 + 4^2 = c^2$$

$$9 + 16 = c^2$$

$$25 = c^2$$

Since $c$ represents a length, we assume that $c$ is positive. Thus, if $c^2$ is 25, $c$ must be 5. The hypotenuse has a length of 5 centimeters.

**E** Exercise Set MyMathLab® ▶

*Find the complement of each angle. See Example 1.*

**1.** 19°

**2.** 65°

**3.** 70.8°

**4.** $45\frac{2}{3}°$

**5.** $11\frac{1}{4}°$

**6.** 19.6°

*Find the supplement of each angle.*

**7.** 150°

**8.** 90°

**9.** 30.2°

**10.** 81.9°

**11.** $79\frac{1}{2}°$

**12.** $165\frac{8}{9}°$

*Solve. See Example 2.*

**13.** If lines *m* and *n* are parallel, find the measures of angles 1 through 7.

**14.** If lines *m* and *n* are parallel, find the measures of angles 1 through 5.

*In each of the following, the measures of two angles of a triangle are given. Find the measure of the third angle. See Example 3.*

**15.** 11°, 79°

**16.** 8°, 102°

**17.** 25°, 65°

**18.** 44°, 19°

**19.** 30°, 60°

**20.** 67°, 23°

*In each of the following, the measure of one angle of a right triangle is given. Find the measures of the other two angles.*

**21.** 45°

**22.** 60°

**23.** 17°

**24.** 30°

**25.** $39\frac{3}{4}°$

**26.** 72.6°

*Given that each of the following pairs of triangles is similar, find the missing length x. See Example 4.*

**27.**

**28.**

**29.**

**30.**

*Use the Pythagorean theorem to find the missing lengths in the right triangles. See Example 5.*

**31.**

**32.**

**33.**

**34.**

# Contents of Student Resources

Student Resources

# Student Resources

## Study Skills Builders

### Attitude and Study Tips

---

#### Study Skills Builder 1

Have You Decided to Complete This Course Successfully?

Ask yourself if one of your current goals is to complete this course successfully.

If it is not a goal of yours, ask yourself why. One common reason is fear of failure. Amazingly enough, fear of failure alone can be strong enough to keep many of us from doing our best in any endeavor.

Another common reason is that you simply haven't taken the time to think about or write down your goals for this course. To help accomplish this, answer the questions below.

**Exercises**

1. Write down your goal(s) for this course.

2. Now list steps you will take to make sure your goal(s) in Exercise 1 are accomplished.

3. Rate your commitment to this course with a number between 1 and 5. Use the diagram below to help.

| High Commitment | | Average Commitment | | Not Committed at All |
|---|---|---|---|---|
| 5 | 4 | 3 | 2 | 1 |

4. If you have rated your personal commitment level (from the exercise above) as a 1, 2, or 3, list the reasons why this is so. Then determine whether it is possible to increase your commitment level to a 4 or 5.

Good luck, and don't forget that a positive attitude will make a big difference.

---

## Study Skills Builder 2

### Tips for Studying for an Exam

To prepare for an exam, try the following study techniques:

- Start the study process days before your exam.
- Make sure that you are up to date on your assignments.
- If there is a topic that you are unsure of, use one of the many resources that are available to you. For example,

  See your instructor.
  View a lecture video on the topic.
  Visit a learning resource center on campus.
  Read the textbook material and examples on the topic.

- Reread your notes and carefully review the Chapter Highlights at the end of any chapter.
- Work the review exercises at the end of the chapter.
- Find a quiet place to take the Chapter Test found at the end of the chapter. Do not use any resources when taking this sample test. This way, you will have a clear indication of how prepared you are for your exam. Check your answers and use the Chapter Test Prep Videos to make sure that you correct any missed exercises.

Good luck, and keep a positive attitude.

### Exercises

*Let's see how you did on your last exam.*

1. How many days before your last exam did you start studying for that exam?

2. Were you up to date on your assignments at that time, or did you need to catch up on assignments?

3. List the most helpful text supplement (if you used one).

4. List the most helpful campus supplement (if you used one).

5. List your process for preparing for a mathematics test.

6. Was this process helpful? In other words, were you satisfied with your performance on your exam?

7. If not, what changes can you make in your process that will make it more helpful to you?

## Study Skills Builder 3

### What to Do the Day of an Exam

Your first exam may be soon. On the day of an exam, don't forget to try the following:

- Allow yourself plenty of time to arrive.
- Read the directions on the test carefully.
- Read each problem carefully as you take your test. Make sure that you answer the question asked.
- Watch your time and pace yourself so that you may attempt each problem on your test.
- Check your work and answers.
- ***Do not turn your test in early.*** If you have extra time, spend it double-checking your work.

Good luck!

### Exercises

*Answer the following questions based on your most recent mathematics exam, whenever that was.*

1. How soon before class did you arrive?

2. Did you read the directions on the test carefully?

3. Did you make sure you answered the question asked for each problem on the exam?

4. Were you able to attempt each problem on your exam?

5. If your answer to Exercise 4 is no, list reasons why.

6. Did you have extra time on your exam?

7. If your answer to Exercise 6 is yes, describe how you spent that extra time.

## Study Skills Builder 4

### Are You Satisfied with Your Performance on a Particular Quiz or Exam?

If not, don't forget to analyze your quiz or exam and look for common errors. Were most of your errors a result of:

- *Carelessness?* Did you turn in your quiz or exam before the allotted time expired? If so, resolve to use any extra time to check your work.

- *Running out of time?* Answer the questions you are sure of first. Then attempt the questions you are unsure of, and delay checking your work until all questions have been answered.

- *Not understanding a concept?* If so, review that concept and correct your work so that you make sure you understand it before the next quiz or the final exam.

- *Test conditions?* When studying for a quiz or exam, make sure you place yourself in conditions similar to test conditions. For example, before your next quiz or exam, take a sample test without the aid of your notes or text.

(For a sample test, see your instructor or use the Chapter Test at the end of each chapter.)

### Exercises

1. Have you corrected all your previous quizzes and exams?

2. List any errors you have found common to two or more of your graded papers.

3. Is one of your common errors not understanding a concept? If so, are you making sure you understand all the concepts for the next quiz or exam?

4. Is one of your common errors making careless mistakes? If so, are you now taking all the time allotted to check over your work so that you can minimize the number of careless mistakes?

5. Are you satisfied with your grades thus far on quizzes and tests?

6. If your answer to Exercise 5 is no, are there any more suggestions you can make to your instructor or yourself to help? If so, list them here and share these with your instructor.

## Study Skills Builder 5

### How Are You Doing?

If you haven't done so yet, take a few moments and think about how you are doing in this course. Are you working toward your goal of successfully completing this course? Is your performance on homework, quizzes, and tests satisfactory? If not, you might want to see your instructor to see if he/she has any suggestions on how you can improve your performance. Reread Section 1.1 for ideas on places to get help with your mathematics course.

### Exercises

*Answer the following.*

1. List any textbook supplements you are using to help you through this course.

2. List any campus resources you are using to help you through this course.

3. Write a short paragraph describing how you are doing in your mathematics course.

4. If improvement is needed, list ways that you can work toward improving your situation as described in Exercise 3.

## Study Skills Builder 6

### Are You Preparing for Your Final Exam?

To prepare for your final exam, try the following study techniques:

- Review the material that you will be responsible for on your exam. This includes material from your textbook, your notebook, and any handouts from your instructor.
- Review any formulas that you may need to memorize.
- Check to see if your instructor or mathematics department will be conducting a final exam review.
- Check with your instructor to see whether final exams from previous semesters/quarters are available to students for review.

- Use your previously taken exams as a practice final exam. To do so, rewrite the test questions in mixed order on blank sheets of paper. This will help you prepare for exam conditions.
- If you are unsure of a few concepts, see your instructor or visit a learning lab for assistance. Also, view the video segment of any troublesome sections.
- If you need further exercises to work, try the Cumulative Reviews at the end of the chapters.

Once again, good luck! I hope you are enjoying this textbook and your mathematics course.

# Organizing Your Work

## Study Skills Builder 7

### Learning New Terms

Many of the terms used in this text may be new to you. It will be helpful to make a list of new mathematical terms and symbols as you encounter them and to review them frequently. Placing these new terms (including page references) on $3 \times 5$ index cards might help you later when you're preparing for a quiz.

### Exercises

1. Name one way you might place a word and its definition on a $3 \times 5$ card.

2. How do new terms stand out in this text so that they can be found?

# Study Skills Builder 8

Are You Organized?

Have you ever had trouble finding a completed assignment? When it's time to study for a test, are your notes neat and organized? Have you ever had trouble reading your own mathematics handwriting? (Be honest—I have.)

When any of these things happens, it's time to get organized. Here are a few suggestions:

- Write your notes and complete your homework assignments in a notebook with pockets (spiral or ring binder).
- Take class notes in this notebook, and then follow the notes with your completed homework assignment.
- When you receive graded papers or handouts, place them in the notebook pocket so that you will not lose them.
- Mark (possibly with an exclamation point) any note(s) that seem extra important to you.
- Mark (possibly with a question mark) any notes or homework that you are having trouble with.
- See your instructor or a math tutor to help you with the concepts or exercises that you are having trouble understanding.
- If you are having trouble reading your own handwriting, *slow down* and write your mathematics work clearly!

## Exercises

1. Have you been completing your assignments on time?

2. Have you been correcting any exercises you may be having difficulty with?

3. If you are having trouble with a mathematical concept or correcting any homework exercises, have you visited your instructor, a tutor, or your campus math lab?

4. Are you taking lecture notes in your mathematics course? (By the way, these notes should include worked-out examples solved by your instructor.)

5. Is your mathematics course material (handouts, graded papers, lecture notes) organized?

6. If your answer to Exercise 5 is no, take a moment and review your course material. List at least two ways that you might better organize it.

## Study Skills Builder 9

Organizing a Notebook

It's never too late to get organized. If you need ideas about organizing a notebook for your mathematics course, try some of these:

- Use a spiral or ring binder notebook with pockets and use it for mathematics only.
- Start each page by writing the book's section number you are working on at the top.
- When your instructor is lecturing, take notes. *Always* include any examples your instructor works for you.
- Place your worked-out homework exercises in your notebook immediately after the lecture notes from that section. This way, a section's worth of material is together.
- Homework exercises: Attempt and check all assigned homework.
- Place graded quizzes in the pockets of your notebook or a special section of your binder.

### Exercises

*Check your notebook organization by answering the following questions.*

1. Do you have a spiral or ring binder notebook for your mathematics course only?

2. Have you ever had to flip through several sheets of notes and work in your mathematics notebook to determine what section's work you are in?

3. Are you now writing the textbook's section number at the top of each notebook page?

4. Have you ever lost or had trouble finding a graded quiz or test?

5. Are you now placing all your graded work in a dedicated place in your notebook?

6. Are you attempting all of your homework and placing all of your work in your notebook?

7. Are you checking and correcting your homework in your notebook? If not, why not?

8. Are you writing in your notebook the examples your instructor works for you in class?

## Study Skills Builder 10

How Are Your Homework Assignments Going?

It is very important in mathematics to keep up with homework. Why? Many concepts build on each other. Often your understanding of a day's concepts depends on an understanding of the previous day's material.

Remember that completing your homework assignment involves a lot more than attempting a few of the problems assigned.

To complete a homework assignment, remember these four things:

- Attempt all of it.
- Check it.
- Correct it.
- If needed, ask questions about it.

### Exercises

*Take a moment and review your completed homework assignments. Answer the questions below based on this review.*

1. Approximate the fraction of your homework you have attempted.

2. Approximate the fraction of your homework you have checked (if possible).

3. If you are able to check your homework, have you corrected it when errors have been found?

4. When working homework, if you do not understand a concept, what do you do?

# MyMathLab and MathXL

## Study Skills Builder 11

Tips for Turning In Your Homework on Time

It is very important to keep up with your mathematics homework assignments. Why? Many concepts in mathematics build upon each other.

Remember these 4 tips to help ensure your work is completed on time:

- Know the assignments and due dates set by your instructor.
- Do not wait until the last minute to submit your homework.
- Set a goal to submit your homework 6–8 hours before the scheduled due date in case you have unexpected technology trouble.
- Schedule enough time to complete each assignment.

Following the tips above will also help you avoid potentially losing points for late or missed assignments.

### Exercises

*Take a moment to consider your work on your homework assignments to date and answer the following questions:*

1. What percentage of your assignments have you turned in on time?

2. Why might it be a good idea to submit your homework 6–8 hours before the scheduled deadline?

3. If you have missed submitting any homework by the due date, list some of the reasons why this occurred.

4. What steps do you plan to take in the future to ensure your homework is submitted on time?

## Study Skills Builder 12

Tips for Doing Your Homework Online

Practice is one of the main keys to success in any mathematics course. Did you know that MyMathLab/MathXL provides you with **immediate feedback** for each exercise? If you are incorrect, you are given hints to work the exercise correctly. You have **unlimited practice opportunities** and can rework any exercises you have trouble with until you master them, and submit homework assignments unlimited times before the deadline.

Remember these success tips when doing your homework online:

- Attempt all assigned exercises.
- Write down (neatly) your step-by-step work for each exercise before entering your answer.
- Use the immediate feedback provided by the program to help you check and correct your work for each exercise.
- Rework any exercises you have trouble with until you master them.
- Work through your homework assignment as many times as necessary until you are satisfied.

### Exercises

*Take a moment to think about your homework assignments to date and answer the following questions:*

1. Have you attempted all assigned exercises?

2. Of the exercises attempted, have you also written out your work before entering your answer—so that you can check it?

3. Are you familiar with how to enter answers using the MathXL player so that you avoid answer entry type errors?

4. List some ways the immediate feedback and practice supports have helped you with your homework. If you have not used these supports, how do you plan to use them with the success tips above on your next assignment?

## Study Skills Builder 13

### Organizing Your Work

Have you ever used any readily available paper (such as the back of a brochure, another course assignment, post-its, etc.) to work out homework exercises before entering the answer in MathXL? To save time, have you ever entered answers directly into MathXL without working the exercises on paper? When it's time to study, have you ever been unable to find your completed work or read and follow your own mathematics handwriting?

When any of these things happen, it's time to get organized. Here are some suggestions:

- Write your step-by-step work for each homework exercise, (neatly) on lined, loose-leaf paper and keep this in a 3-ring binder.

- Refer to your step-by-step work when you receive feedback that your answer is incorrect in MathXL. Double-check against the steps and hints provided by the program and correct your work accordingly.

- Keep your written homework with your class notes for that section.

- Identify any exercises you are having trouble with and ask questions about them.

- Keep all graded quizzes and tests in this binder as well to study later.

If you follow the suggestions above, you and your instructor or tutor will be able to follow your steps and correct any mistakes. You will also have a written copy of your work to refer to later to ask questions and study for tests.

### Exercises

1. Why is it important that you write out your step-by-step work on homework exercises and keep a hard copy of all work submitted online?

2. If you have gotten an incorrect answer, are you able to follow your steps and find your error?

3. If you were asked today to review your previous homework assignments and first test, could you find them? If not, list some ways you might better organize your work.

## Study Skills Builder 14

### Getting Help with Your Homework Assignments

There are many helpful resources available to you through MathXL to help you work through any homework exercises you may have trouble with. It is important that you know what these resources are and know when and how to use them.

Let's review these features found in the homework exercises:

- **Help Me Solve This**—provides step-by-step help for the exercise you are working. You must work an additional exercise of the same type (without this help) before you can get credit for having worked it correctly.

- **View an Example**—allows you to view a correctly worked exercise similar to the one you are having trouble with. You can go back to your original exercise and work it on your own.

- **E-Book**—allows you to read examples from your text and find similar exercises.

- **Video**—your text author, Elayn Martin-Gay, works an exercise similar to the one you need help with. \*\*Not all exercises have an accompanying video clip.

- **Ask My Instructor**—allows you to e-mail your instructor for help with an exercise.

### Exercises

1. How does the "Help Me Solve This" feature work?

2. If the "View an Example" feature is used, is it necessary to work an additional problem before continuing the assignment?

3. When might be a good time to use the "Video" feature? Do all exercises have an accompanying video clip?

4. Which of the features above have you used? List those you found the most helpful to you.

5. If you haven't used the features discussed, list those you plan to try on your next homework assignment.

## Study Skills Builder 15

Tips for Preparing for an Exam

Did you know that you can rework your previous homework assignments in MyMathLab and MathXL? This is a great way to prepare for tests. To do this, open a previous homework assignment and click "similar exercise." This will generate new exercises similar to the homework you have submitted. You can then rework the exercises and assignments until you feel confident that you understand them.

To prepare for an exam, follow these tips:

- Review your written work for your previous homework assignments along with your class notes.

- Identify any exercises or topics that you have questions on or have difficulty understanding.

- Rework your previous assignments in MyMathLab and MathXL until you fully understand them and can do them without help.

- Get help for any topics you feel unsure of or for which you have questions.

### Exercises

1. Are your current homework assignments up to date and is your written work for them organized in a binder or notebook? If the answer is no, it's time to get organized. For tips on this, see Study Skills Builder 13—Organizing Your Work.

2. How many days in advance of an exam do you usually start studying?

3. List some ways you think that practicing previous homework assignments can help you prepare for your test.

4. List two or three resources you can use to get help for any topics you are unsure of or have questions on.

Good luck!

## Study Skills Builder 16

How Well Do You Know the Resources Available to You in MyMathLab?

There are many helpful resources available to you in MyMathLab. Let's take a moment to locate and explore a few of them now. Go into your MyMathLab course, and visit the Multimedia Library, Tools for Success, and E-book.

Let's see what you found.

### Exercises

1. List the resources available to you in the Multimedia Library.

2. List the resources available to you in the Tools for Success folder.

3. Where did you find the English/Spanish Audio Glossary?

4. Can you view videos from the E-Book?

5. Did you find any resources you did not know about? If so, which ones?

6. Which resources have you used most often or found most helpful?

# Additional Help Inside and Outside Your Textbook

## Study Skills Builder 17

How Well Do You Know Your Textbook?

The questions below will help determine whether you are familiar with your textbook. For additional information, see Section 1.1 in this text.

### Exercises

1. What does the ● icon mean?

2. What does the ＼ icon mean?

3. What does the △ icon mean?

4. Where can you find a review for each chapter? What answers to this review can be found in the back of your text?

5. Each chapter contains an overview of the chapter along with examples. What is this feature called?

6. Each chapter contains a review of vocabulary. What is this feature called?

7. There are practice exercises that are contained in this text. What are they and how can they be used?

8. This text contains a student section in the back titled Student Resources. List the contents of this section and how they might be helpful.

9. What exercise answers are available in this text? Where are they located?

## Study Skills Builder 18

Are You Familiar with Your Textbook Supplements?

Below is a review of some of the student supplements available for additional study. Check to see if you are using the ones most helpful to you.

- Chapter Test Prep Videos. These videos provide video clip solutions to the Chapter Test exercises in this text. You will find this extremely useful when studying for tests or exams.

- Interactive DVD Lecture Series. These are keyed to each section of the text. The material is presented by me, Elayn Martin-Gay, and I have placed a ● by the exercises in the text that I have worked on the video.

- The *Student Solutions Manual*. This contains worked-out solutions to odd-numbered exercises as well as every exercise in the Integrated Reviews, Chapter Reviews, Chapter Tests, and Cumulative Reviews.

- Pearson Tutor Center. Mathematics questions may be phoned, faxed, or e-mailed to this center.

- MyMathLab is a text-specific online course. MathXL is an online homework, tutorial, and assessment system.

Take a moment and determine whether these are available to you.

As usual, your instructor is your best source of information.

### Exercises

*Let's see how you are doing with textbook supplements.*

1. Name one way the Lecture Videos can be helpful to you.

2. Name one way the Chapter Test Prep Video can help you prepare for a chapter test.

3. List any textbook supplements that you have found useful.

4. Have you located and visited a learning resource lab located on your campus?

5. List the textbook supplements that are currently housed in your campus's learning resource lab.

## Study Skills Builder 19

Are You Getting All the Mathematics Help That You Need?

Remember that, in addition to your instructor, there are many places to get help with your mathematics course. For example:

- This text has an accompanying video lesson for every section and the CD in this text contains worked-out solutions to every Chapter Test exercise.

- The back of the book contains answers to odd-numbered exercises.

- A *Student Solutions Manual* is available that contains worked-out solutions to odd-numbered exercises as well as solutions to every exercise in the Integrated Reviews, Chapter Reviews, Chapter Tests, and Cumulative Reviews.

- Don't forget to check with your instructor for other local resources available to you, such as a tutoring center.

**Exercises**

1. List items you find helpful in the text and all student supplements to this text.

2. List all the campus help that is available to you for this course.

3. List any help (besides the textbook) from Exercises 1 and 2 above that you are using.

4. List any help (besides the textbook) that you feel you should try.

5. Write a goal for yourself that includes trying everything you listed in Exercise 4 during the next week.

# Bigger Picture— Study Guide Outline

## Solving Equations and Inequalities

### I. Equations

**A. Linear Equations** (Sec. 2.1)

$$5(x - 2) = \frac{4(2x + 1)}{3}$$

$$3 \cdot 5(x - 2) = \cancel{3} \cdot \frac{4(2x + 1)}{\cancel{3}}$$

$$15x - 30 = 8x + 4$$

$$7x = 34$$

$$x = \frac{34}{7}$$

**B. Absolute Value Equations** (Sec. 2.6)

$|3x - 1| = 8$

$3x - 1 = 8$  or  $3x - 1 = -8$

$3x = 9$  or  $3x = -7$

$x = 3$  or  $x = -\dfrac{7}{3}$

$|x - 5| = |x + 1|$

$x - 5 = x + 1$  or  $x - 5 = -(x + 1)$

$\underbrace{-5 = 1}_{\text{No solution}}$  or  $x - 5 = -x - 1$

  or  $2x = 4$

  $x = 2$

**C. Quadratic and Higher-Degree Equations** (Secs. 5.7, 8.1, 8.2, 8.3)

$2x^2 - 7x = 9$

$2x^2 - 7x - 9 = 0$

$(2x - 9)(x + 1) = 0$

$2x - 9 = 0$  or  $x + 1 = 0$

$x = \dfrac{9}{2}$  or  $x = -1$

$2x^2 + x - 2 = 0$

$a = 2, \ b = 1, \ c = -2$

$$x = \frac{-1 \pm \sqrt{1^2 - 4(2)(-2)}}{2 \cdot 2}$$

$$x = \frac{-1 \pm \sqrt{17}}{4}$$

**D. Equations with Rational Expressions** (Sec. 6.4)

$$\frac{7}{x - 1} + \frac{3}{x + 1} = \frac{x + 3}{x^2 - 1}$$

$$\cancel{(x - 1)}(x + 1) \cdot \frac{7}{\cancel{x - 1}} + (x - 1)\cancel{(x + 1)} \cdot \frac{3}{\cancel{x + 1}}$$

$$= \cancel{(x - 1)}\cancel{(x + 1)} \cdot \frac{x + 3}{\cancel{(x - 1)}\cancel{(x + 1)}}$$

$$7(x + 1) + 3(x - 1) = x + 3$$

$$7x + 7 + 3x - 3 = x + 3$$

$$9x = -1$$

$$x = -\frac{1}{9}$$

**E. Equations with Radicals** (Sec. 7.6)

$$\sqrt{5x + 10} - 2 = x$$
$$\sqrt{5x + 10} = x + 2$$
$$(\sqrt{5x + 10})^2 = (x + 2)^2$$
$$5x + 10 = x^2 + 4x + 4$$
$$0 = x^2 - x - 6$$
$$0 = (x - 3)(x + 2)$$
$$x - 3 = 0 \quad \text{or} \quad x + 2 = 0$$
$$x = 3 \quad \text{or} \qquad x = -2$$

Both solutions check.

**F. Exponential Equations** (Secs. 9.3, 9.8)

$$9^x = 27^{x+1}$$
$$(3^2)^x = (3^3)^{x+1}$$
$$3^{2x} = 3^{3x+3}$$
$$2x = 3x + 3$$
$$-3 = x$$

$$5^x = 7$$
$$\log 5^x = \log 7$$
$$x \log 5 = \log 7$$
$$x = \frac{\log 7}{\log 5}$$

**G. Logarithmic Equations** (Sec. 9.8)

$$\log 7 + \log(x + 3) = \log 5$$
$$\log 7(x + 3) = \log 5$$
$$7(x + 3) = 5$$
$$7x + 21 = 5$$
$$7x = -16$$
$$x = -\frac{16}{7}$$

## II. Inequalities

**A. Linear Inequalities** (Sec. 2.4)

$$-3(x + 2) \geq 6$$
$$-3x - 6 \geq 6$$
$$-3x \geq 12$$
$$\frac{-3x}{-3} \leq \frac{12}{-3}$$
$$x \leq -4 \quad \text{or} \quad (-\infty, -4]$$

**B. Compound Inequalities** (Sec. 2.5)

$x \leq 3 \quad \text{and} \quad x < -7$

$(-\infty, -7)$

$x \leq 3 \quad \text{or} \quad x < -7$

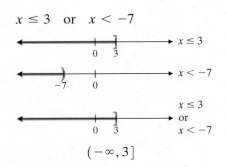

$(-\infty, 3]$

**Study Guide**

### C.  Absolute Value Inequalities (Sec. 2.7)

$$|x - 5| - 8 < -2$$
$$|x - 5| < 6$$
$$-6 < x - 5 < 6$$
$$-1 < x < 11$$
$$(-1, 11)$$

$$|2x + 1| \geq 17$$
$$2x + 1 \geq 17 \quad \text{or} \quad 2x + 1 \leq -17$$
$$2x \geq 16 \quad \text{or} \quad 2x \leq -18$$
$$x \geq 8 \quad \text{or} \quad x \leq -9$$
$$(-\infty, -9] \cup [8, \infty)$$

### D.  Nonlinear Inequalities (Sec. 8.4)

$$x^2 - x < 6$$
$$x^2 - x - 6 < 0$$
$$(x - 3)(x + 2) < 0$$

$$(-2, 3)$$

$$\frac{x - 5}{x + 1} \geq 0$$

$$(-\infty, -1) \cup [5, \infty)$$

# Practice Final Exam

Simplify. If needed, write answers with positive exponents only.

**1.** $\sqrt{216}$

**2.** $\dfrac{(4 - \sqrt{16}) - (-7 - 20)}{-2(1 - 4)^2}$

**3.** $\left(\dfrac{1}{125}\right)^{-1/3}$

**4.** $(-9x)^{-2}$

**5.** $\dfrac{\dfrac{5}{x} - \dfrac{7}{3x}}{\dfrac{9}{8x} - \dfrac{1}{x}}$

**6.** $\dfrac{6^{-1}a^2b^{-3}}{3^{-2}a^{-5}b^2}$

**7.** $\left(\dfrac{64c^{4/3}}{a^{-2/3}b^{5/6}}\right)^{1/2}$

Factor completely.

**8.** $3x^2y - 27y^3$

**9.** $16y^3 - 2$

**10.** $x^2y - 9y - 3x^2 + 27$

Perform the indicated operations and simplify if possible.

**11.** $(4x^3 - 3x - 4) - (9x^3 + 8x + 5)$

**12.** $(6m + n)^2$

**13.** $(2x - 1)(x^2 - 6x + 4)$

**14.** $\dfrac{3x^2 - 12}{x^2 + 2x - 8} \div \dfrac{6x + 18}{x + 4}$

**15.** $\dfrac{2x^2 + 7}{2x^4 - 18x^2} - \dfrac{6x + 7}{2x^4 - 18x^2}$

**16.** $\dfrac{3}{x^2 - x - 6} + \dfrac{2}{x^2 - 5x + 6}$

**17.** $\sqrt{125x^3} - 3\sqrt{20x^3}$

**18.** $(\sqrt{5} + 5)(\sqrt{5} - 5)$

**19.** $(4x^5 - 2x^4 + 4x^2 - 6x + 3) \div (2x - 1)$

Solve each equation or inequality. Write inequality solutions using interval notation.

**20.** $9(x + 2) = 5[11 - 2(2 - x) + 3]$

**21.** $|6x - 5| - 3 = -2$

**22.** $3n(7n - 20) = 96$

**23.** $-3 < 2(x - 3) \le 4$

1. _____

2. _____

3. _____

4. _____

5. _____

6. _____

7. _____

8. _____

9. _____

10. _____

11. _____

12. _____

13. _____

14. _____

15. _____

16. _____

17. _____

18. _____

19. _____

20. _____

21. _____

22. _____

23. _____

24. _____

25. _____

26. _____

27. _____

28. _____

29. _____

30. _____

31. _____

32. _____

33. _____

34. _____

35. _____

36. _____

37. _____

**24.** $|3x + 1| > 5$

**25.** $\dfrac{x^2 + 8}{x} - 1 = \dfrac{2(x + 4)}{x}$

**26.** $y^2 - 3y = 5$

**27.** $x = \sqrt{x - 2} + 2$

**28.** $2x^2 - 7x > 15$

**29.** Solve the system: $\begin{cases} \dfrac{x}{2} + \dfrac{y}{4} = -\dfrac{3}{4} \\ x + \dfrac{3}{4}y = -4 \end{cases}$

*Graph the following.*

**30.** $4x + 6y = 8$

**31.** $2x - y > 5$

**32.** $y = -3$

**33.** $g(x) = -|x + 2| - 1$. Also, find the domain and range of this function.

**34.** $h(x) = x^2 - 4x + 4$. Label the vertex and any intercepts.

**35.** $f(x) = \begin{cases} -\dfrac{1}{2}x \text{ if } x \le 0 \\ 2x - 3 \text{ if } x > 0 \end{cases}$. Also, find the domain and range of this function.

*Write equations of the following lines. Write each equation using function notation.*

**36.** through $(4, -2)$ and $(6, -3)$

**37.** through $(-1, 2)$ and perpendicular to $3x - y = 4$

*Find the distance or midpoint.*

**38.** Find the distance between the points $(-6, 3)$ and $(-8, -7)$.

**39.** Find the midpoint of the line segment whose endpoints are $(-2, -5)$ and $(-6, 12)$.

*Rationalize each denominator. Assume that variables represent positive numbers.*

**40.** $\sqrt{\dfrac{9}{y}}$

**41.** $\dfrac{4 - \sqrt{x}}{4 + 2\sqrt{x}}$

*Solve.*

**42.** The three states where international travelers spend the most money are Florida, California, and New York. International travelers spent $4 billion more money in California than New York, and in Florida they spent $1 billion less than twice the amount spent in New York. If total international spending in these three states is $39 billion, find the amount spent in each state. (*Source:* Travel Industry Association of America)

**43.** Write the area of the shaded region as a factored polynomial.

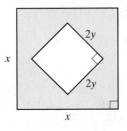

**44.** The product of one more than a number and twice the reciprocal of the number is $\dfrac{12}{5}$. Find the number.

**45.** Suppose that $W$ is inversely proportional to $V$. If $W = 20$ when $V = 12$, find $W$ when $V = 15$.

**46.** Given the diagram shown, approximate to the nearest foot how many feet of walking distance a person saves by cutting across the lawn instead of walking on the sidewalk.

20 ft

$x + 8$

$x$

**47.** A stone is thrown upward from a bridge. The stone's height, $s(t)$, in feet above the water $t$ seconds after the stone is thrown is a function given by the equation

$$s(t) = -16t^2 + 32t + 256$$

**a.** Find the maximum height of the stone.

**b.** Find the time it takes the stone to hit the water. Round the answer to two decimal places.

38. _____

39. _____

40. _____

41. _____

42. _____

43. _____

44. _____

45. _____

46. _____

47. a. _____

b. _____

**48.** _____

**49.** _____

**50.** _____

**51.** _____

**52.** _____

**53.** _____

**54.** _____

**55.** _____

**56.** _____

**57.** _____

**58.** _____

**59.** _____

**60.** _____

**61.** _____

**62.** _____

**48.** The research department of a company that manufactures children's fruit drinks is experimenting with a new flavor. A 17.5% fructose solution is needed, but only 10% and 20% solutions are available. How many gallons of a 10% fructose solution should be mixed with a 20% fructose solution to obtain 20 gallons of a 17.5% fructose solution?

_Perform the indicated operation and simplify. Write the result in the form a + bi._

**49.** $-\sqrt{-8}$

**50.** $(12 - 6i) - (12 - 3i)$

**51.** $(4 + 3i)^2$

**52.** $\dfrac{1 + 4i}{1 - i}$

**53.** If $g(x) = x - 7$ and $h(x) = x^2 - 6x + 5$, find $(g \circ h)(x)$.

**54.** Decide whether $f(x) = 6 - 2x$ is a one-to-one function. If it is, find its inverse.

**55.** Use properties of logarithms to write the expression as a single logarithm.

$$\log_5 x + 3 \log_5 x - \log_5(x + 1)$$

_Solve. Give exact solutions._

**56.** $8^{x-1} = \dfrac{1}{64}$

**57.** $3^{2x+5} = 4$ Give the exact solution and a 4-decimal-place approximation.

**58.** $\log_8(3x - 2) = 2$

**59.** $\log_4(x + 1) - \log_4(x - 2) = 3$

**60.** $\ln \sqrt{e} = x$

**61.** Graph $f(x) = \left(\dfrac{1}{2}\right)^x + 1$

**62.** The prairie dog population of the Grand Forks area now stands at 57,000 animals. If the population is growing at a rate of 2.6% annually, how many prairie dogs will there be in that area 5 years from now?

*Sketch the graph of each equation.*

**63.** $x^2 - y^2 = 36$

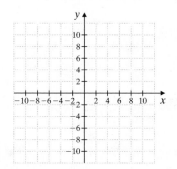

**64.** $16x^2 + 9y^2 = 144$

**65.** $x^2 + y^2 + 6x = 16$

**66.** Solve the system:

$$\begin{cases} x^2 + y^2 = 26 \\ x^2 - 2y^2 = 23 \end{cases}$$

**63.** _____

**64.** _____

**65.** _____

**66.** _____

# Answers to Selected Exercises

## Chapter 1 Real Numbers and Algebraic Expressions
### Section 1.2

**Vocabulary, Readiness & Video Check** **1.** variables **3.** empty **5.** natural numbers **7.** integers **9.** rational number **11.** roster form **13.** Every real number is either a rational number or an irrational number. There are no numbers that are both rational and irrational.

**Exercise Set 1.2** **1.** $\{1, 2, 3, 4, 5\}$ **3.** $\{11, 12, 13, 14, 15, 16\}$ **5.** $\{0\}$ **7.** $\{0, 2, 4, 6, 8\}$ **9.** $3, 0, \sqrt{36}$ **11.** $3, \sqrt{36}$ **13.** $\sqrt{7}$
**15.** $\in$ **17.** $\notin$ **19.** $\notin$ **21.** true **23.** true **25.** false **27.** true **29.** true **31.** false **33.** true **35.** true **37.** false
**39.** $2 \cdot x$ or $2x$ **41.** $x - 10$ **43.** $x + 2$ **45.** $x \div 11$ or $\dfrac{x}{11}$ **47.** $x - 4$ **49.** $x + 2.3$ or $x + 2\dfrac{3}{10}$ **51.** $1\dfrac{1}{3} - x$ **53.** $9 \cdot x$ or $9x$
**55.** $x + 9$ **57.** $2x + 5$ **59.** $12 - 3x$ **61.** $1 + 2x$ **63.** $5x - 10$ **65.** $\dfrac{5}{4 - x}$ **67.** $2(x + 3)$ **69.** answers may vary
**71.** answers may vary **73.** 0 **75.** 1.5 **77.** 2.9 **79.** 1.7 **81.** answers may vary

### Section 1.3

**Vocabulary, Readiness & Video Check** **1.** $<$ **3.** $\neq$ **5.** $\geq$ **7.** $-a$ **9.** commutative **11.** distributive **13.** $=, \neq, <, \leq, >, \geq$
**15.** multiplicative; additive

**Exercise Set 1.3** **1.** $10 + x = -12$ **3.** $2x + 5 = -14$ **5.** $\dfrac{n}{5} = 4n$ **7.** $z - \dfrac{1}{2} = \dfrac{1}{2}z$ **9.** $>$ **11.** $>$ **13.** $<$ **15.** $=$ **17.** $>$
**19.** $<$ **21.** $<$ **23.** $<$ **25.** true **27.** true **29.** false **31.** true **33.** $-9$ **35.** 6.2 **37.** $-\dfrac{4}{7}$ **39.** $\dfrac{5}{11}$ **41.** 0 **43.** $\dfrac{1}{5}$ **45.** $-\dfrac{1}{10}$
**47.** $-4$ **49.** undefined **51.** $\dfrac{16}{19}$ **53.** $-25; \dfrac{1}{25}$ **55.** $-8; -\dfrac{1}{8}$ **57.** $\dfrac{1}{7}; -7$ **59.** 0; undefined **61.** $\dfrac{7}{8}; -\dfrac{7}{8}$ **63.** $y + 7x$ **65.** $w \cdot z$
**67.** $\dfrac{x}{5} \cdot \dfrac{1}{3}$ **69.** $(5 \cdot 7)x$ or $35x$ **71.** $x + (1.2 + y)$ **73.** $14(z \cdot y)$ or $14zy$ **75.** $3x + 15$ **77.** $4z - 24$ **79.** $16a + 8b$ **81.** $6xy - 24x$
**83.** $0.8x + 2y$ **85.** $2x - \dfrac{9}{2}y$ **87.** $12x + 10y + 4z$ **89.** $6 + 3x$ **91.** 0 **93.** 7 **95.** $(10 \cdot 2)y$ or $20y$ **97.** $3x + 12$ **99.** $(5 \cdot 7)y$ or $35y$
**101.** zero; answers may vary **103.** no; answers may vary **105.** answers may vary **107.** $a(b + c) = ab + ac$

### Section 1.4

**Vocabulary, Readiness & Video Check** **1.** b, c **3.** b, d **5.** b **7.** 0 **9.** reciprocal **11.** exponent **13.** square root
**15.** distance; 0; positive; negative **17.** The signs of the numbers determine the sign of the product or quotient. Same signs mean a positive product or quotient, and different signs mean a negative product or quotient **19.** the positive or principal square root

**Exercise Set 1.4** **1.** 2 **3.** 4 **5.** 0 **7.** $-3$ **9.** $-\dfrac{2}{9}$ **11.** 5 **13.** $-24$ **15.** $-11$ **17.** $-3$ **19.** $\dfrac{4}{3}$ **21.** $-2$ **23.** $-21$ **25.** $-\dfrac{1}{2}$
**27.** $-6$ **29.** $-60$ **31.** 80 **33.** 0 **35.** 0 **37.** $-3$ **39.** 3 **41.** $-\dfrac{1}{6}$ **43.** 0.56 **45.** $-7$ **47.** $-8$ **49.** $-49$ **51.** 36 **53.** $-8$
**55.** $-\dfrac{1}{27}$ **57.** 7 **59.** 8 **61.** $\dfrac{1}{3}$ **63.** 4 **65.** $\dfrac{2}{5}$ **67.** 3 **69.** $-12$ **71.** $-18$ **73.** $-72$ **75.** 121 **77.** $-8$ **79.** $-3$ **81.** $\dfrac{1}{21}$ **83.** 7.2
**85.** 15 **87.** $-12.2$ **89.** 0 **91.** undefined **93.** $\dfrac{1}{4}$ **95.** $-\dfrac{3}{7}$ **97.** 4 **99.** 5 **101.** 2 **103.** 2 **105.** $-\dfrac{5}{27}$ **107.** $-1$ **109.** $-\dfrac{3}{2}$
**111.** $-1$ **113.** $-22$ **115.** 35 **117.** 0 **119.** $-256$ **121.** answers may vary **123.** true **125.** $\dfrac{1}{3}$ **127.** 4205 m **129.** b **131.** a
**133.** yes; each player has 4 points **135.** 16.5227 **137.** 4.4272 **139.** 17% **141.** 15% **143.** long term; short term is very volatile

**Integrated Review** **1.** $\{1, 2, 3\}$ **2.** $\{1, 3, 5\}$ **3.** $\{8, 10, 12, \dots\}$ **4.** $\{11, 12, 13, 14\}$ **5.** $2(x - 3)$ **6.** $\dfrac{6}{x + 10}$ **7.** $>$
**8.** $=$ **9.** $<$ **10.** $<$ **11.** $5x = 20$ **12.** $a + 12 = 14$ **13.** $\dfrac{y}{10} = y \cdot 10$ **14.** $x + (-1) = x - 1$ **15.** 3 **16.** 9 **17.** $-28$
**18.** $-220$ **19.** 5 **20.** $-28$ **21.** 5 **22.** $-3$ **23.** $-25$ **24.** 25 **25.** 13 **26.** $-5$ **27.** 0 **28.** undefined **29.** $-24$ **30.** 30 **31.** $-\dfrac{3}{7}$
**32.** $-\dfrac{1}{10}$ **33.** $-\dfrac{1}{2}$ **34.** $-\dfrac{11}{12}$ **35.** $-8$ **36.** $-0.3$ **37.** 8 **38.** 4.4 **39.** $-\dfrac{3}{8}$ **40.** $-\dfrac{4}{7}$ **41.** 7 **42.** 3 **43.** $-4$ **44.** $-8$ **45.** $\dfrac{1}{9}$
**46.** $\dfrac{1}{10}$ **47.** $6; -\dfrac{1}{6}$ **48.** $-4; \dfrac{1}{4}$ **49.** $\dfrac{5}{7}; -\dfrac{5}{7}$ **50.** $-\dfrac{7}{30}; -\dfrac{30}{7}$ **51.** $9m + 45$ **52.** $77 + 11r$ **53.** $6y - 9x$ **54.** $32m - 56n$
**55.** $0.6a + 1.4$ **56.** $1.2n + 3$ **57.** $2x - \dfrac{19}{5}y + 4$ **58.** $5x - \dfrac{19}{2}y + 10$

## Section 1.5

**Vocabulary, Readiness & Video Check**    **1.** like    **3.** combining    **5.** It allows each expression to evaluate to a single number.
**7.** by combining like terms; distributive property

**Exercise Set 1.5**    **1.** 48    **3.** $-1$    **5.** $-9$    **7.** 14.4    **9.** 17    **11.** $-4$    **13.** 40    **15.** $-2$    **17.** 11    **19.** $-\dfrac{3}{4}$    **21.** 7    **23.** 3    **25.** $-11$
**27.** $-2.1$    **29.** $-\dfrac{1}{3}$    **31.** $-\dfrac{79}{15}$    **33.** $-\dfrac{4}{5}$    **35.** $-81$    **37.** $-\dfrac{20}{33}$    **39.** $-235.5$    **41.** 12.25    **43.** 93    **45.** $-12$    **47.** $-\dfrac{23}{18}$    **49.** 5    **51.** $-\dfrac{3}{19}$
**53. a.** 18; 22; 28; 208    **b.** increase; answers may vary    **55. a.** 600; 150; 105    **b.** decrease; answers may vary    **57.** $8293.09    **59.** $8x$
**61.** $18y$    **63.** $-x - 8$    **65.** $-6x + 9$    **67.** $4a - 13b$    **69.** $2x - 2y$    **71.** $0.8x - 3.6$    **73.** $\dfrac{11}{12}b - \dfrac{7}{6}$    **75.** $-3ab - 12.8$    **77.** $6x + 14$
**79.** $2k + 10$    **81.** $-3x + 5$    **83.** $-x^2 + 4xy + 9$    **85.** $4n - 8$    **87.** $-24$    **89.** $-5x + \dfrac{5}{6}y - 1$    **91.** $2x + 10$    **93.** $3a + \dfrac{3}{35}$
**95.** $-0.6a + 3.5$    **97.** $-8t + 12$    **99.** $1.91x + 4.32$    **101.** $15.4z + 31.11$    **103.** $(2 + 7) \cdot (1 + 3)$    **105.** answers may vary
**107.** $(3x + 2) - (5x + 7) = 3x + 2 - 5x - 7 = -2x - 5$    **109.** 20 million    **111.** 70 million    **113.** increasing; answers may vary
**115.** $\dfrac{13}{35}$    **117.** $\dfrac{5}{2}$    **119.** $-0.5876$

## Section 1.6

**Calculator Explorations**    **1.** $6 \times 10^{43}$    **3.** $3.796 \times 10^{28}$

**Vocabulary, Readiness & Video Check**    **1.** $x$    **3.** 3    **5.** $y^7$    **7.** $\dfrac{5}{xy^2}$    **9.** $\dfrac{a^2}{bc^5}$    **11.** $\dfrac{x^4}{y^2}$    **13.** These properties allow us to reorder and
regroup factors to put those with the same bases together so that we may apply the product rule.    **15.** Subtract the exponents on like
bases when applying the quotient rule.    **17.** quotient rule for exponents

**Exercise Set 1.6**    **1.** $4^5$    **3.** $x^8$    **5.** $m^{14}$    **7.** $-20x^2y$    **9.** $-16x^6y^3p^2$    **11.** $-1$    **13.** 1    **15.** $-1$    **17.** 9    **19.** $a^3$    **21.** $-13z^4$    **23.** $x$
**25.** $\dfrac{4}{3}x^3y^2$    **27.** $-6a^4b^4c^6$    **29.** $\dfrac{1}{16}$    **31.** $-\dfrac{1}{27}$    **33.** $\dfrac{1}{x^8}$    **35.** $\dfrac{5}{a^4}$    **37.** $\dfrac{y^2}{x^7}$    **39.** $\dfrac{1}{x^7}$    **41.** $4r^8$    **43.** 1    **45.** $\dfrac{b^7}{9a^7}$    **47.** $\dfrac{6x^{16}}{5}$    **49.** $-140x^{12}$
**51.** $x^{16}$    **53.** $10x^{10}$    **55.** 6    **57.** $\dfrac{1}{z^3}$    **59.** $-2$    **61.** $y^4$    **63.** $\dfrac{13}{36}$    **65.** $\dfrac{3}{x}$    **67.** $r^8$    **69.** $\dfrac{1}{x^9y^4}$    **71.** $24x^7y^6$    **73.** $\dfrac{x}{16}$    **75.** 625    **77.** $\dfrac{1}{8}$
**79.** $\dfrac{a^5}{81}$    **81.** $\dfrac{7}{x^3z^5}$    **83.** $x^{7a+5}$    **85.** $x^{2t-1}$    **87.** $x^{4a+7}$    **89.** $z^{6x-7}$    **91.** $x^{6t-1}$    **93.** $3.125 \times 10^7$    **95.** $1.6 \times 10^{-2}$    **97.** $6.7413 \times 10^4$
**99.** $1.25 \times 10^{-2}$    **101.** $5.3 \times 10^{-5}$    **103.** $7.783 \times 10^8$    **105.** $6 \times 10^{15}$    **107.** $1.24 \times 10^{11}$    **109.** $1.0 \times 10^{-3}$    **111.** 0.0000000036
**113.** 93,000,000    **115.** 1,278,000    **117.** 7,350,000,000,000    **119.** 0.000000403    **121.** 300,000,000    **123.** 0.0095    **125.** 155,000
**127.** answers may vary    **129.** answers may vary

## Section 1.7

**Vocabulary, Readiness & Video Check**    **1.** $x^{20}$    **3.** $x^9$    **5.** $y^{42}$    **7.** $z^{36}$    **9.** $z^{18}$    **11.** The power rule involves a power of a base
raised to a power and exponents are multiplied; the product rule involves a product of like bases and exponents are added.
**13.** multiplied

**Exercise Set 1.7**    **1.** $\dfrac{1}{9}$    **3.** $\dfrac{1}{x^{36}}$    **5.** $9x^4y^6$    **7.** $16x^{20}y^{12}$    **9.** $\dfrac{c^{18}}{64a^{12}b^6}$    **11.** $\dfrac{y^{15}}{x^{35}z^{20}}$    **13.** $-64y^3$    **15.** $\dfrac{1}{a^2}$    **17.** $\dfrac{36}{p^{12}}$    **19.** $-\dfrac{a^6}{512x^3y^9}$    **21.** $\dfrac{64}{27}$
**23.** $4a^8b^4$    **25.** $\dfrac{x^{14}y^{14}}{a^{21}}$    **27.** $\dfrac{1}{y^{10}}$    **29.** $\dfrac{1}{125}$    **31.** $\dfrac{1}{x^{63}}$    **33.** $\dfrac{1}{x^{20}y^{15}}$    **35.** $16x^4$    **37.** $48x^2y^6$    **39.** $\dfrac{x^9}{8y^3}$    **41.** $\dfrac{x^4}{4z^2}$    **43.** $\dfrac{x^4}{16}$    **45.** $\dfrac{1}{y^{15}}$
**47.** $\dfrac{16a^2b^9}{9}$    **49.** $\dfrac{3}{8x^8y^7}$    **51.** $\dfrac{1}{x^{30}b^6c^6}$    **53.** $\dfrac{25}{8x^5y^4}$    **55.** $\dfrac{2}{x^4y^{10}}$    **57.** $x^{9a+18}$    **59.** $x^{16a-1}$    **61.** $b^{10x-4}$    **63.** $y^{15a+3}$    **65.** $16x^{4t+4}$
**67.** $5x^{a+2b}y^{a-2b}$    **69.** $1.45 \times 10^9$    **71.** $8 \times 10^{15}$    **73.** $4 \times 10^{-7}$    **75.** $3 \times 10^{-1}$    **77.** $2 \times 10^1$    **79.** $1 \times 10^1$    **81.** $8 \times 10^{-5}$
**83.** $1.1 \times 10^7$    **85.** $3.5 \times 10^{22}$    **87.** $2.5808 \times 10^{-5}$ sq m    **89.** $\dfrac{8}{x^6y^3}$ cu m    **91.** $2 \times 10^3$ pounds per cubic foot    **93.** yes; $a = \pm 1$
**95.** no; answers may vary    **97.** 90 people per sq mi    **99.** $748 per person

## Chapter 1 Vocabulary Check

**1.** algebraic expression    **2.** opposite    **3.** distributive    **4.** absolute value    **5.** exponent    **6.** variable
**7.** inequality    **8.** reciprocals    **9.** commutative    **10.** associative    **11.** whole    **12.** real

## Chapter 1 Review

**1.** $\dfrac{x}{7}$    **2.** $7x$    **3.** $4(x + 10)$    **4.** $3x - 9$    **5.** $\{-1, 1, 3\}$    **6.** $\{-2, 0, 2, 4, 6\}$    **7.** $\{\ \}$ or $\varnothing$    **8.** $\{\ \}$ or $\varnothing$
**9.** $\{6, 7, 8, \dots\}$    **10.** $\{\dots, -1, 0, 1, 2\}$    **11.** true    **12.** false    **13.** true    **14.** true    **15.** false    **16.** true    **17.** false    **18.** true    **19.** $5, \dfrac{8}{2}, \sqrt{9}$

**20.** $5, \dfrac{8}{2}, \sqrt{9}$  **21.** $5, -\dfrac{2}{3}, \dfrac{8}{2}, \sqrt{9}, 0.3, 1\dfrac{5}{8}, -1$  **22.** $\sqrt{7}, \pi$  **23.** $5, -\dfrac{2}{3}, \dfrac{8}{2}, \sqrt{9}, 0.3, \sqrt{7}, 1\dfrac{5}{8}, -1, \pi$  **24.** $5, \dfrac{8}{2}, \sqrt{9}, -1$  **25.** $12 = -4x$

**26.** $n + 2n = -15$  **27.** $4(y + 3) = -1$  **28.** $6(t - 5) = 4$  **29.** $z - 7 = 6$  **30.** $9x - 10 = 5$  **31.** $x - 5 = 12$  **32.** $-4 = 7y$

**33.** $\dfrac{2}{3} = 2\left(n + \dfrac{1}{4}\right)$  **34.** $t + 6 = -12$  **35.** $\dfrac{3}{4}$  **36.** $-0.6$  **37.** $0$  **38.** $1$  **39.** $-\dfrac{4}{3}$  **40.** $5$  **41.** undefined  **42.** $-1$  **43.** associative property of addition  **44.** distributive property  **45.** additive inverse property  **46.** commutative property of addition  **47.** associative property of multiplication  **48.** multiplicative inverse property  **49.** multiplicative identity property  **50.** commutative property of addition  **51.** additive identity property  **52.** multiplicative identity property  **53.** $5x - 15z$  **54.** $(3 + x) + (7 + y)$, for example

**55.** $2 + (-2)$, for example  **56.** $2 \cdot \dfrac{1}{2}$, for example  **57.** $(3.4)[(0.7)5]$  **58.** $7 + 0$  **59.** $>$  **60.** $>$  **61.** $<$  **62.** $=$  **63.** $<$  **64.** $>$

**65.** $-6x + 21y$  **66.** $-90a - 36b$  **67.** $9m - 4n + \dfrac{1}{2}$  **68.** $4x - 11y + \dfrac{2}{3}$  **69.** $-4$  **70.** $-35$  **71.** $-2$  **72.** $0.31$  **73.** $8$  **74.** $13.3$

**75.** $-4$  **76.** $-22$  **77.** undefined  **78.** $0$  **79.** $4$  **80.** $-5$  **81.** $-\dfrac{2}{15}$  **82.** $2$  **83.** $\dfrac{5}{12}$  **84.** 29,852 ft below sea level  **85.** $9$  **86.** $13$

**87.** $3$  **88.** $54$  **89.** $-\dfrac{32}{135}$  **90.** $-\dfrac{15}{56}$  **91.** $-\dfrac{5}{4}$  **92.** $-\dfrac{5}{2}$  **93.** $\dfrac{5}{8}$  **94.** $-\dfrac{13}{2}$  **95.** $-1$  **96.** $24$  **97.** $1$  **98.** $18$  **99.** $-4$  **100.** $\dfrac{7}{3}$

**101.** $\dfrac{5}{7}$  **102.** $-\dfrac{8}{25}$  **103.** $\dfrac{1}{5}$  **104.** $1$  **105.** $6.28; 62.8; 628$  **106.** increase  **107.** $3x - 13$  **108.** $80y - 1$  **109.** $a + 7y - 6$

**110.** $-23b + 13x - 10$  **111.** $n + 5$  **112.** $y - 4$  **113.** $4$  **114.** $81$  **115.** $-4$  **116.** $-81$  **117.** $1$  **118.** $-1$  **119.** $-\dfrac{1}{16}$  **120.** $\dfrac{1}{16}$

**121.** $-x^2 y^7 z$  **122.** $12x^2 y^3 b$  **123.** $\dfrac{1}{a^9}$  **124.** $\dfrac{1}{a}$  **125.** $\dfrac{1}{x^{11}}$  **126.** $\dfrac{1}{2a^{17}}$  **127.** $\dfrac{1}{y^5}$  **128.** $2y^{x-7}$  **129.** $3.689 \times 10^7$  **130.** $3.62 \times 10^{-4}$

**131.** $0.000001678$  **132.** $410,000$  **133.** $8^{15}$  **134.** $\dfrac{a^2}{16}$  **135.** $27x^3$  **136.** $\dfrac{1}{16x^2}$  **137.** $\dfrac{36x^2}{25}$  **138.** $\dfrac{1}{8^{18}}$  **139.** $\dfrac{9}{16}$  **140.** $-\dfrac{1}{8x^9}$

**141.** $\dfrac{1}{4p^4}$  **142.** $-\dfrac{27y^6}{x^6}$  **143.** $x^{25} y^{15} z^{15}$  **144.** $\dfrac{xz}{4}$  **145.** $\dfrac{x^2}{625y^4 z^4}$  **146.** $\dfrac{2}{27z^3}$  **147.** $27x^{19a}$  **148.** $9x^{4a+2b} y^{-6b}$ or $\dfrac{9x^{4a+2b}}{y^{6b}}$  **149.** $\dfrac{3}{4}; -\dfrac{4}{3}$

**150.** $5; \dfrac{1}{5}$  **151.** $-10x + 6.1$  **152.** $9$  **153.** $-\dfrac{6}{11}$  **154.** $15$  **155.** $-\dfrac{4}{15}$  **156.** $\dfrac{1}{11}$  **157.** $-x + 3y + 1$  **158.** $-\dfrac{1}{25}$  **159.** $\dfrac{8}{x^{30}}$  **160.** $\dfrac{9b^2 c^4}{25a^2}$

**Chapter 1 Test**  **1.** true  **2.** false  **3.** false  **4.** false  **5.** true  **6.** $\{0, 1\}$  **7.** $-3$  **8.** $-225$  **9.** $-2$  **10.** $1$  **11.** $-\dfrac{3}{2}$  **12.** $12$

**13.** $1$  **14. a.** $5.75; 17.25; 57.50; 115.00$  **b.** increase  **15.** $3\left(\dfrac{n}{5}\right) = -n$  **16.** $20 = 2x - 6$  **17.** $-2 = \dfrac{x}{x + 5}$  **18.** distributive property

**19.** associative property of addition  **20.** additive inverse property  **21.** reciprocal: $-\dfrac{11}{7}$; opposite: $\dfrac{7}{11}$  **22.** $6x + 12y + 13.8$

**23.** $-x - 11y$  **24.** $\dfrac{1}{81x^2}$  **25.** $\dfrac{3a^7}{2b^5}$  **26.** $-\dfrac{y^{40}}{z^5}$  **27.** $\dfrac{a^{12}}{4b^{14}}$  **28.** $x^{10w}$  **29.** $6.3 \times 10^8$  **30.** $1.2 \times 10^{-2}$  **31.** $0.000005$  **32.** $5.76 \times 10^4$

**33.** $9 \times 10^{-4}$  **34.** $2.26 \times 10^{10}$

# Chapter 2 Equations, Inequalities, and Problem Solving
## Section 2.1

**Vocabulary, Readiness & Video Check**  **1.** equivalent  **3.** addition  **5.** expression  **7.** equation  **9.** all real numbers  **11.** no solution  **13.** When $x$ is replaced with $-2$ in the given equation, a true statement results.  **15.** distributive property  **17. a.** true  **b.** false

**Exercise Set 2.1**  **1.** yes  **3.** no  **5.** yes  **7.** no  **9.** no  **11.** yes  **13.** $\{6\}$  **15.** $\{-22\}$  **17.** $\{4.7\}$  **19.** $\{10\}$  **21.** $\{-1.1\}$  **23.** $\{-5\}$  **25.** $\{-2\}$  **27.** $\{0\}$  **29.** $\{2\}$  **31.** $\{-9\}$  **33.** $\left\{-\dfrac{10}{7}\right\}$  **35.** $\left\{\dfrac{9}{10}\right\}$  **37.** $\{4\}$  **39.** $\{1\}$  **41.** $\{5\}$  **43.** $\left\{\dfrac{40}{3}\right\}$

**45.** $\{17\}$  **47.** $\{n \mid n$ is a real number$\}$  **49.** $\varnothing$  **51.** $\{x \mid x$ is a real number$\}$  **53.** $\varnothing$  **55.** $\left\{\dfrac{1}{8}\right\}$  **57.** $\{0\}$  **59.** $\{x \mid x$ is a real number$\}$

**61.** $\{4\}$  **63.** $\left\{\dfrac{4}{5}\right\}$  **65.** $\{8\}$  **67.** $\varnothing$  **69.** $\{-8\}$  **71.** $\left\{-\dfrac{5}{4}\right\}$  **73.** $\{-2\}$  **75.** $\{23\}$  **77.** $\left\{-\dfrac{2}{9}\right\}$  **79.** $\dfrac{8}{x}$  **81.** $8x$  **83.** $2x - 5$

**85.** added 19 instead of subtracting; $\{-3\}$  **87.** $0.4 - 1.6 = -1.2$, not 1.2; $\{-0.24\}$  **89. a.** $4x + 5$  **b.** $\{-3\}$  **c.** answers may vary  **91.** answers may vary  **93.** $K = -11$  **95.** $K = -23$  **97.** answers may vary  **99.** $\{5.217\}$  **101.** $\{1\}$

## Section 2.2

**Vocabulary, Readiness & Video Check**  **1.** $>$  **3.** $=$  **5.** $31, 32, 33, 34$  **7.** $18, 20, 22$  **9.** $y, y + 1, y + 2$  **11.** $p, p + 1, p + 2, p + 3$  **13.** distributive property

**Exercise Set 2.2**  **1.** $4y$  **3.** $3z + 3$  **5.** $(65x + 30)$ cents  **7.** $10x + 3$  **9.** $2x + 14$  **11.** $-5$  **13.** $45, 225, 145$  **15.** approximately 1612.41 million acres  **17.** 3533 earthquakes  **19.** 1275 shoppers  **21.** 23%  **23.** 417 employees  **25.** Darlington: 61,000;

Daytona: 159,000   **27.** 40.5 ft; 202.5 ft; 240 ft   **29.** New York: 21.6 million; Los Angeles: 17.2 million; Chicago: 9.8 million
**31.** 29°, 35°, 116°   **33.** 28 m, 36 m, 38 m   **35.** 18 in., 18 in., 27 in., 36 in.   **37.** 494 million; 654 million; 796 million   **39.** biomedical engineer: 8 thousand; event planner: 32 thousand; interpreter: 23 thousand   **41.** AT&T Stadium: 80,000; Levi's Stadium: 68,500; Soldier Field: 61,500
**43.** $430.00   **45.** $461,263   **47.** 75, 76, 77   **49.** 64°, 32°, 84°   **51.** square: 18 cm; triangle: 24 cm   **53.** 72,222 people   **55.** Fallon's ZIP code is 89406; Fernley's ZIP code is 89408; Gardnerville Ranchos' ZIP code is 89410   **57.** 40°, 140°   **59.** incandescent: 1500 bulb hours; fluorescent: 100,000 bulb hours; halogen: 4000 bulb hours   **61.** Encarnacion: 36; Soriano: 34; Longoria: 32   **63.** 208   **65.** $-55$
**67.** 3195   **69.** yes; answers may vary   **71.** 80°   **73.** 2023   **75.** 345   **77.** any three consecutive integers   **79.** 25 skateboards
**81.** company loses money

## Section 2.3

**Vocabulary, Readiness & Video Check**   **1.** $y = 5 - 2x$   **3.** $a = 5b + 8$   **5.** $k = h - 5j + 6$   **7.** That the specified variable will equal some expression and that this expression should not contain the specified variable.

**Exercise Set 2.3**   **1.** $t = \dfrac{d}{r}$   **3.** $r = \dfrac{I}{Pt}$   **5.** $c = P - a - b$   **7.** $y = \dfrac{9x - 16}{4}$ or $y = \dfrac{9}{4}x - 4$   **9.** $l = \dfrac{P - 2w}{2}$ or $l = \dfrac{P}{2} - w$

**11.** $r = \dfrac{E}{I} - R$   **13.** $y = \dfrac{20 - 5x}{4}$ or $y = 5 - \dfrac{5}{4}x$   **15.** $H = \dfrac{S - 2LW}{2L + 2W}$   **17.** $r = \dfrac{C}{2\pi}$   **19.** $L = \dfrac{2s - na}{n}$

**21.**

| $n$ | 1 | 2 | 4 | 12 | 365 |
|---|---|---|---|---|---|
| $A$ | $4703.71 | $4713.99 | $4719.22 | $4722.74 | $4724.45 |

**23. a.** $7313.97   **b.** $7321.14   **c.** $7325.98   **25.** 3.6 hr or 3 hr, 36 min
**27.** 40°C   **29.** 171 packages   **31.** 9 ft   **33.** 2 gal   **35. a.** 1174.86 cu m
**b.** 310.34 cu m   **c.** 1485.20 cu m   **37.** 128.3 mph   **39.** 0.42 ft

**41.** $41.125\pi$ ft; 129.1325 ft   **43.** $1831.96   **45.** $f = \dfrac{C - 4h - 4p}{9}$   **47.** 178 cal   **49.** 1.5 g   **51.** $\{-3, -2, -1\}$

**53.** $\{-3, -2, -1, 0, 1\}$   **55.** answers may vary   **57.** 0.723   **59.** 1.000   **61.** 1.523   **63.** 5.202   **65.** answers may vary

**67.** answers may vary; $g = \dfrac{W}{h - 3t^2}$   **69.** $\dfrac{1}{8}$   **71.** $\dfrac{1}{8}$   **73.** $\dfrac{1}{2}$   **75.** $\dfrac{3}{8}$   **77.** 0   **79.** 0

## Section 2.4

**Vocabulary, Readiness & Video Check**   **1.** d   **3.** b   **5.** $[-0.4, \infty)$   **7.** $(-\infty, -0.4]$   **9.** no   **11.** yes   **13.** The graph of Example 1 is shaded from $-\infty$ to, but not including, $-3$, as indicated by a parenthesis. To write interval notation, write down what is shaded for the inequality from left to right. A parenthesis is always used with $-\infty$, so from the graph, the interval notations is $(-\infty, -3)$.
**15.** same; reverse   **17.** maximum, or less

**Exercise Set 2.4**   **1.** $(-\infty, -3)$   **3.** $[0.3, \infty)$

**5.** $[-7, \infty)$   **7.** $(-2, 5)$   **9.** $(-1, 5]$

**11.** $[-2, \infty)$   **13.** $(-\infty, 1)$   **15.** $(-\infty, 2]$

**17.** $[8, \infty)$   **19.** $(-\infty, -4.7)$   **21.** $(-\infty, -3]$

**23.** $(-\infty, -1]$   **25.** $(-\infty, 11]$   **27.** $(0, \infty)$   **29.** $(-13, \infty)$   **31.** $\left[-\dfrac{79}{3}, \infty\right)$   **33.** $\left(-\infty, -\dfrac{35}{6}\right)$   **35.** $(-\infty, -6)$   **37.** $(4, \infty)$
**39.** $[-0.5, \infty)$   **41.** $(-\infty, 7]$   **43.** $[0, \infty)$   **45.** $(-\infty, -29]$   **47.** $[3, \infty)$   **49.** $(-\infty, -1]$   **51.** $[-31, \infty)$   **53.** $(-\infty, -2]$
**55.** $(-\infty, 9)$   **57.** $\left(-\infty, -\dfrac{11}{2}\right]$   **59.** $(-\infty, -15)$   **61.** $\left[-\dfrac{37}{3}, \infty\right)$   **63.** $(-\infty, 5)$   **65. a.** $\{x \mid x \ge 81\}$   **b.** A final exam grade of
81 or higher will result in an average of 77 or higher.   **67. a.** $\{x \mid x \le 20\}$   **b.** She can move at most 20 whole boxes at one time.
**69. a.** $\{x \mid x \le 1040\}$   **b.** The luggage and cargo must weight 1040 pounds or less.   **71. a.** $\{x \mid x > 200\}$   **b.** If you make more than 200 calls, plan 1 is more economical.   **73.** $\{F \mid F \ge 932°\}$   **75. a.** 2000   **b.** answers may vary   **77.** decreasing; answers may vary
**79.** 28.6 lb   **81.** during 2014   **83.** answers may vary   **85. a.** $\{t \mid t > 22.97\}$   **b.** 2023   **87.** $\{0, 1, 2, 3, 4, 5, 6, 7\}$
**89.** $\{\ldots, -9, -8, -7, -6\}$   **91.** $\{5\}$   **93.** $\{0\}$   **95.** $\{x \mid x \ge 2\}; [2, \infty)$   **97.** $; (-\infty, 0)$

**99.** $\{x \mid -2 < x \le 1.5\};$   **101.** yes   **103.** yes   **105.** $\{4\}$   **107.** $(4, \infty)$   **109.** answers may vary
**111.** answers may vary   **113.** answers may vary

**Integrated Review**   **1.** $\{-5\}$   **2.** $(-5, \infty)$   **3.** $\left[\dfrac{8}{3}, \infty\right)$   **4.** $[-1, \infty)$   **5.** $\{0\}$   **6.** $\left[-\dfrac{1}{10}, \infty\right)$   **7.** $\left(-\infty, -\dfrac{1}{6}\right]$   **8.** $\{0\}$   **9.** $\varnothing$

**10.** $\left[-\dfrac{3}{5}, \infty\right)$   **11.** $\{4.2\}$   **12.** $\{6\}$   **13.** $\{-8\}$   **14.** $(-\infty, -16)$   **15.** $\left\{\dfrac{20}{11}\right\}$   **16.** $\{1\}$   **17.** $(38, \infty)$   **18.** $\{-5.5\}$

**19.** $\left\{\dfrac{3}{5}\right\}$   **20.** $(-\infty, \infty)$   **21.** $\{29\}$   **22.** $\{x \mid x \text{ is a real number}\}$   **23.** $(-\infty, 5)$   **24.** $\left\{\dfrac{9}{13}\right\}$   **25.** $(23, \infty)$   **26.** $(-\infty, 6]$

**27.** $\left(-\infty, \dfrac{3}{5}\right]$   **28.** $\left(-\infty, -\dfrac{19}{32}\right)$

## Section 2.5

**Vocabulary, Readiness & Video Check**   **1.** compound   **3.** or   **5.** $\cup$   **7.** and   **9.** and   **11.** or

**Exercise Set 2.5**   **1.** $\{2, 3, 4, 5, 6, 7\}$   **3.** $\{4, 6\}$   **5.** $\{\ldots, -2, -1, 0, 1, 2, \ldots\}$   **7.** $\{5, 7\}$   **9.** $\{x \mid x$ is an odd integer or $x = 2$ or $x = 4\}$   **11.** $\{2, 4\}$   **13.** $\longleftrightarrow ; (-3, 1)$   **15.** $\longleftrightarrow ; \varnothing$

**17.** $\longleftrightarrow ; (-\infty, -1)$   **19.** $[6, \infty)$   **21.** $(-\infty, -3]$   **23.** $(4, 10)$   **25.** $(11, 17)$   **27.** $[1, 4]$   **29.** $\left[-3, \dfrac{3}{2}\right]$

**31.** $\left[-\dfrac{7}{3}, 7\right]$   **33.** $\longleftrightarrow ; (-\infty, 5)$   **35.** $\longleftrightarrow ; (-\infty, 4] \cup [1, \infty)$

**37.** $\longleftrightarrow ; (-\infty, \infty)$   **39.** $[2, \infty)$   **41.** $(-\infty, -4) \cup (-2, \infty)$   **43.** $(-\infty, \infty)$   **45.** $\left(-\dfrac{1}{2}, \dfrac{2}{3}\right)$

**47.** $(-\infty, \infty)$   **49.** $\left[\dfrac{3}{2}, 6\right]$   **51.** $\left(\dfrac{5}{4}, \dfrac{11}{4}\right)$   **53.** $\varnothing$   **55.** $\left(-\infty, -\dfrac{56}{5}\right) \cup \left[\dfrac{5}{3}, \infty\right)$   **57.** $\left(-5, \dfrac{5}{2}\right)$   **59.** $\left(0, \dfrac{14}{3}\right]$   **61.** $(-\infty, -3]$

**63.** $(-\infty, 1] \cup \left(\dfrac{29}{7}, \infty\right)$   **65.** $\varnothing$   **67.** $\left[-\dfrac{1}{2}, \dfrac{3}{2}\right)$   **69.** $\left(-\dfrac{4}{3}, \dfrac{7}{3}\right)$   **71.** $(6, 12)$   **73.** $-12$   **75.** $-4$   **77.** $\{-7, 7\}$   **79.** $\{0\}$

**81.** $2004, 2005$   **83.** answers may vary   **85.** $(-3, 5)$   **87.** $(2, \infty)$   **89.** $-20.2° \le F \le 95°$   **91.** $67 \le$ final score $\le 94$

## Section 2.6

**Vocabulary, Readiness & Video Check**   **1.** C   **3.** B   **5.** D

**Exercise Set 2.6**   **1.** $\{7, -7\}$   **3.** $\{4.2, -4.2\}$   **5.** $\{7, -2\}$   **7.** $\{8, 4\}$   **9.** $\{5, -5\}$   **11.** $\{3, -3\}$   **13.** $\{0\}$   **15.** $\varnothing$   **17.** $\left\{\dfrac{1}{5}\right\}$

**19.** $\left\{9, -\dfrac{1}{2}\right\}$   **21.** $\left\{-\dfrac{5}{2}\right\}$   **23.** $\{4, -4\}$   **25.** $\{0\}$   **27.** $\varnothing$   **29.** $\left\{0, \dfrac{14}{3}\right\}$   **31.** $\{2, -2\}$   **33.** $\varnothing$   **35.** $\{7, -1\}$   **37.** $\varnothing$   **39.** $\varnothing$

**41.** $\left\{-\dfrac{1}{8}\right\}$   **43.** $\left\{\dfrac{1}{2}, -\dfrac{5}{6}\right\}$   **45.** $\left\{2, -\dfrac{12}{5}\right\}$   **47.** $\{3, -2\}$   **49.** $\left\{-8, \dfrac{2}{3}\right\}$   **51.** $\varnothing$   **53.** $\{4\}$   **55.** $\{13, -8\}$   **57.** $\{3, -3\}$

**59.** $\{8, -7\}$   **61.** $\{2, 3\}$   **63.** $\left\{2, -\dfrac{10}{3}\right\}$   **65.** $\left\{\dfrac{3}{2}\right\}$   **67.** $\varnothing$   **69.** 28%   **71.** $32.4°$   **73.** answers may vary   **75.** answers may vary

**77.** $\varnothing$   **79.** $|x| = 5$   **81.** answers may vary   **83.** $|x - 1| = 5$   **85.** answers may vary   **87.** $|x| = 6$   **89.** $|x - 2| = |3x - 4|$

## Section 2.7

**Vocabulary, Readiness & Video Check**   **1.** D   **3.** C   **5.** A   **7.** The solution set involves "or" and "or" means "union".

**Exercise Set 2.7**   **1.** $\longleftrightarrow ; [-4, 4]$   **3.** $\longleftrightarrow ; (1, 5)$   **5.** $\longleftrightarrow ; (-5, -1)$

**7.** $\longleftrightarrow ; [-10, 3]$   **9.** $\longleftrightarrow ; [-5, 5]$   **11.** $\longleftrightarrow ; \varnothing$

**13.** $\longleftrightarrow ; [0, 12]$   **15.** $\longleftrightarrow ; (-\infty, -3) \cup (3, \infty)$

**17.** $\longleftrightarrow ; (-\infty, -24] \cup [4, \infty)$   **19.** $\longleftrightarrow ; (-\infty, -4) \cup (4, \infty)$

**21.** $\longleftrightarrow ; (-\infty, \infty)$   **23.** $\longleftrightarrow ; \left(-\infty, \dfrac{2}{3}\right) \cup (2, \infty)$   **25.** $\longleftrightarrow ; \{0\}$

**27.** $\longleftrightarrow ; \left(-\infty, -\dfrac{3}{8}\right) \cup \left(-\dfrac{3}{8}, \infty\right)$   **29.** $\longleftrightarrow ; [-2, 2]$

**31.** $\longleftrightarrow ; (-\infty, -1) \cup (1, \infty)$   **33.** $\longleftrightarrow ; (-5, 11)$

**35.** $\longleftrightarrow ; (-\infty, 4) \cup (6, \infty)$   **37.** $\longleftrightarrow ; \varnothing$   **39.** $\longleftrightarrow ; (-\infty, \infty)$

**41.** $\longleftrightarrow ; [-2, 9]$   **43.** $\longleftrightarrow ; (-\infty, -11] \cup [1, \infty)$

**45.** $\longleftrightarrow ; (-\infty, 0) \cup (0, \infty)$   **47.** $\longleftrightarrow ; (-\infty, \infty)$   **49.** $\longleftrightarrow ; \left[-\dfrac{1}{2}, 1\right]$

**51.** $\longleftrightarrow ; (-\infty, -3) \cup (0, \infty)$   **53.** $\longleftrightarrow ; \varnothing$   **55.** $\longleftrightarrow ; \left\{\dfrac{3}{8}\right\}$

**57.** $\longleftrightarrow ; \left(-\dfrac{2}{3}, 0\right)$   **59.** $\longleftrightarrow ; (-\infty, -12) \cup (0, \infty)$   **61.** $\longleftrightarrow ; [-1, 8]$

**63.** $\longleftrightarrow ; \left[-\dfrac{23}{8}, \dfrac{17}{8}\right]$   **65.** $(-2, 5)$   **67.** $\{5, -2\}$   **69.** $(-\infty, -7] \cup [17, \infty)$   **71.** $\left\{-\dfrac{9}{4}\right\}$   **73.** $(-2, 1)$

**75.** $\left\{2, \dfrac{4}{3}\right\}$　**77.** $\varnothing$　**79.** $\left\{\dfrac{19}{2}, -\dfrac{17}{2}\right\}$　**81.** $\left(-\infty, -\dfrac{25}{3}\right) \cup \left(\dfrac{35}{3}, \infty\right)$　**83.** $\dfrac{1}{6}$　**85.** 0　**87.** $\dfrac{1}{3}$　**89.** $-1.5$　**91.** 0　**93.** $|x| < 7$

**95.** $|x| \le 5$　**97.** answers may vary　**99.** $3.45 < x < 3.55$

**Chapter 2 Vocabulary Check**　**1.** compound inequality　**2.** contradiction　**3.** intersection　**4.** union　**5.** identity　**6.** formula
**7.** absolute value　**8.** solution　**9.** consecutive integers　**10.** linear inequality in one variable　**11.** linear equation in one variable

**Chapter 2 Review**　**1.** $\{3\}$　**2.** $\left\{-\dfrac{23}{3}\right\}$　**3.** $\left\{-\dfrac{45}{14}\right\}$　**4.** $\left\{-\dfrac{7}{11}\right\}$　**5.** $\{0\}$　**6.** $\varnothing$　**7.** $\{6\}$　**8.** $\{7.8\}$

**9.** $\{x \mid x \text{ is a real number}\}$　**10.** $\{0\}$　**11.** $\varnothing$　**12.** $\{p \mid p \text{ is a real number}\}$　**13.** $\{-3\}$　**14.** $\{0\}$　**15.** $\left\{\dfrac{96}{5}\right\}$　**16.** $\{-3\}$

**17.** $\{32\}$　**18.** $\{-8\}$　**19.** $\{8\}$　**20.** $\{1\}$　**21.** $\varnothing$　**22.** $\{11\}$　**23.** $-7$　**24.** $140, 145$　**25.** $52$　**26.** $0.12$　**27.** \$35,556
**28.** $10, 11, 12, 13$　**29.** no such integers exist　**30.** width: 40 m; length: 75 m　**31.** 258 mi　**32.** 250 calculators

**33.** $W = \dfrac{V}{LH}$　**34.** $r = \dfrac{C}{2\pi}$　**35.** $y = \dfrac{5x + 12}{4}$　**36.** $x = \dfrac{4y - 12}{5}$　**37.** $m = \dfrac{y - y_1}{x - x_1}$　**38.** $x = \dfrac{y - y_1 + mx_1}{m}$　**39.** $r = \dfrac{E - IR}{I}$

**40.** $g = \dfrac{S - vt}{t^2}$　**41.** $g = \dfrac{T}{r + vt}$　**42.** $P = \dfrac{I}{1 + rt}$　**43. a.** \$3695.27　**b.** \$3700.81　**44.** $32\dfrac{2}{9}°C$　**45.** length: 10 in.; width: 8 in.

**46.** 16 packages　**47.** $(3, \infty)$　**48.** $(-\infty, -4]$　**49.** $(-4, \infty)$　**50.** $(-17, \infty)$　**51.** $(-\infty, 7]$　**52.** $\left(\dfrac{1}{2}, \infty\right)$　**53.** $(-\infty, 1)$

**54.** $[-19, \infty)$　**55.** more than 35 pounds per week　**56.** 9.6　**57.** $\left[2, \dfrac{5}{2}\right]$　**58.** $\left[-2, -\dfrac{9}{5}\right)$　**59.** $\left(\dfrac{1}{8}, 2\right)$　**60.** $\left(-\dfrac{3}{5}, 0\right)$　**61.** $\left(\dfrac{7}{8}, \dfrac{27}{20}\right]$

**62.** $(-5, 2]$　**63.** $\left(\dfrac{11}{3}, \infty\right)$　**64.** $(-\infty, -2] \cup [2, \infty)$　**65.** $260° \le C \le 538°$　**66.** \$1750 to \$3750　**67.** $\{0, -9\}$　**68.** $\left\{-1, \dfrac{11}{3}\right\}$

**69.** $\left\{2, -\dfrac{2}{3}\right\}$　**70.** $\left\{-\dfrac{1}{6}\right\}$　**71.** $\varnothing$　**72.** $\varnothing$　**73.** $\left\{5, -\dfrac{1}{3}\right\}$　**74.** $\{4, 1\}$　**75.** $\left\{7, -\dfrac{8}{5}\right\}$　**76.** $\left\{0, \dfrac{7}{4}\right\}$

**77.** $-\dfrac{8}{5}$ to $2$　$\left(-\dfrac{8}{5}, 2\right)$　**78.** $-4$ to $1$　$(-\infty, -4] \cup [1, \infty)$

**79.** $-3$, $3$　$(-\infty, -3) \cup (3, \infty)$　**80.** $-3$, $3$　$(-3, 3)$　**81.** $0$　$\varnothing$

**82.** $-\dfrac{22}{15}$, $\dfrac{6}{5}$　$\left(-\infty, -\dfrac{22}{15}\right] \cup \left[\dfrac{6}{5}, \infty\right)$　**83.** $-27$, $-9$　$(-\infty, -27) \cup (-9, \infty)$

**84.** $0$　$\varnothing$　**85.** $\{2\}$　**86.** $\left\{\dfrac{37}{8}\right\}$　**87.** China: 137 million; USA: 102 million; France: 93 million

**88.** $B = \dfrac{2A - hb}{h}$　**89.** $h = \dfrac{3V}{\pi r^2}$　**90.** cylinder　**91.** 58 mph　**92.** $(-\infty, 4]$　**93.** $(2, \infty)$　**94.** $\left[-\dfrac{4}{3}, \dfrac{7}{6}\right]$　**95.** $(-\infty, \infty)$

**96.** $(5, \infty)$　**97.** $\{3, -3\}$　**98.** $\varnothing$　**99.** $\left\{-10, -\dfrac{4}{3}\right\}$　**100.** $(-\infty, \infty)$　**101.** $\left(-\dfrac{1}{2}, 2\right)$

**Chapter 2 Test**　**1.** $\{10\}$　**2.** $\{-32\}$　**3.** $\varnothing$　**4.** $\{n \mid n \text{ is a real number}\}$　**5.** $\left\{-\dfrac{80}{29}\right\}$　**6.** $\left\{\dfrac{29}{4}\right\}$　**7.** $\left\{1, \dfrac{2}{3}\right\}$　**8.** $\varnothing$

**9.** $\left\{-4, -\dfrac{1}{3}\right\}$　**10.** $\left\{\dfrac{3}{2}\right\}$　**11.** $y = \dfrac{3x - 8}{4}$　**12.** $g = \dfrac{S}{t^2 + vt}$　**13.** $C = \dfrac{5}{9}(F - 32)$　**14.** $(5, \infty)$　**15.** $\left(-\infty, -\dfrac{11}{3}\right]$　**16.** $\left(\dfrac{3}{2}, 5\right)$

**17.** $(-\infty, -2) \cup \left(\dfrac{4}{3}, \infty\right)$　**18.** $(3, 7)$　**19.** $[5, \infty)$　**20.** $[4, \infty)$　**21.** $\left[1, \dfrac{11}{2}\right)$　**22.** $(-\infty, \infty)$　**23.** 9.6　**24.** 5,055,000 vehicles

**25.** at most 8 hunting dogs　**26.** 58,000　**27.** \$3542.27　**28.** Florida: \$17 billion; California: \$13 billion; New York: \$9 billion

**Cumulative Review**　**1.** $\{101, 102, 103, \ldots\}$; Sec. 1.2, Ex. 2　**2. a.** $\{-2, -1, 0, 1, 2, 3, 4\}$　**b.** $\{4\}$　**3.** $x + 5 = 20$; Sec. 1.3, Ex. 1
**4.** $x - 19 = 4x$　**5.** $\dfrac{z}{9} = 9 + z$; Sec. 1.3, Ex. 3　**6. a.** $>$　**b.** $=$　**c.** $>$　**7.** $-8$; Sec. 1.3, Ex. 16　**8.** 0　**9.** $\dfrac{1}{5}$; Sec. 1.3, Ex. 17　**10.** 7.3

**11.** $-14$; Sec. 1.4, Ex. 6　**12.** $-12.4$　**13.** 5; Sec. 1.4, Ex. 8　**14.** 3　**15.** $-\dfrac{5}{21}$; Sec. 1.4, Ex. 11　**16.** $-7.2$　**17.** $-5$; Sec. 1.4, Ex. 25

**18.** undefined　**19.** 0; Sec. 1.4, Ex. 29　**20.** $-\dfrac{1}{3}$　**21.** $\dfrac{1}{8}$; Sec. 1.4, Ex. 30　**22.** $\dfrac{3}{7}$　**23.** 63; Sec. 1.5, Ex. 1　**24.** $\dfrac{3}{20}$　**25.** $-2x + 4$; Sec. 1.5, Ex. 8

**26.** $-7y - 10$　**27.** $4y$; Sec. 1.5, Ex. 9　**28.** $14x + 8$　**29.** $x^3$; Sec. 1.6, Ex. 11　**30.** $\dfrac{1}{a^2 b}$　**31.** $5x$; Sec. 1.6, Ex. 12　**32.** $\dfrac{z^{20}}{9}$

**33.** $\dfrac{z^2}{9x^4 y^{20}}$; Sec. 1.7, Ex. 14　**34.** $-\dfrac{36}{a^7 b^2}$　**35.** $\dfrac{27a^4 x^6}{2}$; Sec. 1.7, Ex. 15　**36.** $\{-2\}$　**37.** $\{2\}$; Sec. 2.1, Ex. 3　**38.** $\{-4.25\}$

**39.** $\{x \mid x \text{ is a real number}\}$; Sec. 2.1, Ex. 11  **40. a.** $3x + 3$  **b.** $12x + 4$  **41.** $2350; Sec. 2.2, Ex. 4  **42.** 11, 35  **43.** $\dfrac{V}{lw} = h$; Sec. 2.3, Ex. 1

**44.** $x = \dfrac{10 + 4y}{7}$  **45.** 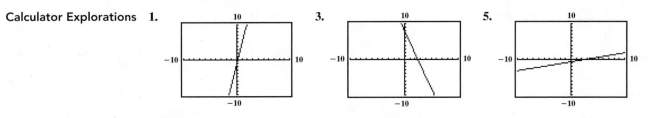 $(-\infty, 6]$; Sec. 2.4, Ex. 7  **46.** $(3, \infty)$  **47.** $(-\infty, \infty)$; Sec. 2.5, Ex. 8  **48.** $\varnothing$

**49.** $\{2, -2\}$; Sec. 2.6, Ex. 1  **50.** $\{2\}$

# Chapter 3 Graphs and Functions

### Section 3.1

**Calculator Explorations  1.**   **3.**  **5.**

**Vocabulary, Readiness & Video Check  1.** origin  **3.** 1  **5.** horizontal  **7.** origin; left (if negative) or right (if positive); up (if positive) or down (if negative)  **9.** The graph of an equation is a picture of its ordered pair solutions; a third point is found as a check to make sure the points line up and a mistake hasn't been made.

**Exercise Set 3.1  1.** $(5, 2)$  **3.** $(3, -1)$  **5.** $(-5, -2)$  **7.** $(-1, 0)$  **9.**

quadrant I;
quadrant II;
quadrant IV;
$y$-axis; quadrant III

**11.** QI  **13.** QII  **15.** QIII  **17.** $y$-axis  **19.** QIII  **21.** $x$-axis  **23.** quadrant IV  **25.** $x$-axis  **27.** quadrant III

**29.**

U.S. Gold Medals in Summer Olympics

**31.** no; yes  **33.** yes; yes  **35.** yes; no  **37.**  **39.**  **41.**

**43.**  **45.**  **47.**  **49.**  **51.**  **53.**  **55.**

**57.**  **59.**  **61.**  **63.**  **65.** $\dfrac{3}{2}$  **67.** 6  **69.** $-\dfrac{6}{5}$  **71.** b  **73.** B  **75.** C

**77.** 1997  **79.** answers may vary

**81. a.** 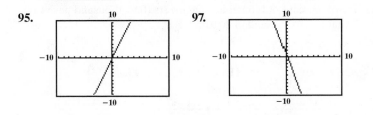  **b.** 14 in.  **83.** $7000  **85.** $500  **87.** depreciation is the same from year to year  **89. a.** $(0, 500)$; 0 tables and 500 chairs are produced  **b.** $(750, 0)$; 750 tables and 0 chairs are produced  **c.** 466 chairs  **91.** answers may vary  **93.** vertical line $x = 0$ has $y$-intercepts

**95.**  **97.**

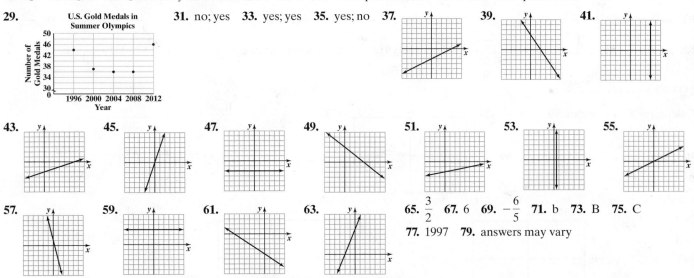

## Section 3.2

**Calculator Explorations**   **1.** 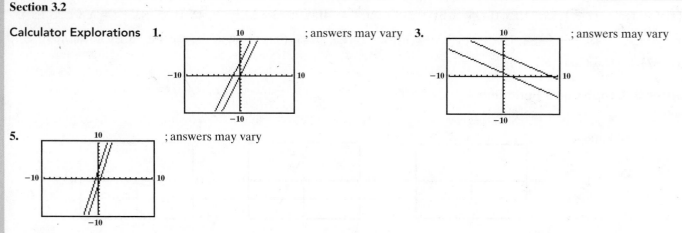 ; answers may vary   **3.** ; answers may vary

**5.** ; answers may vary

**Vocabulary, Readiness & Video Check**   **1.** slope   **3.** $m; (0, b)$   **5.** horizontal   **7.** $-1$   **9.** upward   **11.** horizontal   **13.** $y; x$
**15.** Slope is $0$ indicates $m = 0$ and a horizontal line; undefined slope indicates $m$ is undefined and a vertical line; no slope refers to an undefined slope.   **17.** The slopes of vertical lines are undefined and we can't work mathematically with undefined slopes.

**Exercise Set 3.2**   **1.** $\dfrac{9}{5}$   **3.** $-\dfrac{7}{2}$   **5.** $-\dfrac{5}{6}$   **7.** $\dfrac{1}{3}$   **9.** $-\dfrac{4}{3}$   **11.** $0$   **13.** $m = 2$   **15.** $m = -\dfrac{4}{5}$   **17.** $\dfrac{2}{3}$   **19.** $\dfrac{3}{20}$   **21.** $m = -1; (0, 5)$

**23.** $m = 5; (0, -2)$   **25.** $m = -2; (0, 7)$   **27.** $m = \dfrac{2}{3}; \left(0, -\dfrac{10}{3}\right)$   **29.** $m$ undefined; no $y$-intercept   **31.** $m = 0; (0, -2)$

**33.** $m = \dfrac{1}{2}; (0, 0)$   **35.** $m = \dfrac{1}{3}; \left(0, -\dfrac{8}{3}\right)$   **37.** $m = \dfrac{6}{5}; (0, 6)$   **39.** $m = 7; (0, 0)$   **41.** $m$ undefined; no $y$-intercept   **43.** A   **45.** B

**47.** $l_2$   **49.** $l_2$   **51.** $l_2$   **53.** parallel   **55.** neither   **57.** perpendicular   **59.** neither   **61.** parallel   **63.** perpendicular   **65.** perpendicular

**67.** neither   **69.** $\{9, -3\}$   **71.** $(-\infty, -4) \cup (-1, \infty)$   **73.** $\left[\dfrac{2}{3}, 2\right]$   **75.** incorrect; $m = \dfrac{-14 - 6}{7 - (-2)} = -\dfrac{20}{9}$   **77.** correct   **79.** $-\dfrac{7}{2}$

**81.** $\dfrac{2}{7}$   **83.** $\dfrac{5}{2}$   **85.** $-\dfrac{2}{5}$   **87.** $(6, 20)$   **89.** $-\dfrac{7}{4}$ yd per sec   **91.** $\dfrac{1}{48}$   **93.** answers may vary   **95. a.** $l_1: -2; l_2: -1; l_3: -\dfrac{2}{3}$   **b.** lesser

## Section 3.3

**Calculator Explorations**   **1.** $y = \dfrac{1}{3.5}x$   **3.** $y = \dfrac{-5.78}{2.31}x + \dfrac{10.98}{2.31}$

**5.** $y = x + 3.78$   **7.** $y = 13.3x + 1.5$

**Vocabulary, Readiness & Video Check**   **1.** $m = -4$; $y$-intercept: $(0, 12)$   **3.** $m = 5$; $y$-intercept: $(0, 0)$   **5.** $m = \dfrac{1}{2}$; $y$-intercept: $(0, 6)$
**7.** slope-intercept; $y$-intercept; slope   **9.** Slope-intercept form allows us to see the slope and $y$-intercept so we can discuss them in the context of the application. Slope is a rate of change and tells us the rate of increase of biomedical engineers (0.97 thousand per year). The $y$-intercept tells us the number of biomedical engineers at "zero" year (15.7 thousand in 2010).

**Exercise Set 3.3**

**15.**   **17.**  **19.** C  **21.** D  **23.** $y = -x + 1$  **25.** $y = 2x + \frac{3}{4}$  **27.** $y = \frac{2}{7}x$  **29. a.** 75.3 yr  **b.** $m = 0.19$; the life expectancy of a person born in the United States is approximately 0.19 year more than that of a person born 1 year before.  **c.** $(0, 65.8)$; in year 0, or 1940, the life expectancy of a person born in the U.S. was 65.8 years.  **31. a.** $m = 0.97$; $(0, 15.7)$  **b.** number of people employed as biomedical engineers increases 0.97 thousand for every 1 year  **c.** 15.7 thousand biomedical engineers were employed in 2010.  **33. a.** 598 movies  **b.** 2015  **c.** answers may vary  **35.** $y = 5x + 32$  **37.** $y = 2x - 1$  **39.** incorrect; $y = 3x + 1.7$  **41.** answers may vary  **43.** $y = 42,000x + 2,900,000$, where $y$ is revenue and $x$ is number of years since 2000

## Section 3.4

**Calculator Explorations**  **1.** 18.4  **3.** $-1.5$  **5.** 8.7; 7.6

**Vocabulary, Readiness & Video Check**  **1.** $m = -2$; $(1, 4)$  **3.** $m = \frac{1}{4}$; $(2, 0)$  **5.** $m = 5$; $(3, -2)$  **7.** if one of the two given points is the $y$-intercept  **9.** $y = \frac{1}{3}x - \frac{17}{3}$

**Exercise Set 3.4**  **1.** $y = 3x - 1$  **3.** $y = -2x - 1$  **5.** $y = \frac{1}{2}x + 5$  **7.** $y = -\frac{9}{10}x - \frac{27}{10}$  **9.** $3x - y = 6$  **11.** $2x + y = 1$  **13.** $x + 2y = -10$  **15.** $x - 3y = 21$  **17.** $3x + 8y = 5$  **19.** $x = 2$  **21.** $y = 1$  **23.** $x = 0$  **25.** $y = 4x - 4$  **27.** $y = -3x + 1$  **29.** $y = 4$  **31.** $y = -\frac{3}{2}x - 6$  **33.** $y = -5$  **35.** $y = -4x + 1$  **37.** $2x - y = -7$  **39.** $y = -x + 7$  **41.** $y = -\frac{1}{2}x + \frac{3}{8}$  **43.** $2x + 7y = -42$  **45.** $4x + 3y = -20$  **47.** $x = -2$  **49.** $x + 2y = 2$  **51.** $y = 12$  **53.** $8x - y = 47$  **55.** $x = 5$  **57.** $y = -\frac{3}{8}x - \frac{29}{4}$  **59. a.** $y = 32x$  **b.** 128 ft per sec  **61. a.** $y = -250x + 3500$  **b.** 1625 Frisbees  **63. a.** $y = -600x + 297,000$  **b.** $291,000  **65. a.** $y = 271x + 38,834$  **b.** 40,731  **67.** 31  **69.** $-8.4$  **71.** 4  **73.** $2x + y = 3$  **75.** $2x - 3y = -7$  **77.** true  **79.** answers may vary  **81.**

## Integrated Review

**1.**  **2.**  **3.**  **4.**  **5.** 0  **6.** $-\frac{3}{5}$  **7.** $m = 3$; $(0, -5)$  **8.** $m = \frac{5}{2}$; $\left(0, -\frac{7}{2}\right)$

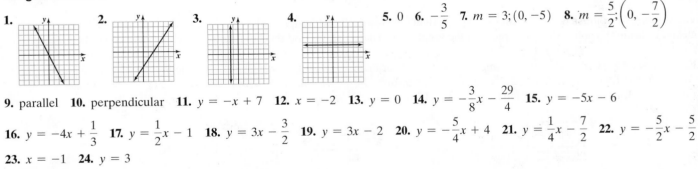

**9.** parallel  **10.** perpendicular  **11.** $y = -x + 7$  **12.** $x = -2$  **13.** $y = 0$  **14.** $y = -\frac{3}{8}x - \frac{29}{4}$  **15.** $y = -5x - 6$  **16.** $y = -4x + \frac{1}{3}$  **17.** $y = \frac{1}{2}x - 1$  **18.** $y = 3x - \frac{3}{2}$  **19.** $y = 3x - 2$  **20.** $y = -\frac{5}{4}x + 4$  **21.** $y = \frac{1}{4}x - \frac{7}{2}$  **22.** $y = -\frac{5}{2}x - \frac{5}{2}$  **23.** $x = -1$  **24.** $y = 3$

## Section 3.5

**Vocabulary, Readiness & Video Check**  **1.** You find the boundary line equation by replacing the inequality symbol with $=$. The points on this line are solutions (line is solid) if the inequality is $\geq$ or $\leq$; the points on this line are not solutions (line is dashed) if the inequality is $>$ or $<$.

**Exercise Set 3.5**

**1.**  **3.**  **5.**  **7.**  **9.**  **11.**  **13.**

**15.** **17.** **19.** **21.** **23.** **25.** **27.**

**29.** D　**31.** A　**33.** yes　**35.** no　**37.** with < or >　**39.**

## Section 3.6

**Vocabulary, Readiness & Video Check** **1.** relation　**3.** domain　**5.** vertical　**7.** An equation is one way of writing down a correspondence that produces ordered pair solutions; a relation between two sets of coordinates, $x$'s and $y$'s.　**9.** If a vertical line intersects a graph two (or more) times, then there's an $x$-value corresponding to two (or more) different $y$-values and is thus not a function.　**11.** No, equations of the form $x = c$, whose graphs are vertical lines, are not functions.

**Exercise Set 3.6** **1.** domain; $\{-1, 0, -2, 5\}$; range: $\{7, 6, 2\}$; function　**3.** domain; $\{-2, 6, -7\}$; range: $\{4, -3, -8\}$; not a function
**5.** domain: $\{1\}$; range: $\{1, 2, 3, 4\}$; not a function　**7.** domain: $\left\{\dfrac{3}{2}, 0\right\}$; range: $\left\{\dfrac{1}{2}, -7, \dfrac{4}{5}\right\}$; not a function　**9.** domain: $\{-3, 0, 3\}$;
range: $\{-3, 0, 3\}$; function　**11.** domain: $\{-1, 1, 2, 3\}$; range: $\{2, 1\}$; function　**13.** domain: $\{1996, 2000, 2004, 2008, 2012\}$;
range: $\{36, 37, 44, 46\}$; function　**15.** domain: $\{32, 104, 212, 50\}$; range: $\{0, 40, 10, 100\}$; function　**17.** domain: $\{2, -1, 5, 100\}$;
range: $\{0\}$; function　**19.** function　**21.** not a function　**23.** function　**25.** not a function　**27.** function　**29.** function
**31.** not a function　**33.** function　**35.** not a function　**37.** not a function　**39.** not a function　**41.** not a function
**43.** not a function　**45.** function　**47.** 15　**49.** 38　**51.** 7　**53.** 3　**55. a.** 0　**b.** 1　**c.** −1　**57. a.** −5　**b.** −5　**c.** −5
**59.** $(1, -10)$　**61.** $(4, 56)$　**63.** −2　**65.** 0　**67.** −4, 0　**69.** 3　**71.** $25\pi$ sq cm　**73.** 2744 cu in.　**75.** 166.38 cm
**77.** 163.2 mg　**79. a.** Agricultural exports to China in 2005 were approximately $7.56 billion.　**b.** $23.76 billion

**81.** **83.** **85.** 　**87.** $(-\infty, 14]$　**89.** $\left[\dfrac{7}{2}, \infty\right)$　**91.** $\left(-\infty, -\dfrac{1}{4}\right)$　**93.** no; answers may vary
**95.** yes; answers may vary　**97.** true　**99.** true　**101.** infinite number
**103.** answers may vary　**105. a.** 0　**b.** $a - 12$　**c.** $-x - 12$
**d.** $x + h - 12$

## Section 3.7

**Vocabulary, Readiness & Video Check** **1.** left; right; up; down　**3.** Although $f(x) = x + 3$ isn't defined for $x = -1$, we need to clearly indicate the point where this piece of the graph ends. Therefore, we find this point and graph it as an open circle.

**Exercise Set 3.7** **1.** domain; $[0, \infty)$; range: $(-\infty, \infty)$　**3.** domain: $(-\infty, \infty)$; range: $[0, \infty)$　**5.** domain: $(-\infty, \infty)$; range:
$(-\infty, -3] \cup [3, \infty)$　**7.** domain: $[2, 7]$; range: $[1, 6]$　**9.** domain: $\{-2\}$; range: $(-\infty, \infty)$　**11.** domain: $(-\infty, \infty)$; range:
$(-\infty, 3]$　**13.** domain: $(-\infty, \infty)$; range: $(-\infty, 3]$　**15.** domain: $[2, \infty)$; range: $[3, \infty)$

**17.** **19.** **21.** **23.** **25.** domain: $(-\infty, \infty)$; range: $[0, \infty)$

**27.** domain: $(-\infty, \infty)$; range: $(-\infty, 5)$　**29.** domain: $(-\infty, \infty)$; range: $(-\infty, 6]$　**31.** domain: $(-\infty, 0] \cup [1, \infty)$; range: $\{-4, -2\}$

**33.** A　**35.** D　**37.** answers may vary　**39.**

## Section 3.8

**Vocabulary, Readiness & Video Check** **1.** C **3.** D **5.** $f(x) = |x|$ **7.** $x$-axis

### Exercise Set 3.8

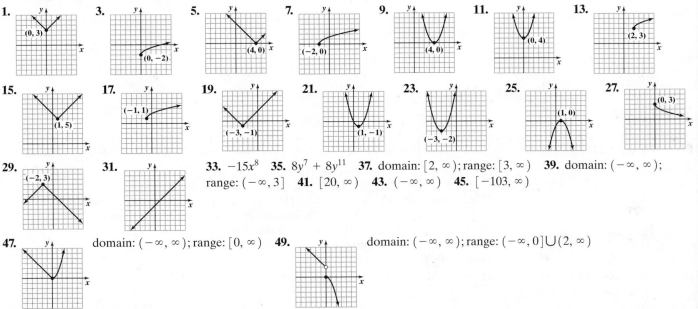

**1.** (0, 3) **3.** (0, −2) **5.** (4, 0) **7.** (−2, 0) **9.** (4, 0) **11.** (0, 4) **13.** (2, 3)

**15.** (1, 5) **17.** (−1, 1) **19.** (−3, −1) **21.** (1, −1) **23.** (−3, −2) **25.** (1, 0) **27.** (0, 3)

**29.** (−2, 3) **31.** **33.** $-15x^8$ **35.** $8y^7 + 8y^{11}$ **37.** domain: $[2, \infty)$; range: $[3, \infty)$ **39.** domain: $(-\infty, \infty)$; range: $(-\infty, 3]$ **41.** $[20, \infty)$ **43.** $(-\infty, \infty)$ **45.** $[-103, \infty)$

**47.** domain: $(-\infty, \infty)$; range: $[0, \infty)$ **49.** domain: $(-\infty, \infty)$; range: $(-\infty, 0] \cup (2, \infty)$

**Chapter 3 Vocabulary Check** **1.** relation **2.** line **3.** linear inequality **4.** Standard **5.** range **6.** Parallel **7.** Slope-intercept **8.** function **9.** slope **10.** perpendicular **11.** $y$ **12.** domain **13.** linear function **14.** $x$ **15.** point-slope

### Chapter 3 Review

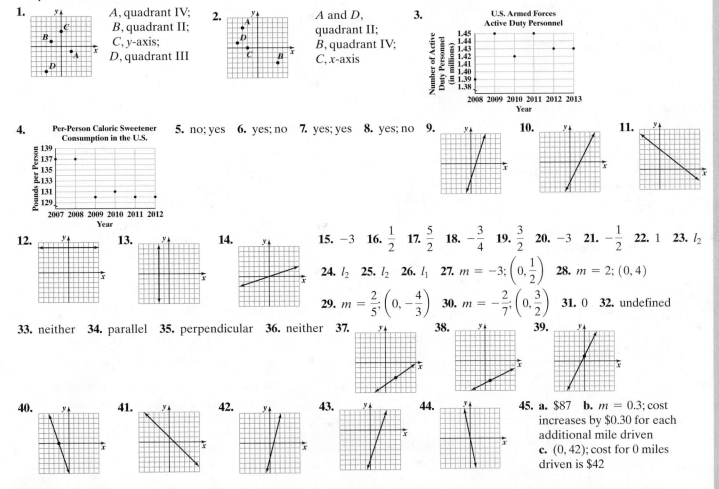

**1.** $A$, quadrant IV; $B$, quadrant II; $C$, $y$-axis; $D$, quadrant III **2.** $A$ and $D$, quadrant II; $B$, quadrant IV; $C$, $x$-axis **3.** U.S. Armed Forces Active Duty Personnel

**4.** Per-Person Caloric Sweetener Consumption in the U.S. **5.** no; yes **6.** yes; no **7.** yes; yes **8.** yes; no **9.** **10.** **11.**

**12.** **13.** **14.** **15.** $-3$ **16.** $\dfrac{1}{2}$ **17.** $\dfrac{5}{2}$ **18.** $-\dfrac{3}{4}$ **19.** $\dfrac{3}{2}$ **20.** $-3$ **21.** $-\dfrac{1}{2}$ **22.** $1$ **23.** $l_2$

**24.** $l_2$ **25.** $l_2$ **26.** $l_1$ **27.** $m = -3$; $\left(0, \dfrac{1}{2}\right)$ **28.** $m = 2$; $(0, 4)$

**29.** $m = \dfrac{2}{5}$; $\left(0, -\dfrac{4}{3}\right)$ **30.** $m = -\dfrac{2}{7}$; $\left(0, \dfrac{3}{2}\right)$ **31.** $0$ **32.** undefined

**33.** neither **34.** parallel **35.** perpendicular **36.** neither **37.** **38.** **39.**

**40.** **41.** **42.** **43.** **44.** **45. a.** $87 **b.** $m = 0.3$; cost increases by $0.30 for each additional mile driven **c.** $(0, 42)$; cost for 0 miles driven is $42

**46. a.** $149   **b.** $m = 0.4$; cost increases $0.40 for each additional mile driven   **c.** $(0, 19)$; cost for 0 miles driven is $19.

**47.** $x = -4$   **48.** $y = -1$   **49.** $y = -\dfrac{1}{2}x - 4$   **50.** $y = 3x + 14$   **51.** $y = \dfrac{3}{4}x + \dfrac{7}{2}$   **52.** $y = -2x - 2$

**53. a.** $y = 12{,}000x + 126{,}000$   **b.** $342,000   **54. a.** $y = -1600x + 20{,}700$   **b.** $11,100   **55.**   **56.**

**57.**   **58.**   **59.**   **60.**   **61.**   **62.**

**63.** domain: $\left\{-\dfrac{1}{2}, 6, 0, 25\right\}$; range: $\left\{\dfrac{3}{4}, 0.65, -12, 25\right\}$; function   **64.** domain: $\left\{\dfrac{3}{4}, 0.65, -12, 25\right\}$; range: $\left\{-\dfrac{1}{2}, 6, 0, 25\right\}$; function

**65.** domain: $\{2, 4, 6, 8\}$; range: $\{2, 4, 5, 6\}$; not a function   **66.** domain: $\{$triangle, square, rectangle, parallelogram$\}$; range: $\{3, 4\}$; function   **67.** $-3$   **68.** $0$   **69.** $18$   **70.** $9$   **71.** $-3$   **72.** $0$   **73.** $381$ lb   **74.** $5080$ lb

**75.**   **76.**   **77.** domain: $(-\infty, \infty)$; range: $(-\infty -1] \cup [1, \infty)$; not a function   **78.** domain: $\{-3\}$; range: $(-\infty, \infty)$; not a function   **79.** domain: $(-\infty, \infty)$; range: $\{4\}$; function   **80.** domain: $[-1, 1]$; range: $[-1, 1]$; not a function   **81.** domain: $\{2\}$; range: $(-\infty, \infty)$   **82.** domain: $(-\infty, \infty)$; range: $(-\infty, \infty)$   **83.** domain: $[-4, 4]$; range: $[-1, 5]$   **84.** domain: $(-\infty, \infty)$; range: $\{-5\}$

**85.**   **86.**   **87.**   **88.**   **89.**   **90.**

**91.**   **92.**   **93.**   **94.** $x = -2$   **95.** $y = 5$   **96.** $y = 2x - 12$   **97.** $y = -11x - 52$   **98.** $y = -\dfrac{3}{2}x - 8$   **99.** $y = -\dfrac{3}{2}x - 1$   **100.** domain: $(-\infty, \infty)$; range: $(-\infty, 0]$   **101.** domain: $(-\infty, \infty)$; range: $(-\infty, \infty)$

**102.**   **103.**   **104.**   **105.**

## Chapter 3 Test

**1.**   $A$ is in quadrant IV. $B$ is on the $x$-axis, no quadrant. $C$ is in quadrant II.

**2.**   **3.**   **4.**   **5.**

**6.**   **7.** $m = -\dfrac{3}{2}$   **8.** $m = -\dfrac{1}{4}$; $\left(0, \dfrac{2}{3}\right)$   **9.** C   **10.** A   **11.** B   **12.** D   **13.** $y = -8$   **14.** $x = -4$   **15.** $y = -2$

**16.** $3x + y = 11$   **17.** $5x - y = 2$   **18.** $y = -\dfrac{1}{2}x$   **19.** $y = -\dfrac{1}{3}x + \dfrac{5}{3}$   **20.** $y = -\dfrac{1}{2}x - \dfrac{1}{2}$   **21.** neither

**22. a.** 75 games   **b.** 54 games   **c.** $120 million   **d.** 0.66; Every million dollars spent on payroll increases winnings by 0.66 game.   **23.**   **24.**   **25.**   **26.** domain: $(-\infty, \infty)$; range: $\{5\}$; function   **27.** domain: $\{-2\}$; range: $(-\infty, \infty)$; not a function   **28.** domain: $(-\infty, \infty)$; range: $[0, \infty)$; function   **29.** domain: $(-\infty, \infty)$; range: $(-\infty, \infty)$; function

**30.**  domain: $(-\infty, \infty)$; **31.** **32.** domain: $(-\infty, \infty)$; **33.**
range: $(-3, \infty)$  range: $(-\infty, -1]$

**Cumulative Review   1.** $\{2, 3, 4, 5\}$; Sec. 1.2, Ex. 1   **2.** $\{-1, 0, 1, 2\}$   **3.** $\{101, 102, 103, \dots\}$; Sec. 1.2, Ex. 2   **4. a.** 7   **b.** 0

**c.** $-10$   **5.** $-\dfrac{1}{9}$; Sec. 1.3, Ex. 21   **6.** undefined   **7.** $\dfrac{4}{7}$; Sec. 1.3; Ex. 22   **8. a.** 98   **b.** 98   **9.** 3; Sec. 1.4, Ex. 1   **10.** $\dfrac{7}{15}$

**11.** $-2.7$; Sec. 1.4, Ex. 4   **12.** $-102.86$   **13.** 23; Sec. 1.5, Ex. 5   **14.** 22   **15.** $2^7$; Sec. 1.6, Ex. 1   **16.** $-\dfrac{3}{2}x^{13}$   **17.** $y^7$; Sec. 1.6, Ex. 3

**18.** $4.86a^8b^{13}$   **19.** $125x^6$; Sec. 1.7, Ex. 5   **20.** $\dfrac{y^8}{81}$   **21.** $64y^2$; Sec. 1.7, Ex. 8   **22.** $\dfrac{c^{15}}{125a^9b^6}$   **23.** $\{0.4\}$; Sec. 2.1, Ex. 4

**24.** $\{x \mid x \text{ is a real number}\}$   **25.** 23, 49, 92; Sec. 2.2, Ex. 3   **26.** 69, 71, 73   **27.** \$11,607.55; Sec. 2.3, Ex. 4   **28. a.** $-7$   **b.** $\dfrac{5}{6}$

**c.** $-13$   **29.** $[2, \infty)$; Sec. 2.4, Ex.1   **30.** $\left(-\infty, -\dfrac{3}{2}\right)$   **31.** $(0.5, 3]$

; Sec. 2.4, Ex. 3   **32.** $(-25, \infty)$   **33.** $(-\infty, 4)$; Sec. 2.5, Ex. 2   **34.** $(-\infty, \infty)$   **35.** $\{4\}$; Sec. 2.6, Ex. 9

**36.** $\left\{1, -\dfrac{1}{5}\right\}$   **37.** $(-\infty, -4) \cup (10, \infty)$; Sec. 2.7, Ex. 6   **38.** $[-2, 18]$   **39.**  ; Sec. 3.1, Ex. 5   **40.** $m = -\dfrac{7}{2}$; $(0, 5)$

**41.** $m = 3$; Sec. 3.2, Ex. 3   **42.** $m = 3$   **43.** $y = \dfrac{2}{3}x - 7$; Sec. 3.3, Ex. 3   **44.** $x + y = 3$   **45.** $y = 3$; Sec. 3.4, Ex. 3   **46.** $x = -2$

**47.** 5; Sec. 3.6, Ex. 16   **48.** $m = -2.3$   **49.** $-2$; Sec. 3.6, Ex. 15   **50.** no

# Chapter 4 Systems of Equations and Inequalities

## Section 4.1

**Calculator Explorations   1.** $(2.11, 0.17)$   **3.** $(-8.20, -6.30)$

**Vocabulary, Readiness & Video Check   1.** B   **3.** A   **5.** The ordered pair must be a solution of *both* equations of the system in order to be a solution of the system.   **7.** Solve one equation for a variable. Next be sure to substitute this expression for the variable into the *other* equation.

**Exercise Set 4.1   1.** yes   **3.** no   **5.** yes   **7.** no   **9.** **11.** **13.** ∅ **15.**

**17.** $(2, 8)$   **19.** $(0, -9)$   **21.** $(1, -1)$   **23.** $\{(x, y) \mid x = -3y + 4\}$   **25.** $\left(\dfrac{5}{2}, \dfrac{5}{4}\right)$   **27.** $(1, -2)$   **29.** $(9, 9)$   **31.** $(7, 2)$   **33.** ∅

**35.** $\{(x, y) \mid 3x + y = 1\}$   **37.** $\left(\dfrac{3}{2}, 1\right)$   **39.** $(2, -1)$   **41.** $(-5, 3)$   **43.** ∅   **45.** $\left(\dfrac{1}{2}, \dfrac{1}{5}\right)$   **47.** $(8, 2)$   **49.** $\{(x, y) \mid x = 3y + 2\}$

**51.** $(-5, 3)$   **53.** $(3, 2)$   **55.** $(7, -3)$   **57.** ∅   **59.** $(3, 4)$   **61.** $\left(-\dfrac{1}{4}, \dfrac{1}{2}\right)$   **63.** $(-2, 1)$   **65.** $(1.2, -3.6)$   **67.** true   **69.** false

**71.** $6y - 4z = 25$   **73.** $x + 10y = 2$   **75.** no solution   **77.** infinite number of solutions   **79.** no; answers may vary

**81.** 5000 DVDs; \$21   **83.** supply greater than demand   **85.** $(1875, 4687.5)$   **87.** makes money   **89.** for $x$-values greater than 1875

**91.** answers may vary; one possibility: $\begin{cases} -2x + y = 1 \\ \phantom{-}x - 2y = -8 \end{cases}$   **93. a.** Consumption of bottled water is increasing, while consumption of carbonated soft drinks is decreasing.   **b.** $(18, 33)$   **c.** In the year 2020, per capita bottled water consumption will be about the same as per capita carbonated soft drink consumption.   **95.** $\left(\dfrac{1}{4}, 8\right)$   **97.** $\left(\dfrac{1}{3}, \dfrac{1}{2}\right)$   **99.** $\left(\dfrac{1}{4}, -\dfrac{1}{3}\right)$   **101.** ∅

## Section 4.2

**Vocabulary, Readiness & Video Check   1.** a, b, d   **3.** yes; answers may vary   **5.** Once we have one equation in two variables, we need to get another equation in the *same* two variables, giving us a system of two equations in the same two variables. We solve this new system to find the value of two variables. We then substitute these values into an original equation to find the value of the third variable.

**Exercise Set 4.2**   **1.** $(-1, 5, 2)$   **3.** $(-2, 5, 1)$   **5.** $(-2, 3, -1)$   **7.** $\{(x, y, z) \mid x - 2y + z = -5\}$   **9.** $\varnothing$   **11.** $(0, 0, 0)$
**13.** $(-3, -35, -7)$   **15.** $(6, 22, -20)$   **17.** $\varnothing$   **19.** $(3, 2, 2)$   **21.** $\{(x, y, z) \mid x + 2y - 3z = 4\}$   **23.** $(-3, -4, -5)$   **25.** $\left(0, \dfrac{1}{2}, -4\right)$
**27.** $(12, 6, 4)$   **29.** $\{5\}$   **31.** $\left\{-\dfrac{5}{3}\right\}$   **33.** 15 and 30   **35.** answers may vary   **37.** answers may vary   **39.** $(1, 1, -1)$   **41.** $(1, 1, 0, 2)$
**43.** $(1, -1, 2, 3)$   **45.** answers may vary

## Section 4.3

**Vocabulary, Readiness & Video Check**   **1.** Up to now we've been choosing one variable/unknown and translating to one equation. To solve this example by a system of equations, we'll choose two variables to represent two unknowns and translate to two equations. **3.** The ordered triple still needs to be interpreted in the context of the application. Each value actually represents the angle measure of a triangle, in degrees.

**Exercise Set 4.3**   **1.** 10 and 8   **3. a.** Ford class: 1106 ft; Nimitz class: 1092 ft   **b.** 3.7 football fields   **5.** plane: 520 mph; wind: 40 mph
**7.** 20 qt of 4%; 40 qt of 1%   **9.** United Kingdom: 34,660 students; Italy: 29,645 students   **11.** 9 large frames; 13 small frames
**13.** $-10$ and $-8$   **15. a.** 2014   **b.** answers may vary   **17.** tablets: $0.80; pens: $0.20   **19.** speed of plane: 630 mph; speed of wind: 90 mph
**21. a.** answers may vary   **b.** 2007   **23.** 28 cm; 28 cm; 37 cm   **25.** 600 mi   **27.** $x = 75; y = 105$   **29.** 625 units   **31.** 3000 units
**33.** 1280 units   **35. a.** $R(x) = 450x$   **b.** $C(x) = 200x + 6000$   **c.** 24 desks   **37.** 2 units of Mix A; 3 units of Mix B; 1 unit of Mix C
**39.** 5 in.; 7 in.; 7 in.; 10 in.   **41.** 18, 13, and 9   **43.** free throws: 90; two-point field goals: 161; three-point field goals: 72
**45.** $x = 60; y = 55; z = 65$   **47.** $5x + 5z = 10$   **49.** $-5y + 2z = 2$   **51.** 2010: 1,593,081; 2013: 1,107,699   **53.** $a = 3, b = 4, c = -1$
**55.** $a = 1.6, b = 23.6, c = 1153.8$; 2015: 1867.8 thousand students

**Integrated Review**   **1.** C   **2.** D   **3.** A   **4.** B   **5.** $(1, 3)$   **6.** $\left(\dfrac{1}{5}, \dfrac{4}{5}\right)$   **7.** $(2, -1)$   **8.** $(5, 2)$   **9.** $\left(\dfrac{3}{2}, 1\right)$   **10.** $\left(-2, \dfrac{3}{4}\right)$   **11.** $\varnothing$
**12.** $\{(x, y) \mid 2x - 5y = 3\}$   **13.** $\left(1, \dfrac{1}{3}\right)$   **14.** $\left(3, \dfrac{3}{4}\right)$   **15.** $(-1, 3, 2)$   **16.** $(1, -3, 0)$   **17.** $\varnothing$   **18.** $\{(x, y, z) \mid x - y + 3z = 2\}$
**19.** $\left(2, 5, \dfrac{1}{2}\right)$   **20.** $\left(1, 1, \dfrac{1}{3}\right)$   **21.** 19 and 27   **22.** 70°; 70°; 100°; 120°

## Section 4.4

**Vocabulary, Readiness & Video Check**   **1.** matrix   **3.** row   **5.** false   **7.** true   **9.** Two rows may be interchanged, the elements of any row may be multiplied/divided by the same nonzero number, the elements of any row may be multiplied/divided by the same nonzero number and added to their corresponding elements in any other row; rows were not interchanged in Example 1.

**Exercise Set 4.4**   **1.** $(2, -1)$   **3.** $(-4, 2)$   **5.** $\varnothing$   **7.** $\{(x, y) \mid 3x - 3y = 9\}$   **9.** $(-2, 5, -2)$   **11.** $(1, -2, 3)$   **13.** $(4, -3)$
**15.** $(2, 1, -1)$   **17.** $(9, 9)$   **19.** $\varnothing$   **21.** $\varnothing$   **23.** $(1, -4, 3)$   **25.** function   **27.** not a function   **29.** c   **31. a.** 2006
**b.** no; answers may vary   **c.** no; it has a positive slope   **d.** answers may vary   **33.** answers may vary

## Section 4.5

**Vocabulary, Readiness & Video Check**   **1.** system   **3.** corner   **5.** No; we can choose any test point except a point on the second inequality's own boundary line.

**Exercise Set 4.5**   **1.**    **3.**    **5.**    **7.**    **9.**    **11.**

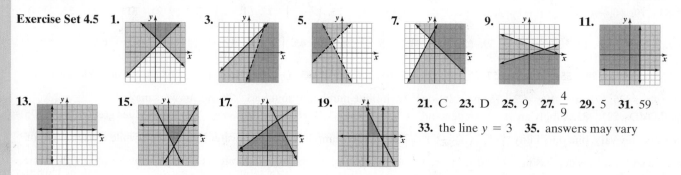

**13.**    **15.**    **17.**    **19.**    **21.** C   **23.** D   **25.** 9   **27.** $\dfrac{4}{9}$   **29.** 5   **31.** 59
**33.** the line $y = 3$   **35.** answers may vary

**Chapter 4 Vocabulary Check**   **1.** system of equations   **2.** solution   **3.** consistent   **4.** triple   **5.** inconsistent   **6.** matrix
**7.** element   **8.** row   **9.** column

**Chapter 4 Review**   **1.** $(-3, 1)$   **2.** $\left(0, \dfrac{2}{3}\right)$   **3.** $\varnothing$   **4.** $\{(x, y) \mid 3x - 6y = 12\}$   **5.** $\left(3, \dfrac{8}{3}\right)$   **6.** 1500 backpacks   **7.** $(2, 0, 2)$
**8.** $(2, 0, -3)$   **9.** $\left(-\dfrac{1}{2}, \dfrac{3}{4}, 1\right)$   **10.** $(-1, 2, 0)$   **11.** $\varnothing$   **12.** $(5, 3, 0)$   **13.** $(1, 1, -2)$   **14.** $(3, 1, 1)$   **15.** 10, 40, and 48

**16.** 63 and 21 **17.** 58 mph; 65 mph **18.** width: 37 ft; length: 111 ft **19.** 20 L of 10% solution; 30 L of 60% solution
**20.** 30 lb of creme-filled; 5 lb of chocolate-covered nuts; 10 lb of chocolate-covered raisins **21.** 17 pennies; 20 nickels; 16 dimes
**22.** larger investment: 9.5%; smaller investment: 7.5% **23.** two sides: 22 cm each; third side: 29 cm **24.** 120, 115, and 60

**25.** $(-3, 1)$ **26.** $\{(x, y) \mid 3x - 6y = 12\}$ **27.** $\left(-\dfrac{2}{3}, 3\right)$ **28.** $\left(\dfrac{1}{3}, \dfrac{7}{6}\right)$ **29.** $\left(\dfrac{5}{4}, \dfrac{5}{8}\right)$ **30.** $(-7, -15)$ **31.** $(1, 3)$ **32.** $(2, 1)$

**33.** $(1, 2, 3)$ **34.** $(2, 0, -3)$ **35.** $(3, -2, 5)$ **36.** $(-1, 2, 0)$ **37.** $(1, 1, -2)$ **38.** $\varnothing$ **39.** **40.**

**47.** $\left(\dfrac{7}{3}, -\dfrac{8}{3}\right)$ **48.** $(10, -4)$ **49.** $\{(x, y) \mid 5x - 2y = 10\}$ **50.** $\varnothing$ **51.** $(-1, 3, 5)$ **52.** 94 and 33 **53.** 28 units, 56 units, 42 units

**54.** **55.** during May

**Chapter 4 Test** **1.** $(1, 3)$ **2.** $\varnothing$ **3.** $(2, -3)$ **4.** $\{(x, y) \mid 10x + 4y = 10\}$ **5.** $(-1, -2, 4)$ **6.** $\varnothing$ **7.** $\left(\dfrac{7}{2}, -10\right)$
**8.** $\{(x, y) \mid x - y = -2\}$ **9.** $(5, -3)$ **10.** $(-1, -1, 0)$ **11.** 53 double rooms; 27 single rooms **12.** 5 gal of 10%; 15 gal of 20%
**13.** 800 packages **14.** $23°, 45°, 112°$ **15.**

**Cumulative Review** **1.** true; Sec. 1.2, Ex. 4 **2.** false **3.** false; Sec. 1.2, Ex. 7 **4.** true **5.** $<$; Sec. 1.3, Ex. 4 **6.** $>$
**7.** $<$; Sec. 1.3, Ex. 7 **8.** $=$ **9.** $<$; Sec. 1.3, Ex. 11 **10.** $<$ **11.** $\dfrac{1}{7}$; Sec. 1.4, Ex. 3 **12.** 7.8 **13.** $-8$; Sec. 1.4, Ex. 5 **14.** $-\dfrac{3}{8}$

**15.** $-1$; Sec. 1.5, Ex. 3 **16. a.** $-5$ **b.** $-24$ **17.** $\dfrac{1}{3x}$; Sec. 1.6, Ex. 16 **18.** $\dfrac{1}{81}$ **19.** $\dfrac{11}{18}$; Sec. 1.6, Ex. 17 **20.** $x^{16}$ **21.** solution; Sec. 2.1, Ex. 1

**22.** $\{0\}$ **23.** \$2350; Sec. 2.2, Ex. 4 **24.** $100°, 100°, 110°, 50°$ **25.** $b = \dfrac{2A - Bh}{h}$; Sec. 2.3, Ex. 3 **26.** $\{m \mid m \text{ is a real number}\}$
**27.** $(-2, \infty)$; Sec. 2.4, Ex. 5 **28.** $(0, \infty)$ **29.** $\{4, 6\}$; Sec. 2.5, Ex. 1 **30.** $\{2, 3, 4, 5, 6, 8\}$ **31.** $\{0\}$; Sec. 2.6, Ex. 5 **32.** $\{1, 9\}$

**33.** ; Sec. 3.1, Ex. 8 **34.** $(-\infty, -3) \cup (2, \infty)$; **35.** slope: $\dfrac{3}{4}$; $y$-intercept: $(0, -1)$; Sec. 3.2, Ex. 4 **36. a.** quadrant III
**b.** quadrant IV **c.** $y$-axis **37.** $y = -3x - 2$; Sec. 3.4, Ex. 1 **38. a.** 75 **b.** 12 **39.** function; Sec. 3.6, Ex. 7
**40.** $m = \dfrac{3}{2}$

**41.** ; Sec. 3.5, Ex. 3 **42.** $y = -2x + 2$ **43.** $(0, -5)$; Sec. 4.1, Ex. 10 **44.** $(-1, 3)$ **45.** $\varnothing$; Sec. 4.2, Ex. 2 **46.** $(-1, 2, 5)$
**47.** 7 and 11; Sec. 4.3, Ex. 2 **48.** $\varnothing$

# Chapter 5 Polynomials and Polynomial Functions

### Section 5.1

**Calculator Explorations** **1.** $x^3 - 4x^2 + 7x - 8$ **3.** $-2.1x^2 - 3.2x - 1.7$ **5.** $7.69x^2 - 1.26x + 5.3$

**Vocabulary, Readiness & Video Check** **1.** coefficient **3.** binomial **5.** trinomial **7.** degree **9.** $6x$ **11.** $2y$ **13.** $7xy^2 - y^2$
**15.** We are finding the degree of the polynomial, which is the greatest degree of any of its terms, so we find the degree of each term first. **17.** combine like terms **19.** evaluating a polynomial at a given replacement value

**Exercise Set 5.1** **1.** 0 **3.** 2 **5.** 3 **7.** 3 **9.** 9 **11.** binomial of degree 1 **13.** trinomial of degree 2 **15.** monomial of degree 3

**17.** degree 4; none of these **19.** $-7y^2 + 6y$ **21.** $x^2y - 5x - \dfrac{1}{2}$ **23.** $6x^2 - xy + 16y^2$ **25.** $18y^2 - 17$ **27.** $3x^2 - 3xy + 6y^2$

**29.** $x^2 - 4x + 8$ **31.** $12x^3y + 8x + 8$ **33.** $4.5x^3 + 0.2x^2 - 3.8x + 9.1$ **35.** $y^2 + 3$ **37.** $-2x^2 + 5x$ **39.** $7y^2 - 12y - 3$

**41.** $7x^3 + 4x^2 + 8x - 10$ **43.** $-20y^2 + 3yx$ **45.** $-3x^2 + 3$ **47.** $2y^4 - 5y^2 + x^2 + 1$ **49.** $5x^2 - 9x - 3$ **51.** $3x^2 + x + 18$

**53.** $4x - 13$ **55.** $-x^3 + 8a - 12$ **57.** $14ab + 10a^2b - 18a^2 + 12b^2$ **59.** $5x^2 + 22x + 16$ **61.** 0 **63.** $8xy^2 + 2x^3 + 3x^2 - 3$

**65.** $3x^2 - 9x + 15$ **67.** $15x^2 + 8x - 6$ **69.** $\dfrac{1}{3}x^2 - x + 1$ **71.** 57 **73.** 499 **75.** 1 **77.** 202 sq in. **79. a.** 284 ft

**b.** 536 ft **c.** 1376 ft **d.** 1296 ft **e.** answers may vary **f.** 19 sec **81. a.** 630 ft **b.** 630.25 ft **c.** 501 ft **d.** 255 ft

**e.** answers may vary **f.** 6 sec **83.** \$80,000 **85.** $-14z + 42y$ **87.** $-15y^2 - 10y + 35$ **89.** a and c

**91.** $(12x - 1.7) - (15x + 6.2) = 12x - 1.7 - 15x - 6.2 = -3x - 7.9$ **93.** answers may vary **95.** answers may vary

**97. a.** \$485.85 billion **b.** \$579.6 billion **c.** \$747.28 billion **d.** answers may vary **99.** $4x^2 - 3x + 6$

**101.** $P(a) = 2a - 3; P(-x) = -2x - 3; P(x + h) = 2x + 2h - 3$ **103.** $12z^{5x} + 13z^{2x} - 2z$ **105.** $(2x^2 + 7xy + 10y^2)$ units

## Section 5.2

**Calculator Explorations** **1.** $x^2 - 16$ **3.** $9x^2 - 42x + 49$ **5.** $5x^3 - 14x^2 - 13x - 2$

**Vocabulary, Readiness & Video Check** **1.** $3x^6$; b **3.** $x^2 - 49$; b **5.** $(a + 1)^2 + 1$; d **7.** distributive property, product rule
**9.** FOIL order or distributive property **11.** Multiply two at a time and multiply as usual, using patterns if you recognize them. Keep multiplying until you are through.

**Exercise Set 5.2** **1.** $-12x^5$ **3.** $86a^5b^8c^3$ **5.** $12x^2 + 21x$ **7.** $-24x^2y - 6xy^2$ **9.** $-4a^3bx - 4a^3by + 12ab$ **11.** $2x^2 - 2x - 12$

**13.** $2x^4 + 3x^3 - 2x^2 + x + 6$ **15.** $15x^2 - 7x - 2$ **17.** $15m^3 + 16m^2 - m - 2$ **19.** $-30a^4b^4 + 36a^3b^2 + 36a^2b^3$ **21.** $x^2 + x - 12$

**23.** $10x^2 - 21xy + 8y^2$ **25.** $16x^2 - \dfrac{2}{3}x - \dfrac{1}{6}$ **27.** $5x^4 - 17x^2y^2 + 6y^4$ **29.** $x^2 + 8x + 16$ **31.** $36y^2 - 1$ **33.** $9x^2 - 6xy + y^2$

**35.** $49a^2b^2 - 9c^2$ **37.** $9x^2 - \dfrac{1}{4}$ **39.** $16b^2 + 32b + 16$ **41.** $4s^2 - 12s + 8$ **43.** $x^4 - 2x^2y^2 + y^4$ **45.** $x^4 - 8x^3 + 24x^2 - 32x + 16$

**47.** $x^4 - 625$ **49.** $-24a^2b^3 + 40a^2b^2 - 160a^2b$ **51.** $36x^2 + 12x + 1$ **53.** $25x^6 - 4y^2$ **55.** $10x^5 + 8x^4 + 2x^3 + 25x^2 + 20x + 5$

**57.** $9x^4 + 12x^3 - 2x^2 - 4x + 1$ **59.** $3x^2 + 8x - 3$ **61.** $9x^6 + 15x^4 + 3x^2 + 5$ **63.** $9x^2 + 6x + 1$ **65.** $9b^2 - 36y^2$ **67.** $49x^2 - 9$

**69.** $9x^3 + 30x^2 + 12x - 24$ **71.** $12x^3 - 2x^2 + 13x + 5$ **73.** $x^2y^2 - 4xy + 4$ **75.** $22a^3 + 11a^2 + 2a + 1$ **77.** $\dfrac{1}{3}n^2 - 7n + 18$

**79.** $45x^3 + 18x^2y - 5x - 2y$ **81.** $a^2 - 3a$ **83.** $a^2 + 2ah + h^2 - 3a - 3h$ **85.** $b^2 - 7b + 10$ **87.** $2x^2$ **89.** $\dfrac{10a^2b^3}{9}$ **91.** $\dfrac{2m^3}{3}$

**93.** $7y(3z - 2) + 1 = 21yz - 14y + 1$ **95.** answers may vary **97.** $30x^2y^{2n+1} - 10x^2y^n$ **99.** $x^{3a} + 5x^{2a} - 3x^a - 15$

**101. a.** $6x + 12$ **b.** $9x^2 + 36x + 35$ **c.** one operation is addition, the other is multiplication **103.** $(8x^2 - 12x + 4)$ sq in.

**105.** $\pi(25x^2 - 20x + 4)$ sq km or $(25\pi x^2 - 20\pi x + 4\pi)$ sq km **107. a.** $a^2 + 2ah + h^2 + 3a + 3h + 2$ **b.** $a^2 + 3a + 2$

**c.** $2ah + h^2 + 3h$ **109.** $5x^2 + 25x$ **111.** $x^4 - 4x^2 + 4$ **113.** $x^3 + 5x^2 - 2x - 10$

## Section 5.3

**Vocabulary, Readiness & Video Check** **1.** the common denominator **3.** The last number is the remainder and the other numbers are the coefficients of the variables in the quotient; the degree of the quotient is one less than the degree of the dividend.

**Exercise Set 5.3** **1.** $2a + 4$ **3.** $3ab + 4$ **5.** $2y + \dfrac{3y}{x} - \dfrac{2y}{x^2}$ **7.** $x + 1$ **9.** $2x - 8$ **11.** $x - \dfrac{1}{2}$ **13.** $2x^2 - \dfrac{1}{2}x + 5$ **15.** $2x^2 - 6$

**17.** $3x^3 + 5x + 4 - \dfrac{2x}{x^2 - 2}$ **19.** $2x^3 + \dfrac{9}{2}x^2 + 10x + 21 + \dfrac{42}{x - 2}$ **21.** $x + 8$ **23.** $x - 1$ **25.** $x^2 - 5x - 23 - \dfrac{41}{x - 2}$

**27.** $4x + 8 + \dfrac{7}{x - 2}$ **29.** $x^6y + \dfrac{2}{y} + 1$ **31.** $5x^2 - 6 - \dfrac{5}{2x - 1}$ **33.** $2x^2 + 2x + 8 + \dfrac{28}{x - 4}$ **35.** $2x^3 - 3x^2 + x - 4$

**37.** $3x^2 + 4x - 8 + \dfrac{20}{x + 1}$ **39.** $3x^2 + 3x - 3$ **41.** $x^2 + x + 1$ **43.** $-\dfrac{5y}{x} - \dfrac{15z}{x} - 25z$ **45.** $3x^4 - 2x$ **47.** 1 **49.** $-133$ **51.** 3

**53.** $-\dfrac{187}{81}$ **55.** $\dfrac{95}{32}$ **57.** $(-9, -1)$ **59.** $(-\infty, -8] \cup [1, \infty)$ **61.** yes **63.** no **65.** a or d **67.** $(x^4 + 2x^2 - 6)$ m **69.** $(3x - 7)$ in.

**71.** $(x^3 - 5x^2 + 2x - 1)$ cm **73.** $x^3 + \dfrac{5}{3}x^2 + \dfrac{5}{3}x + \dfrac{8}{3} + \dfrac{8}{3(x - 1)}$ **75.** $\dfrac{3}{2}x^3 + \dfrac{1}{4}x^2 + \dfrac{1}{8}x - \dfrac{7}{16} + \dfrac{1}{16(2x - 1)}$ **77.** $x^3 - \dfrac{2}{5}x$

**79.** $5x - 1 + \dfrac{6}{x}; x \neq 0$ **81.** $7x^3 + 14x^2 + 25x + 50 + \dfrac{102}{x - 2}; x \neq 2$ **83.** answers may vary **85.** answers may vary

**87.** $(x + 3)(x^2 + 4) = x^3 + 3x^2 + 4x + 12$ **89.** $2x^3 - 5x^2 - 31x + 33$ **91. a.** $m(x) = \dfrac{-71x^3 + 393x^2 - 232x + 1884}{468x^2 - 471x + 8560}$

**b.** $-0.17$

## Section 5.4

**Vocabulary, Readiness & Video Check** **1.** factoring **3.** least **5.** false **7.** false **9.** The GCF is that common variable raised to the smallest exponent in the list. **11.** Look for a GCF other than 1 or $-1$; if you have a four-term polynomial

**Exercise Set 5.4** **1.** 6 **3.** $7x$ **5.** $a^3$ **7.** $3x^2y$ **9.** $5xz^3$ **11.** $6(3x - 2)$ **13.** $4y^2(1 - 4xy)$ **15.** $2x^3(3x^2 - 4x + 1)$
**17.** $4ab(2a^2b^2 - ab + 1 + 4b)$ **19.** $(x + 3)(6 + 5a)$ **21.** $(z + 7)(2x + 1)$ **23.** $(6x^2 + 5)(3x - 2)$ **25.** no common factor other
than 1 **27.** $13x^2y^3(3x - 2)$ **29.** $(a + 2)(b + 3)$ **31.** $(a - 2)(c + 4)$ **33.** $(x - 2)(2y - 3)$ **35.** $(4x - 1)(3y - 2)$
**37.** $3(2x^3 + 3)$ **39.** $x^2(x + 3)$ **41.** $4a(2a^2 - 1)$ **43.** no common factor other than 1 **45.** $4xy(-5x + 4y^2)$ or $-4xy(5x - 4y^2)$
**47.** $5ab^2(2ab + 1 - 3b)$ **49.** $3b(3ac^2 + 2a^2c - 2a + c)$ **51.** $(y - 2)(4x - 3)$ **53.** $(2x + 3)(3y + 5)$ **55.** $(x + 3)(y - 5)$
**57.** $(2a - 3)(3b - 1)$ **59.** $(6x + 1)(2y + 3)$ **61.** $(n - 8)(2m - 1)$ **63.** $3x^2y^2(5x - 6)$ **65.** $(2x + 3y)(x + 2)$
**67.** $(5x - 3)(x + y)$ **69.** $(x^2 + 4)(x + 3)$ **71.** $(x^2 - 2)(x - 1)$ **73.** $x^2 - 3x - 10$ **75.** $x^2 + 5x + 6$ **77.** $y^2 - 4y + 3$
**79.** d **81.** $2\pi r(r + h)$ **83.** $A = 5600(1 + rt)$ **85.** answers may vary **87.** none **89.** answers may vary **91.** $A = P(1 + RT)$
**93.** $x^n(x^{2n} - 2x^n + 5)$ **95.** $2x^{3a}(3x^{5a} - x^{2a} - 2)$ **97. a.** $h(t) = -16t(t - 4)$ **b.** 48 **c.** answers may vary
**99.** $f(x) = 3(-18x^4 + 301x^3 - 1203x^2 + 176x + 30{,}393)$

## Section 5.5

**Vocabulary, Readiness & Video Check** **1.** 5 and 2 **3.** 8 and 3 **5.** Check by multiplying. If you get a middle term of $2x$, not $-2x$, switch the signs of your factors of $-24$. **7.** Write down the factors of the first and last terms. Try various combinations of these factors and look at the sum of the outer and inner products to see if you get the middle term of the trinomial. If not, try another combination of factors. **9.** Our factoring involves a substitution from the original expression, so we need to substitute back in order to get a factorization in terms of the original variable/expression and simplify if necessary.

**Exercise Set 5.5** **1.** $(x + 3)(x + 6)$ **3.** $(x - 8)(x - 4)$ **5.** $(x + 12)(x - 2)$ **7.** $(x - 6)(x + 4)$ **9.** $3(x - 2)(x - 4)$
**11.** $4z(x + 2)(x + 5)$ **13.** $2(x^2 - 12x - 32)$ **15.** $(5x + 1)(x + 3)$ **17.** $(2x - 3)(x - 4)$ **19.** prime polynomial
**21.** $(2x - 3)^2$ **23.** $2(3x - 5)(2x + 5)$ **25.** $y^2(3y + 5)(y - 2)$ **27.** $2x(3x^2 + 4x + 12)$ **29.** $(2x + y)(x - 3y)$
**31.** $2(7y + 2)(2y + 1)$ **33.** $(2x - 3)(x + 9)$ **35.** $(x^2 + 3)(x^2 - 2)$ **37.** $(5x + 8)(5x + 2)$ **39.** $(x^3 - 4)(x^3 - 3)$
**41.** $(a - 3)(a + 8)$ **43.** $(x - 27)(x + 3)$ **45.** $(x - 18)(x + 3)$ **47.** $3(x - 1)^2$ **49.** $(3x + 1)(x - 2)$ **51.** $(4x - 3)(2x - 5)$
**53.** $3x^2(2x + 1)(3x + 2)$ **55.** $(x + 7z)(x + z)$ **57.** $(x - 4)(x + 3)$ **59.** $3(a + 2b)^2$ **61.** prime polynomial
**63.** $(2x + 13)(x + 3)$ **65.** $(3x - 2)(2x - 15)$ **67.** $(x^2 - 6)(x^2 + 1)$ **69.** $x(3x + 1)(2x - 1)$ **71.** $(4a - 3b)(3a - 5b)$
**73.** $(3x + 5)^2$ **75.** $y(3x - 8)(x - 1)$ **77.** $2(x + 3)(x - 2)$ **79.** $(x + 2)(x - 7)$ **81.** $(2x^3 - 3)(x^3 + 3)$ **83.** $2x(6y^2 - z)^2$
**85.** $2xy(x + 3)(x - 2)$ **87.** $(x + 5y)(x + y)$ **89.** $x^2 - 9$ **91.** $4x^2 + 4x + 1$ **93.** $x^3 - 8$ **95.** $\pm 5, \pm 7$ **97.** $x(x + 4)(x - 2)$
**99. a.** 576 ft; 672 ft; 640 ft; 480 ft **b.** answers may vary **c.** $-16(t + 4)(t - 9)$ **101.** $(x^n + 2)(x^n + 8)$ **103.** $(x^n - 6)(x^n + 3)$
**105.** $(2x^n + 1)(x^n + 5)$ **107.** $(2x^n - 3)^2$ **109.** $x^2(x + 5)(x + 1)$ **111.** $3x(5x - 1)(2x + 1)$

## Section 5.6

**Vocabulary, Readiness & Video Check** **1.** $(9y)^2$ **3.** $(8x^3)^2$ **5.** $5^3$ **7.** $(2x)^3$ **9.** $(4x^2)^3$ **11.** See if the first term is a square, say $a^2$, and the last term is a square, say $b^2$. Check to see if the middle term is $2 \cdot a \cdot b$ or $-2 \cdot a \cdot b$. **13.** First rewrite the original binomial so that each term is some quantity cubed. Your answers will then vary, depending on your interpretation.

**Exercise Set 5.6** **1.** $(x + 3)^2$ **3.** $(2x - 3)^2$ **5.** $(2a + 3)^2$ **7.** $3(x - 4)^2$ **9.** $x^2(3y + 2)^2$ **11.** $(4x - 7y)^2$ **13.** $(x + 5)(x - 5)$
**15.** $\left(\dfrac{1}{3} + 2z\right)\left(\dfrac{1}{3} - 2z\right)$ **17.** $(y + 9)(y - 5)$ **19.** $4(4x + 5)(4x - 5)$ **21.** $(x + 2y + 3)(x + 2y - 3)$
**23.** $(x + 8 + x^2)(x + 8 - x^2)$ **25.** $(x + 3)(x^2 - 3x + 9)$ **27.** $(z - 1)(z^2 + z + 1)$ **29.** $(m + n)(m^2 - mn + n^2)$
**31.** $y^2(3 - x)(9 + 3x + x^2)$ **33.** $a(2b + 3a)(4b^2 - 6ab + 9a^2)$ **35.** $2(5y - 2x)(25y^2 + 10xy + 4x^2)$ **37.** $(x - 6)^2$
**39.** $2y(3x + 1)(3x - 1)$ **41.** $(3x + 7)(3x - 7)$ **43.** $(x^2 + 1)(x + 1)(x - 1)$ **45.** $(x^2 - y)(x^4 + x^2y + y^2)$
**47.** $(2x + 3y)(4x^2 - 6xy + 9y^2)$ **49.** $(2x + 1 + z)(2x + 1 - z)$ **51.** $3y^2(x^2 + 3)(x^4 - 3x^2 + 9)$ **53.** $\left(n - \dfrac{1}{3}\right)\left(n^2 + \dfrac{1}{3}n + \dfrac{1}{9}\right)$
**55.** $-16(y + 2)(y - 2)$ **57.** $(x - 5 + y)(x - 5 - y)$ **59.** $(ab + 5)(a^2b^2 - 5ab + 25)$ **61.** $\left(\dfrac{x}{5} + \dfrac{y}{3}\right)\left(\dfrac{x}{5} - \dfrac{y}{3}\right)$
**63.** $(x + y + 5)(x^2 + 2xy + y^2 - 5x - 5y + 25)$ **65.** $\{5\}$ **67.** $\left\{-\dfrac{1}{3}\right\}$ **69.** $\{0\}$ **71.** $\{5\}$ **73.** no; $x^2 - 4$ can be factored further
**75.** yes **77.** $\pi R^2 - \pi r^2 = \pi(R + r)(R - r)$ **79.** $x^3 - y^2x; x(x + y)(x - y)$ **81.** $c = 9$ **83.** $c = 49$
**85. a.** $(x + 1)(x^2 - x + 1)(x - 1)(x^2 + x + 1)$ **b.** $(x + 1)(x - 1)(x^4 + x^2 + 1)$ **c.** answers may vary **87.** $(x^n + 5)(x^n - 5)$
**89.** $(6x^n + 7)(6x^n - 7)$ **91.** $(x^{2n} + 4)(x^n + 2)(x^n - 2)$

**Integrated Review** **1.** $2y^2 + 2y - 11$ **2.** $-2z^4 - 6z^2 + 3z$ **3.** $x^2 - 7x + 7$ **4.** $7x^2 - 4x - 5$ **5.** $25x^2 - 30x + 9$
**6.** $x - 3$ **7.** $2x^3 - 4x^2 + 5x - 5 + \dfrac{8}{x + 2}$ **8.** $4x^3 - 13x^2 - 5x + 2$ **9.** $(x - 4 + y)(x - 4 - y)$ **10.** $2(3x + 2)(2x - 5)$

Answers to Selected Exercises

**11.** $x(x-1)(x^2+x+1)$　**12.** $2x(2x-1)$　**13.** $2xy(7x-1)$　**14.** $6ab(4b-1)$　**15.** $4(x+2)(x-2)$　**16.** $9(x+3)(x-3)$
**17.** $(3x-11)(x+1)$　**18.** $(5x+3)(x-1)$　**19.** $4(x+3)(x-1)$　**20.** $6(x+1)(x-2)$　**21.** $(2x+9)^2$　**22.** $(5x+4)^2$
**23.** $(2x+5y)(4x^2-10xy+25y^2)$　**24.** $(3x-4y)(9x^2+12xy+16y^2)$　**25.** $8x^2(2y-1)(4y^2+2y+1)$
**26.** $27x^2y(xy-2)(x^2y^2+2xy+4)$　**27.** $(x+5+y)(x^2+10x+25-xy-5y+y^2)$
**28.** $(y-1+3x)(y^2-2y+1-3xy+3x+9x^2)$　**29.** $(5a-6)^2$　**30.** $(4r+5)^2$　**31.** $7x(x-9)$　**32.** $(4x+3)(5x+2)$
**33.** $(a+7)(b-6)$　**34.** $20(x-6)(x-5)$　**35.** $(x^2+1)(x+1)(x-1)$　**36.** $5x(3x-4)$　**37.** $(5x-11)(2x+3)$
**38.** $9m^2n^2(5mn-3)$　**39.** $5a^3b(b^2-10)$　**40.** $x(x+1)(x^2-x+1)$　**41.** prime　**42.** $20(x+y)(x^2-xy+y^2)$
**43.** $10x(x-10)(x-11)$　**44.** $(3y-7)^2$　**45.** $a^3b(4b-3)(16b^2+12b+9)$　**46.** $(y^2+4)(y+2)(y-2)$
**47.** $2(x-3)(x^2+3x+9)$　**48.** $(2s-1)(r+5)$　**49.** $(y^4+2)(3y-5)$　**50.** prime　**51.** $100(z+1)(z^2-z+1)$
**52.** $2x(5x-2)(25x^2+10x+4)$　**53.** $(2b-9)^2$　**54.** $(a^4+3)(2a-1)$　**55.** $(y-4)(y-5)$　**56.** $(c-3)(c+1)$
**57.** $9-4x^2$; $(3+2x)(3-2x)$

## Section 5.7

**Calculator Explorations**　**1.** $\{-3.562, 0.562\}$　**3.** $\{-0.874, 2.787\}$　**5.** $\{-0.465, 1.910\}$

**Vocabulary, Readiness & Video Check**　**1.** $\{3, -5\}$　**3.** $\{3, -7\}$　**5.** $\{0, 9\}$　**7.** One side of the equation must be a factored polynomial and the other side must be zero.　**9.** Finding the $x$-intercepts of any equation in two variables means you let $y=0$ and solve for $x$. Doing this with a quadratic equation gives us an equation $=0$, which may be solved by factoring.

**Exercise Set 5.7**　**1.** $\left\{-3, \dfrac{4}{3}\right\}$　**3.** $\left\{-\dfrac{3}{4}, \dfrac{5}{2}\right\}$　**5.** $\{-3, -8\}$　**7.** $\left\{\dfrac{1}{4}, -\dfrac{2}{3}\right\}$　**9.** $\{1, 9\}$　**11.** $\left\{\dfrac{3}{5}, -1\right\}$　**13.** $\{0\}$　**15.** $\{6, -3\}$
**17.** $\left\{\dfrac{2}{5}, -\dfrac{1}{2}\right\}$　**19.** $\left\{\dfrac{3}{4}, -\dfrac{1}{2}\right\}$　**21.** $\left\{-2, 7, \dfrac{8}{3}\right\}$　**23.** $\{0, 3, -3\}$　**25.** $\{-1, 1, 2\}$　**27.** $\left\{-\dfrac{7}{2}, 10\right\}$　**29.** $\{0, 5\}$　**31.** $\{-3, 5\}$
**33.** $\left\{-\dfrac{1}{2}, \dfrac{1}{3}\right\}$　**35.** $\{-4, 9\}$　**37.** $\left\{\dfrac{4}{5}\right\}$　**39.** $\{-5, 0, 2\}$　**41.** $\left\{-3, 0, \dfrac{4}{5}\right\}$　**43.** $\varnothing$　**45.** $\{-7, 4\}$　**47.** $\{4, 6\}$　**49.** $\left\{-\dfrac{1}{2}\right\}$
**51.** $\{-4, -3, 3\}$　**53.** $\{-5, 0, 5\}$　**55.** $\{-6, 5\}$　**57.** $\left\{-\dfrac{1}{3}, 0, 1\right\}$　**59.** $\left\{-\dfrac{1}{3}, 0\right\}$　**61.** $\left\{-\dfrac{7}{8}\right\}$　**63.** $\left\{\dfrac{31}{4}\right\}$　**65.** $\{1\}$　**67.** $-11$ and $-6$,
or 6 and 11　**69.** 75 ft　**71.** 105 units　**73.** 12 cm and 9 cm　**75.** 2 in.　**77.** 10 sec　**79.** width: $7\dfrac{1}{2}$ ft; length: 12 ft　**81.** 10-in. square tier
**83.** E　**85.** F　**87.** B　**89.** $(-3, 0), (0, 2)$　**91.** $(-4, 0), (4, 0), (0, 2), (0, -2)$　**93.** $x-5=0$　or　$x+2=0$
$$x=5 \quad \text{or} \quad x=-2$$
**95.**
$$y(y-5)=-6$$
$$y^2-5y+6=0$$
$$(y-2)(y-3)=0$$
$$y-2=0 \quad \text{or} \quad y-3=0$$
$$y=2 \quad \text{or} \quad y=3$$
**97.** $\left\{-3, -\dfrac{1}{3}, 2, 5\right\}$　**99.** answers may vary　**101.** no; answers may vary
**103.** answers may vary

**Chapter 5 Vocabulary Check**　**1.** polynomial　**2.** Factoring　**3.** degree of a term　**4.** monomial　**5.** trinomial　**6.** quadratic equation　**7.** degree of a polynomial　**8.** binomial　**9.** 0　**10.** FOIL　**11.** synthetic division

**Chapter 5 Review**　**1.** 5　**2.** 1　**3.** $12x-6x^2-6x^2y$　**4.** $-4xy^3-3x^3y$　**5.** $4x^2+8y+6$　**6.** $-4x^2+10y^2$　**7.** $8x^2+2b-22$
**8.** $-4x^3+4x^2+16xy-9x+18$　**9.** $-x+5$　**10.** $12x^2y-7xy+3$　**11.** $x^2-6x+3$　**12.** $x^3+x-2xy^2-y-7$　**13.** 290
**14.** 8　**15.** 58　**16.** 110　**17.** $x^2+4x-6$　**18.** $-x^2+2x+3$　**19.** $(6x^2y-12x+12)$ cm　**20.** $(9x+22)$ in.
**21.** $-24x^3+36x^2-6x$　**22.** $-12a^2b^5-28a^2b^3-4ab^2$　**23.** $2x^2+x-36$　**24.** $9x^2a^2-24xab+16b^2$　**25.** $36x^3-11x^2-8x-3$
**26.** $2x^3+3x^2-12x+5$　**27.** $15x^2+18xy-81y^2$　**28.** $x^2+\dfrac{1}{3}x-\dfrac{2}{9}$　**29.** $x^4+18x^3+83x^2+18x+1$
**30.** $m^4+2m^3-3m^2-4m+4$　**31.** $9x^2-6xy+y^2$　**32.** $16x^2+72x+81$　**33.** $x^2-9y^2$　**34.** $16-9a^2+6ab-b^2$
**35.** $(9y^2-49z^2)$ sq units　**36.** $(121y^2+198y+81)$ sq units　**37.** $1+\dfrac{x}{2y}-\dfrac{9}{4xy}$　**38.** $\dfrac{3}{b}+4b$　**39.** $3x^3+9x^2+2x+6-\dfrac{2}{x-3}$
**40.** $2x^3+6x^2+17x+56+\dfrac{156}{x-3}$　**41.** $2x^3+2x-2$　**42.** $x^2+\dfrac{7}{2}x-\dfrac{1}{4}+\dfrac{15}{8\left(x-\dfrac{1}{2}\right)}$　**43.** $3x^2+2x-1$　**44.** $3x^2+6$
**45.** $3x^2+6x+24+\dfrac{44}{x-2}$　**46.** $4x^2-4x+2-\dfrac{5}{x+\dfrac{3}{2}}$　**47.** $x^4-x^3+x^2-x+1-\dfrac{2}{x+1}$　**48.** $x^2+3x+9-\dfrac{54}{x-3}$
**49.** $3x^3+13x^2+51x+204+\dfrac{814}{x-4}$　**50.** $3x^3-6x^2+10x-20+\dfrac{50}{x+2}$　**51.** $8x^2(2x-3)$　**52.** $12y(3-2y)$
**53.** $2ab(3b+4-2ab)$　**54.** $7ab(2ab-3b+1)$　**55.** $(a+3b)(6a-5)$　**56.** $(x-2y)(4x-5)$　**57.** $(x-6)(y+3)$
**58.** $(a-8)(b+4)$　**59.** $(p-5)(q-3)$　**60.** $(x^2-2)(x-1)$　**61.** $x(2y-x)$　**62.** $4(3y+x)(3y-x)$
**63.** $(x-18)(x+4)$　**64.** $(x-4)(x+20)$　**65.** $2(x-2)(x-7)$　**66.** $3(x+2)(x+9)$　**67.** $x(2x-9)(x+1)$

**68.** $(3x + 8)(x - 2)$  **69.** $(6x + 5)(x + 2)$  **70.** $(15x - 1)(x - 6)$  **71.** $2(2x - 3)(x + 2)$  **72.** $3(x - 2)(3x + 2)$
**73.** $(x + 6)^2(y - 3)(y + 1)$  **74.** $(x + 7)(x + 9)$  **75.** $(x^2 - 8)(x^2 + 2)$  **76.** $(x^2 - 2)(x^2 + 10)$  **77.** $(x + 10)(x - 10)$
**78.** $(x + 9)(x - 9)$  **79.** $2(x + 4)(x - 4)$  **80.** $6(x + 3)(x - 3)$  **81.** $(9 + x^2)(3 + x)(3 - x)$  **82.** $(4 + y^2)(2 + y)(2 - y)$
**83.** $(y + 7)(y - 3)$  **84.** $(x - 7)(x + 1)$  **85.** $(x + 6)(x^2 - 6x + 36)$  **86.** $(y + 8)(y^2 - 8y + 64)$  **87.** $(2 - 3y)(4 + 6y + 9y^2)$
**88.** $(1 - 4y)(1 + 4y + 16y^2)$  **89.** $6xy(x + 2)(x^2 - 2x + 4)$  **90.** $2x^2(x + 2y)(x^2 - 2xy + 4y^2)$  **91.** $(x - 1 + y)(x - 1 - y)$
**92.** $(x - 3 + 2y)(x - 3 - 2y)$  **93.** $(2x + 3)^2$  **94.** $(4a - 5b)^2$  **95.** $\left\{\dfrac{1}{3}, -7\right\}$  **96.** $\left\{-5, \dfrac{3}{8}\right\}$  **97.** $\left\{0, 4, \dfrac{9}{2}\right\}$  **98.** $\left\{-3, -\dfrac{1}{5}, 4\right\}$

**99.** $\{0, 6\}$  **100.** $\{-3, 0, 3\}$  **101.** $\left\{-\dfrac{1}{3}, 2\right\}$  **102.** $\{2, 10\}$  **103.** $\{-4, 1\}$  **104.** $\left\{\dfrac{7}{2}, -5\right\}$  **105.** $\{0, 6, -3\}$  **106.** $\{-21, 0, 2\}$

**107.** $\{0, -2, 1\}$  **108.** $\left\{-\dfrac{3}{2}, 0, \dfrac{1}{4}\right\}$  **109.** $-\dfrac{15}{2}$ or $7$  **110.** $\pi h(R + r)(R - r)$ cu units  **111.** width: 2 m; length: 8 m  **112.** 5 sec

**113.** $3x^3 + 13x^2 - 9x + 5$  **114.** $-2x^2 + 5x - 6.9$  **115.** $7x^2 + 5x + 2 + \dfrac{2}{x - 3}$  **116.** $49a^2b^2 - 7ab + \dfrac{1}{4}$  **117.** $-24$  **118.** $-10$
**119.** $6y^4(2y - 1)$  **120.** $(x^2 - 3)(y + 4)$  **121.** $2(3x + 1)(x - 6)$  **122.** $(4x + 3)^2(y - 20)(y + 1)$  **123.** $z^5(2z + 7)(2z - 7)$
**124.** $(x + 1)(x - 1)(5x^2 + 9)$  **125.** $\{0, 3\}$  **126.** $\{-2, 13\}$

**Chapter 5 Test**  **1.** $-5x^3 - 11x - 9$  **2.** $-12x^2y - 3xy^2$  **3.** $12x^2 - 5x - 28$  **4.** $25a^2 - 4b^2$  **5.** $36m^2 + 12mn + n^2$
**6.** $2x^3 - 13x^2 + 14x - 4$  **7.** $\dfrac{4xy}{3z} + \dfrac{3}{z} + \dfrac{1}{3x}$  **8.** $2x^4 + 2x - 2 + \dfrac{1}{2x - 1}$  **9.** $4x^3 - 15x^2 + 47x - 142 + \dfrac{425}{x + 3}$
**10.** $4x^2y(4x - 3y^3)$  **11.** $(x - 15)(x + 2)$  **12.** $(2y + 5)^2$  **13.** $3(2x + 1)(x - 3)$  **14.** $(2x + 5)(2x - 5)$
**15.** $(x + 4)(x^2 - 4x + 16)$  **16.** $3y(x + 3y)(x - 3y)$  **17.** $2(2y - 1)(4y^2 + 2y + 1)$  **18.** $(x + 3)(x - 3)(y - 3)$
**19.** $\left\{4, -\dfrac{8}{7}\right\}$  **20.** $\{-3, 8\}$  **21.** $\left\{-\dfrac{5}{2}, -2, 2\right\}$  **22.** $(x + 2y)(x - 2y)$  **23.** **a.** 960 ft  **b.** 953.44 ft  **c.** $-16(t - 11)(t + 5)$
**d.** 11 sec

**Cumulative Review**  **1.** $-2x + 1$; Sec. 1.5, Ex. 13  **2.** $-2x - \dfrac{7}{8}$  **3.** $2x + 23$; Sec. 1.5, Ex. 14  **4.** $16.3x - 0.8$  **5.** $6 \times 10^{-5}$; Sec. 1.7,
Ex. 20  **6.** **a.** $\dfrac{a^3}{64}$  **b.** $\dfrac{1}{a^4}$  **c.** $\dfrac{27}{8}$  **d.** $\dfrac{b^{17}}{9a^6}$  **7.** $\{0\}$; Sec. 2.1, Ex. 6  **8.** $\{8\}$  **9.** $y = \dfrac{7 + 2x}{3}$ or $y = \dfrac{7}{3} + \dfrac{2}{3}x$; Sec. 2.3, Ex. 2  **10.** 3 gal
**11.** $\left[\dfrac{5}{2}, \infty\right)$; Sec. 2.4, Ex. 9  **12.** $(-\infty, -5)$  **13.** $\left[-9, -\dfrac{9}{2}\right)$; Sec. 2.5, Ex. 5  **14.** $\left(-1, \dfrac{1}{3}\right]$  **15.** $\left\{\dfrac{3}{4}, 5\right\}$; Sec. 2.6, Ex. 8  **16.** $(-\infty, \infty)$

**17.** undefined; Sec. 3.2, Ex. 5  **18.** $m = -2$  **19.** ; Sec. 3.3, Ex. 1  **20.**  **21.** $y = \dfrac{5}{3}x + \dfrac{13}{3}$; Sec. 3.4, Ex. 6

**22.** $x = -3$  **23.** domain: $\{2, 0, 3\}$; range: $\{3, 4, -1\}$; Sec. 3.6, Ex. 1  **24.** $y = 3$  **25.** $-2$; Sec. 3.6, Ex. 15  **26.** 24  **27.** 5; Sec. 3.6,
Ex. 16  **28.** **a.** $14.1y^8$  **b.** $-6a^5b^3c^4$  **29.** ; Sec. 3.5, Ex. 1  **30.** 37.5 oz of 20% solution; 12.5 oz of 60% solution
**31.** $\left(-4, \dfrac{1}{2}\right)$; Sec. 4.1, Ex. 6  **32.** $(5, 2)$  **33.** $(-4, 2, -1)$; Sec. 4.2, Ex. 1
**34.** $(-3, -2, 5)$  **35.** $(-1, 2)$; Sec. 4.4, Ex. 1  **36.** $(-3, -4, -5)$

**37.** ; Sec. 4.5, Ex. 2  **38.** **a.** $8.25 \times 10^6$  **b.** $3.46 \times 10^{-5}$  **39.** 4; Sec. 5.1, Ex. 9  **40.** $-2x^2 - 5x$  **41.** $10x^2 - 8x$;
Sec. 5.2, Ex. 3  **42.** $10x^3 - 17x^2 + 10x - 2$  **43.** $-7x^3y^2 - 3x^2y^2 + 11xy$; Sec. 5.2, Ex. 5  **44.** $49x^2 - 7x + \dfrac{1}{4}$
**45.** $2x^2 - x + 4$; Sec. 5.3, Ex. 1  **46.** $3xy(2x + y)(2x - y)$  **47.** $(x - 5)(x - 7)$; Sec. 5.5, Ex. 2
**48.** $(5a - 1)(a + 3)$  **49.** $\left\{-5, \dfrac{1}{2}\right\}$; Sec. 5.7, Ex. 2  **50.** $\left\{4, -\dfrac{2}{3}\right\}$

# Chapter 6 Rational Expressions

### Section 6.1

**Calculator Explorations**  **1.** $\{x \mid x \text{ is a real number and } x \neq 6\}$  **3.** $\{x \mid x \text{ is a real number and } x \neq -2, x \neq 2\}$
**5.** $\left\{x \mid x \text{ is a real number and } x \neq -4, x \neq \dfrac{1}{2}\right\}$  **7.** $\{x \mid x \text{ is a real number}\}$

**Vocabulary, Readiness & Video Check   1.** rational   **3.** domain   **5.** 1   **7.** $\dfrac{-a}{b}; \dfrac{a}{-b}$   **9.** $\dfrac{xy}{10}$   **11.** $\dfrac{2y}{3x}$   **13.** $\dfrac{m^2}{36}$   **15.** Rational

expressions are fractions and are therefore undefined if the denominator is zero; the domain of a rational function consists of all real numbers except those for which the rational expression is undefined.   **17.** Yes, multiplying and also simplifying rational expressions often require polynomial factoring. Example 6 involves factoring out a GCF and factoring a difference of squares.   **19.** Since the domain consists of all allowable $x$-values and $x$ is defined as the number of years since the book was published, we do not allow $x$ to be negative for the context of this application.

**Exercise Set 6.1   1.** $\{x \mid x \text{ is a real number}\}$   **3.** $\{t \mid t \text{ is a real number and } t \neq 0\}$   **5.** $\{x \mid x \text{ is a real number and } x \neq 7\}$

**7.** $\left\{ x \middle| x \text{ is a real number and } x \neq \dfrac{1}{3} \right\}$   **9.** $\{x \mid x \text{ is a real number and } x \neq -2, x \neq 0, x \neq 1\}$   **11.** $\{x \mid x \text{ is a real number and } x \neq 2,$

$x \neq -2\}$   **13.** $1 - 2x$   **15.** $x - 3$   **17.** $\dfrac{9}{7}$   **19.** $x - 4$   **21.** $-1$   **23.** $-(x + 7)$   **25.** $\dfrac{2x + 1}{x - 1}$   **27.** $\dfrac{x^2 + 5x + 25}{2}$

**29.** $\dfrac{x - 2}{2x^2 + 1}$   **31.** $\dfrac{1}{3x + 5}$   **33.** $-\dfrac{4}{5}$   **35.** $-\dfrac{6a}{2a + 1}$   **37.** $\dfrac{3}{2(x - 1)}$   **39.** $\dfrac{x + 2}{x + 3}$   **41.** $\dfrac{3a}{5(a - b)}$   **43.** $\dfrac{1}{6}$   **45.** $\dfrac{4x - 3}{3x + 1}$   **47.** $\dfrac{x}{3}$

**49.** $\dfrac{4a^2}{a - b}$   **51.** $\dfrac{4}{(x + 2)(x + 3)}$   **53.** $\dfrac{3x + 7}{6x - 1}$   **55.** $\dfrac{1}{2}$   **57.** $-1$   **59.** $\dfrac{8(a - 2)}{3(a + 2)}$   **61.** $\dfrac{(x + 2)(x + 3)}{4}$   **63.** $\dfrac{2(x + 3)(x - 3)}{5(x^2 - 8x - 15)}$

**65.** $r^2 - rs + s^2$   **67.** $\dfrac{8}{x^2 y}$   **69.** $\dfrac{5x^2 - 2}{(x - 1)^2}$   **71.** $\dfrac{(y + 5)(2x - 1)}{(y + 2)(5x + 1)}$   **73.** $\dfrac{10}{3}, -8, -\dfrac{7}{3}$   **75.** $-\dfrac{17}{48}, \dfrac{2}{7}, -\dfrac{3}{8}$   **77. a.** \$200 million

**b.** \$500 million   **c.** \$300 million   **d.** $\{x \mid x \text{ is a real number}\}$   **79.** $\dfrac{7}{5}$   **81.** $\dfrac{1}{12}$   **83.** $\dfrac{11}{16}$   **85.** b and d   **87.** no; answers may vary

**89.** $\dfrac{5}{x - 2}$ sq m   **91.** $\dfrac{(x + 2)(x - 1)^2}{x^5}$ ft   **93.** answers may vary   **95. a.** 1   **b.** $-1$   **c.** neither   **d.** $-1$   **e.** $-1$   **f.** 1

**97.** answers may vary   **99.** $-1$   **101.** $\dfrac{1}{x^n - 4}$   **103.**

## Section 6.2

**Vocabulary, Readiness & Video Check   1.** a,b   **3.** c   **5.** $\dfrac{12}{y}$   **7.** $\dfrac{35}{y^2}$   **9.** We need to be sure we subtract the entire second numerator—that is, make sure we distribute the subtraction to each term in the second numerator.   **11.** numerator; one; value

**Exercise Set 6.2   1.** $\dfrac{-x + 4}{2x}$   **3.** $\dfrac{16}{y - 2}$   **5.** $-\dfrac{3}{xz^2}$   **7.** $\dfrac{x + 2}{x - 2}$   **9.** $x - 2$   **11.** $\dfrac{-1}{x - 2}$ or $\dfrac{1}{2 - x}$   **13.** $-\dfrac{5}{x}$   **15.** $35x$   **17.** $x(x + 1)$

**19.** $(x + 7)(x - 7)$   **21.** $6(x + 2)(x - 2)$   **23.** $(a + b)(a - b)^2$   **25.** $-4x(x + 3)(x - 3)$   **27.** $\dfrac{17}{6x}$   **29.** $\dfrac{35 - 4y}{14y^2}$

**31.** $\dfrac{-13x + 4}{(x + 4)(x - 4)}$   **33.** $\dfrac{3}{x + 4}$   **35.** 0   **37.** $-\dfrac{x + 2}{x - 1}$ or $\dfrac{x + 2}{1 - x}$   **39.** $\dfrac{-x + 1}{x - 2}$ or $\dfrac{x - 1}{2 - x}$   **41.** $\dfrac{y^2 + 2y + 10}{(y + 4)(y - 4)(y - 2)}$

**43.** $\dfrac{5(x^2 + x - 4)}{(3x + 2)(x + 3)(2x - 5)}$   **45.** $\dfrac{x^2 - 9x - 21}{(x + 3)(x + 1)(x - 2)}$   **47.** $\dfrac{x^2 - x + 20}{(2x - 5)(x + 3)(3x + 2)}$   **49.** $\dfrac{5a + 1}{(a + 1)^2(a - 1)}$   **51.** $\dfrac{3}{x^2 y^3}$

**53.** $-\dfrac{5}{x}$   **55.** $\dfrac{25}{6(x + 5)}$   **57.** $\dfrac{-2x - 1}{x^2(x - 3)}$ or $-\dfrac{2x + 1}{x^2(x - 3)}$   **59.** $\dfrac{b(2a - b)}{(a + b)(a - b)}$   **61.** $\dfrac{2(x + 8)}{(x + 2)^2(x - 2)}$   **63.** $\dfrac{3x^2 + 23x - 7}{(2x - 1)(x - 5)(x + 3)}$

**65.** $\dfrac{5 - 2x}{2(x + 1)}$   **67.** $\dfrac{2(x^2 + x - 21)}{(x + 3)^2(x - 3)}$   **69.** $\dfrac{6x}{(x + 3)(x - 3)^2}$   **71.** $\dfrac{4}{3}$   **73.** 10   **75.** $4 + x^2$   **77.** $\dfrac{2x - 3}{x^2 + 1} - \dfrac{x - 6}{x^2 + 1}$

$= \dfrac{2x - 3 - x + 6}{x^2 + 1} = \dfrac{x + 3}{x^2 + 1}$   **79.** $\dfrac{4x}{x + 5}$ ft; $\dfrac{x^2}{(x + 5)^2}$ sq ft   **81.** answers may vary   **83.** answers may vary   **85.** answers may vary

**87.** $\dfrac{(2x - 3)(x + 6)}{6x^2}$ or $\dfrac{2x^2 + 9x - 18}{6x^2}$   **89.** $\dfrac{4a^2}{9(a - 1)}$   **91.** 4   **93.** $-\dfrac{4}{x - 1}$   **95.** $-\dfrac{32}{x(x + 2)(x - 2)}$   **97.** $\dfrac{3}{2x}$   **99.** $\dfrac{4 - 3x}{x^2}$

**101.**

## Section 6.3

**Vocabulary, Readiness & Video Check** **1.** $\dfrac{7}{1+z}$  **3.** $\dfrac{1}{x^2}$  **5.** $\dfrac{2}{x}$  **7.** $\dfrac{1}{9y}$  **9.** a single fraction in the numerator and in the denominator  **11.** Since a negative exponent moves its base from a numerator to a denominator of the expression only, a rational expression containing negative exponents can become a complex fraction when rewritten with positive exponents.

**Exercise Set 6.3** **1.** 4  **3.** $\dfrac{7}{13}$  **5.** $\dfrac{4}{x}$  **7.** $\dfrac{9(x-2)}{9x^2+4}$  **9.** $2x+y$  **11.** $\dfrac{2(x+1)}{2x-1}$  **13.** $\dfrac{2x+3}{4-9x}$  **15.** $\dfrac{1}{x^2-2x+4}$  **17.** $\dfrac{x}{5(x-2)}$

**19.** $\dfrac{x-2}{2x-1}$  **21.** $\dfrac{x}{2-3x}$  **23.** $-\dfrac{y}{y+1}$  **25.** $-\dfrac{2x^2}{y(x-y)}$  **27.** $\dfrac{2x+1}{y+x^2}$  **29.** $\dfrac{x-3}{9}$  **31.** $\dfrac{1}{x+2}$  **33.** 2  **35.** $\dfrac{xy^2}{x^2+y^2}$  **37.** $\dfrac{2b^2+3a}{b(b-a)}$

**39.** $\dfrac{x}{(x+1)(x-1)}$  **41.** $\dfrac{1+a}{1-a}$  **43.** $\dfrac{x(x+6y)}{2y}$  **45.** $\dfrac{5a}{2(a+2)}$  **47.** $xy(5y+2x)$  **49.** $\dfrac{xy}{5y+2x}$  **51.** $\left\{-\dfrac{5}{6}\right\}$  **53.** {2}  **55.** {54}

**57.** a and c  **59.** a and b  **61.** $\dfrac{770a}{770-s}$  **63.** answers may vary  **65.** $\dfrac{x-3y}{x+3y}$  **67.** $\dfrac{1+x}{2+x}$  **69.** $x(x+1)$  **71. a.** $\dfrac{1}{a+h}$  **b.** $\dfrac{1}{a}$

**c.** $\dfrac{\dfrac{1}{a+h}-\dfrac{1}{a}}{h}$  **d.** $\dfrac{-1}{a(a+h)}$  **73. a.** $\dfrac{3}{a+h+1}$  **b.** $\dfrac{3}{a+1}$  **c.** $\dfrac{\dfrac{3}{a+h+1}-\dfrac{3}{a+1}}{h}$  **d.** $\dfrac{-3}{(a+h+1)(a+1)}$

## Section 6.4

**Vocabulary, Readiness & Video Check** **1.** equation  **3.** expression  **5.** equation  **7.** 14; c  **9.** $(x+4)(x-4)$; a  **11.** Linear and quadratic equations are solved in very different ways, so you need to determine the correct next move to make.

**Exercise Set 6.4** **1.** {72}  **3.** {2}  **5.** {6}  **7.** {2, −2}  **9.** ∅  **11.** $\left\{-\dfrac{28}{3}\right\}$  **13.** {3}  **15.** {−8}  **17.** {3}  **19.** ∅  **21.** {1}

**23.** {3}  **25.** {−1}  **27.** {6}  **29.** $\left\{\dfrac{1}{3}\right\}$  **31.** {−5, 5}  **33.** {3}  **35.** {7}  **37.** ∅  **39.** $\left\{\dfrac{4}{3}\right\}$  **41.** {−12}  **43.** $\left\{1, \dfrac{11}{4}\right\}$

**45.** {−5, −1}  **47.** $\left\{-\dfrac{7}{5}\right\}$  **49.** 5  **51.** length: 15 in.; width: 10 in.  **53.** 12.1%  **55.** 45 to 54  **57.** 21,256 inmates

**59.** answers may vary  **61.** 3000 game disks  **63.** $\left\{\dfrac{1}{16}, \dfrac{1}{3}\right\}$  **65.** $\left\{-\dfrac{1}{5}, 1\right\}$  **67.** {−0.17}  **69.** {0.42}  **71.** {−1, 0}  **73.** {−2}

**75.**  **77.**

**Integrated Review** **1.** $\left\{\dfrac{1}{2}\right\}$  **2.** {10}  **3.** $\dfrac{1+2x}{8}$  **4.** $\dfrac{15+x}{10}$  **5.** $\dfrac{2(x-4)}{(x+2)(x-1)}$  **6.** $-\dfrac{5(x-8)}{(x-2)(x+4)}$  **7.** {4}  **8.** {8}

**9.** {−5}  **10.** $\left\{-\dfrac{2}{3}\right\}$  **11.** $\dfrac{2x+5}{x(x-3)}$  **12.** $\dfrac{5}{2x}$  **13.** {−2}  **14.** $-\dfrac{y}{x}$  **15.** $\dfrac{(a+3)(a+1)}{a+2}$  **16.** $\dfrac{-a^2+31a+10}{5(a-6)(a+1)}$  **17.** $\left\{-\dfrac{1}{5}\right\}$

**18.** $\left\{-\dfrac{3}{13}\right\}$  **19.** $\dfrac{4a+1}{(3a+1)(3a-1)}$  **20.** $\dfrac{-a-8}{4a(a-2)}$ or $-\dfrac{a+8}{4a(a-2)}$  **21.** $\left\{-1, \dfrac{3}{2}\right\}$  **22.** $\dfrac{x^2-3x+10}{2(x+3)(x-3)}$  **23.** $\dfrac{3}{x+1}$

**24.** $\{x \mid x \text{ is a real number and } x \neq 2, x \neq -1\}$  **25.** {−1}  **26.** $\dfrac{22z-45}{3z(z-3)}$  **27. a.** $\dfrac{x}{5}-\dfrac{x}{4}+\dfrac{1}{10}$  **b.** Write each rational expression term so that the denominator is the LCD, 20.  **c.** $\dfrac{-x+2}{20}$  **28. a.** $\dfrac{x}{5}-\dfrac{x}{4}=\dfrac{1}{10}$  **b.** Clear the equation of fractions by multiplying each term by the LCD, 20.  **c.** {−2}  **29.** b  **30.** d  **31.** d  **32.** a  **33.** d

## Section 6.5

**Vocabulary, Readiness & Video Check** **1.** After we multiply through by the LCD, we factor the specified variable out so that we can divide and get the specified variable alone.  **3.** We write the same units in the numerators and the same units in the denominators.  **5.** $x+y$ (or $y+x$)

**Exercise Set 6.5** **1.** $C = \dfrac{5}{9}(F-32)$  **3.** $I = A - QL$  **5.** $R = \dfrac{R_1 R_2}{R_1 + R_2}$  **7.** $n = \dfrac{2s}{a+L}$  **9.** $b = \dfrac{2A - ah}{h}$  **11.** $T_2 = \dfrac{P_2 V_2 T_1}{P_1 V_1}$

**13.** $f_2 = \dfrac{f_1 f}{f_1 - f}$  **15.** $L = \dfrac{n\lambda}{2}$  **17.** $c = \dfrac{2L\omega}{\theta}$  **19.** 1 and 5  **21.** 5  **23.** 4.5 gal  **25.** 5170 women  **27.** 15.6 hr  **29.** 10 min

**31.** 200 mph  **33.** 15 mph  **35.** $-8$ and $-7$  **37.** 36 min  **39.** 45 mph; 60 mph  **41.** 5.9 hr  **43.** 2 hr  **45.** 135 mph  **47.** 12 mi
**49.** $\dfrac{7}{8}$  **51.** $1\dfrac{1}{2}$ min  **53.** 1 hr  **55.** $2\dfrac{2}{9}$ hr  **57.** 2 hr  **59.** $\{-5\}$  **61.** $\{2\}$  **63.** 164 lb  **65.** higher; F  **67.** 6 ohms
**69.** $\dfrac{1}{R} = \dfrac{1}{R_1} + \dfrac{1}{R_2} + \dfrac{1}{R_3}$; $R = \dfrac{15}{13}$ ohms

## Section 6.6

**Vocabulary, Readiness & Video Check**  **1.** direct  **3.** joint  **5.** inverse  **7.** direct  **9.** linear; slope  **11.** $y = ka^2b^5$

**Exercise Set 6.6**  **1.** $k = \dfrac{1}{5}$; $y = \dfrac{1}{5}x$  **3.** $k = \dfrac{3}{2}$; $y = \dfrac{3}{2}x$  **5.** $k = 14$; $y = 14x$  **7.** $k = 0.25$; $y = 0.25x$  **9.** 4.05 lb  **11.** 117,125 tons

**13.** $k = 30$; $y = \dfrac{30}{x}$  **15.** $k = 700$; $y = \dfrac{700}{x}$  **17.** $k = 2$; $y = \dfrac{2}{x}$  **19.** $k = 0.14$; $y = \dfrac{0.14}{x}$  **21.** 54 mph  **23.** 72 amps

**25.** divided by 4  **27.** $x = kyz$  **29.** $r = kst^3$  **31.** $k = \dfrac{1}{3}$; $y = \dfrac{1}{3}x^3$  **33.** $k = 0.2$; $y = 0.2\sqrt{x}$  **35.** $k = 1.3$; $y = \dfrac{1.3}{x^2}$

**37.** $k = 3$; $y = 3xz^3$  **39.** 22.5 tons  **41.** $15\pi$ cu in.  **43.** 8 ft  **45.** $y = kx$  **47.** $a = \dfrac{k}{b}$  **49.** $y = kxz$  **51.** $y = \dfrac{k}{x^3}$  **53.** $y = \dfrac{kx}{p^2}$
**55.** $C = 12\pi$ cm; $A = 36\pi$ sq cm  **57.** $C = 14\pi$ m; $A = 49\pi$ sq m  **59.** 0  **61.** $-1$  **63.** a  **65.** c  **67.** multiplied by 2
**69.** multiplied by 4

## Chapter 6 Vocabulary Check
**1.** complex fraction  **2.** directly  **3.** inversely  **4.** least common denominator  **5.** jointly
**6.** opposites  **7.** rational expression  **8.** equation; expression

## Chapter 6 Review
**1.** $\{x \mid x$ is a real number$\}$  **2.** $\{x \mid x$ is a real number$\}$  **3.** $\{x \mid x$ is a real number and $x \neq 5\}$
**4.** $\{x \mid x$ is a real number and $x \neq 4\}$  **5.** $\{x \mid x$ is a real number and $x \neq 0, x \neq -8\}$  **6.** $\{x \mid x$ is a real number and $x \neq -4, x \neq 4\}$

**7.** $-1$  **8.** $\dfrac{1}{5}$  **9.** $\dfrac{1}{x-1}$  **10.** $\dfrac{1}{x-7}$  **11.** $\dfrac{2(x-3)}{x-4}$  **12. a.** \$119  **b.** \$77  **c.** decrease  **13.** $-\dfrac{3}{2}$  **14.** $\dfrac{2}{5}$  **15.** $\dfrac{a-b}{2a}$  **16.** $\dfrac{1}{6}$

**17.** $\dfrac{12}{5}$  **18.** $\dfrac{(x+4)(x+5)}{3}$  **19.** $\dfrac{a-b}{5a}$  **20.** $\dfrac{3(x+1)}{x-7}$  **21.** $-\dfrac{1}{x}$  **22.** $-\dfrac{x+3}{2(x+2)}$  **23.** $60x^2y^5$  **24.** $2x(x-2)$  **25.** $5x(x-5)$

**26.** $10x^3(x-4)(x+7)(x-3)$  **27.** $\dfrac{4+x}{x-4}$  **28.** $\dfrac{2}{x^2}$  **29.** $\dfrac{3}{2(x-2)}$  **30.** $\dfrac{x-2}{x-10}$  **31.** $\dfrac{-7x-6}{5(x-3)(x+3)}$  **32.** $\dfrac{2x^2-5x-4}{x-3}$

**33.** $\dfrac{5a-1}{(a-1)^2(a+1)}$  **34.** $\dfrac{3x^2-7x-4}{(3x-4)(9x^2+12x+16)}$  **35.** $\dfrac{5-2x}{2(x-1)}$  **36.** $\dfrac{11}{x}$  **37.** $\dfrac{4-3x}{8+x}$  **38.** $\dfrac{x^3}{3(x+1)}$  **39.** $\dfrac{5(4x-3)}{2(5x^2-2)}$

**40.** $\dfrac{y}{x-y}$  **41.** $\dfrac{x(5y+1)}{3y}$  **42.** $\dfrac{x-3}{3}$  **43.** $\dfrac{1+x}{1-x}$  **44.** $-\dfrac{x^2+9}{6x}$  **45.** $\{6\}$  **46.** $\{3\}$  **47.** $\left\{\dfrac{3}{2}\right\}$  **48.** $\{-7, 7\}$

**49.** $\dfrac{2x+5}{x(x-7)}$  **50.** $\dfrac{-5(x+6)}{2x(x-3)}$  **51.** $\left\{-\dfrac{1}{3}, 2\right\}$  **52.** $\left\{\dfrac{3}{2}, 5\right\}$  **53.** $a = \dfrac{2A}{h} - b$  **54.** $R_2 = \dfrac{RR_1}{R_1 - R}$  **55.** $R = \dfrac{E}{I} - r$

**56.** $r = \dfrac{A-P}{Pt}$  **57.** $x = \dfrac{yz}{z-y}$  **58.** $A = \dfrac{HL}{k(T_1-T_2)}$  **59.** 1 and 2  **60.** 7  **61.** $1\dfrac{23}{37}$ hr  **62.** 12 hr  **63.** 8 mph  **64.** 4 mph

**65.** 9  **66.** 3.125 cu m  **67.** 2  **68.** $\dfrac{y-3}{x+2}$  **69.** $\dfrac{3}{5x}$  **70.** $\dfrac{7(x-4)}{2(x-2)}$  **71.** $\dfrac{5(a-2)}{7}$  **72.** $\dfrac{6}{a}$  **73.** $\dfrac{-x+5}{(x+1)(x-1)}$  **74.** $\dfrac{13}{3x}$

**75.** $\dfrac{2(7x-20)}{(x+4)^2(x-4)}$  **76.** $\dfrac{1}{x-2}$  **77.** $\dfrac{2}{15-2x}$  **78.** $\dfrac{2x}{x-2}$  **79.** $\dfrac{2(x-1)}{x+6}$  **80.** $\dfrac{5x}{2}$  **81.** $\{2\}$  **82.** $\left\{\dfrac{5}{3}\right\}$  **83.** $\dfrac{23}{25}$

**84.** $-10$ and $-8$  **85.** 10 hr  **86.** 490 mph  **87.** $63\dfrac{2}{3}$ mph; 45 mph  **88.** $64\pi$ sq in.  **89.** 36  **90.** 4

## Chapter 6 Test
**1.** $\{x \mid x$ is a real number and $x \neq 1\}$  **2.** $\{x \mid x$ is a real number and $x \neq -3, x \neq -1\}$  **3.** $-\dfrac{7}{8}$  **4.** $\dfrac{x}{x+9}$

**5.** $x^2 + 2x + 4$  **6.** $\dfrac{5}{3x}$  **7.** $\dfrac{3}{x^3}$  **8.** $\dfrac{x+2}{2(x+3)}$  **9.** $-\dfrac{4(2x+9)}{5}$  **10.** $-1$  **11.** $\dfrac{1}{x(x+3)}$  **12.** $\dfrac{5x-2}{(x-3)(x+2)(x-2)}$  **13.** $\dfrac{30-x}{6(x-7)}$

**14.** $\dfrac{3}{2}$  **15.** $\dfrac{64}{3}$  **16.** $\dfrac{(x-3)^2}{x-2}$  **17.** $\{7\}$  **18.** $\{8\}$  **19.** $\left\{\dfrac{2}{7}\right\}$  **20.** $\{3\}$  **21.** $x = \dfrac{7a^2+b^2}{4a-b}$  **22.** 5  **23.** $\dfrac{6}{7}$ hr  **24.** 16  **25.** 9  **26.** 256 ft
**27.** 39 earned runs

## Cumulative Review
**1.** true; Sec. 1.2, Ex. 3  **2. a.** $x - \dfrac{1}{3}$  **b.** $5x - 6$  **c.** $8x + 3$  **d.** $\dfrac{7}{2-x}$  **3.** $-1$; Sec. 1.6, Ex. 7  **4.** $x^2y^2$

**5.** 1; Sec. 1.6, Ex. 8  **6.** $\dfrac{8a^7}{9b^{11}}$  **7.** $\varnothing$; Sec. 2.5, Ex. 3  **8.** $\varnothing$  **9.** $\{-1, 1\}$; Sec. 2.6, Ex. 4  **10.** $\left(-\infty, -\dfrac{5}{3}\right] \cup \left[\dfrac{11}{3}, \infty\right)$  **11.** $(0, -12)$,

solution; $(1, 9)$, not a solution; $(2, -6)$, solution; Sec. 3.1, Ex. 3  **12. a.** $-2$  **b.** $-20$  **c.** $-\dfrac{10}{9}$  **13.** parallel; Sec. 3.2, Ex. 8

**14.** 7 **15.** $-5x + 8y = -20$ or $5x - 8y = 20$; Sec. 3.4, Ex. 2 **16.** $f(x) = \frac{1}{2}x + \frac{7}{2}$ **17.** domain: $(-\infty, \infty)$; range: $[0, \infty)$;
Sec. 3.7, Ex. 2 **18.** domain: $(-\infty, \infty)$; range: $(-\infty, \infty)$ **19.** domain: $[-4, 4]$; range: $[-2, 2]$; Sec. 3.7, Ex. 3 **20.** domain: $(-\infty, \infty)$;
range: $(-\infty, -1]$ **21.** $\left(-\frac{21}{10}, \frac{3}{10}\right)$; Sec. 4.1, Ex. 7 **22.** $(3, 4)$ **23.** $\left(\frac{1}{2}, 0, \frac{3}{4}\right)$; Sec. 4.2, Ex. 3 **24.** $(2, 1, 1)$ **25.** 52 mph; 47 mph;
Sec. 4.3, Ex. 3 **26.** Paper, \$3.80; folders, \$5.25 **27.** $(1, -1, 3)$; Sec. 4.4, Ex. 3 **28.** $(0, 5, 4)$ **29.** $\frac{z^2}{9x^4y^{20}}$; Sec. 1.7, Ex. 14

**30. a.** $\frac{7}{12}$ **b.** $-6$ **c.** $\frac{1}{x^3}$ **31.** ; Sec. 4.5, Ex. 1 **32. a.** $48a^{2a}$ **b.** $y^{10b+3}$ **33.** $-5x^2 - 6x$; Sec. 5.1, Ex. 10

**34.** $2x^2 + 6x + 4$ **35.** $8xy - 3x$; Sec. 5.1, Ex. 11 **36.** $-10y + \frac{15}{2}$ **37.** $4x^4 + 8x^3 + 39x^2 + 14x + 56$; Sec. 5.2, Ex. 8
**38.** $16 + 24x - 8y + 9x^2 - 6xy + y^2$ **39.** $2x - 5$; Sec. 5.3, Ex. 3 **40.** $(y + 2)(x - 5)$ **41.** $xy(-3x^2 + 2x - 5)$ or
$-xy(3x^2 - 2x + 5)$; Sec. 5.4, Ex. 6 **42.** $(2x - 5)(3x + 7)$ **43.** $(x + 2)(x + 8)$; Sec. 5.5, Ex. 1
**44.** $(2x - 1 + 3y)(2x - 1 - 3y)$ **45.** $(p^2 + 4)(p + 2)(p - 2)$; Sec. 5.6, Ex. 9 **46. a.** domain: $(-\infty, \infty)$; range: $[-4, \infty)$
**b.** x-intercepts: $(-2, 0), (2, 0)$; y-intercept: $(0, -4)$ **c.** There is no such point. **d.** $(0, -4)$ **e.** $-2, 2$ **f.** between $x = -2$
and $x = 2$ **g.** $\{-2, 2\}$ **47.** $\{-2, 6\}$; Sec. 5.7, Ex. 1 **48.** $\left\{0, -\frac{1}{3}, 3\right\}$ **49.** $\frac{x(x-2)}{2(x+2)}$; Sec. 6.3, Ex. 2 **50.** $\{-4\}$

## Chapter 7 Rational Exponents, Radicals, and Complex Numbers
### Section 7.1

**Vocabulary, Readiness & Video Check** **1.** index; radical sign; radicand **3.** is not **5.** $[0, \infty)$ **7.** $(16, 4)$ **9.** d **11.** d
**13.** Divide the index into each exponent in the radicand. **15.** The square root of a negative number is not a real number, but the
cube root of a negative number is a real number. **17.** For odd roots, $\sqrt[n]{a^n} = a$, whether $a$ is positive, 0, or a negative number, so
absolute value bars aren't needed.

**Exercise Set 7.1** **1.** $2, -2$ **3.** no real number square roots **5.** $10, -10$ **7.** 10 **9.** $\frac{1}{2}$ **11.** 0.01 **13.** $-6$ **15.** $x^5$ **17.** $4y^3$ **19.** 2.646

**21.** 6.164 **23.** 14.142 **25.** 4 **27.** $\frac{1}{2}$ **29.** $-1$ **31.** $x^4$ **33.** $-3x^3$ **35.** $-2$ **37.** not a real number **39.** $-2$ **41.** $x^4$ **43.** $2x^2$

**45.** $9x^2$ **47.** $4x^2$ **49.** 8 **51.** $-8$ **53.** $2|x|$ **55.** $x$ **57.** $|x - 5|$ **59.** $|x + 2|$ **61.** $-11$ **63.** $2x$ **65.** $y^6$ **67.** $5ab^{10}$ **69.** $-3x^4y^3$

**71.** $a^4b$ **73.** $-2x^2y$ **75.** $\frac{5}{7}$ **77.** $\frac{x}{2y}$ **79.** $-\frac{z^7}{3x}$ **81.** $\frac{x}{2}$ **83.** $\sqrt{3}$ **85.** $-1$ **87.** $-3$ **89.** $\sqrt{7}$ **91.** $[0, \infty)$;
**93.** $[3, \infty)$; **95.** $(-\infty, \infty)$; **97.** $(-\infty, \infty)$;

**99.** $-32x^{15}y^{10}$ **101.** $-60x^7y^{10}z^5$ **103.** $\frac{x^9y^5}{2}$ **105.** not a real number **107.** not a real number **109.** answers may vary **111.** b
**113.** b **115.** 1.69 sq m **117.** 11,181 m per sec **119.** answers may vary

### Section 7.2

**Vocabulary, Readiness & Video Check** **1.** true **3.** true **5.** multiply; c **7.** A **9.** C **11.** B **13.** $-\sqrt[5]{3x}$ **15.** denominator;
positive **17.** Write the radical in equivalent fractional exponent form, use rules for exponents to write a single fractional exponent,
simplify the fraction, then write in radical form again.

**Exercise Set 7.2** **1.** 7 **3.** 3 **5.** $\frac{1}{2}$ **7.** 13 **9.** $2\sqrt[3]{m}$ **11.** $3x^2$ **13.** $-3$ **15.** $-2$ **17.** 8 **19.** 16 **21.** not a real number **23.** $\sqrt[5]{(2x)^3}$

**25.** $\sqrt[3]{(7x + 2)^2}$ or $\left(\sqrt[3]{7x + 2}\right)^2$ **27.** $\frac{64}{27}$ **29.** $\frac{1}{16}$ **31.** $\frac{1}{16}$ **33.** not a real number **35.** $\frac{1}{x^{1/4}}$ **37.** $a^{2/3}$ **39.** $\frac{5x^{3/4}}{7}$ **41.** $a^{7/3}$

**43.** $x$ **45.** $3^{5/8}$ **47.** $y^{1/6}$ **49.** $8u^3$ **51.** $-b$ **53.** $\frac{1}{x^2}$ **55.** $27x^{2/3}$ **57.** $\frac{y}{z^{1/6}}$ **59.** $\frac{1}{x^{7/4}}$ **61.** $\sqrt{x}$ **63.** $\sqrt[3]{2}$ **65.** $2\sqrt{x}$ **67.** $\sqrt{x + 3}$

**69.** $\sqrt{xy}$ **71.** $\sqrt[3]{a^2b}$ **73.** $\sqrt[15]{y^{11}}$ **75.** $\sqrt[12]{b^5}$ **77.** $\sqrt[24]{x^{23}}$ **79.** $\sqrt{a}$ **81.** $\sqrt[6]{432}$ **83.** $\sqrt[15]{343y^5}$ **85.** $\sqrt[6]{125r^3s^2}$ **87.** $25 \cdot 3$

**89.** $16 \cdot 3$ or $4 \cdot 12$   **91.** $8 \cdot 2$   **93.** $27 \cdot 2$   **95.** 1509 calories   **97.** 615.8 million   **99.** answers may vary   **101.** $a^{1/3}$   **103.** $x^{1/5}$

**105.** 1.6818   **107.** $\dfrac{t^{1/2}}{u^{1/2}}$

## Section 7.3

**Vocabulary, Readiness & Video Check**   **1.** midpoint; point   **3.** midpoint   **5.** false   **7.** true   **9.** false   **11.** The indexes must be the same.   **13.** The power must be 1. Any even power is a perfect square and will leave no factor in the radicand; any higher odd power can have an even power factored from it, leaving one factor remaining in the radicand.   **15.** average; average

**Exercise Set 7.3**   **1.** $\sqrt{14}$   **3.** 2   **5.** $\sqrt[3]{36}$   **7.** $\sqrt{6x}$   **9.** $\sqrt{\dfrac{14}{xy}}$   **11.** $\sqrt[4]{20x^3}$   **13.** $\dfrac{\sqrt{6}}{7}$   **15.** $\dfrac{\sqrt{2}}{7}$   **17.** $\dfrac{\sqrt[4]{x^3}}{2}$   **19.** $\dfrac{\sqrt[3]{4}}{3}$   **21.** $\dfrac{\sqrt[4]{8}}{x^2}$

**23.** $\dfrac{\sqrt[3]{2x}}{3y^4}$   **25.** $\dfrac{x\sqrt{y}}{13}$   **27.** $\dfrac{x\sqrt{5}}{2y}$   **29.** $-\dfrac{z^2\sqrt[3]{z}}{5x}$   **31.** $4\sqrt{2}$   **33.** $4\sqrt[3]{3}$   **35.** $25\sqrt{3}$   **37.** $2\sqrt{6}$   **39.** $10x^2\sqrt{x}$   **41.** $2y^2\sqrt[3]{2y}$   **43.** $a^2b\sqrt[4]{b^3}$

**45.** $y^2\sqrt{y}$   **47.** $5ab\sqrt{b}$   **49.** $-2x^2\sqrt[5]{y}$   **51.** $x^4\sqrt[3]{50x^2}$   **53.** $-4a^4b^3\sqrt{2b}$   **55.** $3x^3y^4\sqrt{xy}$   **57.** $5r^3s^4$   **59.** $2x^3y\sqrt[4]{2y}$   **61.** $\sqrt{2}$

**63.** 2   **65.** 10   **67.** $x^2y$   **69.** $24m^2$   **71.** $\dfrac{15x\sqrt{2x}}{2}$ or $\dfrac{15x}{2}\sqrt{2x}$   **73.** $2a^2\sqrt[4]{2}$   **75.** $2xy^2\sqrt[5]{x^2}$   **77.** 5 units   **79.** $\sqrt{41}$ units $\approx 6.403$ units

**81.** $\sqrt{10}$ units $\approx 3.162$ units   **83.** $\sqrt{5}$ units $\approx 2.236$ units   **85.** $\sqrt{192.58}$ units $\approx 13.877$ units   **87.** $(4, -2)$   **89.** $\left(-5, \dfrac{5}{2}\right)$

**91.** $\left(\dfrac{5}{2}, 0\right)$   **93.** $\left(-\dfrac{1}{2}, \dfrac{1}{2}\right)$   **95.** $\left(\sqrt{2}, \dfrac{\sqrt{5}}{2}\right)$   **97.** $(6.2, -6.65)$   **99.** $14x$   **101.** $2x^2 - 7x - 15$   **103.** $y^2$   **105.** $x^2 - 8x + 16$

**107.** $\dfrac{\sqrt[3]{64}}{\sqrt{64}} = \dfrac{4}{8} = \dfrac{1}{2}$   **109.** 1.6 m   **111. a.** 2.2 sec   **b.** 1.4 sec   **c.** answers may vary

## Section 7.4

**Vocabulary, Readiness & Video Check**   **1.** Unlike   **3.** Like   **5.** $6\sqrt{3}$   **7.** $7\sqrt{x}$   **9.** $8\sqrt[3]{x}$   **11.** $\sqrt{11} + \sqrt[3]{11}$   **13.** $10\sqrt[3]{2x}$   **15.** Sometimes you can't see that there are like radicals until you simplify, so you may incorrectly think you cannot add or subtract if you don't simplify first.

**Exercise Set 7.4**   **1.** $-2\sqrt{2}$   **3.** $10x\sqrt{2x}$   **5.** $17\sqrt{2} - 15\sqrt{5}$   **7.** $-\sqrt[3]{2x}$   **9.** $5b\sqrt{b}$   **11.** $\dfrac{31\sqrt{2}}{15}$   **13.** $\dfrac{\sqrt[3]{11}}{3}$   **15.** $\dfrac{5\sqrt{5x}}{9}$

**17.** $14 + \sqrt{3}$   **19.** $7 - 3y$   **21.** $6\sqrt{3} - 6\sqrt{2}$   **23.** $-23\sqrt[3]{5}$   **25.** $2a^2b^3\sqrt{ab}$   **27.** $20y\sqrt{2y}$   **29.** $2y\sqrt[3]{2x}$   **31.** $6\sqrt[3]{11} - 4\sqrt{11}$

**33.** $4x\sqrt[4]{x^3}$   **35.** $\dfrac{2\sqrt{3}}{3}$   **37.** $\dfrac{5x\sqrt[3]{x}}{7}$   **39.** $\dfrac{5\sqrt{7}}{2x}$   **41.** $\dfrac{\sqrt[3]{2}}{6}$   **43.** $\dfrac{14x\sqrt[3]{2x}}{9}$   **45.** $15\sqrt{3}$ in.   **47.** $\sqrt{35} + \sqrt{21}$   **49.** $7 - 2\sqrt{10}$

**51.** $3\sqrt{x} - x\sqrt{3}$   **53.** $6x - 13\sqrt{x} - 5$   **55.** $\sqrt[3]{a^2} + \sqrt[3]{a} - 20$   **57.** $6\sqrt{2} - 12$   **59.** $2 + 2x\sqrt{3}$   **61.** $-16 - \sqrt{35}$

**63.** $18 + \sqrt{6} - 24\sqrt{3} - 4\sqrt{2}$   **65.** $3 + 2x\sqrt{3} + x^2$   **67.** $5x - 3\sqrt{15x} - 3\sqrt{10x} + 9\sqrt{6}$   **69.** $-\sqrt[3]{4} + 2\sqrt[3]{2}$

**71.** $\sqrt[3]{x^2} - 4\sqrt[6]{x^5} + 8\sqrt[3]{x} - 4\sqrt{x} + 7$   **73.** $x + 24 + 10\sqrt{x - 1}$   **75.** $2x + 6 - 2\sqrt{2x + 5}$   **77.** $x - 7$   **79.** $\dfrac{7}{x + y}$   **81.** $2a - 3$

**83.** $\dfrac{-2 + \sqrt{3}}{3}$   **85.** $22\sqrt{5}$ ft; 150 sq ft   **87. a.** $2\sqrt{3}$   **b.** 3   **c.** answers may vary   **89.** $2\sqrt{6} - 2\sqrt{2} - 2\sqrt{3} + 6$

## Section 7.5

**Vocabulary, Readiness & Video Check**   **1.** conjugate   **3.** rationalizing the numerator   **5.** $\sqrt{2} - x$   **7.** $5 + \sqrt{a}$   **9.** $-7\sqrt{5} - 8\sqrt{x}$   **11.** to write an equivalent expression without a radical in the denominator   **13.** No, except for the fact you're working with numerators, the process is the same.

**Exercise Set 7.5**   **1.** $\dfrac{\sqrt{14}}{7}$   **3.** $\dfrac{\sqrt{5}}{5}$   **5.** $\dfrac{4\sqrt[3]{9}}{3}$   **7.** $\dfrac{3\sqrt{2x}}{4x}$   **9.** $\dfrac{3\sqrt[3]{2x}}{2x}$   **11.** $\dfrac{3\sqrt{3a}}{a}$   **13.** $\dfrac{3\sqrt[3]{4}}{2}$   **15.** $\dfrac{2\sqrt{21}}{7}$   **17.** $\dfrac{\sqrt{10xy}}{5y}$   **19.** $\dfrac{\sqrt[3]{75}}{5}$

**21.** $\dfrac{\sqrt{6x}}{10}$   **23.** $\dfrac{\sqrt{3z}}{6z}$   **25.** $\dfrac{\sqrt[3]{6xy^2}}{3x}$   **27.** $\dfrac{2\sqrt[4]{9x}}{3x^2}$   **29.** $\dfrac{5\sqrt[5]{4ab^4}}{2ab^3}$   **31.** $-2(2 + \sqrt{7})$   **33.** $\dfrac{7(\sqrt{x} + 3)}{9 - x}$   **35.** $-5 + 2\sqrt{6}$

**37.** $\dfrac{2a + \sqrt{ab} + 2\sqrt{a} + \sqrt{b}}{4a - b}$   **39.** $-\dfrac{8(1 - \sqrt{10})}{9}$   **41.** $\dfrac{x - \sqrt{xy}}{x - y}$   **43.** $\dfrac{5 + 3\sqrt{2}}{7}$   **45.** $\dfrac{5}{\sqrt{15}}$   **47.** $\dfrac{6}{\sqrt{10}}$   **49.** $\dfrac{2x}{7\sqrt{x}}$   **51.** $\dfrac{5y}{\sqrt[3]{100xy}}$

**53.** $\dfrac{2}{\sqrt{10}}$   **55.** $\dfrac{2x}{11\sqrt{2x}}$   **57.** $\dfrac{7}{2\sqrt[3]{49}}$   **59.** $\dfrac{3x^2}{10\sqrt[3]{9x}}$   **61.** $\dfrac{6x^2y^3}{\sqrt{6z}}$   **63.** answers may vary   **65.** $\dfrac{-7}{12 + 6\sqrt{11}}$   **67.** $\dfrac{3}{10 + 5\sqrt{7}}$

**69.** $\dfrac{x - 9}{x - 3\sqrt{x}}$   **71.** $\dfrac{x - 1}{x - 2\sqrt{x} + 1}$   **73.** $\{5\}$   **75.** $\left\{-\dfrac{1}{2}, 6\right\}$   **77.** $\{2, 6\}$   **79.** $r = \dfrac{\sqrt{A\pi}}{2\pi}$   **81.** answers may vary   **83. a.** $\dfrac{y\sqrt{15xy}}{6x^2}$

**b.** $\dfrac{y\sqrt{15xy}}{6x^2}$   **c.** answers may vary   **85.** $\sqrt[3]{25}$

**Integrated Review** **1.** 9 **2.** $-2$ **3.** $\dfrac{1}{2}$ **4.** $x^3$ **5.** $y^3$ **6.** $2y^5$ **7.** $-2y$ **8.** $3b^3$ **9.** 6 **10.** $\sqrt[4]{3y}$ **11.** $\dfrac{1}{16}$ **1**

**13.** $y$ **14.** $16x^{1/2}$ **15.** $x^{5/4}$ **16.** $4^{11/15}$ **17.** $2x^2$ **18.** $\sqrt[4]{a^3b^2}$ **19.** $\sqrt[4]{x^3}$ **20.** $\sqrt[6]{500}$ **21.** $2\sqrt{10}$ **22.** $2xy^2\sqrt[4]{}$

**24.** $-2b^2\sqrt[5]{2}$ **25.** $\sqrt{5x}$ **26.** $4x$ **27.** $7y^2\sqrt{y}$ **28.** $2a^2\sqrt[4]{3}$ **29.** $2\sqrt{5} - 5\sqrt{3} + 5\sqrt{7}$ **30.** $y\sqrt[3]{2y}$ **31.** $\sqrt{15}$

**33.** $4x^2 - 5$ **34.** $x + 2 - 2\sqrt{x+1}$ **35.** $\dfrac{\sqrt{21}}{3}$ **36.** $\dfrac{5\sqrt[3]{4x}}{2x}$ **37.** $\dfrac{13 - 3\sqrt{21}}{5}$ **38.** $\dfrac{7}{\sqrt{21}}$ **39.** $\dfrac{3y}{\sqrt[3]{33y^2}}$ **40.** $\dfrac{}{x + 2\sqrt{x}}$

## Section 7.6

**Graphing Calculator Explorations** **1.** $\{3.19\}$ **3.** $\varnothing$ **5.** $\{3.23\}$

**Vocabulary, Readiness & Video Check** **1.** extraneous solution **3.** $x^2 - 10x + 25$ **5.** Applying the power rule can result in an equation with more solutions than the original equation, so you need to check all proposed solutions in the original equation. **7.** Our answer is either a positive square root of a value or a negative square root of a value. We're looking for a length, which must be positive, so our answer must be the positive square root.

**Exercise Set 7.6** **1.** $\{8\}$ **3.** $\{7\}$ **5.** $\varnothing$ **7.** $\{7\}$ **9.** $\{6\}$ **11.** $\left\{-\dfrac{9}{2}\right\}$ **13.** $\{29\}$ **15.** $\{4\}$ **17.** $\{-4\}$ **19.** $\left\{\dfrac{41}{16}\right\}$ **21.** $\{7\}$

**23.** $\{9\}$ **25.** $\{50\}$ **27.** $\varnothing$ **29.** $\left\{\dfrac{15}{4}\right\}$ **31.** $\{7\}$ **33.** $\{5\}$ **35.** $\{-12\}$ **37.** $\{9\}$ **39.** $\{-3\}$ **41.** $\{1\}$ **43.** $\{1\}$ **45.** $\left\{\dfrac{1}{2}\right\}$

**47.** $\{0, 4\}$ **49.** $\left\{\dfrac{37}{4}\right\}$ **51.** $3\sqrt{5}$ ft **53.** $2\sqrt{10}$ m **55.** $2\sqrt{131}$ m $\approx$ 22.9 m **57.** $\sqrt{100.84}$ mm $\approx$ 10.0 mm **59.** 17 ft **61.** 13 ft

**63.** 14,657,415 sq mi **65.** 100 ft **67.** $\sqrt{625,000}$ ft $\approx$ 790.57 ft **69.** 100 **71.** $\dfrac{\pi}{2}$ sec $\approx$ 1.57 sec **73.** 12.97 ft **75.** answers may vary

**77.** $15\sqrt{3}$ sq mi $\approx$ 25.98 sq mi **79.** answers may vary **81.** 0.51 km **83.** $\dfrac{x}{4x+3}$ **85.** $-\dfrac{4z+2}{3z}$ **87. a.–b.** answers may vary

**89.** $\sqrt{5x-1} + 4 = 7$     **91.** 2743 deliveries
$\quad\quad \sqrt{5x-1} = 3$
$\quad (\sqrt{5x-1})^2 = 3^2$
$\quad\quad\quad 5x - 1 = 9$
$\quad\quad\quad\quad 5x = 10$
$\quad\quad\quad\quad\, x = 2$

## Section 7.7

**Vocabulary, Readiness & Video Check** **1.** complex **3.** $-1$ **5.** real **7.** the product rule for radicals; you need to first simplify each radical separately and have nonnegative radicands before applying the product rule. **9.** the fact that $i^2 = -1$ **11.** $i, i^2 = -1, i^3 = -i, i^4 = 1$

**Exercise Set 7.7** **1.** $9i$ **3.** $i\sqrt{7}$ **5.** $-4$ **7.** $8i$ **9.** $2i\sqrt{6}$ **11.** $-6i$ **13.** $24i\sqrt{7}$ **15.** $-3\sqrt{6}$ **17.** $-\sqrt{14}$ **19.** $-5\sqrt{2}$ **21.** $4i$
**23.** $i\sqrt{3}$ **25.** $2\sqrt{2}$ **27.** $6 - 4i$ **29.** $-2 + 6i$ **31.** $-2 - 4i$ **33.** $2 - i$ **35.** $5 - 10i$ **37.** $8 - i$ **39.** $-12$ **41.** $63$ **43.** $-40$
**45.** $18 + 12i$ **47.** $27 + 3i$ **49.** $18 + 13i$ **51.** $7$ **53.** $12 - 16i$ **55.** $20$ **57.** $2$ **59.** $17 + 144i$ **61.** $-2i$ **63.** $-4i$ **65.** $\dfrac{28}{25} - \dfrac{21}{25}i$
**67.** $-\dfrac{12}{5} + \dfrac{6}{5}i$ **69.** $4 + i$ **71.** $-\dfrac{5}{2} - 2i$ **73.** $-5 + \dfrac{16}{3}i$ **75.** $\dfrac{3}{5} - \dfrac{1}{5}i$ **77.** $\dfrac{1}{5} - \dfrac{8}{5}i$ **79.** $1$ **81.** $i$ **83.** $-i$ **85.** $-1$ **87.** $-64$
**89.** $-243i$ **91.** 5 people **93.** 14 people **95.** 16.7% **97.** $1 - i$ **99.** 0 **101.** $2 + 3i$ **103.** $5 - 4i$ **105.** $2 - i\sqrt{2}$
**107.** answers may vary **109.** $6 - 3i\sqrt{3}$ **111.** yes

**Chapter 7 Vocabulary Check** **1.** conjugate **2.** principal square root **3.** rationalizing **4.** imaginary unit **5.** cube root
**6.** index; radicand **7.** like radicals **8.** complex number **9.** distance **10.** midpoint

**Chapter 7 Review** **1.** 9 **2.** 3 **3.** $-2$ **4.** not a real number **5.** $-\dfrac{1}{7}$ **6.** $x^{32}$ **7.** $-6$ **8.** 4 **9.** $-a^2b^3$ **10.** $4a^2b^6$ **11.** $2ab^2$

**12.** $-2x^3y^4$ **13.** $\dfrac{x^6}{6y}$ **14.** $\dfrac{3y}{z^4}$ **15.** $|x|$ **16.** $|x^2 - 4|$ **17.** $-27$ **18.** $-5$ **19.** $-x$ **20.** $-x$ **21.** $2|2y + z|$ **22.** $5|x - y|$

**23.** $y$ **24.** $|x|$ **25.** 3, 6 **26.** 2, $\sqrt[3]{17}$ **27.** $\dfrac{1}{3}$ **28.** $-\dfrac{1}{3}$ **29.** $-\dfrac{1}{3}$ **30.** $-\dfrac{1}{4}$ **31.** $-27$ **32.** $\dfrac{1}{4}$ **33.** not a real number **34.** $\dfrac{343}{125}$

**35.** $\dfrac{9}{4}$ **36.** not a real number **37.** $x^{2/3}$ **38.** $5^{1/5}x^{2/5}y^{3/5}$ **39.** $\sqrt[5]{y^4}$ **40.** $5\sqrt[3]{xy^2z^5}$ **41.** $\dfrac{1}{\sqrt[3]{x+2}}$ **42.** $\dfrac{1}{\sqrt{x+2y}}$ **43.** $a^{13/6}$

**44.** $\dfrac{1}{b}$ **45.** $\dfrac{1}{a^{9/2}}$ **46.** $\dfrac{y^2}{x}$ **47.** $a^4b^6$ **48.** $\dfrac{1}{x^{11/12}}$ **49.** $\dfrac{b^{5/6}}{49a^{1/4}c^{5/3}}$ **50.** $a - a^2$ **51.** 4.472 **52.** $-3.391$ **53.** 5.191 **54.** 3.826

**55.** $-26.246$ **56.** 0.045 **57.** $\sqrt[6]{1372}$ **58.** $\sqrt[12]{81x^3}$ **59.** $2\sqrt{6}$ **60.** $\sqrt[3]{7x^2yz}$ **61.** $2x$ **62.** $ab^3$ **63.** $2\sqrt{15}$ **64.** $-5\sqrt{3}$ **65.** $3\sqrt[3]{6}$

$2\sqrt[3]{4}$  **67.** $6x^3\sqrt{x}$  **68.** $2ab^2\sqrt[3]{3a^2b}$  **69.** $\dfrac{p^8\sqrt{p}}{11}$  **70.** $\dfrac{y\sqrt[3]{y^2}}{3x^2}$  **71.** $\dfrac{y\sqrt[4]{xy^2}}{3}$  **72.** $\dfrac{x\sqrt{2x}}{7y^2}$  **73.** $\dfrac{5}{\sqrt{\pi}}$ m or $\dfrac{5\sqrt{\pi}}{\pi}$ m  **74.** 5.75 in.

**.** $\sqrt{197}$ units $\approx$ 14.036 units  **76.** $\sqrt{130}$ units $\approx$ 11.402 units  **77.** $\sqrt{73}$ units $\approx$ 8.544 units  **78.** $7\sqrt{2}$ units $\approx$ 9.899 units

**79.** $2\sqrt{11}$ units $\approx$ 6.633 units  **80.** $\sqrt{275.6}$ units $\approx$ 16.601 units  **81.** $(-5, 5)$  **82.** $\left(-\dfrac{15}{2}, 1\right)$  **83.** $\left(-\dfrac{11}{2}, -2\right)$  **84.** $\left(\dfrac{1}{20}, -\dfrac{3}{16}\right)$

**85.** $\left(\dfrac{1}{4}, -\dfrac{2}{7}\right)$  **86.** $(\sqrt{3}, -3\sqrt{6})$  **87.** $-2\sqrt{5}$  **88.** $2x\sqrt{3x}$  **89.** $9\sqrt[3]{2}$  **90.** $3a\sqrt[4]{2a}$  **91.** $\dfrac{15 + 2\sqrt{3}}{6}$  **92.** $\dfrac{3\sqrt{2}}{4x}$  **93.** $17\sqrt{2} - 15\sqrt{5}$

**94.** $-4ab\sqrt[4]{2b}$  **95.** 6  **96.** $x - 6\sqrt{x} + 9$  **97.** $-8\sqrt{5}$  **98.** $4x - 9y$  **99.** $a - 9$  **100.** $\sqrt[3]{a^2} + 4\sqrt[3]{a} + 4$  **101.** $\sqrt[3]{25x^2} - 81$

**102.** $a + 64$  **103.** $\dfrac{3\sqrt{7}}{7}$  **104.** $\dfrac{\sqrt{3x}}{6}$  **105.** $\dfrac{5\sqrt[3]{2}}{2}$  **106.** $\dfrac{2x^2\sqrt{2xy}}{y}$  **107.** $\dfrac{x^2y^2\sqrt[3]{15yz}}{z}$  **108.** $\dfrac{3\sqrt[4]{2x^2}}{2x^3}$  **109.** $\dfrac{3\sqrt{y} + 6}{y - 4}$  **110.** $-5 + 2\sqrt{6}$

**111.** $\dfrac{11}{3\sqrt{11}}$  **112.** $\dfrac{6}{\sqrt{2y}}$  **113.** $\dfrac{3}{7\sqrt[3]{3}}$  **114.** $\dfrac{4x^3}{y\sqrt{2x}}$  **115.** $\dfrac{xy}{\sqrt[3]{10x^2yz}}$  **116.** $\dfrac{x - 25}{-3\sqrt{x} + 15}$  **117.** $\{32\}$  **118.** $\varnothing$  **119.** $\{35\}$  **120.** $\varnothing$

**121.** $\{9\}$  **122.** $\{16\}$  **123.** $3\sqrt{2}$ cm  **124.** $\sqrt{241}$ ft  **125.** 51.2 ft  **126.** 4.24 ft  **127.** $2i\sqrt{2}$  **128.** $-i\sqrt{6}$  **129.** $6i$  **130.** $-\sqrt{10}$

**131.** $15 - 4i$  **132.** $-13 - 3i$  **133.** $-64$  **134.** 81  **135.** $-12 - 18i$  **136.** $1 + 5i$  **137.** $-5 - 12i$  **138.** 87  **139.** $\dfrac{3}{2} - i$

**140.** $-\dfrac{1}{3} + \dfrac{1}{3}i$  **141.** $x$  **142.** $|x + 2|$  **143.** $-10$  **144.** $-x^4y$  **145.** $\dfrac{y^5}{2x^3}$  **146.** 3  **147.** $\dfrac{1}{8}$  **148.** $\dfrac{16}{9}$  **149.** $\dfrac{1}{x^{13/2}}$  **150.** $10x^4\sqrt{2x}$

**151.** $\dfrac{n\sqrt{3n}}{11m^5}$  **152.** $6\sqrt{5} - 11x\sqrt[3]{5}$  **153.** $4x - 20\sqrt{x} + 25$  **154.** $\sqrt{41}$ units  **155.** $(4, 16)$  **156.** $\dfrac{7\sqrt{13}}{13}$  **157.** $\dfrac{2\sqrt{x} - 6}{x - 9}$

**158.** $\{4\}$  **159.** $\{5\}$

**Chapter 7 Test**  **1.** $6\sqrt{6}$  **2.** $-x^{16}$  **3.** $\dfrac{1}{5}$  **4.** 5  **5.** $\dfrac{4x^2}{9}$  **6.** $-a^6b^3$  **7.** $\dfrac{8a^{1/3}c^{2/3}}{b^{5/12}}$  **8.** $a^{7/12} - a^{7/3}$  **9.** $|4xy|$ or $4|xy|$  **10.** $-27$

**11.** $\dfrac{3\sqrt{y}}{y}$  **12.** $\dfrac{8 - 6\sqrt{x} + x}{8 - 2x}$  **13.** $\dfrac{2\sqrt[3]{3x^2}}{3x}$  **14.** $\dfrac{6 - x^2}{8(\sqrt{6} - x)}$  **15.** $-x\sqrt{5x}$  **16.** $4\sqrt{3} - \sqrt{6}$  **17.** $x + 2\sqrt{x} + 1$

**18.** $\sqrt{6} + \sqrt{2} - 4\sqrt{3} - 4$  **19.** $-20$  **20.** 23.685  **21.** 0.019  **22.** $\{2, 3\}$  **23.** $\varnothing$  **24.** $\{6\}$  **25.** $i\sqrt{2}$  **26.** $-2i\sqrt{2}$  **27.** $-3i$

**28.** 40  **29.** $7 + 24i$  **30.** $-\dfrac{3}{2} + \dfrac{5}{2}i$  **31.** $x = \dfrac{5\sqrt{2}}{2}$ in.  **32.** $\sqrt{2}; 5$  **33.** $2\sqrt{26}$ units  **34.** $\sqrt{95}$ units  **35.** $\left(-4, \dfrac{7}{2}\right)$  **36.** $\left(-\dfrac{1}{2}, \dfrac{3}{10}\right)$

**37.** 27 mph  **38.** 360 ft

**Cumulative Review**  **1.** $4x^b$; Sec. 1.7, Ex. 16  **2.** $\dfrac{2}{15}x^{2a}$  **3.** $y^{5a+6}$; Sec. 1.7, Ex. 17  **4.** $-1.4y^{9a}$  **5.** $\left\{\dfrac{3}{4}, 5\right\}$; Sec. 2.6, Ex. 8  **6.** $\left\{\dfrac{2}{3}\right\}$

**7.** $y = -2x + 12$; Sec. 3.4, Ex. 5  **8.** $\dfrac{13}{14}$  **9.** $(-2, 2)$; Sec. 4.1, Ex. 8  **10.** 0  **11.** $\{(x, y, z) \mid x - 5y - 2z = 6\}$; Sec. 4.2, Ex. 4

**12.** $(6, 0)$  **13.** $\varnothing$; Sec. 4.4, Ex. 2  **14. a.** $7x - 3$  **b.** $-3x + 11$  **c.** $6x - 14$  **15.** $13x^3y - xy^3 + 7$; Sec. 5.1, Ex. 14  **16.** $5 \times 10^{-2}$

**17.** $-4$; Sec. 5.1, Ex. 18  **18.** $-5$  **19.** 35; Sec. 5.1, Ex. 19  **20.** $-\dfrac{85}{16}$  **21.** $x^2 + 10x + 25$; Sec. 5.2, Ex. 12  **22.** $3y^2 - 2y - 8$

**23.** $16m^4 - 24m^2n + 9n^2$; Sec. 5.2, Ex. 15  **24.** $6y^3 + 7y^2 - 6y + 1$  **25.** $2x^2 + 5x + 2 + \dfrac{7}{x - 3}$; Sec. 5.3, Ex. 9  **26.** $4y^2 - 1 + \dfrac{9}{y - 3}$

**27.** $(b - 6)(a + 2)$; Sec. 5.4, Ex. 8  **28.** $(x - 1)(x^2 + 4)$  **29.** $2(n^2 - 19n + 40)$; Sec. 5.5, Ex. 4  **30.** $2(x - 2)(x^2 + 2x + 4)$

**31.** $(4x + 3y)^2$; Sec. 5.5, Ex. 8  **32.** $(x + 1 + y)(x + 1 - y)$  **33.** $2(5 + 2y)(5 - 2y)$; Sec. 5.6, Ex. 7  **34.** $(x^2 + 1)(x + 1)(x - 1)$

**35.** $\left\{-\dfrac{1}{2}, 4\right\}$ Sec. 5.7, Ex. 3  **36.** $x^2 + 3$  **37.** $\{-2, 0, 2\}$; Sec. 5.7, Ex. 6  **38. a.** $\dfrac{3(y - 2)}{4(2y + 3)}$  **b.** $\dfrac{x + 4}{16}$  **39.** 1; Sec. 6.1, Ex. 4

**40.** $-a^2 - 2a - 4$  **41.** $\dfrac{1}{5x - 1}$; Sec. 6.1, Ex. 2  **42.** $\dfrac{3}{a + 5}$  **43.** $\dfrac{5 + x}{7z^2}$; Sec. 6.2, Ex. 1  **44. a.** $\dfrac{9x - 2y}{3x^2y^2}$  **b.** $\dfrac{3x(x - 7)}{(x + 3)(x - 3)}$

**c.** $\dfrac{x + 5}{x - 2}$  **45.** $\{-2\}$; Sec. 6.4, Ex. 2  **46.** $\{-1, -5\}$  **47.** $\dfrac{1}{8}$; Sec. 7.2, Ex. 12  **48.** $\dfrac{1}{27}$  **49.** $\dfrac{1}{9}$; Sec. 7.2, Ex. 13  **50.** $\dfrac{1}{25}$

# Chapter 8 Quadratic Equations and Functions

## Section 8.1

**Vocabulary, Readiness & Video Check**  **1.** $\pm\sqrt{b}$  **3.** completing the square  **5.** 9  **7.** We need a quantity shown squared by itself on one side of the equation. The only quantity squared is $x$, so divide both sides by 2 before applying the square root property.  **9.** The coefficient of $y^2$ is 3. To use the completing the square method, the coefficient of the squared variable must be 1, so we first divide through by 3.

**Exercise Set 8.1**  **1.** $\{-4, 4\}$  **3.** $\{-\sqrt{7}, \sqrt{7}\}$  **5.** $\{-3\sqrt{2}, 3\sqrt{2}\}$  **7.** $\{-\sqrt{10}, \sqrt{10}\}$  **9.** $\{-8, -2\}$

**11.** $\{6 - 3\sqrt{2}, 6 + 3\sqrt{2}\}$  **13.** $\left\{\dfrac{3 - 2\sqrt{2}}{2}, \dfrac{3 + 2\sqrt{2}}{2}\right\}$  **15.** $\{-3i, 3i\}$  **17.** $\{-\sqrt{6}, \sqrt{6}\}$  **19.** $\{-2i\sqrt{2}, 2i\sqrt{2}\}$

**21.** $\left\{\dfrac{1 - 4i}{3}, \dfrac{1 + 4i}{3}\right\}$    **23.** $\{-7 - \sqrt{5}, -7 + \sqrt{5}\}$    **25.** $\{-3 - 2i\sqrt{2}, -3 + 2i\sqrt{2}\}$    **27.** 1    **29.** 49    **31.** $x^2 + 16x + 64 = (x + 8)^2$

**33.** $z^2 - 12z + 36 = (z - 6)^2$    **35.** $p^2 + 9p + \dfrac{81}{4} = \left(p + \dfrac{9}{2}\right)^2$    **37.** $r^2 - r + \dfrac{1}{4} = \left(r - \dfrac{1}{2}\right)^2$    **39.** $\{-5, -3\}$

**41.** $\{-3 - \sqrt{7}, -3 + \sqrt{7}\}$    **43.** $\left\{\dfrac{-1 - \sqrt{5}}{2}, \dfrac{-1 + \sqrt{5}}{2}\right\}$    **45.** $\{-1 - \sqrt{6}, -1 + \sqrt{6}\}$    **47.** $\left\{\dfrac{-1 - \sqrt{29}}{2}, \dfrac{-1 + \sqrt{29}}{2}\right\}$

**49.** $\{-4 - \sqrt{15}, -4 + \sqrt{15}\}$    **51.** $\left\{\dfrac{6 - \sqrt{30}}{3}, \dfrac{6 + \sqrt{30}}{3}\right\}$    **53.** $\left\{-4, \dfrac{1}{2}\right\}$    **55.** $\left\{\dfrac{-3 - \sqrt{21}}{3}, \dfrac{-3 + \sqrt{21}}{3}\right\}$    **57.** $\{-1 - i, -1 + i\}$

**59.** $\{-2 - i\sqrt{2}, -2 + i\sqrt{2}\}$    **61.** $\left\{\dfrac{1 - i\sqrt{47}}{4}, \dfrac{1 + i\sqrt{47}}{4}\right\}$    **63.** $\{-5 - i\sqrt{3}, -5 + i\sqrt{3}\}$    **65.** $\{-4, 1\}$

**67.** $\left\{\dfrac{2 - i\sqrt{2}}{2}, \dfrac{2 + i\sqrt{2}}{2}\right\}$    **69.** $\left\{\dfrac{-3 - \sqrt{69}}{6}, \dfrac{-3 + \sqrt{69}}{6}\right\}$    **71.** 20%    **73.** 11%    **75.** 9.63 sec    **77.** 14.32 sec    **79.** 8.29 sec

**81.** 15 ft by 15 ft    **83.** $10\sqrt{2}$ cm    **85.** $3 + 2\sqrt{5}$    **87.** $\dfrac{1 - 3\sqrt{2}}{2}$    **89.** $2\sqrt{6}$    **91.** 5    **93.** complex, but not real numbers

**95.** real numbers    **97.** complex, but not real numbers    **99.** answers may vary    **101.** compound; answers may vary    **103.** $-8x, 8x$
**105.** 6 thousand scissors

## Section 8.2

**Calculator Explorations    1.** $\{-1.27, 6.27\}$    **3.** $\{-1.10, 0.90\}$    **5.** $\varnothing$; answers may vary

**Vocabulary, Readiness & Video Check    1.** $x = \dfrac{-b \pm \sqrt{b^2 - 4ac}}{2a}$    **3.** $-5; -7$    **5.** $1; 0$    **7. a.** Yes, in order to make sure we have
correct values for $a, b,$ and $c$.    **b.** No; clearing fractions makes the work less tedious, but it's not a necessary step.    **9.** With applications,
we need to make sure we answer the question(s) asked. Here we're asked how much distance is saved, so once the dimensions of the
triangle are known, further calculations are needed to answer this question and solve the problem.

**Exercise Set 8.2    1.** $\{-6, 1\}$    **3.** $\left\{-\dfrac{3}{5}, 1\right\}$    **5.** $\{3\}$    **7.** $\left\{\dfrac{-7 - \sqrt{33}}{2}, \dfrac{-7 + \sqrt{33}}{2}\right\}$    **9.** $\left\{\dfrac{1 - \sqrt{57}}{8}, \dfrac{1 + \sqrt{57}}{8}\right\}$

**11.** $\left\{\dfrac{7 - \sqrt{85}}{6}, \dfrac{7 + \sqrt{85}}{6}\right\}$    **13.** $\{1 - \sqrt{3}, 1 + \sqrt{3}\}$    **15.** $\left\{-\dfrac{3}{2}, 1\right\}$    **17.** $\left\{\dfrac{3 - \sqrt{11}}{2}, \dfrac{3 + \sqrt{11}}{2}\right\}$    **19.** $\left\{\dfrac{-5 - \sqrt{17}}{2}, \dfrac{-5 + \sqrt{17}}{2}\right\}$

**21.** $\left\{\dfrac{5}{2}, 1\right\}$    **23.** $\{-3 - 2i, -3 + 2i\}$    **25.** $\{-2 - \sqrt{11}, -2 + \sqrt{11}\}$    **27.** $\left\{\dfrac{3 - i\sqrt{87}}{8}, \dfrac{3 + i\sqrt{87}}{8}\right\}$    **29.** $\left\{\dfrac{3 - \sqrt{29}}{2}, \dfrac{3 + \sqrt{29}}{2}\right\}$

**31.** $\left\{\dfrac{-5 - i\sqrt{5}}{10}, \dfrac{-5 + i\sqrt{5}}{10}\right\}$    **33.** $\left\{\dfrac{-1 - \sqrt{19}}{6}, \dfrac{-1 + \sqrt{19}}{6}\right\}$    **35.** $\left\{\dfrac{-1 - i\sqrt{23}}{4}, \dfrac{-1 + i\sqrt{23}}{4}\right\}$    **37.** $\{1\}$    **39.** $\{3 + \sqrt{5}, 3 - \sqrt{5}\}$

**41.** two real solutions    **43.** one real solution    **45.** two real solutions    **47.** two complex but not real solutions    **49.** two real solutions
**51.** 14 ft    **53.** $(2 + 2\sqrt{2})$ cm, $(2 + 2\sqrt{2})$ cm, $(4 + 2\sqrt{2})$ cm    **55.** width: $(-5 + 5\sqrt{17})$ ft; length: $(5 + 5\sqrt{17})$ ft    **57. a.** $50\sqrt{2}$ m

**b.** 5000 sq m    **59.** 37.4 ft by 38.5 ft    **61.** base: $(2 + 2\sqrt{43})$ cm; height: $(-1 + \sqrt{43})$ cm    **63.** 8.9 sec    **65.** 2.8 sec    **67.** $\left\{\dfrac{11}{5}\right\}$

**69.** $\{15\}$    **71.** $(x^2 + 5)(x + 2)(x - 2)$    **73.** $(z + 3)(z - 3)(z + 2)(z - 2)$    **75.** b    **77.** answers may vary    **79.** $\{0.6, 2.4\}$
**81.** Sunday to Monday    **83.** Wednesday    **85.** $f(4) = 33$; answers may vary    **87. a.** 94.64 billion kilowatts    **b.** 2015    **89. a.** 215 million
**b.** 620 million    **c.** 2018    **91.** $\{0.6, 2.4\}$

## Section 8.3

**Vocabulary, Readiness & Video Check    1.** The values we get for the substituted variable are *not* our final answers. Remember to
always substitute back to the original variable and solve for it if necessary.

**Exercise Set 8.3    1.** $\{2\}$    **3.** $\{16\}$    **5.** $\{1, 4\}$    **7.** $\{3 - \sqrt{7}, 3 + \sqrt{7}\}$    **9.** $\left\{\dfrac{3 - \sqrt{57}}{4}, \dfrac{3 + \sqrt{57}}{4}\right\}$    **11.** $\left\{\dfrac{1 - \sqrt{29}}{2}, \dfrac{1 + \sqrt{29}}{2}\right\}$

**13.** $\{-2, 2, -2i, 2i\}$    **15.** $\{-3, 3, -2i, 2i\}$    **17.** $\left\{-\dfrac{1}{2}, \dfrac{1}{2}, -i\sqrt{3}, i\sqrt{3}\right\}$    **19.** $\{125, -8\}$    **21.** $\left\{-\dfrac{4}{5}, 0\right\}$    **23.** $\left\{-\dfrac{1}{8}, 27\right\}$    **25.** $\left\{-\dfrac{2}{3}, \dfrac{4}{3}\right\}$

**27.** $\left\{-\dfrac{1}{125}, \dfrac{1}{8}\right\}$    **29.** $\{-\sqrt{2}, \sqrt{2}, -\sqrt{3}, \sqrt{3}\}$    **31.** $\left\{\dfrac{-9 - \sqrt{201}}{6}, \dfrac{-9 + \sqrt{201}}{6}\right\}$    **33.** $\{2, 3\}$    **35.** $\{3\}$    **37.** $\{27, 125\}$    **39.** $\{5\}$

**41.** $\left\{\dfrac{1}{8}, -8\right\}$    **43.** $\{-5, 1\}$    **45.** $\{4\}$    **47.** $\{-3\}$    **49.** $\{-\sqrt{5}, \sqrt{5}, -2i, 2i\}$    **51.** $\{6, 12\}$    **53.** $\left\{-\dfrac{1}{3}, \dfrac{1}{3}, -\dfrac{i\sqrt{6}}{3}, \dfrac{i\sqrt{6}}{3}\right\}$

**55.** 5 mph, then 4 mph    **57.** inlet pipe: 15.5 hr; hose: 16.5 hr    **59.** 55 mph; 66 mph    **61.** 8.5 hr    **63.** 12 or $-8$    **65. a.** $(x - 6)$ in.

**b.** $300 = (x - 6) \cdot (x - 6) \cdot 3$    **c.** 16 in. by 16 in.    **67.** 22 ft    **69.** $(-\infty, 3]$    **71.** $(-5, \infty)$    **73.** $\{1, -3i, 3i\}$    **75.** $\left\{-\dfrac{1}{2}, \dfrac{1}{3}\right\}$

**77.** $\left\{-3, \dfrac{3 - 3i\sqrt{3}}{2}, \dfrac{3 + 3i\sqrt{3}}{2}\right\}$    **79.** answers may vary    **81. a.** 51.72 sec    **b.** 54.11 sec    **c.** 106.66 sec    **d.** 53.85 sec

**Integrated Review   1.** $\{-\sqrt{10}, \sqrt{10}\}$   **2.** $\{-2i\sqrt{2}, 2i\sqrt{2}\}$   **3.** $\{1 - 2\sqrt{2}, 1 + 2\sqrt{2}\}$   **4.** $\left\{\dfrac{-5 - 2\sqrt{3}}{2}, \dfrac{-5 + 2\sqrt{3}}{2}\right\}$

**5.** $\{-1 - \sqrt{13}, -1 + \sqrt{13}\}$   **6.** $\{1, 11\}$   **7.** $\left\{\dfrac{-3 - \sqrt{17}}{2}, \dfrac{-3 + \sqrt{17}}{2}\right\}$   **8.** $\left\{\dfrac{-2 - \sqrt{5}}{4}, \dfrac{-2 + \sqrt{5}}{4}\right\}$   **9.** $\left\{\dfrac{2 - \sqrt{2}}{2}, \dfrac{2 + \sqrt{2}}{2}\right\}$

**10.** $\{-3 - \sqrt{5}, -3 + \sqrt{5}\}$   **11.** $\{-2 + i\sqrt{3}, -2 - i\sqrt{3}\}$   **12.** $\left\{\dfrac{-3 - i\sqrt{6}}{5}, \dfrac{-3 + i\sqrt{6}}{5}\right\}$   **13.** $\left\{\dfrac{-3 + i\sqrt{15}}{2}, \dfrac{-3 - i\sqrt{15}}{2}\right\}$

**14.** $\{3i, -3i\}$   **15.** $\{0, -17\}$   **16.** $\left\{\dfrac{1 + \sqrt{13}}{4}, \dfrac{1 - \sqrt{13}}{4}\right\}$   **17.** $\{2 + 3\sqrt{3}, 2 - 3\sqrt{3}\}$   **18.** $\{2 + \sqrt{3}, 2 - \sqrt{3}\}$   **19.** $\left\{-2, \dfrac{4}{3}\right\}$

**20.** $\left\{\dfrac{-5 + \sqrt{17}}{4}, \dfrac{-5 - \sqrt{17}}{4}\right\}$   **21.** $\{1 - \sqrt{6}, 1 + \sqrt{6}\}$   **22.** $\{-\sqrt{31}, \sqrt{31}\}$   **23.** $\{-2\sqrt{3}, 2\sqrt{3}\}$   **24.** $\{-i\sqrt{11}, i\sqrt{11}\}$

**25.** $\{-11, 6\}$   **26.** $\left\{\dfrac{-3 + \sqrt{19}}{5}, \dfrac{-3 - \sqrt{19}}{5}\right\}$   **27.** $\left\{\dfrac{-3 + \sqrt{17}}{4}, \dfrac{-3 - \sqrt{17}}{4}\right\}$   **28.** $\{4\}$   **29.** $\left\{\dfrac{-1 + \sqrt{17}}{8}, \dfrac{-1 - \sqrt{17}}{8}\right\}$

**30.** $10\sqrt{2}$ ft $\approx 14.1$ ft   **31.** Jack: 9.1 hr; Lucy: 7.1 hr   **32.** 5 mph during the first part, then 6 mph

## Section 8.4

**Vocabulary, Readiness & Video Check   1.** $[-7, 3)$   **3.** $(-\infty, 0]$   **5.** $(-\infty, -12) \cup [-10, \infty)$   **7.** We use the solutions to the related equation to divide the number line into regions that either entirely are or entirely are not solution regions; the solutions to the related equation are solutions to the inequality only if the inequality symbol is $\leq$ or $\geq$.

**Exercise Set 8.4   1.** $(-\infty, -5) \cup (-1, \infty)$   **3.** $[-4, 3]$   **5.** $(-\infty, -5] \cup [-3, \infty)$   **7.** $\left(-5, -\dfrac{1}{3}\right)$   **9.** $(2, 4) \cup (6, \infty)$

**11.** $(-\infty, -4] \cup [0, 1]$   **13.** $(-\infty, -3) \cup (-2, 2) \cup (3, \infty)$   **15.** $(-7, 2)$   **17.** $(-1, \infty)$   **19.** $(-\infty, -1] \cup (4, \infty)$

**21.** $(-\infty, 2) \cup \left(\dfrac{11}{4}, \infty\right)$   **23.** $(0, 2] \cup [3, \infty)$   **25.** $(-\infty, 3)$   **27.** $\left[-\dfrac{5}{4}, \dfrac{3}{2}\right]$   **29.** $(-\infty, 0) \cup (1, \infty)$   **31.** $(0, 10)$

**33.** $(-\infty, -4] \cup [4, 6]$   **35.** $\left(-\infty, -\dfrac{2}{3}\right] \cup \left[\dfrac{3}{2}, \infty\right)$   **37.** $(-\infty, -4) \cup [5, \infty)$   **39.** $(-\infty, 1) \cup (2, \infty)$   **41.** $\left(-4, -\dfrac{3}{2}\right) \cup \left(\dfrac{3}{2}, \infty\right)$

**43.** $(-\infty, -5] \cup [-1, 1] \cup [5, \infty)$   **45.** $(-\infty, -6] \cup (-1, 0] \cup (7, \infty)$   **47.** $(-\infty, -8] \cup (-4, \infty)$   **49.** $\left(-\infty, -\dfrac{5}{3}\right) \cup \left(\dfrac{7}{2}, \infty\right)$

**51.** $(-\infty, 0] \cup \left(5, \dfrac{11}{2}\right]$   **53.** $(0, \infty)$   **55.** $0; 1; 1; 4; 4$   **57.** $0; -1; -1; -4; -4$   **59.** answers may vary   **61.** $(-\infty, -1) \cup (0, 1)$

**63.** when $x$ is between 2 and 11   **65.** $(-\infty, -7) \cup (8, \infty)$

## Section 8.5

**Calculator Explorations   1.**

**Vocabulary, Readiness & Video Check   1.** quadratic   **3.** upward   **5.** lowest   **7.** $(0, 0)$   **9.** $(2, 0)$   **11.** $(0, 3)$   **13.** $(-1, 5)$
**15.** Graphs of the form $f(x) = x^2 + k$ shift up or down the $y$-axis $k$ units from $y = x^2$; the $y$-intercept.   **17.** The vertex, $(h, k)$ and the axis of symmetry, $x = h$; the basic shape of $y = x^2$ does not change.   **19.** the coordinates of the vertex, whether the graph opens upward or downward, whether the graph is narrower or wider than $y = x^2$, and the graph's axis of symmetry

**Exercise Set 8.5   1.**

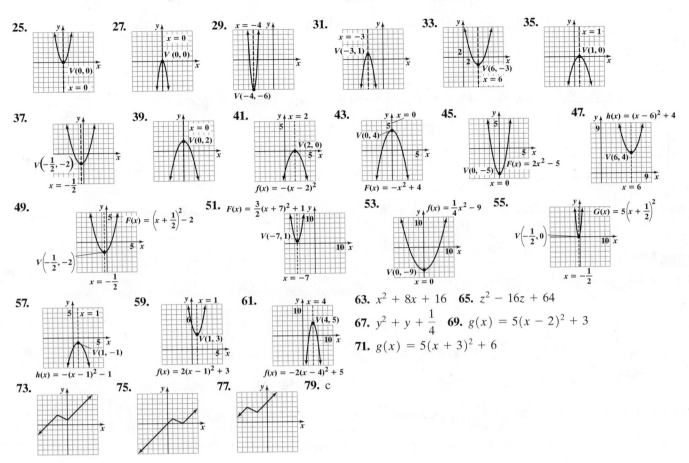

**63.** $x^2 + 8x + 16$   **65.** $z^2 - 16z + 64$

**67.** $y^2 + y + \dfrac{1}{4}$   **69.** $g(x) = 5(x - 2)^2 + 3$

**71.** $g(x) = 5(x + 3)^2 + 6$

**79.** c

## Section 8.6

**Vocabulary, Readiness & Video Check**   **1.** $(h, k)$   **3.** We can immediately identify the vertex $(h, k)$, whether the parabola opens upward or downward, and know its axis of symmetry; completing the square.   **5.** the vertex

**Exercise Set 8.6**   **1.** $0; 1$   **3.** $2; 1$   **5.** $1; 1$   **7.** down   **9.** up   **11.** $(-4, -9)$   **13.** $(5, 30)$   **15.** $(1, -2)$   **17.** $\left(\dfrac{1}{2}, \dfrac{5}{4}\right)$   **19.** D   **21.** B

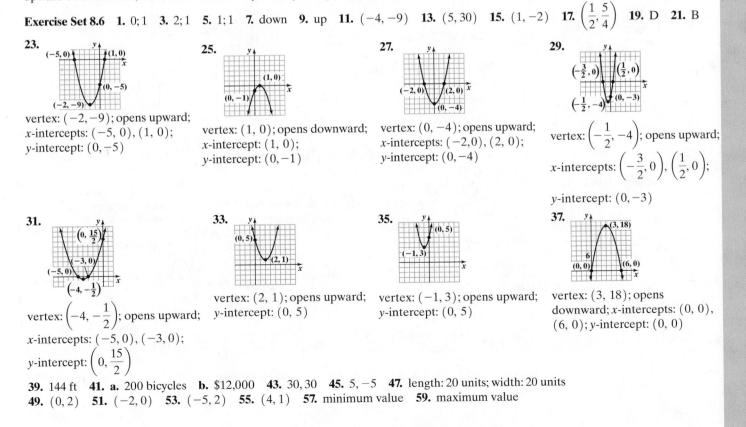

**23.** vertex: $(-2, -9)$; opens upward; $x$-intercepts: $(-5, 0), (1, 0)$; $y$-intercept: $(0, -5)$

**25.** vertex: $(1, 0)$; opens downward; $x$-intercept: $(1, 0)$; $y$-intercept: $(0, -1)$

**27.** vertex: $(0, -4)$; opens upward; $x$-intercepts: $(-2, 0), (2, 0)$; $y$-intercept: $(0, -4)$

**29.** vertex: $\left(-\dfrac{1}{2}, -4\right)$; opens upward; $x$-intercepts: $\left(-\dfrac{3}{2}, 0\right), \left(\dfrac{1}{2}, 0\right)$; $y$-intercept: $(0, -3)$

**31.** vertex: $\left(-4, -\dfrac{1}{2}\right)$; opens upward; $x$-intercepts: $(-5, 0), (-3, 0)$; $y$-intercept: $\left(0, \dfrac{15}{2}\right)$

**33.** vertex: $(2, 1)$; opens upward; $y$-intercept: $(0, 5)$

**35.** vertex: $(-1, 3)$; opens upward; $y$-intercept: $(0, 5)$

**37.** vertex: $(3, 18)$; opens downward; $x$-intercepts: $(0, 0)$, $(6, 0)$; $y$-intercept: $(0, 0)$

**39.** 144 ft   **41. a.** 200 bicycles   **b.** $12,000   **43.** 30, 30   **45.** 5, -5   **47.** length: 20 units; width: 20 units
**49.** $(0, 2)$   **51.** $(-2, 0)$   **53.** $(-5, 2)$   **55.** $(4, 1)$   **57.** minimum value   **59.** maximum value

**61.**    vertex: $(-5, -10)$; opens upward; $y$-intercept: $(0, 15)$; $x$-intercepts: $(-1.8, 0)$, $(-8.2, 0)$

**63.**    **65.** $-0.84$   **67. a.** maximum, answers may vary   **b.** 2030   **c.** 99.9%

**Chapter 8 Vocabulary Check   1.** discriminant   **2.** $\pm\sqrt{b}$   **3.** $\dfrac{-b}{2a}$   **4.** quadratic inequality   **5.** completing the square   **6.** $(0, k)$
**7.** $(h, 0)$   **8.** $(h, k)$   **9.** quadratic formula   **10.** quadratic

**Chapter 8 Review   1.** $\{14, 1\}$   **2.** $\left\{-\dfrac{6}{7}, 5\right\}$   **3.** $\{-7, 7\}$   **4.** $\left\{\dfrac{2-\sqrt{2}}{5}, \dfrac{2+\sqrt{2}}{5}\right\}$   **5.** $\left\{\dfrac{-3-\sqrt{5}}{2}, \dfrac{-3+\sqrt{5}}{2}\right\}$
**6.** $\left\{\dfrac{-3-i\sqrt{7}}{8}, \dfrac{-3+i\sqrt{7}}{8}\right\}$   **7.** 4.25%   **8.** $75\sqrt{2}$ mi; 106.1 mi   **9.** two complex but not real solutions   **10.** two real solutions

**11.** two real solutions   **12.** one real solution   **13.** $\{8\}$   **14.** $\{-5, 0\}$   **15.** $\left\{-\dfrac{5}{2}, 1\right\}$   **16.** $\left\{\dfrac{5-i\sqrt{143}}{12}, \dfrac{5+i\sqrt{143}}{12}\right\}$
**17.** $\left\{\dfrac{1-i\sqrt{35}}{9}, \dfrac{1+i\sqrt{35}}{9}\right\}$   **18.** $\left\{1, \dfrac{9}{4}\right\}$   **19. a.** 20 ft   **b.** $\dfrac{15+\sqrt{321}}{16}$ sec; 2.1 sec   **20.** $(6+6\sqrt{2})$ cm
**21.** $\left\{3, \dfrac{-3+3i\sqrt{3}}{2}, \dfrac{-3-3i\sqrt{3}}{2}\right\}$   **22.** $\{-4, 2-2i\sqrt{3}, 2+2i\sqrt{3}\}$   **23.** $\left\{\dfrac{2}{3}, 5\right\}$   **24.** $\{-5, 5, -2i, 2i\}$   **25.** $\left\{-\dfrac{16}{5}, 1\right\}$
**26.** $\{1, 125\}$   **27.** $\{-1, 1, -i, i\}$   **28.** $\left\{-\dfrac{1}{5}, \dfrac{1}{4}\right\}$   **29.** Jerome: 10.5 hr; Tim: 9.5 hr   **30.** $-5$   **31.** $[-5, 5]$   **32.** $\left(-\dfrac{1}{2}, \dfrac{1}{2}\right)$

**33.** $(-\infty, -4) \cup (-1, 1) \cup (4, \infty)$   **34.** $[-5, -2] \cup [2, 5]$   **35.** $(5, 6)$   **36.** $(-\infty, -6) \cup \left(-\dfrac{3}{4}, 0\right) \cup (5, \infty)$

**37.** $(-\infty, -5] \cup [-2, 6]$   **38.** $(-5, -3) \cup (5, \infty)$   **39.** $(-\infty, 0)$   **40.** $\left(-\dfrac{6}{5}, 0\right) \cup \left(\dfrac{5}{6}, 3\right)$

**41.**   **42.**   **43.**   **44.**   **45.**   **46.**

**47.**   **48.**   **49.** vertex: $(-5, 0)$;   **50.** vertex: $(3, 0)$;   **51.** vertex: $(0, -1)$;
$x$-intercept: $(-5, 0)$;   $x$-intercept: $(3, 0)$;   $x$-intercepts: $\left(-\dfrac{1}{2}, 0\right), \left(\dfrac{1}{2}, 0\right)$;
$y$-intercept: $(0, 25)$   $y$-intercept: $(0, -9)$   $y$-intercept: $(0, -1)$

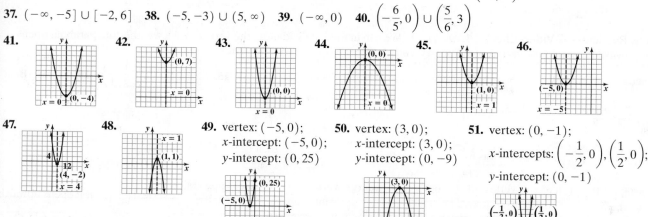

**52.** vertex: $(0, 5)$;
$x$-intercepts: $(-1, 0)$,
$(1, 0)$; $y$-intercept: $(0, 5)$

**53.** vertex: $\left(-\dfrac{5}{6}, \dfrac{73}{12}\right)$; opens downward;   **54. a.** 0.4 sec and 7.1 sec   **b.** answers may vary
$x$-intercepts: $(-2.3, 0)$, $(0.6, 0)$;   **55.** The numbers are both 210.   **56.** 50, $-50$
$y$-intercept: $(0, 4)$   **57.** $\{-5, 6\}$   **58.** $\left\{\dfrac{4}{5}, -\dfrac{1}{2}\right\}$   **59.** $\{-2, 2\}$   **60.** $\left\{-\dfrac{4}{9}, \dfrac{2}{9}\right\}$
**61.** $\left\{\dfrac{-1-3i\sqrt{3}}{2}, \dfrac{-1+3i\sqrt{3}}{2}\right\}$
**62.** $\left\{\dfrac{17-\sqrt{145}}{9}, \dfrac{17+\sqrt{145}}{9}\right\}$

**63.** $\{-i\sqrt{11}, i\sqrt{11}\}$   **64.** $\{-i\sqrt{7}, i\sqrt{7}\}$   **65.** $\left\{\dfrac{21-\sqrt{41}}{50}, \dfrac{21+\sqrt{41}}{50}\right\}$   **66.** $\left\{-\dfrac{8\sqrt{7}}{7}, \dfrac{8\sqrt{7}}{7}\right\}$   **67.** $\{8, 64\}$

**68.** $\left(-\infty, -\dfrac{5}{4}\right] \cup \left[\dfrac{3}{2}, \infty\right)$   **69.** $[-5, 0] \cup \left(\dfrac{3}{4}, \infty\right)$   **70.** $\left(2, \dfrac{7}{2}\right)$   **71. a.** 111,395 thousand passengers   **b.** 2019   **72.** 6.62 sec

**Chapter 8 Test** **1.** $\left\{\dfrac{7}{5}, -1\right\}$ **2.** $\{-1 - \sqrt{10}, -1 + \sqrt{10}\}$ **3.** $\left\{\dfrac{1 + i\sqrt{31}}{2}, \dfrac{1 - i\sqrt{31}}{2}\right\}$ **4.** $\{3 - \sqrt{7}, 3 + \sqrt{7}\}$

**5.** $\left\{-\dfrac{1}{7}, -1\right\}$ **6.** $\left\{\dfrac{3 + \sqrt{29}}{2}, \dfrac{3 - \sqrt{29}}{2}\right\}$ **7.** $\{-2 - \sqrt{11}, -2 + \sqrt{11}\}$ **8.** $\{-3, 3, -i, i\}$ **9.** $\{-1, 1, -i, i\}$ **10.** $\{6, 7\}$

**11.** $\{3 - \sqrt{7}, 3 + \sqrt{7}\}$ **12.** $\left\{\dfrac{2 - i\sqrt{6}}{2}, \dfrac{2 + i\sqrt{6}}{2}\right\}$ **13.** $\left(-\infty, -\dfrac{3}{2}\right) \cup (5, \infty)$ **14.** $(-\infty, -5) \cup (-4, 4) \cup (5, \infty)$

**15.** $(-\infty, -3) \cup (2, \infty)$ **16.** $(-\infty, -3) \cup [2, 3)$

**17.** **18.** **19.** **20.** **21.** 7 ft **22.** $(5 + \sqrt{17})$ hr $\approx 9.12$ hr
**23. a.** 272 ft **b.** 5.12 sec

**Cumulative Review** **1.** $7.3 \times 10^5$; Sec. 1.6, Ex. 28 **2.** $0.002068$ **3.** $\varnothing$; Sec. 2.6, Ex. 6 **4.** $\varnothing$ **5.** not a function; Sec. 3.6, Ex. 8
**6.** vertex: $\left(-\dfrac{1}{2}, -\dfrac{49}{4}\right)$; $y$-intercept: $(0, -12)$; $x$-intercepts: $(3, 0), (-4, 0)$ **7.** $\varnothing$; Sec. 4.1, Ex. 9 **8.** $(-2, -7)$ **9.** $6x^3 - 4x^2 - 2$;
Sec. 5.1, Ex. 16 **10.** $2 - x$ **11.** $6x^2 - 29x + 28$; Sec. 5.2, Ex. 10 **12.** $28a^2 - 29a + 6$ **13.** $15x^4 - x^2y - 2y^2$; Sec. 5.2, Ex. 11
**14.** $6a^2 - 7ab - 5b^2$ **15.** $3x^2 + 2x + 3 + \dfrac{-6x + 9}{x^2 - 1}$; Sec. 5.3, Ex. 6 **16. a.** $\dfrac{a^6}{b^3c^9}$ **b.** $\dfrac{a^8c^6}{b^4}$ **c.** $\dfrac{16b^6}{a^6}$ **17.** $3x^2(2 + x)$; Sec. 5.4, Ex. 3
**18.** $3x(3x^2 + 9x - 5)$ **19.** $(2a - 1)(a + 4)$; Sec. 5.5, Ex. 11 **20.** $(3y - 2)(2x - 5)$ **21.** $3x(a - 2b)^2$; Sec. 5.6, Ex. 4
**22.** $(y + 3)(2x - 1)$ **23.** $\left\{-\dfrac{1}{6}, 3\right\}$; Sec. 5.7, Ex. 5 **24.** $\left\{\dfrac{1}{2}\right\}$ **25.** $\dfrac{n - 2}{n(n - 1)}$; Sec. 6.1, Ex. 9 **26.** $-2 - 6\sqrt{3}$ **27.** $-5x^2$; Sec. 6.1, Ex. 10
**28.** $\dfrac{1}{x - 2}$ **29.** $\dfrac{6x + 5}{3x^3}$; Sec. 6.2, Ex. 6 **30.** $\dfrac{a^2 + 8a - 2}{(a - 2)(a - 4)(a + 4)}$ **31.** $\dfrac{x^2}{y^2}$; Sec. 6.3, Ex. 3 **32.** $\dfrac{2a + b}{a + 2b}$ **33.** $\{-6, -1\}$; Sec. 6.4, Ex. 5
**34.** $\{-1\}$ **35.** $x = \dfrac{yz}{y - z}$; Sec. 6.5, Ex. 1 **36.** $x^2 - 6x + 8$ **37.** $S = krh$; Sec. 6.6, Ex. 5 **38.** $k = 192$; $y = \dfrac{192}{x}$
**39.** $-4$; Sec. 7.1, Ex. 14 **40.** $2a^2b^3\sqrt[3]{3a}$ **41.** $\dfrac{2}{5}$; Sec. 7.1, Ex. 15 **42.** $\dfrac{8\sqrt{x}}{3x}$ **43.** $2$; Sec. 7.3, Ex. 3 **44.** $5 - 2x\sqrt{5} + x^2$
**45.** $\sqrt{\dfrac{2b}{3a}}$; Sec. 7.3, Ex. 5 **46.** $a - b^2$ **47.** $-6\sqrt[3]{2}$; Sec. 7.4, Ex. 5 **48.** $\{2, 6\}$ **49.** $\dfrac{5\sqrt[3]{7x}}{2}$; Sec. 7.4, Ex. 10 **50.** $2x^2\sqrt[4]{3x}$
**51.** $\{-1 + \sqrt{5}, -1 - \sqrt{5}\}$; Sec. 8.1, Ex. 7 **52.** $\{1 + 2\sqrt{6}, 1 - 2\sqrt{6}\}$ **53.** $\left\{\dfrac{2 + \sqrt{10}}{2}, \dfrac{2 - \sqrt{10}}{2}\right\}$; Sec. 8.2, Ex. 2
**54.** $\{2 - 2\sqrt{3}, 2 + 2\sqrt{3}\}$ **55.** $3.9$ sec; Sec. 8.2, Ex. 9 **56.** 15 hr

## Chapter 9 Exponential and Logarithmic Functions
### Section 9.1

**Vocabulary, Readiness & Video Check** **1.** C **3.** F **5.** D **7.** You can find $(f + g)(x)$ and then find $(f + g)(2)$, or you can find
$f(2)$ and $g(2)$ and then add those results.

**Exercise Set 9.1** **1. a.** $3x - 6$ **b.** $-x - 8$ **c.** $2x^2 - 13x - 7$ **d.** $\dfrac{x - 7}{2x + 1}$, where $x \neq -\dfrac{1}{2}$ **3. a.** $x^2 + 5x + 1$ **b.** $x^2 - 5x + 1$
**c.** $5x^3 + 5x$ **d.** $\dfrac{x^2 + 1}{5x}$, where $x \neq 0$ **5. a.** $\sqrt{x} + x + 5$ **b.** $\sqrt{x} - x - 5$ **c.** $x\sqrt{x} + 5\sqrt{x}$ **d.** $\dfrac{\sqrt{x}}{x + 5}$, where $x \neq -5$ **7. a.** $5x^2 - 3x$
**b.** $-5x^2 - 3x$ **c.** $-15x^3$ **d.** $-\dfrac{3}{5x}$, where $x \neq 0$ **9.** 42 **11.** $-18$ **13.** 0 **15.** $(f \circ g)(x) = 25x^2 + 1$; $(g \circ f)(x) = 5x^2 + 5$
**17.** $(f \circ g)(x) = 2x + 11$; $(g \circ f)(x) = 2x + 4$ **19.** $(f \circ g)(x) = -8x^3 - 2x - 2$; $(g \circ f)(x) = -2x^3 - 2x + 4$
**21.** $(f \circ g)(x) = |10x - 3|$; $(g \circ f)(x) = 10|x| - 3$ **23.** $(f \circ g)(x) = \sqrt{-5x + 2}$; $(g \circ f)(x) = -5\sqrt{x} + 2$ **25.** $H(x) = (g \circ h)(x)$
**27.** $F(x) = (h \circ f)(x)$ **29.** $G(x) = (f \circ g)(x)$ **31.** answers may vary **33.** answers may vary **35.** answers may vary **37.** $y = x - 2$
**39.** $y = \dfrac{x}{3}$ **41.** $y = -\dfrac{x + 7}{2}$ **43.** $P(x) = R(x) - C(x)$ **45.** answers may vary

### Section 9.2

**Vocabulary, Readiness & Video Check** **1.** $(2, 11)$ **3.** the inverse of $f$ **5.** vertical **7.** $y = x$ **9.** Every function must have each
$x$-value correspond to only one $y$-value. A one-to-one function must also have each $y$-value correspond to only one $x$-value. **11.** Yes;
by the definition of an inverse function. **13.** Once you know some points of the original equation or graph, you can switch the $x$'s
and $y$'s of these points to find points that satisfy the inverse and then graph it. You can also check that the two graphs (the original
and the inverse) are symmetric about the line $y = x$.

**Exercise Set 9.2**   **1.** not one-to-one   **3.** one-to-one; $h^{-1} = \{(10, 10)\}$   **5.** one-to-one; $f^{-1} = \{(12, 11), (3, 4), (4, 3), (6, 6)\}$
**7.** not one-to-one   **9.** one-to-one;

| Rank in Population (Input) | 1 | 47 | 16 | 25 | 36 | 7 |
|---|---|---|---|---|---|---|
| State (Output) | California | Alaska | Indiana | Louisiana | New Mexico | Ohio |

**11. a.** 3   **b.** 1   **13. a.** 1   **b.** $-1$   **15.** one-to-one   **17.** not one-to-one   **19.** one-to-one   **21.** not one-to-one

**23.** $f^{-1}(x) = x - 4$   **25.** $f^{-1}(x) = \dfrac{x + 3}{2}$   **27.** $f^{-1}(x) = 2x + 2$   **29.** $f^{-1}(x) = \sqrt[3]{x}$   **31.** $f^{-1}(x) = \dfrac{x - 2}{5}$   **33.** $f^{-1}(x) = 5x + 2$

**35.** $f^{-1}(x) = x^3$   **37.** $f^{-1}(x) = \dfrac{\frac{5}{x} - 1}{3}$

**39.** $f^{-1}(x) = \sqrt[3]{x} - 2$

**47.** $(f \circ f^{-1})(x) = x; (f^{-1} \circ f)(x) = x$   **49.** $(f \circ f^{-1})(x) = x; (f^{-1} \circ f)(x) = x$
**51.** 5   **53.** 8   **55.** $\dfrac{1}{27}$   **57.** 9   **59.** $3^{1/2} \approx 1.73$   **61. a.** $(2, 9)$   **b.** $(9, 2)$

**63. a.** $\left(-2, \dfrac{1}{4}\right), \left(-1, \dfrac{1}{2}\right), (0, 1), (1, 2), (2, 5)$

**b.** $\left(\dfrac{1}{4}, -2\right), \left(\dfrac{1}{2}, -1\right), (1, 0), (2, 1) (5, 2)$   **c.**   **d.**   **65.** answers may vary

**67.** $f^{-1}(x) = \dfrac{x - 1}{3};$   **69.** $f^{-1}(x) = x^3 - 3;$

## Section 9.3

**Calculator Explorations**   **1.** 81.98%   **3.** 22.54%

**Vocabulary, Readiness & Video Check**   **1.** exponential; c   **3.** yes   **5.** no; none   **7.** $(-\infty, \infty)$   **9.** In a polynomial function, the base is the variable and the exponent is the constant; in an exponential function, the base is the constant and the exponent is the variable.   **11.** $y = 30(0.996)^{101} \approx 20.0$ lb

**Exercise Set 9.3**   **1.**   **3.**   **5.**   **7.**   **9.**   **11.**

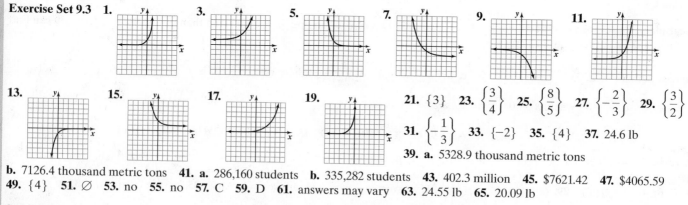

**13.**   **15.**   **17.**   **19.**   **21.** $\{3\}$   **23.** $\left\{\dfrac{3}{4}\right\}$   **25.** $\left\{\dfrac{8}{5}\right\}$   **27.** $\left\{-\dfrac{2}{3}\right\}$   **29.** $\left\{\dfrac{3}{2}\right\}$

**31.** $\left\{-\dfrac{1}{3}\right\}$   **33.** $\{-2\}$   **35.** $\{4\}$   **37.** 24.6 lb

**39. a.** 5328.9 thousand metric tons

**b.** 7126.4 thousand metric tons   **41. a.** 286,160 students   **b.** 335,282 students   **43.** 402.3 million   **45.** $7621.42   **47.** $4065.59
**49.** $\{4\}$   **51.** $\varnothing$   **53.** no   **55.** no   **57.** C   **59.** D   **61.** answers may vary   **63.** 24.55 lb   **65.** 20.09 lb

## Section 9.4

**Vocabulary, Readiness & Video Check**   **1.** The growth rate is given as 5% per year. Since this is "per year," the number of time intervals is the "number of years," or 8.   **3.** time intervals = years/half-life; the decay rate is 50% or $\dfrac{1}{2}$ because half-life is the amount of time it takes half of a substance to decay

**Exercise Set 9.4**   **1.** 451   **3.** 144,302   **5.** 21,231   **7.** 202   **9.** 1470   **11.** 13   **13.** 712,880   **15.** 383   **17.** 333 bison   **19.** 1 g
**21. a.** $\dfrac{14}{7} = 2$; 10; yes   **b.** $\dfrac{11}{7} \approx 1.6$; 13.2; yes   **23.** $\dfrac{500}{152} \approx 3.3$; 2.1; yes   **25.** 4.9 g   **27.** 3   **29.** $-1$   **31.** no; answers may vary

## Section 9.5

**Vocabulary, Readiness & Video Check** **1.** logarithmic; b **3.** yes **5.** no; none **7.** $(-\infty, \infty)$ **9.** First write the equation as an equivalent exponential equation. Then solve.

**Exercise Set 9.5** **1.** $6^2 = 36$ **3.** $3^{-3} = \dfrac{1}{27}$ **5.** $10^3 = 1000$ **7.** $e^4 = x$ **9.** $e^{-2} = \dfrac{1}{e^2}$ **11.** $7^{1/2} = \sqrt{7}$ **13.** $0.7^3 = 0.343$

**15.** $3^{-4} = \dfrac{1}{81}$ **17.** $\log_2 16 = 4$ **19.** $\log_{10} 100 = 2$ **21.** $\log_e x = 3$ **23.** $\log_{10} \dfrac{1}{10} = -1$ **25.** $\log_4 \dfrac{1}{16} = -2$ **27.** $\log_5 \sqrt{5} = \dfrac{1}{2}$ **29.** 3

**31.** $-2$ **33.** $\dfrac{1}{2}$ **35.** $-1$ **37.** 0 **39.** 2 **41.** 4 **43.** $-3$ **45.** $\{2\}$ **47.** $\{81\}$ **49.** $\{7\}$ **51.** $\{-3\}$ **53.** $\{-3\}$ **55.** $\{2\}$ **57.** $\{2\}$

**59.** $\left\{\dfrac{27}{64}\right\}$ **61.** $\{10\}$ **63.** $\{4\}$ **65.** $\{5\}$ **67.** $\left\{\dfrac{1}{49}\right\}$ **69.** 3 **71.** 3 **73.** 1 **75.** $-1$ **77.**  **79.**

**81.**  **83.** **85.** 1 **87.** $\dfrac{x-4}{2}$ **89. a.** $g(2) = 25$ **b.** $(25, 2)$ **c.** $f(25) = 2$ **91.** answers may vary

**93.** **95.** **97.** answers may vary **99.** 0.0827

## Section 9.6

**Vocabulary, Readiness & Video Check** **1.** 36; a **3.** $\log_b 2^7$; b **5.** x; a **7.** No, the product property says the logarithm of a product can be written as a sum of logarithms—the expression in this example is a logarithm of a sum. **9.** Since $\dfrac{1}{x} = x^{-1}$, this gives us $\log_2 x^{-1}$. Using the power property, we get $-1 \log_2 x$ or $-\log_2 x$.

**Exercise Set 9.6** **1.** $\log_5 14$ **3.** $\log_4 9x$ **5.** $\log_6 (x^2 + x)$ **7.** $\log_{10} (10x^2 + 20)$ **9.** $\log_5 3$ **11.** $\log_3 4$ **13.** $\log_2 \dfrac{x}{y}$ **15.** $\log_2 \dfrac{x^2 + 6}{x^2 + 1}$

**17.** $2\log_3 x$ **19.** $-1 \log_4 5 = -\log_4 5$ **21.** $\dfrac{1}{2}\log_5 y$ **23.** $\log_2 5x^3$ **25.** $\log_4 48$ **27.** $\log_5 x^3 z^6$ **29.** $\log_4 4$, or 1 **31.** $\log_7 \dfrac{9}{2}$

**33.** $\log_{10} \dfrac{x^3 - 2x}{x + 1}$ **35.** $\log_2 \dfrac{x^{7/2}}{(x + 1)^2}$ **37.** $\log_8 x^{16/3}$ **39.** $\log_3 4 + \log_3 y - \log_3 5$ **41.** $\log_4 2 - \log_4 9 - \log_4 z$ **43.** $3\log_2 x - \log_2 y$

**45.** $\dfrac{1}{2}\log_b 7 + \dfrac{1}{2}\log_b x$ **47.** $4\log_6 x + 5\log_6 y$ **49.** $3\log_5 x + \log_5 (x + 1)$ **51.** $2\log_6 x - \log_6 (x + 3)$ **53.** 0.2 **55.** 1.2 **57.** 0.35

**59.** 1.29 **61.** $-0.68$ **63.** $-0.125$ **65.** **67.** 2 **69.** 2 **71.** b **73.** true **75.** false **77.** false **79.** because $\log_b 1 = 0$

**Integrated Review** **1.** $x^2 + x - 5$ **2.** $-x^2 + x - 7$ **3.** $x^3 - 6x^2 + x - 6$ **4.** $\dfrac{x - 6}{x^2 + 1}$ **5.** $\sqrt{3x - 1}$ **6.** $3\sqrt{x} - 1$

**7.** one-to-one; $\{(6, -2), (8, 4), (-6, 2), (3, 3)\}$ **8.** not one-to-one **9.** not one-to-one **10.** one-to-one **11.** not one-to-one

**12.** $f^{-1}(x) = \dfrac{x}{3}$ **13.** $f^{-1}(x) = x - 4$ **14.** $f^{-1}(x) = \dfrac{x + 1}{5}$ **15.** $f^{-1}(x) = \dfrac{x - 2}{3}$ **16.** **17.**

**18.** **19.** **20.** $\{3\}$ **21.** $\{7\}$ **22.** $\{-8\}$ **23.** $\{3\}$ **24.** $\{2\}$ **25.** $\left\{\dfrac{1}{2}\right\}$ **26.** $\{32\}$ **27.** $\{4\}$ **28.** $\{5\}$

**29.** $\left\{\dfrac{1}{9}\right\}$ **30.** $\log_2 14x$ **31.** $\log_2 (5^x \cdot 8)$ **32.** $\log_5 \dfrac{x^3}{y^5}$ **33.** $\log_5 x^9 y^3$ **34.** $\log_2 \dfrac{x^2 - 3x}{x^2 + 4}$

**35.** $\log_3 \dfrac{y^4 + 11y}{y + 2}$ **36.** $\log_7 9 + 2\log_7 x - \log_7 y$ **37.** $\log_6 5 + \log_6 y - 2\log_6 z$

**38.** 544,000 mosquitoes

## Section 9.7

**Vocabulary, Readiness & Video Check**   **1.** $10$; c   **3.** $7$; b   **5.** $5$; b   **7.** $\dfrac{\log 7}{\log 2} = \dfrac{\ln 7}{\ln 2}$; a or b   **9.** The understood base of a common logarithm is 10. If you're finding the common logarithm of a known power of 10, then the common logarithm is the known power of 10. **11.** $\log_b b^x = x$

**Exercise Set 9.7**   **1.** $0.9031$   **3.** $0.3636$   **5.** $0.6931$   **7.** $-2.6367$   **9.** $1.1004$   **11.** $1.6094$   **13.** $1.6180$   **15.** $2$   **17.** $-3$   **19.** $2$
**21.** $\dfrac{1}{4}$   **23.** $3$   **25.** $3.1$   **27.** $-4$   **29.** $\dfrac{1}{2}$   **31.** $\{10^{1.3}\}$; $\{19.9526\}$   **33.** $\{e^{1.4}\}$; $\{4.0552\}$   **35.** $\{10^{2.3}\}$; $\{199.5262\}$
**37.** $\{e^{-2.3}\}$; $\{0.1003\}$   **39.** $\left\{\dfrac{10^{1.1}}{2}\right\}$; $\{6.2946\}$   **41.** $\left\{\dfrac{e^{0.18}}{4}\right\}$; $\{0.2993\}$   **43.** $\left\{\dfrac{4 + e^{2.3}}{3}\right\}$; $\{4.6581\}$   **45.** $\left\{\dfrac{10^{-0.5} - 1}{2}\right\}$; $\{-0.3419\}$
**47.** $\$3656.38$   **49.** $\$2542.50$   **51.** $1.5850$   **53.** $0.8617$   **55.** $1.5850$   **57.** $-1.6309$   **59.** $-2.3219$   **61.** $\left\{\dfrac{4}{7}\right\}$   **63.** $x = \dfrac{3y}{4}$
**65.** $\{-6, -1\}$   **67.** answers may vary   **69.**

**71.**

**73.** answers may vary   **75.** $4.2$   **77.** $5.3$

## Section 9.8

**Calculator Explorations**   **1.** 3.67 yr, or 3 yr and 8 mo   **3.** 23.16 yr, or 23 yr and 2 mo

**Vocabulary, Readiness & Video Check**   **1.** $\ln(4x - 2) = \ln 3$ is the same as $\log_e(4x - 2) = \log_e 3$. Therefore, from the logarithm property of equality, we know that $4x - 2 = 3$.   **3.** $2000 = 1000\left(1 + \dfrac{0.07}{12}\right)^{12 \cdot t} \approx 9.9$ yr; as long as the interest rate and compounding are the same, it takes any amount of money the same time to double.

**Exercise Set 9.8**   **1.** $\left\{\dfrac{\log 6}{\log 3}\right\}$; $\{1.6309\}$   **3.** $\left\{\dfrac{\log 5}{\log 9}\right\}$; $\{0.7325\}$   **5.** $\left\{\dfrac{\log 3.8}{2\log 3}\right\}$; $\{0.6076\}$   **7.** $\left\{\dfrac{\ln 5}{6}\right\}$; $\{0.2682\}$
**9.** $\left\{3 + \dfrac{\log 5}{\log 2}\right\}$; $\{5.3219\}$   **11.** $\left\{\dfrac{\log 3}{\log 4} - 7\right\}$; $\{-6.2075\}$   **13.** $\left\{\dfrac{1}{3}\left(4 + \dfrac{\log 11}{\log 7}\right)\right\}$; $\{1.7441\}$   **15.** $\{11\}$   **17.** $\left\{\dfrac{1}{2}\right\}$
**19.** $\left\{\dfrac{3}{4}\right\}$   **21.** $\{-2, 1\}$   **23.** $\{2\}$   **25.** $\left\{\dfrac{1}{8}\right\}$   **27.** $\{4, -1\}$   **29.** $\left\{\dfrac{2}{3}\right\}$   **31.** 103 wolves   **33.** 210.9 million   **35.** 7620   **37.** 9.9 yr
**39.** 1.7 yr   **41.** 8.8 yr   **43.** 55.7 in.   **45.** 11.9 lb per sq in.   **47.** 3.2 mi   **49.** 12 weeks   **51.** 18 weeks   **53.** $-\dfrac{5}{3}$   **55.** $\dfrac{17}{4}$   **57.** 16 yr
**59.** answers may vary   **61.**

; $\{6.93\}$

## Chapter 9 Vocabulary Check
**1.** inverse   **2.** composition   **3.** exponential   **4.** symmetric   **5.** Natural   **6.** Common
**7.** vertical; horizontal   **8.** logarithmic   **9.** Half-life   **10.** exponential

## Chapter 9 Review
**1.** $3x - 4$   **2.** $-x - 6$   **3.** $2x^2 - 9x - 5$   **4.** $\dfrac{2x + 1}{x - 5}$ where $x \neq 5$   **5.** $x^2 + 2x - 1$   **6.** $x^2 - 1$   **7.** 18
**8.** $x^4 - 4x^2 + 2$   **9.** $-2$   **10.** 48   **11.** one-to-one; $h^{-1} = \{(14, -9), (8, 6), (12, -11), (15, 15)\}$   **12.** not one-to-one
**13.** one-to-one;

| Rank in Housing Starts for 2013 (Input) | 4 | 3 | 1 | 2 |
|---|---|---|---|---|
| U.S. Region (Output) | Northeast | Midwest | South | West |

**14.** not one-to-one;

**15.** not one-to-one   **16.** not one-to-one   **17.** not one-to-one   **18.** one-to-one   **19.** $f^{-1}(x) = \dfrac{x - 11}{6}$   **20.** $f^{-1}(x) = \dfrac{x}{12}$
**21.** $f^{-1}(x) = \dfrac{x + 5}{3}$   **22.** $f^{-1}(x) = \dfrac{x - 1}{2}$   **23.** $f^{-1}(x) = -\dfrac{x - 3}{2}$   **24.** $f^{-1}(x) = \dfrac{x + 5}{5}$   **25.** $\{3\}$   **26.** $\left\{-\dfrac{4}{3}\right\}$

**27.** $\left\{\dfrac{3}{2}\right\}$   **28.** $\left\{\dfrac{5}{2}\right\}$

**29.** **30.** **31.** **32.**   **33.** $1131.82 **34.** $2963.11 **35.** 457,393 **36.** 572,165

**37.** 8 players **38.** 636 bears **39.** $\log_7 49 = 2$

**40.** $\log_2 \dfrac{1}{16} = -4$ **41.** $\left(\dfrac{1}{2}\right)^{-4} = 16$

**42.** $0.4^3 = 0.064$ **43.** $\left\{\dfrac{1}{64}\right\}$ **44.** $\{9\}$ **45.** $\{0\}$

**46.** $\{8\}$ **47.** $\{5\}$ **48.** $\{-2\}$ **49.** $\{4\}$ **50.** $\{9\}$ **51.** $\{2\}$ **52.** $\{-8, 1\}$ **53.** **54.** **55.** $\log_3 32$

**56.** $\log_2 18$

**57.** $\log_7 \dfrac{3}{4}$ **58.** $\log_e \dfrac{3}{2}$ **59.** $\log_{11} 4$ **60.** $\log_5 2$ **61.** $\log_5 \dfrac{x^3}{(x+1)^2}$ **62.** $\log_3(x^4 + 2x^3)$ **63.** $3\log_3 x - \log_3(x+2)$

**64.** $\log_4(x+5) - 2\log_4 x$ **65.** $\log_2 3 + 2\log_2 x + \log_2 y - \log_2 z$ **66.** $\log_7 y + 3\log_7 z - \log_7 x$ **67.** 2.02 **68.** $-0.11$

**69.** 0.5563 **70.** $-0.8239$ **71.** 0.2231 **72.** 1.5326 **73.** 3 **74.** $-1$ **75.** $-1$ **76.** 4 **77.** $\{50\}$ **78.** $\left\{\dfrac{e^{1.6}}{3}\right\}$; $\{1.6510\}$

**79.** $\left\{\dfrac{e^{-1}+3}{2}\right\}$; $\{1.6839\}$ **80.** $\left\{\dfrac{e^2 - 1}{3}\right\}$; $\{2.1297\}$ **81.** 0.2920 **82.** 1.2619 **83.** $1957.30 **84.** $1307.51 **85.** $\left\{\dfrac{\log 20}{\log 7}\right\}$; $\{1.5395\}$

**86.** $\left\{\dfrac{\log 7}{2\log 3}\right\}$; $\{0.8856\}$ **87.** $\left\{\dfrac{1}{2}\left(\dfrac{\log 6}{\log 3} - 1\right)\right\}$; $\{0.3155\}$ **88.** $\left\{\dfrac{1}{4}\left(\dfrac{\log 3}{\log 8} + 2\right)\right\}$; $\{0.6321\}$ **89.** $\left\{\dfrac{25}{2}\right\}$ **90.** $\varnothing$

**91.** $\{2\sqrt{2}\}$ **92.** $\{9, -1\}$ **93.** 6.01% **94.** 5.1 yr **95.** 46.2 yr **96.** 115.5 yr **97.** 8.8 yr **98.** 8.5 yr **99.** $\{-2\}$ **100.** $\left\{\dfrac{3}{2}\right\}$

**101.** $\left\{\dfrac{8}{9}\right\}$ **102.** $\left\{\dfrac{7}{2}\right\}$ **103.** $\{3\}$ **104.** $\{3\}$ **105.** $\{-1, 4\}$ **106.** $\left\{\dfrac{17}{3}\right\}$ **107.** $\{e^{-1.2}\}$ **108.** $\left\{\dfrac{9}{10}\right\}$ **109.** $\left\{\dfrac{3e^2}{e^2 - 3}\right\}$ **110.** $\varnothing$

**Chapter 9 Test** **1.** $2x^2 - 3x$ **2.** $-x + 3$ **3.** 5 **4.** $x - 7$ **5.** $x^2 - 6x - 2$ **6.** $f^{-1}(x) = \dfrac{x + 14}{7}$ **7.** one-to-one

**8.** not a function

**9.** one-to-one;

$$f^{-1}(x) = \dfrac{-x + 6}{2}$$

**10.** one-to-one; $f^{-1} = \{(0, 0), (3, 2), (5, -1)\}$ **11.** not one-to-one **12.** $\log_3 24$ **13.** $\log_5 \dfrac{x^4}{x + 1}$

**14.** $\log_6 2 + \log_6 x - 3\log_6 y$ **15.** $-1.53$ **16.** 1.0686 **17.** $\{-1\}$ **18.** $\left\{\dfrac{1}{2}\left(\dfrac{\log 4}{\log 3} - 5\right)\right\}$; $\{-1.8691\}$ **19.** $\left\{\dfrac{1}{9}\right\}$ **20.** $\left\{\dfrac{1}{2}\right\}$

**21.** $\{22\}$ **22.** $\left\{\dfrac{25}{3}\right\}$ **23.** $\left\{\dfrac{43}{21}\right\}$ **24.** $\{-1.0979\}$ **25.** **26.** **27.** $5234.58 **28.** 6 yr **29.** 6.5%; $230

**30.** 100,141 **31.** 64,805 prairie dogs

**32.** 15 yr **33.** 85% **34.** 52%

**Cumulative Review** **1.** 3; Sec. 1.4, Ex. 42 **2.** $4x^2\sqrt[3]{xy^2}$ **3.** 2; Sec. 1.4, Ex. 44 **4.** $3ab\sqrt[4]{2b}$ **5.** 30°, 40°, and 110°; Sec. 4.3, Ex. 6

**6.** $x = 110, y = 70$ **7.** $(x^2 + 5y)(7x - 1)$; Sec. 5.4, Ex. 8 **8.** $\{11\}$ **9.** $\dfrac{5k^2 - 7k + 4}{(k + 2)(k - 2)(k - 1)}$; Sec. 6.2, Ex. 9

**10.** $\dfrac{-x^2 + 23x + 38}{(x - 2)(x + 2)^2}$ **11.** 2; Sec. 7.1, Ex. 30 **12.** $2\sqrt{6}$ **13.** 1; Sec. 7.1, Ex. 32 **14.** $\sqrt[3]{2}$ **15.** $x^{5/6}$; Sec. 7.2, Ex. 14 **16.** $\dfrac{a^4 c^8}{b^6}$

**17.** 32; Sec. 7.2, Ex. 18 **18.** $729x^3$ **19.** $\dfrac{\sqrt{x}}{3}$; Sec. 7.3, Ex. 7 **20.** $2a$ **21.** $\dfrac{\sqrt[4]{3}}{2y}$; Sec. 7.3, Ex. 9 **22.** $9ab^2\sqrt[3]{a^2}$ **23.** $2x - 25$;

Sec. 7.4, Ex. 13 **24.** $\dfrac{1}{4}x^2 - 9$ **25.** $4 - 2\sqrt{3}$; Sec. 7.4, Ex. 14 **26.** **a.** $a - a^{33/4}$ **b.** $x + 2x^{1/2} - 15$ **27.** $\dfrac{7}{3\sqrt{35}}$; Sec. 7.5, Ex. 9

**28.** $\dfrac{3\sqrt[3]{m^2 n}}{m^2 n^3}$ **29.** $\{3\}$; Sec. 7.6, Ex. 4 **30.** $f(x) = \dfrac{x + 20}{3}$ **31.** $-1 + 5i$; Sec. 7.7, Ex. 8 **32.** **a.** 1 **b.** $i$ **c.** $-1$ **d.** $-i$

**33.** $\left\{\dfrac{6 + i\sqrt{5}}{2}, \dfrac{6 - i\sqrt{5}}{2}\right\}$; Sec. 8.1, Ex. 9 **34.** $\left\{\dfrac{-2 + \sqrt{5}}{2}, \dfrac{-2 - \sqrt{5}}{2}\right\}$ **35.** $\{2 + \sqrt{2}, 2 - \sqrt{2}\}$; Sec. 8.2, Ex. 3

**36.** $\left\{\dfrac{3 - \sqrt{5}}{4}, \dfrac{3 + \sqrt{5}}{4}\right\}$ **37.** $\{8, 27\}$; Sec. 8.3, Ex. 5 **38.** $\left\{-\dfrac{7}{5}\right\}$ **39.** $\left(-\dfrac{7}{2}, -1\right)$; Sec. 8.4, Ex. 5 **40.** $4\sqrt{3}$ in.

**41.**   ; Sec. 8.6, Ex. 2   **42.** $x^2 + 2x + 4$   **43. a.** $(f \circ g)(x) = |x - 2|$   **b.** $(g \circ f)(x) = |x| - 2$; Sec. 9.1, Ex. 3

**44. a.** $\dfrac{2a}{a - 1}$   **b.** $\dfrac{-3(a + 6)}{4(a - 3)}$   **c.** $\dfrac{y + x}{x^2 y^2}$   **45.** $f^{-1}(x) = \dfrac{x + 5}{3}$; Sec. 9.2, Ex. 9

**46.** $8x + 4 + \dfrac{1}{x - 2}$   **47.** $\{6\}$; Sec. 9.3, Ex. 6   **48.** $k = \dfrac{1}{24}$; $y = \dfrac{1}{24}x$

**49.** $\{5\}$; Sec. 9.5, Ex. 9   **50. a.** $\dfrac{11\sqrt{5}}{12}$   **b.** $\dfrac{\sqrt[3]{3x}}{6}$   **51.** \$2509.30; Sec. 9.6, Ex. 11   **52.** $\log_3 x^7 y^9$   **53.** $\left\{\dfrac{2}{99}\right\}$; Sec. 9.8, Ex. 4   **54.** $\left\{\dfrac{55}{8}\right\}$

# Chapter 10  Conic Sections

## Section 10.1

### Calculator Explorations

**1.**   **3.**

**Vocabulary, Readiness & Video Check**   **1.** conic sections   **3.** circle; center   **5.** radius   **7.** No, their graphs don't pass the vertical line test.   **9.** Since the standard form of a circle involves a squared binomial for both $x$ and $y$, we need to complete the square on both $x$ and $y$.

**Exercise Set 10.1**   **1.** upward   **3.** to the left   **5.** downward   **7.**   **9.**   **11.**

**13.**   **15.**   **17.**   **19.**   **21.**   **23.**

**25.**   **27.**   **29.**   **31.**   **33.**   **35.**

**37.**   **39.**   **41.**   **43.** $(x - 2)^2 + (y - 3)^2 = 36$   **45.** $x^2 + y^2 = 3$
**47.** $(x + 5)^2 + (y - 4)^2 = 45$

**49.**   **51.**   **53.**   **55.**   **57.**   **59.**

**61.**   **63.**   **65.**   **67.**   **69.**   **71.**

**73.**   **75.**   **77.**   **79.**   **81.**   **83.**

**85.** $\dfrac{\sqrt{3}}{3}$  **87.** $\dfrac{2\sqrt{42}}{3}$  **89.** The vertex is $(1,-5)$.  **91. a.** 16.5 m  **b.** 103.67 m  **c.** 3.5 m  **d.** $(0, 16.5)$  **e.** $x^2 + (y - 16.5)^2 = (16.5)^2$

**93. a.** 125 ft  **b.** 14 ft  **c.** 139 ft  **d.** $(0, 139)$  **e.** $x^2 + (y - 139)^2 = 125^2$  **95.** answers may vary  **97.** 20 m

**99. a.**    **b.** $x^2 + y^2 = 100$  **c.** $x^2 + y^2 = 25$

## Section 10.2

**Calculator Explorations  1.**    **3.**

**Vocabulary, Readiness & Video Check  1.** hyperbola  **3.** focus  **5.** hyperbola; $(0,0)$; $x$; $(a, 0)$ and $(-a, 0)$  **7.** $a$ and $b$ give us the location of 4 intercepts $-(a, 0), (-a, 0), (0, b),$ and $(0, -b)$ for $\dfrac{x^2}{a^2} + \dfrac{y^2}{b^2} = 1$ with center $(0, 0)$. For Example 2, the values of $a$ and $b$ also give us 4 points of the graph, just not intercepts. Here we move a distance of $a$ units horizontally to the left and right of the center and $b$ units above and below the center.

**Exercise Set 10.2**

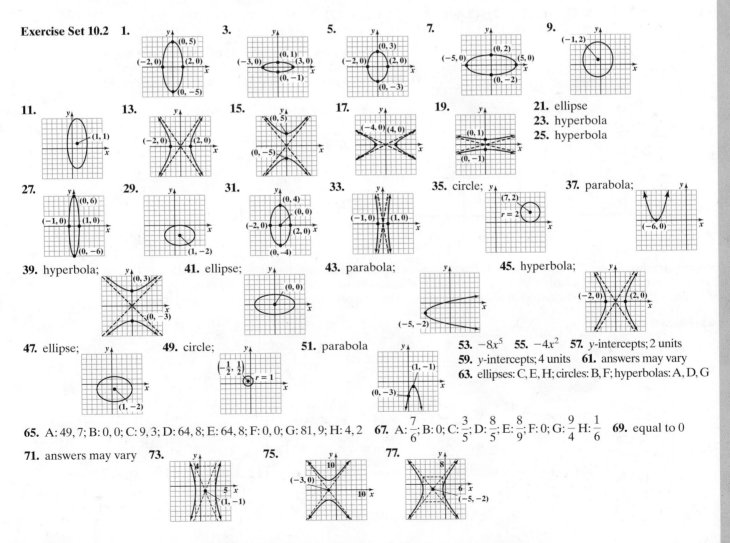

**21.** ellipse  **23.** hyperbola  **25.** hyperbola

**35.** circle;  **37.** parabola;

**39.** hyperbola;  **41.** ellipse;  **43.** parabola;  **45.** hyperbola;

**47.** ellipse;  **49.** circle;  **51.** parabola  **53.** $-8x^5$  **55.** $-4x^2$  **57.** $y$-intercepts; 2 units  **59.** $y$-intercepts; 4 units  **61.** answers may vary  **63.** ellipses: C, E, H; circles: B, F; hyperbolas: A, D, G

**65.** A: 49, 7; B: 0, 0; C: 9, 3; D: 64, 8; E: 64, 8; F: 0, 0; G: 81, 9; H: 4, 2  **67.** A: $\dfrac{7}{6}$; B: 0; C: $\dfrac{3}{5}$; D: $\dfrac{8}{5}$; E: $\dfrac{8}{9}$; F: 0; G: $\dfrac{9}{4}$; H: $\dfrac{1}{6}$  **69.** equal to 0

**71.** answers may vary  **73.**  **75.**  **77.**

**Integrated Review   1.** circle   **2.** parabola   **3.** parabola   **4.** ellipse

**5.** hyperbola   **6.** hyperbola   **7.** ellipse   **8.** circle

**9.** parabola   **10.** parabola   **11.** hyperbola   **12.** ellipse

**13.** ellipse   **14.** hyperbola   **15.** circle

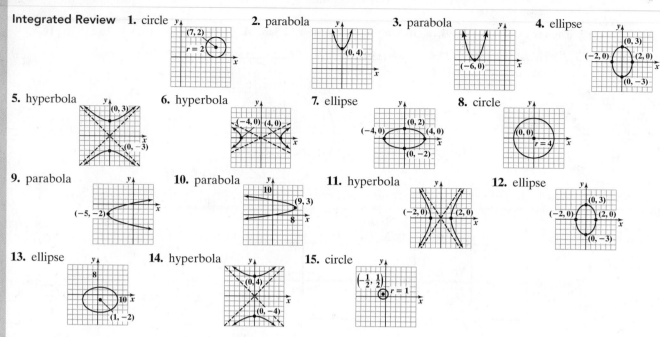

## Section 10.3

**Vocabulary, Readiness & Video Check   1.** Solving for $y$ would either introduce tedious fractions (2nd equation) or square root (1st equation) into the calculations.

**Exercise Set 10.3   1.** $\{(3,-4),(-3,4)\}$   **3.** $\{(\sqrt{2},\sqrt{2}),(-\sqrt{2},-\sqrt{2})\}$   **5.** $\{(4,0),(0,-2)\}$   **7.** $\{(-\sqrt{5},-2),(-\sqrt{5},2),$ $(\sqrt{5},-2),(\sqrt{5},2)\}$   **9.** $\varnothing$   **11.** $\{(1,-2),(3,6)\}$   **13.** $\{(2,4),(-5,25)\}$   **15.** $\varnothing$   **17.** $\{(1,-3)\}$   **19.** $\{(-1,-2),(-1,2),$ $(1,-2),(1,2)\}$   **21.** $\{(0,-1)\}$   **23.** $\{(-1,3),(1,3)\}$   **25.** $\{(\sqrt{3},0),(-\sqrt{3},0)\}$   **27.** $\varnothing$   **29.** $\{(-6,0),(6,0),(0,-6)\}$   **31.** $\{(3,\sqrt{3})\}$   **33.**   **35.**   **37.** $(8x-25)$ in.   **39.** $(4x^2+6x+2)$ m   **41.** answers may vary   **43.** 0, 1, 2, 3, or 4   **45.** $-9$ and 7; 9 and $-7$; 9 and 7; $-9$ and $-7$   **47.** 15 cm by 19 cm   **49.** 15 thousand compact discs; price: $3.75

**51.**   **53.**

## Section 10.4

**Vocabulary, Readiness & Video Check   1.** For both, we graph the related equation to find the boundary and sketch it as a solid boundary for $\le$ or $\ge$ and a dashed boundary for $<$ or $>$; also, we choose a test point not on the boundary and shade the region containing the test point if the test point is a solution of the original inequality or shade the other region if not.

**Exercise Set 10.4   1.**   **3.**   **5.**   **7.**   **9.**   **11.**

**13.**   **15.**   **17.**   **19.**   **21.**   **23.**   **25.**

**27.** 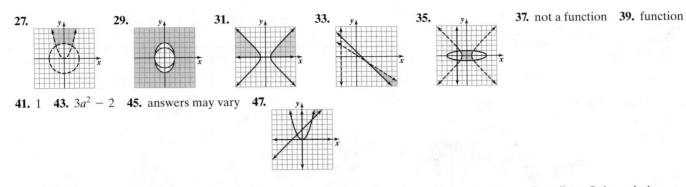 **29.**  **31.**  **33.**  **35.**    **37.** not a function  **39.** function

**41.** 1  **43.** $3a^2 - 2$  **45.** answers may vary  **47.**

**Chapter 10 Vocabulary Check**  **1.** circle; center  **2.** nonlinear system of equations  **3.** ellipse  **4.** radius  **5.** hyperbola
**6.** conic sections  **7.** vertex  **8.** diameter

**Chapter 10 Review**  **1.** $(x + 4)^2 + (y - 4)^2 = 9$  **2.** $(x + 7)^2 + (y + 9)^2 = 11$  **3.** $(x - 5)^2 + y^2 = 25$  **4.** $x^2 + y^2 = \dfrac{49}{4}$

**5.** **6.** **7.** **8.** **9.** **10.** **11.**

**12.** **13.** **14.** **15.** **16.** **17.**

**18.** **19.** **20.** **21.** **22.** **23.**

**24.** **25.** **26.** **27.** **28.** **29.**

**30.** **31.** **32.**  **33.** $\{(1, -2), (4, 4)\}$  **34.** $\varnothing$  **35.** $\{(-1, 1), (2, 4)\}$
**36.** $\{(2, 2\sqrt{2}), (2, -2\sqrt{2})\}$  **37.** $\{(0, 2), (0, -2)\}$
**38.** $\left\{\left(2, \dfrac{5}{2}\right), (-7, -20)\right\}$  **39.** $\{(1, 4)\}$  **40.** $\{(-2, -1), (-2, 1),$
$(2, -1), (2, 1)\}$  **41.** length: 15 ft; width: 10 ft  **42.** 4

**43.** **44.** **45.** **46.** **47.** **48.** **49.**

**50.** **51.** $(x + 7)^2 + (y - 8)^2 = 25$  **52.** **53.** **54.** **55.**

**56.** **57.** **58.** **59.** **60.** **61.**

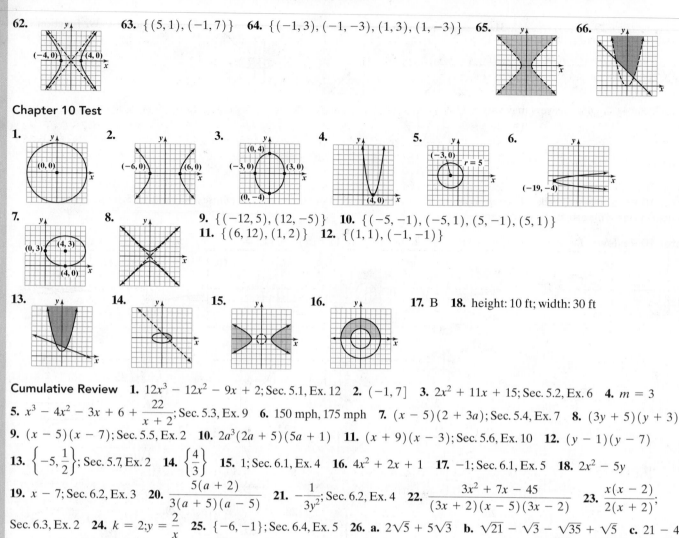

**62.** 

**63.** $\{(5, 1), (-1, 7)\}$ **64.** $\{(-1, 3), (-1, -3), (1, 3), (1, -3)\}$ **65.** **66.**

## Chapter 10 Test

**1.** **2.** **3.** **4.** **5.** **6.**

**7.** **8.** **9.** $\{(-12, 5), (12, -5)\}$ **10.** $\{(-5, -1), (-5, 1), (5, -1), (5, 1)\}$
**11.** $\{(6, 12), (1, 2)\}$ **12.** $\{(1, 1), (-1, -1)\}$

**13.** **14.** **15.** **16.** **17.** B **18.** height: 10 ft; width: 30 ft

**Cumulative Review** **1.** $12x^3 - 12x^2 - 9x + 2$; Sec. 5.1, Ex. 12 **2.** $(-1, 7]$ **3.** $2x^2 + 11x + 15$; Sec. 5.2, Ex. 6 **4.** $m = 3$

**5.** $x^3 - 4x^2 - 3x + 6 + \dfrac{22}{x+2}$; Sec. 5.3, Ex. 9 **6.** 150 mph, 175 mph **7.** $(x - 5)(2 + 3a)$; Sec. 5.4, Ex. 7 **8.** $(3y + 5)(y + 3)$

**9.** $(x - 5)(x - 7)$; Sec. 5.5, Ex. 2 **10.** $2a^3(2a + 5)(5a + 1)$ **11.** $(x + 9)(x - 3)$; Sec. 5.6, Ex. 10 **12.** $(y - 1)(y - 7)$

**13.** $\left\{-5, \dfrac{1}{2}\right\}$; Sec. 5.7, Ex. 2 **14.** $\left\{\dfrac{4}{3}\right\}$ **15.** 1; Sec. 6.1, Ex. 4 **16.** $4x^2 + 2x + 1$ **17.** $-1$; Sec. 6.1, Ex. 5 **18.** $2x^2 - 5y$

**19.** $x - 7$; Sec. 6.2, Ex. 3 **20.** $\dfrac{5(a+2)}{3(a+5)(a-5)}$ **21.** $-\dfrac{1}{3y^2}$; Sec. 6.2, Ex. 4 **22.** $\dfrac{3x^2 + 7x - 45}{(3x+2)(x-5)(3x-2)}$ **23.** $\dfrac{x(x-2)}{2(x+2)}$;

Sec. 6.3, Ex. 2 **24.** $k = 2; y = \dfrac{2}{x}$ **25.** $\{-6, -1\}$; Sec. 6.4, Ex. 5 **26. a.** $2\sqrt{5} + 5\sqrt{3}$ **b.** $\sqrt{21} - \sqrt{3} - \sqrt{35} + \sqrt{5}$ **c.** $21 - 4\sqrt{5}$

**d.** $-7$ **27.** 2; Sec. 6.5, Ex. 2 **28.** $\dfrac{\sqrt{3} - 3}{3}$ **29.** 0; Sec. 7.1, Ex. 5 **30.** $\dfrac{x^3\sqrt{x}}{7}$ **31.** 0.5; Sec. 7.1, Ex. 7 **32.** $3a^2b\sqrt{10a}$ **33.** $\sqrt{x}$; Sec. 7.2,

Ex. 19 **34.** $y\sqrt{x}$ **35.** $\sqrt[3]{rs^2}$; Sec. 7.2, Ex. 21 **36.** $\{6\}$ **37.** $2\sqrt[3]{3}$; Sec. 7.3, Ex. 11 **38.** $12y$ **39.** $2\sqrt[4]{2}$; Sec. 7.3, Ex. 13 **40.** $3x$ **41.** $\dfrac{2\sqrt{5}}{5}$;

Sec. 7.5, Ex. 1 **42. a.** $2\sqrt{2}$ **b.** $2xy^2\sqrt[5]{x^2}$ **c.** $\dfrac{2y^2\sqrt[3]{10}}{5}$ **43.** $\left\{-\dfrac{1}{9}, -1\right\}$; Sec. 7.6, Ex. 2 **44.** $(2, \infty)$ **45.** $13 - 18i$; Sec. 7.7, Ex. 13

**46.** $f^{-1}(x) = 2x - 1$ **47.** $\{-1 + \sqrt{5}, -1 - \sqrt{5}\}$; Sec. 8.1, Ex. 7 **48.** 13 **49.** $\left\{\dfrac{2 + \sqrt{10}}{2}, \dfrac{2 - \sqrt{10}}{2}\right\}$; Sec. 8.2, Ex. 2 **50. a.** $\left\{\dfrac{1}{3}\right\}$

**b.** $\left\{\dfrac{11}{3}\right\}$ **c.** $\{-2\}$ **51.** $\sqrt{2} \approx 1.414$; Sec. 7.3, Ex. 21 **52.** $\{4\}$ **53.** ; Sec. 10.2, Ex. 1 **54. a.** $9x^2 - 21x + 12$ **b.** 90

**c.** $-3x^2 + 9x - 1$ **d.** $-31$ **55.** $\varnothing$; Sec. 10.3, Ex. 3 **56.** $\{(-6, 0), (0, 6)\}$ **57.** $\{(2, \sqrt{3}), (-2, \sqrt{3}), (2, -\sqrt{3}), (-2, -\sqrt{3})\}$;
Sec. 10.3, Ex. 4

## Appendices

**Exercise Set Appendix A** **1.** $\dfrac{21}{30}$ **3.** $\dfrac{4}{18}$ **5.** $\dfrac{16}{20}$ **7.** $\dfrac{1}{2}$ **9.** $\dfrac{2}{3}$ **11.** $\dfrac{3}{7}$ **13.** 1 **15.** 5 **17.** $\dfrac{3}{5}$ **19.** $\dfrac{4}{5}$ **21.** $\dfrac{11}{8}$ **23.** $\dfrac{30}{61}$ **25.** $\dfrac{8}{11}$ **27.** $\dfrac{3}{8}$

**29.** $\dfrac{1}{2}$ **31.** $18\dfrac{20}{27}$ **33.** 37 **35.** $\dfrac{6}{7}$ **37.** 15 **39.** $\dfrac{1}{6}$ **41.** $\dfrac{3}{80}$ **43.** $10\dfrac{5}{11}$ **45.** $2\dfrac{28}{29}$ **47.** 1 **49.** $\dfrac{3}{5}$ **51.** $\dfrac{9}{35}$ **53.** $\dfrac{1}{3}$ **55.** $12\dfrac{1}{4}$ **57.** $1\dfrac{3}{5}$

**59.** $\dfrac{23}{21}$ **61.** $\dfrac{65}{21}$ **63.** $\dfrac{5}{7}$ **65.** $\dfrac{5}{66}$ **67.** $7\dfrac{1}{12}$ **69.** $48\dfrac{1}{15}$ **71.** $\dfrac{7}{5}$ **73.** $\dfrac{17}{18}$ **75.** answers may vary **77.** $\dfrac{1}{5}$ **79.** $\dfrac{3}{8}$ **81.** $10\dfrac{5}{12}$ ft

**Exercise Set Appendix B** **1.** 56 **3.** −32 **5.** 20 **7.** 26 **9.** −19 **11.** 0 **13.** $\dfrac{13}{6}$ **15.** $(1, 2)$ **17.** $\{(x, y) \mid 3x + y = 1\}$
**19.** $(9, 9)$ **21.** $(-3, -2)$ **23.** $(3, 4)$ **25.** 8 **27.** 0 **29.** 15 **31.** 54 **33.** $(-2, 0, 5)$ **35.** $(6, -2, 4)$ **37.** $(-2, 3, -1)$
**39.** $(0, 2, -1)$ **41.** 5 **43.** 0; answers may vary

**Exercise Set Appendix C** **1.** yes **3.** no **5.** answers may vary **7.** answers may vary **9.** answers may vary
**11.** Xmin = −12 Ymin = −12 **13.** Xmin = −9 Ymin = −12 **15.** Xmin = −10 Ymin = −25 **17.** Xmin = −5 Ymin = −15
Xmax = 12 Ymax = 12 Xmax = 9 Ymax = 12 Xmax = 10 Ymax = 25 Xmax = 5 Ymax = 15
Xscl = $\dfrac{6}{5}$ Yscl = $\dfrac{6}{5}$ Xscl = 1 Yscl = 2 Xscl = 2 Yscl = 5 Xscl = 1 Yscl = 3
**19.** Xmin = −20 Ymin = −30 **21.** Setting B **23.** Setting B **25.** Setting B
Xmax = 30 Ymax = 50
Xscl = 5 Yscl = 10

**27.** **29.** **31.** **33.**

**35.** **37.** **39.** **41.**

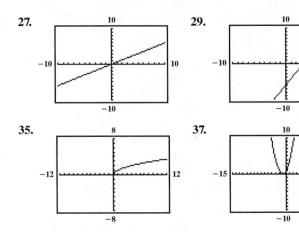

**Exercise Set Appendix D** **1.** **3.** **5.** **7.** **9.** **11.**

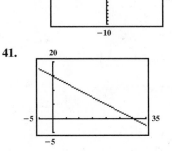

**13.** **15.**

**Exercise Set Appendix E** **1.** 71° **3.** 19.2° **5.** $78\frac{3}{4}°$ **7.** 30° **9.** 149.8° **11.** $100\frac{1}{2}°$ **13.** $m\angle 1 = m\angle 5 = m\angle 7 = 110°$,
$m\angle 2 = m\angle 3 = m\angle 4 = m\angle 6 = 70°$ **15.** 90° **17.** 90° **19.** 90° **21.** 45°, 90° **23.** 73°, 90° **25.** $50\frac{1}{4}°$, 90° **27.** $x = 6$
**29.** $x = 4.5$ **31.** 10 **33.** 12

**Practice Final Exam** **1.** $6\sqrt{6}$ **2.** $-\dfrac{3}{2}$ **3.** 5 **4.** $\dfrac{1}{81x^2}$ **5.** $\dfrac{64}{3}$ **6.** $\dfrac{3a^7}{2b^5}$ **7.** $\dfrac{8a^{1/3}c^{2/3}}{b^{5/12}}$ **8.** $3y(x + 3y)(x - 3y)$
**9.** $2(2y - 1)(4y^2 + 2y + 1)$ **10.** $(x + 3)(x - 3)(y - 3)$ **11.** $-5x^3 - 11x - 9$ **12.** $36m^2 + 12mn + n^2$
**13.** $2x^3 - 13x^2 + 14x - 4$ **14.** $\dfrac{x + 2}{2(x + 3)}$ **15.** $\dfrac{1}{x(x + 3)}$ **16.** $\dfrac{5x - 2}{(x - 3)(x + 2)(x - 2)}$ **17.** $-x\sqrt{5x}$ **18.** −20
**19.** $2x^4 + 2x - 2 + \dfrac{1}{2x - 1}$ **20.** $\{-32\}$ **21.** $\left\{1, \dfrac{2}{3}\right\}$ **22.** $\left\{4, -\dfrac{8}{7}\right\}$ **23.** $\left(\dfrac{3}{2}, 5\right]$ **24.** $(-\infty, -2) \cup \left(\dfrac{4}{3}, \infty\right)$ **25.** $\{3\}$
**26.** $\left\{\dfrac{3 - \sqrt{29}}{2}, \dfrac{3 + \sqrt{29}}{2}\right\}$ **27.** $\{2, 3\}$ **28.** $\left(-\infty, -\dfrac{3}{2}\right) \cup (5, \infty)$ **29.** $\left(\dfrac{7}{2}, -10\right)$
**30.** **31.** **32.** **33.** domain: $(-\infty, \infty)$; range: $(-\infty, -1]$

**34.** **35.** domain: $(-\infty, \infty)$; range: $(-3, \infty)$   **36.** $f(x) = -\dfrac{1}{2}x$   **37.** $f(x) = -\dfrac{1}{3}x + \dfrac{5}{3}$   **38.** $2\sqrt{26}$ units

**39.** $\left(-4, \dfrac{7}{2}\right)$   **40.** $\dfrac{3\sqrt{y}}{y}$   **41.** $\dfrac{8 - 6\sqrt{x} + x}{8 - 2x}$   **42.** Florida: \$17 billion; California: \$13 billion;

New York: \$9 billion   **43.** $(x + 2y)(x - 2y)$   **44.** 5   **45.** 16   **46.** 7 ft   **47. a.** 272 ft

**b.** 5.12 sec   **48.** 5 gal of 10%; 15 gal of 20%   **49.** $0 - 2i\sqrt{2}$   **50.** $0 - 3i$   **51.** $7 + 24i$   **52.** $-\dfrac{3}{2} + \dfrac{5}{2}i$   **53.** $(g \circ h)(x) = x^2 - 6x - 2$

**54.** $f(x)$ is one-to-one; $f^{-1}(x) = \dfrac{-x + 6}{2}$   **55.** $\log_5 \dfrac{x^4}{x + 1}$   **56.** $\{-1\}$   **57.** $\left\{\dfrac{1}{2}\left(\dfrac{\log 4}{\log 3} - 5\right)\right\}$; $\{-1.8691\}$   **58.** $\{22\}$   **59.** $\left\{\dfrac{43}{21}\right\}$

**60.** $\left\{\dfrac{1}{2}\right\}$   **61.**    **62.** 64,805 prairie dogs   **63.**    **64.**    **65.**

**66.** $\{(-5, -1), (-5, 1), (5, -1), (5, 1)\}$

# Index

# Photo Credits